# Introduction to Geography

Tenth Edition

Arthur Getis
San Diego State University

Judith Getis

Jerome D. Fellmann
University of Illinois,
Urbana–Champaign

*With Contributions By*

Victoria L. Getis
Ohio State University

Jon C. Malinowski
United States Military Academy

McGraw Hill Higher Education

Boston Burr Ridge, IL Dubuque, IA Madison, WI New York San Francisco St. Louis
Bangkok Bogotá Caracas Kuala Lumpur Lisbon London Madrid Mexico City
Milan Montreal New Delhi Santiago Seoul Singapore Sydney Taipei Toronto

The McGraw-Hill Companies

Higher Education

INTRODUCTION TO GEOGRAPHY, TENTH EDITION

Published by McGraw-Hill, a business unit of The McGraw-Hill Companies, Inc., 1221 Avenue of the Americas, New York, NY 10020. 

This book is printed on recycled, acid-free paper containing 10% postconsumer waste.

1 2 3 4 5 6 7 8 9 0 QPD/QPD 0 9 8 7 6 5 4

ISBN 0-07-282685-1

Publisher: *Margaret J. Kemp*
Senior Sponsoring Editor: *Daryl Bruflodt*
Senior Developmental Editor: *Lisa A. Bruflodt*
Associate Marketing Manager: *Todd L. Turner*
Lead Project Manager: *Joyce M. Berendes*
Senior Production Supervisor: *Sherry L. Kane*
Lead Media Project Manager: *Judi David*
Senior Media Technology Producer: *Jeffry Schmitt*
Senior Coordinator of Freelance Design: *Michelle D. Whitaker*
Cover/Interior Designer: *Maureen McCutcheon*
(USE) Cover Image: © *Ricardo Azoury /Paulo Fridman Photography;* Spine image: © *Robert Glusic/Getty Images*
Senior Photo Research Coordinator: *Lori Hancock*
Photo Research: *Toni Michaels/PhotoFind, LLC*
Supplement Producer: *Brenda A. Ernzen*
Compositor: *Precision Graphics*
Typeface: *10/12 Palatino*
Printer: *Quebecor World Dubuque, IA*

The views expressed by the authors are theirs alone and do not represent those of the United States Military Academy, the Department of Army, or the Department of Defense.

**Library of Congress Cataloging-in-Publication Data**

Introduction to geography / Arthur Getis . . . [et al.]. — 10th ed.
p. cm.
Rev ed. of: Introduction to geography / Arthur Getis, Judith Getis, Jerome D. Fellmann. 9th ed. ©2004.
Includes bibliographical references and index.
ISBN 0-07-282685-1 (hard copy : alk. paper)
1. Introduction to geography. I. Getis, Arthur, 1934–. II. Getis, Judith. III. Fellmann, Jerome D.
IV. Getis, Victoria L., Malinowski, Jon C.

G128.G45 2006
910—dc22

2004015009
CIP

www.mhhe.com

# Brief Contents

**Preface** ix

**1** Introduction 3

**2** Maps 25

**PART 1**

The Earth Science Tradition 52

**3** Physical Geography: Landforms 55

**4** Physical Geography: Weather and Climate 87

**5** The Geography of Natural Resources 133

**PART 2**

The Culture-Environment Tradition 176

**6** Population Geography 179

**7** Cultural Geography 217

**8** Spatial Interaction 267

**9** Political Geography 299

**PART 3**

The Locational Tradition 338

**10** Economic Geography 341

**11** Urban Geography 389

**12** Human Impact on the Environment 431

**PART 4**

The Area Analysis Tradition 466

**13** The Regional Concept 469

**Appendices** 489

**Glossary** 504

**Index** 517

# Contents


**Preface ix**

**CHAPTER 1**

## Introduction 3

**The Nature of Geography 4**
**What Is Geography? 4**
**Evolution of the Discipline 6**
**Some Core Geographic Concepts 8**
Location, Direction, and Distance 9
*Location 9*
*Direction 10*
*Distance 10*
Size and Scale 11
Physical and Cultural Attributes 13
Attributes of Place Are Always Changing 14
Interrelations between Places 14
Place Similarity and Regions 16
**Geography's Themes and Standards 17**
**Organization of This Book 19**
**Key Words 21**
**For Review & Consideration 22**
**Selected References 22**

**CHAPTER 2**

## Maps 25

**Maps as the Tools of Geography 26**
**Locating Points on a Sphere 27**
The Grid System 27
The Global Positioning System 28
**Map Projections 30**
Area 30
Shape 31
Distance 31
Direction 31
**Scale 33**
**Types of Maps 33**
Topographic Maps and Terrain Representation 34
Thematic Maps and Data Representation 38
*Point Symbols 38*
*Area Symbols 39*
*Line Symbols 40*
Deceptive Practices 40
**Remote Sensing 42**
Aerial Photography 42
Nonphotographic Imagery 42
*Remote-Sensing Techniques 42*
*Satellite Imagery 44*
**Geographic Information Systems 45**
The Geographic Database 47
Applications of GIS 47
**Summary 49**
**Key Words 50**
**For Review & Consideration 50**
**Selected References 51**

PART 1
## The Earth Science Tradition 52

**CHAPTER 3**

## Physical Geography: Landforms 55

**Earth Materials 56**
Igneous Rocks 56
Sedimentary Rocks 57
Metamorphic Rocks 58
**Geologic Time 58**
**Movements of the Continents 58**
**Tectonic Forces 62**
Diastrophism 62
*Broad Warping 62*
*Folding 62*
*Faulting 63*
Volcanism 66
**Gradational Processes 68**
Weathering 68
*Mechanical Weathering 68*
*Chemical Weathering 69*
Mass Movement 70
Erosional Agents and Deposition 71
*Running Water 71*
*Stream Landscapes 72*
*Groundwater 74*
*Glaciers 75*
*Waves, Currents, and Coastal Landforms 76*
*Wind 80*
**Landform Regions 81**
**Summary 83**
**Key Words 83**
**For Review & Consideration 84**
**Selected References 84**

**CHAPTER 4**

## Physical Geography: Weather and Climate 87

**Air Temperature 89**
Earth Inclination 89
Reflection and Reradiation 91
Lapse Rate 92
**Air Pressure and Winds 92**
Pressure Gradient Force 94
The Convection System 95
Land and Sea Breezes 95
Mountain and Valley Breezes 96
The Coriolis Effect 96
The Frictional Effect 97
The Global Air-Circulation Pattern 97
**Ocean Currents 98**
**Moisture in the Atmosphere 99**
Types of Precipitation 101
Storms 102
**Climate, Soils, and Vegetation 104**
Soils and Climate 104
*Soil Formation 104*
*Soil Profiles and Horizons 107*
*Soil Properties 108*
*Soil Classification 110*
Natural Vegetation and Climate 110
*Succession 110*
*Natural Vegetation Regions 110*

**Climate Regions 112**
Tropical Climates 114
*Tropical Rain Forest 114*
*Savanna 114*
Dryland Climates 116
*Hot Deserts 116*
*Midlatitude Deserts and Semideserts 118*
Humid Midlatitude Climates 118
*Mediterranean Climate 119*
*Humid Subtropical Climate 120*
*Marine West Coast Climate 121*
*Humid Continental Climate 122*
Subarctic and Arctic Climates 123
**Climatic Change 123**
Long-Term Climatic Change 124
Short-Term Climatic Change 125
The Greenhouse Effect and Global Warming 126
**Summary 129**
**Key Words 129**
**For Review & Consideration 129**
**Selected References 130**

**CHAPTER 5**

## The Geography of Natural Resources 133

**Resource Terminology 135**
Renewable Resources 135
Nonrenewable Resources 135
Resource Reserves 136
**Energy Resources and Industrialization 137**
**Nonrenewable Energy Resources 137**
Crude Oil 138
Coal 141
Natural Gas 143
Oil Shale and Tar Sands 144
Nuclear Energy 146
*Nuclear Fission 146*
*Nuclear Fusion 147*
**Renewable Energy Resources 148**
Biomass Fuels 148
*Wood 148*
*Waste 148*
Hydropower 148
Solar Energy 150
Other Renewable Energy Resources 152
*Geothermal Energy 152*
*Wind Power 154*
**Nonfuel Mineral Resources 154**
The Distribution of Nonfuel Minerals 156
Copper: A Case Study 157
**Land Resources 159**
Soils 159
Wetlands 163
Forest Resources 164
*U.S. National Forests 166*
*Tropical Rain Forests 167*
**Resource Management 169**
**Summary 172**
**Key Words 173**
**For Review & Consideration 173**
**Selected References 173**

PART 2

## The Culture-Environment Tradition 176

**CHAPTER 6**

## Population Geography 179

**Population Growth 180**
**Some Population Definitions 182**
Birth Rates 182
Fertility Rates 183
Death Rates 186
Population Pyramids 190
Natural Increase 193
Doubling Times 193
**The Demographic Transition 194**
The Western Experience 196
A Divided World Converging 198
**The Demographic Equation 200**
Population Relocation 200
Immigration Impacts 200
**World Population Distribution 200**
**Population Density 204**
Overpopulation 205
Urbanization 207
**Population Data and Projections 207**
Population Data 207
Population Projections 209
**Population Controls 210**
**Population Prospects 212**
**Summary 213**
**Key Words 214**
**For Review & Consideration 214**
**Selected References 215**

**CHAPTER 7**

## Cultural Geography 217

**Components of Culture 218**
**Interaction of People and Environment 220**
Environments as Controls 220
Human Impacts 221
**Subsystems of Culture 221**
The Technological Subsystem 222
The Sociological Subsystem 226
The Ideological Subsystem 228
**Culture Change 229**
Innovation 230
Diffusion 230
Acculturation 233
**Cultural Diversity 235**
**Language 236**
Language Spread and Change 237
Standard and Variant Languages 238
Language and Culture 242
**Religion 246**
Classification and Distribution of Religions 246
The Principal Religions 248
*Judaism 248*
*Christianity 250*
*Islam 253*
*Hinduism 254*
*Buddhism 255*
*East Asian Ethnic Religions 257*
**Ethnicity 257**
**Gender and Culture 259**
**Other Aspects of Diversity 261**
**Summary 263**
**Key Words 264**
**For Review & Consideration 265**
**Selected References 265**

**CHAPTER 8**

## Spatial Interaction 267

**The Definition of *Spatial Interaction* 269**
**Distance and Spatial Interaction 269**
**Barriers to Interaction 269**
**Spatial Interaction and Innovation 270**
**Individual Activity Space 271**
Stage in Life 274
Mobility 275
Opportunities 275

**Diffusion and Innovation 275**
Contagious Diffusion 276
Hierarchical Diffusion 277
**Spatial Interaction and Technology 278**
Automobiles 278
Telecommunications 279
**Migration 281**
Types of Migration 281
Incentives to Migrate 282
Barriers to Migration 286
Patterns of Migration 289
**Globalization 291**
Economic Integration 292
*International Banking 292*
*Transnational Corporations 292*
*Global Marketing 293*
Political Integration 295
Cultural Integration 295
**Summary 296**
**Key Words 296**
**For Review & Consideration 296**
**Selected References 297**

**CHAPTER 9**

## Political Geography 299

**National Political Systems 301**
States, Nations, and Nation-States 301
Evolution of the Modern State 303
Geographic Characteristics of States 303
*Size 303*
*Shape 305*
*Location 305*
*Cores and Capitals 307*
Boundaries: The Limits of the State 310
*Natural and Artificial Boundaries 311*
*Boundaries Classified by Settlement 311*
*Boundaries as Sources of Conflict 312*
Centripetal Forces: Promoting State Cohesion 316
*Nationalism 316*
*Unifying Institutions 316*
*Organization and Administration 317*
*Transportation and Communication 317*
Centrifugal Forces: Challenges to State Authority 318
**Cooperation among States 323**
Supranationalism 323
The United Nations and Its Agencies 324
*Maritime Boundaries 325*
*An International Law of the Sea 325*
*UN Affiliates 326*
Regional Alliances 327
*Economic Alliances 327*
*Military and Political Alliances 329*
**Local and Regional Political Organization 329**
The Geography of Representation: The Districting Problem 330
The Fragmentation of Political Power 333
**Summary 335**
**Key Words 336**
**For Review & Consideration 336**
**Selected References 337**

PART 3
# The Locational Tradition 338

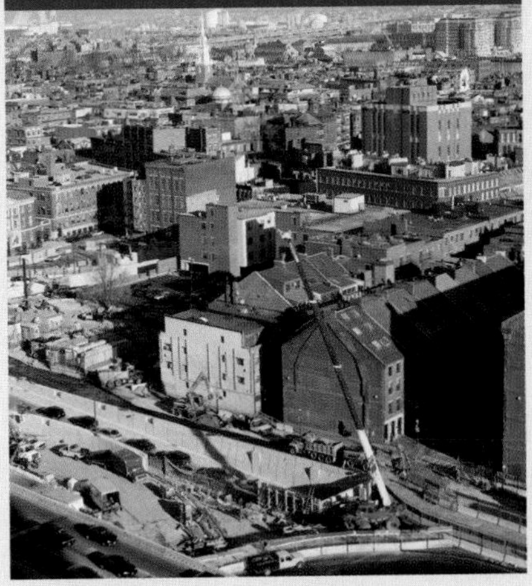

**CHAPTER 10**

## Economic Geography 341

**The Classification of Economic Activity and Economies 342**
Categories of Activity 343
Types of Economic Systems 344
**Primary Activities: Agriculture 345**
Subsistence Agriculture 348
*Extensive Subsistence Agriculture 348*
*Intensive Subsistence Agriculture 350*
Expanding Crop Production 351
*Intensification and the Green Revolution 352*
Commercial Agriculture 353
*Production Controls 354*
*A Model of Agricultural Location 355*
*Intensive Commercial Agriculture 357*
*Extensive Commercial Agriculture 357*
*Special Crops 359*
*Agriculture in Planned Economies 360*
**Other Primary Activities 361**
Fishing 361
Forestry 363
Mining and Quarrying 363
**Trade in Primary Products 364**
**Secondary Activities: Manufacturing 367**
Industrial Locational Models 367
Other Locational Considerations 369
*Transport Characteristics 369*
*Agglomeration Economies 370*
*Just-in-Time and Flexible Production 370*
*Comparative Advantage 371*
*Imposed Considerations 373*
Transnational Corporations (TNCs) 373
World Manufacturing Patterns and Trends 375
High-Tech Patterns 376
**Tertiary and Beyond 378**
Tertiary Services 380
Beyond Tertiary 381
**Services in World Trade 381**
**Summary 384**
**Key Words 385**
**For Review & Consideration 385**
**Selected References 385**

**CHAPTER 11**

## Urban Geography 389

**An Urbanizing World 391**
**The Functions of Urban Areas 394**
**The Location of Urban Settlements 396**
**The Economic Base 398**
**Systems of Urban Settlements 400**
The Urban Hierarchy 400
World Cities 400
Rank-Size and Primacy 401
Urban Influence Zones 402
Network Cities 402
Towns in Agricultural Areas 404
**Inside the City 405**
Competitive Bidding for Land 405
Land Values and Population Density 406
Models of Urban Land Use Structure 408
Social Areas of Cities 409
*Social Status 410*
*Family Status 410*
*Ethnicity 411*
Institutional Controls 411
**Suburbanization in the United States 413**
**Central City Change 414**
Constricted Central Cities 414
Expanding Central Cities 420
**World Urban Diversity 421**
The Anglo American City 421
The West European City 421
The East European City 423
Cities in the Developing World 424

**Summary 427**
**Key Words 427**
**For Review & Consideration 427**
**Selected References 428**

**CHAPTER 12**

## Human Impact on the Environment 431

**Ecosystems 433**
**Impact on Water 433**
- Availability of Water 434
- Modification of Streams 436
- Water Quality 437
- Agricultural Sources of Water Pollution 439
  - *Fertilizers 439*
  - *Biocides 439*
  - *Animal Wastes 439*
- Other Sources of Water Pollution 440
  - *Industry 440*
  - *Mining 441*
  - *Municipalities and Residences 441*
- Controlling Water Pollution 442

**Impact on Air and Climate 442**
- Air Pollutants 442
- Factors Affecting Air Pollution 443
- Acid Rain 444
- Photochemical Smog 446
- Depletion of the Ozone Layer 446
- Controlling Air Pollution 448

**Impact on Landforms 449**
- Landforms Produced by Excavation 449
- Landforms Produced by Dumping 450
- Formation of Surface Depressions 451

**Impact on Plants and Animals 451**
- Habitat Disruption 452
- Hunting and Commercial Exploitation 452
- Exotic Species 453
- Poisoning and Contamination 455

**Solid-Waste Disposal 456**
- Municipal Waste 456
  - *Landfills 457*
  - *Incineration 458*
  - *Source Reduction and Recycling 459*
- Hazardous Waste 460

**Summary 463**
**Key Words 463**
**For Review & Consideration 463**
**Selected References 464**

PART 4

## The Area Analysis Tradition 466

**CHAPTER 13**

## The Regional Concept 469

**The Nature of Regions 470**
**The Structure of This Chapter 472**
**Part I: Regions in the Earth Science Tradition 472**
- Landforms as Regions 472
- Dynamic Regions in Weather and Climate 473
- Natural Resource Regions 475

**Part II: Regions in the Culture–Environment Tradition 476**
- Population as Regional Focus 477
- Language as Region 478
- Mental Regions 479
- Political Regions 480

**Part III: Regions in the Locational Tradition 482**
- Economic Regions 482
- Urban Regions 484
- Ecosystems as Regions 485

**Summary 487**
**Key Words 487**
**For Review & Consideration 488**
**Selected References 488**

**Appendices 489**
**Glossary 504**
**Index 517**

# Preface

**"If you build it, they will come" was the message that inspired the character played by Kevin Costner in the movie *Field of Dreams* to create a baseball field in his Iowa cornfield. A similar hope encouraged us when we first began to think about writing *Introduction to Geography* in 1975. At that time, very few departments of geography in the United States and Canada offered a general introductory course for students—that is, one that sought to acquaint students with the breadth of the entire field. Instead, most departments offered separate courses in physical and human or cultural geography.**

**Recognizing that most students will have only a single college course and textbook in geography, we wanted to develop a book that covers all of the systematic topics that geographers study. Our hope, of course, was that the book would so persuasively identify and satisfy a disciplinary instructional need that more departments would begin to offer a general introductory course to the discipline, a dream that has been realized.**

## APPROACH

Our purpose is to convey concisely and clearly the nature of the field of geography, its intellectual challenges, and the logical interconnections of its parts. Even if students take no further work in geography, we are satisfied that they will have come into contact with the richness and breadth of our discipline and have at their command new insights and understandings for their present and future roles as informed adults. Other students may have the opportunity and interest to pursue further work in geography. For them, we believe, this text will make apparent the content and scope of the subfields of geography, emphasize its unifying themes, and provide the foundation for further work in their areas of interest.

The content is structured around the major research traditions of the discipline. Chapter 1 introduces students to the four organizing traditions that have emerged through the long history of geographic thought and writing: earth science, culture–environment, locational, and area analysis. Each of the four parts of this book centers on one of these geographic perspectives. Within each of the first three parts are chapters devoted to the subfields of geography, each placed with the tradition to which we think it belongs. Thus, the study of weather and climate is part of the earth science tradition; population geography is considered under the culture–environment tradition; and urban geography is included with the locational perspective. The tradition of area analysis—of regional geography—is presented in a single final chapter, which draws on the preceding traditions and themes and is integrated with them by cross-references. A fuller discussion of the book's organization is offered in Chapter 1, pp. 19 to 21.

Of course, our assignment of a topic may not seem appropriate to all users, since each tradition contains many emphases and themes. Some subfields could logically be attached to more than one of the recognized traditions. The rationale for our clustering of chapters is given in the brief introduction to each part of the text.

A useful textbook must be flexible enough in its organization to permit an instructor to adapt it to the time and subject matter constraints of a particular course. Although designed with a one-quarter or one-semester course in mind, this text may be used in a full-year introduction to geography when employed as a point of departure for special topics and amplifications introduced by the instructor or when supplemented by additional readings and class projects.

Moreover, the chapters are reasonably self-contained and need not be assigned in the sequence presented here. The "traditions" structure may be dropped and the chapters rearranged to suit the emphases and sequences preferred by the instructor or found to be of greatest interest to the students. The format of the course should properly reflect the joint contribution of instructor and book, rather than be dictated by the book alone.

## NEW TO THIS EDITION

For the tenth edition, we have made three changes that affect the entire textbook.

- In response to suggestions from reviewers of past editions, we have altered the ordering of the chapters. "The Geography of Natural Resources" is now Chapter 5, in Part One, "The Earth Science Tradition." "Urban Geography" has become Chapter 11. "Human Impact on the Environment" is now Chapter 12, detailing how all the processes, both physical and human, discussed in earlier chapters affect our planet.

- All of the world maps have been put on the Robinson projection, which was designed to show the world in a visually satisfactory manner. It permits some exaggeration of size in the high latitudes in order to improve the shapes of landmasses. Size and shape are most accurate in the temperate and tropical zones, where most people live.

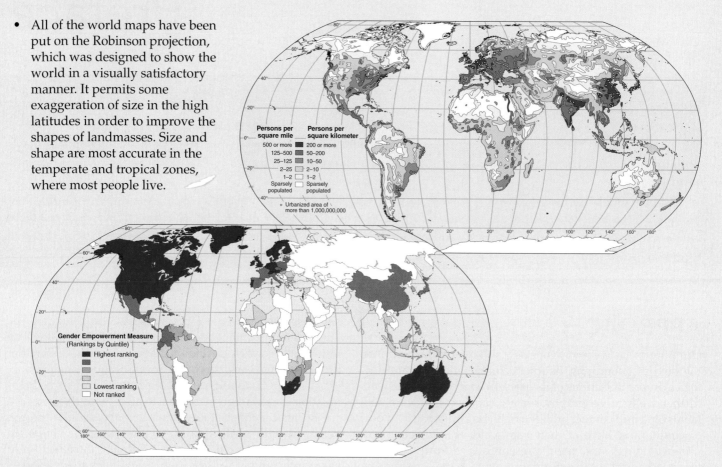

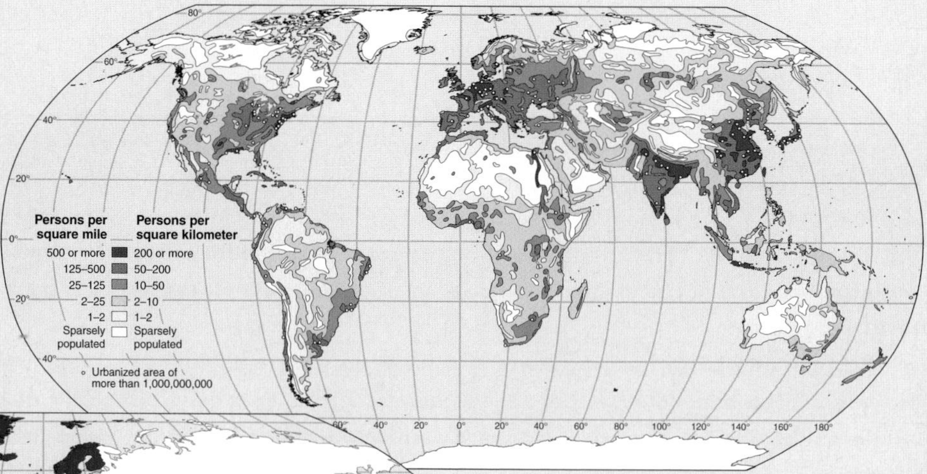


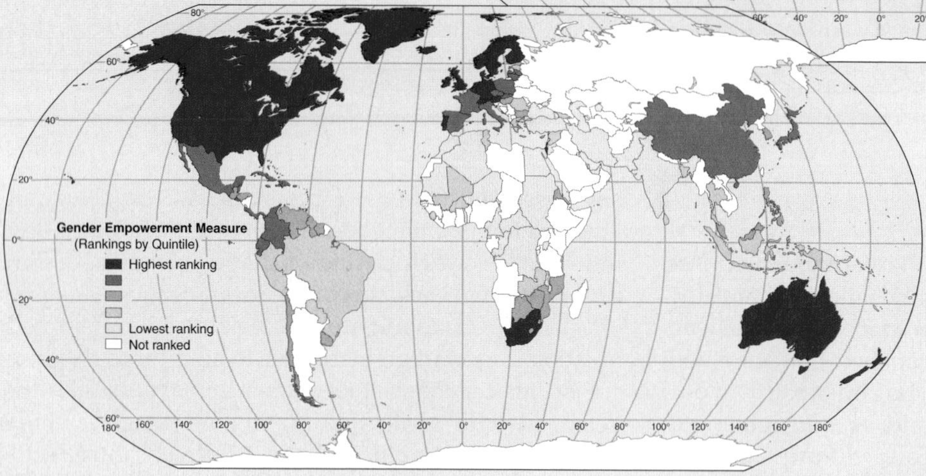


- The colors on the maps have been chosen specifically in order to accommodate colorblind readers. Most of them should be able to distinguish the hues from one another.

As with each new edition, we have added and deleted materials to reflect new research findings in the different topical areas of geography and the spatial consequences of continuing changes in established economic, political, social, and environmental structures and relationships. In addition to the necessary chapter revisions and updating of facts, analyses, and viewpoints mandated by current events, we have made every effort to incorporate in this revision many of the helpful suggestions offered by users. Nearly every chapter contains at least brief text additions or modifications, and four have been altered significantly.

- In response to reviewers who, over the years, have suggested that the topic of map projections is not appropriate for their students, we have moved most of that material to the new Appendix 1. In recognition of the increasing role of the Global Positioning System in our lives, an earlier boxed discussion has been made part of the regular text and expanded upon. The chapter ends with a new section, "Applications of Geographic Information Systems."
- Chapter 4, "Physical Geography: Weather and Climate," concludes with a significantly revised and lengthened discussion of climatic change, with long-term climate changes, short-term climate changes, and the greenhouse effect and global warming as subtopics.
- Extensive text changes in Chapter 10, "Economic Geography," include the addition of material on food resources and nutrition; a new section on expanding crop production with revised "Intensification and the Green Revolution" material; and a consideration of fishing as a primary economic activity. The section on trade in primary products has been completely revised.
- Changes to Chapter 11, "Urban Geography," involve a reorganization and revision of the section "Systems of Urban Settlements" and the addition of a new boxed discussion, "Women in the City." The chapter ends with a new section, "Cities in the Developing World."
- Every table and figure has been reviewed for accuracy and currency and has been replaced, updated, or otherwise revised where necessary. This tenth edition contains 35 new pieces of line art (maps, graphs, and diagrams), 70 new photographs or satellite images, and several new tables. In addition, about 60 figures have been revised and redrawn.

# FEATURES

Every effort has been made to gain and retain student attention, the essential first step in the learning process.

- An outline at the beginning of each chapter clarifies the organization of the chapter.
- Chapter-opening vignettes capture the reader's interest in preparation for the subject matter that follows.
- The text contains more than 450 full-color maps, charts, and photographs, with information and explanations that serve as extensions of the text, not just identification or documentation of the figure.

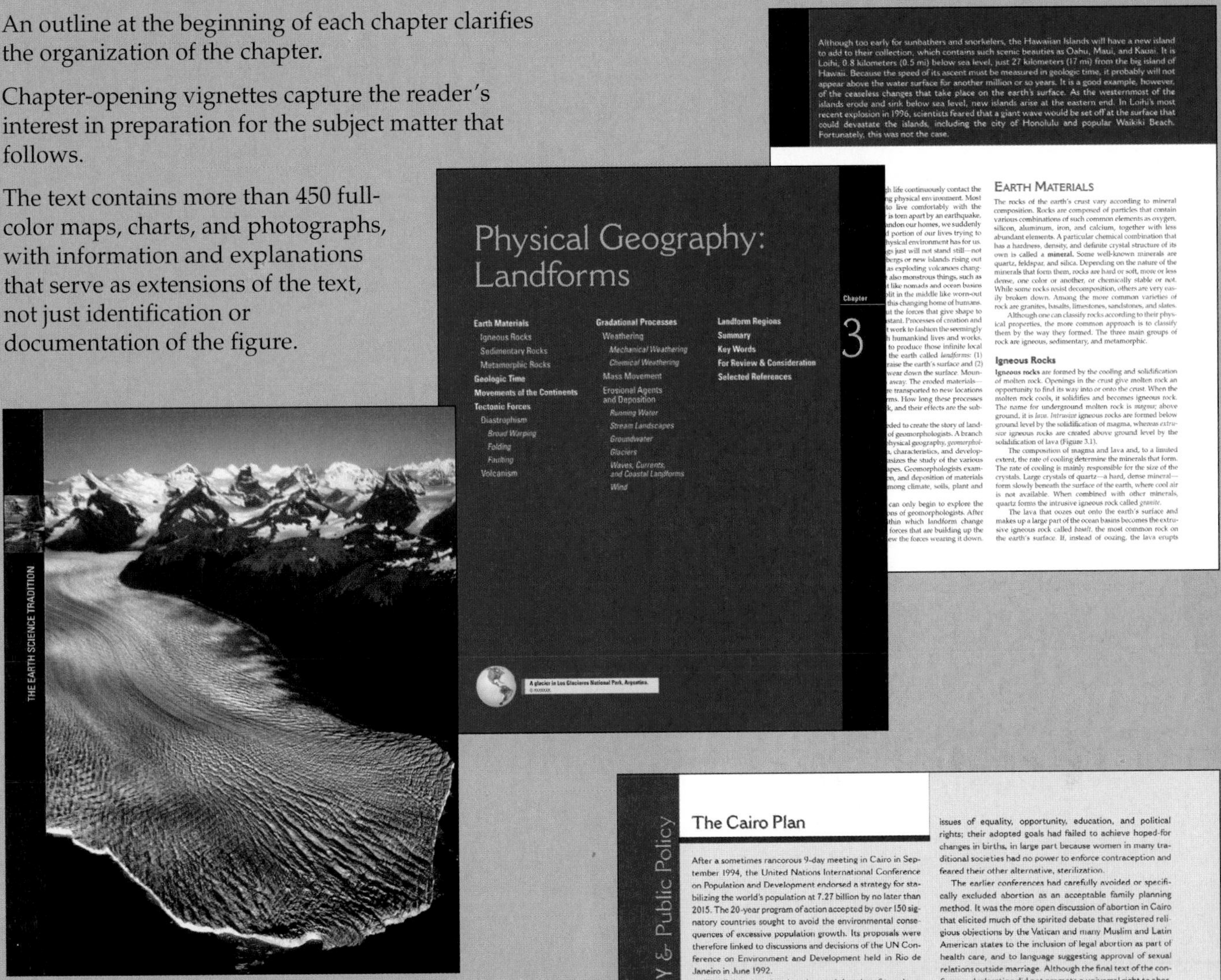

Although too early for sunbathers and snorkelers, the Hawaiian Islands will have a new island to add to their collection, which contains such scenic beauties as Oahu, Maui, and Kauai. It is Loihi, 0.8 kilometers (0.5 mi) below sea level, just 27 kilometers (17 mi) from the big island of Hawaii. Because the speed of its ascent must be measured in geologic time, it probably will not appear above the water surface for another million or so years. It is a good example, however, of the ceaseless changes that take place on the earth's surface. As the westernmost of the islands erode and sink below sea level, new islands arise at the eastern end. In Loihi's most recent explosion in 1996, scientists feared that a giant wave would be set off at the surface that could devastate the islands, including the city of Honolulu and popular Waikiki Beach. Fortunately, this was not the case.

Physical Geography: Landforms

Chapter 3

Earth Materials
Igneous Rocks
Sedimentary Rocks
Metamorphic Rocks
Geologic Time
Movements of the Continents
Tectonic Forces
Diastrophism
*Broad Warping*
*Folding*
*Faulting*
Volcanism
Gradational Processes
Weathering
*Mechanical Weathering*
*Chemical Weathering*
Mass Movement
Erosional Agents and Deposition
*Running Water*
*Stream Landscapes*
*Groundwater*
*Glaciers*
*Waves, Currents, and Coastal Landforms*
*Wind*
Landform Regions
Summary
Key Words
For Review & Consideration
Selected References

A glacier in Los Glaciares National Park, Argentina.

THE EARTH SCIENCE TRADITION

EARTH MATERIALS

The rocks of the earth's crust vary according to mineral composition. Rocks are composed of particles that contain various combinations of such common elements as oxygen, silicon, aluminum, iron, and calcium, together with less abundant elements. A particular chemical combination that has a hardness, density, and definite crystal structure of its own is called a **mineral.** Some well-known minerals are quartz, feldspar, and silica. Depending on the nature of the minerals that form them, rocks are hard or soft, more or less dense, one color or another, or chemically stable or not. While some rocks resist decomposition, others are very easily broken down. Among the more common varieties of rock are granites, basalts, limestones, sandstones, and slates.

Although one can classify rocks according to their physical properties, the more common approach is to classify them by the way they formed. The three main groups of rock are igneous, sedimentary, and metamorphic.

Igneous Rocks

**Igneous rocks** are formed by the cooling and solidification of molten rock. Openings in the crust give molten rock an opportunity to find its way into or onto the crust. When the molten rock cools, it solidifies and becomes igneous rock. The name for underground molten rock is *magma;* above ground, it is *lava. Intrusive* igneous rocks are formed below ground level by the solidification of magma, whereas *extrusive* igneous rocks are created above ground level by the solidification of lava (Figure 3.1).

The composition of magma and lava and, to a limited extent, the rate of cooling determine the minerals that form. The rate of cooling is mainly responsible for the size of the crystals. Large crystals of quartz—a hard, dense mineral—form slowly beneath the surface of the earth, where cool air is not available. When combined with other minerals, quartz forms the intrusive igneous rock called *granite.*

The lava that oozes out onto the earth's surface and makes up a large part of the ocean basins becomes the extrusive igneous rock called *basalt,* the most common rock on the earth's surface. If, instead of oozing, the lava erupts

- Boxed inserts, three to five per chapter, further develop ideas and are written so as to enhance student interest in the material. Except where noted, the authors have written all of the boxes. One box in most chapters explores gender-related issues. See, for example, "100 Million Women Are Missing" in Chapter 6 and "Legislative Women" in Chapter 9.
- A special "Geography and Public Policy" box, which appears in nearly every chapter, highlights an important or controversial issue. The boxes are intended to encourage students to think about the relevance of geography to real-world concerns. Critical-thinking questions at the end of the box, designed to have students reflect on and form an opinion about the issue, can serve as catalysts for class discussion.

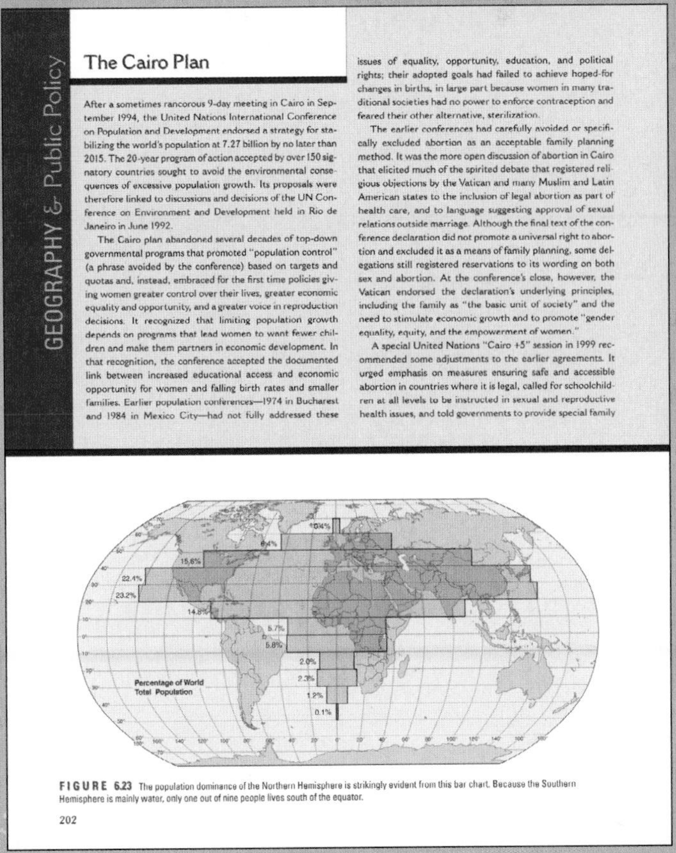


GEOGRAPHY & Public Policy

The Cairo Plan

After a sometimes rancorous 9-day meeting in Cairo in September 1994, the United Nations International Conference on Population and Development endorsed a strategy for stabilizing the world's population at 7.27 billion by no later than 2015. The 20-year program of action accepted by over 150 signatory countries sought to avoid the environmental consequences of excessive population growth. Its proposals were therefore linked to discussions and decisions of the UN Conference on Environment and Development held in Rio de Janeiro in June 1992.

The Cairo plan abandoned several decades of top-down governmental programs that promoted "population control" (a phrase avoided by the conference) based on targets and quotas and, instead, embraced for the first time policies giving women greater control over their lives, greater economic equality and opportunity, and a greater voice in reproduction decisions. It recognized that limiting population growth depends on programs that lead women to want fewer children and make them partners in economic development. In that recognition, the conference accepted the documented link between increased educational access and economic opportunity for women and falling birth rates and smaller families. Earlier population conferences—1974 in Bucharest and 1984 in Mexico City—had not fully addressed these issues of equality, opportunity, education, and political rights; their adopted goals had failed to achieve hoped-for changes in births, in large part because women in many traditional societies had no power to enforce contraception and feared their other alternative, sterilization.

The earlier conferences had carefully avoided or specifically excluded abortion as an acceptable family planning method. It was the more open discussion of abortion in Cairo that elicited much of the spirited debate that registered religious objections by the Vatican and many Muslim and Latin American states to the inclusion of legal abortion as part of health care, and to language suggesting approval of sexual relations outside marriage. Although the final text of the conference declaration did not promote a universal right to abortion and excluded it as a means of family planning, some delegations still registered reservations to its wording on both sex and abortion. At the conference's close, however, the Vatican endorsed the declaration's underlying principles, including the family as "the basic unit of society" and the need to stimulate economic growth and to promote "gender equality, equity, and the empowerment of women."

A special United Nations "Cairo +5" session in 1999 recommended some adjustments to the earlier agreements. It urged emphasis on measures ensuring safe and accessible abortion in countries where it is legal, called for schoolchildren at all levels to be instructed in sexual and reproductive health issues, and told governments to provide special family

FIGURE 6.23 The population dominance of the Northern Hemisphere is strikingly evident from this bar chart. Because the Southern Hemisphere is mainly water, only one out of nine people lives south of the equator.

202

- **End-of-Chapter Material.** *Chapter summaries* bring together and reinforce the major ideas of the chapter. A "Key Words" list contains page references to important terms introduced in the chapter, making it easy for students to verify their understanding of each term. "For Review and Consideration" questions enable readers to check their grasp of chapter material. A limited "Selected References" listing cites important recent or classic considerations of the subject matter of each chapter. We have included both widely available recent books and articles and a few more specialized titles useful to students who want to delve more deeply into particular subfields of geography.

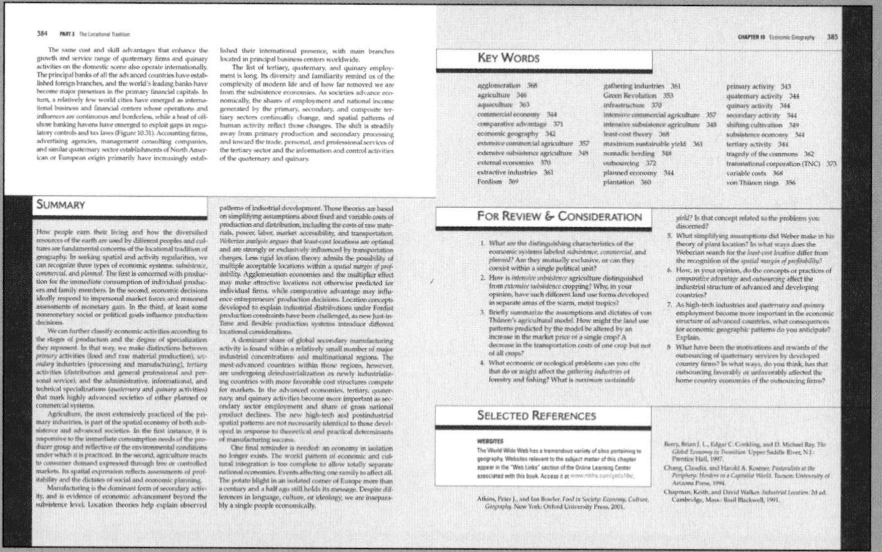


- Websites relevant to the subject matter of each chapter appear in the "Web Links" section of the Online Learning Center associated with this book. It can be accessed at www.mhhe.com/getis10e/.
- As its title indicates, the new Appendix 1, "Map Projections," is a discussion of that topic. It includes a consideration of methods of projection, globe properties and map distortion, and classes of projections.

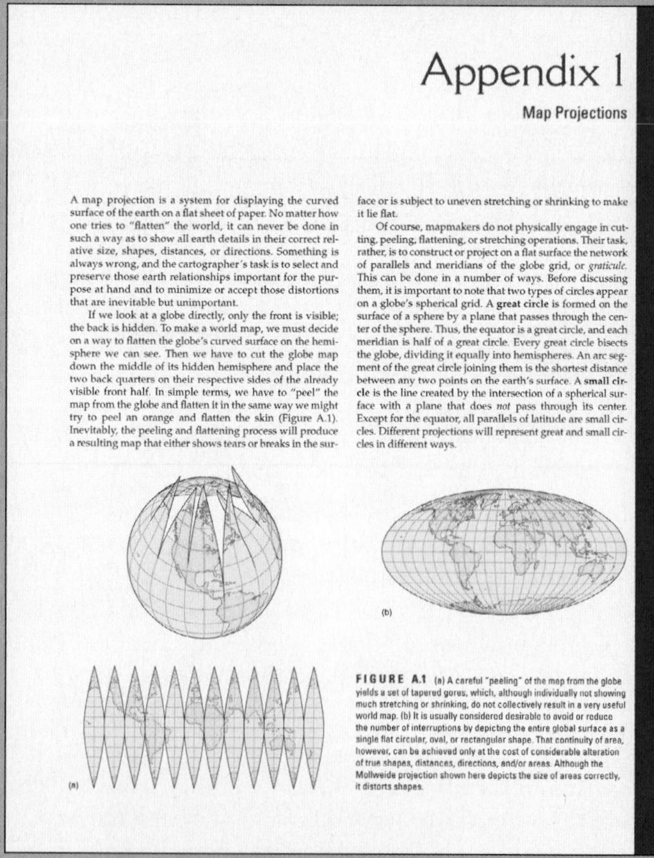


- Appendix 2, a modified version of the 2004 World Population Data Sheet of the Population Reference Bureau, includes basic demographic data and projections for countries, regions, and continents, as well as selected economic and social statistics helpful in national and regional comparisons. The appendix data provide a wealth of useful comparative information for student projects, regional and topical analyses, and study of world patterns.

# SUPPLEMENTS

## The Introduction to Geography Learning/Teaching Package

The tenth edition provides a complete geography program for the student and teacher.

### For the Student

*Online Learning Center at* www.mhhe.com/getis10e

This site gives you the opportunity to further explore topics presented in the book using the Internet. The site contains interactive quizzing with immediate feedback, interactive activities, base maps, animations, flashcards, and critical thinking questions. We have integrated *PowerWeb: Geography's* informative and timely world news, web links, and much more into the site to make these valuable resources easily accessible to students.

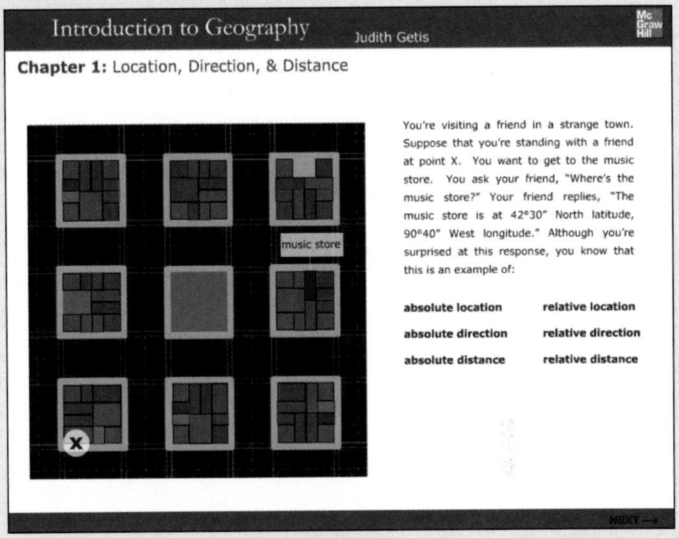


*Interactive World Issues CD-ROM*

Your instructor may require the *Interactive World Issues* CD-ROM. This CD allows you to have hands-on exercises and to see videos of different case studies. The five case studies include Chicago, Oregon, Mexico, China, and South Africa. Since most of us are unable to visit different world regions, this is a good way to understand the issues facing different parts of the world.

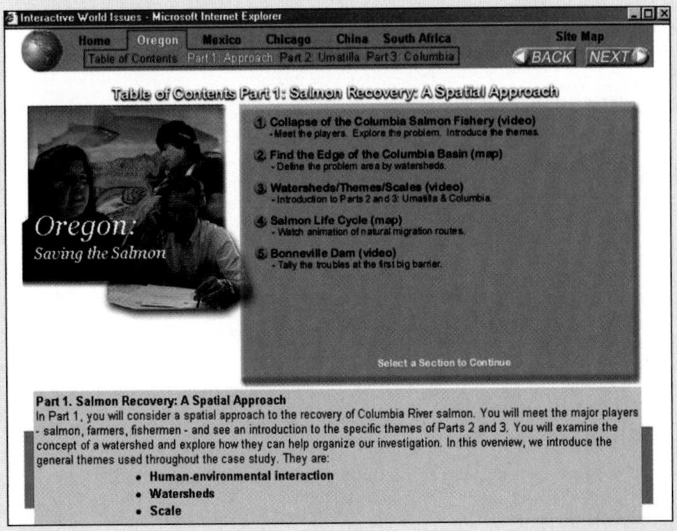


### For the Instructor

*Online Learning Center with PowerWeb: Geography at* www.mhhe.com/getis10e

Take advantage of the classroom activities, PowerPoint slides, and access to PageOut—McGraw-Hill's course management tool. *PowerWeb: Geography* is also available and will provide you with current news updates and articles that are great tools for stimulating class discussion.

*Online Instructor's Manual*

Included in this password-protected section of the Online Learning Center are chapter overviews, key terms, and discussion topics.

*Digital Content Manager CD-ROM*

This CD-ROM contains all of the figures and photographs from the text. The software makes customizing your multimedia presentation easy. You can organize figures in any order you want; add labels, lines, and your own artwork; integrate materials from other sources; edit and annotate lecture notes; and then have the option of placing your multimedia lecture into another presentation program, such as PowerPoint.

*Instructor's Testing and Resource CD-ROM*

This cross-platform CD-ROM provides a wealth of resources for the instructor. Supplements featured on this CD-ROM include computerized testing software that allows instructors to quickly create customized exams. This user-friendly program allows you to sort questions by format; edit existing questions or add new ones; and scramble questions for multiple versions of the same test.

Other assets on the Instructor's Testing and Resource CD-ROM are grouped within easy-to-use folders. The Instructor's Manual and Test Item File are available in both Word and PDF formats. Word files of the test bank are included for those instructors who prefer to work outside of the test-generator software.

*Transparencies*

Included are 100 illustrations from the text, all enlarged for excellent visibility in the classroom.

*Videotape Library*

An extensive array of videotapes is available to qualified adopters. Check with your sales representative for details.

*Interactive World Issues CD-ROM*

This CD allows you to have hands-on exercises and to see videos of different case studies. The five case studies include Chicago, Oregon, Mexico, China, and South Africa. Since most of us are unable to visit different world regions, this is a good way to understand the issues facing different parts of the world.

*Course Management Systems*

Online course content is available for a variety of course management systems, including

- Blackboard
- WebCT
- eCollege
- PageOut

# PACKAGES

*Get a discount when packaging the text with one of these supplements.* McGraw-Hill offers many different packaging options, which not only provide students with valuable geography-related material but will also save them money. Instructors, ask your sales representative for information on the discounts and special ISBNs for ordering a package that contains one or more of the following:

- *Interactive World Issues* CD-ROM
- *Student Atlas of World Geography*
- *New Millennium* CD-ROM by Rand McNally
- *You Can Make a Difference: Be Environmentally Responsible*
- *New York Times* subscription—special 20-week subscription
- *Global Studies* series
- *Student Atlas* series
- *Taking Sides* series
- *Annual Editions* series

# ACKNOWLEDGMENTS

A number of reviewers have greatly improved the content of this and earlier editions of *Introduction to Geography* by their critical comments and suggestions. Although we could not act on every helpful suggestion or adopt every useful observation, all were carefully and gratefully considered. In addition to those acknowledgments of assistance detailed in previous editions, we note the thoughtful advice recently provided by

Derek H. Alderman
*East Carolina University*

Dr. Jeffrey D. Allender
*University of Central Arkansas*

Daniel Block
*Chicago State University*

Jeffrey D. Bradley
*Northwest Missouri State University*

Dr. Bruce E. Davis
*Eastern Kentucky University*

Benjamin Y. Dixon
*State University of New York at Oneonta*

David Gordon
*Bronx Community College (CUNY)*

Anthony F. Grande
*Hunter College–CUNY*

Gregory Haddock
*Northwest Missouri State University*

John C. Jacobs
*Northwest Missouri State University*

Walter Jung
*University of Central Oklahoma*

JoAnne W. Kay
*Brigham Young University–Idaho*

Robert B. Kent
*University of Akron*

Jose Javier Lopez
*Minnesota State University–Mankato*

Chris Lukinbeal
*Southern Connecticut State University*

Paul J. O'Farrell
*Middle Tennessee State University*

Lallie F. Scott
*Northeastern State University*

We gratefully express appreciation to these and unnamed others for their help and contributions and specifically absolve them of responsibility for decisions on content and for any errors of fact or interpretation that users may detect.

We are also indebted to Cynthia A. Brewer of Pennsylvania State University for her research into color palettes accessible by colorblind readers. Many of the maps, graphs, and charts in this edition still reflect the cartographic and design skills of James A. Bier, our close collaborator for many previous editions of the book. We remain grateful for his past contributions.

Finally, we note with deep appreciation and admiration the efforts of the publisher's "book team," separately named on the copyright page, who collectively shepherded this revision to completion. We are grateful for their highly professional interest, guidance, and support.

**Arthur Getis**
**Judith Getis**
**Jerome D. Fellmann**

# Meet the Authors

## Arthur Getis

Arthur Getis received his B.S. and M.S. degrees from The Pennsylvania State University and his Ph.D. from the University of Washington. He is the co-author of several geography textbooks as well as two books dealing with map pattern analysis. He has also published widely in the areas of urban geography, spatial analysis, and geographic information systems. He is co-editor of *Journal of Geographical Systems* and for many years served on the editorial boards of *Geographical Analysis* and *Papers in Regional Science.* He has held administrative appointments at Rutgers University, the University of Illinois, and San Diego State University (SDSU), as well as the Birch Chair of Geographical Studies at SDSU. In 2002, he received the Association of American Geographers Distinguished Scholarship Award. Professor Getis is a member of many professional organizations and has served as an officer in, among others, the Western Regional Science Association and the University Consortium for Geographic Information Science.

## Judith Getis

Judith Getis earned her B.A. and a teaching credential from the University of Michigan and her M.A. from Michigan State University. She has co-authored several geography textbooks and wrote the environmental handbook *You Can Make a Difference.* In addition to numerous articles in the fields of urban geography and geography education, she has written technical reports on topics such as solar power and coal gasification. She and her husband, Arthur Getis, were among the original unit authors of the High School Geography Project, sponsored by the National Science Foundation and the Association of American Geographers. In addition, Mrs. Getis was employed by the Urban Studies Center at Rutgers University; taught at Rutgers; was a social science examiner at Educational Testing Service, Princeton, New Jersey; developed educational materials for Edcom Systems, Princeton, New Jersey; and was a professional associate in the Office of Energy Research, University of Illinois.

## Jerome D. Fellmann

Jerome D. Fellmann received his B.S., M.S., and Ph.D. degrees from the University of Chicago. Except for visiting professorships at Wayne State University, the University of British Columbia, and California State University/Northridge, his professional career has been spent at the University of Illinois at Urbana-Champaign. His teaching and research interests have been concentrated in the areas of human geography in general, and urban and economic geography in particular, in geographic bibliography, the geography of Russia and the CIS, and geographic education. His varied interests have been reflected in articles published in the *Annals of the Association of American Geographers, Professional Geographer, Journal of Geography,* the *Geographical Review,* and elsewhere. He is the co-author of McGraw-Hill's *Human Geography: Landscapes of Human Activity.* In addition to teaching and research, he has held administrative appointments at the University of Illinois and has served as a consultant to private corporations on matters of economic and community development.

# Introduction to Geography

# Introduction

**The Nature of Geography**

**What Is Geography?**

**Evolution of the Discipline**

**Some Core Geographic Concepts**

Location, Direction, and Distance

*Location*

*Direction*

*Distance*

Size and Scale

Physical and Cultural Attributes

Attributes of Place Are Always Changing

Interrelations between Places

Place Similarity and Regions

**Geography's Themes and Standards**

**Organization of This Book**

**Key Words**

**For Review & Consideration**

**Selected References**

**Terraced rice paddies in China have distinctively imprinted human use on a natural landscape formerly dominated by native forests.** © Keren Su/CORBIS.

During the long night of August 7–8, 1975, workers at China's Banqiao Dam struggled in waist-deep water to strengthen the structure. An unusually fierce typhoon, likely to happen only once every 2000 years, had battered the region with three days of nonstop rain. In all, a meter (3.2 ft) of rain fell on parts of central China. Water built up at an alarming rate in the reservoir behind the dam, but when engineers tried to open the dam's sluice gates to relieve the pressure, sediment buildup hindered them and the water continued to rise. With the water now more than 2 meters above the safe level for the dam, workers did what they could to prevent a disaster. It was to no avail. As the dam collapsed early on the morning of August 8, a worker cried out in fear and resignation, "The river dragon has come!"

Immediately, more than 500 million cubic meters of water, enough to fill the Empire State Building almost 500 times, rushed downstream at unsuspecting villages and towns. Six meters (20 ft) high and 12 kilometers (7.5 mi) wide, the wall of water obliterated whole communities in an instant.

In all, more than 60 dams in the region failed during the typhoon and its aftermath. An area three times the size of Rhode Island lay under water and as many as 250,000 people were dead. In the flooded areas, peasants faced the dual struggle of finding food and fighting off disease. Survivors stripped trees bare for food, and more than a million people were stricken with dysentery and other maladies.

## The Nature of Geography

News reports often refer to catastrophes such as this as "natural disasters," but the central China flood was not due to nature alone. Although the storm was certainly a natural event, the failure of the engineers to anticipate it shows that human decisions can affect the impact of nature's extremes. The social and economic actions of humans occur within the context of the environment and may have environmental and human consequences too serious to ignore.

The interaction of humans with the environment works both ways. People can, and often do, inflict irreparable damage on the environment, and the environment can exact a frightening toll from societies that inappropriately exploit it or even innocently occupy it (Figure 1.1).

Should a tragedy such as the Banqiao Dam collapse be called the result of the blind forces of nature, or should it be seen as the logical outcome of human miscalculation and inadequate engineering?

Geographers, as scientists, do not try to make value judgments about such questions. Geography does claim, however, to be a valid and revealing approach to understanding issues of political, economic, social, and ecological concern. Humans and the environment in interaction, the distribution of natural phenomena affecting human use of the earth, the cultural patterns of settlement and exploitation of the physical world—these are the themes of that encompassing discipline called *geography*.

## What Is Geography?

To many people, the word *geography* conjures up memories of learning the names and locations of capitals and rivers. Although geographers value basic knowledge of place names, geography as an academic discipline is much more. Literally, the word *geography* means "writing about the earth," just as *biography* means "writing about a life" and *photography* means "writing with light." But "writing about the earth" is a generic description that could be applied to almost any physical or social science.

*Geography* might better be defined as the study of spatial variation, of how and why things differ from place to place on the surface of the earth. It is, further, the study of how observable spatial patterns evolved through time. Why are the mountains in the eastern United States rounded and those in the West taller and more rugged? Why do you find a concentration of French speakers in Quebec but not in other parts of Canada? Why do Americans in the Northeast call a soft drink a "soda," whereas midwesterners refer to "pop"? If things were the same everywhere, if there were no spatial variation, the kind of curiosity that we call "geographic" would simply not exist. Without the certain conviction that in some interesting and important way landscapes, people, and opportunities differ from place to place, there would be no discipline of geography.

But we do not have to deal in such abstract terms. All of us consciously or subconsciously display geographic

(a)

(b)

**FIGURE 1.1** (a) **Hurricane Mitch,** the fourth-strongest Atlantic hurricane on record, struck Central America in late 1998 with winds in excess of 290 kilometers (180 miles) per hour. But rain rather than wind created devastation when Mitch lingered for days over Honduras, Guatemala, El Salvador, and Nicaragua. Floods and avalanches spawned by torrential downpours killed more than 10,000 people, erased whole villages, destroyed crops, and ruined much of the infrastructure of the countries affected. (b) Similar prolonged, devastating rain and floods during the early months of 2002 killed hundreds and forced thousands of rural and urban Indonesians to flee their homes, including these residents of Jakarta, Indonesia's capital, trying to cross a flooded street.
*(a) Laboratory for Atmospheres/NASA/GSFC. (b) © AP/Tatan Syuflana/Wide World Photos.*

awareness in our own daily lives. You are where you are, doing what you are doing, because of locational choices you have faced and spatial decisions you have made. You cannot be here reading this book and simultaneously be somewhere else—working, perhaps, or at the gym. And should you now want to go to work or take an exercise break, the time involved in going from here to there (wherever "there" is) is time not available for other activities in other locations. Of course, the act of going implies knowing where you are now, where "there" is in relation to "here," and the paths or routes you can take to cover the distance.

These are simple examples of the observation that "space matters" in a very personal way. You cannot avoid the implications of geography in your everyday affairs. Your understanding of your hometown, your neighborhood, or your college campus is essentially a geographic understanding. It is based on your awareness of where things are, of their spatial relationships, and of the varying content of the different areas and places you frequent. You carry out your routine activities in particular places and move on your daily rounds within defined geographic space, following logical paths of connection between different locations. At the same time, those activities and movements are necessarily affected by the physical environment—the terrain features, the weather conditions, and the like—in which they take place.

Just as geography matters in your personal life, so it matters on the larger stage as well. Decisions made by corporations about the locations of manufacturing plants or warehouses in relation to transportation routes and markets

**FIGURE 1.2** Ski development at Whistler Mountain, British Columbia, Canada, clearly shows the interaction of physical environment and human activity. Climate and terrain have made specialized human use attractive and possible. Human exploitation has placed a cultural landscape on the natural environment, thereby altering it. *© Karl Weatherly/Corbis Images.*

are spatially rooted. So, too, are those made by shopping center developers and locators of parks and grade schools. At an even grander scale, judgments about the projection of national power or the claim and recognition of "spheres of influence and interest" among rival countries are related to the implications of distance and area.

Geography, therefore, is about space and the content of space. We think of and respond to places from the standpoint not only of where they are but also, rather more important, of what they contain or what we think they contain. Reference to a place or an area usually calls up images about its physical nature or what people do there and often suggests, without conscious thought, how those physical things and activities are related. "Kansas," "farming," and "tornado" or "Colorado," "mountains," and "snowboarding" are simple examples. The content of an area, that is, has both physical and cultural aspects, and geography is always concerned with understanding both (Figure 1.2).

## Evolution of the Discipline

Geography's combination of interests was apparent even in the work of the early Greek geographers who first gave structure to the discipline. Geography's name was reputedly coined by the Greek scientist Eratosthenes over 2200 years ago from the words *geo,* "the earth," and *graphein,* "to write." From the beginning, that writing focused both on the physical structure of the earth and on the nature and activities of the people who inhabited the various lands of the known world. To Strabo (c. 64 B.C.–A.D. 20), the task of geography was to "describe the several parts of the inhabited world, . . . to write the assessment of the countries of the world [and] to treat the differences between countries." Even earlier, Herodotus (c. 484–425 B.C.) had found it necessary to devote much of his writing to the lands, peoples, economies, and customs of the various parts of the Persian Empire as necessary background to an understanding of the causes and course of the Persian wars.

Greek (and, later, Roman) geographers measured the earth, devised the global grid of parallels and meridians (marking latitudes and longitudes; see p. 7), and drew upon that grid surprisingly sophisticated maps of their known world (Figure 1.3). They explored the apparent latitudinal variations in climate and described in numerous works the familiar Mediterranean basin and the more remote, partly rumored lands of northern Europe, Asia, and equatorial Africa. Employing nearly modern concepts, they described river systems, explored cycles of erosion and patterns of deposition, cited the dangers of deforestation, described

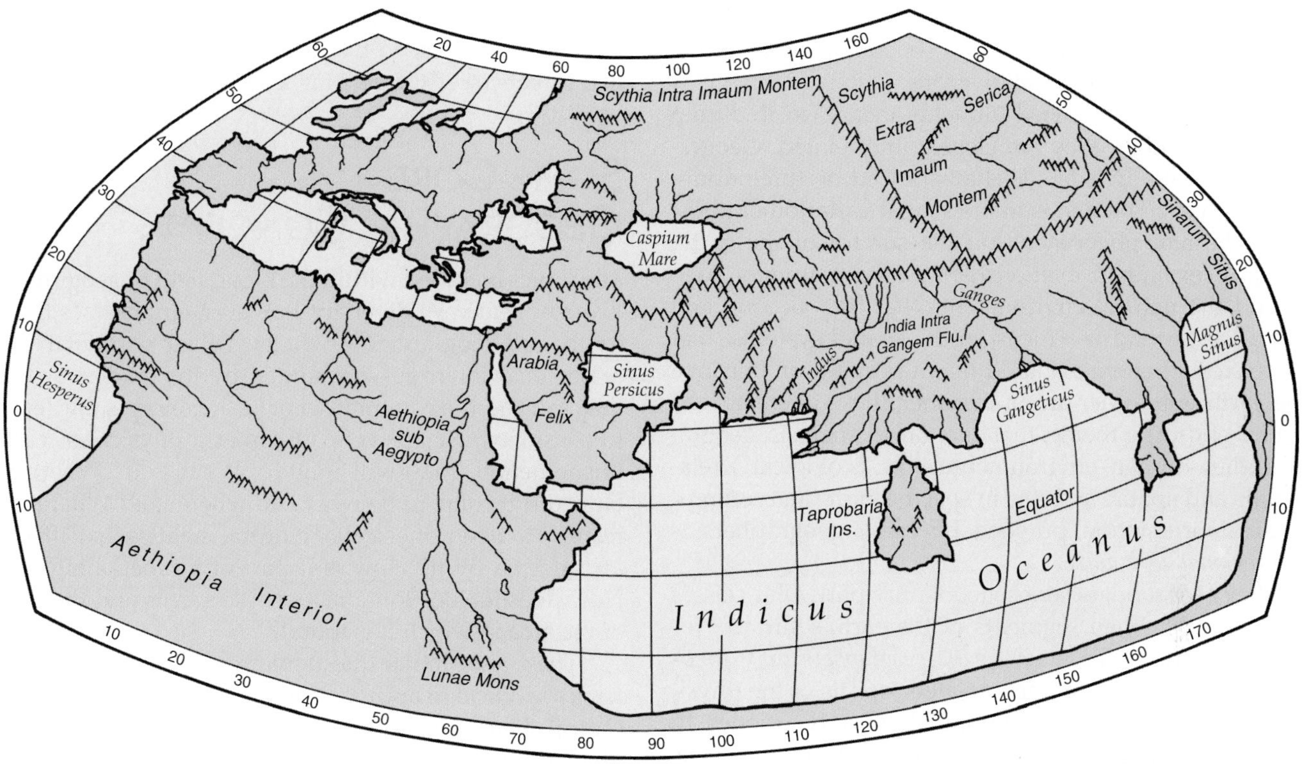


**FIGURE 1.3 World map of A.D. 2d-century Greco-Egyptian geographer-astronomer Ptolemy.** Ptolemy (Claudius Ptolemaeus) adopted a previously developed map grid of latitude and longitude based on the division of the circle into 360°, permitting a precise mathematical location for every recorded place. Unfortunately, errors of assumption and measurement rendered both the map and its accompanying six-volume gazetteer inaccurate. Ptolemy's map, accepted in Europe as authoritative for nearly 1500 years, was published in many variants in the 15th and 16th centuries. The version shown here summarizes the extent and content of the original. Its underestimation of the earth's size convinced Columbus a short westward voyage would carry him to Asia.

areal variations in the natural landscape, and noted the consequences of environmental abuse. Against that physical backdrop, they focused their attention on what humans did in home and distant areas—how they lived; what their distinctive similarities and differences were in language, religion, and custom; and how they used, altered, and perhaps destroyed the lands they inhabited. Strabo, indeed, cautioned against the assumption that the nature and actions of humans were determined by the physical environment they inhabited. He observed that humans were the active elements in a human–environmental partnership.

The interests guiding the early Greek and Roman geographers were and are enduring and universal. The ancient Chinese, for example, were as involved in geography as an explanatory viewpoint as were westerners, though there was no exchange between them. Further, as Christian Europe entered its Middle Ages between A.D. 800 and 1400 and lost its knowledge of Greek and Roman geographical work, Muslim scholars—who retained that knowledge—undertook to describe and analyze their known world in its physical, cultural, and regional variation.

Modern geography had its origins in the surge of scholarly inquiry that, beginning in the 17th century, gave rise to many of the traditional academic disciplines we know today. In its European rebirth, geography from the outset was recognized—as it always had been—as a broadly based integrative study. Patterns and processes of the physical landscape were early interests, as was concern with humans as part of the earth's variation from place to place. The rapid development of geology, botany, zoology, climatology, and other natural sciences by the end of the 18th century strengthened regional geographic investigation and increased scholarly and popular awareness of the intricate interconnections of things in space and between places. By that same time, accurate determination of latitude and longitude and scientific mapping of the earth had made assignment of place information more reliable and comprehensive.

During the 19th century, national censuses, trade statistics, and ethnographic studies gave firmer foundation to human geographic investigation. By the end of the 19th century, geography had become a distinctive and respected discipline in universities throughout Europe and in other regions of the world where European academic examples were followed. The proliferation of professional geographers and geography programs resulted in the development of a whole series of increasingly specialized disciplinary subdivisions, many represented by separate chapters of this book. Political geography, urban geography, and economic geography are examples of some of these subdivisions.

Geography's specialized subfields are not isolated from one another; rather, they are closely interrelated. Geography in all its subdivisions is characterized by three dominating interests. The first is in the spatial variation of physical and human phenomena on the surface of the earth; geography examines relationships between human societies and the natural environments that they occupy and modify. The second is a focus on the spatial systems[1] that link physical phenomena and human activities in one area of the earth with other areas. Together, these interests lead to a third enduring theme, that of regional analysis: geography studies human–environmental (or "ecological") relationships and spatial systems in specific locational settings. This areal orientation pursued by some geographers is called *regional geography.*

Other geographers choose to identify particular classes of things, rather than segments of the earth's surface, for specialized study. These *systematic geographers* may focus their attention on one or a few related aspects of the physical environment or of human populations and societies. In each case, the topic selected for study is examined in its interrelationships with other spatial systems and areal patterns. *Physical geography* directs its attention to the natural environmental side of the human–environmental structure. Its concerns are with landforms and their distribution, with atmospheric conditions and climatic patterns, with soils or vegetation associations, and the like. The other systematic branch of geography is *human geography.* Its emphasis is on people: where they are, what they are like, how they interact over space, and what kinds of landscapes of human use they erect on the natural landscapes they occupy.

A grasp of the broad yet integrated concerns and topics of geography is vital to an understanding of the important national and international problems that dominate daily news reports. Acid rain and the greenhouse effect, economic and social concerns of central cities, international trade imbalances, inadequate food supply and population growth in developing countries, turmoil in Africa and the Middle East—all of these problems occur in a geographic context, and geography helps explain them. To be geographically illiterate is to deny oneself not only the ability to comprehend local and world problems but also the opportunity to contribute meaningfully to the development of policies for dealing with them.

Also important, an understanding of the broad disciplinary subdivisions examined in this book may suggest the great diversity of job opportunities awaiting those who pursue college training in geography. Geographic training can help open the way to wonderfully rewarding and diversified careers (see "Careers in Geography," pp. 10–11).

[1]A "system" is simply a group of elements organized in a way that every element is, to some degree, directly or indirectly interdependent with every other element.

## Some Core Geographic Concepts

The topics included within the broad field of geography are diverse. That very diversity, however, emphasizes the reality that all geographers—whatever their particular topical or regional interests—are united by the similar questions they ask and the common set of basic concepts they employ to consider their answers. Of either a physical or cultural phenomenon, they will inquire: What is it? Where is it? How did it come to be what and where it is? Where is it in relation to other physical or cultural realities that affect it or are affected by it? How is it part of a functioning whole? How does its location affect people's lives and the content of the area in which it is found?

These and similar questions are spatial in focus and systems analytical in approach and are derived from enduring central themes in geography. In answering them, geographers draw upon a common store of concepts, terms, and methods of study that together form the basic structure and vocabulary of geography. Collectively, geographers believe that recognizing spatial patterns is the essential starting point for understanding how people live on and shape the earth's surface. That understanding is not just the task and interest of the professional geographer; it should be, as well, part of the mental framework of all informed persons.

Geographers use the word *spatial* as an essential modifier in framing their questions and forming their concepts. Geography, they say, is a *spatial science.* It is concerned with *spatial behavior* of people, with the *spatial relationships* that are observed between places on the earth's surface, and with the *spatial processes* that create or maintain those behaviors and relationships. The word *spatial* comes, of course, from *space,* and to geographers it always carries the idea of the way things are distributed, the way movements occur, and the way processes operate over the whole or a part of the surface of the earth. The geographer's space, then, is earth space, the surface area occupied or available to be occupied by humans. Spatial phenomena have locations on that surface, and spatial interactions occur among places, things, and people within the earth area available to them. The need to understand those relationships, interactions, and processes helps frame the questions that geographers ask.

Those questions have their starting point in basic observations about the location and nature of places and about how places are similar to or different from one another. Such observations, though simply stated, are profoundly important to our comprehension of the world we occupy.

- Places have location, direction, and distance with respect to other places.

- A place has size; it is large or small. Scale is important.
- A place has both physical structure and cultural content.
- The attributes of places develop and change over time.
- The elements of places interrelate with other places.
- Places may be generalized into regions of similarities and differences.

These are basic notions understandable to everyone. They also are the means by which geographers express fundamental observations about the earth spaces they examine and put those observations into a common framework of reference. Each of the concepts is worth further discussion, for they are not quite as simple as they seem.

## Location, Direction, and Distance

*Location, direction,* and *distance* are everyday ways of assessing the space around us and identifying our position in relation to other things and places of interest. They are also essential in understanding the processes of spatial interaction that are so important in the study of both physical and human geography.

### Location

The location of places and things is the starting point of all geographic study, as well as of our personal movements and spatial actions in everyday life. We think of and refer to location in at least two different senses, *absolute* and *relative.*

**Absolute location** is the identification of place by a precise and accepted system of coordinates; it therefore is sometimes called *mathematical location.* We have several such accepted systems of pinpointing positions. One of them is the global grid of parallels and meridians—that is, latitude and longitude (discussed in Chapter 2, pages 27–28). With it, the absolute location of any point on the earth can be accurately described by reference to its degrees, minutes, and seconds of *latitude* and *longitude.*

Other coordinate systems are also in use. Survey systems such as the township, range, and section description of property in much of the United States give mathematical locations on a regional level, and street address precisely defines a building according to the reference system of an individual town. Absolute location is unique to each described place, is independent of any other characteristic or observation about that place, and has obvious value in the legal description of places, in measuring the distance separating places, or in finding directions between places on the earth's surface.

When geographers—or real estate agents—remark that "location matters," however, their reference is usually not to absolute but to **relative location**—the position of a place or thing in relation to that of other places or things (Figure 1.4). Relative location expresses spatial interconnection and interdependence and may carry social (neighborhood character) and economic (assessed valuations of vacant land) implications. On an immediate and personal level, we think of the location of the school library not in terms of its street address or room number but where it is relative to our classrooms, the cafeteria, or another reference point. On the larger scene, relative location tells us that people, things, and places exist not in a spatial vacuum but in a world of physical and cultural characteristics that differ from place to place.

New York City, for example, may in absolute terms be described as located at (approximately) latitude 40°43′ N (read as 40 degrees, 43 minutes north) and longitude 73°58′ W. We have a better understanding of the *meaning* of its location, however, when reference is made to its spatial relationships: to the continental interior through the Hudson–Mohawk lowland corridor or to its position on the eastern seaboard of the United States. Within the city, we gain understanding of the locational significance of Central Park or the Lower East Side not solely by reference to the street addresses or city blocks they occupy but also by their spatial and functional relationships to the total land use, activity, and population patterns of New York City.

In view of these different ways of looking at location, geographers make a distinction between the *site* and the *situation* of a place. **Site,** an absolute location concept, refers to the physical and cultural characteristics and attributes of the place itself. It is more than mathematical location, for it tells us something about the internal features of that place. **Situation,** on the other hand, refers to the external relations of the locale. It is an expression of relative location with

**FIGURE 1.4** The reality of *relative location* on the globe may be strikingly different from the impressions we form from flat maps. The position of Russia with respect to North America when observed from a polar perspective emphasizes that relative location properly viewed is important to our understanding of spatial relationships and interactions between the two world areas.

## Careers in Geography

Geography admirably serves the objectives of a liberal education. It can make us better informed citizens, more able to understand the important issues facing our communities, our country, and our world and better prepared to contribute solutions.

Can it, as well, be a pathway to employment for those who wish to specialize in the discipline? The answer is "yes," in a number of different types of jobs. One broad cluster is concerned with supporting the field itself through teaching and research. Teaching opportunities exist at all levels, from elementary to university postgraduate. Teachers with some training in geography are in increasing demand in elementary and high schools in the United States, reflecting geography's inclusion as a core subject in the federally adopted *Educate America Act* (Public Law 103-227) and the national determination to create a geographically literate society (see "The National Standards," p. 19). At the college level, specialized teaching and research in all branches of geography have long been established, and geographically trained scholars are prominently associated with urban, community, and environmental studies; regional science; locational economics; and other interdisciplinary programs.

Because of the breadth and diversity of the field, training in geography involves the acquisition of techniques and approaches applicable to a wide variety of jobs outside the academic world. Modern geography is both a physical and social science and fosters a wealth of technical skills. The employment possibilities it presents are as many and varied as are the agencies and enterprises dealing with the natural environment and human activities and with the acquisition and analysis of spatial data.

Many professional geographers work in government, either at the state or local level or in a variety of federal agencies and international organizations. Although many positions do not carry a geography title, physical geographers serve as water and other natural resource analysts, weather and climate experts, soil scientists, and the like. An area of recent high demand is for environmental managers and technicians. Geographers who have specialized in environmental studies find jobs in both public and private agencies. Their work may include assessing the environmental impact of proposed development projects on such things as air and water quality and endangered species, as well as preparing the environmental impact statements required before construction can begin.

Human geographers work in many different roles in the public sector. Jobs include data acquisition and analysis in health care, transportation, population studies, economic development, and international economics. Many geography graduates find positions as planners in local and state governmental agencies concerned with housing and community development, park and recreation planning, and urban and regional planning. They map and analyze land use plans and transportation systems, monitor urban land development, make informed recommendations about the location of public facilities, and engage in basic social science research.

Most of the same specializations are found in the private sector. Geographic training is ideal for such tasks as business planning and market analysis; factory, store, and shopping

particular reference to items of significance to the place in question. Site and situation in the city context are further examined in Chapter 11.

### Direction

Direction is the second universal spatial concept. Like location, it has more than one meaning and can be expressed in absolute or relative terms. **Absolute direction** is based on the cardinal points of north, south, east, and west. These appear uniformly and independently in all cultures, derived from the obvious "givens" of nature: the rising and setting of the sun for east and west, the sky location of the noontime sun and of certain fixed stars for north and south.

We also commonly use **relative,** or *relational,* **directions.** In the United States, we go "out West," "back East," or "down South"; we worry about conflict in the "Near East" or economic competition from the "Far Eastern countries." Despite their reference to cardinal compass points, these directional references are culturally based and locationally variable. The Near East and the Far East locate parts of Asia from the European perspective; they are retained in the Americas by custom and usage, even though one would normally travel westward across the Pacific, for example, to reach the "Far East" from California, British Columbia, or Chile. For many Americans, "back East" and "out West" are reflections of the migration paths of earlier generations for whom home was in the eastern part of the country, to which they might look back. "Up North" and "down South" reflect our accepted custom of putting north at the top and south at the bottom of our maps.

### Distance

*Distance* joins *location* and *direction* as a commonly understood term that has dual meanings for geographers. Like its two companion spatial concepts, distance may be viewed in both an absolute and a relative sense.

center site selection; and community and economic development programs for banks, public utilities, and railroads. Publishers of maps, atlases, news and travel magazines, and the like employ geographers as writers, editors, and mapmakers.

The combination of a traditional, broad-based liberal arts perspective with the technical skills required in geographic research and analysis gives geography graduates a competitive edge in the labor market. These field-based skills include familiarity with geographic information systems (GIS, explained in Chapter 2), cartography and computer mapping, remote sensing and photogrammetry, and competence in data analysis and problem solving. In particular, students with expertise in GIS, who are knowledgable about data sources, hardware, and software, are finding they have ready access to employment opportunities. The following table, based on the booklet "Careers in Geography,"[a] summarizes some of the professional opportunities open to students who have specialized in one (or more) of the various subfields of geography. Also, be sure to read the discussion of geography careers on the homepage of the Association of American Geographers at www.aag.org/Careers/Intro.html. Additional links on the topic of geography careers can be found in the Online Learning Center for this text. The link may be found at the end of this chapter.

[a]"Careers in Geography," by Richard G. Boehm. Washington, D.C.: National Geographic Society, 1996. Previously published by Peterson's Guides, Inc.

| Geographic Field of Concentration | Employment Opportunities |
|---|---|
| Cartography and geographic information systems | Cartographer for federal government (agencies such as Defense Mapping Agency, U.S. Geological Survey, or Environmental Protection Agency) or private sector (e.g., Environmental Systems Research Institute, ERDAS, Intergraph, or Bentley); map librarian; GIS specialist for planners, land developers, real estate agencies, utility companies, local government; remote-sensing analyst; surveyor |
| Physical geography | Weather forecaster; outdoor guide; coastal zone manager; hydrologist; soil conservation/agricultural extension agent |
| Environmental studies | Environmental manager; forestry technician; park ranger; hazardous waste planner |
| Cultural geography | Community developer; Peace Corps volunteer; health care analyst |
| Economic geography | Site selection analyst for business and industry; market researcher; traffic/route delivery manager; real estate agent/broker/appraiser; economic development researcher |
| Urban and regional planning | Urban and community planner; transportation planner; housing, park, and recreation planner; health services planner |
| Regional geography | Area specialist for federal government; international business representative; travel agent; travel writer |
| Geographic education<br>General geography | Elementary/secondary school teacher; college professor; overseas teacher |

**Absolute distance** refers to the spatial separation between two points on the earth's surface measured by an accepted standard unit, such as miles or kilometers for widely separated locales, feet or meters for more closely spaced points. **Relative distance** transforms those linear measurements into other units more meaningful for the space relationship in question.

To know that two competing malls are about equidistant in miles from your residence is perhaps less important in planning your shopping trip than is knowing that, because of street conditions or traffic congestion, one is 5 minutes and the other 15 minutes away (Figure 1.5). Most people, in fact, think of time distance rather than linear distance in their daily activities; downtown is 20 minutes by bus, the library is a 5-minute walk. In some instances, money rather than time is the distance transformation. An urban destination might be estimated to be a $10 cab ride away, information that may affect either the decision to make the trip at all or the choice of travel mode to get there. As a college student, you already know that rooms and apartments are less expensive at a greater distance from campus.

A *psychological* transformation of linear distance is also frequent. A solitary late-night walk back to the car through an unfamiliar or dangerous neighborhood seems far longer than a daytime stroll of the same distance through familiar and friendly territory. A first-time trip to a new destination frequently seems much longer than the return trip over the same path. Nonlinear distance and spatial interaction are further considered in Chapter 8.

## Size and Scale

When we say that a place may be large or small, we speak both of the nature of the place itself and of the generalizations that can be made about it. In either instance, geographers are concerned with **scale,** though we may use that

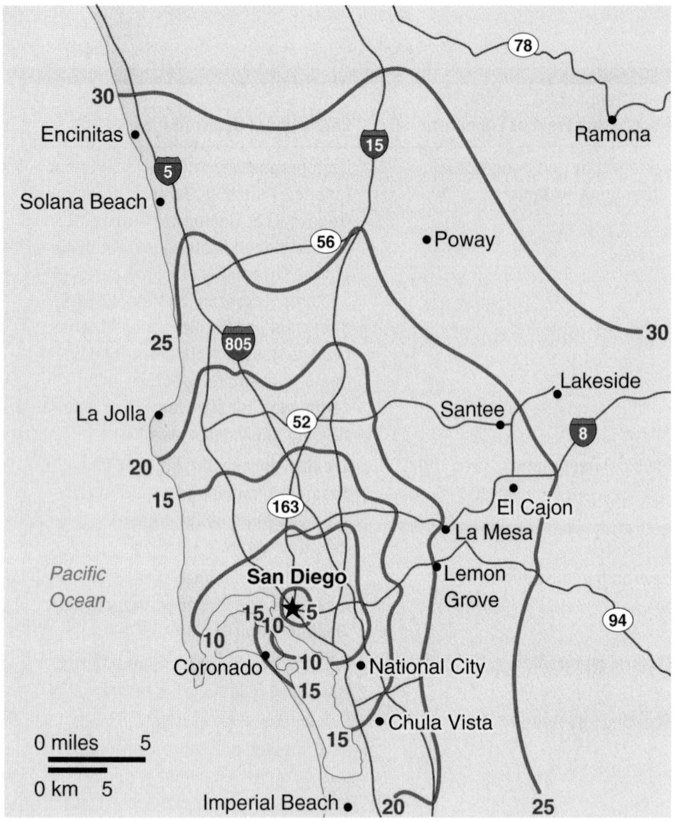


**FIGURE 1.5** Lines of equal travel time (*isochrones:* from Greek, *isos,* equal, and *chronos,* time) mark off the different linear distances accessible within given spans of time from a starting point. The fingerlike outlines of isochrone boundaries reflect variations in road conditions, terrain, traffic congestion, and other aids or impediments to movement. On this map, the areas within 30 minutes' travel time from downtown San Diego are recorded for the year 2002. Note the effect of freeways on travel time.

term in different ways. We can, for example, study a problem such as population or landforms at the local scale or on a global scale. Here, the reference is purely to the size of unit studied. More technically, scale tells us the relationship between the size of an area on a map and the actual size of the mapped area on the surface of the earth. In this sense, as Chapter 2 makes clear, scale is a feature of every map and is essential to recognizing what is shown on that map.

In both senses of the word, *scale* implies the degree of generalization represented (Figure 1.6). Geographic inquiry may be broad or narrow; it occurs at many different size-scales. Climate may be an object of study, but research and generalization focused on climates of the world will differ in degree and kind from study of the microclimates of a city. Awareness of scale is very important. In geographic work, concepts, relationships, and understandings that have meaning at one scale may not be applicable at another.

For example, the study of world agricultural patterns may refer to global climatic regimes, cultural food preferences, levels of economic development, and patterns of world trade. These large-scale relationships are of little concern in the study of crop patterns within single counties of the United States, where topography, soil and drainage conditions, farm size, ownership, and capitalization, or even personal management preferences, may be of greater explanatory significance.

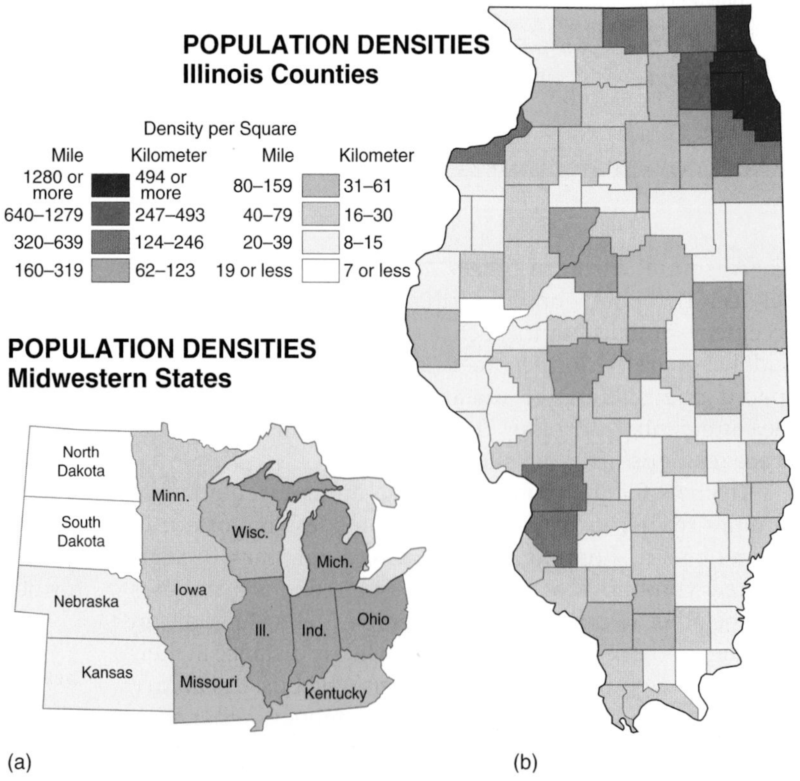


**FIGURE 1.6 Population density and map scale.** "Truth" depends on one's scale of inquiry. Map (a) reveals that the maximum year 2000 population density of Midwestern states was no more than 123 people per square kilometer (319 per sq mi). From map (b), however, we see that population densities in three Illinois counties exceeded 494 people per square kilometer (1280 per sq mi) in 2000. If we were to reduce our scale of inquiry even further, examining individual city blocks in Chicago, we would find densities reaching 2500 or more people per square kilometer (10,000 per sq mi). Scale matters!

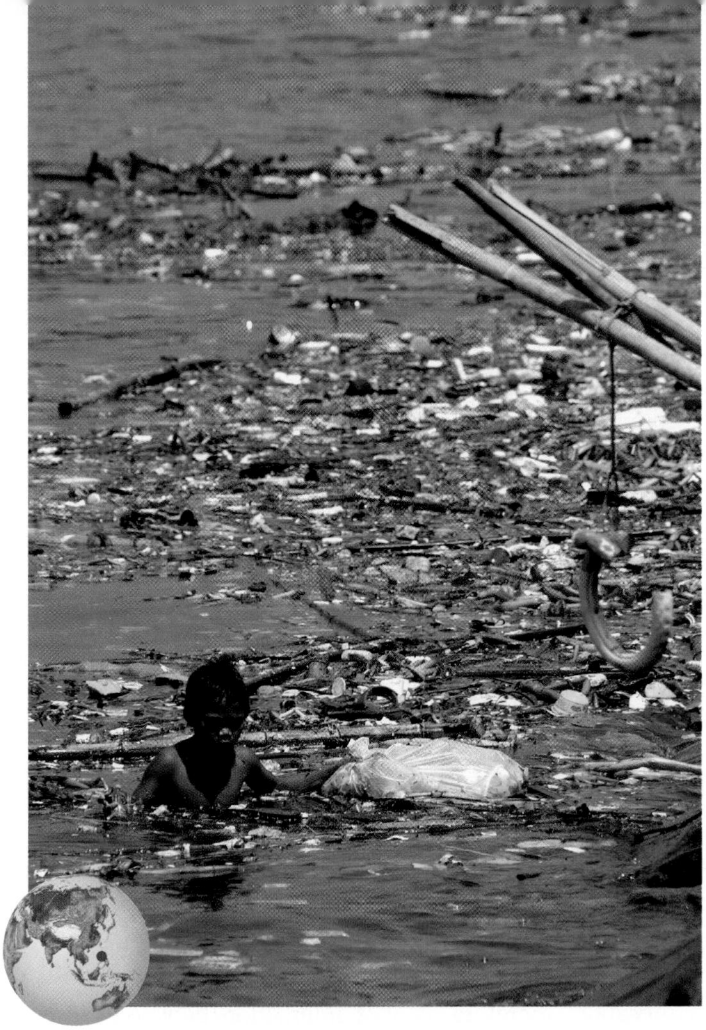

**FIGURE 1.7** Sites (and sights) such as this image from Manila Bay in the Philippines are all-too-frequent reminders of the adverse environmental impacts of humans and their waste products. Many of those impacts are more subtle, hidden in the form of soil erosion, water pollution, increased stream sedimentation, plant and animal extinction, deforestation, and the like. *© David Greedy/Getty Images.*

## Physical and Cultural Attributes

All places have individual physical and cultural attributes distinguishing them from other places and giving them character, potential, and meaning. Geographers are concerned with identifying and analyzing the details of those attributes and, particularly, with recognizing the interrelationship between the physical and cultural components of area: the human–environmental interface.

The physical characteristics of a place are such natural aspects as its climate, soil, water supplies, mineral resources, terrain features, and the like. These **natural landscape** attributes provide the setting within which human action occurs. They help shape—but do not dictate—how people live. The resource base, for example, is physically determined, though how resources are perceived and utilized is culturally conditioned.

Environmental circumstances directly affect agricultural potential and reliability; indirectly, they may affect such matters as employment patterns, trade flows, population distributions, and national diets. The physical environment simultaneously presents advantages and disadvantages with which humans must deal. Thus, the danger of typhoons in central China must be balanced against the agricultural bounty derived from the region's favorable terrain, soil, and moisture conditions. Physical environmental patterns and processes are explored in Chapters 3 and 4 of this book.

At the same time, by occupying a given place, people modify its environmental conditions. The existence of the United States Environmental Protection Agency (and its counterparts elsewhere) is a reminder that humans are the active and frequently harmful agents in the continuing interplay between the cultural and physical worlds (Figure 1.7). Virtually every human activity leaves its imprint on the earth's soil, water, vegetation, animal life, and other resources, as well as on the atmosphere common to all earth space, as Chapter 12 makes clear.

The visible imprint of that human activity is called the **cultural landscape.** It, too, exists at different scales and at different levels of visibility. Contrasts in agricultural practices and land use between Mexico and southern California are evident in Figure 1.8, whereas the signs, structures, and people of Los Angeles's Chinatown leave a smaller, more

**FIGURE 1.8** This NASA image reveals contrasting cultural landscapes along the Mexico-California border. Move your eyes from the Salton Sea (the dark patch at the top of the image) southward to the agricultural land extending to the edge of the picture. Notice how the regularity of the fields and bright colors (representing growing vegetation) give way to a marked break, where irregularly shaped fields and less prosperous agriculture are evident. Above the break is the Imperial Valley of California; below the border is Mexico. *© NASA.*

confined imprint within the larger cultural landscape of the metropolitan area itself.

The physical and human characteristics of places are the keys to understanding both the simple and the complex interactions and interconnections between people and the environments they occupy and modify. Those interconnections and modifications are not static or permanent but are subject to continual change.

## Attributes of Place Are Always Changing

The physical environment surrounding us seems eternal and unchanging but, of course, it is not. In the framework of geologic time, change is both continuous and pronounced. Islands form and disappear; mountains rise and are worn low to swampy plains; vast continental glaciers form, move, and melt away, and sea levels fall and rise in response. Geologic time is long, but the forces that give shape to the land are timeless and relentless.

Even within the short period of time since the most recent retreat of continental glaciers—12,000 or 13,000 years ago—the environments occupied by humans have been subject to change. Glacial retreat itself marked a period of climatic alteration, extending the area habitable by humans to include vast reaches of northern Eurasia and North America formerly covered by thousands of feet of ice. With moderating climatic conditions came changes in vegetation and fauna. On the global scale, these were natural environmental changes; humans were as yet too few in number and too limited in technology to alter materially the course of physical events. On the regional scale, however, even early human societies exerted an impact on the environments they occupied. Fire was used to clear forest undergrowth, to maintain or extend grassland for grazing animals and to drive them in the hunt, and later to clear openings for rudimentary agriculture.

With the dawn of civilizations and the invention and spread of agricultural technologies, humans accelerated their management and alteration of the now no longer "natural" environment. Even the classical Greeks noted how the landscape they occupied differed—for the worse—from its former condition. With growing numbers of people, and particularly with industrialization and the spread of European exploitative technologies throughout the world, the pace of change in the content of area accelerated. The built landscape—the product of human effort—increasingly replaced the natural landscape. Each new settlement or city; each agricultural assault on forests; each new mine, dam, or factory changed the content of regions and altered the temporarily established spatial interconnections between humans and the environment.

Characteristics of places today are the result of constantly changing past conditions. They are the forerunners of differing human-environmental balances yet to be struck. Geographers are concerned with places at given moments of time. But to understand fully the nature and development of places, to appreciate the significance of their relative locations, and to understand the interplay of their physical and cultural characteristics, geographers must view places as the present result of past operation of distinctive physical and cultural processes (Figure 1.9).

You will recall that one of the questions geographers ask about a place or thing is "How did it come to be what and where it is?" This is an inquiry about process and about becoming. The forces and events shaping the physical and explaining the cultural environment of places today are an important focus of geography and are the topics of most of the chapters of this book. To understand them is to appreciate the changing nature of the spatial order of our contemporary world.

## Interrelations between Places

The concepts of relative location and distance that were introduced earlier lead directly to another fundamental spatial reality: places are interrelated with other places in structured and comprehensible ways. In describing the processes and patterns of that **spatial interaction,** geographers add *accessibility* and *connectivity* to the ideas of location and distance.

Tobler's First Law of Geography tells us that, in a spatial sense, everything is related to everything else but relationships are stronger when things are near one another. Our observation, therefore, is that interaction between places diminishes in intensity and frequency as distance between them increases—a statement of the idea of "distance decay," which we explore in Chapter 8. Think about it—are you more likely to go to a fast-food outlet next door or to a nearly identical restaurant across town? Human decision making is strange in many ways, and decisions are sometimes made for unpredictable reasons, but in this case, you can see how the nearer place will probably be frequented more often.

Consideration of distance implies assessment of **accessibility.** How easy or difficult is it to overcome the "friction of distance"? That is, how easy or difficult is it to surmount the barrier of the time and space separation of places? Distance isolated North America from Europe until the development of ships (and aircraft) that reduced the effective distance between the continents. All parts of ancient and medieval cities were accessible by walking; they were "pedestrian cities," a status lost as cities expanded in area and population with industrialization. Accessibility between city districts could be maintained only by the development of public transit systems whose fixed lines of travel increased ease of movement between connected points and reduced it between areas not on the transit lines themselves.

Accessibility, therefore, suggests the idea of **connectivity,** a broader concept implying all the tangible and intangible ways in which places are connected: by physical telephone lines, street and road systems, pipelines and sewers; by unrestrained walking across open countryside; by radio and TV

(a) (b)

**FIGURE 1.9 The process of change in a cultural landscape.** (a) Before the development of the freeway in the 1970s, this portion of suburban Long Island, New York, was largely devoted to agriculture. (b) The construction of the freeway and cloverleaf interchange ramps altered nearby land uses to replace farming with housing developments and new commercial and light industrial activities. Source: *U.S. Geological Survey.*

broadcasts beamed outward uniformly from a central source; and in nature even by movements of wind systems and flows of ocean currents. Where routes are fixed and flow is channelized, *networks*—the patterns of routes connecting sets of places—determine the efficiency of movement and the connectedness of points (Figure 1.10). Demand for universal instantaneous connectivity is common and unquestioned in today's advanced societies. Technologies and devices to achieve it proliferate, as our own lifestyles show. Cell phones, e-mail, broadband wireless Internet, instant messaging, and more have erased time and distance barriers formerly separating and isolating individuals and groups and have reduced our dependence on physical movement and on networks fixed in the landscape.

There is, inevitably, interchange between connected places. **Spatial diffusion** is the process of dispersion of an idea or a thing (a new consumer product or a new song, for example) from a center of origin to more distant points. The rate and extent of that diffusion are affected, again, by the distance separating the origin of the new idea or technology and other places where it is eventually adopted. Diffusion rates are also affected by such factors as population densities, means of communication, obvious advantages of the innovation, and importance or prestige of the originating node. Further discussion of spatial diffusion is found in Chapter 8.

Geographers study the dynamics of spatial relationships. Movement, connection, and interaction are part of the social and economic processes that give character to places and regions. Geography's study of those relationships recognizes that spatial interaction is not just an awkward necessity but a fundamental organizing principle of the physical and social environment. That recognition has become universal, repeatedly expressed in the term *globalization*. **Globalization** implies the increasing interconnection of all societies in all parts of the world as the full range of social, cultural, political, economic, and environmental processes becomes international in scale and effect. Promoted by continuing advances in worldwide accessibility and connectivity, globalization encompasses other core geographic concepts of spatial interaction, accessibility, connectivity, and diffusion. More detailed implications of globalization will be touched on in later chapters of Parts 2 and 3 of this text.

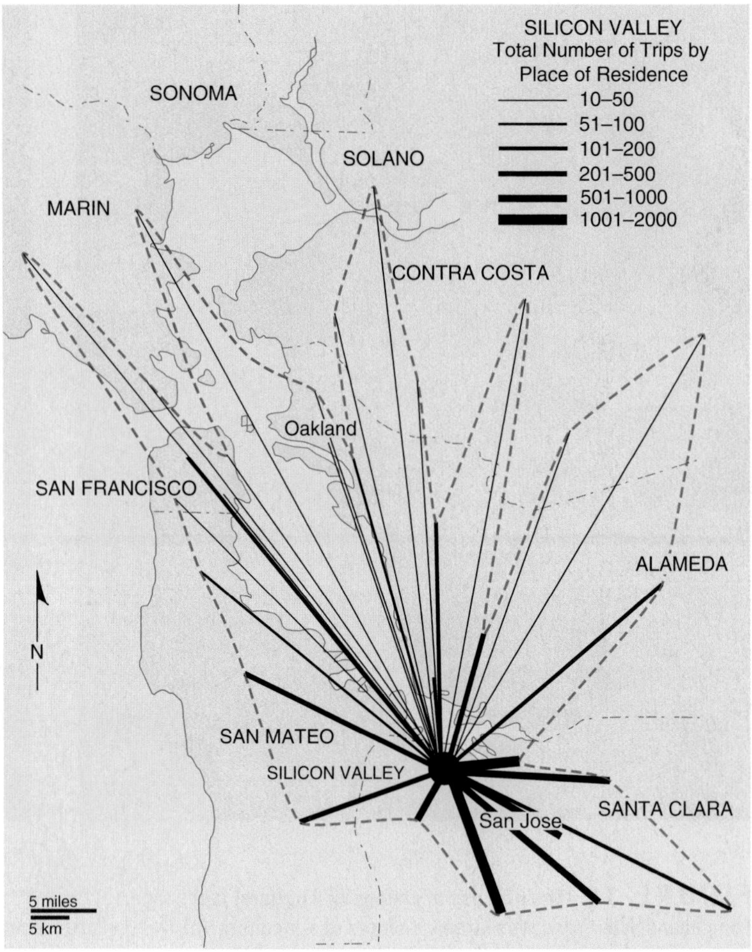


**FIGURE 1.10** An indication of one form of spatial interaction and *connectivity* is suggested by this "desire line" map recording the volume of daily work trips within the San Francisco Bay area to the Silicon Valley employment node. The ends of the desire lines, connected, define the outer reaches of a physical interaction region defined by the network of connecting roads and routes. The region changed in size and shape over time as the network was enlarged and improved, the Valley employment base expanded, and the commuting range of workers increased. The map, of course, gives no indication of the global reach of the Valley's *accessibility* and interaction through other means of communication and interchange.
*Redrawn with permission from Robert Cervero,* Suburban Gridlock. *© 1986 Center for Urban Policy Research, Rutgers, the State University of New Jersey.*

## Place Similarity and Regions

The distinctive characteristics of places—physical, cultural, locational—immediately suggest two geographically important ideas. The first is that no two places on the surface of the earth can be *exactly* the same. Not only do they have different absolute locations, but—as in the features of the human face—the precise mix of physical and cultural characteristics of place is never exactly duplicated. Because geography is a spatial science, the inevitable uniqueness of place would seem to impose impossible problems of generalizing spatial information.

That this is not the case results from the second important idea, that the natural and cultural characteristics of places show patterns of similarity in some areas. For example, a geographer doing fieldwork in France may find that all farmers in one area use a similar, specialized technique to build fences around their fields. Often, such similarities are striking enough for us to conclude that spatial regularities exist. They permit us to recognize and define **regions,** earth areas that display significant elements of internal uniformity and external differences from surrounding territories. Places are, therefore, both unlike and like other places, creating patterns of areal differences and coherent spatial similarity.

The problems of the historian and the geographer are similar. Each must generalize about items of study that are essentially unique. The historian creates arbitrary but meaningful and useful historical periods for reference and study. The "Roaring Twenties" and the "Victorian Era" are shorthand summary names for specific time spans, internally quite complex and varied but significantly distinct from what went before or followed after. The region is the geographer's equivalent of the historian's era: a device to classify the complex reality of the earth's surface into manageable pieces. Just as historians focus on key events to characterize certain historical periods, geographers focus on key unifying elements or similarities to determine the boundaries of regions. By identifying and naming regions, a complex set of interrelated environmental or cultural attributes can easily be conveyed through a simpler construct.

Regions are not "given" in nature any more than "eras" are given in the course of human events. Regions are devised; they are spatial summaries designed to bring order to the infinite diversity of the earth's surface. At their root, they are based on the recognition and mapping of *spatial distributions*—the territorial occurrence of environmental, human, or organizational features selected for study. For example, the location of Welsh speakers in Great Britain is a distribution that can be identified and mapped. As many spatial distributions exist as there are imaginable physical, cultural, or connectivity elements of area to examine (Figure 1.11). Those that are selected for study, however, are those that contribute to the understanding of a specific topic or problem.

Although there are as many individual regions as the objectives of spatial study and understanding demand, two generalized *types* of regions are recognized. A **formal region** is one of essential uniformity in one or a limited number of related physical or cultural features (Figure 1.12). Your home state is a formal political region defined as an area with a common government and laws, whereas the Bible Belt suggests a region based on religious characteristics. The Mississippi Delta and the Ozarks are formal natural regions based on physical features.

A **functional region,** in contrast, may be visualized as a spatial system. Its parts are interdependent, and throughout its extent the functional region operates as a dynamic organizational unit. A functional region has unity not in the sense of static content but in the manner of its operational connectivity. It has a node, or core area, surrounded by the total region defined by the type of control exerted. As the degree and extent of areal control and interaction change, the boundaries of the functional region change in response. Trade areas of towns, national "spheres of influence," and the territories subordinate to the financial, administrative, wholesaling, or retailing centrality exercised by such regional capitals as Chicago, Atlanta, or Minneapolis are cases in point (Figure 1.13).

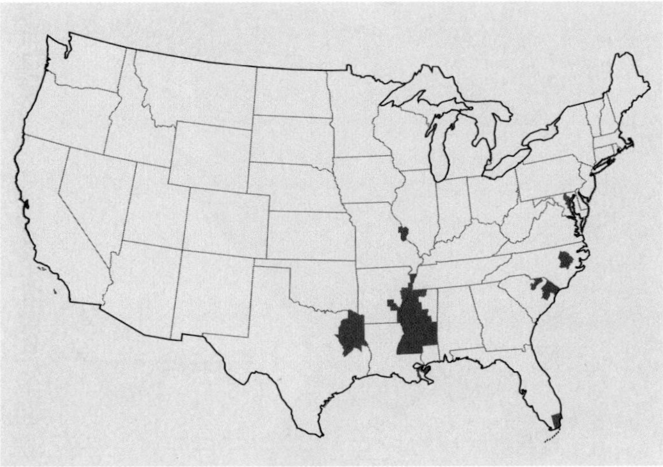

**FIGURE 1.11 Above average bone and joint cancer mortality rates for black females.** All spatial data may be mapped. As this example of the distribution of bone cancer mortality demonstrates, mapped distributions frequently reveal regional patterns that invite analysis. Areas emphasized had above average death rates for the years 1970–1994. The question is, why? Source: *Susan S. Devesa, et al.,* Atlas of Cancer Mortality in the United States 1950–1994. *NIH Publication #99–4564, September 1999. National Institutes of Health, Bethesda, Maryland.*

Of course, all areas are parts of overlapping functional and formal regions at the same time. Consider the New York metropolitan area. On one hand, you can define New York City as that area in which people root for the Yankees or Mets. This would be a formal region defined by a common cultural sports preference. As you moved up the Connecticut coast away from the city, you would find more and more Boston Red Sox fans until eventually you would consider yourself in a different region. But you can also define the New York metro area as a functional region based on the daily commuter rail network. From central New Jersey to the Hudson Valley of New York to southeastern Connecticut, people use a network of trains to commute to the city to work, a classic functional network.

As you read the chapters of this book, notice how many different examples of regions and regionalism are presented in map form and discussed in the text. Note, too, how those depictions and discussions vary between, primarily, formal and functional regions as the subjects and purposes of the examples change. Chapter 13 contains additional special regional studies of both types, illuminating specific topics that are the subjects of Chapters 3 to 12.

# GEOGRAPHY'S THEMES AND STANDARDS

The core geographic concepts discussed so far in this chapter reflect—and are statements of—both the "fundamental themes in geography" and the "National Geography Standards." Together, the "themes" and "standards" have helped organize and structure the study of geography over the past several years at all grade and college levels. Both focus on the development of geographic literacy. The former represent an instructional approach keyed to identification and instruction in the knowledge, skills, and perspectives students should gain from a structured program in geographic education. The latter—"standards"—codify the essential subject matter, skills, and perspectives of geography essential to the mental equipment of all educated adults.

The *five fundamental themes* as summarized by a joint committee of the National Council for Geographic Education and the Association of American Geographers are those basic concepts and topics that recur in all geographic inquiry and at all levels of instruction. They are:

- Location: the meaning of relative and absolute position on the earth's surface;
- Place: the distinctive and distinguishing physical and human characteristics of locales;
- Relationships within places: the development and consequences of human-environmental relationships;
- Movement: patterns and change in human spatial interaction on the earth;
- Regions: how they form and change.

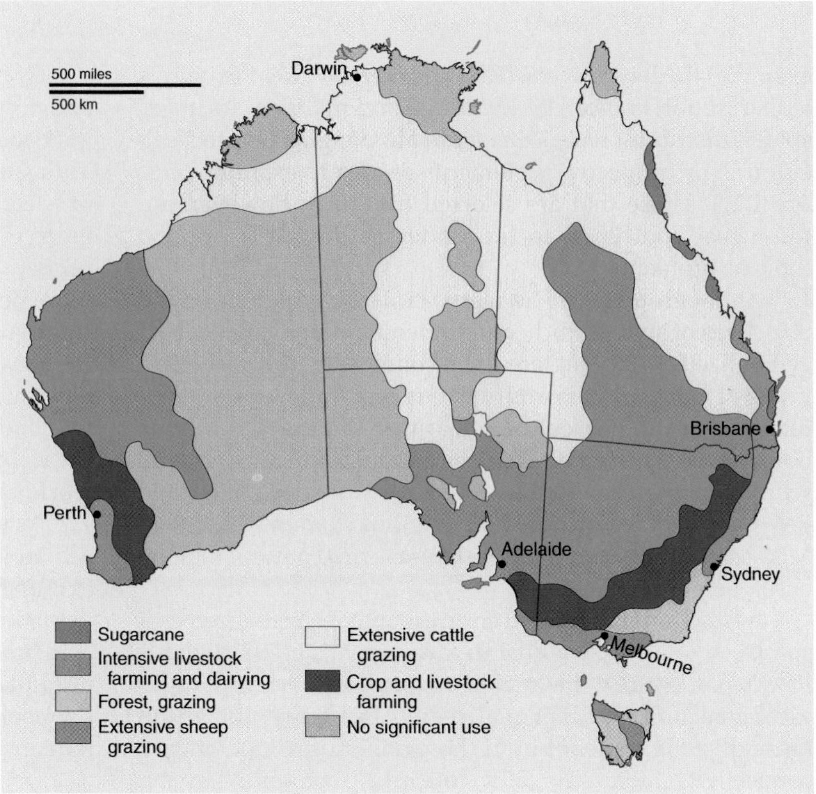


**FIGURE 1.12** This generalized land use map of Australia is composed of *formal regions* whose internal economic characteristics show essential uniformities setting them off from adjacent territories of different condition or use.

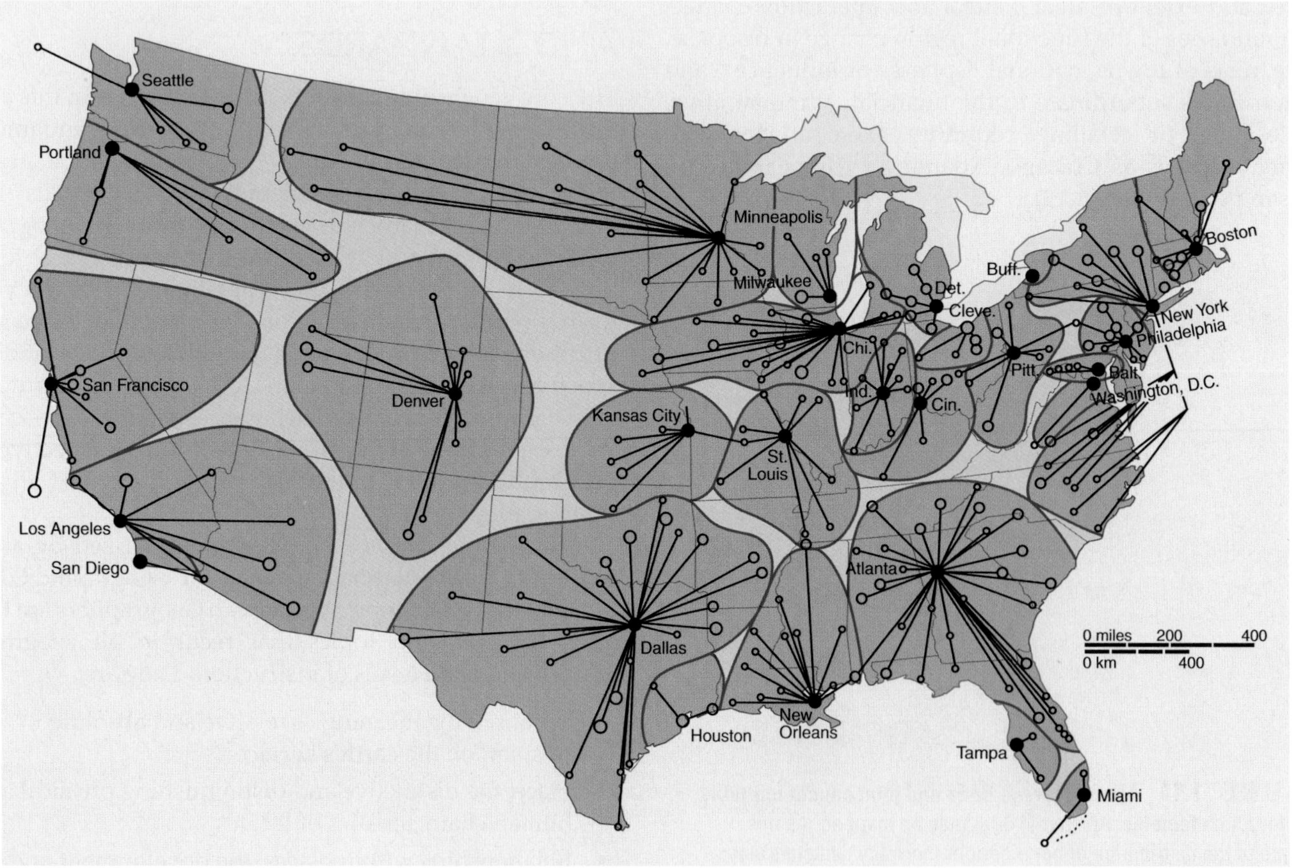


**FIGURE 1.13** The *functional regions* shown on this map were based on linkages between large banks of major central cities and the "correspondent" banks they formerly served in smaller towns. Although the rise of nationwide banks has reduced their role, the regions once defined an important form of *connectivity* between principal cities and locales beyond their own immediate metropolitan area. *Redrawn by permission from* Annals of the Association of American Geographers, *John R. Borchert, Vol. 62, p. 358, Association of American Geographers, 1972.*

## The National Standards

The inclusion of geography in the *Goals 2000* national education program reflects the conviction that a grasp of the skills and understandings of geography is essential in an American educational system "tailored to the needs of productive and responsible citizenship in the global economy." Along with the "basic observations" reviewed in the text, the National Geography Standards 1994 help frame the kinds of understanding we will seek in the following pages and suggest the purpose and benefit of further study of geography.

The 18 geography standards tell us the following:

*The geographically informed person knows and understands:*

### The World in Spatial Terms

1. How to use maps and other geographic tools and technologies to acquire, process, and report information from a spatial perspective.
2. How to use mental maps to organize information about people, places, and environments in a spatial context.
3. How to analyze the spatial organization of people, places, and environments on Earth's surface.

### Places and Regions

4. The physical and human characteristics of places.
5. That people create regions to interpret Earth's complexity.
6. How culture and experience influence people's perceptions of places and regions.

### Physical Systems

7. The physical processes that shape the patterns of Earth's surface.
8. The characteristics and spatial distribution of ecosystems on Earth's surface.

### Human Systems

9. The characteristics, distribution, and migration of human populations on Earth's surface.
10. The characteristics, distribution, and complexity of Earth's cultural mosaics.
11. The patterns and networks of economic interdependence on Earth's surface.
12. The processes, patterns, and functions of human settlement.
13. How the forces of cooperation and conflict among people influence the division and control of Earth's surface.

### Environment and Society

14. How human actions modify the physical environment.
15. How physical systems affect human systems.
16. The changes that occur in the meaning, use, distribution, and importance of resources.

### The Uses of Geography

17. How to apply geography to interpret the past.
18. How to apply geography to interpret the present and plan for the future.

*Source: Geography for Life: National Geography Standards 1994.* Washington, D.C.: National Geographic Research and Exploration, 1994.

The National Geography Standards were established as part of the nationally adopted *Goals 2000: Educate America Act* (see "The National Standards"). Designed specifically as guidelines to the essential geographic literacy to be acquired by students who have gone through the U.S. public school system, the standards address the same conviction underlying this edition of *Introduction to Geography*—that being literate in geography is a necessary part of the mental framework of all informed persons.

## Organization of This Book

The breadth of geographic interest and subject matter, the variety of questions that focus geographic inquiry, and the diversity of concepts and terms geographers employ require a simple, logical organization of topics for presentation to students new to the field. Despite its outward appearance of complex diversity, geography should be seen to have a broad consistency of purpose achieved through the recognition of a limited number of distinct but closely related "traditions." William D. Pattison, who suggested this unifying viewpoint, and J. Lewis Robinson (among others), who accepted and expanded Pattison's reasoning, found that four traditions are logical and inclusive ways of clustering geographic inquiry. Although not all geographic work is confined by the separate traditions, one or more of them are implicit in most geographic studies. The unifying categories—the four traditions within which geographers work—are:

1. the earth science tradition;
2. the culture–environment tradition;
3. the locational (or spatial) tradition;
4. the area analysis (or regional) tradition.

The mutual interdependence of the four traditions is suggested by Figure 1.14, as are their common individual and collective ties to the full range of research and study techniques that geographers employ: maps, of course, but also remote sensing, statistical tools, geographic information systems, and other spatial analytical techniques.

We have used the four traditions as the device for clustering the chapters of this book (from Chapter 3 onward), hoping they will help you recognize the unitary nature of geography while appreciating the diversity of topics studied by geographers. They are, in a sense, introduced by Chapter 2, which views maps (and related tools and techniques) as essential, distinctive, and unifying tools of geographers.

The **earth science tradition** is the branch of the discipline that addresses itself to the earth as the habitat of humans. It is the tradition that in ancient Greece represented the roots of geography, the description of the physical structure of the earth and of the natural processes that give it detailed form. In modern terms, it is the vital environmental half of the study of human–environmental systems, which together constitute geography's subject matter. The earth science tradition prepares the physical geographer to understand the earth and its resources as the common heritage of humankind and to find solutions to the increasingly complex web of pressures placed on the earth by its expanding, demanding human occupants. Consideration of the elements of the earth science tradition constitutes Part 1 (Chapters 3 to 5) of this text.

Part 2 (Chapters 6 to 9) details some of the content of the **culture–environment tradition.** Within this theme of geography, consideration of the earth as a purely physical entity gives way to a primary interest in how people *perceive* the environments they occupy. Its focus is on culture. The landscapes that are explored and the spatial patterns that are central are those that are cultural in origin and expression. People in their numbers, distributions, and diversity; in their patterns of social and political organization; and in their spatial perceptions and behaviors are the orienting concepts of the culture–environment tradition. The theme is distinctive in its thrust but tied to the earth science tradition, because populations exist, cultures emerge, and behaviors occur within the context of the physical realities and patternings of the earth's surface.

The **locational tradition**—or, as it is sometimes called, the *spatial tradition*—is the subject of the chapters of Part 3 (Chapters 10 to 12). It is a tradition that underlies all of geographic inquiry. As Robinson suggested, if we can agree that geology is rocks, that history is time, and that sociology is people, then we can assert that geography is earth space. The locational tradition is primarily concerned with the distribution of cultural phenomena or physical items of significance to human occupance of the earth. It explores as well the spatial patterns of interaction between humans and the ecosystems that sustain them.

Part, but by no means all, of the locational tradition is concerned with distributional patterns. More central are

**FIGURE 1.14** The four traditions of geography do not stand alone. Rather, each is interconnected with the others and all together depend on unifying research skills and tools. As the diagram indicates, the chapters of this text are grouped by reference to the traditions to help you recognize the broad divisions of geography as well as its underlying and unitary nature. To avoid diagram clutter, lines connecting each box to every other have not been drawn but should be understood.

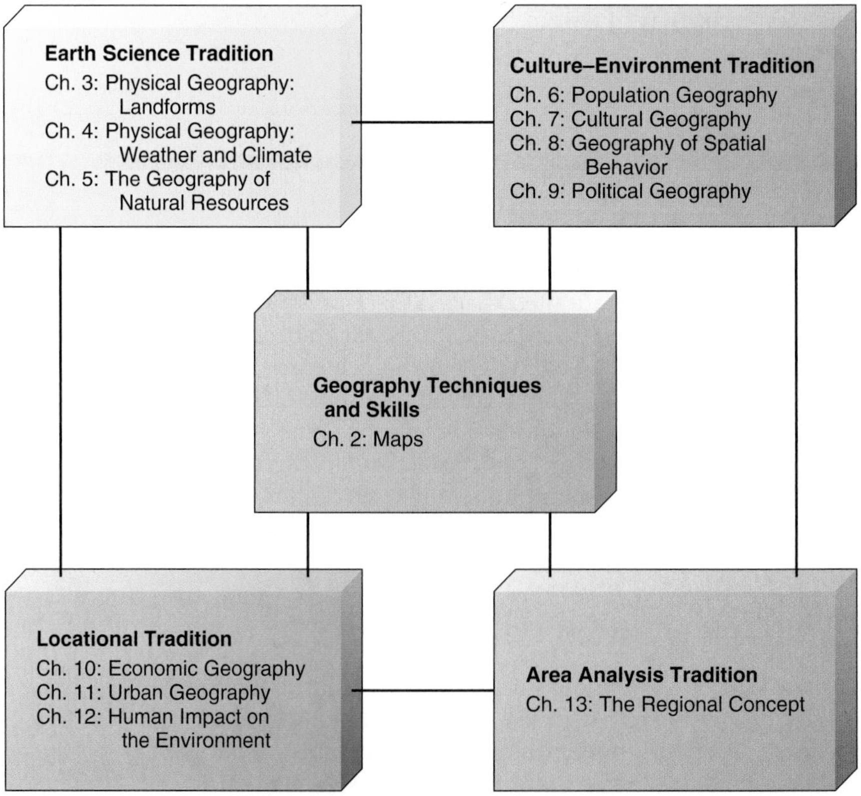

scale, movement, and areal relationships. Map, statistical, geometric, and systems analysis research are among the techniques employed by geographers working within the locational tradition. Irrespective, however, of the analytical tools used or the sets of phenomena studied—economic activities, city systems, or the impact of cultural impositions on the physical landscape—the underlying theme of the locational tradition is the distribution of the phenomenon discussed and the flows and interconnections that unite it to related physical and cultural occurrences.

The **area analysis tradition** is considered in Chapter 13, which makes up Part 4 of *Introduction to Geography.* Again, the roots of this tradition may be traced to antiquity. Strabo's *Geography* was addressed to the leaders of Augustan Rome as a summary of the nature of places in their separate characters and conditions—knowledge deemed vital to the guardians of an empire. Imperial concerns may long since have vanished, but the study of regions and the recognition of their spatial uniformities and differences remain. Such uniformities and differences, of course, grow out of the structure of human-environmental systems and interrelations that are the study of geography. To illustrate the role and diversity of regional studies in geography, much of Chapter 13 consists of special regional investigations and models illuminating specific topics that are the themes of topical Chapters 3 to 12 of this book. References to those special examples are given in Table 1.1 and repeated in the various chapter sections to which they relate.

The identification of the four traditions of geography is not only an organizational convenience but also a recognition that, within that diversity of subject matter called geography, unity of interest is ever-preserved. The traditions, though recognizably distinctive, are intertwined and overlapping. We hope their use as organizing themes—and their further identification in short introductions to the separate sections of this book—will help you grasp the unity in diversity that is the essence of geographic study.

**TABLE 1.1 | Regional Studies Contained in Chapter 13**

| Topic | Page Nos. in Ch. 13 |
|---|---|
| Landforms as regions | 472–473 |
| Dynamic regions in weather and climate | 473–475 |
| Natural resource regions | 475–476 |
| Population as regional focus | 477–478 |
| Language as region | 478–479 |
| Mental regions | 479–480 |
| Political regions | 480–482 |
| Economic regions | 482–483 |
| Urban regions | 484–485 |
| Ecosystems as regions | 485–486 |
| **Chapter Reference** | **Reference Page** |
| Ch. 3: Physical Geography: Landforms | 55 |
| Ch. 4: Physical Geography: Weather and Climate | 87 |
| Ch. 5: The Geography of Natural Resources | 133 |
| Ch. 6: Population Geography | 179 |
| Ch. 7: Cultural Geography | 217 |
| Ch. 8: Spatial Interaction | 267 |
| Ch. 9: Political Geography | 299 |
| Ch. 10: Economic Geography | 341 |
| Ch. 11: Urban Geography | 389 |
| Ch. 12: Human Impact on the Environment | 431 |

## Key Words

absolute direction 10
absolute distance 11
absolute location 9
accessibility 14
area analysis tradition 21
connectivity 14
cultural landscape 13
culture–environment tradition 20
earth science tradition 20
formal region 17
functional region 17
globalization 15
locational tradition 20
natural landscape 13
region 16
relative direction 10
relative distance 11
relative location 9
scale 11
site 9
situation 9
spatial diffusion 15
spatial interaction 14

## For Review & Consideration

1. In what two meanings and for what different purposes do geographers refer to *location?* When geographers say "location matters," what aspect of location commands their interest?
2. What does the term *cultural landscape* imply? Is the nature of the cultural landscape dictated by the physical environment?
3. What kinds of distance transformations are suggested by the term *relative distance?* How is the concept of *psychological distance* related to relative distance?
4. How are the ideas of *distance, accessibility,* and *connectivity* related to processes of *spatial interaction?*
5. Why do geographers concern themselves with *regions?* How are *formal* and *functional* regions different in concept and definition?
6. What are the *four traditions* of geography? Do they represent unifying or divided approaches to geographic understanding?

## Selected References

**WEBSITES**

The World Wide Web has a tremendous variety of sites pertaining to geography. Websites relevant to the subject matter of this chapter appear in the "Web Links" section of the Online Learning Center associated with this book. Access it at www.mhhe.com/getis10e/.

Abler, Ronald F., Melvin G. Marcus, and Judy M. Olson, eds. *Geography's Inner Worlds: Pervasive Themes in Contemporary American Geography.* New Brunswick, N.J.: Rutgers University Press, 1992.

Demko, George J., with Jerome Agel and Eugene Boe. *Why in the World: Adventures in Geography.* New York: Anchor Books/Doubleday, 1992.

Fenneman, Nevin M. "The Circumference of Geography." *Annals of the Association of American Geographers* 9 (1919): 3–11.

*Geography for Life: National Geography Standards.* Washington, D.C.: National Geographic Research and Exploration, 1994.

Holt-Jensen, Arild. *Geography: Its History and Concepts.* 3d ed. Thousand Oaks, Calif.: Sage, 1999.

Lanegran, David A., and Risa Palm. *An Invitation to Geography.* 2d ed. New York: McGraw-Hill, 1978.

Livingstone, David N. *The Geographical Tradition.* Cambridge, Mass.: Blackwell, 1992.

Martin, Geoffrey J., and Preston E. James. *All Possible Worlds: A History of Geographical Ideas.* 3d ed. New York: John Wiley & Sons, 1993.

Massey, Doreen. "Introduction: Geography Matters." In *Geography Matters! A Reader,* ed. Doreen Massey and John Allen, pp. 1–11. New York: Cambridge University Press, 1984.

McDonald, James R. "The Region: Its Conception, Design and Limitations." *Annals of the Association of American Geographers* 56 (1966): 516–28.

Morrill, Richard L. "The Nature, Unity and Value of Geography." *Professional Geographer* 35, no. 1 (Feb. 1983): 1–9.

National Research Council. Rediscovering Geography Committee. *Rediscovering Geography: New Relevance for Science and Society.* Washington, D.C.: National Academy Press, 1997.

Pattison, William D. "The Four Traditions of Geography." *Journal of Geography* 63 (1964): 211–16.

Qing, Dai. *The River Dragon Has Come! The Three Gorges Dam and the Fate of China's Yangtze River and Its People.* Armonk, N.Y.: M.E. Sharpe, 1998.

Robinson, J. Lewis. "A New Look at the Four Traditions of Geography." *Journal of Geography* 75 (1976): 520–30.

Rogers, Alisdair, Heather Viles, and Andrew Goudie. *The Student's Companion to Geography.* 2d ed. Malden, Mass.: Basil Blackwell, 1992.

Sauer, Carl O. "The Education of a Geographer." *Annals of the Association of American Geographers* 46 (1956): 287–99.

White, Gilbert F. "Geographers in a Perilously Changing World." *Annals of the Association of American Geographers* 75 (1985): 10–15.

Wood, Tim F. "Thinking in Geography." *Geography* 72 (1987): 289–99.

Kirkwood
BOUNDARY
Rathbone Lake
Spring
NATIONAL
Ridge
4WD
Red Canyon Creek
Coal Can
Graycroft Ridge
Tepee
Johnson Lake
PACK
Johnson Cr.
Mount Hebgen
2649
Dave
White Lakes
Spring
HEBGEN LAKE
Campground
Grayling
Corey Spring
The Narrows
Grayling Arm
Creek
Spring Creek Campground
Rainbow Point
Campgrounds
Horse Butte Peninsula
Duck
Rumbaugh Campground
Rumbaugh Creek
Campground
Edwards Peninsula
Lookout Tower
South Fork Arm
Madison Arm
Cherry Creek
PACK
Campground
FOREST
MADISON
West Fork
R.3E.
R.4E.
Basin Cabin Spring
Creek
South
4WD
Ski Lift
Denny
DIVIDE
BOUND

# Maps


**Maps as the Tools of Geography**

**Locating Points on a Sphere**

- The Grid System
- The Global Positioning System

**Map Projections**

- Area
- Shape
- Distance
- Direction

**Scale**

**Types of Maps**

- Topographic Maps and Terrain Representation
- Thematic Maps and Data Representation
  - *Point Symbols*
  - *Area Symbols*
  - *Line Symbols*
- Deceptive Practices

**Remote Sensing**

- Aerial Photography
- Nonphotographic Imagery
  - *Remote-Sensing Techniques*
  - *Satellite Imagery*

**Geographic Information Systems**

- The Geographic Database
- Applications of GIS

**Summary**

**Key Words**

**For Review & Consideration**

**Selected References**


**An enlarged portion of a topographic map of the Hebgen Lake, Montana, area.** © USGS.

On May 7, 1999, NATO military pilots mistakenly bombed the Chinese embassy in Belgrade, Yugoslavia, instead of the intended target, a munitions building known as the Federal Directorate for Supply and Procurement some 218 meters (200 yds) away. The missiles killed three people, injured 20 others, and provoked mass protests in China. Expressing deep regret, intelligence officials explained that the out-of-date U.S. military map from which the pilots were working did not show the embassy compound.

Three years later, Americans were riveted by the plight of nine coal miners trapped 73 meters (240 ft) below ground in the flooded Quecreek Mine in southwestern Pennsylvania. On July 24, 2002, a machine operator in the mine broke through to an adjacent, flooded mine that had been abandoned 38 years earlier. As millions of gallons of water rushed in, the nine men looked frantically for a way out, but all the exits were in areas already filled with water. According to the report issued by investigators from the Department of Labor, at one point the miners

> estimated they had about an hour left based on the rate the water was rising. The miners took some time to reflect on their situation and prepared for the worst. Some of the miners tied themselves together so they could be found together in the event they were drowned. They wrote notes to their families and placed them in a plastic bucket. The bucket was closed with a lid, sealed with electrical tape and secured near the roof bolting machine to prevent it from floating away.

Above ground, rescuers drilled holes to pump out the water and sank a shaft 0.8 meters (30 in.) in diameter to reach the miners. After 78 hours underground, all nine were lifted to safety.

This was an accident that should not have happened. The 1957 map the mining companies were using showed the old mine as 138 meters (150 yds) away from Quecreek. But another 421,000 tons of coal had been removed before the mine was closed in 1964, and that additional mining had put one shaft immediately adjacent to Quecreek.

these examples indicate, accurate maps can literally ın the difference between life and death. Governmental ıcies rely on maps of flood-prone areas, of volcanic ɔtions, of earthquake hazard zones, and of areas subject ındslides to develop their long-range plans. Epidemiol-ıts map the occurrence of a disease over time and space, ɔing them identify the source of the outbreak and create an to halt the spread of the disease. Law enforcement ıcies increasingly use maps to identify patterns of spe- types of crime and to help them predict where those ıes are likely to occur in the future. The value of exam-g information in a spatial context cannot be overstated.

## APS AS THE TOOLS GEOGRAPHY

ɔs have a special significance for geographers. They are graphers' primary tools of spatial analysis. For a variety easons, the spatial distributions, patterns, and relations ıterest to geographers usually cannot easily be observed ıterpreted in the landscape itself.

- Many, such as landform or agricultural regions or major cities, are so extensive spatially that they cannot be seen or studied in their totality from one or a few vantage points.
- Many, such as regions of language usage or religious belief, are spatial phenomena but are not tangible or visible.
- Many interactions, flows, and exchanges imparting the dynamic quality to spatial interaction may not be directly observable at all.

Even if all things of geographic interest could be seen and measured through field examination, the infinite variety of tangible and intangible content of an area would make it nearly impossible to isolate for study and interpretation the few topics selected for special investigation.

Therefore, the map has become the essential and distinctive tool of geographers. Only through the map can spatial distributions and interactions of whatever nature be reduced to an observable scale, isolated for individual study, and combined or recombined to reveal relationships not directly measurable in the landscape itself.

The art, science, and technology of making maps is called **cartography.** Modern scientific mapping has its roots in the 17th century, although the earth scientists of ancient Greece are justly famous for their contributions. They recognized the spherical form of the earth and developed map projections and the grid system. Unfortunately, much of the cartographic tradition of Greece was lost to Europe during the Middle Ages and essentially had to be rediscovered. Several developments during the Renaissance gave an impetus to accurate cartography. Among these were the development of printing, the rediscovery of the work of Ptolemy and other Greeks, and the great voyages of discovery.

In addition, the rise of nationalism in many European countries made it imperative to determine and accurately portray boundaries and coastlines, as well as to depict the kinds of landforms contained within the borders of a country. During the 17th century, important national surveys were undertaken in France and England. Many conventions in the way data are presented on maps had their origin in these surveys.

Knowledge of the way information is recorded on maps enables us to read and interpret them correctly. To be on guard against drawing inaccurate conclusions or to avoid being swayed by distorted or biased presentations, we must be able to understand and assess the ways in which facts are represented. Of course, all maps are necessarily distorted because of the need to portray the round earth on a flat surface, to use symbols to represent objects, to generalize, and to record features at a different size than they actually are. This distortion of reality is necessary because the map is smaller than the things it depicts and because its effective communication depends upon selective emphasis of only a portion of reality. As long as map readers know the limitations of the commonly used types of maps and understand what relationships are distorted, they can interpret maps correctly.

# Locating Points on a Sphere

As we noted in Chapter 1, the starting point of all geographic study is the location of places and things, and absolute location is the identification of place by a precise and accepted system of coordinates.

## The Grid System

In order to visualize the basic system for locating points on the earth, think of the world as a sphere with no markings whatsoever on it. There would, of course, be no way of describing the exact location of a particular point on the sphere without establishing a system of reference. We use the **grid system,** which consists of a set of imaginary lines drawn across the face of the earth. The key reference points in that system are the North and South Poles and the equator, which are given in nature, and the prime meridian.

The North and South Poles are the end points of the axis about which the earth spins. The line that encircles the globe halfway between the poles, perpendicular to the axis, is the *equator.* We can describe the location of a point in terms of its distance north or south of the equator, measured as an angle at the earth's center. Because a circle contains 360 degrees, the distance between the two poles is 180 degrees and between the equator and each pole, 90 degrees. **Latitude** is the angular distance north or south of the equator, measured in degrees ranging from 0° (the equator) to 90° (the North and South Poles). As is evident in Figure 2.1a, the lines of latitude, which are parallel to each other and to the equator, run east-west.

The polar circumference of the earth is 24,899 miles; thus, the distance between each degree of latitude equals 24,899 ÷ 360, or about 111 kilometers (69 mi). If the earth were a perfect sphere, all degrees of latitude would be equally long. Due to the slight flattening of the earth in polar regions, degrees of latitude are slightly longer near the poles (111.70 km; 69.41 mi) than near the equator (110.56 km; 68.70 mi).

To record the latitude of a place in a more precise way, degrees are divided into 60 *minutes* (′), and each minute into 60 *seconds* (″), exactly like an hour of time. One minute of latitude is about 1.85 kilometers (1.15 mi), and one second of latitude about 31 meters (101 ft). The latitude of the center of Chicago is written 41°51′50″ North. This system of subdivision of the circle is derived from the sexagesimal (base 60) numerical system of the ancient Babylonians.

Because the distance north or south of the equator is not by itself enough to locate a point in space, we need to specify a second coordinate to indicate distance east or west from an agreed-upon reference line. As a starting point for east-west measurement, cartographers in most countries use as the **prime meridian** an imaginary line passing through the Royal Observatory at Greenwich, England. This prime meridian was selected as the zero-degree longitude by an international conference in 1884. Like all *meridians,* it is a true north-south line connecting the poles of the earth (Figure 2.1b). ("True" north and south vary from magnetic north and south, the direction of the earth's magnetic poles, to which a compass needle points.) Meridians are farthest apart at the equator, come closer and closer together as latitude increases, and converge at the North and South Poles. Unlike *parallels* of latitude, all meridians are the same length.

**Longitude** is the angular distance east or west of the prime (zero) meridian measured in degrees ranging from 0° to 180°. Directly opposite the prime meridian is the 180th meridian, located in the Pacific Ocean. Like parallels of latitude, degrees of longitude can be subdivided into minutes and seconds. However, distance between adjacent degrees of longitude decreases away from the equator because the meridians converge at the poles. With the exception of a few Alaskan islands, all places in North and South America are in the area of west longitude; with the exception of a portion

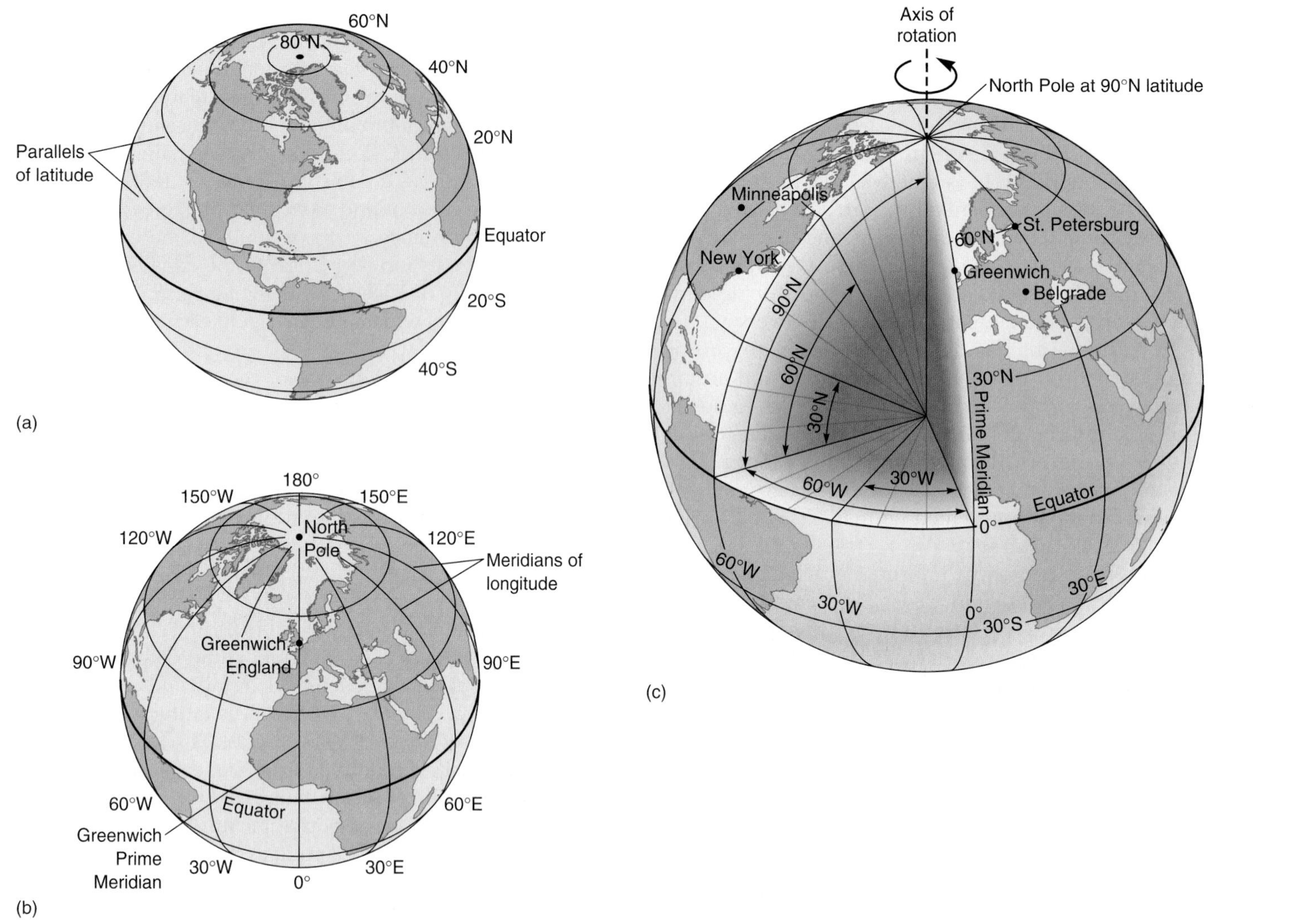


**FIGURE 2.1** (a) **The grid system: parallels of latitude.** Note that the parallels become increasingly shorter closer to the poles. On the globe, the 60th parallel is only half as long as the equator. (b) **The grid system: meridians of longitude.** East-west measurements range from 0° to 180°—that is, from the prime meridian to the 180th meridian in each direction. Because the meridians converge at the poles, the distance between degrees of longitude becomes shorter as one moves away from the equator. (c) **The earth grid, or graticule, consisting of parallels of latitude and meridians of longitude.**

of the Chukchi Peninsula of Siberia, all places in Asia and Australia have east longitude.

Time depends on longitude. The earth, which makes a complete 360-degree rotation once every 24 hours, is divided into 24 time zones roughly centered on meridians at 15-degree intervals. *Greenwich mean time (GMT)* is the time at the prime meridian. The **International Date Line,** where each new day begins, generally follows the 180th meridian. As Figure 2.2 indicates, however, the date line deviates from the meridian in some places in order to avoid having two different dates within a country or an island group. Thus, the International Date Line zigzags so that Siberia has the same date as the rest of Russia and the Aleutian Island and Fiji Island groups are not split. New days begin at the date line and proceed westward, so that west of the line is always 1 day later than east of the line.

By citing the degrees, minutes, and, if necessary, seconds of longitude and latitude, we can describe the location of any place on the earth's surface. To conclude our earlier example, Chicago is located at 41°52′N, 87°40′W. Hong Kong is at 22°15′N, 114°10′E (Figure 2.3).

## The Global Positioning System

In recent years, the **Global Positioning System (GPS)** has made the determination of location significantly easier than it used to be. This navigation and positioning system was conceived in the 1970s and is maintained by the U.S. Department of Defense. The technology uses a string of 24 to 28 Department of Defense satellites that orbit some 20,000 kilometers (12,500 mi) above the earth, passing over the same spot every 24 hours. Each satellite carries four atomic

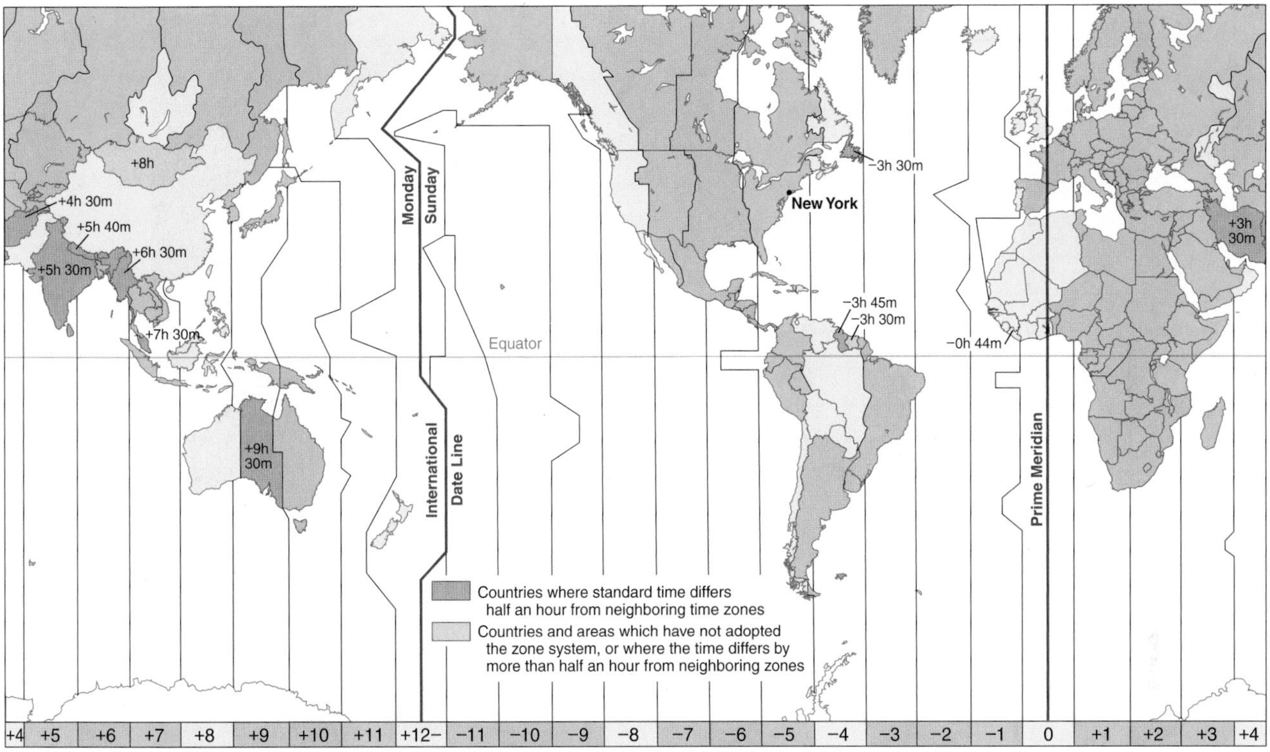


**FIGURE 2.2 World time zones.** Each time zone is about 15° wide, but variations occur to accommodate political boundaries. The figures at the bottom of the map represent the time difference in hours when it is 12 noon in the time zone centered on Greenwich, England. New York is in column −5, so the time there is 7 A.M. when it is noon at Greenwich. Modifications to the universal system of time zones are numerous. Thus, Iceland operates on the same time as Britain, although it is a time zone away. Spain, entirely within the boundaries of the GMT zone, sets its clocks at +1 hour, whereas Portugal conforms to GMT. China straddles five time zones, but the whole country operates on Beijing time (+8 hours). In South America, Chile (in the −5 hour zone) uses the −4 hour designation, whereas Argentina uses the −3 hour zone instead of the −4 hour zone to which it is better suited.

clocks so accurate that they might gain or lose an average of 1 second in 30,000 years.

As they orbit, the satellites continuously transmit their positions, time signals, and other data. The satellites are arranged so that at least four are above the horizon at any time, all over the earth, available for simultaneous measurement. A GPS receiver records the positions of a number of the satellites simultaneously, then determines its latitude, longitude, altitude, and the time (Figure 2.4). To make a precise measurement of a specific location, a receiver needs to pick up signals from at least four satellites.

GPS technology was originally designed for military applications, particularly naval and aerial navigation. Its success was evident in the 1991 Gulf War with Iraq, when U.S. troops used the devices to find their way across the Saudi Arabian desert during Operations Desert Shield and Desert Storm. The technology has also facilitated the development of precision-guided weapons, the so-called smart bombs that home in on a target. Other government applications include the use of GPS receivers for monitoring geologic fault lines and ocean currents, the sensing of global warming in the atmosphere, firefighting, and mapping disaster scenes.

Seeking clues as to why the space shuttle *Columbia* broke apart upon its reentry into the atmosphere on February 1, 2003, for example, federal government investigators used GPS technology to define the debris field, which covered portions of Texas, Louisiana, and several other states. As hundreds of volunteers and law enforcement officers found and collected thousands of pieces of debris, their locations were fed into a computer equipped with special mapping software. Just a few days after the tragedy, the Federal Emergency Management Agency's Disaster Field Office in Lufkin, Texas, was printing more than 1000 maps a day. The maps helped emergency workers focus on the areas they needed to search in order to retrieve more of the shuttle debris.

As GPS receivers have become smaller, lighter, and less expensive, civilian applications have multiplied. Several automobile manufacturers make in-car navigation systems an option in their new cars, and some rental car agencies provide them in their vehicles. The systems tap GPS signals to monitor the car's exact location, comparing it with a computerized atlas stored on a compact disc. The car's location, constantly updated, appears on a computer screen mounted on the dashboard. The navigation systems enable motorists to find out where they are and how to reach their destination. For example, the driver can give a street address or the name of a movie theater, a hospital, or another building, and the system displays it on the screen, indicates how far

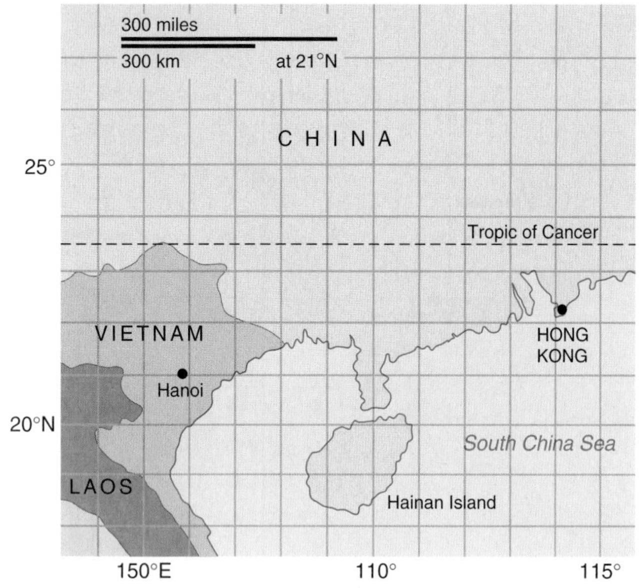


**FIGURE 2.3** The latitude and longitude of Hong Kong are 22°15′N, 114°10′E. What are the coordinates of Hanoi?

**FIGURE 2.4** This handheld GPS receiver shows latitude, longitude, and a map on a small liquid crystal display screen. *Courtesy of Garmin.*

away it is, and how long it should take to drive there, and the system gives turn-by-turn directions on the screen map or directions "spoken" by an electronic voice. Car navigation systems are particularly popular among sales representatives, real estate agents, and repair people, who frequently drive to unfamiliar places.

GPS receivers are also popular with recreational users—hikers, campers, and sailors, for example. In the past few years, systems have been developed for building lightweight, portable GPS receivers into all kinds of things, such as watches, bracelets, belts, and backpacks, in order to ascertain their locations. A number of states use GPS monitoring devices as a surveillance system to track the movements of people on parole and probation, to make sure they do not stray into off-limit areas, such as schools, playgrounds, or victims' houses.

# Map Projections

The earth can be represented with reasonable accuracy only on a globe, but globes are not as convenient as flat maps to store or use, and they cannot depict much detail. For example, if we had a large globe with a diameter of 1 meter, we would have to fit the details of over 100,000 square kilometers of earth surface in an area a few centimeters on a side. Obviously, a globe of reasonable size cannot show the transportation system of a city or the location of very small towns and villages.

In transforming a globe into a map, we cannot flatten the curved surface and keep intact all the properties of the original. **Globe properties** are as follows:

1. all meridians are of equal length; each is one-half the length of the equator;
2. all meridians converge at the poles and are true north-south lines;
3. all lines of latitude (parallels) are parallel to the equator and to each other;
4. parallels decrease in length as one nears the poles;
5. meridians and parallels intersect at right angles;
6. the scale on the surface of the globe is everywhere the same in all directions.

Only the globe grid itself retains all of these characteristics. To project it onto a surface that can be laid flat is to distort some or all of these properties and consequently to distort the reality the map attempts to portray.

The term **map projection** designates the way the curved surface of the globe is represented on a flat map. All flat maps distort, in different ways and to different degrees, some or all of the four main properties of actual earth surface relationships: area, shape, distance, and direction. Figure 2.5 illustrates the distortion inherent in several map projections.

## Area

Some projections enable the cartographer to represent the *areas* of regions in correct or constant proportion to earth reality. That means that any square inch on the map represents an identical number of square miles (or of similar units) anywhere else on the map. As a result, the shape of

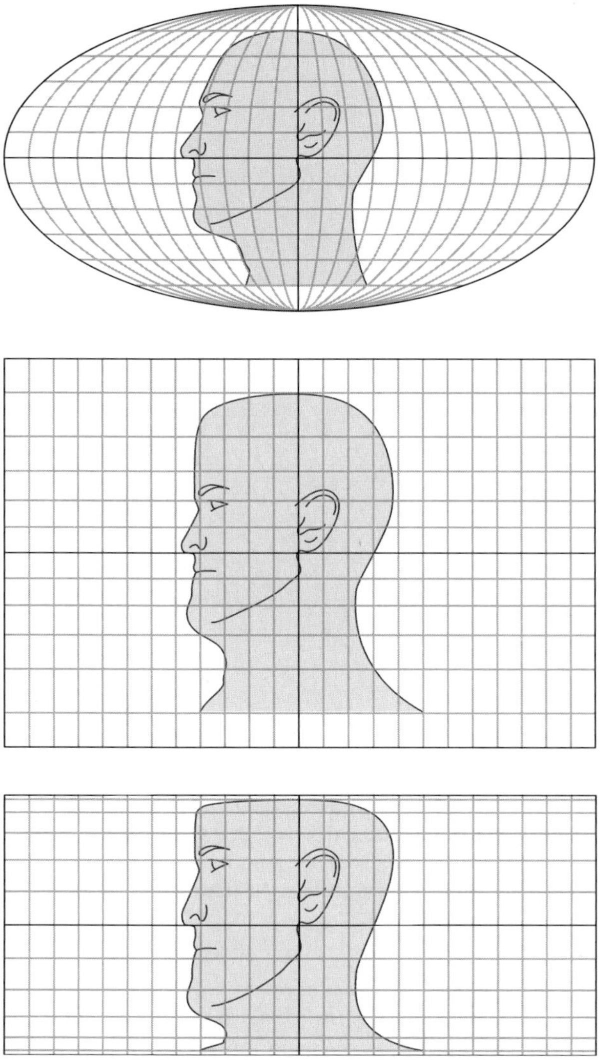

**FIGURE 2.5** This figure illustrates the distortion inherent in three different map projections. A head drawn on one projection has been transferred to two other projections, keeping the latitude and longitude the same as they are found on the first. This does not mean the first projection is the best of the three. The head could have been drawn on any one of them and then plotted on the others. Source: *Arthur Robinson et al.,* Elements of Cartography, *5th ed., Fig. 5.6, p. 85. (New York, Wiley, © 1984.)*

the portrayed area is inevitably distorted. A square on the earth, for example, may become a rectangle on the map, but that rectangle has the correct area. Such projections are called **equal-area** or **equivalent projections** (Figure 2.6a). *A map that shows correct areal relationships always distorts the shapes of regions.*

Equal-area projections are used when a map is intended to show the actual areal extent of a phenomenon on the earth's surface. If we wish to compare the amount of land in agriculture in two parts of the world, for example, it would be very misleading visually to use a map that represented the same amount of surface area at two different scales.

## Shape

Although no projection can provide correct shapes for large areas, some accurately portray the shapes of small areas by preserving correct angular relationships (Figure 2.6b). These true-shape projections are called **conformal projections,** and the importance of *conformality* is that regions and features "look right" and have the correct directional relationships. They achieve these properties for small areas by assuring that parallels of latitude and meridians of longitude cross each other at right angles and that the scale is the same in all directions at any given location. Both these conditions exist on the globe but can be retained for only relatively small areas on maps. Because that is so, the shapes of large regions—continents, for example—are always different from their true earth shapes, even on conformal maps. *A map cannot be both equivalent and conformal.*

## Distance

Distance relationships are nearly always distorted on a map, but some projections maintain true distances in one direction or along certain selected lines. Others, called **equidistant projections,** show true distance in all directions, but only from one or two central points (Figure 2.6c). Distances between all other locations are incorrect and, quite likely, greatly distorted. A planar equidistant map centered on Detroit, for example, shows the correct distance between Detroit and the cities of Boston, Los Angeles, and any other point on the map. But it does *not* show the correct distance between Los Angeles and Boston. *A map cannot be both equidistant and equal-area.*

## Direction

As is true of distances, directions between all points cannot be shown without distortion. On **azimuthal projections,** however, true directions are shown from one central point to all other points. (An *azimuth* is the angle formed at the beginning point of a straight line, in relation to a meridian.) Directions or azimuths from points other than the central point to other points are not accurate. The azimuthal property of a projection is not exclusive—that is, an azimuthal projection may also be equivalent, conformal, or equidistant. The equidistant map shown in Figure 2.6c is, as well, a true-direction map from the same North Pole origin.

Not all maps are equal-area, conformal, or equidistant; most are compromises. One example of such a compromise is the *Robinson projection,* which was designed to show the whole world in a visually satisfactory manner and which is used for most of the world maps in this textbook (Figure 2.7). It does not show true distances or directions and is neither equal-area nor conformal. Instead, it permits some exaggeration of size in the high latitudes in order to improve the shapes of landmasses. Size and shape are most accurate in the temperate and tropical zones, where most people live.

Mapmakers must be conscious of the properties of the projections they use, selecting the one that best suits their

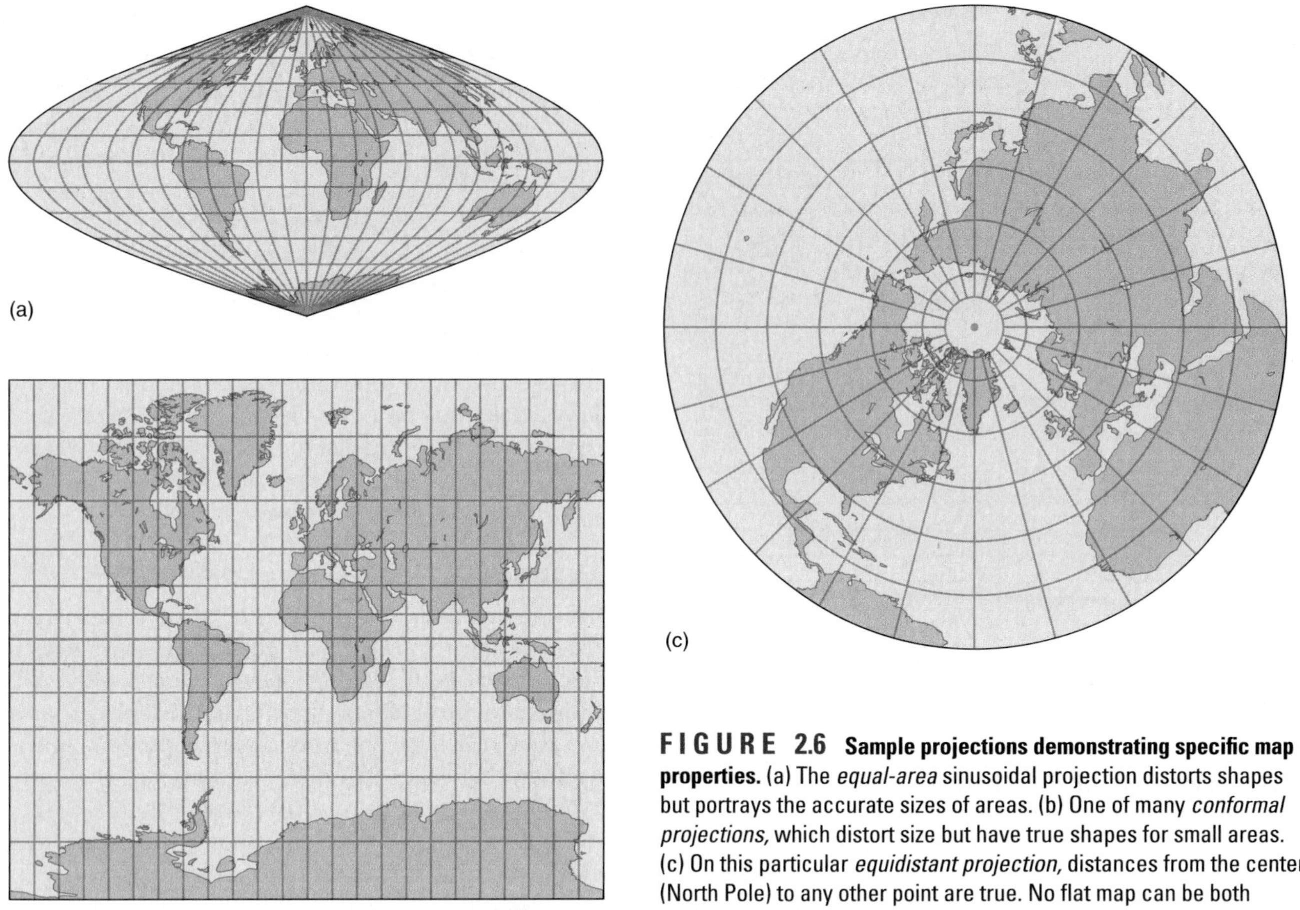


**FIGURE 2.6 Sample projections demonstrating specific map properties.** (a) The *equal-area* sinusoidal projection distorts shapes but portrays the accurate sizes of areas. (b) One of many *conformal projections,* which distort size but have true shapes for small areas. (c) On this particular *equidistant projection,* distances from the center (North Pole) to any other point are true. No flat map can be both equidistant and equal-area.

purposes. If a map shows only a small area, the choice of a projection is not critical—virtually any can be used. The choice is more important when the area to be shown extends over a considerable longitude and latitude; then the selection of a projection depends on the purpose of the map. Some projections are useful for navigation. If numerical data are being mapped, the relative sizes of the areas involved should be correct, so that one of the many equal-area projections is likely to be used. Display maps usually employ conformal projections. Most atlases indicate which projection has been used for each map, thus informing the map reader of the properties of the maps and their distortions. More information about map projections can be found in Appendix 1.

Selection of the map grid, determined by the projection, is the first task of the mapmaker. A second decision involves the scale at which the map is to be drawn.

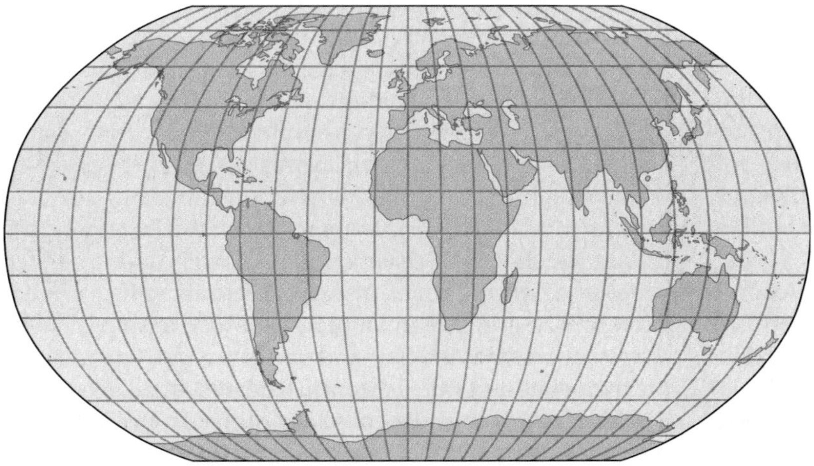

**FIGURE 2.7** The Robinson projection, a compromise between an equal-area and a conformal projection, gives a fairly realistic view of the world. The most pronounced shape distortions are in the less-populated areas of the higher latitudes, such as northern Canada, Greenland, and Russia. On the map, Canada is 21% larger than in reality, while the 48 contiguous states of the United States are 3% smaller than they really are.

# Scale

The *scale* of a map is the ratio between the measurement of something on the map and the corresponding measurement on the earth. Scale is typically represented in one of three ways: verbally, graphically, or numerically as a representative fraction (Figure 2.8). As the name implies, a *verbal* scale is given in words, such as "1 inch to 1 mile" or "10 centimeters to 1 kilometer." A *graphic* scale is a line or bar placed on the map that has been subdivided to show the map lengths of units of earth distance.

A *representative fraction (RF)* scale gives two numbers, the first representing the map distance and the second indicating the ground distance. The fraction may be written in a number of ways. There are 5280 feet in 1 mile and 12 inches in 1 foot; 5280 times 12 equals 63,360, the number of inches in 1 mile. The fractional scale of a map at 1 inch to 1 mile can be written as 1:63,360 or 1/63,360. On the simpler metric scale, 1 centimeter to 1 kilometer is 1:100,000. The units used in each part of the fractional scale are the same; thus, 1:63,360 could also mean that 1 foot on the map represents 63,360 feet on the ground, or 12 miles—which is, of course, the same as 1 inch represents 1 mile. Numerical scales are the most accurate of all scale statements and can be understood in any language.

The map scale, or ratio between the map dimensions and those of reality, can range from very large to very small. A *large-scale map,* such as a plan of a city, shows an area in considerable detail. That is, the ratio of map to ground distance is relatively large, for example 1:600 (1 inch on the map represents 600 inches or 50 feet on the ground) or 1:24,000. At this scale, features such as buildings and highways can be drawn to scale. Figure 2.10 is an example of a large-scale map. *Small-scale maps,* such as those of countries or continents, have a much smaller ratio. Buildings, roads, and other small features cannot be drawn to scale and must be magnified and represented by symbols to be seen. Figures 2.2 and 2.3 are small-scale maps. Although no rigid numerical limits differentiate large-scale from small-scale maps, most cartographers would consider large-scale maps to have a ratio of 1:50,000 or less, and maps with ratios of 1:500,000 or more to be small-scale.

Each of the four maps in Figure 2.9 is drawn at a different scale. Although each is centered on Boston, notice how scale affects both the area that can be shown in a square that is 2 inches on a side and the amount of detail that can be depicted. On map (a), at a scale of 1:25,000, about 2.6 inches represent 1 mile, so that the 2-inch square shows less than 1 square mile. At this scale, one can identify individual buildings, highways, rivers, and other landscape features. Map (d), drawn to a scale of 1 to 1 million (1:1,000,000, or 1 inch represents almost 16 miles), shows an area of almost 1000 square miles. In this map, only major features, such as main highways and the location of cities, can be shown, and even the symbols used for that purpose occupy more space on the map than would the features depicted if they were drawn true to scale.

Small-scale maps like (c) and (d) in Figure 2.9 are said to be very *generalized.* They give a general idea of the relative locations of major features but do not permit accurate measurement. They show significantly less detail than do large-scale maps and typically smooth out such features as coastlines, rivers, and highways.

"1 inch to 1 mile"
"1 centimeter to 5 kilometers"

(a)

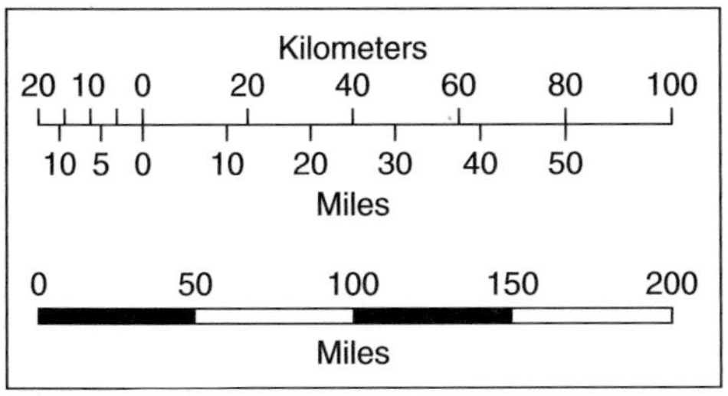


(b)

$\frac{1}{62{,}500}$ 1:62,500

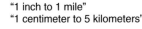

(c)

**FIGURE 2.8** Map scales relate a map distance to a distance on the earth's surface. (a) A verbal scale is given in words. (b) A graphic scale divides a line into units, each unit representing the distance between two points on the earth's surface. Graphic scales help the map user measure earth distance from the map. A graphic scale is automatically enlarged or reduced when the map is photographed and reproduced at a different size. (c) A representative fraction (RF) scale is a simple fraction or ratio. The units of distance on both sides of the scale must be the same; they need not be stated.

# Types of Maps

The properties of the globe grid and of various projections are the concern of the cartographer. In contrast, geographers are more interested in the depiction of spatial data and in the analysis of the patterns those data present. Out of the many items constituting the content of an area, the geographer must first select those that are relevant to the problem at hand and then decide how to display them. In that effort, geographers can choose from among different types of maps.

*General-purpose, reference,* or *location* maps make up one major class of maps familiar to everyone. Their purpose is simply to display without analysis one or more natural and/or human-made features of an area or of the world as a whole. Common examples of the natural features shown on maps are water features (coastlines, rivers, lakes, and so on) and the shape and elevation of terrain. Cultural features include transportation routes, settlements, property-ownership lines, political boundaries, and names.

The other major type of map is called the *thematic,* or *special-purpose,* map, one of any scale that shows a specific spatial distribution or category of data. Again, the phenomena being mapped may be physical (climate, vegetation, soils, and so on) and/or cultural (e.g., the distribution of population, religions, diseases, or crime).

## Topographic Maps and Terrain Representation

As we noted, some general-purpose maps depict the shape and elevation of the terrain. These are called **topographic maps.** They portray the surface features of relatively small areas, often with great accuracy (Figure 2.10). They not only show the elevations of landforms, streams, and other water bodies but may also display features that people have added to the natural landscape. These include transportation routes, buildings, and such land uses as orchards, vineyards, and cemeteries. Boundaries of all kinds, from state boundaries to field or airport limits, are also depicted on topographic maps.

The U.S. Geological Survey (USGS), the chief federal agency for topographic mapping in this country, produces several map series, each on a standard scale. Complete topographic coverage of the United States is available at scales of 1:250,000 and 1:100,000. Maps are also available at various other scales. A single map in one of these series is called a *quadrangle.* Topographic quadrangles at the scale of 1:24,000 exist for the entire area of the 48 contiguous states, Hawaii, and territories, a feat that requires about 57,000 maps. Each map covers a rectangular area that is 7.5 minutes of latitude by 7.5 minutes of longitude. As is evident from Figure 2.10, these 7.5-minute quadrangle maps provide detailed information about the natural and cultural features of an area.

Because of Alaska's large size and sparse population, the primary scale for mapping that state is 1:63,360 (1 inch represents 1 mile). The Alaska quadrangle series consists of more than 2900 maps.

In Canada, the responsibility for national mapping lies with Survey and Mapping and Remote Sensing, Natural Resources (NRCAN). Maps at a scale of 1:250,000 are available for the entire country; the more heavily populated southern part of Canada is covered by 1:50,000-scale maps. Provincial mapping agencies produce detailed maps at even larger scales.

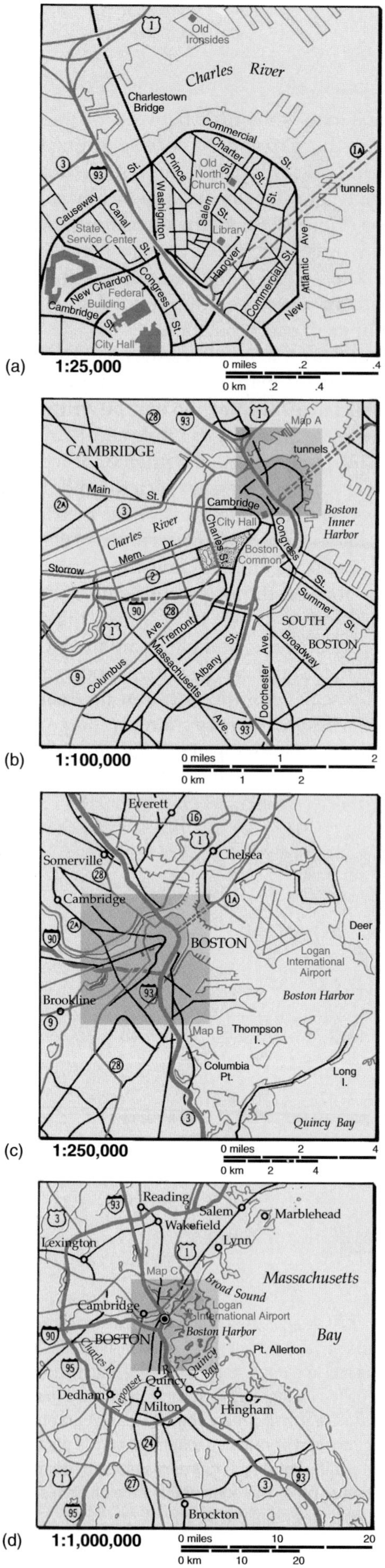


**FIGURE 2.9 The effect of scale on area and detail.** The larger the scale, the greater the number and kinds of features that can be included because the map shows a more restricted area than does a map at a smaller scale.

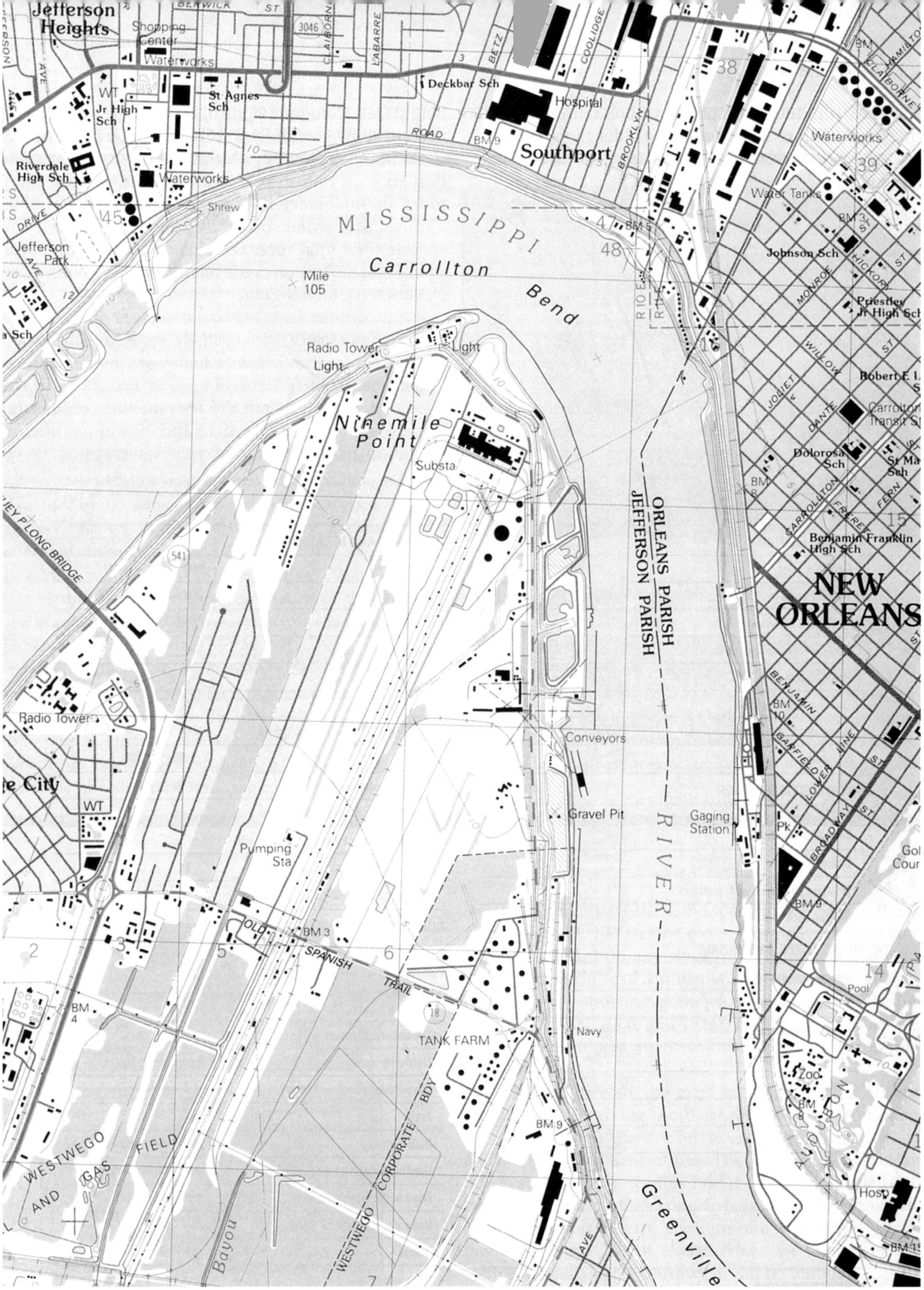


**FIGURE 2.10 A portion of the "New Orleans West Quadrangle, Louisiana," 7.5-minute series of U.S. Geological Survey topographic maps.** The fractional scale is 1:24,000 (1 inch equals about one-third of a mile), allowing considerable detail to be shown. The gray tint denotes built-up areas, in which only schools, churches, cemeteries, parks, and other public facilities are shown. Source: *U.S. Geological Survey.*

The tremendous amount of information on topographic maps makes them useful to engineers, regional planners, land use analysts, and developers, as well as to hikers and casual users. Given such a wealth of information, the experienced map reader can make deductions about both the physical character of the area and the economic and cultural use of the land.

## Thematic Maps and Data Representation

The study of the spatial patterns and interrelationships of things, whether people, crops, or traffic flows, is the essence of geography. Various kinds of symbols are used to record the location or numbers of these phenomena on thematic maps. The symbols and maps may be either *qualitative* or *quantitative.*

The principal purpose of the qualitative map is to show the distribution of a particular class of information. The world location of producing oil fields, the distribution of national parks, and the pattern of areas of agricultural specialization within a country are examples. The interest is in where these things are, and nothing is reported about, for example, barrels of oil extracted, number of park visitors, or value of crops produced.

In contrast, quantitative thematic maps show the spatial characteristics of numerical data. Usually, a single variable, such as population, income, or land value, is chosen, and the map displays the variation of that feature from place to place.

### Point Symbols

Features that occur at a particular point in space are represented on maps by *point symbols.* Thousands of types of such features exist on the earth: churches, schools, cemeteries, and historical sites, to name a few. Symbols used to represent them include dots, crosses, triangles, and other shapes. On a qualitative thematic map, each such symbol records merely the location of that feature at a particular point on Earth.

Sometimes, however, our interest is in showing the variation in the number of things that exist at several points—for example, the population of selected cities, the tonnage handled at certain terminals, or the number of passengers at given airports.

There are two chief means of symbolizing such distributions, as Figures 2.13 and 2.14 indicate. One method is to choose a symbol, usually a dot, to represent a given quantity of the mapped item (such as 50 people) and to repeat that symbol as many times as necessary. Such a map is easily understood because the dots give the map reader a visual impression of the pattern. Sometimes pictorial symbols—for example, human figures or oil barrels—are used instead of dots.

If the range of the data is great, the geographer may find it inconvenient to use a repeated symbol. For example, if one port handles 50 or 100 times as much tonnage

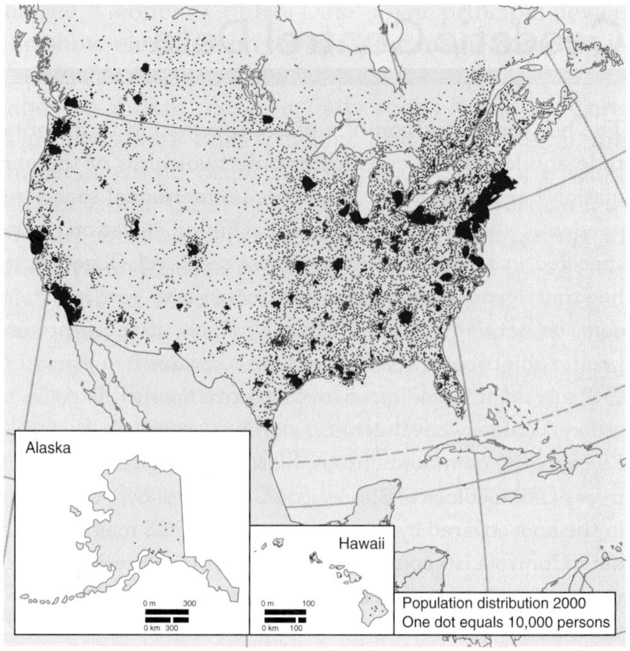


**FIGURE 2.13 Population distribution in the United States and Canada, 2000.** The dot map provides a good visual impression of the distribution and relative density of a phenomenon—in this case, how the number of people varies from place to place.

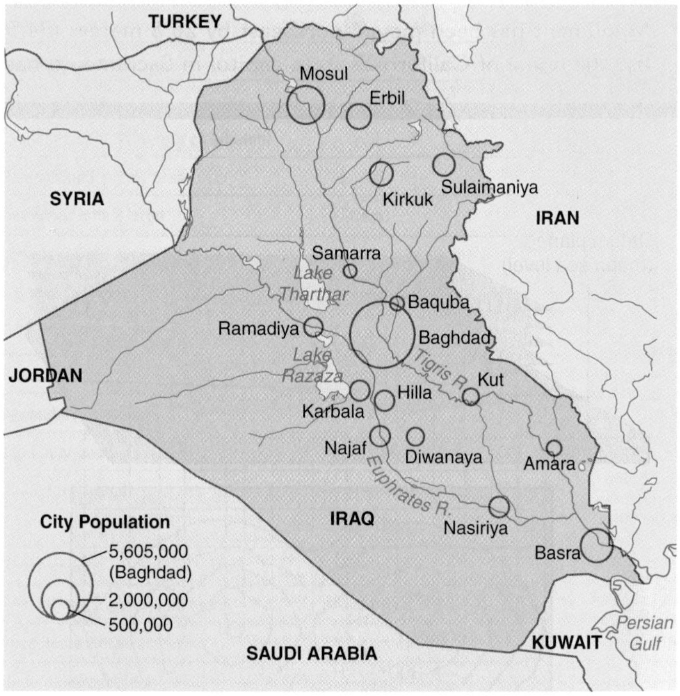


**FIGURE 2.14 Urban population, Iraq.** On this map, the area of the circle is proportional to the size of a city. Cities with fewer than 200,000 people are not shown. The scale in the bottom left corner helps the reader interpret the map.

as another, that many more dots would have to be placed on the map. To circumvent this problem, the cartographer can choose a second method and use proportional symbols. The size of the symbol is varied according to the quantities represented. Thus, if bars are used, they can be shorter or longer as necessary. If squares or circles are used, the *area* of the symbol ordinarily is proportional to the quantity shown (Figure 2.14).

There are occasions, however, when the range of the data is so great that even circles or squares would take up too much room on the map. In such cases, three-dimensional symbols, usually spheres or cubes, are used, and their *volume* is proportional to the data. Unfortunately, many map readers fail to perceive the added dimension implicit in volume, and most cartographers do not recommend the use of such symbols.

## Area Symbols

Features found within defined areas of the earth's surface are represented on maps by *area symbols.* Such maps fall into two general categories: those showing differences in kind and those showing differences in quantity. Atlases contain numerous examples of the first category, such as patterns of religions, languages, political entities, vegetation, or types of rock. Normally, different colors or patterns are used for different areas, as shown in Figure 2.15.

Maps that show how the *amount* of a phenomenon varies from area to area are called **choropleth maps.** The term is derived from the Greek words *choros* ("place") and *pleth* ("magnitude" or "value"). The quantities shown may be absolute numbers (e.g., the population of counties) or derived values, such as percentages, ratios, rates, and densities (e.g., population density by county). The data are grouped into a limited number of classes, each represented by a distinctive color, shade, or pattern. Figure 2.16 is an example of a choropleth map. In this case, the areal units are states. Other commonly used subdivisions are counties, townships, cities, and census divisions.

As Figures 2.15 and 2.16 reveal, three main problems characterize maps (whether qualitative or quantitative) that show the distribution of a phenomenon in an area:

1. they give the impression of uniformity to areas that may actually contain significant variations;
2. boundaries attain unrealistic precision and significance, implying abrupt changes between areas when, in reality, the changes may be gradual;
3. unless colors are chosen wisely, some areas may look more important than others.

A special type of area map is the **area cartogram,** or **value-by-area map,** in which the areas of units are proportional to the data they represent (Figure 2.17). Population, income, cost, or another variable becomes the standard of measurement. Depending on the idea that the mapmaker

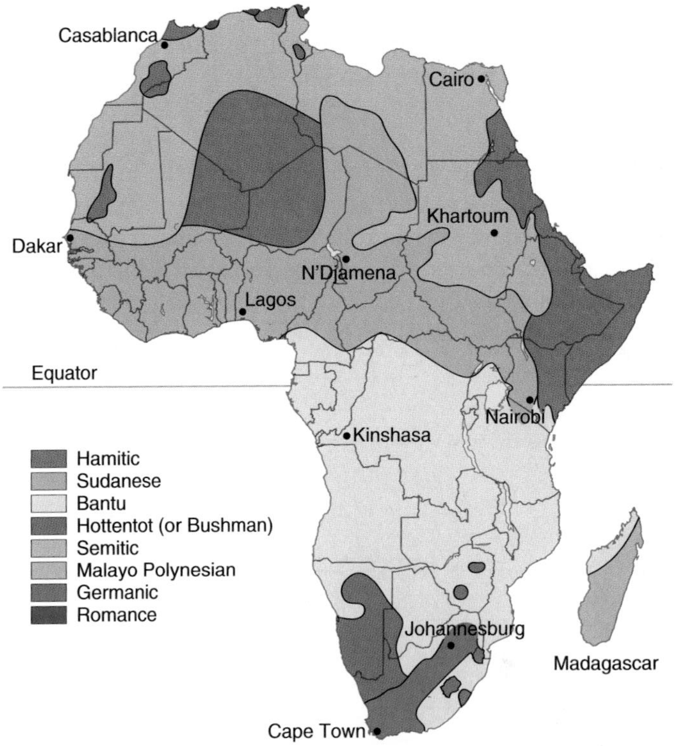


**FIGURE 2.15 Language regions of Africa.** Maps such as this one may give the false impression of uniformity within a given area—for example, that only Bantu is spoken over much of southern Africa. Such maps are intended to represent only the predominant language in an area.

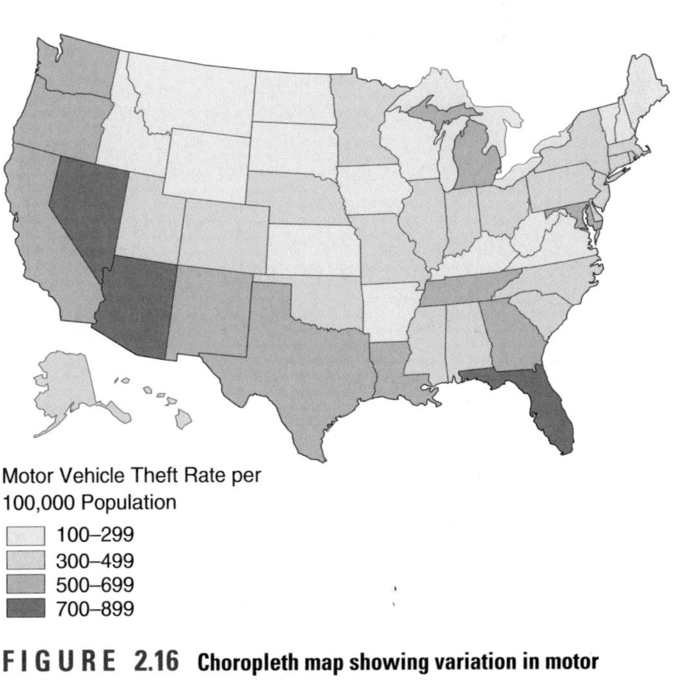


**FIGURE 2.16 Choropleth map showing variation in motor vehicle theft rates by state, 1998.** Quantitative variation by area is more easily visualized in map than in tabular form. Source: *Redrawn from* Crime and Justice Atlas 2000, *U.S. Department of Justice, p. 55.*

wishes to convey, the sizes and shapes of areas may be altered, distances and directions may be distorted, and contiguity may or may not be preserved.

### Line Symbols

As the term suggests, *line symbols* represent features that have length but insignificant width. Some lines on maps do not have numerical significance. The lines representing rivers, political boundaries, roads, and railroads, for example, are not quantitative. They are indicated on maps by such standardized symbols as the ones shown here and in Figure 2.11.

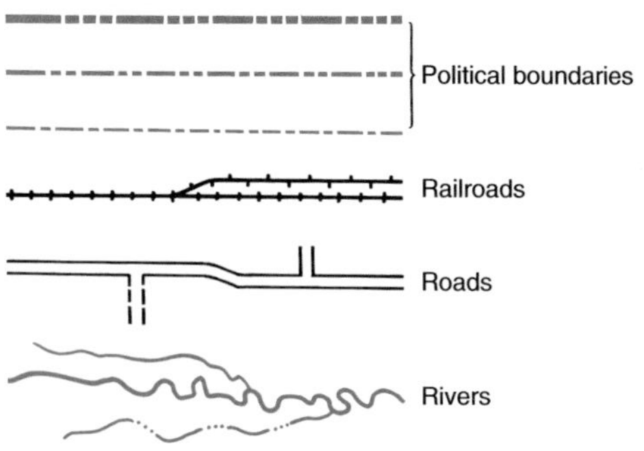


Often, however, lines on maps do denote specific numerical values. Contour lines that connect points of equal elevation above mean sea level are a kind of **isoline,** or line of constant value. Other examples of isolines are *isohyets* (equal rainfall), *isotherms* (equal temperature), and *isobars* (equal barometric pressure).

**Flow-line maps** are used to portray linear movement between places. They may be qualitative or quantitative. Examples of *qualitative* flow maps are those showing ocean currents or airline routes. The lines are of uniform thickness and generally have arrowheads to denote direction of movement. On *quantitative* flow maps, on the other hand, the flow lines are scaled so that their widths are proportional to the amounts they represent. Migration, traffic, and commodity flows are usually portrayed this way. The location of the route taken, the direction of movement, and the volume of flow can all be depicted. The amount shown may be either an absolute or a derived value—for example, the actual traffic flows or a per mile figure. In Figure 2.18, the width of the flow lines is proportional to the number of interstate migrants in the United States. Another flow-line map appears in Figure 5.7.

## Deceptive Practices

Most people have a tendency to believe what they see in print. Maps are particularly persuasive because of the implied precision of their lines, scales, symbol placement, and information content. It is useful to be reminded that the printed image doesn't always mirror reality, that the

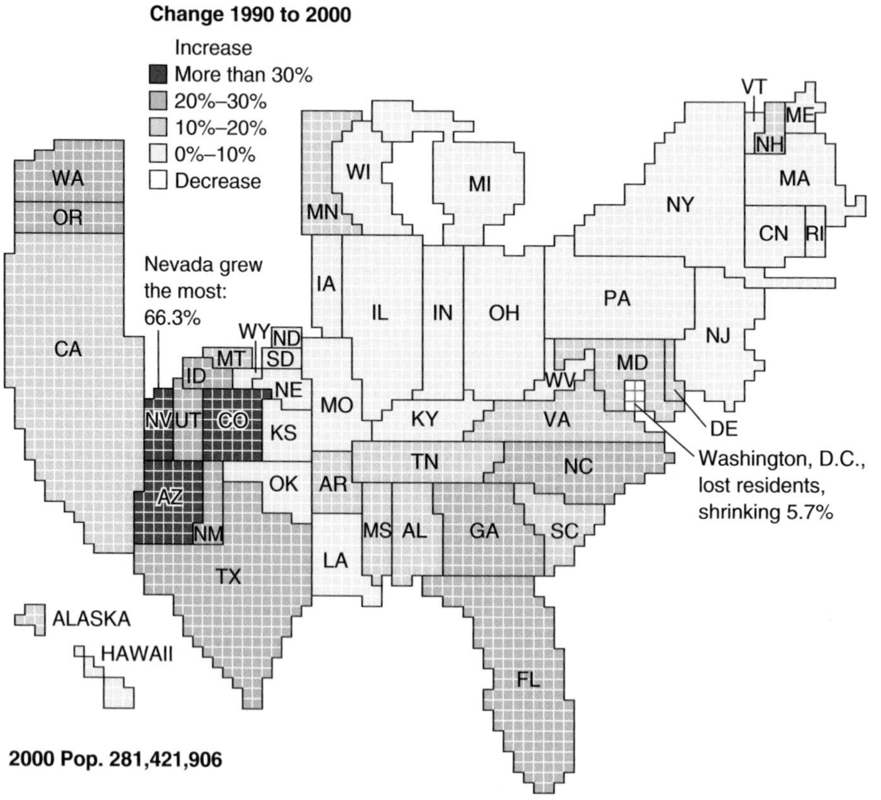


**FIGURE 2.17** A cartogram in which each state is sized according to its number of residents in the year 2000 as counted by the U.S. Census Bureau. The cartogram also shows the percent change in population between 1990 and 2000.
Source: *U.S. Bureau of the Census.*

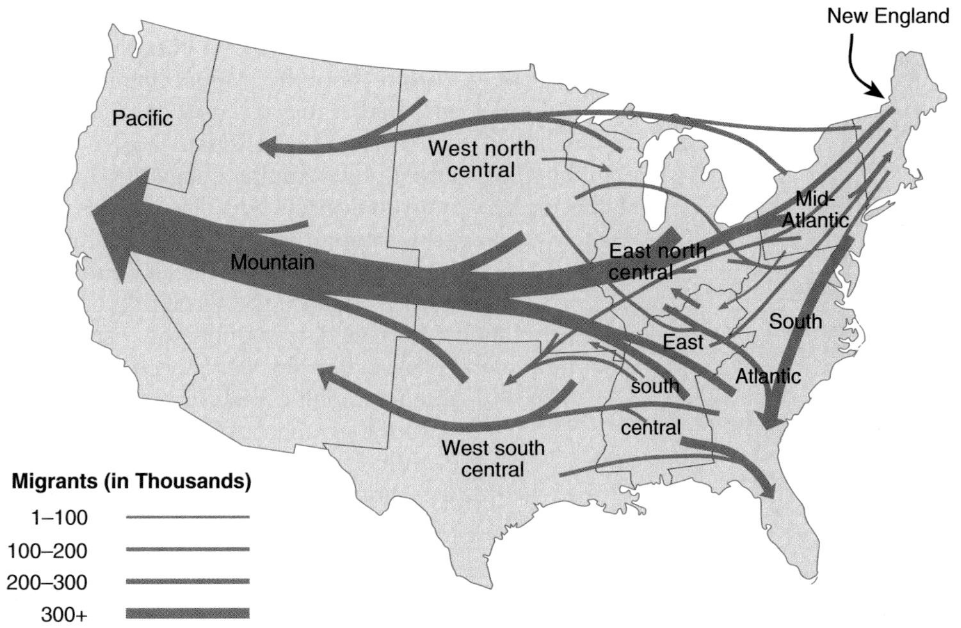


**FIGURE 2.18 A quantitative flow-line map of migration patterns in the United States in the 1950s.**

truth can be distorted. As in all forms of communication, the message conveyed by a map reflects the intent and, perhaps, the biases of its author. Maps can subtly or blatantly manipulate the message they impart or contain intentionally false information.

Sometimes the cause of cartographic distortion has been ignorance, as when the cartographers of the Middle Ages filled the unknown interiors of continents with mythical beasts. Other times the motivation for distortion has been propaganda; some of the maps of Nazi Germany fall into this category. Yet another reason for falsification has been to thwart foreign military and intelligence operations.

The chief cartographer of the USSR acknowledged in 1988 that for 50 years the policy of the Soviet Union had been to deliberately falsify almost all publicly available maps of the country. The policy apparently stemmed from the belief that cartography should serve military needs and began when the government mapping agency was put under the control of the security police. The types of cartographic distortions on Soviet maps included the displacement and omission of features and the use of incorrect grid coordinates (Figure 2.19). The routes of highways, rivers, and railroads were sometimes altered by as much as 10 kilometers (6 mi). A city or town might be shown on the east bank of a river, when in fact it is on the west bank. Even when features were shown correctly, the latitude and longitude grid might be misplaced. The distortion was not consistent over time. Thus, a town shown on one side of a river on one map might appear on the opposite bank on a later map, move several kilometers on a subsequent edition, and disappear altogether on yet another.

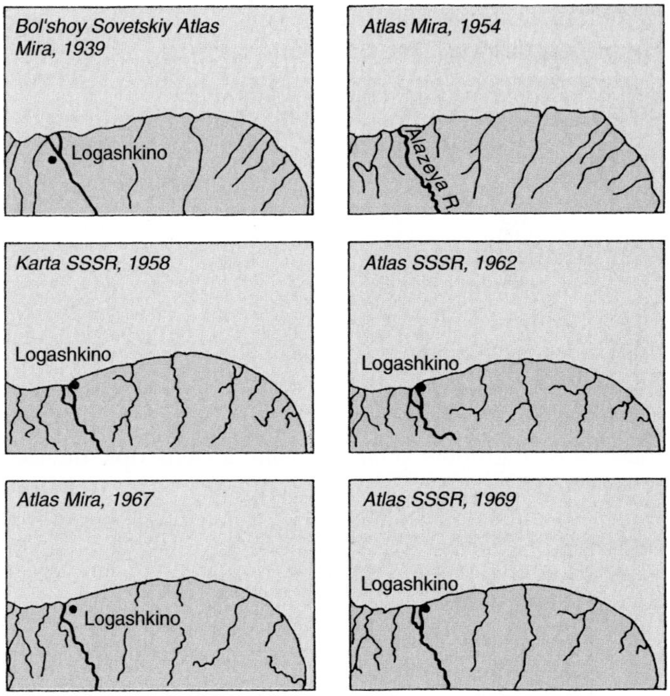


**FIGURE 2.19 Representation of Logashkino and vicinity as portrayed on various Soviet maps.** Deliberate cartographic disinformation was a Cold War tactic of the Soviet Union. We usually trust maps to tell us exactly where things are located. On the maps shown, however, the town of Logashkino migrates from west of the river away from the coast to east of the river on the coast, while the river itself gains and loses a distributary and, in 1954, the town itself disappears. The changing misinformation was intended to obscure the precise location of possible military targets from potential enemies.

In summary, maps can distort and lie as readily as they can convey verifiable spatial data or scientifically valid analyses. The more that map users are aware of those possibilities and the more understanding they possess of map projections, symbolization, and common forms of thematic and reference mapping standards, the more likely they are to reasonably question and clearly understand the messages maps communicate.

# Remote Sensing

When topographic maps were first developed, it was necessary to obtain the data for them through fieldwork, a slow and tedious process, which involved relating a given point on the earth's surface to other points by measuring its distance, direction, and altitude. The technological developments that have taken place in aerial photography since the 1930s have made it possible to speed up production and greatly increase the land area represented on topographic maps. Aerial photography is only one of a number of remote sensing techniques now employed.

**Remote sensing** is a relatively new term, but the process it describes—detecting the nature of an object from a distance—has been taking place for well over a century. Soon after the development of the camera, photographs were made from balloons and kites. Even carrier pigeons wearing miniature cameras that took exposures automatically at set intervals were used to take aerial photographs of Paris. The airplane, first used for mapping in the 1930s, provided a platform for the camera and the photographer so that it was possible to take photographs from planned positions.

## Aerial Photography

Although there is now a variety of sensing devices, aerial photography employing cameras with returned film remains a widely used remote sensing technique. Mapping from the air has certain obvious advantages over surveying from the ground, the most evident being the bird's-eye view that the cartographer obtains. Using stereoscopic devices, the cartographer can determine the exact slope and size of features, such as mountains, rivers, and coastlines. Areas that are otherwise hard to survey, such as mountains and deserts, can be mapped easily from the air. Furthermore, millions of square miles can be surveyed in a very short time. Aerial photographs must, of course, be interpreted by using such clues as the size, shape, tone, and color of the recorded objects before maps can be made from them. Maps based on aerial photographs can be made quickly and revised easily so that they are kept up-to-date. With aerial photography, the earth can be mapped more accurately, more completely, and more rapidly than ever before.

In 1975, the U.S. Department of the Interior instituted the National Mapping Program to improve the collection and analysis of cartographic data and to prepare maps that would assist decision makers who deal with resource and environmental problems. One of the first goals of the program was to achieve complete coverage of the country with orthophotographic imagery for all areas not already mapped at the scale of 1:24,000.

An **orthophotomap** is a multicolored, distortion-free aerial photographic image to which certain supplementary information (such as place names, a locational grid, boundaries, contour lines, and other symbols) has been added. The prefix *ortho* (from the Greek word *orthos,* meaning "correct") is used because the aerial photographs have been rectified, or corrected, to remove the displacement caused by differences in terrain elevation and camera tilt. An orthophotomap combines the image characteristics of a photograph with the geometric characteristics of a map. Note in Figure 2.20 that, in contrast to the conventional topographic map, the photograph is the chief means of representing information. Orthophotomaps have a variety of uses, which include forest management, soil erosion assessment, flood hazard and pollution studies, and city planning.

Standard photographic film detects reflected energy within the visible portion of the electromagnetic spectrum (Figure 2.21). Although they are invisible, *near-infrared* wavelengths can be recorded on special sensitized infrared film. Discerning and recording objects that are not visible to the human eye, infrared film has proved particularly useful for the classification of vegetation and hydrographic features. Color-infrared photography yields what are called **false-color images,** "false" because the film does not produce an image that appears natural. For example, leaves of healthy vegetation have a high infrared reflectance and are recorded as red on color-infrared film, whereas unhealthy or dormant vegetation appears as blue, green, or gray. Clear water appears as black, but sediment-laden water is light blue.

## Nonphotographic Imagery

For wavelengths longer than 1.2 micrometers (a micrometer is 1 one-millionth of a meter) on the electromagnetic spectrum, sensing devices other than photographic film must be used.

### Remote-Sensing Techniques

**Thermal scanners,** which sense the energy emitted by objects on Earth, are used to produce images of thermal radiation (Figure 2.22). That is, they record the longwave radiation (which is proportional to surface temperature) emitted by water bodies, clouds, and vegetation, as well as by buildings and other structures. Unlike conventional photography, thermal sensing can be employed during nighttime as well as daytime, giving it military applications. It is widely used for studying various aspects of water resources, such as ocean currents, water pollution, surface energy budgets, and irrigation scheduling.

Operating in a different band of the electromagnetic spectrum, **radar** (short for *ra[dio] d[etecting] a[nd] r[anging]*) systems can also be used during the day or night. They transmit pulses of energy toward objects and sense the

(a)

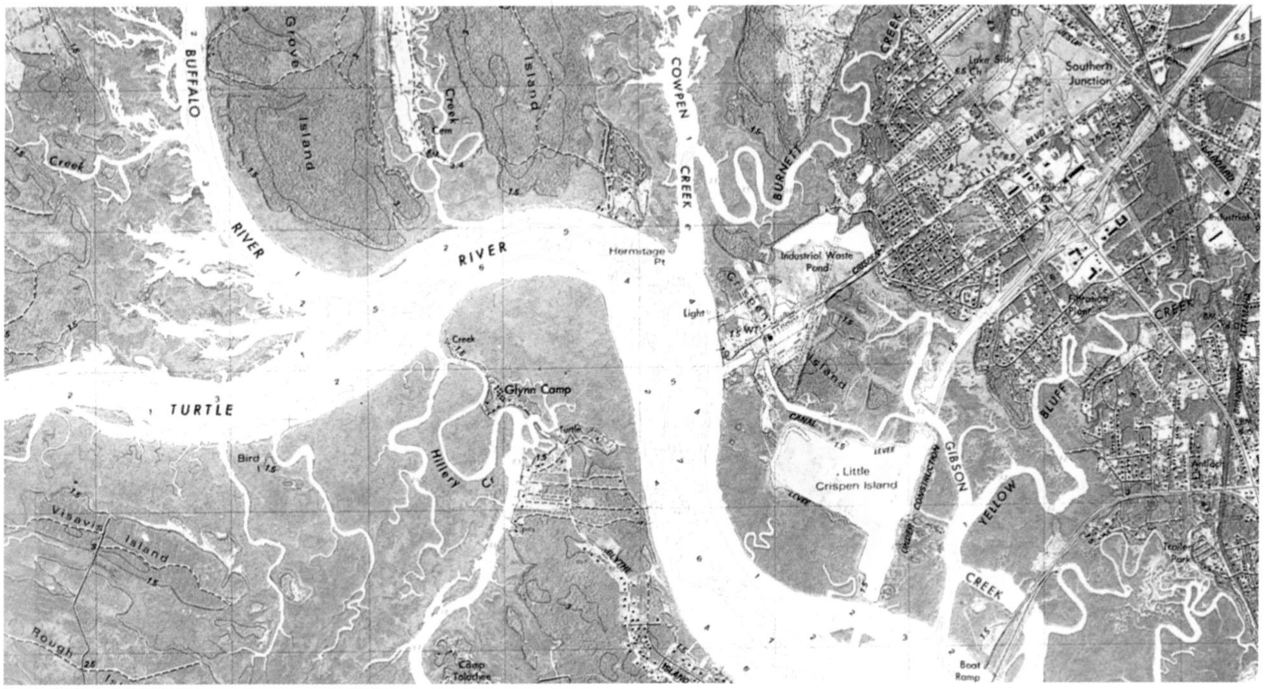


(b)

**FIGURE 2.20** (a) **Topographic map and** (b) **orthophotomap of a portion of the Brunswick West quadrangle in southeastern Georgia.** Because aerial photographs show subtle topographic detail in areas of very low relief, orthophotomaps are especially well suited for portraying marshlands and coastal zones. A useful complement to the standard USGS topographic maps, orthophotomaps are used to update and revise existing maps.
Source: *U.S. Geological Survey.*

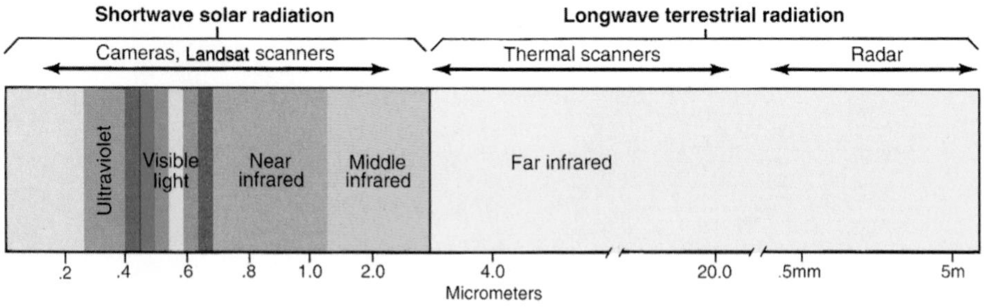


**FIGURE 2.21 Wavelengths of the electromagnetic spectrum in micrometers.** One micrometer equals 1 one-millionth of a meter. Sunlight is made up of different wavelengths. The human eye is sensitive to only some of these wavelengths, the ones we see in the colors of the rainbow. Although invisible, near-infrared wavelengths can be recorded on special sensitized film and by scanners on satellites. The scanners measure reflected light in both the visible and near-infrared portions of the spectrum. Wavelengths longer than 4.0 micrometers characterize terrestrial radiation.

(a)

(b)

**FIGURE 2.22 Thermal radiation images of the World Trade Center site in New York City, 2001.** Following the collapse of the twin towers on September 11, 2001, fire-fighting and rescue teams relied on daily thermal images of the site to detect fire patterns in the rubble and underground. Based on the patterns revealed, the teams decided where to work that day. (a) A few days after the attack, a nearly constant field of heat (shown in red) covered most of the 16-acre site. (b) A month later, the underground blaze was confined largely to where the two towers once stood.
*The New York State Office for Technology (OFT) © 2001.*

energy reflected back. The data are used to create images such as that shown in Figure 2.23, which was produced by radar equipment mounted on an airplane. Because radar can penetrate clouds and vegetation as well as darkness, it is particularly useful for monitoring the locations of airplanes, ships, and storm systems and for mapping parts of the world that are perpetually hazy or cloud-covered, such as the Amazon Basin.

### Satellite Imagery

For more than 30 years, both manned and unmanned spacecraft have supplemented the airplane as the vehicle for imaging the terrain. Concurrently, many steps have been taken to automate mapping, including the use of electronic mapping techniques, automatic plotting devices, and automatic data processing. Many images are now taken either from continuously orbiting satellites, such as those in the U.S. Landsat series and the French SPOT series, or from manned spacecraft flights, such as those of the *Apollo* and *Gemini* missions. Among the advantages of satellites are the speed of coverage and the fact that views of large regions can be obtained.

In addition, because they are equipped to record and report back to the earth information from parts of the electromagnetic spectrum that are outside the range of human eyesight, these satellites enable us to map the invisible. A number of agencies in the United States, Japan, and Russia have launched satellites specifically to monitor the weather. Data obtained by satellites have greatly improved weather forecasting and the tracking of major storm systems and, in the process, have saved countless lives. The satellites are one of the sources of the weather maps shown daily on television and in newspapers.

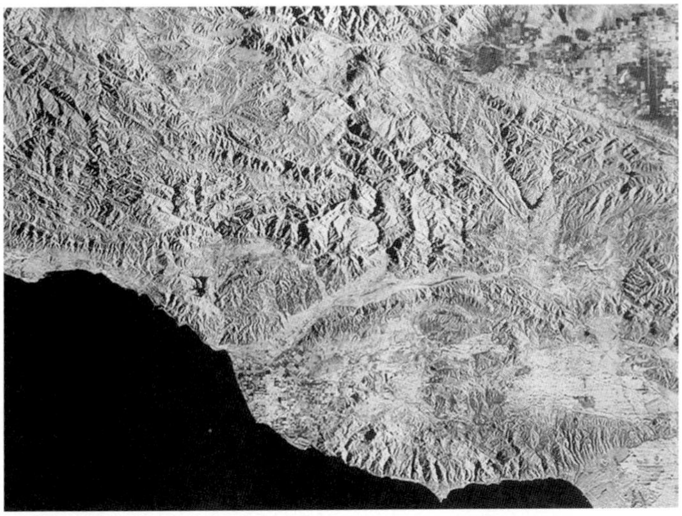

**FIGURE 2.23 A radar image of Los Angeles, California.** This mosaic was compiled from many radar image strips. Radar images can be collected regardless of the weather or time of day.
Source: *U.S. Geological Survey.*

Mapping is only one of the applications of remote sensing, which has also proven to be an effective method of conducting resource surveys and monitoring the natural environment. Geologists have found remote sensing to be particularly useful in conducting resource surveys in desert and remote areas. For example, information about vegetation or folding patterns of rocks can be used to help identify likely sites for mineral or oil prospecting. Remote sensing imagery has been used to monitor a variety of environmental phenomena, including water pollution, the effects of acid rain, and rain forest destruction. Because remotely sensed images can be used to calculate such factors as rates of transpiration and photosynthesis, they are invaluable for modeling relationships between the atmosphere and the earth's surface.

Military applications of remotely sensed images include better aircraft navigation, improved weapons targeting, and enhanced battlefield management and tactical planning, which raises the question of who should have access to the information (see "Civilian Spy Satellites").

Perhaps the best known remote sensing spacecraft are the **Landsat satellites,** the first of which was launched in 1972. They take about 1 hour and 40 minutes to orbit the earth and provide repetitive coverage of the planet every 16 days. The satellites carry scanning instruments that pick up sunlight reflected by foliage, water, rocks, and other objects. Rather than recording data photographically, the Landsat satellites relay electronic signals to receiving stations, where computers convert them into photo-like images that can be adjusted to fit special map projections. Composite images can be made by combining information from different wavelengths of light energy. Landsat 7, launched in 1999, acquires data in the visible, near infrared, shortwave infrared, and thermal infrared spectrums.

Landsat's sensors are capable of resolving objects larger than 15 meters (50 ft) apart. The Landsat satellites are in a polar orbit; as they fly from south to north, the earth turns below them, so that each orbit covers a strip of surface adjacent to the previous one. The satellites image the earth in 183-kilometer (115-mi) swaths at the same local time on consecutive passes, repeating the pattern of successive ground tracks at 16-day intervals.

Landsat was the first civilian program of the National Aeronautics and Space Administration (NASA) devoted to studying the earth and how the global environment is changing. Its primary aim is to create a long-term collection of consistently calibrated earth imagery to aid researchers investigating the complex interactions of all the components—air, water, land, and plant and animal life—of the earth's system. The satellites have provided a continuous set of data to users since 1972 with images that are consistent in terms of data acquisition format, geometry, spatial resolution, calibration, coverage, and spectral characteristics.

Landsat images have a variety of research applications, including the following:

- tracing ocean currents
- assessing water quality in lakes
- mapping snow cover, glaciers, and polar ice sheets
- analyzing soil and vegetation conditions
- monitoring global deforestation
- monitoring strip mining reclamation
- identifying geologic structures and associated mineral deposits
- mapping population changes in metropolitan areas.

Some Landsat data are utilized not for long-term scientific research but for immediately monitoring, mapping, and responding to natural and human-caused disasters, such as storms, floods, earthquakes, volcanoes, wildfires, and oil spills (Figure 2.24).

## Geographic Information Systems

A major development in cartography is the use of computers to assist in making maps. Within the past 25 years, computers have become an integral part of almost every stage of the cartographic process, from the collection and recording of data to the production and revision of maps. Although the initial cost of equipment is high, the investment is repaid in the more efficient and more accurate production and revision of maps.

Computers are at the heart of what has come to be known as a **geographic information system (GIS).** A GIS is a computer-based set of procedures for assembling, storing,

# Civilian Spy Satellites

Although remote sensing satellites have been orbiting the earth for about three decades, only since 1999 have detailed satellite images such as that shown here become available to the general public. Until recently, images from commercial satellites were considerably fuzzier and less detailed than those from military satellites. That is no longer the case. In 1994, the federal government lifted technical restrictions on private companies, permitting them to build a new generation of civilian "spy" satellites and to sell the images to any buyer. Space Imaging Corporation, based in Denver, Colorado, launched the first high-resolution commercial satellite, *Ikonos I* (after the Greek word for "image") in 1999. Two other U.S. firms launched similar satellites the following year. The new imaging technology is so powerful that it can detect and record objects on the ground as small as 1 meter wide: cars, houses, even hot tubs.

The new imagery has been welcomed by clients such as geologists, city planners, and disaster-relief officials. Nongovernmental public interest groups use the images for such activities as pressuring governments to live up to environmental laws and treaties, tracking refugee movements, detecting illegal waste dumps, and monitoring arms control agreements. At the same time, the sharp visual detail and widespread availability of the images provided by the satellites have raised fears about national security. "Any time a powerful new technology is introduced, there's a battle over the uses to which it's put," one intelligence official noted.

**A satellite image of San Francisco International Airport.** Only recently have detailed, accurate, high-quality images such as this one been taken from commercial satellites and made available to the general public. 

"Here, the potential for beneficial uses is very high, on balance. But the potential for abuse certainly exists and we'll no doubt see some of that."

One of the chief concerns is that the images, which rival military reconnaissance in sharpness and accuracy, can be purchased by those who would threaten the national welfare. In wartime, for example, foes could use them to track the location of military troops and equipment. Terrorists could use the imagery to plan surprise attacks. General Richard B. Myers, head of the U.S. Space Command, warns that the government must decide what to do in times of armed conflict. "There's going to be risk in operations like this when you're selling what can be used against you."

American companies, however, operate under a number of security restrictions. They are forbidden from selling images to customers in several countries, including Cuba and North Korea. And in the interests of national security, the government can put any area off limits. Soon after September 11, 2001, for example, the Pentagon purchased exclusive rights to all satellite images of Afghanistan and Pakistan from Space Imaging.

Some experts believe that the civilian satellites can promote peace by reducing military secrecy and false governmental claims. "It provides an independent check on what the government is saying," one scientist noted. The U.S. military, for example, claimed in 1999 that a top-secret missile base in North Korea posed a serious threat to the United States and argued for the construction of a multibillion-dollar missile shield, in violation of the antiballistic missile treaty. But the Federation of American Scientists (FAS), a nonprofit group, used imagery from *Ikonos I* to contradict the claim. Given the absence of transportation links, paved roads, propellant storage, and staff housing, the FAS concluded the facility was "incapable of supporting the extensive test program that would be needed to fully develop a reliable missile system."

**QUESTIONS TO CONSIDER**

1. Do you think that the availability of the new, detailed satellite images is a potential threat to national security? Might access to images of their enemies make belligerent countries more dangerous than they already are? Why or why not?
2. In what ways might access to satellite imagery stem the tides of environmental and social destabilization and foster peace?
3. Should the federal government, which licenses the satellites, be allowed to exercise "shutter control," cutting off image sales during wartime? Defend your answer.

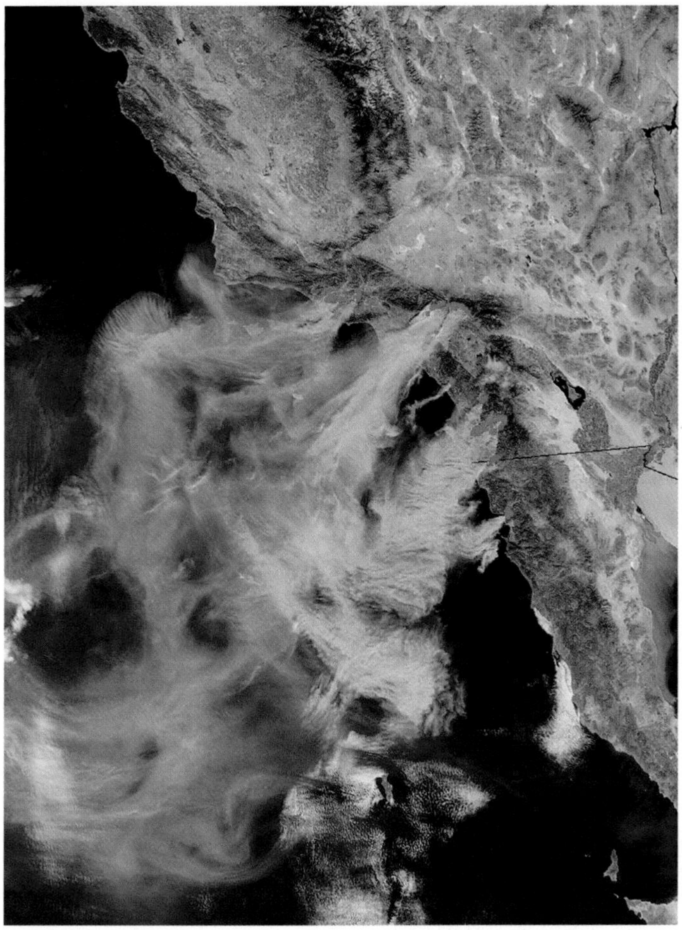

**FIGURE 2.24** Three major clusters of fires are evident in this satellite image of southern California taken on October 26, 2003: near Los Angeles, in the San Bernardino Mountains, and in San Diego. Data from remotely sensed images can be used to keep track of the extent and intensity of fires and to update maps of active fires several times a day. They assist fire managers on the ground in determining where best to position firefighters to contain a blaze, to assess damage after a fire is contained, and to plan recovery efforts. *© Jacques Descloitres, MODIS Rapid Response Team, NASA/GSFC.*

manipulating, analyzing, and displaying geographically referenced information. Any data that can be located spatially can be entered into a GIS. The five major components of a GIS are the following:

1. a data input component that converts maps and other data from their existing form into digital or computer-readable form;
2. a data management component used to store and retrieve data;
3. data manipulation functions that allow data from disparate sources to be used simultaneously;
4. analysis functions that enable the extraction of useful information from the data;
5. a data output component that makes it possible to visualize maps and tables on the computer monitor or as hard copy (such as paper).

## The Geographic Database

The first step in developing a GIS is to create a **geographic database,** a digital record of geographic information from such sources as maps, field surveys, aerial photographs, and satellite imagery. As long as the data are geographically referenced, a GIS can use information from many different sources and in many different forms. The purpose of the study will determine the data to be entered into the database. For a physical geographer studying the sensitivity of wetlands to damage in a particular area, the source data might include maps of rainfall amounts at different points, soil types, vegetative cover, originating points of water pollution, contours, and direction of stream flow. An urban geographer or regional planner, on the other hand, might use a GIS data set that contains the great amount of place-specific information collected and published by the U.S. Census Bureau, including political boundaries, census tracts, population distribution, a building inventory, race, ethnicity, income, housing, employment, and so on.

Once geographic information is in the computer in digital form, the data can be manipulated, analyzed, and displayed with a speed and precision not otherwise possible. Because computers can process millions of facts in seconds, they are particularly useful for researchers who need to analyze many variables simultaneously. The development of geographic information systems has deemphasized the use of maps to store information and has enabled researchers to concentrate on using maps for analyzing and communicating spatial information. With the appropriate software, a computer operator can display any combination of data, showing the relationships among variables almost instantly (Figure 2.25). In this sense, a GIS allows an operator to generate maps that were virtually impossible to create only a few decades ago.

GIS operations can produce several types of output: displays on a computer monitor, listings of data, or hard copy. When a map is to be produced, the specialist can quickly call up the desired data. Geographic information systems are particularly useful for revising existing maps because outdated data—for example, population sizes—can be modified or replaced easily.

## Applications of GIS

Who uses geographic information systems? Tens of thousands of people in a variety of fields use them for a variety of purposes. Every issue of *ArcNews,* a monthly publication of the Environmental Systems Research Institute, Inc., describes numerous examples of "GIS in Action" from many fields. In human geography, the vast and growing array of spatial data has encouraged the use of GIS to explore models of regional economic and social structure, to examine transportation systems and urban growth patterns, to study patterns of voting behavior, and so on. For physical geographers, the analytic and modeling capabilities of GIS are fundamental to the

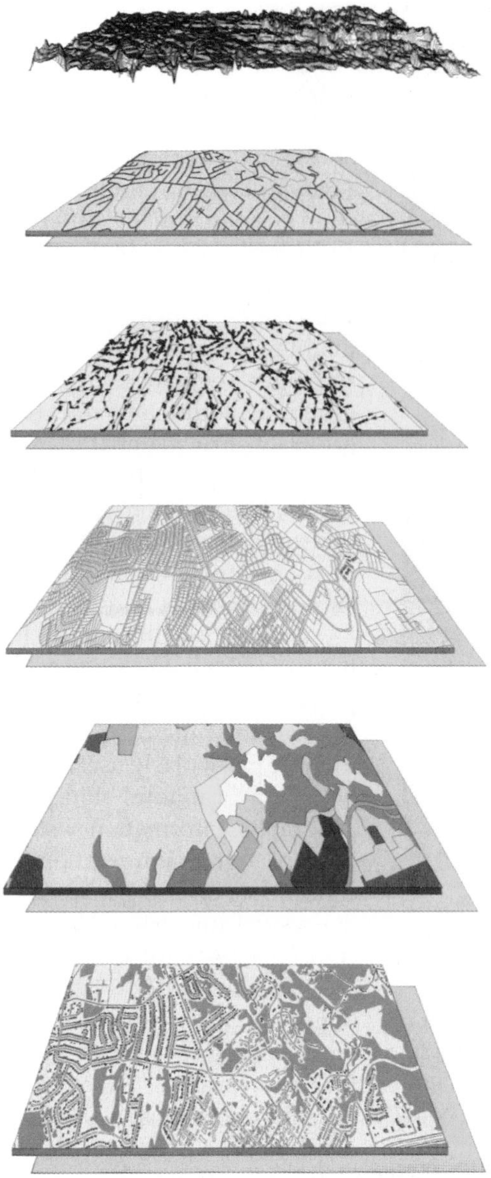

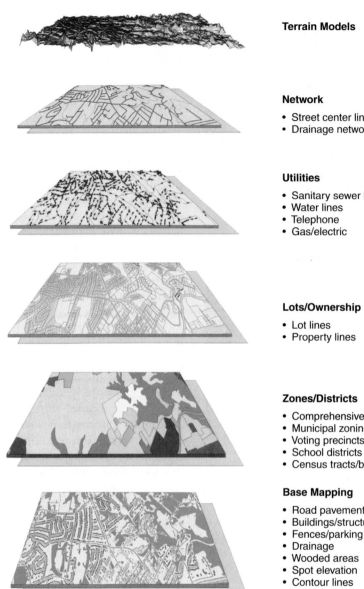

**FIGURE 2.25** Information layering is the essence of a GIS. Map information that has been converted to digital data is stored in the computer in different data "layers." A GIS enables the user to combine just the layers that are desired to produce a composite map and to analyze how those variables relate to each other. *Reprinted by permission of Shaoli Huang.*

understanding of processes and interrelations in the natural environment.

In addition to geographers, researchers in fields ranging from archaeology to zoology use geographic information systems, as a few examples indicate.

- Biologists and ecologists use GIS to study numerous environmental problems, including air and water pollution, landscape conservation, wildlife management, and the protection of endangered species.
- Epidemiologists need accurately mapped information to study both the diffusion of diseases, such as malaria, SARS, AIDS, and dengue fever, and entomological risk factors.
- GIS software has made it possible for political scientists to evaluate existing legislative districts using criteria such as compactness and contiguity and to suggest ways the boundaries of the districts might be redrawn.
- Sociologists have used GIS to identify clusters of segregation and to examine the changing structures of segregation over time.

Many companies in the private sector also use computerized mapmaking systems. Among others, oil and gas companies, restaurant chains, soft-drink bottlers, and car rental companies rely on GIS systems to perform such diverse tasks as identifying drilling sites, picking locations for new franchises, analyzing sales territories, and calculating optimal driving routes.

Many bureaus and agencies at the local, regional, state, and national levels of government employ geographic information systems. These include such departments as highway and traffic control, public utilities, and planning. Law enforcement agencies use spatially based software packages to analyze patterns of crime, identify "hot spots" of criminal activity, and redeploy police resources accordingly.

Government officials also use GIS to help plan emergency response to natural and human-induced disasters, such as tornadoes, hurricanes, earthquakes, floods, and wildfires. Because disasters usually occur suddenly, threatening people and structures, they create chaos and panic. Increasingly, GIS technology is being used to help communities prepare for disasters and create response plans. By merging information on the types of roads, the locations of fire stations, the anticipated response times of fire and rescue squads, and other data, for example, planners can produce maps depicting evacuation zones, evacuation routes, and shelter locations. When disaster strikes, whether it is a wildfire in Arizona or a tornado in Oklahoma, maps produced using GIS have proved invaluable in tasks such as locating houses, identifying property ownership, helping responders decide where to send field crews or rescue workers, and siting field hospitals and paramedic bases. In the wake of a disaster, maps have been used to locate damaged structures, assess property damage, and prepare for debris removal.

One of the most dramatic examples of GIS applications occurred in the days immediately following the attacks on the World Trade Center on September 11, 2001. Response and recovery teams needed to know such things as the stability of the rubble and the remaining buildings, which subway lines were damaged, where watermains were located, and where there were utility outages. With fires still burning at the site of the attack, and the mayor's planning office destroyed, GIS experts immediately began gathering the data needed to provide highly accurate maps of the site. The maps, continually updated, helped emergency managers track the expansion or abatement of the underground fires and enabled them to determine how rescue equipment could gain access to the site, where to safely position large recovery equipment, and by what route debris could be removed (Figure 2.26).

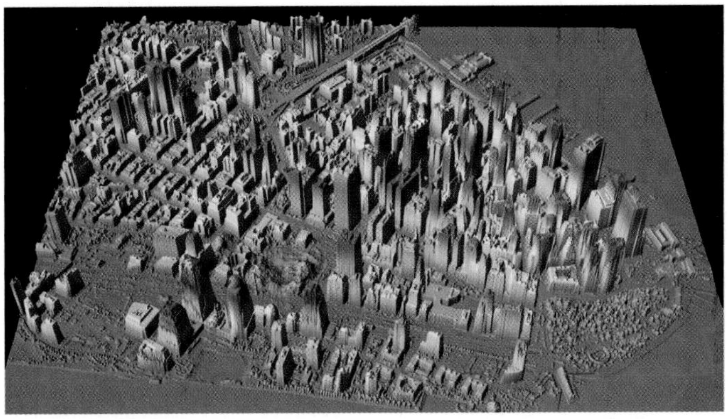

**FIGURE 2.26 A three-dimensional LIDAR image of the site of the World Trade Center.** Accurately mapping the wreckage of the World Trade Center using a combination of remote sensing, GPS, and GIS was invaluable in recovery and cleanup efforts after the attacks of 9/11. The GPS system was used to position both ground and airborne mapping sensors. Aircraft outfitted with three kinds of sensors collected high-resolution aerial photography, thermal imagery, and LIDAR (Light Detection and Ranging) data. Within hours after the data were collected, GIS professionals from government, industry, and academia merged the digital data to produce large, high-resolution images of the building structures and the surrounding area. The three-dimensional models created by the LIDAR system enabled engineers to calculate the volume of the rubble piles, track their movement and change, and determine the reach needed by cranes to remove it. *© NOAA/U.S. Army JPSD.*

Because of the growing importance of GIS in all sorts of public and private spatial inquiries, demand in the job market is growing for those skilled in its techniques. Most university courses in GIS are taught in geography departments, and "GIS/remote sensing" is a primary occupational specialty for which many geography undergraduate and graduate majors seek preparation.

## Summary

In this chapter, we did not attempt to discuss all aspects of the field of cartography. Mapmaking and map design, systems of land survey, map compilation, and techniques of map reproduction are among the topics that were deliberately omitted. The intent of the chapter was to introduce those aspects of map study that will aid in map reading and interpretation and to hint at new techniques of map creation and of the design and purpose of geographic information systems.

Maps are one of the oldest and most basic means of communication. They are as indispensable to the geographer as words, photographs, and quantitative techniques of analysis. Geographers are not unique in their dependence on maps. People involved in the analysis and solution of many of the critical issues of our time also rely on them. Global warming, pollution, globalization, energy supplies, national security, crime, and public health are just a few of the issues that call for the accurate representation of elements on the earth's surface.

The roots of modern mapping lie in the 17th century. Three key developments that made it both desirable and possible to map the surface features of the earth were the rediscovery of the works of earth scientists of ancient Greece, transoceanic voyages, and the invention of the printing press.

The grid system of longitude and latitude is used to locate points on the earth's surface. Latitude is the measure of distance north and south of the equator, while longitude is the angular distance east or west of the prime meridian. Both latitude and longitude are given in degrees, which for greater precision can be subdivided into minutes and seconds. Global positioning devices enable users to locate their positions on the planet.

All systems of representing the curved earth on a flat map distort one or more earth features. Any given projection will distort area, shape, distance, and/or direction. Cartographers select the projection that best suits their purpose, and they may elect to use an equal-area, conformal, or equidistant projection or one that shows correctly the directions from a single point to all others. Many useful projections have none of these qualities, however.

Among the most accurate and most useful large-scale maps are the topographic quadrangles produced by a country's chief mapping agency. They contain a wealth of information about both the physical and cultural landscape and are used for a variety of purposes.

In recent years, remote sensing techniques have resulted in the faster and more accurate mapping of the world. Both aerial photography and satellite imagery can discern reflected energy in the visible and near-infrared portions of the spectrum. Applications of remote sensing include mapping, monitoring the environment, and conducting resource surveys. The need to store, process, and retrieve the vast amounts of data generated by remote sensing has spurred the development of geographic information systems. The growing power and flexibility of computers and associated software have made GIS mapping invaluable in the search for solutions to a variety of problems.

As you read the remainder of this book, note the many different uses of maps. For example, notice in Chapter 3 how important maps are to your understanding of the theory of continental drift; in Chapter 7, how maps aid geographers in identifying cultural regions; and in Chapter 8, how behavioral geographers use maps to record people's perceptions of space.

## Key Words

area cartogram 39
azimuthal projection 31
cartography 27
choropleth map 39
conformal projection 31
contour interval 36
contour line 36
equal-area projection 31
equidistant projection 31
equivalent projection 31
false-color image 32
flow-line map 40
geographic database 47
geographic information system (GIS) 45
Global Positioning System (GPS) 28
globe properties 30
grid system 27
International Date Line 28
isoline 40
Landsat satellite 45
latitude 27
longitude 27
map projection 30
orthophotomap 42
prime meridian 27
radar 42
remote sensing 42
scale 33
thermal scanner 42
topographic map 34
value-by-area map 39

## For Review & Consideration

1. What important map and globe reference purpose does the *prime meridian* serve? Is the prime, or any other, meridian determined in nature or devised by humans? How is the prime meridian designated or recognized?
2. What happens to the length of a degree of longitude as one approaches the poles? What happens to a degree of latitude between the equator and the poles?
3. From a world atlas, determine, in degrees and minutes, the locations of New York City; Moscow, Russia; Sydney, Australia; and your hometown.
4. List at least five properties of the globe grid.
5. Briefly make clear the differences in properties and purposes of *conformal, equivalent,* and *equidistant* projections. Give one or two examples of the kinds of map information that would best be presented on each type of projection. Give one or two examples of how misunderstandings might result from data presented on an inappropriate projection.

6. In what different ways can *map scale* be presented? Convert the following map scales into their verbal equivalents.

   1:1,000,000 1:63,360 1:12,000

7. What is the purpose of a *contour line?* What is a *contour interval?* What landscape feature is implied by closely spaced contours?

8. What kinds of data acquisition are suggested by the term *remote sensing?* Describe some ways in which different portions of the spectrum can be sensed. To what uses are remotely sensed images put?

9. What are the basic components of a *geographic information system?* How is spatial information recorded in a *geographic database?* What are some of the uses of computerized mapmaking systems?

# Selected References

### WEBSITES

The World Wide Web has a tremendous variety of sites pertaining to geography. Websites relevant to the subject matter of this chapter appear in the "Web Links" section of the Online Learning Center associated with this book. Access it at www.mhhe.com/getis10e/.

American Cartographic Association. Committee on Map Projections. *Choosing a World Map: Attributes, Distortions, Classes, Aspects.* Special Publication No. 2. Falls Church, Va.: American Congress on Surveying and Mapping, 1988.

———. *Matching the Map Projection to the Need.* Special Publication No. 3. Falls Church, Va.: American Congress on Surveying and Mapping, 1991.

———. *Which Map Is Best? Projections for World Maps.* Special Publication No. 1. Falls Church, Va.: American Congress on Surveying and Mapping, 1986.

Barnes, Scottie B. "GPS Comes Down to Earth." *Mercator's World* 4, no. 3 (May/June 1999): 62–63.

Brown, Lloyd A. *The Story of Maps.* Boston: Little, Brown, 1949; reprint ed., New York: Dover, 1977.

Campbell, James B. *Introduction to Remote Sensing.* 2d ed. New York: Guilford, 1996.

Campbell, John. *Map Use and Analysis.* 4th ed. New York: McGraw-Hill, 2001.

Dent, Borden. *Cartography: Thematic Map Design.* 5th ed. Dubuque, Iowa: WCB/McGraw-Hill, 1998.

Easterbrook, Don, and Dori Kovanen. *Interpretation of Landforms from Topographic Maps and Air Photographs.* Englewood Cliffs, N.J.: Prentice Hall, 1998.

Lillesand, Thomas M., and Ralph W. Kiefer. *Remote Sensing and Image Interpretation.* 4th ed. New York: John Wiley & Sons, 1999.

MacEachren, Alan M. *How Maps Work.* New York: Guilford, 1995.

Monmonier, Mark. *Drawing the Line: Tales of Maps and Cartocontroversy.* New York: Henry Holt, 1995.

———. *How to Lie with Maps.* 2d ed. Chicago: University of Chicago Press, 1996.

———. "Perfectly Flat: Orthophotos and Orthophotoquad Maps." *Mercator's World* 6, no. 1 (Jan./Feb. 2001): 50–53.

Muehrcke, Phillip C. *Map Use: Reading, Analysis, Interpretation.* Revised 4th ed. Madison, Wis.: JP, 2001.

Robinson, Arthur H., et al. *Elements of Cartography.* 6th ed. New York: John Wiley & Sons, 1995.

Thrower, Norman J. W. *Maps and Civilization: Cartography in Culture and Society.* 2d ed. Chicago: University of Chicago Press, 1999.

———. "A Primer on Projections." *Mercator's World* 5, no. 6 (Nov./Dec. 2000): 32–38.

Tyner, Judith. *Introduction to Thematic Cartography.* Englewood Cliffs, N.J.: Prentice Hall, 1992.

Wilford, John Noble. "Revolutions in Mapping." *National Geographic* 193, no. 2 (Feb. 1998): 6–39.

———. *The Mapmakers.* Revised ed. New York: Alfred A. Knopf, 2000.

PART 1

# The Earth Science Tradition

# The Earth Science Tradition

## PART 1

**For nearly a month the mountain had rumbled, emitting puffs of steam and flashes of fire. Within the past week, on its slopes and near its base, deaths had been recorded from floods, mud slides, and falling rock. A little after 8:00 on the morning of May 8, 1902, the climax came for volcanic Mount Pelée and the thriving port of Saint Pierre on the island of Martinique. To the roar of one of the biggest explosions the world has ever known and to the clanging of church bells aroused in swaying steeples, a fireball of gargantuan size burst forth from the upper slope of the volcano. Lava, ash, steam, and superheated air engulfed the town, and 29,933 people met their death.**

**More selective, but for those victimized just as deadly, was the sudden "change in weather" that struck central Illinois on December 20, 1836. Within an hour, preceded by winds gusting to 70 miles per hour, the temperature plunged from 40°F to –30°F. On a walk to the post office through slushy snow, Mr. Lathrop of Jacksonville, Illinois, found, just as he passed the Female Academy, that "the cold wave struck me, and as I drew my feet up the ice would form on my boots until I made a track . . . more like that of [an elephant] than a No. 7 boot." Two young salesmen were found frozen to death along with their horses; one "was partly in a kneeling position, with a tinderbox in one hand, a flint in the other, with both eyes open as though attempting to light the tinder in the box." Others died, too—inside of horses that had been disemboweled and used as makeshift shelters, in fields, woods, and on roads both a short distance and an eternity from the travelers' destinations.**

**Fortunately, few human–environmental encounters are as tragic as these.** Rather, the physical world in all its spatial variation provides the constant and generally passive background against which the human drama is played. It is to that background that physical geographers, acting within the earth science tradition of the discipline, direct their attention. Their primary concern is with the natural rather than the cultural landscape and with the interplay between the encompassing physical world and the activities of humans.

The interest of those working within the earth science tradition of geography, therefore, is not necessarily in the physical sciences as an end in themselves but in physical processes as they create landscapes and environments of significance to humankind. The objective is not solely to trace the physical and chemical reactions that have produced a piece of igneous rock or to describe how a glacier has scoured the rock; it is the relationship of the rock to people. Geographers are interested in what the rock can tell us about the evolution of the earth as our home or in the significance of certain types of rock for the distribution of mineral resources and fertile soil.

The questions that physical geographers ask do not usually deal with the catastrophic, though catastrophe is an occasional element of human–earth-surface relationships. Rather, they ask questions that go to the root of understanding the earth as the home and the workplace of humankind. What does the earth as our home look like? How have its components been formed? How are they changing, and what changes are likely to occur in the future through natural or human causes? How are environmental features, in their spatially distinct combinations, related to past, present, and prospective human use of the earth? These are some of the basic questions that geographers who follow the earth science tradition attempt to answer.

The three chapters in Part 1 of our review of geography deal, in turn, with landforms, weather and climate, and natural resources. Although the subject matter of each clearly identifies it with the fundamental earth science background of the discipline, a brief word about the content and purpose of each chapter is appropriate.

The treatment of so vast a field as landforms (Chapter 3) must be highly selective within an introductory text. The aim here is simply to summarize the great processes by which landforms are created and to depict the general classes of features resulting from those processes without becoming overly involved in scientific reasoning and technical terms.

In Chapter 4, "Physical Geography: Weather and Climate," the major elements of the atmosphere—temperature, precipitation, and air pressure—are discussed, and their regional generalities and associations in patterns of world climates are presented. The study of weather and climate adds coherence to our understanding of the earth as the home of people. Frequent reference will be made to the climatic background of human activities in later sections of our study.

Part 1 concludes with Chapter 5, "The Geography of Natural Resources." Here, we first explore the renewable and nonrenewable energy and nonfuel mineral resources that are the bases of today's diversified industrial and service economies. Later in the chapter, we discuss the land resources supporting all human life and activity and review the strategies of resource management to achieve environmentally sustainable economies.

The focus of Part 1, therefore, is on the earth as the environment and habitat of humans. Both the diversity and reciprocal nature of human-environmental interactions are recurring themes not only of the following three chapters but in all geographic inquiry.

**A storm brewing at Pictured Rocks National Lakeshore, Michigan.** © Konrad Zobel/Corbis Images.

Although too early for sunbathers and snorkelers, the Hawaiian Islands will have a new island to add to their collection, which contains such scenic beauties as Oahu, Maui, and Kauai. It is Loihi, 0.8 kilometers (0.5 mi) below sea level, just 27 kilometers (17 mi) from the big island of Hawaii. Because the speed of its ascent must be measured in geologic time, it probably will not appear above the water surface for another million or so years. It is a good example, however, of the ceaseless changes that take place on the earth's surface. As the westernmost of the islands erode and sink below sea level, new islands arise at the eastern end. In Loihi's most recent explosion in 1996, scientists feared that a giant wave would be set off at the surface that could devastate the islands, including the city of Honolulu and popular Waikiki Beach. Fortunately, this was not the case.

Humans on their trip through life continuously contact the ever-changing, active, moving physical environment. Most of the time, we are able to live comfortably with the changes, but when a freeway is torn apart by an earthquake, or floodwaters force us to abandon our homes, we suddenly realize that we spend a good portion of our lives trying to adapt to the challenges the physical environment has for us.

For the geographer, things just will not stand still—not only little things, such as icebergs or new islands rising out of the sea, or big ones, such as exploding volcanoes changing their shape and form, but also monstrous things, such as continents that wander about like nomads and ocean basins that expand, contract, and split in the middle like worn-out coats. It is a fascinating story, this changing home of humans.

Geologic time is long, but the forces that give shape to the land are timeless and constant. Processes of creation and destruction are continually at work to fashion the seemingly eternal structure upon which humankind lives and works. Two types of forces interact to produce those infinite local variations in the surface of the earth called *landforms:* (1) forces that push, move, and raise the earth's surface and (2) forces that scour, wash, and wear down the surface. Mountains rise and are then worn away. The eroded materials—soil, sand, pebbles, rocks—are transported to new locations and help create new landforms. How long these processes have worked, how they work, and their effects are the subject of this chapter.

Much of the research needed to create the story of landforms results from the work of geomorphologists. A branch of the fields of geology and physical geography, *geomorphology* is the study of the origin, characteristics, and development of landforms. It emphasizes the study of the various processes that create landscapes. Geomorphologists examine the erosion, transportation, and deposition of materials and the interrelationships among climate, soils, plant and animal life, and landforms.

In a single chapter, we can only begin to explore the many and varied contributions of geomorphologists. After discussing the contexts within which landform change takes place, we consider the forces that are building up the earth's surface and then review the forces wearing it down.

## Earth Materials

The rocks of the earth's crust vary according to mineral composition. Rocks are composed of particles that contain various combinations of such common elements as oxygen, silicon, aluminum, iron, and calcium, together with less abundant elements. A particular chemical combination that has a hardness, density, and definite crystal structure of its own is called a **mineral.** Some well-known minerals are quartz, feldspar, and silica. Depending on the nature of the minerals that form them, rocks are hard or soft, more or less dense, one color or another, or chemically stable or not. While some rocks resist decomposition, others are very easily broken down. Among the more common varieties of rock are granites, basalts, limestones, sandstones, and slates.

Although one can classify rocks according to their physical properties, the more common approach is to classify them by the way they formed. The three main groups of rock are igneous, sedimentary, and metamorphic.

### Igneous Rocks

**Igneous rocks** are formed by the cooling and solidification of molten rock. Openings in the crust give molten rock an opportunity to find its way into or onto the crust. When the molten rock cools, it solidifies and becomes igneous rock. The name for underground molten rock is *magma;* above ground, it is *lava. Intrusive* igneous rocks are formed below ground level by the solidification of magma, whereas *extrusive* igneous rocks are created above ground level by the solidification of lava (Figure 3.1).

The composition of magma and lava and, to a limited extent, the rate of cooling determine the minerals that form. The rate of cooling is mainly responsible for the size of the crystals. Large crystals of quartz—a hard, dense mineral—form slowly beneath the surface of the earth, where cool air is not available. When combined with other minerals, quartz forms the intrusive igneous rock called *granite.*

The lava that oozes out onto the earth's surface and makes up a large part of the ocean basins becomes the extrusive igneous rock called *basalt,* the most common rock on the earth's surface. If, instead of oozing, the lava erupts

**FIGURE 3.1 Various rock types.** Source: *Hubbard Scientific Company.*

from a volcano crater, it may cool very rapidly. Some of the igneous rocks formed in this manner contain cavities and are light, such as *pumice.* Some may be dense, even glassy, as is *obsidian.* The glassiness occurs when lava meets standing water and cools suddenly.

## Sedimentary Rocks

Some **sedimentary rocks** are composed of particles of gravel, sand, silt, and clay that were eroded from already existing rocks. Surface waters carry the sediment to oceans, marshes, lakes, or tidal basins. Compression of these materials by the weight of additional deposits on top of them and a cementing process brought on by the chemical action of water and certain minerals cause sedimentary rock to form.

Sedimentary rocks evolve under water in horizontal beds called *strata* (Figure 3.2). Usually one type of sediment collects in a given area. If the particles are large and rounded—for instance, the size and shape of gravel—a gravelly rock called *conglomerate* forms. Sand particles are the ingredient for *sandstone,* whereas silt and clay form *shale* or *siltstone.*

Sedimentary rocks also derive from organic material, such as coral, shells, and marine skeletons. These materials settle into beds in shallow seas, forming *limestone.* If the organic material forms mainly from decomposing vegetation, it can develop into a sedimentary rock called *bituminous coal. Petroleum* is also a biological product, formed during the millions of years of burial by chemical reactions that transform some of the organic material into liquid and gaseous compounds. The oil and gas are light; therefore, they rise through the pores of the surrounding rock to places where dense rocks block their upward movement. Sedimentary rocks vary considerably in color (from coal black to chalk white), hardness, density, and resistance to chemical decomposition.

Large parts of the continents contain sedimentary rocks. For example, nearly the entire eastern half of the United States is overlain with these rocks. Such formations indicate that in the geologic past, seas covered even larger proportions of the earth than they do today.

**FIGURE 3.2** The sedimentary rocks of the Grand Canyon in Arizona are evident in this photograph. *© Robert N. Wallen.*

### Metamorphic Rocks

**Metamorphic rocks** are formed from igneous and sedimentary rocks by earth forces that generate heat, pressure, or chemical reaction. The word *metamorphic* means "changed shape." The internal earth forces may be so great that heat and pressure change the mineral structure of a rock, forming new rocks. For example, under great pressure, shale, a sedimentary rock, becomes *slate,* a rock with different properties. Limestone, under certain conditions, may become *marble,* and granite may become *gneiss* (pronounced nice). Materials metamorphosed at great depths and exposed only after overlying surfaces have been slowly eroded away are among the oldest rocks known on Earth. Like igneous and sedimentary rocks, however, their formation is a continuing process.

Rocks are the constituent ingredients of most landforms. The strength or weakness, permeability, and mineral content control the way rocks respond to the forces that shape and reshape them. Two principal processes alter rocks: (1) the forces that tend to build landforms up and (2) the gradational processes that wear landforms down. All rocks are part of the *rock cycle* through which old rocks are continually transformed into new ones by these processes. No rocks have been preserved unaltered throughout the earth's history.

## Geologic Time

The earth was formed about 4.7 billion years ago. When we think of a person who lives to be 100 years old as having had a long life, it becomes clear that the earth is incredibly old indeed. Because our usual concept of time is dwarfed when we speak of billions of years, it is useful to compare the age of the earth with something more familiar.

Imagine that the height of the Sears Tower in Chicago represents the age of the earth. The tower is 110 stories, or 412 meters (1447 ft), tall. In relative terms, even the thickness of a piece of paper laid on the rooftop would be too great to represent an average person's lifetime. Of the total building height, only 4.8 stories represent the 200 million years that have elapsed since the present ocean floors began to form.

At this moment, the landforms on which we live are ever so slightly being created and destroyed. The processes involved have been in operation for so long that any given location most likely was the site of ocean and land at a number of different times in its past. Many of the landscape features on Earth today can be traced back millions of years. The processes responsible for building up and tearing down those features are occurring simultaneously, but usually at different rates.

In the past 40 years, scientists have developed a useful framework within which one can best study our constantly changing physical environment. Their work is based on the early 20th-century geologic studies of Alfred Wegener, who proposed the theory of **continental drift.** He believed that all landmasses were once united in one supercontinent, which Wegener named Pangaea ("all Earth"), and that over many millions of years the continents broke away from each other, slowly drifting to their current positions. Although Wegener's theory was initially rejected outright, new evidence and new ways of rethinking old knowledge have led to wide acceptance in recent years by earth scientists of the idea of moving continents. Wegener's ideas were a forerunner of the broader **plate tectonics** theory, which is explained in the next section, "Movements of the Continents."

## Movements of the Continents

The landforms mapped by cartographers are only the surface features of a thin cover of rock, the earth's *crust* (Figure 3.3). Above the earth's interior is a partially molten layer called the **asthenosphere.** It supports a thin but strong solid shell of rocks called the **lithosphere,** of which the outer, lighter portion is the earth's crust. The crust consists of one set of rocks found below the oceans and another set that makes up the continents.

The lithosphere is broken into about 12 large and many small, rigid plates, each of which, according to the theory of plate tectonics, slides or drifts very slowly over the heavy, semimolten asthenosphere. A single plate often contains both oceanic and continental crust. Figure 3.5 shows that the North American plate, for example, contains the northwest Atlantic Ocean and most, but not all, of North America. The peninsula of Mexico (Baja California) and part of California are on the Pacific plate.

Scientists are not certain why lithospheric plates move. One reasonable theory suggests that heat and heated material from the earth's interior rise by convection into particular crustal zones of weakness. These zones are sources for the divergence of the plates. The cooled materials then sink downward in subduction zones. In this way, the plates are thought to be set in motion. Strong evidence indicates that, about 225 million years ago, the entire continental crust was connected in one supercontinent, which was broken into

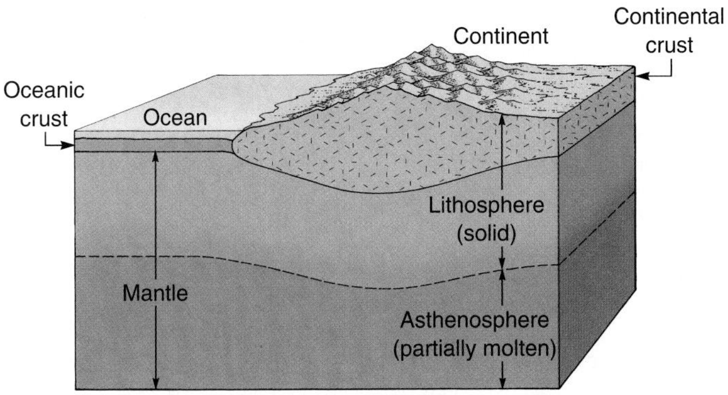


**FIGURE 3.3 The outer zones of the earth (not to scale).** The *lithosphere* includes the crust. The *asthenosphere* lies below the lithosphere.

plates as the seafloor began to spread. The divergence came from the widening of what is now the Atlantic Ocean. Figure 3.4 shows four stages of the drifting of the continents.

Materials from the asthenosphere have been rising along the mid-Atlantic Ocean fracture and, as a result, the seafloor has continued to spread. The Atlantic Ocean is now 6920 kilometers (4300 mi) wide at the equator. If it diverges by a bit less than 2.5 centimeters (1 in.) per year, as scientists have estimated, one could calculate that the separation of the continents did, in fact, begin about 225 million years ago. Notice on Figures 3.5a and 3.6 how the ridge line that makes up the axis of the ocean runs parallel to the eastern coast of North and South America and the western coast of Europe and Africa.

Boundaries where plates move away from each other are called *divergent plate boundaries. Transform boundaries* occur where one plate slides horizontally past another plate, whereas at *convergent boundaries* two plates move toward each other (Figure 3.5b). Collisions sometimes occur

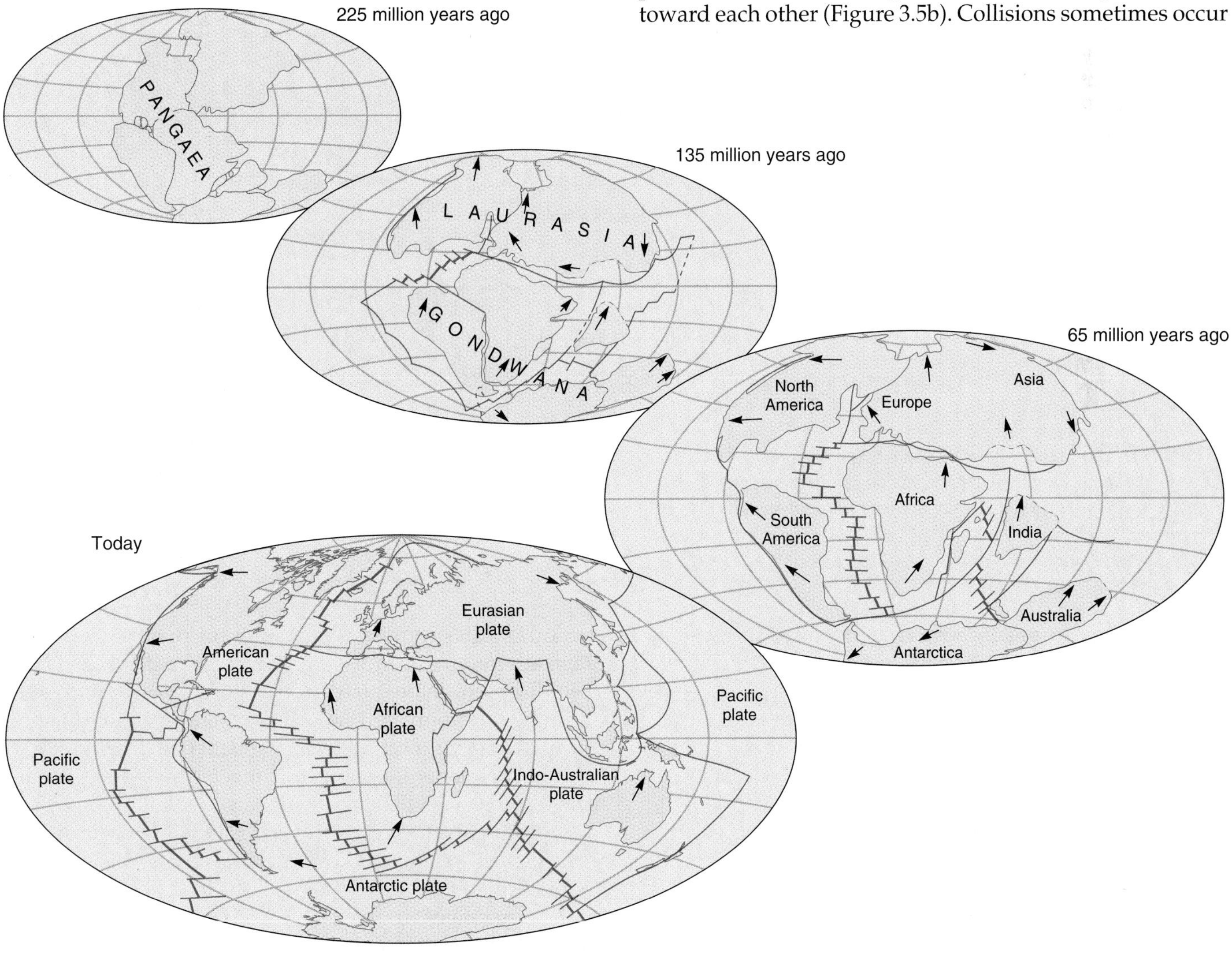


**FIGURE 3.4 Reconstruction of plate movements during the past 225 million years.** The northern and southern portions of Pangaea are called Laurasia and Gondwana, respectively. Some 225 million years ago, the continents were connected as one large landmass. After they split apart, the continents gradually moved to their present positions. Notice how India broke away from Antarctica and collided with the Eurasian landmass. The Himalayas were formed at the zone of contact. Source: *American Petroleum Institute.*

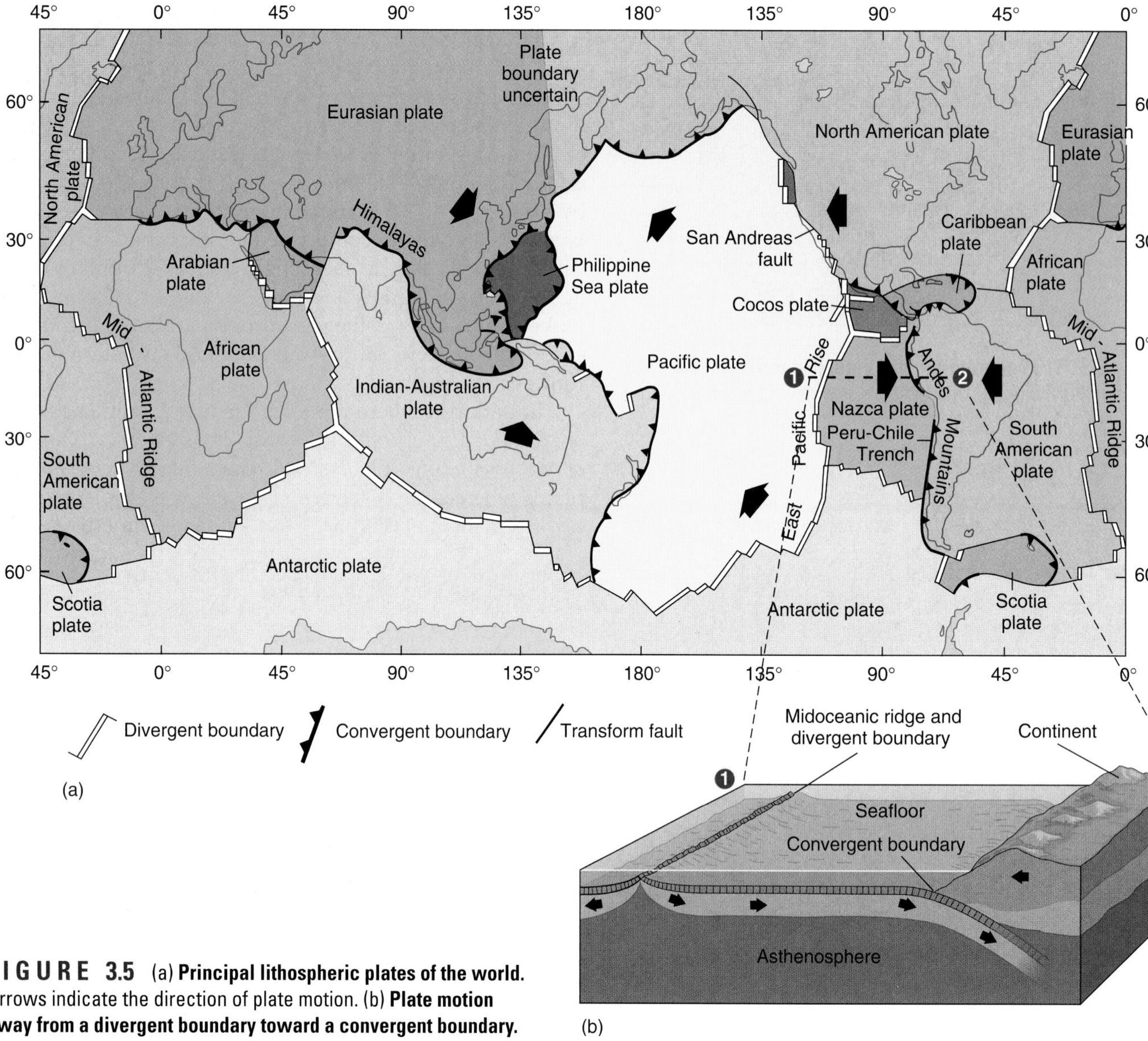


**FIGURE 3.5** (a) **Principal lithospheric plates of the world.** Arrows indicate the direction of plate motion. (b) **Plate motion away from a divergent boundary toward a convergent boundary.**

as the lithospheric plates move. The pressure exerted at the intersections of plates can cause earthquakes, which over periods of many years change the shape and features of landforms. Figure 3.7 shows the location of near-surface earthquakes for a recent time period. Comparison with Figure 3.5a illustrates that the areas of greatest earthquake activity are at plate boundaries.

The famous San Andreas fault in California is part of a long fracture separating two lithospheric plates, the North American and the Pacific. Earthquakes occur along **faults** (fractures in rock along which there has been movement) when the tension or compression at the junction becomes so great that only an earth movement can release the pressure.

Despite the availability of scientific knowledge about earthquake zones, the general disregard for this danger is a difficult cultural phenomenon with which to deal (see the section on diastrophism, beginning on p. 62). Every year there are hundreds and sometimes thousands of casualties resulting from inadequate preparation for earthquakes. In some highly populated areas, the chances that damaging earthquakes will occur are very great. The distribution of earthquakes shown in Figure 3.7 reveals the potential dangers to densely settled areas of Japan, the Philippines, parts of Southeast Asia, and the western rim of the Americas.

Convergent movement of the lithospheric plates results in the formation of deep-sea trenches and continental-scale mountain ranges, as well as in the occurrence of earthquakes. The continental crust is made up of lighter rocks than is the oceanic crust. Where plates with different types of crust at their edges converge, there is a tendency for the denser but thinner oceanic crust to be forced down into the asthenosphere. Deep trenches form below the ocean at these

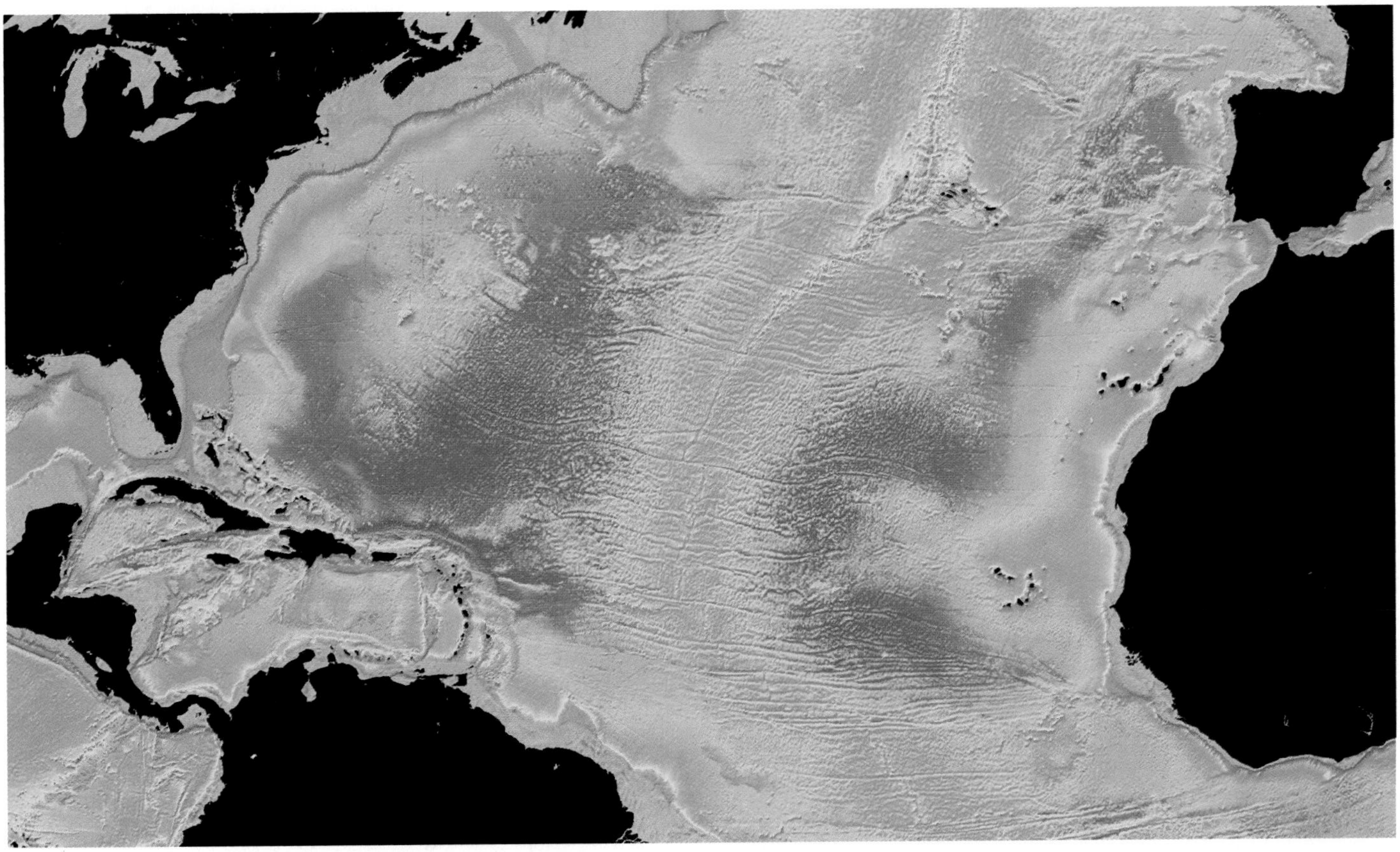

**FIGURE 3.6 An accurate map of the North Atlantic Ocean seafloor created by the National Oceanic and Atmospheric Administration using gravity measurements taken from satellite readings.** The configuration of the seafloor is evidence of the dynamic processes shaping continents and ocean basins. Source: *David T. Sandwell, 1995. Scripps Institution of Oceanography.*

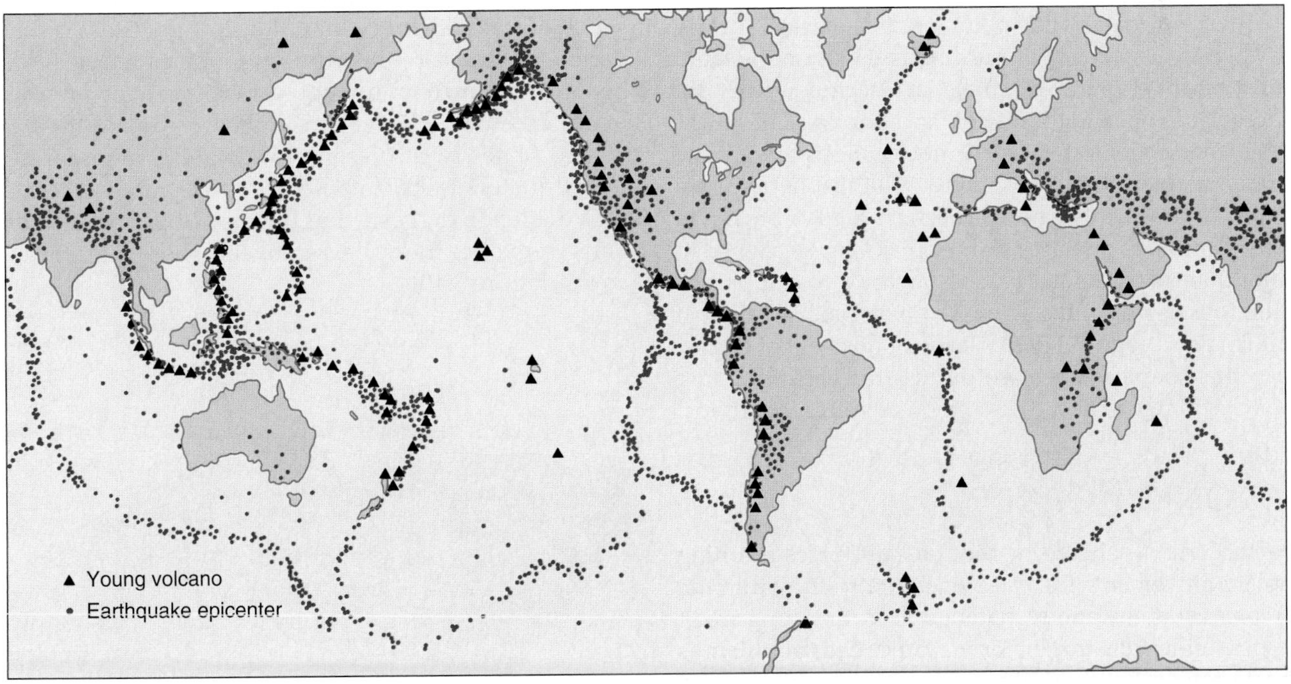


**FIGURE 3.7 Locations of young volcanoes and the epicenters of earthquakes.** Notice that they are concentrated at the margins of the lithospheric plates, as comparison with Figure 3.5 reveals. Volcanoes can also grow in the middle of a plate. The volcanoes of Hawaii, for example, are located in the middle of the Pacific plate. *Map plotted by the Environmental Data and Information Service of NOAA; earthquakes from U.S. Coast and Geodetic Survey.*

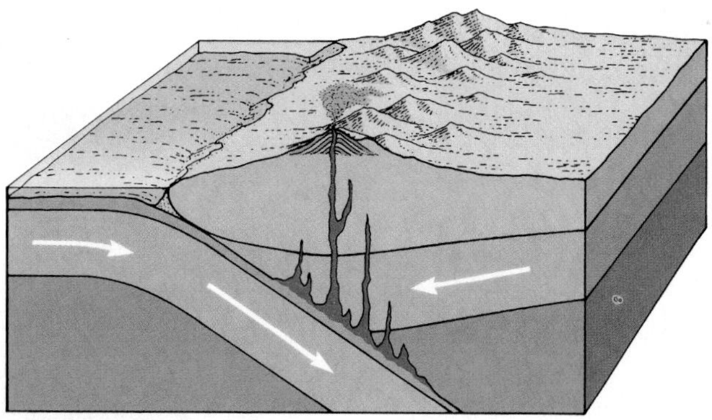

**FIGURE 3.8 The process of subduction.** When lithospheric plates collide, the heavier oceanic crust is usually forced beneath the lighter continental material. See Figure 3.5a for the subduction zones of the world.

convergent boundaries. This type of collision is termed **subduction** (Figure 3.8). The subduction zones of the world are shown in Figure 3.5a.

Most of the Pacific Ocean is underlain by a plate that, like the others, is constantly pushing and being pushed. The continental crust on adjacent plates is being forced to rise and fracture, making an active volcano zone of the Pacific Ocean rim (sometimes called the "ring of fire"). The tremendous explosion that rocked Mount St. Helens in the state of Washington in 1980 is an example of continuing volcanic activity along the Pacific rim. A number of damaging earthquakes have occurred along the San Andreas fault in recent years. Their epicenter (the point on the earth's surface directly above the focus of an earthquake) was on the fault. The most recent strong earthquake, in December 2003, did considerable damage to the city of Pasa Robles, California.

Plate intersections are not the only locations susceptible to readjustments in the lithosphere. As lithospheric plates have moved, the earth's crust has been cracked or broken in virtually thousands of places. Some breaks are weakened to the point that they become *hot spots,* areas of volcanic eruption due to a rising plume of molten material. The molten material may explode out of a volcano or ooze out of cracks. Later in this chapter, when we discuss the earth-building forces, we will return to the discussion of volcanic activity.

# TECTONIC FORCES

The earth's crust is altered by the constant forces resulting from plate movement. *Tectonic* (generated from within the earth) forces shaping and reshaping the earth's crust are of two types, either diastrophic or volcanic. **Diastrophism** is the great pressure acting on the plates that deforms them by folding, twisting, warping, breaking, or compressing rock. **Volcanism** is the force that transports heated material to or toward the surface of the earth. When particular places on the continents are subject to diastrophism or volcanism, the changes that take place can be as simple as the bowing or cracking of rock or as dramatic as lava exploding from the crater and sides of Mount St. Helens.

## Diastrophism

In the process of plate tectonics, pressures build in various parts of the earth's crust, and slowly, usually over thousands of years, the crust is transformed (see "Mount Everest: The Jewel in the Crown," p. 74). By studying rock formations, geologists are able to trace the history of the development of a region. Over geologic time, most continental areas have been subjected to both tectonic and gradational activity—building up and tearing down. They usually have a complex history of broad warping, folding, faulting, and leveling. Some flat plains in existence today may hide a history of great mountain development in the past.

### Broad Warping

Great tectonic forces resulting from the movement of continents may bow an entire continent. Also, the changing weight of a large region may result in the **warping** of the surface. For example, the down-warping of the eastern United States is evident in the many irregularly shaped stream estuaries. As the coastal area warped downward, the sea advanced, forming estuaries and underwater canyons.

### Folding

When the compressional pressure caused by plate movements is great, layers of rock are forced to buckle. The result may be a warping or bending effect, and a ridge or series of parallel **folds** may develop. Figure 3.9 shows a variety of structures resulting from folding. The folds can be thrust upward many thousands of feet and laterally for many miles. The Ridge and Valley region of the eastern United States is, at present, low parallel mountains—300 to 900 meters (1000 to 3000 ft) above sea level—but the rock evidence suggests that the tops of the present mountains were once the valleys between 9100-meter (30,000-ft) crests (Figure 3.10).

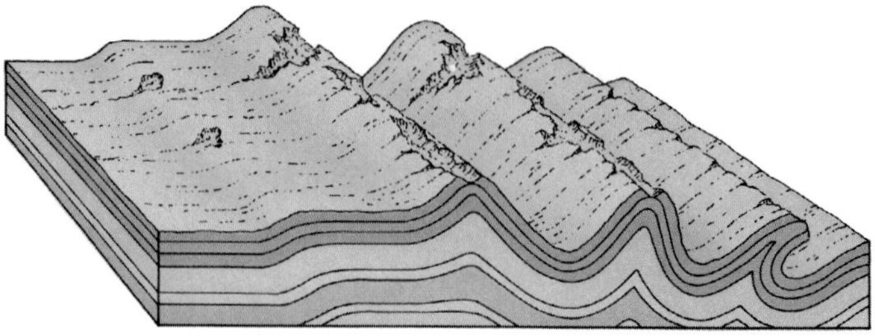

**FIGURE 3.9 Stylized forms of folding.** Degrees of folding vary from slight undulations of strata with little departure from the horizontal to highly compressed or overturned beds.

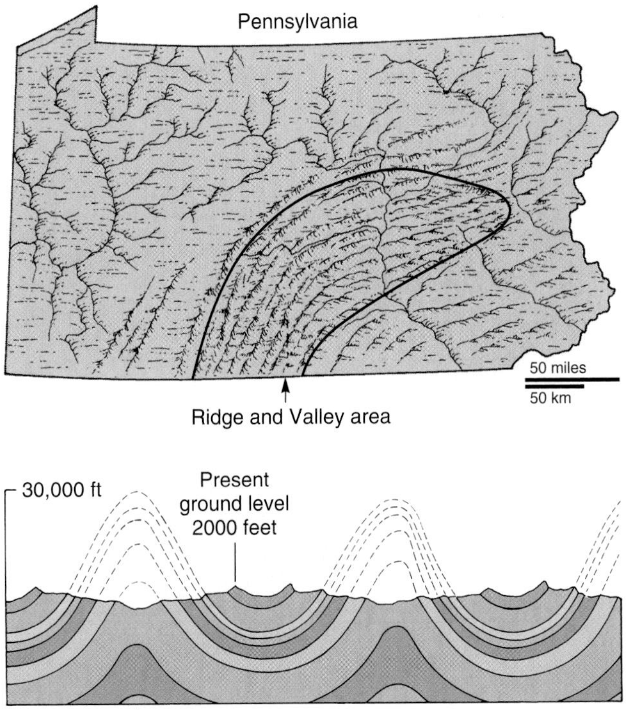

**FIGURE 3.10** (a) The Ridge and Valley region of Pennsylvania, now eroded to hill lands, is the relic of 9100-meter (30,000-ft) folds that were reduced to form *synclinal* (downarched) hills and *anticlinal* (uparched) valleys. The rock in the original troughs, having been compressed, was less susceptible to erosion. (b) **A syncline in Maryland on the border of Pennsylvania, exposed by a road cut.** *(b) © Mark C. Burnett/Photo Researchers, Inc.*

## Faulting

A fault is a break or fracture in rock along which movement has taken place. The stress causing a fault results in displacement of the earth's crust along the fracture zone. Figure 3.11 depicts examples of fault types. There may be uplift on one side of the fault or downthrust on the other. In some cases, a steep slope known as a fault *escarpment,* which may be several hundred feet high and several hundred miles long, is formed. The stress can push one side up and over the other side, or a separation away from the fault may cause the sinking of land, creating a *rift valley* (Figure 3.12).

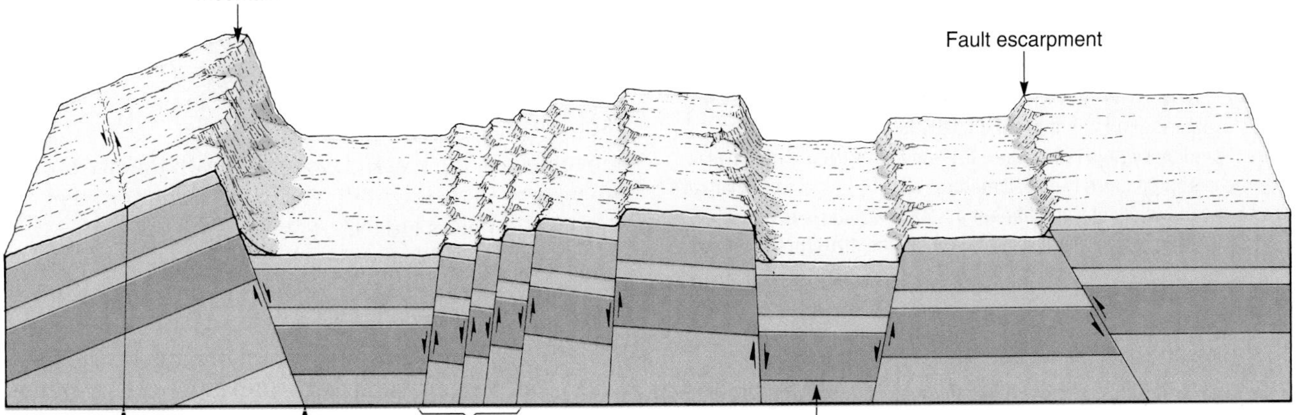


**FIGURE 3.11** Faults, in their great variation, are common features of mountain belts where deformation is great. The different forms of faulting are categorized by the direction of movement along the plane of fracture.

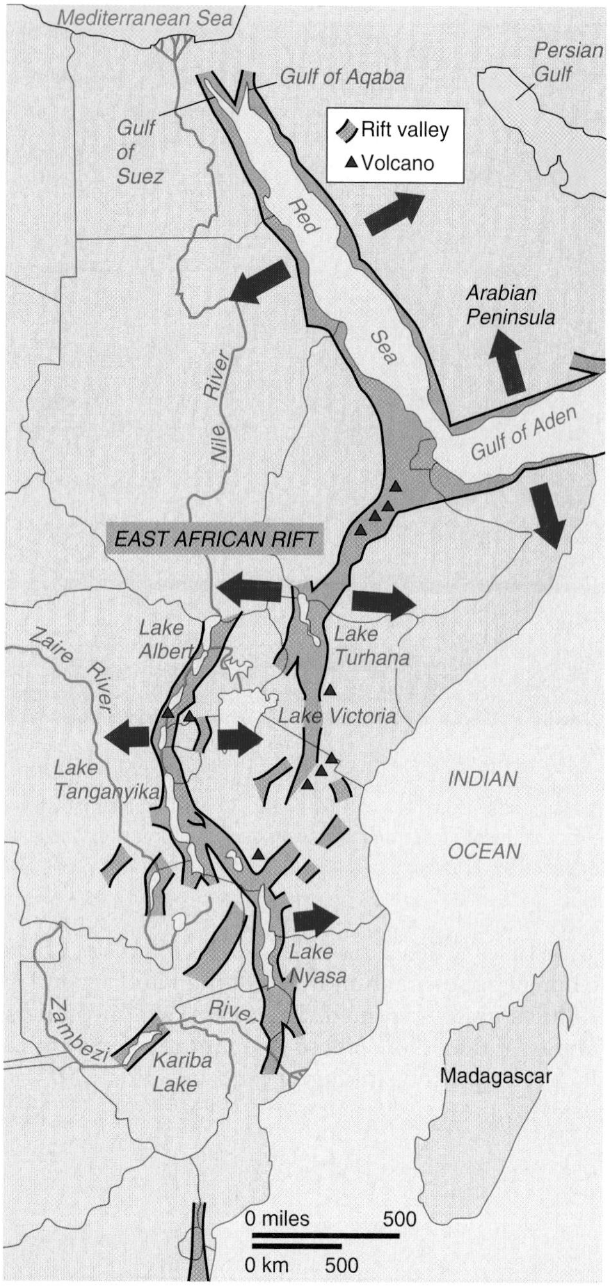


**FIGURE 3.12 The rift valleys of East Africa.** Great fractures in the earth's crust resulted in the creation, through subsidence, of an extensive rift valley system (see Figure 3.11) in East Africa. The parallel faults, some reaching more than 610 meters (2000 ft) below sea level, are bordered by steep walls of the adjacent plateau, which rises to 1500 meters (5000 ft) above sea level and from which the structure dropped.

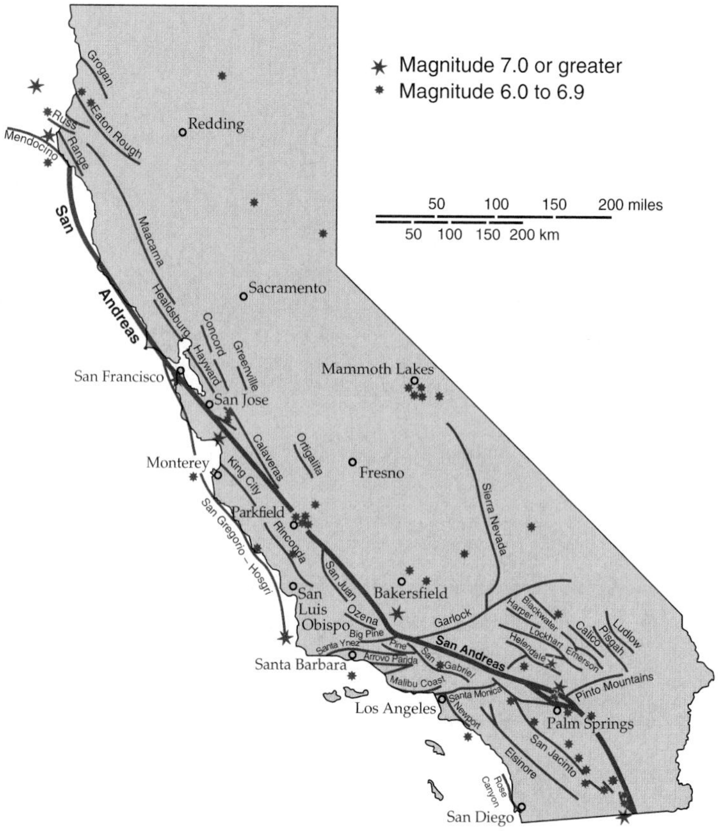


**FIGURE 3.13 The San Andreas fault system in California, with the epicenters of magnitude 6.0 and greater earthquakes of the 20th century.** Source: *Map updated from "The San Andreas Fault System, California," ed. By Robert E. Wallace, U.S. Geological Survey Professional Paper 1515, 1990.*

Many fractures are merely cracks (called *joints*) with little noticeable movement along them. In other cases, however, mountains such as the Sierra Nevadas of California have risen as the result of faulting. Sometimes, the movement has been horizontal along the surface rather than upward or downward. The San Andreas transform fault, shown in Figures 3.13 and 3.14, is such a case.

Whenever movement occurs along a fault, or at another point of weakness, an earthquake results. The greater the movement, the greater the magnitude of the earthquake (see "Scaling Earthquakes," p. 65). Stress builds in rock as tectonic forces are applied and, when a critical point is finally reached, an earthquake occurs and tension is reduced.

The earthquake that occurred in Alaska on Good Friday in 1964 was one of the strongest measured, with a magnitude of 8.2 on the Richter scale. Although the stress point of that earthquake was below ground 121 kilometers (75 mi) away from Anchorage, vibrations called *seismic waves* caused earth movement in the weak clays underneath the city. Sections of Anchorage slid downhill, and part of the business district dropped 3 meters (10 ft).

Table 3.1 indicates the characteristic effects associated with earthquakes of different magnitudes. Figure 3.15 shows earthquake-induced damage.

If an earthquake or a volcanic eruption occurs below an ocean, the movement can cause **tsunami,** very large sea waves (see "Tsunami," p. 70). Though not noticeable on the open sea, tsunami may become 9 or more meters (30 ft) high as they approach land, sometimes thousands of kilometers from the earthquake focus. The devastation in Hilo, Hawaii, in 1946 prompted the development of a tsunami warning system.

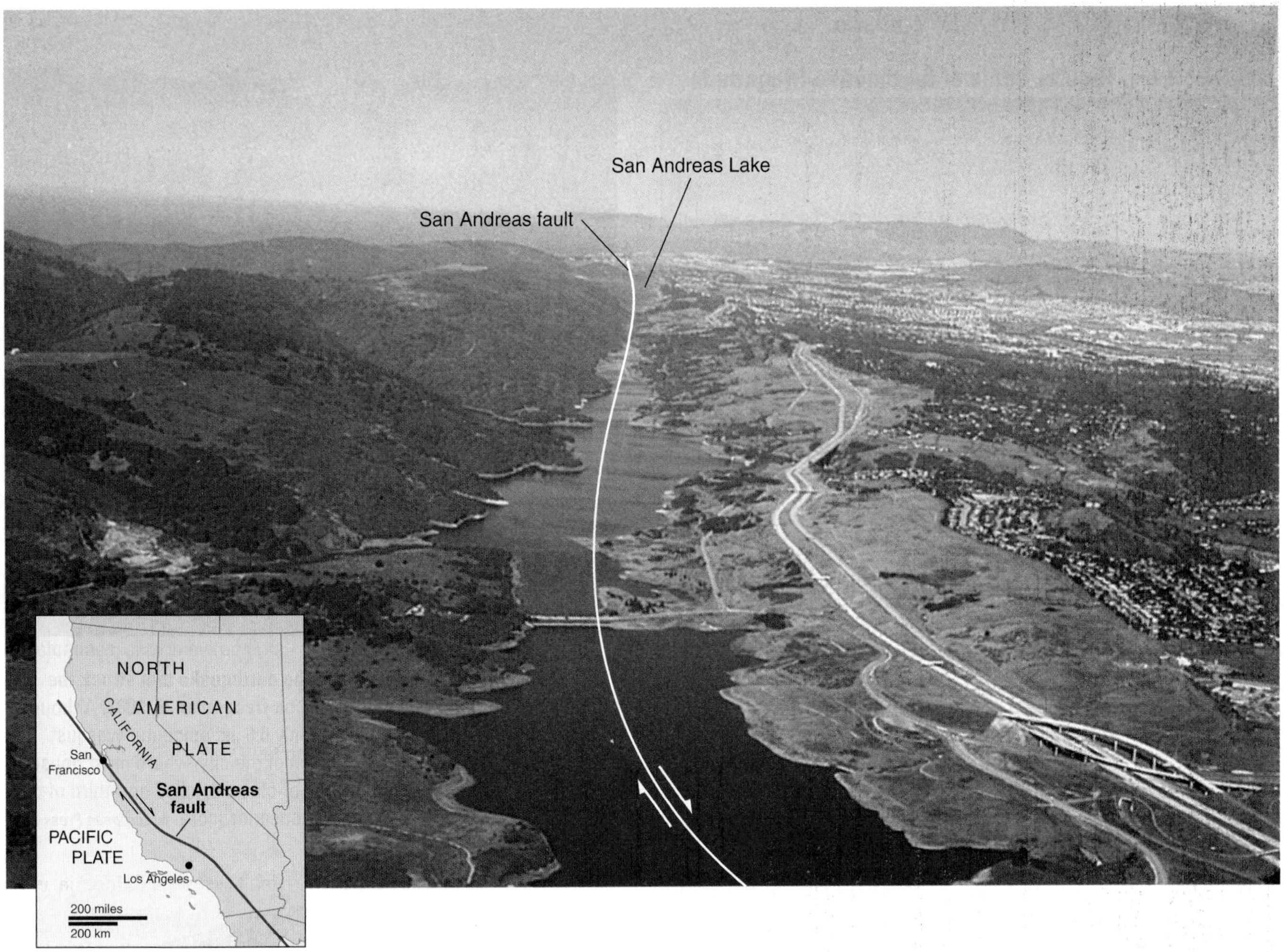


**FIGURE 3.14 View along part of the San Andreas fault, looking northward toward San Francisco.** Here, lakes occupy the fault zone. A *transform fault,* (see Figure 3.11), the San Andreas marks a part of the slipping boundary between the Pacific and the North American plates. The inset map shows the relative southward movement of the American plate; dislocation averages about 1 centimeter (0.4 in.) per year.
*(Photo) © Kevin Shafer/Tom Stack & Associates.*

## Scaling Earthquakes

In 1935, C.F. Richter devised a scale of earthquake *magnitude,* a measure of the energy released during an earthquake. An earthquake is really a form of energy expressed as wave motion passing through the surface layer of the earth. Radiating in all directions from the earthquake focus, seismic waves gradually dissipate their energy at increasing distances from the *epicenter.* On the Richter scale, the amount of energy released during an earthquake is estimated by measurement of the ground motion that occurs. Seismographs record earthquake waves, and by comparison of wave heights, the relative strength of quakes can be determined. Although Richter scale numbers run from 0 to 9, there is no absolute upper limit to earthquake severity. Presumably, nature could outdo the magnitude of the most intense earthquakes recorded so far, which reached 8.5–8.6.

Because magnitude, as opposed to intensity (a measure of an earthquake's size by its effect on people and buildings), can be measured accurately, the Richter scale has been widely adopted. Nevertheless, it is still only an approximation of the amount of energy released in an earthquake. In addition, the height of the seismic waves can be affected by the rock materials under the seismographic station, and some seismologists believe that the Richter scale underestimates the magnitude of major tremors.

In recent years, seismologists have used a measure called *moment magnitude.* This measure accounts for the movement of the earth's surface during an earthquake. If the movement (slip) is great, compared with the size (length) of the fault, then the moment magnitude will be great. A small slip on a large fault is considered a minor earthquake. The Northridge, California, earthquake of 1994 had similar Richter and moment values, but the disastrous "Good Friday" quake in Alaska 30 years earlier had an 8.2 Richter value and a 9.0 moment magnitude.

(a)

(b)

(c)

**FIGURE 3.18** (a) **Mount St. Helens, Washington, before it erupted on May 18, 1980.** (b) The eruption sent a cloud of steam and hot ash out of the cone 15 kilometers (9 mi) into the atmosphere, and a landslide carried ash and coarser debris down the slopes of the mountain. (c) The explosion blew away the top and much of one side of the volcano, lowering its summit by nearly 400 meters (1300 ft). *(a) David Muench/Getty Images. (b) © 1980 Keith Ronnholm. (c) © U.S. Geological Survey, CVO.*

On continents, the Deccan Plateau of India and the Columbia Plateau of the Pacific Northwest in the United States are examples of this type of process (Figure 3.19).

# GRADATIONAL PROCESSES

**Gradational processes** are responsible for the reduction of land surface. If a land surface where a mountain once stood is now a low, flat plain, gradational processes have been at work. The worn, scraped, or blown away material is deposited in new places and, as a result, new landforms are created. In terms of geologic time, the Rocky Mountains are a recent phenomenon; gradational processes are active there, just as on all land surfaces, but they have not yet had time to reduce these huge mountains.

Three kinds of gradational processes occur: *weathering, mass movement,* and *erosion.* Both mechanical and chemical weathering processes play a role in preparing bits of rock for the creation of soils and for movement to new sites by means of gravity or erosion. Mass movement transfers downslope by gravity any loosened, higher-lying material, including rock debris and soil, and the agents of running water, moving ice, wind, waves, and currents erode and carry these loose materials to other areas, where landforms are created or changed.

## Weathering

The breakdown and decomposition of rocks and minerals at or near the earth's surface in response to atmospheric factors (water, air, and temperature) is called **weathering.** It occurs as a result of both mechanical and chemical processes.

### Mechanical Weathering

**Mechanical weathering** is the physical disintegration of earth materials at or near the surface; that is, larger rocks are broken into smaller pieces. A number of processes cause mechanical weathering. The three most important are frost action, the development of salt crystals, and root action.

If water that soaks into a rock (between particles or along joints) freezes, ice crystals grow and exert pressure on the rock. When the process is repeated—freezing, thawing, freezing, thawing, and so on—the rock begins to disintegrate. Salt crystals act similarly in dry climates, where groundwater is drawn to the surface by *capillary action* (water rising because of surface tension). This action is similar to the process in plants whereby liquid plant nutrients move upward through the stem and leaf system. Evaporation leaves behind salt crystals that form, expand, and disintegrate rocks. Roots of trees and other plants may also find their way into rock joints; as they grow, they break and disintegrate the rock. These are all mechanical processes because they are physical in nature and do not alter the chemical composition of the material upon which they act.

**FIGURE 3.19 Fluid basaltic lavas created the Columbia Plateau, covering an area of 130,000 square kilometers (50,000 sq mi).** Some individual flows were more than 100 meters (300 ft) thick and spread up to 60 kilometers (40 mi) from their original fissures. *© Wolfgang Kaehler.*

## Chemical Weathering

A number of **chemical weathering** processes cause rock to decompose rather than disintegrate. In other words, the minerals composing rocks separate into component parts by chemical reaction rather than fragmentation. The three most important processes are oxidation, hydrolysis, and carbonation. Because each of these depends on the availability of water, less chemical weathering occurs in dry and cold areas than in moist and warm ones. Chemical reactions are accelerated in the presence of moisture and heat, so less chemical weathering occurs in cold, dry areas than in warm, moist ones.

*Oxidation* occurs when oxygen combines with mineral components, such as iron, to form oxides. As a result, some rock areas in contact with oxygen begin to decompose. Decomposition also results when water comes into contact with certain rock minerals, such as aluminosilicates. The chemical change that occurs is called *hydrolysis.* When carbon dioxide gas from the atmosphere dissolves in water, a weak carbonic acid forms. The action of the acid, called *carbonation,* is particularly evident on limestone because the calcium bicarbonate salt created in the process readily dissolves and is removed by groundwater and surface water.

Weathering, either mechanical or chemical, does not itself create distinctive landforms. Nevertheless, it prepares rock particles for erosion and for the creation of soil. After the weathering process decomposes rock, the force of gravity and the erosional agents of running water, wind, and moving ice are able to carry the weathered material to new locations.

Mechanical and chemical weathering create *soil,* the thin layer of fine material containing organic matter, air, water, and weathered rock materials that rests on the solid rock below it. Soil type is much more a function of the

## Tsunami

Tsunami follow submarine earthquakes that cause fissures, or cracks, in the earth's surface. Water rushes in to fill the depression caused by the falling away of the ocean bottom. The water then moves outward, building in momentum and rhythm, as swells of tremendous power. The waves that hit Hilo, Hawaii, following the April 1, 1946, earthquake off of Dutch Harbor, Alaska, were moving at approximately 640 kilometers per hour (400 mph) with a crest-to-crest spacing of some 130 kilometers (80 mi).

The long swells of tsunami are largely unnoticed in the open ocean. Only when the wave trough (the low point of a wave) scrapes sea bottom in shallow coastal areas or when the narrowed topography of inlets and harbors focuses the waves into smaller spaces does the water pile up into precipitous peaks. The tsunami at Hilo on the exposed northeast part of the island of Hawaii were estimated at between 14 and 30 meters (45 and 100 ft). The water smashed into the city, deposited more than 4 meters (14 ft) of silt in its harbor, left fish stranded in palm trees, created millions of dollars in damage, and caused 175 deaths—many of these casualties were people who had gone to the shore to see the giant waves arrive.

Far more lethal were the following tsunami.

| Date | Quake Site | Est. Wave Height | Est. Deaths |
|---|---|---|---|
| June 1896 | Japan | 29 m | 27,000 |
| May 1960 | Chile | 10 m | 1,250 |
| July 1998 | Papua New Guinea | 15 m | 3,000 |

In an attempt to minimize loss of life in Pacific coastal communities, the U.S. Coast and Geodetic Survey in 1948 established a Tsunami Early Warning System based in Hawaii. Any major earthquake in the Pacific region that is capable of generating tsunami is reported to the Tsunami Warning Center in Honolulu. If tsunami are detected, the center relays information about their source, speed, and estimated time of arrival to low-lying coastal communities in danger.

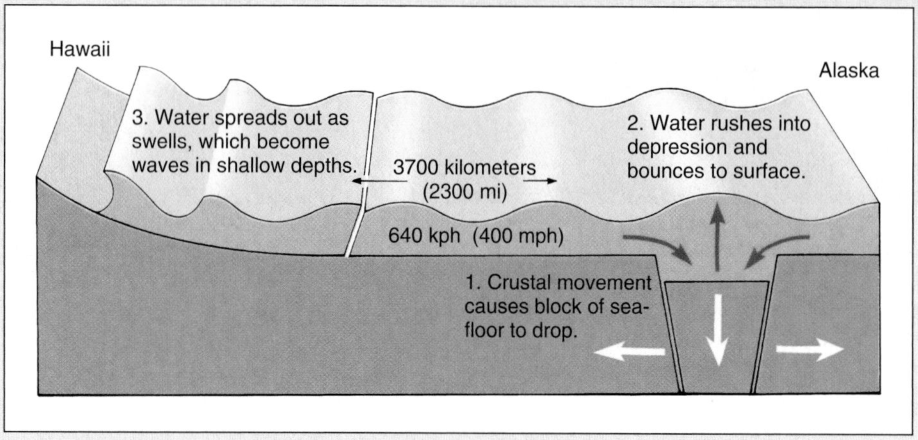


climate of the region in which it occurs than it is of the kind of bedrock below it. Temperature and rainfall act on minerals, in conjunction with the decaying of overriding vegetation, to form soils. The topic of soils is discussed in more detail in Chapter 4.

### Mass Movement

The force of gravity—that is, the attraction of the earth for bodies at or near its surface—is constantly pulling on all materials. The downslope movement of material due to gravity is called *mass wasting* or **mass movement.** Because it is more descriptive, this book uses the latter term. Small particles or huge boulders, if not held back by solid rock or other stable material, will fall down slopes. Spectacular acts of mass movement include avalanches and landslides. More widespread, but less noticeable, are mass movements such as soil-creep and the flow of mud down hillsides (Figure 3.20).

Especially in dry areas, a common but very dramatic landform created by the accumulation of rock particles at the base of hills and mountains is *talus,* pictured in Figure 3.20a. As pebbles, particles of rock, or even larger stones break away from exposed bedrock on a mountainside because of weathering, the masses fall and accumulate, producing large, conelike landforms. The larger rocks travel farther than the fine-grained sand particles, which remain near the top of the slope.

(a)

(b)

**FIGURE 3.20** (a) Rockfall from this butte has created talus, an accumulation of broken rock at the base of a cliff. (b) Creeping soil has caused trees to tilt. *(a) © Robert N. Wallen. (b) Courtesy of Arthur Getis.*

## Erosional Agents and Deposition

**Erosional agents,** such as wind, water, and glaciers, carve already existing landforms into new shapes. Fast-moving agents carry debris, and slow-moving agents drop it. The material that has been worn, scraped, or blown away is deposited in new places, and new landforms are created. Each erosional agent is associated with a distinctive set of landforms.

### Running Water

Running water is the most important erosional agent. Water, whether flowing across land surfaces or in stream channels, plays an enormous role in wearing down and building up landforms.

Running water's ability to erode depends upon several factors: (1) the amount of precipitation; (2) the length and steepness of the slope; and (3) the kind of rock and vegetative cover. Steeper slopes and faster-flowing water results, of course, in more rapid erosion. Vegetative cover sometimes slows the flow of water. When this vegetation is reduced, perhaps because of farming or livestock grazing, erosion can be severe, as shown in Figure 3.21.

**FIGURE 3.21** Gullying can result from heavy rain and poor farming techniques, including overgrazing by livestock or many years of continuous row crops. Surface runoff removes topsoil easily when vegetation is too thin to protect it. *© Grant Heilman/Grant Heilman Photography, Inc.*

Even the impact of precipitation—heavy rain or hail—can cause erosion. After hard rain dislodges soil, the surface becomes more compact; therefore, further precipitation fails to penetrate the soil. The result is that more water, prevented from seeping into the ground, becomes available for surface erosion. Soil and rock particles in the water are carried to streams, leaving behind gullies and small stream channels.

Both the force of water and the particles contained in the stream are agents of erosion. Abrasion, or wearing away, takes place when particles strike against stream channel walls and along the streambed. Because of the force of the current, large particles, such as gravel, slide along the streambed, grinding rock on the way.

Floods and rapidly moving water are responsible for dramatic changes in channel size and configuration, sometimes forming new channels. In cities where paved surfaces cover soil that would otherwise have absorbed or held water, runoff is accentuated so that nearby rivers and streams rapidly increase in size and velocity after heavy rains. Oftentimes, flash floods and severe erosion result.

Small particles, such as clay and silt, are suspended in water and constitute—together with material dissolved in the water or dragged along the bottom—the *load* of a stream. Rapidly moving floodwaters carry huge loads. As high water or floodwater recedes, and stream velocity decreases, sediment contained within the stream no longer remains suspended, and particles begin to settle. Heavy, coarse materials drop the quickest; finer particles are carried longer and transported farther. The decline in velocity and the resulting deposition are especially pronounced and abrupt when streams meet slowly moving water in bays, oceans, and lakes. Silt and sand accumulate at the intersections, creating *deltas,* as pictured in Figure 3.22.

A great river, such as the Chang Jiang (Yangtze) in China, has a large, growing delta, but less prominent deltas exist at the mouths of many streams. Until the recent completion of the Aswan Dam, the huge delta of the Nile River had been growing. Now much of the silt is being dropped in Lake Nasser behind the dam.

In plains adjacent to streams, land is sometimes built up by the deposition of *stream load.* If the deposited material is rich, it may be a welcome and necessary part of farming activities, such as that historically known in Egypt along the Nile. Should the deposition be composed of sterile sands and boulders, however, formerly fertile bottomland may be destroyed. By drowning crops or inundating inhabited areas, the floods themselves, of course, may cause great human and financial loss. More than 900,000 lives were lost in the floods of the Huang He (Yellow River) of China in 1887.

## Stream Landscapes

A landscape is in a particular state of balance between the uplift of land and its erosion. Rapid uplift is not followed by nicely ordered stages of erosion. Recall that uplift and erosion take place simultaneously. At a given location, one force may be greater than the other at a given time, but as yet there is no way to predict accurately the "next" stage of landscape evolution.

**FIGURE 3.22** The delta of the Mississippi River. Notice the ongoing deposition of silt and the effect that both river and gulf currents have on the movement of the silt. Source: *NASA.*

Perhaps the most important factor differentiating the effect of streams on landforms is whether the recent climate (for example, the past several million years) has tended to be humid or arid.

***Stream Landscapes in Humid Areas*** Perhaps weak surface material or a depression in rock allows the development of a stream channel. In its downhill run in mountainous regions, a stream may flow over precipices, forming *falls* in the process. The steep downhill gradient allows streams to flow rapidly, cutting narrow, V-shaped channels in the rock (Figure 3.23a). Under these conditions, the erosional process is greatly accelerated. Over time, the stream may have worn away sufficient rock for the falls to become rapids, and the stream channel becomes incised below the height of the surrounding landforms. This is evident in the upper reaches of the Delaware, Connecticut, and Tennessee rivers.

In humid areas, the effect of stream erosion is to round landforms. Streams flowing down moderate gradients tend to carve valleys that are wider than those in mountainous areas. Surrounding hills become rounded, and valleys eventually become **floodplains** as they broaden and flatten. Streams work to widen the floodplain. Their courses meander, constantly carving out new river channels. The channels left behind as new ones are cut become *oxbow* lakes, hundreds of which are found in the Mississippi River floodplain (Figure 3.23b). An oxbow lake is crescent-shaped and occupies the abandoned channel of a stream meander.

In nearly flat floodplains, the highest elevations may be the banks of rivers, where *natural levees* are formed by the deposition of silt at river edges during floods. Floods that breach levees are particularly disastrous because flood waters fill the floodplain as they equalize its elevation with that of the swollen river. The U.S. Army Corps of Engineers has augmented the natural levees in particularly susceptible areas, such as the banks of the lower Mississippi River.

***Stream Landscapes in Arid Areas*** A distinction must be made between the results of stream erosion in humid regions and those in arid areas. The lack of vegetation in arid regions greatly increases the erosional force of running water. Water originating in mountainous areas sometimes never reaches the sea if the channel runs through a desert. In fact, stream channels may be empty except during rainy periods, when water rushes down hillsides to collect and form temporary lakes called *playas.* In the process, **alluvium** (sand and mud) builds up in the lakes and at lower elevations, and *alluvial fans* are formed along hillsides (Figure 3.24). The fan is produced by the deposition of silt, sand, and gravel outward as the stream reaches lowlands at the base of the slope it traverses. If the process has been particularly long-standing, alluvial deposits may bury the eroded mountain masses. In desert regions in Nevada, Arizona, and California, it is not unusual to observe partially buried mountains poking through alluvium.

(a)

(b)

**FIGURE 3.23** (a) **V-shaped valley of a rapidly downcutting stream, the Yellowstone River in Wyoming.** (b) **Oxbow-shaped lakes adjoining a meandering stream in Alaska.** *(a) © Robert N. Wallen. (b) © B. Anthony Stewart/National Geographic Image Collection.*

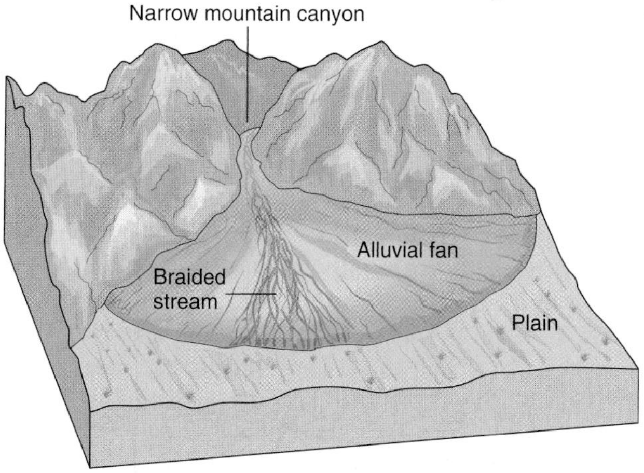


**FIGURE 3.24** Alluvial fans are built where the velocity of streams is reduced as they flow out upon the more level land at the base of the mountain slope. The abrupt change in slope and velocity greatly reduces the stream's capacity to carry its load of coarse material. Deposition occurs, choking the stream channel and diverting the flow of water. With the canyon mouth fixing the head of the alluvial fan, the stream sweeps back and forth, building and extending a broad area of deposition. *Redrawn from Charles C. Plummer and David McGeary,* Physical Geology, *8th ed.*

## Mount Everest: The Jewel in the Crown

© Corbis/R-F Website.

The fastest growing mountain range on Earth happens also to contain the world's tallest set of mountains. Nearly in the center of the Himalayas stands the world's highest peak, Mount Everest. Currently, Mount Everest is measured as standing 8850 meters (29,028 ft) above sea level. Recent measurements, however, indicate that Mount Everest and many of the other peaks, such as K-2, are growing at about 1 centimeter (1/2 inch) a year.

As the mountains build ever higher, their great weight softens underlying materials, causing the mountains to settle. In other words, there are two forces at work. The force building the mountains is that of the Indian plate moving northward, crashing into the Eurasian plate. The Himalayas, at the edge of the Eurasian plate, react to the great force by pushing higher and higher. But apparently, a second force that works against the first keeps mountains on Earth from rising to heights of 15,000 or 18,000 meters (roughly 50,000 or 60,000 ft). One may think of these great mountains as being in a kind of equilibrium—the bigger they get, the heavier they get, and the more likely they are to sag.

The battle between the two plates began about 45 million years ago. Usually, one plate is forced under the more stationary plate (subduction). In this case, however, the rocks of the two plates are similar in weight and density. Thus, subduction has not taken place, and what normally would have been a modest crinkle on the earth's surface turned into the tallest, most rugged mountain range on Earth. Particularly interesting is one view of the Himalayas. The Indian plate contains the relatively low-elevation Indian subcontinent and, as a result, the view of the Himalayas from the plains of northern India is one of the great sights on Earth. No wonder Mount Everest is the jewel in the mountain climber's crown. Edmund Hillary and Tenzing Norgay were the first to reach the summit, in 1953. Since then, more than 1500 climbers have reached the top, most of them since 1990; 176 have died attempting to scale Mount Everest.

Because streams in arid areas have only a temporary existence, their erosional power is less consistent than that of the freely flowing streams of humid areas. In some instances, they barely mark the landscape; in other cases, swiftly moving water may carve deep, straight-sided *arroyos.* Water may rush onto an alluvial plain in a complicated pattern resembling a multistrand braid, leaving in its wake an alluvial fan. The channels resulting from this rush of water are called *washes.* The erosional power of unrestricted running water in arid regions is dramatically illustrated by the steep-walled configuration of *buttes* and *mesas* (large buttes), such as those in Utah shown in Figure 3.25.

### Groundwater

Some of the water supplied by rain and snow sinks underground into pores and cracks in rocks and soil, not in the form of an underground pond or lake but simply as subsurface material. When underground water accumulates, a zone of saturation called an *aquifer* forms, through which water can move readily. As indicated in Figure 3.26, the upper level of this zone is the **water table;** below it, the soils and rocks are saturated with water. A well must be drilled into an aquifer to ensure a supply of water. Groundwater moves constantly but very slowly (usually only centimeters a day). Most remains underground, seeking the lowest level. When the surface of the land dips below the water table, however, ponds, lakes, and marshes form. Some water finds its way to the surface by capillary action in the ground or in vegetation. When the ground surface extends below the level of the water table, the most common feature to develop is a stream.

Groundwater, particularly when combined with carbon dioxide, dissolves soluble materials by a chemical process called *solution.* Although groundwater tends to decompose many types of rocks, its effect on limestone is most spectacular. Many of the great caves of the world have been created by the underground movement of water through limestone regions. Water sinking through the overlying rock leaves carbonate deposits as it drips into empty spaces. The deposits hang from cave roofs (*stalactites*) and build upward

**FIGURE 3.25 Canyonlands National Park, Utah.** The resistant caprock of the mesa protects softer, underlying strata from downward erosion. Where the caprock is removed, lateral erosion lowers the surface, leaving the mesa as an extensive and pronounced relic of the former higher-lying landscape. *© Carr Clifton/Minden Pictures.*

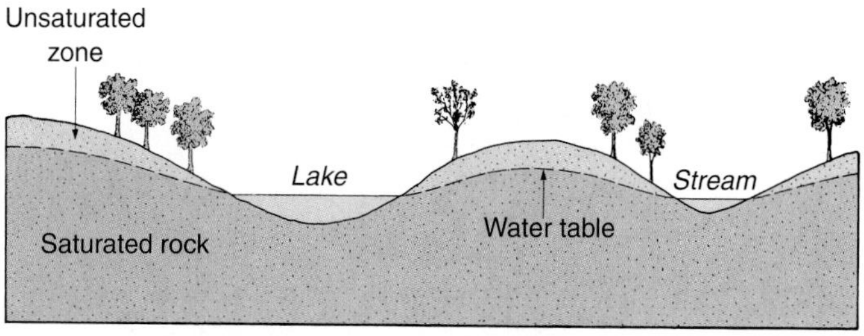


**FIGURE 3.26** The groundwater table generally follows surface contours but in subdued fashion. Water flows slowly through the saturated rock, emerging at earth depressions that are lower than the level of the water table. During a drought, the table is lowered and the stream channel becomes dry.

from cave floors (*stalagmites*). In some areas, the uneven effect of groundwater erosion on limestone leaves a landscape pockmarked by a series of *sinkholes,* surface depressions in an area of collapsing caverns.

**Karst topography** refers to a large limestone region marked by sinkholes, caverns, and underground streams, as shown in Figure 3.27. East central Florida, a karst area, has suffered considerable damage from the creation and widening of sinkholes. This type of topography gets its name from a region on the Adriatic Sea at the Italy-Slovenia border. The Mammoth Cave region of Kentucky, another karst area, has many kilometers of interconnected limestone caves.

## Glaciers

Another agent causing erosion and deposition to occur is glaciers. Although they are much less extensive today, glaciers covered a large part of the earth's land area as recently as 10,000 to 15,000 years ago. Many landforms were created by the erosional or depositional effects of glaciers.

Glaciers form only in very cold places with short or nonexistent summers, where annual snowfall exceeds annual snowmelt and evaporation. The weight of the snow causes it to compact at the base and form ice. When the snowfall reaches a thickness of about 100 meters (328 ft), ice at the bottom becomes plasticlike and begins to move slowly. A **glacier,** then, is a large body of ice moving slowly down a slope or spreading outward on a land surface (Figure 3.28). Some glaciers appear to be stationary simply because the melting and evaporation at the glacier's edge equal the speed of the ice advance. Glaciers can, however, move as much as a meter per day.

Most theories of glacial formation concern earth climatic cooling. Perhaps a combination of the following theories explains the evolution of glaciers. The first theory attributes cooling to periods when there may have been excessive amounts of volcanic dust in the atmosphere. The argument is that the dust, by reducing the amount of solar energy reaching the earth, effectively lowered temperatures at the surface. A second theory attributes the ice ages to known changes in the shape, tilt, and seasonal positions of the earth's orbit around the sun over the last half-million years. Such changes alter the amount of solar radiation received by the earth and its distribution over the earth. A recent theory suggests that when large continental plates drift over polar regions, temperatures on Earth become more extreme and, as a result, induce the development of glaciers. This theory, of course, cannot explain the most recent ice ages.

Today, continental-size glaciers exist on Antarctica, Greenland, and Baffin Island in Canada, but mountain glaciers are found in many parts of the world. About 10% of the earth's land area is under ice. During the most recent advance of ice, the continental ice of Greenland was part of an enormous glacier that covered nearly all of Canada (Figure 3.29) and the northernmost portions of the United States and Eurasia. The giant glacier reached thicknesses of 3000 meters (10,000 ft) (the depth in Greenland today), enveloping entire mountain systems. Another feature of the last ice age was the development of *permafrost,* a permanently frozen layer of ground that can be as much as 300 meters (1000 ft) deep. Because the permafrost layer prevents the downward percolation of moisture, the surface soil may become saturated with water during the brief summer season, when only a thin surface thaws (see "Permafrost," p. 79).

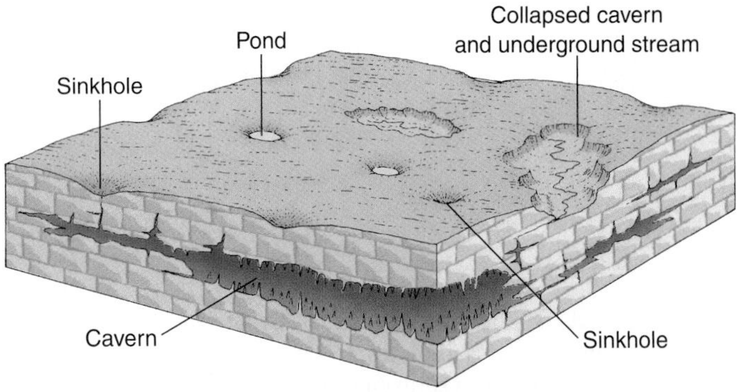


(a)

(b)

**FIGURE 3.27** Limestone erodes easily in the presence of water. (a) *Karst* topography, such as that shown here, occurs in humid areas where limestone in flat beds is at the surface. (b) This satellite photo of east central Florida shows the many round lakes formed in the sinkholes of a karst landscape. *(b) NASA.*

The weight of glaciers breaks up underlying rock and prepares it for transportation by the moving mass of ice. Consequently, glaciers change landforms by erosion. Glaciers scour the land as they move, leaving surface scratches, or striations, on rocks that remain. Much of eastern Canada has been scoured by glaciers that left little soil but many ice-gouged lakes and streams. The erosional forms created by glacial scourings have a variety of names. A *glacial trough* is a deep, U-shaped valley visible only after the glacier has receded. If the valley is below sea level today, as in Norway or British Columbia, *fiords,* or arms of the sea, are formed. Some of the landforms created by scouring are shown in Figure 3.28. Figure 3.30 shows *tarns,* small lakes in the hollowed-out depressions of a *cirque,* and arêtes, sharp ridges that separate adjacent glacially carved valleys. Cirques are formed by ice erosion at the head of a glacial valley.

Glaciers create landforms when they deposit the debris they have transported. These deposits, called *glacial till,* consist of rocks, pebbles, and silt. As the great tongues of ice move forward, debris accumulates in parts of the glacier. The ice that scours valley walls and the ice at the tip of the advancing tongue are particularly filled with debris. As a glacier melts, it leaves behind hills of glacial till of different sizes and shapes, such as moraines, eskers, and drumlins (Figure 3.31).

Many other landforms have been formed by glaciers. The most important is the *outwash plain,* a gently sloping area in front of a melting glacier. The melting along a broad front sends thousands of small streams running out from the glacier in braided fashion, streams that deposit neatly stratified glacial till. Outwash plains, which are essentially great alluvial fans, cover a wide area and provide new, rich parent-material for soil formation. Most of the midwestern part of the United States owes some of its soil fertility to relatively recent glacial deposition.

Before the end of the most recent ice age, at least three previous major advances occurred during the 1.5 million years of the Pleistocene period. Firm evidence is not available on whether we have emerged from the cycle of ice advance and retreat. Factors concerning the earth's changing temperature, which are discussed in Chapter 4, must be considered before assessing the likelihood of a new ice advance. For the first half of the 20th century, the world's glaciers were melting faster than they were building up. Current trends are not clear, although there is fear that the greenhouse effect (discussed in Chapter 4) is warming the earth and will cause the seas to rise.

### Waves, Currents, and Coastal Landforms

Whereas glacial action is intermittent in earth history, the breaking of ocean waves on continental coasts and islands is unceasing and causes considerable change in coastal landforms. As waves reach shallow water close to shore, they are forced by friction along the seafloor to heighten until a breaker is formed, as shown in Figure 3.32. The uprush of water not only carries sand for deposition but also erodes the landforms at the coast, while the backwash carries the eroded material away. This type of action results in different kinds of landforms, depending on conditions.

If land at the coast is well above sea level, the wave action causes cliffs to form. Cliffs then erode at a rate dependent on the rock's resistance to the constant assault of salt water. During storms, a great deal of power is released by the forward thrust of waves, and much erosion takes place. Landslides are a hazard during coastal storms, and they occur particularly in areas where weak sedimentary rock or glacial till exists.

**FIGURE 3.28 Alpine glacial landforms.** Frost shattering and ice movement carve *cirques,* the irregular bottoms of which may contain lakes (*tarns*) after glacial melt. Where cirque walls adjoin from opposite sides, knifelike ridges called *arêtes* are formed, interrupted by overeroded passes, or *cols.* The intersection of three or more creates a pointed peak, or *horn.* Rock debris falling from cirque walls is carried along by the moving ice. *Lateral moraines* form between the ice and the valley walls; *medial moraines* mark the union of such debris where two valley glaciers join. *Recessional moraines* form where the end of the ice remains stationary long enough for an accumulation of sediment to form, while a *terminal moraine* marks the glacier's farthest advance. Small conical hills of sediment are called *kames.*

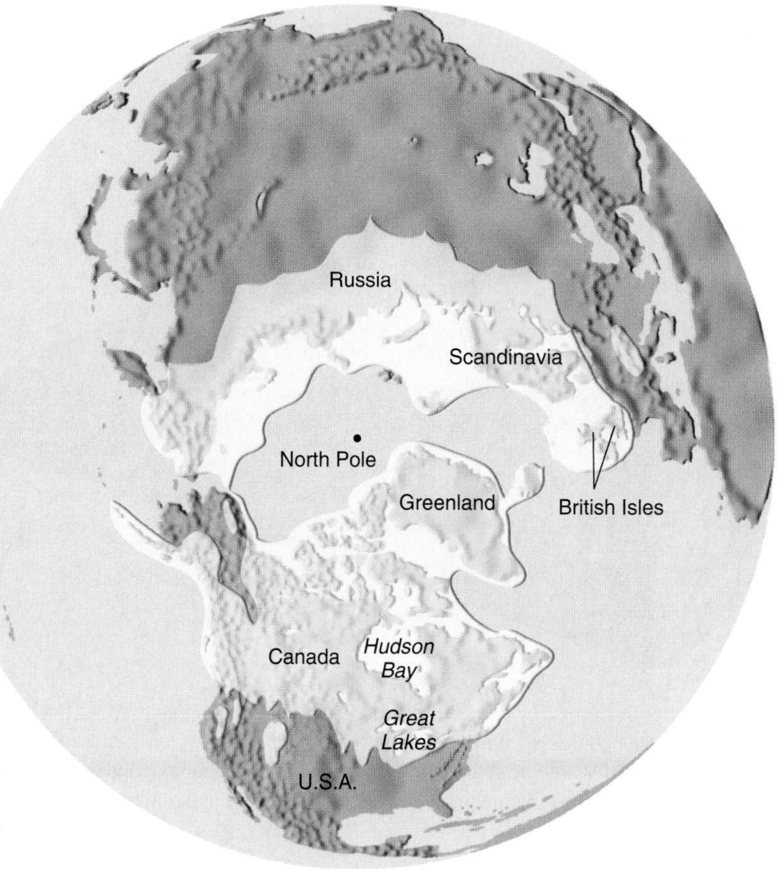


**FIGURE 3.29 Maximum extent of continental glaciation in the Northern Hemisphere (about 15,000 years ago).** Because sea level was lower than at present, due to the large volume of moisture trapped as ice on the land, glaciation extended beyond present continental shorelines. Separate centers of snow accumulation and ice formation developed. Large lakes were created between the western mountains of North America and the advancing ice front. To the south, huge rivers carried away glacial meltwaters.

**FIGURE 3.30 Tarns in glacial cirques, with arêtes behind.** The glacially formed lakes shown here are in the Glacier Peak Wilderness Area, Washington. *© Bob & Ira Spring.*

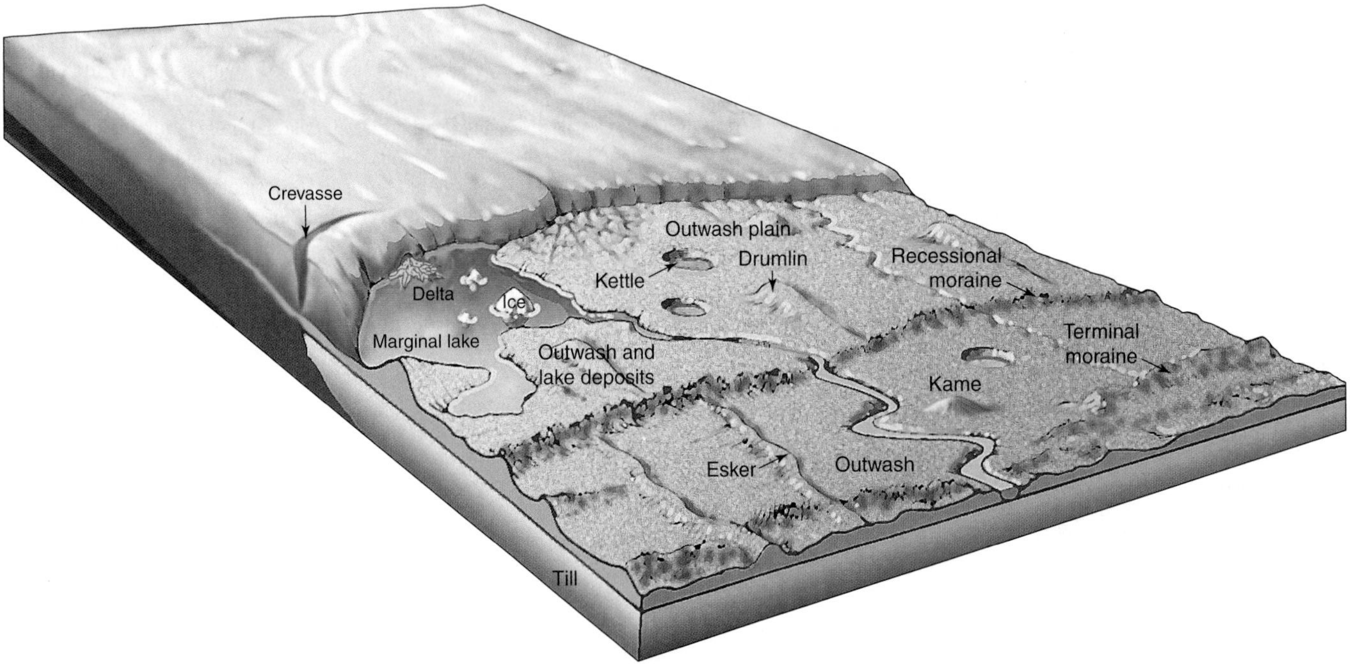


**FIGURE 3.31 Depositional features formed by ice sheets.** The debris carried by continental glaciers and left behind as they retreat creates various landforms. Moraines of *till,* unsorted glacial drift, form at the retreating edge. Streams carry silt to form *outwash plains. Kames* are small, conical hills formed of outwash deposits; *drumlins* are elongated hills made of till and oriented in the direction of ice movement; and *eskers* are long ridges of sediment deposited by glacial meltwater. Enclosed depressions caused by the melting of a stagnant block of ice that was surrounded and buried by sediment are called *kettles.*

## Permafrost

In 1577, on his second voyage to the New World in search of the Northwest Passage, Sir Martin Frobisher reported finding ground in the far north that was frozen to depths of "four or five fathoms, even in summer" and that the frozen condition "so combineth the stones together that scarcely instruments with great force can unknit them." The permanently frozen ground, now termed *permafrost*, underlies perhaps a fifth of the earth's land surface (a). In the lands surrounding the Arctic Ocean, its maximum thickness is about 600 meters (2000 ft).

For almost 300 years after Frobisher's discovery, little attention was paid to this frost phenomenon. But in the 19th and 20th centuries, during the building of the Trans-Siberian railroad, the construction of buildings associated with the discovery of gold in Alaska and the Yukon Territory, and the development of the oil pipeline that now connects Prudhoe Bay in northern Alaska with Valdez in southern Alaska, attention focused on the unique nature of permafrost.

Uncontrolled construction activities lead to the thawing of permafrost. This, in turn, produces unstable ground susceptible to soil movement and landslides, ground subsidence, and frost heaving (b). Scientists have found that successful use requires that the least possible disturbance of the frozen ground occur. The Alaska pipeline was built above the earth's surface in order to allow the passage of migrating wildlife and to reduce the likelihood that the relatively warm oil would disturb the permafrost and thus destroy the pipeline. It was necessary to preserve the insulating value of the ground surface as much as possible, so the vegetation mat was not removed. Instead, a coarse gravel fill was added along the surface below the pipeline.

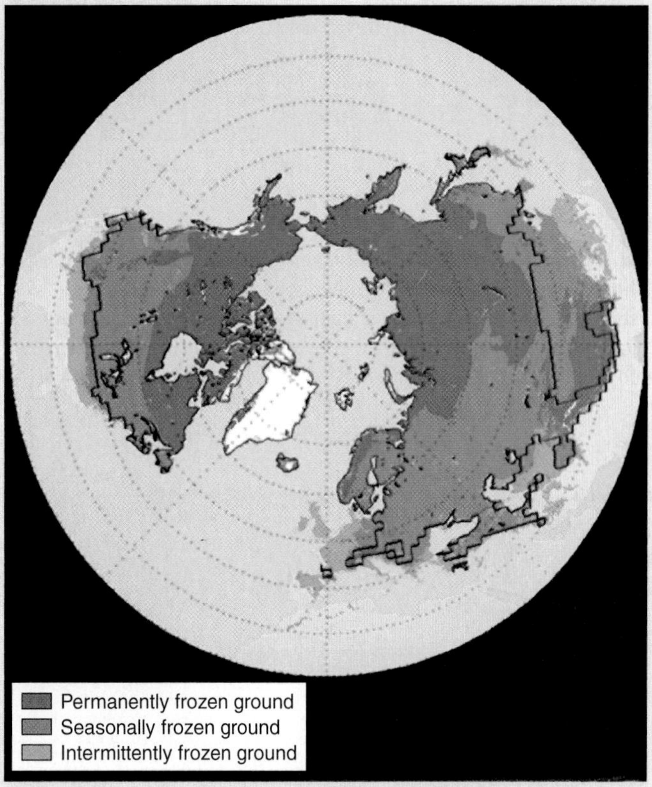


(a)

(b)

(a) **Regions of modern permafrost.** (b) These Alaskan railroad tracks have been warped by the effect of melting permafrost. *(a) NASA. (b) O. J. Ferrains, U.S. Geological Survey/L. A. Yehle photo.*

Beaches are formed by the deposition of sand grains contained in the water. The sand originates from the vast amount of coastal erosion and streams (Figure 3.33). *Longshore currents,* which move roughly parallel to the shore, transport the sand, forming beaches and *spits.* A more sheltered area increases the chances of a beach being built.

The backwash of waves, however, takes sand away from beaches if no longshore current exists. As a result, *sandbars* can develop a short distance away from the shoreline. If sandbars become extensive, they can eventually close off the shore, creating a new coastal configuration that encloses *lagoons* or *inlets. Salt marshes* very often develop in

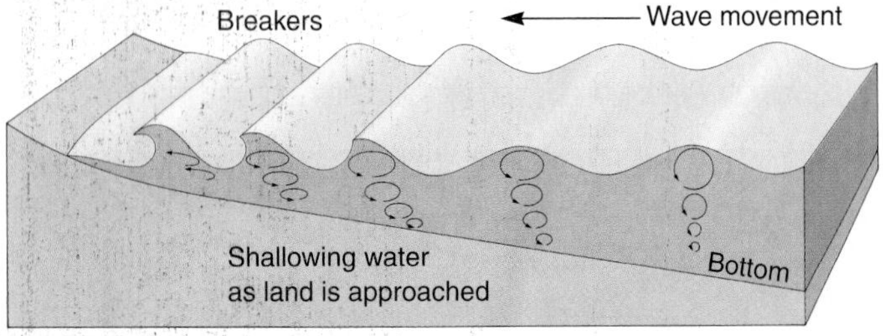


(a)

(b)

**FIGURE 3.32 Formation of waves and breakers.** (a) As the offshore swell approaches the gently sloping beach bottom, sharp-crested waves form, build up to a steep wall of water, and break forward in plunging surf. (b) Evenly spaced breakers form as successive waves touch bottom along a regularly sloping shore. *(b) © George Lepp/Getty Images.*

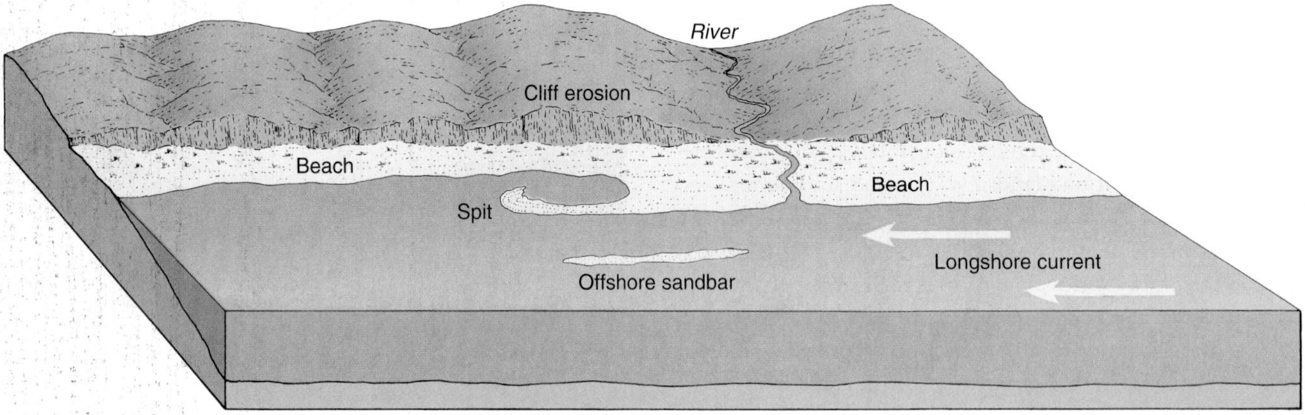


**FIGURE 3.33** The cliffs behind the shore are eroded by waves during storms and high water. Sediment from the cliff and river forms the beach deposit; the longshore current moves some sediment downcurrent to form a *spit. Offshore sandbars* are created from material removed from the beach and deposited by retreating waves. Generally, not all the features shown here would occur in a single setting.

and around these areas. For example, the Outer Banks of North Carolina, composed of long ribbons of sand, are constantly shifting as a result of currents and storms.

*Coral reefs*, made not from sand but from coral organisms growing in shallow tropical water, are formed by the secretion of calcium carbonate in the presence of warm water and sunlight. Reefs, consisting of millions of colorful coral skeletons, develop short distances offshore. Off the coast of northeastern Australia lies the most famous coral reef, the Great Barrier Reef. *Atolls,* found in the South Pacific, are reefs formed in shallow water around a volcano that has since been covered or nearly covered by water.

### Wind

In humid areas, vegetation cover confines the effect of wind mainly to sandy beach areas, but in dry climates, wind is a powerful agent of erosion and deposition. Limited vegetation in dry areas leaves exposed particles of sand, clay, and silt subject to movement by wind. Thus, many of the sculptured features found in dry areas result from mechanical weathering, that is, from the abrasive action of sand and dust particles as they are blown against rock surfaces. Sand and dust storms occurring in a drought-stricken farm area may make it unusable for agriculture. Inhabitants of Oklahoma, Texas, and Colorado suffered greatly in the 1930s when their farmlands became the "Dust Bowl" of the United States.

Several types of landforms are produced by wind-driven sand. Figure 3.34 depicts one of these. Although sandy deserts are much less common than gravelly deserts, also called *desert pavement,* their characteristic landforms are better known. Most of the Sahara, the Gobi, and the western U.S. deserts are covered with rocks, pebbles, and gravel, not sand. Each also has a small portion (and the Saudi Arabian Desert has a large area) covered with sand blown by wind into a series of waves, or *dunes.* Unless vegetation stabilizes them, the dunes move as sand is blown from their windward faces onto and over their crests. One of the most distinctive sand desert dunes is the crescent-shaped *barchan.* Along seacoasts and inland lakeshores, in both wet and dry climates, wind can create sand ridges that reach a height of 90 meters

**FIGURE 3.34** The prevailing wind from the left has given these transverse dunes a characteristic gentle windward slope and a steep, irregular leeward slope. *Courtesy of James A. Bier.*

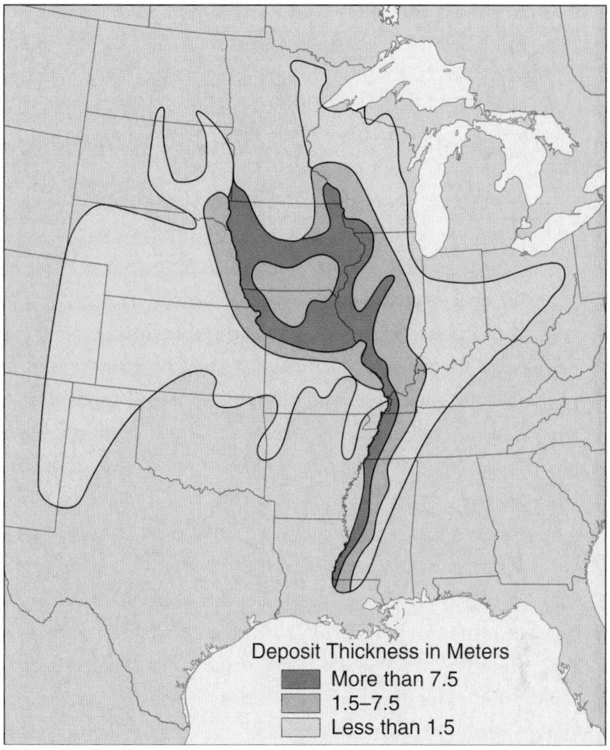


**FIGURE 3.35 Location of windblown silt deposits, including loess, in the United States.** The thicker layers, found in the upper Mississippi Valley area, are associated with the wind movement of glacial debris. Farther west, in the Great Plains, wind-deposited materials are sandy in texture, not loessial. *Adapted from* Geology of Soils *by Charles B. Hunt, copyright 1972 W. H. Freeman and Company. Reprinted by permission of the author.*

(300 ft). Sometimes, coastal communities and farmlands are threatened or destroyed by moving sand (see "Beaches on the Brink," p. 82).

Another kind of wind-deposited material, silty in texture and pale yellow or buff in color, is called **loess.** Encountered usually in midlatitude westerly wind belts, it covers extensive areas in the United States (Figure 3.35), central Europe, central Asia, and Argentina. It has its greatest development in northern China, where loess covers hundreds of thousands of square miles, often to depths of more than 30 meters (100 ft). The windborne origin of loess is confirmed by its typical occurrence downwind from extensive desert areas, though major deposits are assumed to have resulted from wind erosion of nonvegetated sediment deposited by meltwater from retreating glaciers. Because rich soils usually form from loess deposits, if climatic circumstances are appropriate, these areas are among the most productive agricultural lands in the world.

# Landform Regions

Every piece of land not covered by buildings and other structures contains clues as to how it has changed over time. Geomorphologists interpret these clues, studying such things as earth materials and soils, the availability of water, drainage patterns, evidences of erosion, and glacial history. The scale of analysis may be as small as a stream or as large as a *landform region,* a large section of the earth's surface where a great deal of homogeneity occurs among the types of landforms that characterize it.

Most atlases have a map showing, in a general way, the kinds of landform regions found in different parts of the world. Note how the *mountain* belts generally coincide with convergent plate boundaries not found beneath the sea (see Figure 3.5a) and with the earthquake-prone areas (see Figure 3.7). Vast *plains* exist in North and South America, Europe, Asia, and Australia. Many of these regions were created under former seas and appeared as land when seas contracted. These, and the smaller plains areas, are the drainage basins for some of the great rivers of the world, such as the Mississippi-Missouri, Amazon, Volga, Nile, Ganges, and Tigris-Euphrates. The valleys carved by these rivers and the silt deposited by them are among the most agriculturally productive areas in the world. The *plateau* regions are many and varied. The African plateau region is the largest. Much of the African landscape is characterized by low mountains and hills whose base is about 700 meters (2300 ft) above sea level. Generally quiet from the standpoint of tectonic activity, Africa is largely made up of geologically ancient continental blocks that have been in an advanced stage of erosion for millions of years.

Humans affect and are affected by the landscape, landforms, moving continents, and earthquakes; however, except at times of natural disaster, these elements of the physical world are, for most of us, quiet, accepted background. More immediate in affecting our lives and fortunes are the great patterns of climate. Climate helps define the limits of the economically possible with present levels of technology, the daily changes of weather that affect the success of picnics and crop yield alike, and the patterns of vegetation and soils. We turn our attention to these elements of the natural environment in Chapter 4.

# Beaches on the Brink

Headlines such as "Beaches on the Brink," "Storm-Lashed Cape Is a Fragile Environment," "Fighting the Development Tide," and "State Looks for Money to Restore the Shore" signal a growing concern with the condition of coastlines. In addition, they raise the central question of how we can utilize coastlines without, at the same time, destroying them.

Because many of the world's people live or vacation on coasts, and because such areas are often densely populated, coastal processes have a considerable impact on humans. Nature's forces are continually shaping and reshaping the coasts; they are dynamic environments, always in a state of flux. Some processes are dramatic and induce rapid change: tropical cyclones (hurricanes and typhoons), tsunami, and floods can wreak havoc, take thousands of lives, and cause millions of dollars in damage in a matter of hours. A less hazardous process is beach erosion, although it tends to magnify the effects of storms.

Some beach erosion is caused by natural processes, both marine and land. Waves carry huge loads of suspended sand, and longshore currents constantly move sand along the shoreline. The weathering and erosion of sea cliffs produce sediment, and rivers carry silt from mountains to beaches.

Human activities affect both erosion and deposition, however. Dams, for example, decrease the flow of sand to the water's edge by trapping sediment upstream. We fill marshes, construct dikes, and bulldoze dunes or remove their natural stabilizing vegetation. We hasten erosion when we build roads, houses, and other structures on cliffs and dunes or when we plant trees and lawns on top of them.

Especially vulnerable to erosion are barrier islands, narrow strips of sand parallel to the mainland. Under natural conditions, they are not stable places; their ends typically migrate, and during storms, waves can wash right over them. Some barrier islands are highly developed and densely populated, including Atlantic City and Miami Beach.

Once hotels and condominiums, railroads, and highways have been built along the waterfront, people attempt to protect their investments by preventing beaches from eroding. Breakwaters, offshore structures built to absorb the force of large, breaking waves and to provide quiet water near shore, are designed to trap sand and thus retard erosion, but they are not always successful. Although some are locally beneficial by forming new areas of deposition, they almost invariably accelerate erosion in adjacent areas. The efforts of one community are often negated by those of nearby towns and cities, underscoring the need for agencies to coordinate efforts and develop comprehensive land use plans for an extensive length of shoreline.

An alternative to erecting artificial structures is beach replenishment: adding sand to beaches to replace that lost by erosion. This provides recreational opportunities; at the same time, it buffers property from damage by storms. Sand may be dredged from harbors or pumped from offshore sandbars. The disadvantages of this technique are that dredging disrupts marine organisms, sand of the right texture may be hard to obtain, and replenished beaches can be short-lived. For example, more than $5 million was spent replenishing the beach at Ocean City, New Jersey, in 1982. The beach disappeared within 3 months, after a series of northeasters hit the area. Similarly, a massive dredging project that cost more than $17 million replenished 8 kilometers (5 mi) of beaches along the San Diego coast in 2000. Within months, after the largest waves of the year pounded the coast, more than half of the sand was gone.

The expense of maintaining coastal zones raises two basic questions: Who benefits? Who should pay? Some people contend that the interests of those who own coastal property are not compatible with the public interest, and it is unwise to expend large amounts of public funds to protect the property of only a few. People argue that shorefront businesses and homeowners are the chief beneficiaries of shore protection measures, that they often deny the public access to the beaches in front of their property, and therefore they should pay for the majority of the cost of maintaining the shoreline. At present, however, this is rarely the case; costs are typically shared by communities, states, and the federal government. Indeed, 51 federal programs subsidize coastal development and redevelopment. The largest is the National Flood Insurance Program, which offers low-cost insurance to homeowners in flood-prone areas. People have used the protection of that insurance to build anywhere, even in high-risk coastal areas.

This beach house in Kitty Hawk, North Carolina, was destroyed by Hurricane Isabel in September 2003. *© AP/Wide World Photos.*

**QUESTIONS TO CONSIDER**

1. From 1994 to the present, the Army Corps of Engineers has spent millions of dollars annually to pump sand from the ocean bottom onto eroded beaches in New Jersey. This ongoing project is based on the assumption that additional replenishment will be required every 5 or 6 years as beaches erode. Currently, the federal government pays 65% of the cost, the state government 25%, and local governments 10%. New Jersey Senator Frank Lautenberg contends that beach replenishment is critical for the region's future. "People's lives and property are at stake. Jersey's beaches bring crucial tourist dollars to the state." But James Tripp of the Environmental Defense Fund argues that beach rebuilding is simply throwing taxpayers' money into the ocean. "Pumping all the sand in the world is not going to save the day." Do you think the beach replenishment project is a wise use of taxpayers' money? Does the federal government have an obligation to protect or rebuild storm-damaged beaches? Why or why not?
2. Coastal erosion is not a problem for beaches, only for people who want to use them. Do you believe that we should learn to live with erosion, not building in the coastal zone unless we are prepared to consider our structures temporary and expendable? Should communities adopt zoning plans that prohibit building on undeveloped lands within, say, 50 meters (164 ft) of the shore?
3. Should the federal government curtail programs that provide inexpensive storm insurance for oceanfront houses and businesses, as well as speedy grants and loans for storm repairs not covered by insurance? Should people be allowed to rebuild storm-damaged buildings even if they are vulnerable to future damage? Why or why not?
4. Coastal erosion will become more serious if the current rise in sea level—about 2.5 centimeters (1 in.) every 12 years—continues or even increases as global warming causes seawater to expand or the polar ice caps to melt. Many of the world's major cities would be threatened by this rise in sea level. What are some of those cities? How might they protect themselves?

## Summary

Rocks, the materials that constitute the earth's surface, are classified as igneous, sedimentary, and metamorphic. In the most recent 200 million of the earth's 4.7 billion years, continental plates drifted on the asthenosphere to their present positions. At or near plate intersections, tectonic activity is particularly in evidence in two forms. Diastrophism, such as faulting, results in earthquakes and, on occasion, tsunami. Volcanism moves molten materials toward the earth's surface.

The building up of the earth's surface is balanced by three gradational processes—weathering, mass movement, and erosion. Weathering, both mechanical and chemical, prepares materials for transport by disintegrating rocks. It is also instrumental in the development of soils. Talus slopes and soil-creep are examples of mass movement. The erosional agents of running water, groundwater, glaciers, waves and currents, and wind move materials to new locations. Examples of landforms created by the collection of eroded materials are alluvial fans, deltas, natural levees, moraines, and sand dunes.

## Key Words

alluvium 73
asthenosphere 58
chemical weathering 69
continental drift 58
diastrophism 62
erosional agents 71
faults 60
floodplain 72
folds 62
glacier 75
gradational processes 68
igneous rocks 56
karst topography 75
lithosphere 58
loess 81
mass movement 70
mechanical weathering 68
metamorphic rocks 58
mineral 56
plate tectonics 58
sedimentary rocks 57
subduction 62
tsunami 64
volcanism 62
warping 62
water table 74
weathering 68

# For Review & Consideration

1. How can rocks be classified? List three classes of rocks according to their origin. In what ways can they be distinguished from one another?
2. What evidence makes the theory of *plate tectonics* plausible?
3. What is the meaning and name of the process that occurs when two plates collide?
4. Explain what is meant by *gradation* and *volcanism.*
5. What is meant by *folding, joint,* and *faulting?*
6. Draw a diagram indicating the varieties of ways *faults* occur.
7. With what earth movements are earthquakes associated? What are *tsunami* and how do they develop?
8. What is the distinction between *mechanical* and *chemical weathering?* Is weathering responsible for landform creation? In what ways do glaciers engage in mechanical weathering?
9. Explain the origin of the various landforms one usually finds in desert environments.
10. How do *glaciers* form? What landscape characteristics are associated with glacial erosion? With glacial deposition?
11. How are alluvial fans, deltas, natural levees, and moraines formed?
12. How is groundwater erosion differentiated from surface water erosion?
13. How are the processes that bring about change due to waves and currents related to the processes that bring about change by the force of wind?
14. What processes account for the landform features of the area in which you live?

# Selected References

**WEBSITES**

The World Wide Web has a tremendous variety of sites pertaining to geography. Websites relevant to the subject matter of this chapter appear in the "Web Links" section of the Online Learning Center associated with this book. Access it at www.mhhe.com/getis10e/.

Ahnert, Frank. *Introduction to Geomorphology.* London: Edward Arnold, 1998.

Benn, Douglas I. *Glaciers and Glaciation.* London: Edward Arnold, 1998.

Briggs, David, et al., eds. *Fundamentals of the Physical Environment.* 2d ed. New York: Routledge, 1997.

Chorley, Richard J. *Water, Earth, and Man.* London: Methuen, 1969.

Easterbrook, Don J. *Surface Processes and Landforms.* 2d ed. Upper Saddle River, N.J.: Prentice Hall, 1998.

French, Hugh M. *The Periglacial Environment.* 2d ed. Harlow, Essex, England: Addison Wesley Longman, 1996.

Goudie, Andrew. *The Changing Earth: Rates of Geomorphological Processes.* Oxford, England: Blackwell, 1995.

Haslett, Simon. *Coastal Systems.* New York: Routledge, 2001.

Huggett, Richard J. *Fundamentals of Geomorphology.* New York: Routledge, 2002.

Knighton, David. *Fluvial Forms and Processes.* Oxford, England: Oxford University Press, 1988.

Kump, Lee, James Kesting, and Robert Crane. *The Earth System.* Upper Saddle River, N.J.: Prentice Hall, 1999.

Livingston, Ian, and Andrew Warren. *Aeolian Geomorphology: An Introduction.* Harlow, Essex, England: Addison Wesley Longman, 1996.

Montgomery, Carla W. *Environmental Geology.* 6th ed. Boston: McGraw-Hill, 2003.

Ollier, Cliff, and Colin Pain. *The Origin of Mountains.* New York: Routledge, 2000.

Plummer, Charles, David McGeary, and Diane Carlson. *Physical Geology.* 9th ed. Boston: McGraw-Hill, 2003.

*Progress in Physical Geography.* London: Edward Arnold. Various issues.

Rolls, David, and Will J. Bland. *Weathering: An Introduction to Basic Principles.* London: Edward Arnold, 1998.

Spencer, Edgar W. *Earth Science.* Boston: McGraw-Hill, 2003.

Strahler, Alan, and Arthur Strahler. *Physical Geography: Science and Systems of the Human Environment.* 2d ed. New York: John Wiley & Sons, 2002.

Trenhaile, Alan S. *Coastal Dynamics and Landforms.* Oxford, England: Clarendon Press, 1997.

TRUCKS
OVER
5 TONS
EXCLUDED
STOP

Chapter 4

# Physical Geography: Weather and Climate


**Air Temperature**
- Earth Inclination
- Reflection and Reradiation
- Lapse Rate

**Air Pressure and Winds**
- Pressure Gradient Force
- The Convection System
- Land and Sea Breezes
- Mountain and Valley Breezes
- The Coriolis Effect
- The Frictional Effect
- The Global Air-Circulation Pattern

**Ocean Currents**

**Moisture in the Atmosphere**
- Types of Precipitation
- Storms

**Climate, Soils, and Vegetation**
- Soils and Climate
  - *Soil Formation*
  - *Soil Profiles and Horizons*
  - *Soil Properties*
  - *Soil Classification*
- Natural Vegetation and Climate
  - *Succession*
  - *Natural Vegetation Regions*

**Climate Regions**
- Tropical Climates
  - *Tropical Rain Forest*
  - *Savanna*
- Dryland Climates
  - *Hot Deserts*
  - *Midlatitude Deserts and Semideserts*
- Humid Midlatitude Climates
  - *Mediterranean Climate*
  - *Humid Subtropical Climate*
  - *Marine West Coast Climate*
  - *Humid Continental Climate*
- Subarctic and Arctic Climates

**Climatic Change**
- Long-Term Climatic Change
- Short-Term Climatic Change
- The Greenhouse Effect and Global Warming

**Summary**

**Key Words**

**For Review & Consideration**

**Selected References**


**Damage from the massive ice storm that struck New York, New England, and the province of Quebec in January 1998.** © Syracuse Newspapers/Image Works.

It started as a small disturbance in the Atlantic Ocean, east of the Caribbean islands in early September. But as is often the case, the disturbance soon became a tropical storm, and then, after its sustained wind speeds were clocked at greater than 120 kilometers per hours (74 mph), it met the definition of a hurricane and was named Isabel. Fed by warm tropical waters, Isabel advanced northwestward at even greater speeds (Figure 4.1).

After two weeks of churning its way toward the Atlantic coast, the hurricane hit land with deadly force on September 18, 2003. The residents of North Carolina's Outer Banks were the first to feel the storm's might as Isabel carved a new inlet into Cape Hatteras, eroded beaches, and washed away part of Highway 12, as well as houses, mobile homes, and even entire motels. Soon, all of eastern North Carolina was affected by Isabel. Wind-driven high water flooded coastal regions, and winds as high as 170 kilometers per hour (105 mph) downed trees and power lines and severely damaged roads and bridges.

Within hours, the hurricane was wreaking havoc in Virginia and Maryland. Storm-swollen waters from Chesapeake Bay that reached 2 meters (6 ft) in places flooded streets and buildings in Baltimore, while the Potomac River coursed into Alexandria, Virginia, and Washington, D.C., causing the federal government to shut down. Trees toppled onto houses and cars and blocked roads. Millions of houses and businesses were without electricity for a week or more.

The winds weakened as Isabel passed over Pennsylvania and West Virginia, and it was downgraded to a tropical storm. Nevertheless, the winds were strong enough to down trees and power lines, and heavy rains caused flash floods. By the time Isabel had spun itself out over Canada, it had left at least 26 people dead and caused billions of dollars worth of damage.

The power of hurricanes is concentrated in a narrow path. Whether such meteorological events occur in Asia or North America, they do great damage. The lives of all in the paths of these storms are affected. Tropical storms are one extreme type of weather phenomenon. Most people are "weather watchers"—they watch television forecasts with great interest and plan their lives around weather events. In this chapter, we review the subsection of physical geography concerned with weather and climate. It deals with normal, patterned phenomena from which such extreme weather events as Hurricane Isabel occasionally emerge.

A weather forecaster describes current conditions for a limited region, such as a metropolitan area, and predicts future weather conditions. If the elements that make up the **weather,** such as temperature, wind, and precipitation, are recorded at specified moments in time, such as every hour, an inventory of weather conditions can be developed. By finding trends in data that have been gathered over an extended period of time, we can speak about typical conditions. These characteristic circumstances describe the **climate** of a region. Weather is a moment's view of the lower atmosphere, whereas climate is a description of typical weather conditions in an area or at a place over a period of time. Geographers analyze the differences in weather and climate from place to place in order to understand how climatic elements affect human occupance of the earth.

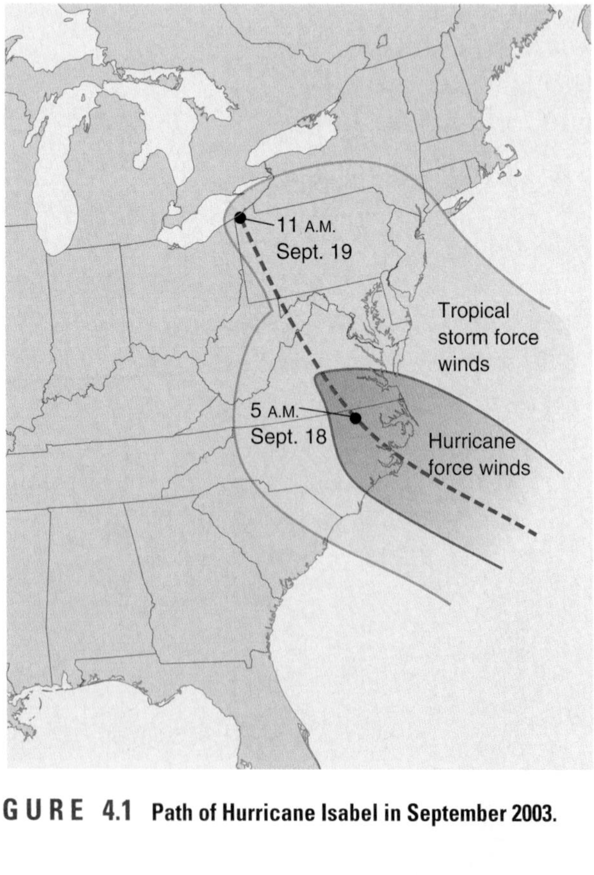


**FIGURE 4.1** Path of Hurricane Isabel in September 2003.

In geography, we are particularly interested in the physical environment that surrounds us. That is why the **troposphere,** the lowest layer of the earth's atmosphere, attracts our attention. This layer, extending about 10 kilometers (6 mi) above the ground, contains virtually all of the air, clouds, and precipitation of the earth.

In this chapter, we try to answer the questions usually raised regarding characteristics of the lower atmosphere. By discussing these answers from the viewpoint of averages or average variations, we attempt to give a view of the earth's climatic differences, a view held to be very important for understanding the way people use the land. Climate is a key to understanding, in a broad way, the distribution of world population. People have great difficulty living in areas that are, on average, very cold, very hot, very dry, or very wet. They are also negatively affected by huge storms or flooding. In this chapter, we first discuss the elements that constitute weather conditions and then describe the various climates of the earth.

# AIR TEMPERATURE

Perhaps the most fundamental question about weather is "Why do temperatures vary from place to place?" The answer to this question requires the discussion of a number of concepts to help focus on the way heat accumulates on the earth's surface.

Energy from the sun, called *solar energy,* is transformed into heat, primarily at the earth's surface and secondarily in the atmosphere. Not every part of the earth or its overlying atmosphere receives the same amount of solar energy. At any given place, the amount of incoming solar radiation, or **insolation,** available depends on the intensity and duration of radiation from the sun. These are determined by both the angle at which the sun's rays strike the earth and the number of daylight hours. These two fundamental factors, plus the following five modifying variables, determine the temperature at any given location:

1. the amount of water vapor in the air;
2. the degree of cloud cover (or cover in general);
3. the nature of the surface of the earth (land or water);
4. the elevation above sea level;
5. the degree and direction of air movement.

Let us look at these factors briefly.

## Earth Inclination

The axis of the earth—that is, the imaginary line connecting the North Pole to the South Pole—always remains in the same position. It is tilted about 23.5° away from the perpendicular (Figure 4.2). Every 24 hours, the earth rotates once on that axis, as shown in Figure 4.3. While rotating, the earth is slowly revolving around the sun in a nearly circular annual orbit (Figure 4.4). If the earth were not tilted

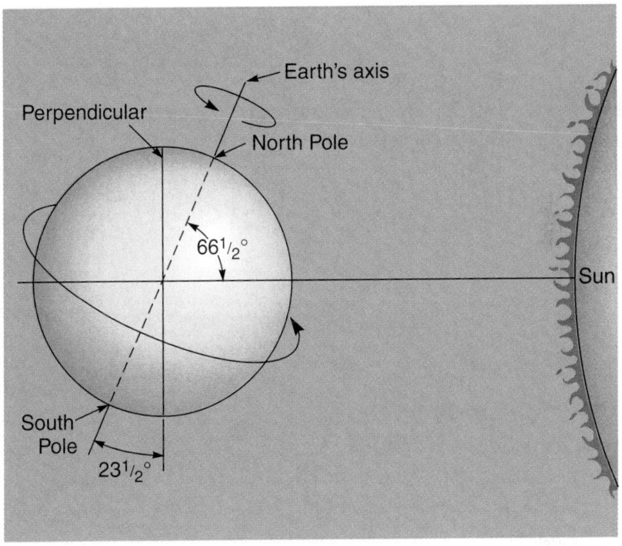


**FIGURE 4.2 The earth's position relative to the sun.** The earth is in its summer position in the Northern Hemisphere (winter in the Southern Hemisphere).

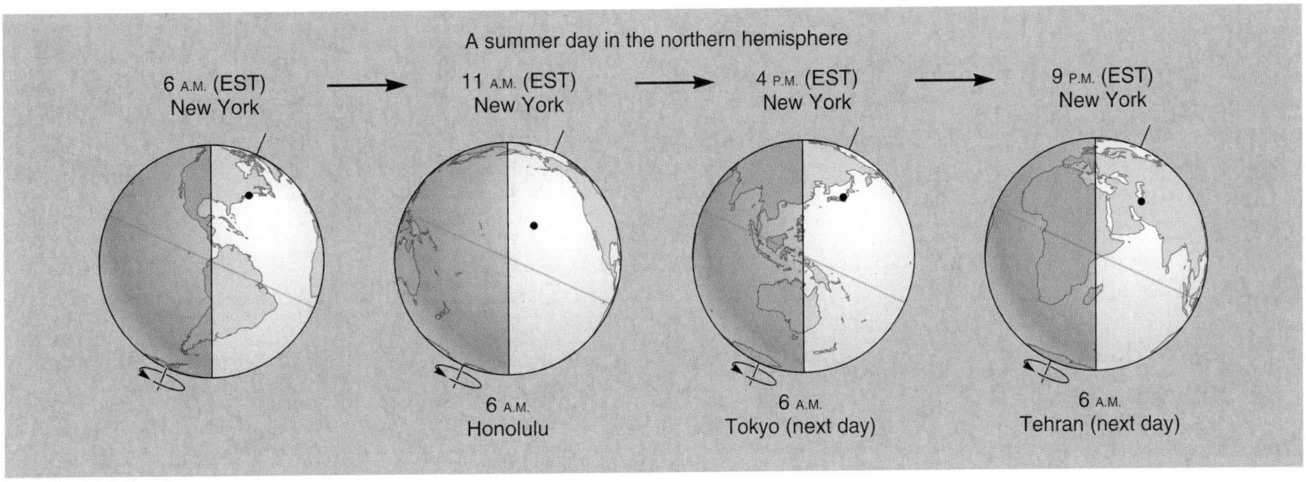


**FIGURE 4.3 The process of the 24-hour rotation of the earth on its axis.**

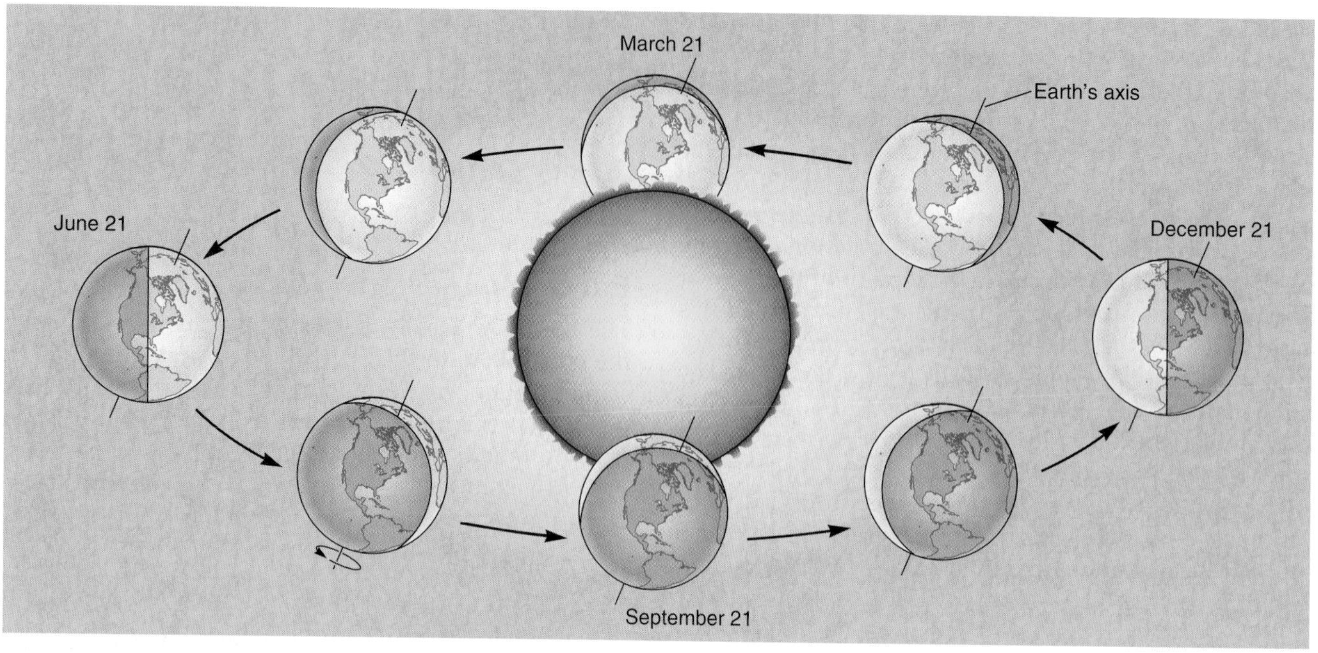


**FIGURE 4.4 The process of the yearly revolution of the earth around the sun.** The sun, which is about 93 million miles from Earth, is not drawn to scale; it is much larger relative to the size of the earth.

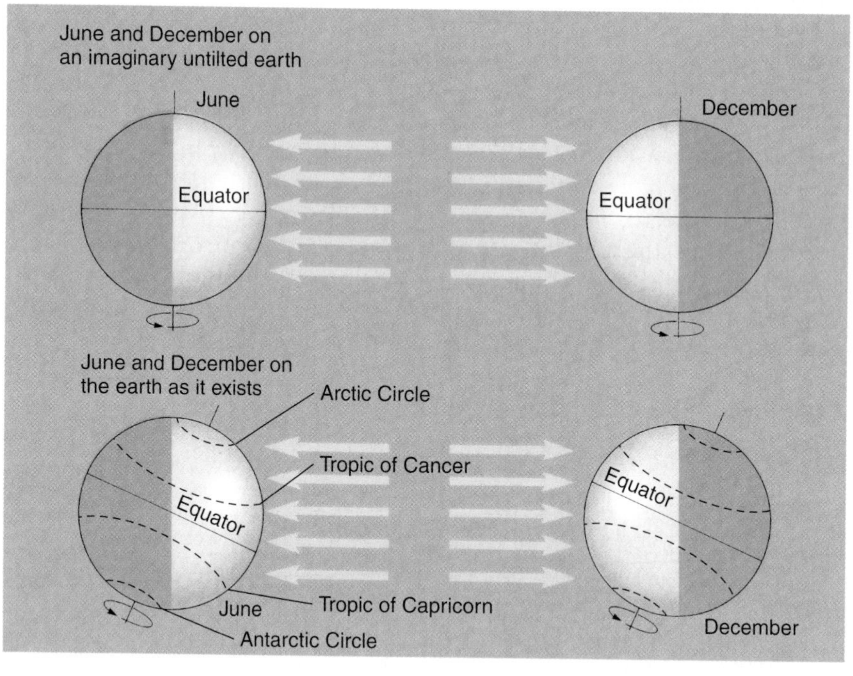


**FIGURE 4.5** Notice on the bottom two diagrams that, as the earth revolves, the north polar area in June is bathed in sunshine for 24 hours, while the south polar area is dark. The most intense of the sun's rays are felt north of the equator in June and south of the equator in December. None of this is true in the untilted examples shown on the upper two diagrams.

from the perpendicular, the solar energy received *at a given latitude* would not vary during the course of the year. The rays of the sun would directly strike the equator, and as distance away from the equator became greater, rays would strike the earth at ever-increasing angles, therefore diminishing the intensity of the energy and giving climates a latitudinal standardization (Figures 4.5 and 4.6).

Because of the inclination, however, the location of highest incidence of incoming solar energy varies during the course of the year. When the Northern Hemisphere is tilted directly toward the sun, the vertical rays of the sun are felt as far north as 23.5°N latitude (Tropic of Cancer). This position of the earth occurs about June 21, the summer *solstice* for the Northern Hemisphere and the winter solstice for the Southern Hemisphere. About December 21, when the vertical rays of the sun strike near 23.5°S latitude (Tropic of Capricorn), it is the beginning of summer in the Southern Hemisphere and the onset of winter in the Northern Hemi-

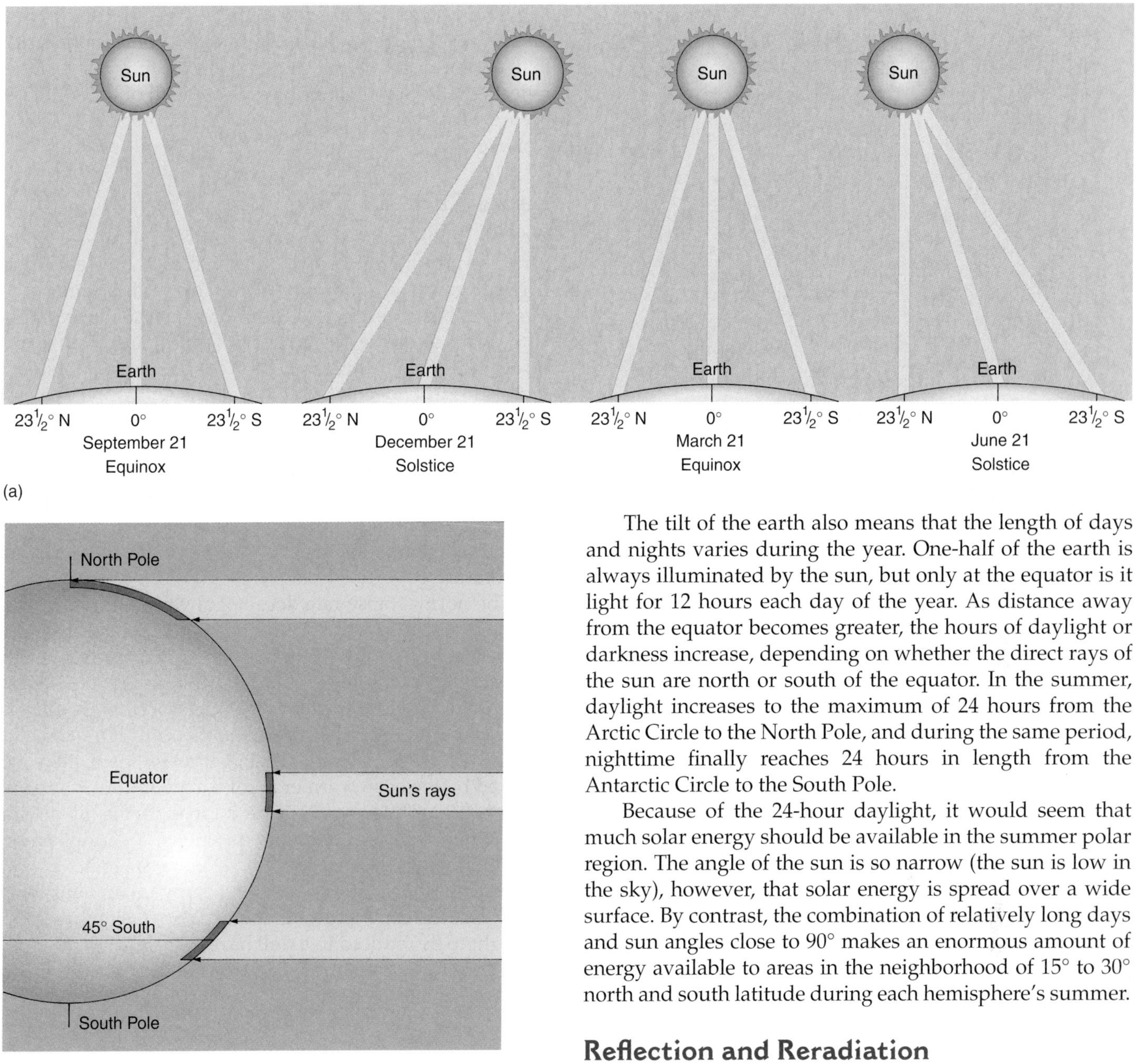


**FIGURE 4.6** **(a) Imaginary rays from the sun at spring and autumn equinoxes and summer and winter solstices.** (b) Three equal, imaginary rays from the sun are shown striking the earth at different latitudes at the time of the equinox. As distance away from the equator increases, the rays become more diffused, showing how the sun's intensity is diluted in the high latitudes.

sphere. During the rest of the year, the position of the earth relative to the sun results in direct rays migrating from about 23.5°N to 23.5°S and back again. On about March 21 and September 21 (the spring and autumn *equinoxes*), the vertical rays of the sun strike the equator.

The tilt of the earth also means that the length of days and nights varies during the year. One-half of the earth is always illuminated by the sun, but only at the equator is it light for 12 hours each day of the year. As distance away from the equator becomes greater, the hours of daylight or darkness increase, depending on whether the direct rays of the sun are north or south of the equator. In the summer, daylight increases to the maximum of 24 hours from the Arctic Circle to the North Pole, and during the same period, nighttime finally reaches 24 hours in length from the Antarctic Circle to the South Pole.

Because of the 24-hour daylight, it would seem that much solar energy should be available in the summer polar region. The angle of the sun is so narrow (the sun is low in the sky), however, that solar energy is spread over a wide surface. By contrast, the combination of relatively long days and sun angles close to 90° makes an enormous amount of energy available to areas in the neighborhood of 15° to 30° north and south latitude during each hemisphere's summer.

## Reflection and Reradiation

Much of the potentially receivable solar radiation is, in fact, sent back to outer space or diffused in the troposphere in a process known as **reflection.** Clouds, which are dense concentrations of suspended, tiny water or ice particles, reflect a great deal of energy. Light-colored surfaces, especially snow cover, also reflect large amounts of solar energy.

Energy is lost through reradiation as well as reflection. In the **reradiation** process, the earth's surface acts as a communicator of energy. As indicated in Figure 4.7, the energy that is absorbed into the land and water is returned to the atmosphere in the form of terrestrial radiation. On a clear night, when no clouds can block or diffuse movement, temperatures continually decrease, as the earth reradiates as heat the energy it has received and stored during the course of the day.

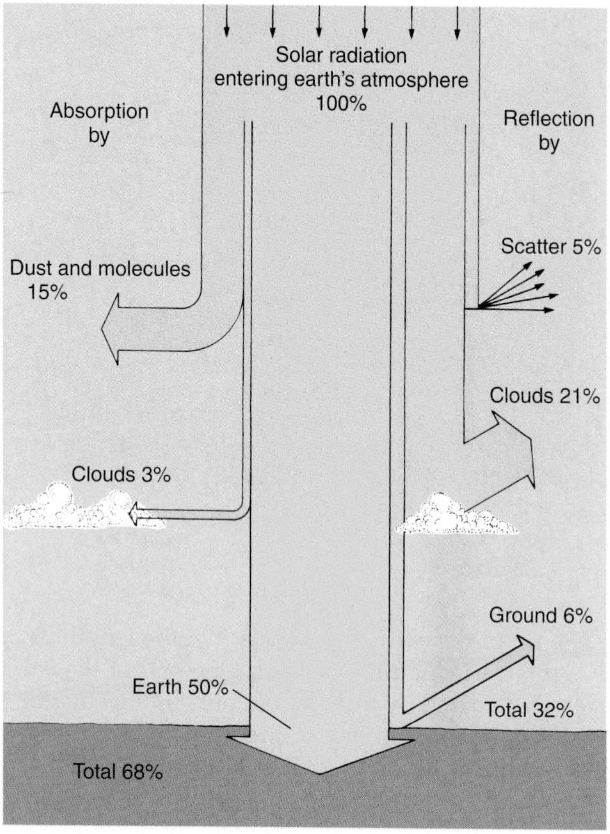


**FIGURE 4.7** Consider the incoming solar radiation as 100%. The portion that is absorbed into the earth (50%) is eventually released to the atmosphere and then reradiated into space.

Some kinds of earth surface material, especially water, store solar energy more effectively than others. Because water is transparent, solar rays can penetrate a great distance below its surface. If water currents are present, heat is distributed even more effectively. On the other hand, land surfaces are opaque, so all of the energy received from the sun is concentrated at the surface. Land, having more heat available at the surface, reradiates its energy faster than water. Air is heated by the process of reradiation from the earth and not directly by energy from the sun passing through it. Thus, because land heats and cools more rapidly than water, hot and cold temperature extremes recorded on Earth occur on land and not the sea.

Temperatures are moderated by the presence of large bodies of water near land areas. Note in Figure 4.8 that coastal areas have lower summer temperatures and higher winter temperatures than those places at the same distance from the equator, excluding seacoasts. Land areas affected by the moderating influences of water are considered *marine* environments; those areas not affected by nearby water are *continental* environments.

Temperatures vary in a cyclical way each day. In the course of a day, as incoming solar energy exceeds energy lost through reflection and reradiation, temperatures begin to rise. The ground stores some heat, and temperatures continue to rise until the angle of the sun becomes so narrow that energy received no longer exceeds that lost by the reflection and reradiation processes. Not all of the heat loss occurs during the night, but long nights appreciably deplete stored energy.

## Lapse Rate

We may think that, as we move vertically away from the earth toward the sun, temperatures increase. However, this is not true within the troposphere. The earth absorbs and reradiates heat; therefore, temperatures are usually warmest at the earth's surface and lower as elevation increases. Note on Figure 4.9 that this temperature **lapse rate** (the rate of temperature change with altitude in the troposphere) averages about 6.4°C per 1000 meters (3.5°F per 1000 ft). For example, the difference in elevation between Denver and Pikes Peak is about 2700 meters (9000 ft), which normally results in a 17°C (32°F) difference in temperature. Jet planes flying at an altitude of 9100 meters (30,000 ft) are moving through air that is about 56°C (100°F) colder than ground temperatures.

The normal lapse rate does not always hold, however. Rapid reradiation sometimes causes temperatures to be higher above the earth's surface than at the surface itself. This particular condition, in which air at lower altitudes is cooler than air aloft, is called a **temperature inversion.** An inversion is important because of its effect on air movement. Warm air at the surface, which normally rises, may be blocked by the even warmer air of a temperature inversion (Figure 4.10). Thus, surface air is trapped; if it is filled with automobile exhaust emissions or smoke, a serious smog condition may develop (see "The Donora Tragedy," p. 95). Because of the configuration of nearby mountains, Los Angeles often experiences temperature inversions, causing sunlight to be reduced to a dull haze (Figure 4.11).

The effect of air movement on temperature is made clear in the following section, "Air Pressure and Winds."

# AIR PRESSURE AND WINDS

The second fundamental question about weather and climate concerns **air pressure:** How do differences in air pressure from place to place affect weather conditions? The answer to this question first requires that we explain why differences in air pressure occur.

Air is a gaseous substance whose weight affects air pressure. If it were possible to carve out 16.39 cubic centimeters (1 cu in.) of air at the earth's surface and weigh it, along with all the other cubic centimeters of air above it, under normal conditions the total weight would equal approximately 6.67 kilograms (14.7 lbs) of air as measured at sea level. Actually, this is not very heavy when you consider the dimensions of the column of air: 2.54 centimeters by 2.54 centimeters (1 in. by 1 in.) by about 9.7 kilometers (6

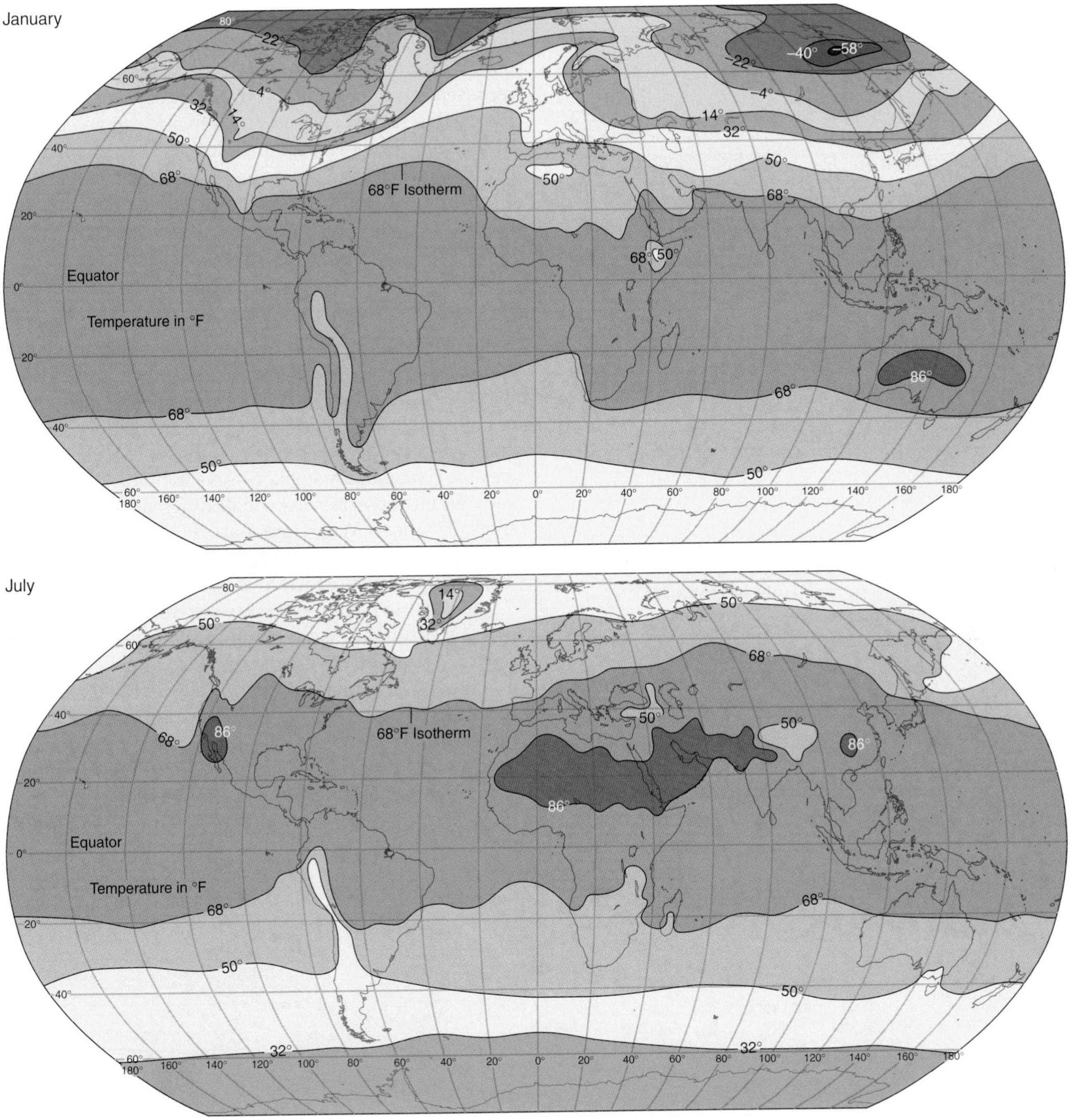


**FIGURE 4.8** At a given latitude, water areas are warmer than land areas in winter and cooler in summer. *Isotherms* are lines of equal temperature.

mi), or about 6.2 cubic meters (220 cu ft). The weight of air 4.8 kilometers (3 mi) above the earth's surface, however, is considerably less than 6.67 kilograms (14.7 lbs) because there is correspondingly less air above it. Thus, it is clear that air is heavier and air pressure is higher closer to the earth's surface.

It is a physical law that, for equal amounts of cold and hot air, the cold air is denser. This law exemplifies why hot-air balloons, filled with lighter air, can rise into the atmosphere. A cold morning is characterized by relatively heavy air, but as afternoon temperatures rise, air becomes lighter.

*Barometers* of various types are used to record changes in air pressure. Barometric readings in inches of mercury or millibars are a normal part, along with recorded temperatures, of every weather report. Air pressure at a given location changes as surfaces heat or cool. Barometers record a drop in atmospheric pressure when air heats and a rise in pressure when air cools.

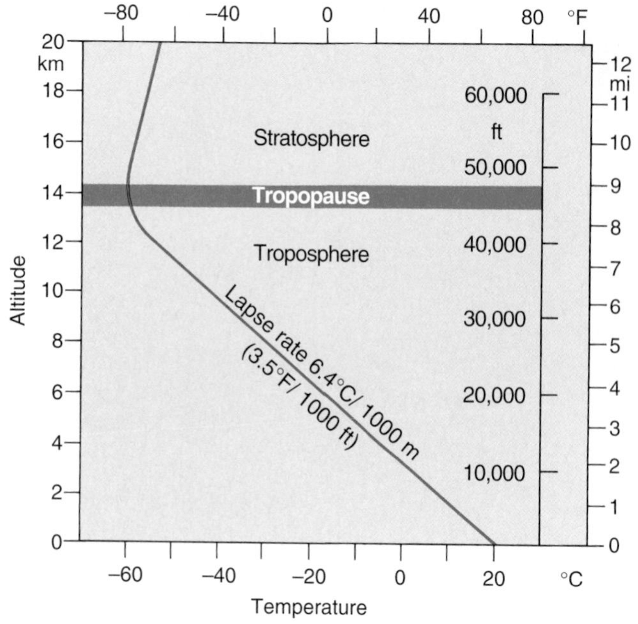


**FIGURE 4.9 The temperature lapse rate under typical conditions.** The *tropopause* is a transition zone between the troposphere and the stratosphere. It marks the level at which temperature ceases to fall with altitude.

In order to visualize the effect of air movements on weather, it is useful to think of air as a liquid made up of two fluids with different densities (representing light air and heavy air)—for example, water and gasoline. If the fluids are put into a tank at the same time, the lighter liquid will move to the top and the heavier air will move to the bottom, representing the vertical motion of air. The heavier liquid spreads out horizontally along the bottom of the tank, becoming the same thickness everywhere. This flow represents the horizontal movement of the air or wind on the earth's surface. Air attempts to achieve an equilibrium by evening out pressure imbalances that result from the heating and cooling processes. Air races from heavy (cold) air locations to light (warm) air locations. Thus, the greater the differences in air pressure between places, the greater the wind.

## Pressure Gradient Force

Because of differences in the nature of the earth's surface—water, snow cover, dark green forests, cities, and so on—and the other factors that affect energy receipt and retention, zones of high and low air pressure develop. Sometimes, these high- and low-pressure zones cover entire continents, but usually they are considerably smaller—several hundred miles wide—and within these regions, small differences are noted over short distances. When pressure differences exist between areas, a **pressure gradient force** causes air to blow from an area of high pressure toward an area of low pressure.

In order to balance pressure differences that have developed, air from the heavier high-pressure areas flows to low-pressure zones. Heavy air stays close to the earth's surface as it moves, producing winds, and forces the upward movement of warm air. The velocity, or speed, of the wind is in direct proportion to pressure differences. Winds are caused by pressure differences that induce airflow from zones of high pressure to zones of low pressure. If distances between high- and low-pressure zones are short, pressure gradients

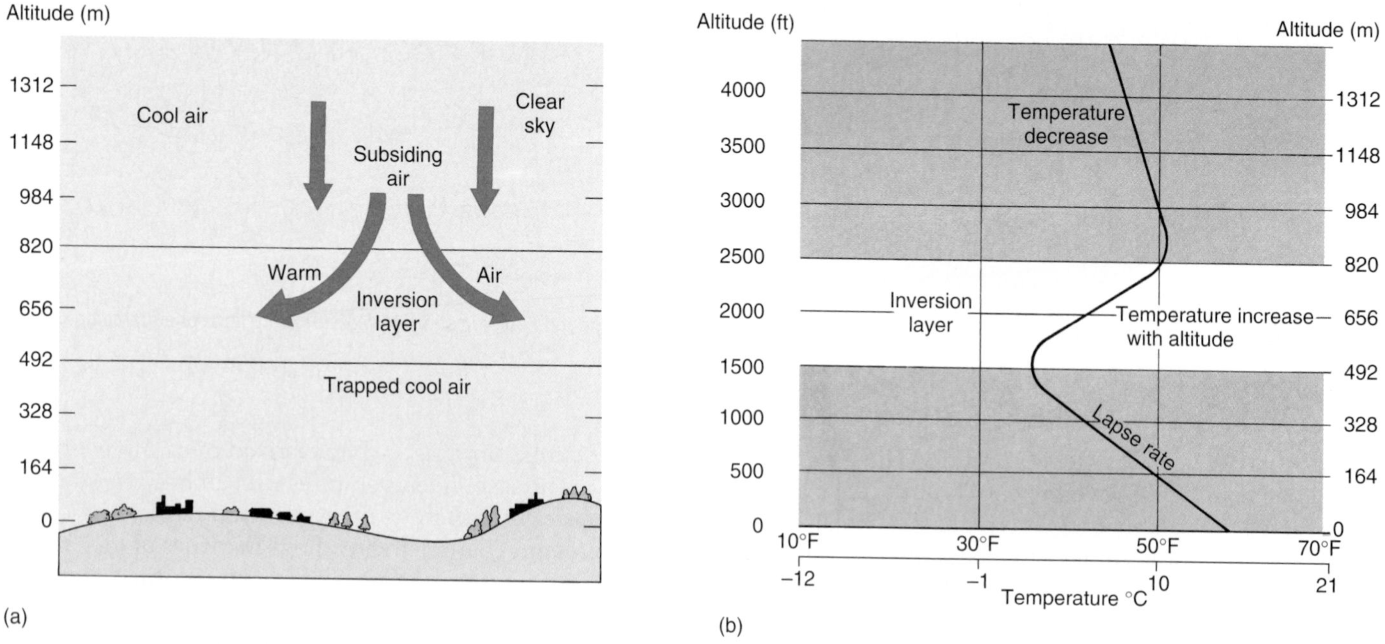


**FIGURE 4.10 Temperature inversion.** (a) A layer of warm, subsiding air acts as a cap, temporarily trapping cooler air close to the ground. (b) Note that air temperature decreases with distance from the ground until the warm inversion layer is reached, at which point the temperature increases.

## The Donora Tragedy

A heavy fog settled over the valley town of Donora, Pennsylvania, in late October 1948. Stagnant, moisture-filled air was trapped in the valley by surrounding hills and by a temperature inversion that held cooler air, gradually filling with smoke and fumes from the town's zinc works, against the ground under a lid of lighter, warmer upper air. For 5 days, the smog increased in concentration; the sulfur dioxide emitted from the zinc works continually converted to deadly sulfur trioxide by contact with the air.

Both old and young, with and without histories of respiratory problems, reported to doctors and hospitals a difficulty in breathing and unbearable chest pains. Before the rains washed the air clean nearly a week after the smog buildup, 20 died and hundreds of others were hospitalized. A normally harmless, water-saturated inversion had been converted to deadly poison by a tragic union of natural weather processes and human activity.

are steep and wind velocities are great. More gentle air movements occur when zones of different pressure are far apart and the degree of difference is not great.

### The Convection System

A room's temperature is lower near the floor than the ceiling because warm air rises and cool air descends. The circulatory motion of descending cool air and ascending warm air is known as **convection** (Figure 4.12). A convectional wind system results when surface-heated warm air rises and is replaced by cool air from above.

### Land and Sea Breezes

A good example of a convectional system is **land** and **sea breezes** (Figure 4.13a). Close to a large body of water, the differential daytime heating between land and water is great. As a result, warmer air over the land rises vertically, only to be replaced by cooler air from over the sea. At night, just the opposite occurs; the water is warmer than the land, which has reradiated much of its heat, and the result is a land breeze toward the sea. These two winds make seashore locations in warm climates particularly comfortable.

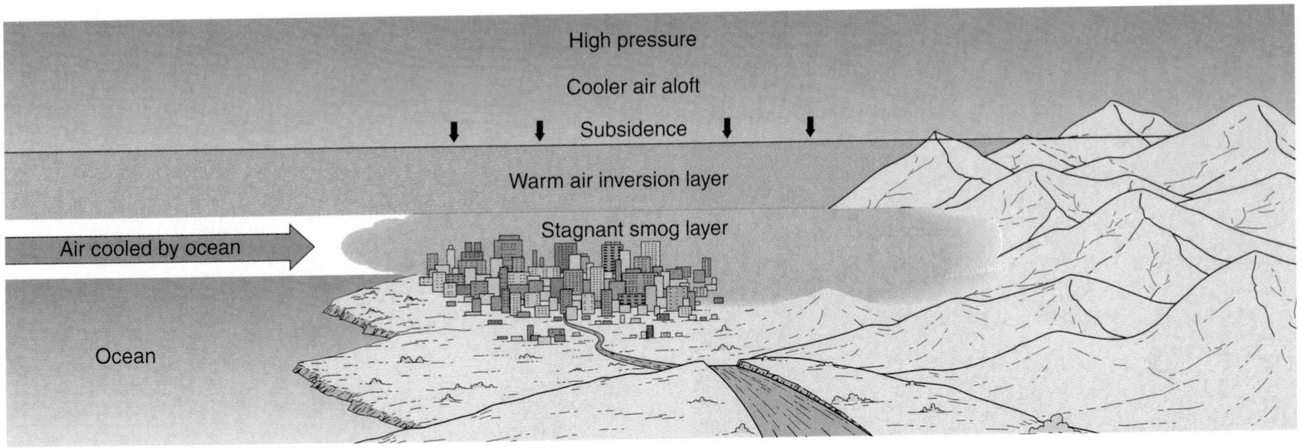


**FIGURE 4.11 Smog in the Los Angeles area.** Below the inversion layer, stagnant air holds increasing amounts of pollutants, caused mainly by automobile exhausts. See also Figure 12.14.

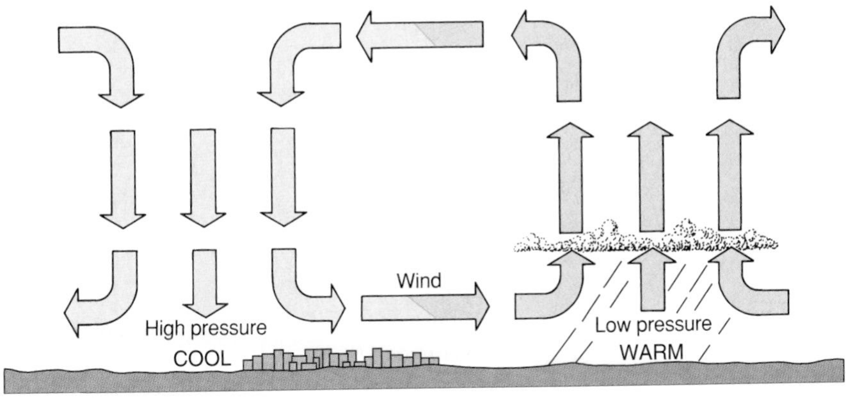


**FIGURE 4.12 A convection system.** Descending cool air flows toward low pressure. Precipitation very often occurs in low-pressure zones. As warm air rises, it cools and can become supersaturated, resulting in precipitation.

## Mountain and Valley Breezes

Gravitational force causes the heavy cool air that accumulates over snow in mountainous areas to descend into lower valley locations, as suggested in Figure 4.13b. Consequently, valleys can become much colder than the slopes, and a temperature inversion occurs. Slopes are the preferred sites for agriculture in mountainous regions because cold air from **mountain breezes** can cause freezing conditions in the valleys. In densely settled narrow valleys where industry is concentrated, air pollution can become particularly dangerous. Mountain breezes usually occur during the night; **valley breezes**—caused by warm air moving up slopes in mountainous regions—are usually a daytime phenomenon. The canyons of southern California are the scenes of strong mountain and valley breezes. In addition, during the dry season, they become dangerous areas for the spread of brush and forest fires.

## The Coriolis Effect

In the process of moving from high to low pressure, wind veers toward the right of the direction of travel in the Northern Hemisphere and toward the left in the Southern Hemisphere. This deflection is called the **Coriolis effect.** Were it not for this effect, winds would move in exactly the direction specified by the pressure gradient.

To illustrate the impact of the Coriolis effect upon winds, a familiar example may be helpful. Imagine a line of ice skaters holding hands while skating in a circle, with one skater nearest the center of the circle. This skater turns slowly, while the outermost skater must skate very rapidly in order to keep the line straight. In a similar way, the equatorial regions of the earth are rotating at a much faster rate than the areas around the poles.

Next, suppose that the center skater threw a ball directly toward the skater at the end of the line. By the time the ball arrived, it would pass behind the outside skater. If the skaters are going in a counterclockwise direction—as the earth appears to be moving viewed from the position of the North Pole—the ball appears to the person at the North Pole to pass to the right of the outside skater. If the skaters are going in a clockwise direction—as the earth appears to be moving viewed from the South Pole—the deflection is to the left. Because air (like the ball) is not firmly attached to the earth, it, too, will appear to be deflected. The air maintains its direction of movement, but the earth's surface

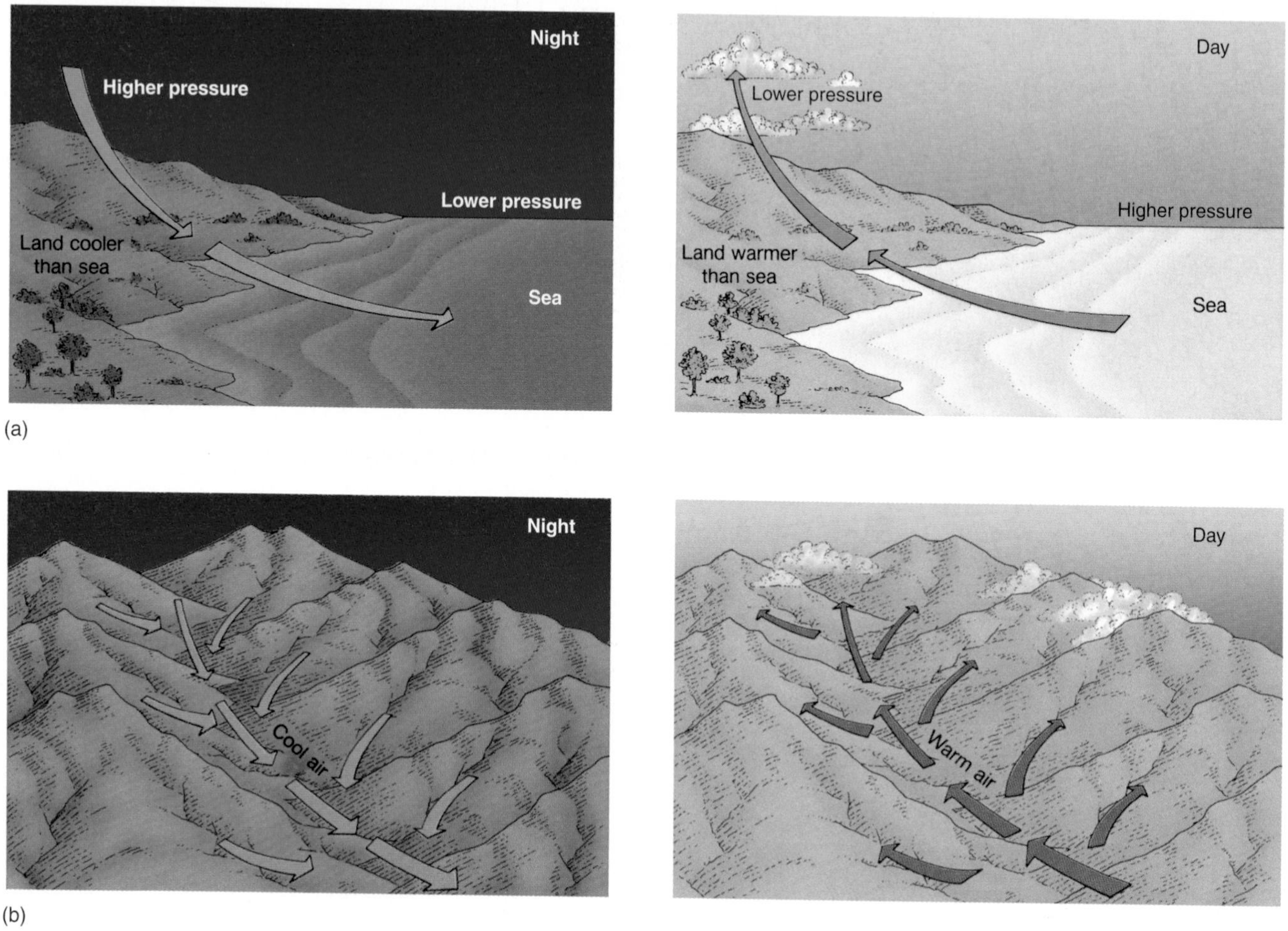


**FIGURE 4.13 Convectional wind effects due to differential heating and cooling.** (a) **Land and sea breezes.** (b) **Mountain and valley breezes.**

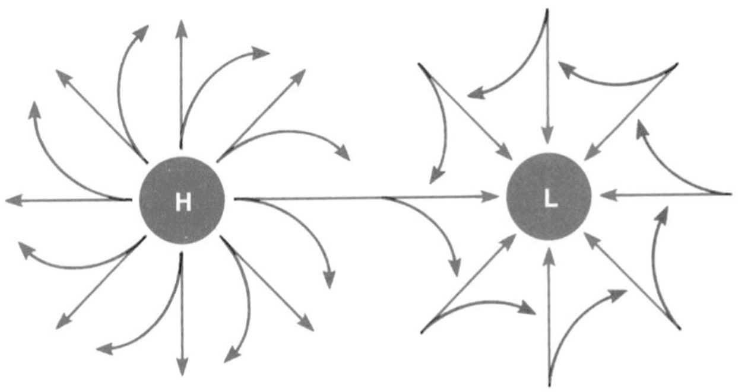


**FIGURE 4.14 The effect of the Coriolis force on flowing air in the Northern Hemisphere.** The straight arrows indicate the paths that winds would follow flowing out of an area of high (H) pressure or into one of low (L) pressure were they to follow the paths dictated by pressure differentials. The curved arrows represent the apparent deflecting effect of the Coriolis force. Wind direction—indicated on the diagram by curved arrows—is always given by the direction *from* which the wind is coming.

moves out from under it. Since the position of the air is measured relative to the earth's surface, the air appears to have diverged from its straight path.

The Coriolis effect and the pressure gradient force produce spirals rather than simple straight-line patterns of wind, as indicated in Figure 4.14. The spiral of wind is the basic form of the many storms that are so important to the earth's air circulation system. These storm patterns are discussed later in this chapter.

## The Frictional Effect

Wind movement is slowed by the frictional drag of the earth's surface. The effect is strongest at the surface and declines until it becomes ineffective at about 1500 meters (about 1 mi) above the surface. Not only is wind speed decreased, but wind direction is changed as well. Instead of following a course exactly dictated by the pressure gradient force or by the Coriolis effect, the **frictional effect** causes wind to follow an intermediate path.

## The Global Air-Circulation Pattern

Equatorial areas of the earth are zones of low pressure. Intense solar heating in these areas is responsible for a convectional effect. Note in Figure 4.15 how the warm air rises and tends to move away from the *equatorial low* pressure in both northerly and southerly directions. As equatorial air rises, it cools and eventually becomes dense. The lighter air near the surface cannot support the cool, heavy air. The heavy air falls, forming surface zones of high pressure. These areas of *subtropical high* pressure are located at about 30°N and 30°S of the equator.

When this cooled air reaches the earth's surface, it, too, moves in both northerly and southerly directions. The Coriolis effect, however, modifies wind direction and creates, in the Northern Hemisphere, belts of winds called the *northeast trades* in the tropics and the *westerlies* (really the southwesterlies) in the midlatitudes. The names refer to the direction from which the winds come. Most of the United States lies within the belt of westerlies; that is, the air usually moves across the country from southwest to northeast. A series of ascending air cells also exists over the oceans to the north of the westerlies, called the *subpolar low.* These areas tend to be cool and rainy. The *polar easterlies* connect the subpolar low areas to the *polar high.* The general global air-circulation pattern is modified by local wind conditions.

It should be clear that these belts move in unison as the vertical rays of the sun change position. For example, equatorial low conditions are evident in the area just north of the equator during the Northern Hemisphere summer and just south of the equator during the Southern Hemisphere summer. Air circulation will be discussed in more detail in the section "Types of Precipitation."

The strongest flows of upper air winds, 9 to 12 kilometers (30,000 to 40,000 ft), are the **jet streams.** These air streams, moving at 160 to 320 kilometers per hour (100 to 200 mph), from west to east in both the Northern and Southern Hemispheres, circle the earth in an undulating pattern, first north then south as they move westward. There are three to six undulations at any one time in the Northern Hemisphere, but the waves are not always continuous. These undulations, or waves, control the flow of air masses on the earth's surface. More stable undulations are likely to create similar day-to-day weather conditions. These waves tend to separate cold polar air from warm tropical air. When a wave dips far to the south in the Northern Hemisphere, cold air is taken equatorward and warm air moves poleward, bringing severe weather changes to the midlatitudes. The jet stream is more pronounced in the winter than in the summer.

Nowhere does the manner in which the seasonal shift takes place have such a profound effect on humanity as in the densely populated areas of southern and eastern Asia. The wind, which comes from the southwest during summer in India, reaches the landmass after picking up a great deal of moisture over the warm Indian Ocean. As it crosses the coast mountains and foothills of the Himalayas, the monsoon rains begin. A **monsoon** wind is one that changes direction seasonally. The summer monsoon wind brings heavy showers over most of South Asia.

In the southern and eastern parts of Asia, the farm economy, and particularly the rice crop, is totally dependent on summer monsoon rainwater. If, for any of several possible reasons, the wind shift is late or the rainfall is significantly more or less than optimum, crop failure may result. The undue prolongation of the summer monsoon rains in 1978 caused disastrous flooding, crop failure, and the loss of lives in eastern India and Southeast Asia.

The transition to dry northerly monsoon winter winds occurs gradually across the region, first becoming noticeable in the north in September. By January, most of the subcontinent is dry. Then, beginning in March in southern areas, the yearly cycle repeats itself.

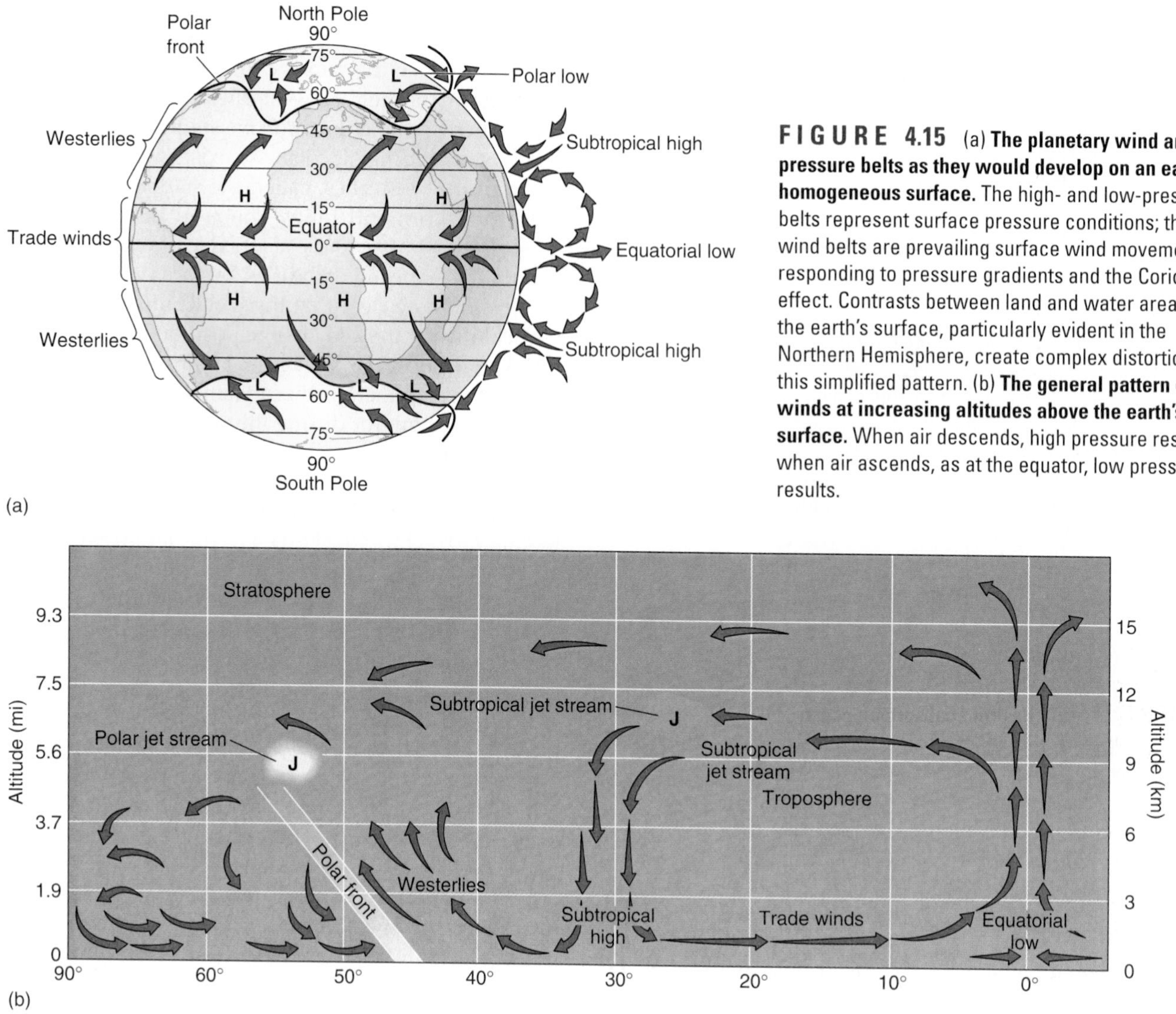


**FIGURE 4.15** (a) **The planetary wind and pressure belts as they would develop on an earth of homogeneous surface.** The high- and low-pressure belts represent surface pressure conditions; the wind belts are prevailing surface wind movements responding to pressure gradients and the Coriolis effect. Contrasts between land and water areas on the earth's surface, particularly evident in the Northern Hemisphere, create complex distortions of this simplified pattern. (b) **The general pattern of winds at increasing altitudes above the earth's surface.** When air descends, high pressure results; when air ascends, as at the equator, low pressure results.

# Ocean Currents

Surface ocean currents correspond roughly to global wind direction patterns because the winds of the world set ocean currents in motion. In addition, just as differences in air pressure cause wind movements, so do differences in the density of water cause water movement. When water evaporates, residues of salt and other minerals that will not evaporate are left behind, making water denser. High-density water exists in areas of high pressure, where descending dry air readily picks up moisture. In areas of low pressure, where rainfall is plentiful, ocean water is low in density. Wind direction (including the Coriolis effect) and the differences in density cause water to move in wide paths from one part of the ocean to another (Figure 4.16).

There is an important difference between surface air movements and surface water movements. Landmasses are barriers to water movement, deflecting currents and sometimes forcing them to move in a direction opposite to the main current. Air, on the other hand, moves freely over both land and water.

The shape of an ocean basin also has an important effect on ocean current patterns. For example, the north Pacific current, which moves from west to east, strikes the western coast of Canada and the United States. The current is then forced to move both north and south, although the major movement is the cold ocean current that moves south along the California coast. In the Atlantic Ocean, however, as Figure 4.16 indicates, the current is deflected in a northeasterly direction by the shape of the coast (Nova Scotia and Newfoundland jut far into the Atlantic). It then moves freely across the ocean, past the British Isles and Norway, finally reaching the extreme northwest coast of Russia. This massive movement of warm water to northerly lands, called the **North Atlantic drift,** has enormous significance to inhabitants of those areas. Without it, northern Europe would be much colder.

Ocean currents affect not only the temperature but also the precipitation on land areas adjacent to the ocean. A cold

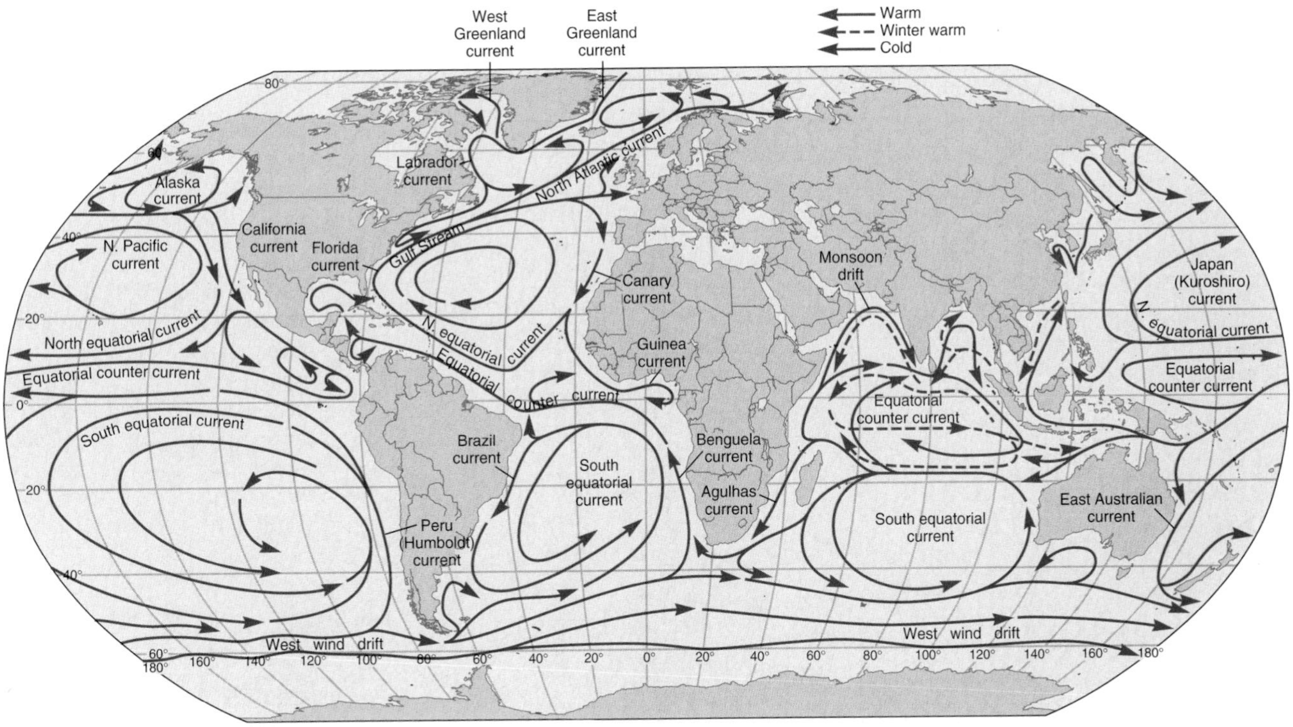


**FIGURE 4.16 The principal surface ocean currents of the world.** Notice how the warm waters of the Gulf of Mexico, the Caribbean, and the tropical Atlantic Ocean drift to northern Europe.

ocean current near land causes the air just above the water to be cold while the air above is warm. There is very little opportunity for convection, thus denying moisture to nearby land. Coastal deserts of the world usually border cold ocean currents. On the other hand, warm ocean currents—such as those off the coast of India—take moisture to the adjacent land area, especially when prevailing winds are landward (see "El Niño," p. 108).

The earlier question about ways in which differences in air pressure affect weather conditions is now answered, in terms of warm and cool air movement over various surfaces at different times of the year and different times of the day. A more complete answer regarding the causes of different types of weather conditions, however, requires an explanation of the susceptibility of places to receive precipitation, because rainfall and wind patterns are highly related.

## Moisture in the Atmosphere

Air contains water vapor (what we feel as humidity), which is the source of all precipitation. **Precipitation** is any form of water particles—rain, sleet, snow, or hail—that fall from the atmosphere and reach the earth's surface. Ascending air can expand easily because less pressure is on it. When heat from the lower air spreads through a larger volume in the troposphere, the mass of air becomes cooler. Cool air is less able to hold water vapor than warm air is (Figure 4.17).

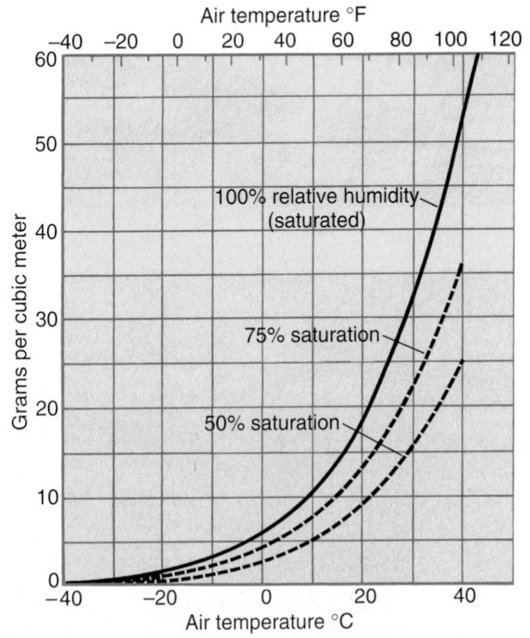


**FIGURE 4.17 The water-carrying capacity of air and relative humidity.** The actual water in the air (water vapor) divided by the water-carrying capacity (×100) equals the relative humidity. The solid line represents the maximum water-carrying capacity of air at different temperatures.

Air is said to be *supersaturated* when it contains so much water vapor that the vapor condenses (changes from a gas to a liquid) and forms droplets if fine particles, called *condensation nuclei,* are present. These particles, mostly dust, pollen, smoke, and salt crystals, nearly always exist. At first, the tiny water droplets are usually too light to fall. As many droplets coalesce into larger drops, and become too heavy to remain suspended in air, they fall as rain. When temperatures below the freezing point cause water vapor to form ice crystals instead of water droplets, snow is created (Figure 4.18).

Large numbers of rain droplets or ice crystals form clouds, which are supported by slight upward movements of air. The form and altitude of clouds depend on the amount of water vapor in the air, the temperature, and wind movement. Descending air in high-pressure zones usually yields cloudless skies. Whenever warm, moist air rises, clouds form. The most dramatic cloud formation is probably the *cumulonimbus,* pictured in Figure 4.19. This is the anvil-head cloud that often accompanies heavy rain. Low, gray *stratus* clouds appear more often in cooler seasons than

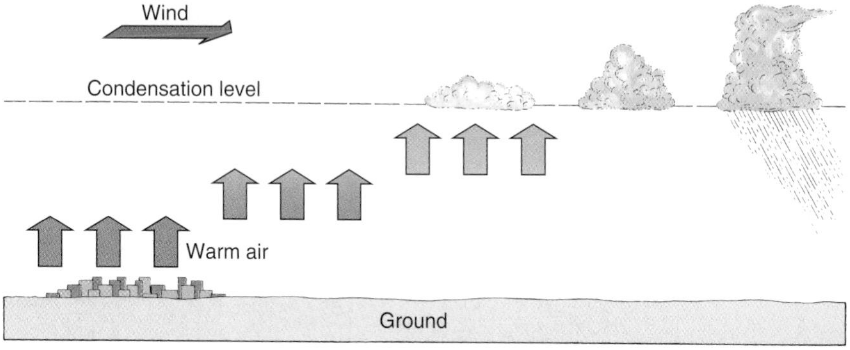


**FIGURE 4.18** As warm air rises, it cools. As it cools, its water vapor condenses and clouds form. If the air becomes supersaturated, precipitation occurs.

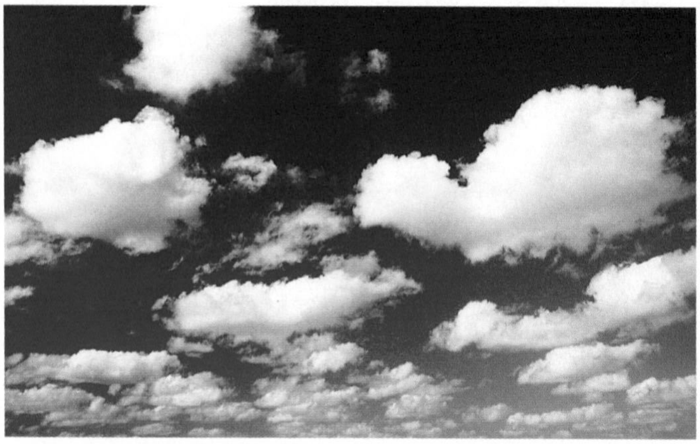

(a)

(b)

(c)

(d)

**FIGURE 4.19 Cloud types:** (a) **fair-weather cumulus,** (b) **cumulonimbus,** (c) **stratus,** (d) **cirrus.** *(a) © Doug Armand/Getty Images. (b) © Dr. Barry Barker. (c) A. Copley/VU. (d) © NOAA.*

in warmer months. The very high, wispy *cirrus* clouds that appear in all seasons are made entirely of ice crystals.

**Relative humidity** is a percentage measure of the moisture content of the air, expressed as the amount of water vapor present relative to the maximum that can exist at the current temperature. As air gets warmer, the amount of water vapor it can contain increases. If the relative humidity is 100%, the air is completely saturated with water vapor. A relative humidity value of 60% on a hot day means the air is extremely humid and very uncomfortable. A 60% reading on a cold day, however, indicates that, although the air contains relatively large amounts of water vapor, it holds, in absolute terms, much less vapor than on a hot, muggy day. This example demonstrates that relative humidity is meaningful only if we keep air temperature in mind.

Dew on the ground in the morning means that nighttime temperatures dropped to the level at which condensation took place (see Figure 4.17). The critical temperature for condensation is called the **dew point.** Foggy or cloudy conditions on the earth's surface imply that the dew point has been reached and that relative humidity is valued at 100%.

## Types of Precipitation

When large masses of air rise, precipitation may take place in one of three types: (1) convectional, (2) orographic, or (3) cyclonic, or frontal.

The first type, **convectional precipitation,** results from rising, heated, moisture-laden air. As air rises, it cools. When its dew point is reached, condensation and precipitation occur, as Figure 4.20 shows. This process is typical of summer storms or showers in tropical and continental climates. Usually, the ground is heated during the morning and early afternoon. Warm air that accumulates begins to rise, forming first cumulus clouds and then cumulonimbus clouds. Finally, lightning, thunder, and heavy rainfall occur, which may affect each part of the ground for only a brief period when the storm is moving. It is common for these convectional storms to occur in late afternoon or early evening.

If quickly rising air currents rapidly circulate air within a cloud, ice crystals may form near the top of the cloud. When these ice crystals are large enough to fall, a new updraft containing water can force them back up, enlarging the pieces of ice. This process may occur repeatedly until the updrafts can no longer sustain the ice pieces as they fall to the ground in the form of hail.

**Orographic precipitation,** the second type and depicted in Figure 4.21, occurs as warm air is forced to rise because hills or mountains block moisture-laden winds. This type of precipitation is typical in areas where mountains and hills are downwind from oceans or large lakes. Saturated air from over the water blows onshore, rising as the land rises. Again, the processes of cooling, condensation, and precipitation take place. The *windward* side—the side exposed to the prevailing wind—of the hills and mountains receives a great deal of precipitation. The opposite side, called the *leeward* side or rain shadow, and the adjoining regions downwind are very often dry. The air that passes over the mountains or hills descends and warms. As we have seen, descending air does not produce precipitation, and warming air absorbs moisture from surfaces it passes over. A graphic depiction of the great differences in rainfall over very short distances is shown on the map of the state of Washington in Figure 4.22.

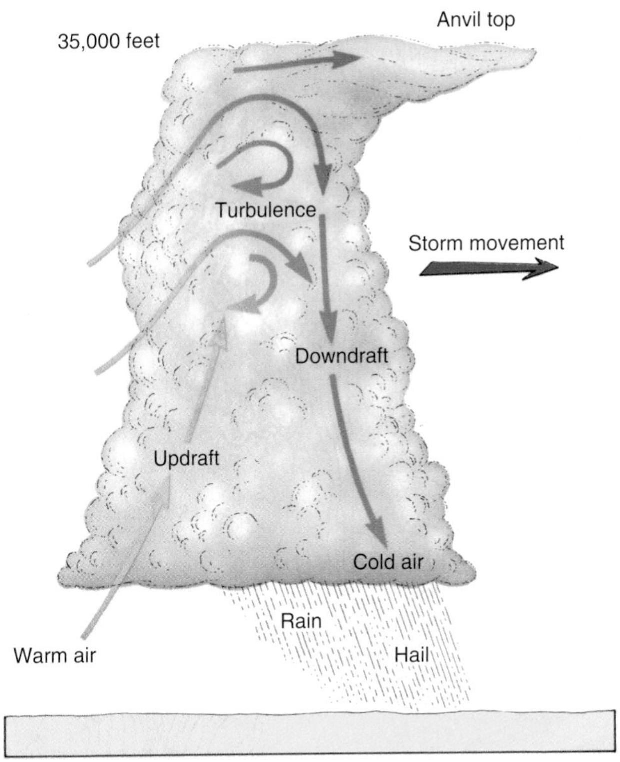


**FIGURE 4.20** When warm air laden with moisture rises, a cumulonimbus cloud may develop and convectional precipitation may occur. The falling particles within the system create a downdraft of cold upper-altitude air.

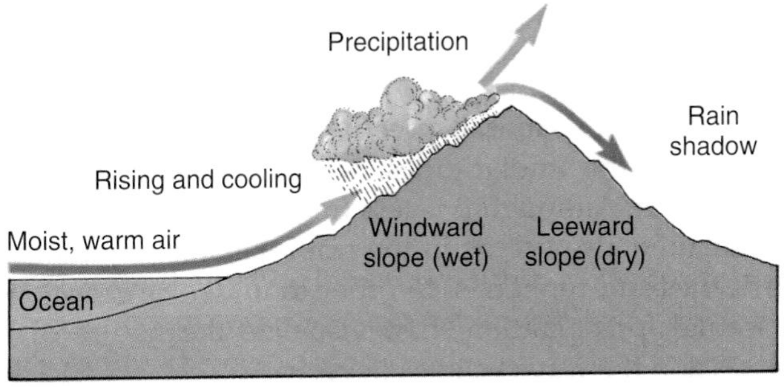


**FIGURE 4.21 Orographic precipitation.** Surface winds may be raised to higher elevations by hills or mountains lying in their paths. If such orographically lifted air is sufficiently cooled, precipitation occurs. Descending air on the leeward side of the upland barrier becomes warmer, its capacity to retain moisture is increased, and water absorption rather than release takes place.

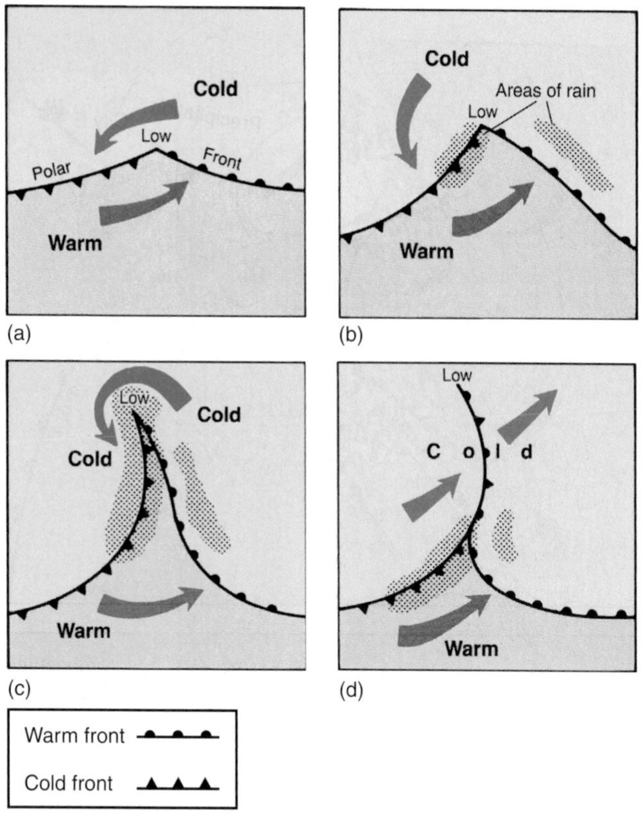


**FIGURE 4.25** When warm and cold air come into contact along a low-pressure trough in the Northern Hemisphere midlatitudes, the possibility of cyclonic storm formation occurs. (a) A wave begins to form along the polar front. (b) Cold air begins to turn in a southerly direction, while warm air moves north. (c) Cold air, generally moving faster than warm air, begins to overtake the warm air, forcing it to rise and, in the process, the storm deepens. (d) Eventually, two sections of cold air join; the warm air forms a pocket overhead, removing it from its energy and moisture source. The cyclonic storm dissipates as the cold front is reestablished.

additional stream of moist air from the Atlantic Ocean. In the meantime, a high-pressure system was moving south with Arctic air from Canada. These movements converged on the northeast United States. The result was nearly a meter (2 to 3 ft) of crippling snow that brought cities such as Washington, D.C.; Baltimore; Philadelphia; New York; and Boston to a standstill (Figure 4.27a).

The most violent of all storms is the **tornado.** It is also the smallest storm (Figure 4.27b), typically measuring less than 30 meters (100 ft) in diameter. Tornadoes are spawned in the huge cumulonimbus clouds that sometimes travel in advance of a cold front along a squall line. During the spring or fall, when adjacent air masses differ the most, the central United States is prone to many of these funnel-shaped killer clouds. Although winds can reach 500 kilometers per hour (about 300 mph), these storms are small and usually travel on the ground for less than a mile, so only limited areas are affected, though they may be devastated (Figure 4.27c).

The *Fujita scale* of tornado intensity links reported damage to wind speed. It ranges from F0, a "weak" tornado with wind speeds up to 116 kilometers per hour (72 mph), to F5, a "violent" one, with winds as great as 512 kilometers per hour (318 mph). Most (74%) tornadoes are either F0 or F1, whereas 25% are classified as F2 or F3, "strong" tornadoes capable of causing major structural damage. Only 1% are in the violent (F4 and F5) categories.

# Climate, Soils, and Vegetation

We have traced some of the causes of weather changes that occur as air from high-pressure zones flows toward low-pressure areas, fronts pass and waves develop, dew points are reached, and sea breezes arise. Some parts of the world experience these changes more rapidly and more often than do other parts.

Day-to-day weather conditions can be explained by the principles explained in this chapter. However, the effect of weather elements—temperature, precipitation, and air pressure and winds—cannot be understood unless a person is conscious of the earth's surface features. Weather forecasters in each location on Earth must deal with weather elements in the context of their local physical and built environments.

The complexities of daily weather conditions may be summarized by statements about climate. The climate of an area is a generalization based on daily and seasonal weather conditions. Are summers warm, on the average? Is heavy snow in the winter likely? Are winds normally from the southeast? Are climatic averages typical of daily weather conditions, or are the day-to-day or week-to-week variations so great that one should speak of average variations rather than just averages? These are the questions we must ask in order to form an intelligent description of the differences in conditions from place to place.

Before beginning a discussion of climate regions of the world, it is useful to describe two major characteristics of the earth's surface: soils and vegetation. These two elements of the physical geography of our planet are intimately tied to climatic variations. When we observe climatic differences, we immediately see that soil types and various forms of vegetation go hand in hand with temperature levels and the seasonal distribution of precipitation.

## Soils and Climate

Soil is one of the most important components of the physical environment. Life as we know it could not exist without it. Soils play an important role by storing and purifying water and are essential to plant, and thus animal and human, life.

### Soil Formation

*Soil* can be defined as a layer of fine material containing *organic matter* (dead plant and animal material), *inorganic matter* (weathered rock materials), air, and water that rests on the bedrock underlying it. The physical and chemical

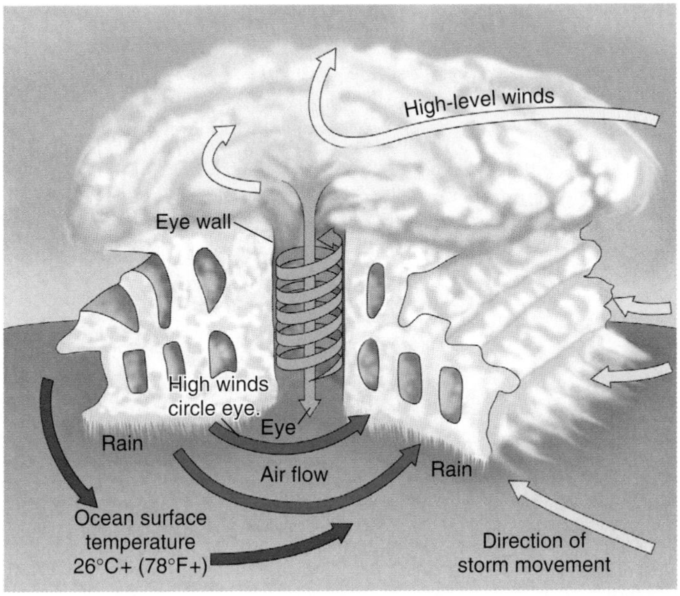


(a)

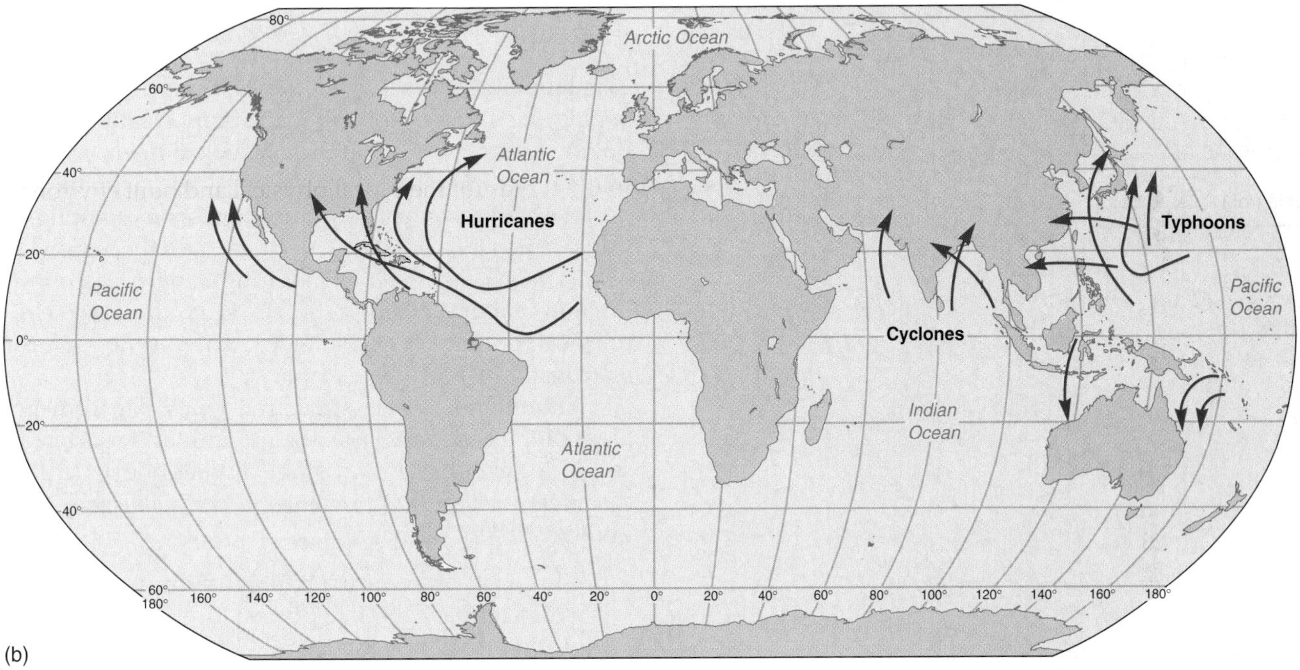


(b)

**FIGURE 4.26** (a) **Characteristics of a mature hurricane.** Spiral bands of cumulonimbus clouds spawn heavy rainfall. Air ascends in the clouds near the center of the storm. The descending, warming air at the center creates an *eye,* a small area of calm on the surface. Intense convectional circulation creates strong winds away from the eye.
(b) **General hurricane paths.** *(a) and (b) From Michael Bradshaw and Ruth Weaver,* Physical Geography: An Introduction to Earth Environment, *pp. 177, 179. Reprinted by permission of The McGraw-Hill Companies, Inc.*

## TABLE 4.1 | The Force of Hurricanes

| Category | Barometric Pressure (Inches) | Wind (mph) | Potential Damage |
|---|---|---|---|
| 1 | Over 28.94 | 74–95 | Damage mainly to trees, shrubbery, unanchored mobile homes; storm surge damage for all categories |
| 2 | 28.50–28.94 | 96–110 | Some trees blown down; major damage to exposed mobile homes; some damage to roofs |
| 3 | 27.91–28.49 | 111–130 | Trees stripped of foliage, large trees blown down; mobile homes destroyed; some structural damage to small buildings |
| 4 | 27.17–27.90 | 131–155 | All signs blown down; extensive damage to windows, doors, and roofs; flooding inland as far as 10 kilometers (6 mi); major damage to lower floors of structures near shore |
| 5 | Less than 27.17 | Over 155 | Severe damage to windows, doors, and roofs; small buildings overturned and blown away; major damage to structures less than 4.6 meters (15 ft) above sea level within 458 meters (500 yds) of shore |

(a)

(b)

(c)

**FIGURE 4.27 Storms.** (a) Blizzards bring the transportation systems of cities to a halt. (b) In the United States, tornadoes occur most frequently in the central and south central parts of the country (especially in Oklahoma, Kansas, and the Texas Panhandle), where cold polar air very often meets warm, moist Gulf air. (c) The Oklahoma City tornado of May 3, 1999, was an F5; it leveled neighborhoods such as this one. *(a) Leland Bobbe/Getty Images. (b) © Sheila Beougher/Getty Images. (c) Jeff Mitchell/Hulton Archive/Getty Images.*

disintegration of rock, called *weathering* (see Chapter 3), begins the process of soil formation. Weathering breaks down solid rock, ultimately producing finely fragmented mineral particles. These particles, together with the decomposed organic matter that comes to rest on top of them, are transformed into soils by water, heat, and the various living agents (such as bacteria and fungi) that decompose the organic matter. Soil formation is a dynamic process. The physical, chemical, and biological activities that produce it are constantly at work.

Although broad classes of soil types extend over large areas of the earth's surface, the individual characteristics of soil in a given place may differ markedly over very short distances. Variation is due chiefly to the five major factors involved in soil formation.

1. The *geologic factor* is the parent (underlying) rock, which influences the depth, texture, drainage, and nutrient content of a soil.
2. The *climatic factor* refers to the effects of temperature and precipitation on soil. Temperature affects the length of the growing season, the speed of vegetation decay, and the rate of evaporation. Precipitation totals and intensity influence the type of vegetation that grows in an area and thus the supply of *humus,* decomposed organic matter.
3. The *topographic factor* refers to the height of the land, the *aspect* of a slope (which direction it faces), and the angle of slope. All of these affect the amounts of precipitation, cloud cover, and wind; temperature; and the surface runoff of water, drainage, and rate of soil erosion.
4. The *biological factor* refers to both living and dead plants and animals, which add organic matter to the soil and interact in the nutrient cycle. Plants take up mineral nutrients from the soil and return them to it when they die. Microorganisms, such as bacteria and

fungi, assist in the decomposition of dead organic matter, whereas larger organisms, such as ants and worms, mix and aerate the soil.

5. The *chronological factor* indicates the length of time the preceding four factors have been interacting to create a particular soil. Recall that soil formation is an ongoing but gradual process. Young soils that have been forming for hundreds of—or even a few thousand—years retain many of the characteristics of their parent-material, whereas fully mature soils have evolved over tens of thousands of years.

### Soil Profiles and Horizons

Over time, soils tend to develop into layers of various thicknesses. These layers, called **soil horizons,** differ from one another in their structure, texture, color, and other characteristics. A **soil profile** is a vertical cross section of the soil, from the earth's surface down into the underlying parent-material, showing its different horizons (Figure 4.28).

- The surface layer, or *O-horizon* (*O* for *organic*), consists largely of fresh and decaying organic matter: leaves, twigs, animal droppings, dead insects, and so on.

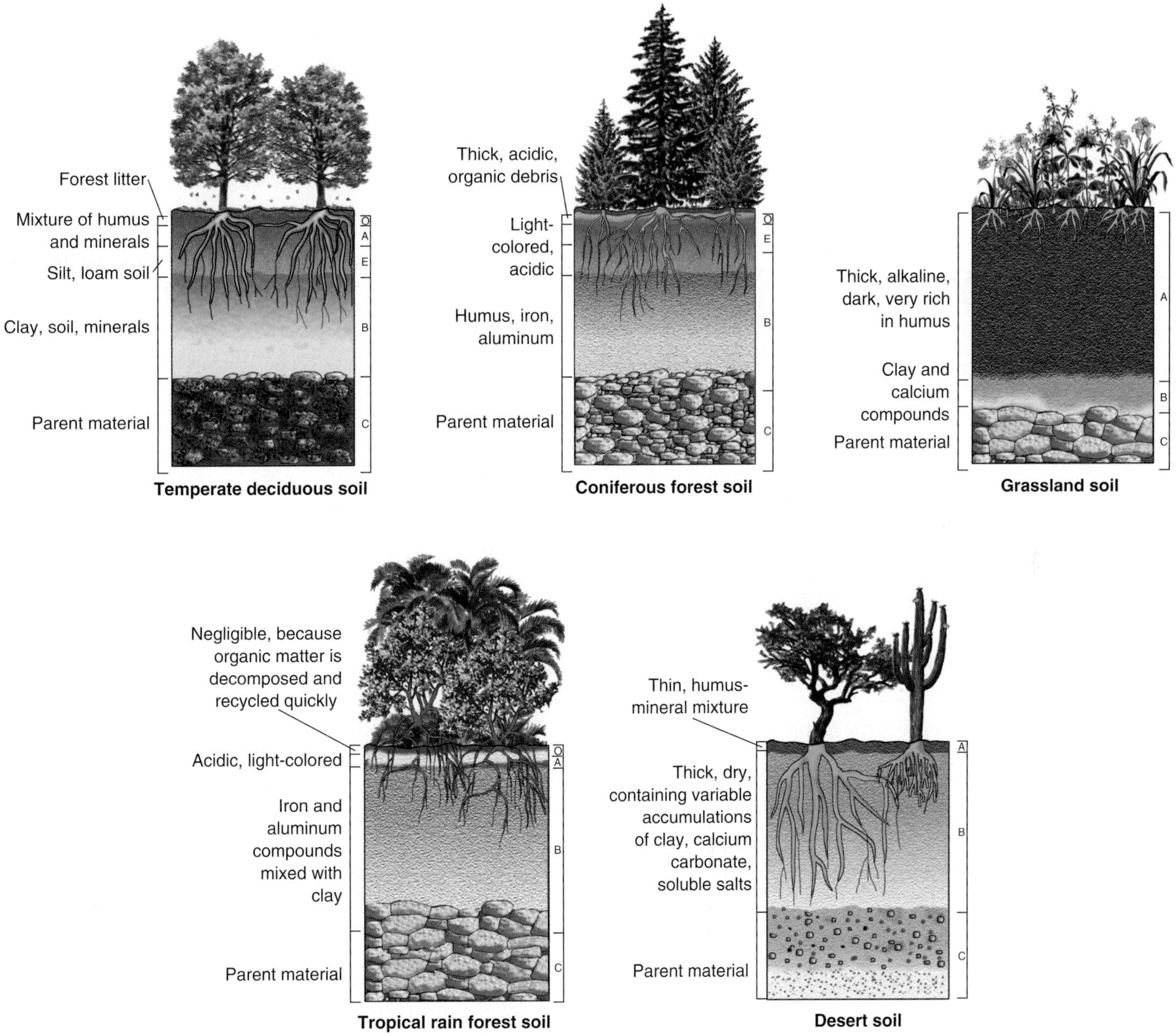


**FIGURE 4.28 Generalized soil profiles found in five major ecosystems.** The number, composition, and thickness of the soil horizons vary depending on the type of soil. Not shown on these profiles is the lowest layer, the unaltered bedrock, or R-horizon. *From* Biosphere 2000: Protecting Our Global Environment, *3d ed. by Donald G. Kaufman and Cecilia M. Franz. (New York: HarperCollins Publishers, 2000), Fig. 16.3, p. 313.*

## El Niño

**El Niño** is a term coined years ago by fishermen who noticed that the normally cool waters off the coasts of Ecuador and Peru were considerably warmer every 3 or 4 years around Christmastime, hence the name El Niño, Spanish for "the child," referring to the infant Jesus. The fish catch was significantly reduced during these periods. If fishermen had been able to identify the scientific associations that present-day oceanographers and climatologists make, they would have recognized a host of other effects that follow from El Niño.

During the winter of 1997–1998, an unusually severe El Niño caused enormous damage and hundreds of deaths. The west coast of the United States, especially California, was inundated with rainfall amounts double, triple, and even quadruple normal. For the November to March winter period, San Francisco received 102.24 centimeters (40.25 in.) of rain—normal is 41.63 centimeters (16.39 in.). The 38 centimeters (15 in.) in February 1998 was the most for that month in the 150 years of record keeping in San Francisco. The resort city of Acapulco, Mexico, was badly battered by torrential rains and high, wind-blown tides. Parts of South America, especially Ecuador, Peru, and Chile, were ravaged by floods and mud slides, while droughts and fires scorched eastern South America, Australia, and parts of Asia, especially Indonesia. A stronger than usual southern branch of the jet stream generated by El Niño spawned dozens of tornadoes, which killed more than 100 people in Alabama, Georgia, and Florida.

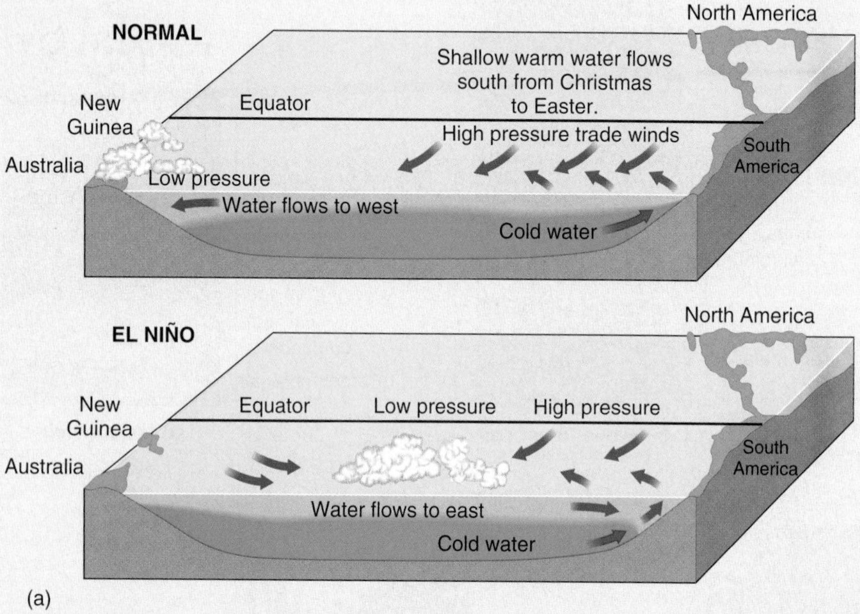


(a) The top diagram shows normal circumstances in the southern Pacific Ocean. Trade winds blow warm surface water westward and allow cold water to come to the surface along the South American coast. The bottom diagram shows that, during El Niño, winds from near Australia blow warm water eastward to the coast of South America. (b) Sea surface temperatures in degrees Celsius are shown for La Niña, normal, and El Niño conditions. Notice how the extent of the warm water (red, orange colors) changes, particularly in the eastern Pacific Ocean. Source: *(a) From Michael Bradshaw and Ruth Weaver,* Physical Geography: An Introduction to Earth Environments, *p. 211. Reprinted by permission of the McGraw-Hill Companies, Inc. (b) Richard W. Reynolds, National Centers for Environmental Prediction, National Weather Service, National Oceanic and Atmospheric Administration (NOAA).*

- Beneath the O-horizon lies the fertile *A-horizon,* representing topsoil. Plant nutrients abound, and biological activity and humus content are at their maximum. The humus helps give this horizon a dark color.
- Water percolating through the soil removes some of the organic and mineral matter from the bottom of the A-layer in a process called *eluviation* (outwashing), yielding the lighter-colored *E-horizon.*
- The material removed from the E-horizon is deposited in the *B-horizon,* or zone of illuviation (inwashing). Because it contains little organic matter, it is less fertile than the A-horizon. Depending on the types of minerals that have been deposited in this layer, it may be darker or more brightly colored than the E-horizon.
- The *C-horizon* is where weathering is slowly transforming bedrock into soil particles. The older the soil becomes, and the warmer and wetter the climate, the deeper and more discernible the C-horizon will be.
- The lowest layer, the unaltered bedrock, is designated the *R-horizon* (*R* for *regolith*).

### Soil Properties

The four major components of soil—minerals, organic matter, water, and air—interact to produce distinctive soils. **Soil properties** are the characteristics that enable us to distinguish one type of soil from another.

Soils contain both *organic* and *inorganic matter.* The latter, produced by weathering, consists of minerals, such as quartz,

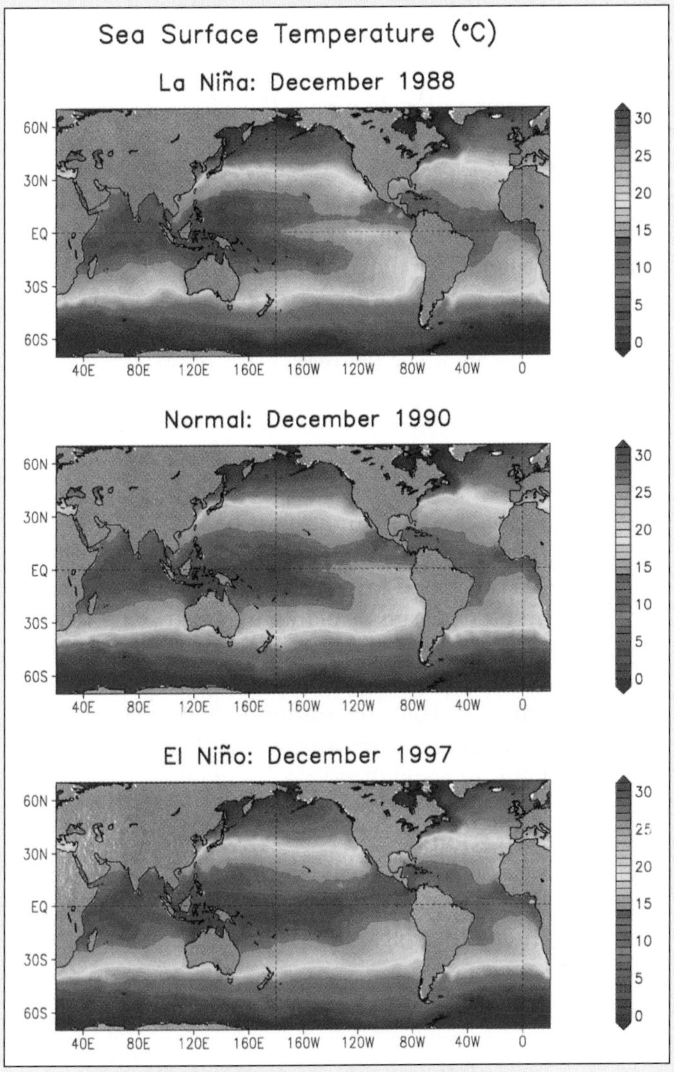


(b)

During periods of El Niño, winds that usually blow from east to west over the central Pacific Ocean, from the cold ocean current to the warm waters of East Asia, slow or even reverse. This phenomenon occurs every 2 to 7 years, but with different degrees of intensity. For example, an El Niño occurred in 1986–1987 and again in 1991–1992, but the modest warm-water buildup did not cause extraordinary circumstances, whereas the 1982–1983 and 1997–1998 events were among the most extreme on record. The cold-water peak between El Niño occurrences is called **La Niña.** The last major La Niña occurred in 1988, a year marked by drought for large portions of North America.

The El Niño condition is an example of the interaction of atmospheric pressure and ocean temperature. The atmospheric and oceanic states encourage each other. Under normal conditions, the contrast in temperatures across the ocean helps drive winds that, in turn, keep pushing water to the west, maintaining the contrast in water temperature. But when a condition called the *southern oscillation* occurs, there is warming in the eastern Pacific, enhancing the usual temperature contrasts between the equator and the earth's poles. Atmospheric pressure rises near Australia, the wind falters, and El Niño is created off the coast of South America. The greater the temperature disparity combined with moisture available from the Pacific Ocean, the more severe the weather.

iron, and aluminum. When weathering breaks down rocks into soil particles, the minerals are released to nourish plant growth.

*Texture* refers to the size of the mineral matter in the soil and is determined by the proportion of sand, silt, and clay particles. Sand is the largest particle type, followed by silt and then clay. The most agriculturally productive soil, called *loam,* is about 40% sand, 40% silt, and 20% clay.

Texture influences *soil structure,* which is defined by the way individual particles aggregate into larger clumps. The size, shape, and alignments of clumps affect the capacity of the soil to hold water, air, and plant nutrients.

Soils vary considerably in the *nutrients* they contain. These chemical elements—such as carbon, hydrogen, and oxygen—are essential for plant growth and to maintain the fertility of the soil. Soils deficient in nutrients can be made productive by adding fertilizer artificially.

As we have noted, organic matter, or humus, is derived mainly from dead and decaying plant and animal matter. Humus holds water and supplies nutrients to plants. The highest amounts of humus are found in the fertile prairies of North America, the Argentinian pampas, and the treeless grasslands (steppes) of Russia.

High humus content gives the soil a dark brown or black *color,* another soil property. In tropical and subtropical regions, iron compounds can give soil a yellow or reddish color. Light colors (gray or white) often indicate highly leached soils in wet areas and alkaline soils in dry areas. Leached soils are those in which groundwater has dissolved and removed the minerals.

The pH scale, discussed in Chapter 5, measures the *acidity* or *alkalinity* of a soil. The most agriculturally productive soils tend to be balanced between being very acidic and very alkaline.

### Soil Classification

Over the years, scientists have devised a number of ways of classifying soils. One of the most commonly used classifications is that developed by the U.S. Department of Agriculture, known simply as *Soil Taxonomy,* which is based on the present-day characteristics of the soil (see "Soil Taxonomy"). It divides soils into 11 orders, which in turn are subdivided into suborders, great groups, subgroups, families, and, finally, thousands of soil series. A **soil order** is a very general grouping of soils with similar composition, horizons, weathering, and leaching processes. The *sol* at the end of each soil order is from the Latin *solum,* meaning "soil." Figure 4.29 shows the world distribution of soil orders.

## Natural Vegetation and Climate

Each climate is typified by a particular mixture of **natural vegetation**—that is, the plants that would exist in an area if people did not modify or interfere with the growth process. Natural vegetation, little of which remains today in areas of human settlement, has close interrelationships not only with climate but also with soils, landforms, groundwater, and other features of habitat, including animals.

### Succession

The natural vegetation of a particular area develops in a sequence of stages known as **succession** until a final stage of equilibrium has been reached with the natural environment. Succession usually begins with a relatively simple *pioneer* plant community, the first organisms (e.g., lichens) to colonize bare rock. They begin the process of soil formation. Over time, the pioneer community alters the environment; as the alterations become more pronounced, plants appear that could not have survived under the original conditions and eventually dominate the pioneer community. For example, lichens may be replaced by mosses and ferns (Figure 4.30).

This vegetative evolutionary process continues as each succeeding community prepares the way for the next by the changes it has made in the topsoil, soil structure, the ability of the soil to retain moisture, and so on. In general, each successive community shows an increase in the number of species and the height of the plants. To continue our earlier example, mosses and ferns may be replaced by grasses and herbs, and once sufficient soil has accumulated, they will be succeeded by low shrubs, which in turn are replaced by trees. The idealized plant succession depicted in Figure 4.30 can take hundreds or even thousands of years to develop.

The final step in the succession of plant communities in a specific area is referred to as the **climax community,** a self-perpetuating assemblage of plants that is in balance with the climate and soils of an area. Climax communities are not permanent, however. They change as environmental conditions change. Volcanic eruptions, forest fires, floods, droughts, and other disturbances alter the environment, forcing changes in vegetation.

### Natural Vegetation Regions

Figure 4.31 shows the general pattern of the earth's natural vegetation regions. In the hotter parts of the world, where rainfall is heavy and well scattered throughout the year, the

## Soil Taxonomy

| Soil Order | Brief Description |
|---|---|
| Oxisols | Red, orange, and yellow; highly weathered, leached, and acidic; low fertility; found in the humid tropics of South America and Africa |
| Ultisols | Red and yellow; highly weathered and leached; less acidic than oxisols; low fertility; develop in warm, wet or dry tropics and subtropics |
| Alfisols | Gray-brown; moderately weathered and leached; fertile, extremely rich in nutrients; found in humid midlatitudes |
| Spodosols | Light-colored sandy A-horizon, red-brown B-horizon; moderately weathered, leached, and acidic; form beneath coniferous forests |
| Mollisols | Dark brown to black; moderately weathered and leached; extremely rich in nutrients; world's most fertile soil; form beneath grasslands in midlatitudes |
| Aridisols | Light in color; sandy; usually saline or alkaline; dry and low in organic matter but agriculturally productive if irrigated properly |
| Inceptisols | Immature, poorly developed soils; form in high-latitude cold climates, especially tundra and mountain areas; limited agricultural potential except in river valleys where seasonal floods deposit fresh layers of sediment |
| Vertisols | Dark in color; high in clay content; form under grasses in tropical and subtropical areas with pronounced wet and dry periods; fertile but difficult to cultivate |
| Entisols | Thin and sandy; immature, poorly developed; low in nutrients; found on tundra, mountain slopes, and recent floodplains |
| Histosols | Black; acidic; composed primarily of organic matter in various stages of decay; waterlogged all or part of the year; found in poorly drained areas (peat bogs, swamps, meadows) of the upper midlatitudes and tundra; can be fertile if drained |
| Andisols | Young, immature soils developing on parent-materials of volcanic origin, such as ash and basalt deposits; high organic content; acidic |

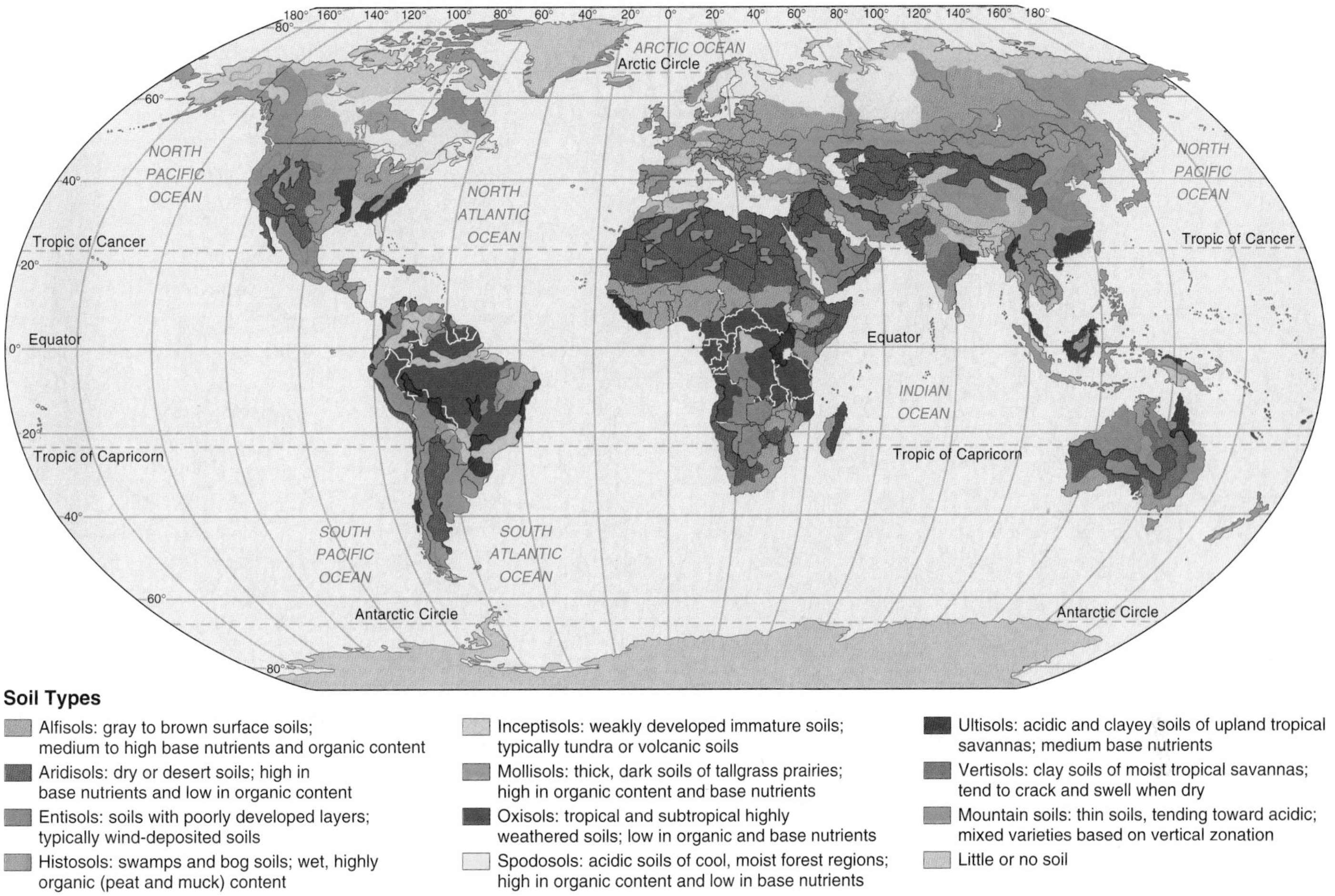


**FIGURE 4.29** **World soils map.**

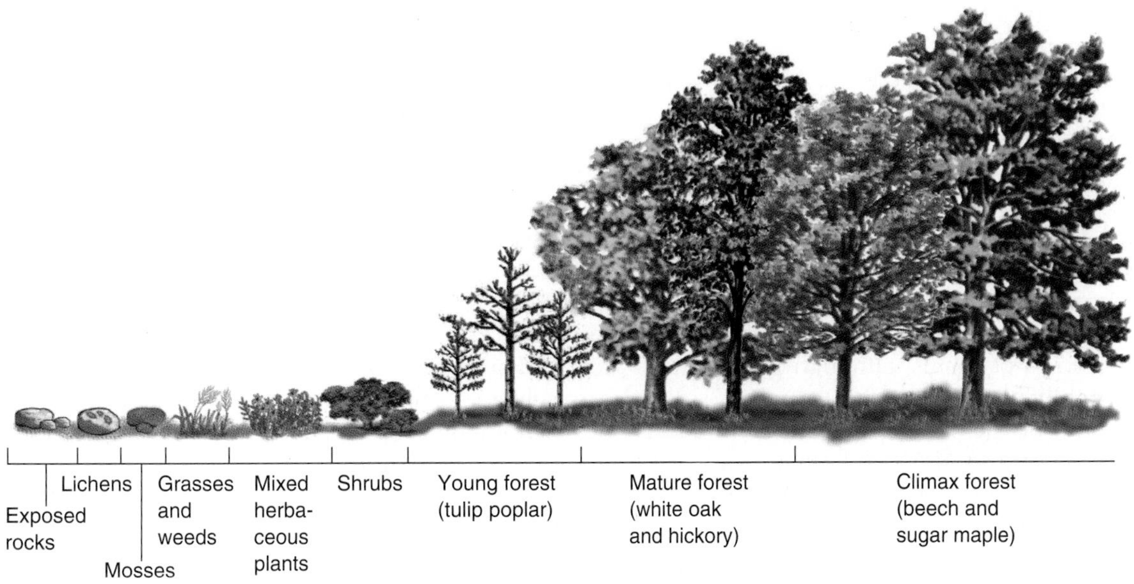


**FIGURE 4.30** **An idealized plant succession in a temperate deciduous region.** An increasing number of species is usually found at each stage of the succession. The particular species in any one location are determined by local differences in such factors as bedrock, elevation, temperature, sunlight, and rainfall. *From* Biosphere 2000: Protecting Our Global Environment, *3d ed. by Donald G. Kaufman and Cecilia M. Franz. (New York: HarperCollins Publishers, 2000), Fig. 5.3, p. 86.*

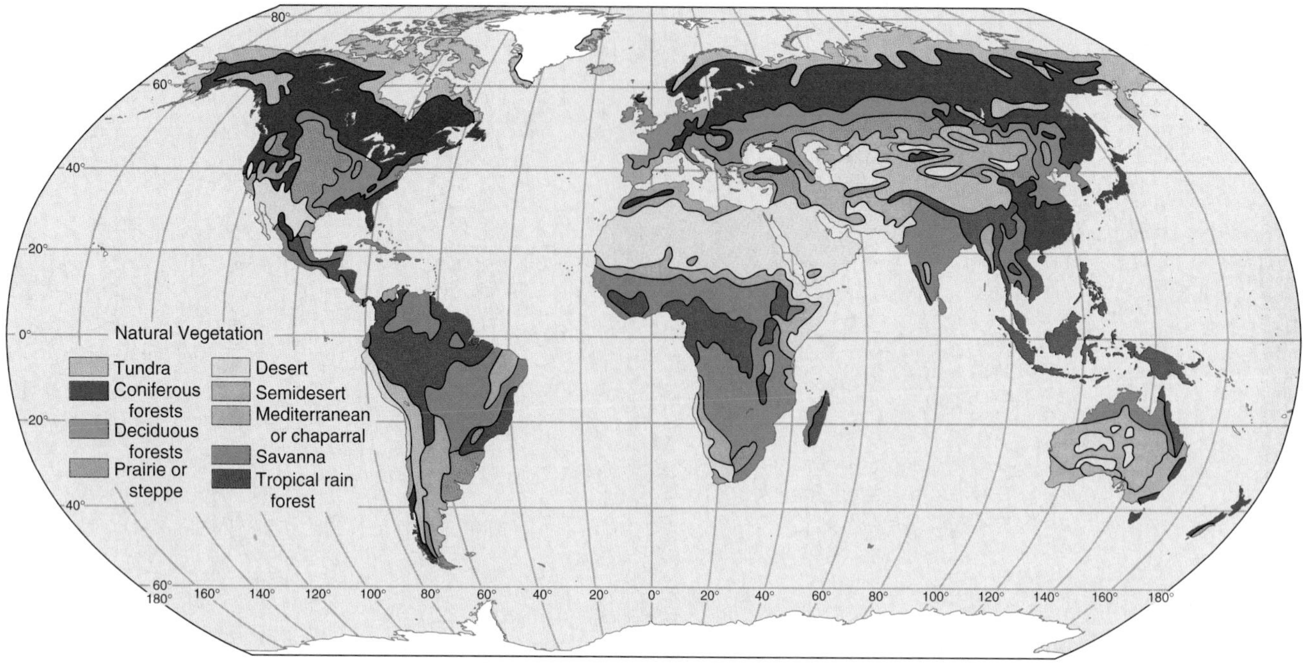


**FIGURE 4.31 World map of natural vegetation.**

vegetation type is *tropical rain forest. Forests,* in general, are made up of trees growing closely together, creating a continuous and overlapping leaf canopy. In the tropics, the forest consists of hundreds of tree species in any small area. Because the canopy blocks the sun's rays, only sparse undergrowth exists. When tropical rainfall is seasonal, *savanna* vegetation occurs, characterized by a low grassland with occasional patches of forests or individual trees. The high evaporation rate denies the savanna region sufficient moisture for dense vegetation.

*Mediterranean,* or *chaparral,* vegetation is found in the hot summer and mild, damp winter midlatitudes characterizing California, Australia, Chile, South Africa, and the Mediterranean Sea regions. This type of vegetation consists mainly of shrubs and trees of limited size, such as the live oak. Together, they form a low, dense vegetation that is green during the wet season and brown during the dry season. Most dryland areas support some vegetation. *Semidesert* and *desert* vegetation is made up of dwarf trees, shrubs, and various types of cactus, although in gravelly and sandy areas, virtually no plants exist.

In temperate parts of the world with modest year-round rainfall, such as central North America, southern South America, and south-central Asia, the most prevalent type of vegetation is *prairie,* or *steppe.* These are extensive grasslands, usually growing on high humus content soils. When rainfall is higher in temperate areas, natural vegetation turns to *deciduous woodlands.* These types of trees, such as oak, elm, and sycamore, lose their leaves during the cold season.

Beyond temperate zones, in northerly regions that have mild summers and very cold winters, *coniferous forests* are common. Evaporation rates are low. Usually, only a few species of trees predominate, such as pines and spruces. Still farther north, forests yield to *tundra* vegetation, a complex mix of very low growing shrubs, mosses, lichens, and grasses.

## Climate Regions

The two most important elements that differentiate weather conditions are temperature and precipitation. Although air pressure is also an important weather element, differences in air pressure are hardly noticeable without the use of a barometer. Thus, we may regard warm, moderate, cold, or very cold temperatures as characteristic of a place or region. In addition, high, moderate, and low precipitation are good indicators of the degree of humidity or aridity in a place or region. For these two scales, we will define the terms more precisely and will map the areas of the world having various combinations of temperatures and precipitation.

Because extreme seasonal changes occur, two global climatic maps are shown in Figure 4.32, one for winters and one for summers. Maps could have been developed for each of the four seasons, or for the 12 months of the year. Rather, these two maps give a good, though brief, description of climatic differences. Keep in mind that a summer map of the world is a combination of Northern Hemisphere climates from June 21 to September 21 and Southern Hemisphere cli-

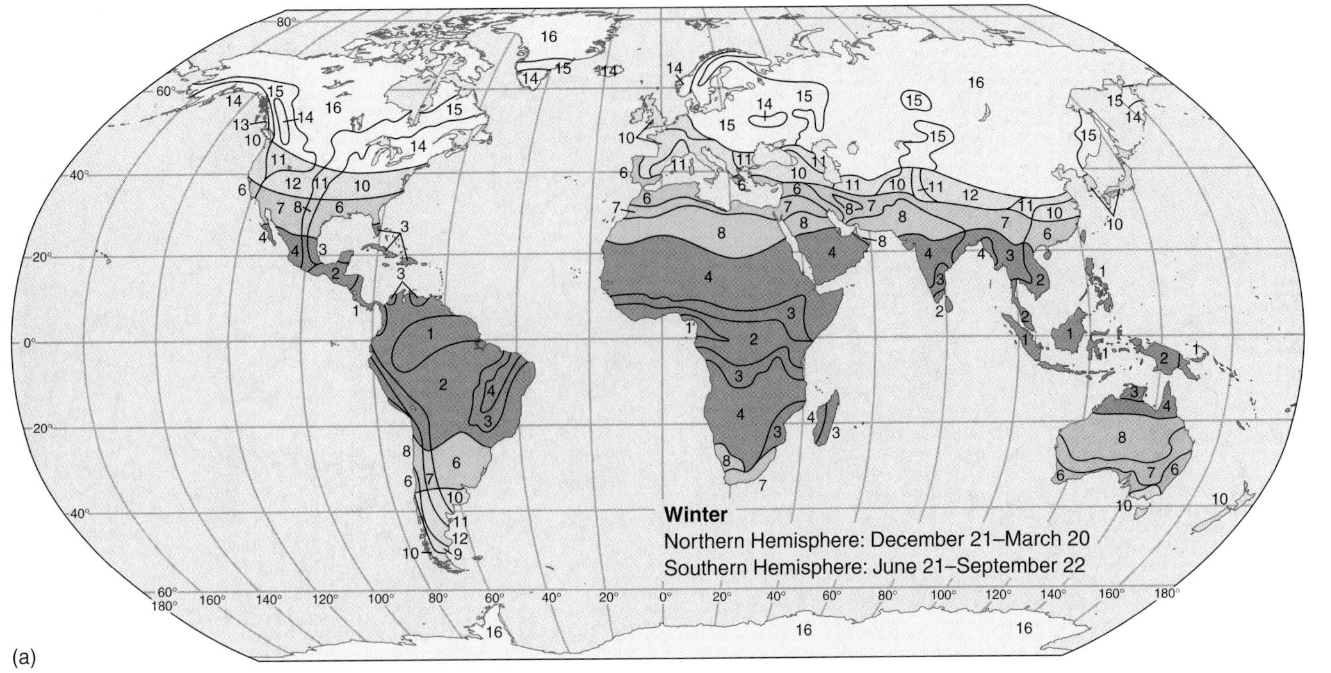


(a)

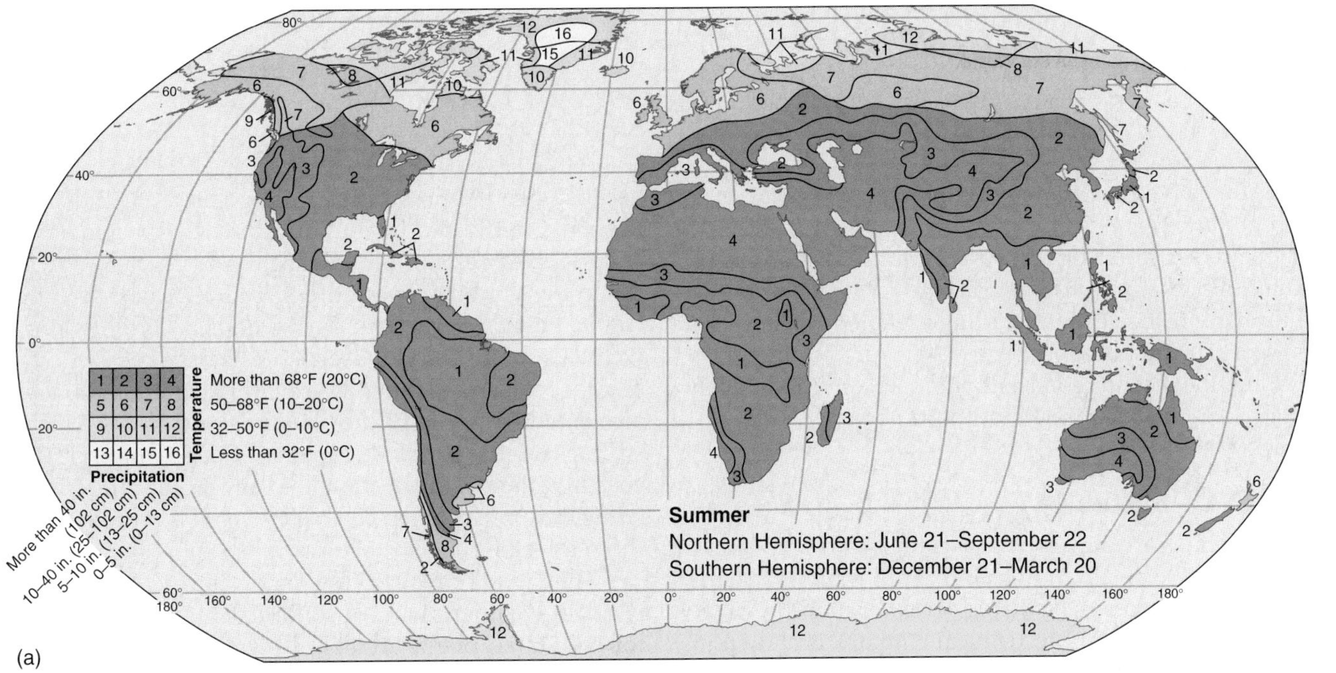


(a)

**FIGURE 4.32** These maps combine temperature and precipitation data to display seasonal variations in the basic components of climate. In reading the maps, remember that the winter climate shown in (a) represents the winter season for both the Northern and Southern Hemispheres. This means that December, January, and February data were used for the north latitudes, and June, July, and August data were used for the south latitudinal parts of the world. The result is a winter view of the world, one of greatly varying temperatures and small amounts of precipitation. (b) The summer view shows a world of nearly universally warm temperatures and large amounts of precipitation.

mates from December 21 to March 21, because the seasons are reversed in the two hemispheres.

Figure 4.33 depicts the various climates of the world and is based on the type of information presented in Figure 4.32. Called the Köppen system, it is the best known of a number of similar climate classification schemes. Developed in 1918, it is based on natural vegetation in addition to temperature and precipitation criteria.

Table 4.2 shows the multilevel system developed by Köppen. There are six broad categories, designated as A, B, C, D, E, and H. The A climates are tropical, B are dry, C are mild climates in the midlatitudes, D climates of the

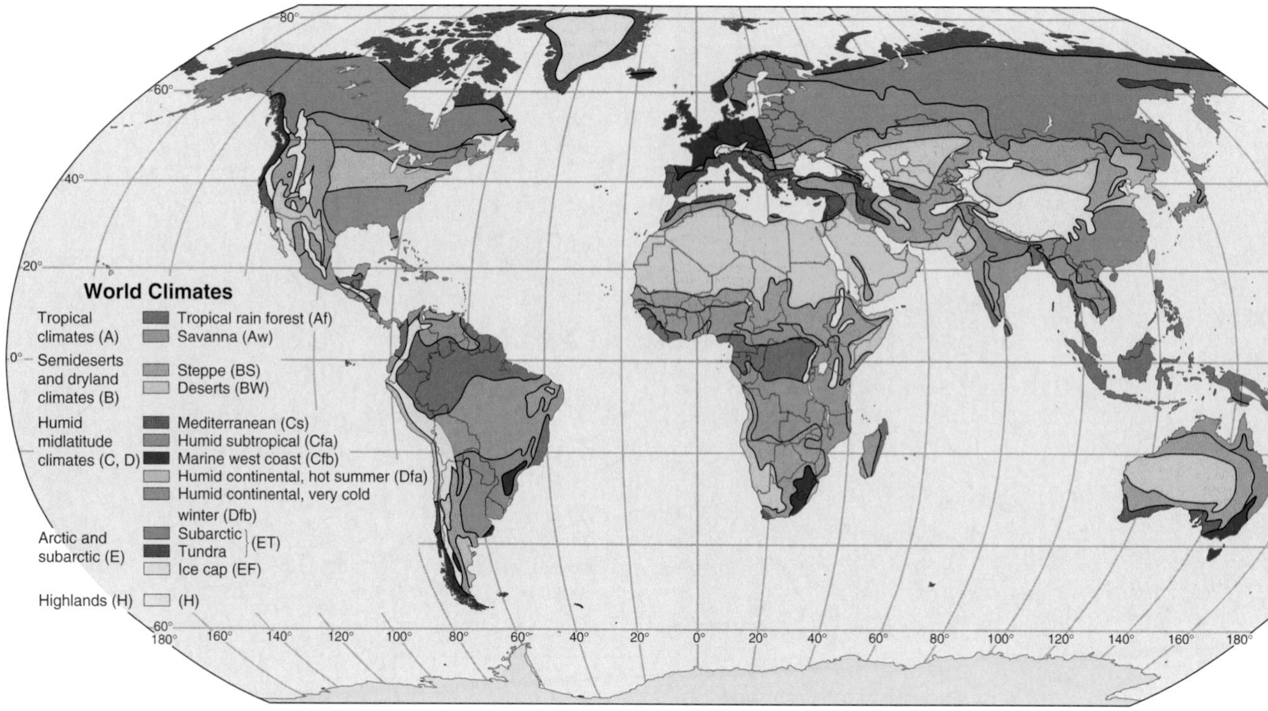


**FIGURE 4.33** **Climates of the world.**

midlatitudes have severely cold winters, E are polar, and H are highland climates.

The letters of the Köppen system in the section headings that follow refer to those in Figure 4.33 and Table 4.2. The numbers in the section headings refer to the keys on Figure 4.32. The first number represents typical winter conditions, and the second represents summer conditions. Each number represents idealized conditions.

## Tropical Climates (A)

Tropical climates are generally associated with earth areas lying between the northernmost and southernmost lines of the sun's vertical rays—the *Tropic of Cancer* and the *Tropic of Capricorn.* The location of tropical climates is shown in Figure 4.34.

### Tropical Rain Forest (Af: 1, 1)

The areas that straddle the equator are generally located within the equatorial low-pressure zone. These regions are called **tropical rain forest.** They are warm, wet climates in both the winter and summer (Figure 4.35). Rainfall usually comes from daily convectional thunderstorms, and although most days are sunny and hot, by afternoon, cumulonimbus clouds form and convectional rain falls. The caption for Figure 4.35 explains how to interpret a *climagraph.*

Tropical rain forests are typically filled with natural vegetation, which is still present but declining rapidly because of intentional fires in large areas of the Amazon Basin of South America and the Zaire River Basin of Africa. Tall, dense forests of broadleaf trees and heavy vines predominate. Among the hundreds of species of trees found in tropical rain forests, both dark woods and light woods, as well as spongy softwoods, such as balsa, and hardwoods, such as teak and mahogany, exist (Figure 4.36). Rain forests also extend away from the equator along coasts where prevailing winds supply a constant source of moisture to coastal uplands. In addition, the orographic effect provides sufficient precipitation for heavy vegetation to develop in these forests.

The typical soils of these regions are *oxisols.* As a result of rapid weathering, most soil nutrients that are necessary for cultivated plants are absent. Only with large inputs of fertilizers can the soils be made to sustain continued agricultural use.

### Savanna (Aw, Am: 3, 1)

As the sun's vertical rays extend farther from the equator in the summer, the equatorial low-pressure zone follows the sun's path. Thus, areas to the north and south of the rain forest are wet in the summer months, although still hot, but are dry the remainder of the year because the moist equatorial low has been replaced by the dry air of subtropical highs. These areas are known as **savanna** lands because of the kind of natural vegetation that grows here.

The natural vegetation of savannas resembles a form of scrub forest; however, these areas are now recognized as a

**TABLE 4.2 | Climate Characteristics**

| Climate Type | Köppen Classification | Temperature and Precipitation | Soil, Vegetation, Wildlife |
|---|---|---|---|
| ***Tropical*** | A | | |
| Tropical rain forest (1;1) | Af | Constant high temperatures<br>Rainfall heavy all year; convectional<br>High amount of cloud cover<br>High humidity | Dense, many species of trees<br>Jungle where light penetrates<br>Many small animals, insects<br>Oxisols |
| Savanna (3;1) | Aw | High temperatures<br>Rainfall heavy in summer high-sun period; convectional<br>Dry in winter low-sun period | Forests to grassland, depending on rainfall amount<br>Large animals<br>Ultisols, vertisols, and aridisols |
| | Am | Monsoon: highest temperature just before rainy season | |
| ***Semidesert and Dryland*** | B | | |
| Hot deserts (7;4) | BWh | Extremely high temperatures in summer; warm winters<br>Very little rainfall<br>Low humidity | Shrubs in gravelly or sandy environments<br>Reptiles<br>Aridisols |
| Steppe (10;4) and desert (4;4) | BS<br>BWk | Warm to hot summers<br>Cold winters<br>Some convectional rainfall in summer<br>Some frontal snowfall in winter | Grass and desert shrubs<br>Mollisols in grasslands<br>Aridisols in deserts |
| ***Humid Midlatitude*** | C and D | | |
| Mediterranean (6;3) | Cs | Warm to hot summers<br>Mild to cool winters<br>Dry summers<br>Frontal precipitation in winter<br>Generally low humidity | Chapparal vegetation (scrub oak trees and bushes)<br>Alfisols, aridisols |
| Humid subtropical (6;2) | Cfa | Hot summers<br>Mild winters<br>Convectional showers in summer<br>Frontal precipitation in winter | Deciduous forests<br>Coniferous forests, especially in sandy soils<br>Mainly alfisols |
| Marine west coast (10;6) | Cfb | Westerly winds year-round<br>Mild summers<br>Cool to cold winters<br>Low rainfall in summer<br>Frontal rainfall in winter | Vast coniferous forests in orographic regions<br>Deciduous forest in plains<br>Spodosols |
| Humid continental (10;2:14;2 and 15;6) | Dfa and Dfb | Hot to mild summers<br>Cool to very cold winters<br>Convectional showers in summer<br>Frontal snowfall in winter | Coniferous forests<br>Spodosols |
| ***Arctic and Subarctic*** | E | | |
| Subarctic (16;7), tundra (16;16), (16;12) | ET | Cool to cold, short summers<br>Extremely cold winters | Stunted coniferous forests to mosses and lichens to permanent ice |
| Ice cap (16;16) | EF | Dry climates with some summer and winter precipitation | Inceptisols |
| ***Highlands*** | H | Great variety of conditions based on elevation, prevailing winds, sun or nonsun-facing slopes, latitude, valley or nonvalley, ruggedness | |

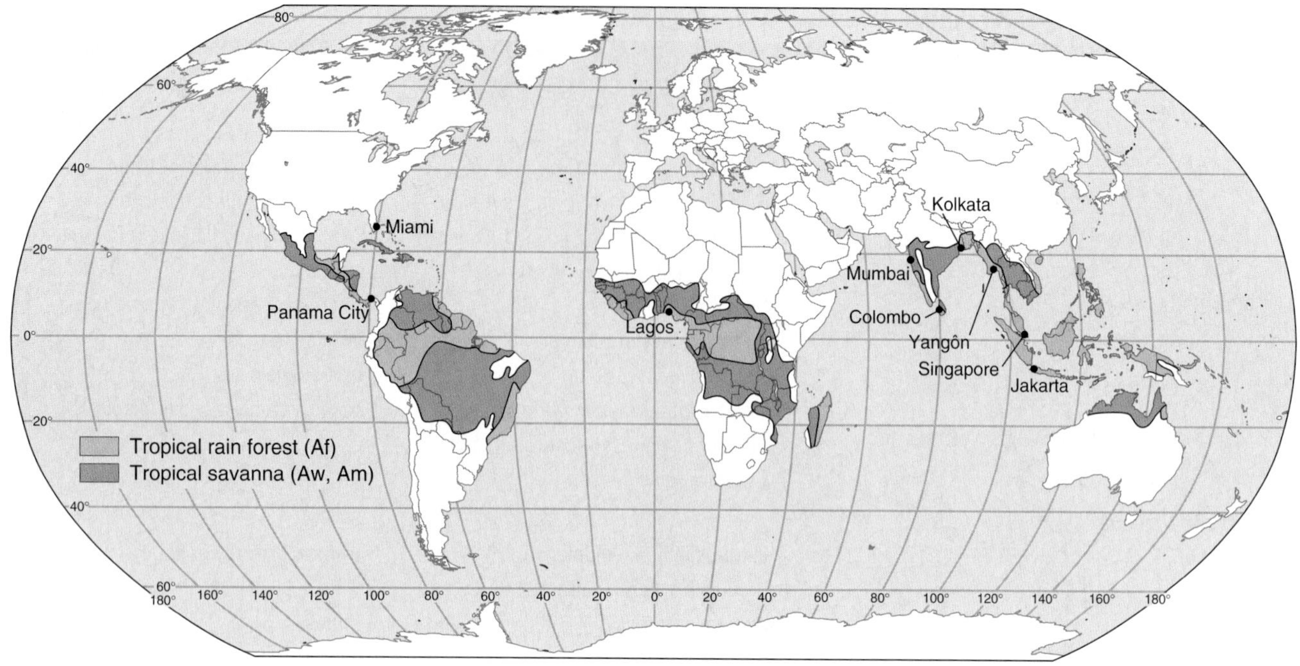


**FIGURE 4.34 The location of tropical climates.**

grassland with widely dispersed trees. The natural tendency toward a more forested cover has been reduced by the periodic clearing by burning that local agriculturalists and hunters engage in. Savannas sometimes seem to have been purposely designed because of their parklike appearance, as indicated in Figure 4.37a. The East African region of Kenya and Tanzania contains well-known grasslands—Serengeti National Park, for example—and fire-resisting species of trees, where large animals, such as giraffes, lions, and elephants, roam. The *campos* and *llanos* of South America are other huge savanna areas.

The wetter portions of the savanna lands are often underlain by *ultisols,* soils that develop in warm, wet-dry regions beneath forest vegetation (Figure 4.37b). These soils are weak in nutrients for cultivated plants but respond favorably when lime and fertilizers are applied. In drier parts of the savanna, the characteristic soils are *vertisols,* which form under grasses in warm climates. When it rains, the surface becomes plasticlike and some of the soil slides into the cracks formed during the dry season. Consequently, vertisols are difficult to till and are used productively as grazing land.

Generally, breaks between different climates are hard to differentiate and distinguish. Instead, zones of transition are apparent. These transitional zones are typical of plains and plateaus—and not so gradual in mountainous regions. Less dense forests exist between the thick tropical rain forest and the savanna.

A special case in Asia needs mentioning, however. When summer monsoon winds carry water-laden air to the mainland, a significant increase in rainfall on the hills, mountains, and adjacent plains occurs. The amount of precipitation in the savanna region is not as abundant—notice the pattern of precipitation in Figure 4.38. As a result, vegetation is dense, even though the winters are dry. Jungle growth and large forests are the natural vegetation. Much of this vegetation, however, has ceased to exist because people have been using the land for rice and tea production for many generations.

## Dryland Climates (B)

The location of these climates is shown in Figure 4.39. In the interior of continents where mountains block west winds, or in lands far from the reaches of moist tropical air, extensive regions of desert and *semidesert* conditions appear.

### Hot Deserts (BWh:7,4)

On the poleward side of the savannas, grasses begin to shorten, and desert shrubs become evident. This is where we approach the belt of subtropical high pressure that brings considerable sunshine, hot summer weather, and very little precipitation. Note the minute amount of rainfall shown in Figure 4.40. The precipitation that does fall is convectional but sporadic. As conditions become drier, fewer and fewer drought-resistant shrubs appear and, in some areas, only gravelly and sandy deserts exist, as suggested in Figure 4.41.

The great, hot deserts of the world, such as the Sahara, the Arabian, the Australian, and the Kalahari, are all the products of high-pressure zones (see Figure 4.39). Often, the

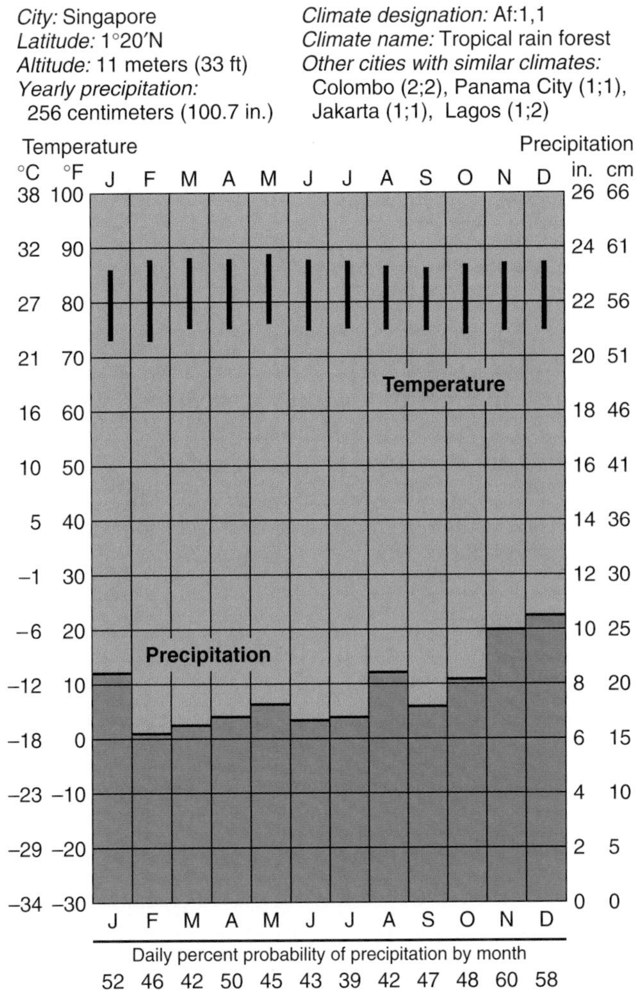


**FIGURE 4.35** This and succeeding climate charts (*climagraphs*) show average daily high and low temperatures for each month, the average precipitation for each month, and the probability of precipitation on any particular day in a designated month. For Singapore, the average daily high temperature in August is 30.5°C (87°F); the low is 24°C (75°F). The rainfall for the month, on average, is 21 centimeters (8.4 in.), and on a given day in August, there is a 42% chance of rainfall. The numbers in parentheses following a city's name refer to winter and summer temperature and precipitation data as shown on Figure 4.32 (a) and (b).

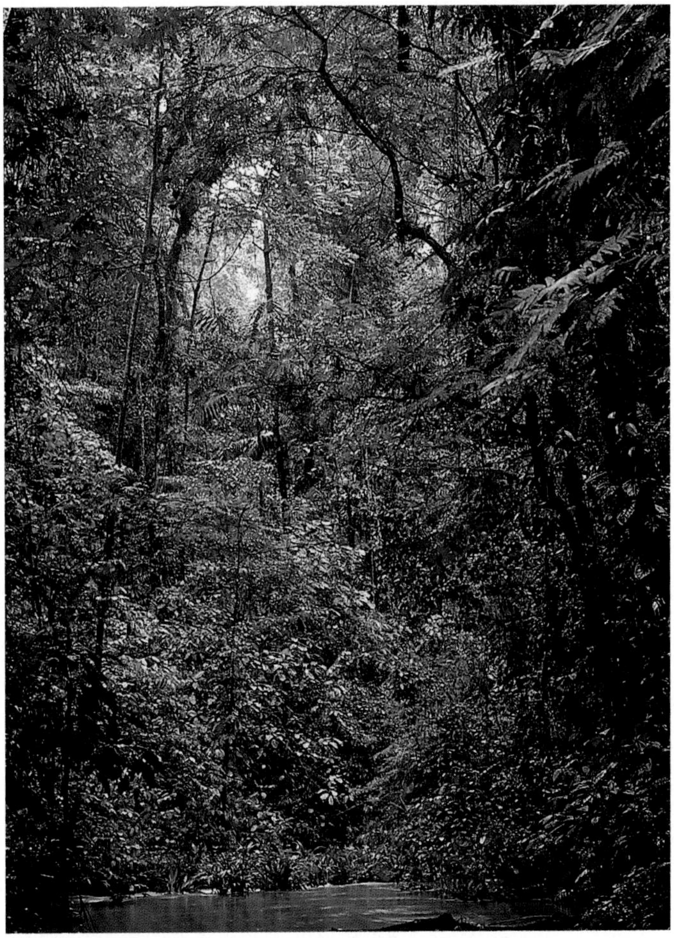

**FIGURE 4.36 Tropical rain forest.** The vegetation is characterized by tall, broadleaf, hardwood trees and vines.

(a)

(b)

**FIGURE 4.37 The parklike landscapes of grasses and trees characteristic of the (a) drier and (b) wetter tropical savanna.** *(a) © Aubrey Lang/Valan Photos. (b) © Thomas J. Bassett, Dept. of Geography, University of Illinois.*

*City:* Yangôn, Myanmar
*Latitude:* 16°46′N
*Altitude:* 5.5 meters (18 ft)
*Yearly precipitation:* 252 centimeters (99.2 in.)
*Climate designation:* Am:3,1
*Climate name:* Savanna (monsoon type)
*Other cities with similar climates:* Mumbai (4;1), Calcutta (3;1), Miami (3;1)

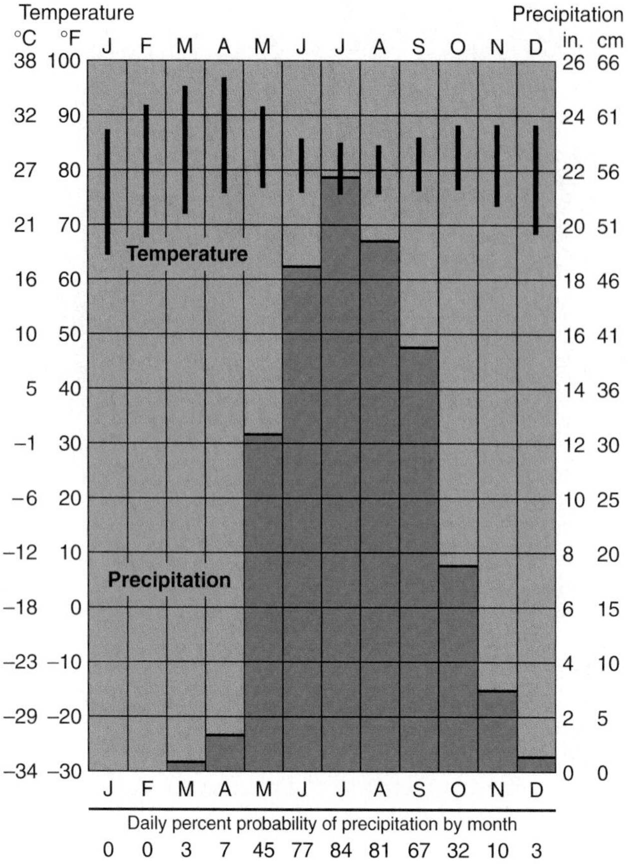


**FIGURE 4.38 Climagraph for Yangôn, Myanmar.** (See Figure 4.35 for explanation.)

driest parts of these deserts are along the western coasts, where cold ocean currents are found. The soils, called *aridisols,* respond well to cultivation when irrigation is made available. Earlier, mention was made of the relationship between cold ocean currents and deserts.

#### Midlatitude Deserts and Semideserts (BWk:4,4;BS:10,4)

Figure 4.42 illustrates typical temperature and precipitation patterns in these midlatitude drylands. Occasionally, a summer convectional storm or a frontal system with some moisture occurs. The extreme dry areas are known as cold deserts. The moderately dry lands are called **steppes.** The natural vegetation is grass, although desert shrubs, pictured in Figure 4.43, are found in drier portions of the steppes. Rain is not plentiful, but soils are rich because the grasses return nutrients to the soil. The soils, called *mollisols,* have a dark brown to black A-horizon and are among the most naturally fertile soils in the world. As a result, humans have made the steppes of the United States, Canada, Ukraine, and China some of the most productive agricultural regions on Earth. The steppes are also known for their hot, dry summers and biting winter winds, which sometimes bring blizzards.

### Humid Midlatitude Climates (C, D)

Figure 4.44 shows the location of several climate types that are all humid—that is, not having desert conditions in the winter, summer, or both. In addition, winter temperatures well below those of the tropical climates are characteristic of the humid midlatitudes. These climate types would be neatly defined, paralleling the lines of latitude, were it not

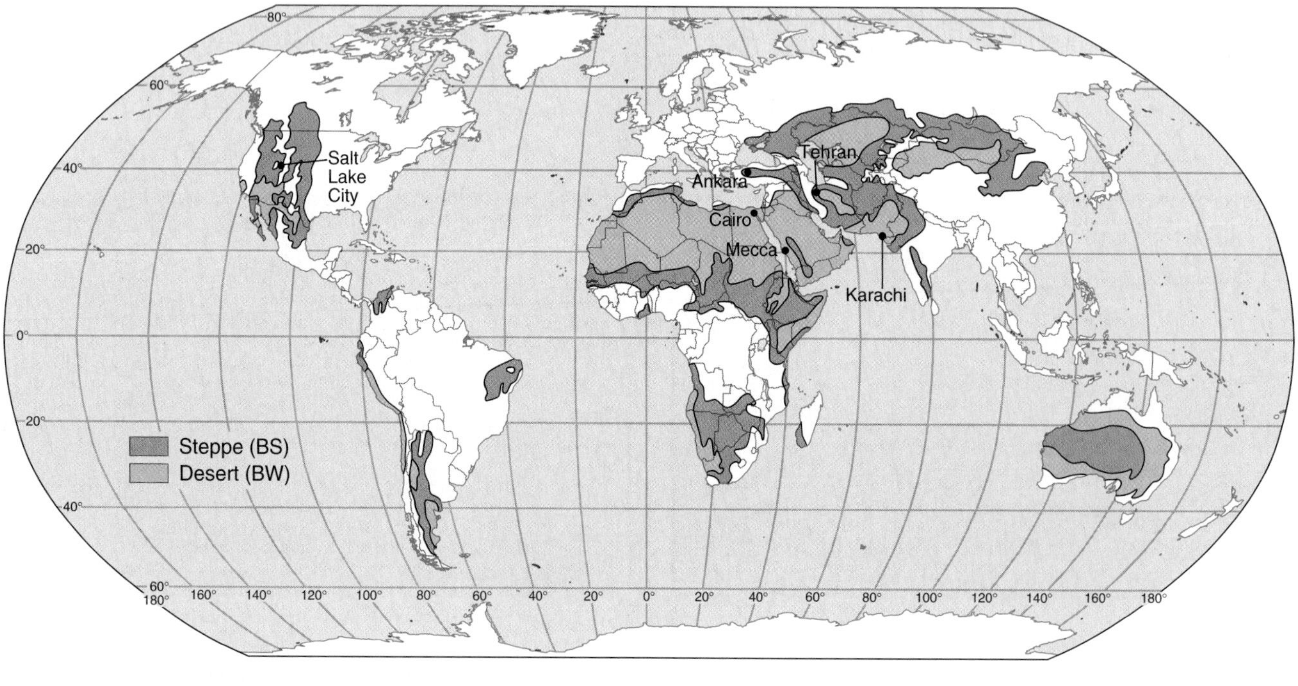


**FIGURE 4.39 The location of steppe and desert climates.**

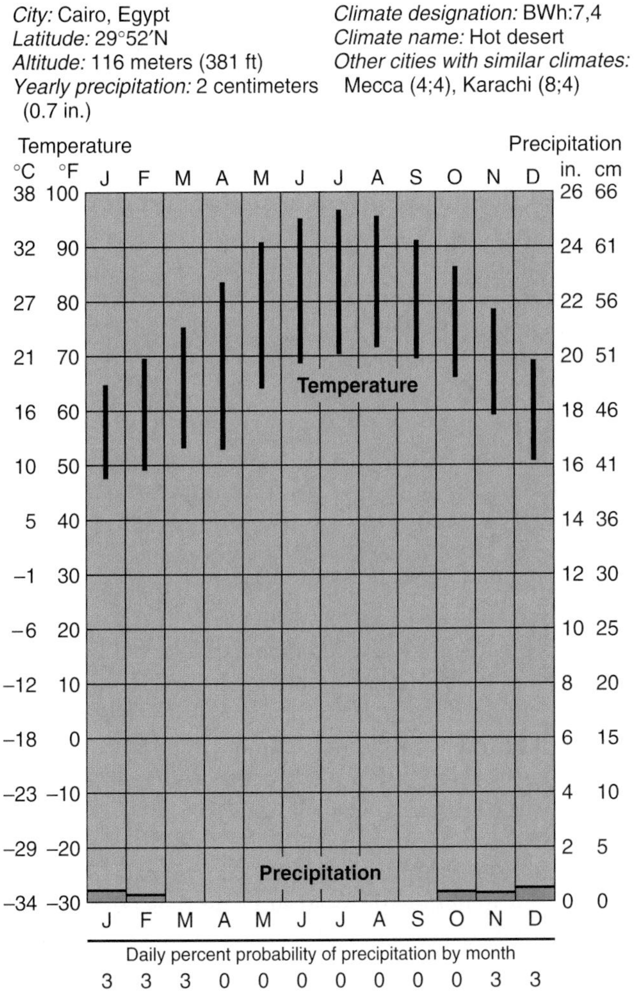


**FIGURE 4.40 Climagraph for Cairo, Egypt.** (See Figure 4.35 for explanation.)

for mountain ranges, warm or cold ocean currents, and, particularly, land-water configurations. These factors cause the greatest variations in the middle latitudes.

### Mediterranean Climate (Cs:6,3)

Midlatitude winds generally blow from the west in both the Northern and Southern Hemispheres, and a significant amount of the precipitation is produced from frontal systems. Thus, it is important to know if the water is cold or warm near land areas. Several climatic zones are noticeable in the middle latitudes, all marked by warm summer temperatures except those in areas cooled by westerly winds from the ocean.

To the poleward side of the hot deserts, a transition zone occurs between the subtropical high and the moist westerlies zones. Here, cyclonic storms bring rainfall only in the winter, when the westerlies shift toward the equator. Summers are dry and hot as the subtropical highs shift slightly poleward (Figure 4.45). Winters are not cold. These conditions describe the **Mediterranean climate,** which is often found on the

**FIGURE 4.41 Death Valley, California.** Devoid of stabilizing vegetation, desert sands are constantly rearranged in complex dune formations. *© Dietrich Stock Photos, Inc.*

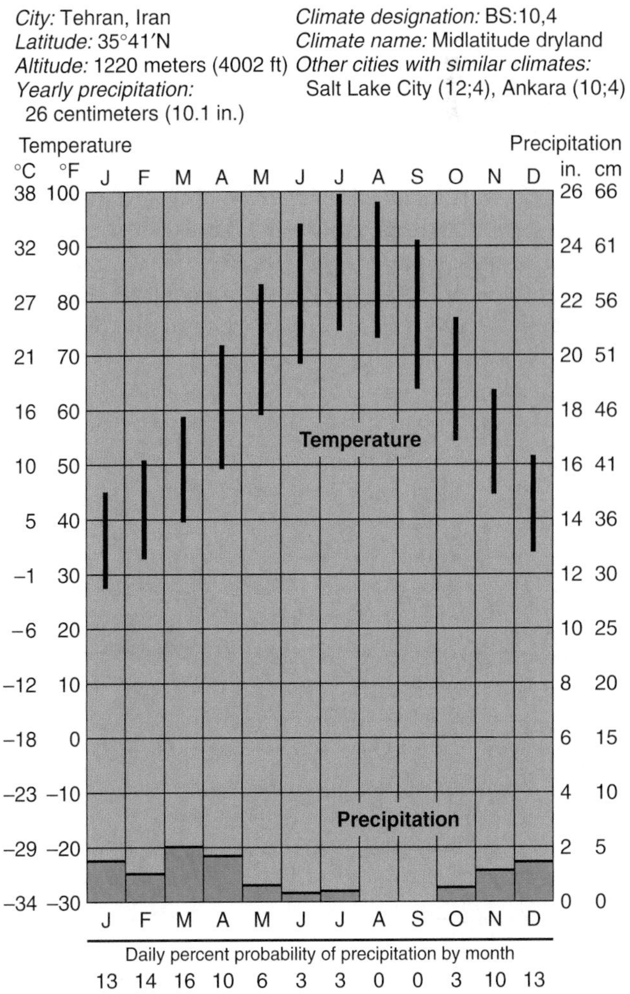


**FIGURE 4.42 Climagraph for Tehran, Iran.** (See Figure 4.35 for explanation.)

**FIGURE 4.43 Desert shrubs in the midlatitude drylands of northern Mexico.** *© Leonard Lee Rue, Jr./Photo Researchers.*

western coasts of continents in the middle latitudes. Southern California, the Mediterranean area itself, western Australia, the tip of South Africa, and central Chile in South America are characterized by this type of climate. In these areas, where precipitation is sufficient, shrubs and small deciduous trees, such as the scrub oak, grow (Figure 4.46).

The Mediterranean climate area—long and densely settled in southern Europe, the Near East, and North Africa—has more moisture and a greater variety of vegetation and soil types than the desert. Clear, dry air predominates; winters are relatively short and mild; and plants and flowers grow year-round. Even though summers are hot, nights are usually cool and clear. Much of the vegetation in this area is now in the form of crops.

### Humid Subtropical Climate (Cfa:6,12)

On the eastern coasts of continents, the transition is from the equatorial climate to the **humid subtropical climate.** Convectional summer showers and winter cyclonic storms are the sources of precipitation. As illustrated in Figure 4.47, this climate is characterized by hot, moist summers and moderate, moist winters. In the fall, on occasion, hurricanes that develop in tropical waters strike the coastal areas.

The generally even distribution of rainfall allows for the presence of deciduous forests containing hardwood trees, such as oak and maple, whose leaves turn orange and red before falling in autumn. In addition, conifers become mixed with deciduous trees as a second-growth forest.

The poleward transition to the continental climates is accompanied by increasingly colder winters and shorter summers. In this direction as well, cyclonic storms become more responsible for rainfall than are convectional showers. The region can no longer be characterized as humid subtropical; rather, it is described as *humid continental* (see the section "Humid Continental Climate," p. 122). Southern Brazil, the southeastern United States, and southern China all have a humid subtropical climate.

Beneath the deciduous forests of both the humid subtropical and the humid continental climates are *alfisols.* The A-horizon is generally gray-brown, and these soils are usually rich in plant nutrients. Derived from the strongly basic humus created by fallen leaves of deciduous trees, alfisols retain moisture during the hot days of summer, allowing for productive agriculture.

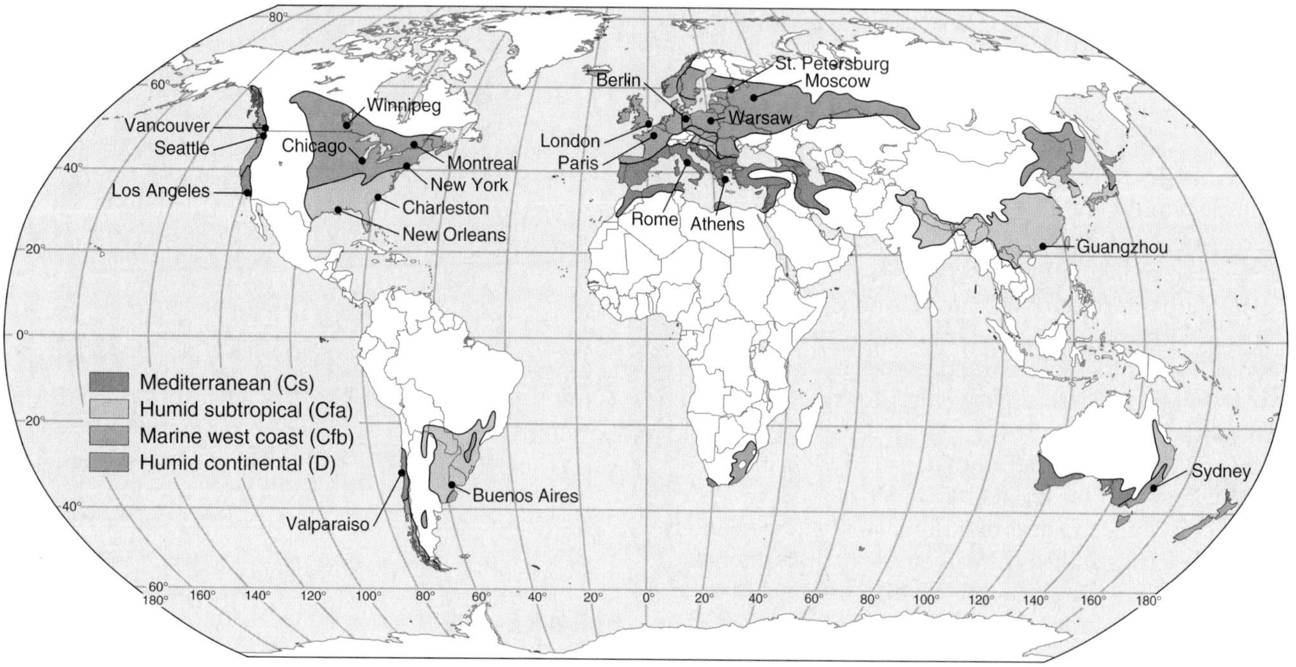


**FIGURE 4.44 The location of humid midlatitude climates.**

*City:* Rome, Italy
*Latitude:* 41°48′N
*Altitude:* 115 meters (377 ft)
*Yearly precipitation:* 85 centimeters (33.3 in.)
*Climate designation:* Cs:6,3
*Climate name:* Mediterranean
*Other cities with similar climates:* Athens (6;3), Los Angeles (6;4), Valparaiso (6;4)

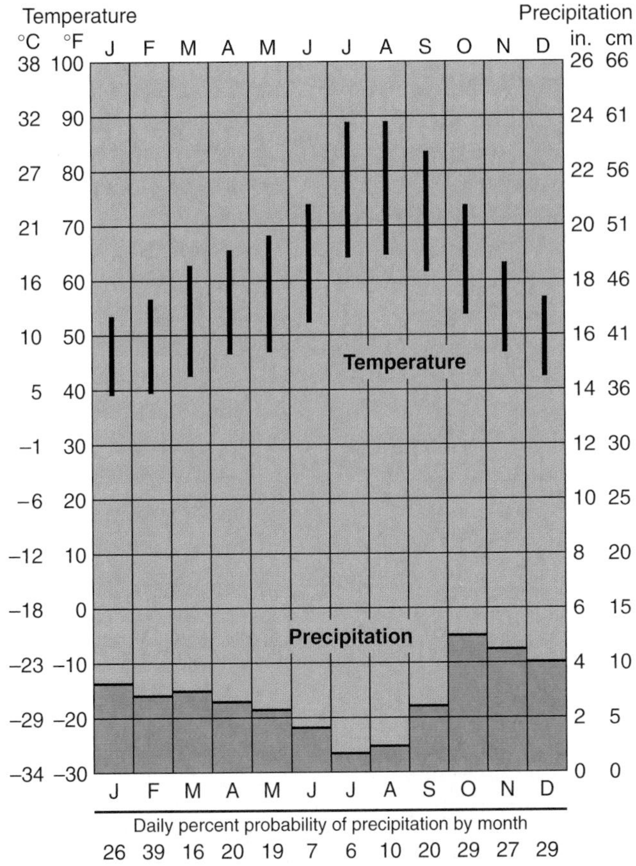


**FIGURE 4.45 Climagraph for Rome, Italy.** (See Figure 4.35 for explanation.)

## Marine West Coast Climate (Cfb:10,6)

Closer to the poles, but still within the westerly wind belt, are areas of **marine west coast climate.** Here, cyclonic storms and orographic precipitation play a relatively large role. In the winter, more rainfall and cooler temperatures prevail than in the Mediterranean zones. Compare the patterns in Figures 4.45 and 4.48. In the transitional zone just poleward of the Mediterranean climate, little rainfall occurs during the summer. Closer to the poles, however, rainfall increases appreciably in the summer and even more so in the winter. Marine winds from the west moderate both summer and winter temperatures. Thus, summers are pleasantly cool, and winters, though cold, do not normally produce freezing temperatures.

This climate affects relatively small land areas in all but one region. Because northern Europe contains no great mountain belt to thwart the west-to-east flow of moist air, the marine west coast climate stretches well across the continent to Poland. In Poland, cyclonic storms originating in the Arctic regions are noticeable. Northern Europe's moderate climate also owes its existence to a relatively warm

**FIGURE 4.46 Vegetation typical of an area with a Mediterranean climate.** Trees such as scrub oak are short and scattered. *© Carr Clifton/Minden Pictures.*

*City:* Sydney, Australia
*Latitude:* 33°58′S
*Altitude:* 9 meters (29 ft)
*Yearly precipitation:* 116 centimeters (46.5 in.)
*Climate designation:* Cfa:6,2
*Climate name:* Humid subtropical
*Other cities with similar climates:* Guangzhou (6;2), Charleston (6;2), New Orleans (6;2)

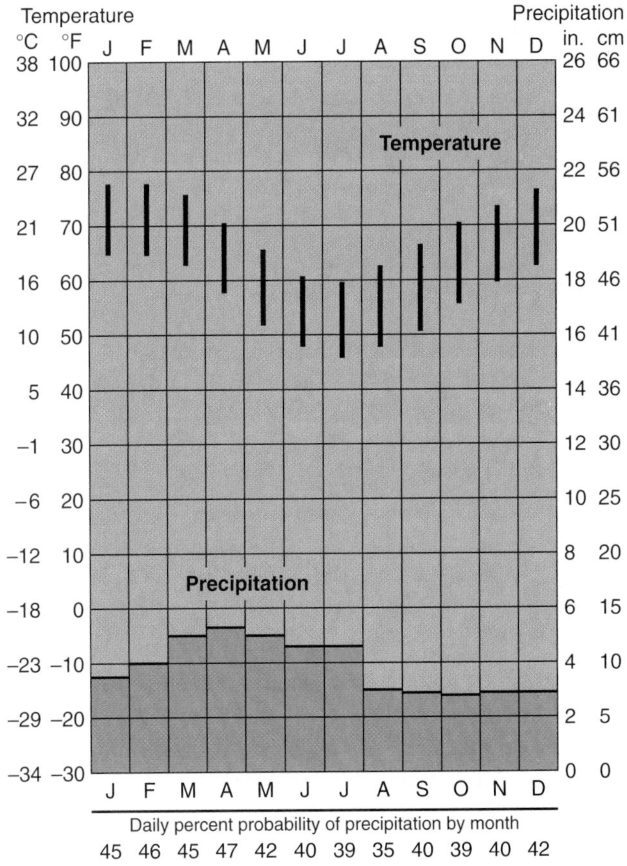


**FIGURE 4.47 Climagraph for Sydney, Australia.** (See Figure 4.35 for explanation.) Because Sydney is in the Southern Hemisphere, the warmest days are in January and the coldest are in July.

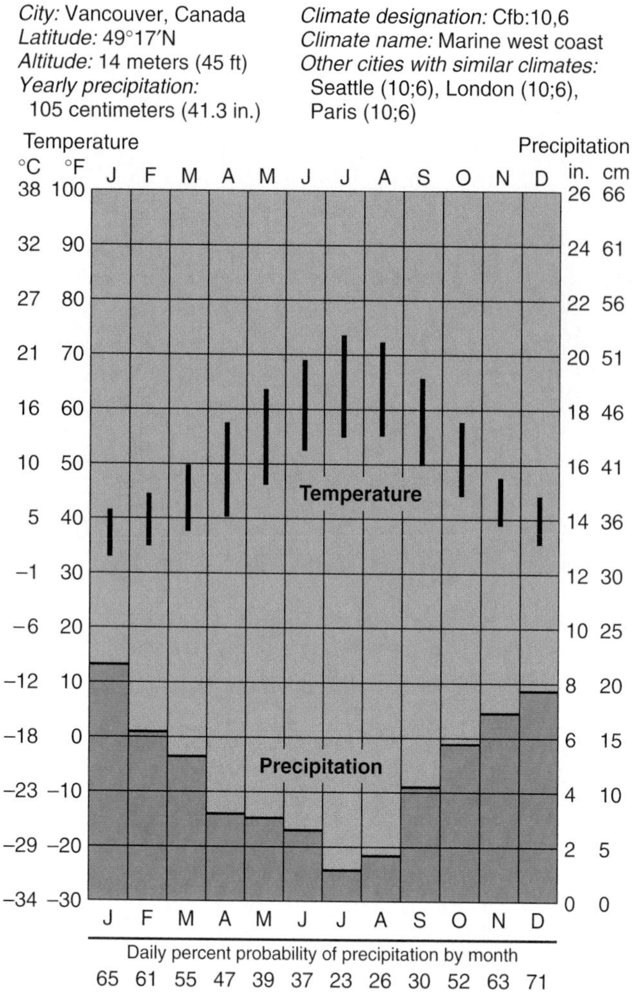


**FIGURE 4.48 Climagraph for Vancouver, Canada.** (See Figure 4.35 for explanation.)

ocean current whose influence is felt for nearly 1600 kilometers (1000 mi), from Ireland to central Europe.

The orographic effect from mountains in areas such as the northwestern United States, western Canada, and southern Chile produces enormous amounts of precipitation, often in the form of snow on the windward side (see Figure 4.22). Vast coniferous forests—needle-leaf trees, such as pines, spruces, and firs—cover the mountains' lower elevations. Because the mountains prevent moist air from continuing to the leeward side, midlatitude deserts are found to the east of these marine west coast areas.

The main soils of marine west coast areas are *spodosols*. They are strongly acidic and low in plant nutrients, a result of the acidic humus that develops from the fallen needles of coniferous trees. Fertilizers must be used for farming in order to neutralize the acidity.

### Humid Continental Climate (Dfa, Dfb: 10, 2; 14, 2 and 15, 6)

Air masses that originate close to the poles and drift toward the equator and other air masses that drift toward the poles from the tropics produce frontal precipitation. Whenever warmer air or marine air blocks cold continental air masses, or vice versa, frontal storms develop. The climates that these air masses influence are described as **humid continental climates.** Figures 4.49 and 4.50 show the range and the dominance of winter temperatures within this climatic type.

The continental climate may be contrasted to marine west coast climates in that the former has prevailing winds from the land, the latter from the sea. Coniferous forests become more plentiful in the direction of the poles, until temperatures become so low that trees are denied an adequate growing season; therefore, their growth is stunted (Figure 4.51). Along with coniferous forests are the infertile spodosols. The transition from the very cold air masses of the winter to the occasional convectional storms of the summer means that four distinct seasons are apparent.

Three huge areas of the world are characterized by a humid continental climate: (1) the northern and central United States and southern Canada, (2) most of the European portion of Russia, and (3) northern China. Because

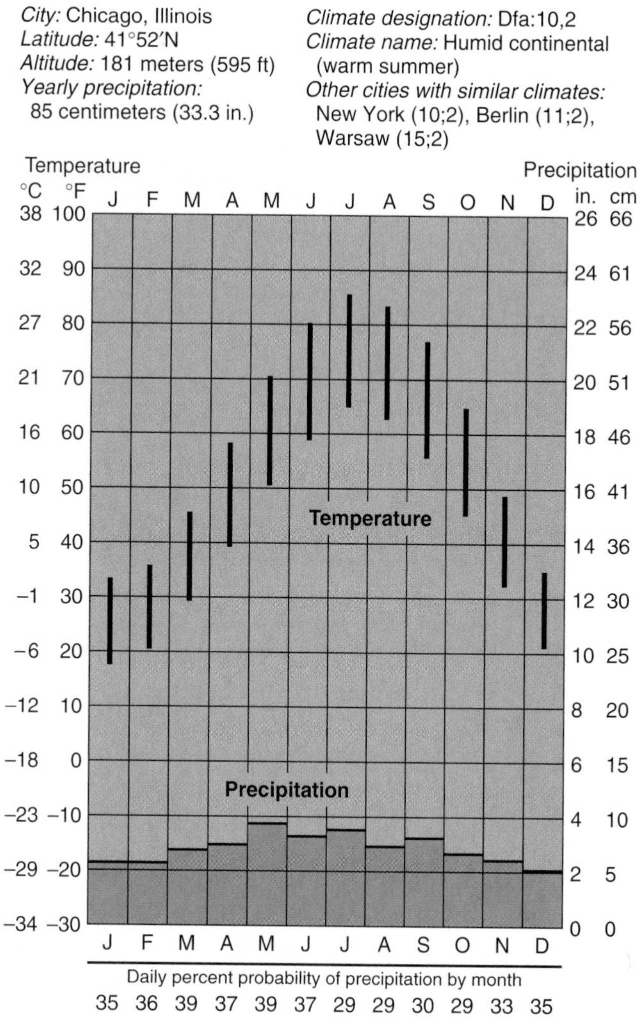


**FIGURE 4.49 Climagraph for Chicago, Illinois.** (See Figure 4.35 for explanation.)

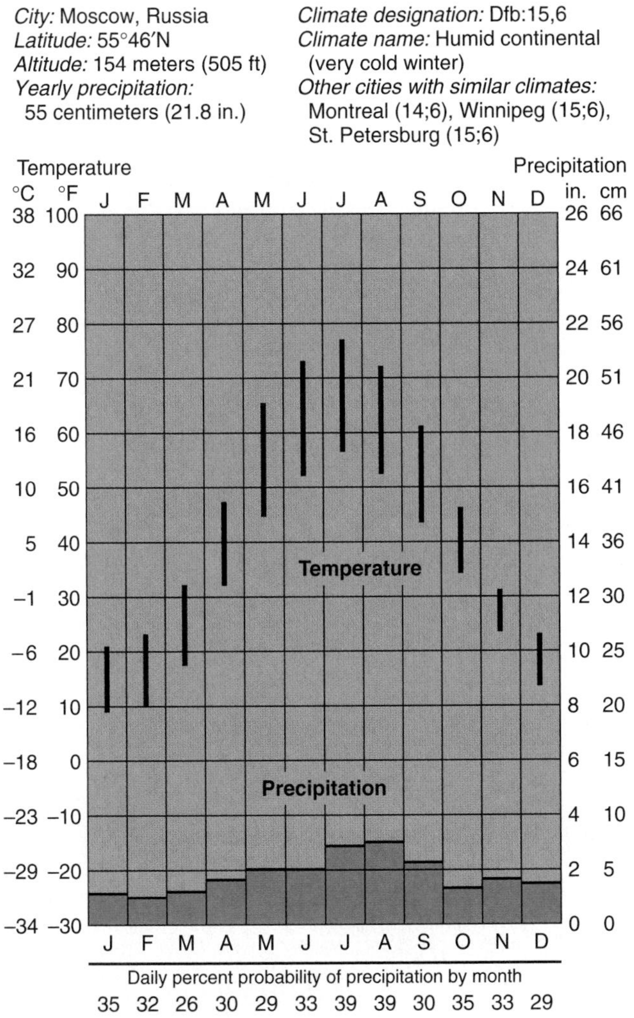


**FIGURE 4.50 Climagraph for Moscow, Russia.** (See Figure 4.35 for explanation.)

**FIGURE 4.51** In the extensive region of east central Canada and the area around Moscow, Russia, the summers are long and warm enough to support a dense coniferous forest. Farther north, growth is less luxuriant. *© Vol. 36/PhotoDisc.*

there are no land areas at a comparable latitude in the Southern Hemisphere, this climate is not represented there. In fact, the only nonmountain cold climate in the Southern Hemisphere is the polar climate of Antarctica.

## Subarctic and Arctic Climates (E:16, 7)

Toward northern areas and into the interior parts of the North American and Eurasian landmasses, increasingly colder temperatures prevail (Figures 4.52 and 4.53). Trees become stunted, and eventually only mosses and other cool-weather plants of the type shown in Figure 4.54 will grow.

The word **tundra** is often used to describe the northern boundary zone beyond these treed subarctic regions. Because long, cold winters predominate, the ground is frozen most of the year. A few cool summer months, with an abundant supply of mosquitoes, break up the monotony of extreme cold. Although very cold temperatures characterize the tundra, snowfall is not very abundant. Strong easterly winds blow snow, which, combined with ice fogs and little winter sunlight, contributes to a very bleak climate. Alaska, northern Canada, and northern Russia are covered with either the stunted trees of the subarctic climate or by the bleak, treeless expanse of the tundra. Antarctica and Greenland, however, are icy deserts.

Soils in these vast arctic regions are varied. Perhaps the most characteristic soils are *histosols*. They tend to be peat or muck, consisting of plant remains that accumulate in water. In forested areas, they tend to resemble spodosols, which form in areas of poor drainage both in the Arctic and the midlatitudes.

These thumbnail sketches of climatic conditions throughout the world give us the basic patterns of large regions. On any given day, conditions may be quite different from those discussed or mapped in this chapter. However, the physical climatological processes, in general, are what concern us. We can deepen our understanding of climates by applying our knowledge of the elements of weather.

# Climatic Change

We have stressed that climates are only averages of, perhaps, greatly varying day-to-day conditions. Figure 4.55 illustrates the global variation in yearly precipitation. Temperatures are less changeable than precipitation on a year-to-year basis, but they too vary. How can we account for these variations? Scientists in research stations all around the world are investigating this question. The data they use range from daily temperature and precipitation records to calculations concerning the position of the earth in relation to the sun. Because day-to-day records for most places date back only 50 to 100 years, scientists look for additional information about past climates in rock formations, the chemical composition of earth materials, ice cores, lake-floor sediments, tree rings, and other sources.

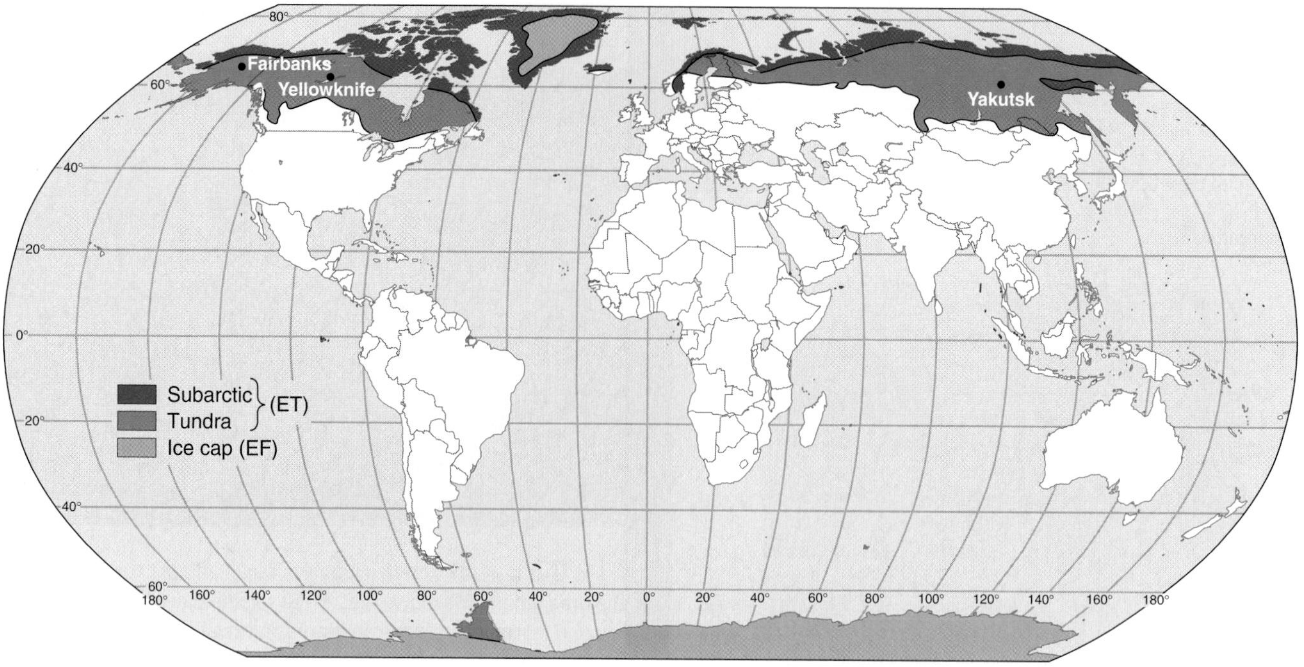


**FIGURE 4.52 The location of Arctic and subarctic climates.**

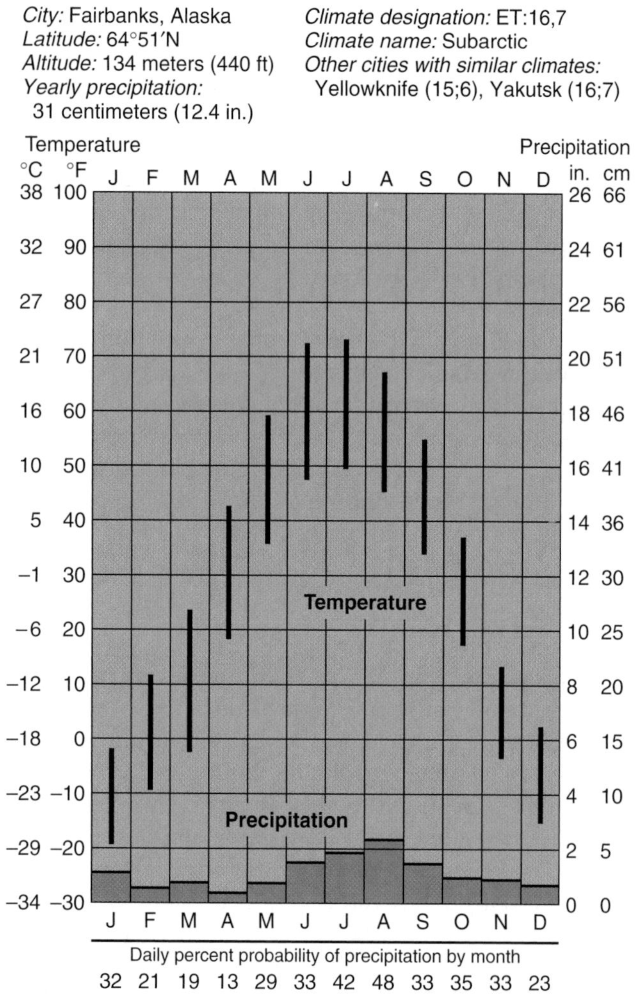


**FIGURE 4.53 Climagraph for Fairbanks, Alaska.** (See Figure 4.35 for explanation.)

**FIGURE 4.54 Tundra vegetation in Canada.** *© John Shaw/Tom Stack & Associates.*

## Long-Term Climatic Change

Significant variations in climate have occurred over geologic time. For example, approximately 65 million years ago, at the end of the Cretaceous period, there was a sudden cooling of the earth's climate. This cooldown is thought to have caused the extinction of some 75% of all existing plant and animal species, including most dinosaurs. To take another example, cycles of ice sheet formation and breakup occurred at least five times during the last ice age, which lasted 100,000 years and ended only 11,000 years ago.

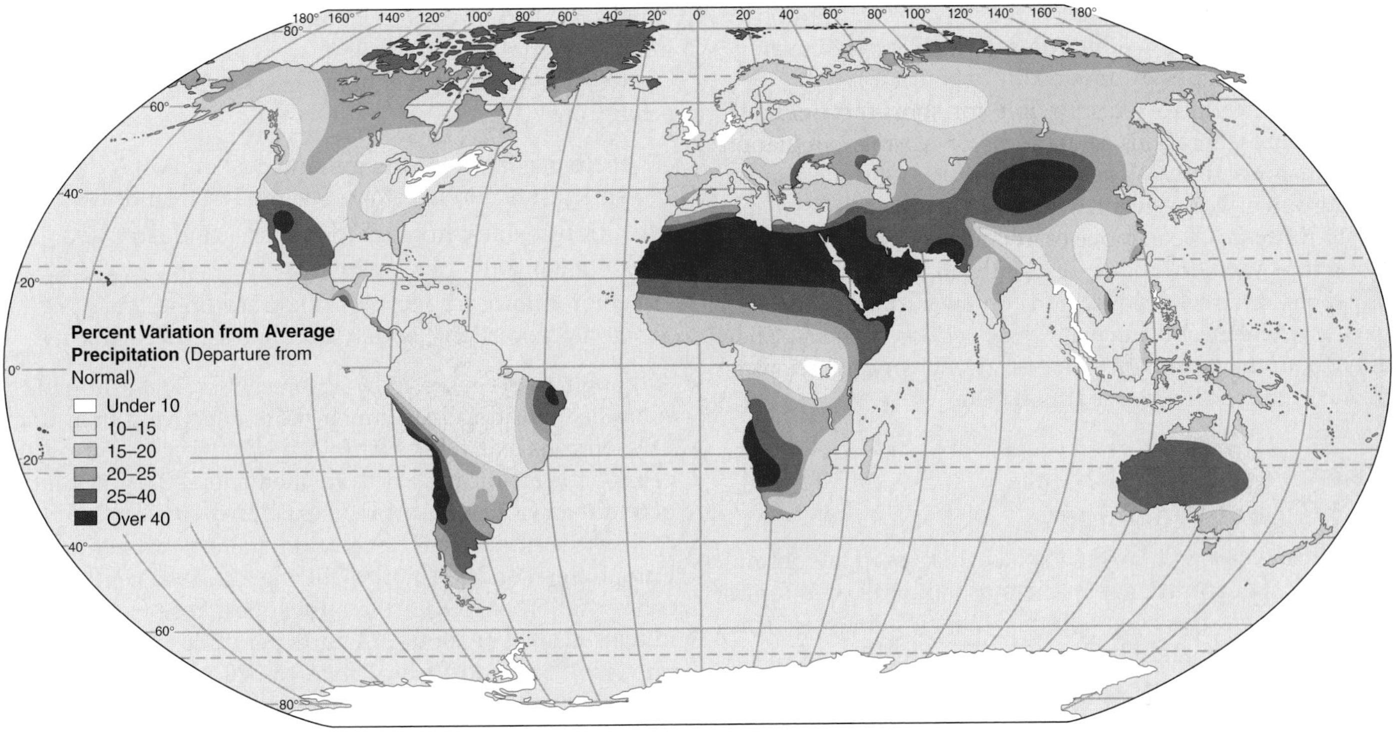


**FIGURE 4.55 The world pattern of precipitation variability.** Regions of low total precipitation tend to have high variability. In general, the drier the climate, the greater is the probability that there will be considerable differences in rainfall and/or snowfall from one year to the next.

Climatologists have identified two major climatic periods just in the past 1000 years: a medieval warm period and a "little ice age." Between about A.D. 800 and 1200, during the medieval warm period, temperatures were as warm or warmer than they are now. Settlement and farming expanded northward and to higher altitudes, the Vikings colonized Iceland and Greenland, and vineyards flourished in Britain. During the little ice age, which lasted from about A.D. 1300 to 1850, Arctic ice expanded, glaciers advanced, drier areas of the earth were desiccated, and crop failures and starvation were common.

Scientists have suggested several explanations for such periodic changes in climate. Some of the climatic variations are thought to be due to three aspects of the earth's motion, all of which affect the amount of solar radiation reaching the planetary surface. First is the shape of the earth's orbit around the sun, which varies from nearly circular to more elliptical over a period of about 100,000 years. When the orbit is nearly circular, the earth experiences relatively cold temperatures. When it is elliptical, as it is now, the earth is closer to the sun for several months, is exposed to more total solar radiation, and thus it has higher temperatures.

Another cycle corresponds to the tilt of the earth's axis relative to the orbital plane. The tilt varies from 21.5° to 24.5° every 41,000 years. The amount of solar radiation striking polar regions changes as the angle of tilt changes. A low tilt position—that is, a more perpendicular position of the earth—is accompanied by periods of colder climate. Cooler climates are thought to be critical in the formation of ice sheets.

Finally, like an unbalanced spinning top, the earth wobbles slightly as it rotates, changing the earth's orientation to the sun. The gyration of the rotation axis repeats every 23,000 years. When the tilt of the axis is greatest, the polar regions receive less solar radiation than they do at other times and become colder.

## Short-Term Climatic Change

Climate can change more quickly and irregularly than the earth's cycles suggest. Great volcanic eruptions can alter climates for several years. They spew enormous amounts of ash, water vapor, sulfur dioxide, and other gases into the upper atmosphere. As these solid and liquid particles spread over much of the planet, they block some of the incoming solar radiation that normally would reach the earth's surface, producing a cooling effect. The famous "year without a summer"—1816—in New England, when snow fell in June and frost came in July, probably was caused by the eruption a year earlier of the Indonesian volcano Tambora. The explosion ejected an estimated 200 million tons of gaseous aerosols and 50 cubic kilometers (30 cu mi) of dust and ash into the atmosphere. The reflective cooling effect lasted for years. A similar decline in temperatures occurred after the 1883 volcanic eruption of Krakatoa, also in Indonesia. A less extreme drop in temperatures in the early 1990s was attributed to the July 1991 eruption of Mount Pinatubo in the Philippines, which lowered average global temperatures by about 0.5°C (about 1°F).

Two other factors responsible for short-term climatic changes are alterations in patterns of oceanic circulation and sunspot activity. As described on page 108, for example, during an El Niño event, warm surface waters from the western Pacific Ocean move eastward, changing the climate along the western coasts of South and North America. Sunspots, relatively cool regions on the surface of the sun, vary in number and intensity over periods of years. They affect both the output of solar energy and concentrations of ozone in the earth's upper atmosphere.

## The Greenhouse Effect and Global Warming

All of the cycles and factors we have discussed are natural processes. In contrast, one of the most hotly debated topics in recent years has been whether human beings are contributing to climatic change through what is popularly termed the **greenhouse effect.** Put simply, the theory is that certain gases concentrate in the atmosphere, where they function as an insulating barrier, trapping infrared radiation that would otherwise be radiated back into the upper atmosphere and reradiating it earthward. In other words, like glass in a greenhouse, the gases admit incoming solar radiation but retard its reradiation back into space. You have experienced such a greenhouse effect if you have gotten into a car on a cold but sunny day; the car's interior is warmer than the outside air.

The earth has a *natural* greenhouse effect, provided mainly by water vapor that has evaporated from the ocean or evapotranspired from land. The water vapor remains a constant, but during the past 150 years or so, human activities have increased the amount of other greenhouse gases in the atmosphere, augmenting its heat-trapping ability. Many scientists fear that an *enhanced* greenhouse effect could result in a gradual increase in the earth's average surface temperature, with significant impacts on the earth's ecosystems, a process called **global warming.** *That* greenhouse effect is far less benign and nurturing than the name implies.

Carbon dioxide ($CO_2$) is the primary greenhouse gas whose amount has been increased by human activities. Although it occurs naturally, excessive quantities of it are released by burning fossil fuels. Beginning with the Industrial Revolution in the mid-1700s, large amounts of coal, petroleum, and natural gas have been burned to power industry, to heat and cool cities, and to drive vehicles. Their combustion has turned fuels into carbon dioxide and water vapor. At the same time, much of the world's forests have been destroyed by logging and to clear land for agriculture. Deforestation adds to the greenhouse effect in two ways: it means there are fewer trees to capture carbon dioxide and produce oxygen, and burning the wood sends $CO_2$ back into the atmosphere at an accelerated rate. The relative contribution of carbon dioxide to the potential for global warming is thought to be about 55%.

Other important greenhouse gases influenced by human activity are

1. methane, from natural gas and coal mining, agriculture and livestock, swamps, and landfills;
2. nitrous oxides, from motor vehicles, industry, and nitrogen-containing fertilizers;
3. chlorofluorocarbons, hydrofluorocarbons, and perfluorocarbons, widely used industrial chemicals.

Although these gases may be present in small amounts, some of them trap heat much more effectively than does $CO_2$. Nitrous oxide, for example, has 360 times the capacity of $CO_2$ to trap heat, and even methane is 24 times more potent than $CO_2$ in absorbing heat close to the earth.

As the Industrial Revolution gained momentum in Europe and North America during the 19th century, the concentration of $CO_2$ in the atmosphere rose from its preindustrial level of about 274 parts per million (ppm) to 315 ppm in 1958; it rose since then to 370 ppm in 2004. The methane concentration in the lower atmosphere has *already* more than doubled from its preindustrial level and is currently increasing by just over 1% per year.

Proponents of the greenhouse effect theory believe that human activity has already put enough of the various gases into the atmosphere to affect climate. They fear that the accelerated warming trend of the past 50 years may have exceeded typical climate shifts and cite evidence such as the following to support their beliefs:

- The 20th century was the warmest century of the past 600 years, and most of its warmest years were concentrated near its end. The world's average surface temperature rose about 0.6°C (a bit over 1°F) in the 20th century, and the 1990s were the hottest decade of the century (Figure 4.56).

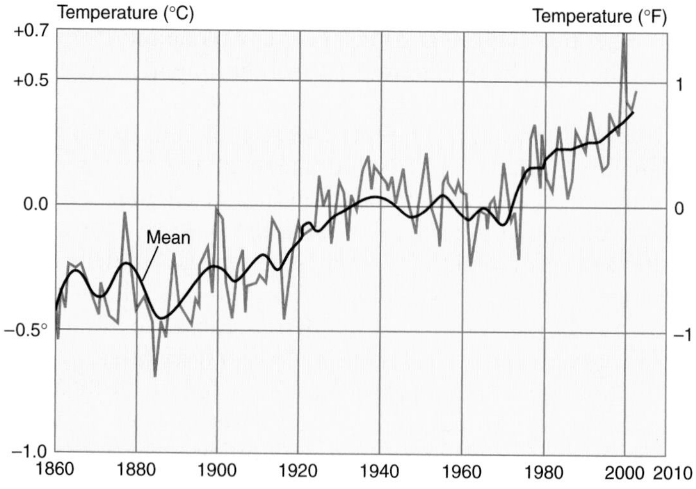


**FIGURE 4.56 Global average temperature change, 1860–2002.**
Source: *NOAA.*

- Winter temperatures in the Arctic have risen about 4°C (7°F) since the 1950s. The Arctic as a whole is losing its ice cap. Between 1978 and 2000, the coverage of Arctic sea ice in winter decreased by 6%, and the average thickness of Arctic ice declined by 42%, from 3.1 meters to 1.8 meters (10.2 to 5.9 ft). Similarly, the sea ice west of the Antarctic Peninsula has diminished more than 20% since 1970.
- On every continent, glaciers are thinning and retreating. For example, glaciers atop Mount Kilimanjaro and Mount Kenya in Africa shrank by 70% or more during the 20th century, and glaciers in the Swiss Alps are estimated to have lost half their volume since 1850. These patterns of glacial thinning and retreat are repeated in Alaska, Peru, Russia, India, China, Irian Jaya, New Zealand, and elsewhere. In some places (e.g., portions of Montana's Glacier National Park and the eastern Himalayas), glaciers melted and disappeared altogether in the 20th century. Although glaciers have grown and retreated for thousands of years, many scientists believe that the rate of melt has accelerated in the past few decades and now exceeds anything in recent centuries.

Whatever the attributable causes of global warming, most climatologists agree on certain of its general consequences should it continue (Figure 4.57). Increases in sea temperatures would cause ocean waters to expand slightly (thermal expansion) and the polar ice caps to melt at least a bit. More serious consequences would result from the melting of the Greenland ice sheet and the rapid retreat or total melting of glaciers throughout the world. Although melting sea ice has no effect on sea levels, water melted from continental sources adds to ocean volumes. Inevitably, sea levels would rise, perhaps 1 meter (3.3 ft) in a hundred years, with devastating impacts, especially in the tropics and warm temperate regions, where many coastlines are heavily settled. As Figure 4.58 shows, the most vulnerable areas are along the north and west coasts of Africa, South and Southeast Asia, and low-lying coral atolls in the Pacific and Indian oceans.

Other problems would result from changes in precipitation patterns. The warming of lakes and oceans would speed evaporation, causing more active convection currents in the atmosphere. It is important to note that changes in precipitation would be regional, with some areas receiving more and others less precipitation than they do now. Polar and equatorial regions might get heavier rainfall, whereas the continental interiors of the midlatitudes could become drier and suffer at least periodic drought. More northerly agricultural regions, such as parts of Canada, Scandinavia, and Russia, might benefit from general temperature rises; the longer growing season would make them more productive. Changes in temperature and precipitation would affect soils and vegetation. The composition of forests would change, as some areas would become less favorable for certain species of plants but more hospitable to others.

Several cautionary notes are needed. Some analysts point out that increases in temperatures recorded at land-based observation stations were inevitable but misleading. Many weather stations originally sited in rural areas have, through urban expansion, now been made part of city "heat islands." The result, they maintain, has been a distortion of officially recorded long-term temperature trends and a false suggestion of steadily rising temperatures.

Further, many of those who accept that the global climate has been warming point out that climate prediction is not an exact science, and they disagree with the scenarios of change just presented. Temperature differences are the engine driving the global circulation of winds and ocean currents and help create conditions inducing or inhibiting winter and summer precipitation and daily weather conditions. Exactly how those vital climate details would express themselves locally and regionally is uncertain.

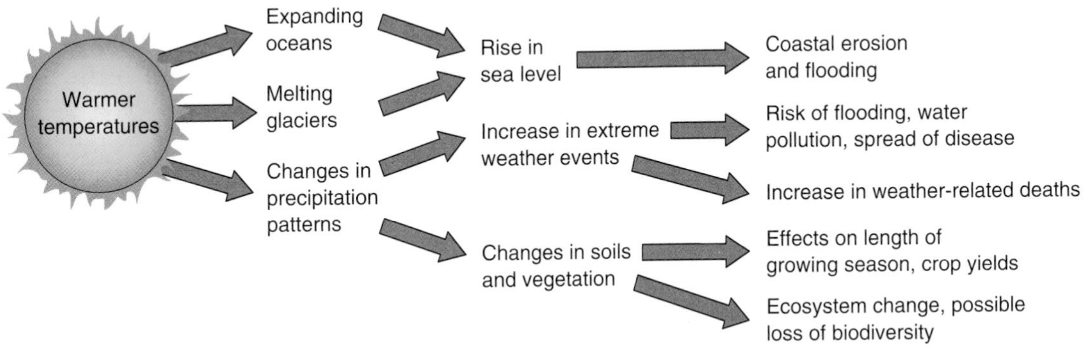


**FIGURE 4.57 Potential effects of global warming.** Global warming is projected to cause an increase in the frequency and severity of extreme weather events and weather-related disasters: heavy downpours, floods, hurricanes, heat waves, drought, and wildfires. Secondary effects of short-term extremes of weather would be shifts in plant and animal ranges; an increase in water pollution; and the spread of infectious diseases as warmer, wetter weather conditions widen the range of disease-carrying insects.

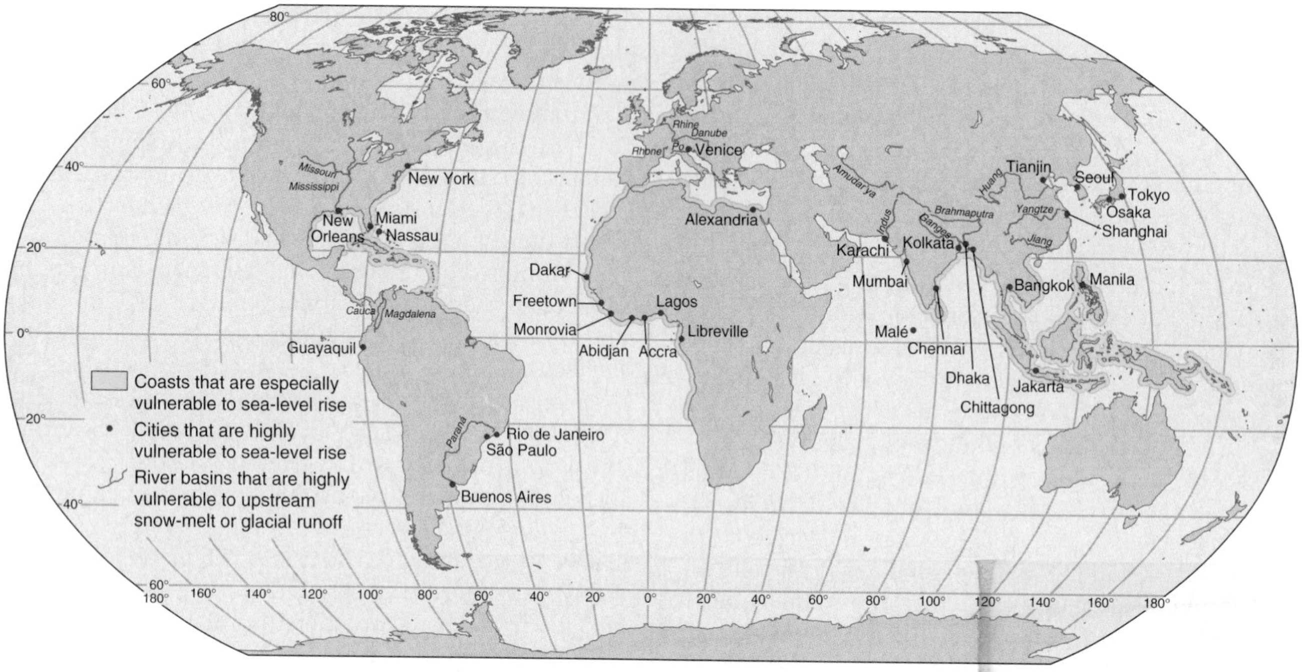


**FIGURE 4.58** The acceleration of global warming could produce significant changes in sea level and patterns of precipitation. Approximately one-sixth of the world's people live at or just a few meters above sea level, and hundreds of millions live in cities downstream from mountains where accelerated melting of glaciers or snow could contribute to severe flooding, especially if the river basins have been heavily deforested. Even a 1-meter (3-ft) rise in sea level would be enough to cover the Maldives and other low-lying island countries. The homes of between 50 and 100 million people would be inundated, a fifth of Egypt's arable land in the Nile delta would be flooded, and the impact on the people of the Bangladesh who live on thousands of alluvial islands known as "chars" would be catastrophic. Source: *Redrawn from "The Melting of the World's Ice," from* WorldWatch, *Nov./Dec. 2000, pp. 5–7. Reprinted by permission of Worldwatch Institute, www.worldwatch.org.*

Some critics believe that climate models don't adequately account for clouds and argue that global temperatures might stabilize or even decrease as the concentration of greenhouse gases increases. A hotter atmosphere, they say, would increase evaporation, sending up more water vapor, which could condense into clouds. The increased cloud cover might reflect so much sunlight that it would slow the rate at which the earth would be heated. Others contend that the increased evaporation would produce more rainfall. As it fell, the rain would cool the land and subsequently cool the air over the land.

Some scientists have gone so far as to suggest that global warming could, paradoxically, bring about the onset of another ice age. They reason that greenhouse-related temperature increases would likely warm upper latitudes sufficiently to permit heavy snowfall to commence in regions—such as northern Greenland—where snow now rarely falls. An increase in upper-latitude precipitation would yield a dramatic influx of fresh water and ice subject to melting in polar oceans. That influx would dilute the salt content of oceans; major ocean currents, such as the Gulf Stream (which provides northern Europe with a relatively warm climate), would slow and so would their warming effect on northern regions. An abrupt drop in polar temperatures can bring about significant changes in world climatic patterns. Recent evidence, it is claimed, indicates that polar regions did, in fact, grow slightly warmer before the onset of the previous glacial period, just as they are warming now.

## SUMMARY

In this chapter, we examined various concepts and terms that are useful in understanding weather and climate. We identified solar energy as the great generator of the main weather elements of temperature, moisture, and atmospheric pressure. Spatial variation in these elements is caused by two elements: the earth's broad physical characteristics, such as greater solar radiation at the equator than at the poles; and local physical characteristics, such as the effect of water bodies or mountains on local weather conditions.

Climate regions help us simplify the complexities that arise from such special conditions as monsoon winds in Asia or cold ocean currents off the western coast of South America. In just a few descriptive sentences, it is possible to identify the essence of the wide variations in weather conditions that may exist at a place over the course of a year. When a person says that Seattle is in a region of marine west coast climate, for example, images of the average daily conditions communicate not only the facts of weather but also the reasons for conditions as they are. In addition, knowledge of climate tells us about the conditions under which one carries out life's daily tasks.

Weather and climate are among the building blocks that help us understand more clearly the role of humans on the surface of the earth. Although the following chapters focus mainly on the characteristics of human cultural landscapes, one should keep in mind that the physical landscape significantly affects human behavior.

## KEY WORDS

air mass 102
air pressure 92
blizzard 103
climate 88
climax community 110
convection 95
convectional precipitation 101
Coriolis effect 96
cyclone 103
cyclonic (frontal) precipitation 102
dew point 101
El Niño 108
frictional effect 97
front 102
global warming 126
greenhouse effect 126
humid continental climate 122
humid subtropical climate 120
hurricane 103
insolation 89
jet stream 97
land breeze 95
La Niña 109
lapse rate 92
marine west coast climate 121
Mediterranean climate 119
monsoon 97
mountain breeze 96
natural vegetation 110
North Atlantic drift 98
orographic precipitation 101
precipitation 99
pressure gradient force 94
reflection 91
relative humidity 101
reradiation 91
savanna 114
sea breeze 95
soil horizon 107
soil order 110
soil profile 107
soil properties 108
source region 102
steppe 118
succession 110
temperature inversion 92
tornado 104
tropical rain forest 114
troposphere 89
tundra 123
typhoon 103
valley breeze 96
weather 88

## FOR REVIEW & CONSIDERATION

1. What is the difference between *weather* and *climate?*
2. What determines the amount of *insolation* received at a given point? Does all potentially receivable solar energy actually reach the earth? If not, why?
3. How is the atmosphere heated? What is the *lapse rate* and what does it indicate about the atmospheric heat source? Describe a *temperature inversion.*
4. What is the relationship between atmospheric pressure and surface temperature? What is a *pressure gradient* and of what concern is it in weather forecasting?

5. In what ways do land and water areas respond differently to equal insolation? How are these responses related to atmospheric temperatures and pressures?
6. Draw and label a diagram of the planetary wind and pressure system. Account for the occurrence and character of each wind and pressure belt. Why are the belts latitudinally ordered?
7. What is *relative humidity?* How is it affected by changes in air temperatures? What is the *dew point?*
8. What are the three types of large-scale *precipitation?* How does each occur?
9. What are *air masses?* What is a *front?* Describe the development of a cyclonic storm, showing how it relates to air masses and fronts.
10. What factors were chiefly responsible for today's weather?
11. Describe the five factors that affect soil formation. How do the different soil horizons vary from one another? How can we distinguish one type of soil from another? To what soil order does the soil in your area belong?
12. Describe the vegetative evolutionary process known as *succession.* What is the difference between a pioneer plant community and a climax community?
13. Summarize the distinguishing temperature, moisture, vegetation, and soil characteristics of each type of climate.
14. What is the climate at Tokyo, London, São Paulo, Leningrad, and Bangkok?
15. What causes the greenhouse effect? What impact might it have on the environment?

# Selected References

**WEBSITES**

The World Wide Web has a tremendous variety of sites pertaining to geography. Websites relevant to the subject matter of this chapter appear in the "Web Links" section of the Online Learning Center associated with this book. Access it at www.mhhe.com/getis10e/.


Aguado, Edward, and James E. Burt. *Understanding Weather and Climate.* 2d ed. Upper Saddle River, N.J.: Prentice Hall, 2000.

Bair, Frank E. *Climates of States.* 5th ed. Detroit: Visible Ink Press, 2001.

Barry, R. G., and R. J. Chorley. *Atmosphere, Weather, and Climate.* 7th ed. New York: Methuen, 1998.

Danielson, E., J. Levin, and Elliot Abrams. *Meteorology.* 2d ed. Boston: McGraw-Hill, 2003.

Geiger, Rudolf, Robert Aron, and Paul Todhunter. *The Climate Near the Ground.* 6th ed. Lanham, Md.: Rowman and Littlefield, 2003.

Glantz, Michael H. *Currents of Change: El Niño's Impact on Climate and Society.* Cambridge, England: Cambridge University Press, 1996.

Gordon, Adrian, et al. *Dynamic Meteorology: A Basic Course.* 2d ed. London: Edward Arnold, 1997.

Hoyt, Douglas V., and Kenneth H. Shatten. *The Role of the Sun in Climate Change.* Oxford, England: Oxford University Press, 1997.

McGregor, Glenn R., and Simon Nieuwolt. *Tropical Climatology: An Introduction to the Climates of the Low Latitudes.* 2d ed. New York: Wiley, 1998.

Miller, Raymond, and Duane T. Gardiner. *Soils in Our Environment.* 9th ed. Upper Saddle River, N.J.: Prentice Hall, 2000.

Pielke, Roger A. *Hurricanes: Their Nature and Impact on Society.* New York: Wiley, 1998.

Schneider, Stephen H., ed. *Encyclopedia of Climate and Weather.* Oxford, England: Oxford University Press, 1996.

Simmons, I. G. *Biogeographical Processes.* London: Allen & Unwin, 1982.

Strahler, Alan H., and Arthur N. Strahler. *Physical Geography: Science and Systems of the Human Environment.* New York: Wiley, 2002.

Tarbuck, Edward J., and Frederick Lutgens. *Earth Science.* 9th ed. Upper Saddle River, N.J.: Prentice Hall, 1999.

Thompson, Russell D., and Allen Perry, eds. *Applied Climatology.* New York: Routledge, 1997.

U.S. Global Change Research Program, National Assessment Synthesis Team. *Climate Change Impacts on the United States: The Potential Consequences of Climate Variability and Change.* New York: Cambridge University Press, 2000.

The Weather Channel (television) provides information on and explanations of current weather conditions throughout the world.

*Weatherwise.* Issued six times a year by Weatherwise, Inc., 230 Nassau St., Princeton, N.J., 08540.

Wells, Neil. *The Atmosphere and the Ocean: A Physical Introduction.* 2d ed. New York: Wiley, 1997.

# The Geography of Natural Resources

**Resource Terminology**
- Renewable Resources
- Nonrenewable Resources
- Resource Reserves

**Energy Resources and Industrialization**

**Nonrenewable Energy Resources**
- Crude Oil
- Coal
- Natural Gas
- Oil Shale and Tar Sands
- Nuclear Energy
  - *Nuclear Fission*
  - *Nuclear Fusion*

**Renewable Energy Resources**
- Biomass Fuels
  - *Wood*
  - *Waste*
- Hydropower
- Solar Energy
- Other Renewable Energy Resources
  - *Geothermal Energy*
  - *Wind Power*

**Nonfuel Mineral Resources**
- The Distribution of Nonfuel Minerals
- Copper: A Case Study

**Land Resources**
- Soils
- Wetlands
- Forest Resources
  - *U.S. National Forests*
  - *Tropical Rain Forests*

**Resource Management**

**Summary**

**Key Words**

**For Review & Consideration**

**Selected References**

**A nuclear power plant in Nottingham, England.**
© David Noton/Masterfile.

The world's richest nation" was how an article in *National Geographic* described Nauru in 1976. A single-island country in the South Pacific Ocean, halfway between Hawaii and Australia, Nauru today is a wasteland whose barren and inhospitable landscape suggests a creation of science fiction. What happened between 1976 and now is that Nauru was almost totally stripped of the one natural resource that gave it value—high-grade phosphate rock laid down over millions of years—and most of the island is no longer inhabitable.

Located just south of the equator, Nauru is tiny, with an area of just 21 square kilometers (8 sq mi). A narrow coastal belt surrounds a central plateau that rises to 65 meters (213 ft) above sea level, covers 80% of the land area, and contains valuable deposits of phosphate. There is just one paved road, a loop that circles the island.

For several thousand years, Nauruans lived within the island's limits despite frequent droughts. A population of about 1000 depended on food from the sea and from the tropical vegetation. The plunder of the island began after Nauru became a colony of Germany in 1888. At the end of World War I, the League of Nations gave Great Britain, Australia, and New Zealand the right to administer the island and set up the British Phosphate Commission to run the phosphate industry. Most of the phosphate was shipped to Australia for fertilizer. When Nauru received independence in 1968, the people chose to continue the extraction, which brought in tens of millions of dollars annually and gave the country one of the highest per capita incomes in the world.

Now, however, most of the phosphate has been mined. Deposits are expected to be exhausted in just a few years. Apart from some tropical fruits, no other resources exist; virtually everything must be imported for the 12,000 inhabitants: foodstuffs, fuels, manufactured goods, machinery, building and construction materials, and even water when the desalination plant fails. Poor investments and mismanagement of a billion-dollar trust fund have plunged the country into debt. The environmental damage from strip mining has been severe. The lush tropical forests were removed to facilitate mining. For thousands of years, they had provided refuge for native and migratory birds. The plateau is now a dry, jagged wasteland of tall fossilized coral pinnacles, all that remain after the phosphate that lies between them is removed (Figure 5.1).

Nauru exemplifies the difference between a sustainable society and one that squanders resources that have accumulated over millions of years. Population numbers and economic development have magnified the extent and the intensity of human depletion of the treasures of the earth. Resources of land, of ores, and of most forms of energy are finite, but the resource demands of an expanding, economically advancing population appear to be limitless. This imbalance between resource availability and use has been a concern for more than a century, at least since the time of Malthus and Darwin, but it wasn't until the 1970s that the rate of resource depletion and the environmental degradation associated with it became a major and controversial issue.

Resources are unevenly distributed in kind, amount, and quality and do not match uneven distributions of population and demand. In this chapter, we survey the natural resources on which societies depend, their patterns of production and consumption, and the problem of managing those resources in light of growing demands and shrinking reserves.

We begin our discussion by defining some commonly used terms.

**FIGURE 5.1 Environmental devastation.** Intensive phosphate mining has left most of Nauru a wasteland. These tall coral spires stand in stark contrast to the tropical forests that once grew above them. *© Don Brice Photography.*

# RESOURCE TERMINOLOGY

A **resource** is a naturally occurring, exploitable material that a society perceives to be useful to its economic and material well-being. Willing, healthy, and skilled workers constitute a valuable resource, but without access to materials such as fertile soil or petroleum, human resources are limited in their effectiveness. In this chapter, we devote our attention to physically occurring resources, or, as they are more commonly called, *natural* resources.

The availability of natural resources is a function of two things: the physical characteristics of the resources themselves, and human economic and technological conditions. The physical processes that govern the formation, distribution, and occurrence of natural resources are determined by physical laws over which people have no direct control. We take what nature gives us. To be considered a resource, however, a given substance must be *understood* to be a resource. This is a cultural, not purely a physical, circumstance. Native Americans may have viewed the resource base of Pennsylvania as composed of forests for shelter and fuel, as well as the habitat of the game animals (another resource) on which they depended for food. European settlers viewed the forests as the unwanted covering of the resource that *they* perceived to be of value: soil for agriculture. Still later, industrialists appraised the underlying coal deposits, ignored or unrecognized as a resource by earlier occupants, as the item of value for exploitation (Figure 5.2)

Natural resources are usually recognized as falling into one of two broad classes: renewable and nonrenewable.

**FIGURE 5.2** The original hardwood forest covering these Virginia hills was removed by settlers who saw greater resource value in the underlying soils. The soils, in turn, were stripped away for access to the still more valuable coal deposits below. Resources are as a culture perceives them, though exploitation may consume them and destroy the potential of an area for alternate uses. *© Corbis/R-F Website.*

## Renewable Resources

**Renewable resources** are materials that are replaced or replenished by natural processes. They can be used over and over again; the supplies are not depleted. A distinction can be made, however, between those that are perpetual and those that are renewable only if carefully managed (Figure 5.3). **Perpetual resources** come from sources that are virtually inexhaustible, such as the sun, wind, waves, tides, and geothermal energy.

**Potentially renewable resources** are renewable if left to nature but can be destroyed if people use them carelessly. These include groundwater, soil, plants, and animals. If the rate of exploitation exceeds that of regeneration, these renewable resources can be depleted. Groundwater extracted beyond the replacement rate in arid areas may be as permanently dissipated as if it were a nonrenewable ore. Soils can be totally eroded, and an animal species may be completely eliminated. Forests are a renewable resource only if people are planting at least as many trees as are being cut.

## Nonrenewable Resources

**Nonrenewable resources** exist in finite amounts or are generated in nature so slowly that for all practical purposes the

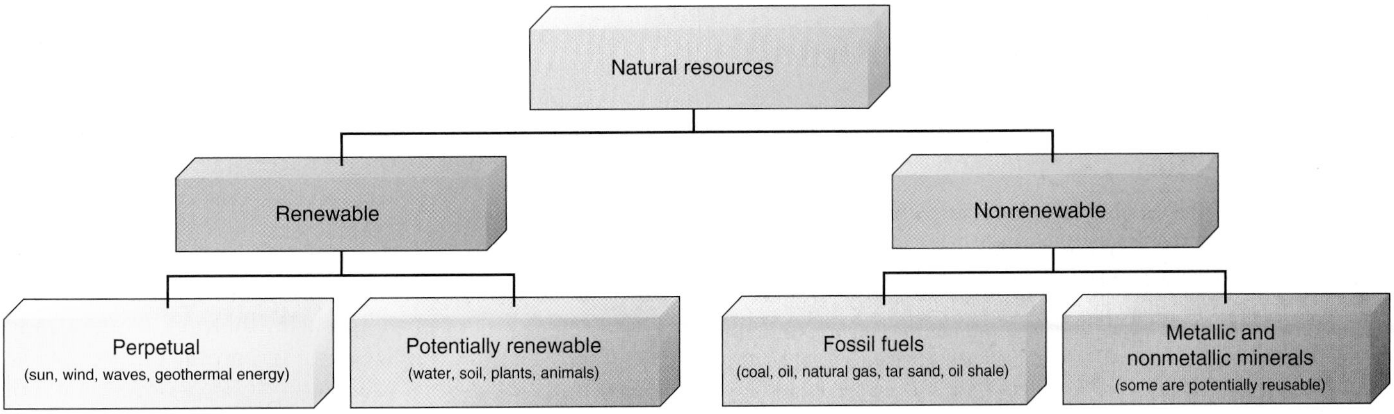


**FIGURE 5.3 A classification of natural resources.** Renewable resources can be depleted if the rate of use exceeds that of regeneration.

supply is finite. They include the fossil fuels (coal, crude oil, natural gas, oil shales, and tar sands), the nuclear fuels (uranium and thorium), and a variety of nonfuel minerals, both metallic and nonmetallic. Although the elements of which these resources are composed cannot be destroyed, they can be altered to less useful or available forms, and they are subject to depletion. The energy stored in a unit volume of the fossil fuels may have taken eons to concentrate in usable form; it can be converted to heat in an instant and be effectively lost forever.

Fortunately, many minerals can be *reused* even though they cannot be *replaced*. If they are not chemically destroyed—that is, if they retain their original chemical composition—they are potentially reusable. Aluminum, lead, zinc, and other metallic resources, plus many of the nonmetallics, such as diamonds and petroleum by-products, can be used time and time again. However, many of these materials are used in small amounts in any given object, so that recouping them is economically unfeasible. In addition, many materials are now being used in manufactured products, so that they are unavailable for recycling unless the product is destroyed. Consequently, the term *reusable resource* must be used carefully. At present, all mineral resources are being mined much faster than they are being recycled.

## What Is Energy?

People have built their advanced societies by using inanimate energy resources. **Energy**—the ability to do work—is either potential or kinetic. *Potential* energy is stored energy; when released, it is in a form that can be harnessed to do work. *Kinetic* energy is the energy of motion; all moving objects possess kinetic energy.

Assume that a reservoir contains a large amount of stored water. The water is a storehouse of potential energy. When the gates of the dam holding back the water are opened, water rushes out. Potential energy has become kinetic energy, which can be harnessed to do such work as driving a generator of electricity. No energy has been lost, it has simply been converted from one form to another.

Unfortunately, energy conversions are never complete. Not all of the potential energy of the water can be converted into electrical energy. Some potential energy is always converted to heat and dissipated to the surroundings. *Energy efficiency* is the measure of how well we can convert one form of energy into another without waste—that is, the ratio of energy that is produced to the amount of energy consumed in the production process.

## Resource Reserves

Some regions contain many resources, others relatively few. No industrialized country, however, has all the resources it needs to sustain itself. The United States has abundant deposits of many minerals, but it depends on other countries for such items as tin and manganese. The actual or potential scarcity of key nonrenewable resources makes it desirable to predict their availability in the future. We want to know, for example, how much petroleum remains in the earth and how long we will be able to continue using it.

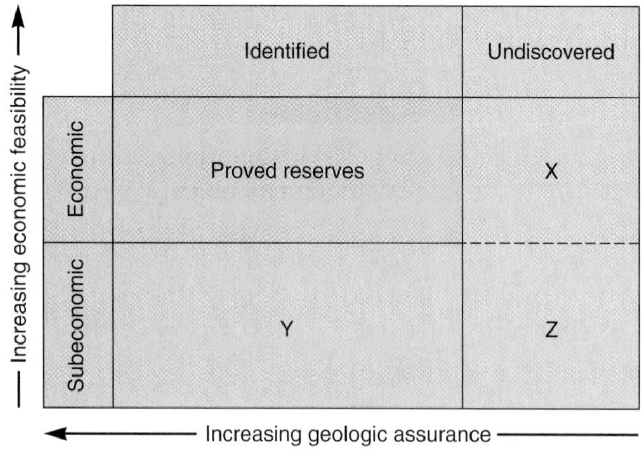


**FIGURE 5.4 The variable definition of *reserves*.** Proved or usable reserves consist of amounts that have been identified and can be recovered at current prices and with current technology. *X* denotes amounts that would be attractive economically but have not yet been discovered. Identified but not economically attractive amounts are labeled *Y*, and *Z* represents undiscovered amounts that would not be attractive now even if they were discovered. Source: *General classification of resources by the U.S. Geological Survey.*

Any answer will be only an estimate, and for a variety of reasons such estimates are difficult to make. Exploration has revealed the existence of certain deposits, but we have no sure way of knowing how many remain undiscovered. Further, our definition of what constitutes a usable resource depends on current *economic* and *technological* conditions. If they change—if, for example, it becomes possible to extract and process ores more efficiently—our estimate of reserves also changes. Finally, the answer depends in part on the rate at which the resource is being used, but it is impossible to predict future rates of use with any certainty. The current rate could drop if a substitute for the resource in question were discovered, or it could increase if population growth or industrialization placed greater demands on it.

A useful way of viewing reserves is illustrated by Figure 5.4. Assume that the large rectangle includes the total stock of a particular resource, all that exists of it in or on the earth. Some deposits of that resource have been discovered; they are shown in the left-hand column as "identified" amounts. Deposits that have not been located are called "undiscovered" amounts. Deposits that are economically recoverable with current technology are at the top of the diagram, whereas those labeled "subeconomic" are not attractive for any of a number of reasons (the concentration is not rich enough, it would require expensive treatment after mining, it is not accessible, and so on).

We can properly term **proved,** or **usable, reserves**—quantities of a resource that can be extracted profitably from known deposits—only the portion of the rectangle indicated by the pink tint. These are the amounts that have been identified and that can be recovered under existing economic and operating conditions. If new deposits of the resource are discovered, the reserve category will shift to the right; improved technology or increased prices for the product can shift the reserve boundary downward. An ore that was not considered a reserve in 1950, for example, may become a reserve in 2010 if ways are found to extract it economically.

## Energy Resources and Industrialization

Although people depend on a wide range of resources contained in the biosphere, energy resources are the "master" natural resources. We use energy to make all other resources available (see "What Is Energy?" p. 136). Without the energy resources, all other natural resources would remain in place, unable to be mined, processed, and distributed. When water becomes scarce, we use energy to pump groundwater from greater depths or to divert rivers and build aqueducts. Likewise, we increase crop yields in the face of poor soil management by investing energy in fertilizers, herbicides, farm implements, and so on. By the application of energy, the conversion of materials into commodities and the performance of services far beyond the capabilities of any single individual are made possible. Further, the application of energy can overcome deficiencies in the material world that humans exploit. High-quality iron ore may be depleted, but by massive applications of energy, the iron contained in rocks of very low iron content can be extracted and concentrated for industrial use.

Energy can be extracted in a number of ways. Humans themselves are energy converters, acquiring their fuel from the energy contained in food. Our food is derived from the solar energy stored in plants via photosynthesis. In fact, nearly all energy sources are really storehouses for energy originally derived from the sun. Among them are wood, water, the wind, and the fossil fuels. People have harnessed each of these energy sources, to a greater or lesser degree. Preagricultural societies depended chiefly on the energy stored in wild plants and animals for food, although people developed certain tools (such as spears) and customs to exploit the energy base. For example, they added to their own energy resources by using fire for heating, cooking, and clearing forest land.

Sedentary agricultural societies developed the technology to harness increasing amounts of energy. The domestication of plants and animals, the use of wind to power ships and windmills, and the use of water to turn waterwheels all expanded the energy base. For most of human history, wood was the predominant source of fuel, and even today at least half the world's people depend largely on fuel wood for cooking and heating.

However, it was the shift from renewable resources to those derived from nonrenewable minerals, chiefly fossil fuels, that sparked the Industrial Revolution, made possible the population increases discussed in Chapter 6, and gave population-supporting capacity to areas far in excess of what would be possible without inanimate energy sources. The enormous increase in individual and national wealth in industrialized countries has been built in large measure on an economic base of coal, oil, and natural gas. They are used to provide heat, to generate electricity, and to run engines.

Energy consumption goes hand in hand with industrial production and with increases in personal wealth. In general, the greater the level of energy consumption, the higher the gross national income per capita. This correlation of energy consumption with economic development points out a basic disparity between societies. Countries that can afford to consume great amounts of energy continue to expand their economies and to increase their levels of living. Those without access to energy, or those unable to afford it, see the gap between their economic prospects and those of the developed countries growing ever greater.

## Nonrenewable Energy Resources

Crude oil, natural gas, and coal have formed the basis of industrialization. Figure 5.5 shows past energy consumption patterns in the United States. Burning wood supplied most energy needs until about 1885, by which time coal had risen to prominence. The proportion of energy needs satisfied by burning coal peaked about 1910; from then on, oil and natural gas were increasingly substituted for coal. The graph shows the absolute dominance of the fossil fuels as

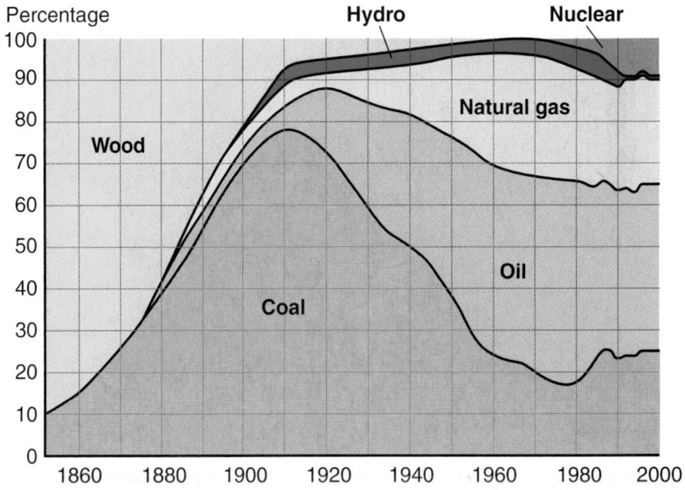


**FIGURE 5.5 Sources of energy in the United States, 1850–2000.** The fossil fuels provided about 90% of the energy supply in 2000.

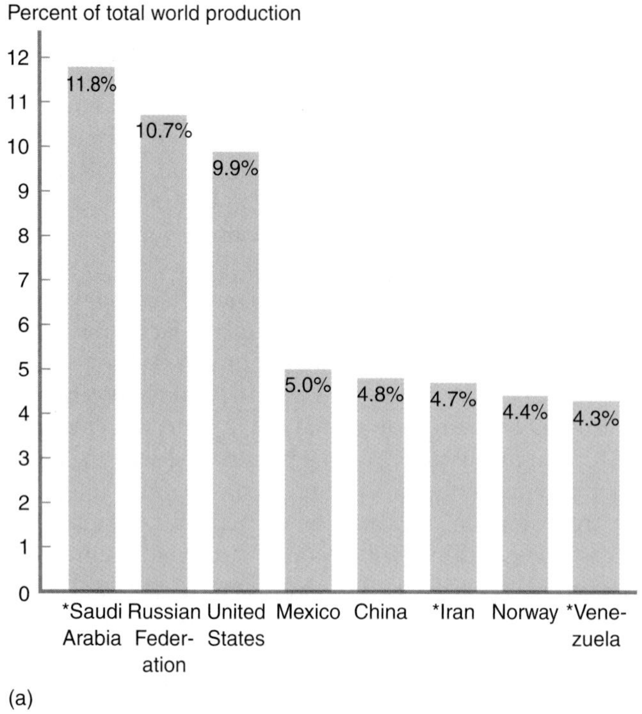


**FIGURE 5.6** (a) **Leading producers of oil.** These eight countries produced 56% of the world's oil in 2002. Members of OPEC are marked with an asterisk. All told, OPEC countries accounted for 38% of oil production in 2002. (b) **Leading consumers of oil.** Only the five countries shown here consumed 3% or more of the world's oil in 2002. Sources: *Data from* The BP Statistical Review of World Energy, *June 2003.*

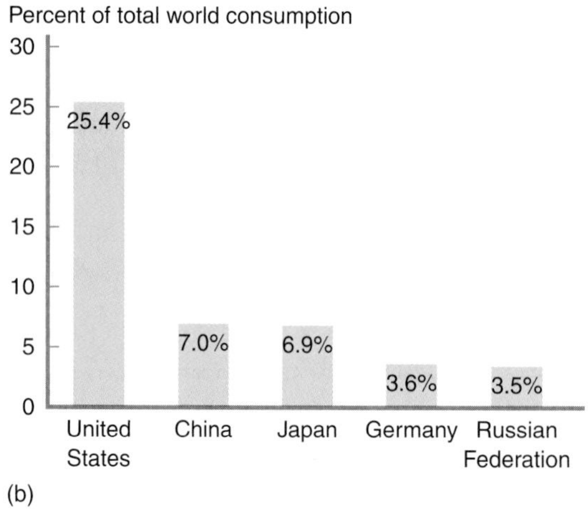


energy sources during the last 100 years. In 2000, they accounted for about 90% of national energy consumption.

## Crude Oil

Today, crude oil and its by-products account for almost 40% of the commercial energy (excluding wood and other traditional fuels) consumed in the world. Some world regions and industrial countries have a far higher dependency. Figure 5.6 shows the main producers and consumers of crude oil (also called petroleum).

After it is extracted from the ground, crude oil must be refined. The hydrocarbon compounds are separated and distilled into waxes and tars (for lubricants, asphalt, and many other products) and various fuels. Petroleum rose to importance because of its combustion characteristics and its adaptability as a concentrated energy source for powering moving

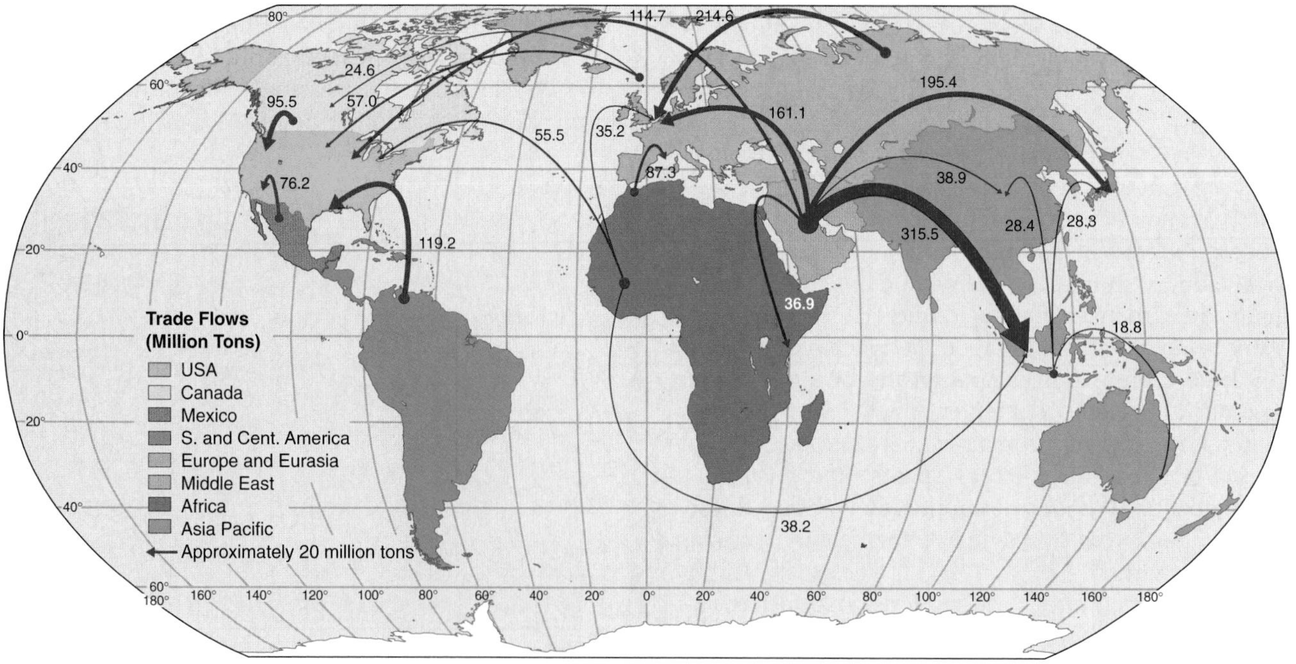


**FIGURE 5.7** **International crude oil flow by sea, 2002.** Note the dominant position of the Middle East in terms of oil exports. The arrows indicate origin and destination, not specific routes. The line widths are proportional to the volume of movement. In 2002, the United States imported three-fifths of the oil it used. Source: *Data from* The BP Statistical Review of World Energy, *June 2003.*

vehicles. Although there are thousands of oil-based products, fuels such as home heating oil, diesel and jet fuels, and gasoline are the major output of refineries. In the United States, transport fuels account for two-thirds of all oil consumption.

As Figure 5.7 shows, oil from a variety of production centers flows, primarily by water, to the industrially advanced countries. Note that the United States imports oil from a number of regions. The other major importers, Western Europe and Japan, import chiefly Middle Eastern oil.

The efficiency of pipelines, supertankers, and other modes of transport and the low cost of oil helped create a world dependence on that fuel, even though coal was still generally and cheaply available. The pattern is aptly illustrated by American reliance on foreign oil. For many years, United States oil production had remained at about the same level, 8 to 9 million barrels per day. Between 1970 and 1977, however, as domestic supplies became much more expensive to extract, consumption of oil from foreign sources increased dramatically, until almost half of the oil consumed nationally was imported. The dependence of the United States and other advanced industrial economies on imported oil gave the oil-exporting countries tremendous power, reflected in the soaring price of oil in the 1970s. During that decade, oil prices rose dramatically, largely as a result of the strong market position of the Organization of Petroleum Exporting Countries (OPEC).

Among the side effects of the oil "shocks" of 1973–1974 and 1979–1980 were worldwide recessions, large net trade deficits for oil importers, a reorientation of world capital flows, and a depreciation of the U.S. dollar against many other currencies. On the positive side, the soaring oil prices of the 1970s triggered oil exploration in non-OPEC countries, improvements in oil-drilling technology, and a search for alternative energy sources. Perhaps most important, for a number of years, they diminished total energy demand, partly because of the recession and partly because the high prices fostered conservation. Industrial countries have learned to use much less oil for each unit of output. In general, cars, planes, and other machines are more energy efficient than they were in the 1970s, as are industries and buildings constructed in recent years.

Since 1985, however, both global production and consumption of oil have increased steadily. And the United States, which satisfied 69% of its oil needs by domestic production in each year between 1982 and 1986, has since then relied increasingly on imports. By 2002, the United States depended on foreign sources for 60% of the oil it consumed annually (see "Fuel Economy and CAFE Standards," p. 140).

It is particularly difficult to estimate the size of oil reserves. Not only are estimates constantly revised as oil is extracted and new reserves are located, but many governments tend to maintain some secrecy about the sizes of reserves, understating official national estimates. Nonetheless, it is clear that oil is a finite resource and that oil reserves are very unevenly distributed among the world's countries (Figure 5.8). Slightly more than 1000 billion barrels are classified as proved reserves, and another 900 billion are thought to exist in undiscovered reservoirs. If all the oil

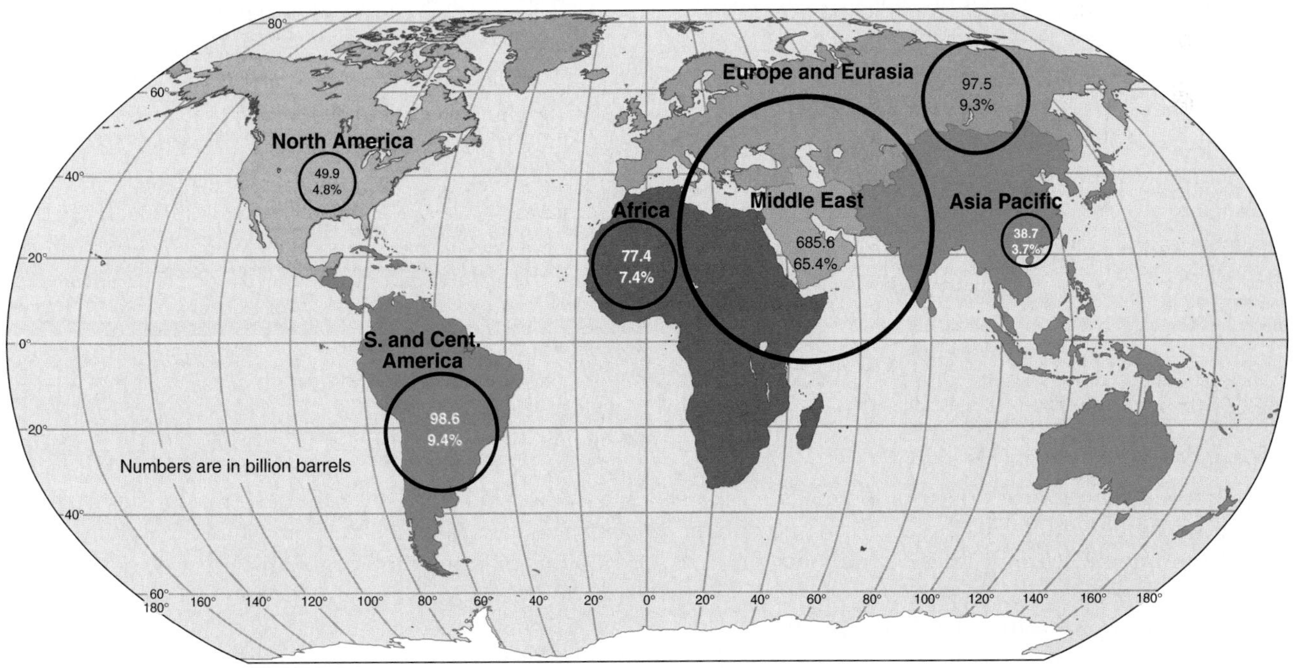


**FIGURE 5.8 Regional shares of proved oil reserves, in billions of barrels, January 1, 2003.** Oil supplies are finite, and some countries are likely to deplete their reserves in the foreseeable future. The United States consumes about one-quarter of the world's oil supply but possesses less than 3% of the world's reserves. Middle Eastern countries contain about two-thirds of the proved reserves (Saudi Arabia alone has 25% of the world's oil reserves). At the beginning of 2003, the world's proved oil reserves were estimated to be 1048 billion barrels, a figure that tends to rise over time with new methods of locating and extracting oil deposits and changes in price. In 2002, global consumption of oil was about 28 billion barrels per year.

Source: *Data from* The BP Statistical Review of World Energy, *June 2003.*

GEOGRAPHY & Public Policy

## Fuel Economy and CAFE Standards

The United States is a crude-oil junkie, dependent on daily fixes of petroleum. On average, Americans consume almost 19 million barrels of oil a day. That is equivalent to 2.9 gallons per person each day, or about 1000 gallons per person annually. What are some of the implications of this dependence?

Consider the data in the following table.

| Country | 2002 Proved Reserves (Billion Barrels) | 2002 Production (Billion Barrels) | 2002 Consumption (Billion Barrels) |
|---|---|---|---|
| United States | 30.0 | 2.8 | 7.2 |
| Canada | 6.5 | 1.1 | 0.7 |
| Mexico | 26.9 | 1.3 | 0.6 |

Notice the imbalance between yearly production and consumption for the United States. In contrast to its hemispheric neighbors, Canada and Mexico, the United States consumes far more oil than it produces. At current rates of consumption, and assuming no imports, the proved reserves would meet domestic demand for only five years. Americans continue to drive their cars, and their manufacturing plants continue to turn out a wide range of petroleum-based products, only because the country imports almost 10 million barrels of oil a day. That is, the United States relies on foreign sources to meet more than half its crude-oil needs.

About 40% of the oil used in the United States is burned as fuel in private automobiles and light trucks. When the price of oil tripled in the mid-1970s, Congress set new standards for fuel efficiency, and Americans began buying smaller cars, helping bring down the per capita consumption of gasoline. Between 1973 and 1985, the average fuel efficiency of American automobiles increased from 13.1 miles per gallon (mpg) to 17.9 mpg. That increase alone cut gasoline consumption by 20 billion gallons per year, lowering oil imports by 1.3 million barrels of oil per day.

As the price of gasoline declined in the 1990s, however, Americans reverted to past practices, buying large, non-fuel-efficient vehicles, such as sport utility vehicles (SUVs), minivans, and light trucks. They have become so popular in the United States that they now account for more than half of all passenger vehicles sold in the country. Unfortunately, big SUVs are gas guzzlers, and their popularity has driven down the average gas mileage (fuel-economy) ratings of vehicles driven in America. Despite advances in fuel-saving technology, the average fuel economy of new vehicles is now lower than it was in 1988.

The fuel efficiency of automobiles varies widely. The most efficient vehicles get more than 60 miles per gallon, while the least efficient average only 10. The high fuel consumption of large SUVs not only exacerbates U.S. dependence on foreign sources of oil but also has negative environmental impacts. If the average fuel efficiency of the vehicle fleet increased just 3 mpg, the United States would save 1 million barrels of oil per day, out of the roughly 10 million barrels it imports. Automobiles account for about 20% of the country's carbon dioxide emissions, and the more fuel a vehicle consumes, the greater its emissions. A car that gets 26 miles to the gallon has only half the emissions of one that averages 13 mpg. Every gallon of gasoline burned puts about 13 pounds of carbon dioxide into the atmosphere, contributing to the greenhouse effect and global warming. The nitrogen oxides and hydrocarbons emitted by vehicles contribute to acid rain and ozone smog (topics discussed in Chapter 12), respectively.

One way to address these environmental concerns and simultaneously reduce American dependence on imported oil would be to increase energy efficiency. Current federal *Corporate Average Fuel Economy (CAFE)* standards require each automaker's fleet of new passenger cars to average 27.5 miles per gallon and its light truck fleet (pickups, SUVs, and minivans) 20.7 mpg. The less-stringent fuel economy standard for light trucks originally was intended to avoid penalizing builders, farmers, and others who relied on pickup trucks for their work, but it opened the way for automakers to

could be extracted from the known reserves, and if the current rate of production holds, the proved reserves would last only about 40 years. The ratio of production to reserves, however, gives some Middle Eastern countries more than a century of pumping at current rates before their oil fields run dry.

For more than 40 years, there have been predictions that the world would soon run out of oil. Pessimists still believe that global oil production could peak by 2010, whereas optimists think we will rely on oil for most if not all of this century. They argue that technological advances in exploration and production, such as deep-water drilling and enhanced recovery, will so significantly increase the amount of oil that can be recovered from beneath the ground that we needn't worry about gas pumps running dry.

Whereas it was once thought that offshore oil existed only in shallow waters, current thinking is that enormous amounts of oil lie thousands of feet below sea level off the Gulf of Mexico, Brazil, and West Africa. For several years now, a number of oil companies have pumped oil from the Gulf of Mexico at depths exceeding 1000 meters (3280 ft).

Even more promising is the possibility of recovering more oil from existing reservoirs. Currently, on average, only 30% to 35% of the oil in a reservoir is brought to the surface; most of the oil remains unrecovered. Oil industry optimists believe that enhanced recovery techniques (injecting

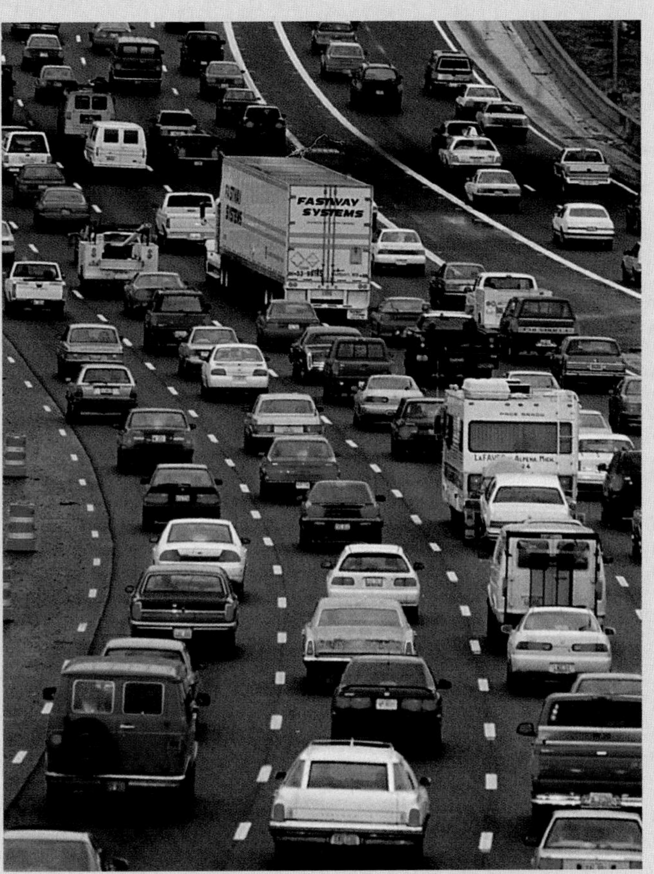

© J. Crawford/The Image Works.

replace station wagons and sedans with vehicles that met the definition of a light truck: SUVs and minivans.

Proponents of raising the CAFE standards note that the technologies to boost automobile efficiency significantly without sacrificing safety already exist. They point out that Japanese standards mandate 30.3 mpg, and European regulations an even higher 33.0 mpg. Increased energy efficiency is the quickest and cheapest path to reduced fuel consumption and a cleaner and healthier world, they argue. Additionally, they contend that SUVs endanger other motorists because they limit visibility and because they cause more damage in a crash.

Automobile industry officials, however, point out that they offer a variety of vehicles, some more fuel efficient, some less, and that people are free to choose the model they like. Most Americans, they say, are more concerned about automotive safety, comfort, and performance than about fuel economy. Some buyers cite the safety of SUVs. As one SUV owner noted, "It's a very safe vehicle. I bought this thing because it was a big iron tanker that wouldn't bend. You get hit by something and don't feel it." Others value the space a minivan or an SUV offers: "When you have a boat to haul and a family to transport, you don't have a lot of choice; you need large capacity."

**QUESTIONS TO CONSIDER**

1. With which, if any, of the following statements do you agree or disagree? Why?
   "Allowing the American free-market system to meet America's energy needs hasn't worked. The automobile industry cannot be trusted to regulate itself. Automakers won't improve the fuel economy of their fleets unless the government forces them to."
   "Government regulation isn't the way to go. Maybe $3-per-gallon gasoline will close the SUV/minivan mileage loophole."
   "We can't force people to buy more efficient vehicles. We have to balance environmental concerns with what customers want, and many of them favor bigger engines and vehicles more than fuel efficiency."
2. Do you think Congress should raise CAFE standards, requiring automakers to improve vehicle mileage?
3. Should SUVs and minivans continue to be subject to less stringent regulations than other passenger cars? Why or why not?
4. Large SUVs that weigh from 8,500 to 10,000 pounds, such as the Hummer H2, Ford Excursion, and Chevrolet Suburban, do not have to comply with the CAFE standards, despite the fact that they are used primarily as passenger vehicles. Do you think they should be exempt from the fuel standards? Why or why not?

water, gases, or chemicals into wells to force out more oil) will make 60% to 70% of the oil in a reservoir recoverable.

Finally, although most geologists agree that there are few huge oil fields left to be discovered, some of these have yet to be tapped in a major way. Industry analysts think significant increases in production are likely to come from Russia and a number of the countries of the former Soviet Union (Kazakhstan, Tajikistan, Uzbekistan, Turkmenistan, Kyrgyzstan, and Azerbaijan). Kazakhstan, for example, is developing large new oil fields and constructing new pipelines out of the Caspian basin, and Russia has begun to develop huge offshore oil fields in its far east, near Sakhalin Island in the Sea of Okhotsk.

## Coal

Coal was the fuel basis of the Industrial Revolution. From 1850 to 1910, the proportion of U.S. energy supplied by coal rose from 10% to almost 80%. Although the consumption of coal declined as the use of petroleum expanded, coal remained the single most important domestic energy source until 1950 (see Figure 5.5, p. 137).

Although coal is a nonrenewable resource, world supplies are so great that its resource life expectancy may be measured in centuries, not in the much shorter spans usually cited for oil and natural gas. The United States alone possesses more than 240 billion tons of coal considered potentially mineable on an economic basis with existing technology. At current production levels, these demonstrated reserves would be sufficient to meet the domestic demand for coal for another 2.5 centuries.

Worldwide, the most extensive deposits are concentrated in the middle latitudes of the Northern Hemisphere, as shown in Figure 5.9. Two countries, China and the United States, have dominated world coal production in recent years, accounting for half the coal produced in the world. Since 1990, the use of coal worldwide has remained steady, but distinct regional variations in its use are evident. Coal production has risen in the United States but has fallen dramatically in Europe and the countries of the former Soviet Union, as governments have discontinued subsidies to the industry. The use of coal continues to grow in many Asian countries (India, Indonesia, South Korea, Taiwan, and Japan), but in China both the production and consumption of coal peaked in 1996 and have declined each year since then. In the United States and other industrialized countries, coal is used chiefly for electric power generation and to make coke for steel production. In less developed countries, coal is widely used for home heating and cooking, as well as to generate electricity and fuel factories.

Coal is not a resource of constant quality. It ranges from lignite (barely compacted from the original peat) through bituminous coal (soft coal) to anthracite (hard coal), each *rank* reflecting the degree to which organic material has been transformed. Anthracite has a fixed carbon content of about 90% and contains very little moisture. Conversely, lignite has the highest moisture content and the lowest amount of elemental carbon, and thus the lowest heat value. About half of the demonstrated reserve base in the United States is bituminous coal, concentrated primarily in the states east of the Mississippi River.

Besides rank, the *grade* of a coal, which is determined by its content of waste materials (particularly ash and sulfur), helps to determine its quality. Good-quality bituminous coals with the caloric content and the physical properties suitable for producing coke for the steel industry are decreasingly available readily and are increasing in cost. Anthracite, formerly a dominant fuel for home heating, is now much more expensive to mine and finds no ready industrial market. The Schuylkill anthracite deposits of eastern Pennsylvania are discussed as a special type of resource region in Chapter 13.

The value of a coal deposit is determined not only by its rank and grade but also by its accessibility, which depends on the thickness, depth, and continuity of the coal seam and its inclination to the surface. Much coal can be mined relatively cheaply by open pit (surface) techniques, in which huge shovels strip off surface material and remove the exposed seams. Much coal, however, is available only by expensive and more dangerous shaft mining, as in Appalachia and most of Europe. In spite of their generally lower heating value, western U.S. coals are now attractive because of their low sulfur content. They do, however, require expensive transportation to markets or high-cost transmission lines if they are used to generate electricity for distant consumers (Figure 5.10).

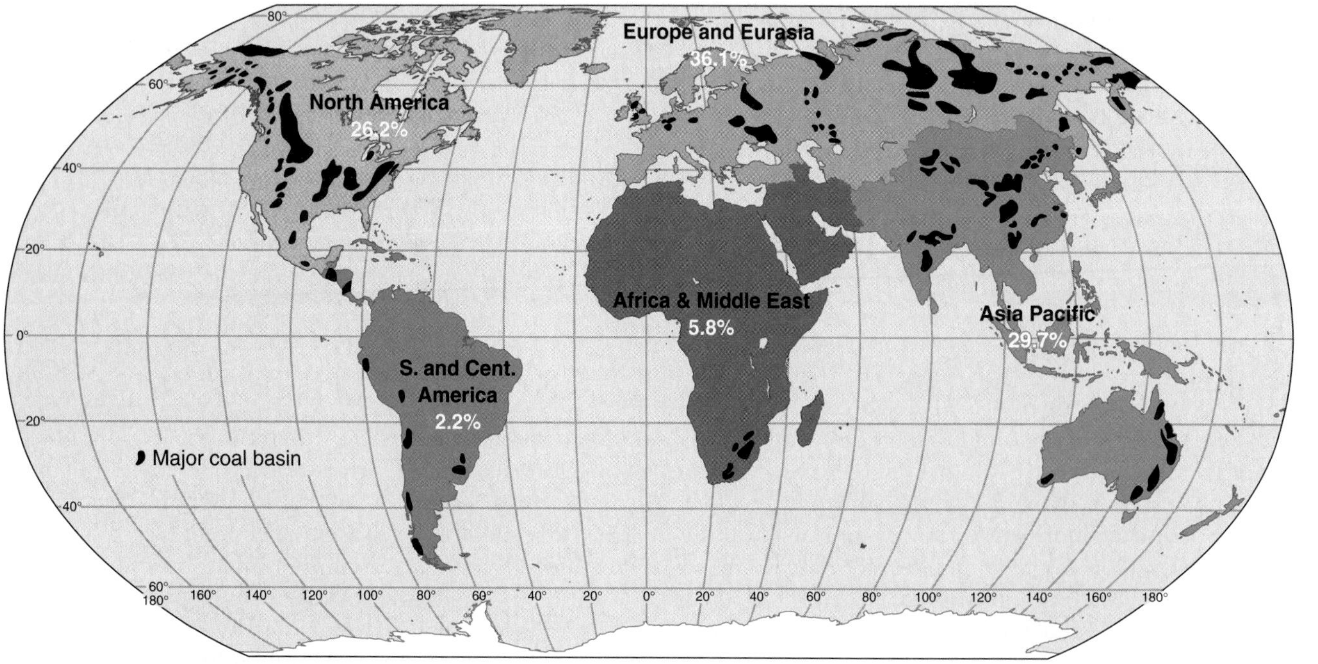


**FIGURE 5.9 Regional shares of proved coal reserves, January 1, 2003.** Major coal basins are concentrated in the Northern Hemisphere. Despite their large size, Africa and South America have very little coal. In 2002, the United States was the world's largest coal producer and consumer, followed by China. Nearly all of the reserves in the "Asia Pacific" region are in just three countries: China (11.6%), Australia (8.3%), and India (8.6%).
Source: *Reserve data from* The BP Statistical Review of World Energy, *June 2003.*

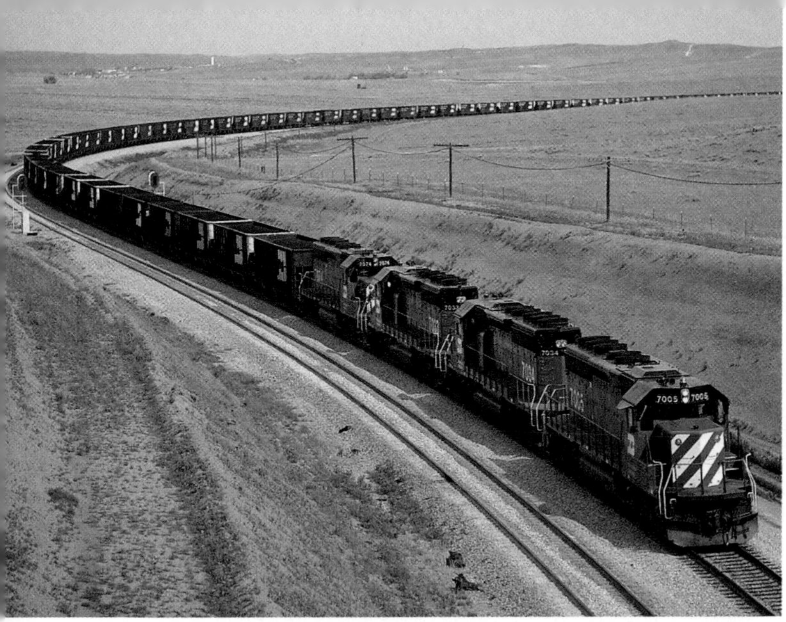

**FIGURE 5.10** Long-distance transportation adds significantly to the cost of low-sulfur western U.S. coals because they are remote from eastern U.S. markets. To minimize these costs, unit trains carrying only coal engage in a continuous shuttle movement between western strip mines and eastern utility companies. *© Bruce McAllister/Getty Images.*

The ecological, health, and safety problems associated with the mining and combustion of coal must also be figured into its cost. The mutilation of the original surface and the acid contamination of lakes and streams associated with the strip mining and burning of coal are partially controlled by environmental protection laws, but these measures add to the costs. Eastern U.S. coals have a relatively high sulfur content, and costly techniques for the removal of sulfur and other wastes from stack gases are now required by most industrial countries, including the United States.

The cost of moving coal influences its patterns of production and consumption. Coal is bulky and is not as easily transported as nonsolid fuels. As a rule, coal is usually consumed in the general vicinity of the mines. Indeed, the high cost of transporting coal induced the development of major heavy industrial centers directly on coal fields—for example, Pittsburgh, the Ruhr, the English Midlands, and the Donets district of Ukraine.

## Natural Gas

Coal is the most abundant fossil fuel, but natural gas has been called the nearly perfect energy source. It is a highly efficient, versatile fuel that requires little processing and is environmentally benign. Of the fossil fuels, natural gas (which is mainly methane) has the least impact on the environment. It burns cleanly. The chemical products of burned methane are carbon dioxide and water vapor, which are not pollutants, although they are greenhouse gases.

As Figure 5.5 indicates, the 20th century saw an appreciable growth in the proportion of U.S. energy supplied by gas. In 1900, it accounted for about 3% of the national energy supply. By 1980, the figure had risen to 30%, but then it declined to about 25% by 2000. The trend in the rest of the world has been in the opposite direction. Global production and consumption increased significantly after the oil shock of 1973–1974 and by 2002 had nearly doubled, accounting for almost 25% of global energy consumption.

Most gas is used directly for industrial and residential heating. In fact, gas has overtaken both coal and oil as a house-heating fuel, and more than half of the homes in the United States are now heated by gas. A portion is also used in electricity-generating plants, and some is chemically processed into products as diverse as motor fuels, plastics, synthetic fibers, and insecticides.

Very large natural gas fields were discovered in Texas and Louisiana as early as 1916. Later, additional large deposits were found in the Kansas–Oklahoma–New Mexico region. At that time, the south-central United States was too sparsely settled to make use of the gas, and in any case, it was oil, not gas, that was being sought. Many wells that produced only gas were capped. Gas found in conjunction with oil was vented or burned at the wellhead as an unwanted by-product of the oil industry. The situation changed only in the 1930s, when pipelines were built to link the southern gas wells with customers in Chicago, Minneapolis, and other northern cities.

Like oil, natural gas flows easily and cheaply by pipeline. Unlike oil, however, gas does not move as freely in international trade by sea (Figure 5.11). Transoceanic shipment involves costly equipment for liquefying the gas by cooling it to –126°C (–260°F), for vessels that can contain the liquid under appropriate temperature conditions, and for regasification facilities at the destination port. **Liquefied natural gas (LNG)** is extremely hazardous, because the mixture of methane and air is explosive. Although the United States imports some LNG, chiefly from Venezuela, most gas is transferred by pipeline. In the United States, the pipeline system is more than 1.6 million kilometers (1 million mi) long.

Like other fossil fuels, natural gas is nonrenewable; its supply is finite. Estimates of reserves are difficult to make because they depend on what customers are willing to spend for the fuel, and estimates have risen as the price of gas has increased. Further complicating the estimate of supplies is uncertainty about potential gas resources in unusual types of geologic formations. These include tight sandstone formations, deep (below 6000 meters, or 20,000 ft) geologic basins, and shale and coal beds.

Worldwide, just two regions contain two-thirds of the proved gas reserves: Russia (31%) and the Middle East (36%) (Figure 5.12). The remaining 33% is divided roughly equally among North America, Western Europe, Africa, Asia, and Latin America, each of which has from 4% to 8% of the total. The gas in the proved reserves would last about 65 more years at current production rates, but developing countries, particularly in South and Southeast Asia, may well have undiscovered deposits that could add significantly to the life expectancy of world reserves if they were developed.

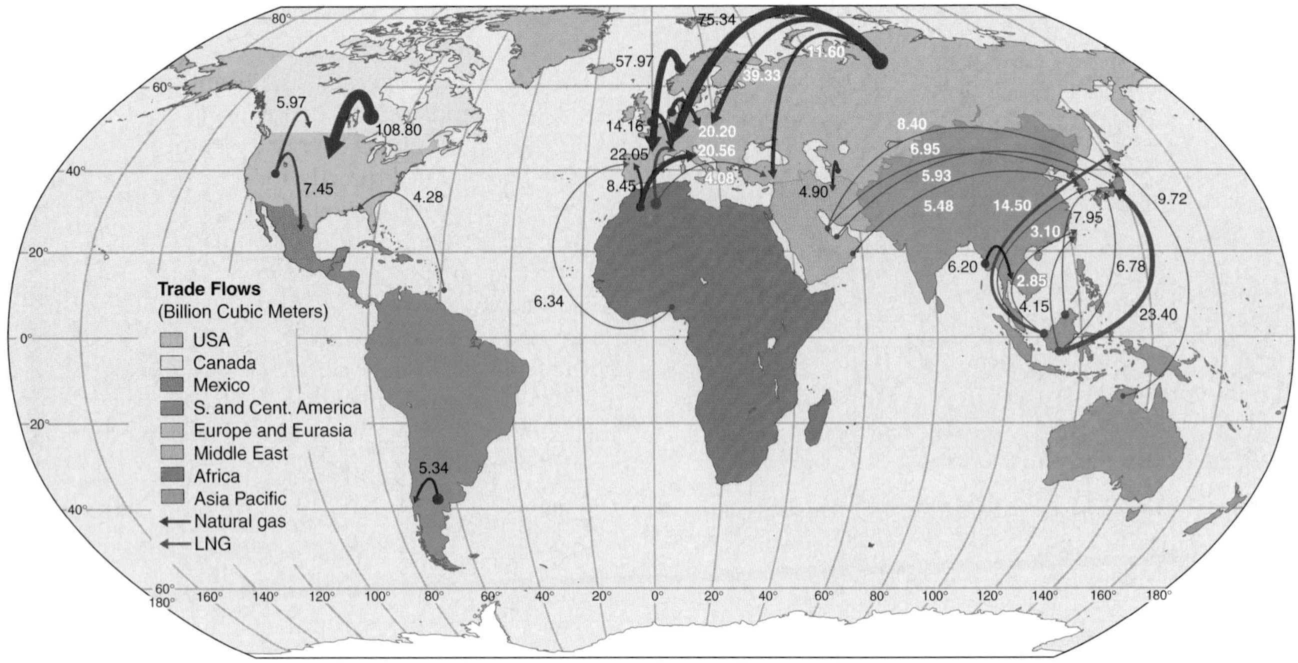


**FIGURE 5.11** (a) **Worldwide trade flows of natural gas, 2002.** Although most natural gas flows by pipeline, diminishing supplies of natural gas in many developed countries, the discovery of large reserves of natural gas in remote regions, and a reduction in the costs associated with trade in liquefied natural gas (LNG) are combining to make LNG more attractive. In 2002, Japan, South Korea, and Taiwan accounted for almost 75% of the world trade in LNG, but that pattern may shift during the next decade. Source: *Data from* The BP Statistical Review of World Energy, *June 2003.*

In the United States, the Texas-Louisiana and Kansas–Oklahoma–New Mexico regions account for about 90% of the domestic natural gas output, but there are thought to be gas deposits beneath almost all states. In addition, many offshore areas are known to contain gas. Potential Alaskan reserves are estimated to be at least twice as large as today's proved reserves in the rest of the country, containing enough gas to heat all of the houses in the country for a decade. If the technology necessary to produce gas from unconventional sources is developed, gas reserves may be sufficient for another century. Of course, these less accessible supplies will be more costly to develop, and hence more expensive.

## Oil Shale and Tar Sands

A similar situation affects the prospects of the extraction of oil from **oil shale,** fine-grained rock containing organic material called *kerogen.* A tremendous potential reserve of hydrocarbon energy, the rocks involved are not shales but calcium and magnesium carbonates more similar to limestone than to shale, and the hydrocarbon, kerogen, is not oil but a waxy, tarlike substance that adheres to the grains of carbonate material. The crushed rock is heated to a temperature high enough (more than 480°C, or 900°F) to decompose the kerogen, releasing a liquid oil product, *shale oil.*

World reserves of oil shale are enormous. Known deposits estimated to contain billions of barrels of shale oil are found in the United States, Brazil, Russia, China, and Australia (Figure 5.13). The richest deposits in the United States are in the Green River Formation near the junction of Colorado, Utah, and Wyoming. They contain enough oil to supply the needs of the United States for another century, and in the 1970s were thought to be the answer to national energy self-sufficiency. Several billion dollars were invested in oil shale research and development operations in the Piceance Basin near Grand Junction, Colorado, but as oil prices fell in the 1980s, interest in the projects waned. The last plant, that at Parachute Creek, Colorado, was abandoned in 1991.

Another potential source of petroleum liquids is **tar sand,** sandstone saturated with a viscous, high-carbon petroleum called *bitumen.* Global tar-sand resources are thought to be many times larger than conventional oil resources, containing several trillion barrels of oil, most of it in Canada. Four major deposits in Alberta are estimated to contain as much as 1 trillion barrels of bitumen. The Athabasca deposit is the largest; it supplies about 10% of Canada's annual oil demand (Figure 5.14). Tar-sand deposits are also found in Venezuela, Trinidad, Russia, and Utah.

Producing oil from either oil shale or tar sands requires high capital outlays and carries substantial environmental costs. It disturbs large areas of land and produces enormous amounts of waste. Nevertheless, because they exist in vast amounts, oil shale and tar sands could be plentiful sources of gaseous and liquid fuels. Unlike nuclear energy or most renewable energy sources (such as solar or hydroelectric

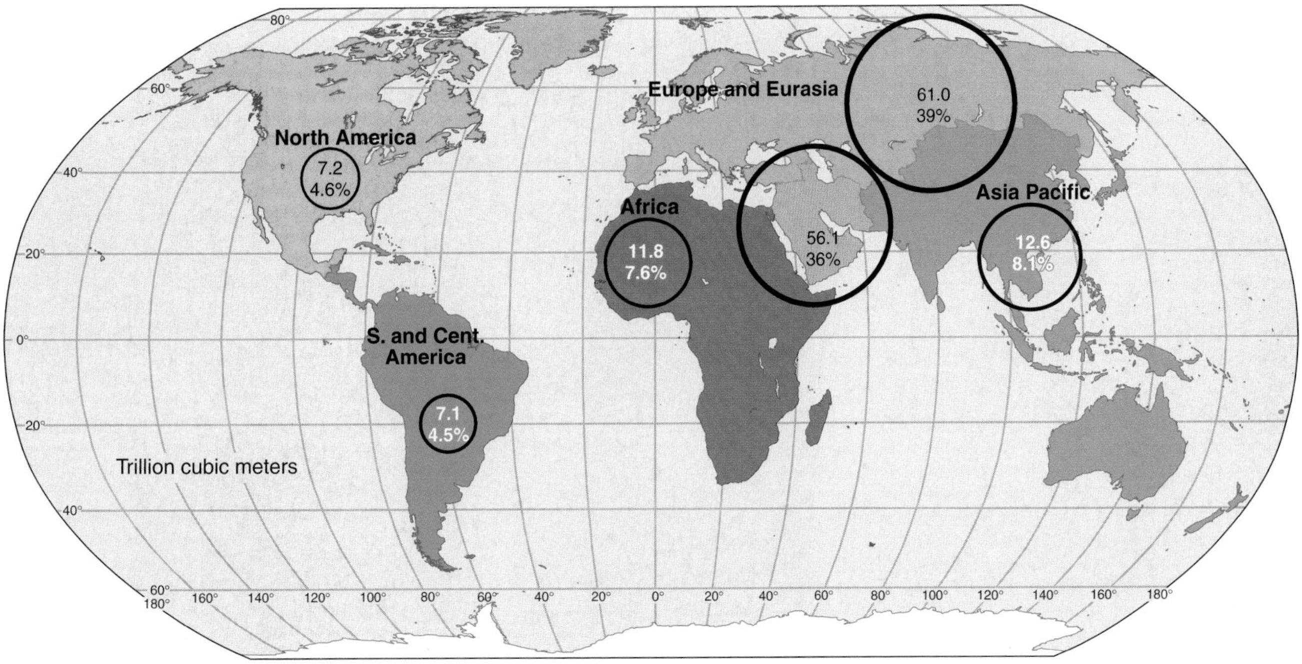


**FIGURE 5.12 Proved natural gas reserves, January 1, 2003.** Russia contains the largest natural gas reserves, about 30% of the world total (and more than 9 times the size of U.S. reserves). Major reserves also exist in the countries of the Middle East. Source: *Data from* The BP Statistical Review of World Energy, *June 2003.*

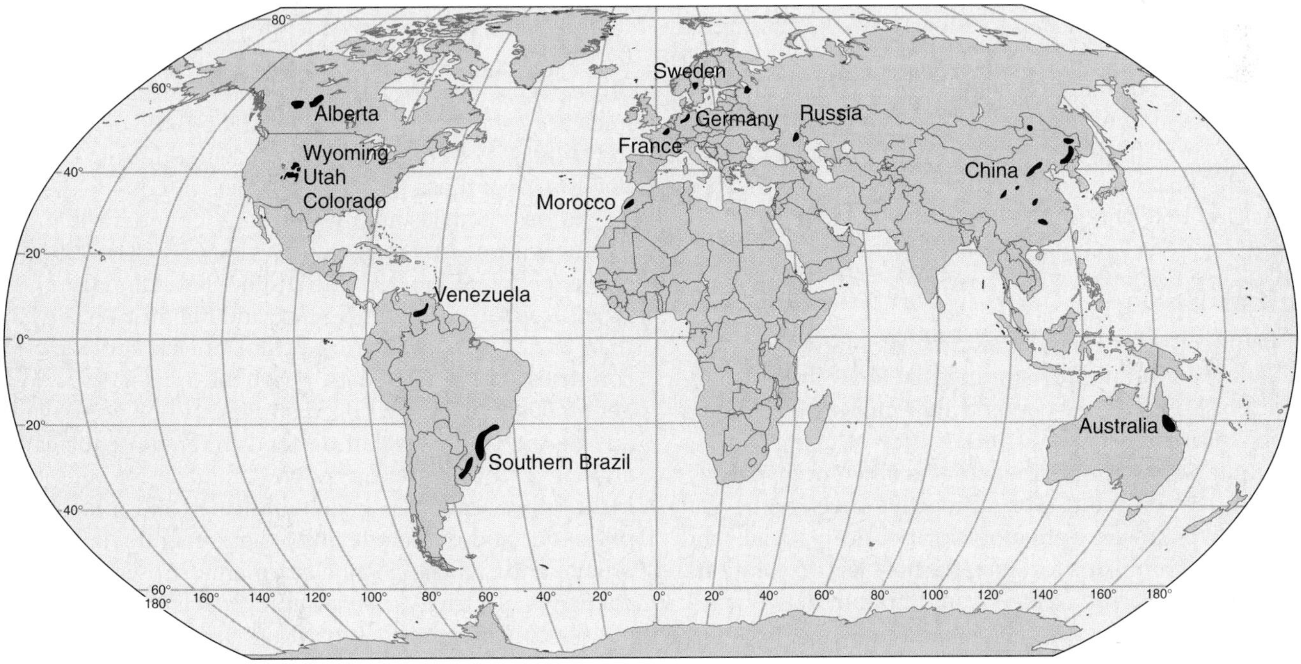


**FIGURE 5.13 Oil shale deposits.** The United States contains about two-thirds of the world's known supply of oil shale, with the richest deposits located in the Green River formation. Oil shale is unlikely to become an important resource until ways are found to process it economically and to solve the waste-disposal and land-reclamation problems its mining poses.

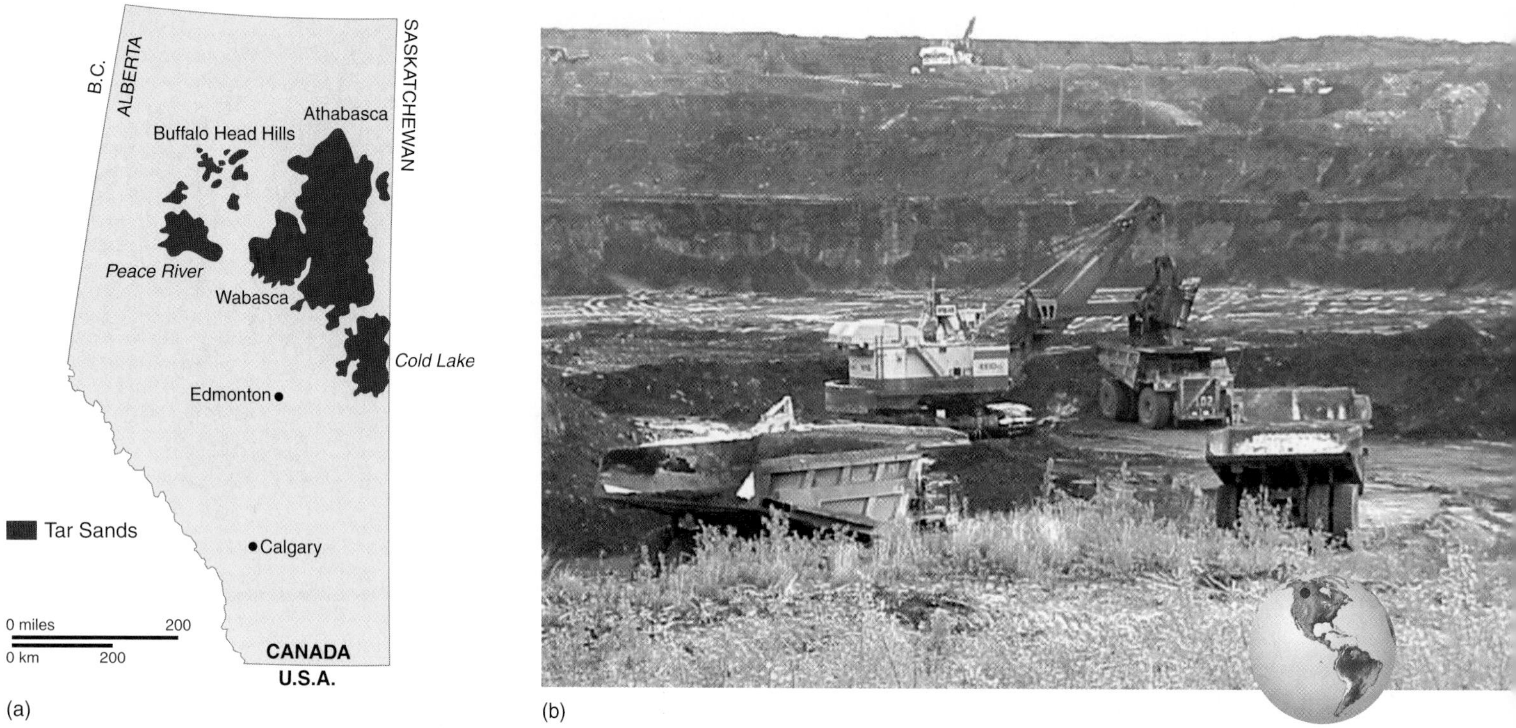


**FIGURE 5.14** (a) **Tar sand deposits in Alberta, Canada.** (b) Production of synthetic oil from the tar sands involves four steps: (1) the removal of overburden; (2) mining and transportation to extraction units; (3) the addition of steam and hot water to separate the bitumen from the tailings residue; and (4) the refinement of the bitumen into coke and distillates. *(b) Courtesy of Suncor.*

power), they could provide the gasoline, jet, and other fuels on which industrialized societies depend. The lowering of oil and gas prices in the 1980s and a decline in government support put a temporary halt to efforts to make alternative fuels economically competitive with oil. As reserves of oil and natural gas become depleted, however, at some point their prices are bound to rise. When that occurs, countries are likely to turn to these more unconventional sources of fuel. In the meantime, technologies employing nuclear energy as a source of nonfuel power are already developed.

## Nuclear Energy

Proponents consider nuclear power a major long-term solution to energy shortages. Assuming that the technical problems can be solved, they contend that nuclear fuels could provide a virtually inexhaustible source of energy. Other commentators, pointing to the dangers inherent in any system dependent on the use of radioactive fuels, argue that nuclear power poses technological, political, social, and environmental problems for which society has no solutions. Basically, energy can be created from the atom in two ways: nuclear fission and nuclear fusion.

### Nuclear Fission

The conventional form of **nuclear fission** for power production involves the controlled "splitting" of an atomic nucleus of uranium-235, the only naturally occurring fissile isotope. When U-235 atoms are split, about one-thousandth of the original mass is converted to heat. The released heat is transferred through a heat exchanger to create steam, which drives turbines to generate electricity.

A single pound of U-235 contains the energy equivalent of nearly 5500 barrels of oil. More than 400 commercial nuclear reactors around the world tap that energy, generating about 17% of the world's electricity (Figure 5.15). About one-fourth of those plants are in the United States.

Some countries are much more dependent than others on nuclear power. It accounts for 15% to 20% of the electric power generated in both the United States and Canada, but two European countries (France and Belgium) receive more than 65% of their electricity from nuclear power. Several countries have decided to reject the nuclear option altogether. Denmark, Greece, Australia, and New Zealand are among the countries that decided to remain "nuclear-free" and never built nuclear power plants. A few countries, including Germany, Sweden, and the Philippines, plan to phase out and dismantle all of their reactors. Japan, on the other hand, whose 53 nuclear plants provide 36% of the country's electricity, is continuing to build new plants, as are South Korea and China. Russian and Ukraine companies have announced their intention of resuming construction of several new reactors.

In both the United States and the world as a whole, however, nuclear capacity is likely to peak in the near future. Although a few countries continue to build new nuclear

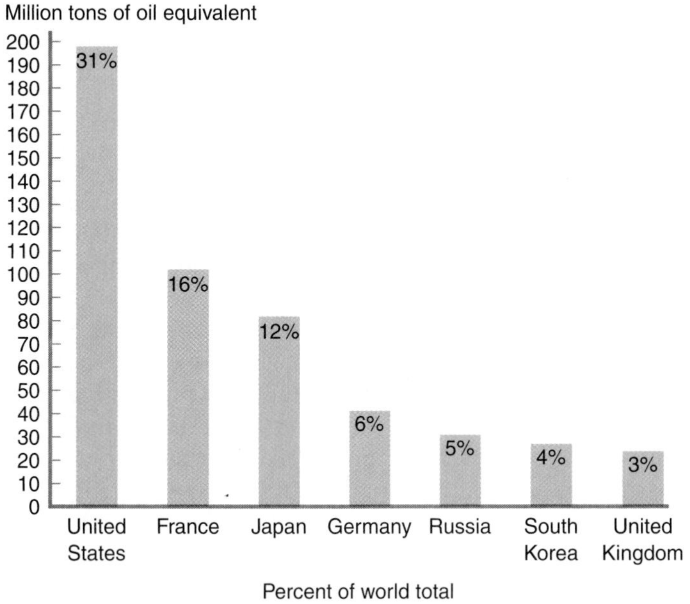


**FIGURE 5.15 Leading producers and consumers of nuclear energy, January 1, 2003.** These seven countries produce and use almost 80% of the world's nuclear energy. The percent of world total nuclear energy consumption is indicated for each country. Few nuclear power plants exist in developing countries. Source: *Data from* The BP Statistical Review of World Energy, *June 2003.*

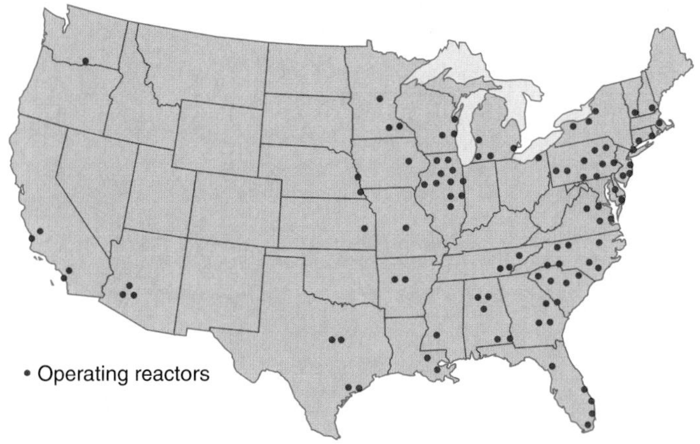


**FIGURE 5.16 Operating nuclear power plants in the United States as of January 1, 2002.** Concentrated east of the Mississippi River, the plants produce about 20% of the country's electricity. Source: *Data from U.S. Nuclear Regulatory Commission.*

plants, many older plants will be closed in the coming years as they exhaust their usefulness. The average lifespan of a nuclear plant is estimated to be only 20 to 30 years. After that time, steam generators corrode and steel pressure vessels become too brittle to operate safely. In the United States, no new plants have been ordered since 1978, for a variety of reasons (Figure 5.16). The high costs of constructing, licensing, and operating the plants has meant that, instead of being "too cheap to meter," nuclear power is about twice as expensive as energy derived from other sources, such as coal. Heightened concerns about safety and the disposal of radioactive waste have also eroded public support for nuclear energy.

Uranium-235 is a raw material that is itself in short supply. The economically exploitable reserves are expected to be depleted by about the year 2035. In the 1970s, the anticipated shortage of U-235 for conventional fission plants spurred interest in the *fast breeder reactor* as a "bridge" between the fission plants of today and the nuclear fusion plants of the long-term future. Breeder reactors could stretch uranium supplies to more than 700 years.

If a core of fissionable U-235 is wrapped in a blanket of U-238 (a more abundant but nonfissionable isotope), the U-235 both generates electricity and converts its wrapping to fissionable plutonium. This is the breeder reactor, so named because it can "breed" its own fuel. Once made, plutonium can replace U-235 as the breeder's core to fuel chain reactions. In theory, a breeder could produce more plutonium fuel than it consumes.

France and Russia operate demonstration fast breeder reactors. Japan shut down its prototype reactor in 1995 after it was damaged in an accident. Many analysts doubt that the reactor will ever be reactivated. Funding for the first U.S. demonstration project was halted in 1982. The project had faced continuous opposition from those who argued that the breeder reactor is unnecessary, uneconomical, and unsafe.

### Nuclear Fusion

Unlike a fission reaction, which splits an atom, a **nuclear fusion** reaction forces certain lightweight atoms, such as deuterium and tritium (forms of hydrogen), to combine, releasing tremendous amounts of energy. Fusion is the process that makes the sun and other stars burn. It is also the basis of the hydrogen bomb, which uses a brief, uncontrolled thermonuclear fusion reaction. The problem, still not solved for commercial use despite 50 years of research, is to control fusion for a usable release of its energy.

More difficult to achieve than fission, nuclear fusion requires heating the atoms to extremely high temperatures, until their nuclei collide and fuse. One of the technological problems facing fusion researchers is finding material for the containment vessel that is resistant to radiation and temperatures 100 million °C or higher.

If the developmental problems associated with nuclear fusion are solved, electricity requirements would presumably be satisfied for millions of years. One cubic kilometer of ocean water, the source of deuterium atoms, contains as much potential energy as that available from the world's entire known oil reserves. Fusion also offers the advantages of the total absence of radioactive wastes and greater inherent safety in energy generation than fission plants. Like conventional nuclear power, however, fusion energy could be accompanied by problems we haven't conceived of yet.

# Renewable Energy Resources

The problems posed by the use of nuclear energy, the threatened depletion of some finite fossil fuels, and the desire to be less dependent on foreign sources of energy have increased interest in industrialized countries in the *renewable* resources. One of the advantages of such resources is their ubiquity. Most places on Earth have an abundance of sunlight, rich plant growth, strong wind, or heavy rainfall. Another advantage is ease of use. It doesn't take advanced technology to utilize many of the renewable resources, one reason for their widespread use in developing countries. The most common renewable source of energy is plant matter.

## Biomass Fuels

More than half of the people in the world depend on wood and other forms of biomass for their daily energy requirements. **Biomass fuel** is any organic material produced by plants, animals, or microorganisms that can be burned directly as a heat source or converted into a liquid or gas. In addition to wood, biomass fuels include leaves, crop residues, peat, manure, and other plant and animal material. In Ethiopia and Bangladesh, biomass supplies 90% or more of the total energy consumed; in India and Pakistan, the figure is about 40%. In contrast, energy from the conversion of wood, grasses, and other organic matter is insignificant in the developed world.

There are two major sources of biomass: (1) trees, grain and sugar crops, and oil-bearing plants such as sunflowers, and (2) wastes, including crop residues, animal wastes, garbage, and human sewage. Biomass can be transformed into fuel in many ways, including direct combustion, gasification, and anaerobic digestion. Further, conversion processes can be designed to produce, in addition to electricity, solid (wood and charcoal), liquid (oils and alcohols), and gaseous (methane and hydrogen) fuels that can be easily stored and transported.

### Wood

The great majority of the energy that is produced from biomass comes from *wood.* In 1850, the United States obtained about 90% of its energy needs from wood. Although wood now contributes only 3% to the energy mix of the country as a whole, the percentage varies by region, with wood fuel providing about 15% of the energy used in Maine and Vermont. In developing countries, wood is a key source of energy, used for space heating, cooking, water heating, and lighting. This dependence on wood is leading in places to severe depletion of forests, a subject discussed later in this chapter (see "Forest Resources," p. 164).

A second biomass contribution to energy systems is *alcohol,* which can be produced from a variety of plants. After the oil shortages of the 1970s, Brazil, which is poorly endowed with fossil fuels, embarked on an effort to develop its indigenous energy supplies in order to reduce the country's dependence on imported oil. All gasoline sold in Brazil contains 26% ethanol made from sugarcane, a mixture on which conventional automobile engines can run without modifications.

### Waste

*Waste,* including crop residues and animal and human refuse, represents the second broad category of organic fuels. Particularly in rural areas, energy can be obtained by fermenting such wastes to produce methane gas (also called *biogas*) in a process known as *anaerobic digestion* (Figure 5.17). A number of countries, including India, South Korea, and Thailand, have national biogas programs, but the largest effort to generate substantial quantities of methane gas for rural households has been undertaken in China. There, backyard fermentation tanks (biodigesters) supply as many as 35 million people with fuel for cooking, lighting, and heating. The technology has been kept intentionally simple. A stone fermentation tank is fed with wastes, which can include straw and other crop residues in addition to manure. These are left to ferment under pressure, producing methane gas that is later drawn through a hose into the farm kitchen. After the gas is spent, the remaining waste is pumped out and used in the fields for fertilizer.

## Hydropower

Biomass, particularly wood, is the most commonly used source of renewable energy. The second most common is **hydropower,** which exploits the energy present in falling water. Hydropower is generated when water falls from one level to another, either naturally or over a dam. The falling water can then be used to turn waterwheels, as it was in

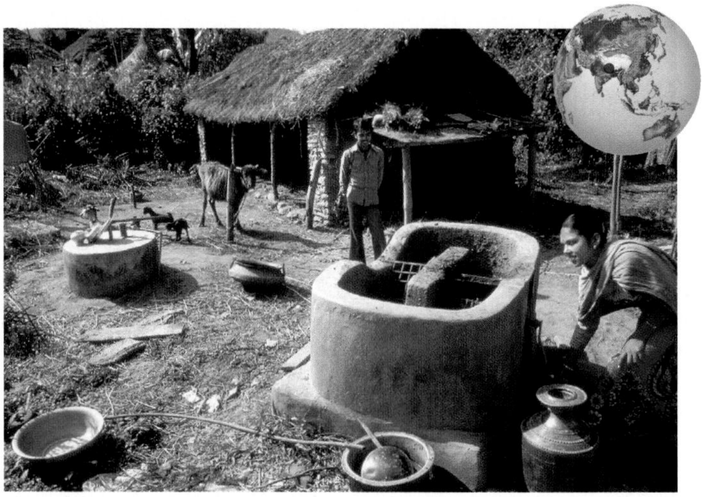

**FIGURE 5.17 A biogas generator in Nepal.** Animal and vegetable wastes are significant sources of fuel in countries such as Nepal, Pakistan, India, and China. Wastes inserted into the tank in the foreground are mixed with water and decaying organic material. As the wastes decompose, gases are emitted. The large unit in the background is the gas-collection chamber. Tubes lead from it to family kitchens.

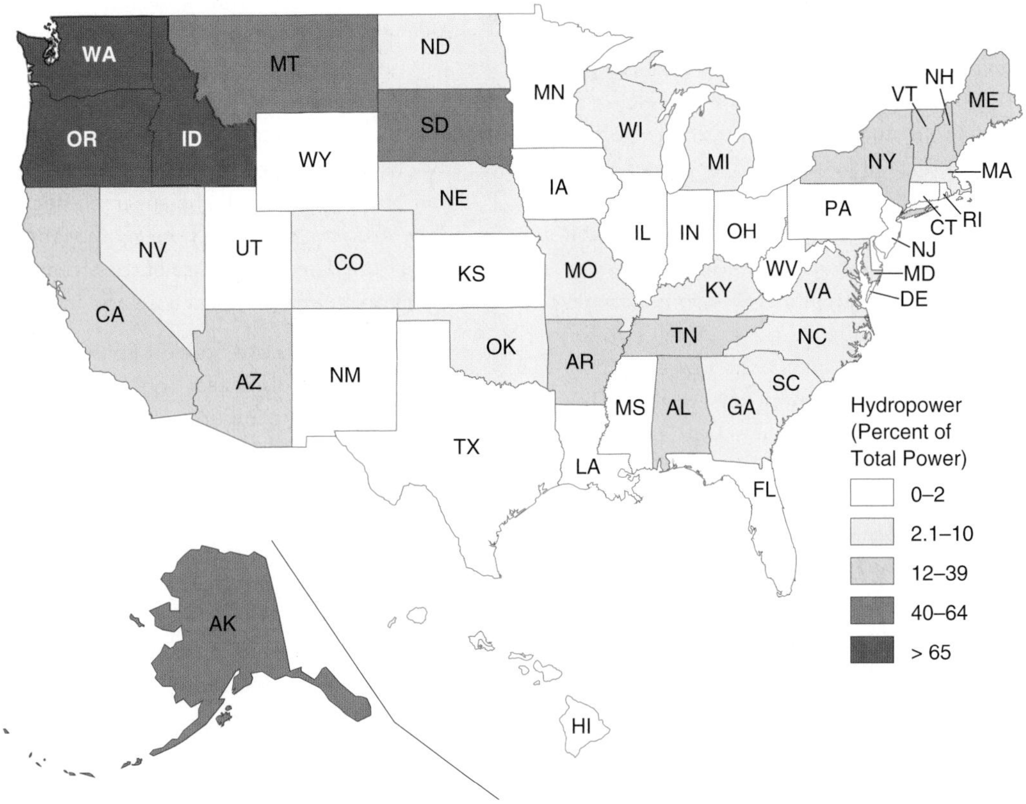


**FIGURE 5.18 Percentage of total power produced coming from hydroelectric sources in the United States.** The United States has developed about half of its hydroelectric capacity.

ancient Egypt, or modern turbine blades, powering a generator to produce electricity. Hydropower is a clean source of energy. The water is neither polluted nor consumed during power generation, although in arid areas some water in the reservoir may be lost to evaporation. Generally speaking, as long as a stream continues to flow, hydropower is renewable.

Tied as it is to a source of water, hydropower production is location-specific with regard to generation. In the United States, it is generated at more than 1900 sites in 47 states. Still, more than 60% of the country's developed capacity is concentrated in just three areas: the Pacific states (Washington, Oregon, and California), the multistate Tennessee River valley area of the Southeast, and the Northeast (Figure 5.18). That pattern is a result both of the location of the resource base and the role that agencies such as the Tennessee Valley Authority (TVA) have played in hydropower development.

Transmitting hydroelectricity over long distances is costly, so it is generally consumed in the region where it is produced. This fact helps account for variations in the pattern of consumption and for the energy mix used in specific regions. Thus, although hydropower provides the United States as a whole with one-tenth of its electricity, it supplies all of Idaho's needs and about 90% of the electricity used in Oregon and Washington.

Consumption patterns around the world are shown in Figure 5.19. The contribution of hydropower to a country's energy supplies varies greatly. Falling water provides Norway with 99% of its electricity (and 50% of all its energy needs). Canada, the world's leading producer, gets about 70% of its electricity from hydropower, a share roughly comparable to that found in Brazil, Switzerland, and New Zealand. Almost 40 countries in Asia, Africa, and South America obtain more than half of their electricity from hydropower. The Itaipu hydroelectric power plant on the Parana River in South America, currently the largest such facility in the world, provides Paraguay with 80% of its electricity and Brazil with 20%.

In addition to supplying electrical power, dams provide flood protection and irrigation water. Despite these and other advantages, the exploitation of hydropower has significant environmental and social costs (see "Dammed Trouble," p. 150). Flooding valleys for water storage submerges forests, farmlands, and villages, sometimes resulting in the displacement of tens of thousands of people. The reservoirs flood natural wetlands and river habitats, alter streamflow patterns, and trap silt that would otherwise flow downstream to be deposited on agricultural lands, leading to long-term declines in soil fertility. The disruption of downstream ecosystems that evolved to take advantage of a river's sporadic flood pattern typically reduces the diversity of aquatic species.

GEOGRAPHY & Public Policy

## Dammed Trouble

Dams are built for three primary purposes: to generate power, to prevent floods, and to provide a reliable source of water for agriculture and municipalities. They offer many advantages. They exploit a free and endlessly renewable source of energy, water. After the initial capital investment, the costs of operation are relatively low. Many suitable sites exist in less developed countries. Dams permit large numbers of people to inhabit and cultivate arid regions—to make the desert bloom. Some open up shipping passages on formerly wild rivers. Dams on the Columbia and Lower Snake rivers, for example, allow barge traffic to penetrate almost 800 kilometers (500 mi) into the interior, making Lewiston, Idaho, the West Coast's most inland seaport.

The United States built its largest dams in the 1930s and 1940s, when the Tennessee, Colorado, Columbia, and other rivers were dammed. It was not a time when environmentalists held much sway, and the negative consequences of dam construction went largely unrecognized. The benefits, on the other hand, were evident. The folksinger Woody Guthrie's tribute to the damming of the Columbia River ran, in part:

> And up on the river is Grand Coulee Dam,
> The mightiest thing ever built by a man,
> To run the great factories and water the land;
> It's roll on, Columbia, roll on![a]

The best sites for water development in the United States have already been exploited, but this is not the case in much of the rest of the world. Large dam construction projects are underway on the Narmada River in India and the Tigris and Euphrates Rivers in Turkey, among other places. In 1994, the first concrete was poured for what is intended to be the world's largest hydroelectric project, the Three Gorges Dam. The enormous structure will block the Chang Jiang (Yangtze) River in China's scenic Three Gorges area, honored by painters and poets, where the river flows through a series of sheer chasms. Although the dam itself will not be the world's biggest, its hydroelectric output will be more than 18,000 megawatts (the equivalent of 18 large coal or nuclear power plants), by far the largest in the world.

The Three Gorges Dam has elicited intense opposition from Chinese and foreign critics. Scheduled to be completed in 2009, the dam will have significant social and environmental costs. They are representative of the disadvantages of hydropower, although the specific effects of any single dam depend on such factors as the geomorphology of the area, the size of the dam, the nature of the stream the dam impounds, the climate, and the local flora and fauna.

- When valleys are flooded for water storage, people and wildlife are displaced, forests are destroyed. About the length of Lake Superior, the huge reservoir—600 kilometers (375 mi) long—now rising behind the Three Gorges Dam is submerging the houses, farmland, and villages of an estimated 1.9 million people. Critics of the project argue that the successful resettlement of so many people is impossible. Moving them up on the surrounding hillsides is unrealistic; the slopes cannot support new farming. Resettling people far from their homes hasn't worked in other regions. Many drift back to their native homes; others live as impoverished refugees, jobless and landless.
- Although its proponents embrace hydropower as a source of clean energy, the flooding that accompanies big dams often submerges large forests. As the vegetation decays, it releases methane, a potent greenhouse gas, thus contributing to global warming.
- Upstream, the water will slow down and drop an enormous amount of silt on the bottom of the reservoir; as it fills with sediment, upstream flooding is possible.
- Within a few decades, silting-up can reduce the original generating capacity of a dam by as much as 70%; in the very long run, reservoirs will fill with silt and the river will flow over the dam.
- There will be a loss of nutrients to river life because the sediment trapped behind the dam and turbines contains organic matter that nourishes downriver food webs.
- Although they provide water for irrigating crops, dams also concentrate salts in the topsoil, particularly in arid areas. Long desert rivers, such as the Nile and Colorado, are especially salty, and irrigating a hectare of cropland can leave several tons of salt dispersed in the soil.

[a] *Roll On, Columbia* Words by Woody Guthrie. Music based on "Goodnight Irene" by Huddie Ledbetter and John A. Lomax. ™ © 1936 (renewed), 1957 (renewed), and 1963 (renewed) Ludlow Music, Inc., New York, NY. Used by permission.

### Solar Energy

Each year, the earth intercepts solar energy that is equivalent to many thousands of times the energy people currently use. Inexhaustible and nonpolluting, **solar energy** is the ultimate origin of most forms of utilized energy: fossil fuels and plant life, waterpower and wind power. It is, however, the direct capture of solar energy that is seen by many as the best hope of satisfying a large proportion of future energy needs with minimal environmental damage and maximum conservation of earth resources. The chief drawback to solar power is its diffuse and intermittent nature. Being diffuse, it must be collected over a large area to make it practical to use, and because it is intermittent, it requires some means of storage.

- Many places have experienced an increase in the incidence of waterborne diseases in reservoirs, irrigation canals, and rivers following dam construction. In Egypt, the Aswan High Dam expanded the breeding areas for snails that transmit schistosomiasis, a debilitating disease that is almost impossible to cure. Malaria typically increases following dam construction in tropical regions because hundreds of square kilometers of ideal mosquito breeding area are exposed during periods of reservoir drawdown.
- The reduced water flow during the dry season will disrupt the Chang Jiang's estuary, causing seawater to move upstream. Because sediment is no longer deposited by floodwaters, the shoreline may erode. As estuaries, beaches, and wetlands shrink, wildlife habitat will be diminished.
- Large dams affect the temperature and oxygen content of downstream waters, altering the mix of aquatic species. Rivers become less livable for some species, better for others. Some analysts predict a dramatic reduction in the number of fish in the Chang Jiang because the lower water temperature and the curtailing of spring floods will affect spawning behavior. The dams of the U.S. Pacific Northwest have proved lethal to the region's wild salmon, which need to migrate from the rivers out to the Pacific Ocean and then back again to spawn where they were born. The Army Corps of Engineers has struggled for 25 years with little success to find ways for salmon and the dams to coexist.
- Finally, there is a possibility of a catastrophic dam failure, the result of age, poor design and construction, or an earthquake. Some have charged that the Three Gorges Dam is so poorly constructed that it is already cracking and could collapse.

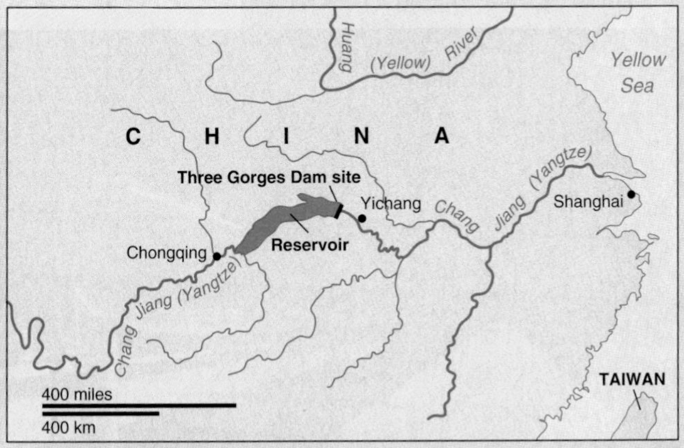


Options to the construction of huge dams do exist. The efficiency of existing dams can be increased. Replacing old, inefficient generators with newer, highly efficient ones can triple a dam's electrical output. Small dams can deliver many of the benefits of hydropower at a fraction of the cost and without displacing people. Many developing countries are building small dams and installing small generators to supply power to areas far from electrical utility lines.

**QUESTIONS TO CONSIDER**

1. The advantages of hydropower are manifold, yet one observer charges that water development has amounted to a Faustian bargain between people and the natural world. What does he mean?
2. If you were the Minister of Water Resources in China, would you have recommended that the Three Gorges Dam be built? What questions would you want answered in order to assess the likely benefits and costs of the dam? How would you respond to critics who argue that China could more quickly and cheaply meet its energy and flood control needs by reinforcing dikes and building smaller dams along the Chang Jiang and its tributaries?
3. In the United States, there is an emerging belief that old dams that have outlived their usefulness and whose environmental costs now outweigh their benefits should be dismantled, returning rivers to good health and restoring their free-flowing nature. By 2002, nearly 500 dams had been decommissioned and torn down in Maine, Vermont, North Carolina, Wisconsin, California, and elsewhere. About one-quarter of the dams in the United States are at least 50 years old, and many are in dire need of repair. Would you favor dismantling them? Why or why not?
4. About 100,000 dams regulate American rivers, yet according to the U.S. Geological Survey, flooding continues to be the most destructive and costly type of natural disaster in the country, there has been no reduction in the average number of flood deaths each year, and even when adjusted for inflation, flood damage to property has almost tripled since 1951. Can you think of some reasons this might be so?

The technology for using solar energy for domestic uses such as hot-water heating and space heating is well known. In the United States, both passive and active solar-heating technologies have secured a permanent foothold in the marketplace. More than half a million homes use solar radiation for water and space heating. Japan is reported to have some 4 million solar hot-water systems, and in Israel, two-thirds of all the homes heat water with solar energy. The use of solar panels for individual homes is best in climates that are warm and sunny, with not too much cloud cover and not too many hours of darkness in winter, when the demand for energy is at its highest.

A second type of solar-energy use involves converting concentrated solar energy into thermal energy to generate

Although relatively few places have geothermal steam that can be exploited to generate electricity, geothermal energy can also be used directly for heating and cooling. Geothermal heat pumps (also known as ground-source heat pumps) take advantage of the constant temperature found in soil below the frost line to heat or cool air pumped through a building. Loops of piping are buried in the ground; an electric compressor circulates refrigerant through them and then cools or heats the air, which is distributed throughout a building. Energy-efficient and environmentally clean, geothermal heating systems have grown in popularity in the United States in recent years, particularly for new construction.

### Wind Power

Although wind power was used for centuries to pump water, grind grain, and drive machinery, its contribution to the energy supply in the United States virtually disappeared more than a century ago, when windmills were replaced first by steam and later by the fossil fuels. Windmills offer many advantages as sources of electrical power. They can turn turbines directly, do not use any fuel, and can be built and erected rather quickly. They need only strong, steady winds to operate, and these exist at many sites. Furthermore, wind turbine generators do not pollute the air or water and do not deplete scarce natural resources. Technological advances in design have lowered the cost of using wind turbines to generate electricity, so they are becoming increasingly competitive with conventional power plants. Wind power now costs from 3 to 6 cents per kilowatt-hour, about the same as power plants fired by fossil fuels.

In the 1980s, California dominated the worldwide development of wind energy, spurred by the gasoline shortages of previous years, federal and state tax incentives, and favorable long-term utility contracts. By 1987, some 15,000 turbines had been installed in three parts of the state, representing 90% of the world's wind capacity. That percentage declined steadily during the 1990s as other states and a number of European countries began to invest in **wind farms,** clusters of wind-powered turbines producing commercial electricity. At the beginning of 2002, the United States accounted for about 17% of the world operating wind capacity, with California responsible for roughly half of that amount (Figure 5.22)

In the 1990s, a commitment to reducing dependence on fossil fuels and developing renewable resources stimulated the growth of wind-power installations in several European countries, particularly Germany and Denmark. Germany is the world's leading producer of wind-powered energy, accounting in 2002 for more than one-third of global capacity. Denmark has the highest per capita output of wind energy in the world, with wind power generating about 16% of the country's electricity. With the exception of India, Asian countries have been somewhat slower to build wind farms, but sizable projects are under construction in China and Japan.

Offshore wind turbine installations are expected to play a growing role in the coming years, particularly in northern Europe. The world's biggest offshore windmill park, at the entrance to Copenhagen's harbor, supplies 3% of that Danish city's electricity, or enough for 32,000 households. The Netherlands and Sweden also have offshore windmill parks, and planned or under construction are parks off the coasts of Britain, Ireland, Belgium, and Germany.

The chief disadvantage of wind power is that it is unreliable and intermittent; because its energy cannot be easily stored, it requires a backup system. Detractors point out that it takes thousands of wind turbines to produce the same amount of electricity as a single nuclear power plant. Environmental concerns include the aesthetic impact of wind farms (they are very visible, often covering entire hillsides and dominating the landscape) and the hazard they pose to migrating birds.

## Nonfuel Mineral Resources

The mineral resources already discussed provide the energy that enables people to do their work. Equally important to our economic well-being are the *nonfuel* minerals, for they can be processed into steel, aluminum, and other metals, and into glass, cement, and other products. Our buildings, tools, and weapons are chiefly mineral in origin.

Virtually all the resources we deem essential, including metals, nonmetallic minerals, rocks, and the fuels, are contained in the earth's crust, the thin outer skin of the planet. Of the 92 natural elements, just 8 account for more than 98% of the mass of the earth's crust (Figure 5.23). They can be thought of as geologically abundant, and all others as geologically scarce. In most places, minerals exist in concentrations too low to make their exploitation profitable. If the concentration is high enough to make mining feasible, the mineral deposit is called an **ore.** Thus what is or is not an ore depends on demand, price, and technology and changes over time.

Exploitation of a mineral resource typically involves six steps:

1. *exploration* (finding concentrated deposits of the material);
2. *extraction* (removing it from the earth);
3. *concentration* (separating the desired material from the ore);
4. *smelting* and/or *refining* (breaking down the mineral to the desired pure material);
5. *transporting* it to where it will be used;
6. *manufacturing* the ore into a finished product.

Each step requires inputs of energy and materials.

Five factors help determine the practicality and profitability of mining a specific deposit of a mineral: its value, the quantity available, the richness of the ore in a particular

(a)

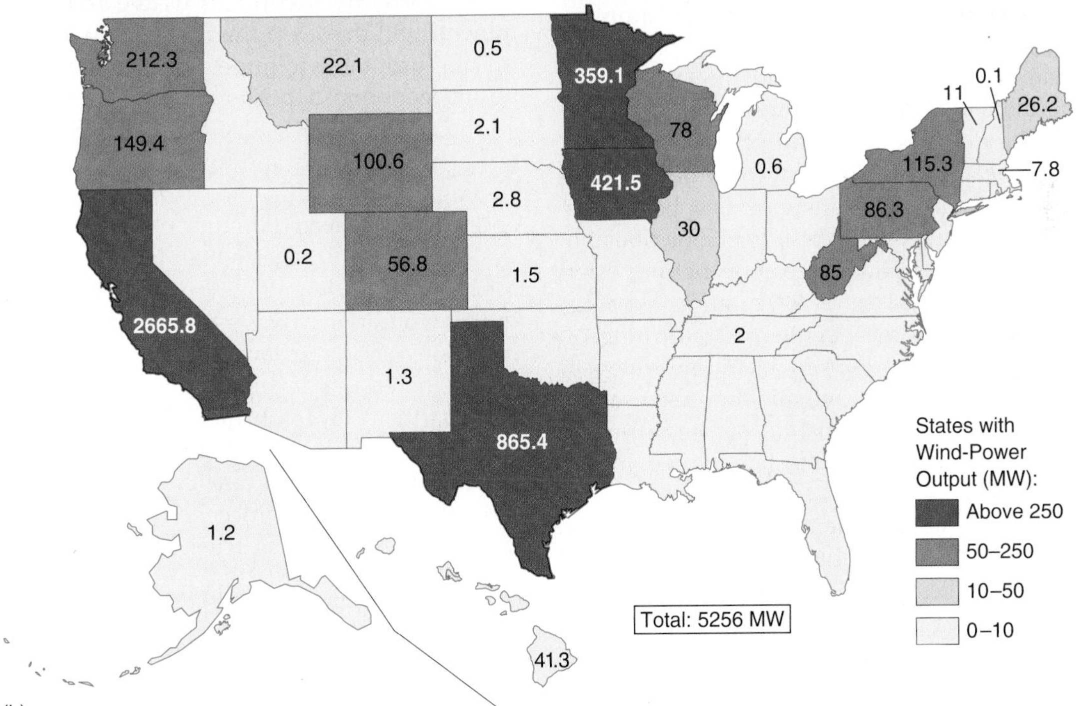


(b)

**FIGURE 5.22** (a) **A wind farm in the San Gorgonio Pass, near Palm Springs, California.** The wind turbine generators harness wind power to produce electricity. California's wind farms produce enough electricity to supply the residential power demands of about 1 million people. (b) **Wind-generating capacity by state, in megawatt output, 2002.** One thousand megawatts meet the energy needs of about 50,000 U.S. households. Although only a small portion of the country's wind potential has been tapped, new wind-power projects are underway in a number of states. Wind turbine technology advanced dramatically during the 1990s; modern turbines are significantly more powerful and reliable than earlier ones. In recent years, wind energy has been the fastest growing source of electricity in both the United States and the world as a whole. *(a) © The McGraw-Hill Companies, Inc./Dog Sherman, photographer.*

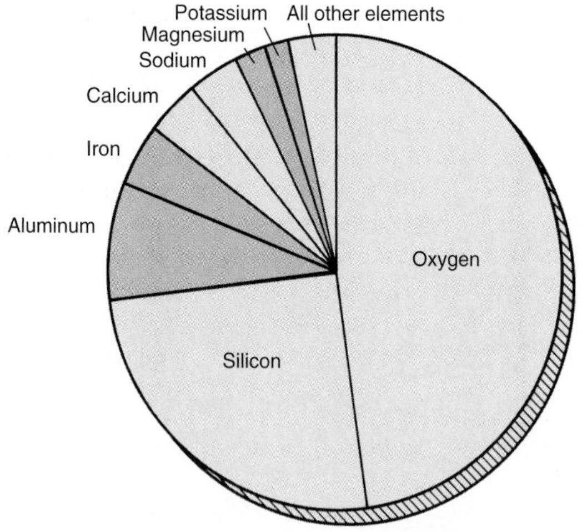


**FIGURE 5.23 The relative abundance, by weight, of elements in the earth's crust.** Only four of the economically important elements—those shown in purple—are geologically abundant, accounting for more than 1% of the total weight of the earth's crust. Fortunately, these and other commercially valuable minerals have been concentrated in specific areas within the crust. Were they uniformly disseminated throughout the crust, their exploitation would not be feasible.

**TABLE 5.1 Projected Lifetimes of Reserves of Selected Minerals**

| | Years Remaining of Proved Reserves | |
|---|---|---|
| **Mineral** | **World** | **United States** |
| Silver | 18 | 7 |
| Lead | 21 | 4 |
| Zinc | 25 | 15 |
| Copper | 27 | 15 |
| Tin | 37 | 0 |
| Nickel | 40 | 0 |
| Phosphate | 87 | 22 |
| Manganese | 101 | 0 |
| Cobalt | 159 | 0 |
| Bauxite | 203 | 4 |
| Chromium | 281 | 0 |
| Potash | 333 | 19 |

*Source: Mineral Commodities Summaries 2000*, U.S. Geological Survey.

Note: These figures reflect the approximate number of years the proved reserves of selected minerals will last, based on current rates of production and consumption. Such figures are only suggestive, because reserve totals and consumption rates fluctuate over time. A decline in the costs of exploitation, increase in value of the material, and/or new discoveries of deposits will extend the lifetimes shown here.

deposit, the distance to market, and land acquisition and royalty costs. Even if these conditions are favorable, mines may not be developed or even remain operating if supplies from competing sources are available more cheaply. In the 1980s, more than 25 million tons of iron ore–producing capacity was permanently shut down in the United States and Canada. Similar declines occurred in North American copper, nickel, zinc, lead, and molybednum mining as market prices fell below domestic production costs. Beginning in the early 1990s, as a result of both resource depletion and low-cost imports, the United States became a net importer of nonfuel minerals for the first time. Although increases in mineral prices may lead to the opening or reopening of mines that have been deemed unprofitable, the developed industrial countries with market economies find themselves at a competitive disadvantage against developing country producers with lower-cost labor and state-owned mines with abundant reserves.

Natural processes produce minerals so slowly that they fall into the category of nonrenewable resources, existing in finite deposits. The supplies of some, however, are so abundant that a ready supply will exist far into the future. These include coal, sand and gravel, potash, and magnesium. The supply of others, such as tin and mercury, is small and getting smaller as industrial societies place ever-greater demands on them. Table 5.1 gives one estimate of "years remaining" for some important metals. It should be taken as suggestive rather than definitive, because mineral reserves are difficult to estimate. As we noted in the case of fossil fuels, such estimates are based on economic and technological conditions, and we cannot predict either future prices for minerals or improvements in technology. The depletion of the currently identified, usable reserves of a valuable mineral will drive up the price of the mineral, which will make it profitable to mine some of what are now classified as subeconomic deposits (see Figure 5.4). Those deposits would then be reclassified as proved reserves. The discovery of new deposits and/or improvements in mineral-processing technology would also increase the reserve figure and thus its projected lifetime.

Although human societies began to use metals as early as 3500 B.C., world demand remained small until the Industrial Revolution. It was not until after World War II that increasing shortages and rising prices (and in the United States, increasing dependence on foreign sources) began to impress themselves on the general consciousness. Worldwide technological development has established ways of life in which minerals are the essential constituent. That industrialization has proceeded so rapidly and so cheaply is the direct result of the earlier ready availability of rich and accessible deposits of the requisite materials. Economies grew fat by skimming the cream. The question, yet unanswered, is whether the remaining supplies of scarce minerals will limit the expansion of industrialized and developing economies or whether, and how, people will find a way to cope with shortages.

## The Distribution of Nonfuel Minerals

Because the distribution of mineral resources is the result of long-term geologic processes that concentrated certain elements into commercially exploitable deposits, it follows that the larger the country, the more likely it is to contain such deposits. And in fact, Russia, China, Canada, the United States, Brazil, and Australia possess abundant and diverse mineral resources. As Figure 5.24 indicates, these are the leading mining countries. They contain roughly half of the nonfuel mineral resources and produce the bulk of the metals (e.g., iron, manganese, and nickel) and nonmetals (e.g., potash and sulfur).

Many types of nonfuel minerals are concentrated in a small number of countries, and some scarce elements occur in just a few regions of the world. Thus, extensive deposits of cobalt and diamonds are largely confined to Russia and central-southern Africa. South Africa has nearly half of the world's gold ore and more than three-quarters of the chromium and platinum-group metals. Some countries contain only one or two exploitable minerals—Morocco has phosphates, for example, and New Caledonia, nickel. Several countries with large populations are at a disadvantage with respect to mineral reserves. They include industrialized countries such as France and Japan, which are able to import the resources, as well as developing countries such as Nigeria and Bangladesh, which are less able to afford imports.

It is important to note that no country contains all of the economically important mineral resources. Some, such as the United States, which were bountifully supplied by nature, have spent much of their assets and now depend on foreign sources. Although the United States was virtually self-sufficient in mineral supplies in the 1940s and 1950s, it is not today. Because of its history of use of domestic reserves and its continually expanding economy, the United States now depends on other countries for more than 50% of its supply of a number of essential minerals, some of which are shown in Table 5.2.

The increasing costs and declining availability of metals encourage the search for substitutes. The fact that industrial chemists and metallurgists have been so successful in the search for new materials that substitute for the traditional resources has tended to allay fears of possible resource depletion. But it must be understood that no adequate replacements have been found for some minerals, such as cobalt and chromium. Other substitutes are frequently synthetics, often employing increasingly scarce and costly hydrocarbons in their production. Many, in their use or disposal, constitute environmental hazards, and all have their own high and increasing price tags.

## Copper: A Case Study

Table 5.1 indicates that the world reserves of copper will last only another 27 years or so, based on current rates of production and consumption and assuming that no new extractable reserves appear. Copper is a relatively scarce mineral, and its importance to industrialized societies is evidenced by the fact that more copper is mined annually than any other nonferrous metal except aluminum.

Three properties make copper desirable: it conducts both heat and electricity extremely well, it can be hammered or drawn into thin films or wires, and it resists corrosion. More than half of all copper produced is used in electrical wire and equipment. About one-third of the copper produced in the United States is used to make bronze and brass.

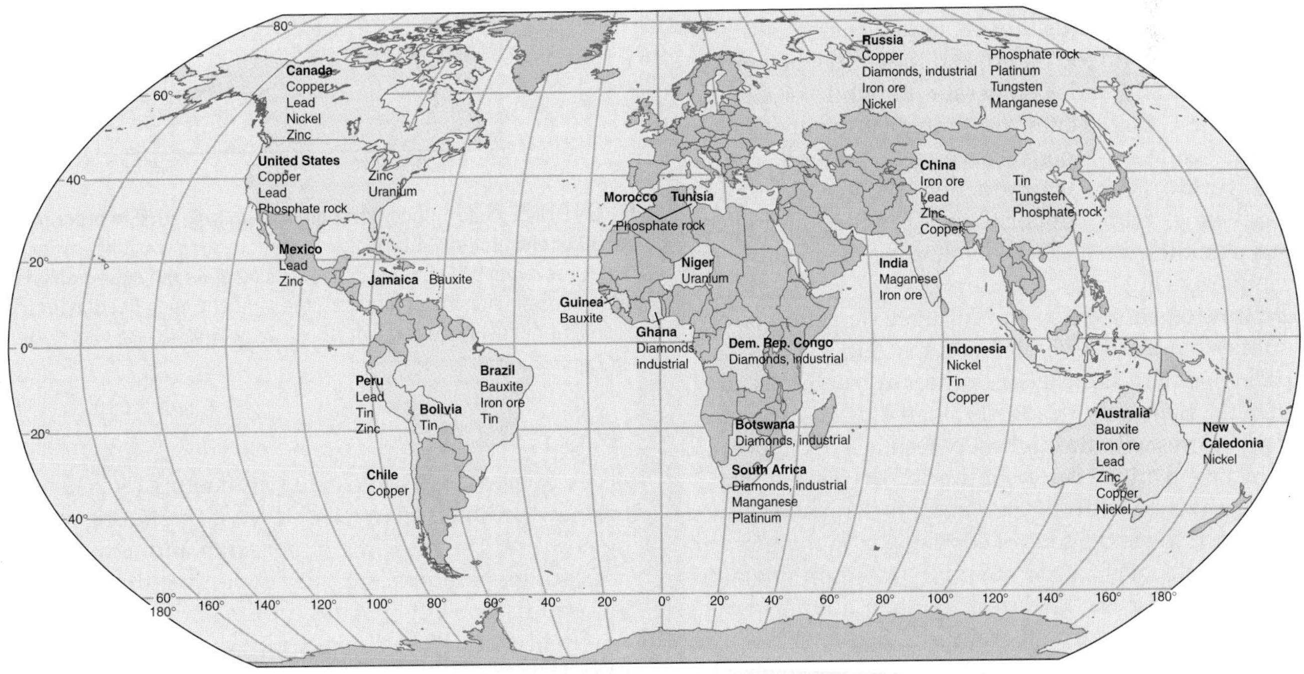


**FIGURE 5.24 Leading producers of selected minerals.** The countries shown are not necessarily those with the largest deposits. India, for example, contains reserves of bauxite, China of manganese, and South Africa of phosphates, but none is yet a major producer of those materials.
*Data from World Resources Institute.*

**TABLE 5.2 | U.S. Reliance on Foreign Supplies of Selected Minerals**

| Mineral | % Imported in 2000 | Major Sources | Major Uses |
|---|---|---|---|
| Bauxite and alumina | 100% | Australia, Guinea, Jamaica | Aluminum production |
| Columbium | 100 | Brazil, Canada, Germany | Additive in steelmaking and superalloys |
| Manganese | 100 | S. Africa, Gabon, Australia | Steelmaking, batteries, agricultural chemicals |
| Mica | 100 | India, Belgium, Germany | Electronic and electrical equipment |
| Quartz crystal | 100 | Brazil, Germany, Madagascar | Electronics, optical applications |
| Strontium | 100 | Mexico, Germany | TV picture tubes, ferrite magnets, pyrotechnics |
| Platinum | 83 | S. Africa, Russia, United Kingdom | Catalysts in the automotive, chemical industries |
| Tantalum | 80 | Australia, China, Thailand | Capacitors, superalloys, cemented carbide tools |
| Chromium | 78 | S. Africa, Russia, Kazakhstan | Steel, chemicals, refractories |
| Cobalt | 74 | Norway, Finland, Zambia | Superalloys for jet engines, cutting tools, magnets, chemicals |
| Barite | 71 | China, India, Mexico | Oil and gas well drilling fluids, chemicals |
| Potash | 70 | Canada, Russia, Belarus | Fertilizers, chemicals |
| Tungsten | 68 | China, Russia, Bolivia | Electrical and electronic components, tool steel |
| Zinc | 60 | Canada, Mexico, Peru | Galvanizing, zinc-base alloys, brass and bronze |
| Nickel | 58 | Canada, Norway, Russia | Stainless steel, superalloys for the chemical and aerospace industries |
| Silver | 52 | Canada, Mexico, Peru | Photographic materials, electrical products, catalysts |

*Source:* U.S. Geological Survey.

The remainder is used in construction, in industrial and farm machinery, for transportation, and for plumbing pipes.

Like most minerals, copper is unevenly distributed in the earth's crust. The largest copper deposits are found at convergent tectonic margins in western North America, western South America, and Australia. Copper deposits in sedimentary basins include those extending across northern Europe, from England to Poland, and the copper belt of central Africa (Zambia and Democratic Republic of Congo). Chile and the United States are thought to contain about 40% of the world's copper reserves. Those two countries lead the world in copper production. Because the United States consumes more copper than it produces, it imports significant amounts to meet its demand for the metal. Other major importers are Japan and Germany.

The threatened scarcity and ultimate depletion of copper supplies have had several effects that suggest how societies will cope with shortages of other raw materials. First, in the United States, the grade of mined ores has decreased steadily. The ores with the highest percentage of copper (2% and above) were mined early (Figure 5.25). Now, ore of 0.5% grade is the average. Thus, 1000 tons of rock must be mined and processed to yield 5 tons of copper—or, in more practical terms, 3 tons of rock are necessary to equip one automobile with the copper used in its radiator and its various electrical components. The remaining 2.985 tons is waste, generated at the mine, the concentrator, and the smelter.

Second, the recovery of copper by recycling has increased. Unlike some minerals, copper lends itself to recycling because much of it is used in pure form in sizeable pieces. It is less expensive to remanufacture recycled copper into new products than to produce it by mining and refining new ore. The recycling of old scrap and new scrap from fabricating operations supplies about one-third of the copper used in the United States each year.

Finally, price rises have spurred the search for substitutes. In many of its applications, copper is being replaced by other, less expensive materials. Aluminum is replacing copper in some electrical applications and in heat exchangers. Plastics are supplanting copper in plumbing pipes and

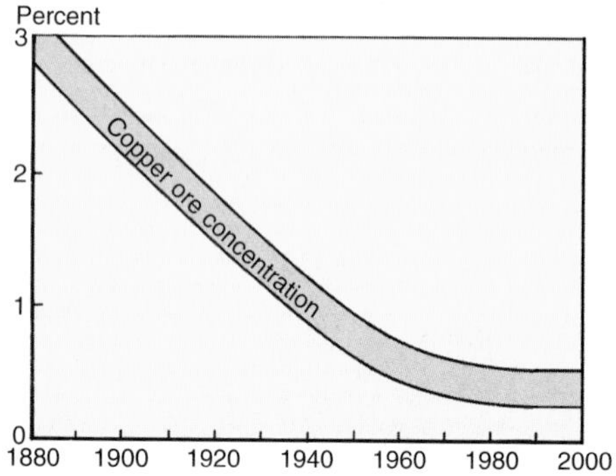


**FIGURE 5.25 Concentration of copper needed in order to be mined economically.** In 1880, 3% copper ore rock was necessary, but today, rock with 0.5% or less copper is mined. As the supply of a metal decreases and its price increases, the concentration needed for economic recovery goes down. Source: *Data from the U.S. Bureau of Mines, U.S. Department of the Interior.*

building materials. Glass fibers are employed in many telephone transmission lines, and steel can be used in shell casings and coinage. Due to its increased use in motors and electronic equipment, however, the amount of copper used in automobiles actually has increased by 40% since 1980.

# Land Resources

The mineral resources we have just discussed are nonrenewable. We turn our attention now to land resources that are potentially renewable, examining the distribution and status of three of those resources: soils, wetlands, and forests. Because they support or are living things, they are sometimes called *biological resources.*

## Soils

By design or by accident, people have brought about many changes in the physical, chemical, and biochemical nature of the soil and altered its structure, fertility, and drainage characteristics. The exact nature of the changes in any area depends on past practices as well as on the original nature of the land.

Over much of the earth's surface, the thin layer of topsoil upon which life depends is only a few inches deep, usually less than 30 centimeters (1 ft). Below it, the lithosphere is a complex mixture of rock particles, inorganic mineral matter, organic material, living organisms, air, and water. Under natural conditions, soil is constantly being formed by the physical and chemical decomposition of rock material and by the decay of organic matter. It is simultaneously being eroded, for **soil erosion**—the removal of soil particles, usually by wind or running water—is as natural a process as soil formation, and it occurs even when land is totally covered by forests or grass. Under most natural conditions, however, the rate of soil formation equals or exceeds the rate of soil erosion, so that soil depth and fertility tend to increase with time.

When land is cleared and planted to crops, or when the vegetative cover is broken by overgrazing or other disturbances, the process of erosion accelerates. When its rate exceeds that of soil formation, the topsoil becomes thinner and eventually disappears, leaving behind only sterile subsoil or barren rock. At that point, the renewable soil resource has been converted through human impact into a nonrenewable and dissipated asset. Carried to the extreme of bare rock hillsides or wind-denuded plains, erosion spells the total end of agricultural use of the land.

Such massive destruction of the soil resource could endanger the survival of the civilization it has supported. For the most part, however, farmers devise ingenious ways to preserve and even improve the soil resource upon which their lives and livelihoods depend. Farming skills have not declined in recent years, but pressures upon farmlands have increased with population growth. Farming has been forced higher up onto steeper slopes, more forest land has been converted to cultivation, grazing and crops have been pushed farther and more intensively into semiarid areas, and existing fields have had to be worked more intensively and less carefully. Many traditional agricultural systems and areas that were ecologically stable and secure as recently as 1950, when world population stood at 2.5 billion people, are disintegrating under the pressures of more than 6 billion people.

The pressure of growing population numbers is having an especially destructive effect on tropical rain forests. Expanded demand for fuel and commercial wood, and a midlatitude market for beef that can be satisfied profitably by replacing tropical forest with cleared grazing land are responsible for some of the loss, but the major cause of *deforestation* is clearing the land for crops. Extending across parts of Asia, Africa, and Latin America, the tropical rain forests are the most biologically diverse places on Earth, but vast expanses are being destroyed every year. About 45% of their original expanse has already been cleared or degraded. Deforestation is discussed in more detail on pages 168–169, but it is important to note here that accelerated soil erosion quickly removes tropical forest soils from deforested areas. Lands cleared for agriculture almost immediately become unsuitable for that use, partially because of soil loss (Figure 5.26).

The tropical rain forests can succumb to deliberate massive human assaults and be irretrievably lost. With much less effort, and with no intent to destroy or alter the environment, humans are similarly affecting the arid and semiarid regions of the world. The process is called **desertification,** the expansion or intensification of areas of degraded or destroyed soil and vegetation cover; it usually occurs in

**FIGURE 5.26** The tropical rain forest was cleared on this tract in Amazonia, Brazil, to make room for a tin mining operation. Exposed soils quickly deteriorate in structure and fertility and are easily eroded. *© Mark Edwards/Still Pictures.*

arid and semiarid environments. Climatic change—unpredictable cycles of rainfall and drought—is often a contributing cause, but desertification accelerates because of human activity, mainly overgrazing, deforestation for fuelwood, clearing of original vegetation for cultivation, and burning. Desertification implies a continuum of ecological alteration from slight to extreme (Figure 5.27).

Whatever its degree of development, when the process results from human rather than climatic change, it begins in the same fashion: the disruption or removal of the native cover of grasses and shrubs through farming or overgrazing (Figure 5.28). If the disruption is severe enough, the original vegetation cannot reestablish itself, and the exposed soil is made susceptible to erosion during the brief, heavy rains that dominate precipitation patterns in semiarid regions. Water runs off the land surface instead of seeping in, carrying soil particles with it. When the water is lost through surface flow rather than seepage downward, the water table is lowered. Eventually, even deep-rooted bushes are unable to reach groundwater, and all natural vegetation is lost. The process is accentuated when too many grazing animals pack the earth down with their hooves, blocking the passage of air and water through the soil. When both plant cover and soil moisture are lost, desertification has occurred.

It happens with increasing frequency in many areas of the earth as pressures upon the land continue. Africa is most at risk; the United Nations has estimated that 40% of that continent's nondesert land is in danger of human-induced desertification. But nearly a third of Asia and a fifth of Latin America's land are similarly endangered. In countries where desertification is particularly extensive and severe (Algeria, Ethiopia, Iraq, Jordan, Lebanon, Mali, and Niger), per capita food production declined by some 40% between 1950 and the late 1990s. The resulting threat of starvation spurs populations of the affected areas to increase their farming and livestock pressures on the denuded land, further contributing to their desertification.

Desertification is but one expression of land deterioration leading to accelerated soil erosion. The evidence of that deterioration is found in all parts of the world. In Guatemala, for example, some 40% of the productive capacity of the land has been lost through erosion, and several areas of the country have been abandoned because agriculture has become economically impracticable. The figure is 50% in El Salvador, and Haiti has no high-value soil left at all. In Turkey, about half of the land is severely or very severely eroded. A full one-quarter of India's total land area has been significantly eroded.

In recent years, soil erosion in the United States has been at an all-time high (Figure 5.29) (see "Maintaining Soil Productivity," p. 162). Wind and water are blowing and washing soil off croplands in Iowa and Missouri, pasturelands in the Great Plains, and ranches in Texas. America's croplands lose almost 2 billion tons of soil per year to erosion, an average

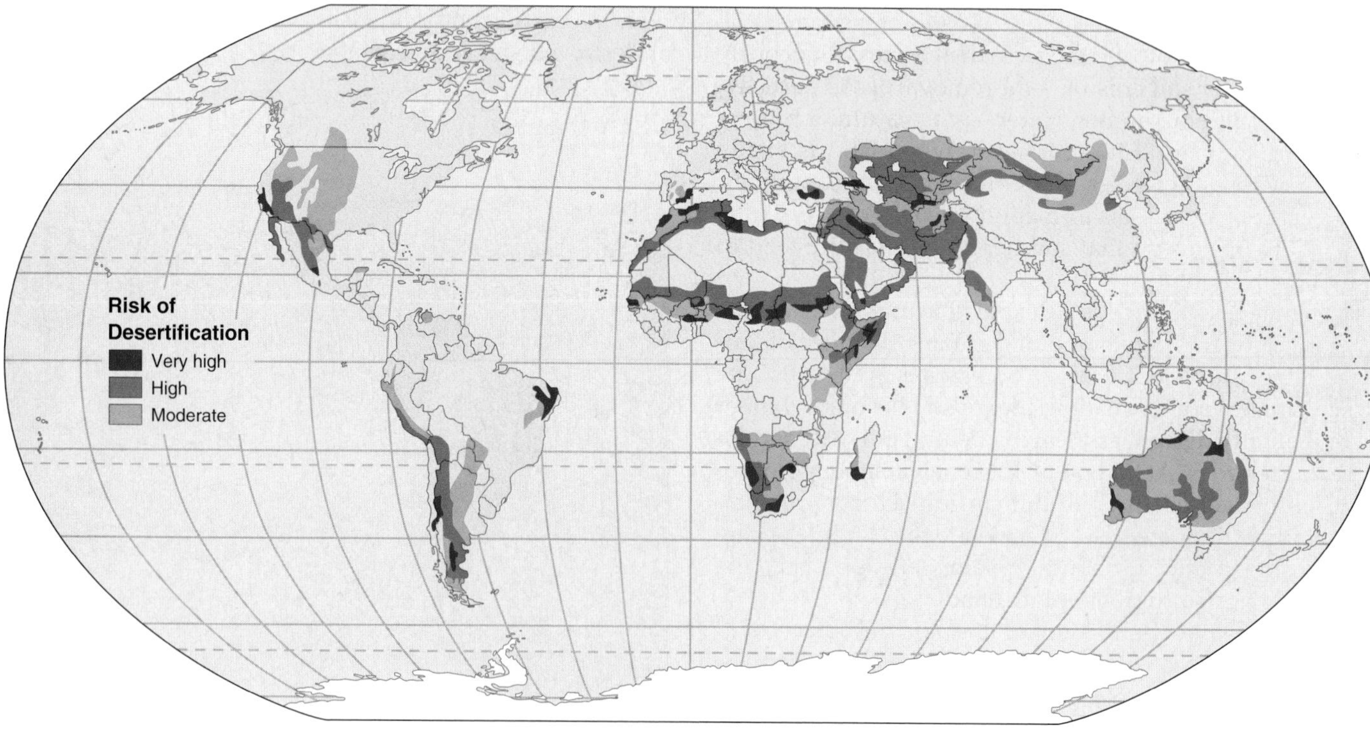


**FIGURE 5.27** Desertification affects about 1 billion people in more than 100 countries. According to the United Nations, one-quarter of the world's land surface now qualifies as degraded semidesert. Desertification is usually understood to imply the steady advance of the margins of the world's deserts into their bordering drylands, converting through human mistreatment formerly usable pastures and croplands into barren and sterile landscapes. In reality, the process may result from natural climatic fluctuations as much as from human abuse.

**FIGURE 5.28** Windblown dust is engulfing the scrub forest in this drought-stricken area of Mali, near Timbuktu. The district is part of the Sahel region of Africa, where desertification has been accelerated by both the climate and human pressures on the land. The cultivation of marginal land, overgrazing by livestock, and recurring droughts have led to the destruction of native vegetation, erosion, and land degradation. *© Wolfgang Kaehler.*

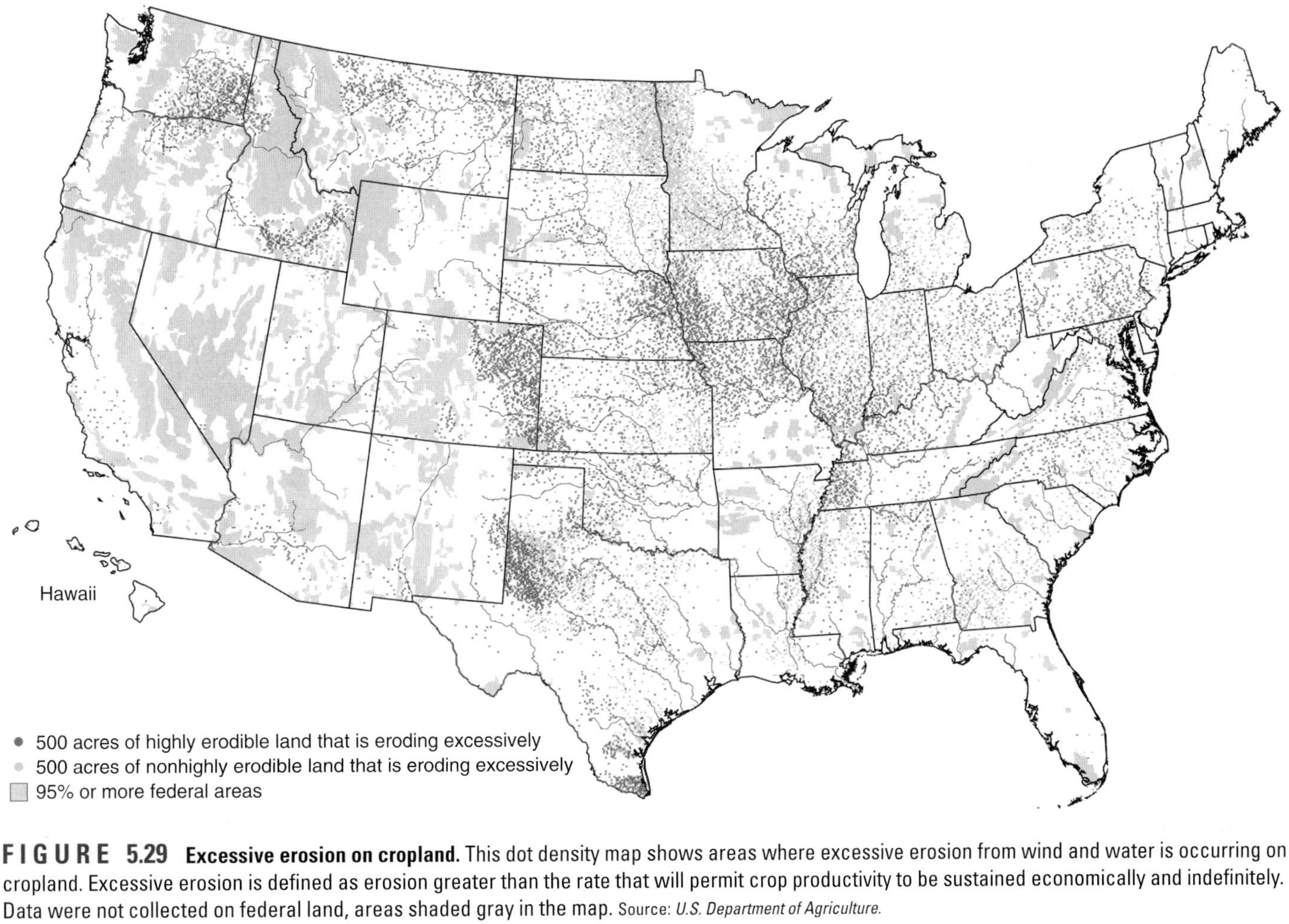


**FIGURE 5.29 Excessive erosion on cropland.** This dot density map shows areas where excessive erosion from wind and water is occurring on cropland. Excessive erosion is defined as erosion greater than the rate that will permit crop productivity to be sustained economically and indefinitely. Data were not collected on federal land, areas shaded gray in the map. Source: *U.S. Department of Agriculture.*

## Maintaining Soil Productivity

In much of the world, increasing population numbers are largely responsible for accelerated soil erosion. In the United States, economic conditions rather than population pressures have often contributed to excessive rates of erosion. Federal tax laws and the high farmland values of the 1970s encouraged farmers to plow virgin grasslands and to tear down windbreaks to increase their cultivable land and yields. The secretary of agriculture exhorted farmers to plant all of their land, "from fencerow to fencerow," to produce more grain for export. Land was converted from cattle grazing to corn and soybean production as livestock prices declined.

When prices of both land and agricultural products declined in the 1980s, farmers felt impelled to produce as much as they could in order to meet their debts and make any profit at all. To maintain or increase their productivity, many neglected conservation practices, plowing under marginal lands and using fields for the same crops every year.

Conservation techniques were not forgotten, of course. They are practiced by many and persistently advocated by farm organizations and soil conservation groups. Techniques to reduce erosion by holding the soils in place are well known. They include contour plowing, terracing, strip-cropping and crop rotation, erecting windbreaks and constructing water diversion channels, and practicing no-till or reduced tillage farming (the practice of sowing seeds without turning over the soil, thus allowing plants to grow amid the stubble of the previous year's crop).

Another method of reducing erosion and reclaiming land that is badly damaged is to pay farmers to stop producing crops on highly erodible farmland. After the mid-1980s, federal farm programs attempted to reverse some of the damage resulting from past economic pressures and farming practices. One objective has been to retire for conservation purposes some 18.6 million hectares (46 million acres) of croplands that were eroding faster than three times the natural rate of soil formation. The Conservation Reserve Program of 1985 pays farmers to take highly erodible land out of production; in return, farmers agree to plant the land in trees or a grass or legume cover and then retire it from further use for 10 years.

In 1996, however, Congress passed the Federal Agriculture Improvement and Reform Act, which contained as its cornerstone a program known as "Freedom to Farm." The premise of the program is that the market and not the federal government should determine which crops are planted. Almost immediately, farmers began expanding their acreage, planting crops with a high market value on land that had been set aside. It remains to be seen how much land that was previously protected by conservation measures will be cultivated using more destructive farming methods.

Only by reducing the economic pressures that lead to abuse of farmlands and by continuing to practice known soil conservation techniques can the country maintain the long-term productivity of soil, the resorce base upon which all depend.

**Contour plowing and strip-cropping.** *© Jerry Irwin/Photo Researchers.*

**No-till farming.** Source: *Gene Alexander/Soil Conservation Service, USDA.*

annual loss of more than 4 tons per acre. In some areas, the average is 15 to 20 tons per acre. Of the roughly 167 million hectares (413 million acres) of land that are intensively cropped in the United States, more than one-third are losing topsoil faster than it can be replaced naturally (it can take up to a century to replace an inch of topsoil). In parts of Iowa and Illinois where the topsoil was once a foot deep, less than half of it remains.

Like most processes, soil erosion has secondary effects. As the soil quality and quantity decline, croplands become less productive and yields drop. Streams and reservoirs experience accelerated siltation. In countries where the topsoil is heavily laden with agricultural chemicals, erosion-borne silt pollutes water supplies. The danger of floods increases as bottomlands fill with silt, and the costs of maintaining navigation channels grow.

Accelerated erosion is a primary cause of agricultural soil deterioration, but in arid and semiarid areas, salt accumulation can be a contributing factor. **Salinization** is the concentration of salts in the topsoil as a result of the evaporation of surface water. It occurs in poorly drained soils in dry climates, where evaporation exceeds precipitation. As water evaporates, some of the salts are left behind to form a white crust on the surface of the soil (Figure 5.30).

Like erosion, salinization is a natural process that has been accelerated by human activities. Poorly drained irrigation systems are the primary culprit, because irrigation water tends to move slowly and thus to evaporate rapidly. All irrigation water contains dissolved salts, which are left behind on the surface when water evaporates. Mild or moderate salinity makes soil less productive and lowers crop yields; extreme salinity ultimately can render the land unsuitable for agriculture.

Thousands of once fertile acres have been abandoned in Iran and Iraq; over 25% of the irrigated areas of India, Pakistan, Syria, and Egypt are affected by salinization. Approximately 1.6 million hectares (4 million acres) of cultivated soils in the Canadian provinces of Saskatchewan and Alberta are classified as overly saline. Areas of serious salinization also appear in the U.S. Southwest, particularly in the Colorado River drainage basin and in the Central Valley of California. Ironically, the irrigation water that transformed that arid, 430-kilometer-long (270-mi-long) valley into one of the country's most productive farming regions now threatens to make portions of it worthless again.

**FIGURE 5.30** Salinization has left a white crust of salt on the surface of this soil. The problem is most severe in hot, dry areas where irrigation is practiced. *© Mark Gibson.*

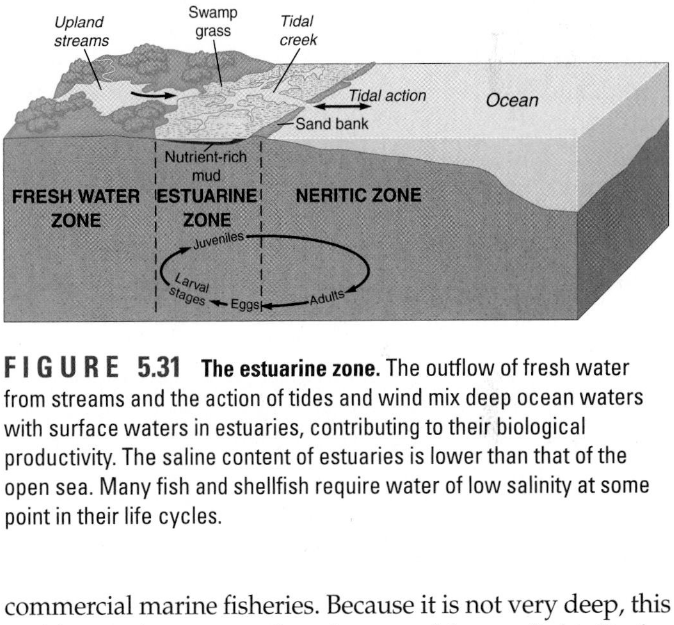


**FIGURE 5.31 The estuarine zone.** The outflow of fresh water from streams and the action of tides and wind mix deep ocean waters with surface waters in estuaries, contributing to their biological productivity. The saline content of estuaries is lower than that of the open sea. Many fish and shellfish require water of low salinity at some point in their life cycles.

## Wetlands

Vegetated land surfaces that are periodically or permanently covered by or saturated with standing water are called **wetlands.** Transitional zones between land and water, wetlands take a variety of forms, including grassy marshes, wooded swamps, tidal flats, and estuaries, and are found on all continents. Some are permanently wet; others have standing water for only part of the year. In North America, wetlands range in size from the small prairie potholes of the Midwest and Alberta, Canada, to areas as large as Florida's Everglades. Among the best known wetlands of the United States are the Okefenokee Swamp in Georgia and the bayous of Louisiana and Mississippi. Alaska has large expanses of wetlands, mostly peatlands.

The two major categories of wetlands are *inland* and *coastal.* In the United States, most wetlands are freshwater inland wetlands. They include bogs, marshes, swamps, and floodplains adjacent to rivers. Either fresh or salt water covers coastal wetlands, which are crucial to marine ecosystems. The part of the sea lying above the continental shelf is the most productive of all ocean waters, supporting the major commercial marine fisheries. Because it is not very deep, this *neritic zone* is penetrated and warmed by sunlight. It also receives the nutrients flowing into oceans from streams and rivers, so that vegetation and a great variety of aquatic life can flourish. However, the neritic zone depends to a considerable extent on the continued functioning of the **estuarine zone,** the relatively narrow area of wetlands along coastlines where salt water and fresh water meet and mix (Figure 5.31).

Extremely valuable ecological systems, wetlands perform a number of vital functions. Trapping and filtering the silt, pollutants, and nutrients that rivers bring downstream, wetlands improve water quality and help replenish underground aquifers. They absorb floodwaters and help stabilize shorelines by providing barriers to coastal erosion. Among the most diverse and productive of all ecosystems, wetlands provide habitat and food for a wide variety of plants and animals. Indeed, wetlands are essential to the survival of many species of fish and shellfish by serving as their spawning grounds. Wetlands are also major breeding, feeding, nesting, and wintering grounds for many types of birds (Figure 5.32). Not only are these areas extraordinarily productive themselves, but they also contribute to the productivity of the neritic zone, where fish feed on the life that flows from wetlands into the sea.

People have not always recognized or appreciated the value of wetlands, tending to view them as swampy, smelly areas that provide breeding grounds for mosquitoes and impeded settlement—wastelands that should be reclaimed for productive uses such as agriculture and commercial development. Indeed, in the United States, Congress in the mid-1800s passed the Swamp Land Acts, which made it national policy to drain and fill wetlands. Scientists estimate that half the wetlands of the world have been destroyed. The United States is thought to have lost about 50% of its original wetlands, Europe at least 60%, and Australia and New Zealand about 90%.

Wetlands have been drained, dredged, filled and built upon, converted to cropland, and used as garbage dumps. They are polluted by chemicals, excess nutrients, and other waterborne wastes. Natural shorelines have been bulldozed, and artificial levees and breakwaters interfere with the flooding that nurtures wetlands with fresh infusions of sediment and water. Habitat alteration or destruction inevitably disrupts the intricate ecosystems of the wetlands.

Growing awareness of the importance of wetlands and of how much wetland acreage has been damaged or destroyed has led in the United States and elsewhere to efforts to preserve and protect them. In the United States, the Clean Water Act of 1972 and subsequent amendments gave wetlands a measure of federal protection. The act prohibits the filling of wetlands without a permit issued by the U.S. Army Corps of Engineers (USCOE), a provision critics have likened to putting a fox in charge of the hen house, since the corps has long emphasized dredging, stream straightening, and dike building. Federal protection does not mean a wetland cannot be developed, and the USCOE has issued thousands of permits letting homeowners and developers fill in hundreds of thousands of acres of wetlands. Thus, although the rate of wetlands loss has slowed in recent years, many of the wetlands that remain are in danger of degradation or loss.

**FIGURE 5.32 A salt marsh in Louisiana.** Tidal marshlands have been subjected to dredging and filling for residential and industrial development. The loss of such areas reduces the essential habitat of waterfowl, fish, crustaceans, and mollusks. Many waterfowl breed and feed in coastal marshes and use them for rest during long migrations. *© Franke Keating/Photo Researchers, Inc.*

## Forest Resources

Wetlands are only one of the renewable resources in danger of irreparable damage by human action. In many parts of the world, forests are similarly endangered.

After the retreat of continental glaciers some 12,000 years ago, and before the rise of agriculture, the world's forests and woodlands probably covered some 45% of the earth's land area exclusive of Antarctica. They were a sheltered and productive environment for earlier societies that subsisted on gathered fruits, nuts, berries, leaves, roots, and fibers collected from trees and woody plants. Few such cultures remain, though the gathering of forest products is still an important supplemental activity, particularly among subsistence agricultural societies.

Even after millennia of land clearance for agriculture and, more recently, commercial lumbering, cattle ranching, and fuelwood gathering, forests still cover about one-third of the world's land area. As an industrial raw material source, however, forests are more restricted in area. Although forests of some type reach discontinuously from the equator northward to beyond the Arctic Circle and southward to the tips of the southern continents, *commercial forests* are restricted to two very large global belts (Figure 5.33). One, nearly continuous, occupies the upper-middle latitudes of the Northern Hemisphere. The second straddles the equatorial zones of South and Central America, Central Africa, and Southeast Asia. These forest belts differ in the types of trees they contain and in the type of market or use they serve.

The *northern coniferous*, or softwood, forest is the largest and most continuous stand, circling the globe below the polar regions. Its pine, spruce, fir, and other types of conifers are used for construction lumber and to produce pulp for paper, rayon, and other cellulose products. To its south are the *temperate hardwood* forests, containing deciduous species such as oak, hickory, maple, and birch. These and the trees of the mixed forest lying between the hardwood and softwood belts have been greatly reduced in areal extent by centuries of agricultural and urban settlement and development, though they still are commercially important for hardwood applications: furniture, veneers, railroad ties, and so on.

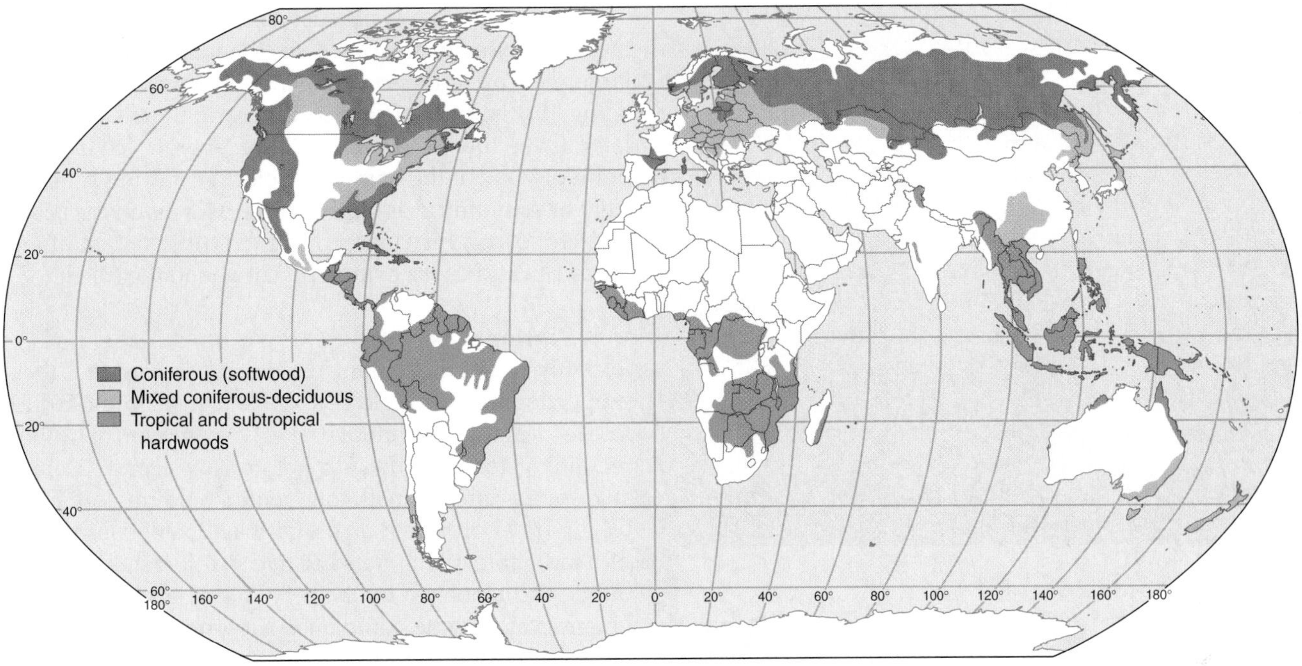


**FIGURE 5.33 Major commercial forest regions.** Much of the original forest, particularly in midlatitude regions, has been cut over. Many treed landscapes that remain do not contain commercial stands. Significant portions of the northern forests are not readily accessible and at current prices cannot be considered commercial.

The *tropical lowland hardwood* forests are exploited primarily for fuelwood and charcoal, on which the populations of developing countries are heavily dependent. About 90% of world fuelwood production comes from the forests of Africa, Asia, Oceania, and Latin America. An increasing quantity of special-quality woods from the tropical forests is cut for export as lumber, however. Southeast Asian countries such as Myanmar and Indonesia now account for much of the world's hardwood log exports (Figure 5.34).

The adage about not being able to see the forest for the trees is applicable to those who view forests only for the commercial value of the trees they contain. Forests are more than trees, and timbering is only one purpose that forests serve. Chief among the other purposes are soil and watershed conservation, the provision of a habitat for wildlife, and recreation. Forests also play a vital role in the global recycling of water, carbon, and oxygen.

Because forests serve a variety of purposes, the kind of management techniques employed in any one area depend on the particular use(s) to be emphasized. Thus, if the goal is to maintain a diversity of native plant species in order to provide a maximum number of ecological niches for wildlife, the forest will be managed differently than if it is designed for public recreation or the protection of watersheds. Even if the use to be emphasized is timber production, different management approaches may be taken. Logging techniques for the production of plywood or wood chips, for example, differ from those used for the production of high-quality lumber.

Commercial forests can be considered a renewable resource only if sustained-yield techniques are practiced—that is, if harvesting is balanced by new growth (see the term *maximum sustainable yield* in Chapter 10). Timber companies employ a number of methods of tree harvesting and regeneration. Two quite different practices, clear cutting and selective cutting, illustrate the diversity of such approaches (Figure 5.35).

*Clear cutting* is one of the most controversial logging practices. As the name implies, all the trees are removed

**FIGURE 5.34 Teak logs for export stacked near Mandalay, Myanmar.** © Larry Tackett/Tom Stack & Associates.

(a)

(b)

**FIGURE 5.35** (a) **Clear cutting in the Gifford Pinchot National Forest, Washington.** Cutting every tree, regardless of species or size, drives out wildlife, damages watersheds, disrupts natural regeneration, and removes protective ground cover, exposing slopes to erosion. (b) **Selective cutting in eastern Canada.** Older, mature specimens are removed at first cutting. Younger trees are left for later harvesting.
*(a) © Milton Rand/Tom Stack & Associates. (b) © R. Moller/Valan Photos.*

from a given area at one time. The site is then left to regenerate naturally or is replanted, often with fast-growing seedlings of a single species. Excessive clear cutting, particularly on steep slopes, destroys wildlife habitats, accelerates soil erosion and water pollution, replaces a mixed forest with a wood plantation of no great genetic diversity, and reduces or destroys the recreational value of the area.

*Selective cutting* is practiced in mixed-forest stands containing trees of varying ages, sizes, and species. Medium and large trees are cut either singly or in small groups, encouraging the growth of younger trees that will be harvested later. Over time, the forest will regenerate itself. From the point of view of the harvester, selective cutting is not as efficient or economical as clear cutting. Moreover, the practice is often followed very loosely, and the construction of logging roads can still cause extensive damage to the forest.

### U.S. National Forests

Roughly one-third of the United States is forested, the same proportion for the world as a whole. Only some 40% of those forests provide the annual harvest of commercial timber. The remaining forests either do not contain economically valuable species; are in small, fragmented holdings; are inaccessible; or are in protected areas. Of that 40% of commercial forest land, almost half is in 171 national forests owned by the public and managed by the U.S. Forest Service (Figure 5.36). Logging by private companies is permitted; timber companies pay for the right to cut designated amounts of timber. Currently, the Forest Service is at the center of debates over how the forests should be managed. Among the issues are methods of harvesting, the cutting of very old tree stands, road building, and rates of reforestation.

By law—the 1960 Multiple Use Sustained Yield Act—the national forests are to be managed for four purposes: recreation, timber production, watershed protection, and wildlife habitat preservation. Although no use is to be particularly favored over others, conservationists charge that the Forest Service increasingly supports commercial logging and that the forests are being cut at an unprecedented rate.

In recent years, billions of board feet of timber have been taken from the national forests. Environmentalists are especially concerned that nearly half of this has come from national forests in Oregon and Washington, most of it irreplaceable "old growth." These virgin forests contain trees that are among the tallest and oldest in the world, indeed that were alive when Pilgrims set foot on Plymouth Rock.

Old-growth forests include trees of every age and size, both living and dead. Some ancient trees are immense, capable of growing 90 meters (300 ft) high, and they may live for more than 1000 years. They include the Douglas fir, Western red cedar, sequoia, and redwood. Tons of dead and decaying logs carpet the forest floor, where, sodden with moisture, they help control erosion and protect the forest from fire. As they decay, the logs release nutrients back into the soil. Such forests provide a habitat for hundreds of types of insects and animals, some of them threatened or endangered species.

The only large expanses of old-growth forest remaining in the United States are in the Pacific Northwest, most of them owned by the federal government. These ancient forests once covered about 60% of the forested areas between the Cascade Mountains and the Pacific Ocean, stretching 3200 kilometers (2000 mi) from California to Alaska. Today, only 10% of the old-growth forests remain, and they are

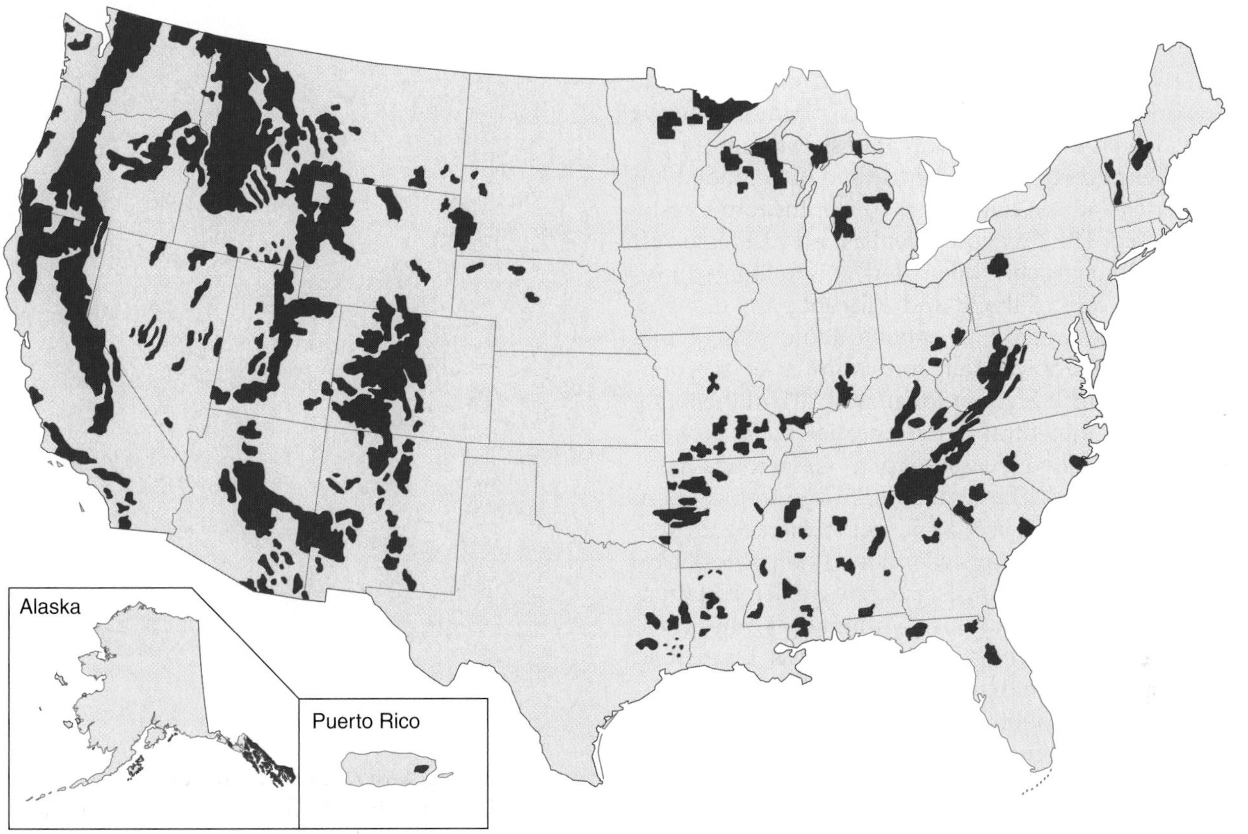


**FIGURE 5.36 National forests of the United States.** Every year, more than 180,000 hectares (450,000 acres) of trees are cut down within these forests—about 4 square kilometers (2 sq mi) of deforestation every day. To accommodate the cutting, 547,000 kilometers (340,000 mi) of logging roads (10 times the length of the U.S. interstate highway system) had been built in formerly pristine areas by 2001.

being logged at the rate of 25,000 hectares (60,000 acres) per year. If logging continues at the present rate, they will be gone in about 20 years, ceasing to exist as they are today. Although companies plant new seedlings to replace those they cut, timber is being harvested twice as fast as new trees can replace it. Further, traditional management practices, including clear cutting, road building, and harvesting after decades, not centuries, of regeneration prevent the development of a true old-growth forest ecosystem.

It is ironic that many Americans condemn the burning of the tropical rain forests while the U.S. government not only permits the destruction of forests just as ecologically precious but, in fact, subsidizes that destruction. The federal government annually loses more than $1 billion on timber sales, because building and maintaining the logging roads costs far more than the timber companies pay for the wood.

Perhaps the worst abuse has occurred in the Tongass National Forest, North America's largest temperate rain forest, which stretches along 800 kilometers (500 mi) of Alaska's southeastern coastline (Figure 5.37). It is thought to be one of the last remaining places in the United States containing every plant and animal species that existed in it before European colonization. A storehouse of biodiversity, the Tongass is home to such threatened species as grizzly bears and bald eagles; its waters teem with aquatic life, from salmon to whales. Clear cutting and road building, however, are endangering the viability of those wildlife habitats. The federal government has spent hundreds of millions of dollars to build some 7400 kilometers (4600 mi) of access roads and to promote commercial timbering and has received a pittance in return. Five-hundred-year-old trees 3 meters (10 ft) in diameter and more than 32 meters (105 ft) tall have been sold for $3 each and ground into pulp that is sent overseas and converted into products such as rayon and cellophane.

### Tropical Rain Forests

It is not only in the United States that government economic policies accelerate the rate of forest destruction. Much of the deforestation occurring in tropical areas also has governmental sanction. Brazil, Indonesia, and the Philippines are among the countries where governments subsidize projects aimed at converting forests to other uses, such as farming, cattle ranching, and mining. Their economic policies are driven by the pressure of growing population numbers, the need for more agricultural land, expanded demand for fuel and commercial wood, a midlatitude market for beef that can be satisfied profitably by replacing tropical forest with cleared grazing land, and an increasing demand in China for soybeans, soy oil, and soy meal by-products.

The tropical forests extend across parts of Asia, Africa, and Latin America (Figure 5.38). Millions of acres are being completely cleared every year, and almost half their original expanse has already been either cleared or degraded. In Central America and the Caribbean, 70% of the rain forests have disappeared. The Amazon River Basin, a vast region of more than 4 million square kilometers (1.5 million sq mi) extending across part of Brazil and adjacent parts of eight other countries, is thought to contain about half of the world's remaining tropical rain forests. Most of the rain forests of Africa, which account for around 30% of the total, now exist mainly in central Africa because those of west Africa, from Sierra Leone eastward to Cameroon, have been largely destroyed. The remaining 20% of tropical rain forests are found in the Asia/Pacific region. Here again, the picture is bleak. India, Malaysia, and the Philippines have already lost much of their forests, and rates of deforestation recently have risen sharply in Myanmar, Kampuchea (Cambodia), Thailand, Vietnam, and Indonesia. Overall, it is estimated that nearly half of Asia's natural forest is gone.

Although no one doubts that the tropical rain forests are being cleared, there is considerable uncertainty about the rate at which that is happening. The United Nations estimates that on average about 40,000 square kilometers (15,000 sq mi) have been cleared annually in recent years, or slightly less than 1% per year of the remaining forests. Satellite imagery of the Brazilian forests in the Amazon yields a similar figure. In Brazil, this means that, on average, an area roughly the size of Connecticut is deforested each year.

Deforestation in Brazil has become the focus of international attention because it has both the largest area of tropical rain forests *and* one of the highest rates of clearing. No country better illustrates the fact that deforestation is a complicated issue that pits the governments of developing countries against those of industrialized countries, international development organizations against environmental groups, and multinational corporations against advocates of human rights. To critics of its policy of deliberately developing the Amazon Basin, Brazilians respond that it is justified because it relieves high population densities in its crowded northeast; gives farmers and ranchers land on which to build a livelihood; permits the country to tap unexploited natural resources; and helps Brazil repay its huge foreign debt. Pointing out that most of the forests in Western Europe and the United States were cleared decades if not centuries ago, and that subsequent resource exploitation led to wealth and prosperity, Brazilians ask why they shouldn't be able to use their natural resources in what they deem to be their own best interest.

There are good reasons, however, why North Americans should care what happens to the tropical rain forests. Their destruction raises three principal global concerns and a host of local ones. First, all forests play a major role in maintaining the oxygen and carbon balance of the earth.

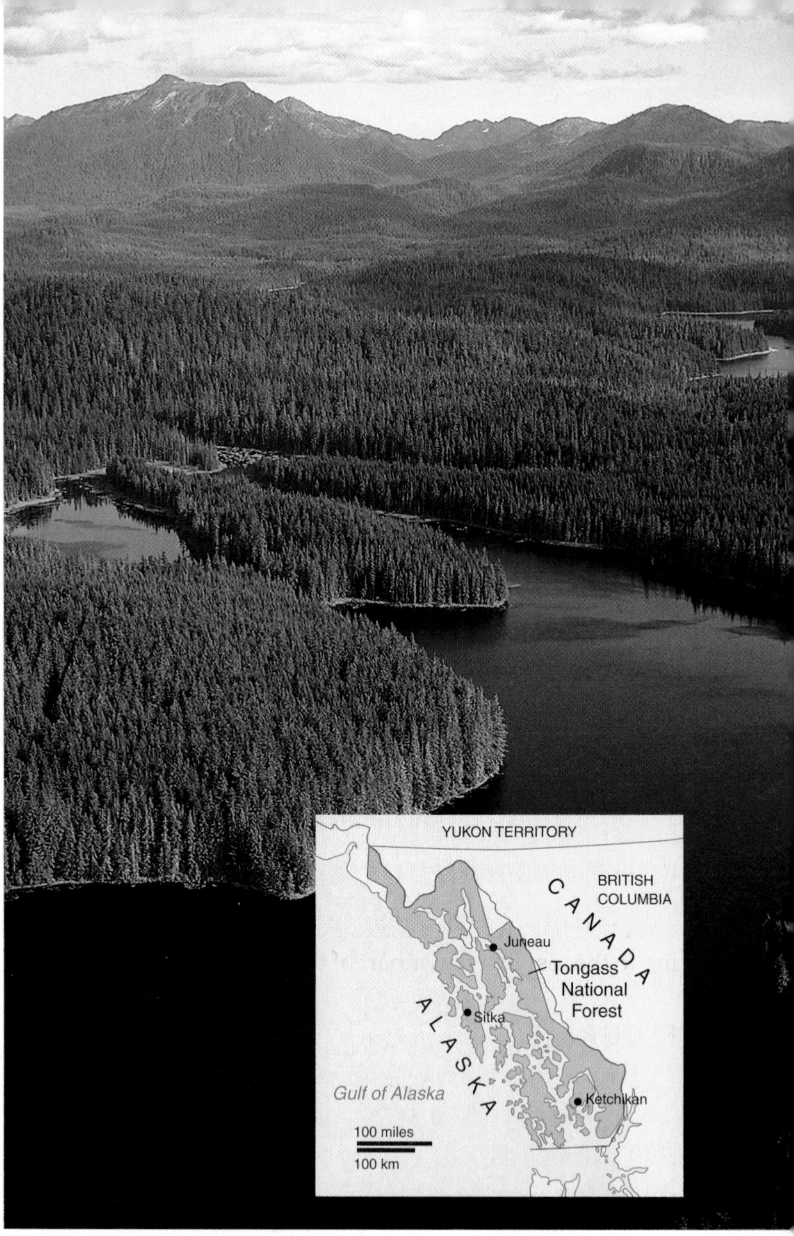


**FIGURE 5.37 The Tongass National Forest,** the largest national forest in North America, covers approximately 7 million hectares (17 million acres). The innumerable islands, inlets, and fiords of the Tongass are set against a backdrop of coastal mountains. Since the late 1950s, the federal government has subsidized the timber industry by promising companies a long-term supply of cheap wood from the Tongass. Conservationists contend that old-growth trees are not needed for forest products and that the Tongass should be managed for wildlife, fishing, and tourism, as well as logging. *Photo: © Don Pitcher/Alaska Stock.*

People and their industries consume oxygen; vegetation both extracts the carbon from atmospheric carbon dioxide and releases oxygen back into the atmosphere. Indeed, the forests of the Amazon have been called the "lungs of the world" for their major contribution to the oxygen breathed by humankind. When the tropical forest is cleared, its role both as a carbon "sink" and as an oxygen-replenisher is lost.

A second global concern is the contribution of forest clearing to air pollution and climate change. Deforestation by burning releases vast quantities of carbon dioxide into the atmosphere. Brazilian scientists estimate that the thousands

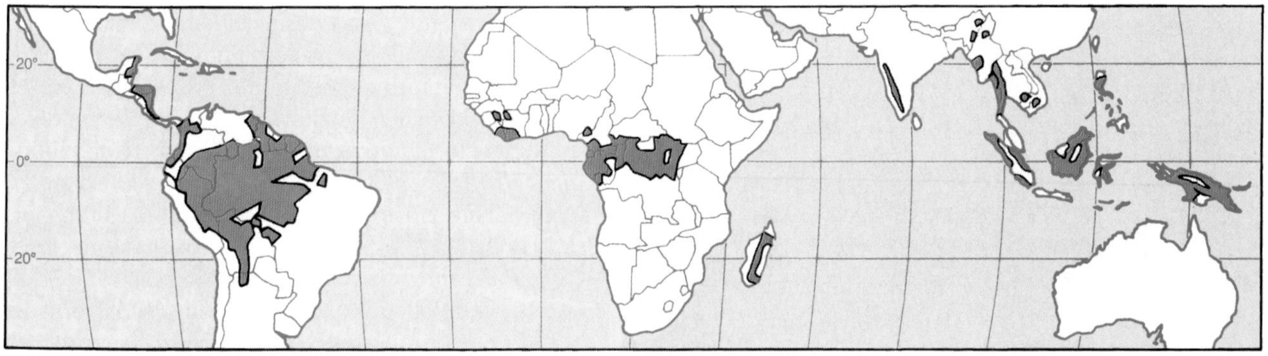


**FIGURE 5.38** Tropical rain forests exist in tropical latitude regions with high temperatures and high levels of humidity year-round. The Amazon River Basin has the world's largest continuous area of rain forests. Three countries contain more than half of the tropical rain forests: Brazil, the Democratic Republic of Congo, and Indonesia. Large tracts are being cleared to make way for farming, cattle ranching, commercial logging, and development projects.

of fires that are set to clear the Amazon forest account for one-tenth of the global production of carbon dioxide, contributing to the warming of the atmosphere (Figure 5.39). In addition, the fires generate gases (nitrogen oxides and methane) that create acid rain and contribute to the depletion of the ozone layer, topics discussed in Chapter 12.

Finally, the eradication of tropical forests is already leading to the loss of a major part of the biological diversity of the planet. The forests are one component in an intricate ecosystem that has developed over millions of years. The trees, vines, flowering plants, animals, and insects depend on one another for survival. The destruction of the habitat by clearance annually causes the extinction of thousands of plant and animal species that exist nowhere else. Although the tropical rain forests now occupy less than 10% of the earth's land surface, they are thought to contain anywhere from 50% to 70% of all the species of plants, animals, and microorganisms in the world. Many of the plants have become important world staple food crops, among them rice, corn, cassava, squash, banana, pineapple, and sugarcane. Unknown additional potential food species remain as yet unexploited. In addition, the tropical forests yield an abundance of industrial products (oils, gums, latexes, and turpentines) and are the world's main storehouse of medicinal plants (see "Tropical Forests and Medical Resources," p. 172).

Deforestation also incurs heavy environmental, economic, and social costs on a more local basis. All forests anchor topsoil and absorb excess moisture. In a vicious cycle, forest clearance accelerates soil erosion and siltation of streams and irrigation channels, leaving the area vulnerable to flooding and drought, leading in turn to future shortages of food and fuelwood. Within a matter of years, land that has been cleared for agriculture can become unsuitable for that use. In the Himalayan watershed, in the Ethiopian highlands, and in numerous other places, deforestation, erosion, and rainfall runoff have aggravated floods that have killed tens of thousands of people and left millions of others homeless.

# Resource Management

The destruction of the rain forests is a tragedy that yields no long-term benefits. The world is approaching the end of a period in which resources were cheap, readily available, and lavishly used. Over the centuries, the earth has been viewed as an almost inexhaustible storehouse of resources for humans to exploit and, simultaneously, as a vast repository for the waste products of society. Now there is a growing realization that resources can be depleted, even renewable ones, such as forests, that many have lifespans measured only in decades, and that the air and water—also resources—cannot absorb massive amounts of pollutants and yet retain their life-supporting abilities.

That realization was reflected in June 1992, at the Earth Summit in Rio de Janeiro, when the world's governments agreed to form the UN Commission on Sustainable Development. Since that time, more than 70 countries, including the United States, have launched efforts to chart a path toward **sustainable development,** which is generally defined as development that satisfies current needs without jeopardizing the ability of future generations to meet their own needs. The principles of sustainability are straightforward. The sustainable use of renewable resources means using them at rates within their capacity for regeneration. Over the long term,

- soil erosion cannot exceed soil formation;
- forest destruction cannot exceed forest regeneration;
- species extinction cannot exceed species evolution;
- fish catches cannot exceed the regenerative capacity of fisheries; and
- pollutants cannot exceed the capacity of the system to absorb them.

A society can violate the principles of sustainability in the short run but not in the long run and still endure.

"Zero, possibly even negative [population] growth" was the 1972 slogan proposed by the prime minister of Singapore, an island country in Southeast Asia. His nation's population, which stood at 1 million at the end of World War II (1945), had doubled by the mid-1960s. To avoid the overpopulation he foresaw, the government decreed "Boy or girl, two is enough" and refused maternity leaves and access to health insurance for third or subsequent births. Abortion and sterilization were legalized, and children born fourth or later in a family were to be discriminated against in school admissions policy. In response, by the mid-1980s, birth rates had fallen below the level necessary to replace the population, and abortions were terminating more than one-third of all pregnancies.

"At least two. Better three. Four if you can afford it" was the national slogan proposed by that same prime minister in 1986, reflecting fears that the stringencies of the earlier campaign had gone too far. From concern that overpopulation would doom the country to perpetual Third World poverty, Prime Minister Lee Kuan Yew was moved to worry that population limitation would deprive it of the growth potential and national strength implicit in a youthful, educated workforce adequate to replace and support the present aging population. His 1990 national budget provided for sizable long-term tax rebates for second children born to mothers under 28. Not certain that financial inducements alone would suffice to increase population, the Singapore government annually renewed its offer to take 100,000 Hong Kong Chinese who might choose to leave when China took over that territory in 1997.

The policy reversal in Singapore reflects an inflexible population reality: the structure of the present controls the content of the future. The size, characteristics, growth trends, and migrations of today's populations help shape the well-being of peoples yet unborn but whose numbers and distributions are now being determined. The numbers, age, and sex distribution of people; patterns and trends in their fertility and mortality; and their density of settlement and rate of growth all affect and are affected by the social, political, and economic organization of a society. Through population data we begin to understand how the people in a given area live, how they may interact with one another, how they use the land, what pressure on resources exists, and what the future may bring.

**Population geography** provides the background tools and understandings of those interests. It focuses on the number, composition, and distribution of human beings in relation to variations in the conditions of earth space. It differs from **demography,** the statistical study of human population, in its concern with *spatial* analysis—the relationship of numbers to area. Regional circumstances of resource base, type of economic development, level of living, food supply, and conditions of health and well-being are basic to geography's population concerns. They are, as well, fundamental expressions of the human–environmental relationships that are the substance of all human geographic inquiry.

## Population Growth

Sometime early in 2004, a human birth raised Earth's population to about 6.4 billion people. At the start of 1991, the count was nearly 5.4 billion. That is, over the 13 years between those dates, the world population grew on average by about 77 million people annually, or some 211,000 per day. The average, however, conceals the reality that annual increases have been declining over the years. During the early 1990s, the U. S. Census Bureau and the United Nations Population Division regularly reported yearly growth at 85 to 90 million. Even with the slower pace of estimated increase, the United Nations early this century still projected that the world would likely contain almost 9 billion inhabitants in 2050 and grow to perhaps 9.1 billion around the year 2100. Many demographers, however, impressed by dramatic birth rate reductions reported by 2002 for many developing and populous countries—India, significantly—lowered their estimates to predict end-of-century world totals peaking at between 8 and 9 billion followed by numerical decline, not stability. All do agree, however, that essentially all of any future growth will occur in countries now considered "developing" (Figure 6.1). We will return to these projections and to the difficulties and disagreements inherent in making them later in this chapter.

Just what is implied by numbers in the millions and billions? With what can we compare the 2003 population of Guinea-Bissau in Africa (about 1.3 million) or of China (about 1.3 billion)? Unless we have some grasp of their scale and meaning, our understanding of the data and data manipulations of the population geographer can at best be superficial. It is difficult to appreciate a number as vast as 1 million or 1 billion, and the great distinction between them. Some examples offered by the Population Reference Bureau may help in visualizing their immensity and implications.

- A 2.5-centimeter (1-in.) stack of U.S. paper currency contains 233 bills. If you had a *million* dollars in

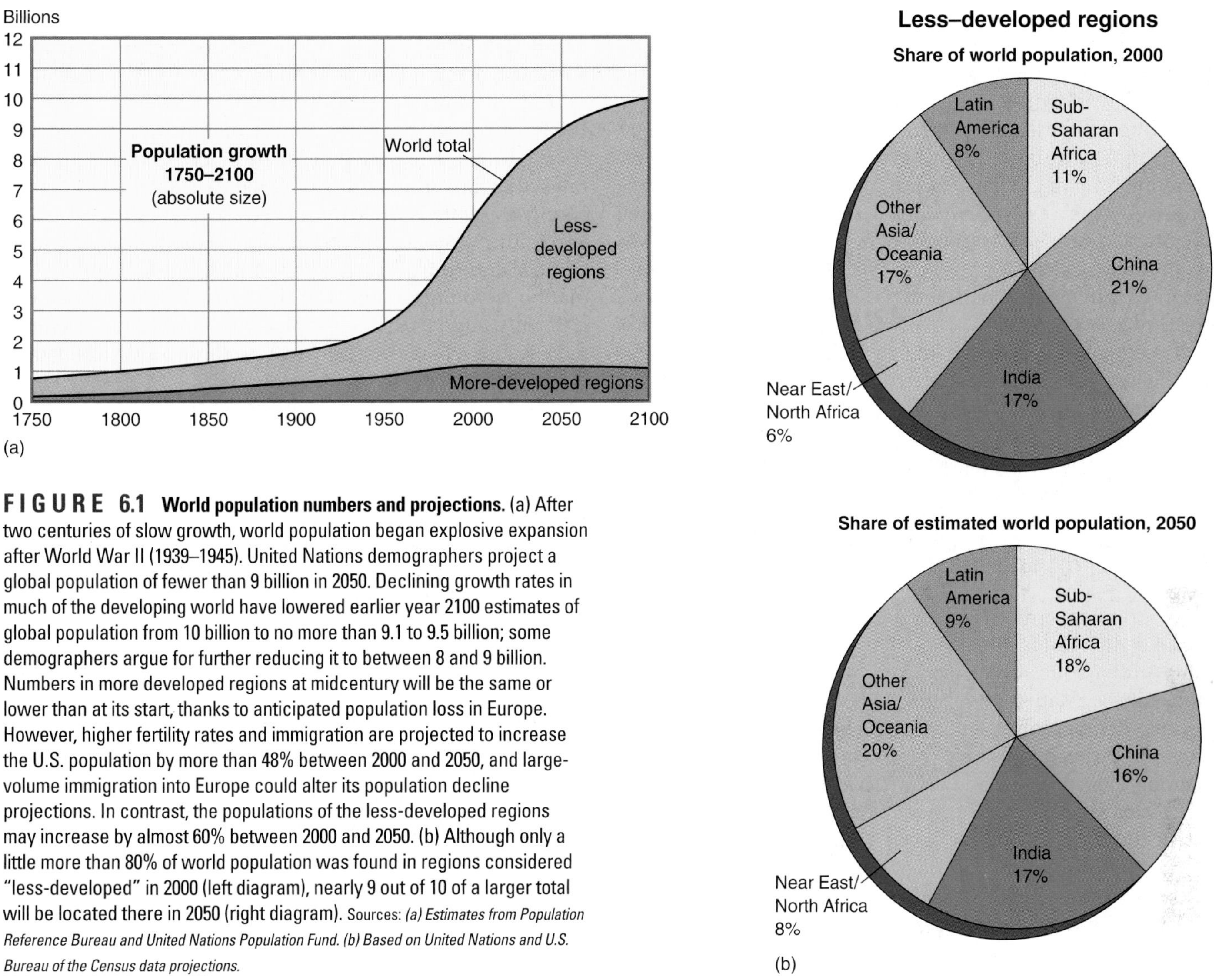


**FIGURE 6.1 World population numbers and projections.** (a) After two centuries of slow growth, world population began explosive expansion after World War II (1939–1945). United Nations demographers project a global population of fewer than 9 billion in 2050. Declining growth rates in much of the developing world have lowered earlier year 2100 estimates of global population from 10 billion to no more than 9.1 to 9.5 billion; some demographers argue for further reducing it to between 8 and 9 billion. Numbers in more developed regions at midcentury will be the same or lower than at its start, thanks to anticipated population loss in Europe. However, higher fertility rates and immigration are projected to increase the U.S. population by more than 48% between 2000 and 2050, and large-volume immigration into Europe could alter its population decline projections. In contrast, the populations of the less-developed regions may increase by almost 60% between 2000 and 2050. (b) Although only a little more than 80% of world population was found in regions considered "less-developed" in 2000 (left diagram), nearly 9 out of 10 of a larger total will be located there in 2050 (right diagram). Sources: *(a) Estimates from Population Reference Bureau and United Nations Population Fund. (b) Based on United Nations and U.S. Bureau of the Census data projections.*

thousand-dollar bills, the stack would be 11 centimeters (4.3 in.) high. If you had a *billion* dollars in thousand-dollar bills, your pile of money would reach 109 meters (358 ft)—about the length of a football field.

- You had lived a *million* seconds when you were 11.6 days old. You won't be a *billion* seconds old until you are 31.7 years of age.
- When still flying, the supersonic airplane, the *Concorde,* could theoretically circle the globe in only 18.5 hours at its cruising speed of 2150 kilometers (1336 mi) per hour. It would take 31 days for a passenger to journey a *million* miles on the Concorde, whereas a trip of a *billion* miles would last 85 years.

The implications of the present numbers and the potential increases in population are of vital current social, political, and ecological concern. Population numbers were much smaller some 12,000 years ago when continental glaciers began their retreat, people spread to formerly unoccupied portions of the globe, and human experimentation with food sources initiated the Agricultural Revolution. The 5 or 10 million people who then constituted all of humanity obviously had considerable potential to expand their numbers. In retrospect, we see that the natural resource base of the earth had a population-supporting capacity far in excess of the pressures exerted on it by early hunting and gathering groups.

Some observers maintain that despite present numbers or even those we can reasonably anticipate for the future, the adaptive and exploitative ingenuity of humans is in no danger of being taxed. Others, however, compare Earth to a self-contained spaceship and declare with chilling conviction that a finite vessel cannot bear an ever-increasing number of passengers. They point to recurring problems of malnutrition and starvation (though these are realistically more a matter of failures of distribution than of inability to produce enough foodstuffs worldwide). They cite dangerous conditions of air and water pollution, the loss of forest and farmland, the apparent nearing exhaustion of many minerals and fossil fuels, and other evidences of strains on world resources as foretelling the discernible outer limits of population growth.

On a worldwide basis, populations grow only one way: the number of births in a given period exceeds the number of deaths. Ignoring for the moment regional population changes resulting from migration, we can conclude that observed and projected increases in population must result from the failure of natural controls to limit the number of births or to increase the number of deaths, or from the success of human ingenuity in circumventing such controls when they exist. In contrast, current estimates of slowing world population growth and eventual stability or decline clearly indicate that humans, by their individual and collective decisions, may effectively limit growth and control global population numbers. The implications of these observations will become clearer after we define some terms important in the study of world population and explore their significance.

## Some Population Definitions

Demographers employ a wide range of measures of population composition and trends, though all their calculations start with a count of events: of individuals in the population, births, deaths, marriages, and so on. To those basic counts, demographers add refinements that make the figures more meaningful and useful in population analysis. Among them are *rates* and *cohort* measures.

**Rates** simply record the frequency of occurrence of an event during a given time frame for a designated population—for example, the marriage rate as the number of marriages performed per 1000 population in the United States last year. **Cohort** measures refer data to a population group unified by a specified common characteristic—the age cohort of 1–5 years, perhaps, or the college class of 2008 (Figure 6.2). Basic numbers and rates useful in the analysis of world population and population trends have been reprinted in this book with the permission of the Population Reference Bureau as Appendix 2. Examination of them will help illustrate the discussion that follows.

### Birth Rates

The **crude birth rate** (CBR), often referred to simply as the *birth rate*, is the annual number of live births per 1000 population. It is "crude" because it relates births to total population without regard to the age or sex composition of that population. A country with a population of 2 million and with 40,000 births a year has a crude birth rate of 20 per 1000.

$$\frac{40{,}000}{2{,}000{,}000} = 20 \text{ per } 1000$$

The birth rate of a country is, of course, strongly influenced by the age and sex structure of its population, by the customs and family size expectations of its inhabitants, and by its adopted population policies. Because these conditions vary widely, recorded national birth rates vary—early in the 21st century, from a high of 45 to 50 or more in some West African states to lows of 9 or 10 per 1000 in 20 or more European countries. Although birth rates of 30 or above per 1000 are considered *high*, almost one-fifth of the world's people (down from one-half in 1990) live in countries with rates that are that high or higher (Figure 6.3). In these countries—chiefly in Africa, western and southern Asia, and Latin America—the population is predominantly agricultural and rural, and a high proportion of the female population is young. In many of them, birth rates may be significantly higher than official records indicate. Available data suggest that every year around 50 million births go unregistered and therefore uncounted.

Birth rates of less than 18 per 1000 are reckoned *low* and are characteristic of industrialized, urbanized countries. All European countries, including Russia, as well as Anglo America, Japan, Australia, and New Zealand have low rates, as, importantly, do an increasing number of developing states. Some of these, such as China (see "China's Way—and Others"), have adopted effective family planning programs. In others, changed cultural norms have reduced desired family size. *Transitional* birth rates (between 18 and 30 per 1000) characterize some, mainly smaller developing and newly industrializing countries, although giant India entered that group in 1994.

As the recent population histories of Singapore and China indicate, birth rates are subject to change. The decline to current low birth rates of European countries and of some of the areas that they colonized is usually ascribed to industrialization, urbanization, and in recent years, maturing populations. While restrictive family planning policies in China rapidly reduced the birth rate from over 33 per 1000 in 1970 to 18 per 1000 in 1986, industrializing Japan experienced a comparable 15-point decline in the decade 1948–1958 with little governmental intervention. Indeed,

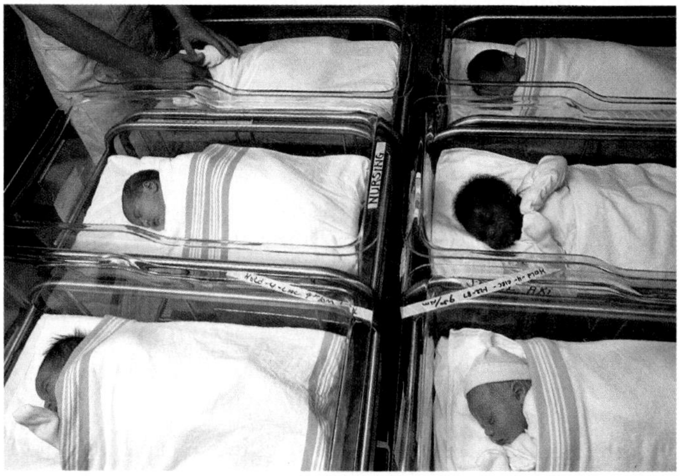

**FIGURE 6.2** Whatever their differences by race, sex, or ethnicity, these babies will forever be clustered demographically into a single *birth cohort*. © Herb Snitzer/Stock, Boston.

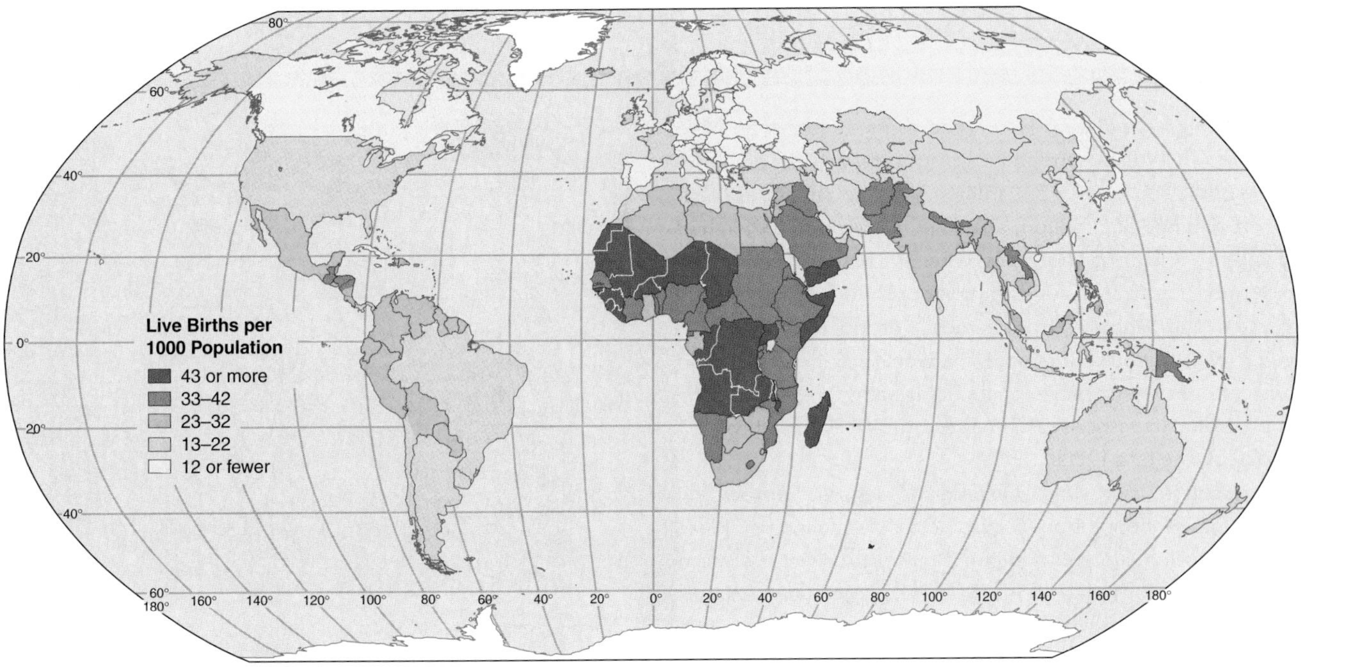


**FIGURE 6.3 Crude birth rates.** The map suggests a degree of precision that is misleading in the absence of reliable, universal registration of births. The pattern shown serves, however, as a generally useful summary of comparative reproduction patterns if class divisions are not taken too literally. Reported or estimated population data vary annually, so this and other population maps may not agree in all details with the figures recorded in Appendix 2. Source: *Data from Population Reference Bureau.*

the stage of economic development appears closely related to variations in birth rates among countries, although rigorous testing of this relationship proves it to be imperfect (Figure 6.3). As a group, the more developed states of the world showed a crude birth rate of 11 per 1000 in the early years of the 21st century; less-developed countries (excluding China) registered about 28 per 1000 (down from 35 in 1990).

Religious and political beliefs can also affect birth rates. The convictions of many Roman Catholics and Muslims that their religion forbids the use of artificial birth control techniques often lead to high birth rates among believers. However, dominantly Catholic Italy has nearly the world's lowest birth rate, and Islam itself does not prohibit contraception. Similarly, some European governments—concerned about birth rates too low to sustain present population levels—subsidize births in an attempt to raise those rates. Regional variations in projected percentage contributions to world population growth are summarized in Figure 6.4.

## Fertility Rates

Crude birth rates may display such regional variability because of differences in age and sex composition or disparities in births among the reproductive-age, rather than total, population. The rate is "crude" because its denominator contains persons who have no chance at all of giving birth—males, young girls, and old women. The **total fertility rate** (TFR) is a more refined and thus more satisfactory statement than the crude birth rate for showing the rate and

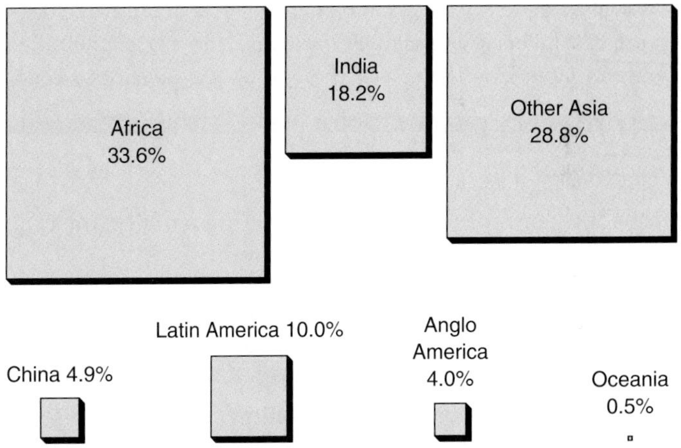


**FIGURE 6.4 Projected percentage contributions to world population growth by region, 2000–2050.** Birth rate changes recorded by differently sized regional populations with differing age structures are altering the world pattern of population increase. Africa, containing 13% of world population in 2000, will probably account for more than one-third of total world increase between 2000 and 2050. Between 1965 and 1975, China's contribution to world growth was 2.5 times that of Africa; between 2000 and 2050, Africa's numerical growth will be more than 8 times that of China. India, which reached the 1 billion level in 2000, is projected to grow by more than 50% over the first half of the 21st century and have by far the world's largest population. In contrast to the growth within the world regions shown, Europe's population is expected to decrease by 90 million over the same half-century period, according to UN 2002 projections. Sources: *Projections based on World Bank and United Nations figures.*

## China's Way—and Others

An ever larger population is "a good thing," Chairman Mao announced in 1965 when China's birth rate was 37 per 1000 and its population totaled 540 million. At Mao's death in 1976, numbers reached 852 million, although the birth rate then had dropped to 25. During the 1970s, when it became evident that population growth was consuming more than half the annual increase in the country's gross domestic product, China introduced a well-publicized campaign advocating the "two-child family" and providing services, including abortions, supporting that program. In response, China's birth rate dropped to 19.5 per 1000 by the late 1970s.

"One couple, one child" became the slogan of a new and more vigorous population control drive launched in 1979, backed by both incentives and penalties to assure its success in China's tightly controlled society. Late marriages were encouraged; free contraceptives, cash awards, abortions, and sterilizations were provided to families limited to a single child. Penalties, including steep fines, were levied for second births. At the campaign's height in 1983, the government ordered the sterilization of either husband or wife for couples with more than one child. Infanticide—particularly the exposure or murder of female babies—was a reported means both of conforming to a one-child limit and of increasing the chances that the one child would be male. By 1986, China's officially reported crude birth rate had fallen to 18 per 1000, far below the 37 per 1000 then registered among the rest of the world's less-developed countries. The one-child policy was effectively dropped in 1984 to permit two-child limits in rural areas where 70% of Chinese population still resides, but in 2002 it was reinstated as nationwide law following documentation of extensive underreporting of rural births.

© Owen Franken/Corbis Images.

In contrast, newly prosperous urbanites have voluntarily reduced their fertility to well below replacement levels, with childless couples increasingly common. Nationally, past and

probability of reproduction among fertile females, the only segment of population bearing children.

The TFR (Figure 6.5) tells us the average number of children that would be born to each woman if, during her childbearing years, she bore children at the current year's rate for women that age. The fertility rate minimizes the effects of fluctuation in the population structure and summarizes the demonstrated and expected reproductive behavior of women. It is thus a more useful and reliable figure for regional comparative and predictive purposes than is the crude birth rate.

Although a TFR of 2.0 would seem sufficient exactly to replace present population (one baby to replace each parent), in reality, replacement levels are reached only with TFRs of 2.1 to 2.3. The fractions over 2.0 are required to compensate for infant and childhood mortalities, childless women, and unexpected deaths in the general population. On a worldwide basis, the TFR in 2003 was 2.8, down from 5.0 in 1965 and from 3.7 in the mid-1980s. The more-developed countries recorded a 1.5 rate early in the 21st century, down from a near-replacement 2.0 in 1985. That decrease has been dwarfed in amount and significance by the rapid changes in reproductive behavior in much of the developing world. Since 1960, the average TFR in the less-developed world has fallen by half from the traditional 6 or more to near 3.0 today. Between the early 1960s and the end of the 20th century, the largest fertility declines occurred in Latin America and Asia (down by 55% and 52% respectively). The smallest decrease—of 15%—came in sub-Saharan Africa.

The recent fertility declines in developing states, however, have been more rapid and widespread than anyone expected. Indeed, the TFRs for so many less-developed countries have dropped so dramatically since the early 1960s (Figure 6.6) that earlier, widely believed world population projections anticipating 10 billion or more at the end of this century are now generally discounted and rejected. Indeed, worldwide in 2003, 65 countries and territories containing some 45% of global population had fertility rates less than 2.1, with more poised to join their ranks. China's decrease from a TFR of 5.9 births per woman in the period 1960–1965 to (officially) about 1.7 in 2003 and comparable drops in TFRs of Bangladesh, Brazil, Mozambique, and other states demonstrate that fertility reflects cultural values, not biological imperatives. If those values now favor current population controls have been so successful that by 2001 serious concerns were being expressed by demographers and government officials that population decrease, not increase, is the problem next to be confronted. Projections suggest that by 2042, because of lowered fertility rates, China's population numbers will actually start falling. The country is already beginning to face a pressing social problem: a declining proportion of working-age persons and an absence of an adequate welfare network to care for a rapidly growing number of senior citizens.

Concerned with their own increasing numbers, many developing countries have introduced their own less extreme programs of family planning, stressing access to contraception and sterilization. International agencies have encouraged these programs, buoyed by such presumed success as the 21% fall in fertility rates in Bangladesh from 1970 to 1990 as the proportion of married women of reproductive age using contraceptives rose from 3% to 40% under intensive family planning encouragement and frequent adviser visits. The costs per birth averted, however, were reckoned at more than the country's $160 per capita gross domestic product.

Research suggests that fertility falls because women decide they want smaller families, not because they have unmet needs for contraceptive advice and devices. Nineteenth-century northern Europeans without the aid of science, it is observed, had lower fertility rates than their counterparts today in middle-income countries. With some convincing evidence, improved women's education has been proposed as a surer way to reduce births than either encouraged contraception or China's coercive efforts. Studies from individual countries indicate that 1 year of female schooling can reduce the fertility rate by between 5% and 10%, yet the fertility rate of uneducated Thai women is only two-thirds that of Ugandan women with secondary education. Obviously, the demand for babies is not absolutely related to educational levels.

Instead, that demand seems closely tied to the use value placed on children by poor families in some parts of the developing world. Where those families share in such communal resources as firewood, animal fodder, grazing land, and fish, the more of those collective resources that can be converted to private family property and use, the better off is the family. Indeed, the more communal resources that are available for "capture," the greater are the incentives for a household to have more children to appropriate them. Some population economists conclude that only when population numbers increase to the point of total conversion of communal resources to private property—and children have to be supported and educated rather than employed—will poor families in developing countries want fewer children. If so, coercion, contraception, and education may be less effective as checks on births than the economic consequence of population increase itself.

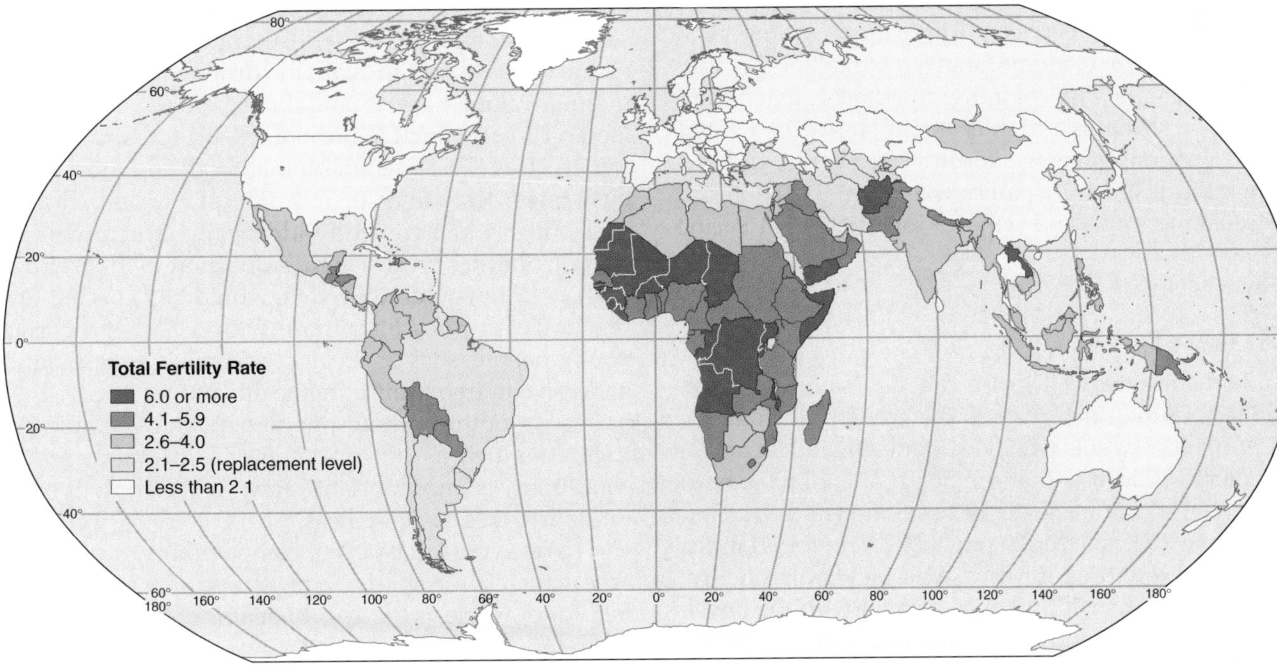


**FIGURE 6.5** **Total fertility rate (TFR)** indicates the average number of children that would be born to each woman if, during her childbearing years, she bore children at the same rate as women of those ages actually did in a given year. Since the TFR is age-adjusted, two countries with identical birth rates may have quite different fertility rates and therefore different prospects for growth. Depending on mortality conditions, a TFR of 2.1 to 2.5 children per woman is considered the "replacement level," at which a population will eventually stop growing. Source: *Data from Population Reference Bureau.*

fewer children than formerly, population projections based on earlier, higher TFR rates must be adjusted.

In fact, demographers have long assumed that recently observed developing country—and therefore global—fertility rate declines to the replacement level would continue and, in the long run, lead to stable population numbers. However, nothing in logic or history requires population stability at any level. Indeed, rather than assume, as in the past, a fertility decline to a constant continuing rate of 2.1, the 2002 United Nations' world population projection predicts a long-term fertility rate of 1.9 for most less-developed countries—below the replacement level. Should the UN's new assessment of fertility prove correct, world population would not just stop growing, as its past projections envisioned; it would inevitably decline (see "A Population Implosion?" p. 187). Of course, should cultural values change to again favor children, growth would resume. Different TFR estimates imply conflicting population projections and vastly different regional and world population concerns.

Individual country projections based on current fertility rates, it should be noted, may not accurately anticipate population levels even in the near future. Massive international population movements are occurring in response to political instabilities and, particularly, to differentials in perceived economic opportunities. Resulting migration flows may cause otherwise declining national populations to stabilize or even grow. For example, the European Union in recent years has had a negative rate of natural increase, yet since 2000 it has experienced essentially a constant population solely because of immigrant influx from Eastern Europe, Asia, and Africa.

World regional and national fertility rates reported in Appendix 2 and other sources are summaries that conceal significant variations between population groups. The Caribbean region, for example, showed a total fertility rate of 2.7 in 2003, but the TFRs of individual states ranged from a low of 1.6 in Cuba to a high of 4.7 in Haiti. The United States 2002 national average fertility rate of 2.1 did not reveal that the TFR for Hispanics was about 3.0, about 2.2 for African Americans, and only 1.9 for Asians and Pacific Islanders.

## Death Rates

The **crude death rate** (CDR), also called the **mortality rate,** is calculated in the same way as the crude birth rate: the annual number of events per 1000 population. In the past, a valid generalization was that the death rate, like the birth rate, varied with national levels of development. Characteristically, highest rates (over 20 per 1000) were found in the less-developed countries of Africa, Asia, and Latin America; lowest rates (less than 10) were associated with the developed states of Europe and Anglo America. That correlation became decreasingly valid as dramatic reductions in death rates occurred in developing countries in the years following World War II. Infant mortality rates and life expectancies improved as antibiotics, vaccinations, and pesticides to treat diseases and control disease carriers were made available in almost all parts of the world and as increased attention was paid to funding improvements in urban and rural sanitary facilities and safe water supplies.

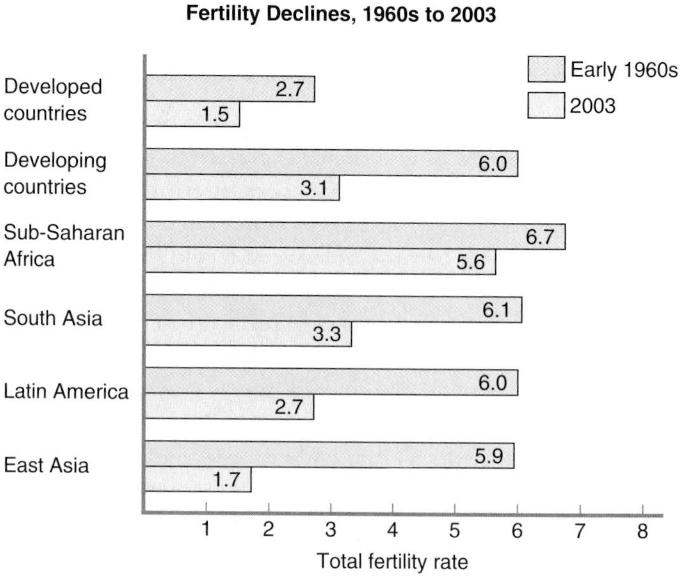


**FIGURE 6.6 Differential fertility declines.** Fertility has declined most rapidly in Latin America and Asia and much more slowly in sub-Saharan Africa. Developed countries as a group now have below-replacement-level fertility. Europe is far below with a 2003 TFR of 1.4; the United States, however, with a TFR of 2.0, is nearly at the replacement point of 2.1. Sources: *Population Reference Bureau and United Nations Population Fund.*

Distinctions between more-developed and less-developed countries in mortality (Figure 6.7), indeed, have been so reduced that by 1994 death rates for less-developed countries as a group actually dropped below those for more-developed states and have remained lower since. Notably, that reduction did not extend to maternal mortality rates (see "The Risks of Motherhood," p. 189). Like crude birth rates, death rates are meaningful for comparative purposes only when we study identically structured populations. Countries with a high proportion of elderly people, such as Denmark and Sweden, would be expected to have higher death rates than those with a high proportion of young people, such as Iceland, assuming equality in other national conditions affecting health and longevity. The pronounced youthfulness of populations in developing countries, as much as improvements in sanitary and health conditions, is an important factor in the recently reduced mortality rates of those areas.

To overcome that lack of comparability, death rates can be calculated for specific age groups. The *infant mortality rate,* for example, is the ratio of deaths of infants aged 1 year or under per 1000 live births:

$$\frac{\text{deaths age 1 year or less}}{\text{1000 live births}}$$

# A Population Implosion?

For much of the last half of the 20th century, demographers and economists focused on a "population explosion" and its implied threat of a world with too many people and too few resources of food and minerals to sustain them. By the end of the century, those fears for some observers were being replaced by a new prediction of a world with too few rather than too many people.

That possibility was suggested by two related trends. The first had become apparent by 1970, when it was noted that the total fertility rates (TFRs) of 19 countries, almost all of them in Europe, had fallen below the **replacement level**—the level of fertility at which populations replace themselves—of 2.1. Simultaneously, Europe's population pyramid began to become noticeably distorted, with a smaller proportion of young and a growing share of middle-aged and retirement-aged inhabitants. The decrease in native working-age cohorts had already, by 1970, encouraged the influx of non-European "guest workers" whose labor was needed to maintain economic growth and to sustain the generous security provisions guaranteed to what was becoming the oldest population of any continent.

Many countries of Western and Eastern Europe sought to reverse their birth rate declines by adopting pronatalist policies. The communist states of the East rewarded pregnancies and births with generous family allowances, free medical and hospital care, extended maternity leaves, and child care. France, Italy, the Scandinavian countries, and others gave similar bonuses or awards for first, second, and later births. Despite those inducements, however, reproduction rates continued to fall. By 2003, every one of the 43 European countries and territories had fertility rates below replacement levels, and the populations of Spain and Italy, for example, are projected to contract by a quarter between 2000 and 2050; Europe as a whole is forecast to shrink by 90 million people by midcentury. "In demographic terms," France's prime minister remarked, "Europe is vanishing."

Europe's experience soon was echoed in other societies of advanced economic development on all continents. By 1995, the United States, Canada, Australia, New Zealand, Japan, Taiwan, South Korea, Singapore, and other older and newly industrializing countries (NICs) had registered fertility rates below the replacement level. As they have for Europe, simple projections foretold their aging and declining population. Japan's numbers, for example, will begin to decline in 2006, when its population will be older than Europe's; Taiwan forecasts negative growth by 2035.

The second trend indicating to many that world population numbers should stabilize and even decline during the lifetimes of today's college cohort is a simple extension of the first: TFRs are being reduced to or below the replacement levels in countries at all stages of economic development in all parts of the world. While only 18% of total world population in 1975 lived in countries with a fertility rate below replacement level, nearly 45% did so by the end of the century. By 2015, demographers estimate, half the world's countries and about two-thirds of its population will show TFRs below 2.1 children per woman. Exceptions to the trend are and still will be found in Africa, especially sub-Saharan Africa, and in some areas of South, Central, and West Asia; however, even in those regions, fertility rates have been decreasing in recent years. "Powerful globalizing forces [are] at work pushing toward fertility reduction everywhere" was a 1997 observation from the French National Institute of Demographic Studies.

That conclusion is plausibly supported by assumptions of the United Nations' 2002 forecast of a decline of long-term fertility rates of most less-developed states to an average of 1.9. The same UN assessment envisions that those countries will reach those below-replacement fertilities by or before 2050. Should those assumptions prove valid, global depopulation could commence before midcentury. Between 2040 and 2050, one projection indicates, world population would fall by about 85 million (roughly the amount of its annual growth during much of the 1990s) and would shrink further by about 25% with each successive generation.

If the UN scenario is realized in whole or in part a much different worldwide demographic and economic future is promised than that prophesied so recently by "population explosion" forecasts. Declining rather than increasing pressure on world food and mineral resources would be in our future, along with shrinking rather than expanding world, regional, and national economies. Even the achievement of **zero population growth** (ZPG), a condition for individual countries when births plus immigration equal deaths plus emigration, has social and economic consequences not always perceived by its advocates. These inevitably include an increasing proportion of older citizens, fewer young people, a rise in the median age of the population, and a growing old-age dependency ratio with ever-increasing pension and social services costs borne by a shrinking labor force. Actual population decline, now the common European condition, would only exaggerate those consequences on a worldwide basis.

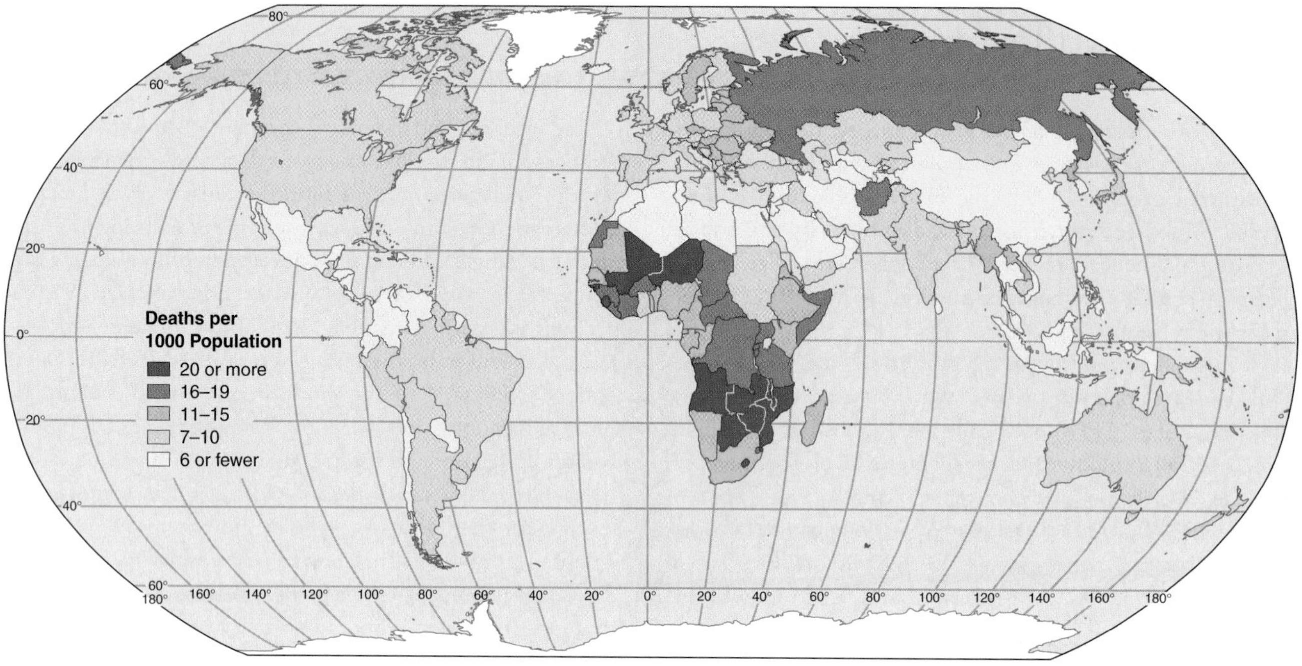


**FIGURE 6.7 Crude death rates** show less worldwide variability than do the birth rates displayed in Figure 6.3. The widespread availability of at least minimal health protection measures and a generally youthful population in the developing countries yield death rates frequently lower than those recorded in "old-age" Europe. Source: *Data from Population Reference Bureau.*

Infant mortality rates are significant because it is at these ages that the greatest declines in mortality have occurred, largely as a result of the increased availability of health services. The drop in infant mortality accounts for a large part of the decline in the general death rate in the last few decades, because mortality during the first year of life is usually greater than in any other year.

Two centuries ago, it was not uncommon for 200 to 300 infants per 1000 to die in their first year. Even today, despite significant declines in those rates over the last 60 years in many countries (Figure 6.8), striking world regional and national variations remain. For all of Africa, infant mortality rates are near 90 per 1000, and individual African states (for example, Angola, Liberia, Niger, and Sierra Leone) showed rates above 150 early in this century. Nor are rates uniform within single countries. The former Soviet Union reported a national infant mortality rate of 23 (1991), but it registered above 110 in parts of its Central Asian region. In contrast, infant mortality rates in Anglo America and Western and Northern Europe are more uniformly in the 4 to 7 range.

Modern medicine and sanitation have increased life expectancy and altered age-old relationships between birth and death rates. In the early 1950s, only five countries, all in northern Europe, had life expectancies at birth of more than 70 years. By the end of the 20th century, some 60 countries outside Europe and North America—although none in sub-Saharan Africa—were on that list. The availability and employment of modern methods of health and sanitation have varied regionally, and the least-developed countries

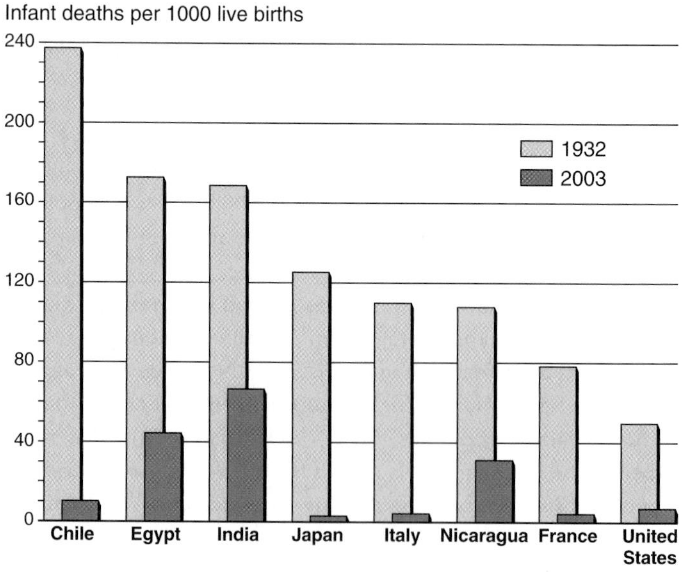


**FIGURE 6.8 Infant mortality rates for selected countries.** Dramatic declines in the rate have occurred in all countries, a result of international programs of health care delivery aimed at infants and children in developing states. Nevertheless, the decreases have been proportionately greatest in the urbanized, industrialized countries, where sanitation, safe water, and quality health care are widely available. Sources: *Data from U.S. Bureau of the Census and Population Reference Bureau.*

## The Risks of Motherhood

The worldwide leveling of crude death rates does not apply to pregnancy-related deaths. In fact, the maternal mortality ratio—maternal deaths per 100,000 live births—is the single greatest health disparity between developed and developing countries. According to the World Health Organization, approximately 515,000 women die each year from causes related to pregnancy or childbirth; 99% of them live in less-developed states where, as a group, the maternal mortality ratio is some 40 times greater than in the more-developed countries. Complications of pregnancy, childbirth, and abortions are the leading slayers of women of reproductive age throughout the developing world, although the incidence of maternal mortality is by no means uniform, as the charts indicate. In Africa, the risk is around 1 death in 16 pregnancies, compared with 1 in 65 in Asia and 1 in 3700 in North America. Country-level differences are even more striking: in Ethiopia, for example, 1 out of every 9 women dies from pregnancy-related complications, compared with 1 in 8700 in Switzerland.

Excluding China, less-developed countries as a group in the late 1990s had a maternal mortality ratio of 580, and 10% of all deaths were due to perinatal and maternal causes. Although 42% of all maternal deaths occurred in Asia (which accounts for about 60% of the world's births), sub-Saharan African women, burdened with 53% of world maternal mortality, were at greatest statistical risk. There, maternal death ratios reached above 2000 in Rwanda and Sierra Leone, and 1 in 13 women in sub-Saharan Africa risked dying of maternal causes. In contrast, the maternal mortality ratio in developed countries as a group (including Russia and Eastern Europe) is 21, and in some—Ireland, Sweden, and Switzerland, for example—it is as low as 5 to 7 (it was 6 in Canada and 8 in the United States in the late 1990s).

The vast majority of maternal deaths in the developing world are preventable. Most result from causes rooted in the social, cultural, and economic barriers confronting females in their home environment throughout their lifetimes: malnutrition, anemia, lack of access to timely basic maternal health care, physical immaturity due to stunted growth, and unavailability of adequate prenatal care or trained medical assistance at birth. Part of the problem is that women are considered expendable in societies where their status is low, although the correlation between women's status and maternal mortality is not exact. In those cultures, little attention is given to women's health or their nutrition, and pregnancy, although a major cause of death, is simply considered a normal condition warranting no special consideration or management. To alter that perception and increase awareness of the affordable measures available to reduce maternal mortality worldwide, 1998 was designated "The Year of Safe Motherhood" by a United Nations interagency group.

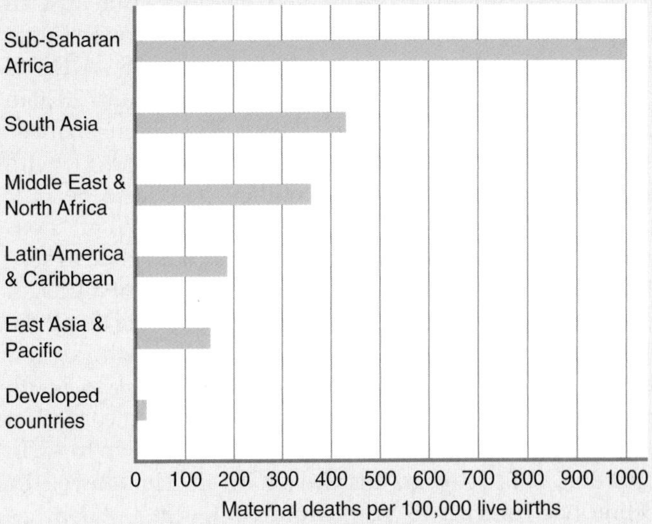


Regional Ratios of Maternal Deaths

(a)

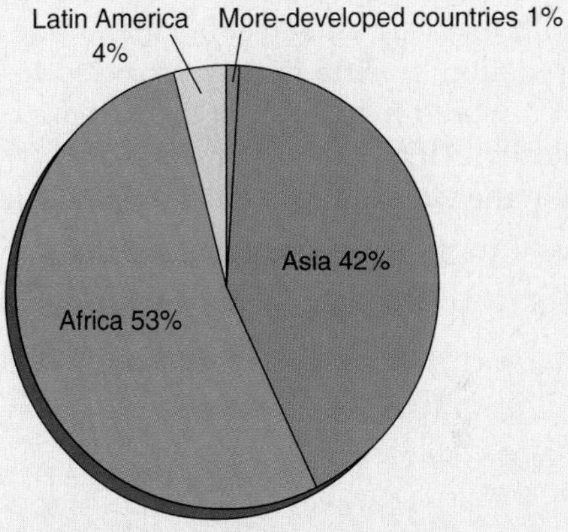


Regional Shares of Maternal Deaths

(b)

Source: *Graph data from WHO.*

have least benefited from them. In such underdeveloped and impoverished areas as much of sub-Saharan Africa, the chief causes of death other than HIV/AIDS are those no longer of immediate concern in more-developed lands: diseases such as malaria; intestinal infections; typhoid; cholera; and especially among infants and children, malnutrition and dehydration from diarrhea.

HIV/AIDS is the tragic and, among developing regions particularly, widespread exception to observed global improvements in life expectancies and reductions in adult death rates and infant and childhood mortalities. AIDS has become the fourth most common cause of death worldwide and is forecast to surpass the Black Death of the 14th century—which caused an estimated 25 million deaths in Europe and 13 million in China—as history's worst-ever epidemic. According to a report by UNAIDS, AIDS is expected to kill 68 million people between 2000 and 2020 in the 45 most affected countries; about 55 million of those deaths will occur in sub-Saharan Africa. The World Health Organization has estimated 42 million people to be HIV positive early in the 21st century. Some 95% of those infected live in developing countries, and 70% reside in sub-Saharan Africa, where women account for 60% of all cases. In that hardest-hit region, as much as one-fourth of the adult population in some countries is HIV positive, and average life expectancy has been cut drastically. In South Africa, the life expectancy of a baby born in the early 21st century should be 66 years; AIDS has cut that down to 47. In Botswana, it is 36 years instead of 70; in Zimbabwe, the decline has been to 43 years from 69. Overall, AIDS has killed some 15 million Africans since the 1950s, when HIV (originally a disease of monkeys) appears to have established itself in Africa as a virulent human epidemic strain. Along with the deep cuts in sub-Saharan life expectancies, total population by 2015 is now projected to be 60 million less than it would have been in the absence of the disease. Economically, AIDS will cut an estimated 8% off national incomes in the worst-hit sub-Saharan countries by 2010.

Nonetheless, because of their high fertility rates, populations in all sub-Saharan countries except South Africa are still expected to grow significantly between 2000 and 2050, adding nearly 1 billion to the continent's total. Indeed, despite high mortality rates due to HIV/AIDS, the population of the world's 48 least-developed countries as a group will, according to UN projections, almost triple between 2000 and 2050, the consequence of their high fertility levels. However, warnings of the rapid spread of the AIDS epidemic in Russia, Ukraine, and South and East Asia—particularly China and India—raise new global demographic concerns even as more hopeful reports of declining infection and mortality rates in some African and Southeast Asian countries are appearing.

## Population Pyramids

Another means of comparing populations is through the **population pyramid,** a graphic device that represents a population's age and sex composition. The term *pyramid* describes the diagram's shape for many countries in the 1800s, when the display was created: a broad base of younger age groups and a progressive narrowing toward the apex as older populations were thinned by death. Now, many different shapes are formed, each reflecting a different population history (Figure 6.9) and some suggest "population profile" is a more appropriate label. By grouping several generations of people, the pyramids, or profiles,

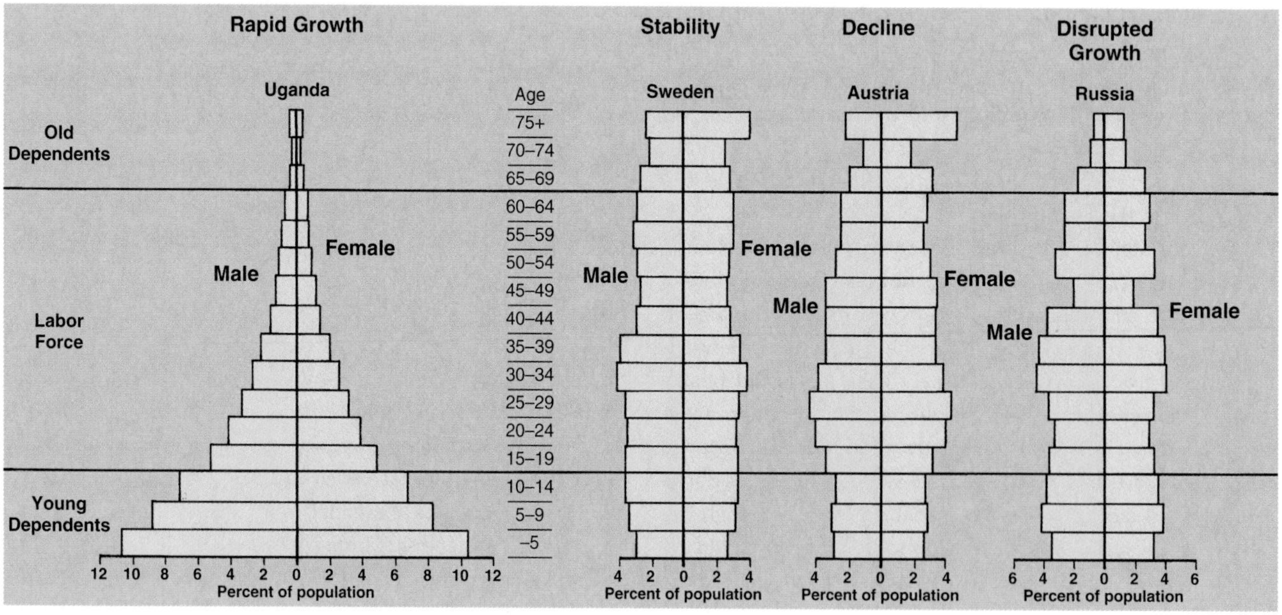


**FIGURE 6.9 Four patterns of population structure.** These diagrams show that population "pyramids" assume many shapes. The age distribution of national populations reflects the past, records the present, and foretells the future. In countries such as Uganda, social costs related to the young are important and economic expansion is vital to provide employment for new entrants into the labor force. Austria's negative growth means a future with fewer workers to support a growing demand for social services for the elderly. The 1992 pyramid for Russia reported the sharp decline in births during World War II as a "pinching" of the 45–49 cohort and showed in the large deficits of men above age 65 the heavy male mortality of the war and late-Soviet period sharp reductions in Russian male longevity. Sources: *The World Bank; the United Nations; Population Reference Bureau; and Carl Haub, "Population Change in the Former Soviet Republics,"* Population Bulletin *49, no. 4 (1994).*

highlight the impact of "baby booms," population-reducing wars, birth rate reductions, and external migrations.

A rapidly growing country such as Uganda has most people in the lowest age cohorts; the percentage in older age groups declines successively, yielding a pyramid with markedly sloping sides. Typically, female life expectancy is reduced in older cohorts of less-developed countries, so that for Uganda the proportion of females in older age groups is lower than in, for example, Sweden. Female life expectancy and mortality rates may also be affected by cultural rather than economic developmental causes (see "100 Million Women Are Missing"). In Sweden, a wealthy country with a very slow rate of growth, the population is nearly equally divided among the age groups, giving a "pyramid" with almost vertical sides. Among older cohorts, as Austria shows, there may be an imbalance between men and women because of the greater life expectancy of the latter. The impacts of war, as Russia's pyramid vividly shows, were evident in the early 1990s in that country's depleted age cohorts and male-female disparities. The sharp contrasts between the composite pyramids of sub-Saharan Africa and Western Europe summarize the differing population concerns of the developing and developed regions of the world; the projection for Botswana suggests the degree to which accepted pyramid shapes can quickly change (Figure 6.10).

The population profile provides a quickly visualized demographic picture of immediate practical and predictive value. For example, the percentage of a country's population in each age group strongly influences demand for goods and services within that national economy. A country with a high proportion of young has a high demand for educational facilities and certain types of health delivery services. In addition, of course, a large portion of the population is too young to be employed (Figures 6.10 and 6.11). On the other hand, a population with a high percentage of elderly people also requires medical goods and services specific to that age group, and these people must be supported by a smaller proportion of workers. As the profile of a national population changes, differing demands are placed on a country's social and economic systems (Figure 6.12). The **dependency ratio** is a simple measure of the number of economic dependents, old or young, that each 100 people in the productive years (usually, 15–64) must

## 100 Million Women Are Missing

Worldwide, some 100 million females are missing, victims of nothing more than their sex. In China, India, Pakistan, New Guinea, and many other developing countries, a traditional preference for boys has meant neglect and death for girls, millions of whom are killed at birth, deprived of adequate food, or denied the medical attention provided to sons favored as old-age and wealth-gathering insurance for parents. In both China and India, ultrasound and amniocentesis tests are employed, often against government directives, to determine the sex of a fetus, so that it can be aborted if it is a female.

The evidence for the missing women starts with one fact: about 106 males are conceived and born for every 100 females. Normally, girls are hardier and more resistant to disease than boys, and in populations where the sexes are treated equally in matters of nutrition and health care, there are about 105 to 106 females for every 100 males. However, the 2001 census of India found just 93.2 females for every 100 males, whereas in China nearly 10% of all girls of the 1990s birth cohorts are "missing," and in 2000 there were 120 boys under age 5 for every 100 girls. China's 2000 census recorded a national average disparity in births of 117 boys for every 100 girls—a deepening imbalance from the 1990 census ratio of 111 newborn boys to 100 newborn girls. Even higher year 2000 differentials were registered in Hainan and Guangdong provinces in southeastern China, with newborn ratios of between 130 and 140 males to 100 females.

Ratio deviations are most striking for second and subsequent births. In China, South Korea, Taiwan, and Hong Kong, for example, the most recent figures for first-child sex ratios are near normal but rise to 121 boys per 100 girls for a second Chinese child and to 185 per 100 for a third Korean child. On that evidence, the problem of missing females is getting worse. Conservative calculations suggest there are more than 60 million females missing in China alone, almost 5% of the national population and more than are unaccounted for in any other country.

The problem is seen elsewhere. In much of South and West Asia and North Africa, there are only some 94 females for every 100 males, a shortfall of about 12% of normal (Western) expectations. A 2000 United Nations report on South Asia suggests the "100 million" world total of missing females is an understatement. It declares that abortions of female fetuses along with infanticide and the food favoratism shown boys have meant that 79 million lost females are attributed to discrimination in South Asia, including some 40 million in India alone.

But not all poor countries show the same disparities. In sub-Saharan Africa, where poverty and disease are perhaps more prevalent than on any other continent, there are 102 females for every 100 males and, in Latin America and the Caribbean, there are equal numbers of males and females. Cultural norms and practices, not poverty or underdevelopment, seem to determine the fate and swell the numbers of the world's 100 million missing women.

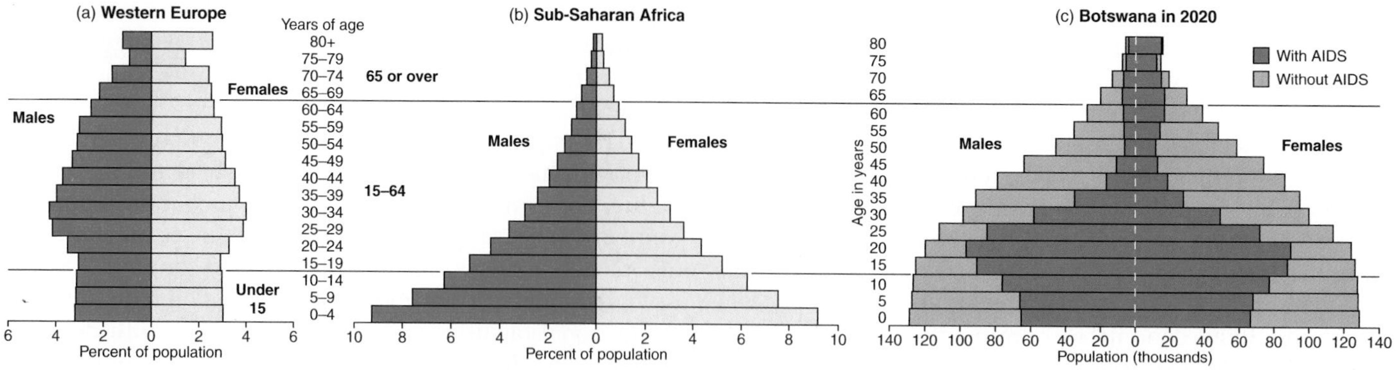


**FIGURE 6.10** **Summary population pyramids.** The mid-1990s pyramids for (a) Western Europe and (b) sub-Saharan Africa show the sharp contrasts in the age structure of older developed regions with their characteristic lowered birth and total fertility rates and that of the much more youthful developing sub-Saharan states. Even in the early 2000s, nearly 45% of the sub-Saharan population was below age 15. That percentage, however, was smaller than it had been in the early and middle 1990s and would have been lower still if not for AIDS-related population loss among young adults. A continuing regionwide decline in relative size of young cohorts may come as a result of economic development and changing family size decisions, but for some countries (c) and perhaps for the region as a whole, tremendous pyramid distortions will result from the demographic impact of AIDS. By 2020, the otherwise expected "normal" pyramid of Botswana may well be distorted into a "population chimney" in which there are more adults in their sixties and seventies than adults in their forties and fifties. Sources: *(a) and (b) Lori S. Ashford, "New Perspectives on Population: Lessons from Cairo,"* Population Bulletin *50, no. 1 (1995), Figure 3; (c) U.S. Bureau of the Census,* World Population Profile 2000.

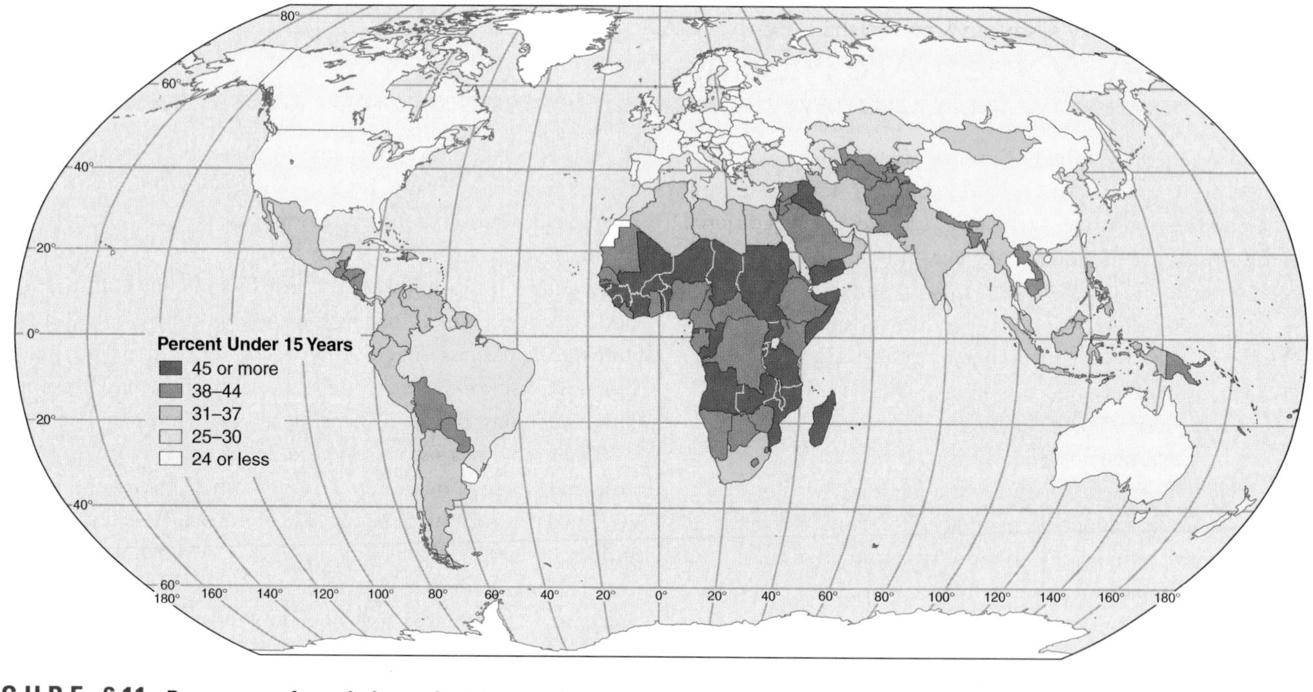


**FIGURE 6.11** **Percentage of population under 15 years of age.** A high proportion of a country's population under 15 increases the dependency ratio of that state and promises future population growth as the youthful cohorts enter childbearing years. Source: *Data from Population Reference Bureau.*

support. Population pyramids give quick visual evidence of that ratio.

They also foretell future problems resulting from present population policies or practices. The strict family-size rules and widespread preferences for sons in China, for example, skews the pyramid in favor of males. On current evidence, about 1 million excess males a year will enter an imbalanced marriage market in China beginning about 2010. Even now, the Chinese population pyramid shows that never-married men ages 20–44 outnumber their female counterparts by nearly two to one. The 40 million bachelors China is likely to have in 2020, unconnected to society by wives and children, may pose threats to social order and, perhaps, national stability not foreseen or planned when family control programs were put in place but clearly suggested when made evident by population pyramid distortions.

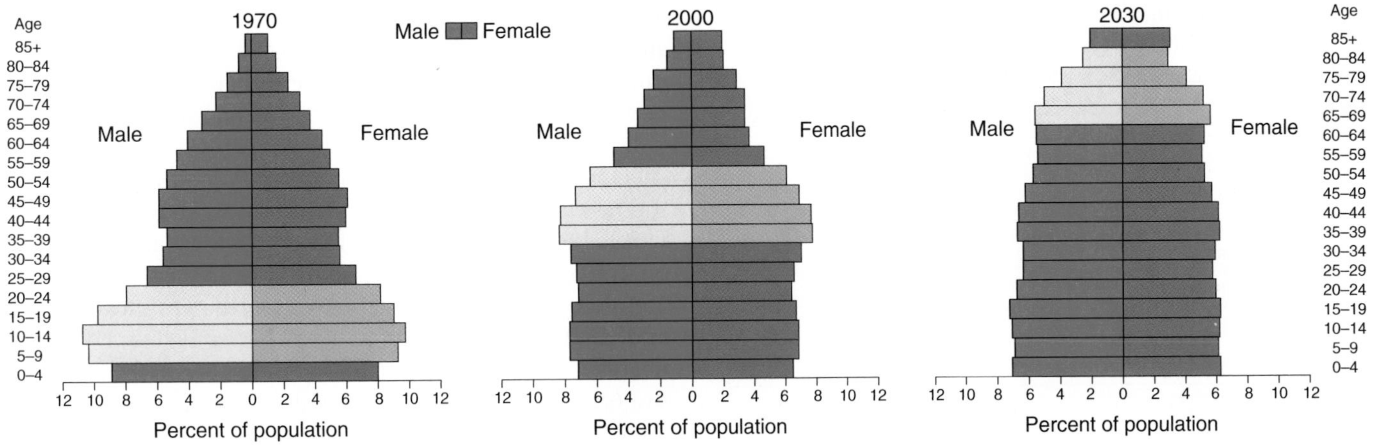


**FIGURE 6.12 The progression of the "boomers."** The baby-boom cohort born between 1946 and 1964—through the U.S. population pyramid—has been associated with changing American lifestyles and expenditure patterns. In 1970, national priorities focused on childhood and young adult interests and the needs, education, and support of younger age groups. At the start of the 21st century, boomers formed the largest share of the working-age adult population, and their wants and spending patterns shaped the national culture and economy. By 2030, the pyramid foretells, their desires and support needs—now for retirement facilities and old-age care—will again be central concerns. *Graphs from "Elderly Americans" by Christine L. Himes in* Population Bulletin *56, No. 4, December 2001, Figure 1. Redrawn by permission of Population Reference Bureau.*

## Natural Increase

Knowledge of a country's sex and age distributions also enables demographers to forecast its future population levels, though the reliability of projections decreases with increasing length of forecast (Figure 6.13). Thus, a country with a high proportion of young people will experience a high rate of natural increase unless there is a very high mortality rate among infants and juveniles or fertility and birth rates change materially. The **rate of natural increase** of a population is derived by subtracting the crude death rate from the crude birth rate. *Natural* means that increases or decreases due to migration are not included. If a country had a birth rate of 22 per 1000 and a death rate of 12 per 1000 for a given year, the rate of natural increase would be 10 per 1000. This rate is usually expressed as a percentage—that is, as a rate per 100 rather than per 1000. In the example given, the annual increase would be 1%.

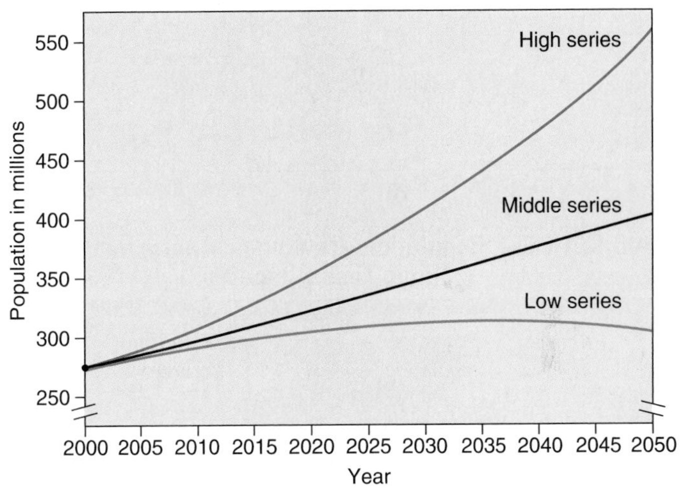


**FIGURE 6.13 Possible population futures for the United States.** As these population projections to 2050 illustrate, expected future numbers vary greatly because the fertility, birth and death rates, and immigration flow assumptions they are based on are different. Depending on those assumptions, 2000 Census Bureau projections of U.S. population in 2050 ranged from 313.5 million (low series) to 552.7 million (high series). The middle series estimate of 403.6 million is the one most often cited. The folly of basing planning decisions on very long-range projections is made clear by the Bureau's extension of its trends for a full century. By 2100, it calculated, the U.S. population could range from a low series 282.7 million to a high series 1.18 billion. Source: *U.S. Bureau of the Census.*

## Doubling Times

The rate of increase can be related to the time it takes for a population to double if the current growth rate remains constant—that is, the **doubling time.** Table 6.1 shows that it would take 70 years for a population with a rate of increase of 1% (approximately the rate of growth of Thailand or Argentina at the turn of the 21st century) to double. A 2% rate of increase—recorded in 2003 by Kenya and Peru—means that the population would double in only 35 years. (Population doubling time can be estimated by dividing the current growth rate into the number 70. Thus, $70 \div 2 = 35$ years.) How could adding only 20 people per 1000 cause a population to grow so quickly? The principle is the same as that used to compound interest in a bank. Table 6.2 shows the number yielded by a 2% rate of increase at the end of successive 5-year periods.

Until recently, for the world as a whole, the rates of increase have risen over the span of human history. Therefore, the doubling time has steadily decreased (Table 6.3). Growth rates vary regionally, and in countries with high rates of increase (Figure 6.14), the doubling time is less than the 54 years projected for the world as a whole (at early 21st-century

**TABLE 6.1 Doubling Time in Years at Different Rates of Increase**

| Annual Percentage Increase | Doubling Time (Years) |
|---|---|
| 0.5 | 140 |
| 1.0 | 70 |
| 2.0 | 35 |
| 3.0 | 24 |
| 4.0 | 17 |
| 5.0 | 14 |
| 10.0 | 7 |

**TABLE 6.2 Population Growth Yielded by a 2% Rate of Increase**

| Year | Population |
|---|---|
| 0 | 1000 |
| 5 | 1104 |
| 10 | 1219 |
| 15 | 1345 |
| 20 | 1485 |
| 25 | 1640 |
| 30 | 1810 |
| 35 | 2000 |

**TABLE 6.3 Population Growth and Approximate Doubling Times since A.D. 1**

| Year | Estimated Population | Doubling Time (Years) |
|---|---|---|
| 1 | 250 million | |
| 1650 | 500 million | 1650 |
| 1804 | 1 billion | 154 |
| 1927 | 2 billion | 123 |
| 1974 | 4 billion | 47 |
| *World population may reach:* | | |
| 2030 | 8 billion | 56[a] |

*Source:* United Nations.

[a]The final estimate of doubling time reflects assumptions of decreasing and stabilizing fertility rates. No current projections contemplate a further doubling to 16 billion people.

growth rates). Should world fertility rates decline (as they have in recent years), theoretical population doubling time would correspondingly increase, as it has since 1990.

Here, then, lies the answer to the question posed earlier. Even small annual additions accumulate to large total increments because we are dealing with geometric or exponential (1, 2, 4, 8) rather than arithmetic (1, 2, 3, 4) growth (Figure 6.15). The ever-increasing base population has reached such a size that each additional doubling would, if actually achieved, result in an astronomical increase in the total. A simple mental exercise suggests the inevitable consequences of such doubling, or **J-curve,** growth. Take a very large sheet of the thinnest paper you can find and fold it in half. Fold it in half again. After seven or eight folds, the sheet will have become as thick as a book—too thick for further folding by hand. If you could make 20 folds, the stack would be nearly as high as a football field is long. From then on, the results of further doubling would be astounding. At 40 folds, the stack would be well on the way to the moon and, at 70 it would reach twice as far as the distance to the nearest star. Rounding the bend on the J-curve, which Figure 6.16 suggests world population did around 1900, fostered after 1950 dire predictions of inevitable unsupportable pressures on the planet's population support capabilities.

By 2000, however, it had become apparent that few developed countries, particularly in Europe, are likely in the foreseeable future or ever to double their population size if their growth is projected—as usually done—solely on current rates of natural increase. But individual country growth is also dependent on patterns of immigration and emigration and of changes in life expectancy. That is, a country's "natural" population growth based solely on births and deaths may yield significantly lower population projections and longer doubling times than does the same country's "overall" growth taking migration into account. The contrast may be striking. The United States in the early 21st century had a 0.6% rate of natural increase and a doubling time of 117 years; it had, however, an *overall* growth rate of 1.2% with a doubling time of 58 years.

With immigration largely absent, declining fertility rates in most of the developing world and the incidence of AIDS in sub-Saharan Africa and parts of populous Asia cast doubt on the utility or applicability of long-term doubling time projections even for present high growth rate countries. Although the United Nations estimates that the population of the 48 least-developed countries will possibly almost triple between 2000 and 2050, we have learned that doubling time assumptions are inherently misleading and population increases are limited.

## THE DEMOGRAPHIC TRANSITION

The theoretical consequence of exponential population growth cannot be realized. Some form of braking mechanism must necessarily operate to control totally unregulated population growth. If voluntary population limitation is not undertaken, involuntary controls of an unpleasant nature may be set in motion.

One attempt to summarize a historically observed voluntary limitation of population growth—and relating that control to economic development—is the **demographic transition** model. It traces the changing levels of human fertility and mortality presumably associated with industrialization and urbanization. Over time, the model assumes, high birth

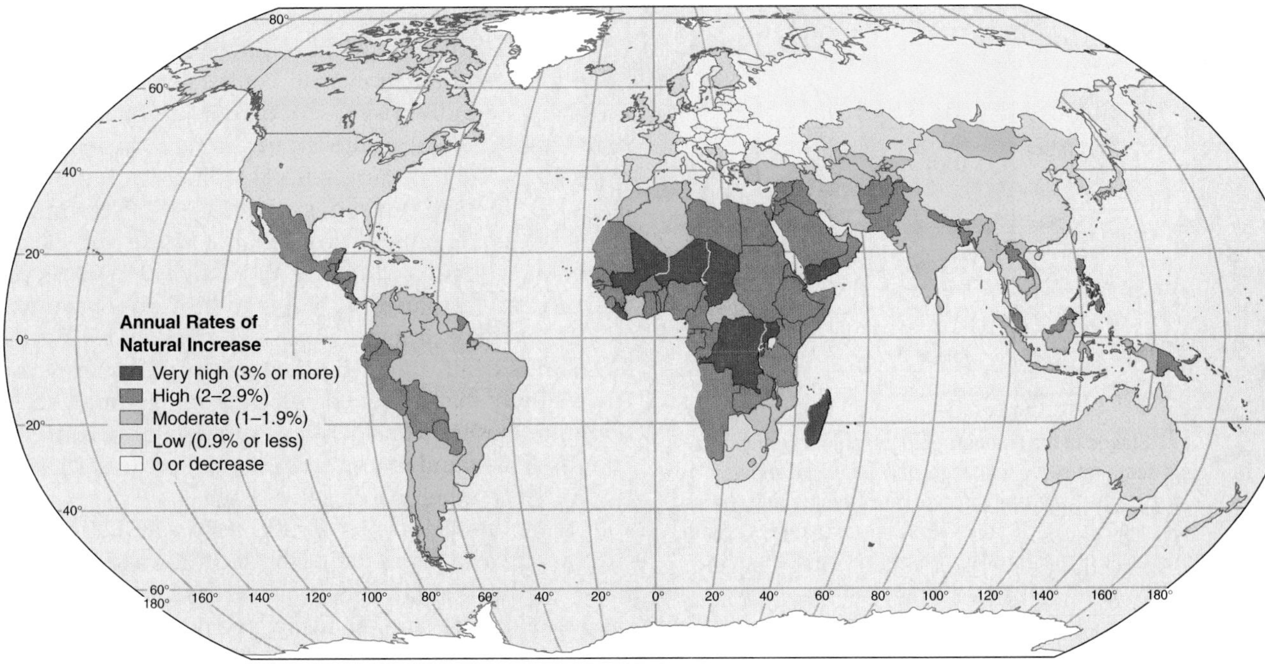


**FIGURE 6.14 Annual rates of natural increase.** The world's 2003 rate of natural increase (1.3%) would mean a doubling of population in 54 years. Since demographers now anticipate that world population—currently above 6 billion—will stabilize at around 10 billion (in about A.D. 2100) and perhaps actually decline after that, the "doubling" implication and time frame of current rates of natural increase reflect mathematical, not realistic, projections. Many individual continents and countries, of course, deviate widely from the global average rate of growth and have vastly different doubling times. Africa as a whole has the highest rates of increase, followed by Central America and western Asia. Europe as a whole (including Russia) had negative growth early in the 21st century, with some individual countries showing increases so small that their doubling times must be measured in millennia. Source: *Data from Population Reference Bureau.*

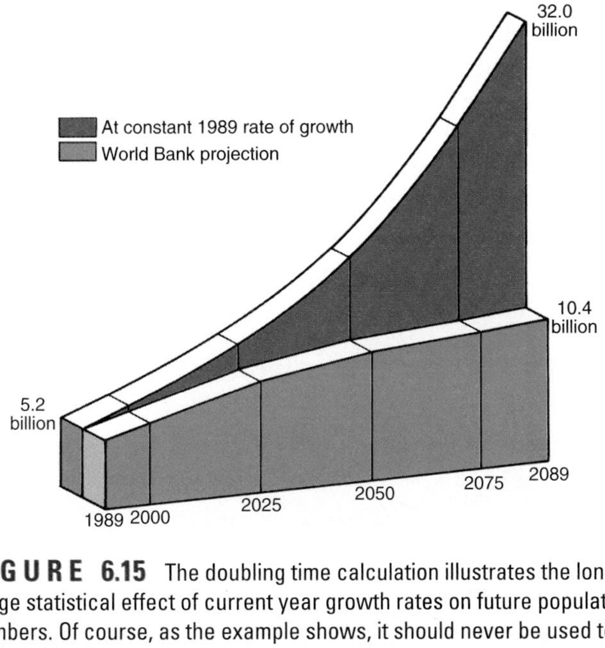


**FIGURE 6.15** The doubling time calculation illustrates the long-range statistical effect of current year growth rates on future population numbers. Of course, as the example shows, it should never be used to suggest a realistic prediction of future population size, for population growth reflects not just birth rates but also death rates, age structure, collective family-size decisions, and migration. Because of lower recorded fertility rates, recent projections of 2089 world totals are far below the World Bank's 1989 estimates. Source: *Population Reference Bureau,* 1989 World Population Data Sheet.

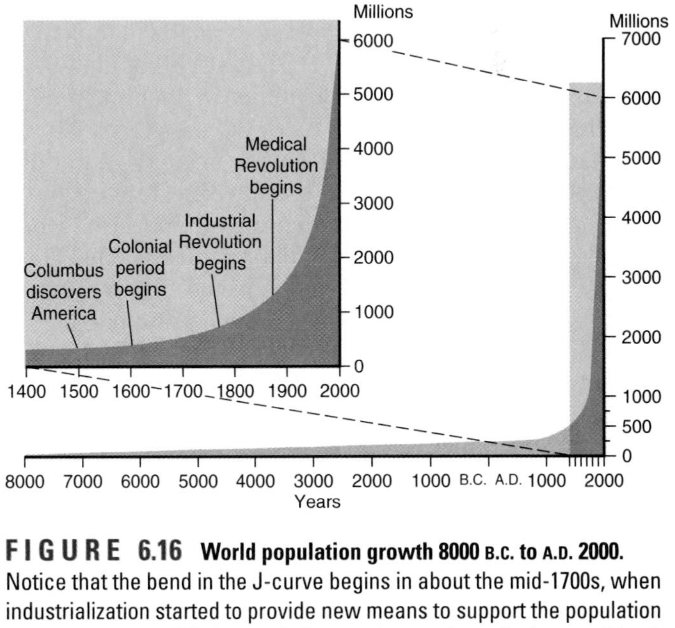


**FIGURE 6.16 World population growth 8000 B.C. to A.D. 2000.** Notice that the bend in the J-curve begins in about the mid-1700s, when industrialization started to provide new means to support the population growth made possible by revolutionary changes in agriculture and food supply. Improvements in medical science, sanitation, and nutrition reduced death rates near the opening of the 20th century in the industrializing countries.

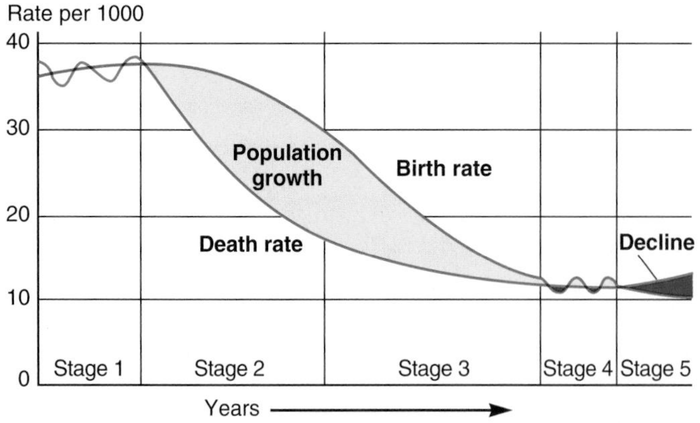


**FIGURE 6.17 Stages in the demographic transition.** During the first stage, birth and death rates are both high, and population grows slowly. When the death rate drops and the birth rate remains high, there is a rapid increase in numbers. During the third stage, birth rates decline and population growth is less rapid. The fourth stage is marked by low birth and death rates and, consequently, by a low rate of natural increase or even by decrease if death rates should exceed those of births. Indeed, the negative growth rates of many European countries have suggested to some that a fifth stage, one of population decline, is regionally—and ultimately worldwide—a logical extension of the transition model.

and death rates will gradually be replaced by low rates (Figure 6.17). The *first stage* of that replacement process—and of the demographic transition model—is characterized by high birth and high but fluctuating death rates.

As long as births only slightly exceed deaths, even when the rates of both are high, the population will grow only slowly. This was the case for most of human history until about A.D. 1750. Demographers think that it took from approximately A.D. 1 to A.D. 1650 for the population to increase from 250 million to 500 million, a doubling time of more than a millennium and a half.

Growth was not steady, of course. There were periods of regional expansion that were usually offset by sometimes catastrophic decline. Wars, famine, and other disasters took heavy tolls. For example, the bubonic plague (the Black Death), which swept across Europe in the 14th century, is estimated to have killed between one-third and one-half of the population of that continent, and epidemic diseases brought by Europeans to the Western Hemisphere are believed to have reduced New World native populations by 95% within a century or two of first contact. The first stage of the demographic transition model is no longer found in any country. At the end of the 20th century, few countries—even in poorer regions of sub-Saharan Africa—had death rates as high as 20 per 1000. However, in several African states, birth rates approached or were above 50 per 1000.

## The Western Experience

The demographic transition model was developed to explain the population history of Western Europe. That area entered a *second stage* with the industrialization that began about 1750. Its effects—declining death rates accompanied by continuing high birth rates—have been dispersed worldwide even without universal conversion to an industrial economy. Rapidly rising population during the second demographic stage results from dramatic increases in life expectancy. That, in turn, reflects falling death rates due to advances in medical and sanitation practices, improved foodstuff storage and distribution, a rising per capita income, and the urbanization that provides the environment in which sanitary, medical, and food distributional improvements are concentrated (Figure 6.18). Birth rates do not fall as soon as death rates; ingrained cultural patterns change more slowly than technologies. In many agrarian societies, large families are considered advantageous. Children contribute to the family by starting to work at an early age and by supporting their parents in old age.

Many countries in southern Asia and Latin America display the characteristics of the second stage in the population model. Bhutan, with a birth rate of 34 and a death rate of 9, and Nicaragua, with respective rates of 32 and 5 (2003 estimates), are typical. The annual rates of increase of such countries are near or above 25 per 1000, and their populations will double in about 25 years. Such rates, of course, do not mean that the full impact of the Industrial Revolution has been worldwide; they do mean that the underdeveloped societies have been beneficiaries of the life preservation techniques associated with it.

The *third stage* follows when birth rates decline as people begin to control family size. The advantages of having many children in an agrarian society are not as evident in urbanized, industrialized cultures. In fact, such cultures may view children as economic liabilities rather than assets. When the birth rate falls and the death rate remains low, the population size begins to level off. Many countries are now registering the low death rates and transitional birth rates of the third stage.

**FIGURE 6.18** Paris, France, at the end of the 19th century. A modernizing Europe experienced improved living conditions and declining death rates during that century of progress. *Library of Congress.*

The classic demographic transition model ends with a *fourth* and final stage, characterized by very low birth and death rates. This stage yields at best only very slight percentage increases in population and doubling times stretch to a thousand years or more. A significant and irreversible aging of the world's population is a direct and profound consequence of the worldwide transition from high to low levels of fertility and mortality associated with the fourth stage of the model (see p. 213). In a few countries, death rates have begun to equal or exceed birth rates and populations are actually declining.

This extension of the fourth stage into a *fifth* of population decrease has so far been largely confined to the rich, industrialized world—notably, Europe and Japan—but increasingly promises to affect much of the rest of the world as well. Even now, the dramatic decline in fertility recorded in almost all countries since the 1980s suggests that by 2010 at the latest a majority of the world's population will reside in areas where the only significant population growth will result from demographic momentum (see p. 212), not from second-stage expansion.

The original transition model was devised to describe the experience of northwest European countries as they went from rural-agrarian societies to urban-industrial ones. It may not fully reflect the prospects of contemporary developing countries. In Europe, church and municipal records, some dating from 16th century, show that people tended to marry late or not at all. In England before the Industrial Revolution, as many as half of all women in the 15 to 50 age cohort were unmarried. Infant mortality was high; life expectancy was low. With the coming of industrialization in the 18th and 19th centuries, immediate factory wages instead of long apprenticeship programs permitted earlier marriage and more children. Since improvements in sanitation and health came only slowly, death rates remained high. Around 1800, 25% of Swedish infants died before their first birthday. Population growth rates remained below 1% per year in France throughout the 19th century.

Beginning about 1860, first death rates and then birth rates began their significant, though gradual, decline. This "mortality revolution" came first, as an *epidemiologic transition* echoed the demographic transition with which it is associated. Many formerly fatal epidemic diseases became endemic—that is, essentially continual within a population—and mortality patterns showed a shift from communicable to noncommunicable diseases. As people developed partial immunities, mortalities associated with them declined. Improvements in animal husbandry, crop rotation and other agricultural practices, and new foodstuffs (the potato was an early example) from overseas colonies raised the level of health of the European population in general.

At the same time, sewage systems and sanitary water supplies became common in larger cities, and general levels of hygiene improved everywhere (Figure 6.19). Deaths due to infectious, parasitic, and respiratory diseases and to malnutrition declined, while those related to chronic illnesses associated with a maturing and aging population increased. Western Europe passed from a first stage "Age of Pestilence and Famine" to a presumed ultimate "Age of Degenerative and Human-Origin Diseases." However, recent increases in drug- and antibiotic-resistant diseases, pesticide resistance of disease-carrying insects, and such new scourges of both the less-developed and more-developed countries as AIDS cast doubt on the finality of that "ultimate" stage (see "Our Delicate State of Health," p. 202). Nevertheless, even the resurgence of old and the emergence of new scourges, such as malaria, tuberculosis, and AIDS (which together caused an estimated 150 million deaths between 1945 and 2000), are unlikely to have decisive demographic consequences on the global scale.

In Europe, the striking reduction in death rates was echoed by similar declines in birth rates as societies began to alter their traditional concepts of ideal family size. In cities, child labor laws and mandatory schooling meant that children no longer were important contributors to family economies. As "poor-relief" legislation and other forms of public welfare substituted for family support structures, the insurance value of children declined. Family consumption patterns altered as the Industrial Revolution made more widely available goods that served consumption desires, not just basic living needs. Children hindered rather than aided the achievement of the age's promise of social mobility and lifestyle improvement. Perhaps most important, and by some measures preceding and independent of the implications of the Industrial Revolution, were changes in the status of women and in their spreading conviction that control over childbearing was within their power and to their benefit.

**FIGURE 6.19** Pure piped water replacing individual or neighborhood wells, and sewers and waste treatment plants instead of privies, became increasingly common in urban Europe and North America during the 19th century. Their modern successors, such as the Las Vegas, Nevada, treatment plant shown here, helped complete the *epidemiologic transition* in developed countries. *Lynn Betts, USDA, Natural Resources Conservation Center.*

## Our Delicate State of Health

Death rates have plummeted, and the benefits of modern medicines, antibiotics, and sanitary practices have enhanced both the quality and expectancy of life in the developed and much of the developing world. Far from being won, however, the struggle against infectious and parasitic diseases is growing in intensity and is, perhaps, unwinnable. More than a half-century after the discovery of antibiotics, the diseases they were to eradicate are on the rise, and both old and new disease-causing microorganisms are emerging and spreading all over the world. Infectious and parasitic diseases kill between 17 and 20 million people each year; they officially account for one-quarter to one-third of global mortality and, because of poor diagnosis, certainly are responsible for far more. And their global incidence is rising.

The five leading infectious killers are acute respiratory infections such as pneumonia, diarrheal diseases, tuberculosis, malaria, and measles. In addition, AIDS was killing 3 million or more persons yearly early in this century, far more than measles or malaria. The incidence of infection, of course, is far greater than the occurrence of deaths. Nearly 30% of the world's people, for example, are infected with the bacterium that causes tuberculosis, but only 2 to 3 million are killed by the disease each year. More than 500 million people are infected with such tropical diseases as malaria, sleeping sickness, schistosomiasis, and river blindness, with perhaps 3 million annual deaths. Newer pathogens are constantly appearing, such as those causing Lassa fever, Rift Valley fever, Ebola fever, Hanta fever, West Nile encephalitis, hepatitis C, and severe acute respiratory syndrome (SARS), incapacitating and endangering far more than they kill. In fact, at least 30 previously unknown infectious diseases have appeared since the mid-1970s.

The spread and virulence of infectious diseases are linked to the dramatic changes so rapidly occurring in the earth's physical and social environments. Climate warming permits temperature-restricted pathogens to invade new areas and claim new victims. Deforestation, water contamination, wetland drainage, and other human-induced alterations to the physical environment disturb ecosystems and simultaneously disrupt the natural system of controls that keeps infectious diseases in check. Rapid population growth and explosive urbanization, increasing global tourism, population-dislocating wars and migrations, and expanding world trade all increase interpersonal disease-transmitting contacts and the mobility and range of disease-causing microbes, including those brought from previously isolated areas by newly opened road systems and air routes. Add in poorly planned or executed public health programs, inadequate investment in sanitary infrastructures, and inefficient distribution of medical personnel and facilities, and the causative role of humans in many of the current disease epidemics is clearly visible.

In response, a worldwide Program for Monitoring Emerging Diseases (ProMED) was established in 1993 and developed a global online infectious disease network linking health workers and scientists in more than 100 countries to battle what has been called a growing "epidemic of epidemics." The most effective weapons in that battle are already known. They include improved health education; disease prevention and surveillance; research on disease vectors and incidence areas (including GIS and other mapping of habitats conducive to specific diseases); careful monitoring of drug therapy; mosquito control programs; provision of clean water supplies; and distribution of such simple and cheap remedies and preventatives as childhood immunizations, oral rehydration therapy, and vitamin A supplementation. All, however, require expanded investment and attention to those spreading infectious diseases—many with newly developed antibiotic-resistant strains—so recently thought to be no longer of concern.

## A Divided World Converging

The demographic transition model described the presumed inevitable course of population events from the high birth and death rates of premodern (underdeveloped) societies to the low and stable rates of advanced (developed) countries. The model failed to anticipate, however, that the population history of Europe was apparently not relevant to all developing countries of the middle and late 20th century. Many developing societies seemingly remained locked in the second stage of the model, unable to realize the economic gains and social changes necessary to progress to the third stage of falling birth rates.

The introduction of Western technologies of medicine and public health, including antibiotics, insecticides, sanitation, immunization, infant and child health care, and eradication of smallpox, quickly and dramatically lowered the death rates in developing countries. Such imported technologies and treatments accomplished in a few years what it took Europe 50 or 100 years to experience. Sri Lanka, for example, sprayed extensively with DDT to combat malaria; life expectancy jumped from 44 years in 1946 to 60 only 8 years later. With similar public health programs, India also experienced a steady reduction in its death rate after 1947. Simultaneously, with international sponsorship, food aid cut the death toll of developing states during drought and other disasters. The dramatic decline in mortality that had emerged only gradually throughout the European world occurred with startling speed in developing countries after 1950.

Corresponding reductions in birth rates did not immediately follow, and world population totals soared: from 2.5 billion in 1950 to 3 billion by 1960 and 5 billion by the middle 1980s. Alarms about the "population explosion" and its predicted devastating impact on global food and mineral resources were frequent and strident. In demographic terms, the world was viewed by many as permanently divided between developed regions that had made the demographic transition to stable population numbers and the underdeveloped, endlessly expanding ones that had not.

Birth rate levels, of course, unlike life expectancy improvements, depend less on supplied technology and assistance than they do on social acceptance of the idea of fewer children and smaller families (Figure 6.20). That acceptance began to grow broadly but unevenly worldwide even as regional and world population growth seemed uncontrollable. In 1984, only 18% of world population lived in countries with fertility rates at or below replacement levels (that is, countries that had achieved the demographic transition). By 2000, however, 44% lived in such countries, and early in the 21st century it is increasingly difficult to distinguish between developed and developing societies on the basis of their fertility rates. Those rates in many separate Indian states (Kerala and Tamil Nadu, for example) and in such countries as Sri Lanka, Thailand, South Korea, and China are below those of the United States and some European countries. Significant decreases to near the replacement level have also occurred in the space of a single generation in many other Asian and Latin American states with high recent rates of economic growth. Increasingly, it appears, low fertility is becoming a feature of both rich and poor, developed and developing states.

Despite this general substantial convergence in fertility, there still remains a significant minority share of the developing world with birth rates averaging 1.5 to 2 times or more above the replacement level. Indeed, early in the 21st century, almost 1.4 billion persons live in countries or regions where total fertility is still 3.5 or greater (a level not considered high, of course, in the early 1950s, when only a quarter of the world's population had a TFR below that mark). For the most part, these current high-fertility countries and areas are in sub-Saharan Africa and the northern parts of the Indian subcontinent. High TFR regions collectively, United Nations demographers predict, will provide the majority of world population growth to at least 2050.

The established patterns of both high- and low-fertility regions tend to be self-reinforcing. Low growth permits the expansion of personal income and the accumulation of capital

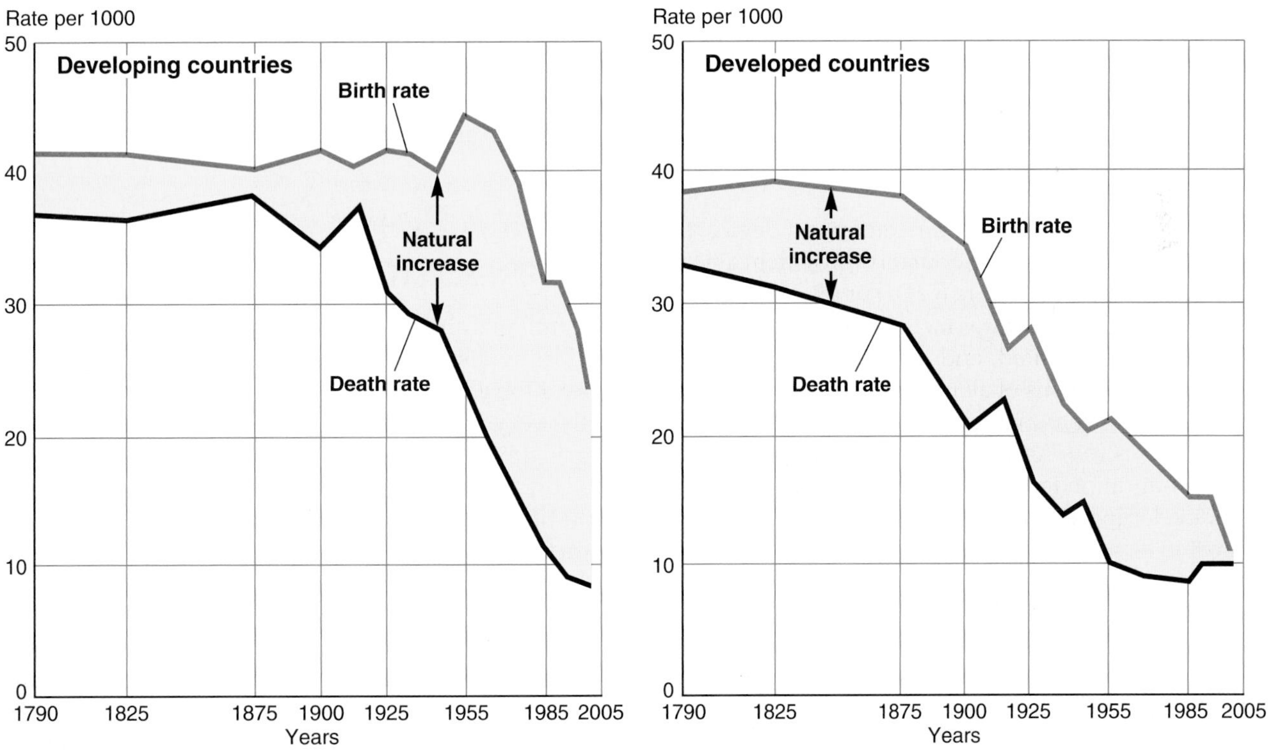


**FIGURE 6.20 World birth and death rates.** The "population explosion" after World War II (1939–1945) reflected the effects of drastically reduced death rates in developing countries without simultaneous and compensating reductions in births. By the end of the 20th century, however, three interrelated trends had appeared in many developing world countries: (1) fertility had overall dropped further and faster than had been predicted 25 years earlier, (2) contraceptive acceptance and use had increased markedly, and (3) age at marriage was rising. In consequence, the demographic transition had been compressed from a century to a generation in some developing states. In others, fertility decline began to slacken in the mid-1970s but continued to reflect the average number of children—four or more—still desired in many societies. Source: *Revised and redrawn from Elaine M. Murphy,* World Population: Toward the Next Century, *revised ed. (Washington, D.C.: Population Reference Bureau, 1989).*

that enhance the quality and security of life and make large families less attractive or essential. In contrast, in high birth rate regions, population growth consumes in social services and assistance the investment capital that might promote economic expansion. Increasing populations place ever-greater demands on limited soil, forest, water, grassland, and cropland resources. As the environmental base deteriorates, productivity declines and population-supporting capacities are so diminished as to make difficult or impossible the economic progress on which the demographic transition depends, an apparent equation of increasing international concern (see "The Cairo Plan," p. 202).

# The Demographic Equation

Births and deaths within a region's population—natural increases or decreases—tell only part of the story of population change. Migration involves the long-distance movement of people from one residential location to another. When that relocation occurs across political boundaries, it affects the population structure of both the origin and destination jurisdictions. The **demographic equation** summarizes the contribution made to regional population change over time by the combination of *natural change* (difference between births and deaths) and *net migration* (difference between in-migration and out-migration).[1] On a global scale, of course, all population change is accounted for by natural change. The impact of migration on the demographic equation increases as the population size of the areal unit studied decreases.

## Population Relocation

In the past, emigration proved an important device for relieving the pressures of rapid population growth in at least some European countries (Figure 6.21). For example, in one 90-year span, 45% of the natural increase in the population of the British Isles emigrated, and between 1846 and 1935, some 60 million Europeans of all nationalities left that continent. Despite recent massive movements of economic and political refugees across Asian, African, and Latin American boundaries, emigration today provides no comparable relief valve for developing countries. Total population numbers are too great to be much affected by migrations of even millions of people. In only a few countries—Afghanistan, Cuba, El Salvador, and Haiti, for example—have as many as 10% of the population emigrated in recent decades. A more detailed treatment of the processes and patterns of international and intranational migrations as expressions of spatial interaction is presented in Chapter 8.

## Immigration Impacts

Where cross-border movements are massive enough, migration may have a pronounced impact on the demographic equation and result in significant changes in the population structures of both the origin and destination regions. Past European and African migrations, for example, not only altered but substantially created the population structures of new, sparsely inhabited lands of colonization in the Western Hemisphere and Australasia. In some decades of the late 18th and early 19th centuries 30% to more than 40% of the population increase in the United States was accounted for by immigration. Similarly, eastward-moving Slavs colonized underpopulated Siberia and overwhelmed native peoples.

Migrants are rarely a representative cross section of the population group they leave, and they add an unbalanced age and sex component to the group they join. A recurrent research observation is that emigrant groups are heavily skewed in favor of young singles. Whether males or females dominate the outflow varies with circumstances. Although males traditionally have far exceeded females in international flow, in recent years females have accounted for between 40% and 60% of all transborder migrants.

At the least, then, the receiving country will have its population structure altered by an outside increase in its younger age and, probably, unmarried cohorts. The results are both immediate in a modified population pyramid and potential in their future impact on reproduction rates and excess of births over deaths. The origin area will have lost a portion of its young, active members of childbearing years. It perhaps will have suffered distortion in its young adult sex ratios, and it certainly will have recorded a statistical aging of its population. The destination society will likely experience increases in births associated with the youthful newcomers and, in general, have its average age reduced.

# World Population Distribution

The millions and billions of people of our discussion are not uniformly distributed over the earth. The most striking feature of the world population distribution map (Figure 6.22) is the very unevenness of the pattern. Some land areas are nearly uninhabited, others are sparsely settled, and still others contain dense agglomerations of people. A little more than half of the world's people are found—unevenly concentrated—in rural areas. Nearly half are urbanites, however, and a constantly growing proportion are residents of very large cities of 1 million or more.

Earth regions of apparently very similar physical makeup show quite different population numbers and densities, perhaps the result of differently timed settlement or of settlement by different cultural groups. Northern and Western Europe, for example, inhabited thousands of years before North America, contain as many people as the United States on 70% less land; the present heterogeneous population of the Western Hemisphere is vastly more dense overall than was that of earlier Native Americans.

We can draw certain generalizing conclusions from the uneven but far from irrational distribution of population shown in Figure 6.22. First, almost 90% of all people live

[1]See the Glossary definition for the calculation of the equation.

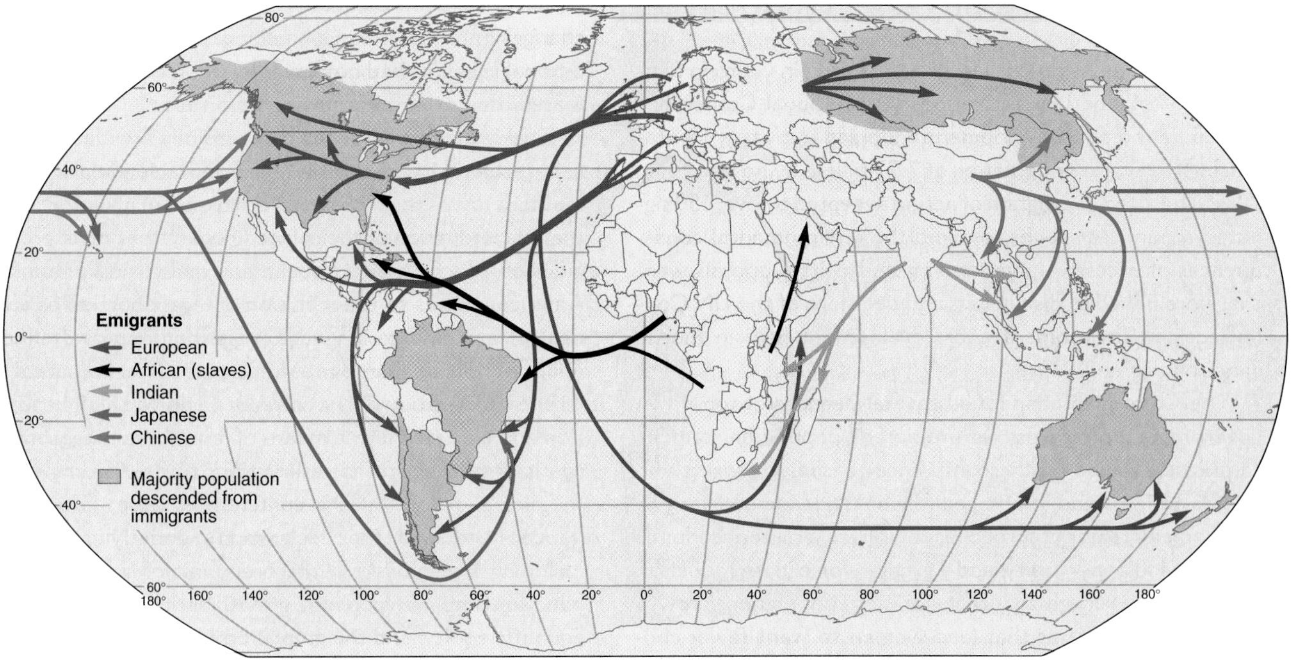


**FIGURE 6.21 Principal migrations of recent centuries.** The arrows suggest the major free and forced international population movements since about 1700. The shaded areas on the map are regions whose present population is more than 50% descended from the immigrants of recent centuries. Source: *Shaded zones after Daniel Noin,* Géographie de la Population, *p. 85, 1979, © Masson, Paris.*

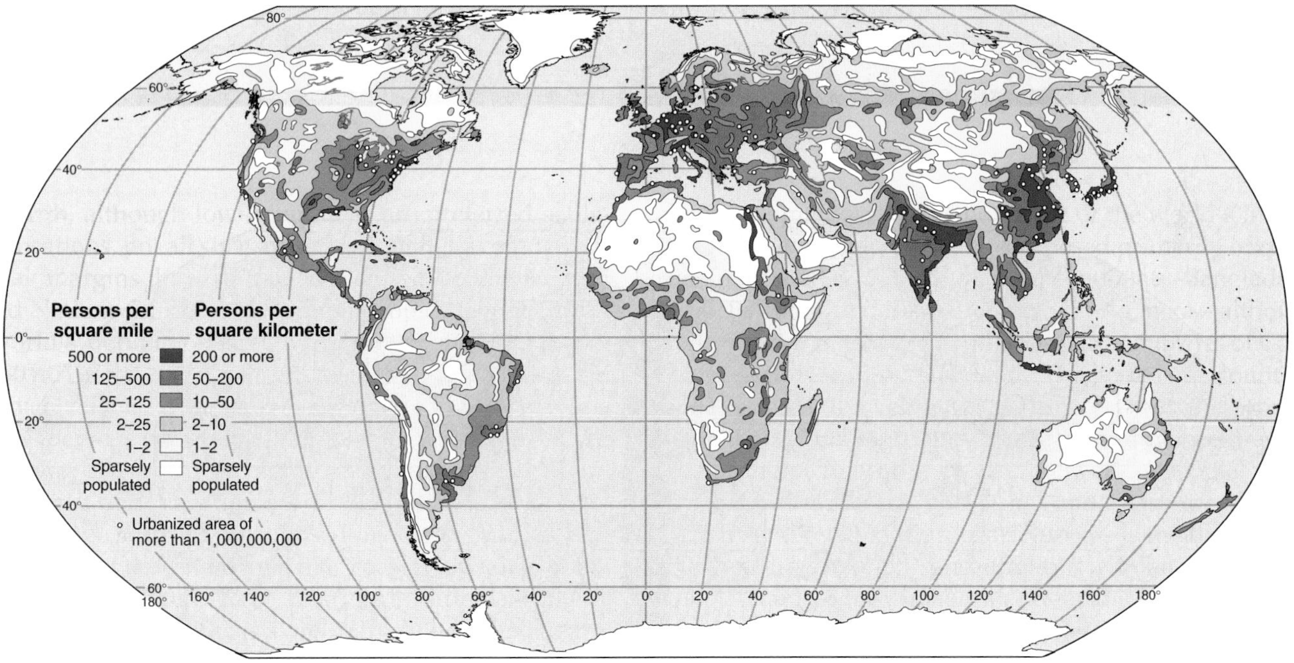


**FIGURE 6.22 World population density.**

north of the equator, and two-thirds of the total dwell in the midlatitudes between 20° and 60° North (Figure 6.23). Second, a large majority of the world's inhabitants occupy only a small part of its land surface. More than half the people live on about 5% of the land, two-thirds on 10%, and almost nine-tenths on less than 20%. Third, people congregate in lowland areas; their numbers decrease sharply with increases in elevation. Temperature, length of growing season, slope and erosion problems, even oxygen reductions at very high altitudes all appear to limit the habitability of higher elevations. One estimate is that between 50% and 60% of all people live below 200 meters (650 ft), a zone containing less than 30% of total land area. Nearly 80% reside below 500 meters (1650 ft).

Clearly, natural conditions are less restrictive than Greek geographers believed. Both ancient and modern technologies have rendered habitable the areas that natural conditions make forbidding. Irrigation, terracing, diking, and draining are among the methods devised to extend the ecumene locally (Figure 6.24).

At the world scale, the ancient observation of habitability appears remarkably astute. The **nonecumene,** or *anecumene,* the uninhabited or very sparsely occupied zone, does include the permanent ice caps of the Far North and Antarctica and large segments of the tundra and coniferous forest of northern Asia and North America. But the nonecumene is not continuous, as the ancients supposed. It is discontinuously encountered in all portions of the globe and includes parts of the tropical rain forests of equatorial zones, midlatitude deserts of both the Northern and Southern Hemispheres, and high mountain areas.

Even parts of these unoccupied or sparsely occupied districts have localized, dense settlement nodes or zones based on irrigation agriculture, mining and industrial activities, and the like. Perhaps the most striking case of settlement in an environment elsewhere considered part of the nonecumene world is that of the dense population in the Andes Mountains of South America and the plateau of Mexico. There, Native Americans found temperate conditions away from the dry coast regions and the hot, wet Amazon Basin. The fertile high basins have served a large population for more than a thousand years.

Even with these locally important exceptions, the nonecumene portion of the earth is extensive; 35% to 40% of all the world's land surface is inhospitable and without significant settlement. This is, admittedly, a smaller proportion of the earth than would have qualified as uninhabitable in ancient times or even during the 19th century. Since the end of the Ice Age some 12,000 years ago, humans have steadily expanded their areas of settlement.

## Population Density

Margins of habitation could only be extended, of course, as humans learned to support themselves from the resources of new settlement areas. The numbers that could be sustained in old or new habitation zones were and are related to the

**FIGURE 6.24** Terracing of hillsides is one device to extend a naturally limited productive area. The technique is effectively used here in Wakayama Prefecture, southern Honshu, Japan. *© Robert Essel/Corbis Images.*

resource potential of those areas and the cultural levels and technologies possessed by the occupying populations. The term **population density** expresses the relationship between the number of inhabitants and the area they occupy.

Density figures are useful, if sometimes misleading, representations of regional variations of human distribution. The **crude density** or **arithmetic density** of population is the most common and least satisfying expression of that variation. It is the calculation of the number of people per unit area of land, usually within the boundaries of a political entity. It is an easily reckoned figure. All that is required is information on total population and total land area, both commonly available for national or other political units. The figure can, however, be misleading and may obscure more of reality than it reveals. The calculation is an average that blankets a country's largely undevelopable or sparsely populated regions, along with its intensively settled and developed districts. A national average density figure reveals nothing about either class of territory. In general, the larger the political unit for which crude or arithmetic population density is calculated, the less useful is the figure.

Various modifications may be made to refine density as a meaningful abstraction of distribution. Its descriptive precision is improved if the area in question can be subdivided into comparable regions or units. Thus, it is more revealing to know that, in 2000, New Jersey had a density of 438 and Wyoming of 3.5 persons per square kilometer (1134 and 9 per sq mi) of land area than to know only that the figure for the conterminous United States (48 states) was 36 per square kilometer (94 per sq mi). If Hawaii and large, sparsely populated Alaska are added, the U.S. density figure drops to 31 per square kilometer (80 per sq mi). The calculation may also be modified to provide density distinctions between classes of population—rural versus urban, for example. Rural densities in the United States rarely exceed 115 per square kilometer (300 per sq mi), while portions of major cities can have thousands of people in equivalent space.

Another revealing refinement of crude density relates population not simply to total national territory but to the area of a country that is or may be cultivated—that is, to *arable* land. When total population is divided by arable land area alone, the resulting figure is the **physiological density,** which is, in a sense, an expression of population pressure exerted on agricultural land. Table 6.4 makes evident that countries differ in physiological density and that the contrasts between crude and physiological densities of countries point up actual settlement pressures that are not revealed by arithmetic densities alone. The calculation of physiological density, however, depends on uncertain definitions of arable and cultivated land, assumes that all arable land is equally productive and comparably used, and includes only one part of a country's resource base.

**Agricultural density** is still another useful variant. It simply excludes city populations from the physiological density calculation and reports the number of rural residents per unit of agriculturally productive land. It is, therefore, an estimate of the pressure of people on the rural areas of a country.

## Overpopulation

It is an easy and common step from concepts of population density to assumptions about overpopulation or overcrowding. It is wise to remember that **overpopulation** is a

**TABLE 6.4 | Comparative Densities for Selected Countries**

| | Crude Density | | Physiological Density[a] | | Agricultural Density[b] | |
|---|---|---|---|---|---|---|
| **Country** | sq mi | km² | sq mi | km² | sq mi | km² |
| Argentina | 35 | 14 | 384 | 149 | 44 | 17 |
| Australia | 6 | 2 | 100 | 38 | 10 | 4 |
| Bangladesh | 2552 | 985 | 4083 | 1577 | 3129 | 1208 |
| Canada | 9 | 3 | 176 | 68 | 36 | 14 |
| China | 352 | 136 | 2654 | 1025 | 1829 | 706 |
| Egypt | 178 | 69 | 6350 | 2452 | 3362 | 1298 |
| India | 873 | 337 | 1605 | 620 | 1176 | 454 |
| Iran | 107 | 41 | 1213 | 468 | 414 | 160 |
| Japan | 910 | 348 | 7321 | 2827 | 1562 | 603 |
| Nigeria | 355 | 137 | 1131 | 437 | 653 | 252 |
| United Kingdom | 643 | 248 | 2634 | 1017 | 272 | 105 |
| United States | 80 | 31 | 404 | 156 | 93 | 36 |

*Sources:* World Bank, *World Development Indicators 2003;* and Population Reference Bureau, *World Population Data Sheet.*

[a]Total population divided by area of arable land.

[b]Rural population divided by area of arable land.

*Note:* Rounding may produce apparent conversion discrepancies.

value judgment reflecting an observation or a conviction that an environment or a territory is unable adequately to support its present population. (A related but opposite concept of *underpopulation* refers to the circumstance of too few people to sufficiently develop the resources of a country or region to improve the level of living of its inhabitants.)

Overpopulation is not the necessary and inevitable consequence of high density of population. Tiny Monaco, a principality in southern Europe about half the size of New York's Central Park, has a crude density of nearly 17,500 people per square kilometer (45,000 people per sq mi). Mongolia, a sizeable state of 1,565,000 square kilometers (604,000 sq mi) between China and Siberian Russia, has 1.6 persons per square kilometer (4.1 per sq mi); Iran, only slightly larger, has 41 per square kilometer (107 per sq mi). Macao, a former island possession of Portugal off the coast of China, has some 26,000 persons per square kilometer (67,000 per sq mi); the Falkland Islands off the Atlantic coast of Argentina count at most 1 person for every 5 square kilometers (2 sq mi) of territory. No conclusions about conditions of life, levels of income, adequacy of food, or prospects for prosperity can be drawn from these density comparisons.

Overcrowding is a reflection not of numbers per unit area but of the **carrying capacity** of land—the number of people an area can support on a sustained basis, given the prevailing technology. A region devoted to efficient, energy-intensive commercial agriculture that makes heavy use of irrigation, fertilizers, and biocides can support more people at a higher level of living than one engaged in the slash-and-burn agriculture described in Chapter 10. An industrial society that takes advantage of resources such as coal and iron ore and has access to imported food will not feel population pressure at the same density levels as a country with rudimentary technology.

Since carrying capacity is related to the level of economic development, maps such as Figure 6.22, displaying present patterns of population distribution and density, do not suggest a correlation with conditions of life. Many industrialized, urbanized countries have lower densities and higher levels of living than do less-developed ones. Densities in the United States, where there is a great deal of unused and unsettled land, are considerably lower than those in Bangladesh, where essentially all land is arable and which, with 1020 people per square kilometer (2640 per sq mi), is the most densely populated nonisland state in the world. At the same time, many African countries have low population densities and low levels of living, whereas Japan combines both high densities and wealth.

Overpopulation can be equated with levels of living or conditions of life that reflect a continuing imbalance between numbers of people and carrying capacity of the land. One measure of that imbalance might be the unavailability of food supplies sufficient in caloric content to meet individual daily energy requirements or so balanced as to satisfy normal nutritional needs. Unfortunately, dietary insufficiencies—with long-term adverse implications for life expectancy, physical vigor, and mental development—are most likely to be encountered in the developing countries, where much of the population is in the younger age cohorts (see Figure 6.11).

If those developing countries simultaneously have rapidly increasing population numbers dependent on domestically produced foodstuffs, the prospects must be for continuing undernourishment and overpopulation. Much of sub-Saharan Africa finds itself in this circumstance. Its per capita food production decreased during the 1990s, with continuing decline predicted over the following quarter-century as the population–food gap widens (Figure 6.25). The countries of North Africa are similarly strained. Egypt already must import well over half the food it consumes. Africa is not alone. The international Food and Agriculture Organization (FAO) estimates that, early in the 21st century, at least 65 countries with over 30% of the population of the developing world were unable to feed their inhabitants adequately from their own national territories at the low level of agricultural technology and inputs employed. Even rapidly industrializing China, an exporter of grain until 1994, now in most years is a net grain importer.

In the contemporary world, insufficiency of domestic agricultural production to meet national caloric requirements cannot be considered a measure of overcrowding or poverty. Only a few countries are agriculturally self-sufficient. Japan, a leader among the advanced states, is the world's biggest food importer and supplies from its own

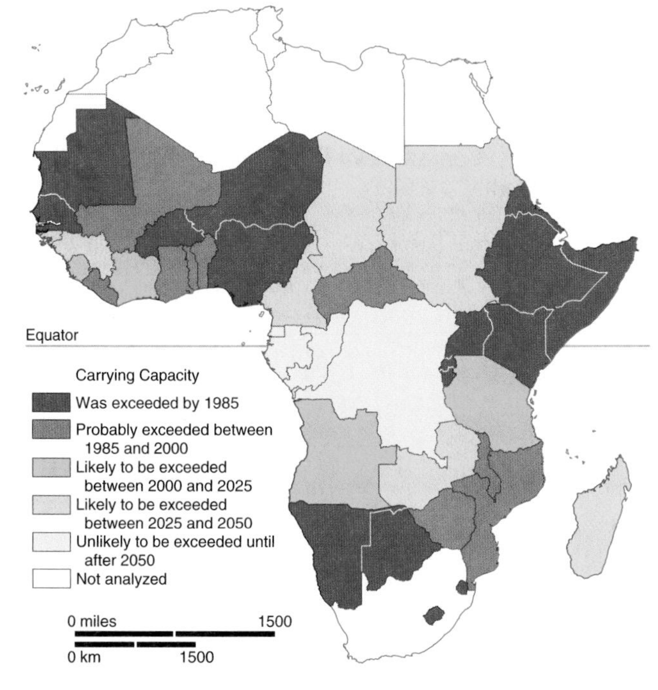


**FIGURE 6.25 Carrying capacity and potentials in sub-Saharan Africa.** The map assumes that (1) all cultivated land is used for growing food; (2) food imports are insignificant; and (3) agriculture is conducted by low-technology methods. Sources: *World Bank; United Nations Development Programme; Food and Agriculture Organization (FAO); and Bread for the World Institute.*

production only 40% of the calories its population consumes. Its physiological density is high, as Table 6.4 indicated, but it obviously does not rely on an arable land resource for its present development. Largely lacking in either agricultural or industrial resources, it nonetheless ranks well on all indicators of national well-being and prosperity. For countries such as Japan, South Korea, Malaysia, and Taiwan—all of which currently import more than 70% of the grain they consume—a sudden cessation of the international trade that permits the exchange of industrial products for imported food and raw materials would be disastrous. Domestic food production could not maintain the dietary levels now enjoyed by their populations and they, more starkly than many underdeveloped countries, would be "overpopulated."

### Urbanization

Pressures on the land resource of countries are increased not just by their growing populations but also by the reduction of arable land caused by such growth. More and more of the world population increase must be accommodated not in rural areas but in cities that hold the promise of jobs and access to health, welfare, and other public services. As a result, the *urbanization* (transformation from rural to urban status according to individual countries' definition of *urban*) of population in developing states is increasing dramatically. Since the 1950s, cities have grown faster than rural areas in nearly all developing countries. Indeed, because of the now rapid flow of migrants from countrysides to cities, population growth in the rural areas of the developing world has essentially stopped. Although Latin America, for example, has experienced substantial overall population increase, the size of its rural population is actually declining.

On UN projections, essentially all world population increase between 2000 and 2030 will be in urban areas and almost entirely within the developing regions and countries, continuing a pattern established by 1950 (Figure 6.26). In those areas collectively, cities are growing by over 3% a year, and the poorest regions are experiencing the fastest growth. By 2020, the UN anticipates, a majority of the population of less-developed countries will live in urban areas. In East, West, and Central Africa, for example, cities are expanding by 5% a year, a pace that can double their population every 14 years. Global urban population, just 750 million in 1950, grew to nearly 3 billion early in the 21st century and is projected to rise to 5.1 billion by 2030. The uneven results of past urbanization are summarized in Figure 6.27.

The sheer growth of cities in people and territory has increased pressures on arable land and adjusted upward both arithmetic and physiological densities. Urbanization consumes millions of hectares of cropland each year. In Egypt, for example, urban expansion and new development between 1965 and 1985 took out of production as much fertile soil as the massive Aswan Dam on the Nile River made newly available through irrigation with the

Urban and rural population growth, 1950–2030

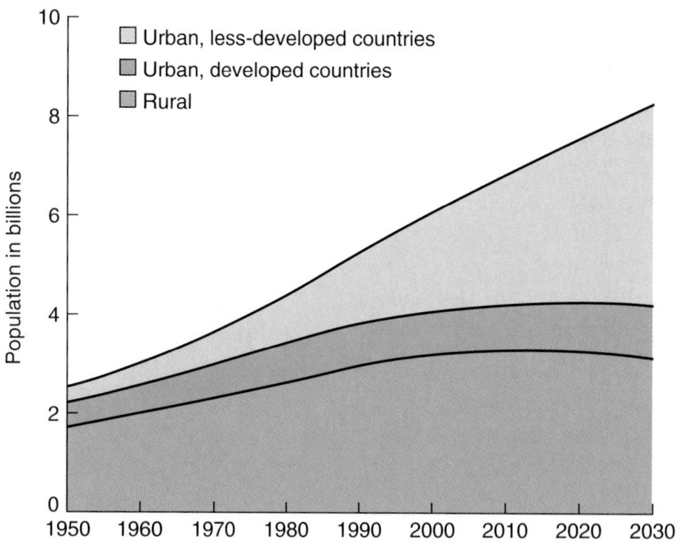


**FIGURE 6.26 Past and projected urban and rural population growth.** According to UN projections, some 65% of the world's total population may be urbanized by 2030. No universal definition of *urban* exists; the Population Reference Bureau, however, notes that "typically the population living in towns of 2000 or more . . . is classified as 'urban.'"
*Redrawn from* Population Bulletin *vol. 53, no. 1, Figure 3, page 12 (Population Reference Bureau, 1998).*

water it impounded. And during much of the 1990s, China lost close to 1 million hectares (2.5 million acres) of farmland each year to urbanization, road construction, and industrialization; the pace of such loss continued into this century. By themselves, some of the developing world cities, often surrounded by concentrations of people living in uncontrolled settlements, slums, and shantytowns (Figure 6.28), are among the most densely populated areas in the world. They face massive problems in trying to provide housing, jobs, education, and adequate health and social services for their residents. These and other matters of urban geography are the topics of Chapter 11.

## Population Data and Projections

Population geographers, demographers, planners, governmental officials, and a host of others rely on detailed population data to make their assessments of present national and world population patterns and to estimate future conditions. Birth rates and death rates, rates of fertility and of natural increase, age and sex composition of the population, and other items are all necessary ingredients for their work.

### Population Data

The data that students of population employ come primarily from the United Nations Statistical Office, the World

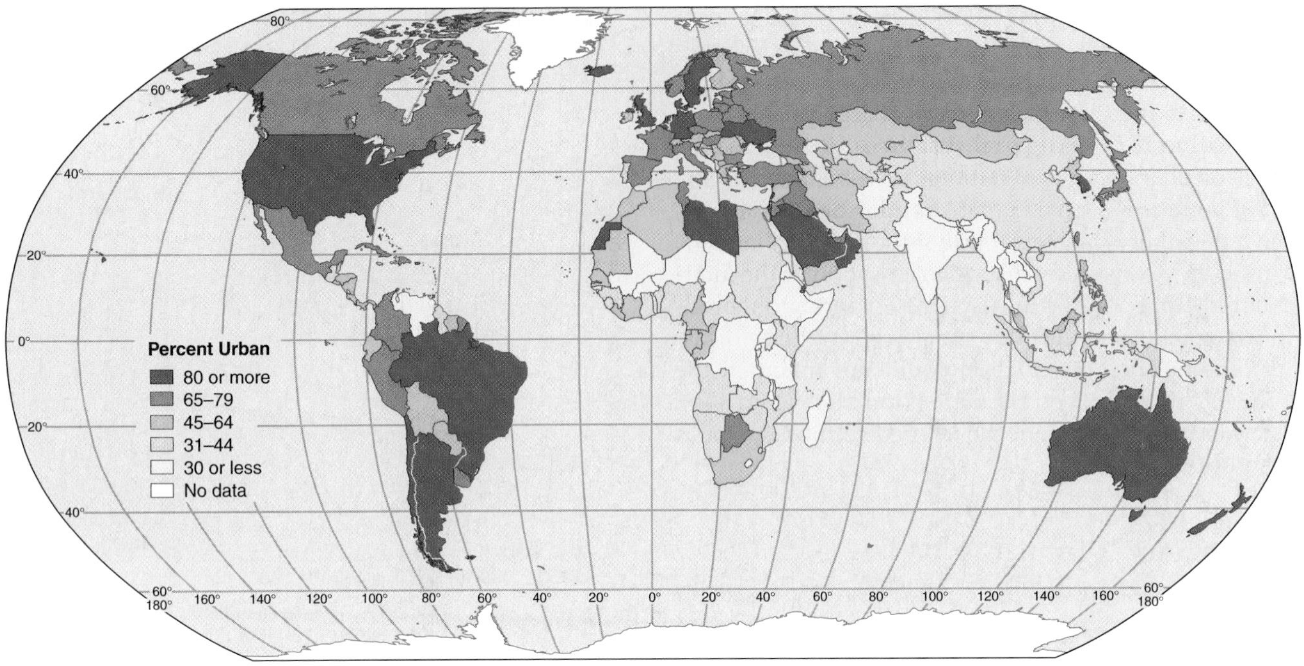


**FIGURE 6.27** **Percentage of national population classified as urban.** Urbanization has been particularly rapid in the developing continents. In 1950, only 17% of Asians and 15% of Africans were urban; by 2003, one-third of Africans and nearly 40% of Asians were city dwellers, and collectively the less-developed areas contained 70% of the world's city population. Source: *Data from Population Reference Bureau.*

**FIGURE 6.28** Millions of people of the developing world live in shantytowns on the fringes of large cities, without benefit of running water, electricity, sewage systems, or other public services. The UN reports that up to 40% of all urban dwellers worldwide live in such squatter settlements and slums. The hillside slum pictured here is one of the many *favelas* that are home for nearly half of Rio de Janeiro's more than 10 million residents. *© Luiz Claudio Marigo/Peter Arnold, Inc.*

Bank, the Population Reference Bureau, and ultimately from national censuses and sample surveys. Unfortunately, the data as reported may be more misleading than informative. For much of the developing world, a national census is a massive undertaking. Isolation and poor transportation, insufficiency of funds and trained census personnel, high rates of illiteracy limiting the type of questions that can be asked, and populations suspicious of all things governmental restrict the frequency, coverage, and accuracy of population reports.

However derived, detailed data are published by the major reporting agencies for all national units, even when those figures are poorly based on fact or are essentially fictitious. For example, for years, data on the total population, birth and death rates, and other vital statistics for Somalia were regularly reported and annually revised. The fact was, however, that Somalia had never had a census and had no system whatsoever for recording births. Seemingly precise data were regularly reported as well for Ethiopia. When that country had its first-ever census in 1985, at least one data source had to drop its estimate of the country's birth rate by 15% and increase its figure for Ethiopia's total population by more than 20%. And a disputed 1992 census of Nigeria officially reported a population of 88.5 million, still the largest in Africa but far below the then generally accepted and widely cited estimates of between 110 and 120 million Nigerians.

Fortunately, census coverage on a world basis is improving. Almost every country has now had at least one census of its population, and most have been subjected to periodic sample surveys (Figure 6.29). However, only about 10% of the developing world's population live in countries with anything approaching complete systems for registering births and deaths. Estimates are that 40% or less of live births in Indonesia, Pakistan, India, and the Philippines are officially recorded; sub-Saharan Africa has the highest percentage of unregistered births (71%), according to UNICEF. Apparently, deaths are even less completely reported than births throughout Asia. And whatever the deficiencies of Asian states, African statistics are still less complete and reliable. It is, of course, on just these basic birth and death data that projections about population growth and composition are founded.

**FIGURE 6.29** By the early 21st century, most countries of the developed and developing worlds had conducted a relatively recent census, although some were of doubtful completeness or accuracy. The photo shows an enumerator interviewing a Quito resident during the Ecuador census of 2001. *© Guillermo Granja/Reuters America.*

## Population Projections

For all their inadequacies and imprecisions, current data reported for country units form the basis of **population projections,** estimates of future population size, age, and sex composition based on current data. Projections are not forecasts, and demographers are not the social science equivalent of meteorologists. Weather forecasters work with a myriad of accurate observations applied against a known, tested model of the atmosphere. Demographers, in contrast, work with sparse, imprecise, out-of-date, and missing data applied to human actions that will be unpredictably responsive to stimuli not yet evident.

Population projections, therefore, are based on assumptions for the future applied to current data which are, themselves, frequently suspect. Since projections are not predictions, they can never be wrong. They are simply the inevitable result of calculations about fertility, mortality, and migration rates applied to each age cohort of a population now living, as well as the making of birth rate, survival, and migration assumptions about cohorts yet unborn. Of course, the perfectly valid *projections* of future population size and structure resulting from those calculations may be dead wrong as *predictions.*

Because those projections are invariably treated as scientific expectations by a public that ignores their underlying qualifying assumptions, agencies such as the UN that estimate the population of, say, Africa in the year 2025, do so by not one but three or more projections: high, medium, and low, for example (see "World Population Projections," p. 210). For areas as large as Africa, a medium projection is assumed to benefit from compensating errors and statistically predictable behaviors of very large populations. For individual African countries and smaller populations, the medium projection may be much less satisfying. The usual tendency in projections is to assume that something like current conditions will be applicable in the future. Obviously, the more distant the future, the less likely is that assumption to remain true. The resulting observation should be that the further into the future the population structure of small areas is projected, the greater is the implicit and inevitable error (see Figure 6.13).

## World Population Projections

Population projections can be useful and instructive but, of necessity, are based on current evaluations of future events. In making them, demographers are guided by controlling assumptions about the future levels of birth and death rates and, in some cases, the amount of migration. These assumptions are based on answers to such questions as: What are the present levels of fertility, of literacy, and of education? Does the government have a policy to influence population growth? What is the status of women? What might be the impact of, for example, HIV/AIDS on life expectancies?

Along with these questions must be weighed the likelihood of socioeconomic change, for it is generally assumed that as a country "develops," a preference for smaller families will cause fertility to fall to the replacement level of about two children per woman. But when can one expect this to happen in less-developed countries? And for the majority of more-developed countries with fertility currently below replacement level, can one assume that fertility will rise to avert the eventual disappearance of the population and, if so, when?

Predicting the pace of fertility decline is most important, as illustrated by one earlier set of United Nations long-range projections for Africa. As with many projections, these were issued in a "series" to show the effects of different assumptions. The "low" projection for Africa assumed that replacement-level fertility would be reached in 2030, which would put the continent's population at 1.4 billion in 2100. If the attainment of replacement-level fertility were delayed to 2065, the population would reach 4.4 billion in 2100. That difference of 3 billion serves as a cautionary note on both the acceptability of distant projections and the need to consider not only the possibilities but also the probabilities of future events.

Unfortunately, demographers usually cast their projections in an environmental vacuum, ignoring the realities of soils, vegetation, water supplies, and climate that ultimately determine feasible or possible levels of population support. Inevitably, different analysts present different assessments of the absolute carrying capacity of the earth. At an unrealistically low level, the World Hunger Project in the 1990s calculated that the world's ecosystem could, with then present agricultural technologies and with equal distribution of food supplies, support on a sustained basis no more than 5.5 billion people, a number already far exceeded. Many agricultural economists, in contrast—citing present trends and prospective increases in crop yields, fertilizer efficiencies, and intensification of production methods—are confident that the earth can readily feed 10 billion or more on a sustained basis. Nearly all observers, however, agree that physical environmental realities make unrealistic purely demographically based projections of a world population two or three times its present size.

# Population Controls

All population projections include an assumption that at some point in time population growth will cease and plateau at the replacement level. Without that assumption, future numbers become unthinkably large. For the world at unchecked present growth rates, there would be 1 trillion people three centuries from now, 4 trillion four centuries in the future, and so on. Although there is reasonable debate about whether the world is now overpopulated and about what either its optimum or maximum sustainable population should be, totals in the trillions are beyond any reasonable expectation.

Population pressures do not come from the amount of space humans occupy. It has been calculated, for example, that the entire human race could easily be accommodated within the boundaries of the state of Delaware. The problems stem from the food, energy, and other resources necessary to support the population and from the impact on the environment of the increasing demands and the technologies required to meet them. Rates of growth currently prevailing in many countries make it nearly impossible for them to achieve the kind of social and economic development they would like.

Clearly, at some point, population will have to stop increasing as fast as it has been. That is, either the self-induced limitations on expansion implicit in the demographic transition will be adopted or an equilibrium between population and resources will be established in more dramatic fashion. Recognition of this eventuality is not new. "[P]estilence, and famine, and wars, and earthquakes have to be regarded as a remedy for nations, as the means of pruning the luxuriance of the human race," was the opinion of the theologian Tertullian during the 2d century A.D.

Thomas Robert **Malthus** (1766–1834), an English economist and demographer, put the problem succinctly in a treatise published in 1798: all biological populations have a potential for increase that exceeds the actual rate of increase, and the resources for the support of increase are limited. In later publications, Malthus amplified his thesis by noting the following:

1. Population is inevitably limited by the means of subsistence.

2. Populations invariably increase with increase in the means of subsistence unless prevented by powerful checks.
3. The checks that inhibit the reproductive capacity of populations and keep it in balance with means of subsistence are either "private" (moral restraint, celibacy, and chastity) or "destructive" (war, poverty, pestilence, and famine).

The deadly consequences of Malthus's dictum that unchecked population increases geometrically while food production can increase only arithmetically[2] have been reported throughout human history, as they are today. Starvation, the ultimate expression of resource depletion, is no stranger to the past or present. By conservative estimate, more than 40 people worldwide will die of causes related to hunger and malnutrition (11 million per year) during the 2 minutes it takes you to read this page; half will be children under 5. They will, of course, be more than replaced numerically by new births during the same 2 minutes. Losses are nearly always recouped. All battlefield deaths, perhaps 70 million, in all of humankind's wars over the past 300 years equal less than a 1-year replacement period at present rates of natural increase.

Yet, inevitably—following the logic of Malthus, the apparent evidence of history, and observations of animal populations—equilibrium must be achieved between numbers and support resources. When overpopulation of any species occurs, a population dieback is inevitable. The madly ascending leg of the J-curve is bent to the horizontal, and the J-curve is converted to an **S-curve.** It has happened before in human history, as Figure 6.30 summarizes. The top of the S-curve represents a population size consistent with and supportable by the exploitable resource base. When the population is equivalent to the carrying capacity of the occupied area, it is said to have reached a **homeostatic plateau.**

In animals, overcrowding and environmental stress apparently release an automatic physiological suppressant of fertility. Although famine and chronic malnutrition may reduce fertility in humans, population limitation usually must be either forced or self-imposed. The demographic transition to low birth rates matching reduced death rates is cited as evidence that Malthus's first assumption was wrong: human populations do not inevitably grow geometrically. Fertility behavior, it was observed, is conditioned by social determinants, not solely by biological or resource imperatives.

Although Malthus's ideas had been discarded as deficient by the end of the 19th century in light of the European population experience, the concerns he expressed were revived during the 1950s. Observations of population growth in underdeveloped countries and the strain that growth placed on their resources inspired the viewpoint that improvements in living standards could be achieved only by raising investment per worker. Rapid population growth was seen as a serious diversion of scarce resources away from capital investment and into unending social welfare programs. In order to lift living standards, the existing national efforts to lower mortality rates had to be balanced by governmental programs to reduce birth rates. **Neo-Malthusianism,** as this viewpoint became known, has been the underpinning of national and international programs of population limitation primarily through birth control and family planning (Figure 6.31).

Neo-Malthusianism has had a mixed reception. Asian countries, led by China and India, have in general—though

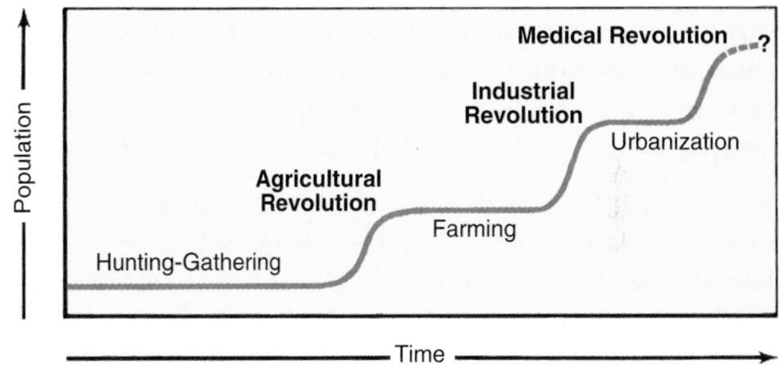


**FIGURE 6.30** The steadily higher *homeostatic plateaus* (states of equilibrium) achieved by humans are evidence of their ability to increase the carrying capacity of the land through technological advance. Each new plateau represents the conversion of the J-curve into an S-curve. *Medical Revolution* implies the range of modern sanitary and public health technologies and disease preventative and curative advances that materially reduced morbidity rates.

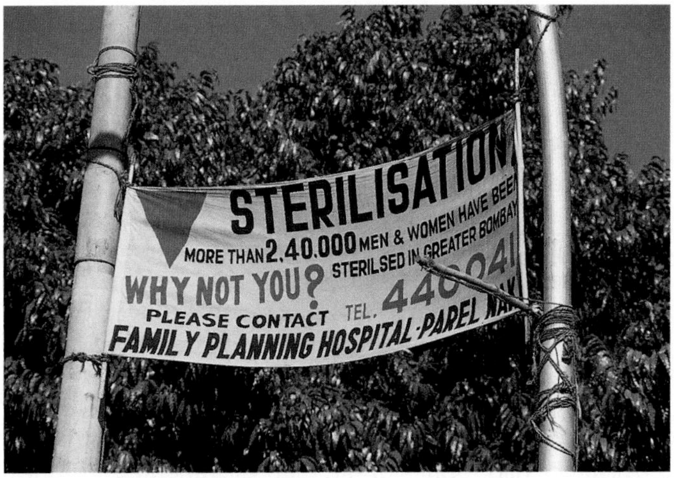


**FIGURE 6.31** A Mumbai, India, sign promoting the government's continuing program to reduce the country's high fertility rate. Female sterilization is the world's most popular form of birth prevention, and in India, Brazil, and China, a reported one-third or more of all married women have been sterilized. The comparable worldwide married male sterilization rate, in contrast, is 4%. © *Carl Purcell.*

[2]"Within a hundred years or so, the population can increase from fivefold to twentyfold, while the means of subsistence . . . can increase only from three to five times," was the observation of Hung Liangchi of China, a spatially distant early 19th-century contemporary of Malthus.

with differing successes—adopted family planning programs and policies. In some instances, success has been declared complete. Singapore established its Population and Family Planning Board in 1965, when its fertility rate was 4.9 lifetime births per woman. By 1986, that rate had declined to 1.7, well below the 2.1 replacement level for developed countries, and the board was abolished as no longer necessary. Caribbean and South American countries, even the poorest and most agrarian, have also experienced declining fertility rates, though often these reductions have been achieved despite pronatalist views of governments influenced by the Roman Catholic church.

Africa and the Middle East have generally been less responsive to the neo-Malthusian arguments because of ingrained cultural convictions among people, if not in all governmental circles, that large families—six or seven children—are desirable. Although total fertility rates have begun to decline in most sub-Saharan African states, they still remain above replacement levels nearly everywhere. Islamic fundamentalism opposed to birth restrictions also is a cultural factor in the Near East and North Africa. However, the Muslim theocracy of Iran has endorsed a range of contraceptive procedures and developed one of the world's more aggressive family planning programs.

Other barriers to fertility control exist. When first proposed by Western states, neo-Malthusian arguments that family planning is necessary for development were rejected by many less-developed countries. Reflecting both nationalistic and Marxist concepts, they maintained that remnant colonial-era social, economic, and class structures rather than population increase hinder development. Some governmental leaders think there is a correlation between population size and power and pursue pronatalist policies, as did Mao's China during the 1950s and early 1960s. And a number of American economists called *cornucopians* expressed the view, beginning in the 1980s, that population growth is a stimulus, not a deterrent, to development and that human minds and skills are the world's ultimate resource base. Since the time of Malthus, they observe, world population has grown from 900 million to over 6 billion without the predicted dire consequences—proof that Malthus failed to recognize the importance of technology in raising the carrying capacity of the earth. Still higher population numbers, they suggest, are sustainable, perhaps even with improved standards of living for all.

A third view, modifying cornucopian optimism, admits that products of human ingenuity such as the Green Revolution (see page 353) increases in food production have managed to keep pace with rapid population growth since 1970. But its advocates argue that scientific and technical ingenuity to enhance food production does not automatically appear; both complacency and inadequate research support have hindered continuing progress in recent years. And even if further advances are made, they observe, not all countries or regions have the social and political will or capacity to take advantage of them. Those that do not, third-view advocates warn, will fail to keep pace with the needs of their populace and will sink into varying degrees of poverty and environmental decay, creating national and regional—though not necessarily global—crises.

# Population Prospects

Regardless of population philosophies, theories, or cultural norms, the fact remains that many or most developing countries are showing significantly declining population growth rates. Global fertility and birth rates appear to be falling to an extent not anticipated by pessimistic Malthusians and at a pace that suggests a peaking of world population numbers sooner—and at lower totals—than previously projected (see "A Population Implosion?" p. 187). In all world regions, steady and continuous fertility declines have been recorded over the past years, reducing fertility from global five children per woman levels in the early 1950s to less than three per woman early in the present century.

Reducing fertility levels even to the replacement level of about 2.1 births per woman does not mean an immediate end to population growth. Because of the age composition of many societies, numbers of births will continue to grow even as fertility rates per woman decline. The reason is to be found in **demographic** (or **population**) **momentum,** and the key to that is the age structure of a country's population.

When a high proportion of the population is young, the product of past high fertility rates, larger and larger numbers enter the childbearing age each year; that is the case for major parts of the world early in the 21st century. The populations of developing countries are far younger than those of the established industrially developed regions (see Figure 6.11), with about one-third (in Asia and Latin America) to over 40% (in Africa) below the age of 15. The consequences of the fertility of these young people are yet to be realized. A population with a greater number of young people tends to grow rapidly, regardless of the level of childbearing. The results will continue to be felt until the now youthful groups mature and work their way through the population pyramid.

Inevitably, while this is happening, even the most stringent national policies limiting growth cannot stop it entirely. A country with a large present population base will experience large numerical increases despite declining birth rates. Indeed, the higher fertility is to begin with and the sharper its drop to low levels, the greater will be the role of momentum even after rates drop below replacement. A simple comparison of South Korea and the United Kingdom may demonstrate the point. In 2002, the two countries had the same level of fertility, with women averaging about 1.6 children each. Between that year and 2025, the larger population of the U.K. (without considering immigration or the births associated with newcomers) was projected to decline by 2 million persons, whereas smaller, more youthful South Korea was expected to continue growing, adding 2 million people.

Eventually, of course, young populations grow older, and even the youthful developing countries are beginning to face the consequences of that reality. The problems of a rapidly aging population that already confront the industrialized economies are now being realized in the developing world as well. Globally, there will be more than 1 billion persons 60 years of age or older by 2025 and nearly 2 billion by 2050, when the world will contain more people aged 60 or above than children under the age of 15. That momentous reversal in relative proportions of young and old already occurred in 1998 in the more-developed regions. The progression toward older populations is considered irreversible, the result of the now global demographic transition from high to low levels of fertility and mortality. The youthful majorities of the past are unlikely to occur again, for globally the population of older persons early in this century is growing by 2% per year—much faster than the population as a whole—and between 2025 and 2030, the 60+ growth rate will reach 2.8% per year. By 2050, the UN projects, one out of five persons worldwide will be 60 years old or older.

Three-fourths of the midcentury elderly folk will live in the less-developed world, for the growth rate of older people is three times as high in developing countries as in the developed ones. In the developing world, older persons are projected to make up 20% of the population by 2050, in contrast to the 8% over age 60 there in 2000. Since the pace of aging is much faster in the developing countries, they will have less time than the developed world did to adjust to the consequences of that aging. And those consequences will be experienced at lower levels of personal and national income and economic strength.

In both rich and poor countries, the working-age populations will face increasing burdens and obligations. The potential support ratio (PSR), the number of persons aged 15–64 years per one citizen aged 65 or older, has steadily fallen. Between 1950 and 2000, it dropped from 12 to 9 workers for each older person; by midcentury, the PSR is projected to drop to 4. The implications for social security schemes and social support obligations are obvious and are made more serious because the older population itself is aging. By the middle of the century, one-fifth of older persons will be 80 years old or older and, on average, will require more support expenditures for health and long-term care than do younger seniors. The consequences of population aging appear most intractable for the world's poorest developing states that generally lack health, income, housing, and social service support systems adequate for the needs of their older citizens. To the social and economic implications of their present population momentum, therefore, developing countries must add the aging consequences of past patterns and rates of growth (Figure 6.32).

**FIGURE 6.32** These senior residents of a Moroccan nursing home are part of the rapidly aging population of many developing countries. Worldwide, the over-60 cohort will number 22% of total population by 2050 and will be larger than the number of children less than 15 years of age. But by 2020, a third of Singapore citizens will be 55 or older, and China will have as large a share of its population over 60—about one in four—as will Europe. Already, the numbers of old people in the world's poorer countries are beginning to dwarf those in the rich world. At the start of the 21st century, there are nearly twice as many persons over 60 in developing countries as in the advanced ones, but most are without the old-age assistance and welfare programs developed countries have put in place. *© Nathan Benn/Corbis Images.*

# Summary

Birth, death, fertility, and growth rates are important in understanding the numbers, composition, distribution, and spatial trends of population. Recent "explosive" increases in human numbers and the prospects of continuing population expansion can be traced to sharp reductions in death rates, increases in longevity, and the impact of demographic momentum on a youthful population largely concentrated in the developing world. Control of population numbers historically was accomplished through a demographic transition first experienced in European societies that adjusted their fertility rates downward as death rates fell and life expectancies increased. The introduction of advanced technologies of preventative and curative medicine, pesticides, and famine relief have reduced mortality rates in developing countries without, until recently, always a compensating reduction in birth rates. Recent fertility declines in many developing regions suggest the demographic transition is no longer limited to the advanced industrial countries and promise world population stability earlier and at lower numbers than envisioned just a few years ago.

Even with the advent of more widespread fertility declines, the 6 billion human beings present at the end of the 20th century will still likely grow to almost 9 billion by the middle of the 21st century. That growth will largely reflect increases unavoidable because of the size and youth of populations in developing countries. Eventually, a new balance between population numbers and the carrying capacity of the world will be reached, as it has always been following past periods of rapid population increase.

People are unevenly distributed over the earth. The ecumene, or permanently inhabited portion of the globe, is discontinuous and marked by pronounced differences in population concentrations and numbers. East Asia, South Asia, Europe, and northeastern United States/southeastern Canada represent the world's greatest population clusters, though smaller areas of great density are found in other regions and continents. Because growth rates are highest and population doubling times generally shorter in world regions outside these four present main concentrations, new patterns of population localization and dominance are taking form.

A respected geographer once commented that "population is the point of reference from which all other elements [of geography] are observed." Certainly, population geography is the essential starting point of the human component of the human–environmental concerns of geography. But human populations are not merely collections of numerical units; nor are they to be understood solely through statistical analysis. Societies are distinguished not just by the abstract data of their numbers, rates, and trends but also by experiences, beliefs, understandings, and aspirations which collectively constitute that human spatial and behavioral variable called *culture.* It is to that fundamental human diversity that we next turn our attention.

## Key Words

agricultural density 205
arithmetic density 205
carrying capacity 206
cohort 182
crude birth rate (CBR) 182
crude death rate (CDR) 186
crude density 205
demographic equation 200
demographic (population) momentum 212
demographic transition 194
demography 180
dependency ratio 191
doubling time 193
ecumene 203
homeostatic plateau 211
J-curve 194
Malthus 210
mortality rate 186
neo-Malthusianism 211
nonecumene 204
overpopulation 205
physiological density 205
population density 205
population geography 180
population projection 209
population pyramid 190
rate of natural increase 193
rates 182
replacement level 187
S-curve 211
total fertility rate (TFR) 183
zero population growth (ZPG) 187

## For Review & Consideration

1. How do the *crude birth rate* and the *fertility rate* differ? Which measure is the more accurate statement of the amount of reproduction occurring in a population?
2. How is the *crude death rate* calculated? What factors account for the worldwide decline in death rates since 1945?
3. How is a *population pyramid* constructed? What shape of "pyramid" reflects the structure of a rapidly growing country? Of a population with a slow rate of growth? What can we tell about future population numbers from those shapes?
4. What variations do we discern in the spatial pattern of the *rate of natural increase* and, consequently, of population growth? What rate of natural increase would double population in 35 years?
5. How are population numbers projected from present conditions? Are projections the same as predictions? If not, in what ways do they differ?
6. Describe the stages in the *demographic transition.* Where has the final stage of the transition been achieved? What appears to be the applicability of the demographic transition to other parts of the world?
7. Contrast *crude population density, physiological density,* and *agricultural density.* For what different purposes might each be useful? How is *carrying capacity* related to the concept of density?

8. What was Malthus's underlying assumption concerning the relationship between population growth and food supply? In what ways do the arguments of *neo-Malthusians* differ from the original doctrine? What governmental policies are implicit in neo-Malthusianism?

9. Why is *demographic momentum* a matter of interest in population projections? In which world areas are the implications of demographic momentum most serious in calculating population growth, stability, or decline?

# Selected References

**WEBSITES**

The World Wide Web has a tremendous variety of sites pertaining to geography. Websites relevant to the subject matter of this chapter appear in the "Web Links" section of the Online Learning Center associated with this book. Access it at www.mhhe.com/getis10e/.


Brea, Jorge A. "Population Dynamics in Latin America." *Population Bulletin* 58, no. 1. Washington, D.C.: Population Reference Bureau, 2003.

Brown, Lester R., Gary Gardner, and Brian Halweil. *Beyond Malthus: Nineteen Dimensions of the Population Challenge.* New York: Norton, 1999.

Bulatao, Rodolfo A., and John B. Casterline, eds. *Global Fertility Transition.* Supplement to *Population and Development Review* 27 (2001).

Castles, Stephen, and Mark J. Miller. *The Age of Migration: International Population Movements in the Modern World.* 3d ed. New York: Guilford Press, 2003.

De Souza, Roger-Mark, John S. Williams, and Frederick A.B. Meyerson, "Critical Links: Population, Health, and the Environment." *Population Bulletin* 58, no. 3. Washington, D.C.: Population Reference Bureau, 2003.

Gelbard, Alene, Carl Haub, and Mary M. Kent. "World Population Beyond Six Billion." *Population Bulletin* 54, no. 1. Washington, D.C.: Population Reference Bureau, 1999.

Harrison, Paul, and Fred Pearce. *AAAS Atlas of Population and Environment,* ed. Victoria D. Markham. Berkeley, Calif.: American Association for the Advancement of Science and the University of California Press, 2001.

Haupt, Arthur, and Thomas Kane. *Population Handbook.* 4th ed. Washington, D.C.: Population Reference Bureau, 1997.

Hines, Christine L. "Elderly Americans." *Population Bulletin,* 56, no. 4. Washington, D.C.: Population Reference Bureau, 2001.

Martin, Philip, and Jonas Widgren. "International Migration: Facing the Challenge." *Population Bulletin* 57, no. 1. Washington, D.C.: Population Reference Bureau, 2002.

McFalls, Joseph A., Jr. "Population: A Lively Introduction." 4th ed. *Population Bulletin* 58, no. 4. Washington, D.C.: Population Reference Bureau, 2003.

Newbold, K. Bruce. *Six Billion Plus: Population Issues in the Twenty-First Century.* Lanham, Md.: Rowman and Littlefield, 2002.

Olshansky, S. Jay, Bruce Carnes, Richard G. Rogers, and Len Smith. "Infectious Diseases—New and Ancient Threats to World Health." *Population Bulletin* 52, no. 2. Washington, D.C.: Population Reference Bureau, 1997.

O'Neill, Brian, and Deborah Balk. "World Population Futures." *Population Bulletin* 56, no. 3. Washington, D.C.: Population Reference Bureau, 2001.

Peters, Gary L., and Robert P. Larkin. *Population Geography: Problems, Concepts, and Prospects.* 7th ed. Dubuque, Iowa: Kendall/Hunt, 2002.

Population Reference Bureau staff. "Transitions in World Population." *Population Bulletin* 59, no. 1. Washington, D.C.: Population Reference Bureau, 2004.

United Nations Population Fund. *The State of World Population.* New York: United Nations, annual.

U.S. Census Bureau. *The AIDS Pandemic in the 21st Century.* International Population Reports WP/02-2. Washington, D.C.: Government Printing Office, 2004.

U.S. Census Bureau. *Global Population at a Glance: 2002 and Beyond.* International Population Reports WP 02-1. Washington, D.C.: U.S. Census Bureau, 2004.

U.S. Census Bureau. *Global Population Profile.* International Population Reports WP/02. Washington, D.C.: Government Printing Office, 2004.

**FIGURE 7.4 The physical and cultural landscapes in juxtaposition.** Advanced societies are capable of so altering the circumstances of nature that the cultural landscapes they create become controlling environments. The city of Cape Town, South Africa, is a "built environment" largely unrelated to its physical surroundings. *© Charles O'Rear/Corbis Images.*

that a culture can be viewed as a three-part structure composed of subsystems that he termed *technological, sociological,* and *ideological.* In a similar classification, the biologist Julian Huxley identified three components of culture: *artifacts, sociofacts,* and *mentifacts.* Together, according to these interpretations, the subsystems—recognized by their separate components—constitute the structure of culture as a whole. But they are integrated; each reacts to the others and is affected by them in turn.

The **technological subsystem** is composed of the material objects and the techniques of their use by means of which people are able to live. Such objects are the tools and other instruments that enable us to feed, clothe, house, defend, transport, and amuse ourselves. The **sociological subsystem** of a culture is the sum of those expected and accepted patterns of interpersonal relations that find their outlet in economic, political, military, religious, kinship, and other associations. The **ideological subsystem** consists of the ideas, beliefs, and knowledge of a culture and of the ways in which they are expressed in speech or other forms of communication.

## The Technological Subsystem

Examination of variations in culture and in the manner of human existence from place to place centers on a series of commonplace questions: How do the people in an area make a living? What resources and what tools do they use to feed, clothe, and house themselves? Is a larger percentage of the population engaged in agriculture than in manufacturing? Do people travel to work in cars, on bicycles, or on foot? Do they shop for food or grow their own?

These questions concern the adaptive strategies used by different cultures in "making a living." In a broad sense, they address the technological subsystems at the disposal of those cultures—the instruments and tools people use in the daily cycle of existence. Huxley termed the material objects we employ to carry on our activities *artifacts.* For most of human history, people lived by hunting and gathering, taking the bounty of nature with only minimal dependence on weaponry, implements, and the controlled use of fire. Their adaptive skills were great, but their technological level was low. They had few specialized tools, could exploit only a limited range of potential resources, and had little or no

## Chaco Canyon Desolation

© Jon Malinowski.

It is not certain when they first came, but by A.D. 1000 the Anasazi people were building a flourishing civilization in what are now the states of Arizona and New Mexico. In the Chaco Canyon alone, they erected as many as 75 towns, all centered around pueblos, huge stone-and-adobe apartment buildings as tall as five stories and with as many as 800 rooms. These were the largest and tallest buildings of North America prior to the construction of iron-framed "cloud scrapers" in major cities at the end of the 19th century. An elaborate network of roads and irrigation canals connected and supported the pueblos. About A.D. 1200, the settlements were abruptly abandoned. The Anasazi, advanced in their skills of agriculture and communal dwelling, were—according to some scholars—forced to move on by the ecological disaster their pressures had brought to a fragile environment.

They needed forests for fuel and for the hundreds of thousands of logs used as beams and bulwarks in their dwellings. The pinyon-juniper woodland of the canyon was quickly depleted. For larger timbers needed for construction, they first harvested stands of ponderosa pine found some 40 kilometers (25 mi) away. As early as A.D. 1030, these, too, were exhausted, and the community switched to spruce and Douglas fir from mountaintops surrounding the canyon. When they were gone by 1200, the Anasazi's fate was sealed—not only by the loss of forest but also by the irreversible ecological changes deforestation and agriculture had occasioned. With forest loss came erosion that destroyed the topsoil. The surface water channels that had been built for irrigation were deepened by accelerated erosion, converting them into enlarging arroyos, useless for agriculture.

The material roots of their culture destroyed, the Anasazi turned on themselves and warfare convulsed the region. Smaller groups sought refuge elsewhere, re-creating on a reduced scale their pueblo way of life, but now in nearly inaccessible, highly defensible mesa and cliff locations. The destruction they had wrought destroyed the Anasazi in turn.

control of nonhuman sources of energy. Their impact on the environment was small, but at the same time the "carrying capacity" of the land discussed in Chapter 6 was low everywhere, for technologies and artifacts were essentially the same among all groups.

The retreat of the last glaciers about 12,000 to 13,000 years ago marked the start of a period of unprecedented cultural development. It led from primitive hunting and gathering economies at the outset through the evolution of agriculture and animal husbandry to, ultimately, urbanization, industrialization, and the intricate complexity of the modern technological subsystem. Since not all cultures passed through all stages at the same time, or even at all, *cultural divergence* between human groups became evident.

Cultural diversity among ancient societies reflected the proliferation of technologies that followed a more assured food supply and made possible a more intensive and extensive utilization of resources. Different groups in separate environmental circumstances developed specialized tools and behaviors to exploit resources they recognized. Beginning with the Industrial Revolution of the 18th century, however, a reverse trend—toward commonality of technology—began.

Today, advanced societies are nearly indistinguishable in the tools and techniques at their command. They have experienced *cultural convergence*—the sharing of technologies, organizational structures, and even culture traits and artifacts that is so evident among widely separated societies in a modern world united by instantaneous communication and efficient transportation. Those differences in technological traditions that still exist between developed and underdeveloped societies reflect, in part, national and personal wealth, stage of economic advancement and complexity, and, importantly, the level and type of energy used (Figure 7.5).

In technologically advanced countries, many people are employed in manufacturing or allied service trades. Per capita incomes tend to be high, as do levels of education and nutrition, life expectancies, and medical services. These countries wield great economic and political power. In contrast, technologically less-advanced countries have a high percentage of people engaged in farming, with much of the agriculture at a subsistence level (Figure 7.6). The gross national income or GNI (which measures the total domestic and foreign value added of all goods and services claimed

(a) (b)

**FIGURE 7.5** (a) This Balinese farmer, working with draft animals, employs tools typical of the low technological levels of subsistence economies. (b) Cultures with advanced technological subsystems use complex machinery to harness inanimate energy for productive use. *(a) © Dave G. Houser/Hillstrom Stock Photos. (b) Scott Bauer/USDA Agricultural Research Service.*

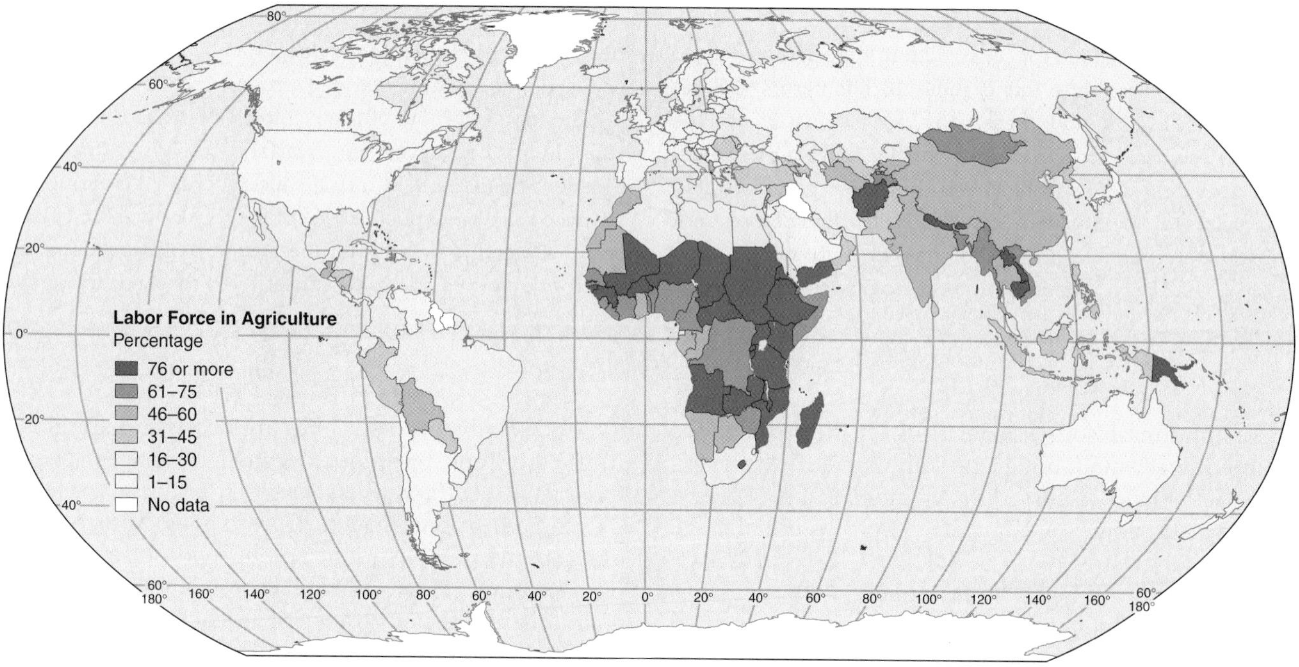


**FIGURE 7.6 Percentage of labor force engaged in agriculture.** For the world as a whole, agricultural workers make up slightly less than half of the total labor force. Highly developed economies usually have relatively low proportions of their labor forces in the agricultural sector, but the contrast between advanced and underdeveloped countries in the agricultural labor force measure is diminishing. Rapid population growth in developing countries has resulted in increased rural landlessness and poverty, from which escape is sought by migration to cities. The resulting reduction in the agricultural labor force percentage is an expression of relocation of poverty and unemployment, not of economic advancement.
Sources: *Data from C.I.A.* The World Factbook *and United Nations Development Programme.*

by residents of a country during a year) of these countries are much lower than those of industrialized states. Per capita incomes (Figure 7.7), life expectancies, and literacy rates also tend to be low.

Labels such as *advanced–less advanced, developed–underdeveloped,* or *industrial–nonindustrial* can mislead us into thinking in terms of "either–or." They may also be misinterpreted to mean, in general cultural terms, "superior–inferior." This belief is totally improper because the terms relate solely to economic and technological circumstances and bear no qualitative relationship to such vital aspects of culture as music, art, or religion.

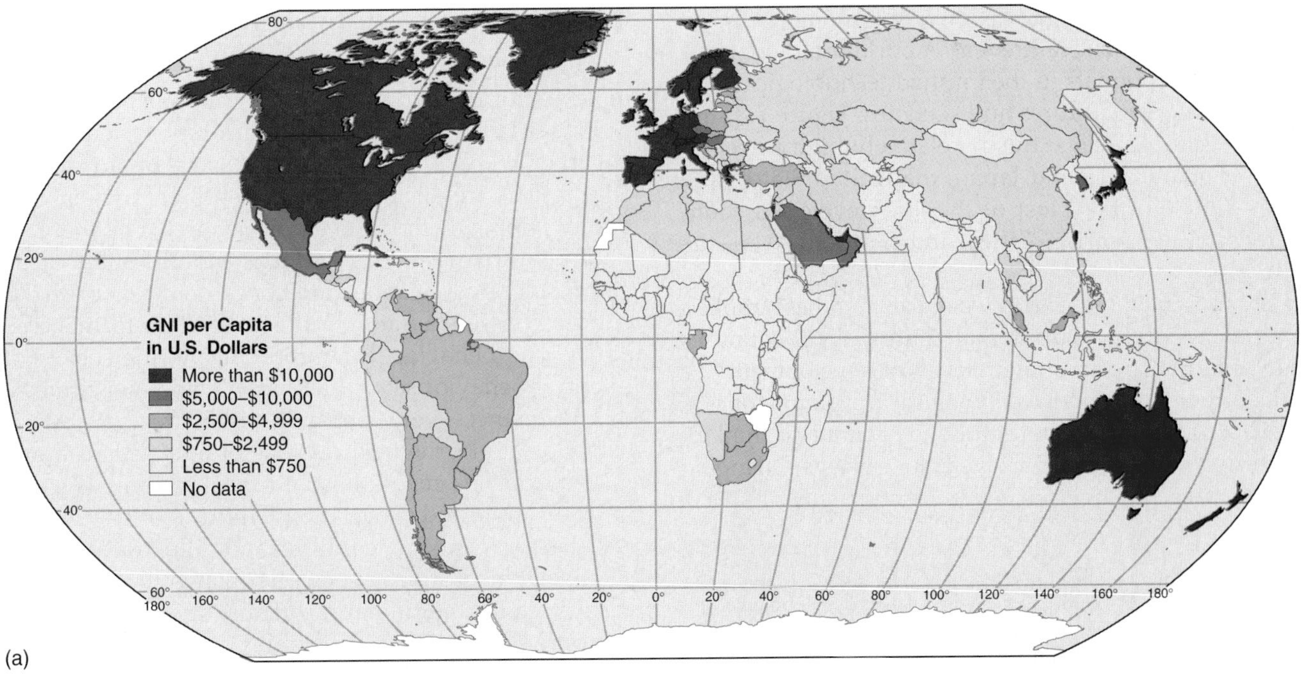


(a)

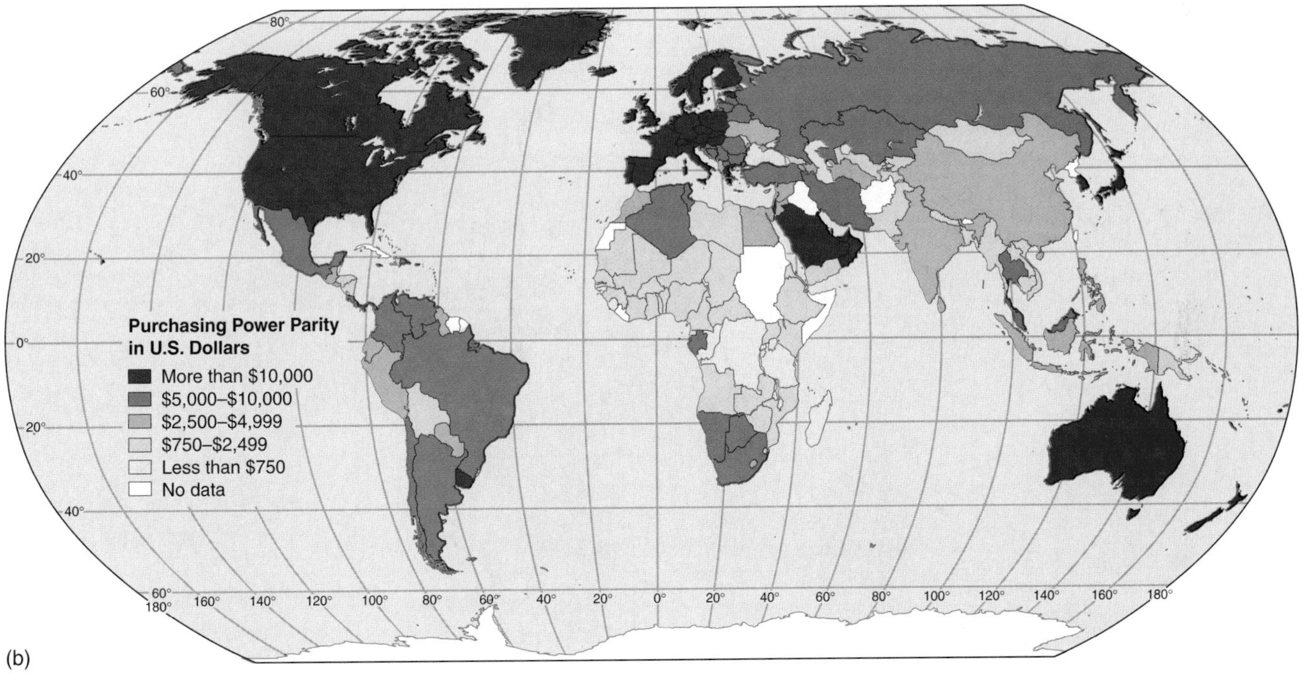


(b)

**FIGURE 7.7 Two contrasting views of income.** Both assume that total national income is evenly divided among all citizens, but each takes a different view of the meaning of the per capita income that division implies. (a) **Gross national income per capita.** GNI per capita expresses the individual's presumed income converted into U.S. dollars at official market exchange rates. Showing great contrasts between more- and less-advanced economies, it is a frequently employed summary of degree of technological development, though high incomes in sparsely populated, oil-rich countries may not have the same meaning in subsystem terms as do comparable per capita values in industrially advanced states. The map implies an unrealistic precision. For many states, when uncertain GNI is divided by unreliable population totals, the resulting GNI per capita is at best a rough approximation. (Gross national product is now called "gross national income" by the World Bank). (b) **Purchasing Power Parity (PPP)** takes into account variations in price levels to measure the relative size of economies. It is based on the idea that identical baskets of traded goods should cost the same in all countries and thus attempts to measure the relative domestic purchasing power of local currencies. By this more realistic measure, the abject poverty suggested by per capita GNI is seen to be much reduced in many developing countries.

Sources: *Data from the World Bank and United Nations.*

Properly understood, however, terms and measures of economic development can reveal important national and world regional contrasts in the implied technological subsystems of different cultures and societies. Figure 7.8 suggests that technological status is relatively high in nearly all European countries and Japan, the United States, and Canada—the "North." Most of the less developed countries are in Latin America, Africa, and southern Asia—the "South."[3]

It is important, however, to recognize that these implied national averages conceal internal contrasts. All countries include areas that are at different levels of development. We must also remember that technological development is a dynamic concept. It is most useful and accurate to think of the countries of the world as arrayed along an ever-changing continuum of technological levels and subsystems.

[3]The terms "North" and "South" to describe relative developmental levels were introduced in *North–South: A Programme for Survival* (sometimes called the *Brandt Report*), published in 1980 by the Independent Commission on International Development Issues. The former Soviet Union was at that time included within the North, and its successor states retain that association.

## The Sociological Subsystem

Continuum and change also characterize the religious, political, formal and informal educational, and other institutions that constitute the sociological subsystem of culture. Together, these *sociofacts* define the social organizations of a culture. They regulate how the individual functions relative to the group, whether it be family, church, or state.

There are no "givens" as far as patterns of interaction in any of these associations are concerned, except that most cultures possess a variety of formal and informal ways of structuring behavior. The importance to the society of the differing behavior sets varies among, and constitutes obvious differences between, cultures. Differing patterns of behavior are learned expressions of culture and are transmitted from one generation to the next by formal instruction or by example and expectation (Figure 7.9). The story of the Gauda's son that opened this chapter illustrates the point.

Social institutions are closely related to the technological system of a culture group. Thus, hunter-gatherers have one set of institutions; industrial societies have quite different ones. Preagricultural societies tended to be composed of small bands based on kinship ties, with little social differentiation or specialization of function in the band; the

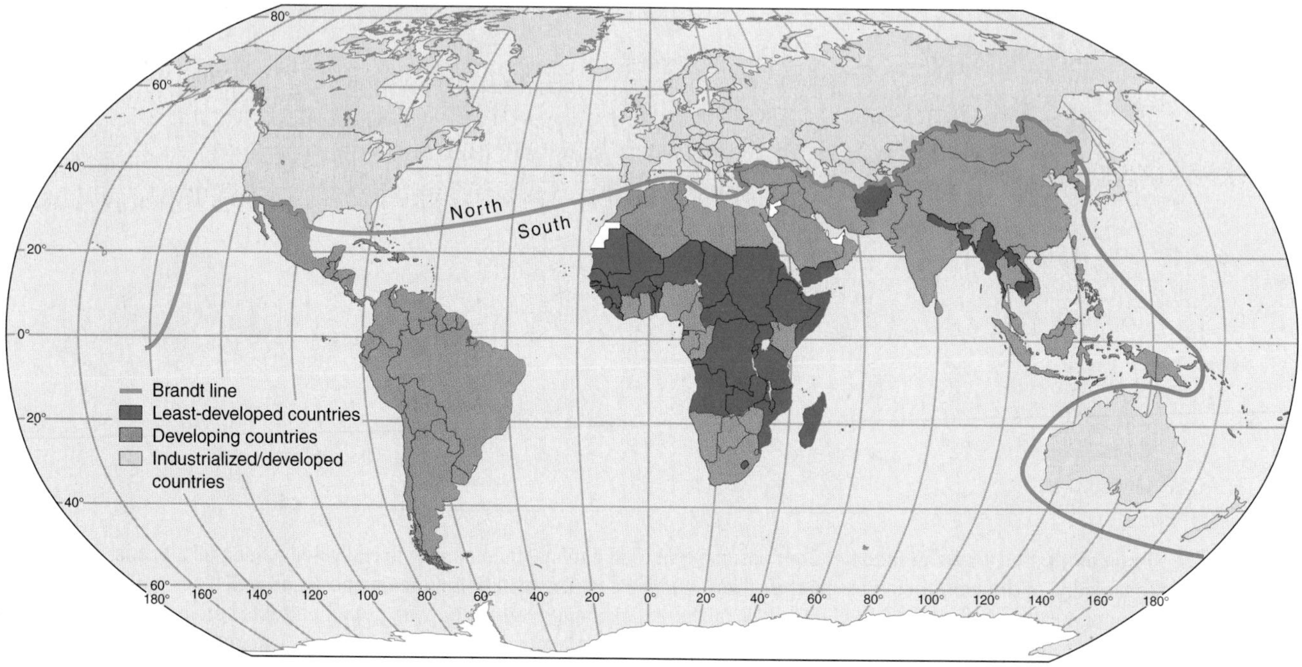


**FIGURE 7.8 Comparative developmental levels.** The "North–South" line of the 1980 *Brandt Report* suggested a simplified world contrast of development and underdevelopment based largely on degree of industrialization and per capita wealth registered then. In 2004, the United Nations Economic and Social Council and the UN Conference on Trade and Development (UNCTAD) recognized 50 "least-developed countries." That recognition reflects low ratings in three criteria: gross domestic product per capita, human resources as measured by a series of "quality of life" indicators, and level of economic diversification. The inclusive category of "developing countries" ignores recent significant economic and social gains in several Asian and Latin American states, raising them now to "industrialized/developed" status. Some "least-developed" states are small island countries not shown at this map scale. Sources: *UNCTAD and United Nations Development Programme.*

(a)

(b)

(c)

(d)

**FIGURE 7.9** All societies prepare their children for membership in the culture group. In each of these settings, certain values, beliefs, skills, and proper ways of acting are being transmitted to young people. *(a) © Cary Wolinsky/Stock, Boston. (b) © John Eastcott/Momatiuk/The Image Works. (c) © Yigol Pardo/Hillstrom Stock Photo. (d) © Stephanie Maze/Corbis Images.*

San (Bushmen) of arid southern Africa and isolated rain forest groups in Amazonia might serve as modern examples (Figure 7.10).

The revolution in food production occasioned by plant and animal domestication beginning around 10,000 years ago touched off a social transformation that included increases in population, urbanization, work specialization, and structural differentiation within the society. Politically, the rules and institutions by which people were governed changed with the formation of sedentary, agricultural societies. Loyalty was transferred from the kinship group to the state; resources became possessions rather than the common property of all. Equally far-reaching changes occurred after the 18th-century Industrial Revolution, leading to the complex of human social organizations that we experience and are controlled by today in "developed" states and that increasingly affect all cultures everywhere.

Culture is a complexly intertwined whole. Each organizational form or institution affects, and is affected by, related culture traits and complexes in intricate and variable ways.

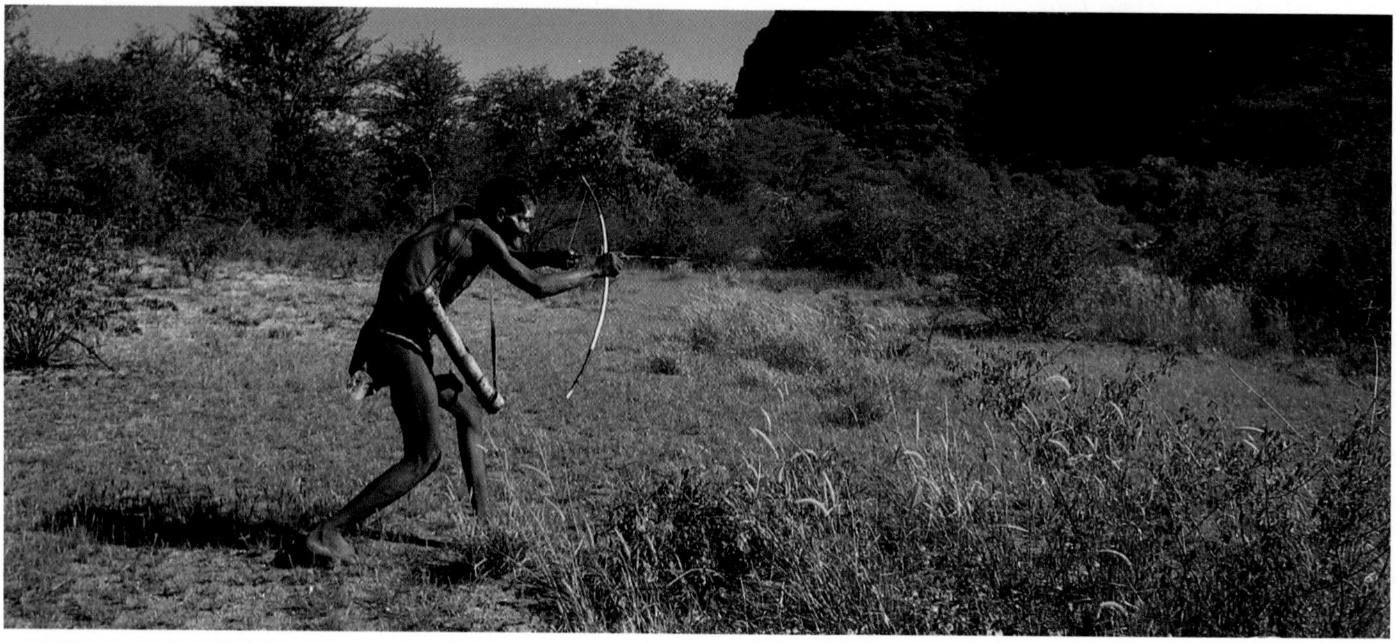

**FIGURE 7.10** Hunter-gatherers practiced the most enduring lifestyle in human history, trading it for the more arduous life of farmers under the necessity to provide larger quantities of less diversified foodstuffs for a growing population. Unlike their settled farmer rivals and successors, among hunter-gatherers, age and sex differences, not caste or economic status, have been the primary basis for interpersonal relations and the division of labor. This San (Bushman) hunter of Botswana is a member of one of the very few groups in Africa or the world still at least partially pursuing this ancient lifestyle. *© Aubrey Lang/Valan Photos.*

Systems of land and property ownership and control, for example, are institutional expressions of the sociological subsystem. They are simultaneously explicitly central to the classification of economies and to the understanding of spatial and structural patterns of economic development, as Chapter 10 will examine. Again, for each country the adopted system of laws and justice is a cultural variable identified with the sociological subsystem but extending its influence to all aspects of economic and social organization, including the political geographic systems discussed in Chapter 9.

## The Ideological Subsystem

The third class of elements defining and identifying a culture is the ideological subsystem. This subsystem consists of ideas, beliefs, and knowledge, as well as the ways we express these things in our speech and other forms of communication. Mythologies, theologies, legend, literature, philosophy, folk wisdom, and commonsense knowledge make up this category. Passed on from generation to generation, these abstract belief systems, or *mentifacts*, tell us what we ought to believe, what we should value, and how we ought to act. Beliefs form the basis of the socialization process.

Often, we know—or think we know—what the beliefs of a group are from written sources. Sometimes, however, we must depend on the actions or objectives of a group to tell us what its true ideas and values are. "Actions speak louder than words" and "Do as I say, not as I do" are commonplace recognitions of the fact that actions and words do not always coincide. The values of a group cannot be deduced from the written record alone.

Nothing in a culture stands totally alone. Changes in the ideas that a society holds may affect the sociological and technological systems, just as, for example, changes in technology force changes in the social system. The abrupt alteration after World War I (1914–1918) of the ideological structure of Russia from a monarchical, agrarian, capitalistic system to an industrialized, communistic society involved sudden, interrelated alteration in all facets of that country's culture system. The equally abrupt disintegration of Russian communism in the early 1990s was similarly disruptive of all its established economic, social, and administrative organizations. The interlocking nature of all aspects of a culture is termed **cultural integration.**

The recognition of three distinctive subsystems of culture, although helping us appreciate its structure and complexity, can at the same time obscure the many-sided nature of individual elements of culture. *Cultural integration* means that any cultural object or act may have a number of meanings. A dwelling, for example, is an artifact providing shelter for its occupants. It is, simultaneously, a sociofact reflecting the nature of the family or kinship group it is designed to house, and a mentifact summarizing a culture group's convictions about appropriate design, orientation, and building materials of dwelling units. In the same vein, clothing serves as an artifact of bodily protection appropriate to climatic conditions, available materials and techniques, or the activity in which the wearer is engaged. But garments also may be sociofacts, identifying an individual's role in the social structure of the community or culture, and mentifacts, evoking larger community value systems (Figure 7.11).

## Culture Change

The recurring theme of cultural geography is change. No culture is, or has been, characterized by a permanently fixed set of material objects, systems of organization, or even ideologies, although all of these may be long-enduring within a stable, isolated society. Such isolation and stability have always been rare. On the whole, although cultures are essentially conservative, they are always in a state of flux.

Many individual changes, of course, are so slight that they initially are almost unnoticed, though collectively they

(a)

(b)

(c)

**FIGURE 7.11** (a) When clothing serves primarily to cover, protect, or assist in activities, it is an *artifact.* (b) Some garments are *sociofacts,* identifying a role or position within the social structure: the distinctive "uniforms" of a soldier, a cleric, or a beribboned ambassador immediately proclaim their respective roles in a culture's social organizations. (c) The sometimes mandatory burkas or chadors worn by Muslim women are *mentifacts,* indicative not specifically of the role of the wearer but of the values of the culture the wearer represents.

*(a) © Corbis/Vol. #94. (b) © W. Marc Bernsau/The Image Works. (c) © Corbis/Vol. #145.*

may substantially alter the affected culture. Think of how the culture of the United States differs today from what it was in 1940—not in essentials, perhaps, but in the innumerable electric, electronic, and transportational devices that have been introduced and in the recreational, social, and behavioral adjustments they and other technological changes have wrought. Among the latter have been shifts in employment patterns to include greater participation by women in the waged workforce and associated changes in attitudes toward the role of women in the society at large. Such cumulative changes occur because the culture traits of any group are not independent; they are clustered in a coherent and integrated pattern. Change on a small scale will have wide repercussions as associated traits also change to accommodate the adopted adjustment. Change, both major and minor, within cultures is induced by *innovation, spatial diffusion,* and *acculturation.*

## Innovation

**Innovation** implies changes to a culture that result from ideas created within the social group itself and adopted by the culture. The novelty may be an invented improvement in material technology, such as the bow and arrow or the jet engine. It may involve the development of nonmaterial forms of social structure and interaction: feudalism or Christianity, for example.

Premodern and traditional societies characteristically are not innovative or receptive to change. In cultures at equilibrium with their environment and with no unmet needs, change has no adaptive value and has no reason to occur. Indeed, all societies have an innate resistance to change, since innovation inevitably creates tensions between the new reality and other established socioeconomic conditions. Those tensions can be resolved only by adaptive changes elsewhere in the total system. Complaints about youthful fads or the glorification of times past are familiar examples of reluctance to accept or adjust to change. However, when a social group is unreasonably resistant or unresponsive—mentally, psychologically, or economically—to changing circumstances and innovation, it is said to exhibit *cultural lag.* For example, for at least 1500 years, most California Indians were in contact with cultures utilizing both maize and pottery, yet they failed to accept either innovation.

Innovation—invention—frequently under stress, has marked the history of humankind. An expanded food base accompanied the pressures of growing populations at the end of the Ice Age. Domestication of plants and animals appears to have occurred independently in several recognizable areas of "invention" of agriculture, shown in Figure 7.12. From them, presumably, came a rapid diffusion of food types, production techniques, and new modes of economic and social organization as the majority of humankind changed from hunting–gathering to sedentary farming by no later than 2000 years ago. All innovation has a radiating impact upon the web of culture; the more basic the innovation, the more pervasive its consequences.

Few innovations in human history have been more basic than the Agricultural Revolution. It affected every aspect of society. Culture altered at an accelerating pace, and change itself became a way of life. Humans learned the arts of spinning and weaving plant and animal fibers. They learned to use the potter's wheel, to fire clay, and to make utensils. They developed techniques of brick making, mortaring, and building construction. They discovered the skills of mining, smelting, and casting metals. Special local advantages in resources or products promoted the development of long-distance trading connections. On the foundation of such technical advancements a more complex, exploitative culture appeared, including a stratified society to replace the rough equality of hunting and gathering economies.

The source regions of such social and technical revolutions were initially spatially confined. The term **culture hearth** is used to describe those restricted areas of innovation from which key culture elements diffused to exert an influence on surrounding regions. The hearth may be viewed as the "cradle" of any culture group whose developed systems of livelihood and life created a distinctive cultural landscape. Most of the thousands of hearths that evolved across the world in all regions and natural settings remained at low levels of social and technical development. Only a few produced the trappings of *civilization,* which are usually assumed to include writing (or other forms of record keeping), metallurgy, long-distance trade connections, astronomy and mathematics, social stratification and labor specialization, formalized governmental systems, and a structured urban society. Several major culture hearths emerged, some as early as 7000 to 8000 years ago, following the initial revolution in food production. Prominent centers of early creativity were located in Egypt, Mesopotamia, the Indus Valley of the Indian subcontinent, northern China, southeastern Asia, and in several locations in Africa, the Americas, and elsewhere (Figure 7.13).

In most modern societies, innovative change has become common, expected, and inevitable, though it may be rejected by some of their separate culture groups (see "Folk and Popular Culture"). The rate of invention, at least as measured by the number of patents granted, has steadily increased, and the period between idea conception and product availability has been decreasing. A general axiom is that the more ideas available and the more minds able to exploit and combine them, the greater the rate of innovation. The spatial implication is that larger urban centers of advanced economies tend to be centers of innovation, not just because of their size but also because of the number of ideas interchanged. Indeed, ideas not only stimulate new ideas but also create circumstances in which new solutions must be developed to maintain the forward momentum of the society (Figure 7.14).

## Diffusion

**Spatial diffusion** is the process by which a concept, a practice, an innovation, or a substance spreads from its point of origin to new territories. Diffusion can assume a variety of

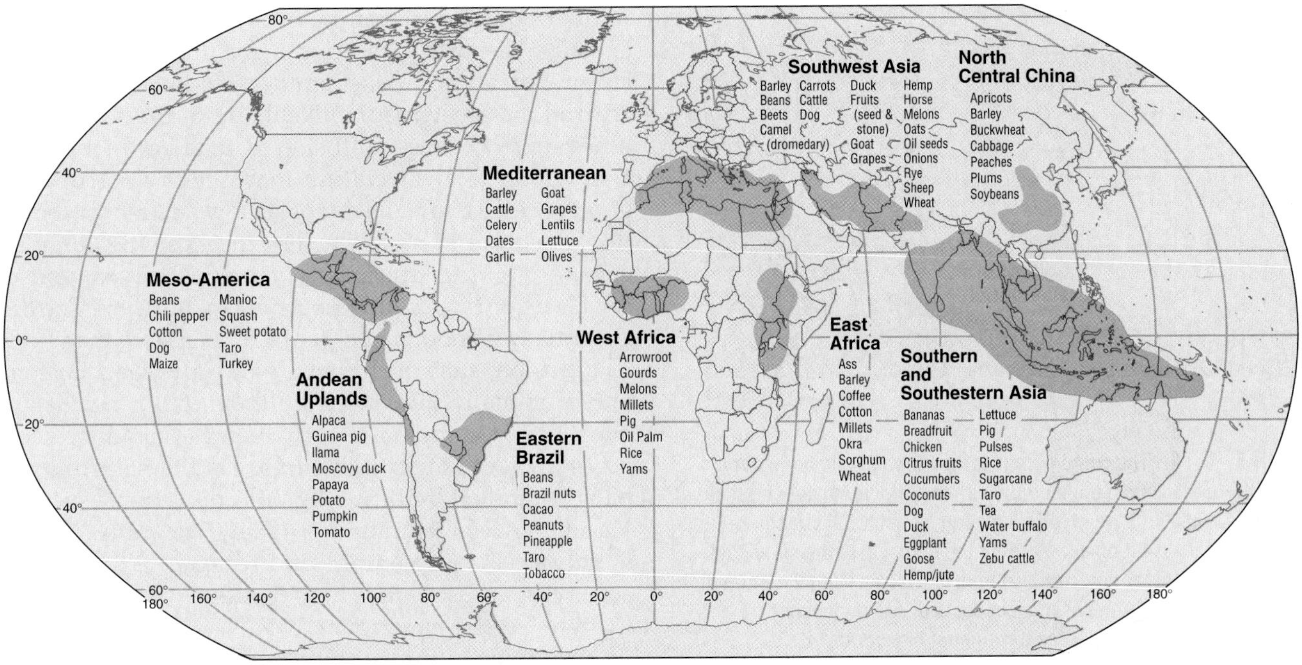


**FIGURE 7.12 Chief centers of plant and animal domestication.** The Southern and Southeastern Asian center was characterized by the domestication of plants, such as taro, that are propagated by the division and replanting of existing plants (vegetative reproduction). Reproduction by the planting of seeds (e.g., maize and wheat) was more characteristic of Meso-America and Southwest Asia. The African and Andean areas developed crops reproduced by both methods. The lists of crops and livestock associated with the separate origin areas are selective, not exhaustive.

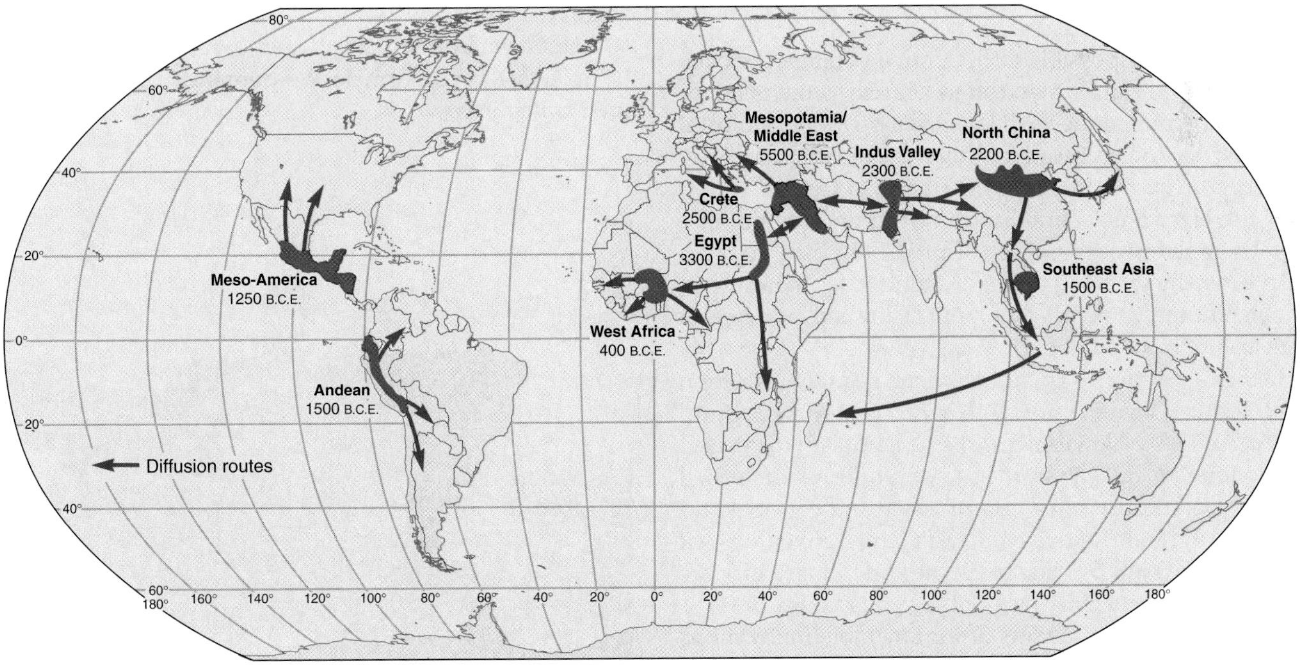


**FIGURE 7.13 Early culture hearths of the Old World and the Americas.** The B.C.E. (Before the Common Era) dates approximate times when the hearths developed complex social, intellectual, and technological bases and served as cultural diffusion centers.

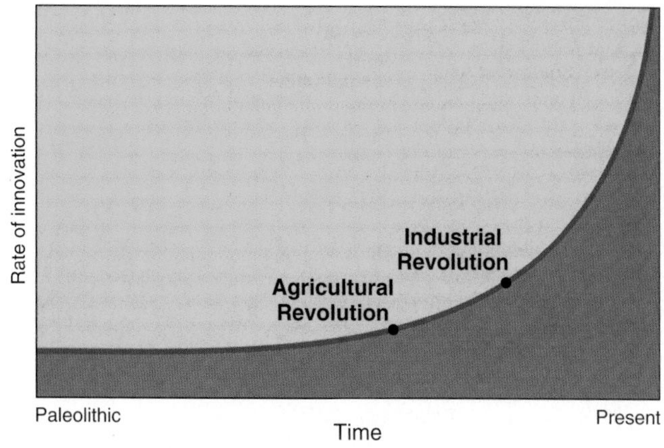


**FIGURE 7.14 The trend of innovation through human history.** Hunter-gatherers, living in equilibrium with their environment and their resource base, had little need for innovation and cultural change. The Agricultural Revolution accelerated the diffusion of the ideas and techniques of domestication, urbanization, and trade. With the Industrial Revolution, dramatic increases in all aspects of socioeconomic innovation began to alter cultures throughout the world.

forms, but basically two processes are involved. Either people move to a new area and take their culture with them (as the immigrants to the American colonies did), or information about an innovation (like barbed wire or hybrid corn) can spread throughout a culture. In either case, new ideas are transferred from their source region to new areas and to different culture groups. Spatial diffusion will be discussed in more detail in Chapter 8.

It is not always possible to determine whether the existence of a culture trait in two different areas is the result of diffusion or of independent (or *parallel*) innovation. Cultural similarities do not necessarily prove that spatial diffusion has occurred. The pyramids of Egypt and of Central America most likely were separately conceived and are not evidence, as some have proposed, of pre-Columbian voyages from the Mediterranean to the Americas. A monument-building culture, after all, has only a limited number of shapes from which to choose.

Historical examples of independent, parallel invention are numerous: logarithms by Napier (1614) and Burgi (1620), calculus by Newton (1672) and Leibnitz (1675), and the telephone by Elisha Gray and Alexander Graham Bell (1876) are commonly cited examples. It appears beyond doubt that agriculture was independently invented not only both in the New World and in the Old but also in more than one culture hearth in each of the hemispheres.

All cultures are amalgams of innumerable innovations spread spatially from their points of origin and integrated into the structure of the receiving societies. It has been estimated that no more than 10% of the cultural items of any society are traceable to innovations created by its members and that the other 90% come to the society through diffusion (see "A Homemade Culture," p. 234).

Barriers to diffusion do exist, of course, as Chapter 8 explains. Generally, the closer and the more similar two cultural areas are to one another, the lower those barriers are and the greater is the likelihood of the adoption of an innovation, because diffusion is a selective process. Of course, the receiver culture may selectively adopt some goods or ideas from the donor society and reject others. The decision to adopt is governed by the receiving group's own culture. Political restrictions, religious taboos, and other social customs are cultural barriers to diffusion. The French Canadians, although geographically close to many centers of diffusion, such as Toronto, New York, and Boston, are only minimally influenced by them. Both their language and culture complex govern French Canadian selective acceptance of Anglo influences. Traditional groups, perhaps controlled by firm religious conviction—such as the Hasidic Jewish community of New York City—may very largely reject culture traits and technologies of the larger society in whose midst they live (see "Folk and Popular Culture," p. 233 and Figure 7.15).

Adopting cultures do not usually accept intact items originating from the outside. Diffused ideas and artifacts commonly undergo some alteration of meaning or form that makes them acceptable to a borrowing group. The process of the fusion of the old and new, called **syncretism,** is a major feature of cultural change. It can be seen in alterations to religious ritual and dogma made by convert societies seeking acceptable conformity between old and new beliefs; the mixture of Catholic rites and voodooism in Haiti is an example. On a more familiar level, syncretism is reflected in subtle or blatant alterations of imported cuisines to make them conform to the demands of America's fast-food franchises.

**FIGURE 7.15** Motivated by religious conviction that the "good life" must be reduced to its simplest forms, the Amish community of east central Illinois shuns all modern luxuries of the majority, secular society around them. Children use horse and buggy, not school bus or automobile, on the daily trip to their rural school. *Courtesy of Jean Fellmann.*

## Folk and Popular Culture

Not all culture groups, even in "developed" societies, readily adopt or adjust to cultural change. In general understanding, *culture* is taken to mean "our way of life"—how we act (and why), what we eat and wear, how we amuse ourselves, what we believe, whom we admire. There are distinctions to be made, however, on the universality of the "way of life" that is accepted.

Folk groups exist in either spatial or self-imposed social isolation from the common culture of the larger societies of which they are presumably a part. **Folk culture** connotes the traditional and nonfaddish way of life characteristic of a homogeneous, cohesive, largely self-sufficient group that is essentially isolated from or resistant to outside influences. Tradition controls folk culture, and resistance to change is strong. The homemade and handmade dominate in tools, food, music, story, and ritual. *Folk life* is a cultural whole composed of both tangible and intangible elements. **Material culture** is the tangible part, made up of physical, visible things: everything from musical instruments to furniture, tools, and buildings. In folk societies, these things are products of the household or community itself, not of commercial mass production. Their intangible **nonmaterial culture** comprises the mentifacts and sociofacts expressed in oral tradition, folk song and folk story, and customary behavior; ways of speech, patterns of worship, outlooks and philosophies are passed to following generations by teachings and examples.

Within Anglo America, true folk groups are few and dwindling. Their earlier, larger numbers were based on the customs and beliefs brought to the New World by immigrant groups distinguished by their different languages, religious beliefs, and areas of origin. With time, many of their imported *ethnic* characteristics became transmuted into American "folk" features. For example, the traditional songs of western Virginia can be considered both nonmaterial folk expressions of the Upland South or evidence of an immigrant ethnic heritage derived from rural English forebears.

In that respect, each of us bears the evidence of ethnic origin and folk life. Each of us uses proverbs traditional to our family or culture; each is familiar with childhood nursery rhymes and fables. We rap wood for luck, have heard how to plant a garden by phases of the moon, and know what is the "right" way to celebrate a holiday or prepare a favorite dish. For most, however, such evidences of folk culture are minor elements in our life, and only a few groups—such as the Old Order Amish, with their rejection of electricity, the internal combustion engine, and other "worldly" accouterments in favor of buggy, hand tools, and traditional dress—remain in the United States as reminders of the folk cultural distinctions formerly widely recognizable. Canada, on the other hand, has retained a greater number of clearly recognizable ethnically unique folk and decorative arts traditions.

By general understanding, **popular culture** stands in opposition—and as the replacement for—folk culture. *Popular* implies the general mass of people, not the small-group distinctiveness and individuality of folk groups. It suggests a process of constantly adopting, conforming to, and quickly abandoning ever-changing fads and common modes of behavior. In that process, locally distinctive lifestyles and material and nonmaterial folk culture traits are largely replaced and lost; uniformity is substituted for variety, and small-group identity is eroded. For most of us, it is a sought-after uniformity. In the 1750s, George Washington wrote to his British agent to request ". . . two pair of Work'd Ruffles . . . ; if work'd Ruffles shou'd be out of fashion send such as are not . . ." and "whatever goods you may send me . . . you will let them be fashionable." His desire, echoed today, was to fit in with the peer group and larger social milieu of which he was a part.

Popular culture may be seen as both a leveling and a liberating force. On the one hand, it obliterates those locally distinctive folk culture lifestyles that emerge when groups remain isolated and self-sufficient. At the same time, however, individuals are exposed to a broader range of available opportunities—in clothing, foods, tools, recreations, and lifestyles—than ever are available to them in a cultural environment controlled by the restrictive and limited choices imposed by custom and isolation. Broad areal uniformity—in the form of the repetitive content of national discount stores, duplicate retailers in identical shopping malls, or the familiarity of ubiquitous fast-food chains—may displace folk cultural localisms. But it is a cultural uniformity vastly richer in content, diversity, and possibilities than any it replaces—though not necessarily an improvement in the social or religious values it contains or promotes.

### Acculturation

**Acculturation** is the process by which one culture group undergoes a major modification by adopting many of the characteristics of another, usually dominant, culture group. In practice, acculturation may involve changes in the original cultural patterns of either or both of two groups involved in prolonged firsthand contact. Such contact and subsequent cultural alteration may occur in a conquered or colonized region. Very often, the subordinate or subject population is forced to acculturate or does so voluntarily, overwhelmed by the superiority in numbers or the technical level of the conqueror.

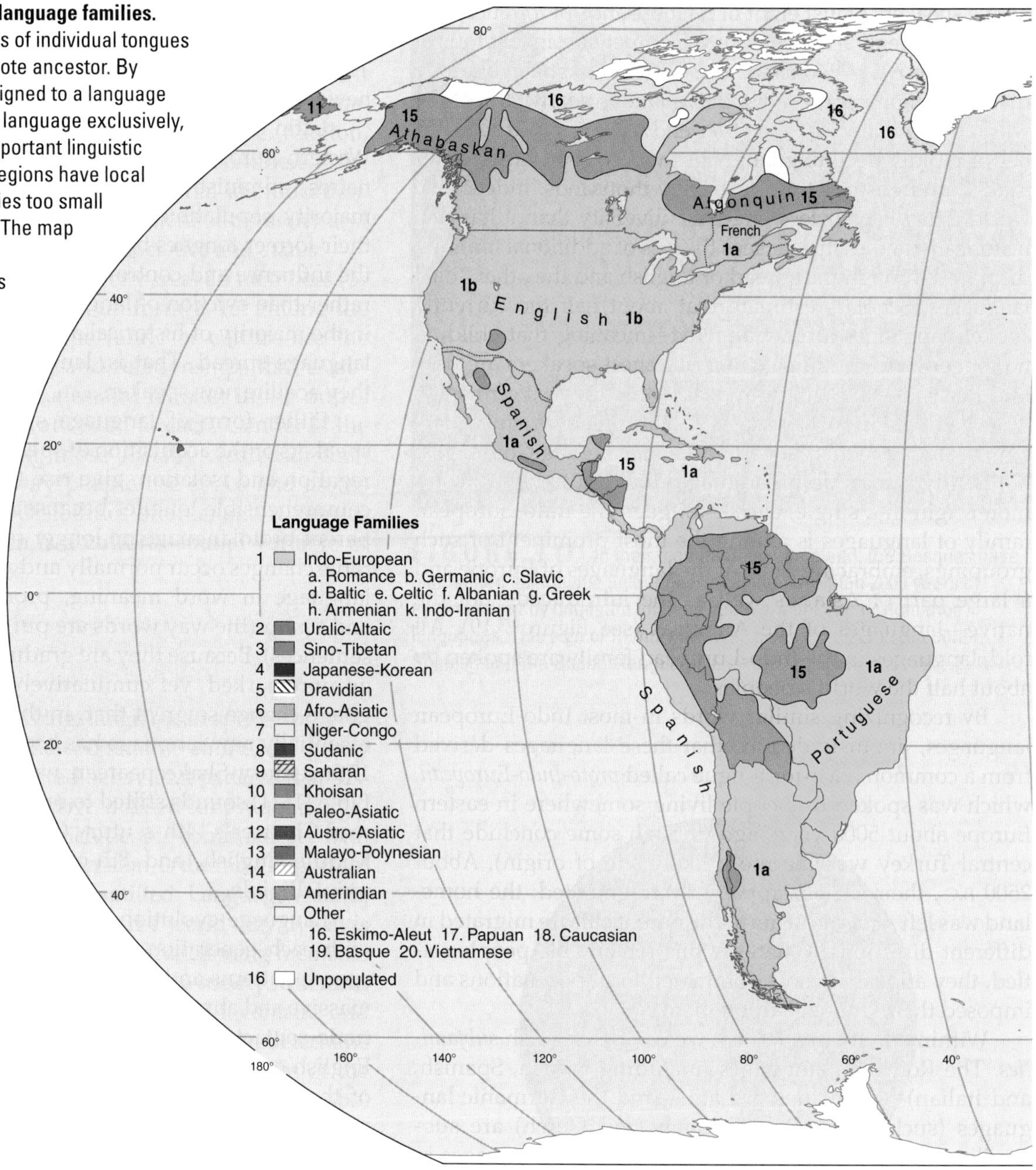


**FIGURE 7.19 World language families.** Language families are groups of individual tongues that have a common but remote ancestor. By suggesting that the area assigned to a language or language family uses that language exclusively, the map pattern conceals important linguistic detail. Many countries and regions have local languages spoken in territories too small to be recorded at this scale. The map also fails to report that the population in many regions is fluent in more than one language, or that a second language serves as the necessary vehicle of commerce, education, or government. Nor is important information given about the number of speakers of different languages; the fact that there are more speakers of English in India or Africa than in Australia is not even hinted at by a map at this scale.

were adopted secondhand from Spanish variants of South American native words: *cigar, potato, chocolate, tomato, tobacco, hammock.*

A worldwide diffusion of the language resulted as English colonists carried it to the Western Hemisphere and Australasia and as trade, conquest, and territorial claim took it to Africa and Asia. In that areal spread, English was further enriched by its contacts with other languages. By becoming the accepted language of commerce and science, it contributed in turn to the common vocabularies of other tongues. Within some 400 years, English has developed from a localized language of 7 million islanders off the European coast to a truly international language with some 375 million native speakers, perhaps 175 million who use it as a second language, and another 500 million who have reasonable competence in English as a foreign language. English serves as an official language of some 60 countries (Figure 7.21), far exceeding in that role French (32), Arabic (25), or Spanish (21), the other leading current international languages. At the start of the 21st century, over 78% of Internet web pages used English. No other language in history has assumed so important a role on the world scene.

## Standard and Variant Languages

People who speak a common language such as English are members of a *speech community,* but membership does not necessarily imply linguistic uniformity. A speech commu-

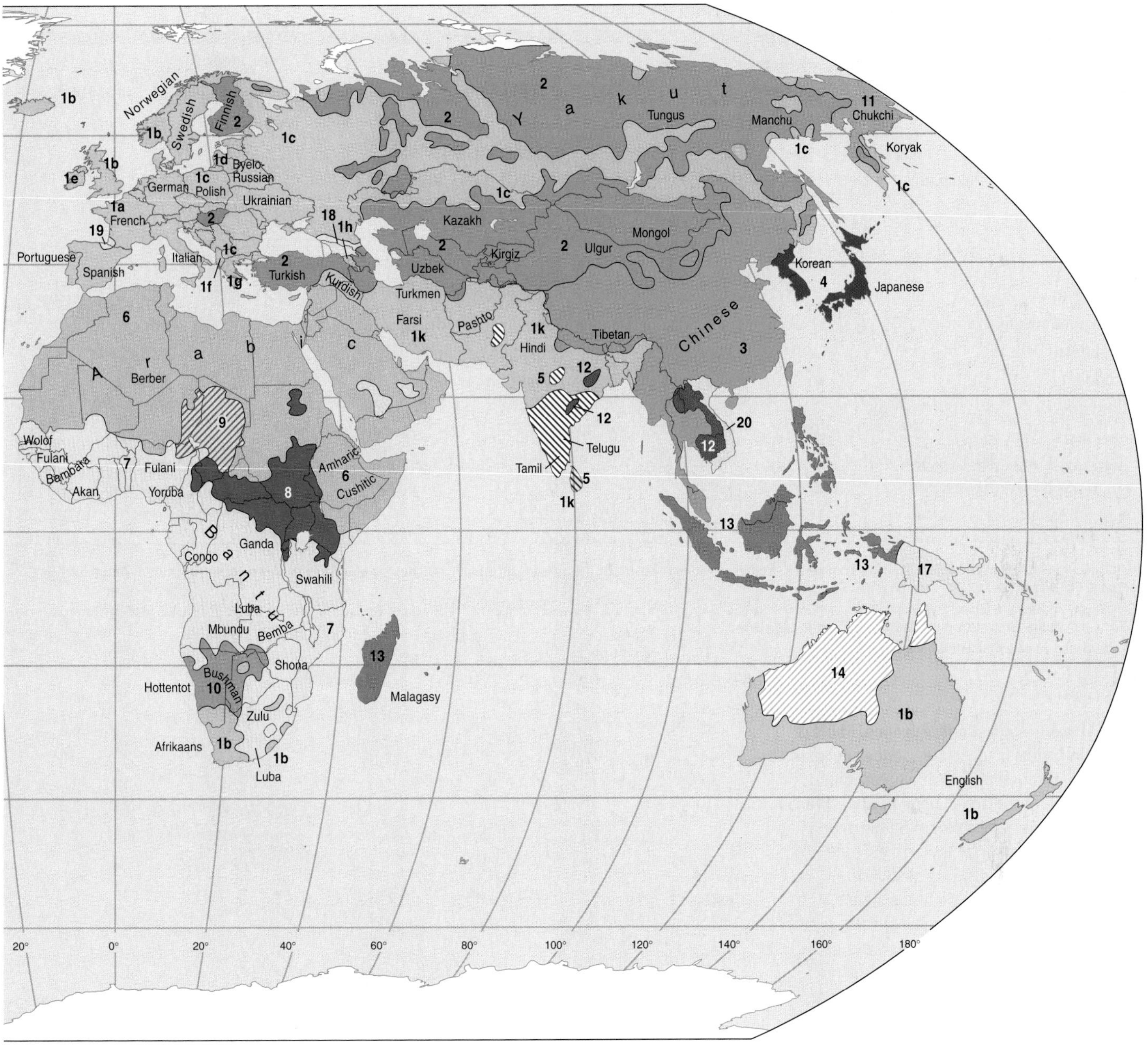


nity usually possesses both a **standard language** comprising the accepted community norms of syntax, vocabulary, and pronunciation and a number of more or less distinctive **dialects,** the ordinary speech of areal, social, professional, or other subdivisions of the general population.

An official or unofficial standard language is the form carrying governmental, educational, or societal sanction. In Arab countries, for example, classical Arabic is the language of the mosque, of education, and of the newspapers and is standardized throughout the Arabic-speaking world. Colloquial Arabic is used at home, in the street, and at the market—and in its regional variants, it may be as widely different as are, for example, Portuguese and Italian. On the other hand, the United States, English-speaking Canada, Australia, and the United Kingdom all have only slightly different forms of standard English.

Just as no two individuals talk exactly the same, all but the smallest and most closely knit speech communities display recognizable speech variants called dialects. Vocabulary, pronunciation, rhythm, and the speed at which the language is spoken may clearly set groups of speakers apart from one another. Dialects may coexist in space. Cockney and cultured English share the streets of London; black English ("Ebonics") and Standard American are heard in the same schoolyards throughout the United States. In many societies, *social dialects* denote social class and educational levels, with speakers of higher socioeconomic status or educational achievement most likely to follow the norms of

## TABLE 7.1 | Languages Spoken by More Than 40 Million People, 2004

| Language | Millions of Speakers (Native Plus Second Language Users) | Language | Millions of Speakers (Native Plus Second Language Users) |
|---|---|---|---|
| Mandarin (China) | 1076 | Tamil (India, Sri Lanka) | 76 |
| English | 551 | Marathi (India) | 73 |
| Hindi/Urdu[a] (India, Pakistan) | 498 | Panjabi (India, Pakistan) | 72 |
| Spanish | 427 | Yue (Cantonese) (China) | 72 |
| Russian | 267 | Vietnamese | 69 |
| Bengali (Bangladesh, India) | 215 | Italian | 67 |
| Portuguese | 195 | Turkish | 63 |
| Malay-Indonesian | 176 | Swahili (East Africa) | 50 |
| Japanese | 132 | Egyptian Arabic[b] | 48 |
| French | 131 | Min (China) | 48 |
| German | 128 | Jinyu (China) | 47 |
| Javanese | 78 | Ukrainian | 47 |
| Korean (Korea, China, Japan) | 78 | Gujarati (India, Pakistan) | 46 |
| Wu (China) | 78 | Polish | 44 |
| Telugu (India) | 77 | | |

[a]Hindi and Urdu are basically the same language: Hindustani. Written in the Devangari script, it is called *Hindi,* the official language of India; in the Arabic script, it is called *Urdu,* the official language of Pakistan.

[b]Collectively, the many, often mutually unintelligible versions of colloquial Arabic are used by at least 260 million native speakers. Classical or literary Arabic, the language of the Koran, is uniform and standardized but is restricted to formal usage as a spoken tongue.

*Source:* Based on data from *Ethnologue* and others.

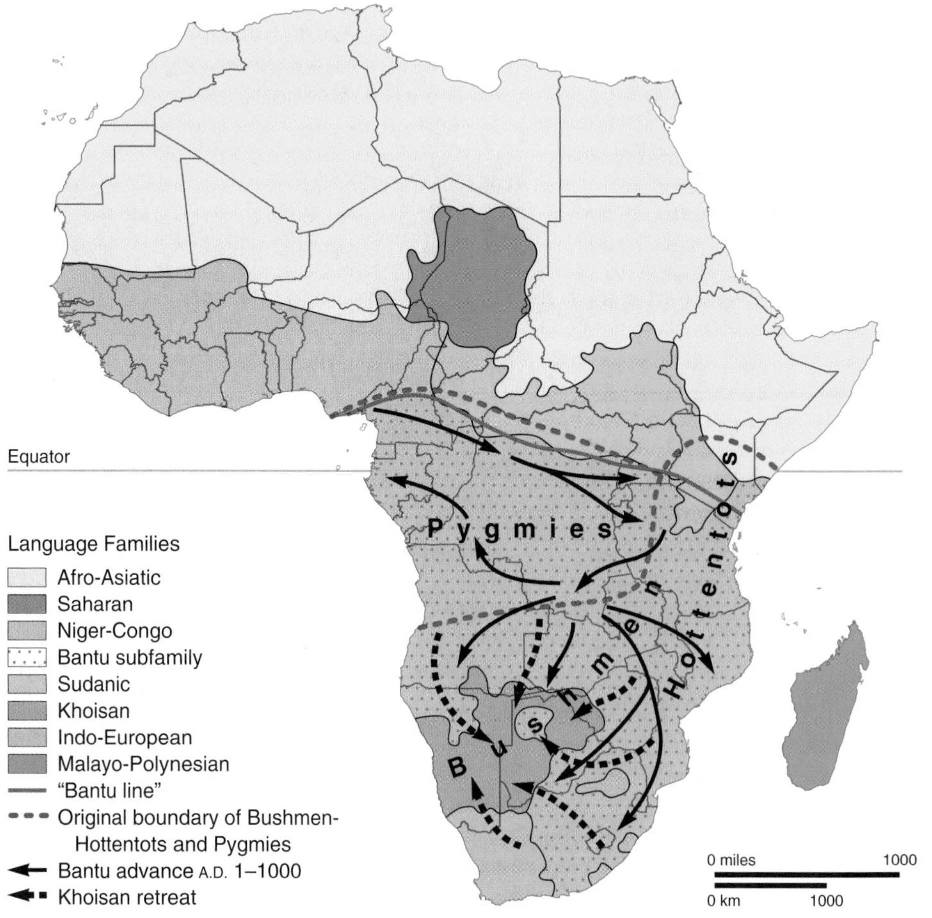


**FIGURE 7.20 Bantu advance, Khoisan retreat in Africa.** Linguistic evidence suggests that proto-*Bantu* speakers originated in the region of the Cameroon-Nigeria border, spread eastward across the southern Sudan, then turned southward to Central Africa. From there, they dispersed slowly eastward, westward, and, against slight resistance, southward. The earlier *Khoisan*-speaking occupants of sub-Saharan Africa were no match against the advancing metal-using Bantu agriculturalists. Pygmies, adopting a Bantu tongue, retreated deep into the forests. Bushmen and Hottentots retained their distinctive Khoisan "click" language but were forced out of forests and grasslands into the dry steppes and deserts of the southwest.

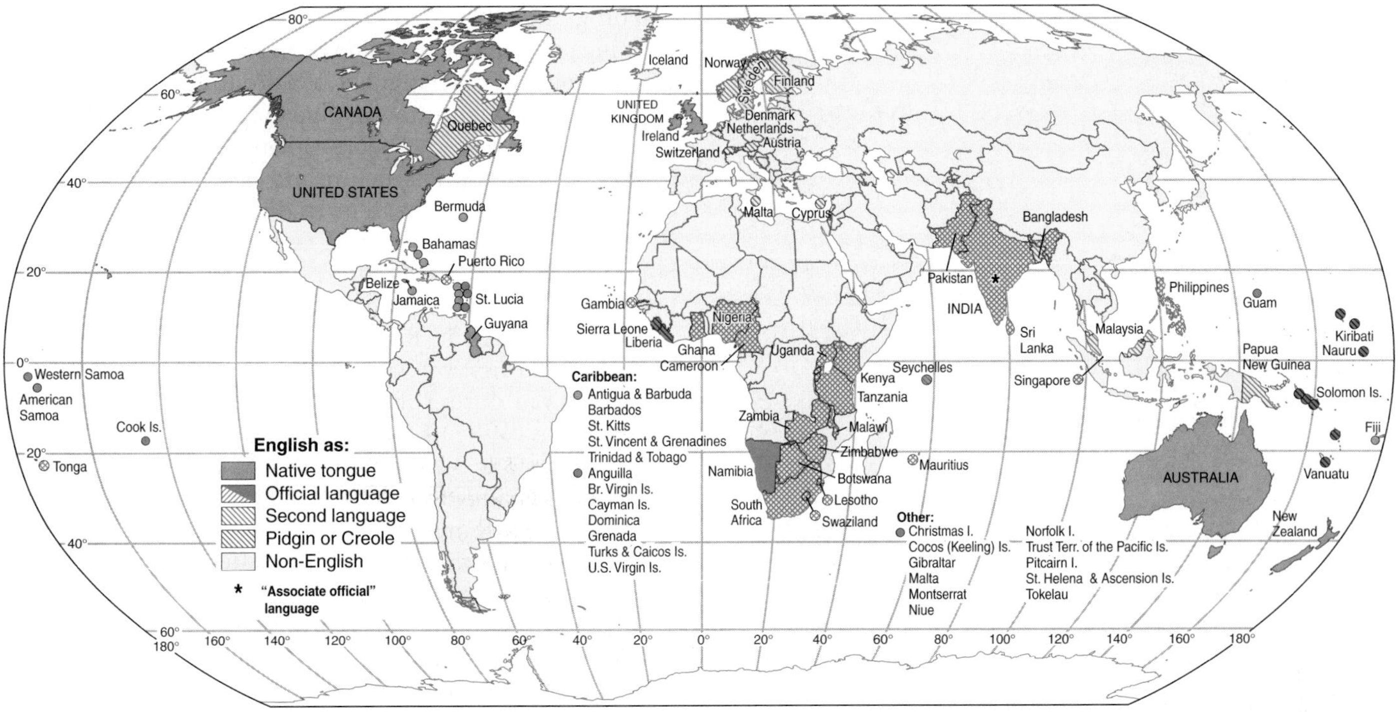


**FIGURE 7.21 International English.** In worldwide diffusion and acceptance, English has no past or present rivals. Along with French, it is one of the two working languages of the United Nations, two-thirds of all scientific papers are published in it, making it the first language of scientific discourse. In addition to being the accepted language of international air traffic control, English is the sole or joint official language of more nations and territories, some too small to be shown here, than any other tongue. It also serves as the effective unofficial language of administration in other multilingual countries with different formal official languages. "English as a second language" is indicated for countries with near-universal or mandatory English instruction in public schools. The full extent of English penetration of Continental Europe, where over 80% of secondary school students study it as a second language and more than one-third of European Union residents can easily converse in it, is not evident on this map.

their standard language. Less-educated or lower-status people or groups consciously distinguishing themselves from the mainstream culture are more apt to use the *vernacular*—nonstandard language or dialect adopted by the social group. In some instances, however, as in Germany and German-Switzerland, local dialects are preserved and prized as badges of regional identity.

More commonly, we think of dialects in spatial terms. Speech is a geographic variable. Each locale is likely to have its own, perhaps slight, language differences from neighboring places. Such differences in pronunciation, vocabulary, word meanings, and other language characteristics help define the *linguistic geography*—the study of the character and spatial pattern of *geographic,* or *regional, dialects*—of a generalized speech community. Figure 7.22 records the variation in usage associated with just one phrase. In the United States, Southern English and New England speech are among the regional dialects that are most easily recognized by their distinctive accents. In some instances, there is so much variation among geographic dialects that some are almost foreign tongues to other speakers of the same language. Effort is required for Americans to understand Australian English or that spoken in Liverpool, England, or in Glasgow, Scotland (see "World Englishes," p. 243). An interesting United States example is discussed in the regional study "Gullah as Language," in Chapter 13.

Local dialects and accents do not display predictable patterns of consistency or change. In ethnically and regionally complex United States, for example, mixed conclusions concerning local speech patterns have been drawn by researchers examining the linguistic results of an increasingly transient population, immigration from other countries and cultures, and the pervasive and presumed leveling effects of the mass media. The distinct evidence of increasing contrasts among the speech patterns and accents of Chicago, New York, Birmingham, St. Louis, and other cities is countered by reports of decreasing local dialect pronunciations in such centers as Dallas and Atlanta, which have experienced major influxes of northerners. And other studies find that some regional accents are fading in small towns and rural areas, presumably because mass-media standardization is more influential than local dialect reinforcement as areal populations decline.

Language is rarely a total barrier to communication between peoples. Bilingualism or multilingualism may permit skilled linguists to communicate in a jointly understood third language, but long-term contact between less-able populations may require the creation of a new language—a

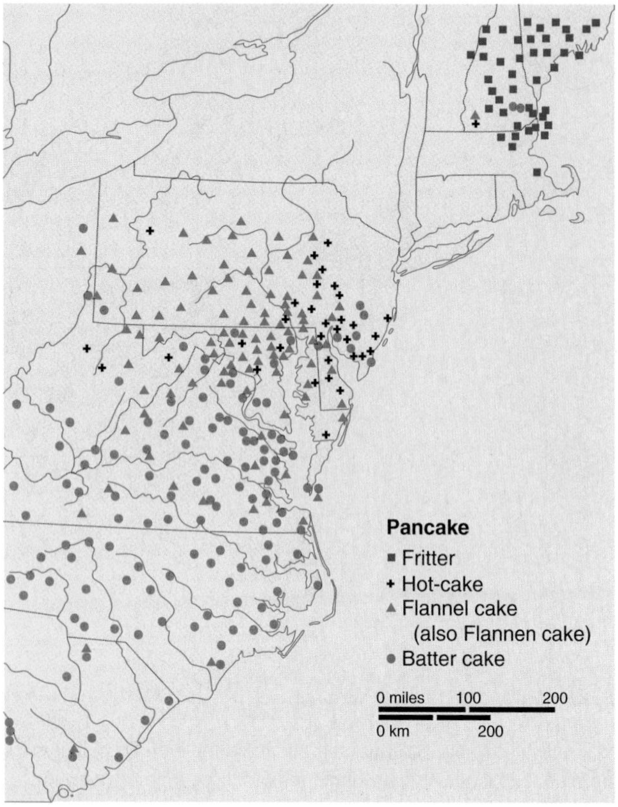


**FIGURE 7.22** Maps such as this are used to record variations over space and among social classes in word usage, accent, and pronunciation. The differences are due not only to initial settlement patterns but also to more recent large-scale movements of people—for example, from rural to urban areas and from the South to the North. Despite the presumed influence of national radio and television programs in promoting a "general" or "standard" American accent and usage, regional and ethnic language variations persist.

*Redrawn with permission from* A Word Geography of the Eastern United States *by Hans Kurath. Copyright © 1949 by the University of Michigan Press.*

pidgin—learned by both parties. A **pidgin** is an amalgam of languages, usually a simplified form of one of them, such as English or French, with borrowings from another, perhaps non-European local language. In its original form, a pidgin is not the mother tongue of any of its speakers; it is a second language for everyone who uses it, one generally restricted to such specific functions as commerce, administration, or work supervision.

Pidgins are characterized by a highly simplified grammatical structure and a sharply reduced vocabulary adequate to express basic ideas but not complex concepts. For example, fanagalo, a pidgin created earlier in South Africa's gold mines to allow spoken communication between workers of different tribes and nationalities and between workers and Afrikaner bosses, is being largely abandoned. Since the mid-1990s, workers have increasingly been schooled in basic English as fanagalo—which lacks the vocabulary to describe how to operate new, automated mining machinery and programmable winches with their multiple sensors and warnings in English—became decreasingly useful. If a pidgin becomes the first language of a group of speakers—who may have lost their former native tongue through disuse—a **creole** has evolved. Creoles invariably acquire a more complex grammatical structure and enhanced vocabulary.

Creole languages have proved useful integrative tools in linguistically diverse areas; several have become symbols of nationhood. Swahili, a pidgin formed from a number of Bantu dialects with major vocabulary additions from Arabic, originated in the coastal areas of East Africa and spread inland first by Arab ivory and slave caravans and later by trade during the period of English and German colonial rules. When Kenya and Tanzania gained independence, they made Swahili the national language of administration and education. Other examples of creolization are Afrikaans (a pidginized form of 17th-century Dutch used in the Republic of South Africa); Haitian Creole (the language of Haiti, derived from the pidginized French used in the slave trade); and Bazaar Malay (a pidginized form of the Malay language, a version of which is the official national language of Indonesia).

A **lingua franca** is an established language used habitually for communication by people whose native tongues are mutually incomprehensible. For them, it is a *second language,* one learned in addition to the native tongue. Lingua franca (literally, "Frankish tongue") was named from the French dialect adopted as a common language by the Crusaders at war in the Holy Land. Later, Latin became the lingua franca of the Mediterranean world until, finally, it was displaced by vernacular European tongues. Arabic followed Muslim conquest as the unifying language of that international religion after the 7th century. Mandarin Chinese and Hindi in India have traditionally had a lingua franca role in their linguistically diverse countries. The immense linguistic complexity of Africa has made regional lingua francas there necessary and inevitable—Swahili in East Africa, for example, and Hausa in parts of West Africa.

## Language and Culture

Language embodies the culture complex of a people, reflecting both environment and technology. Arabic has 80 words related to camels, an animal on which a regional culture relied for food, transport, and labor, and Japanese contains more than 20 words for various types of rice. Russian is rich in terms for ice and snow, indicative of the prevailing climate of its linguistic cradle; and the 15,000 tributaries and subtributaries of the Amazon River have obliged the Brazilians to enrich Portuguese with words that go beyond "river." Among them are *paraná* (a stream that leaves and reenters the same river), *igarapé* (an offshoot that runs until it dries up), and *furo* (a waterway that connects two rivers).

Most—perhaps all—cultures display subtle or pronounced differences in ways males and females use language. Most have to do with vocabulary and with gram-

## World Englishes

Non-native speakers of English far outnumber those for whom English is the first language. Most of the more than 1 billion people who speak and understand at least some English as a second language live in Asia; they are appropriating the language and remaking it in regionally distinctive fashions to suit their own cultures, linguistic backgrounds, and needs.

It is inevitable that widely spoken languages separated by distance, isolation, and cultural differences will fragment into dialects, which in turn evolve into new languages. Latin splintered into French, Spanish, Italian, and other Romance languages; the many national variants of spoken Arabic are effectively different tongues. English is similarly experiencing that sort of regional differentiation, shaped by the variant needs and inputs of its far-flung community of speakers and following the same path to mutual unintelligibility. Although Standard English may be one of or the sole official language of their countries of birth, millions of people around the world claiming proficiency in English or English as their national language cannot understand each other. Even teachers of English from India, Malaya, Nigeria, or the Philippines, for example, may not be able to communicate in their supposedly common tongue—and find cockney English of London utterly alien.

The splintering of spoken English is a fact of linguistic life, and its offspring—called "World Englishes" by linguists—defy frequent attempts by various governments to remove localisms and encourage adherence to international standards. Singlish (Singapore English) and Taglish (a mixture of English and Tagalog, the dominant language of the Philippines) are commonly cited examples of the multiplying World Englishes but equally distinctive regional variants have emerged in India, Malaysia, Hong Kong, Nigeria, the Caribbean, and elsewhere. One linguist suggest that, beyond an "inner circle" of countries where English is the first and native language—e.g., Canada, Australia, the United States—lies an "outer circle" where English is a second language (Bangladesh, Ghana, India, Kenya, Pakistan, Zambia, and many others) and where the regionally distinctive World Englishes are most obviously developing. Even farther out is an "expanding circle" of such as China, Egypt, Korea, Nepal, and Saudi Arabia where English is a foreign language and distinctive local variants in common usage have not yet developed.

Although the constant modern world electronic and literary interaction between the variant regional Englishes make it likely that the common language will remain universally intelligible, it also seems probable that mutually incomprehensible forms of English will become entrenched as the language is taught, learned, and used in world areas far removed from contact with first-language. "Our only revenge," said a French official deploring the declining role of French within the European Union, "is that the English language is being killed by all these foreigners speaking it so badly."

matical forms peculiar to individual cultures. For example, among the Caribs of the Caribbean, the Zulu of Africa, and elsewhere, men have words that women through custom or taboo are not permitted to use, and the women have words and phrases that the men never use "or they would be laughed to scorn," an informant reports. Evidence from English and many other unrelated tongues indicates that, as a rule, female speakers use forms considered to be "better" or "more correct" than do males of the same social class. The greater and more inflexible the difference in the social roles of men and women in a particular culture, the greater and more rigid are the observed linguistic differences between the sexes.

A common language fosters unity among people. It promotes a feeling for a region; if it is spoken throughout a country, it fosters nationalism. For this reason, languages often gain political significance and serve as a focus of opposition to what is perceived as foreign domination. Although nearly all people in Wales speak English, many also want to preserve Welsh because they consider it an important aspect of their culture. They think that, if the language is forgotten, their entire culture may also be threatened. French Canadians received government recognition of their language and established it as the official language of Quebec Province; Canada itself is officially bilingual. In India, with 18 constitutional languages and 1652 other tongues, serious riots have occurred by people expressing opposition to the imposition of Hindi as the single official national language.

Bilingualism or multilingualism complicates national linguistic structure. Areas are considered bilingual if more than one language is spoken by a significant proportion of the population. In some countries—Belgium and Switzerland, for example—there is more than one official language. In many others, such as the United States, only one language may have implicit or official governmental sanction, although several others are spoken (see "An Official U.S. Language?" p. 244). Speakers of one of these may be concentrated in restricted areas (e.g., most speakers of French in Canada live in Quebec Province). Less often, they are distributed fairly evenly throughout the country. In some countries, the language in which instruction, commercial transactions, and government business take place is not a domestic language at all. In linguistically complex sub-Saharan Africa, nearly all countries have selected a European tongue—usually that of their former colonial governors—as an official language (Figure 7.23).

GEOGRAPHY & Public Policy

## An Official U.S. Language?

Within recent years in Lowell, Massachusetts, public school courses were offered in Spanish, Khmer, Lao, Portuguese, and Vietnamese, and all messages from schools to parents were translated into five languages. Polyglot New York City gave bilingual programs in Spanish, Chinese, Haitian Creole, Russian, Korean, Vietnamese, French, Greek, Arabic, and Bengali. In most states, it is possible to get a high school equivalency diploma without knowing English, because tests are offered in French and Spanish. In at least 39 states, driving tests are available in foreign languages; California provides 30 varieties, New York 23, and Michigan 20, including Arabic and Finnish. And as required by the 1965 federal Voting Rights Act, multilingual ballots are provided in some 300 electoral jurisdictions in 30 states.

These, and innumerable other evidences of governmentally sanctioned linguistic diversity, may come as a surprise to the many Americans who assume that English is the official language of the United States. It isn't; nowhere does the Constitution provide for an official language, and no federal law specifies one. The country was built by a great diversity of cultural and linguistic immigrants, who nonetheless shared an eagerness to enter mainstream American life. At the start of the 21st century, a reported 18% of all American residents spoke a language other than English in the home. In California public schools, one out of three students uses a non-English tongue within the family. In Washington, D.C., schools, students speak 127 languages and dialects, a linguistic diversity duplicated in other major city school systems.

Nationwide bilingual teaching began as an offshoot of the civil rights movement in the 1960s, was encouraged by a Supreme Court opinion authored by Justice William O. Douglas, and has been actively promoted by the U.S. Department of Education under the Bilingual Education Act of 1974 as an obligation of local school boards. Its purpose has been to teach subject matter to minority-language children in the language in which they think while introducing them to English, with the hope of achieving English proficiency in 2 or 3 years. Disappointment with the results led to a successful 1998 California anti-bilingual education initiative, Proposition 227, to abolish the program. Similar rejection elsewhere has followed California's lead.

Opponents of the implications of governmentally encouraged multilingual education, bilingual ballots, and ethnic separatism argue that a common language is the unifying glue of the United States and all countries; without that glue, they fear, the process of "Americanization" and *acculturation*—the adoption by immigrants of the values, attitudes, behavior, and speech of the receiving society—will be undermined. Convinced that early immersion and quick proficiency in English are the only sure ways for minority newcomers to gain necessary access to jobs, higher education, and full integration into the economic and social life of the country, proponents of "English only" use in public education, voting, and state and local governmental agencies successfully passed Official English laws and constitutional amendments in 27 states from the late 1980s to 2002.

Although the amendments were supported by sizeable majorities of the voting population, resistance to them—and to their political and cultural implications—was, in every instance, strong and persistent. Ethnic groups—particularly Hispanics, who are the largest of the linguistic groups affected—charged that they were evidence of blatant Anglocentric racism, discriminatory and repressive in all regards. Some educators argued persuasively that all evidence proved that, although immigrant children eventually acquire English proficiency in any event, they do so with less harm to their self-esteem and subject-matter acquisition when initially taught in their own language. Businesspeople with strong minority labor and customer ties and political leaders—often themselves members of ethnic communities or with sizeable minority constituencies—argued against "discriminatory" language restrictions.

*Toponyms*—place-names—are language on the land, the record of past and present cultures whose namings endure as reminders of their passing and their existence. **Toponymy,** the study of place-names, therefore, is a revealing tool of historical cultural geography, because place-names become a part of the cultural landscape that remains long after the name givers have passed from the scene.

In England, for example, place-names ending in *chester* (as in Winchester and Manchester) evolved from the Latin *castra,* meaning "camp." Common Anglo-Saxon suffixes for tribal and family settlements were *ing* (people or family) and *ham* (hamlet or, perhaps, meadow), as in Birmingham and Gillingham. Norse and Danish settlers contributed place-names ending in *thwaite* (meadow) and others denoting such landscape features as *fell* (an uncultivated hill) and *beck* (a small brook). The Arabs, sweeping out from Arabia across North Africa and into Iberia, left their imprint in place-names to mark their conquest and control. *Cairo* means "victorious," *Sudan* is "the land of the blacks," and *Sahara* is "wasteland" or "wilderness." In Spain, a corrupted version of the Arabic *wadi,* "watercourse," is found in *Guadalajara* and *Guadalquivir.*

In the New World, not one people but many placed names on landscape features and new settlements. In doing so, they remembered their homes and homelands, honored their monarchs and heroes, borrowed and mispronounced from rivals, adopted and distorted Amerindian names, followed fads, and recalled the Bible. Homelands were honored in New England, New France, and New Holland; settlers' hometown memories brought Boston, New Bern, and New Rochelle from England, Switzerland, and France.

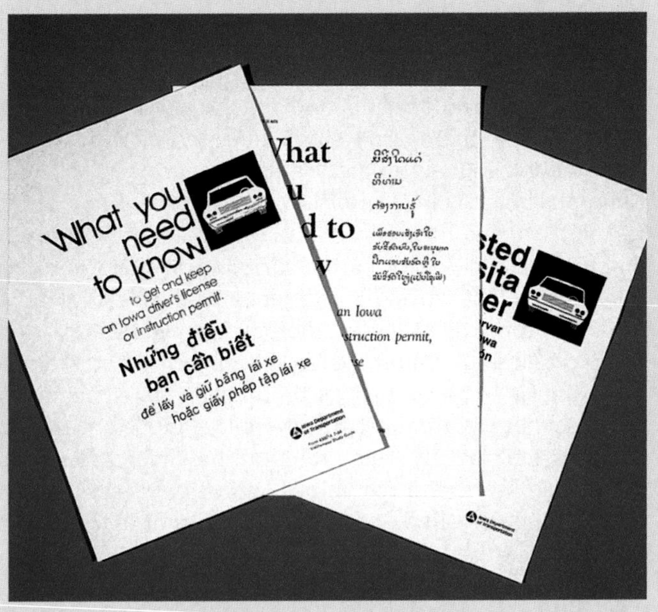


© Bob Coyle and courtesy of Iowa Department of Transportation.

And historians noted that it had all been tried unsuccessfully before. The anti-Chinese Workingmen's Party in 1870s California led the fight for English-only laws in that state. The influx of immigrants from central and southeastern Europe at the turn of the 20th century led Congress to make oral English a requirement for naturalization, and anti-German sentiment during and after World War I led some states to ban any use of German. The Supreme Court struck down those laws in 1923, ruling that the "protection of the Constitution extends to all, to those who speak other languages as well as to those born with English on their tongue." Following suit, some of the recent state language amendments have also been voided by state courts. In ruling its state's English-only law unconstitutional, Arizona's Supreme Court in 1998 noted it "chills First Amendment rights."

To counter those judicial restraints and the possibility of an eventual multilingual, multicultural United States in which English and, likely, Spanish would have coequal status and recognition, U.S. English—an organization dedicated to the belief that "English is, and ever must remain, the only official language of the people of the United States"—actively supports the proposed U.S. constitutional amendment first introduced in Congress by former senator S. I. Hayakawa in 1981 and resubmitted by him and others in subsequent years. The proposed amendment would simply establish English as the official national language but would impose no duty on people to learn English and would not infringe upon any right to use other languages. Whether or not these modern attempts to designate an official U.S. language eventually succeed, they represent a divisive subject of public debate affecting all sectors of American society.

### QUESTIONS TO CONSIDER

1. Do you think that multiple languages and ethnic separatism represent a threat to America's cultural unity that can be avoided only by viewing English as a necessary unifying force? Or do you think making English the official language might divide its citizens and damage its legacy of tolerance and diversity?
2. Do you believe that immigrant children would learn English faster if bilingual classes were reduced and immersion in English were more complete? Or do you think that a slower pace of English acquisition is acceptable if subject-matter comprehension and cultural self-esteem are enhanced?
3. Do you think Official English laws inflame prejudice against immigrants or provide all newcomers with a common standard of admission to the country's political and cultural mainstream?

Monarchs were remembered in Virginia for the Virgin Queen Elizabeth, Carolina for one English king, Georgia for another, and Louisiana for a king of France. Washington, D.C.; Jackson, Mississippi and Michigan; Austin, Texas; and Lincoln, Illinois, memorialized heroes and leaders.

Names given by the Dutch in New York were often distorted by the English; Breukelyn, Vlissingen, and Haarlem became Brooklyn, Flushing, and Harlem. French names underwent similar twisting or translation, and Spanish names were adopted, altered, or later put into such bilingual combinations as Hermosa Beach. Amerindian tribal names—the Yenrish, Maha, Kansa—were modified, first by French and later by English speakers, to Erie, Omaha, and Kansas. A faddish classical revival after the American Revolution gave us Troy, Athens, Rome, Sparta, and other ancient town names and later spread them across the country. Bethlehem, Ephrata, Nazareth, and Salem came from the Bible.

Of course, European colonists and their descendants gave place names to a physical landscape already adequately named by indigenous peoples. Those names were sometimes adopted but often shortened, altered, or—certainly—mispronounced. The vast territory that local Amerindians called "Mesconsing," meaning "the long river," was recorded by Lewis and Clark as "Quisconsing," later to be further distorted into "Wisconsin." Milwaukee, Winnipeg, Potomac, Niagara, Adirondack, Chesapeake, Shenandoah, and Yukon; the names of 28 of the 50 United States; and the present identity of thousands of North American places and features, large and small, had their origin in Native American languages.

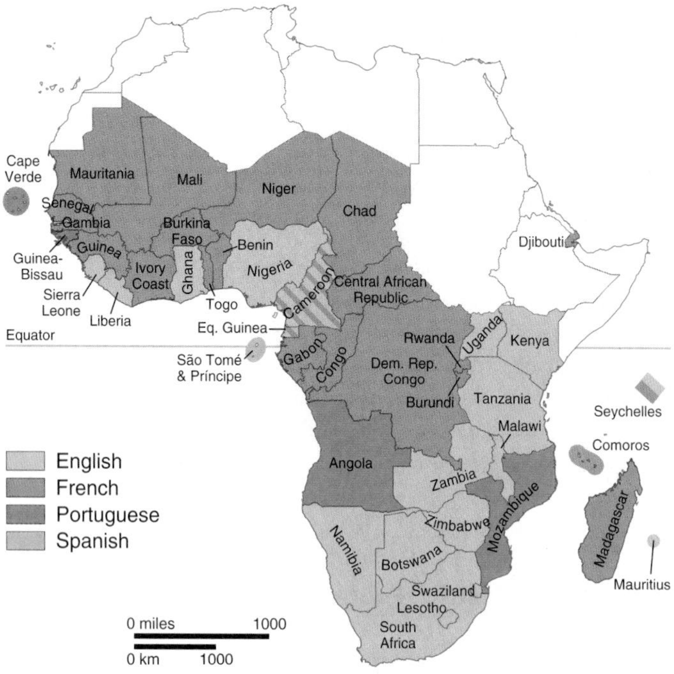


**FIGURE 7.23 Europe in Africa through official languages.** Both the linguistic complexity of sub-Saharan Africa and the colonial histories of its present political units are implicit in the designation of a European language as the sole or joint "official" language of the various countries.

# Religion

Enduring place-names are only one measure of the importance of language as a powerful, unifying thread in the culture complex of people and as a fiercely defended symbol of the history and individuality of a distinctive social group. But language is not alone in that role. In some ways, it yields to religion as a cultural rallying point. French Catholics slaughtered French Huguenots (Protestants) in the name of religion in the 16th century. English Roman Catholics were hounded from the country after the establishment of the Anglican church. Religious enmity between Muslims and Hindus forced the partition of the Indian subcontinent after the departure of the British in 1947. And recent years have witnessed continuing religious confrontations between, for example, Catholic and Protestant Christian groups in Northern Ireland; Muslim sects in Lebanon, Iran, and Iraq; Muslims and Jews in Palestine; Christians and Muslims in the Balkans, the Philippines, and Lebanon; and Buddhists and Hindus in Sri Lanka.

However, unlike language, which is an attribute of all people, religion varies in its cultural role—dominating in some societies, unimportant or rejected in others. All societies have value systems—common beliefs, understandings, expectations, and controls—which unite their members and set them off from other, different culture groups. Such a value system is termed a *religion* when it involves systems of formal or informal worship and faith in the sacred and divine. In a more inclusive sense, religion may be viewed as a unified system of beliefs and practices that join all those who adhere to them into a single moral community.

Religion may intimately affect all facets of a culture. Religious belief is, by definition, an element of the ideological subsystem; formalized and organized religion is an institutional expression of the sociological subsystem. And religious beliefs strongly influence attitudes toward the tools and rewards of the technological subsystem.

Nonreligious value systems—humanism or Marxism, for example—can be just as binding on the societies that espouse them as are more traditional religious beliefs. Even societies that largely reject religion, however, are strongly influenced by traditional values and customs set by predecessor religions—in days of work and rest or in legal principles, for example.

Because religions are formalized views about the relation of the individual to this world and to the hereafter, each carries a distinct conception of the meaning and value of this life, and most contain strictures about what must be done to achieve salvation (Figure 7.24). These beliefs become interwoven with the traditions of a culture. One cannot understand India without a knowledge of Hinduism, or Israel without an appreciation of Judaism.

Economic patterns may be intertwined with past or present religious beliefs. Traditional restrictions on food and drink may affect the kinds of animals that are raised or avoided, the crops that are grown, and the importance of those crops in the daily diet. Occupational assignment in the Hindu caste system is, in part, religiously supported. In many countries, there is a state religion; that is, religious and political structures are intertwined. Buddhism, for example, has been the state religion in Myanmar, Laos, and Thailand. By their official names, the Islamic Republic of Pakistan and the Islamic Republic of Iran proclaim their identity of church and government. Despite the country's overwhelming Muslim majority, Indonesia sought and formerly found domestic harmony by recognizing five official religions and a state ideology—*pancasila*—whose first tenet is belief in one god.

## Classification and Distribution of Religions

Religions are cultural innovations. They may be unique to a single culture group, closely related to the faiths professed in nearby areas, or derived from or identical to belief systems spatially far removed. Although interconnections and derivations among religions can frequently be discerned—as Christianity and Islam can trace descent from Judaism—family groupings are not as useful in classifying religions as they are in studying languages. A distinction between *monotheism*, belief in a single deity, and *polytheism*, belief in many gods, is frequent but not particularly spatially relevant. It is more useful for the spatial interests of geographers to categorize religions as *universalizing*, *ethnic*, or *tribal (traditional)*.

**FIGURE 7.24** Worshipers gathered during *hajj,* the pilgrimage to Mecca. The black structure is the Ka'bab, the symbol of Allah's (God's) oneness and the unity of God and humans. Many rules concerning daily life are given in the Koran, the holy book of the Muslims. All Muslims are expected to observe the five pillars of the faith: (1) repeated saying of the basic creed; (2) prayers five times daily, facing Mecca; (3) a month of daytime fasting (Ramadan); (4) almsgiving; and (5) if possible, a pilgrimage to Mecca. *© Topham/*The Observer/*The Image Works.*

Christianity, Islam, and Buddhism are the major world **universalizing religions,** faiths that claim applicability to all humans and that seek to transmit their beliefs to all lands through missionary work and conversion. Membership in universalizing religions is open to anyone who chooses to make a symbolic commitment, such as baptism in Christianity. No one is excluded because of nationality, ethnicity, or previous religious belief.

**Ethnic religions** have strong territorial and cultural group identification. One usually becomes a member of an ethnic religion by birth or by adoption of a complex lifestyle and cultural identity, not by a simple declaration of faith. These religions do not usually proselytize (convert nonbelievers), and their members form distinctive closed communities identified with a particular ethnic group, region, or political unit. An ethnic religion—for example, Judaism, Indian Hinduism, or Japanese Shinto—is an integral element of a specific culture. To be part of the religion is to be immersed in the totality of the culture.

**Tribal** (or *traditional*) **religions** are special forms of ethnic religions distinguished by their small size, their unique identity with localized culture groups not yet fully absorbed into modern society, and their close ties to nature. *Animism* is the name given to their belief that life exists in all objects, from rocks and trees to lakes and mountains, or that such objects are the abode of the dead, of spirits, and of gods. *Shamanism* is a form of tribal religion that involves community acceptance of a *shaman* who, through special powers, can intercede with and interpret the spirit world.

The nature of the different classes of religions is reflected in their distributions over the world (Figure 7.25) and in their number of adherents. Universalizing religions tend to be expansionary, carrying their message to new peoples and areas. Ethnic religions, unless their adherents are dispersed, tend to be regionally confined or to expand only slowly and over long periods. Tribal religions tend to contract spatially as their adherents are incorporated increasingly into modern society and converted by proselytizing faiths.

As we expect in cultural geography, the map records only the latest stage of a constantly changing reality. While established religious institutions tend to be conservative and resistant to change, religion as a culture trait is dynamic. Personal and collective beliefs may alter in response to developing individual and societal needs and challenges. Religions may be imposed by conquest, adopted by conversion, or defended and preserved in the face of surrounding hostility or indifference.

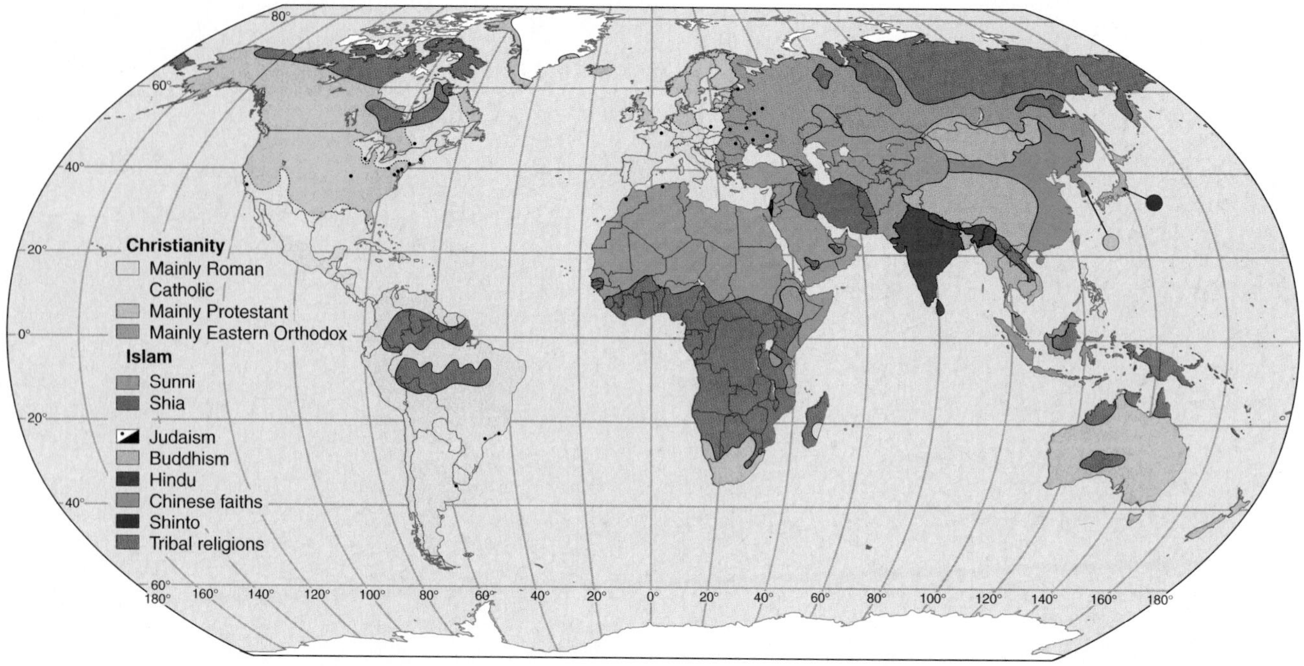


**FIGURE 7.25 Principal world religions.** The assignment of individual countries to a single religious category conceals a growing intermixture of faiths in European and other Western countries that have experienced recent major immigration flows. In some instances, those influxes are altering the effective, if not the numerical, religious balance. In nominally Christian, Catholic France, for example, low church-going rates suggest that now more Muslims than practicing Catholics reside there and that, considering birth rate differentials, someday Islam may be the country's predominant religion, as measured by the number of practicing adherents. Secularism—rejection of religious belief—is common in many countries but is not locationally indicated on this map.

Nor does the map present a full picture even of current religious regionalization or affiliation. Few societies are homogeneous, and most modern ones contain a variety of faiths or, at least, variants of the dominant professed religion. Some of those variants in many religions are intolerant or antagonistic toward other faiths or toward the sects and members of their own faith deemed insufficiently committed or orthodox (see "Militant Fundamentalism," p. 249).

Frequently, members of a particular religion show areal concentration within a country. Thus, in urban Northern Ireland, Protestants and Catholics reside in separate areas whose boundaries are clearly understood and respected. The "Green Line" in Beirut, Lebanon, marked a guarded border between the Christian east and the Muslim west sides of the city, whereas within the country as a whole, regional concentrations of adherents of different faiths and sects are clearly recognized (Figure 7.26). Religious diversity within countries may reflect the degree of toleration a majority culture affords minority religions. In dominantly (90%) Muslim Indonesia, Christian Bataks, Hindu Balinese, and Muslim Javanese lived in peaceful coexistence for many years. By contrast, the fundamentalist Islamic regime in Iran has persecuted and executed those of the Baha'i faith.

One cannot assume that all people within a mapped religious region are adherents of the designated faith, nor can it be assumed that membership in a religious community means active participation in its belief system. *Secularism*, an indifference to or rejection of religion and religious belief, is an increasing part of many modern societies, particularly of the industrialized countries and those now or recently under communist regimes. In England, for example, the state Church of England claims only 20% of the British as communicants, but only 2% of the population attends its Sunday services. Even in devoutly Roman Catholic South American states, low church attendance attests to the rise of at least informal secularism. In Colombia, only 18% of the people attend Sunday services; in Chile, the figure is 12%; in Mexico, 11%; and in Bolivia, 5%.

## The Principal Religions

Each of the major religions has its own unique mix of cultural values and expressions, each has had its own pattern of innovation and spatial diffusion (Figure 7.27), and each has had its own impact on the cultural landscape. Together, they contribute importantly to the worldwide pattern of human diversity.

### Judaism

We can begin our review of world faiths with *Judaism*, whose belief in a single God laid the foundation for both Christianity and Islam. Unlike its universalizing offspring, Judaism is closely identified with a single ethnic group and with a complex and restrictive set of beliefs and laws. It

## Militant Fundamentalism

The term *fundamentalism* entered the social science vocabulary in the late 20th century to describe any religious orthodoxy that is revivalist and ultraconservative. Originally, it designated an American Christian movement named after a set of volumes—*The Fundamentals: A Testimony of the Truth*—published between 1910 and 1915 and embracing both absolute religious orthodoxy and a commitment to inject its beliefs into the political and social arenas. More recently, fundamentalism has become a generic description for all religious movements that seek to regain and publicly institutionalize traditional social and cultural values that are usually rooted in the teachings of a sacred text or written dogma.

Springing from rejection of the secularist tendencies of modernity, fundamentalism is now found in every dominant religion wherever a Western-style society has developed, including Islam, Christianity, Hinduism, Judaism, Sikhism, Buddhism, Confucianism, and Zoroastrianism. Fundamentalism is, therefore, a reaction to the modern world; it represents an effort to draw upon a "golden age" religious tradition in order to cope with and counteract an already changing society that is denounced as trying to erase the true faith and traditional religious values. The near-universality of fundamentalist movements is seen by some as another expression of a widespread rebellion against the presumed evils fostered by secular globalization.

Fundamentalists always place a high priority on doctrinal conformity and its necessity for achieving salvation. Further, they are convinced of the unarguable correctness of their beliefs and the necessity of the unquestioned acceptance of those beliefs by the general society. Since the truth is knowable and indisputable, fundamentalists hold, there is no need to discuss it or argue it in open forum. To some observers, therefore, fundamentalism is, by its nature, undemocratic and states controlled by fundamentalist regimes combining politics and religion of necessity stifle debate and punish dissent. In the modern world that rigidity seems most apparent in Islam where, it is claimed, "all Muslims believe in the absolute inerrancy of the Quran . . ." (*The Islamic Herald,* April 1995) and several countries—for example, the Islamic Republic of Iran and the Islamic Republic of Pakistan—proclaim by offical name their administrative commitment to religious control.

In most of the modern world, however, such commitment is not overt or official, and fundamentalists often believe that they and their religious convictions are under mortal threat. They view modern secular society—with its assumption of equality of competing voices and values—as trying to eradicate the true faith and religious verities. Initially, therefore, every fundamentalist movement begins as an intrareligious struggle directed against its own coreligionists and countrymen in response to a felt assault by the liberal or secular society they inhabit. At first, the group may blame its own weakness and irresolution for the oppression they feel and the general social decay they perceive. To restore society to its idealized standards, the aroused group may exhort its followers to ardent prayer, ascetic practices, and physical or military training.

If it is unable to impose its beliefs on others peacefully, the fundamentalist group—seeing itself as the savior of society—may justify other, more extreme actions against perceived oppressors. Initial protests and nonviolent actions may escalate to attacks on corrupt public figures thwarting their vision and to outright domestic guerrilla warfare. That escalation is advanced and gains willing supporters when inflexible fundamentalism is combined with the unending poverty and political impotence felt in many, particularly Middle Eastern, societies today. When an external culture or power—commonly, a demonized United States—is seen as the unquestioned source of the pollution and exploitation frustrating their social vision, some fundamentalists have been able to justify any extreme action and personal sacrifice for their cause. In their struggle, it appears an easy progression from domestic dispute and disruption to international terrorism.

emerged some 3000 to 3500 years ago in the Near East, one of the ancient culture hearth regions (see Figure 7.13).

Judaism is a distinctively *ethnic* religion, the determining factors of which are descent from Israel (the patriarch Jacob), the Torah (law and scripture), and the traditions of the culture and the faith. Early military success gave the Jews a sense of territorial and political identity to supplement their religious self-awareness. Later conquests by nonbelievers led to their dispersion (*diaspora*) to much of the Mediterranean world and farther east into Asia by A.D. 500 (Figure 7.28).

During the 13th and 14th centuries, many Jews sought refuge in Poland and Russia from persecution in western and central Europe; during the later 19th and early 20th centuries, Jews were important elements of the European immigrant stream to the Western Hemisphere. The establishment of the state of Israel in 1948 was a fulfillment of the goal of *Zionism,* the belief in the need to create an autonomous Jewish state in Palestine. It demonstrated a determination that Jews not lose their identity by absorption into alien cultures and societies. It also spatially united two earlier separated Jewish communities: the Sephardim, who were expelled from Iberia in the

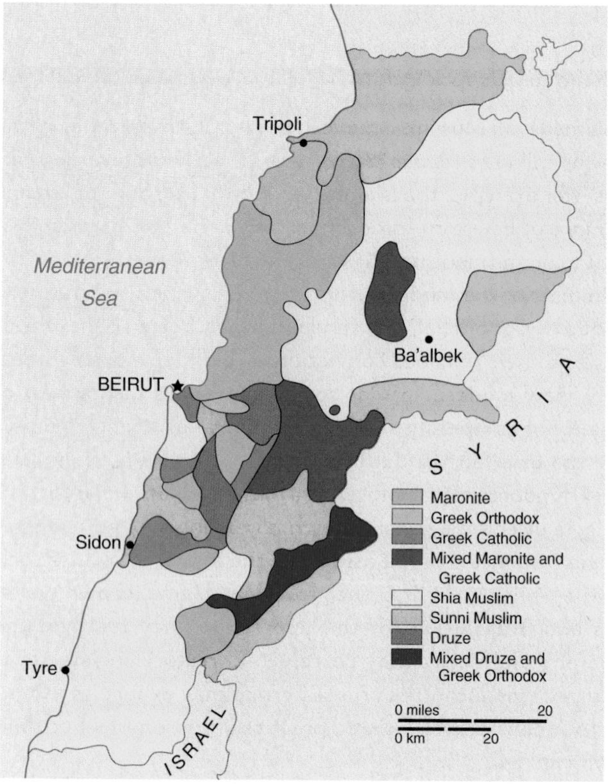


**FIGURE 7.26 Religious regions of Lebanon.** Long-standing religious territoriality and rivalry led, in the 1960s and 1970s, to open conflict between Muslims and Christians and among various branches of each major faith in this eastern Mediterranean country.

late 15th century, fleeing initially to North Africa and the Near East, and the Ashkenazim, who, between the 13th and 16th centuries, sought refuge in eastern Europe from persecutions in western and central Europe.

Judaism's imprint upon the cultural landscape has been subtle and unobtrusive. The Jewish community reserves space for the practice of communal burial; the spread of the cultivated citron in the Mediterranean area during Roman times has been traced to Jewish ritual needs; and the religious use of grape wine assured the cultivation of the vine in their areas of settlement. The synagogue as place of worship has tended to be less elaborate than its Christian counterpart. The essential for religious service is a community of at least 10 adult males, not a specific structure.

## Christianity

*Christianity* had its origin in the life and teachings of Jesus, a Jewish preacher of the 1st century of the modern era, whom his followers believed was the messiah promised by God. The new covenant he preached was not a rejection of traditional Judaism but a promise of salvation to all humankind, rather than to just a chosen people.

Christianity's mission was conversion. As a universal religion of salvation and hope, it spread quickly among the underclasses of both the eastern and western parts of the Roman Empire, carried to major cities and ports along the excellent system of Roman roads and sea lanes (Figure 7.29). In A.D. 313, Emperor Constantine proclaimed Christianity the state religion. Much later, of course, the faith

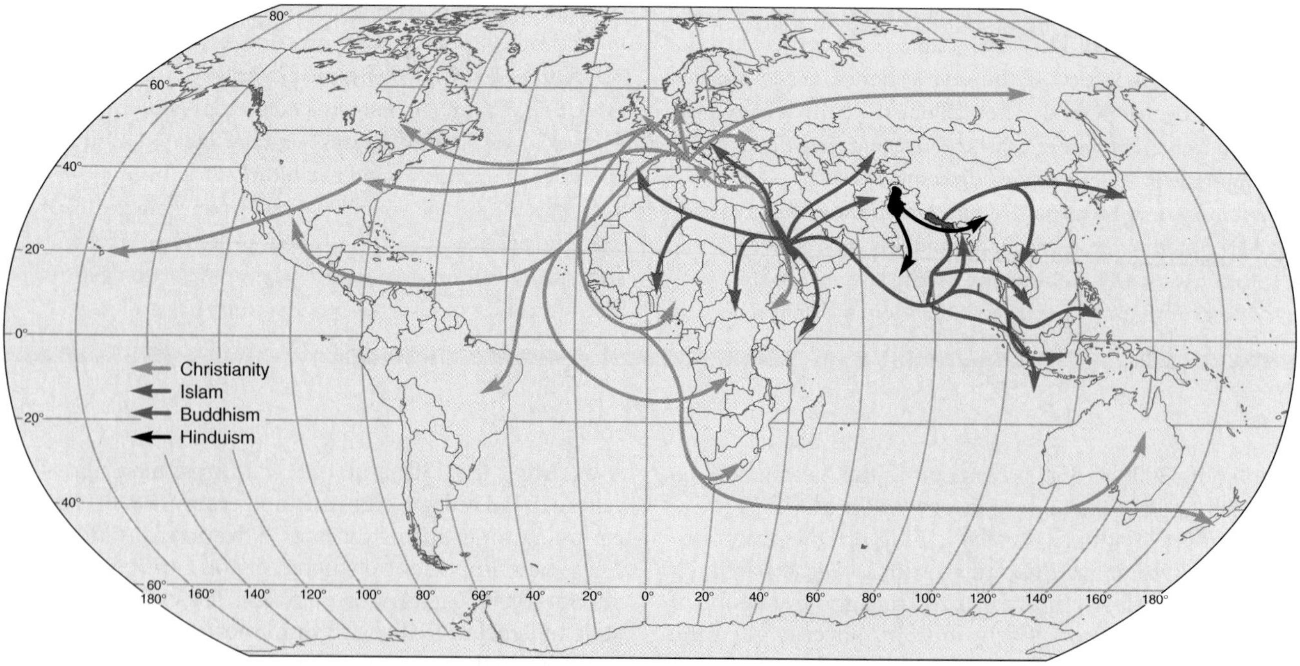


**FIGURE 7.27 Innovation areas and diffusion routes of major world religions.** The monotheistic (single-deity) faiths of Judaism, Christianity, and Islam arose in southwestern Asia, Judaism and Christianity in Palestine in the eastern Mediterranean region and Islam in western Arabia near the Red Sea. Hinduism and Buddhism originated within a confined hearth region in the northern part of the Indian subcontinent. Their rates, extent, and directions of spread are suggested here and detailed on later maps.

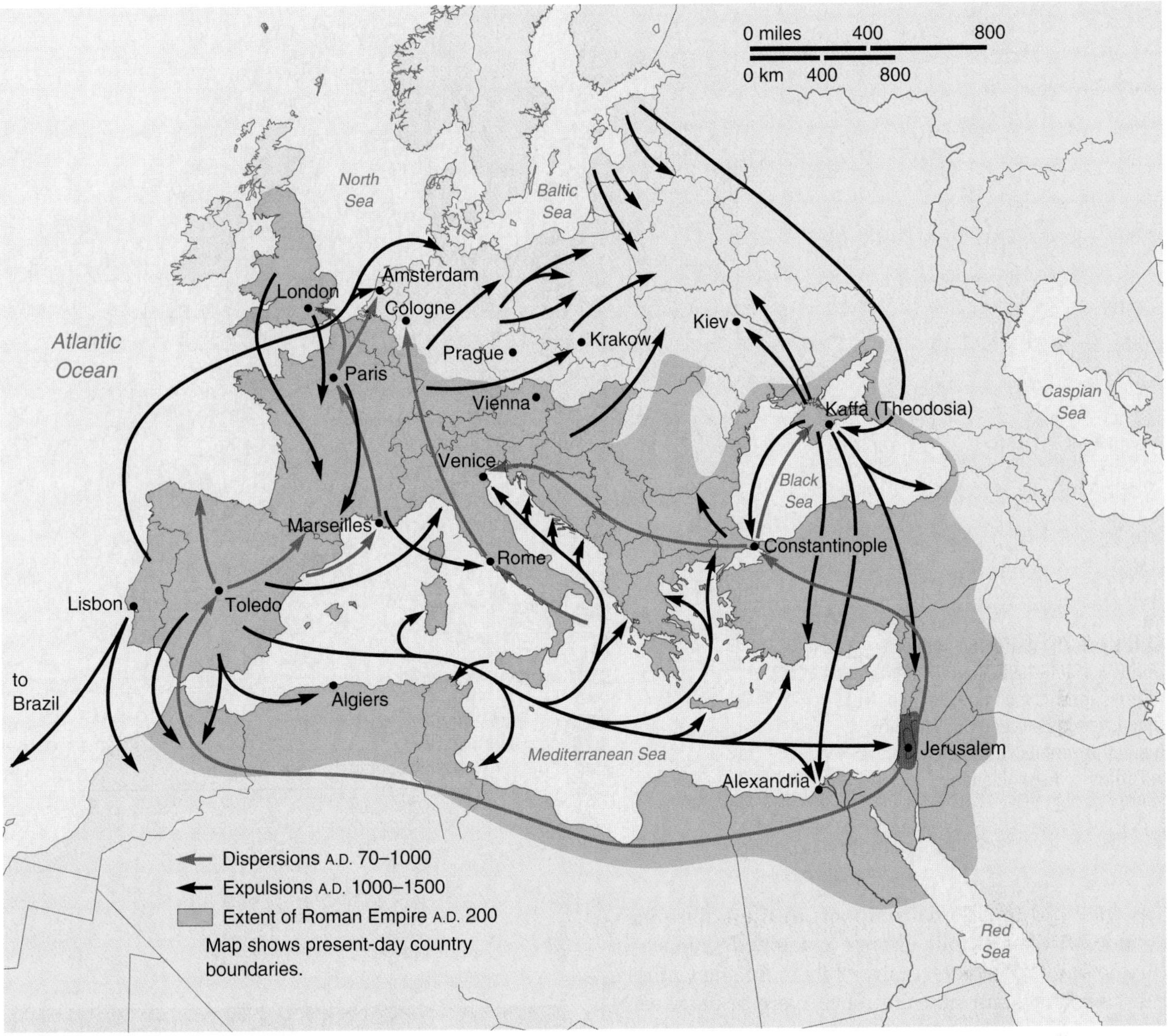


**FIGURE 7.28 Jewish dispersions, A.D. 70–1500.** A revolt against Roman rule in A.D. 66 was followed by the destruction of the Jewish Temple years later and an imperial decision to Romanize the city of Jerusalem. Judaism spread from the hearth region, carried by its adherents dispersing from their homeland to Europe, Africa, and eventually in great numbers to the Western Hemisphere. Although Jews established themselves and their religion in new lands, they did not lose their sense of cultural identity or seek to attract converts to their faith.

was brought to the New World with European settlement (Figure 7.27).

The dissolution of the Roman Empire into a western and eastern half after the fall of Rome also divided Christianity. The Western Church, based in Rome, was one of the very few stabilizing and civilizing forces uniting western Europe during the Dark Ages. Its bishops became the civil as well as ecclesiastical authorities over vast areas devoid of other effective government. Parish churches were the focus of rural and urban life, and the cathedrals replaced Roman monuments and temples as the symbols of the social order.

Secular imperial control endured in the eastern empire, whose capital was Constantinople. Thriving under its protection, the Eastern Church expanded into the Balkans, eastern Europe, Russia, and the Near East. The fall of the eastern empire to the Turks in the 15th century opened eastern Europe temporarily to Islam, though the Eastern Orthodox Church (the direct descendant of the Byzantine state church) remains, in its various ethnic branches, a major component of Christianity.

The Protestant Reformation of the 15th and 16th centuries split the church in the west, leaving Roman Catholicism supreme in southern Europe but installing a variety of Protestant denominations and national churches in western and northern Europe. The split was reflected in the subsequent worldwide dispersion of Christianity. Catholic Spain and Portugal colonized Latin America, bringing both their languages and the Roman church to that area (Figure

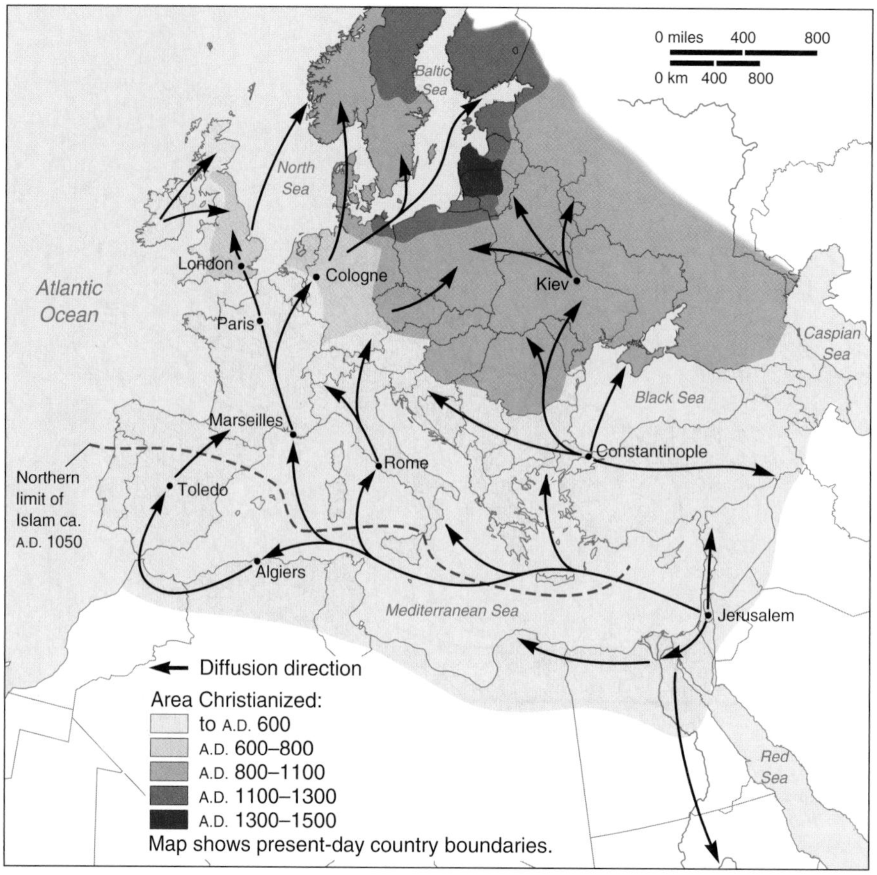


**FIGURE 7.29 Diffusion paths of Christianity, A.D. 100–1500.** Routes and dates are for Christianity as a composite faith. No distinction is made between the Western Church and the various subdivisions of the Eastern Orthodox denominations.

7.27), as they did to colonial outposts in the Philippines, India, and Africa. Catholic France colonized Quebec in North America. Protestants, many of them fleeing Catholic or repressive Protestant state churches, were primary early settlers of Anglo America, Australia, New Zealand, Oceania, and South Africa.

Although religious intermingling rather than rigid territorial division is characteristic of the contemporary American scene (Figure 7.30), the beliefs and practices of various immigrant groups and the innovations of domestic congregations have created a particularly varied spatial patterning of "religious regions" in the United States (Figure 7.31).

The mark of Christianity on the cultural landscape has been conspicuous and enduring. In pre-Reformation Catholic Europe, the parish church formed the center of life for small neighborhoods of every town, the village church was the centerpiece of every rural community, and in larger cities the central cathedral served simultaneously as a glorification of God, a symbol of piety, and the focus of religious and secular life (Figure 7.32a).

Protestantism placed less importance on the church as a monument and symbol, although in many communities—colonial New England, for example—the churches of the principal denominations were at the village center (Figure 7.32b). Many were adjoined by a cemetery, because Chris-

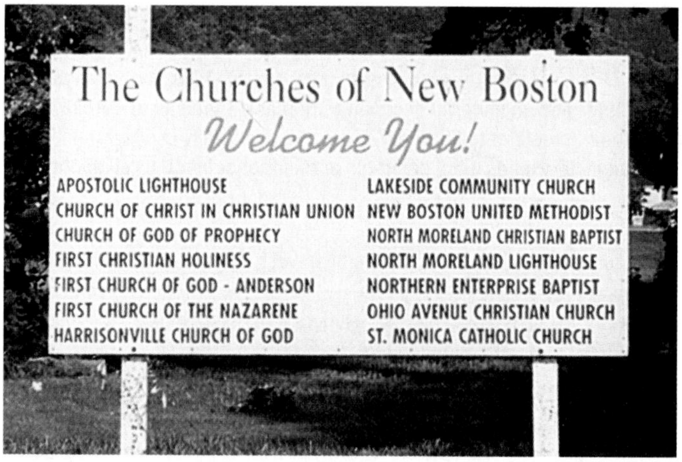


**FIGURE 7.30** Advertised evidence of religious diversity in the United States. The sign details only a few Christian congregations. In reality, the United States has become the most religiously diverse country in the world, with essentially all of the world's faiths represented within its borders. Welcoming signs for other, particularly larger, towns might also announce Muslim, Hindu, Buddhist, and many other congregations in their varied religious mix. *Photo courtesy of Susan Reisenweaver.*

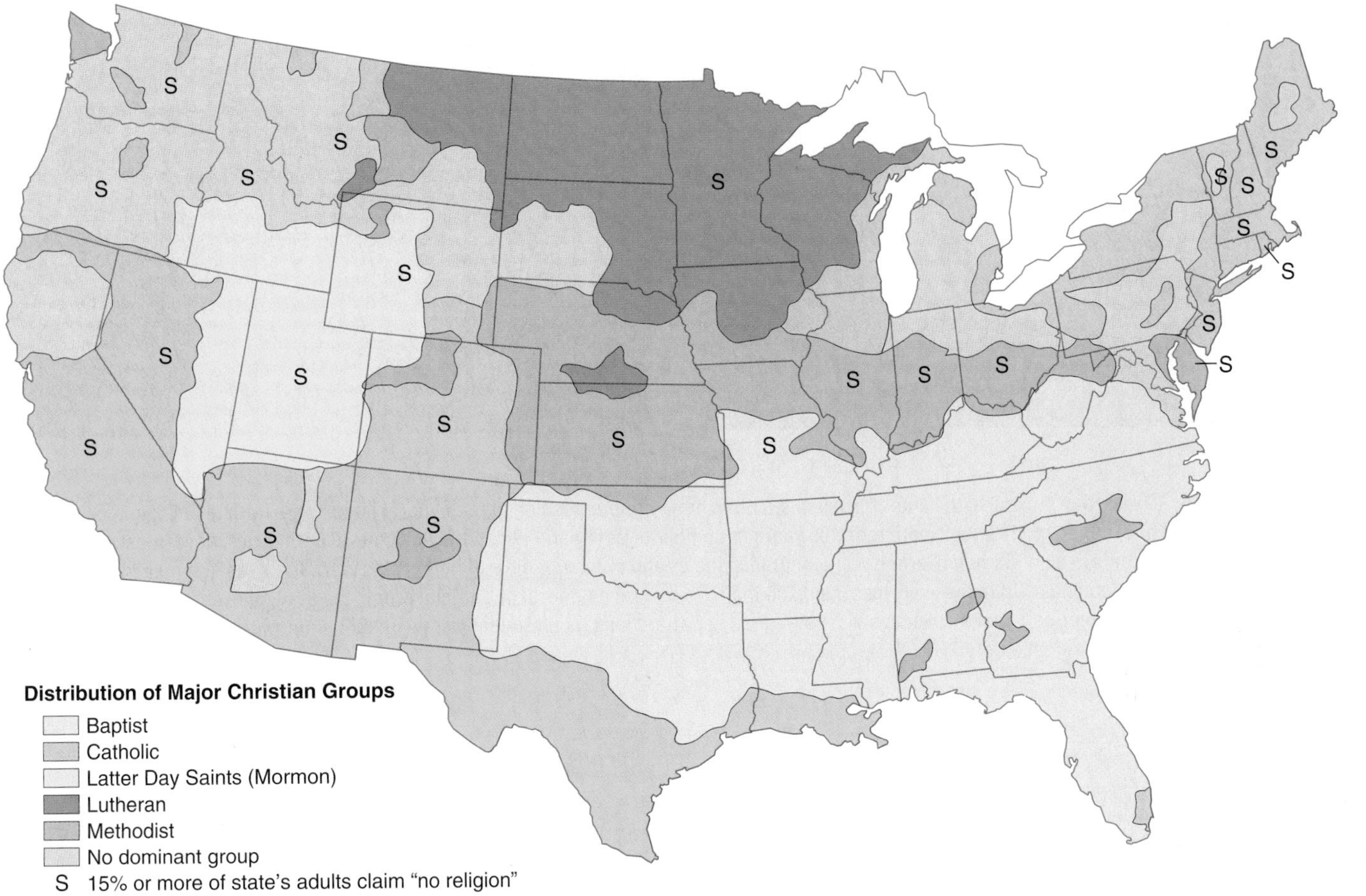


**FIGURE 7.31 Religious affiliation in the conterminous United States.** The greatly generalized areas of religious dominance shown conceal the reality of immense diversity of church affiliations throughout the United States. A sizeable number of Americans claim to have "no religion." Secularism (marked by *S* on the map) is particularly prominent in the western states, in the industrial Midwest, and in the Northeast. Sources: *Based on data or maps from: the 2001 "American Religious Identity Survey" by the Graduate School at City University of New York; religious denomination maps prepared by Ingolf Vogeler of the University of Wisconsin, Eau-Claire, based on data compiled by the Roper Center for Public Research; and* Churches and Church Membership in the United States *(Atlanta, Georgia: Glenmary Research Center, 1992).*

tians—in common with Muslims and Jews—practice burial in areas reserved for the dead. In Christian countries, particularly, the cemetery—whether connected to the church, separate from it, or unrelated to a specific denomination—has traditionally been a significant land use within urban areas.

### Islam

*Islam* springs from the same Judaic roots as Christianity and embodies many of the same beliefs: there is only one God, who can be revealed to humans through prophets; Adam was the first human; Abraham was one of his descendants. Mohammed is revered as the prophet of *Allah* (God), succeeding and completing the work of earlier prophets of Judaism and Christianity, including Moses, David, and Jesus. The Koran, the word of Allah revealed to Mohammed, contains not only rules of worship and details of doctrine but also instructions on the conduct of human affairs. For fundamentalists, it thus becomes the unquestioned guide to matters both religious and secular. Observance of the "five pillars" (see Figure 7.24) and surrender to the will of Allah unites the faithful into a community that has no concern with race, color, or caste.

It was that law of community that unified an Arab world sorely divided by tribes, social ranks, and multiple local deities. Mohammed was a resident of Mecca but fled in A.D. 622 to Medina, where the Prophet proclaimed a constitution and announced the universal mission of the Islamic community. That flight—*Hegira*—marks the starting point of the Islamic (lunar) calendar. By the time of Mohammed's death in A.H. 11 (*Anno*—in the year of—*Hegira*, or A.D. 632), all of Arabia had joined Islam. The new religion swept quickly outward from that source region over most of Central Asia and, at the expense of Hinduism, into northern India (Figure 7.33). Its advance westward was particularly rapid and inclusive in North Africa. Later, Islam dispersed into Indonesia, southern Africa, and the Western Hemisphere. It continues its spatial spread as the fastest-growing major religion at the present time.

(a)

(b)

**FIGURE 7.32** In Christian societies, the church assumes a prominent, central position in the cultural landscape. (a) The building of Notre Dame Cathedral in Paris, France, begun in 1163, took more than 100 years to complete. Between 1170 and 1270, some 80 cathedrals were constructed in France alone. The cathedrals in all of Catholic Europe were located in the centers of major cities. Their plazas were the sites of markets, public meetings, and religious ceremonies. (b) Individually less imposing than the central cathedral of Catholic areas, the several Protestant churches common in small and large American towns collectively constitute an important land use frequently sited in the center of the community. The church shown here is in Waitsfield, Vermont. *(a) © David A. Burney. (b) © David Brownell.*

Disagreements over the succession of leadership after the Prophet led to a division between two primary groups, Sunnis and Shi'ites. Sunnis, the majority (80% to 85% of Muslims) recognize the first four *caliphs* (originally, "successor" and later the title of the religious and civil head of the Muslim state) as Mohammed's rightful successors. The Shi'ites reject the legitimacy of the first three and believe that Muslim leadership rightly belonged to the fourth caliph, the Prophet's son-in-law, Ali, and his descendants. At the start of the 21st century, Sunnis constitute the majority of Muslims in all countries except Iran, Iraq, Bahrain, and perhaps Yemen.

The mosque—place of worship, community clubhouse, meeting hall, and school—is the focal point of Islamic communal life and the primary imprint of the religion on the cultural landscape. Its principal purpose is to accommodate the Friday communal service, mandatory for all male Muslims. It is the congregation rather than the structure that is important; small or poor communities are as well served by a bare, whitewashed room as are larger cities by architecturally splendid mosques. With its perfectly proportioned, frequently gilded or tiled domes; its graceful, soaring towers and minarets (from which the faithful are called to prayer); and its delicately wrought parapets and cupolas, the carefully tended mosque is frequently the most elaborate and imposing structure of the town (Figure 7.34).

### Hinduism

*Hinduism* is the world's oldest major religion. Though it has no datable founding event or initial prophet, some evidence traces its origin back 4000 or more years. Hinduism is an ethnic religion, an intricate web of religious, philosophical, social, economic, and artistic elements constituting a distinctive Indian civilization. Its estimated 1 billion adherents are primarily Asian and largely confined to India, where it claims 80% of the population.

From its cradle area in the valley of the Indus River, Hinduism spread eastward down the Ganges River and southward throughout the subcontinent and adjacent regions by amalgamating, absorbing, and eventually supplanting earlier native religions and customs. Its practice eventually spread throughout Southeast Asia, into Indonesia, Malaysia, Cambodia, Thailand, Laos, and Vietnam, as well as into neighboring Myanmar and Sri Lanka. The largest Hindu temple complex is in Cambodia, not India, and Bali remains a Hindu pocket in dominantly Islamic Indonesia.

There is no common creed, single doctrine, or central ecclesiastical organization defining the Hindu. A Hindu is one born into a caste, a member of a complex social and economic—as well as religious—community. Hinduism accepts and incorporates all forms of belief; adherents may believe in one god or many or none. The *caste* (meaning "birth") structure of society is an expression of the eternal transmigration of souls. For Hindus, the primary aim of this life is to conform to prescribed social and ritual duties and to the rules of conduct for the assigned caste and profession. Those requirements constitute that individual's *dharma*—law and duties. Traditionally, each craft or profession is the property of a particular caste.

The practice of Hinduism is rich with rites and ceremonies, festivals and feasts, processions and ritual gatherings of literally millions of celebrants. It involves careful observance of food and marriage rules and the performance of duties within the framework of the caste system. Worship

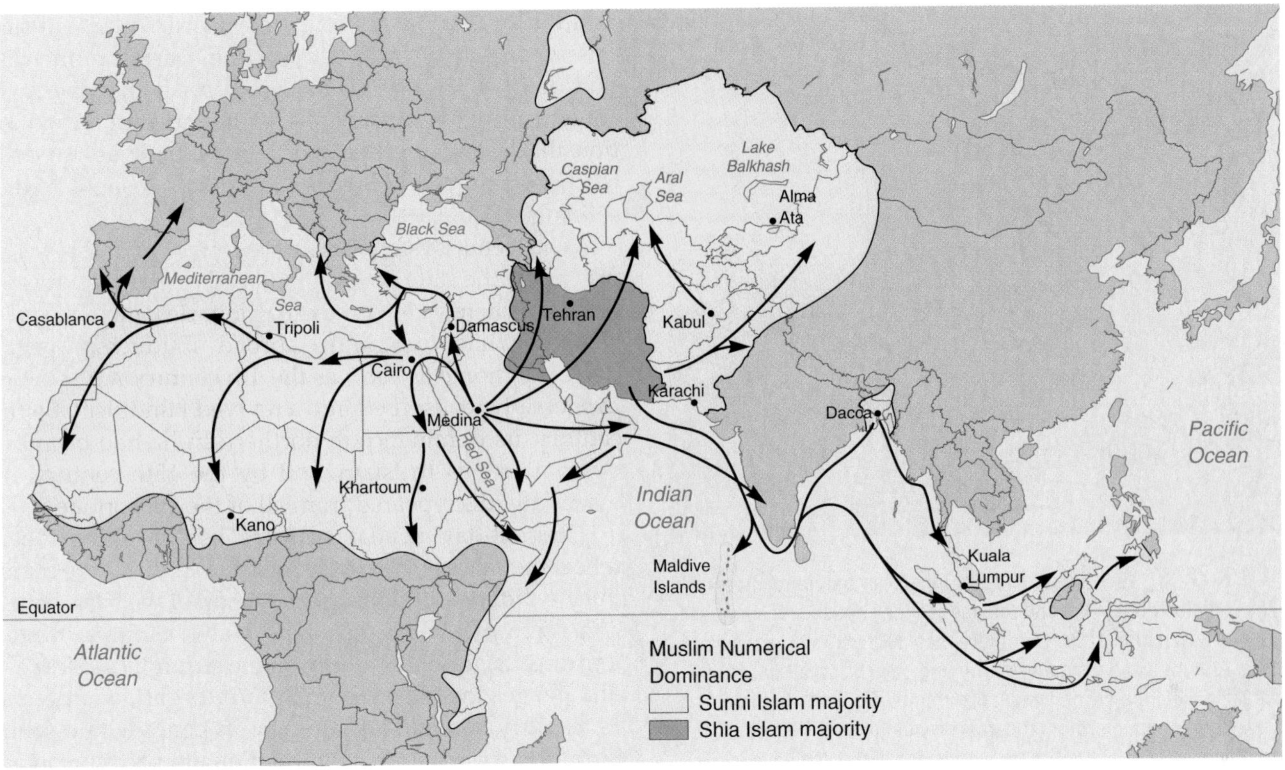


**FIGURE 7.33 The spread and extent of Islam.** Islam predominates in more than 35 countries along a band across northern Africa to Central Asia and the northern part of the Indian subcontinent. Still farther east, Indonesia has the largest Muslim population of any country. Islam's greatest development is in Asia, where it is second only to Hinduism, and in Africa, where, some observers suggest, it may be the leading faith. Current Islamic expansion is particularly rapid in the Southern Hemisphere.

**FIGURE 7.34** The Blue Mosque in Istanbul, Turkey. The common architectural features of the mosque make it an unmistakable landscape evidence of the presence of Islam in any local culture. *© Royalty-Free Getty Images.*

in the temples and shrines that are found in every village (Figure 7.35) and the leaving of offerings to secure merit from the gods are required. The temples, shrines, daily rituals and worship, numerous specially garbed or marked holy men and ascetics, and ever-present sacred animals mark the cultural landscape of Hindu societies, a landscape infused with religious symbols and sights that are part of a total cultural experience.

### Buddhism

Numerous reform movements have derived from Hinduism over the centuries, some of which have endured to the present day as major religions on a regional or world scale. For example, *Sikhism* developed in the Punjab area of northwestern India in the late 15th century A.D., rejecting the formalism of both Hinduism and Islam and proclaiming a gospel of universal toleration. The great majority of some 20 million Sikhs still live in India, mostly in the Punjab, though others have settled in Malaysia, Singapore, East Africa, the United Kingdom, and North America.

The largest and most influential of the dissident movements is *Buddhism,* a universalizing faith founded in the 6th century B.C. in northern India by Siddhartha Gautama, the Buddha ("Enlightened One"). The Buddha's teachings were more a moral philosophy that offered an explanation for evil and human suffering than a formal religion. He viewed the road to enlightenment and salvation to lie in understanding the "four noble truths": existence involves suffering; suffering is the result of desire; pain ceases when desire is destroyed; the destruction of desire comes through knowledge of correct behavior and correct thoughts. The

**FIGURE 7.35** The Hindu temple complex at Khajuraho in central India. The creation of temples and the images they house has been a principal outlet of Indian artistry for more than 3000 years. At the village level, the structure may be simple, containing only the windowless central cell housing the divine image, a surmounting spire, and the temple porch or stoop to protect the doorway of the cell. The great temples, of immense size, are ornate extensions of the same basic design. *© Fred Bruemmer/Valan Photos.*

Buddha instructed his followers to carry his message as missionaries of a doctrine open to all castes, for no distinction among people was recognized. In that message, all could aspire to ultimate enlightenment, a promise of salvation that raised the Buddha in popular imagination from teacher to savior and Buddhism from philosophy to universalizing religion.

The belief system spread throughout India, where it was made the state religion in the 3d century B.C. It was carried elsewhere into Asia by missionaries, monks, and merchants. While expanding abroad, Buddhism began to decline at home as early as the 4th century A.D., slowly but irreversibly reabsorbed into a revived Hinduism. By the 8th century, its dominance in northern India had been broken by conversions to Islam, and by the 15th century, it had essentially disappeared from all of the subcontinent.

Present-day spatial patterns of Buddhist adherence reflect the schools of thought, or *vehicles,* that were dominant during different periods of dispersion of the basic belief system (Figure 7.36). In all of its many variants, Buddhism imprints its presence vividly on the cultural landscape. Buddha images in stylized human form began to appear in the 1st century A.D. and are common in painting and sculpture throughout the Buddhist world. Equally widespread are the three main types of buildings and monuments: the *stupa,* a

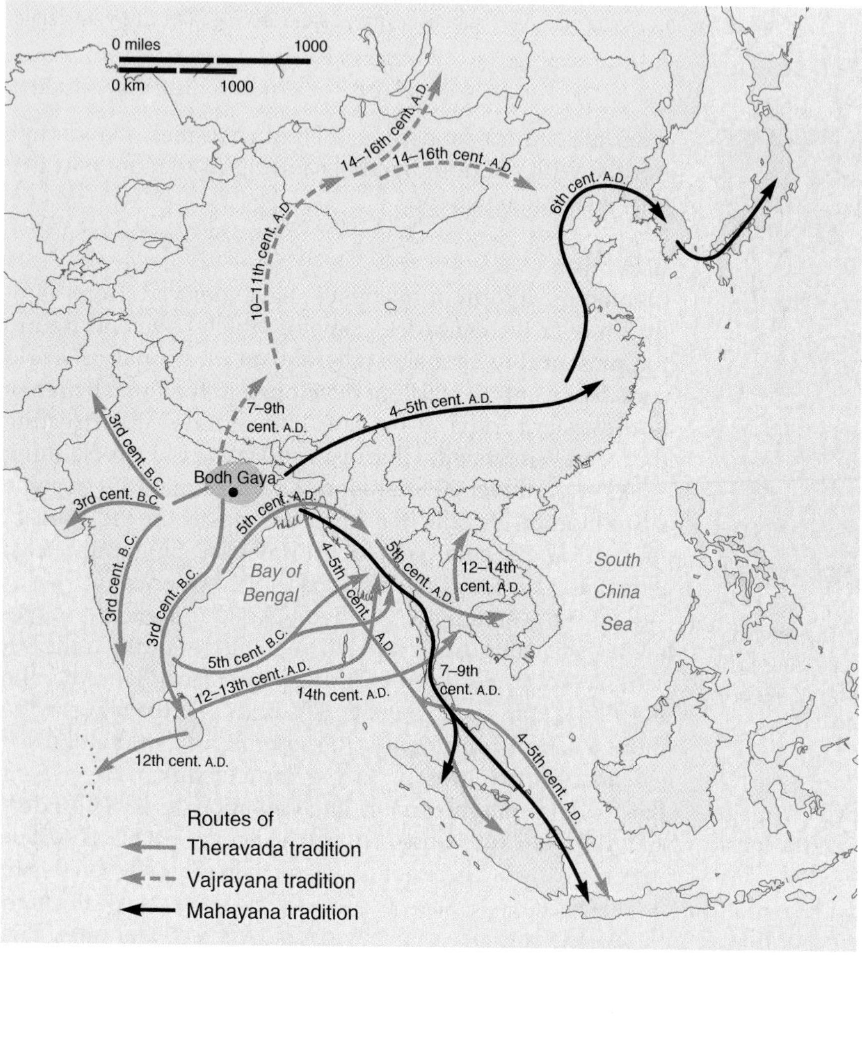


**FIGURE 7.36** **Diffusion paths, times, and vehicles of Buddhism.**

commemorative shrine; the temple or pagoda enshrining an image or a relic of the Buddha (Figure 7.37); and the monastery, some of them the size of small cities.

### East Asian Ethnic Religions

When Buddhism reached China from the south some 1500 to 2000 years ago and was carried to Japan from Korea in the 6th century, it encountered and later amalgamated with already well-established ethical belief systems. The Far Eastern ethnic religions are syncretisms, combinations of different forms of belief and practice. In China, the union was with Confucianism and Taoism, themselves becoming intermingled by the time of Buddhism's arrival, and in Japan, it was with Shinto, a polytheistic animism and shamanism.

Chinese belief systems address not so much the hereafter as the achievement of the best possible way of life in the present existence. They are more ethical or philosophical than religious in the pure sense. Confucius (K'ung Fu-tzu), a compiler of traditional wisdom who lived about the same time as Gautama Buddha, emphasized the importance of proper conduct between ruler and subjects and between family members. The family was extolled as the nucleus of the state, and filial piety was the loftiest of virtues. There are no churches or clergy in *Confucianism,* though its founder believed in a heaven seen in naturalistic terms, and the Chinese custom of ancestor worship as a mark of gratitude and respect was encouraged.

Confucianism was joined by, or blended with, *Taoism,* an ideology that, according to legend, was first taught by Lao Tsu in the 6th century B.C. Its central theme is *Tao* (the Way), a philosophy teaching that eternal happiness lies in total identification with nature and deploring passion, unnecessary invention, unneeded knowledge, and governmental interference in the simple life of individuals. Buddhism, stripped by Chinese pragmatism of much of its Indian otherworldliness and defining a *nirvana* achievable in this life, was easily accepted as a companion to these traditional Chinese belief systems. Along with Confucianism and Taoism, Buddhism became one of the honored Three Teachings, and to the average person there was no distinction in meaning or importance between a Confucian temple, Taoist shrine, or Buddhist stupa.

Buddhism also joined and influenced Japanese Shinto, the traditional religion of Japan, which developed out of nature and ancestor worship. *Shinto*—the Way of the Gods—is basically a structure of customs and rituals rather than an ethical or a moral system. It observes a complex set of deities, including deified emperors, family spirits, and the divinities residing in rivers, trees, certain animals, mountains and, particularly, the sun and moon. At first resisted, Buddhism was later amalgamated with traditional Shinto. Buddhist deities were seen as Japanese gods in a different form, and Buddhist priests formerly but no longer assumed control of most of the numerous Shinto shrines in which the gods are believed to dwell and which are approached through ceremonial *torii,* or gateway arches (Figure 7.38).

## ETHNICITY

Any discussion of cultural diversity would be incomplete without the mention of **ethnicity.** Based on the root word *ethnos,* meaning "people" or "nation," the term is usually

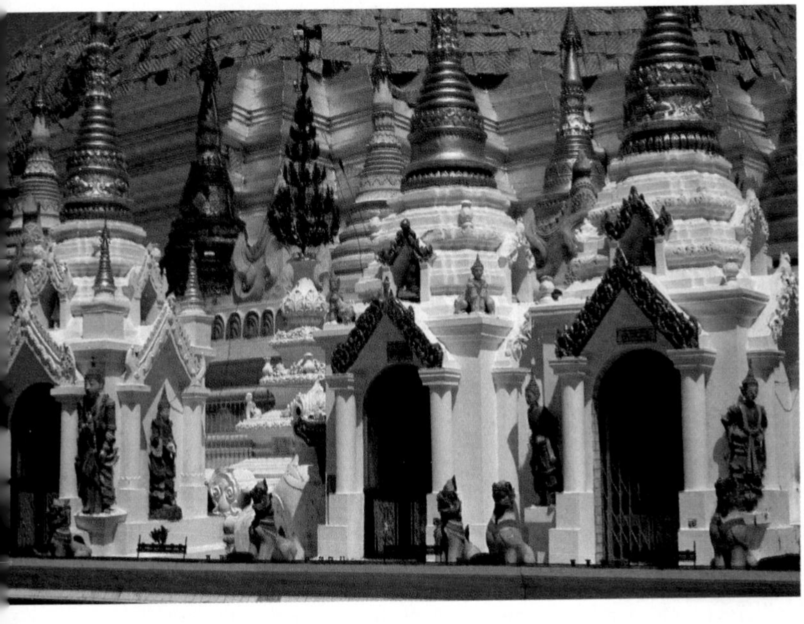

**FIGURE 7.37** The golden stupas of the Swedagon pagoda, Yangon, Myanmar (Rangoon, Burma). *© Wolfgang Kaehler.*

**FIGURE 7.38** A Shinto shrine, Nikko, Honshu Island, Japan. *© Charlotte Kahler.*

used to refer to the ancestry of a particular people who have in common distinguishing characteristics associated with their heritage. No single trait denotes ethnicity. Recognition of ethnic communities may be based on language, religion, national origin, unique customs, or, improperly, an ill-defined concept of "race" (see "The Matter of Race," p. 259). Whatever the unifying thread, ethnic groups may strive to preserve their special shared ancestry and cultural heritage through the collective retention of language, religion, festivals, cuisines, traditions, and in-group work relationships, friendships, and marriages. Those preserved associations are fostered by and support *ethnocentrism,* the feeling that one's own ethnic group is superior.

Normally, reference to ethnic communities is recognition of their minority status within a country or region dominated by a different, majority culture group. We do not identify Koreans living in Korea as an ethnic group because theirs is the dominant culture in their own land. Koreans living in Japan, however, constitute a discerned and segregated group in that foreign country. Ethnicity, therefore, is an evidence of areal cultural diversity and a reminder that culture regions are rarely homogeneous in the characteristics displayed by all of their occupants.

*Territorial segregation* is a strong and sustaining trait of ethnic identity, one that assists groups to retain their distinction. On the world scene, indigenous ethnic groups have developed over time in specific locations and have established themselves in their own and others' eyes as distinctive peoples with defined homeland areas. The boundaries of most countries of the world encompass a number of racial or ethnic minorities (Figure 7.39). Their demands for special territorial recognition have sometimes increased with advances in economic development and self-awareness, as Chapter 9 points out. Where clear territorial separation does not exist but ethnic identities are distinct and animosities bitter, tragic conflict within single political units can erupt. Recent histories of deadly warfare between Tutsi and Hutu in Rwanda, or Serb and Croat in Bosnia, make vivid the often continuing reality of ethnic discord and separatism.

Increasingly in a world of movement, ethnicity is less a matter of indigenous populations and more one of outsiders in an alien culture. Immigrants, legal and illegal, and refugees from war, famine, or persecution are a growing presence in countries throughout the world. Immigrants to a country typically have two choices. They may hope for *assimilation* by giving up many of their past culture traits, losing their distinguishing characteristics and merging into the mainstream of the dominant culture, as reviewed earlier on p. 234. Or they may try to retain their distinctive cultural heritage. In either case, they usually settle initially in an area where other members of their ethnic group live, as a place of refuge and learning (Figure 7.40). With the passage of time, they may leave their protected community and move out among the general population.

The Chinatowns, Little Havanas, and Little Italys of Anglo American cities have provided the support systems essential to new immigrants in an alien culture region. Japanese, Italians, Germans, and other ethnics have formed agricultural colonies in Brazil in much the same spirit. Such ethnic enclaves may provide an entry station, allowing both

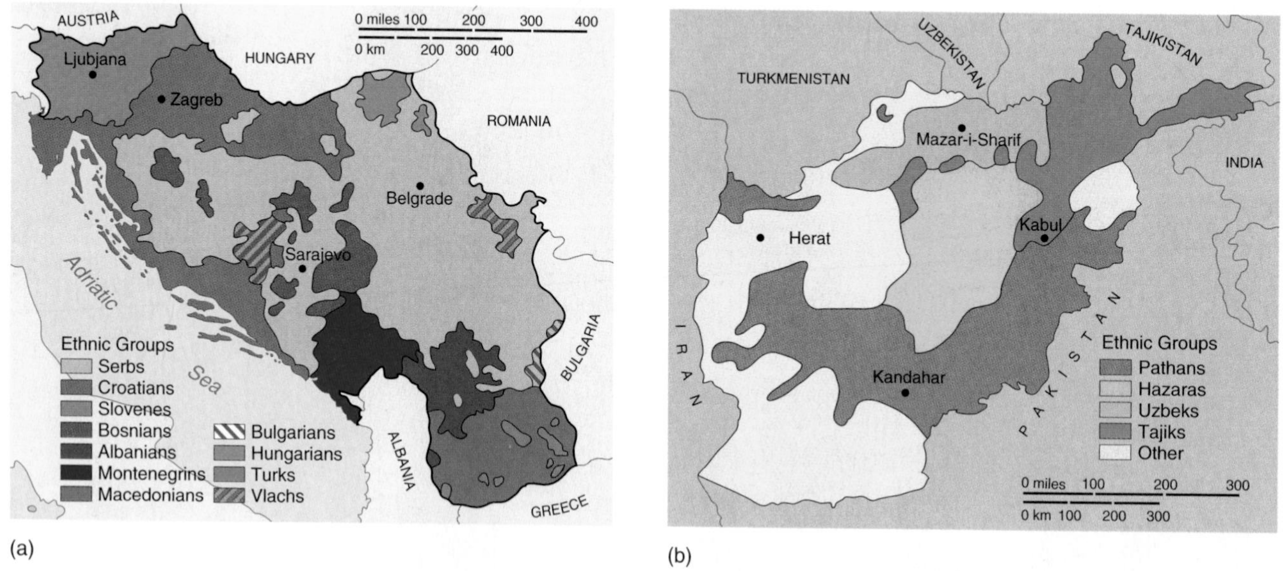


**FIGURE 7.39** (a) **Ethnicity in former Yugoslavia.** Yugoslavia was formed after World War I (1914–1918) from a patchwork of Balkan states and territories, including the former kingdoms of Serbia and Montenegro, Bosnia and Herzegovina, Croatia-Slavonia, and Dalmatia. The authoritarian central government created in 1945 began to disintegrate in 1991 as non-Serb minorities voted for regional independence and national separation. Religious differences between Eastern Orthodox, Roman Catholic, and Muslim adherents compounded resulting conflicts rooted in nationality and rival claims to ethnic homelands. (b) **Afghanistan** houses Pathan, Tajik, Uzbek, and Hazara ethnic groups (among others), speaking Pashto, Dari Persian, Uzbek, and several minor languages, and split between majority Sunni and minority Shia Muslim believers.

## The Matter of Race

Human populations can be differentiated from one another on any number of bases: gender, nationality, stage of economic development, and so on. One common form of differentiation is based on recognizable inherent physical characteristics, or *race*.

A **race** is usually understood to be a population subset whose members have in common some hereditary biological characteristics that set them apart physically from other human groups. The spread of human beings over the earth and their occupation of different environments were accompanied by the development of physical variations in skin pigmentation, hair texture, hair and eye color, facial characteristics, blood composition, and other traits largely related to soft tissue. Some subtle skeletal differences among peoples also exist.

Such differences have formed the basis for the segregation, by some anthropologists, of humanity into different racial groups, although recent research indicates that the greatest genetic variation between any racial groups ever identified is far less than the variation within any given population. Caucasoid, Negroid, Mongoloid, Amerindian, Australoid and other races have been recognized in a process of arbitrary classification invention, modification, and refinement that began at least two centuries ago. Racial differentiation, as commonly understood—based largely on surface appearance—is old and can reasonably be dated at least to the Paleolithic (100,000 to about 11,000 years ago) spread and isolation of population groups.

Although racial classifications vary by author, most are derived from recognized geographic variations of populations. Thus, Mongoloids are associated with northern and eastern Asia; Australoids are the aboriginal people of Australasia; Amerindians developed in the Americas, and so on. If all of humankind belongs to a single species that can freely interbreed and produce fertile offspring, how did this areal differentiation by race occur? Why is it that, despite millennia of mixing and migration, people with distinct combinations of physical traits appear to be clustered in particular areas of the world?

Two causative forces appear to be most important. First, through evolutionary **natural selection** or **adaptation**, characteristics are transmitted that enable people to adapt to particular environmental conditions, such as climate. Studies have suggested a plausible relationship between, for example, solar radiation and skin color and between temperature and body size. In tropical climates, for example, it presumably is advantageous to be short, since that means a greater bodily surface area for sweat to evaporate. In frigid Arctic regions, it is suggested, Inuits and other native populations developed round heads and bodies with increased bodily volume and decreased evaporative surface area. The second force, **genetic drift**, refers to a heritable trait (such as flatness of face) that appears by chance in one group and becomes accentuated through inbreeding. If two populations are too separated spatially for much interaction to occur (*isolation*), a trait may develop in one but not in the other. Unlike natural selection, genetic drift differentiates populations in nonadaptive ways.

Natural selection and genetic drift promote racial differentiation. Countering them is **gene flow** via interbreeding (also called *admixture*), which acts to homogenize neighboring populations. Genetically, it has been observed, there is no such thing as "pure" race, since people breed freely outside their local groups. Opportunities for interbreeding, always part of the spread and intermingling of human populations, have accelerated with the growing mobility and migrations of people in the past few centuries. Although we may have an urge to group humans "racially," we cannot use biology to justify it, and anthropologists have largely abandoned, and geneticists have dismissed, the idea of race as a scientific concept.

Nor does race have meaningful application to any human characteristics that are culturally acquired. That is, race is *not* equivalent to ethnicity or nationality and has no bearing on differences in religion or language. There is no "Irish" or "Hispanic" race, for example. Such groupings are based on culture, not genes. Culture summarizes the way of life of a group of people, and members of the group may adopt it irrespective of their individual genetic heritage, or race.

individuals and the groups to which they belong to undergo cultural and social modifications sufficient to enable them to operate effectively in the new, majority society. Sometimes, of course, settlers have no desire to assimilate or are not allowed to assimilate, so that they and their descendants form a more or less permanent subculture in the larger society. The Chinese in Malaysia belong to this category. Ethnicity in the context of nationality is discussed more fully in Chapter 9.

## Gender and Culture

**Gender** refers to socially created—not biologically based—distinctions between femininity and masculinity. Because gender relationships and role assignments differ among societies, the status of women is a cultural spatial variable and becomes, therefore, a topic of geographic interest and inquiry. Gender distinctions are complex, and the role and reward assignments of males and females differ from soci-

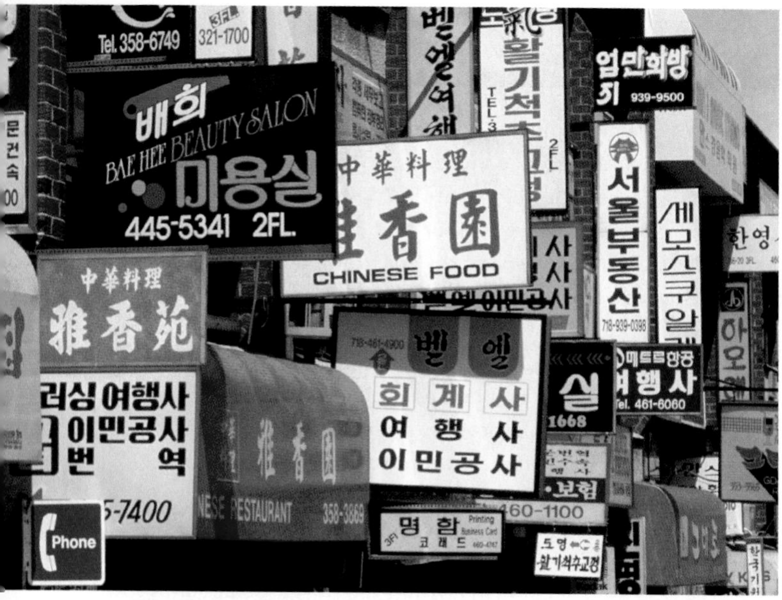


**FIGURE 7.40** The advertising signs along Union Street in the Flushing, Queens, area of New York City proclaim unmistakably the existence of a distinctive, well-established mixed Korean-Chinese community. Chinatowns, Little Italys, and other ethnic enclave counterparts in cities throughout the United States and Western Europe provide both the spatial refuge and the support systems essential to new arrivals in an alien culture realm. *© Lee Snider/Corbis Images.*

ety to society. In many fundamental ways those assignments are conditioned by areally different levels of economic development. Therefore, we might well assume a close similarity between the economic roles and production assignments of males and females—and the status of women—in different cultures that are at the same level of technological advancement. Indeed, it has been observed that modern African or Asian subsistence agricultural groups and those of 18th-century frontier America show more similarities in gender relationships than do pioneer rural and contemporary postindustrial American society.

It may further be logical to believe that advancement in the technological sense would be reflected in an enhancement of the status and rewards of both men and women in all developing societies. The pattern that we actually observe is not quite that simple or straightforward, however. In addition to a culture's economic stage, religion and custom play their important roles in determining gender relationships and female prestige. Further, it appears that, at least in the earlier phases of technological change and development, women generally lose rather than gain in standing and rewards. Only recently and only in the most-developed countries have gender contrasts been reduced within and between societies.

Hunting and gathering cultures observed a general egalitarianism; each sex had a respected, productive, coequal role in the kinship group (see Figure 7.10). Gender is more involved and changeable in agricultural societies (see "Women and the Green Revolution," Chapter 10). The Agricultural Revolution—a major change in the technological subsystem—altered the earlier structure of gender-related responsibilities. In the hoe agriculture that was the first advance over hunting and gathering and is today found in much of sub-Saharan Africa and in South and Southeast Asia, women became responsible for most of the actual fieldwork, while retaining their traditional duties in child rearing, food preparation, and the like; their economic role and status remained equivalent to males. Plow agriculture, on the other hand, tended to subordinate the role of women and diminish their level of equality. Women might have hoed, but men plowed, and female participation in farmwork was drastically reduced. This is the case today in Latin America and, increasingly, in sub-Saharan Africa, where women are often more visibly productive in the market than in the field (Figure 7.41). As women's agricultural productive role declined, they were afforded less domestic authority, less control over their own lives, and few, if any, property rights independent of male family members.

At the same time, considering all their paid and unpaid labor, women spend more hours per day working than do men. In developing countries, the UN estimates, when unpaid agricultural work and housework are considered along with wage labor, women's work hours exceed men's by 30% and may involve at least as arduous—or heavier—physical labor. The UN's Food and Agriculture Organization reports that "rural women in the developing world carry 80 tons or more of fuel, water and farm produce for a distance of 1 km during the course of a year. Men carry much less. . . ." Everywhere, women are paid less than men for comparable employment.

Western industrial—"developed"—society emerged directly from the agricultural tradition of the subordinate female who was not considered an important element in the economically active population, no matter how arduous or essential the domestic tasks assigned. With the growth of cities and industry in 19th-century America, for example, as women began to enter the workforce in increasing numbers, a "cult of true womanhood" developed as a reaction to the competitive pressures of the marketplace and factory floor. It held that women were morally superior to men; their role was private, not public. A woman's job was to rear children, attend church, and above all to keep a sober, virtuous, and cultured home, a place that offered the male breadwinner refuge, security, and privacy. This Victorian ideal fostered in America and much of Western Europe both social and economic discrimination against working women.

Only within the later 20th century, and then only in the more-developed countries, did that subordinate role pattern change. Women became increasingly economically active and placed themselves as never before in direct competition with men for similar occupations and wages. The feminist movement in modern industrialized societies was the direct response to the barriers that formerly restricted favored economic and legal positions to men. Even though

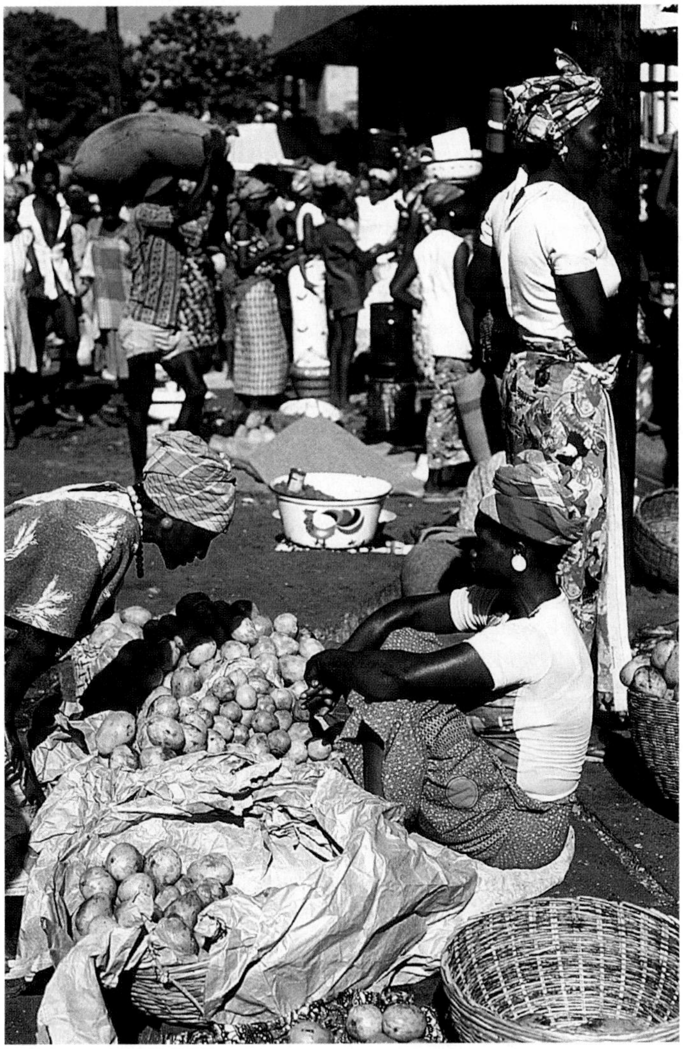

**FIGURE 7.41** Women dominate the once-a-week *periodic* markets in nearly all developing countries. There, they sell produce from their gardens or the family farm and frequently offer for sale processed goods to which their labor has added value: oil pressed from seeds or—in Niger, for example—from peanuts grown on their own fields; cooked, dried, or preserved foods; simple pottery, baskets, or decorated gourds. The market shown here is in the West African country of Ghana. More than half the economically active women in sub-Saharan Africa and southern Asia and about one-third in northern Africa and the rest of Asia are self-employed, working primarily in the informal sector. In the developed world, only about 14% of active women are self-employed. *© Liba Taylor/Corbis Images.*

women made up over 46% of the total employed labor force in the United States in the late 1990s, their representation in higher-paying, more-prestigious positions was much lower, despite increasing acceptance of the standard of equal pay for equal work. The elimination of remaining legal, social, and economic discriminations has been a primary objective of the North American "feminist revolution."

Such a revolution is much less likely or possible in strongly conservative and traditional economies and societies (see "Leveling the Field," p. 262). The present world pattern of gender-related institutional and economic role assignments is influenced not only by a country's level of economic development but also by the persistence of religious and customary restrictions its culture imposes on women and by the nature of its economic—particularly, agricultural—base. The first control is reflected in contrasts between the developed and developing world; the second and third are evidenced in variations within the developing world itself.

A distinct gender-specific regionalization has emerged. Among the Arab or Arab-influenced Muslim areas of western Asia and North Africa, the proportion of the female population that is economically active is low; religious tradition restricts women's acceptance in economic activities outside the home. The same cultural limitations do not apply under the different rural economic conditions of Muslims in southern and southeastern Asia, where labor force participation by women in Indonesia and Bangladesh, for example, is much higher than it is among the western Muslims. In Latin America, well known for a patriarchal social structure, women have been overcoming cultural restrictions on their growing employment outside the home.

Sub-Saharan Africa, highly diverse culturally and economically, in general is highly dependent on female farm labor and market income. The traditional role of strongly independent, property-owning women formerly encountered under traditional agricultural and village systems, however, has increasingly been replaced by subordination of women with modernization of agricultural techniques and introduction of formal, male-dominated financial and administrative agricultural institutions. For the developing world in general, a series of indicators has been combined to establish a "gender empowerment measure" and ranking (Figure 7.42); it clearly displays regional differentials in the position of women in different cultures and world areas.

Regional and cultural contrasts in gender relationships are also encountered in the advanced economies and the industrial components of developing countries. In modernizing Eastern Asian states, for example, women have yet to achieve the status they enjoy in most Western economies. In China, although it ranks high in gender empowerment (see Figure 7.42), women are generally not effectively competitive with men and are largely absent from the highest managerial and administrative levels; in Japan, males nearly exclusively run the huge industrial and political machinery of the country. In contrast, economic and social gender equality is more advanced in the Scandinavian countries than perhaps in any other portion of the industrialized world.

# Other Aspects of Diversity

Culture is the sum total of the way of life of a society. It is misleading to isolate, as we have done in this chapter, only a few elements of the technological, sociological, and ideological subsystems and imply by that isolation that

## Leveling the Field

The Fourth World Conference on Women held in Beijing during September 1995 produced the strongest and broadest declaration of women's rights and equality ever forthcoming from a series of international meetings on the subject beginning in Mexico City in 1975. Noting that its conclusions as adopted by delegates from 185 countries were in full conformity with the Charter of the United Nations and with international law, the Beijing Conference called on all governments to formulate strategies, programs, and laws designed to assure women of their full human rights to equality and development. Its final declaration detailed recommended policies in the areas of sexuality and childbearing, violence against women, discrimination against girls, female inheritance rights, and family protection. Its particular emphasis focused on efforts to "ensure women's equal access to economic resources including land, credit, science, and technology, vocational training, information, communication and markets as a means to further advancement and empowerment of women and girls." In the opinion of one United States delegate, "This is a document that guarantees the same rights for women that have been long enjoyed by men."

The conference did not reach its decisions and recommendations without spirited debate clearly rooted in the contrasting ways that different major cultures assign gender roles. Those contrasts, and the relationship between human rights and national customs they imply, were put in international perspective in the preamble to the conference platform: "While the significance of national and regional particularities and various historical, cultural and religious background must be borne in mind, it is the duty of states regardless of their political, economic and cultural systems to promote and protect all human rights and fundamental freedoms."

All the declarations of basic human rights of women articulated at the Beijing Conference were preserved at the "Beijing Plus Five" conference held at the United Nations in June 2000. In addition, delegates to that meeting agreed on principles calling for the punishment of domestic violence and for the criminalization of the cross-border trafficking of women and girls for sexual or wage slavery purposes. Again, reservations to many conference proposals were voiced by different delegations that typically split into fluctuating camps of conservative countries, largely Islamic and Roman Catholic and more secular states, including most of Europe and much of Latin America. Despite the formally adopted resolutions of "Beijing Plus Five," those strongly ingrained reservations suggest that the playing field remains uneven.

they are identifying characteristics differentiating culture groups. Economic developmental levels, language, religion, ethnicity, and gender all are important and common distinguishing culture traits, but they tell only a partial story. Other suggestive, though perhaps less pervasive, basic elements exist.

Architectural styles in public and private buildings are evocative of region of origin, even when they are indiscriminately mixed together in American cities. The Gothic and New England churches, the neoclassical bank, and the skyscraper office building suggest not only the functions they house but also the culturally and regionally variant design solutions that gave them form. The Spanish, Tudor, French Provincial, or ranch-style residence may not reveal the ethnic background of its American occupant, but it does constitute a culture statement of the area and the society from which it diffused.

Music, food, games, and other evidence of the joys of life, too, are cultural indicators associated with particular world or national areas. Music is an emotional form of communication found in all societies, but, being culturally patterned, it varies among them. Instruments, scales, and types of composition are technical forms of variants; the emotions aroused and the responses evoked among peoples to musical cues are learned behaviors. The Christian hymn means nothing emotionally to a pagan New Guinea clan. The music of a Chinese opera may be simply noise to the European ear. Where there is sufficient similarity between musical styles and instrumentation, blending (syncretism) and transferral may occur. American jazz represents a blend; calypso and flamenco music have been transferred to the Anglo American scene. Foods identified with other culture regions have similarly been transferred to become part of the culinary environment of the American "melting pot."

These are but a few additional minor statements of the variety and the intricate interrelationships of that human mosaic called culture. Individually and collectively, however, they are in their areal expressions and variations only part of the subject matter of the cultural geographer. Patterns and controls of spatial interaction and behavior, national and international political structures, economic activities and orientations, and levels and patterns of urbanization—all fundamental aspects of contemporary culture—are the subjects of the following chapters.

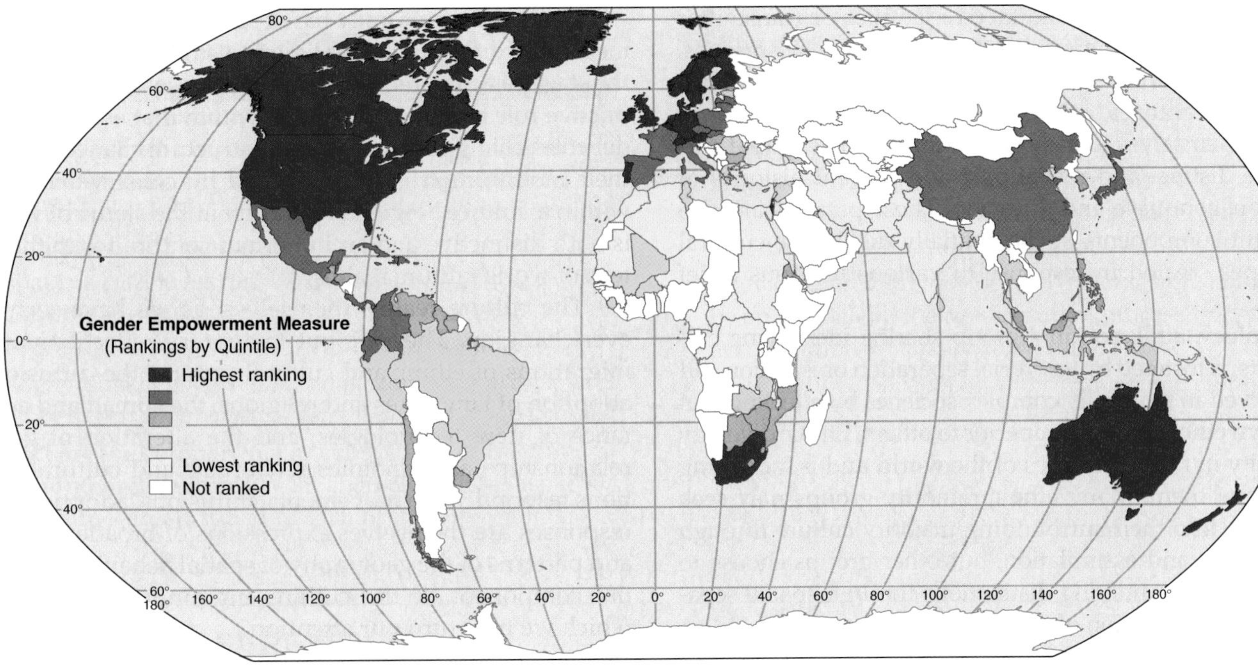


**FIGURE 7.42 The gender empowerment measure (GEM)** summarizes women's access to political and economic power based on three variables: female share of parliamentary seats; share of professional and technical jobs; and share of administrative and managerial positions. The UN's GEM rankings are heavily biased toward measures that technologically advanced, career-oriented Western cultures consider indicative of gender equality and progress; they do not consider the standards and values accepted in other cultural settings. As calculated, the GEM shows that gender equality in political, economic, and professional activities is not necessarily related to level of national wealth or development. According to this measure, some developing countries—China, for example, where women are afforded a large share of legislative seats and political administrative positions—outperform industrialized South Korea. Only 102 countries are ranked in this report from the end of the 20th century; in most, women are in a distinct minority in the exercise of economic power and decision making. Source: *Rankings from United Nations Development Programme.*

# SUMMARY

Culture is the learned behaviors and beliefs of distinctive groups of people. Culture traits, the smallest differentiating items of culture, are the building blocks of integrated culture complexes. Together, traits and complexes in their spatial patterns create human—"cultural"—landscapes, define culture regions, and distinguish culture groups. Those landscapes, regions, and group characteristics change through time as human societies interact with their environments, develop new solutions to collective needs, or are altered through innovations adopted from outside the group.

The detailed complexities of culture can be simplified by recognition of its component subsystems. The technological subsystem is composed of the material objects (artifacts) and techniques of livelihood. The sociological subsystem comprises the formal and informal institutions (sociofacts) that control the social organization of a culture group. The ideological subsystem consists of the ideas and beliefs (mentifacts) a culture expresses in speech and through belief systems.

The presumed cultural uniformity of a preagricultural world was lost as domestication of plants and animals in many world areas led to the emergence of culture hearths of wide-ranging innovation and to a cultural divergence between different groups. Although modern-day shared technologies contribute to cultural convergence throughout the world, many elements of cultural distinction remain to identify and separate social groups. Among the most prominent of the differentiating culture traits are language, religion, ethnicity, and gender.

Language and religion are both transmitters of culture and identifying traits of culture groups. Both have distinctive spatial patterns, reflecting past and present processes of interaction and change. Although languages can be grouped by origins and historical development, their world distributions depend as much on the movements of peoples and histories of conquest and colonization as they do on linguistic evolution. Toponymy, the study of place-names, helps document that history of movement. Linguistic geography studies spatial variations in languages, variations that may be minimized by encouragement of standard languages or overcome by pidgins, creoles, and lingua francas.

"For a brief 18-month period between April 1860 and October 1861, an undying blend of courage and endurance was created by the Pony Express riders. Racing through Nebraska, Wyoming, Utah, and Nevada, these horsemen carried letters written on lightweight paper (postage ranged from $2 to $10 an ounce) from one relay station to another. They covered the 1,966 miles from St. Joseph, Missouri, to Sacramento, California, in 11 days. The nature of the work is implied in the newspaper advertisements seeking riders who weighed less than 135 pounds, did not drink or carouse, and were 'daring young men—preferably orphans.'"[1] The twice-weekly trip took longer if riders had to avoid Indians or a herd of a million buffalo (Figure 8.1).

The $10 charge for an ounce of mail in 1860 is equivalent to $220 in today's money. Only such things as diamonds are as expensive to ship today. It is interesting to note that there are six nonstop flights leaving St. Louis every day for San Francisco (cities near St. Joseph and Sacramento, respectively), and an average one-way fare is about $200 per person (the average weight of a person with baggage is about 180 pounds). It takes approximately 4 hours for a plane to make the trip. The time differential averages out to one flight every 6 hours versus one pony express ride every 84 hours, and the cost for an ounce is 7 cents by plane versus 22,000 cents by pony express.

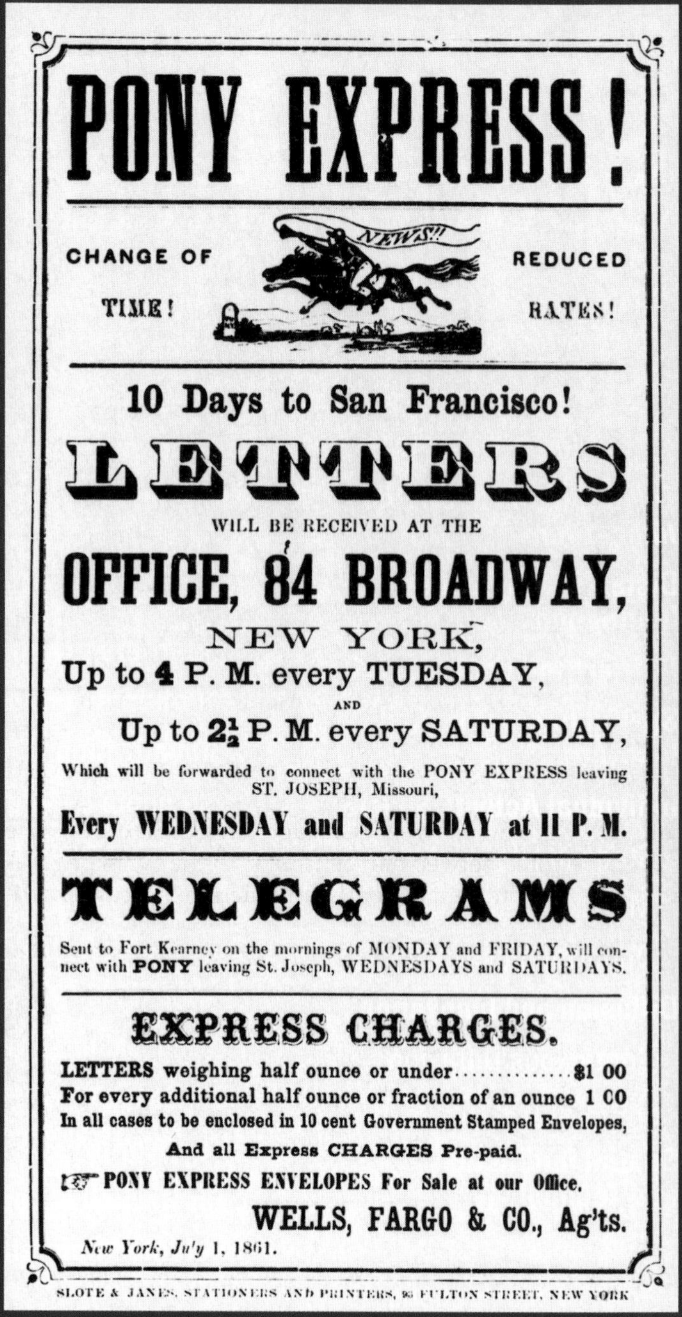


**FIGURE 8.1** **A poster advertising the services of the Pony Express, whose riders sped across the West on horses bred to run fast.** In operation from only 1860 to 1861, the Pony Express was rendered obsolete by railroads and the telegraph. In 1861, a telegraph wire strung from New York City to San Francisco reduced the time for communication between the coasts from days to seconds. *Courtesy of Pony Express Museum, St. Joseph, MO. © Bettmann/Corbis Images.*

This comparison shows unmistakably that the interaction between the Midwest part of the United States and the West has grown enormously over the past 144 years. The level of interaction is a function of the demand for travel, its speed, and its cost. All of these factors are conditioned by technology. In this case, the technology changed from fresh horses spaced at 16- to 24-kilometer (10- to 15-mi) intervals in the Pony Express days to jet planes. Today, with populations in each area in the millions and the cost of travel low, it is no wonder that six planes leave daily from each city for the other.

[1] *The Story of America*, The Reader's Digest Association, Pleasantville, New York, 1975, p. 199.

# THE DEFINITION OF *SPATIAL INTERACTION*

**Spatial interaction** is the term geographers use to represent the interdependence between geographic areas. Spatial interaction can be the movement of people between places, the flow of goods from one region to another, the diffusion of ideas from a center of knowledge to other areas, or the spread of a communicable disease from a group of people living in one area to those living in another area. What all of these examples have in common is that there is some sort of a flow over a distance separating people. Spatial interaction is the geographic counterpart of human interaction. The difference is that the location of those involved in the interaction can be clearly represented on maps.

If there is no one at a site (for example, on an iceberg), there can be no spatial interaction between the site and any other site. On the other hand, if there are a great number of people at one site—for example, Chicago—and a great number of people at another site—say, New York—there will be a great deal of spatial interaction between them. But if the second site is a city as far away as Tokyo is from Chicago, there will be fewer interactions between Chicago and Tokyo than between Chicago and New York. The attractive force between areas is akin to the force of gravity. Thus, the amount of spatial interaction is a function of the size of the interdependent populations and the distance between them.

# DISTANCE AND SPATIAL INTERACTION

Because people make many more short-distance trips than long ones, there is greater human interaction over short distances than long distances. This is the principle of **distance decay,** the decline of an activity, a function, or an amount of interaction with increasing distance from the point of origin. The tendency is for the frequency of trips to decrease very rapidly beyond an individual's **critical distance,** the distance beyond which cost, effort, means, and perception play an overriding role in our willingness to travel. Figure 8.2 illustrates this principle with regard to journeys from a homesite. The critical distance is different for each person. The variables of a person's age, mobility, and opportunity, together with an individual's interests and demands, help define how much and how far a person will travel. Because distance retards spatial interaction, we can say there is a *friction of distance* in our lives and activities.

A small child, for example, will make many trips up and down the block, but he or she will be inhibited by parental admonitions from crossing the street. Different but equally effective constraints control adult behavior. Daily or weekly shopping may be within the critical distance of an individual, and little thought may be given to the cost or effort involved. Shopping for special goods, however, is relegated to infrequent trips, and cost and effort are considered. The majority of our social contacts tend to be at a short distance within our own neighborhoods or with friends who live relatively close at hand; longer social trips to visit relatives are less frequent. In all such trips, however, the distance decay function is clearly at work.

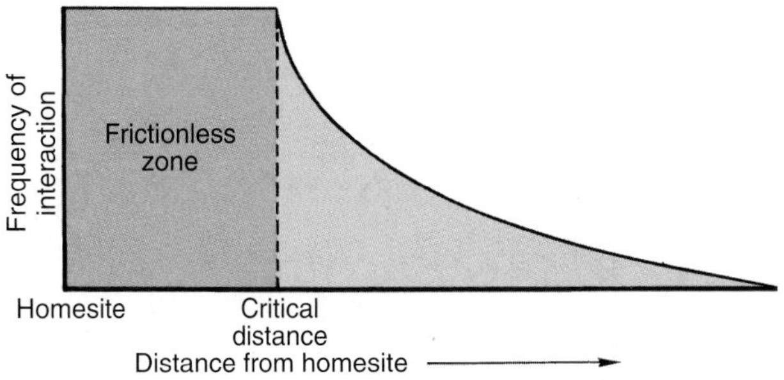


**FIGURE 8.2** This general diagram indicates how most people observe distance. For each activity, there is a distance beyond which the intensity of contact declines. This is called the *critical distance.* The distance up to the critical distance is a frictionless zone, in which time or distance considerations do not effectively figure into the trip decision.

Effort may be measured in terms of *time-distance*—that is, the time required to complete the trip. For the journey to work, time rather than cost often plays the major role in determining the critical distance. When significant differences between our cognition of distance and real distance are evident, we use the term *psychological distance* to describe our perception of distance. A number of studies show that people tend to psychologically consider known places as nearer than they really are and little-known places as farther than true distance. A humorous example of this is seen in Figure 8.3, a more serious one in Figure 8.4. Also, see "Mental Maps," p. 272.

We gain information about the world from many sources. Although information obtained from radio, television, the Internet, and newspapers is important to us, face-to-face contact is assumed to be the most effective means of communication. The distance decay principle implies that as the distance away from the home or workplace increases, the number of possible face-to-face contacts usually decreases. We expect more spatial interaction at short distances than at long distances. Where population densities are high, such as in cities (particularly central business districts during business hours), the spatial interaction between individuals can be at a very high level, which is one reason these centers of commerce are often also centers for the development of new ideas.

# BARRIERS TO INTERACTION

Recent changes in technology permit us to travel farther than ever before, with greater safety and speed, and to communicate without physical contact more easily and completely than previously possible. This intensification

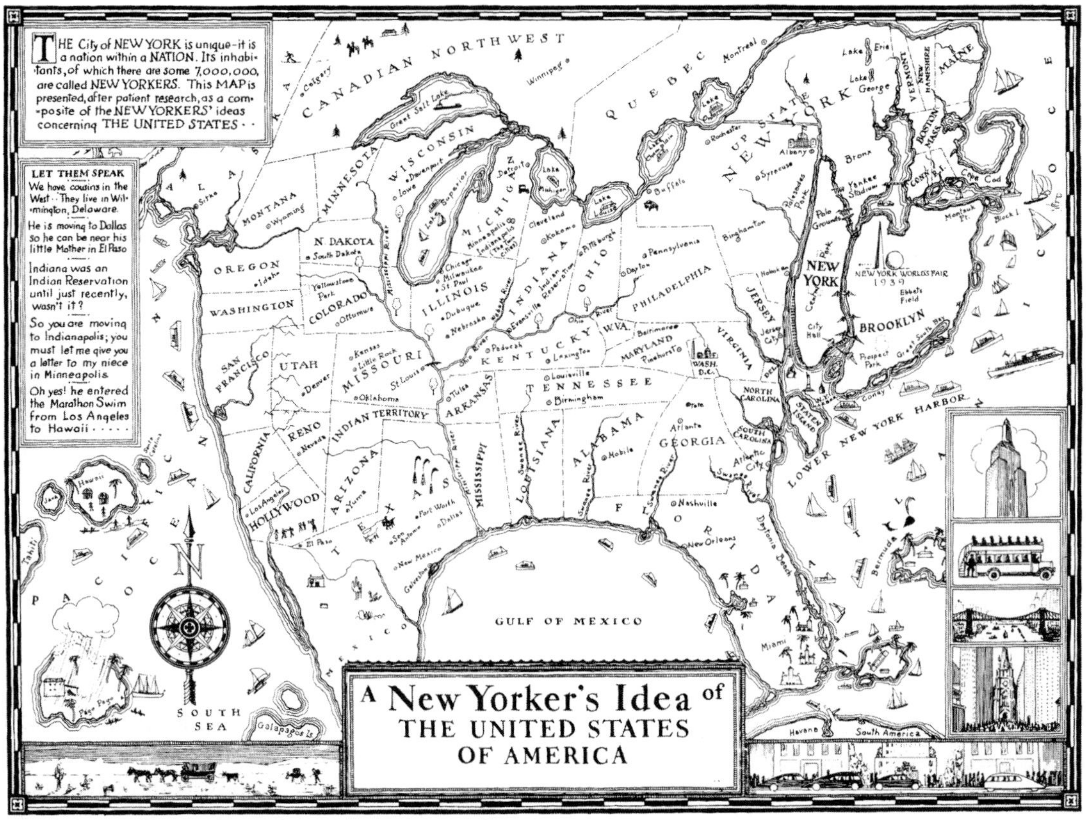


**F I G U R E 8.3** **Psychological distance: an artist's conception of a New Yorker's view of the United States.** *Copyright Florence Thierfeldt, Milwaukee, Wisconsin.*

of contact has resulted in an acceleration of innovation and in the rapid spread of goods and ideas. Several millennia ago, innovations such as the smelting of metals took hundreds of years to spread. Today, worldwide diffusion may be almost instantaneous.

The fact that the possible number of interactions is high, however, does not necessarily mean that the effective occurrence of interactions is high. That is, a number of barriers to interaction exist. Such barriers are any conditions that hinder either the flow of information or the movement of people and thus retard or prevent the acceptance of an innovation. Distance itself is a barrier to interaction. Generally, the farther two areas are from each other, the less likely is interaction. The concept of distance decay says that, all else being equal, the amount of interaction decreases as the distance between two areas increases.

Cost represents another barrier to interaction. Relatives, friends, and associates living long distances apart may find it difficult to afford to see one another. The frequency and time allocated to telephone and e-mail communication, relatively inexpensive forms of interaction, are very much a function of the location of friends and relatives—which, of course, favors short-distance interaction.

Interregional contact may also be hindered by the physical environment and by the cultural barriers of differing religions, languages, ideologies, genders, and political systems. Mountains and deserts, oceans and rivers can, and have, acted as physical barriers slowing or impeding interaction. Cultural barriers may be equally impenetrable. Those nearby who practice religions differently or speak another language may not be in touch with their neighbors. Governments (such as that of North Korea) that interfere with radio reception, control the flow of foreign literature, and discourage contact between their citizens and foreign nationals may overtly impede cultural contact.

In crowded areas, people commonly set psychological barriers around themselves, so that only a limited number of interactions take place. The barriers are raised in defense against information overload and for psychological well-being. We must have a sense of privacy in order to filter out information that does not directly concern us. As a result, we tend to reduce our interests to a narrow range when we find ourselves in crowded situations, allowing our wider interests to be satisfied by our use of the communications media.

## Spatial Interaction and Innovation

The probability that new ideas will be generated out of old ideas is a function of the number of available old ideas in contact with one another. People who specialize in a particular field of interest seek out others with whom they wish to interact. Crowded central cities are characteristically com-

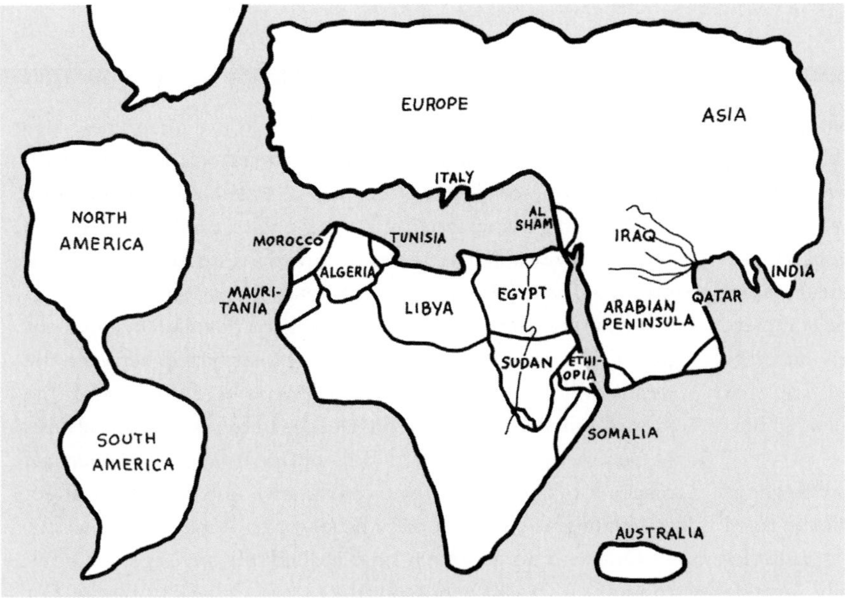


**FIGURE 8.4** **A mental map of the world.** This map was drawn by a Palestinian high school student from Gaza. The map reflects the secondary school education the author is receiving, which conforms to the Egyptian national school curriculum and thus is influenced by the importance of the Nile River and pan-Arabism. Al Sham is the old, but still used, name for the area including Syria, Lebanon, and Palestine. The map might be quite different if the Gaza school curriculum were designed by Palestinians or if an Israeli drew it.

posed of specialists in very narrow fields of interest. Consequently, under short-distance, high-density circumstances, the old ideas are given a hearing and new ideas are generated by the interaction. New inventions and new social movements usually arise in circumstances of high spatial interaction. An exception, of course, is the case of intensely traditional societies—Japan in the 17th and 18th centuries, for example—where the culture rejects innovation and clings steadfastly to customary ideas and methods.

The culture hearths of an earlier day (see Chapter 7) were the most densely settled, high-interaction centers of the world. At present, the great national and regional capital cities attract people who want or need to interact with others in special-interest fields. The association of population concentrations and the expression of human ingenuity have long been noted. The home addresses recorded for patent applicants by the U.S. Patent Office over the past century indicate that inventors were typically residents of major urban centers, presumably people in close contact and able to exchange ideas with those in shared fields of interest. It still appears that the metropolitan centers of the world attract those who are young and ambitious and that face-to-face or word-of-mouth contact is important in the creation of new ideas and products. The relatively recent revolution in communications, which allows for inexpensive interaction through a variety of phone and Internet services, has suggested to some that the traditional importance of cities as collectors of creative talent may decline in the future.

## Individual Activity Space

We will see in Chapter 9 that groups and countries draw boundaries around themselves to divide space into territories that are defended if necessary. The concept of **territoriality**—the emotional attachment to, and the defense of, home ground—has been seen by some as a root explanation of many human actions and responses. It is true that some collective activity appears to be governed by territorial defense responses: the conflict between street groups in claiming and protecting their "turf" (and the fear for their lives when venturing beyond it) and the sometimes violent rejection by ethnic urban neighborhoods of any different encroaching population group it considers threatening.

But for most of us, our personal sense of territoriality is a tempered one. We regard our homes and property as defensible private domains but open them to innocent visitors, known or unknown, or to those on private or official business. Nor do we confine our activities so exclusively within controlled home territories as street-gang members do within theirs. Rather, we have a more or less extended home range, an **activity space** or area within which we move freely on our rounds of regular activity, sharing that space with others who are also about their daily affairs.

Figure 8.5, p. 274 suggests probable activity spaces for a family of five for 1 day. Note that the activity space for each individual for 1 day is rather limited, even though two members of the family use automobiles. If 1 week's activity were shown, more paths would have to be added to the map, and in a year's time, several long trips would probably have to be noted. Long trips are usually taken irregularly.

The kind of activities individuals engage in can be classified according to type of trip: journeys to work, school, shopping center, recreational area, and so on. People in nearly all parts of the world make the same types of journeys, though the spatially variable requirements of culture, economy, and personal circumstances dictate their frequency, duration, and significance in the time budget of an individual.

# Mental Maps

Spatial interaction is affected by the way people perceive places. When information about a place is sketchy, blurred pictures develop. These influence the impression we have of places. Our willingness or ability to interact with what might be thought of as strange or unknown places cannot be discounted. We may say that each individual has a **mental map** of the world. No single person, of course, has a true and complete image of the world; therefore, having a completely accurate mental map is impossible. In fact, the best mental map that most individuals have is that of their own residential neighborhood, the place where they spend the most time.

Whenever individuals think about a place or how to get to a place, they produce a mental map. What are believed to be unnecessary details are left out, and only the important elements are incorporated. Those elements usually include awareness that the object or the destination does indeed exist, some conception of the distances separating the starting point and the named object(s), and a feeling for the directional relationships between points. A mental route map may also include reference points to be encountered on the chosen path of connection or on alternate lines of travel. Although mental maps are highly personalized, people with similar experiences tend to give similar answers to questions about the environment and to produce roughly comparable sketch maps.

Awareness of places is usually accompanied by opinions about them, but there is no necessary relationship between the depth of knowledge and the perceptions held. In general, the more familiar we are with a locale, the better is the factual basis of our mental image of it. But we form firm impressions of places totally unknown by us personally, and these may affect travel, migration decisions, or other forms of spatial interaction.

One way to ascertain how individuals envisage the environment is to ask them what they think of various places. For instance, they might be asked to rate places according to desirability—perhaps residential desirability—or to make a list of the best and worst places in a region such as the United

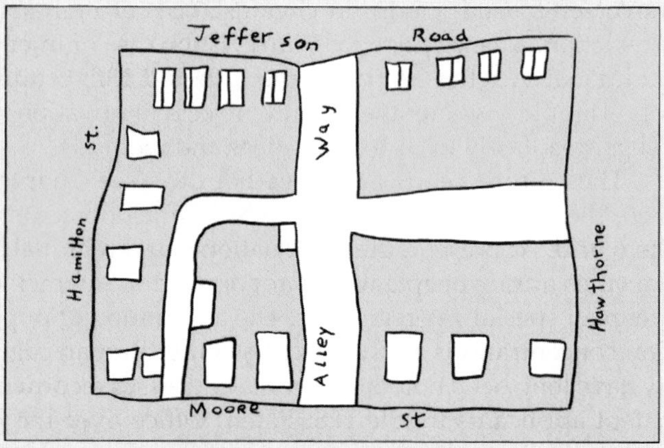


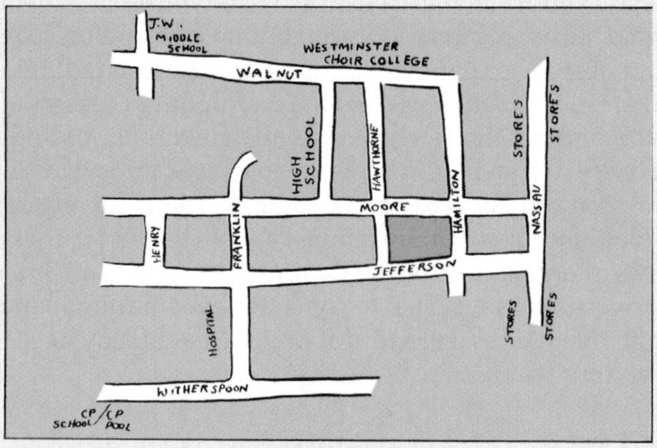


Three children, aged 6, 10, and 13, who lived in the same house, were asked to draw maps of their neighborhood. No further instructions were given. Notice how perspectives broaden and neighborhoods expand with age. For the 6-year-old, the neighborhood consisted of the houses on each side of her own. The square block on which she lived was the neighborhood for the 10-year-old. The wider activity space of the 13-year-old is also evident. The square block that the 10-year-old drew is shaded in the 13-year-old's sketch.

States. Certain regularities appear in such studies. The accompanying figure presents some residential desirability data elicited from college students in three provinces of Canada. These and comparable mental maps suggest that near places are preferred to far places unless much information is available about the far places. Places with similar cultural forms are preferred, as are places with high standards of living. Individuals tend to be indifferent to unfamiliar places and to dislike unfamiliar areas that have competing cultural interests (such as disliked political and military activities) or a physical environment known to be unpleasant.

On the other hand, places perceived to have superior climates or landscape amenities are rated highly in mental map studies and favored in tourism and migration decisions. The southern and southwestern coast of England is attractive to citizens of generally wet and cloudy Britain, and holiday tours to Spain, the south of France, and the Mediterranean islands are heavily booked by the English. A U.S. Census Bureau study indicates that "climate" is, after work and family proximity, the most often reported reason for interstate moves by adults of all ages. International studies reveal a similar migration motivation based not only on climate but also on concepts of natural beauty and amenities.

Each of these maps shows the residential preference of a sampled group of Canadians from the provinces of Quebec, Ontario, and British Columbia. Note that each group of respondents prefers its own area but that all like the Canadian and U.S. west coasts.

*Redrawn with permission from Herbert A. Whitney, "Preferred Locations in North America: Canadians, Clues, and Conjectures," in* Journal of Geography, *Vol. 83, No. 5, p. 222. Copyright © 1984 National Council for Geographic Education, Indiana, PA.*

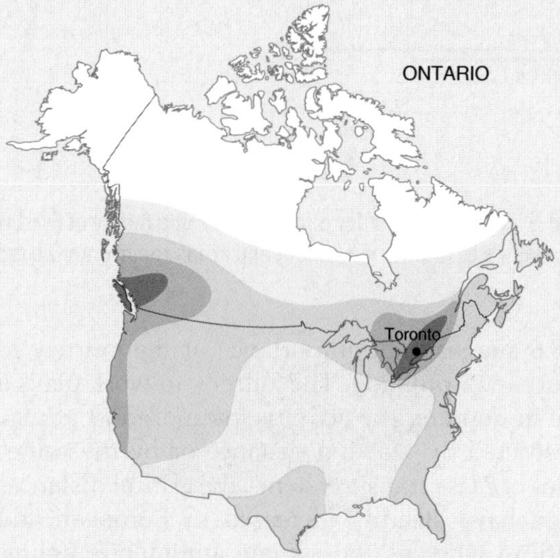

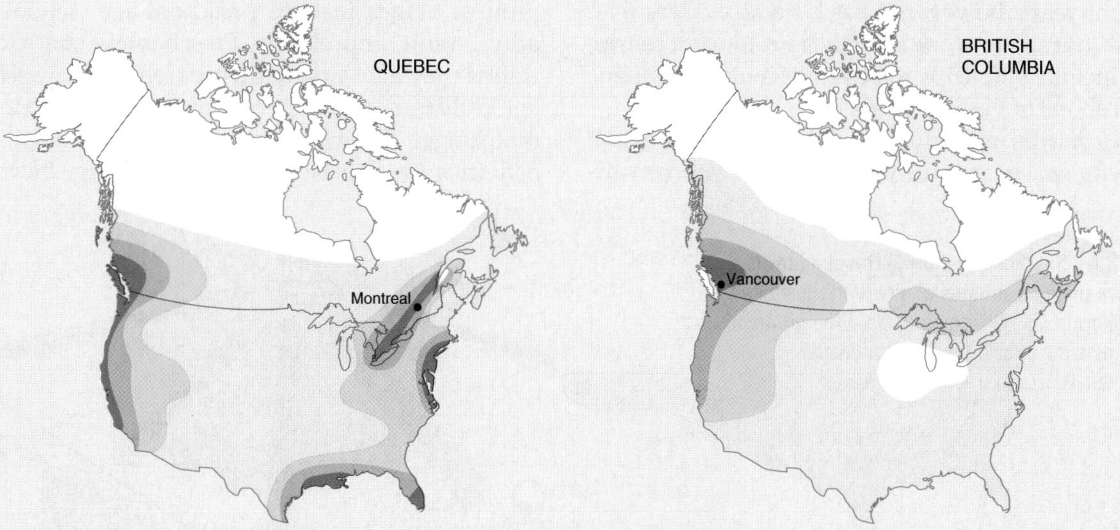

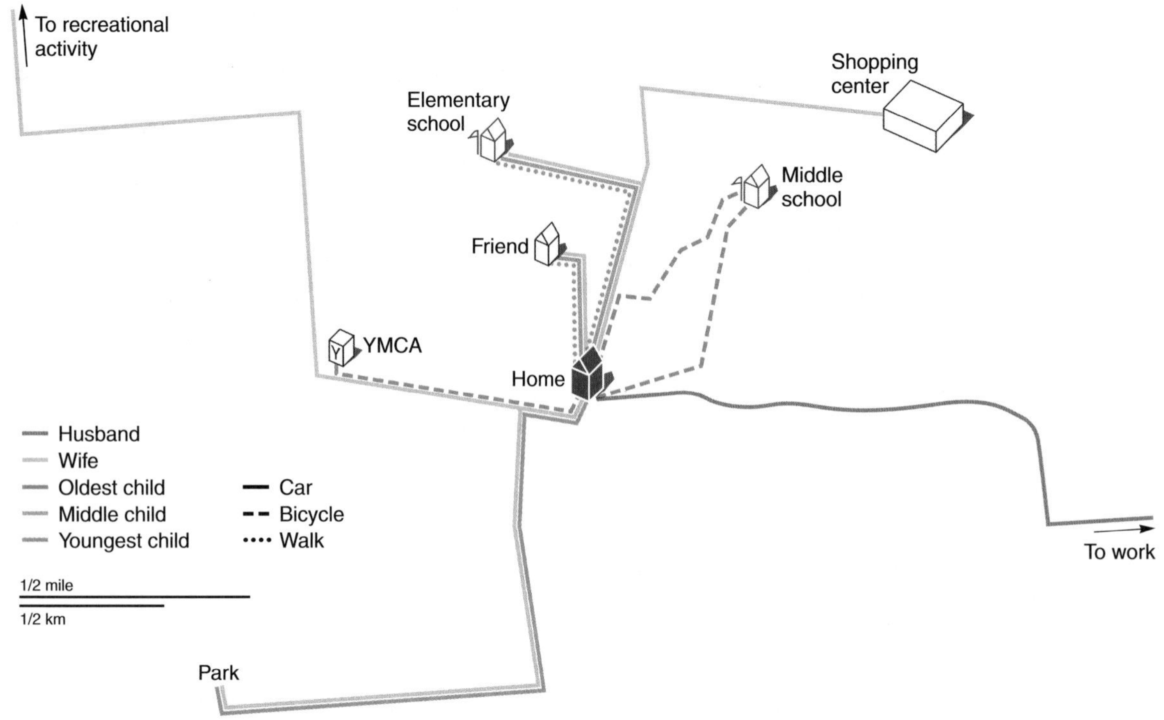


**FIGURE 8.5 Activity space for each member of a family of five for a typical weekday.** One parent commutes to work, while the other parent works at home. Routes of regular movement and areas recurrently visited help to foster a sense of territoriality and to affect one's perceptions of space.

Figure 8.6 suggests the importance of the journey to work in an urban population. The journey to work plays a decisive role in defining the activity space of most adults. Formerly restricted by walking distance or by the routes and schedules of mass transit systems, the critical distances of work trips have steadily increased in European and Anglo American cities as the private automobile figures more importantly in the movement of workers (Figure 8.7). In more recent years, however, it has become evident that, for many, the journey to work is really a multipurpose trip, which may include side trips to day-care centers, cleaners, schools, and shops of various kinds.

The types of trips individuals make, and thus the extent of their activity space, are partly determined by three variables: people's stage in life (age); the means of mobility at their command; and the opportunities implicit in their daily activities.

## Stage in Life

The first variable determining the types of trips individuals make, **stage in life,** refers to membership in specific age groups. Stages include preschool-age, school-age, young adult, adult, and elderly. Preschoolers stay close to home unless they accompany their parents or caregivers. School-age children usually travel short distances to lower schools and longer distances to upper-level schools. After-school activities tend to be limited to walking, bicycle, or auto-

**FIGURE 8.6 Southern California travel patterns.** The numbers are the percentages of all urban trips taken in southern California on a typical weekday. The greatest single movement is the journey to and from work.
Source: *Data from 1991 Association of Governments Survey.*

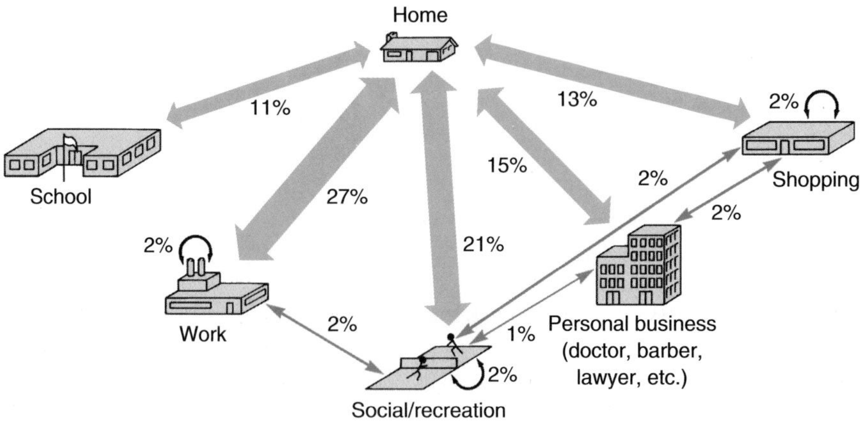

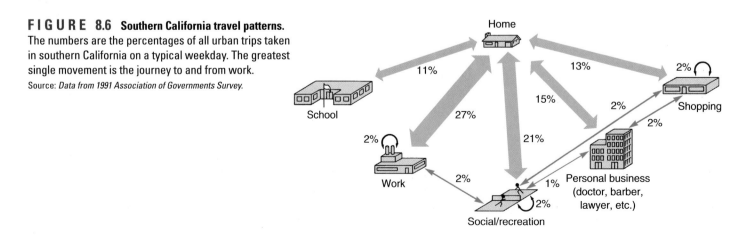

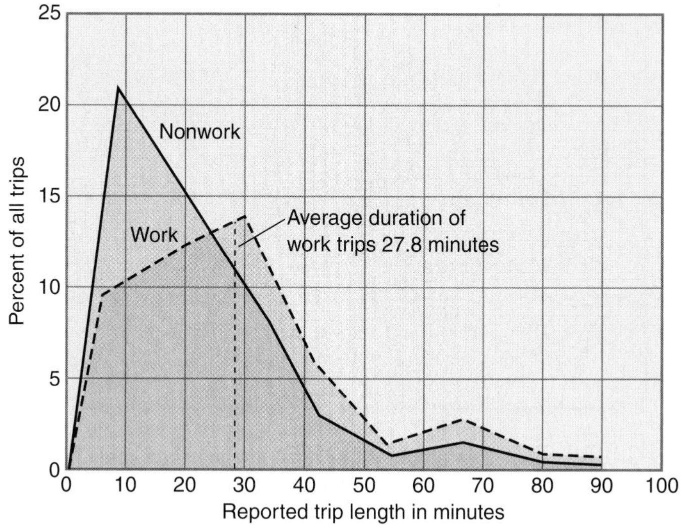


**FIGURE 8.7 The frequency distribution of work and nonwork trip lengths in minutes in Toronto.** Studies in various metropolitan areas support the conclusions documented by this graph: work trips are usually longer than other recurring journeys. In the United States in the early 1990s, the average work trip covered 17.1 kilometers (10.6 mi), and half of all trips to work took under 22 minutes; for suburbanites commuting to the central business district, the journey to work involved between 30 and 45 minutes. By 2000, increasing sprawl had lengthened average commuting distances and, because of growing traffic congestion, had increased the average work trip commuting time to 25 minutes; 15% of workers had commutes of more than 45 minutes. The situation is similar elsewhere; in the middle 1990s, the average British communting distance was 12.5 kilometers (7.8 mi). Most nonwork trips in all countries are relatively short. Source: *Maurice Yeates,* Metropolitan Toronto and Region Transportation Study, *figure 42. The Queen's Printer, Toronto: 1966.*

mobile trips provided by parents to nearby locations. High school students and other young adults are usually more mobile and take part in more activities than do younger children. They engage in more spatial interaction. Adults responsible for household duties make shopping trips and trips related to child care, as well as journeys away from home for social, cultural, and recreational purposes. Wage-earning adults usually travel farther from home than other family members. Elderly people may, through infirmity or interests, have less extensive activity spaces.

## Mobility

The second variable that affects the extent of activity space is *mobility,* or the ability to travel. An informal consideration of the cost and effort required to overcome the friction of distance is implicit. Where incomes are high, automobiles are available, and the cost of fuel is a minor item in the family budget, mobility may be great and individual action space can be large. In societies where cars are not a standard means of personal conveyance, the daily activity space may be limited to the shorter range afforded by bicycles or walking. Obviously, both intensity of purpose and the condition of the roadway affect the execution of movement decisions.

The mobility of individuals in countries or in sections of countries with high incomes is relatively great; people's activity space horizons are broad. These horizons, however, are not limitless. There is a fixed number of hours in a day, most of them consumed in performing work, preparing and eating food, and sleeping. In addition, there is a fixed number of road, rail, and air routes, so even the most mobile individuals are constrained in the amount of activity space they can use. No one can easily claim the world as his or her activity space. An example of this limitation is that of women living in suburban communities who must balance family obligations, such as preparing meals and caring for children, with their workforce activities. In this case, women's mobility is restricted; as a result, their occupational opportunities are limited.

## Opportunities

The third factor limiting activity space is the individual assessment of the availability of possible activities or *opportunities.* In the teeming cities of Asia, for example, the very poor satisfy their daily needs nearby; the impetus for journeys away from the residence is minimal. In impoverished countries and neighborhoods, low incomes limit the inducements, opportunities, destinations, and necessity of travel. Similarly, if one lives in a remote, sparsely settled area, with few or no roads, schools, factories, or stores, one's expectations and opportunities are limited, and activity space is therefore reduced. Opportunities plus mobility conditioned by life stage bear heavily on the amount of spatial interaction in which individuals engage.

# Diffusion and Innovation

As we noted in Chapter 7, **spatial diffusion** is the process by which a concept, practice, or substance spreads from its point of origin to new territories. A *concept* is an idea or invention, such as a new way of thinking—for example, deciding that shopping on the Internet is worth doing. A *practice* might be the actual process of shopping on the Internet. A *substance* is a tangible thing, such as the goods bought by means of the Internet. Diffusion is at the heart of the geography of spatial interaction.

Ideas generated in a center of activity will remain there unless some process is available for their spread. *Innovations,* the changes to a culture that result from the adoption of new ideas, spread in various ways. Some new inventions are so obviously advantageous that they are put to use quickly by those who can afford and profit from them. A new development in petroleum extraction may promise such material reward as to lead to its quick adoption by all major petroleum companies, irrespective of their distance from the point of introduction. The new strains of wheat and rice that significantly increased agricultural yields in much of the world and that were part of the Green Revolution (discussed in Chapter 10) were quickly made known to agronomists in all cereal-producing countries. However, they were more slowly taken up in poor countries, which could benefit most from them, partly because farmers had difficulty paying for them and spatial interaction was limited.

Many innovations are of little consequence by themselves, but sometimes the widespread adoption of seemingly inconsequential innovations bring about large changes when viewed over a period of time. A new musical tune, "adopted" by a few people, may lead many individuals to fancy that tune plus others of a similar sound. This, in turn, may have a bearing on dance routines, which may then bear on clothing selection, which in turn may affect retailers' advertising campaigns and consumers' patterns of expenditures. Eventually, a new cultural form is identified that may have an important impact on the thinking processes of the adopters and on those who come into contact with the adopters. Notice that a broad definition of *innovation* is used, but notice also that what is important is whether or not innovations are adopted.

In spatial terms, we can identify a number of processes for the diffusion of innovations. Each is based on the way innovations spread from person to person and, therefore, from place to place. These processes are discussed in the following sections, "Contagious Diffusion" and "Hierarchical Diffusion."

## Contagious Diffusion

Let us suppose that a scientist develops a gasoline additive that noticeably improves the performance of his or her car. Assume further that the person shows friends and associates the invention and that they, in turn, tell others. This process is similar to the spread of a contagious disease. The innovation will continue to diffuse until barriers (that is, people not interested in adopting the new idea) are met or until the area is saturated (that is, all available people have adopted the innovation). This **contagious diffusion** process follows the rules of distance decay spatial interaction at each step. Short-distance contacts are more likely than long-distance contacts, but over time the idea may have spread far from the original site. Figure 8.8 illustrates a theoretical contagious diffusion process.

A number of characteristics of this kind of diffusion are worth noting. If an idea has merit in the eyes of potential adopters, and they become adopters, the number of contacts of adopters with potential adopters will compound. Consequently, the innovation will spread slowly at first and then more and more rapidly, until saturation occurs or a barrier is reached. The incidence of adoption is represented by the S-shaped curve in Figure 8.9. The area in which those affected are located will be small at first, and then the area will enlarge at a faster and faster rate. The spreading process will slow as the available areas and/or people decrease. Figure 8.10 shows the contagious diffusion of a disease.

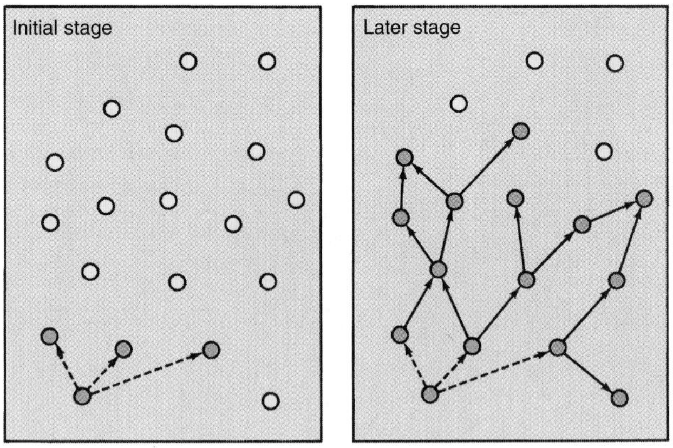


**FIGURE 8.8 Contagious diffusion.** A phenomenon spreads from one place to neighboring locations, but in the process it remains and is often intensified in the place of origin. *Reproduced by permission from* Resource Publications for College Geography, Spatial Diffusion, *by Peter R. Gould, page 4. Association of American Geographers, 1969.*

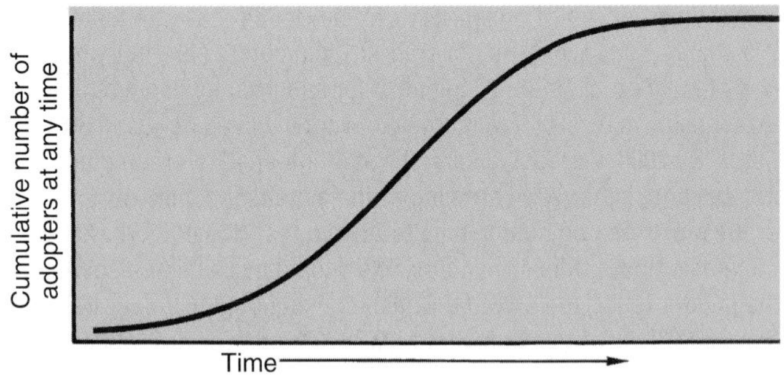


**FIGURE 8.9 The diffusion of innovations over time.** The number of adopters of an innovation rises at an increasing rate until the point at which about one-half the total who ultimately decide to adopt the innovation have made the decision. At this point, the number of adopters increases at a decreasing rate.

If an inventor's idea falls into the hands of a commercial distributor, the diffusion process may follow a somewhat different course than that just discussed. The distributor might "force" the idea into the minds of individuals by using the mass media. If the media were local in impact, such as newspapers, then the pattern of adoptions would be similar to that just described (Figure 8.11). If, however, a nationwide television, newspaper, or magazine advertising campaign were undertaken, the innovation would become known in numbers roughly corresponding to the population density. Where more people live, there would, of course, be more potential adopters. Economic or other barriers may also affect the diffusion. One immediately sees, however, why large television markets are so valuable and why national advertising is so expensive.

This type of contagious diffusion process may act together with the distance-decay process. Many of those who accept the innovation after learning of it in the mass media will tell others, so that a locally contagious effect will

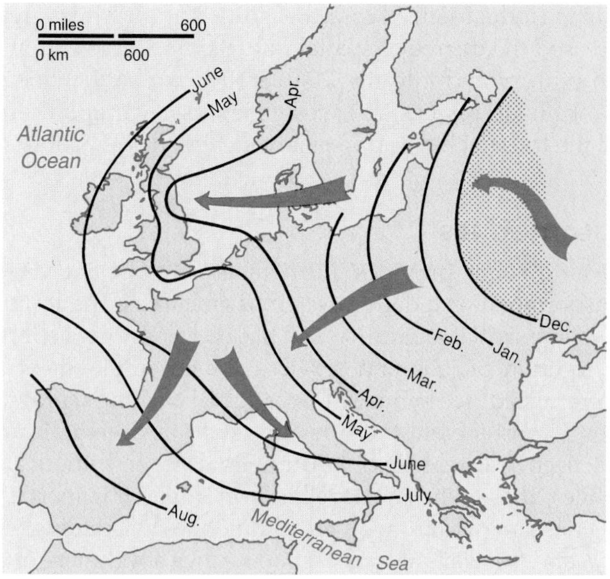


**FIGURE 8.10** The process of contagious diffusion is sensitive to both time and distance, as suggested by the diffusion of the European influenza pandemic of 1781. The flu began in Russia and moved westward, covering Europe in about 8 months. Source: *Based on Gerald F. Pyle and K. David Patterson,* Ecology of Disease *2, no. 3 (1984): 179.*

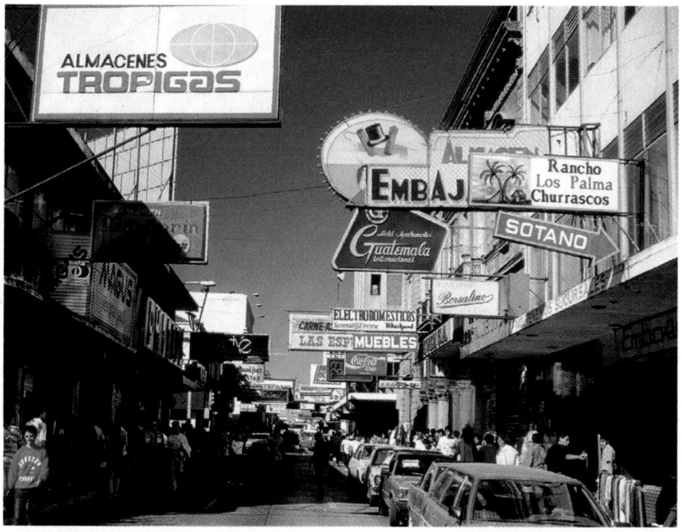


**FIGURE 8.11 A street scene in Guatemala City.** In modern society, advertising is a potent force for diffusion. Advertisements over radio and television, in newspapers and magazines, and on billboards and signs communicate information about many different products and innovations. *© Laurence Fordyce; Eye Ubiquitous/Corbis Images.*

begin to take over soon after the original contact is made. Each type of medium has its own level of effectiveness. Advertisers have found they must repeat messages time and again before they are accepted as important information. This fact says something about the effectiveness of the mass media as opposed to, say, face-to-face contact.

## Hierarchical Diffusion

A second way innovations are spread combines some aspects of contagious diffusion with the inclusion of a new element: a hierarchy. A *hierarchy* is a classification of objects into categories, so that categories are increasingly complex or have increasingly higher status. Hierarchies are found in many systems of organization, such as governmental offices (an organization chart), universities (instructors, professors, deans, and the president), and cities (villages, towns, regional centers, and metropolises). **Hierarchical diffusion** is the spread of innovation up or down a hierarchy of places.

As an example, let us suppose that a new way of automobile traffic control is adopted in a major city. Information on the innovation is spread, but only officials in comparably sized cities are in a position to accept the idea at first. It may be that the quality of information diffused to larger cities is better or that larger cities are more financially able to adopt the idea than smaller cities. Eventually, the innovation is adopted in smaller cities, and so on down the hierarchy, as it becomes better known or more financially feasible. A hypothetical scheme showing how a four-level hierarchy may be connected in the flow of information is presented in Figure 8.12. Note that the lowest-level centers are connected to higher-level centers but not to each other. Observe, too, that connections may bypass intermediate levels and link only with the highest-level center.

Many times, hierarchical diffusion takes place simultaneously with contagious diffusion. One might expect variations when the density of high-level centers is great and when distances between centers are short. A quick and inexpensive way to spread an idea is to communicate information about it at high-order hierarchical levels. Then the three types of diffusion processes can be used most effectively; even while an idea is diffusing through a high level in the hierarchy, it is also spreading outward from high-level centers. Consequently, low-level centers that are a short distance from high-level centers may be apprised of the innovation before more distant medium-level centers. People living in suburbs and small towns near a large city are privy to much that is new in the large city, as are individuals in other large cities half a continent away. Figures 8.13 and 8.14 show these patterns for a case taken from Japan.

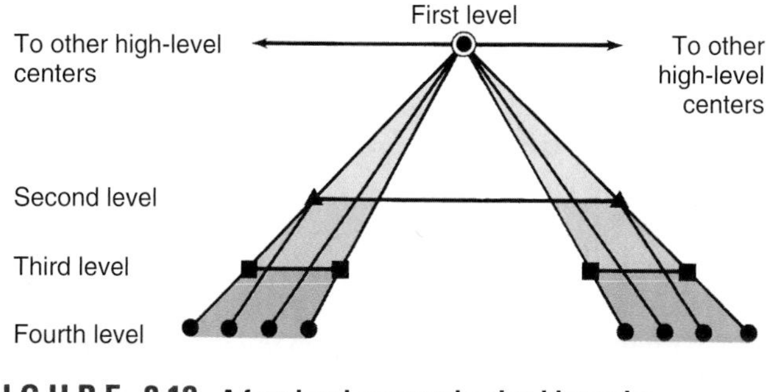


**FIGURE 8.12 A four-level communication hierarchy.**

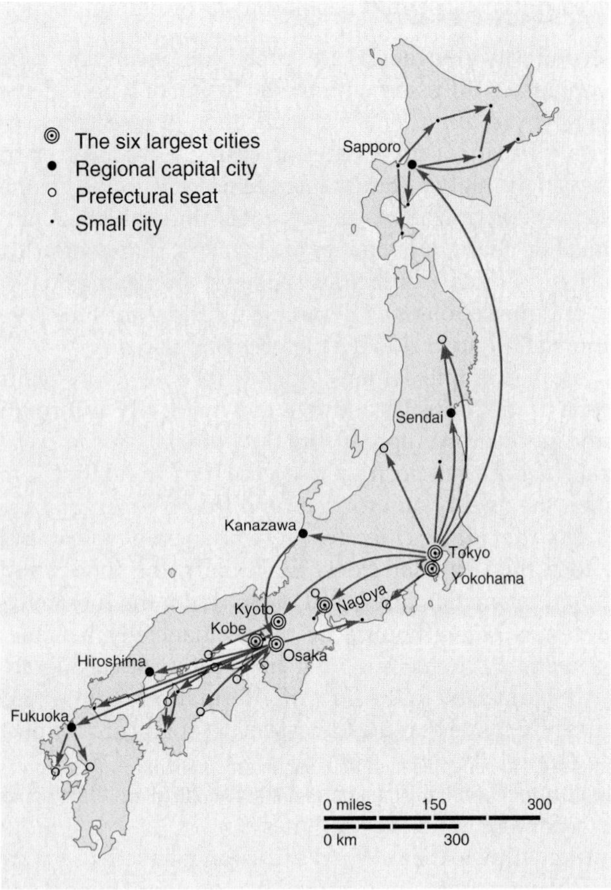


**FIGURE 8.13** Rotary Clubs, members of the international service association, were established in the large cities of Japan during the 1920s. New clubs were established under the sponsorship of the original clubs. This map shows both a hierarchical and a contagious pattern of diffusion. *Redrawn with permission from Yoshio Sugiura, "Diffusion of Rotary Clubs in Japan, 1920–1940: A Case of Non-Profit Motivated Innovation Diffusion under a Decentralized Decision-Making Structure," in* Economic Geography, *Vol. 62, No. 2, p. 128. Copyright © 1986 Clark University, Worcester, MA.*

These forms of diffusion operate in the spread of culture. The consequences are the spatial interaction and innovation already discussed (see pages 270 to 271). Also, recall from Chapter 7 that migration, invasions, selective cultural adoptions, and cultural transference aid the diffusion of innovation. These broader movements and exchanges represent interactions of people beyond their usual activity spaces (see "Documenting Diffusion," p. 279).

# Spatial Interaction and Technology

When opportunities for spatial interaction abound, spatial interaction becomes a major part of people's lives. Opportunities to interact are based not only on the monetary ability to engage in spatial interaction but also on the means of interaction. In the 20th century, the ability of average wage earners in the industrialized countries to own automobiles greatly increased the degree of spatial interaction. At the end of the 20th century and into the 21st century, we have witnessed how cellular phones, low-cost telephone communication, and the Internet have appreciably increased interaction.

## Automobiles

The automobile provides individuals with fast and flexible transportation on a daily basis. It has increased the ability to overcome spatial separation and has had a profound effect on the location of jobs and services. Much employment has decentralized to suburban locations, creating sprawling cities. Unfortunately, the side effects of this decentralization have been a decrease in opportunities for those without automobiles, those who must depend on public transportation. Societies have accommodated automobiles by building highways and freeways, as opposed to further developing public transportation systems. Those with automobiles are able to

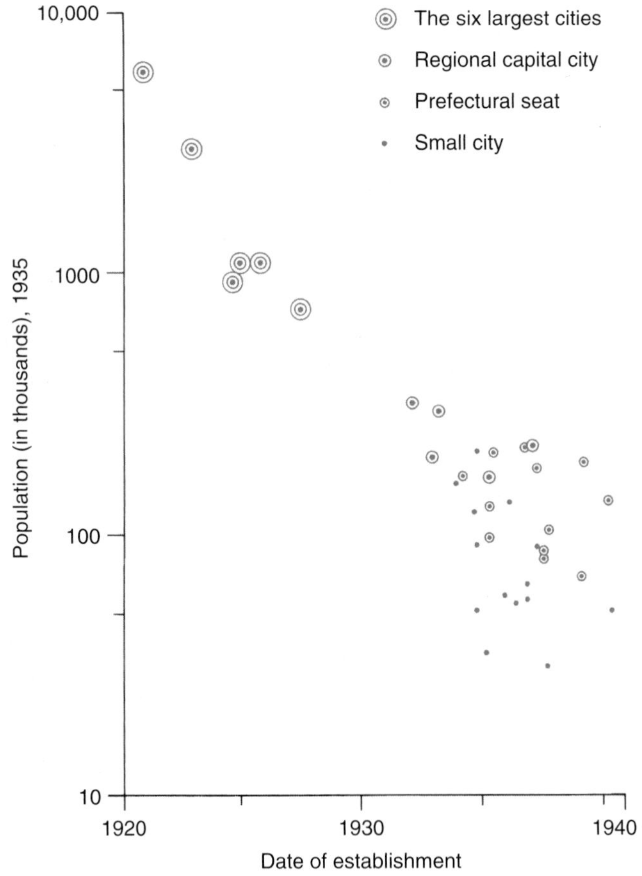


**FIGURE 8.14** This diagram shows the hierarchical diffusion component of the spread of Rotary Clubs in Japan. The largest cities were the first centers of Rotary Club activity, followed by cities at lower and lower levels of urban population and city function. *Redrawn with permission from Yoshio Sugiura, "Diffusion of Rotary Clubs in Japan, 1920–1940: A Case of Non-Profit Motivated Innovation Diffusion under a Decentralized Decision-Making Structure," in* Economic Geography, *Vol. 62, No. 2, p. 128. Copyright © 1986 Clark University, Worcester, MA.*

## Documenting Diffusion

The places of origin of many ideas, items, and technologies important in contemporary cultures are only dimly known or supposed, and their routes of diffusion are speculative at best. Gunpowder, printing, and spaghetti are presumed to be the products of Chinese inventiveness; the lateen sail has been traced to the Near Eastern culture world. The moldboard plow is ascribed to 6th-century Slavs of northeastern Europe. The sequence and routes of the diffusion of these innovations have not been documented.

In other cases, such documentation exists, and the process of diffusion is open to analysis. Clearly marked is the diffusion path of the custom of smoking tobacco, a practice that originated with Amerindians. Sir Walter Raleigh's Virginia colonists, returning home to England in 1586, introduced smoking in English court circles, and the habit very quickly spread among the general populace. England became the source region of the new custom for northern Europe; English medical students introduced smoking to Holland in 1590. Dutch and English together spread the habit to the Baltic and Scandinavian areas and overland through Germany to Russia. The innovation continued its eastward diffusion, and within 100 years, tobacco spread across Siberia and was, in the 1740s, reintroduced to the American continent at Alaska by Russian fur traders. A second route of diffusion for tobacco smoking can be traced from Spain through the Mediterranean area into Africa, the Near East, and Southeast Asia.

In more recent times, hybrid corn was originally adopted by imaginative farmers of northern Illinois and eastern Iowa in the mid-1930s. By the late 1930s and early 1940s, the new seeds were being planted as far east as Ohio and north to Minnesota, Wisconsin, and northern Michigan. By the late 1940s, all commercial corn-growing districts of the United States and southern Canada were cultivating hybrid varieties.

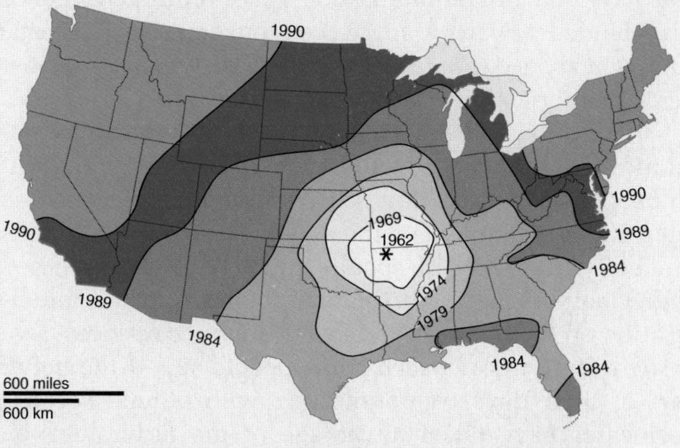


Redrawn from data in Thomas O. Graff and Dub Ashton, "Spatial Diffusion of Wal-Mart," *The Professional Geographer,* 46, No. 1, pp. 19–29. Association of American Geographers, 1994. Reprinted by permission of Blackwell Publishers.

A similar pattern of contagious diffusion marked the expansion of the Wal-Mart stores chain. From its origin in northwest Arkansas in 1962, the discount chain had dispersed throughout the United States by the 1990s to become the country's largest business in sales volume. In its expansion, Wal-Mart displayed a "reverse hierarchical" diffusion, initially spreading by way of small towns before opening its first stores in larger cities and metropolitan areas.

commute, shop, see friends and family, and engage in group activities with little inconvenience. As automobiles become more comfortable and high technology oriented, they further encourage people to seek more opportunities for interaction. Unless local governments control urban growth and development, the result is a sprawling urban environment where people appear to be constantly on the move from one place to another. This process that increases spatial interaction has been ongoing since the automobile became affordable to large numbers of people.

### Telecommunications

For information flows, space has a different meaning than it does for the movement of people or commodities. Communication, for example, does not necessarily imply the time-consuming physical relocation of things (though in the case of letters and print media, it usually does). Indeed, in modern telecommunications, the process of information flow can be instantaneous regardless of distance. The result is space-time convergence to the point of the obliteration of space. A Bell System report tells us that in 1920, putting through a transcontinental telephone call took 14 minutes and eight operators and cost more than $15.00 for a 3-minute call. By 1940, the call completion time had been reduced to less than $1^1/_2$ minutes and the cost had fallen to $4.00. In the 1960s, direct distance dialing allowed a transcontinental connection in less than 30 seconds, and electronic switching has now reduced the completion time to that involved in dialing a number and answering a phone. The price of long-distance conversation essentially disappeared with the advent of voice communication over the Internet in the late 1990s.

The Internet and communication satellites have made worldwide personal and mass communication immediate and data transfers instantaneous. The same technologies that have led to communication space-time convergence have tended toward a space-cost convergence. Domestic mail, which once charged a distance-based postage, is now carried nationwide or across town for the same price.

It is conceivable that the current revolution in telecommunications technology will have a more profound effect on people's lives, and thus social and industrial structure, than the automobile. For those with telecommunications capability, the level of spatial interaction has increased appreciably (Figure 8.15). Cellular phones, e-mail, communication on the Internet, and low-cost phone services have created lifestyles in which some people spend the better part of their days communicating with others. Because businesses have taken advantage of the technology by offering goods and services online, the number of shopping trips of all sorts has declined. In addition, a number of individuals make a living in a telecommuting environment. That is, they conduct business on the Internet and therefore do not take part in the morning and evening journey to and from work. The implication is that many people will find it less compelling to live in a crowded urban environment, so that sprawl is likely to increase at an accelerated pace.

Many types of industries will become footloose; that is, they won't need to be tied spatially to other industries or an urban environment. Most likely, it will be low-cost wage locations that will prosper. This is due to the fact that if industrialists and service providers have their choice of any location, they will seek one where wages are low, skills are high, and amenities, such as the existence of a warm tropical environment, are plentiful.

While automobiles foster long-distance commuting to work, telecommuting reduces the need to commute at all. Both, however, encourage urban sprawl. Because the telecommunications revolution began in the mid-1990s and is still in its infancy, we are not yet in a position to predict how it will affect all aspects of our lifestyles.

**FIGURE 8.15 The Internet.** This map was created by sending test data from a computer in Murray Hill, New Jersey, to each of the approximately 168,000 autonomous networks registered in the Internet routing databases kept by Merit Network, Inc., and other authoritative sources and graphing the resulting connections. The data were collected on June 1, 2001, and include more than 135 countries. The colors show the Internet domains where network switches (routers) were registered. Source: *Map design by Gregory Staple and Irena Slage © 2001 Peacock Maps, Inc. and Lumeta Corporation.*

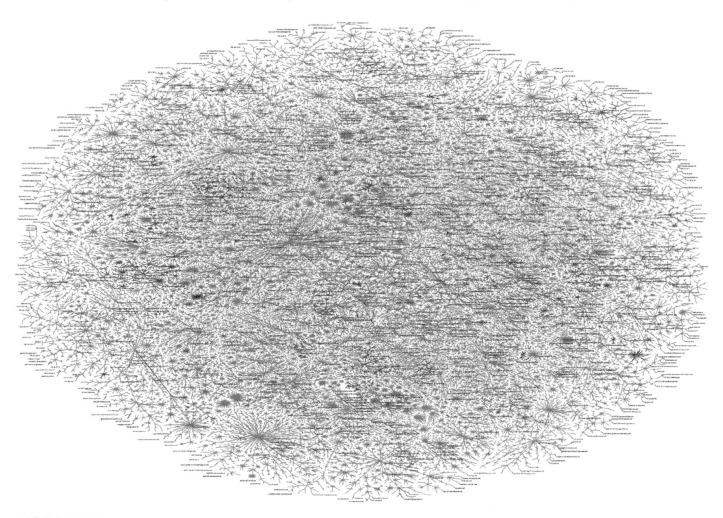

# MIGRATION

An important aspect of human history has been the **migration** of peoples—that is, the permanent relocation of both place of residence and activity space. It has contributed to the evolution of separate cultures, to the relocation diffusion of those cultures, and to the complex mix of peoples and cultures found in various parts of the world. The settlement of North America, Australia, and New Zealand involved great long-distance movements of peoples. The flight of refugees from past and recent wars, the settlement of Jews in Israel, the current migration of workers to the United States from Mexico and Central America, and innumerable other examples of mass movement come quickly to mind. In all cases, societies transplanted their cultures to the new areas, their cultures therefore diffused and intermixed, and history was altered.

Massive movements of people within countries, across national borders, and between continents have emerged as a pressing concern of recent decades. They affect national economic structures; determine population density and distribution patterns; alter traditional ethnic, linguistic, and religious mixtures; and inflame national debates and international tensions. Migration patterns and conflicts touch many aspects of social and economic relations and have become an important part of current geographic realities. Our interest in this chapter is with migration as an unmistakable, recurring, and near-universal expression of human spatial behavior.

## Types of Migration

Migration flows can be discussed at different scales, from massive intercontinental torrents to individual decisions to move to a new house or apartment within the same metropolitan area. At each level, although the underlying controls on spatial behavior remain constant, the immediate motivating factors influencing the spatial interaction are different, with differing impacts on population patterns and cultural landscapes.

Naturally, the length of a move and its degree of disruption of people's lives raise distinctions important in the study of migration. A change of residence from the central city to the suburbs certainly changes the activity space of schoolchildren and of adults in many of their nonworking activities, but the workers may still retain the city—indeed, the same place of employment there—as an activity space. On the other hand, migration from Europe to the United States and the massive farm-to-city movements of rural Americans late in the 19th and early 20th centuries meant a total change of all aspects of behavioral patterns.

At the broadest scale, *intercontinental* movements range from the earliest peopling of the habitable world to the most recent flight of Asian and African refugees to countries of Europe and the Western Hemisphere. The population structure of the United States, Canada, Australia and New Zealand, Argentina, Brazil, and other South American countries is a reflection and result of massive intercontinental flows of immigrants that began as a trickle during the 16th and 17th centuries and reached a flood during the 19th and early 20th centuries. Later in the 20th century, World War II (1939–1945) and its immediate aftermath involved more than 25 million permanent population relocations, all of them international but not all intercontinental.

*Intracontinental* and *interregional* migrations involve movements between countries and within countries, most commonly in response to individual and group assessments of improved economic prospects, but often reflecting flight from difficult or dangerous environmental, military, economic, or political conditions. The millions of refugees leaving their homelands following the dissolution of Eastern European communist states, including the former USSR and Yugoslavia, exemplify that kind of flight. Between 1980 and 2000, Europe received some 20 million newcomers, often refugees, who joined the 15 million labor migrants ("guest workers") already in Western European countries by the early 1990s. About 175 million people—3% of the world population—live in a country other than the country of their birth in the early 2000s, and migration had become a world social, economic, and political issue of first priority.

Migrations may be forced or voluntary, or, in many instances, reluctant relocations imposed on the migrants by circumstances. In *forced* or *involuntary migrations*, the relocation decision is made solely by people other than the migrants themselves (Figure 8.16). Between 10 and 12 million Africans were forcibly transferred as slaves to the Western Hemisphere from the late 16th to the early 19th centuries. Half or more were destined for the Caribbean and most of the remainder for Central and South America, though nearly a million arrived in the United States. Australia owed its earliest European settlement to convicts transported after the 1780s to the British penal colony established in southeastern Australia (New South Wales). More recent involuntary migrants include millions of Soviet citizens forcibly relocated from countryside to cities and from the western areas to labor camps in Siberia and the Russian Far East beginning in the late 1920s.

Less than fully voluntary relocation—*reluctant migration*—of some 8 million Indonesians has taken place under an aggressive governmental campaign begun in 1969 to move people from densely settled Java to other islands and territories of the country in what has been called the "biggest colonization program in history." International refugees from war and political turmoil or repression numbered some 15 million in 2003, according to the World Refugee Survey—1 out of every 415 people on the planet. In the past, refugees sought asylum mainly in Europe and other developed areas. More recently, the flight of people is primarily from developing countries to other developing regions, and many countries with the largest refugee populations are among the world's poorest. Between 1990 and 2003, Iran became host to about 2.4 million Afghans and an equal number of Iraqis fleeing repression, persecution, and war. Sub-Saharan Africa alone houses more than 3 million refugees (Figure 8.17).

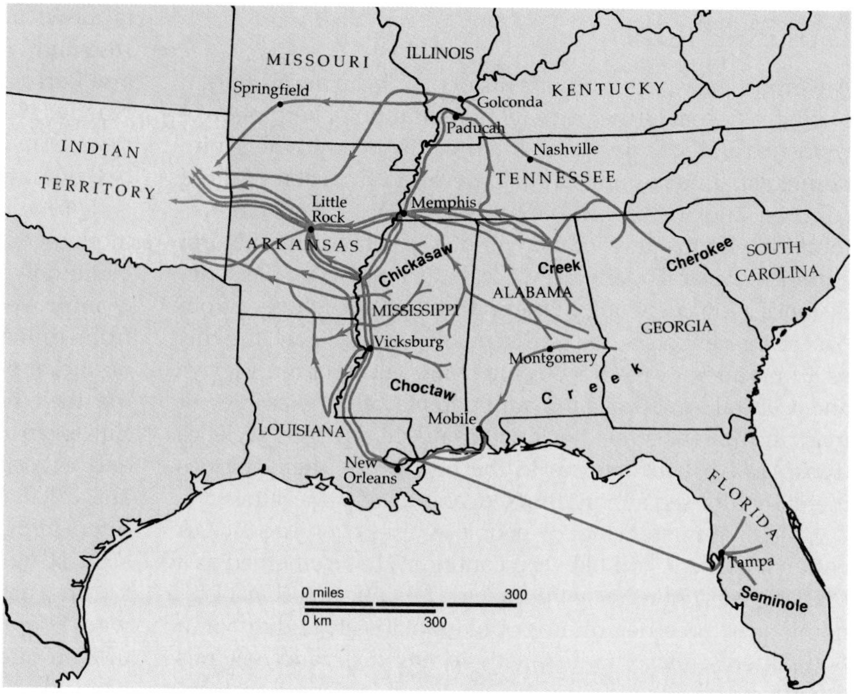


**FIGURE 8.16 Forced migrations: the Five Civilized Tribes.** Between 1825 and 1840, some 100,000 southeastern Amerindians were removed from their homelands and transferred by the army across the Mississippi River to "Indian Territory" in present-day Oklahoma. By far, the largest number were members of the Five Civilized Tribes of the South: Cherokees, Choctaws, Chickasaws, Creeks, and Seminoles. Settled, Christianized, literate small-farmers, their forced eviction and arduous journey—particularly along what the Cherokees named their "Trail of Tears" in the harsh winter of 1837–1838—resulted in much suffering and death.

Worldwide, an additional 22 million people were "internally displaced," effectively internal refugees within their own countries. In a search for security or sustenance, they have left their home areas but not crossed an international boundary.

The great majority of migratory movements, however, are *voluntary*, representing individual response to the factors influencing all spatial interaction decisions. At root, migrations take place because the migrants believe that their opportunities and lives will be better at their destination than they are at their present location.

## Incentives to Migrate

The decision to move is a cultural and temporal variable. Nomads fleeing famine and spreading deserts in the Sahel of Africa obviously are motivated by considerations different from those of an executive receiving a job transfer to Chicago, a resident of Appalachia seeking factory employment in the city, or a retired couple searching for sun and sand. In general, people who voluntarily decide to migrate are seeking better economic, political, or cultural conditions or certain amenities. For many, the reasons for migration are frequently a combination of several of these categories.

Negative home conditions that impel the decision to migrate are called **push factors.** They include loss of job, lack of professional opportunity, overcrowding, and a variety of other influences, including poverty, war, and famine. The presumed positive attractions of the migration destination are known as **pull factors.** They include all the attractive attributes perceived to exist at the new location—safety and food, perhaps, or job opportunities, better climate, lower taxes, more room, and so forth. Very often, migration is a result of both perceived push and pull factors (Figure 8.18). It is the perception of the opportunities and want satisfaction that is important, whether or not the perception is supported by objective reality.

*Economic considerations* have impelled more migrations than any other single incentive. If migrants face unsatisfactory conditions at home (e.g., unemployment or famine) and believe that the economic opportunities are better elsewhere, they will be attracted to the thought of moving. Poverty is the great motivator. Some 30% of the world's population—nearly 2 billion people—have less than $1.00 per day income. Many additionally are victims of drought, floods, and other natural catastrophes, or of wars and terrorism. Poverty in developing countries is greatest in the countryside; rural areas are home to around 750 million of the world's poorest people. Of these, 20 to 30 million move each year to towns and cities—many as "environmental refugees," abandoning land so eroded or exhausted it can no longer support them. In the cities, they join the 40% or more of the labor force that is unemployed or underemployed in their home country and seek legal or illegal entry into the more promising economies of the developed world. All, rural or urban, respond to the same basic forces: the push of poverty and the pull of perceived or hoped-for opportunity.

The desire to escape war and persecution at home and to pursue the promise of freedom in a new location is a *political incentive* for migration. Americans are familiar with the history of settlers who migrated to North America seeking religious and political freedom (Figure 8.19). In more recent times, the United States has received hundreds of thousands of refugees from countries such as Hungary, following the uprising of 1956; Cuba, after its takeover by Fidel Castro; and Vietnam, after the fall of South Vietnam. The massive

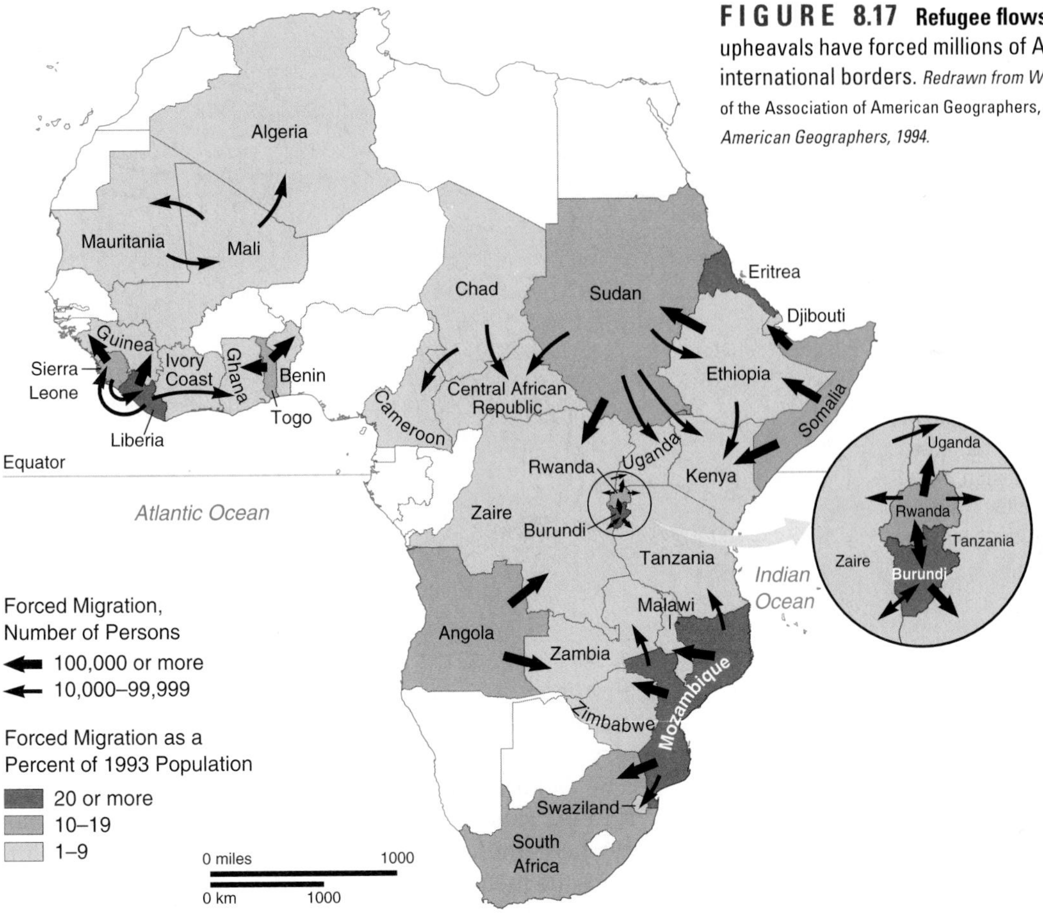


**FIGURE 8.17 Refugee flows in Africa.** In recent years, political upheavals have forced millions of Africans from their homes and across international borders. *Redrawn from William B. Wood, "Forced Migration,"* The Annals of the Association of American Geographers, *84, no. 4, Figure 3, p. 619. Association of American Geographers, 1994.*

movements of Hindus and Muslims across the Indian subcontinent in 1947, when Pakistan and India were established as governing entities, and the exodus of Jews fleeing persecution in Nazi Germany in the 1930s are other examples of politically inspired moves. More recently, about 1 million Hutus fled into neighboring African countries after ethnic Tutsis took over Rwanda's government; ethnic Muslims living in Bosnia were pushed out of their ancestral homes by Serbs; and many Haitians, under severe economic privation during a political crisis, have left for the United States.

Migration normally but not always involves a hierarchy of decisions. Once people have decided to move and have selected a general destination (e.g., America or the Sunbelt), they must still choose a particular site in which to settle. At this scale, *cultural variables* can be important pull factors. Migrants tend to be attracted to areas where the language, religion, and racial or ethnic background of the inhabitants are similar to their own. This similarity can help migrants feel at home when they arrive at their destination, and it may make it easier for them to find a job and to become assimilated into the new culture. The Chinatowns and Little Italys of large cities attest to the drawing power of cultural factors, as discussed in Chapter 7.

Another set of inducements is grouped under the heading *amenities,* the particularly attractive or agreeable features

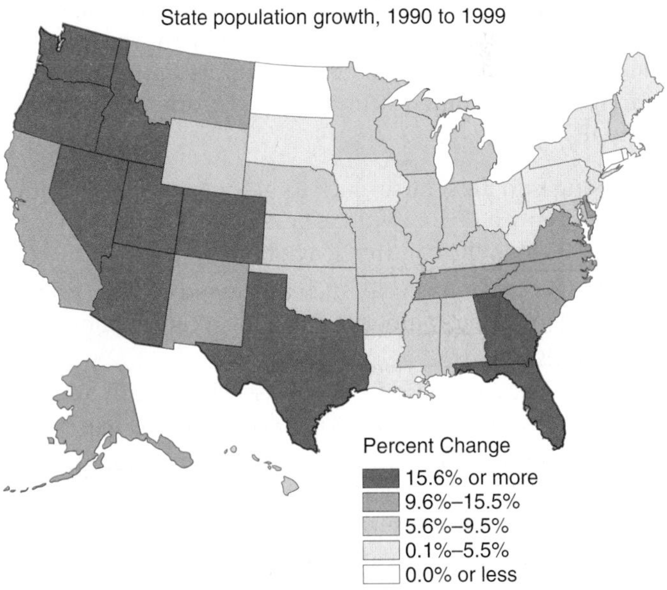


**FIGURE 8.18 Migration in the United States.** Although birth and death rates have a strong bearing on population growth, the growth pattern in the United States from 1990 to 1999 was largely a result of net migration to the West, Southwest, and Southeast. The *pull factors* included job opportunities and mild climates, the *push factors* included loss of job opportunities and harsh winters. Source: *Population Reference Bureau.*

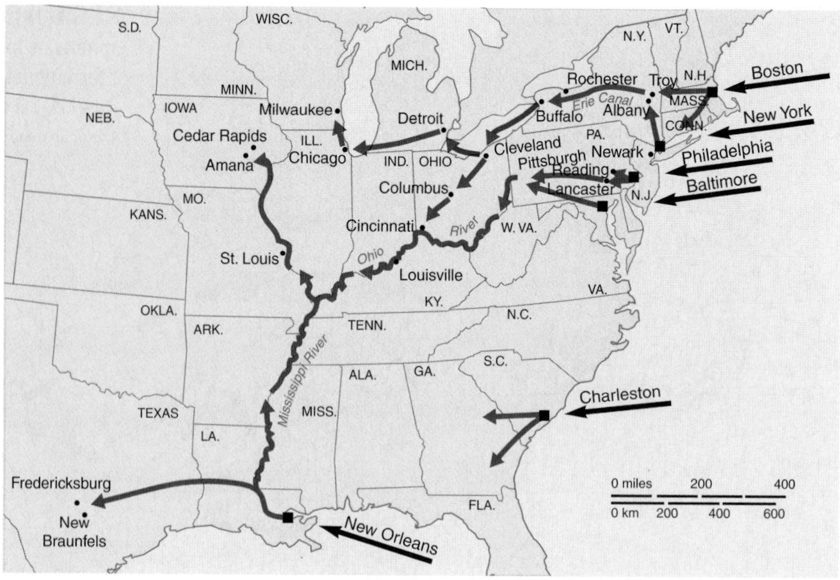


**FIGURE 8.19 The major paths of the early migration of Germans to America.** Most emigrants left Germany because of religious and political persecution. They chose the United States not only because immigrants were made welcome but also because labor was in demand and farmland was available. The first immigrants landed, and many settled, in Boston, New York, Philadelphia, Baltimore, Charleston, and New Orleans. The migrants carried with them such aspects of their culture as religion, language, and food preferences.

that are characteristic of a place. Amenities may be natural (mountains, oceans, climate, and the like) or cultural (e.g., the arts and music opportunities available in large cities). They are particularly important to relatively affluent people seeking "the good life." Amenities help to account for the attractiveness of the so-called Sunbelt states in the United States for retirees; a similar movement to the southern coast has also been observed in countries such as the United Kingdom and France.

The significance of the various incentives varies according to the age, sex, education, and economic status of the migrants (see "Gender and Migration," p. 285). For the modern American, reasons to migrate have been summarized into a limited number of categories that are not mutually exclusive:

1. changes in life course, such as getting married, having children, getting a divorce, or needing less dwelling space when the children leave home;
2. changes in career cycle, such as leaving college, getting a first job or a promotion, receiving a career transfer, or retiring;
3. forced or reluctant migrations associated with urban development, construction projects, and the like;
4. neighborhood changes from which there is flight, perhaps pressures from new and unwelcome ethnic groups, building deterioration, street gangs, or similar rejected alterations in activity space;
5. changes of residence associated with individual personality (chronic mobility).

Some people simply tend to move often for no easily discernible reasons, whereas others, *stayers*, settle into a community permanently. Of course, for a country such as China, with its limitations on emigration and severe housing shortages, a totally different set of summary migration factors are present. By *emigration* we mean leaving one's country or region to settle elsewhere.

The factors that contribute to mobility tend to change over time. However, in most societies, one group has always been the most mobile: young adults (Figure 8.20). They are the members of society who are launching careers and making initial decisions about occupation and location. They have the fewest responsibilities of all adults; thus, they are not as strongly tied to family and institutions as older people are. Most of the major voluntary migrations have been composed primarily of young people who suffered from a lack of opportunities in the home area and who were easily able to take advantage of opportunities elsewhere.

The concept of **place utility** helps us understand the decision-making process that potential voluntary migrants undergo. Place utility is the value that an individual puts on a given residential site. The decision to migrate is a reflection of the appraisal—the perception—by the prospective migrant of the current homesite as opposed to other sites of which something is known or hoped for. The individual may adjust to conditions at the homesite and decide not to migrate.

In evaluating comparative place utility, the decision maker considers not only the perceived value of the present location but also the expected place utility of each potential destination. The evaluations are matched with the individual's *aspiration level*—that is, the level of accomplishment or ambition that the individual sees for himself or herself. Aspirations tend to be adjusted to what an individual considers attainable. If the person is satisfied with present circumstances, then he or she does not initiate search behavior. If, on the other hand, the person is dissatisfied with the home location, he or she assigns a utility to each possible new site. The utility is based on past or expected future

## Gender and Migration

Gender is involved in migration at every level. In a household or family, women and men are likely to play different roles regarding decisions or responsibilities for activities such as childcare. These differences, and the inequalities that underlie them, help determine who decides whether the household moves, which household members migrate, and the destination for the move. Outside the household, societal norms about women's mobility and independence often restrict their ability to migrate.

The economies of both sending and receiving areas play a role as well. If jobs are available for women in the receiving area, women have an incentive to migrate, and families are more likely to encourage the migration of women as necessary and beneficial. Thousands of women from East and Southeast Asia have migrated to the oil-rich countries of the Middle East, for example, to take service jobs.

The impact of migration is also likely to be different for women and men. Moving to a new economic or social setting can affect the regular relationships and processes that occur within a household or family. In some cases, women might remain subordinate to the men in their families. A study of Greek-Cypriot immigrant women in London and of Turkish immigrant women in the Netherlands found that although these women were working for wages in their new societies, these new economic roles did not affect their subordinate standing in the family in any fundamental way.

In other situations, however, migration can give women more power in the family. In Zaire, women in rural areas move to towns to take advantage of job opportunities there, and gain independence from men in the process.

One of the keys to understanding the role of gender in migration is to disentangle household decision-making processes. Many researchers see migration as a family decision or strategy, but some family members will have more sway than others, and some members will benefit more than others will from those decisions.

For many years, men predominated in the migration streams flowing from Mexico to the United States. Women played an important role in this migration stream, even when they remained in Mexico. Mexican women influenced the migration decisions of other family members; they married migrants to gain the benefits from and opportunity for migration; and they resisted or accepted the new roles in their families that migration created.

In the 1980s, Mexican women began to migrate to the United States in increasing numbers. Economic crises in Mexico and an increase in the number of jobs available for women in the United States, especially in factories, domestic service, and service industries, have changed the backdrop of individual migration decisions. Now, women often initiate family moves or resettlement efforts.

Mexican women have begun to build their own migration networks, which are key to successful migration and resettlement in the United States. Networks provide migrants with information about jobs and places to live and have enabled many Mexican women to make independent decisions about migrating.

In immigrant communities in the United States, women are often the vital links to social institution services and to other immigrants. Thus, women have been instrumental in the way that Mexican immigrants have settled and become integrated into new communities.



rewards at the various sites. Because the new places are unfamiliar to the individual, the information received about them acts as a substitute for the personal experience of the homesite. The decision maker can do no more than sample information about the new sites, and, of course, there may be errors in both information and interpretation.

One goal of the potential migrant is to minimize uncertainty. Most decision makers either elect not to migrate or postpone the decision unless uncertainty can be lowered sufficiently. That objective may be achieved either by going through a series of transitional relocation stages or by following the example of known predecessors. **Step migration** involves the place transition from, for example, rural to central city residence through a series of less extreme locational changes—from farm to small town to suburb and, finally, to the major central city itself. The term **chain migration** indicates that the mover is part of an established migrant flow from a common origin to a prepared destination. An advance group of migrants, having established itself in a new home area, is followed by second and subsequent migrations originating in the same home district and frequently united by kinship or friendship. Public and private services for legal migrants and informal service networks for undocumented or illegal migrants become established and contribute to the continuation or expansion of the chain migration flow. Ethnic and foreign-born enclaves in major cities and rural areas in a number of countries are the immediate result.

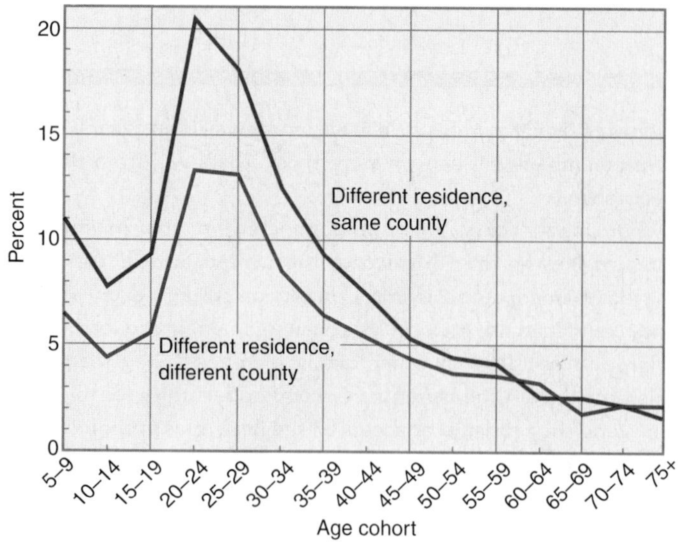


**FIGURE 8.20 Percentage of 2000 population over 5 years of age with a different residence than in 1999.** Young adults figure most prominently in both short- and long-distance moves in the United States, an age-related pattern of mobility that has remained constant over time. For the sample year shown, 33% of people in their twenties moved, whereas fewer than 5% of those 65 and older did so. Short-distance moves predominate; 56% of the 43 million U.S. movers between March 1999 and March 2000 relocated within the same county, and another 20% moved to another county in the same state. Some two-thirds of intracounty (mobility) moves were made for housing-related reasons; long-distance moves (migration) are likely to be made for work-related reasons. Source: *U.S. Bureau of the Census.*

Sometimes the chain migration is specific to occupational as well as ethnic groups. For example, nearly all newspaper vendors in New Delhi, in the north of India, are reported to come from one small district in Tamil Nadu, in the south of India. Most construction workers in New Delhi come either from Orissa, in the east of India, or Rajasthan, in the northwest, and taxicab drivers originate in the Punjab area. A network of about 250 related families who come from a small town several hundred miles to the north dominates the diamond trade of Mumbai (Bombay), India.

A similar domination of businesses by a single ethnic group occurs in the United States. Familiar examples of these *ethnic niche businesses* include fruit stores owned by Koreans and diners owned by Greeks, but there are many others. Female immigrants from Vietnam dominate the manicure trade, accounting for more than half the women in that profession. A large proportion of female immigrants from the Philippines find employment as nurses, particularly in New York City. About 30% of the Filipinos in New York City and its suburbs work as nurses or other health practitioners, the result of their aggressive recruitment by American hospitals and the fact that U.S. immigration authorities have made it easy for nurses to obtain work visas and green cards giving them permanent resident status. Immigrants from India, mostly from the state of Gujarat, now own more than one-third of the hotels in the United States, most of them budget and midpriced franchises such as Holiday Inns, Days Inns, and Ramadas.

Another goal of the potential migrant is to avoid physically dangerous or economically unprofitable outcomes in the final migration decision. Place utility evaluation, therefore, requires assessments not only of perceived pull factors of new sites but also of the potentially negative economic and social reception the migrant might experience at those sites. An example of some of these observations can be seen in the case of the large numbers of young people from the Caribbean, Mexico, and Central America who have migrated both legally and illegally to the United States over the past 30 years (Figure 8.21).

Faced with poverty at home, these young adults regard the place utility in their own country as minimal. Their space-searching ability is limited, however, by both the lack of money and the lack of alternatives in the land of their birth. With a willingness to work and with aspirations for success—perhaps wealth—in the United States, they learn from friends and relatives of job opportunities north of the border, low-paying though they may be. Hundreds of thousands quickly place high utility on perhaps a temporary relocation (maybe 5 or 10 years) to the United States. Many know that dangerous risks are involved if they attempt to enter the country illegally, but even legal immigrants face legal restrictions designed to reduce the pull attractions of the United States (see "Backlash," p. 288). The arrival of those who consider the rewards worth the risk indicated their assignment of higher utility to the new site than to the old one.

In the 20th century, nearly all countries experienced a great movement of people from agricultural areas to the cities, continuing the pattern of *rural-to-urban migration* that first became prominent during the 18th and 19th-century Industrial Revolution in advanced economies. The migration presumably paralleled the number of perceived opportunities within cities and convictions of absence of place utility in the rural districts. Perceptions, of course, do not necessarily accord with reality. Rapid increases in impoverished rural populations of developing countries put increasing and unsustainable pressures on land, fuel, and water in the countryside. Landlessness and hunger, as well as the loss of social cohesion that the growing competition for declining resources induces, help force migration to cities. As a result, while the rate of urban growth is decreasing in the more-developed countries, urbanization in the developing world continues apace, as will be discussed more fully in Chapter 11.

## Barriers to Migration

Paralleling the incentives to migration is a set of disincentives, or barriers, to migration. They help account for the fact that many people do not choose to move even when conditions are bad at home and are known to be better elsewhere. Migration depends on knowledge of the opportuni-

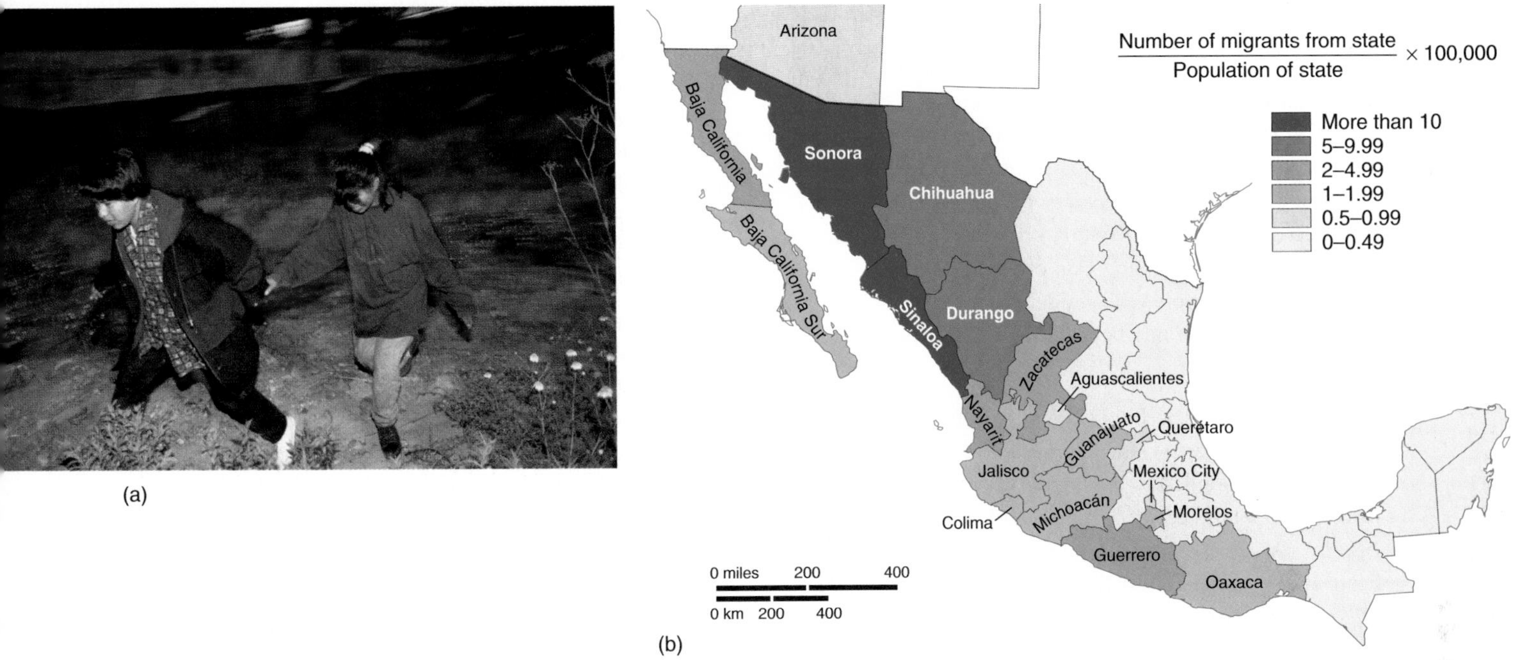


(a) (b)

**FIGURE 8.21** (a) **Illegal Mexican immigrants running from the Border Patrol.** (b) **Undocumented migration rate to Arizona.** The Arizona region and nearby Mexican states have historical ties that go back to the early 1800s. In many respects, the international border cuts through a culture region. Note that distance plays a large role in the decision to migrate to the United States, with over half the migrants coming from four nearby Mexican states: Sonora, Sinaloa, Durango, and Chihuahua. *(a) © AP/Wide World Photos. (b) Redrawn from John P. Harner, "Continuity Amidst Change,"* The Professional Geographer, *47, No. 4, Fig. 2, p. 403. Copyright © Association of American Geographers, 1995. Reprinted by permission of Blackwell Publishers.*

ties in other areas. People with a limited knowledge of the opportunities elsewhere are less likely to migrate than are those who are better informed. Other barriers include physical features, the costs of moving, ties to individuals and institutions in the original activity space, and governmental regulations.

*Physical barriers* to travel include seas, mountains, swamps, deserts, and other natural features. In prehistoric times, physical barriers played an especially significant role in limiting movement. Thus, the spread of the ice sheets across most of Europe in Pleistocene times was a barrier to both migration and human habitation. Physical barriers to movement have probably assumed less importance only within the past 400 years. The developments that made possible the great age of exploration, beginning about A.D. 1500, and the technological developments associated with industrialization have enabled people to conquer space more easily. With industrialization came improved forms of transportation, which made travel faster, easier, and cheaper. Even then, as the conditions the Pony Express riders experienced only a century and a half ago show, travel could be arduous, and in some parts of the world, it remains so today.

*Economic barriers* to migration include the cost of both traveling and establishing a residence elsewhere. Frequently, the additional expense of maintaining contact with those left behind is also a pertinent cost factor. Normally, all of these costs increase with the distance traveled and are a more significant barrier to travel for the poor than for the rich. Many immigrants to the United States were married men who came alone; when they had acquired enough money, they sent for their families to join them. This phenomenon is still evident among recent immigrants from the Caribbean area to the United States and among Turks, Yugoslavs, and West Indians who have settled in northern and Western European countries. The *cost factor* limits long-distance movement, but the larger the differential between present circumstances and perceived opportunities, the more individuals are willing to spend on moving. For many, especially older people, the differentials must be extraordinarily high for movement to take place.

*Cultural factors* also contribute to decisions not to migrate. Family, religious, ethnic, and community relationships defy the principle of differential opportunities. Many people will not migrate under any but the most pressing of circumstances. The fear of change and human inertia—the fact that it is easier not to move than to do so—may be so great that people consider, but reject, a move. Ties to one's own country, culture group, neighborhood, or family may be so strong as to compensate for the disadvantages of the home location. Returning migrants may convince potential leavers that the opportunities are not, in fact, better elsewhere—or that, even if they are better, they are not worth entering an alien culture or sacrificing home and family.

Restrictions on immigration and emigration constitute *political barriers* to migration. Many governments frown on movements into or outside their own borders and restrict

GEOGRAPHY & Public Policy

## Backlash

Migrants can enter a country legally—with a passport, a visa, a working permit, or another authorization—or illegally. Some aliens enter a country legally but on a temporary basis (as a student or tourist, for example) but then remain after their departure date. Others may arrive claiming the right of political asylum but actually seeking economic opportunity. Recent years have seen a rising tide of emotion against the estimated 12 million people who reside illegally in the United States, a sentiment that has been reflected in a number of actions.

- Security fears since the September 11, 2001, assaults on the World Trade Center and Pentagon have led to more stringent visa applicant background checks, greater restrictions on admitting refugees and asylum seekers, tighter border controls, and stricter enforcement of Immigration and Naturalization Service (INS) rules on alien residency reports and visa time restrictions.
- Greater efforts are being made to deter illegal crossings along the Mexican border by increasing the number of Border Patrol agents and by building steel fences near El Paso, Texas; Nogales, Arizona; San Ysidro, California; and elsewhere.
- Four states—Florida, Texas, Arizona, and California—are suing the federal government to win reimbursement for the costs of illegal immigration.
- The U.S.-Mexico Border Counties Coalition, composed of representatives from the 24 counties in the United States that abut Mexico, is demanding that the federal government reimburse local administrations for money spent on legal and medical services for undocumented aliens. These services include the detention, prosecution, and defense of immigrants; emergency medical care; ambulance service; and even autopsies and burials for those who die while trying to cross the border.

In the 1990s, concern over the estimated 2 million illegal immigrants residing in California led the voters of that state to approve a trio of ballot initiatives aimed at curbing what their proponents saw as unwarranted privileges for undocumented immigrants. Proposition 187, passed in 1994, denied certain public services to illegal aliens. It prohibited state and local governmental agencies from providing publicly funded education, nonemergency health care, welfare benefits, and social services to any person whom they could not verify to be either a U.S. citizen or a person legally admitted to the country. The measure also required state and local agencies to report suspected illegal immigrants to the Immigration and Naturalization Service and to certain state officials.

Proponents of Proposition 187 argued that California could no longer support the burden of high levels of immigration, especially if the immigrants could not enter the more skilled professions. They contended that welfare, medical, and educational benefits are magnets that draw illegal aliens into the state. These unauthorized immigrants are estimated to cost California taxpayers more than $3.5 billion per year and result in overcrowded schools and public health clinics, as well as the reduction of services to legal residents. Why should the latter pay for benefits for people who are breaking the law, the proposition's supporters asked?

© Bettmann/Corbis Images.

out-migration. These restrictions may make it impossible for potential migrants to leave, and they certainly limit the number who can do so. On the other hand, countries suffering from an excess of workers often encourage emigration. The huge migration of people to the Americas in the late 19th and early 20th centuries is a good example of perceived opportunities for economic gain far greater than in the home country. Many European countries were overpopulated, and their political and economic systems stifled domestic economic opportunity at a time when people were needed by American entrepreneurs hoping to increase their wealth in untapped resource-rich areas.

The most developed countries where per capita incomes are high, or perceived as high, are generally the most desired international destinations. In order to protect themselves against overwhelming migration streams, such countries as the United States, Australia, France, and Germany have restrictive policies on immigration. In addition to absolute

Those opposed to proposition 187 contended that projected savings would be illusory because the proposition collided with federal laws that guarantee access to public education for all children in the United States. It also violated federal Medicaid laws, so California would be in danger of losing all regular Medicaid funding. Forcing an estimated 300,000 children out of school and onto the streets would increase the risk of juvenile crime. Forbidding doctors from giving immunizations or basic medical care to anyone suspected of being an illegal immigrant would encourage the spread of communicable diseases throughout the state, putting everyone at risk. Educators, doctors, and other public service officials would be turned into immigration officers, a task for which they are ill suited. Finally, opponents argued that the proposition would not stop the flow of illegal aliens because it did nothing to increase enforcement at the border or to punish employers who hire undocumented workers.

A week after the passage of Proposition 187, a federal judge issued a temporary restraining order blocking the enforcement of most of its provisions pending the resolution of legal issues. Shortly thereafter, another federal judge struck down portions of the proposition, declaring them unconstitutional. Finally, 5 years after the passage of the proposition, the U.S. Court of Appeals for the Ninth Circuit in 1999 permanently voided its core provisions, including those that prevent illegal immigrants from attending public schools and receiving social services and health care. It also voided the requirement that local law enforcement authorities, school administrators, and social and medical workers turn in suspected illegal immigrants to federal and state authorities. "Today's announcement is the final shovel of dirt on the grave of Proposition 187," said the director of the American Civil Liberties Union of southern California. "Hopefully it brings to a close what has been a very ugly chapter in California politics," added a spokesman for the state attorney general's office. The original sponsors of Proposition 187, however, warned that "the will of the people has been frustrated" and predicted that "the battle may not be over."

Proposition 209, passed in 1996, banned state and local governmental preferences based on race and gender in hiring and school admissions. No "positive" discrimination for racial minorities is allowed, and affirmative action programs were discontinued. Although opponents of the proposition appealed the initiative, the U.S. Supreme Court in 1997 declined to hear the appeal, allowing the ban on racial and gender preferences to stand.

Finally, in 1998, California voters overwhelmingly approved Proposition 227, characterized by a spokesman for the Mexican American Legal Defense and Education Fund as the third in a row of anti-Latino measures. The proposition scraps the system of bilingual education, in which non-English-speaking children were taught in their native language until they learned English well enough to be mainstreamed into regular classrooms. Instead, students will receive 1 year of instruction in English. While opponents of the measure called it immigrant-bashing, its supporters argued that bilingual education has been a failure; few children graduate into English-speaking classes each year, and many leave school unable to speak, read, or write well in the language of their adopted country.

**QUESTIONS TO CONSIDER**

1. What do you think are the magnets that draw immigrants across the border: jobs or benefits? Would the denial of services be likely to lessen the perceived place utility of the United States and thus reduce illegal immigration?
2. People who believe that states should receive full federal payment for all costs associated with illegal immigrants argue that "state taxpayers should not bear the burden of the federal government's failure to control the border." Do you think the federal government has an obligation to fully or partially reimburse states for the costs of education, medical care, incarceration, and other legal services for unauthorized immigrants? Why or why not?
3. Should the United States require citizens to have a national identification card? Why or why not?
4. If you had been able to vote on Proposition 187, how would you have voted? Why?
5. Is it good policy not to educate or give basic medical care to any people, even those not legally in the country? If so, under what circumstances?

quotas on the number of immigrants (usually classified by country of origin), a country may impose other requirements, such as the possession of a labor permit or sponsorship by a recognized association.

## Patterns of Migration

Several geographic concepts deal with patterns of migration. The first of these is the **migration field,** an area or areas that dominate a locale's in- and out-migration patterns. For any single place, the origin of its in-migrants and the destination of its out-migrants remain fairly stable spatially over time. As would be expected, areas near the point of origin make up the largest part of the migration field (Figure 8.22). However, places far away, especially large cities, may also be prominent. These characteristics of migration fields are functions of the hierarchical movement to larger places and the fact that so many people live in large metropolitan areas

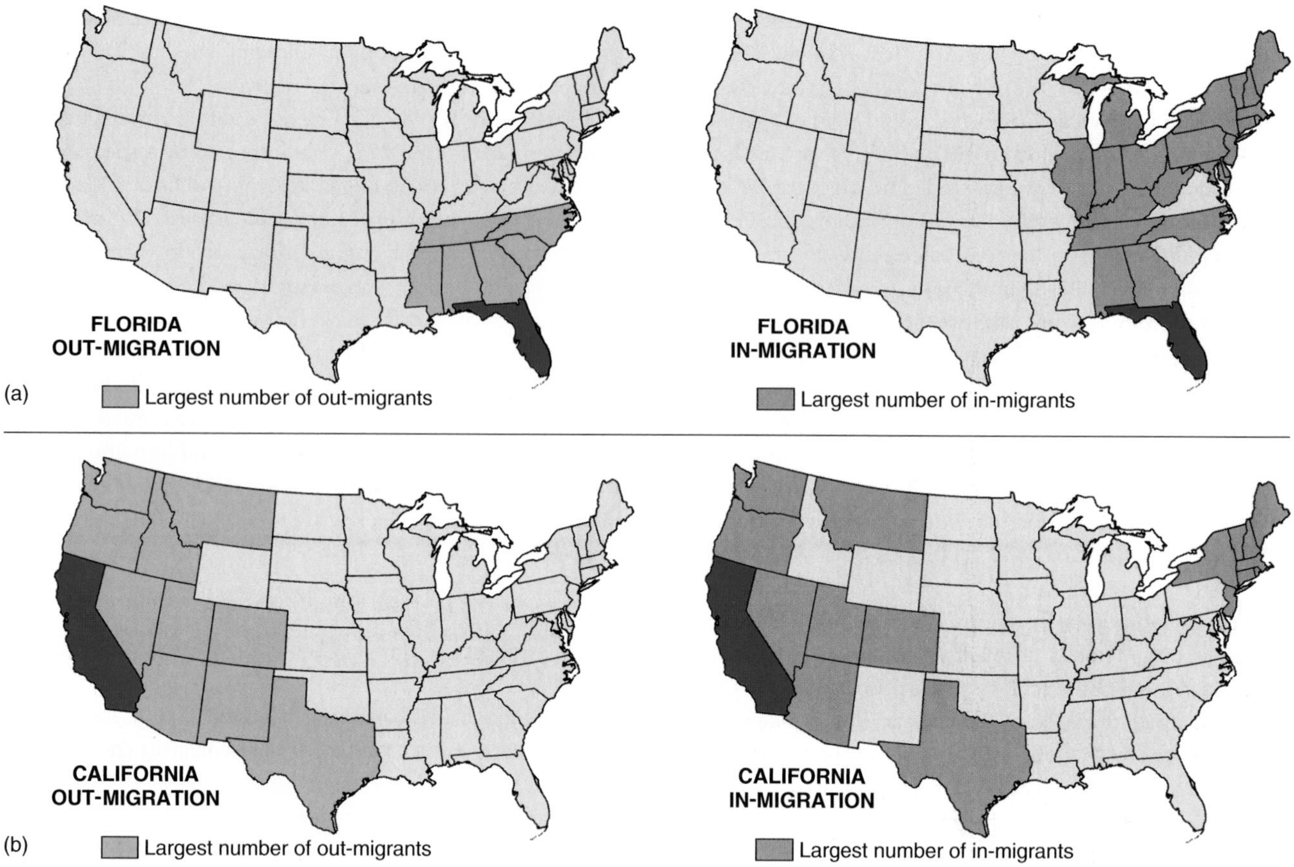


**FIGURE 8.22 The migration fields of Florida and California in 1980.** (a) For Florida, nearby southern states receive most out-migrants, but in-migrants, especially retirees, originate from much of the eastern United States. (b) For California, the out-migration areas are western states. The in-migration origins include both western and heavily populated northeastern states. *From Kavita Pandit, "Differentiating between Subsystems and Typologies in the Analysis of Migration Regions,"* The Professional Geographer, *46, No. 3, Figs. 5 & 6, pp. 342–343. Association of American Geographers, 1994.*

that one may expect some migration into and out of them from most areas within a country.

Migration fields do not conform exactly to the diffusion concepts mentioned earlier in the section "Diffusion and Innovation." As Figure 8.23 shows, some migration fields reveal a distinctly *channelized* pattern of flow. The channels link areas that are socially and economically tied to one another by past migration patterns, economic trade considerations, or some other affinity. As a result, flows of migration along these channels are greater than would otherwise be the case. The former movements of blacks from the southern United States to the North; of Scandinavians to Minnesota and Wisconsin; of Mexicans to such border states as California, Texas, and New Mexico; and of retirees to Florida and Arizona are all examples of **channelized migration** flows.

Of course, not all immigrants stay permanently at their first destination. Of the approximately 80 million newcomers to the United States between 1900 and 1980, some 10 million returned to their homelands or moved to another country. Estimates for Canada indicate that perhaps 40 of each 100 immigrants eventually leave, and about 25% of newcomers to Australia also depart permanently. A corollary of all out-migration flows is, therefore, **return migration,** or **countermigration,** the return of migrants to the regions from which they had earlier emigrated (Figure 8.24).

Within the United States, return migration—moving back to one's state of birth—makes up about 20% of all domestic moves. That figure varies dramatically among states, however. More than a third of recent in-migrants to West Virginia, for example, were returnees, as were over 25% of those moving to Pennsylvania, Alabama, Iowa, and a few other states. Such widely different states as New Hampshire, Maryland, California, Florida, Wyoming, and Alaska were among those that found returnees were fewer than 10% of their in-migrants. Interviews suggest that states deemed attractive draw new migrants in large numbers, whereas those with high proportions of returnees in the migrant stream are not perceived as desirable destinations by other than former residents.

If freedom of movement is not restricted, it is not unusual for as many as 25% of all migrants to return to their place of origin. Unsuccessful migration is sometimes due to

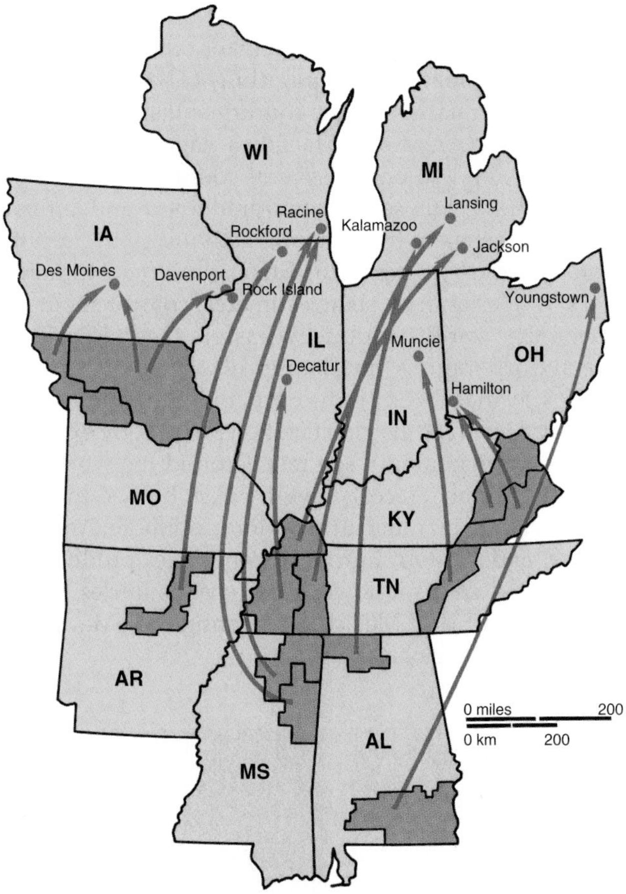


**FIGURE 8.23 Channelized migration flows from the rural South to Midwestern cities of medium size.** Distance is not necessarily the main determinant of flow direction. Perhaps through family and friendship links, the rural southern areas are tied to particular Midwestern destinations. *Redrawn by permission from* Proceedings of the Association of American Geographers, *C. C. Roseman, Vol. 3, p. 142. Copyright © Association of American Geographers. Reprinted by permission of Blackwell Publishers.*

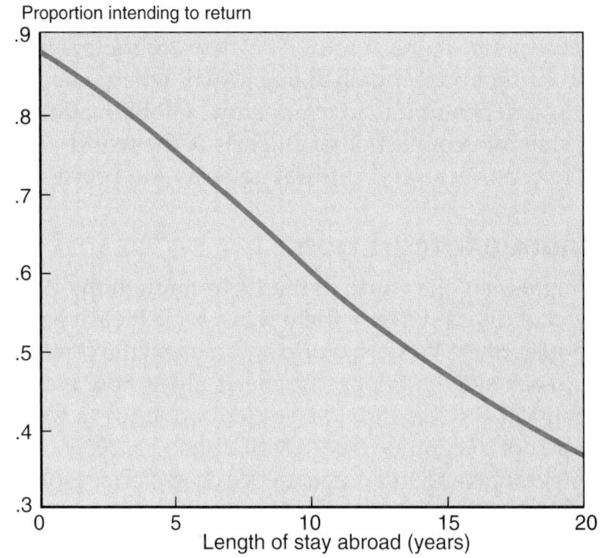


**FIGURE 8.24 Migrants originally from the former Yugoslavia intending to return home from Germany.** As the length of stay in Germany increases, the proportion intending to return decreases, but even after 10 years abroad, more than half intend to return. *Redrawn from B. Waldorf, "Determinants of International Return Migration Intentions,"* The Professional Geographer, *47, no. 2, Fig. 2, p. 132. Association of American Geographers, 1995.*

an inability to adjust to the new environment. More often, it is the result of false expectations based on distorted mental images of the destination at the time of the move. Myths, secondhand and false information, and people's own exaggerations contribute to what turn out to be a mistaken decision to move. Although return migration often represents the unsuccessful adjustment of individuals to a new environment, it does not necessarily mean that negative information about a place returns with the migrant. It usually means a reinforcement of the channel, as communication lines between the unsuccessful migrant and would-be migrants take on added meaning and understanding.

In addition to channelization, the influence of large cities causes migration fields to deviate from the distance-decay pattern. The concept of **hierarchical migration** assists in understanding the nature of migration fields. Earlier we noted that sometimes information diffuses according to a hierarchical rule—that is, from city to city at the highest level in the hierarchy and then to lower levels. Hierarchical migration, in a sense, is a response to that flow. The tendency is for individuals in domestic relocations to move up the level in the hierarchy, from small places to larger ones. Very often, levels are skipped on the way up; only in periods of general economic decline is there considerable movement down the hierarchy. The suburbs of large cities are considered part of the metropolitan area, so the movement from a town to a suburb is considered a move up the hierarchy. From this hierarchical pattern, we can envisage information flowing down to small places from cities and metropolitan areas, and migration flowing up from rural to urban regions.

# Globalization

We have seen how the cost of communication affects the degree of spatial interaction between places. In the past 20 years, we have witnessed the development of the Internet and have benefited from relatively low transportation costs. Increased computerization of transactions has made it easy to buy goods from abroad and to travel abroad. During this period, there also has been a strong movement throughout the world to reduce the barriers to trade and foreign investment and ownership. The European Union is a good case in point. Its currency, the *euro,* now makes it possible to conduct financial transactions in a single currency, much as is the case in the United States. Computer technology enables investors to buy stock on foreign stock exchanges or to buy mutual funds that represent firms whose headquarters are in many parts of the world. The new technologies have helped bring about a globalizing world where people are more interdependent than ever before. **Globalization**—the increasing interconnection of all parts of the world—affects economic, political, and cultural patterns and processes.

## Economic Integration

One might view the world of the 1950s through the 1980s as a period of division, when there was a wide breach between the peoples of the Western world and those of the communist world. Each side had opposing views about how economic and political systems should be organized. It was a world of division, not integration (see "DOSCapital," p. 293).

Integration and interdependence characterize globalization. The fact that Eastern and Western Europe are coming together as a single economic entity, and that the East and South Asian countries' economies are being integrated into those of Europe and North America, is as much a function of the revolution in communication and computer technologies as it is the will of the world's political and financial leaders. Low-cost, high-speed computers; communications satellites; fiber-optic networks; and the Internet are the main technologies of the revolution, but other technologies, such as robotics, microelectronics, electronic mail, fax machines, and cellular phones also play an important role (Figure 8.25). The fact that a consumer in Athens, Greece, can order a book from Amazon.com or clothes from Lands' End, obtain news from CNN, or engage in a transaction on the London Stock Exchange while talking on a cellular phone to a colleague in Tokyo is revolutionary. The forces at work on Americans, the Japanese, and the British have the same effect on Greek consumers as they go about their daily activities. Thus, globalization brings about greater integration and a great deal more spatial interaction.

### International Banking

International banking has become a complex system of integrated and interdependent financial markets. It is no wonder that the various stock exchanges of the world tend to rise and fall together. There are exceptions, of course, but when a pharmaceutical company such as Pfizer buys Warner-Lambert Pharmaceuticals, it affects pharmaceutical stock trading throughout the world. The financial balances between companies are monitored closely by brokers and traders around the world. As a result, split-second changes in all markets are possible. Within minutes of the 9/11/01 attacks on the World Trade Center and Pentagon, stock markets everywhere went down as investors sensed that the international marketplace was in jeopardy of losing its stability.

The internationalization of finance is also reflected in the immense amount of money in foreign investments. Many Americans, for example, own foreign stocks and bonds, either directly or through mutual funds and pension plans. Similarly, people outside the United States have significant holdings in U.S. companies.

### Transnational Corporations

The past 20 years have seen a tremendous increase in the number of **transnational corporations (TNCs),** companies that have headquarters in one country and subsidiary companies, factories, and other facilities (laboratories, offices, warehouses, and so on) in several countries. As many as 65,000 TNCs—with several hundred thousand affiliates worldwide—engage in economic activities that are international in scope (Figure 8.26). They account for trillions of dollars in sales and, by some estimates, control about one-third of the world's productive assets (see also "Transnational Corporations" in Chapter 10).

The way that TNCs produce and provide goods and services is also a part of the globalization process. By exploiting the large differential in wage rates around the world, they keep their production costs down, which has led to the decentralization of manufacturing. More and more, American, Japanese, and Western European companies produce their manufactured goods in lower labor cost countries, such as China, Thailand, and Mexico, integrating these developing countries into the global economy.

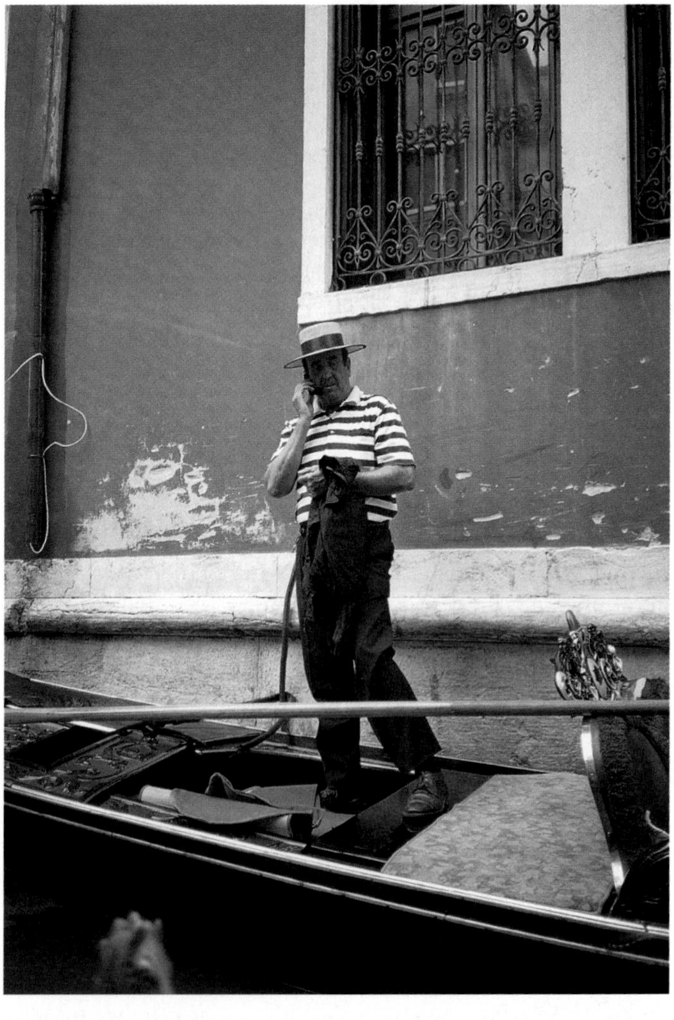

**FIGURE 8.25 The old and the new: a gondolier in Venice, Italy, conducting business on a cellular phone.** *Courtesy of Arthur Getis.*

## DOSCapital

In the following excerpt, Thomas Friedman contrasts the Cold War and globalization.

> A new international system has now clearly replaced the Cold War: globalization. That's right, globalization—the integration of markets, finance, and technologies in a way that is shrinking the world from a size medium to a size small and enabling each of us to reach around the world farther, faster, and cheaper than ever before. It's not just an economic trend, and it's not just some fad. Like all previous international systems, it is directly or indirectly shaping the domestic politics, economic policies, and foreign relations of virtually every country.
>
> As an international system, the Cold War had its own structure of power: the balance between the United States and the USSR, including their respective allies. The Cold War had its own rules: In foreign affairs, neither superpower would encroach on the other's core sphere of influence. The Cold War had its own dominant ideas: the clash between communism and capitalism, as well as detente, nonalignment, and perestroika.
>
> The Cold War system was characterized by one overarching feature: division. The world was chopped up, and both threats and opportunities tended to grow out of whom you were divided from. Appropriately, that Cold War system was symbolized by a single image: the Wall. The globalization system also has one overarching characteristic: integration. Today, both the threats and opportunities facing a country increasingly grow from whom it is connected to. This system is also captured by a single symbol: the World Wide Web. So in the broadest sense, we have gone from a system built around walls to a system increasingly built around networks.
>
> Integration has been driven in large part by globalization's defining technologies: computerization, miniaturization, digitization, satellite communications, fiber optics, and the Internet. And that integration, in turn has led to many other differences between the Cold War and globalization systems.
>
> Whereas the defining measurement of the Cold War was weight, particularly the throw-weight of missiles, the defining measurement of the globalization system is speed—the speed of commerce, travel, communication, and innovation. The Cold War was about Einsteins's mass-energy equation, $e = mc^2$. Globalization is about Moore's Law, which states that the performance power of microprocessors will double every 18 months.
>
> If the Cold War were a sport, it would be sumo wrestling, says Johns Hopkins University professor Michael Mandelbaum. "It would be two big fat guys in a ring, with all sorts of posturing and rituals and stomping of feet, but actually very little contact until the end of the match, when there is a brief moment of shoving and the loser gets pushed out of the ring, but nobody gets killed." By contrast, if globalization were a sport, it would be the 100-meter dash, over and over and over. No matter how many times you win, you have to race again the next day. And if you lost by just one-hundredth of a second, it can be as if you lost by an hour.



In the United States, the service sector of employment has risen dramatically as the manufacturing sector has declined. Instead of producing heavy manufactured goods, the U.S. economy has moved toward the production of high-technology goods and services. The service sector has blossomed as the number of financial transactions has increased. With the increased wealth resulting from the effects of new technologies on production processes, Americans are able to travel more and to stay at hotels and eat out more often. This has a great bearing on the types of jobs people hold, not only in the United States but throughout the business and tourist world. These developments are discussed in more detail in Chapter 10.

### Global Marketing

The greater wealth of those who have benefited from globalization has created a huge new market for goods and services. TNCs market their products around the world, whether they be goods such as Swiss watches, Italian shoes, or Coca-Cola or the services provided by, for example, worldwide hotel chains or cellular phone companies. Japanese cars are found in all parts of the world, just as are fast-food chains. Some of the ramifications of this are discussed in the section "Cultural Integration," p. 295.

It is important to remember, however, that globalization is a fairly recent development, and at present its benefits accrue to a minority of the world's 6 billion people. Only one-tenth of those people, about 600 million, are affluent enough to live comfortably and to purchase the goods and services alluded to earlier. According to the United Nations Development Fund, one-quarter of humanity lives on less than $1 per day. An illiterate farmer in a remote village in Tibet more than likely lacks access to a telephone, much less the new technologies of the revolution in communications.

Nokia—Finland (Pakistan)

Nestlé—Switzerland (Egypt)

IBM—United States (Vietnam)

Sony—Japan (Vietnam)

British Petroleum—U.K. (China)

Ford—United States (China)

**FIGURE 8.26** The world's transnational corporations increased in number from about 7,000 in 1970 to nearly 65,000 in 2003. Ninety of the top 100 TNCs are headquartered in the European Union, the United States, and Japan. Their recognition and impact, however, are global, as suggested by these billboards advertising just a sample of leading TNCs in distant settings. Corporate names and headquarters countries are followed by billboard locations in parentheses. ***Nokia mobile billboard:*** *© Chris Stowers/Panos Pictures;* ***Nestle billboard:*** *© Mark Henley/Panos Pictures;* ***IBM billboard:*** *© Deborah Harse/Image Works;* ***Sony billboard:*** *© Mark Henley/Panos Pictures;* ***BP sign:*** *© Caroline Penn/Panos Pictures;* ***Ford Motor Company billboard:*** *© Steven Harris/Newsmakers/Getty News.*

## Political Integration

The flow of money (capital), goods, ideas, and information around the globe links people in ways that transcend national boundaries. One effect has been to stimulate the formation of new, supranational alliances or the restructuring of older ones. To enhance commerce, countries are signing free trade agreements and joining economic organizations such as the North American Free Trade Agreement (1994) and the World Trade Organization (1995). As will be discussed in Chapter 9, many other international alliances—military, political, and cultural—have been created since 1980, and some that are older, such as NATO, are enlarging their membership.

Another effect of globalization has been the enormous increase in the number of international *nongovernmental organizations (NGOs)*. Their number more than quadrupled during the 1990s, from about 6000 to more than 26,000. As the names of such well-known organizations as Amnesty International, Doctors without Borders, and Greenpeace indicate, the concerns of international NGOs range widely, from human rights and acid rain to famine relief and resource depletion. What they have in common is that they bring together people in far-flung parts of the world in pursuit of mutual goals.

The transmission of news has never been wider or faster. *Time Magazine* can be bought as easily in New Delhi, India, as in New York City, and CNN broadcasts around the world, informing people about current events and, sometimes, helping bring about government intervention in places where it might not have occurred. Coverage of the wars in Bosnia and Kosovo, for example, stimulated the United Nations and NATO to send in peacekeeping troops to stop the carnage. More recently, in 2002, satellite television's graphic imagery of the havoc wrought by Palestinian suicide bombers and Israeli retaliation fueled support for the Palestinian cause throughout the Arab world.

Finally, the Internet has given people a powerful tool to affect governmental policies. To give just a few examples, in 1989, prodemocracy students in Beijing, China, used the Internet to publicize what was happening in that city and to rally support for their cause. The woman who won the Nobel Peace Prize in 1997 for her contribution to the international campaign to ban landmines, Jody Williams, used e-mail to organize 1000 human rights and arms control groups on six continents. Similarly, NGOs used the Internet to coordinate a massive protest against the 1999 meeting of the World Trade Organization in Seattle (Figure 8.27).

**FIGURE 8.27** Some of the approximately 70,000 people who demonstrated against the World Trade Organization (WTO) and its policies in Seattle, Washington, in 1999. Among the protesters were those representing nearly 800 NGOs. The scene was repeated at subsequent meetings of the WTO. *© Steven Rubin/The Image Works.*

## Cultural Integration

Imagine this scene: wearing a Yankees baseball cap, a GAP shirt, Levis, and Reebok shoes, a teenager in Lima, Peru, goes with her friends to see the latest thriller. After the movie, they plan to eat at a nearby McDonald's. Meanwhile, her brother sits at home, listening to Britney Spears on his Sony Walkman while playing a video game on the family's computer. Both children are evidence of the globalization of culture, particularly pop culture. That culture is Western in origin and chiefly American. U.S. movies, television shows, software, music, food, and fashion are marketed worldwide. They influence the beliefs, tastes, and aspirations of people in virtually every country, although their effect is most pronounced on young people. They, rather than their elders, are the ones who want to emulate the stars they see in movies and on MTV and to adopt what they think are Western lifestyles, manners, and modes of dress.

Another indication of cultural integration is the worldwide spread of the English language. It has become *the* medium of communication in economics, technology, and science.

Both the dominance of English and the globalization of popular culture are resented by many people and rejected by some. Iran, Singapore, and China attempt to restrict the programming that reaches their people, although their citizens' access to satellite dishes and the Internet likely mean those governments will not succeed in stopping the spread of Western culture. French ministers of culture try to keep the French language pure, unadulterated by English. Other people decry what they see as the homogenization of culture, and still others find the spread of what they define as Western cultural values abhorrent. Whether or not movies, music, and other communications media accurately reflect Western culture, critics argue that they reflect values such as materialism, innovation, self-indulgence, sexuality, spontaneity, and defiance of authority and tradition. Many movies and television shows appear to glorify violence and rebellion and to promote the new over the old, leisure over work, and wealth over self-fulfillment.

## Summary

In this discussion of spatial interaction, we emphasized the factors that influence how individuals move about and use space. At the beginning of the chapter, we examined variables such as the nature of the information available, people's age, their past experiences, and their values. The concept of distance decay shows how distance plays a role in the degree of spatial interaction that takes place. The question of differences in the extent of space used led us to the concept of activity space. The age of an individual, his or her degree of mobility, and the availability of opportunities all play a role in defining the limits of individual activity space. We saw, however, that as distance increases, familiarity with the environment decreases. The concept of critical distance identifies the distance beyond which the decrease in familiarity with places begins to be significant. The new telecommunications technologies have stimulated great increases in interaction and our knowledge of the world.

Certain factors having to do with the diffusion of innovations indicate what opportunities exist for individuals living in various places. Contagious and hierarchical diffusion influence the geographic direction that cultural change will take. Of course, barriers to diffusion, such as effort and cost, do exist.

A special type of human interaction is migration. When strong enough, the various push and pull forces motivate a long-distance, permanent move. Migration fosters the spread of culture by means of diffusion processes. The decision to migrate is a function of the utility people assign to places and the opportunities at those places.

In recent years, complex world banking systems and modern technology have fostered globalization. Even though the globalization process has brought wealth to many, large numbers of poor people around the world feel left out or are critical of the values engendered by globalization. In all of the processes mentioned in this chapter, there was a strong emphasis on information flow and the effect of distance decay.

## Key Words

activity space 271
chain migration 285
channelized migration 290
contagious diffusion 276
countermigration 290
critical distance 269
distance decay 269
globalization 292
hierarchical diffusion 277
hierarchical migration 291
mental map 272
migration 281
migration field 289
place utility 284
pull factor 282
push factor 282
return migration 290
spatial diffusion 275
spatial interaction 269
stage in life 274
step migration 285
territoriality 271
transnational corporations (TNCs) 292

## For Review & Consideration

1. What is the role of distance in helping us understand spatial interaction?
2. Think of the various conflicts that have taken place in the world in the past century. Does the concept of distance decay bear on the location of adversaries? Why?
3. On a blank piece of paper, and without any maps to guide you, draw a map of the United States, putting in state boundaries wherever possible; this is your mental map of the country. Compare it with a standard atlas map. What conclusions can you reach?
4. What is meant by *activity space?* What factors affect the spatial extent of the activity space of an individual?
5. Recall the places you have visited in the past week. In your movements, were the distance-decay and critical-distance rules operative? What variables affect an individual's critical distance?
6. Briefly distinguish between *contagious diffusion* and *hierarchical diffusion.* In what ways, if any, were these forms of diffusion in operation in the culture hearths discussed in Chapter 7?
7. What considerations affect a decision to migrate? What is *place utility* and how does its perception induce or inhibit migration?

8. What common barriers to migration exist? Why do most people migrate within their own country?
9. Define the term *migration field*. Some migration fields show a *channelized* flow of people. Select a particular channelized migration flow (such as the movement of Scandinavians to the United States, people from the Great Plains to California, or southern blacks to the North) and explain why a channelized flow developed.
10. Identify the effects of globalization on your lifestyle and the patterns of trade in your urban area.

# Selected References

**WEBSITES**

The World Wide Web has a tremendous variety of sites pertaining to geography. Websites relevant to the subject matter of this chapter appear in the "Web Links" section of the Online Learning Center associated with this book. Access it at www.mhhe.com/getis10e/.


Boyle, Paul, and Keith Halfacre, eds. *Migration and Gender in the Developed World.* New York: Routledge, 1999.

Brown, Lawrence A. *Innovation Diffusion: A New Perspective.* New York: Methuen, 1981.

Brunn, Stanley, and Thomas Leinbach. *Collapsing Space and Time: Geographic Aspects of Communication and Information.* Winchester, Mass.: Unwin Hyman, 1991.

Castles, Stephen, and Mark J. Miller. *The Age of Migration: International Population Movements in the Modern World.* 3d ed. New York: Guilford Press, 2003.

Clark, William A. V. *The California Cauldron: Immigration and the Fortunes of Local Communities.* New York: Guilford Press, 1998.

Clark, William A. V. *Immigrants and the American Dream.* New York: Guilford Press, 2003.

Fotheringham, A. Stewart, and Morton E. O'Kelly. *Spatial Interaction Models: Formulations and Applications* (Studies in Operational Regional Science). Dordrecht, Netherlands: Kluwer, 1989.

Golledge, Reginald G., and Robert J. Stimson. *Spatial Behavior: A Geographic Perspective.* New York: Guilford Press, 1996.

Gould, Peter R. *The Slow Plague.* Oxford, England: Blackwell, 1993.

Gould, Peter R., and Rodney White. *Mental Maps.* 2d ed. New York: Routledge, 1993.

Hägerstrand, Torsten. *Innovation Diffusion as a Spatial Process.* Chicago: University of Chicago Press, 1967.

Haines, David W., and Karen E. Rosenblum, eds. *Illegal Immigration in America: A Reference Handbook.* Westport, Conn.: Greenwood, 1999.

Held, David, and Anthony McGrew. *Globalization/Anti-Globalization.* Cambridge, England: Polity Press, 2003.

Janelle, Don G., and David C. Hodge, eds. *Information, Place, and Cyberspace: Issues in Accessibility.* Berlin, Germany: Springer, 2000.

Marcuse, Peter, and Ronald van Kempen, eds. *Globalizing Cities: A New Spatial Order?* Oxford, England: Blackwell, 2000.

Morrill, Richard L., Gary L. Gaile, and Grant Ian Thrall. *Spatial Diffusion.* Scientific Geography Series; vol. 10. Newbury Park, Calif.: Sage, 1988.

Pooley, Colin G., and Ian D. Whyte, eds. *Migrants, Emigrants and Immigrants: A Social History of Migration.* New York: Routledge, 1991.

Sassen, Saskia, ed. *Global Networks: Linked Cities.* New York: Routledge, 2002.

United Nations High Commissioner for Refugees. *The State of the World's Refugees.* New York: Oxford University Press, annual.

Zonn, Leo E., ed. *Place Images in the Media: A Geographic Appraisal.* Savage, Md.: Rowman and Littlefield, 2000.

Zook, Matthew. *The Geography of the Internet Industry.* Malden, Mass.: Blackwell, 2003.

*country* or *nation*. That is, a nation can also be defined as (1) an independent political unit holding sovereignty over a territory (e.g., a member of the United Nations). But it can also be used to describe (2) a community of people with a common culture and territory (e.g., the Kurdish nation). The second definition is *not* synonymous with *state* or *country*.

To avoid confusion, we shall define a **state** on the international level as an independent political unit occupying a defined, permanently populated territory and having full sovereign control over its internal and foreign affairs. We will use *country* as a synonym for the territorial and political concept of "state." Not all recognized territorial entities are states. Antarctica, for example, has neither an established government nor a permanent population; it is, therefore, not a state. Nor are *colonies* or *protectorates* recognized as states. Although they have defined extent, permanent inhabitants, and some degree of separate governmental structure, they lack full control over all of their internal and external affairs.

We use *nation* in its second sense, as a reference to people, not to political structure. A **nation** is a group of people with a common culture occupying a particular territory, bound together by a strong sense of unity arising from shared beliefs and customs. Language and religion may be unifying elements, but even more important are an emotional conviction of cultural distinctiveness and a sense of ethnocentrism. For example, the Cree nation exists because of its cultural uniqueness, not by virtue of territorial sovereignty.

The composite term **nation-state** properly refers to a state whose territorial extent coincides with that occupied by a distinct nation or people or, at least, whose population shares a general sense of cohesion and adherence to a set of common values (Figure 9.4a). That is, a nation-state is an entity whose members feel a natural connection with each other by virtue of sharing language, religion, or some other cultural characteristic strong enough both to bind them together and to give them a sense of distinction from all others outside the community. In reality, very few countries can claim to be true nation-states, since few are or have ever been wholly uniform ethnically or culturally. Iceland, Slovenia, Poland, and the two Koreas are often cited as acceptable examples.

A *binational* or *multinational state* is one that contains more than one nation (Figure 9.4b). Often, no single ethnic group dominates the population. In the constitutional structure of the Soviet Union before 1988, one division of the legislative branch of the government was termed the Soviet of Nationalities. It was composed of representatives from civil divisions of the Soviet Union populated by groups of officially recognized "nations": Ukrainians, Kazakhs, Estonians, and others. In this instance, the concept of nationality was territorially less than the extent of the state.

Alternatively, a single nation may be dispersed across and be predominant in two or more states. This is the case with a *part-nation state* (Figure 9.4c). Here, a people's sense of nationality exceeds the areal limits of a single state. An example is the Arab nation, which dominates 17 states.

Finally, there is the special case of the *stateless nation*, a people without a state. The Kurds, for example, are a nation

(a)

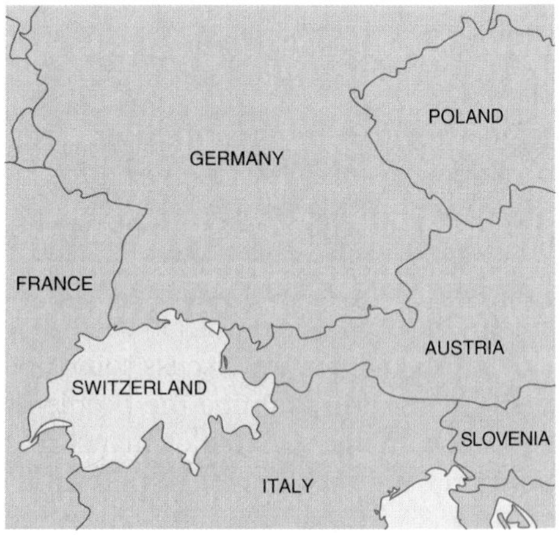


(b)

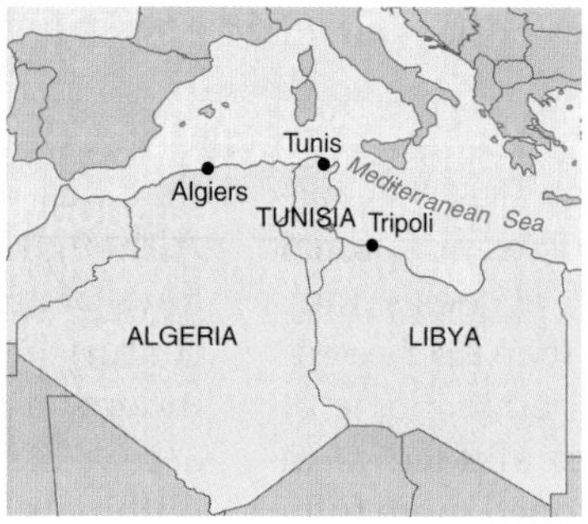


(c)

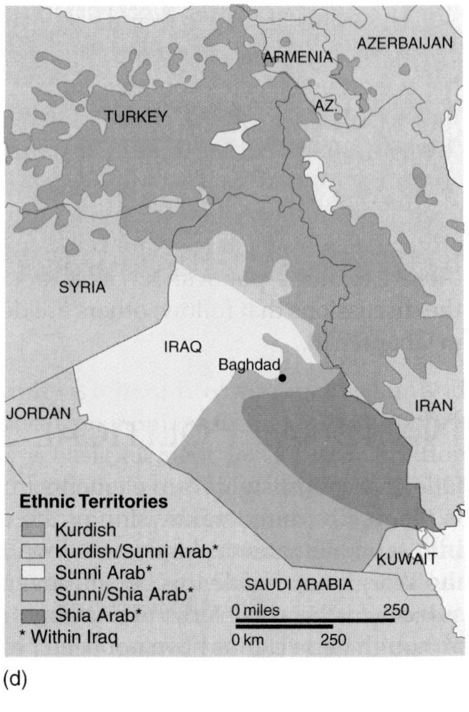


(d)

**FIGURE 9.4 Types of relationships between states and nations.** (a) **Nation-states.** Poland and Slovenia are examples of states occupied by a distinct nation, or people. (b) **A multinational state.** Switzerland shows that a common ethnicity, language, or religion is not necessary for a strong sense of nationalism. (c) **A part-nation state.** The Arab nation extends across and dominates many states in northern Africa and the Middle East. (d) **A stateless nation.** An ancient group with a distinctive language, Kurds are concentrated in Turkey, Iran, and Iraq. Smaller numbers live in Syria, Armenia, and Azerbaijan.

of approximately 20 million people divided among six states and dominant in none (Figure 9.4d). Kurdish nationalism has survived over the centuries, and many Kurds nurture a vision of an independent Kurdistan. Other stateless nations are the Roma (gypsies), Basques, and Palestinians.

## Evolution of the Modern State

The concept and practice of the political organization of space and people arose independently in many parts of the world. Our Western orientations and biases may incline us to trace ideas of spatial political organization through their Near Eastern, Mediterranean, and Western European expressions. Mesopotamian and classical Greek city-states, the Roman Empire, and European colonizing kingdoms and warring principalities were, however, not unique. Southern, southeastern, and eastern Asia had their counterparts, as did sub-Saharan Africa and the Western Hemisphere. Although Western European models and colonization strongly influenced the forms and structures of modern states around the world, the cultural roots of statehood run deeper and reach further back in many parts of the world than the European example alone suggests.

The now universal idea of the modern state was developed by European political philosophers in the 18th century. Their views advanced the concept that people owe allegiance to a state and the people it represents, rather than to its leader, such as a king or a feudal lord. The new concept coincided in France with the French Revolution and spread over Western Europe, to England, Spain, and Germany.

Many states are the result of European expansion during the 17th, 18th, and 19th centuries, when much of Africa, Asia, and the Americas was divided into colonies. Usually these colonial claims were given fixed and described boundaries where none had earlier been formally defined. Of course, precolonial native populations had relatively fixed home areas of control within which there was recognized dominance and border defense and from which there were, perhaps, raids of plunder or conquest of neighboring "foreign" territories. Beyond understood tribal territories, great empires arose, again with recognized outer limits of influence or control: Mogul and Chinese; Benin and Zulu; Incan and Aztec. Upon them where they still existed, and upon the less formally organized spatial patterns of effective tribal control, European colonizers imposed their arbitrary new administrative divisions of the land. In fact, groups that had little in common were often joined in the same colony (Figure 9.5). The new divisions, therefore, were not usually based on meaningful cultural or physical lines. Instead, the boundaries simply represented the limits of the colonizing empire's power.

As these former colonies have gained political independence, they have retained the idea of the state. They have generally accepted—in the case of Africa, by a conscious decision to avoid precolonial territorial or ethnic claims that could lead to war—the borders established by their former European rulers. The problem that many of the new countries face is "nation-building"—developing feelings of loyalty to the state among their arbitrarily associated citizens. Julius Nyerere, president of Tanzania, noted in 1971, "These new countries are artificial units, geographical expressions carved on the map by European imperialists. These are the units we have tried to turn into nations."

The idea of separate statehood grew slowly at first and more recently has accelerated rapidly. At the time of the Declaration of Independence of the United States in 1776, there were only about 35 empires, kingdoms, and countries in the entire world. By the beginning of World War II in 1939, their number had only doubled to about 70. Following that war, the end of the colonial era brought a rapid increase in the number of sovereign states. By 2003, independent states totaled nearly 200. The number has grown in recent years due to the disintegration of the USSR, Czechoslovakia, and Yugoslavia, which created more than 20 countries, where only 3 had existed before (Figure 9.6).

## Geographic Characteristics of States

Every state has certain geographic characteristics by which it can be described and that set it apart from all other states. A look at the world political map inside the cover of this book confirms that every state is unique. The size, shape, and location of any one state combine to distinguish it from all others. These characteristics also affect the power and stability of states.

### Size

The area that a state occupies may be large, as is true of China, or small, as is Liechtenstein. The world's largest country, Russia, occupies more than 17 million square kilometers (6.5 million sq mi), or some 11% of the land surface of the world. It is more than a million times as large as Nauru, one of the ministates found in all parts of the world (see "The Ministates," p. 306).

One might assume that the larger a state's area, the greater is the chance that it will have useful resources, such as fertile soil and minerals. In general, that assumption is valid, but much depends upon accidents of location. Mineral resources are unevenly distributed, and size alone does not guarantee their presence within a state. And Australia, Canada, and Russia, though large, have relatively small areas capable of supporting productive agriculture. Great size, in fact, may be a disadvantage. A very large country may have vast areas that are inaccessible, sparsely populated, and hard to integrate into the mainstream of economy and society. Small states are more apt than large ones to have a culturally homogeneous population. They find it easier to develop transportation and communication systems to link the sections of the country, and, of course, they have shorter boundaries to defend against invasion. Size alone, then, is not critical in determining a country's stability and strength, but it is a contributing factor.

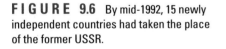

**FIGURE 9.5 The discrepancies between ethnic groups and national boundaries in Africa.** Cultural boundaries were ignored by European colonial powers. The result was significant ethnic diversity in nearly all African countries and conflicts between countries over borders. *Modified from* World Regional Geography: A Question of Place *by Paul Ward English, with James Andrew Miller. Copyright © 1977 Harper & Row. Used by permission of the author.*

FIGURE 9.6 By mid-1992, 15 newly independent countries had taken the place of the former USSR.

## Shape

Like size, a country's shape can affect the well-being of a state by fostering or hindering effective organization. Assuming no major topographical barriers, the most efficient form would be a circle, with the capital located in the center. In such a country, all places could be reached from the center in a minimal amount of time and with the least expenditure for roads, railway lines, and so on. It would also have the shortest possible borders to defend. Uruguay and Poland have roughly circular shapes, forming a **compact state** (Figure 9.7).

**Prorupt states** are nearly compact but possess one or sometimes two narrow extensions of territory. Proruption may simply reflect peninsular elongations of land area, as in the case of Myanmar and Thailand. In other cases, the extensions have an economic or strategic significance, having been designed to secure state access to resources or to establish a buffer zone between states that would otherwise adjoin. Whatever their origin, proruptions tend to isolate a portion of a state.

The least efficient shape administratively is represented by countries like Norway and Chile, which are long and narrow. In such **elongated states,** the parts of the country far from the capital are likely to be isolated because great expenditures are required to link them to the core. These countries are also likely to encompass more diversity of climate, resources, and peoples than compact states, perhaps to the detriment of national cohesion or, perhaps, to the promotion of economic strength.

The fourth class of shapes, that of **fragmented states,** includes countries composed entirely of islands (e.g., the Philippines and Indonesia), countries that are partly on islands and partly on the mainland (Italy and Malaysia), and those that are chiefly on the mainland but whose territory is separated by another state (the United States). Fragmentation makes it harder for the state to impose centralized control over its territory, particularly when the parts of the state are far from one another. This is a problem in Indonesia, which is made up of more than 13,000 islands, stretched out along a 5100-kilometer (3200-mi) arc. Fragmentation helped lead to the disintegration of Pakistan. It was created in 1947 as a fragmented state, but East and West Pakistan were 1610 kilometers (1000 mi) from one another. That distance exacerbated economic and cultural differences between the two, and when the eastern part of the country seceded in 1971 and declared itself the independent state of Bangladesh, West Pakistan was unable to impose its control.

A special case of fragmentation occurs when a territorial outlier of one state, an **exclave,** is located within another state. Before German unification, West Berlin was an outlier of West Germany within East Germany (the German Democratic Republic). Europe has many such outlying bits of one country inside another. Kleinwalsertal, for example, is a piece of Austria accessible only from Germany. Baarle-Hertog is a fragment of Belgium inside Holland. Llivia is a Spanish town just inside France. Exclaves are not limited to Europe, of course. African examples include Cabinda, an exclave of Angola, and Melilla and Ceuta, two Spanish exclaves in Morocco (Figure 9.8).

The counterpart of an exclave, an **enclave,** helps to define the fifth class of shapes, the **perforated state.** A perforated state completely surrounds a territory that it does not rule, as the Republic of South Africa surrounds Lesotho. The enclave, the surrounded territory, may be independent or may be part of another state. Two of Europe's smallest independent states, San Marino and Vatican City, are enclaves that perforate Italy. As an *exclave* of West Germany, West Berlin perforated the national territory of former East Germany and was an *enclave* in it. The stability of the perforated state can be weakened if the enclave is occupied by people whose value systems differ from those of the surrounding country (see "The Gnarled Politics of an Enclave," p. 309).

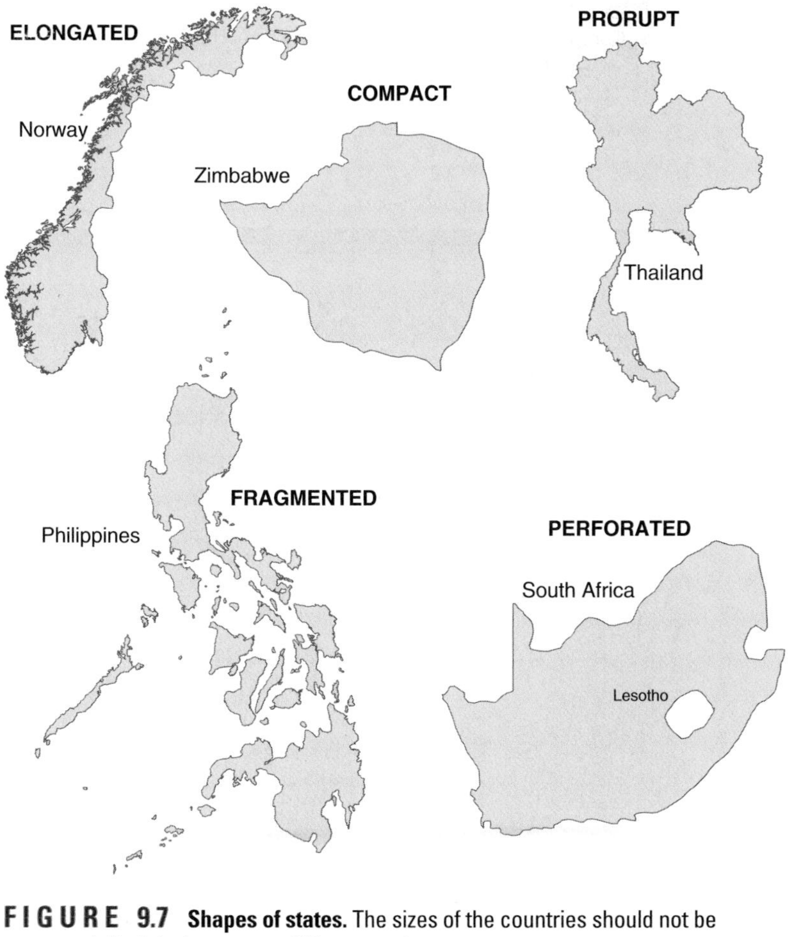


**FIGURE 9.7 Shapes of states.** The sizes of the countries should not be compared. Each is drawn on a different scale.

## Location

The significance of size and shape as factors in national well-being can be modified by a state's

## The Ministates

Totally or partially autonomous political units that are small in area and population pose some intriguing questions. Should size be a criterion for statehood? What is the potential of ministates to cause friction among the major powers? Under what conditions are they entitled to representation in international assemblies like the United Nations?

Of the world's growing number of small countries, more than 40 have less than 1 million people, the population size adopted by the United Nations as the upper limit defining "small states," though not too small to be members of that organization. Nauru has about 12,000 inhabitants on its 21 square kilometers (8.2 sq mi). Other areally small states, such as Singapore, covering 580 square kilometers (224 sq mi), have populations (4.2 million) well above the UN criterion. Many are island territories located in the Caribbean and the Pacific Ocean (such as Grenada and Tonga Islands), but Europe (Vatican City and Andorra), Asia (Bahrain and Brunei), and Africa (Djibouti and Equatorial Guinea) have their share.

Many ministates are vestiges of colonial systems that no longer exist. Some of the small countries of West Africa and the Arabian peninsula fall into this category. Others, such as Mauritius, served primarily as refueling stops on transoceanic voyages. However, some occupy strategic locations (such as Bahrain, Malta, and Singapore), and others contain valuable minerals (Kuwait and Trinidad). The possibility of claiming 370-kilometer-wide (200-nautical-mile) zones of adjacent seas (see p. 325) adds to the attraction of yet others.

Their strategic or economic value can expose small islands and territories to unwanted attention from larger neighbors. The 1982 war between Britain and Argentina over the Falkland Islands (called the Islas Malvinas by Argentina) and the Iraqi invasion of Kuwait in 1990 demonstrate the ability of such areas to bring major powers into conflict and to receive world attention that is out of proportion to their size and population.

The proliferation of tiny countries raises the question of their representation and their voting weight in international assemblies. Should there be a minimum size necessary for participation in such bodies? Should countries receive a vote proportional to their population? New members accepted into the United Nations in 1999 and 2000 included four small Pacific island countries, all with populations under 100,000: Nauru, Tonga, Kiribati, and Tuvalu. Within the United Nations, the Alliance of Small Island States (AOSIS) has emerged as a significant power bloc, controlling more than one-fifth of UN General Assembly votes.

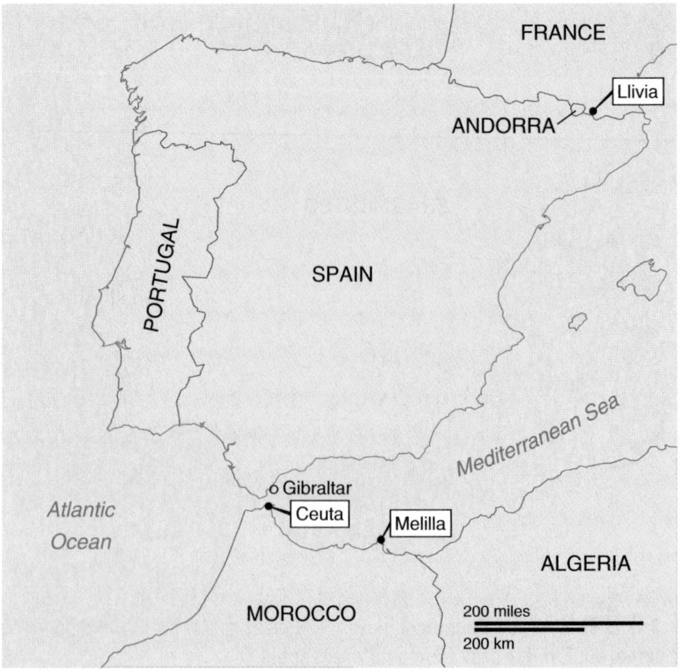


**FIGURE 9.8 Spanish exclaves in North Africa and France.** Although Spanish troops seized the garrison towns of Melilla and Ceuta almost 500 years ago, and a majority of the exclaves' residents are of Spanish descent, Morocco still claims sovereignty over the towns. In recent years, Ceuta and Melilla have become the temporary stopping points for thousands of would-be migrants to Spain. Emigrants from Mali, Nigeria, and as far away as Kashmir and Iraq enter Ceuta and Melilla requesting political asylum and seeking work permits or visas to enter Europe. Llivia became an exclave in 1660 when Spain ceded the surrounding area to France in the Treaty of the Pyrenees. Gibraltar is a British colony, and Andorra is an independent ministate.

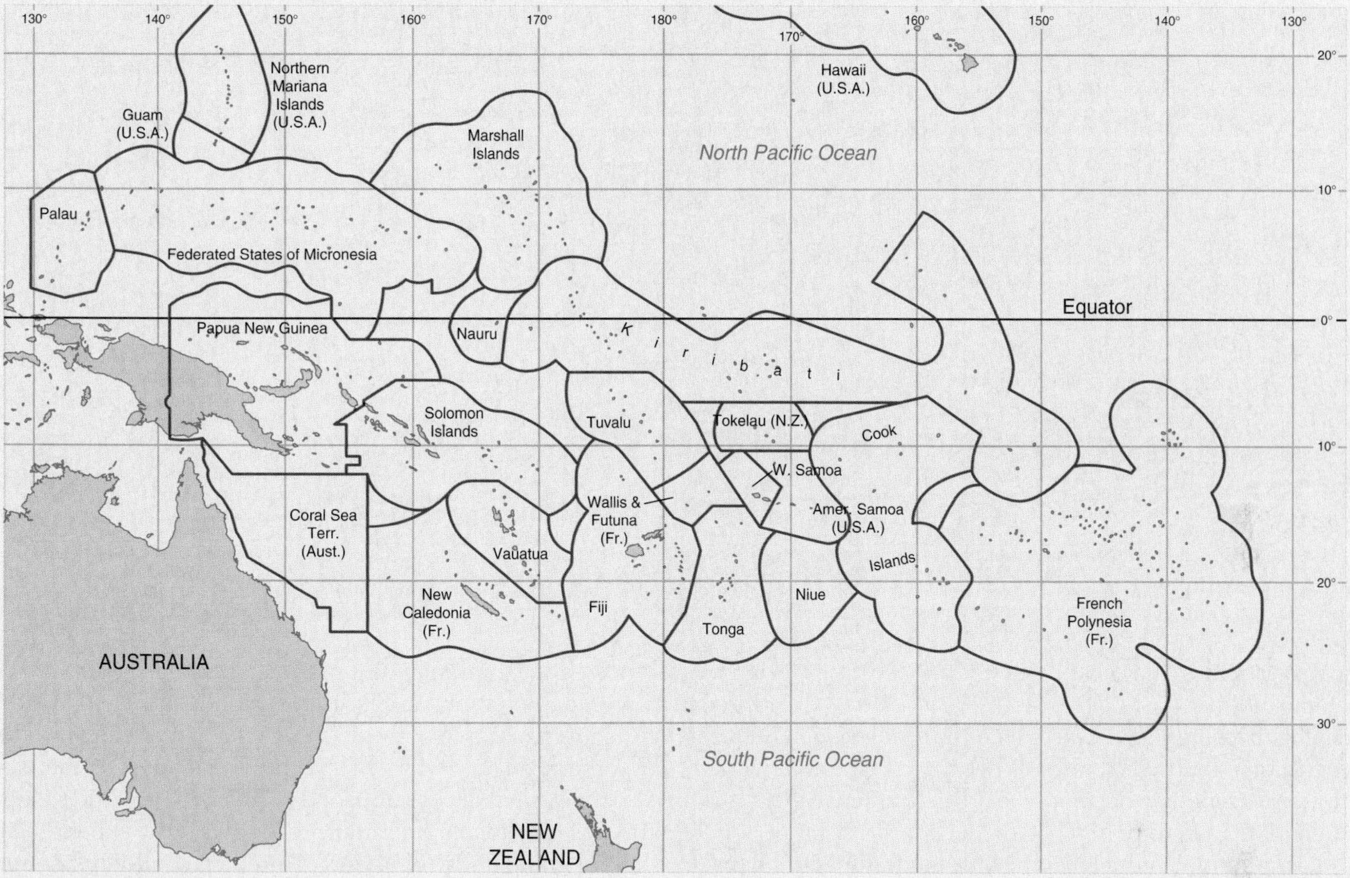


Pacific Ocean ministates.

location, both absolute and relative. Although both Canada and Russia are extremely large, their *absolute* location in the upper-middle latitudes reduces their size advantages when agricultural potential is considered. For another example, Iceland has a reasonably compact shape, but its location in the North Atlantic Ocean, just south of the Arctic Circle, means that most of the country is barren. Settlement is confined to the rims of the island.

A state's *relative* location, its position compared to that of other countries, is as important as its absolute location. *Landlocked* states, those lacking ocean frontage and surrounded by other states, are at a geographic disadvantage (Figure 9.9). They lack easy access to maritime (seaborne) trade and to the resources found in coastal waters and submerged lands. Bolivia gained 480 kilometers (300 mi) of sea frontier along with its independence in 1825, but lost its ocean frontage by conquest to Chile in 1879. Its annual Day of the Sea ceremony reminds Bolivians of their loss and of continuing diplomatic efforts to secure an alternate outlet. The number of landlocked states—about 40—increased greatly with the dissolution of the Soviet Union and the creation of new, smaller countries out of such former multinational countries as Yugoslavia and Czechoslovakia.

In a few instances, a favorable relative location constitutes the primary resource of a state. Singapore, a state of only 580 square kilometers (224 sq mi) and 4.2 million people, is located at a crossroads of world shipping and commerce. Based on its port and commercial activities, and buttressed by its more recent industrial development, Singapore has become a notable Southeast Asian economic success. In general, history has shown that countries benefit from a location on major trade routes, not only from the economic advantages such a location carries, but also because they are exposed to the diffusion of new ideas and technologies.

### Cores and Capitals

Many states have come to assume their present shape, and thus the location they occupy, as a result of growth over centuries. They grew outward from a central region, gradually expanding into surrounding territory. The original nucleus, or **core area,** of a state usually contains its densest population

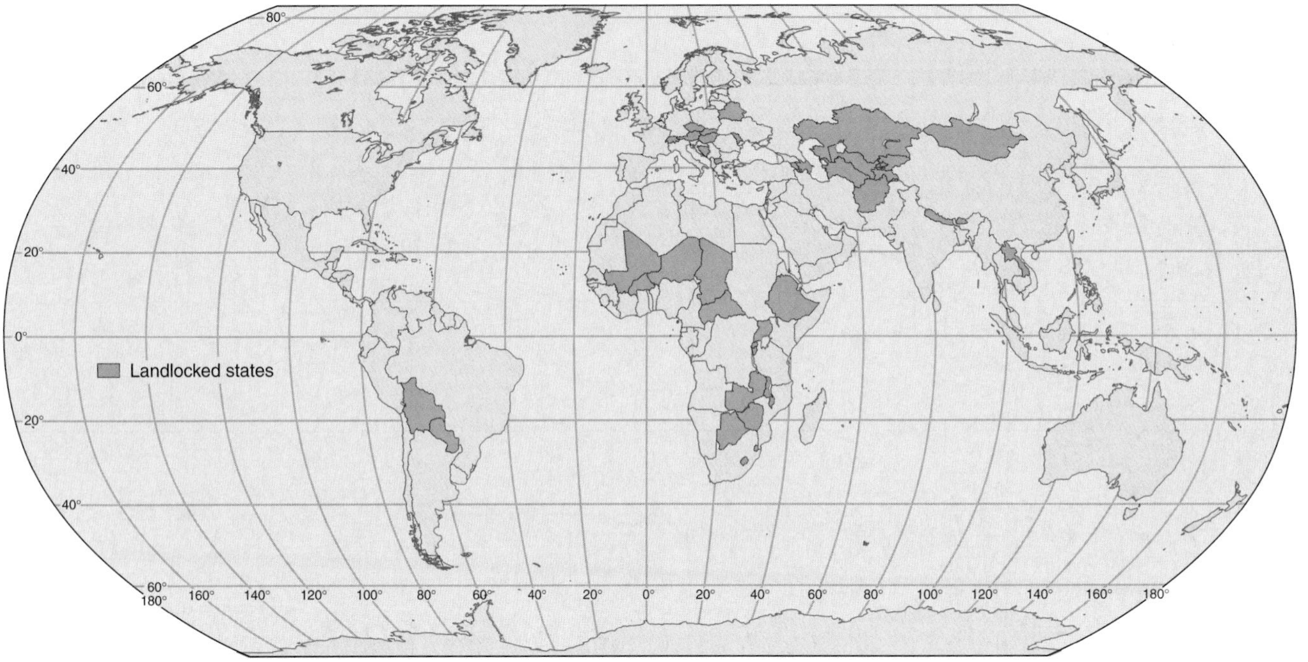


**FIGURE 9.9 Landlocked states.** Landlocked states are at a commercial and strategic disadvantage, compared to countries that have ocean frontage.

and largest cities, the most highly developed transportation system, and the most-developed economic base. All of these elements become less intense away from the national core. Urbanization ratios and city sizes decline, transport networks thin, and economic development is less intensive on the periphery than in the core.

Easily recognized and unmistakably dominant national cores include the Paris Basin of France; London and southeastern England; Moscow and the major cities of European Russia; northeastern United States and southeastern Canada; and the Buenos Aires megalopolis in Argentina. Not all countries have such clearly defined cores, and some may have two or more rival core areas. Chad, Mongolia, and Saudi Arabia have no clearly defined core, for instance, whereas Ecuador, Nigeria, Democratic Republic of the Congo, and Vietnam are examples of multicore states.

The capital city of a state is usually within its core region and frequently is the very focus of it, dominant not only because it is the seat of central authority but because of the concentration of population and economic functions as well. That is, in many countries, the capital city is also the largest or *primate* city, dominating the structure of the entire country. Paris in France, London in the United Kingdom, and Mexico City are all examples of that kind of political, cultural, and economic primacy.

This association of capital with core is common in what have been called the *unitary states,* countries with highly centralized governments, relatively few internal cultural contrasts, a strong sense of national identity, and borders that are clearly cultural as well as political boundaries. Most European cores and capitals are of this type. This association is also found in many newly independent countries whose former colonial occupiers established a primary center of exploitation and administration and developed a functioning core in a region that lacked an urban structure or organized government. With independence, the new states retained the established infrastructure, added new functions to the capital, and, through lavish expenditures on governmental, public, and commercial buildings, sought to create prestigious symbols of nationhood.

In *federal states,* associations of more or less equal provinces or states with strong regional governmental responsibilities, the capital city may have been newly created to serve as the administrative center. Although part of a generalized core region of the country, the designated capital was not its largest city and acquired few of the additional functions to make it so. Ottawa, Canada; Washington, D.C.; and Canberra, Australia, are examples (Figure 9.10).

A new form of state organization—*regional* government, or *asymmetric federalism*—is emerging in Europe as formerly strong unitary states acknowledge the autonomy aspirations of their several subdivisions and grant to them varying degrees of local administrative control while retaining in central hands authority over matters of nationwide concern, such as monetary policy, defense, and foreign relations. That new form of federalism involves the recognition of regional capitals, legislative assemblies, administrative bureaucracies, and the like. The asymmetric federalism of the United Kingdom, for example, now involves separate status for Scotland, Wales, and Northern Ireland, with their own capitals at Edinburgh, Cardiff, and Belfast. That of Spain recognizes Catalonia and the Basque country with capitals in Barcelona and Vitoria, respectively.

## The Gnarled Politics of an Enclave

The collapse of the Soviet Union was accompanied by a resurgence of ethnic and nationalist feelings that had been long suppressed in its tightly controlled republics. Nowhere was this more evident than in Nagorno-Karabakh, a mountainous enclave in Azerbaijan that is 80% Armenian. Armenians are Christians and Azeris are Shiite Muslims, but religious differences are only partly to blame for the conflict that began to consume the region in 1988.

Both Armenia and Azerbaijan were incorporated into the Soviet Union during the early 1920s. Bowing to pressure from Turkey, which had a history of discord with its own Armenian minority group and did not want a larger than necessary Armenian presence along its border, Stalin awarded the region of Nagorno-Karabakh to Azerbaijan. Although it was accorded a separate status within that republic as an autonomous district, Armenians were resentful. The Azeris are ethnically Turkish, and Armenians claim that more than 1 million of their ethnic kin were slaughtered by Turks during World War I.

The Soviets believed that nationalism would eventually disappear under communism, but that did not happen. Although vastly outnumbered by more than 7 million Azerbaijanis, the 180,000 Karabakh Armenians retained a strong sense of national identity buttressed by their own language and adherence to the Armenian Christian Church.

In 1987, encouraged by Mikhail Gorbachev's policy of liberalization and his stated willingness to rectify wrongs of the past, Karabakh Armenians agitated for union with Armenia. Banners displayed at mass demonstrations proclaimed, "Karabakh is a test case for *perestroika*" (restructuring). Rallies were held in Yerevan, the Armenian capital, to support Karabakh Armenians. There was talk of creating a corridor to link Nagorno-Karabakh with Armenia; less than 16 kilometers (10 mi) separate the two at the enclave's southwestern tip.

Azerbaijan responded by tightening its control of the enclave. Officials banned the teaching of Armenian history in the schools of Nagorno-Karabakh and discouraged the use of the Armenian language.

Violence against Armenians in the city of Sumqayit in February 1988 touched off open warfare between Armenians and Azeris. Soviet troops were called in to stop the violence,

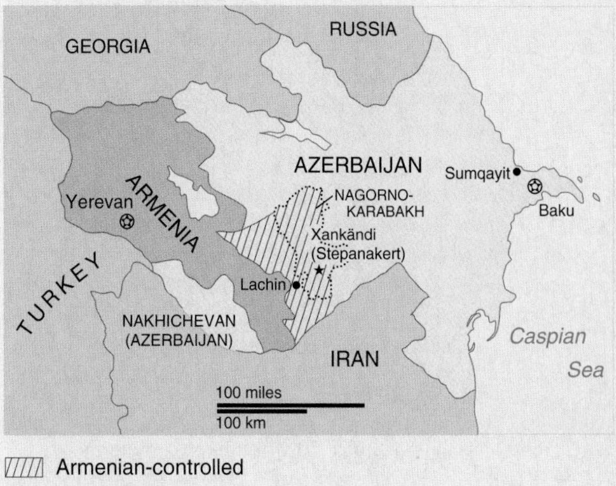


but they were unable to restore order and were largely withdrawn by 1992, by which time both Armenia and Azerbaijan had become independent states. Thousands of people died; tens of thousands of both Azeris and Armenians became refugees, seeking sanctuary in their respective homelands. In 1994, Azerbaijan agreed to a truce.

Nagorno-Karabakh is now Azeri-free. Villages and towns once populated by Azeris have been resettled, mainly with Armenian refugees from Azerbaijan. Although the territory still is recognized internationally as part of Azerbaijan, it is controlled by its resident Armenians, as is the Lachin region west of Nagorno-Karabakh. Among the proposals put forward at recent international peace talks were granting Nagorno-Karabakh and the Lachin region self-governing status and compensating Azerbaijan for its loss of territory with an internationally protected road linking it to its own exclave, Nakhichevan.

Like other enclaves, Nagorno-Karabakh tests the stability of the surrounding state. It demonstrates how boundary decisions made in the past may reverberate in the present and how forces of disruption strengthen when central authority wanes. Finally, the conflict in the enclave raises the recurring political-geographic question of how to accommodate peacefully the demands of a minority when they conflict with the wishes of the majority.

All other things being equal, a capital located in the center of the country provides equal access to the government, facilitates communication to and from the political hub, and enables the government to exert its authority easily. Many capital cities, such as Washington, D.C., were centrally located when they were designated as seats of government, but lost their centrality as the state expanded.

Some capital cities have been relocated outside of peripheral national core regions, at least in part to achieve the presumed advantages of centrality. Two examples of such relocation are from Karachi inland to Islamabad in Pakistan, and from Istanbul to Ankara, in the center of Turkey's territory. A particular type of relocated capital is the *forward-thrust capital* city, one that has been deliberately

**FIGURE 9.10** Canberra, the planned capital of Australia, was deliberately sited away from the country's two largest cities, Sydney and Melbourne. Planned capitals are often architectural showcases, providing a focus for national pride. *© Chris Groenhout.*

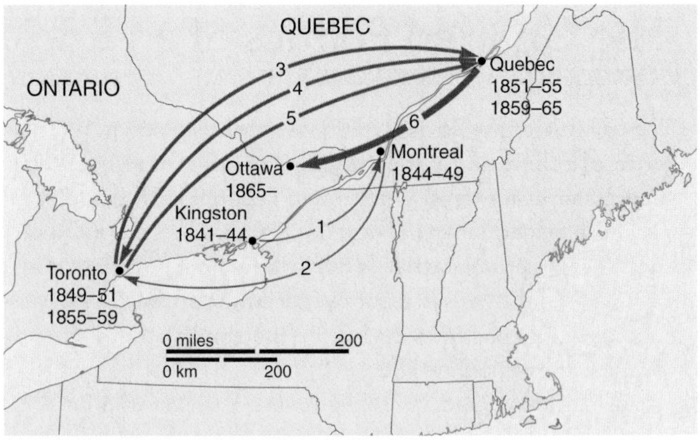


**FIGURE 9.11 Canada's migratory capital.** Kingston was chosen as the first capital of the united Province of Canada in preference to either Quebec, capital of Lower Canada, or Toronto, that of Upper Canada. In 1844, governmental functions were relocated to Montreal, where they remained until 1849, after which they shifted back and forth—as the map indicates—between Toronto and Quebec. An 1865 session of the provincial legislature was held in Ottawa, the city that became the capital of the Confederation of Canada in 1867. *Redrawn with permission from David B. Knight,* A Capital for Canada. *Chicago: University of Chicago, Department of Geography, Research Paper No. 182, 1977, Fig. 1, p. vii.*

sited in a state's frontier zone to signal the government's awareness of regions away from the core and its interest in encouraging more uniform development. In the late 1950s, Brazil moved its capital from Rio de Janeiro to the new city of Brasília to demonstrate its intent to develop the vast interior of the country. The West African country of Nigeria has been building the new capital of Abuja near its geographic center since the late 1970s, with the relocation there of government offices and foreign embassies in the early 1990s. The British colonial government relocated Canada's capital six times between 1841 and 1865, in part seeking centrality to the mid-19th century population pattern and in part seeking a location that bridged that country's cultural divide (Figure 9.11).

## Boundaries: The Limits of the State

Recall that no portion of the earth's land surface is outside the claimed control of a national unit, that even uninhabited Antarctica has had territorial claims imposed upon it (see Figure 9.3). Each of the world's states is separated from its neighbors by *international boundaries,* or lines that establish the limit of each state's jurisdiction and authority. Boundaries indicate where the sovereignty of one state ends and that of another begins.

Within its own bounded territory, a state administers laws, collects taxes, provides for defense, and performs other such governmental functions. Thus, the location of the boundary determines the kind of money people in a given area use, the legal code to which they are subject, the army they may be called upon to join, and the language and perhaps the religion children are taught in school. These examples suggest how boundaries serve as powerful reinforcers of cultural variation over the earth's surface.

Territorial claims of sovereignty are three-dimensional. International boundaries mark not only the outer limits of a state's claim to land (or water) surface but also are projected downward to the center of the earth in accordance with international consensus allocating rights to subsurface resources. States also project their sovereignty upward, but with less certainty because of a lack of agreement on the upper limits of territorial airspace. Properly viewed, then, an international boundary is a line without breadth; it is a vertical interface between adjacent state sovereignties.

Before boundaries were delimited, nations or empires were likely to be separated by *frontier zones,* ill-defined and fluctuating areas marking the effective end of a state's authority. Such zones were often uninhabited or only sparsely populated and were liable to change with shifting settlement patterns. Many present-day international boundaries lie in former frontier zones, and in that sense, the boundary line has replaced the broader frontier as a marker of a state's authority.

### Natural and Artificial Boundaries

Geographers have traditionally distinguished between "natural" and "artificial" boundaries. **Natural** (or *physical*) **boundaries** are those based on recognizable physiographic features such as mountains, rivers, and lakes. Although they might seem to be attractive as borders because they actually exist in the landscape and are visible dividing elements, many natural boundaries have proved to be unsatisfactory. That is, they do not effectively separate states.

Many international boundaries lie along mountain ranges—for example, in the Alps, Himalayas, and Andes. Some have proved to be stable, others have not. Mountains are rarely total barriers to interaction. Although they do not invite movement, they are crossed by passes, roads, and tunnels. High pastures may be used for seasonal grazing, and a mountain region may be a source of water for hydroelectric power. Nor is the definition of a boundary along a mountain range a simple matter. Should it follow the crests of the mountains or the *water divide* (the line dividing two drainage areas)? The two are not always the same. Border disputes between China and India are, in part, the result of the failure of mountain crests and headwaters of major streams to coincide (Figure 9.12).

Rivers can be even less satisfactory as boundaries. In contrast to mountains, rivers foster interaction. River valleys are likely to be agriculturally or industrially productive and to be densely populated. For example, for hundreds of miles, the Rhine River serves as an international boundary in Western Europe. It is also a primary traffic route lined by chemical plants, factories, and power stations, and dotted by the castles and cathedrals that make it one of Europe's major tourist attractions. It is more a common, intensively used resource than a barrier in the lives of the states it borders.

**FIGURE 9.12** Several international borders run through the jumble of the Himalayas. The mountain boundary between India and China has long been in dispute. Source: *© Fred Bavendam/Peter Arnold, Inc.*

The alternative to natural boundaries are **artificial** or **geometric boundaries.** Frequently delimited as sections of parallels of latitude or meridians of longitude, they are found chiefly in Africa, Asia, and the Americas. The western portion of the United States–Canada border, which follows the 49th parallel, is an example of a geometric boundary. Many such boundaries were established when the areas in question were colonies, the land was only sparsely settled, and detailed geographic knowledge of the frontier region was lacking.

### Boundaries Classified by Settlement

Boundaries can also be classified according to whether they were laid out before or after the principal features of the cultural landscape developed. An **antecedent boundary** is one drawn across an area before it is well populated, that is, before most of the cultural landscape features developed. To continue our earlier example, the western portion of the United States–Canada boundary is such an antecedent line, having been established by a treaty between the United States and Great Britain in 1846.

Boundaries drawn after the development of the cultural landscape are termed **subsequent boundaries.** One type of subsequent boundary is a **consequent** (also called *ethnographic*) **boundary,** a border drawn to accommodate existing religious, linguistic, ethnic, or economic differences between countries. An example is the boundary drawn between Northern Ireland and Eire (Ireland). Subsequent **superimposed boundaries** may also be forced upon existing cultural landscapes, a country, or a people by a conquering or colonizing power that is unconcerned about preexisting cultural patterns. The colonial powers in 19th-century Africa superimposed boundaries upon established African cultures without regard to the tradition, language, religion, or ethnic affiliation of those whom they divided (see Figure 9.5).

When Great Britain prepared to leave the Indian subcontinent after World War II, it was decided that two independent states would be established in the region: India and Pakistan. The boundary between the two countries, defined in the partition settlement of 1947, was thus both a *subsequent* and a *superimposed* line. As millions of Hindus migrated from the northwestern portion of the subcontinent to seek homes in India, millions of Muslims left what would become India for Pakistan. In a sense, they were attempting to ensure that the boundary would be *consequent*—that is, that it would coincide with a division based on religion. This boundary example is more fully discussed in "Political Regions in the Indian Subcontinent" in Chapter 13.

If a former boundary line that no longer functions as such is still marked by some landscape features or differences on the two sides, it is termed a *relic boundary*. The abandoned castles dotting the former frontier zone between Wales and England are examples of a relic boundary. They are also evidence of the disputes that sometimes attend the process of boundary making. A more recent example is the Berlin Wall, built by communist East Germany in 1961 to seal off the border between East and West Berlin. The dismantling of the wall in 1990 marked the reunification of Germany; any part of the wall that remains standing as a historic monument is a relic boundary.

### Boundaries as Sources of Conflict

Boundaries create many possibilities for conflict between countries. At the start of the 21st century, about 80 countries are involved in disputes with one or more of their neighbors. Although the causes of conflict are varied, geographic considerations underlie many of them. Figure 9.13 shows for an imaginary state, Hypothetica, the spatial conditions that could give rise to conflict with its neighbors. Each condition is identified by number, and each is illustrated by real-world examples in the discussion that follows.

#### *Landlocked States (Potential Trouble Spot #1)*

Hypothetica is a landlocked country, as are about one-fifth of the world's states (see Figure 9.9). To trade with overseas markets, landlocked states have to import and export their goods by land-based modes of transportation. They must cooperate with neighboring states and arrange for the goods to travel across a foreign country, and cooperation may be difficult.

Access to the sea usually is gained in one of two ways. Typically, the landlocked country arranges to use facilities at a foreign port plus to have the right to travel to that port, but such arrangements are not without their problems. For centuries, landlocked countries have had to contend with restrictions, tolls, high fees for transit and storage, complicated customs formalities, the risk that the goods will be lost or damaged, and other obstructions to the easy movement of goods to and from the sea. In addition, they have had little or no control over the availability and efficiency of the transport and port facilities outside their borders, and they face the possibility that war will close their access to the sea. However, the situation has improved in recent decades for countries that have signed international conventions permitting the movement of goods across intervening territories without discriminatory taxes, tolls, or freight charges. Bolivia, for example, has access to the Chilean port of Arica, the Peruvian port of Ilo, and the Argentinian city of Rosario on the Parana River (Figure 9.14).

Rather than depend on another country's port and goodwill, some landlocked states have gained access to the sea through a narrow corridor of land that reaches either the sea or a navigable river. Examples include the Congo Corridor of the Democratic Republic of Congo and the Caprivi Strip of Namibia, which was designed by the Germans to give what was then their colony of Southwest Africa access to the Zambezi River and Indian Ocean. Although these corridors have endured, others—such as the Polish and Finnish Corridors established after World War I—were short-lived.

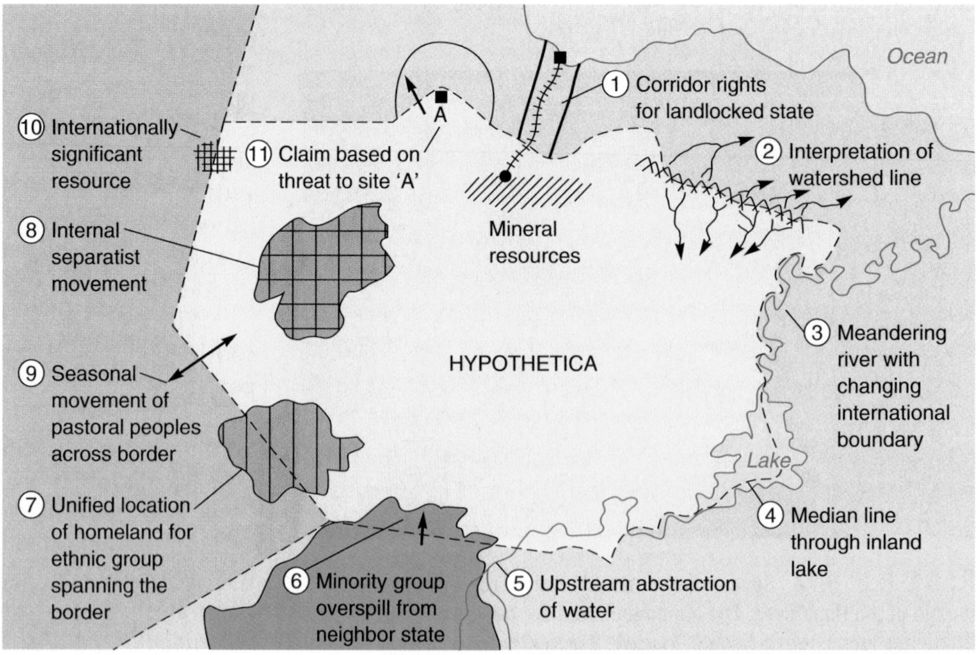


**FIGURE 9.13 Geographic sources of international stress.** To illustrate the conditions that can give rise to conflicts between states, eminent British geographer Peter Haggett drew this map of a hypothetical state and identified potential trouble spots. Real-world examples of the stress points and disputes shown in this map are discussed in the text. *Redrawn from Peter Haggett,* Geography: A Global Synthesis *(Prentice Hall, 2001). Figure 17.10, p. 522.*

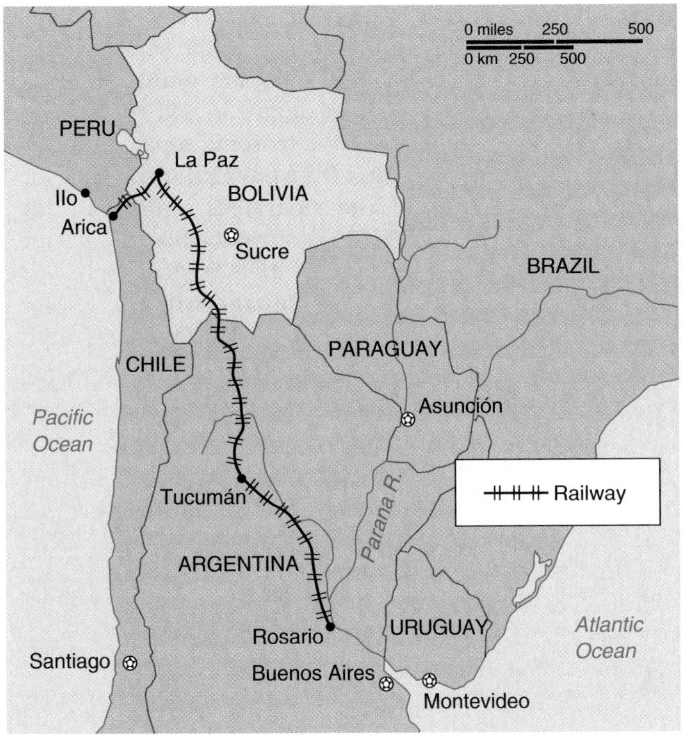


**FIGURE 9.14** Like many other landlocked countries, Bolivia has gained access to the sea through arrangements with neighboring states. Unlike most landlocked countries, however, Bolivia can access ports on two oceans.

***Waterbodies as National Boundaries (#2, #3, #4, #5)***

As we noted earlier (p. 311), although rivers and lakes form parts of many national borders, they create many opportunities for conflict. Any body of water that forms part of a border requires agreement on where the boundary line should lie: along the right or left bank or shore, along the center of the waterway, or perhaps along the middle of the navigable channel. Soviet insistence that its sovereignty extended to the Manchurian (Dongbei) bank of the Amur and Ussuri rivers was a long-standing matter of dispute and border conflict between the USSR and the People's Republic of China, only resolved with Russian agreement in 1987 that the boundary should pass along the main channel of the rivers. Even an agreement in accordance with international custom that the boundary be drawn along the main channel may be impermanent if the river changes its course, floods, or dries up.

Potential trouble spot #2 relates to the use of a watershed boundary in defining an international dividing line—that is, one that runs along a ridge or crest dividing two drainage areas. Disputes occur when states disagree about the interpretation of documents that define a boundary and/or the way the boundary was delimited. The boundary between Argentina and Chile, originally defined during Spanish colonial rule and then established by treaty in 1881, was to follow "the most elevated crests of the Andean Cordillera dividing the waters" (Figure 9.15). Because the southern Andes had not been adequately explored and mapped, it wasn't apparent that the crestlines (highest peaks) and the watershed divides between east- and west-flowing rivers do not always coincide. In some places, the water divide is many miles east of the highest peaks, leaving a long, narrow area of about 52,000 square kilometers (20,000 sq mi) in dispute. The discrepancies in claims made for difficult relations between Argentina and Chile for nearly a century.

Pressure point #3 shows a meandering river (one that changes its course). If the river constitutes part of the international boundary, as it does here, the border will change over time. The boundary between the United States and Mexico, for example, which runs along the main channel of the Rio Grande, has changed as the river has altered its course. Similarly, a lake (#4) requires agreement on where the boundary between two countries lies. In the case of the

**FIGURE 9.15 The disputed boundary between Argentina and Chile in the southern Andes.** The treaty establishing the boundary between the two countries preceded adequate exploration and mapping of the area, leaving its precise location in doubt.

United States and Canada, the two countries agreed that a line equidistant from the shores of Lake Erie and Ontario would form part of the international border.

Another trouble spot (#5) relates to Hypothetica's use of the river that flows downstream into it from another country. Water use is an increasing source of conflict among countries, particularly in arid or semiarid regions. When water is scarce, its abstraction, diversion, or pollution by one country can significantly affect the quantity and quality of water available to those downstream. Growing shortages of fresh water are leading to tensions along many rivers, including the Jordan, Tigris and Euphrates, Nile, Indus, Ganges, and Brahmaputra.

### *Minority Group Identification (#6, #7, #8)*

Like nearly all countries, Hypothetica contains more than one culture group. In the real world, the locations of minority groups have led to international tensions, civil wars, wars of liberation, and international strife around the globe. As one of the most difficult problems with which countries must deal, minority group identification is discussed in more detail in "Centrifugal Forces," but brief examples of pressure points #6, #7, and #8 are provided in the following paragraphs.

Conflicts can arise if the people of one state claim and seek to acquire a territory whose population is historically or ethnically related to that of the state but is now subject to a foreign government. This condition is represented by pressure point #6, minority group overspill from a neighboring state. Under these conditions, the desire to expand the country's borders is called **irredentism,** from the Italian word for "unredeemed." In 1938, Hitler used the existence of German minorities in a part of Czechoslovakia known as Sudetenland to justify Germany's occupation of the region, which had been awarded to Czechoslovakia after World War I. Sudetenland was restored to Czechoslovakia after World War II.

Hungary's claims to Transylvania, a Romanian province, are based on both historical and ethnic ties. The two countries have quarreled over it for centuries. Transylvania was under Hungarian control from 1649 until 1920, when—as part of the reordering of the political map of Europe that followed World War I—it became part of Romania. In 1940, Germany and Italy forced Romania to give the province back to Hungary, but the country had to surrender it again after World War II.

We earlier discussed the case of the stateless nation (p. 302), a people without a state, and cited as examples Kurds, Roma, Basques, and Palestinians. Pressure point #7 shows a distinct ethnic group or nation located in both Hypothetica and a neighboring state. Conflict occurs when nations seek to govern themselves in their own state and try to carve out a new nation-state from portions of existing countries. As the example of the Basques indicates, they need not represent a majority of their residents in order to foment discord.

Basques live in a region overlapping France and Spain (Figure 9.16). In an attempt to dampen the separatist fires that had burned since the 1960s, Spain granted its three Basque provinces a significant degree of self-rule in 1978, but that has failed to satisfy the extreme separatist movement Euskadi ta Askatasuna (ETA), which means "Basque Homeland and Liberty." The separatists contend that the Spanish state has attempted to destroy the Basques' unique cultural identity and to suppress their language, Euskadi, which is unrelated to any other language in the world. They demand an independent, unified Basque state, not only for the Basque region of Spain but for a portion of southern France as well, and the ETA continues to wage urban guerrilla warfare against Spain, with much of the staging for the attacks taking place in France. The nationalist political party in the region rejects the violent tactics of the ETA separatists and favors peaceful moves toward independence. Members of yet other political parties are satisfied with the autonomy the region now possesses, but the ETA keeps alive an atmosphere of division and tension.

In the case of pressure point #8, an internal separatist movement, the group seeking independence is totally contained within Hypothetica. The Civil War in the United States in the 19th century is one example of a secessionist conflict, but wars of secession were numerous in the 20th century, particularly in Africa and Asia. The 1967–1970 civil war in Nigeria is one example.

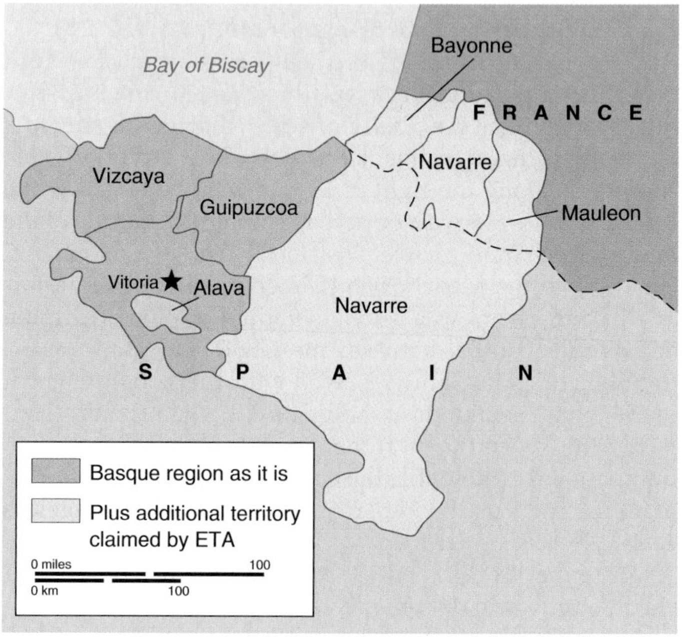


**FIGURE 9.16** The Basque region straddles the border between Spain and France. Although the Basques were granted a measure of self-rule for their region, militant separatists in the Euskadi ta Askatasuna (ETA) want to see the establishment of an independent state for the Basque region of Spain and a portion of southern France.

### *Resource Disputes (#9, #10, #11)*

Neighboring states are likely to covet the resources—whether they be valuable mineral deposits, rich fishing grounds, or a cultural resource, such as a site of religious significance—lying in border areas and to disagree over their use. In recent years, for example, the United States has been involved in resource disputes with both its immediate neighbors: with Mexico over the shared resources of the Colorado River and Gulf of Mexico and with Canada over the Georges Bank fishing grounds in the Atlantic Ocean.

Conflicts arise when neighboring states disagree over policies to be applied along a border. Such policies may concern the movement of traditionally nomadic groups (#9), immigration, customs regulations, and the like. U.S. relations with Mexico, for example, have been affected by the increasing number of illegal aliens and the flow of drugs entering the United States from Mexico (Figure 9.17). Russia and Finland have entered into an agreement that allows Laplanders to herd reindeer irrespective of the border between the two countries.

The location of an internationally significant resource in a border region (#10) provides another opportunity for conflict. One of the causes of the 1990–1991 war in the Persian Gulf was the huge oil reservoir known as the Rumaila field, which lies mainly in Iraq, with a small extension into Kuwait (Figure 9.18). Because the two countries had been unable to agree on percentages of ownership, or a formula for sharing production costs and revenues, Kuwait pumped oil from Rumaila without any international agreement. Iraq helped justify its invasion of Kuwait by contending that the latter had been stealing Iraqi oil in what amounted to economic warfare.

A final potential trouble spot is represented by site A on the map in Figure 9.11 (#11), the location of a resource that the state believes is crucial to its survival and that must be defended, even if it means claiming an adjacent piece of land in a neighboring state. The resource might be physical, such as a military post, or cultural, such as a holy city. Syria and Israel, for example, dispute the ownership of the Golan Heights, which contains water and is high ground. It allows the country that controls it to look down upon and listen to the country on the other side—Israelis can literally look at

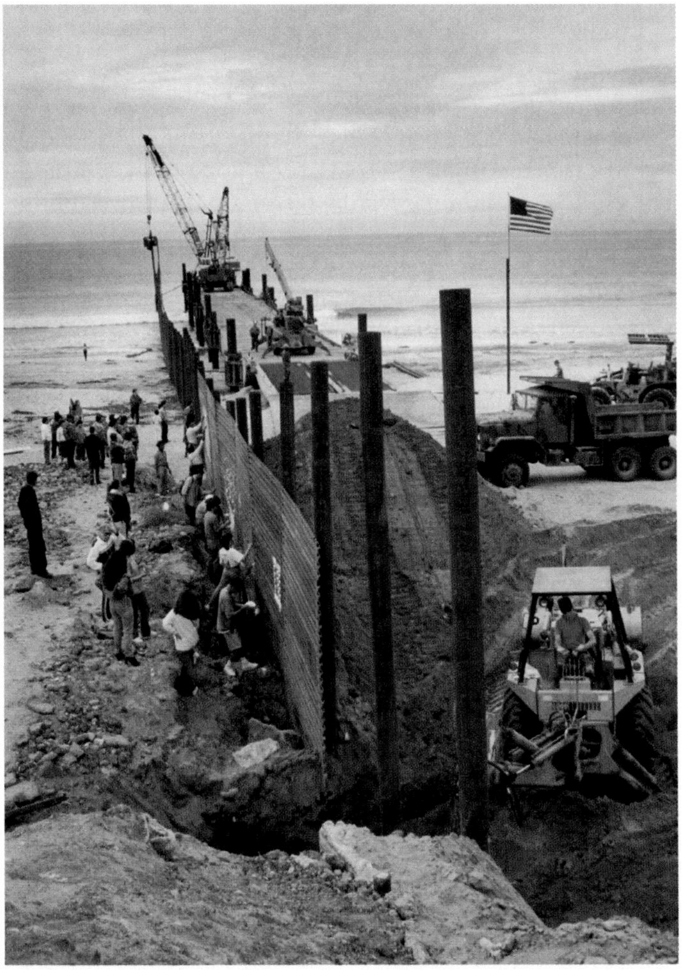

**FIGURE 9.17** To stem the flow of undocumented migrants entering California from Baja California, the United States, in 1993, constructed a fence 3 meters (10 ft) high along the border. *© San Diego Union Tribune/John Nelson.*

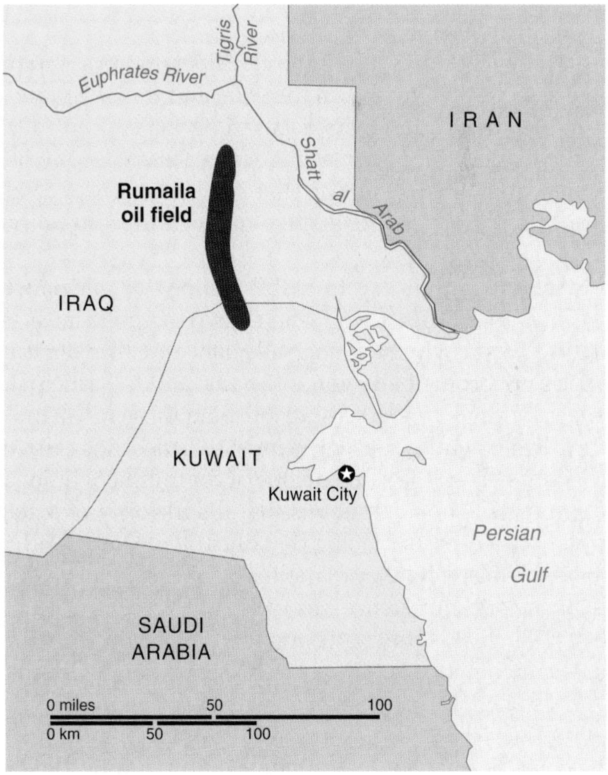


**FIGURE 9.18 The Rumaila oil field.** One of the world's largest oil reservoirs, Rumaila straddles the Iraq-Kuwait border. Iraqi grievances over Kuwaiti drilling were partly responsible for Iraq's invasion of Kuwait in 1990.

Damascus from Mount Hermon. A disputed cultural resource is represented by Jerusalem, a site of great religious significance to Christians, Jews, and Muslims. It has been the source of conflicts since at least the beginning of the first crusade in A.D. 1096. Believing that Jerusalem is vital to its own identity, Israel has in recent years effectively annexed much of Muslim East Jerusalem. Currently, one of the principal points of contention between the Israeli government and Palestinians is access to and the control of holy sites in the city.

## Centripetal Forces: Promoting State Cohesion

At any moment in time, a state is characterized by forces that promote unity and national stability and by others that disrupt them. Political geographers refer to the former as **centripetal forces.** These unifying factors bind together the people of a state, enable it to function, and give it strength. **Centrifugal forces,** on the other hand, destabilize and weaken a state. If centrifugal forces are stronger than those promoting unity, the very existence of the state will be threatened. In the sections that follow, we examine four forces—nationalism, unifying institutions, effective organization and administration of government, and systems of transportation and communication—to see how they can promote cohesion.

### Nationalism

One of the most powerful of the centripetal forces is **nationalism,** an identification with the state and the acceptance of national goals. Nationalism is based on the concept of allegiance to a single country and the ideals and the way of life it represents. It is an emotion that provides a sense of identity and loyalty and of collective distinction from all other peoples and lands.

States purposely try to instill feelings of allegiance in their constituents, for such feelings give the political system strength. People who have such allegiance are likely to accept the rules governing behavior in the area and to participate in the decision-making process establishing those rules. In light of the divisive forces present in most societies, not everyone, of course, will feel the same degree of commitment or loyalty. The important consideration is that the majority of a state's population accept its ideologies, adhere to its laws, and participate in its effective operation. For many countries, such acceptance and adherence has come only recently and partially; in some, it is frail and endangered.

Recall that true nation-states are rare; in only a few countries do the territory occupied by the people of a particular nation and the territorial limits of the state coincide. Most countries have more than one culture group that considers itself separate in an important way from other citizens. In a multicultural society, nationalism helps integrate different groups into a unified population. This kind of consensus nationalism has emerged in countries such as the United States and Switzerland, where different culture groups have joined together to create political entities commanding the loyalties of all of their citizens.

States promote nationalism in a number of ways. *Iconography* is the study of the symbols that help unite people. National anthems and other patriotic songs, flags, national sports teams, rituals, and holidays are all developed as symbols of a state in order to promote nationalism and attract allegiance (Figure 9.19) (see "Sports and National Identity," p. 317). By ensuring that all citizens, no matter how diverse the population, will have at least these symbols in common, they impart a sense of belonging to a political entity, called, for example, Japan or Canada. In some countries, certain documents, such as the Magna Charta in England or the Declaration of Independence in the United States, serve the same purpose. Royalty may fill the need: in Sweden, Japan, and Great Britain, the monarchy functions as the symbolic focus of allegiance. Symbols and beliefs are major components of every culture. When a society is very heterogeneous, composed of people with different customs, religions, and languages, belief in the national unit can help weld them together.

### Unifying Institutions

Institutions as well as symbols help develop the sense of commitment and cohesiveness essential to the state. Schools, particularly elementary schools, are among the most important of these. Children learn the history of their own country and relatively little about other countries. Schools are expected to instill the society's goals, values, and traditions, to teach the common language that conveys them, and to guide youngsters to identify with their country.

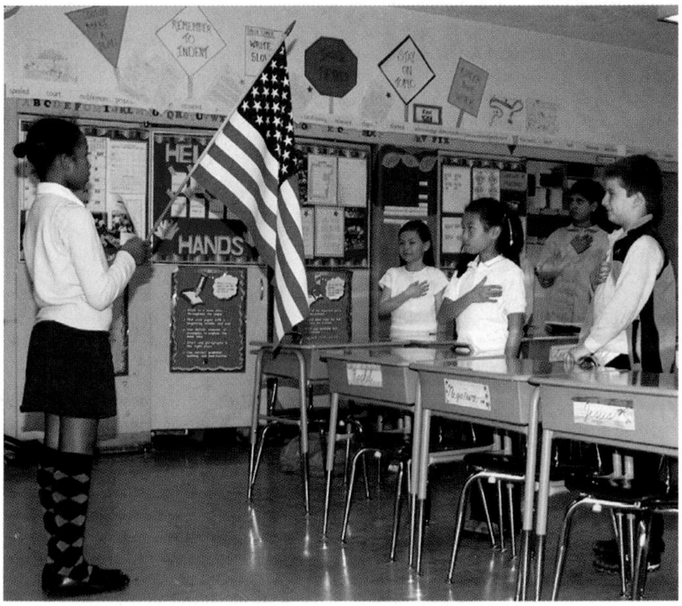


**FIGURE 9.19** The ritual of the Pledge of Allegiance is just one way in which schools in the United States seek to instill a sense of national identity in students. *© The McGraw-Hill Companies.*

## Sports and National Identity

In 2003, two of the states of the former Yugoslavia, Serbia and Montenegro, were renamed, becoming "The State Union of Serbia and Montenegro." Residents agree that the new name is awful. "It takes an eternity just to pronounce it," scoffed Maja Jovanovic, a 28-year-old hairdresser in Belgrade. "I can't exactly imagine anyone falling into a state of frenzy chanting it at a soccer match." There is no national flag, there is no national anthem, and the name is so long that people have taken to calling it Sam. One economist remarked, "Nobody screams for 'Serbia and Montenegro' at soccer matches."

The Serbians and Montenegrans are not obsessed with soccer, but vitally concerned about centripetal forces that might bring unity to a war-torn land. Sports bring people together. At the close of the 1976 summer Olympics, an article in the *New York Times* remarked:

> As the curtain drops on the Games of the XXI Olympiad, this image emerges: Remove nationalism from the Olympics and there can be no Olympics. Let the purists moralize and the editorial writers agonize, but the sad fact is—for better or for worse—nationalism is what the Olympics are all about.
>
> The truth of this was written in every quivering rafter of the Forum on Tuesday night when the United States team won the basketball gold medal by defeating Yugoslavia. American flags waved against a backdrop of faces screaming "U-S-A, U-S-A" and the scene was one of rampant nationalism.
>
> There were voices and flags for the Yugoslavs, too, and earlier for the Russians and the Canadians. Had these been teams without a country there would have been neither cheers nor flags—because there would have been no people.
>
> The people were there to cheer not individuals and individual performances but to vent the strong feelings of patriotism and to see their national team win. . . .
>
> It is fashionable and chic and one-worldish to expound the joys of competition for its own sake, but the human spirit needs something more to cling to.
>
> Eliminate the majesty of the opening Olympic ceremony, the parade of nations, the flags, the ethnic costumes. Let the athletes march only as individuals, under one generic flag with only an Olympic anthem to sing. Should these things happen, the Olympic movement would soon die of financial malnutrition.

This article was written in 1976. Have things changed now, more than a quarter of a century later? Are sports any more or less important to national feeling?



Other institutions that promote nationalism are the armed forces and, sometimes, a state church. The armed forces are, of necessity, taught to identify with the state. They see themselves as protecting the state's welfare from what are perceived to be its enemies.

In about one-quarter of the world's countries, the religion of the majority of the people has by law been designated a state church. In such cases, the church sometimes becomes a force for cohesion, helping unify the population. This is true of Islam in Pakistan, Judaism in Israel, Buddhism in Thailand, and Hinduism in Nepal. In countries such as these, the religion and the church are so identified with the state that belief in one is transferred to allegiance to the other.

The schools, the armed forces, and the church are just three of the institutions that teach people what it is like to be members of a state. As institutions, they operate primarily on the level of the sociological subsystem of culture, helping structure the outlooks and behaviors of the society. But by themselves, they are not enough to give cohesion, and thus strength, to a state. Indeed, each of the institutions we have discussed can be a destabilizing centrifugal force.

### Organization and Administration

A further bonding force is public confidence in the effective organization of the state. Can it provide security from external aggression and internal conflict? Are its resources distributed and allocated in such a way as to be perceived to promote the economic welfare of all its citizens? Are all citizens afforded equal opportunity to participate in governmental affairs (see "Legislative Women," p. 318)? Do institutions that encourage consultation and the peaceful settlement of disputes exist? How firmly established are the rule of law and the power of the courts? Is the system of decision making responsive to the people's needs?

The answers to these questions and the relative importance of the answers vary from country to country, but they and similar ones are implicit in the expectation that the state will, in the words of the Constitution of the United States, "establish justice, insure domestic tranquillity, provide for the common defense, (and) promote the general welfare. . . ." If those expectations are not fulfilled, the loyalties promoted by national symbols and unifying institutions may be weakened or lost.

### Transportation and Communication

A state's transportation network fosters political integration by promoting interaction between areas and by joining them economically and socially. The role of a transportation network in uniting a country has been recognized since ancient times. The saying that all roads lead to Rome had its

## Legislative Women

Women, a majority of the world's population, in general fare poorly in the allocation of such resources as primary and higher education, employment opportunities and income, and health care. That their lot is improving is encouraging. In nearly every developing country, women have been closing the gender gap in literacy, school enrolment, and acceptance in the job market.

But in the political arena, where power ultimately lies, women's share of influence is increasing only slowly and selectively. In 2003, fewer than 15 countries out of a world total of some 200 had women as heads of government: presidents or prime ministers. Nor did they fare much better as members of parliaments. In that year, women held just 15% of all the seats in the world's legislatures.

Only in 25 countries did women occupy one-quarter or more of the seats in the lower or single legislative house in 2003. Of these 25, 10 were European and 5 were African. Sweden was the most feminist, with 45% of its members female. In no country were women a legislative majority, and a number of countries had no female representatives at all. Although 10 of the countries with women making up one quarter of the legislature are in Europe, in many other European countries, women occupy a tiny minority of legislative seats. This includes both established democracies of Northern and Western Europe and virtually all of the countries of Southern and Eastern Europe. Typical percentages are those for France (12%), Hungary (10%), and Greece (9%).

Many countries are witnessing increased discontent with the proportion of women in legislatures. In the 1990s, women's legislative representation began to expand materially in many developed and developing democracies, and their "fair share" of political power began to be formally recognized or enforced. In Western countries, particularly, improvement in female parliamentary participation has become a matter of plan and pride for political parties and, occasionally, for governments themselves. Political parties from Mexico to China have tried to correct female underrepresentation, usually by setting quotas for women candidates, and a few governments—including Belgium and Italy—have tried to require their political parties to improve their balance.

France went further than any other country in acknowledging the right of women to equal access to elected office when, in 1999, it passed a constitutional amendment requiring *parité*—parity, or equality. A year later, the National Assembly enacted legislation requiring the country's political parties to fill 50% of the candidacies in all elections in the country (municipal, regional, and European Parliament) with women or lose a corresponding share of their campaign funding, which, in France, the state provides. That is, all political parties must put forward as many female as male candidates.

Quotas are controversial, however, and often are viewed with disfavor, even by avowed feminists. Some argue that quotas are demeaning because they imply that women cannot match men on merit alone. Others fear that other groups (e.g., religious groups, ethnic minorities) would also seek quotas to assure their fair representation in legislatures.

Notice that the United States was not among the 25 countries where women held 25% or more of national legislative seats in 2003. In the 108th Congress (2003–2005), only 14 women served in the Senate and 62 in the House of Representatives, making up 14% of both House and Senate, an all-time high.

American women have made greater electoral gains in state legislatures, where their percentage rose steadily over the past several decades, from 4% in 1969 to 22% in 2003. Wide dis-

origin in the impressive system of roads that linked Rome to the rest of the empire. Centuries later, a similar network was built in France, linking Paris to the various departments of the country. Often, the capital city is better connected to other cities than the outlying cities are to one another. In France, for example, it can take less time to travel from one city to another by way of Paris than by direct route.

Roads and railroads have played a historically significant role in promoting political integration. In the United States and Canada, they not only opened up new areas for settlement, but also increased interaction between rural and urban areas. Because transportation systems play a major role in a state's economic development, it follows that, the more economically advanced a country is, the more extensive its transport network is likely to be. At the same time, the higher the level of development, the more money there is to be invested in building transport routes. In other words, the two reinforce one another.

Transportation and communication, although encouraged within a state, are frequently curtailed or at least controlled between states as a conscious device for promoting state cohesion through limitation on external spatial interaction (Figure 9.20). The mechanisms of control include restrictions on trade through tariffs and embargoes, legal barriers to immigration and emigration, and limitations on travel through passports and visa requirements.

### Centrifugal Forces: Challenges to State Authority

State cohesion is not easily achieved or, once gained, invariably retained. Destabilizing centrifugal forces are

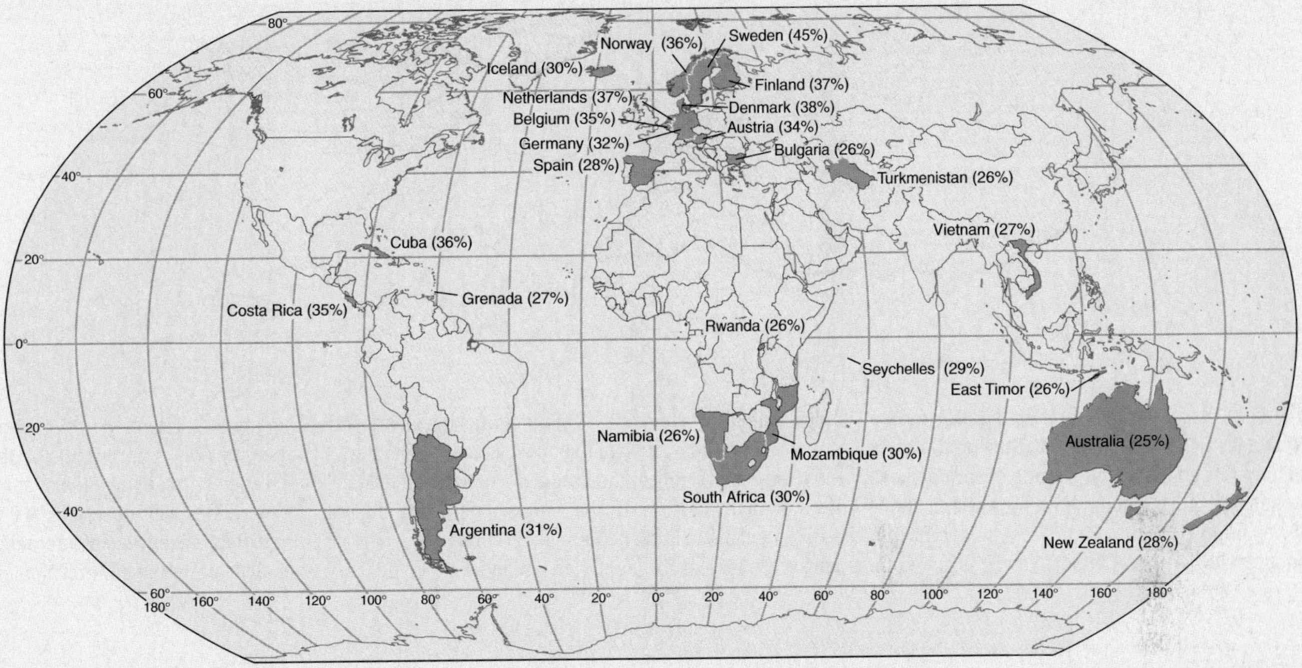


**Countries where women held 25% or more of the seats in the lower or single legislative house in 2003.**

parities exist among the states, however. In 2003, there were 7382 state legislators, of which 1648, or 22.3%, were women. In Washington, 37% of state lawmakers were women in 2003; in Colorado and Maryland, 33%; and in Oregon, 31%. At the other end of the spectrum was South Carolina, where women constituted just 9% of the state legislature.

A significant presence of women in legislative bodies makes a difference in the kinds of bills that get passed and the kinds of programs that receive governmental emphasis. Regardless of party affiliation, women tend to have somewhat different priorities than male lawmakers. For example, women are more apt than their male counterparts to sponsor bills and vote for measures in such policy areas as child care, long-term care for the elderly, affordable health insurance, women's health issues, and women's rights, including divorce and spousal-abuse laws.

ever-present, sowing internal discord and challenges to the state's authority (see "Terrorism and Political Geography," p. 322). Transportation and communication may be hindered by a country's shape or great size, leaving some parts of the country not well integrated with the rest. A state that is not well organized or administered stands to lose the loyalty of its citizens. Institutions that in some states promote unity can be a divisive force in others.

*Organized religion,* for example, can be a potent centrifugal force. It may compete with the state for people's allegiance—one reason the former USSR and other communist governments suppressed religion and promoted atheism. Conflict between majority and minority faiths within a country—as between Catholics and Protestants in Northern Ireland or Hindus and Muslims in Kashmir and Gujarat State in India—can destabilize social order. Opposing sectarian views within a single, dominant faith can also promote civil conflict. Recent years have seen Muslim militant groups attempt to overturn official or constitutional policies of secularism or replace a government deemed insufficiently ardent in its imposition of religious laws and regulations. Islamic fundamentalism led to the 1979 overthrow of the shah of Iran; more recently, Islamic militancy has been a destabilizing force in, among other countries, Afghanistan, Algeria, Tunisia, Egypt, and Saudi Arabia.

*Nationalism,* in contrast to its role as a powerful centripetal agency, is also a potentially very disruptive centrifugal force. We previously identified four types of relationships between states and nations: a nation-state, a multinational state, a part-nation state, and a stateless nation (see Figure 9.4). The idea of the nation-state is that states are formed around and coincide with nations. It is a

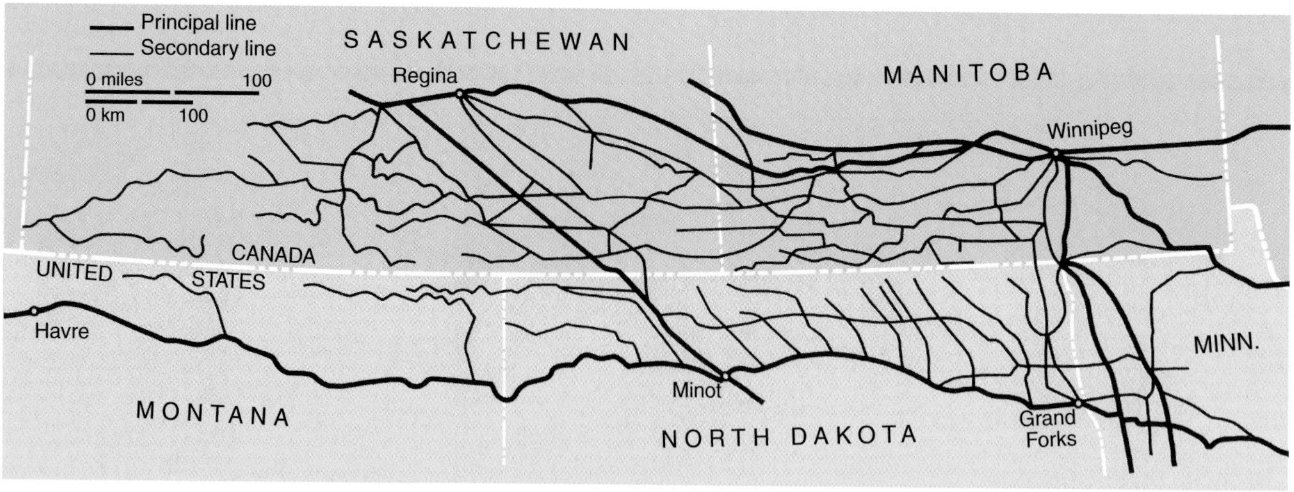


**FIGURE 9.20 Canadian–United States railroad discontinuity.** Canada and the United States developed independent railway systems connecting their respective prairie regions with their separate national cores. Despite extensive rail construction during the 19th and early 20th centuries, the pattern that emerged even before recent track abandonment was one of discontinuity at the border. Note how the political boundary restricted the ease of spatial interaction between adjacent territories. Many branch lines approached the border, but only eight crossed it. In fact, for more than 480 kilometers (300 mi), no railway bridged the boundary line. The international border—and the cultural separation it represents—inhibits other expected degrees of interaction. Telephone calls between Canadian and U.S. cities, for example, are far less frequent than would be expected if distance alone were the controlling factor.

small step from that to the notion that every nation has the right to its own state or territory.

Centrifugal forces are particularly strong in countries containing multiple nationalities and unassimilated minorities, racial or ethnic conflict, contrasting cultures, and a multiplicity of languages or religions. Such states are susceptible to nationalist challenges from within their borders: a country whose population is not bound by a shared sense of nationalism but is split by several local primary allegiances suffers from **subnationalism.** That is, many people give their primary allegiance to traditional groups or nations that are smaller than the population of the entire state.

Any country that contains one or more important national minorities is susceptible to challenges from within its borders if the minority group has an explicit territorial identification and believes that its right to *self-determination*—the right of a group to govern itself in its own state or territory—has not been satisfied. In its intense form, **regionalism,** a strong minority group self-awareness and identification with a region rather than with the state, can be expressed politically as a desire for more autonomy (self-government) or even separation from the rest of the country. It is prevalent in many parts of the world today and has created currents of unrest within many countries, even long-established ones.

Canada, for example, houses a powerful secessionist movement in French-speaking Quebec, the country's largest province. In October 1995, a referendum to secede from Canada and become a sovereign country failed by a scant margin (49% yes, 51% no). Quebec's nationalism is fueled by strong feelings of collective identity and distinctiveness, as well as by a desire to protect its language and culture. Additionally, separatists believe that the province, which has ample resources and one of the highest standards of living in the industrialized world, could manage as well or better on its own than within Canada.

In Western Europe, five countries (the United Kingdom, France, Belgium, Italy, and Spain) contain political movements whose members reject total control by the existing sovereign state and who claim to be the core of a separate national entity (Figure 9.21). Some separatists would be satisfied with *regional autonomy,* usually in the form of self-government, or "home rule"; others seek complete independence for their regions.

In an effort to defuse these separatist movements and to accommodate politically or culturally diverse peoples within their own borders, several European governments have moved in the direction of regional recognition and **devolution** (decentralization) of political control. Britain, France, Spain, Portugal, and Italy are among the countries that have recognized the need for administrative structures that reflect regional concerns and have granted a degree of political autonomy to recognized political subunits, giving them a measure of self-rule short of complete independence. In 1999, for example, voters in Scotland and Wales elected representatives to two newly created legislatures—the Scottish Parliament and the National Assembly of Wales. The legislatures have authority over such domestic matters as local government, housing, health, education, culture, transportation, and the environment. The British Parliament kept its powers over broad national policy concerns: defense, foreign policy, the economy, and monetary policy.

Nationalist challenges to state authority affect many countries outside of Western Europe, of course, and indeed

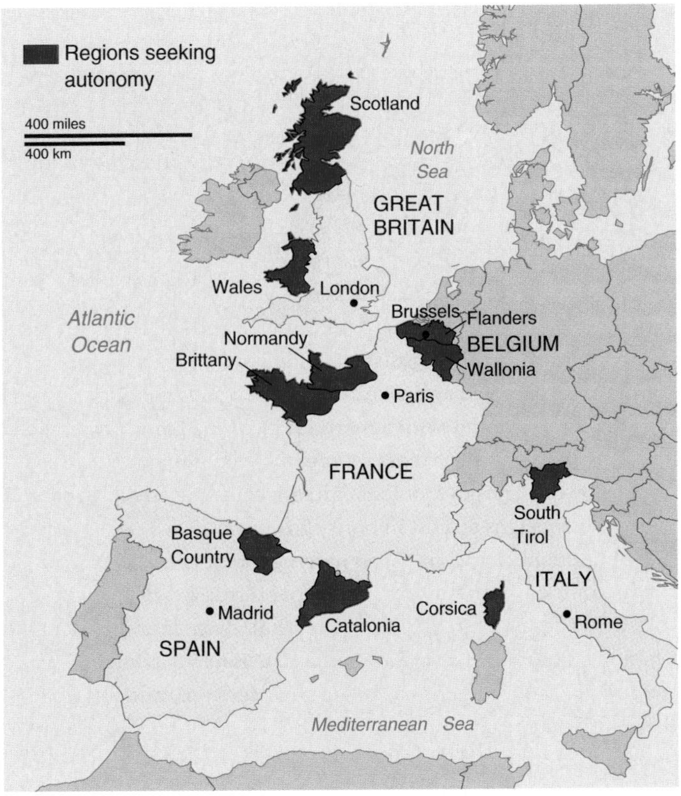


(a)

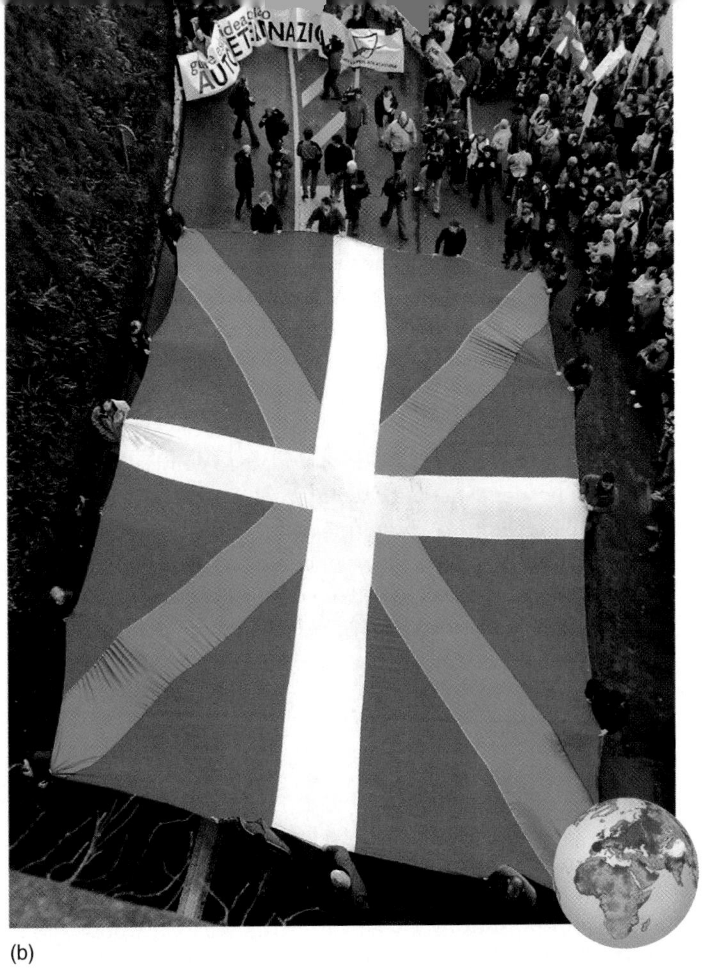

(b)

**FIGURE 9.21** (a) **Regions in Western Europe seeking autonomy.** Despite long-standing state attempts to culturally assimilate these historic nations, each contains a political movement that has recently sought or is currently seeking a degree of self-rule that recognizes its separate identity. Separatists on the island of Corsica, for example, want to secede from France, and separatists in Catalonia demand independence from Spain. The desires of nationalist parties in both Wales and Scotland were partially accommodated by the creation in 1999 of their own parliaments and a degree of regional autonomy, an outcome labeled "separation but not a divorce" from the United Kingdom. (b) Demonstrators carry a giant Basque flag during a march to call for independence for the Basque region. Spain has granted limited independence to the Basque region in an example of asymmetric federalism. *(b) © AP/Wide World Photos.*

are more characteristic of developing countries, especially those formed since the end of World War II and containing disparate groups more motivated by enmity than affinity. The Basques of Spain and the Bretons of France have their counterparts in the Palestinians in Israel, the Sikhs in India, the Tamils in Sri Lanka, the Moros in the Philippines, and many others.

The countries of Eastern Europe and the republics of the former Soviet Union have seen many instances of regionally rooted nationalist feelings. Now that the forces of ethnicity, religion, language, and culture are no longer suppressed by communism, ancient rivalries are more evident than at any time since World War II. The end of the Cold War aroused hopes of decades of peace. Instead, the collapse of communism and the demise of the USSR spawned many smaller wars. Numerous ethnic groups, large and small, are asserting their identities and what they perceive to be their right to determine their own political status.

The national independence claimed in the early 1990s by the 15 former Soviet republics did not assure the satisfaction of all separatist movements within them (see Figure 9.6). Many of the new individual countries are subject to strong destabilizing forces that challenge their territorial integrity and survival. The Russian Federation itself, the largest and most powerful remnant of the former USSR, has 89 components, including 21 "ethnic republics" and a number of other nationality regions. Many are rich in natural resources, have non-Russian majorities, and seek greater autonomy within the federation. Some, indeed, want total independence. One, the predominantly Muslim republic of Chechnya, in 1994 claimed the right of self-determination and attempted to secede from the federation, provoking a bloody war that escalated again in 1996 and 1999.

As the USSR declined and eventually disbanded, it lost control of its communist satellites in Eastern Europe. That loss and resurgent nationalism led to a dramatic reordering of the region's political map. East Germany was reunified with West Germany in 1990, and 3 years later, the people of Czechoslovakia agreed to split their country into two separate, ethnically based states: the Czech Republic and Slovakia. More violently, Yugoslavia shattered into five pieces in 1991–1992, but with the exception of Slovenia, the

## Terrorism and Political Geography

"Where were you when the world stopped turning?" asks Alan Jackson in his song about the September 11, 2001, terrorist attacks on the United States. You probably know the answer to his question, and you probably always will. Of course, the world didn't really stop turning, but that's how it felt to millions of Americans with no previous exposure to terrorism.

What is terrorism? How does it relate to political geography? Do all countries experience terrorism? Is terrorism new? Is there a way to prevent it? Attempting to answer these questions, difficult as they are, may help us understand the phenomenon.

**Terrorism** is the calculated use of violent acts against civilians and symbolic targets to publicize a cause, intimidate or coerce a civilian population, or affect the conduct of a government. *International terrorism,* such as the attacks of September 11, 2001, includes acts that transcend national boundaries. International terrorism is intended to intimidate people in other countries. *Domestic terrorism* consists of acts by individuals or groups against the citizens or government of their own country. *State terrorism* is committed by the agents of a government. *Subnational terrorism* is committed by groups outside a government.

Terrorism, thus, is a weapon. It is a weapon whose aim is intimidation and whose victims are usually civilians.

State terrorism is probably as old as the concept of a state. As early as 146 B.C., for example, Roman forces sacked and completely destroyed the city of Carthage, burning it to the ground, killing men, women, and children, and sowing salt on the fields so that no crops could grow. Governments have used systematic policies of violence and intimidation to further dominate and control their own populations. Nazi Germany, the Pol Pot regime in Cambodia, and Stalinist Russia are 20th-century examples of state terrorism. Heads of state ordered the murder, imprisonment, or exile of enemies of the state—politicians, intellectuals, dissidents—anyone who dared to criticize the government. In Rwanda, the former Yugoslavia, and Saddam Hussein's Iraq, state terrorism aimed against ethnic and religious minorities provided the government with a method of consolidating power; in each case, genocide, or mass murder of ethnic minority groups was the result. The government or its agencies waged full-fledged military campaigns against minority groups.

Subnational terrorism began much later, at the same time as the rise of the nation-state. Subnational terrorism can be perpetrated by those who feel wronged by their own or another government. For example, ethnic groups in a minority who feel that the national government has taken their territory and absorbed them into a larger political entity, such as the Basques in Spain, have used terrorist acts to resist the government. Ethnic and religious groups that have been split by national boundaries imposed by others, such as Palestinian Arabs in the Middle East, have used terrorism to make governance impossible. Political, ethnic, or religious groups that feel oppressed by their own government, such as the Oklahoma City bombers in the United States, have committed acts of domestic terrorism.

boundaries of the five new republics did not match the territories occupied by nationalities, a situation that plunged the region into war as nations fought to redefine the boundaries of their countries. One tactic used to transform a multinational area into one containing only one nation is **ethnic cleansing,** the killing or forcible relocation of less powerful minorities. It occurred in Croatia, Bosnia-Herzegovina, and the Kosovo province of southern Serbia. An uneasy truce now prevails in each of these regions, enforced in Bosnia and Kosovo by NATO peacekeeping forces. Few doubt that, if the soldiers were withdrawn, fighting would erupt again.

It is too early to tell whether these new states and others will be viable political entities. We can, however, make some generalizations about nationalist challenges to state authority.

The two preconditions common to all separatist movements are *territory* and *nationality.* First, the group must be concentrated in a core region that it claims as a national homeland. It seeks to regain control of land and power that it believes were unjustly taken by the ruling party. Second, certain cultural characteristics must provide a basis for the group's perception of separateness, identity, and cultural unity. These might be language, religion, or distinctive group customs, which promote feelings of group identity at the same time that they foster exclusivity. Normally, these cultural differences have persisted over several generations and have survived despite strong pressures toward assimilation.

Other characteristics common to many separatist movements are a *peripheral location* and *social and economic inequality.* Troubled regions tend to be peripheral, often isolated in rural pockets, and their location away from the seat of central government engenders feelings of alienation, exclusion, and neglect. Second, the dominant culture group is often seen as an exploiting class that has suppressed the local language, controlled access to the civil service, and taken more than its share of wealth and power. Poorer regions complain that they have lower incomes and greater unemployment than prevail in the rest of the state and that "outsiders" control key resources and industry. Separatists in relatively rich regions believe that they could exploit their resources for themselves and do better economically without the constraints imposed by the central state.

Nearly every country has experienced some form of terrorism at some point since the mid-19th century. These acts have been as various as the anarchist assassinations of political leaders in Europe during the 1840s and in the United States in the late 19th century, the abduction of Canadian government officials by the Front Liberation du Quebec (FLQ) in 1970, and the release of sarin gas in the Tokyo subways in 1995 by the group Aum Shinrikyo.

The political and religious aims of these attackers, however, can cause confusion on the world stage. In 2001, the Reuters News Agency told its reporters to stop using the word *terrorism*, because "one person's terrorist is another's freedom fighter." The definition of *terrorism* rests on the ability to identify motives.

Although it may be difficult to distinguish among types of terrorism, it is even more difficult to prevent it. Generally speaking, there are four common responses to terrorism on the part of governments and international bodies:

1. Reducing or addressing the causes of terrorism. In some cases, political change can reduce a terrorist threat. For example, the 1998 Good Friday Agreement in Northern Ireland led to a reduction in terrorist acts; the Spanish government's granting of some regional autonomy to the Basques helped quiet the actions of the ETA and reduced the support of many Basque people for such acts.
2. Increasing international cooperation in the surveillance of subnational groups. Spurred by terrorist crimes in Bahrain and Saudi Arabia, the Arab Gulf States agreed in 1998 to exchange intelligence regarding terrorist groups, to share intelligence regarding the prevention of an anticipated terrorist act, and to assist each other in investigating terrorist crimes.
3. Increasing security measures in a country. In the United States, following September 11, 2001, the government organized a Department of Homeland Security, federalized air traffic screening, and increased efforts to reduce financial support for foreign terrorist organizations. In concert, the European Union froze the assets of any group on its list of terrorist organizations.
4. Using military means, either unilaterally or multilaterally against terrorists or governments that sponsor terrorists. Following the September 11 attacks, the United States led a coalition of countries in attacking the government of Afghanistan, which had harbored Osama bin Laden's al-Queda terrorist organization.

Each response to terrorism is expensive, politically difficult, and/or potentially harmful to the life and liberty of civilians. Governments must decide which response or combination of responses is likely to have the most beneficial effect.

# Cooperation Among States

The strivings of groups with distinct territorially based identities and a desire for fuller control of that territory are a reminder of the fragility of the modern state. In many ways, countries are now weaker than ever before. Many are economically frail, others are politically unstable, and some are both. Strategically, no state is safe from military attack, for technology now enables countries to shoot weapons halfway around the world. Some people believe that no national security is possible in the nuclear age.

The recognition that a country cannot by itself guarantee either its prosperity or its own security has led to increased cooperation among states. These cooperative ventures are proliferating quickly, and they involve countries everywhere. They are also adding a new dimension to the concept of political boundaries, since the associations of states have borders of a higher spatial order than those between individual states. Such boundaries as the former Iron Curtain or the current division between NATO (North Atlantic Treaty Organization) and non-NATO states, represent a different scale of the political ordering of space.

## Supranationalism

Associations among states represent a new dimension in the ordering of national power and national independence. Recent trends in economic globalization and international cooperation suggest to some that the sovereign state's traditional responsibilities and authorities are being diluted by a combination of forces and partly delegated to higher-order political and economic organizations. Corporations and even nongovernmental agencies often operate in controlling ways outside of nation-state jurisdiction.

The rise of transnational corporations dominant in global markets, for example, limits the economic influence of individual countries. Cyberspace and the Internet are controlled by no one and are largely immune to the state restrictions on the flow of information exerted by many governments. And increasingly, individual citizens of any country have their lives and actions shaped by decisions not only of local and national authorities but also of regional economic associations (e.g., the North American Free Trade Agreement), of multiparty military alliances (e.g., NATO), and of global political agencies (e.g., the United Nations).

The roots of such multistate cooperative systems are ancient—for example, the leagues of city-states in the ancient Greek world or the Hanseatic League of free German cities in Europe's medieval period. New cooperative systems have proliferated since the end of World War II. They represent a world trend toward a **supranationalism** composed of associations of three or more states created for mutual benefit and the achievement of shared objectives. Although many individuals and organizations decry the loss of national independence that supranationalism entails, the many supranational associations in existence early in the 21st century are evidence of their attraction and pervasiveness. Nearly all countries, in fact, are members of at least one—and most are members of many—supranational groupings.

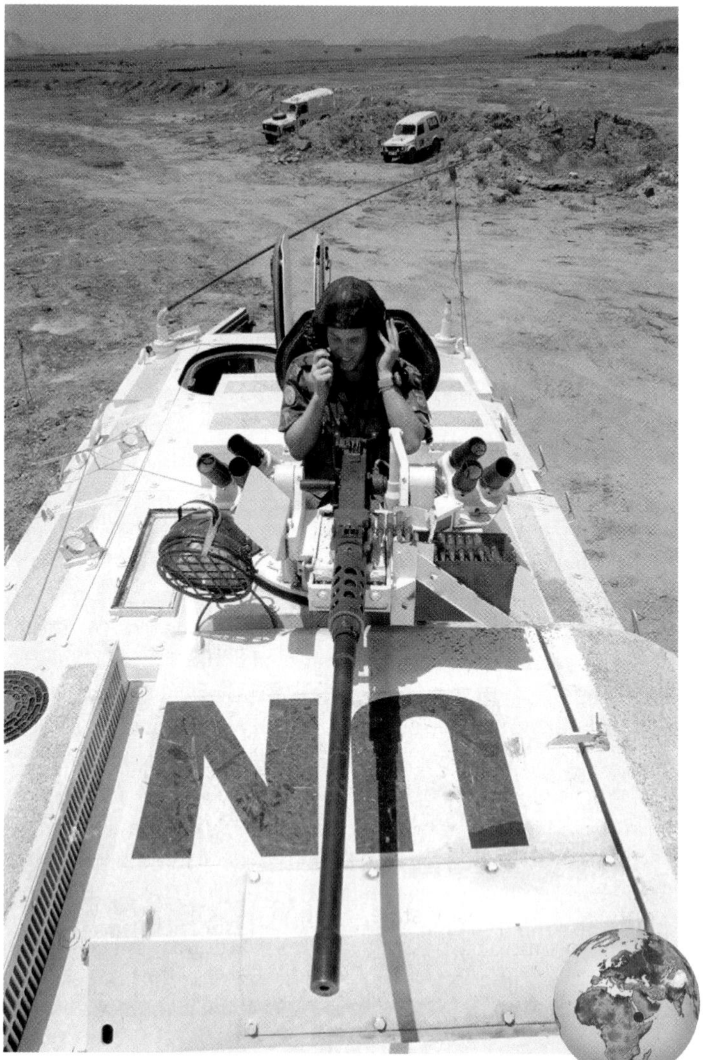

**FIGURE 9.22 United Nations peacekeeping forces on duty in Eritrea.** Under the auspices of the UN, soldiers from many countries staff peacekeeping forces and military observer groups in many world regions in an effort to halt or mitigate conflicts. After two years of conflict in a border dispute between Ethiopia and Eritrea, the UN agreed to deploy troops to monitor the ceasefire and patrol the border. The demand for peacekeeping and observer operations is indicated by the recent deployment of UN forces in Bosnia, Croatia, Cyprus, East Timor, Haiti, Iraq/Kuwait, Israel, Kosovo, Lebanon, Pakistan/India, Sierra Leone, Somalia, and elsewhere. *© UN/DPI Photo.*

## The United Nations and Its Agencies

The United Nations (UN) is the only organization that tries to be universal. Its membership has expanded from 51 countries in 1945 to 191 in 2002. Switzerland and Timor-Leste joined the UN in 2002.

The UN is the most ambitious attempt ever undertaken to bring together the world's countries in international assembly and to promote world peace. Stronger and more representative than its predecessor, the League of Nations, it provides a forum where countries can discuss international problems and regional concerns and a mechanism, admittedly weak but still significant, for forestalling disputes or, when necessary, for ending wars (Figure 9.22). The United Nations also sponsors 40 programs and agencies aimed at fostering international cooperation with respect to specific goals. Among these are the World Health Organization (WHO), the Food and Agriculture Organization (FAO), and the United Nations Educational, Scientific, and Cultural Organization (UNESCO). Many other UN agencies and much of the UN budget are committed to assisting member states with matters of economic growth and development.

Member states have not surrendered sovereignty to the UN, and the world body is legally and effectively unable to make or enforce a world law. Nor is there a world police force. Although there is recognized international law adjudicated by the International Court of Justice, rulings by this body are sought only by countries agreeing beforehand to abide by its arbitration. Finally, the United Nations has no authority over the military forces of individual countries.

A pronounced change both in the relatively passive role of the United Nations and in traditional ideas of international relations has begun to emerge, however. Long-established rules of total national sovereignty that allowed governments to act internally as they saw fit, free of outside interference, are fading as the United Nations increasingly applies a concept of "interventionism." The Persian Gulf War of 1991 was UN-authorized under the old rules prohibiting one state (Iraq) from violating the sovereignty of another (Kuwait) by attacking it. After the war, the new interventionism sanctioned UN operations within Iraq to protect Kurds within that country. Later, the UN intervened with troops and relief agencies in Somalia, Bosnia, and elsewhere, invoking an "international jurisdiction over inalienable human rights" that prevails without regard to state frontiers or sovereignty considerations.

Whatever the long-term prospects for interventionism replacing absolute sovereignty, for the short term the UN remains the only institution where the vast majority of the world's countries can collectively discuss international political and economic concerns and attempt peacefully to resolve their differences. It has been particularly influential in formulating a law of the sea.

### Maritime Boundaries

Boundaries define political jurisdictions and areas of resource control, but claims of national authority are not restricted to land areas alone. Water covers about two-thirds of the earth's surface, and increasingly countries have been projecting their sovereignty seaward to claim adjacent maritime areas and resources. A basic question involves the right of states to control water and the resources it contains. The inland waters of a country, such as rivers and lakes, have traditionally been regarded as being within the sovereignty of that country. Oceans, however, are not within any country's borders. Are they, then, to be open to all states to use, or may a single country claim sovereignty and limit access and use by other countries?

For most of human history, the oceans remained effectively outside individual national control or international jurisdiction. The seas were a common highway for those daring enough to venture on them, an inexhaustible larder for fishermen, and a vast refuse pit for the muck of civilization. By the end of the 19th century, however, most coastal countries claimed sovereignty over a continuous belt 3 or 4 nautical miles wide (a *nautical mile,* or *nm,* equals 1.15 statute miles, or 1.85 kilometers). At the time, the 3-nm limit represented the farthest range of artillery and thus the effective limit of control by the coastal state. Though recognizing the rights of others to innocent passage, such sovereignty permitted the enforcement of quarantine and customs regulations, allowed national protection of coastal fisheries, and made claims of neutrality effective during other people's wars. The primary concern was with security and unrestricted commerce. No separately codified law of the sea existed, however, and none seemed to be needed until after World War I.

A League of Nations Conference for the Codification of International Law, convened in 1930, inconclusively discussed maritime legal matters and identified areas of concern that were to become increasingly pressing after World War II. Important among these was an emerging shift from interest in commerce and national security to a preoccupation with the resources of the seas, an interest fanned by the *Truman Proclamation* of 1945. Motivated by a desire to exploit offshore oil deposits, the U.S. federal government, under this doctrine, laid claim to all resources on the continental shelf contiguous to its coasts. Other states, many claiming even broader areas of control, hurried to annex marine resources. Within a few years, a quarter of the earth's surface was appropriated by individual coastal countries.

### An International Law of the Sea

Unrestricted extensions of jurisdiction and territorial disputes over proliferating claims to maritime space and resources led to a series of United Nations conferences on the Law of the Sea. Meeting over a period of years, delegates from more than 150 countries attempted to achieve consensus on a treaty that would establish an internationally agreed-upon "convention dealing with all matters relating to the Law of the Sea." The meetings culminated in a draft treaty in 1982, the **United Nations Convention on the Law of the Sea (UNCLOS).**

The convention delimits territorial boundaries and rights by defining four zones of diminishing control (Figure 9.23).

1. A *territorial sea* of up to 12 nm (19 km) in breadth, over which coastal states have sovereignty, including exclusive fishing rights. Vessels of all types normally have the right of innocent passage through the territorial sea, although under certain circumstances noncommercial vessels (primarily military and research) can be challenged.
2. A *contiguous zone* of up to 24 nm (38 km). Although a coastal state does not have complete sovereignty in this zone, it can enforce its customs, immigration, and sanitation laws and has the right of hot pursuit out of its territorial waters.
3. An **exclusive economic zone (EEZ)** of up to 200 nm (370 km), in which the state has recognized rights to explore, exploit, conserve, and manage the natural resources, both living and nonliving, of the seabed and waters (Figure 9.24). Countries have exclusive rights to the resources lying within the continental shelf when

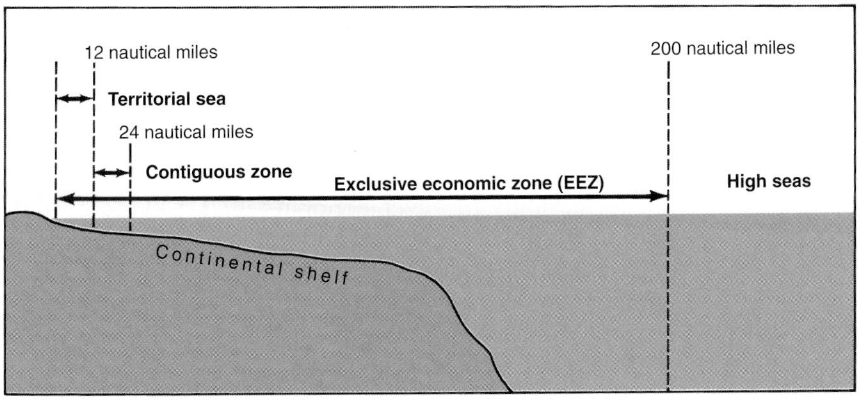


**FIGURE 9.23** **Territorial claims permitted by the 1982 United Nations Convention on the Law of the Sea (UNCLOS).**

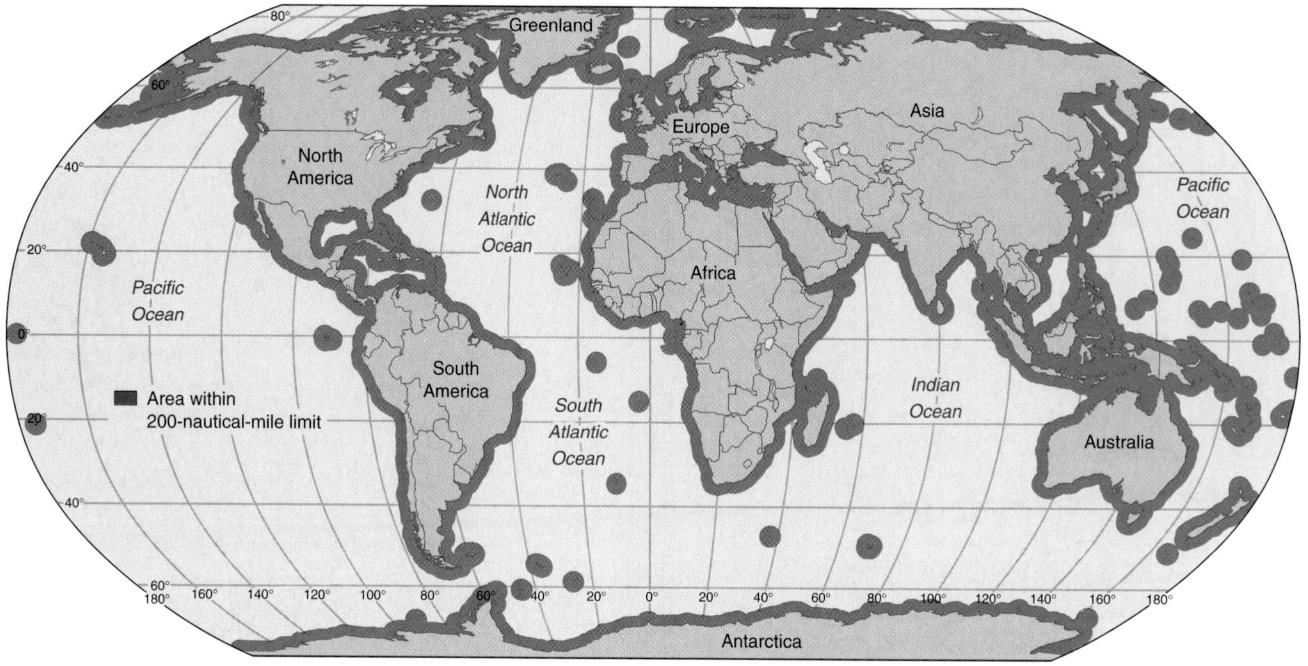


**FIGURE 9.24 The 200-nautical-mile exclusive economic zone (EEZ) claims of coastal states.** The provisions of the Law of the Sea Convention, in effect, changed the maritime map of the world. Three important consequences flowed from the 200-nm EEZ concept: (1) islands gained a new significance; (2) countries gained a host of new neighbors; and (3) the EEZ lines resulted in overlapping claims. EEZ lines are drawn around a country's possessions as well as around the country itself. Every island, no matter how small, has its own 200-nm EEZ. This means that although the United States shares continental borders only with Canada and Mexico, it has maritime boundaries with countries in Asia, South America, and Europe. All told, the United States may have to negotiate some 30 maritime boundaries, which is likely to take decades. Other countries, particularly those with many possessions, will have to engage in similar lengthy negotiations.

this extends farther, up to 350 nm (560 km) beyond their coasts. The traditional freedoms of the high seas are to be maintained in this zone.

4. The *high seas* beyond the EEZ. Outside any national jurisdiction, they are open to all states, whether coastal or landlocked. Freedom of the high seas includes the right to sail ships, fish, fly over, lay submarine cables and pipelines, and pursue scientific research. Mineral resources in the international deep seabed area beyond national jurisdiction are declared the common heritage of humankind, to be managed for the benefit of all the peoples of the earth.

By the end of the 1980s, most coastal countries, including the United States, had used the UNCLOS provisions to proclaim and reciprocally recognize jurisdiction over 12-nm (19-km) territorial seas and 200-nm (370-km) economic zones. Despite reservations held by the United States and a few other industrial countries about the deep seabed mining provisions, the convention received the necessary ratification by 60 states and became international law in 1994.

## UN Affiliates

Other fully or essentially global supranational organizations with influences on the economic, social, and cultural affairs of states and individuals have been created. Most are specialized international agencies, autonomous and with their own memberships but with affiliated relationships with the United Nations and operating under its auspices. Among them are the Food and Agriculture Organization (FAO), the World Bank, the International Labor Organization (ILO), the United Nations Children's Fund (UNICEF), the World Health Organization (WHO), and—of growing economic importance—the World Trade Organization (WTO).

The World Trade Organization, which came into existence in 1995, has become one of the most significant of the global expressions of supranational economic control. It is charged with enforcing the global trade accounts that grew out of years of international negotiations under the terms of the General Agreement on Tariffs and Trade (GATT). The basic principle behind the WTO is that the 146 (as of 2003) member countries should work to cut tariffs, dismantle nontariff barriers to trade, liberalize trade in services, and treat all other countries uniformly in matters of trade. Any preference granted to one should be available to all.

Increasingly, however, regional rather than global trade agreements are being struck, and free-trade areas are proliferating. Only a few WTO members are not already part of another regional trade association. Such regional alliances—about 80 of them by 2003—it can be argued, make world trade less free by scrapping tariffs on trade among member states but retaining them on exchanges with nonmembers.

## Regional Alliances

In addition to their membership in international agencies, countries have shown themselves willing to relinquish some of their independence to participate in smaller, multinational systems. These groupings can be economic, military, or political. Cooperation in the economic sphere seems to come more easily to states than does political or military cooperation.

### Economic Alliances

Among the most powerful and far-reaching of the economic alliances are those that have evolved in Europe, particularly the **European Union (EU)** and its several forerunners. Shortly after the end of World War II, the Benelux countries (Belgium, the Netherlands, and Luxembourg) formed an economic union to create a common set of tariffs and to eliminate import licenses and quotas. Formed at about the same time were the Organization for European Cooperation (1948), which coordinated the distribution and use of Marshall Plan funds, and the European Coal and Steel Community (1952), which integrated the development of that industry in the member countries. A few years later, in 1957, the *European Economic Community (EEC)*, or *Common Market*, was created, composed at first of only six states: France, Italy, West Germany, and the Benelux countries.

To counteract these Inner Six, as they were called, other countries in 1960 formed the European Free Trade Association (EFTA). Known as the Outer Seven, they were the United Kingdom, Norway, Denmark, Sweden, Switzerland, Austria, and Portugal (Figure 9.25). Between 1973 and 1986, three members (the United Kingdom, Denmark, and Portugal) left EFTA for membership in the Common Market and were replaced by Iceland and Finland. Other Common Market additions were Greece in 1981 and Spain and Portugal in 1986. Austria, Finland, and Sweden became members of the European Union, as the organization embracing the Common Market is now called, in 1995.

Between 1997 and 2000, the EU opened entry negotiations with 12 countries. In 2004, it admitted 8 former Soviet bloc nations, from Estonia in the north to Slovenia in the south, and added the island states of Malta and Cyprus (Figure 9.26). These 10 additions increased the EU's landmass by 23%, raised its total population to more than 450 million people, and expanded its economy to rival that of the United States. They make the EU the world's largest and richest bloc of countries.

Over the years, members of the European Union have taken many steps to integrate their economies and coordinate their policies in such areas as transportation, agriculture, and fisheries. A council of ministers, a commission, a European parliament, and a court of justice give the European Union supranational institutions with effective ability to make and enforce laws. By January 1, 1993, the EU had abolished most remnant barriers to free trade and the free movement of capital and people among its members, creating a single European market. In another step toward economic and monetary union, the EU's single currency, the *euro*, replaced separate national currencies in 1999. Notes and coins in national currencies—such as the Portuguese escudo and the deutsche mark—were withdrawn after July 2002. A few countries within the EU such as Sweden and the United Kingdom have elected to delay moving to the single currency.

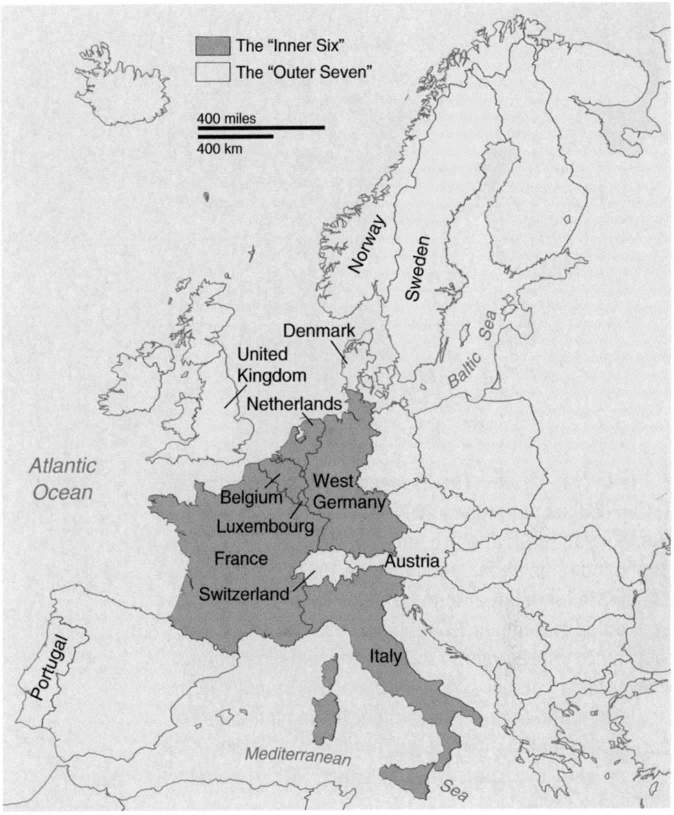


**FIGURE 9.25 The original "Inner Six" and "Outer Seven" of Europe.**

We have traced this European development process, not because it is important to remember all the forerunners or the present structure of the European Union but to illustrate the fluid process by which regional alliances are made. Countries come together in an association, some drop out, and others join. New treaties are made, and new coalitions emerge. Indeed, a number of such regional economic and trade associations have been added to the world supranational map. None are as encompassing in power and purpose as the EU, but all represent a cession of national independence to achieve broader regional goals.

NAFTA, the North American Free Trade Agreement, launched in 1994 and linking Canada, the United States, and Mexico in an economic community aimed at lowering or removing trade and movement restrictions among the countries, is perhaps the best known to North American students. The Americas as a whole, however, have other, similar associations with comparable trade-enhancement objectives, though frequently they also have social, political,

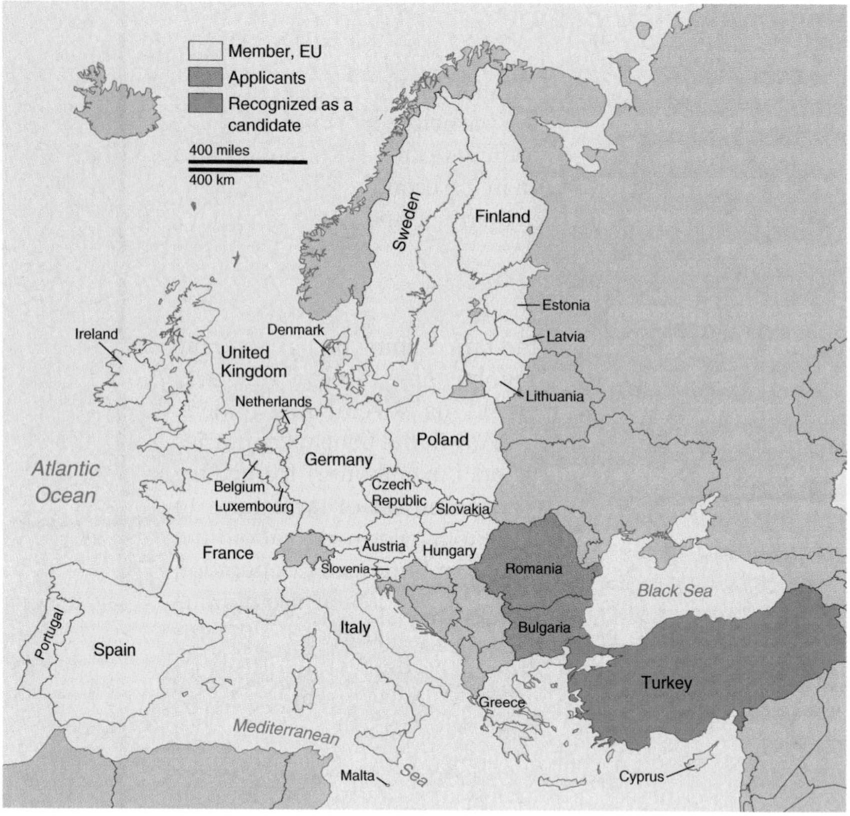


**FIGURE 9.26 The 25 members of the European Union (EU) as of January 2004.** Accession talks with Romania and Bulgaria are underway. Turkey, whose application has been pending since 1987, is also a candidate for membership; formal talks were scheduled to begin in 2004. The EU has stipulated that in order to join, a country must have stable institutions guaranteeing democracy, the rule of law, human rights, and protection of minorities; a functioning market economy; and the ability to accept the obligations of membership, including the aims of political, economic, and monetary union.

and cultural interests in mind. CARICOM (Caribbean Community and Common Market), for example, was established in 1974 to further cooperation among its 15 members in economic, health, cultural, and foreign policy arenas. MERCOSUR (the Common Market of the Southern Cone), which unites Brazil, Argentina, Uruguay, and Paraguay (and associate members Peru, Bolivia, and Chile) in the creation of a customs union to eliminate levies on goods moving between them, is a South American example.

A similar interest in promoting economic, social, and cultural cooperation and development among its members underpins the Association of Southeast Asian Nations (ASEAN). A less-wealthy African example is the Economic Community of West African States (ECOWAS). The Asia Pacific Economic Cooperation (APEC) forum includes China, Japan, Australia, Canada, and the United States among its 18 members and has a grand plan for "free trade in the Pacific" by 2020. More restricted bilateral and regional preferential trade arrangements have also proliferated, numbering more than 400 in 2003 and creating a maze of rules, tariffs, and commodity agreements that result in trade restrictions and preferences contrary to the free-trade intent of the World Trade Organization.

Economic interests, then, motivate the establishment of most international alliances, though political, social, and cultural objectives also figure largely in many. Although the alliances themselves will change, the idea of supranational associations appears to have been permanently added to the national political and global realities of the 21st century. The world map pattern those alliances create must be recognized to understand the current international order.

Three further points about regional international alliances are worth noting. The first is that the formation of a coalition in one area often stimulates the creation of another alliance by countries left out of the first. Thus, the union of the Inner Six gave rise to the treaty among the Outer Seven. Similarly, a counterpart of the Common Market was the Council of Mutual Economic Assistance (CMEA), also known as Comecon, which linked the former communist countries of Eastern Europe and the USSR through trade agreements.

Second, the new economic unions tend to be composed of contiguous states (Figure 9.27). This was not the case with the recently dissolved empires, which included far-flung territories. Contiguity facilitates the movement of people and goods. Communication and transportation are simpler and more effective among adjoining countries than among those far removed from one another, and common cultural, linguistic, and political traits and interests are more to be expected in countries adjacent to one another.

Finally, it does not seem to matter whether countries are alike or distinctly different in their economies, as far as joining economic unions is concerned. There are examples of both. If the countries are dissimilar, they may complement each other. This was one basis for the European Common Market. Dairy products and furniture from Denmark

**FIGURE 9.27 Western Hemisphere economic unions in 2004.** The number of international trade organizations has risen rapidly since the 1960s.

are sold in France, freeing that country to specialize in the production of machinery and clothing. On the other hand, countries that produce the same raw materials hope that, by joining together in an economic alliance, they might be able to enhance their control of markets and prices for their products. The Organization of Petroleum Exporting Countries (OPEC) is a case in point. Other attempts to form commodity cartels and price agreements between producing and consuming countries are represented by the International Tin Agreement, the International Coffee Agreement, and others.

### Military and Political Alliances

As we have seen, countries form alliances for other than economic reasons. Strategic, political, and cultural considerations may also foster cooperation. *Military alliances* are based on the principle that unity brings strength. Such pacts usually provide for mutual assistance in the case of aggression. Once again, action breeds reaction when such an association is created. The formation of the North Atlantic Treaty Organization (NATO), a defensive alliance of many European countries and the United States, was countered by the establishment of the Warsaw Treaty Organization, which joined the USSR and its satellite countries of Eastern Europe. Both pacts allowed the member states to base armed forces in one another's territories, a relinquishment of a certain degree of sovereignty uncommon in the past.

Military alliances depend on the perceived common interests and political goodwill of the countries involved. As political realities change, so, too, do the strategic alliances. NATO was created to defend Western Europe and North America against the Soviet military threat. When the dissolution of the USSR and Warsaw Pact removed that threat, the purpose of the NATO alliance became less clear and, during the 1990s, its relationships with Eastern European states and Russia were under review. Most of those countries sought ways to foster cooperation with NATO, and three of them (Poland, the Czech Republic, and Hungary) joined the alliance in 1999 (Figure 9.28).

All international alliances recognize communities of interest. In economic and military associations, common objectives are clearly seen and described, and joint actions are agreed upon with respect to the achievement of those objectives. More generalized mutual concerns or appeals to historical interest may be the basis for primarily *political alliances*. Such associations tend to be rather loose, not requiring their members to yield much power to the union. Examples are the Commonwealth of Nations (formerly the British Commonwealth), composed of many former British colonies and dominions, and the Organization of American States (OAS), both of which offer economic as well as political benefits.

There are many examples of abortive political unions that have foundered because the individual countries could not agree on questions of policy and were unwilling to subordinate individual interests to make the union succeed. The United Arab Republic, the Central African Federation, the Federation of Malaysia and Singapore, and the Federation of the West Indies fall within this category.

Although many such political associations have failed, observers of the world scene speculate about the possibility that "superstates" will emerge from one or more of the international alliances that now exist. Will a "United States of Europe," for example, under a single common government, be the logical outcome of the successes of the EU? No one knows, but as long as the individual state is regarded as the highest form of political and social organization (as it is now) and as the body in which sovereignty rests, such total unification is unlikely.

## Local and Regional Political Organization

The most profound contrasts in cultures tend to occur between, rather than within, states, one reason political geographers traditionally have been interested primarily in country units. The emphasis on the state, however, should not obscure the fact that, for most of us, it is at that local level that we find our most intimate and immediate contact

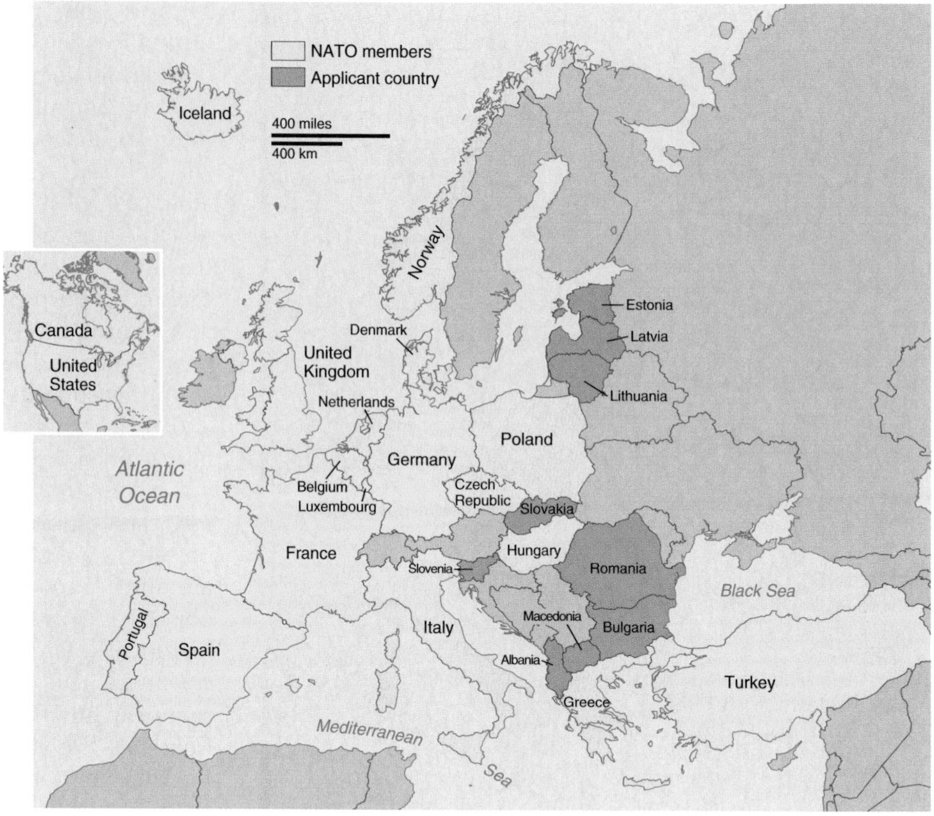


**FIGURE 9.28** The NATO military alliance as of 2004 had 19 members. Three countries (Poland, the Czech Republic, and Hungary) joined the alliance in 1999, the first enlargement since Spain became a member in 1982. Nine other countries have applied for membership. Proponents of expansion argue that it is necessary in order to create a zone of stability and security throughout Europe. Opponents contend that enlargement is a divisive move that will cast a shadow over the future of relations with Russia, which is opposed to expansion so close to its borders.

with government and its influence on the administration of our affairs. In the United States, for example, an individual is subject to the decisions and regulations made by the local school board, the municipality, the county, the state, and perhaps, a host of special-purpose districts—all in addition to the laws and regulations issued by the federal government and its agencies. Among other things, local political entities determine where children go to school, the minimum size lot on which a person may build a house, and where one may legally park a car. Adjacent states of the United States may be characterized by sharply differing personal and business tax rates; differing controls on the sale of firearms, alcohol, and tobacco; variant administrative systems for public services; and different levels of expenditures for them (Figure 9.29).

All of these governmental entities are *spatial systems.* Because they operate within defined geographic areas, and because they make behavior-governing decisions, they are topics of interest to political geographers. In the concluding sections of this chapter, we will examine two aspects of political organization at the local and regional level. Our emphasis will be on the United States and Canadian scene simply because their local political geography is familiar to most of us. We should remember, however, that the North American structure of municipal governments, minor civil divisions, and special-purpose districts has counterparts in other regions of the world.

## The Geography of Representation: The Districting Problem

There are more than 85,000 local governmental units in the United States. Slightly more than half of these are municipalities, townships, and counties. The remainder are school districts, water-control districts, airport authorities, sanitary districts, and other special-purpose bodies. Boundaries have been drawn around each of these districts. Although the number of districts does not change greatly from year to year, many boundary lines are redrawn in any single year. Such *redistricting,* or *reapportionment,* is made necessary by shifts in population, as areas gain or lose people.

Every 10 years, following the United States census, updated figures are used to redistribute the 435 seats in the House of Representatives among the 50 states. Redrawing the Congressional districts to reflect population changes is required by the Constitution, the intention being to make sure that each legislator represents roughly the same number of people. Since 1964, Canadian provinces and territories have entrusted redistricting for federal offices to independent electoral boundaries commissions. Although a few states in the United States also have independent, nonpartisan boards or commissions draw district boundaries, most rely on state legislatures for the task. Across the United States, the decennial census data are also used to redraw the boundaries of legislative districts within each

**FIGURE 9.29 The Four Corners Monument, marking the meeting of Utah, Colorado, New Mexico, and Arizona.** Jurisdictional boundaries within countries may be precisely located but are usually not highly visible on the landscape. At the same time, those boundaries may be very significant in citizens' personal affairs and in the conduct of economic activities. *© Cameramann International, Ltd.*

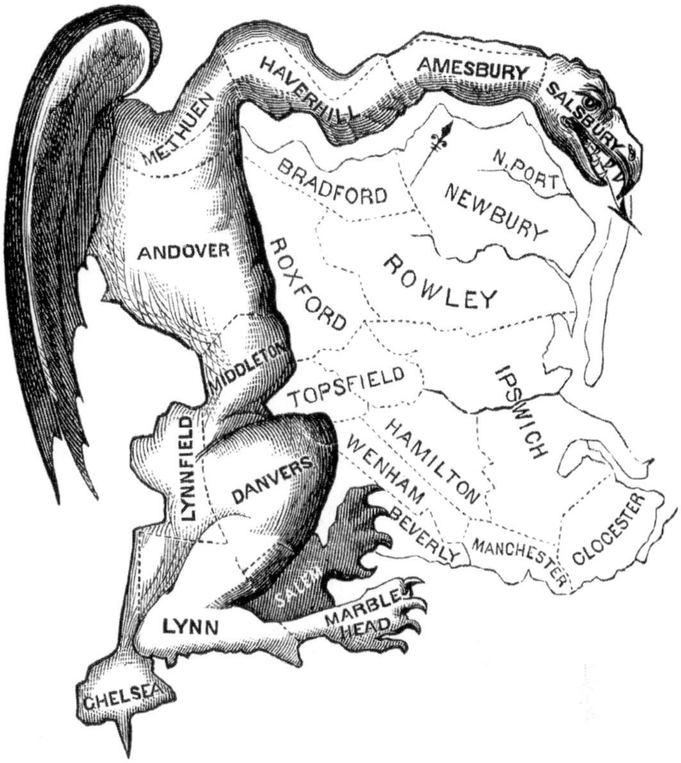


THE GERRY-MANDER. (Boston, 1811.)

**FIGURE 9.30 The original gerrymander.** The term *gerrymander* originated in 1811 from the shape of an electoral district formed in Massachusetts while Elbridge Gerry was governor. When an artist added certain animal features, the district resembled a salamander and quickly came to be called a gerrymander. Source: *© Bettmann/Corbis Images.*

state as well as those for local offices, such as city councils and county boards.

The analysis of how boundaries are drawn around voting districts is one aspect of **electoral geography,** which also addresses the spatial patterns yielded by election results and their relationship to the socioeconomic characteristics of voters. In a democracy, it might be assumed that election districts should contain roughly equal numbers of voters, that electoral districts should be reasonably compact, and that the proportion of elected representatives should correspond to the share of votes cast by members of a given political party. Problems arise because the way in which the boundary lines are drawn can maximize, minimize, or effectively nullify the power of a group of people.

**Gerrymandering** is the practice of drawing the boundaries of legislative districts so as to unfairly favor one political party over another, to fragment voting blocs, or to achieve other nondemocratic objectives (Figure 9.30). A number of strategies have been employed over the years for that purpose. *Stacked* gerrymandering involves drawing circuitous boundaries to enclose pockets of strength or weakness of the group in power; it is what we usually think of as gerrymandering. The *excess vote* technique concentrates the support of the opposition in a few districts, which it can win easily, but leaves it few potential seats elsewhere. Conversely, the *wasted vote* strategy dilutes the opposition's strength by dividing its votes among a number of districts.

Assume that X and O represent two groups with an equal number of voters but different policy preferences. Although there are equal numbers of Xs and Os, the way electoral districts are drawn affects voting results. In Figure 9.31a, the Xs are concentrated in one district and will probably elect only one representative of four. The power of the Xs is maximized in Figure 9.31b, where they may control three of the four districts. The voters are evenly divided in Figure 9.31c, where the Xs have the opportunity to elect two of the four representatives. Finally, Figure 9.31d shows how both political parties may agree to delimit the electoral districts to provide "safe seats" for incumbents. Such a partitioning offers little chance for change.

Figure 9.31 depicts a hypothetical district, compact in shape with an even population distribution and only two groups competing for representation. In actuality, American municipal voting districts are often oddly shaped because of such factors as the city limits, historic settlement patterns, current population distribution, and transportation routes—as well as past gerrymandering. Further, in any large area, many groups vie for power. Each electoral interest group promotes its version of "fairness" in the way boundaries are delimited. Minorities seek representation in proportion to their numbers, so that they will be able to elect representatives who are concerned about and responsive to their needs (see "Voting Rights and Race," p. 332).

## Voting Rights and Race

Irregularly shaped Congressional voting districts, such as those shown here, were created by several state legislatures after the 1990 census to make minority representation in Congress more closely resemble minority presence in each state's voting-age population. Most were devised to contain a majority of black votes, but what opponents called extreme examples of racial gerrymandering was, in a few cases, utilized to accommodate Hispanic majorities. All represented a deliberate attempt to balance voting rights and race; all were specifically intended to comply with the federal Voting Rights Act of 1965, which provides that members of racial minorities shall not have "less opportunity than other members of the electorate . . . to elect representatives of their choice."

Because at least some of the newly created districts had very contorted boundaries, on appeal by opponents they have at least in part been ruled unconstitutional by the Supreme Court. The state legislatures' attempts at fairness and adherence to the Congressional mandate contained in the Voting Rights Act were held not to meet such other standards as rough equality of district population size, reasonably compact shape, and avoidance of disenfranchisement of any class of voters. The conflicts reflected the uncertainty of exactly what were the controlling requirements in voting district creation.

In North Carolina, for example, although 24% of the 1990 population of that state was black, past districting had divided black voters among a number of districts, with the result that blacks had not elected a single Congressional representative in the 20th century. In 1991, the Justice Department ordered North Carolina to redistrict so that at least two districts would contain black majorities. Because of the way the black population is distributed, the only way to form black-majority districts was to string together cities, towns, and rural areas in elongated, sinuous belts. The two newly created districts had slim (53%) black majorities.

The redistricting in North Carolina and other states had immediate effects. Black membership in the House of Representatives increased from 26 in 1990 to 39 in 1992; blacks constituted nearly 9% of the House as against 12% of blacks in the total population. Within a year, those electoral gains were threatened as lawsuits challenging the redistricting were filed in a number of states. The chief contention of the plaintiffs

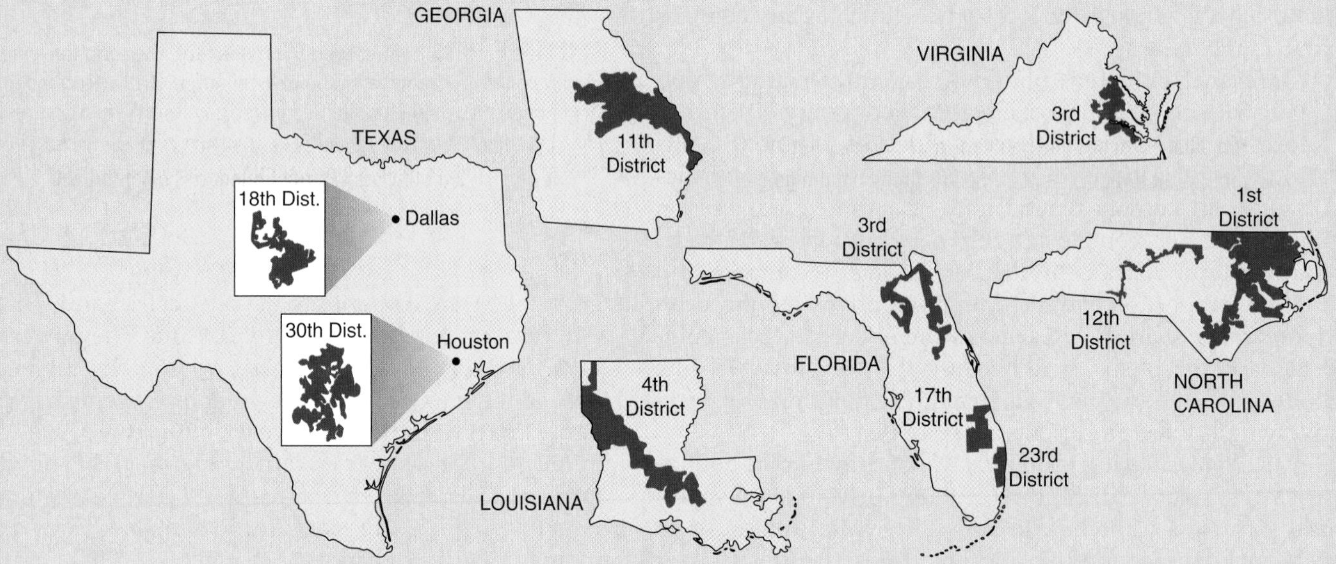


**FIGURE 9.31**
**Alternative districting strategies.** Xs and Os might represent Republicans and Democrats, urban and rural voters, blacks and whites, or any other distinctive groups.

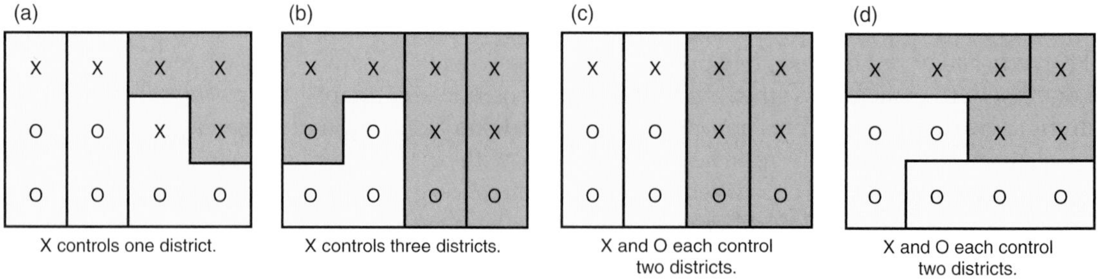

was that the irregular shapes of the districts were a product of racial gerrymandering and amounted to reverse discrimination against whites.

In June 1993, a sharply divided Supreme Court ruled in *Shaw* v. *Reno* that North Carolina's 12th Congressional District might violate the constitutional rights of white voters and ordered a district court to review the case. The 5–4 ruling gave evidence that the country had not yet reached agreement on how to comply with the Voting Rights Act. It raised a central question: should a state maximize the rights of racial minorities or not take racial status into consideration? A divided court provided answers in 1995, 1996, and 1997 rulings that rejected congressional redistricting maps for Georgia, Texas, North Carolina, and Virginia on the grounds that "race cannot be the predominant factor" in drawing election district boundaries, nor can good-faith efforts to comply with the Voting Rights Act insulate redistricting plans from constitutional attack.

The difficulty in interpreting and complying with the Voting Rights Act is illustrated by the fact that, although the Supreme Court in 1996 declared North Carolina's 12th Congressional District unconstitutional, federal courts in 1998 and 1999 rejected alternative district designs. Finally, in *Easley* v. *Cromartie*, the Court held in 2001 that a redrawn 12th District met constitutional requirements and added that race was a legitimate consideration in redistricting as long as it was not "the dominant and controlling" one. Some observers hailed the decision as providing guidance to state legislatures and lower federal courts in creating and judging new districts. Others contended that the ruling left unresolved the basic tension between the mandate of the Voting Rights Act and the Court's admonition against using race as a predominant factor in redistricting.

**QUESTIONS TO CONSIDER**

1. Do you believe that race should be a consideration in the electoral process? Why or why not? If so, should voting districts be drawn to increase the likelihood that representatives of racial or ethnic minorities will win elections? If not, how can one be certain that the voting power of minorities will not be diluted in white-majority districts?
2. With which of the following arguments in *Shaw* v. *Reno* do you agree? Why? ". . . Racial gerrymandering, even for remedial purposes, may balkanize us into competing racial factions; it threatens to carry us further from the goal of a political system in which race no longer matters" (Justice Sandra Day O'Connor). ". . . Legislators will have to take race into account in order to avoid dilution of minority voting strength" (Justice David Souter).
3. One of the candidates in North Carolina's 12th Congressional District said, "I love the district because I can drive down I-85 with both car doors open and hit every person in the district." Given a good transportation and communication network, how important is it that voting districts be compact?
4. Blacks face difficult hurdles in being elected in districts that do not have a black majority, as witnessed by their numerical underrepresentation in most legislative bodies. But critics of "racial gerrymandering" contend that blacks have been and can continue to be elected in white-majority districts and, further, that white politicians can and do adequately represent the needs of all, including black, voters in their districts. Do you agree? Why or why not?
5. Given partisanship and the desire of incumbent legislators to protect their own seats, is there an inherent conflict in having legislators draw district boundaries? Defend your answer.
6. After the 1990 census, more than 130 suits were filed in 40 states challenging either the states' overall redistricting plans or individual districts, the result being that, in many cases, courts ended up determining district boundaries. Does it seem democratic for a judge elected in one district to draw up a redistricting plan for an entire state? How might such a situation be avoided?

Gerrymandering is not automatically successful. First, a districting arrangement that appears to be unfair may be appealed to the courts. In addition, many factors other than political party affiliation influence voting decisions. Key issues may cut across party lines, scandal may erode (or personal charm increase) votes unexpectedly, and the amount of candidate financing or the number of campaign workers may determine election outcomes if compelling issues are absent.

## The Fragmentation of Political Power

Boundary drawing at any electoral level is never easy, particularly when political groups want to maximize their representation and minimize that of opposition groups. Furthermore, the boundaries that we may want for one set of districts may *not* be those that we want for another. For example, sewage districts must take natural drainage features into account, whereas police districts may be based on the distribution of the population or the number of miles of street to be patrolled. And school attendance zones must consider the numbers of school-aged children and the capacities of individual schools.

As these examples suggest, the United States is subdivided into great numbers of political administrative units whose areas of control are spatially limited. The 50 states are partitioned into more than 3000 counties ("parishes" in Louisiana), most of which are further subdivided into townships, each with a still lower level of governing power. This political fragmentation is further increased by the existence of nearly innumerable special-purpose districts whose boundaries rarely coincide with the standard major and minor civil divisions of the country, or even with each other (Figure 9.32). Each district represents a form of political allocation of territory to achieve a specific aim of local need or legislative intent.

Canada, a federation of 10 provinces and 3 territories, has a similar pattern of political subdivision. Each of the provinces contains minor civil divisions—municipalities—under provincial control, and all (cities, towns, villages, and rural municipalities) are governed by elected councils. Ontario and Quebec also have counties that group smaller

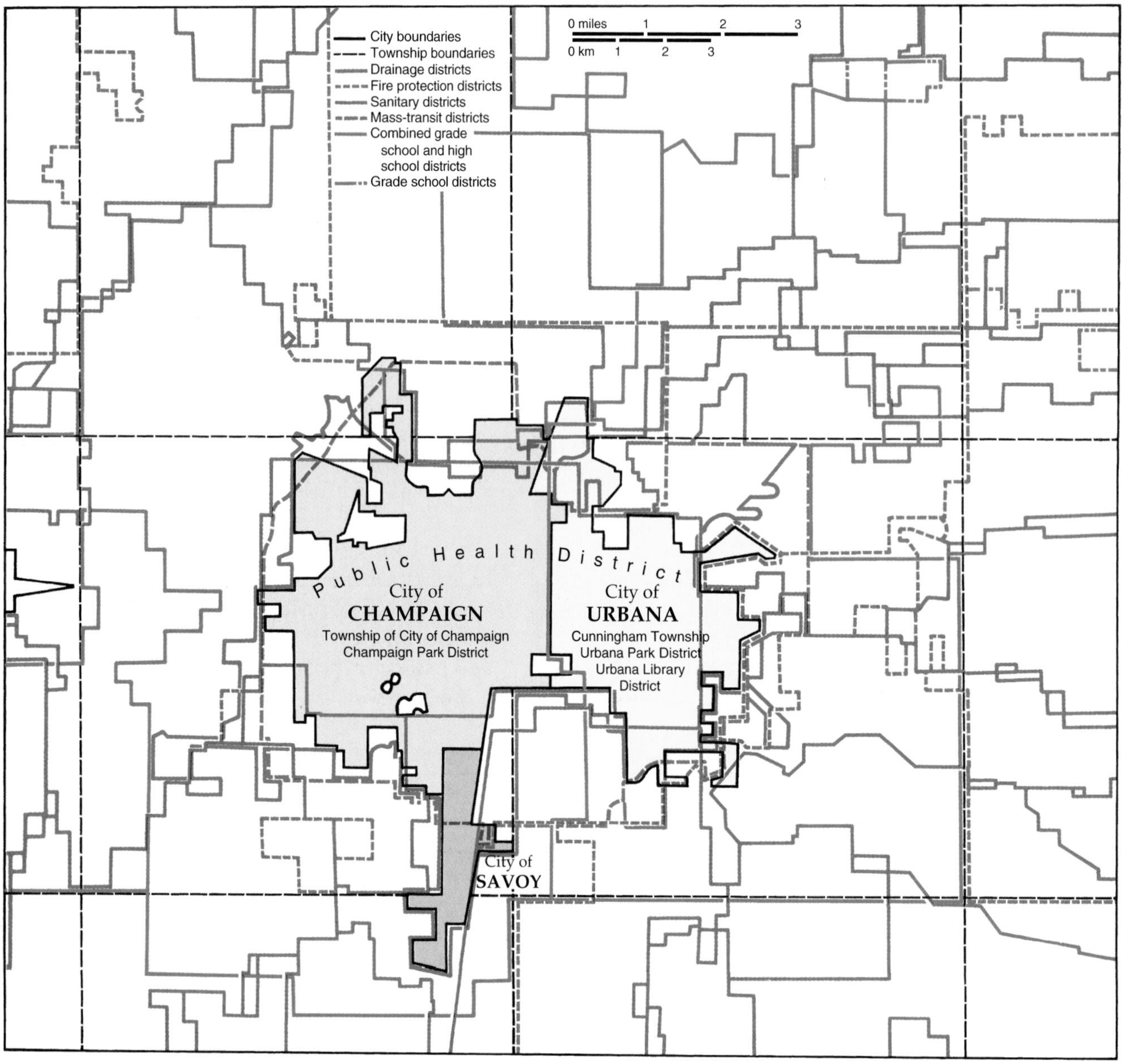


**FIGURE 9.32 Political fragmentation in Champaign County, Illinois.** The map shows a few of the independent administrative agencies with separate jurisdictions, responsibilities, and taxing powers in a portion of a single Illinois county. Among the other such agencies forming the fragmented political landscape are Champaign County itself, a forest preserve district, a public health district, a mental health district, the county housing authority, and a community college district.

municipal units for certain purposes. In general, municipalities are responsible for police and fire protection, local jails, roads and hospitals, water supply and sanitation, and schools, duties that are discharged either by elected agencies or by appointed commissions.

Most North Americans live in large and small cities. In the United States, these, too, are subdivided, not only into wards or precincts for voting purposes but also into special districts for such functions as fire and police protection, water and electricity supply, education, recreation, and sanitation. These districts almost never coincide with one another, and the larger the urban area, the greater the proliferation of small, special-purpose governing and taxing units. Although no Canadian community has quite the multiplicity of governmental entities plaguing many U.S. urban areas, major Canadian cities may find themselves with complex and growing systems of a similar nature. Even before its major expansion on January 1, 1998, for example, metropolitan Toronto had more than 100 authorities that could be classified as "local governments."

The existence of such a great number of districts in metropolitan areas may cause inefficiency in public services and hinder the orderly use of space. *Zoning ordinances,* for example, are determined by each municipality. They are intended to allow the citizens to decide how land is to be used and, thus, are a clear example of the effect of political decisions on the division and development of space. Zoning policies dictate the areas where light and heavy industries may be located, the sites of parks and other recreational areas, the location of business districts, and the types and location of housing. Unfortunately, in large urban areas, the efforts of one community may be hindered by the practices of neighboring communities. Thus, land zoned for an industrial park in one city may abut land zoned for single-family residences in an adjoining municipality. Each community pursues its own interests, which may not coincide with those of its neighbors or of the larger region.

Inefficiency and duplication of effort characterize not just zoning but many of the services provided by local governments. The efforts of one community to avert air and water pollution may be, and often are, counteracted by the rules and practices of other towns in the region, although state and national environmental protection standards are now reducing such potential conflicts. Social as well as physical problems spread beyond city boundaries. Thus, nearby suburban communities are affected when a central city lacks the resources to maintain high-quality schools or to attack social ills. The provision of health care facilities, electricity and water, transportation, and recreational space affects the whole region and, many professionals think, should be under the control of a single, unified metropolitan government.

The growth in the number and size of metropolitan areas has increased awareness of the problems of their administrative fragmentation. In response, new approaches to the integration of those areas have been proposed and adopted. The aims of all plans of metropolitan government are to create or preserve coherence in the management of areawide concerns and to assure that both the problems and the benefits of growth are shared without regard to their jurisdictional locations.

## SUMMARY

The sovereign state is the dominant entity in the political subdivision of the world. It constitutes an expression of cultural separation and identity as pervasive as that inherent in language, religion, or ethnicity. A product of 18th-century political philosophy, the idea of the state was diffused globally by colonizing European powers. In most instances, the colonial boundaries they established have been retained as their international boundaries by newly independent countries.

The greatly varying geographic characteristics of states contribute to national strength and stability. Size, shape, and relative location influence countries' economies and international roles, while national cores and capitals are the heartlands of states. Boundaries, the legal definition of a state's size and shape, determine the limits of its sovereignty. They may or may not reflect preexisting cultural landscapes and, in any given case, may or may not prove to be viable. Whatever their nature, boundaries are at the root of many international conflicts. Maritime boundary claims, particularly as reflected in the United Nations Convention on the Law of the Sea, add a new dimension to traditional claims of territorial sovereignty.

State cohesiveness is promoted by a number of centripetal forces. Among these are national symbols, a variety of institutions, and confidence in the aims, organization, and administration of government. Also helping foster political and economic integration are transportation and communication connections. Destabilizing centrifugal forces, particularly ethnically based separatist movements, threaten the cohesion and stability of many states.

Although the state remains central to the partitioning of the world, a broadening array of political entities affects people individually and collectively. Recent decades have seen a significant increase in supranationalism, in the form and variety of global and regional alliances, to which states have surrendered some sovereign powers. At the other end of the spectrum, expanding Anglo American urban areas and governmental responsibilities raise questions of fairness in districting procedures and of effectiveness when political power is fragmented.

## Key Words

antecedent boundary 311
artificial boundary 311
centrifugal force 316
centripetal force 316
compact state 305
consequent boundary 311
core area 307
devolution 320
electoral geography 331
elongated state 305
enclave 305
ethnic cleansing 322
European Union (EU) 327
exclave 305
exclusive economic zone (EEZ) 325
fragmented state 305
geometric boundary 311
gerrymandering 331
irredentism 314
nation 302
nationalism 316
nation-state 302
natural boundary 311
perforated state 305
political geography 300
prorupt state 305
regionalism 320
state 302
subnationalism 320
subsequent boundary 311
superimposed boundary 311
supranationalism 324
terrorism 322
United Nations Convention on the Law of the Sea (UNCLOS) 325

## For Review & Consideration

1. What are the differences among a *state,* a *nation,* and a *nation-state?* Why is a colony not a state? How can one account for the rapid increase in the number of states since World War II?
2. What attributes differentiate states from one another? How do a country's size and shape affect its power and stability? How can a piece of land be both an *enclave* and an *exclave?*
3. How can boundaries be classified?
4. How do borders create opportunities for conflict? Describe and give examples of three types of such conflicts.
5. Distinguish between *centripetal* and *centrifugal* political forces. Why is nationalism both a centripetal and a centrifugal force? What are some of the ways national cohesion and identity are achieved?
6. What characteristics are common to all or most autonomist movements? Where are some of these movements active? Why do they tend to be located on the periphery rather than at the national core?
7. What types of international organizations and alliances can you name? What were the purposes of their establishment? What generalizations can you make regarding economic alliances?
8. How does the *United Nations Convention on the Law of the Sea* define zones of diminishing national control? What are the consequences of the concept of the 200-nm *exclusive economic zone?*
9. Why does it matter how boundaries are drawn around electoral districts? Theoretically, is it always possible to delimit boundaries "fairly"? Support your answer.
10. What reasons can you suggest for the great political fragmentation of the United States? What problems stem from such fragmentation?

# Selected References

**WEBSITES**

The World Wide Web has a tremendous variety of sites pertaining to geography. Websites relevant to the subject matter of this chapter appear in the "Web Links" section of the Online Learning Center associated with this book. Access it at www.mhhe.com/getis10e/.

Agnew, John. *Making Political Geography.* London: Oxford University Press, 2002.

Berdun, M. Montserrat Guibernau. *Nations without States.* Cambridge, England: Blackwell, 1999.

Blacksell, Mark. *Political Geography.* New York: Routledge, 2004.

Blake, Gerald H. *Maritime Boundaries.* London: Routledge, 2002.

Boyd, Andrew. *An Atlas of World Affairs.* 10th ed. New York: Routledge, 1998.

Chinn, Jeff, and Robert Kaiser. *Russians as the New Minority: Ethnicity and Nationalism in the Soviet Successor States.* Boulder, Colo.: Westview Press, 1996.

Croft, Stuart, et al., eds. *The Enlargement of Europe.* New York: St. Martin's Press, 1999.

Demko, George J., and William B. Wood, eds. *Reordering the World: Geopolitical Perspectives on the Twenty-first Century.* 2nd ed. Boulder, Colo.: Westview Press, 1999.

Gellner, Martin I. *Encounters with Nationalism.* Oxford, England: Basil Blackwell, 1995.

Gibb, Richard, and Mark Wise. *The European Union.* London: Edward Arnold, 2000.

Glassner, Martin I. *Political Geography.* 3d ed. New York: Wiley, 2003.

Graham, Brian, ed. *Modern Europe: Place, Culture, and Identity.* London: Edward Arnold, 1998.

Hooson, David, ed. *Geography and National Identity.* Oxford, England: Basil Blackwell, 1994.

Jones, Martin, Rhys Jones, and Michael Woods. *Introduction to Political Geography: Place, Space, and Politics.* London: Routledge, 2003.

Muir, Richard. *Political Geography: A New Introduction.* New York: Macmillan, 1999.

Newhouse, John. "Europe's Rising Regionalism." *Foreign Affairs* 76 (Jan./Feb. 1997): 67–84.

Pinder, David. *The New Europe: Economy, Society, and Environment.* New York: Wiley, 1998.

Renner, Michael. *The Anatomy of Resource Wars.* Worldwatch Paper 162. Washington, D.C.: Worldwatch Institute, 2002.

"The Rise of Europe's Little Nations." *The Wilson Quarterly* 18, no. 1 (1994): 50–81.

Shelley, Fred M., et al. *Political Geography of the United States.* New York: Guilford Press, 1996.

Spencer, Metta, ed. *Separatism: Democracy and Disintegration.* Lanham, Md.: Rowman and Littlefield, 1998.

Taylor, Peter J., and Colin Flint. *Political Geography: World-Economy, Nation-State and Locality.* 4th ed. Upper Saddle River, N.J.: Prentice Hall, 2000.

Williams, Allan M. *The European Community: The Contradiction of Integration.* 2d ed. Oxford, England: Basil Blackwell, 1997.

The crop bloomed luxuriantly that summer of 1846. The disaster of the preceding year seemed over and the potato, the sole sustenance of some 8 million Irish peasants, would again yield in the bounty needed. Yet within a week, wrote Father Mathew, "I beheld one wide waste of putrefying vegetation. The wretched people were seated on the fences of their decaying gardens . . . bewailing bitterly the destruction that had left them foodless." Colonel Gore found that "every field was black," and an estate steward noted that "the fields . . . look as if fire has passed over them." The potato was irretrievably gone for a second year; famine and pestilence were inevitable.

Within 5 years, the settlement geography of the most densely populated country in Europe was forever altered. The United States received a million immigrants, who provided the cheap labor needed for the canals, railroads, and mines that it was creating in its rush to economic development. New patterns of commodity flows were initiated as American maize for the first time found an Anglo-Irish market—as part of Poor Relief—and then entered a wider European market that had also suffered general crop failure in that bitter year. Within days, a microscopic organism, the cause of the potato blight, had altered the economic and human geography of two continents.

That alteration resulted from a complex set of intertwined causes and effects that demonstrates once again our repeated observation that apparently separate physical and cultural geographic patterns are really interconnected parts of a single reality. Central among those patterns are the ones the economic geographer isolates for special study.

Simply stated, **economic geography** is the study of how people earn their living, how livelihood systems vary by area, and how economic activities are spatially interrelated and linked. Of course, we cannot really comprehend the totality of the economic pursuits of more than 6 billion human beings. We cannot examine the infinite variety of production and service activities found everywhere on the earth's surface; nor can we trace all their innumerable interrelationships, linkages, and flows. Even if that level of understanding were possible, it would be valid for only a fleeting instant of time, for economic activities are constantly undergoing change.

Economic geographers seek consistencies. They attempt to develop generalizations that will aid in the comprehension of the maze of economic variations characterizing human existence. From their studies emerges a deeper awareness of the dynamic, interlocking diversity of human enterprise, of the impact of economic activity on all other facets of human life and culture, and of the increasing interdependence of differing national and regional economic systems (see "Economic Regions" in Chapter 13). The potato blight, although it struck only one small island, ultimately affected the economies of continents. In like fashion, the depletion of America's natural resources and the "deindustrialization" of its economy and conversion to postindustrial service and knowledge-based activities are altering the relative wealth of countries, flows of international trade, domestic employment and income patterns, and more (Figure 10.1).

## The Classification of Economic Activity and Economies

The search for an understanding of livelihood patterns is made more difficult by the complex environmental and cultural realities controlling the economic activities of humans. Many production patterns are rooted in the spatially variable circumstances of the *physical environment.* The staple crops of the humid tropics, for example, are not part of the agricultural systems of the midlatitudes; livestock types that thrive in American feedlots or on western ranges are not adapted to the Arctic tundra or the margins of the Sahara Desert. The unequal distribution of useful mineral deposits gives some regions and countries economic prospects and employment opportunities denied to others. Forestry and fishing depend on still other natural resources unequal in occurrence, type, and value.

Within the bounds of the environmentally possible, economic or production decisions may be conditioned by *cultural considerations.* For example, culturally based food preferences rather than environmental limitations may dictate the choice of crops or livestock. Maize is a preferred grain in Africa and the Americas; wheat in North America, Australia, Argentina, southern Europe, and Ukraine; and rice in much of Asia. Pigs are not produced in Muslim areas. The level of *technological development* of a culture will affect its recognition of resources or its ability to exploit them. Preindustrial societies do not know of, or need, iron ore or coking coal underlying their hunting, gathering, or gardening grounds. *Political decisions* may encourage or discourage—through subsidies, protective tariffs, or production restrictions—patterns of economic activity. And,

**FIGURE 10.1** These Japanese cars unloading at Seattle were forerunners of a continuing flow of imported goods capturing a share of the domestic market traditionally held by American manufacturers. Established patterns of production and exchange are constantly subject to change in a world of increasing economic and cultural interdependence and of changing relative competitive strengths. *© John Maher/EKM-Nepenthe.*

ultimately, production is controlled by *economic factors* of demand, whether that demand is expressed through a free market mechanism, through government intervention, or through the consumption requirements of a single family producing for its own needs.

## Categories of Activity

Such regionally varying environmental, cultural, technological, political, and market conditions add spatial details to more generalized ways of categorizing the world's productive work. One approach to that categorization is to view economic activity as ranged along a continuum of both increasing complexity of product or service and increasing distance from the natural environment. Seen from that perspective, a small number of distinctive stages of production and service activities can be distinguished (Figure 10.2).

**Primary activities** are those that harvest or extract something from the earth. They are at the beginning of the production cycle, where humans are in closest contact with the resources and potentialities of the environment. Such primary activities involve basic foodstuffs and raw material production. Hunting and gathering, grazing, agriculture, fishing, forestry, and mining and quarrying are examples.

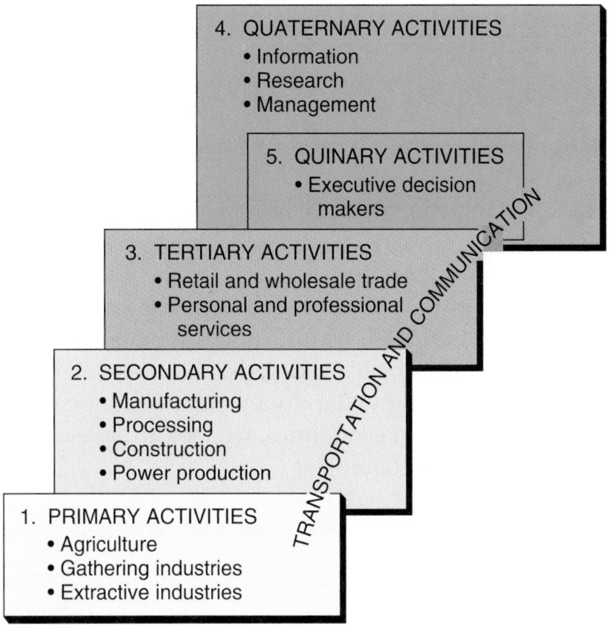


**FIGURE 10.2 The categories of economic activity.** The five main sectors of the economy do not stand alone. They are connected and integrated by transportation and communication services and facilities not assigned to any single sector but common to all.

**Secondary activities** are those that add value to materials by changing their form or combining them into more useful, and therefore more valuable, commodities. That provision of *form utility* ranges from the simple handicraft production of pottery or woodenware to the delicate assembly of electronic goods or space vehicles (Figure 10.3). Copper smelting, steelmaking, metalworking, automobile production, the textile and chemical industries—indeed, the full array of *manufacturing* and *processing industries*—are included in this phase of the production process. Also included are the production of *energy* (the "power company") and the *construction* industry.

**Tertiary activities** consist of those business and labor specializations that provide *services* to the primary and secondary sectors and *goods* and *services* to the general community and to the individual. These include professional, clerical, and personal services. They constitute the vital link between producer and consumer, for tertiary occupations importantly include the wholesale and retail *trade* activities—including "dot-com" Internet sales—necessary in highly interdependent societies.

In economically advanced societies, a growing number of individuals and entire organizations are engaged in the processing and dissemination of information and in the administration and control of their own or other enterprises. The term **quaternary** is applied to this fourth class of economic activities, which is composed entirely of services rendered by white collar professionals working in education, government, management, information processing, and research. Sometimes, a subdivision of these management functions—**quinary activities**—is distinguished to recognize high-level decision-making roles in all types of large organizations, public or private. The distinctions between tertiary, quaternary, and quinary activities are further developed later in this chapter under the section "Tertiary and Beyond." As Figure 10.2 further suggested, transportation and communication services cut across the general activity categories, unite them, and make possible the spatial interactions that all human enterprise requires.

These categories of production and service activities help us see an underlying structure to the nearly infinite variety of things people do to earn a living and to sustain themselves. But they tell us little about the organization of the larger economy of which the individual worker or enterprise is a part. For that wider organizational understanding of world and regional economies, we look to *systems* rather than *components* of economies.

## Types of Economic Systems

Broadly viewed, national economies in the early 21st century fall into one of three major types of system: *subsistence, commercial,* or *planned.* None of these economic systems is or has been "pure." That is, none exists in isolation in an increasingly interdependent world. Each, however, displays certain underlying characteristics based on its distinctive forms of resource management and economic control.

**FIGURE 10.3** These logs entering a lumber mill are products of *primary production.* Processing them into boards, plywood, or prefabricated houses is a *secondary activity* that increases their value by altering their form. The products of many secondary industries—sheet steel from steel mills, for example—constitute raw materials for other manufacturers. *© Mark Gibson.*

In a **subsistence economy,** goods and services are created for the use of the producers and their kinship groups. Therefore, there is little exchange of goods and only limited need for markets. In the **commercial economies** that have become dominant in nearly all parts of the world, producers or their agents in theory freely market their goods and services, the laws of supply and demand determine price and quantity, and market competition is the primary force shaping production decisions and distributions. In the extreme form of **planned economies** associated with the communist-controlled societies that have now collapsed in nearly every country where they were formerly created or imposed, producers or their agents disposed of goods and services through governmental agencies that controlled both supply and price. The quantities produced and the locational patterns of production were tightly programmed by central planning departments.

Rigidly planned economies no longer exist in their classical form; they have been modified or dismantled in favor of free market structures or are only partially retained in the lesser degree of economic control associated with governmental supervision or ownership of selected sectors of increasingly market-oriented economies. Nevertheless, their landscape evidence lives on. The physical structures, patterns of production, and imposed regional interdependencies they created remain to influence or distort the economic decisions of successor societies.

In actuality, few people are members of only one of these systems, although one may be dominant. A farmer in India may produce rice and vegetables primarily for the family's consumption but also save some of the produce to sell. In addition, members of the family may market cloth or other handicrafts they make. With the money derived from those sales, the Indian peasant is able to buy, among other things, clothes for the family, tools, or fuel. Thus, that Indian farmer is a member of at least two systems: subsistence and commercial.

In the United States, governmental controls on the production of various types of goods and services (such as growing wheat or tobacco, producing alcohol, constructing and operating nuclear power plants, or engaging in licensed personal and professional services) mean that the country does not have a purely commercial economy. To a limited extent, its citizens participate in a controlled and planned as well as in a free market environment. Many African, Asian, and Latin American market economies have been decisively shaped by governmental policies encouraging or demanding the production of export commodities rather than domestic foodstuffs or promoting through import restrictions the development of domestic industries not readily supported by the national market alone. Example after example would show that there are very few people in the world who are members of only one type of economic system (Figure 10.4).

Inevitably, spatial patterns, including those of economic systems and activities, are subject to change. For example, the commercial economies of Western European countries, some with sizeable infusions of planned economy controls, are being restructured by both increased free market competition and supranational regulation under the World Trade Organization and the European Union (see pp. 326 and 327). Many of the countries of Latin America, Africa, Asia, and the Middle East that traditionally were dominated by subsistence economies are now benefiting from technology transfer from advanced economies and integration into expanding global production and exchange patterns.

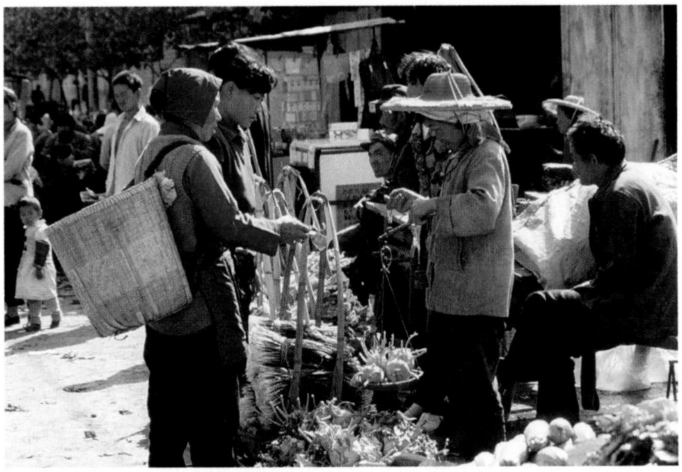

**FIGURE 10.4** Independent street merchants, shop owners, and peddlers in modern China are members of both a planned and a market system. Free markets and private vendors multiplied after government price controls on most food items were removed in May 1985. Increasingly, nonfood trade and manufacturing, too, are being freed of central governmental control and thriving in the private sector. As state-run companies shrank and laid off workers, privately owned businesses in 2003 accounted for over half of China's gross domestic product and employed 130 million workers—only one-fifth of the total workforce but a large majority of the industrial workers. The photo shows a portion of the crowded produce market in Kunming, Yunnan Province. *© Liba Taylor/Corbis Images.*

No matter what economic system locally prevails, in all systems transportation is a key variable. No advanced economy can flourish without a well-connected transport network. All subsistence societies—and subsistence areas of developing countries—are characterized by their isolation from regional and world routeways (Figure 10.5), and that isolation restricts their progression to more-advanced forms of economic structure.

Former sharp contrasts in economic organization are becoming blurred and national economic orientations are changing as globalization reduces structural contrasts in national economies. Still, both approaches to economic classification—by type of activity and by organization of economies—help us visualize and understand world economic geographic patterns. In this chapter, our path to that understanding leads through the successive categories of economic activity, from primary to quinary, with emphasis on the technologies, spatial patterns, and organizational systems that are involved in each category.

## Primary Activities: Agriculture

Humankind's basic economic concern is producing or securing food resources sufficient in caloric content to meet individual daily energy requirements and so balanced as to satisfy normal nutritional needs. Those supplies may be acquired by the consumer directly through the primary economic activities of hunting, gathering, farming and fishing (a form of "gathering") or indirectly through the performance of other primary, secondary, or higher-level economic endeavors that yield income sufficient to provide the earner and the earner's family with income to obtain needed daily sustenance.

Since the middle of the 20th century, a recurring but unrealized fear has been that the world's steadily increasing population would inevitably and quickly exceed available or potential food supplies. Instead, although global population has more than doubled since 1950, the total amount of human food produced worldwide since then has also more than doubled. The Food and Agriculture Organization (FAO) of the United Nations has set the minimum daily requirement for caloric intake per person at 2350. By that measure, annual food supplies are more than sufficient to meet world needs. That is, if total food resources were evenly distributed, everyone would have access to amounts sufficient for adequate daily nourishment. In reality, however, nearly one-seventh of the world's population—nearly

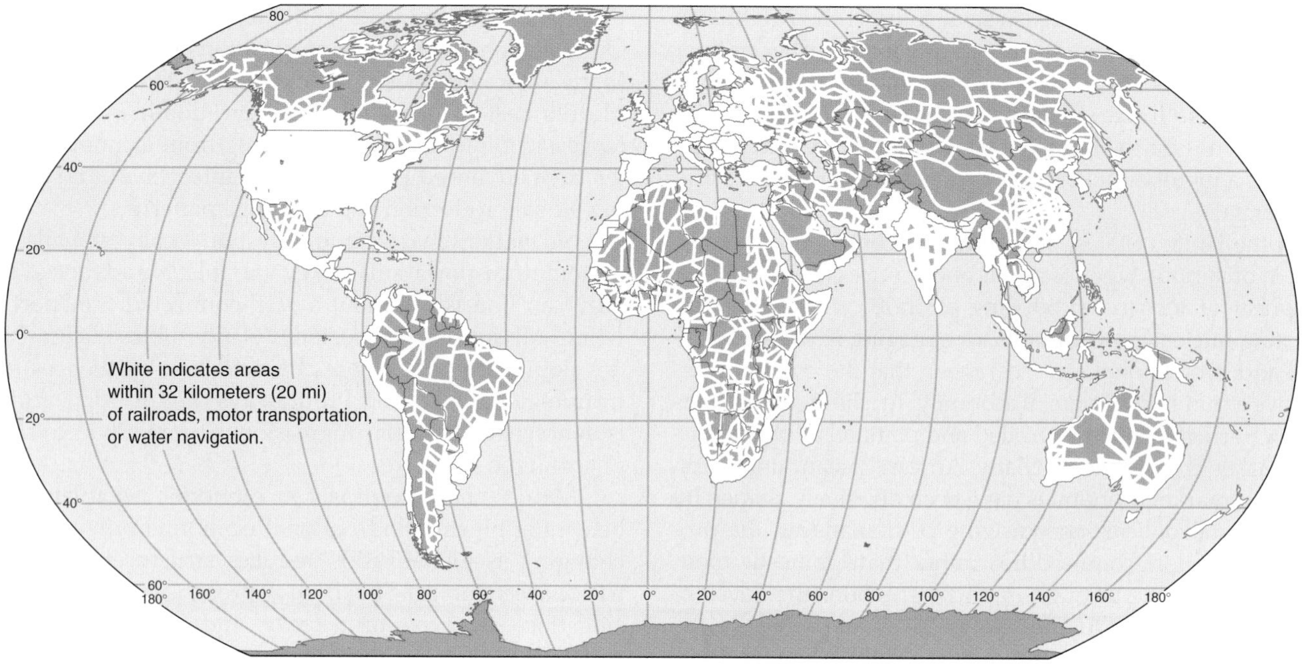


**FIGURE 10.5 Patterns of access and isolation.** Accessibility is a key measure of economic development and of the degree to which a world region can participate in interconnected market activities. Isolated areas of countries with advanced economies suffer a price disadvantage because of high transportation costs. Lack of accessibility in subsistence economic areas slows their modernization and hinders their participation in the world market. Source: *Copyright Permission: Hammond Incorporated, Maplewood, NJ 07040.*

all in the less-developed countries—are inadequately supplied with food and nutrients (Figure 10.6). Conservatively, 54 countries fall below achieving the FAO minimum per capita requirement; they do not produce enough food to supply their populations or have the economic means to close the gap through imports.

This stark contradiction between sufficient worldwide food supplies and widespread malnutrition reflects inequalities in national and personal incomes; population growth rates; lack of access to fertile soils, credit, and education; local climatic conditions or catastrophes; and lack of transportation and storage facilities, among other reasons. By mid-century, the increasingly interconnected world population will expand to nearly 9 billion and concerns with individual states' food supplies will inevitably continue and remain, as well, a persistent international issue.

Before there was farming, *hunting* and *gathering* were the universal forms of primary production. These preagricultural pursuits are now practiced by at most a few thousand persons worldwide, primarily in isolated and remote pockets within the low latitudes and among the sparse populations of very high latitudes. The interior of New Guinea, rugged areas of interior Southeast Asia, diminishing segments of the Amazon rain forest, a few districts of tropical Africa and northern Australia, and parts of the Arctic regions still contain such preagricultural people. Their numbers are few and declining, and wherever they are brought into contact with more-advanced cultures, their way of life is eroded or lost.

**Agriculture,** defined as the growing of crops and the tending of livestock, whether for the subsistence of the producers or for sale or exchange, has replaced hunting and gathering as economically the most significant of the primary activities. It is spatially the most widespread, found in all world regions where environmental circumstances—including adequate moisture, good growing season length, and productive soils—permit (Figure 10.7). Crop farming alone covers some 15 million square kilometers (5.8 million sq mi) worldwide, about 10% of the earth's total land area. The United Nations estimates that more than one-third of the world's land (excluding Greenland and Antarctica) is in agricultural use, including permanent pastureland. In many developing economies, at least two-thirds of the labor force is directly involved in farming and herding. In some, such as Bhutan in Asia or Burkina Faso and Burundi in Africa, the figure is more than 90%. Overall, however, employment in agriculture is steadily declining in developing countries, echoing but trailing the trend in highly developed commercial economies, where direct employment in agriculture involves only a small fraction of the labor force (Figure 7.6).

It has been customary to classify agricultural societies on the twin bases of the importance of off-farm sales and the level of mechanization and technological advancement. *Subsistence, traditional* (or *intermediate*), and *advanced* (or *modern*) are usual terms employed to recognize both aspects. These are not mutually exclusive but, rather, are recognized stages along a continuum of farm economy variants. At one end

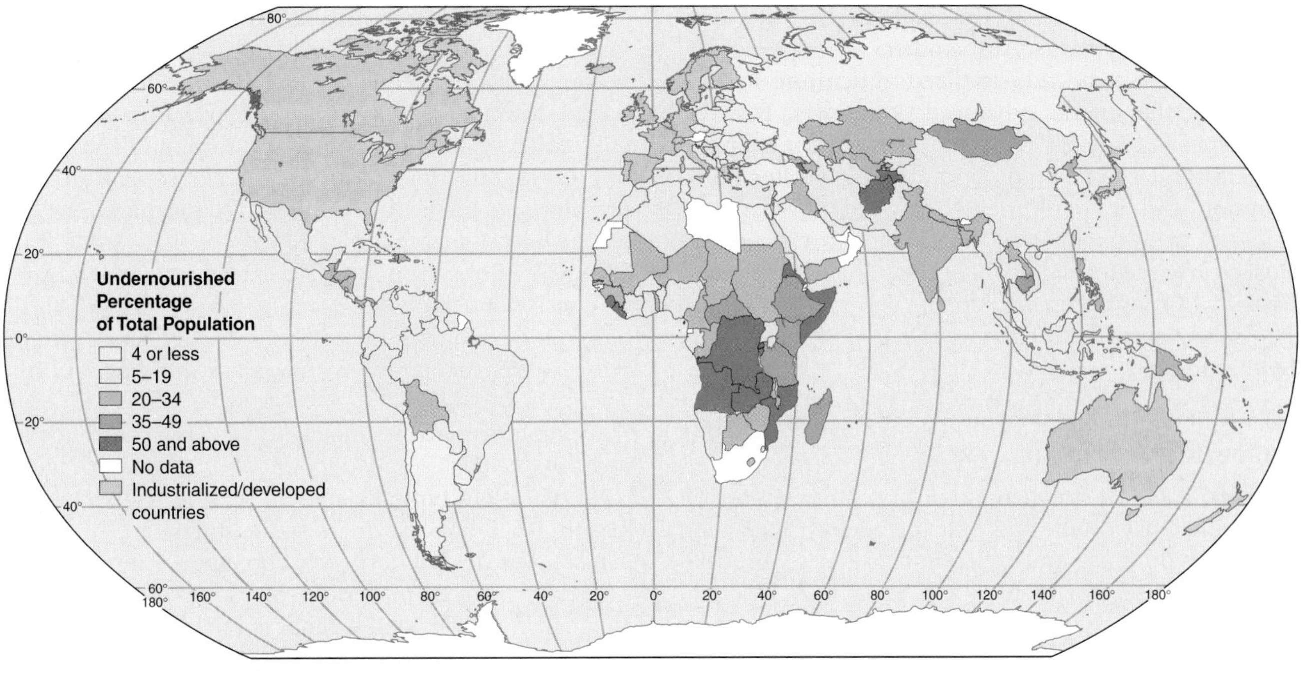


**FIGURE 10.6 Percent of national population that is undernourished, 2001.** Early in the 21st century, there were about 840 million undernourished people worldwide facing chronic hunger or starvation, undernutrition, and deficiencies of essential iron, iodine, vitamin A, and other micronutrients. For many, sickness and parasites take the nutritive value from what little food is eaten. The world's nutritional levels have proportionally improved in the past several decades. More than one-third of people living in poor countries in 1970 were undernourished; by 2001, that figure had fallen to 16%. Numerically, however, the number of malnourished people across the developing world grew by an average of 4.5 million a year between 1995 and 2001. Sub-Saharan Africa's incidence of undernourishment has remained constant, a reflection of the region's continuing poverty and progressive drop in per capita food production since the 1960s. In contrast, the FAO indicates that all industrialized countries have average daily per capita caloric intake above 110% of physiological requirements, although that generalization masks troubling incidences of areal and household hunger and malnutrition.

Sources: *FAO,* State of Food Insecurity in the World; *and Bread for the World Institute.*

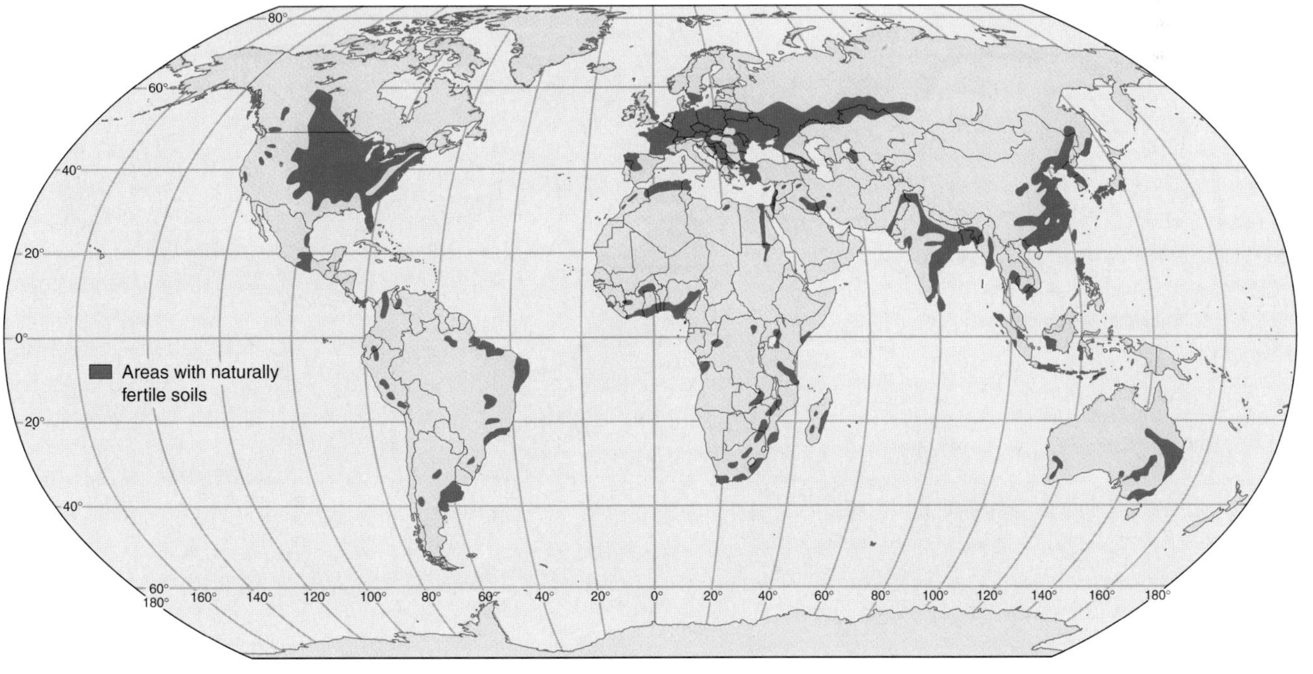


**FIGURE 10.7 Areas with naturally fertile soils** account for much of the world's grain production and include the corn and wheat belts of North America, Ukraine, and southern Siberia and the rice-producing regions of India and Southeast Asia. A comparison of this map with Figure 10.6 will help explain the prevalence of undernourishment in sub-Saharan Africa.

lies production solely for family sustenance, using rudimentary tools and native plants. At the other is the specialized, highly capitalized, near-industrialized agriculture for off-farm delivery that marks advanced economies. Between these extremes is the middle ground of traditional agriculture, where farm production is in part destined for home consumption and in part oriented toward off-farm sale either locally or in national and international markets. We can most clearly see the variety of agricultural activities and the diversity of controls on their spatial patterns by examining the "subsistence" and "advanced" ends of the agricultural continuum.

## Subsistence Agriculture

By definition, a *subsistence* economic system involves nearly total self-sufficiency on the part of its members. Production for exchange is minimal, and each family or close-knit social group relies on itself for its food and other most essential requirements. Farming for the immediate needs of the family is, even today, the predominant occupation of humankind. In most of Africa, much of Latin America, and most of southern and eastern Asia, the majority of the people are primarily concerned with feeding themselves from their own land and livestock.

Two chief types of subsistence agriculture may be recognized: *extensive* and *intensive.* Although each type has several variants, the essential contrast between them is realizable yield per unit of area used and, therefore, population-supporting potential. **Extensive subsistence agriculture** involves large areas of land and minimal labor input per hectare. Both product per land unit and population densities are low. **Intensive subsistence agriculture** involves the cultivation of small land holdings through the expenditure of great amounts of labor per acre. Yields per unit area and population densities are both high (Figure 10.8).

### Extensive Subsistence Agriculture

Of the several types of *extensive subsistence* agriculture—varying one from another in their intensities of land use—two are of particular interest: nomadic herding and shifting cultivation.

**Nomadic herding,** the wandering but controlled movement of livestock solely dependent on natural forage, is the most extensive type of land use system (see Figure 10.8). That is, it requires the greatest amount of land area per person sustained. Over large portions of the Asian semidesert and desert areas, in certain highland areas, and on the fringes of and within the Sahara, a relatively small number of people graze animals for consumption by the herder group, not for market sale. Sheep, goats, and camels are most common, while cattle, horses, and yaks are locally important. The reindeer of Lapland were formerly part of the same system.

Whatever the animals involved, their common characteristics are hardiness, mobility, and an ability to subsist on sparse forage. The animals provide a variety of products: milk, cheese, and meat for food; hair, wool, and skins for

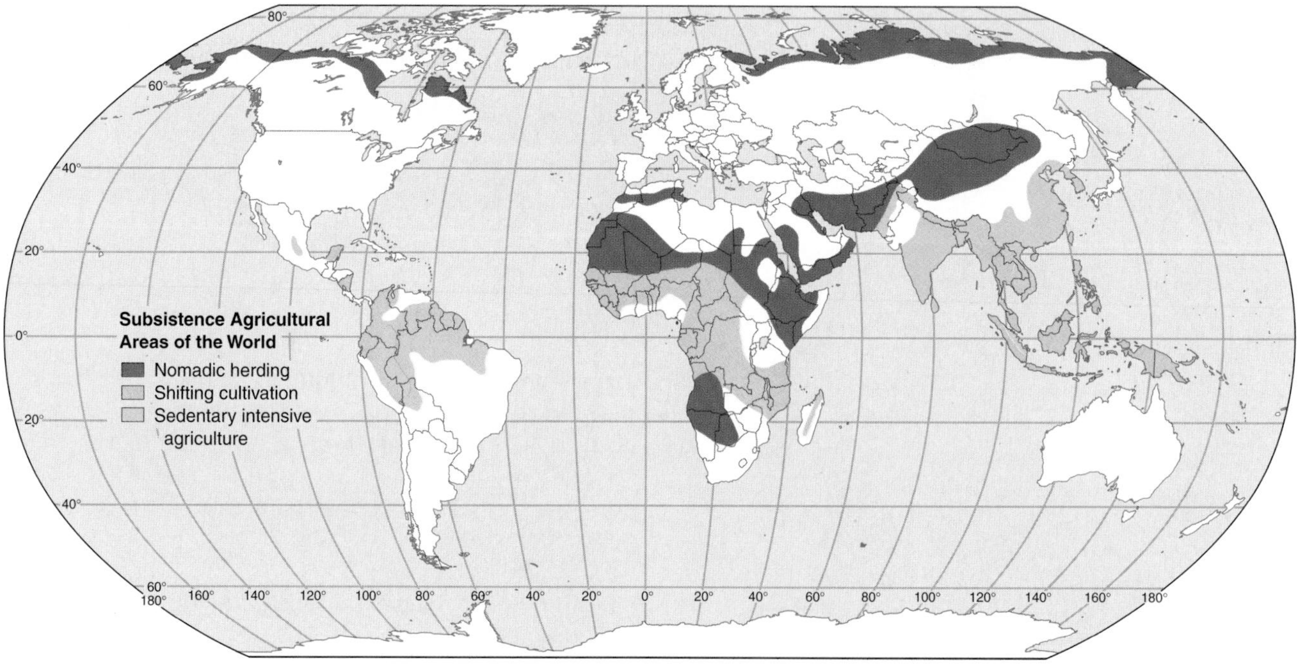


**FIGURE 10.8 Subsistence agricultural areas of the world.** Nomadic herding, supporting relatively few people, was the age-old way of life in large parts of the dry and cold world. Shifting or swidden agriculture maintains soil fertility by tested traditional practices in tropical wet and wet-and-dry climates. Large parts of Asia support millions of people engaged in sedentary intensive cultivation, with rice and wheat the chief crops.

clothing; skins for shelter; and excrement for fuel. For the herder, they represent primary subsistence. Nomadic movement is tied to sparse and seasonal rainfall or to cold temperature regimes and to the areally varying appearance and exhaustion of forage. Extended stays in a given location are neither desirable nor possible. *Transhumance* is a special form of seasonal movement of livestock to exploit specific locally varying pasture conditions. Used by permanently or seasonally sedentary pastoralists and pastoral farmers, transhumance involves either the regular vertical alteration from mountain to valley pastures between summer and winter months or horizontal movement between established lowland grazing areas to reach pastures temporarily lush from monsoonal (seasonal) rains.

As a type of economic system, nomadic herding is declining. Many economic, social, and cultural changes are causing nomadic groups to alter their way of life or to disappear entirely. On the Arctic fringe of Russia, herders under communism were made members of state or collective herding enterprises; in post-Soviet years, extensive oil and natural gas exploration and extraction are damaging or destroying large portions of tundra habitat. In northern Scandinavia, Lapps (Saami) are engaged in commercial more than in subsistence livestock farming. In the Sahel region of Africa on the margins of the Sahara, oases once controlled by herders have been taken over by farmers, and the great droughts of recent decades have forever altered the formerly nomadic way of life of thousands.

A much differently based and distributed form of extensive subsistence agriculture is found in all of the warm, moist, low-latitude areas of the world. There, many people engage in a kind of nomadic farming. Through clearing and use, the soils of those areas lose many of their nutrients (as soil chemicals are dissolved and removed by surface and groundwater—*leaching*—or nutrients are removed from the land in the vegetables picked and eaten), and farmers cultivating them need to move on after harvesting several crops. In a sense, they rotate fields rather than crops to maintain productivity. This type of **shifting cultivation** has a number of names, the most common of which are *swidden* (an English localism for "burned clearing") and *slash-and-burn.* Each region of its practice has its own name—for example, *milpa* in Middle and South America, *chitemene* in Africa, and *ladang* in Southeast Asia.

Characteristically, the farmers hack down the natural vegetation, burn the cuttings, and then plant such crops as maize (corn), millet (a cereal grain), rice, manioc or cassava, yams, and sugarcane (Figure 10.9). Increasingly included in many of the crop combinations are such high-value, labor-intensive commercial crops as coffee, providing the cash income that is evidence of the growing integration of all peoples into exchange economies. Initial yields—the first and second crops—may be very high, but they quickly become lower with each successive planting on the same plot. As that occurs, cropping ceases, native vegetation is allowed to reclaim the clearing, and gardening shifts to another newly prepared site. The first clearing will ideally not be used again for crops until, after many years, natural fallowing replenishes its fertility (see "Swidden Agriculture," p. 350).

**FIGURE 10.9** An African swidden plot being fired. Stumps and trees left in the clearing will remain after the burn. *© Wolfgang Kaehler.*

Less than 3% of the world's people are still predominantly engaged in tropical shifting cultivation on about one-seventh of the world's land area (see Figure 10.8). Because the essential characteristic of the system is the intermittent cultivation of the land, each family requires a total occupance area equivalent to the garden plot in current use plus all land left fallow for regeneration. Population densities are characteristically low, because much land is needed to support few people. It may be argued that shifting cultivation is a highly efficient cultural adaptation where land is abundant in relation to population, and levels of technology and capital availability are low. As those conditions change, the system becomes less viable.

## Swidden Agriculture

The following account describes shifting cultivation among the Hanunóo people of the Philippines. Nearly identical procedures are followed in all swidden farming regions.

> When a garden site of about one-half hectare (a little over one acre) has been selected, the swidden farmer begins to remove unwanted vegetation. The first phase of this process consists of slashing and cutting the undergrowth and smaller trees with bush knives. The principal aim is to cover the entire site with highly inflammable dead vegetation so that the later stage of burning will be most effective. Because of the threat of soil erosion the ground must not be exposed directly to the elements at any time during the cutting stage. During the first months of the agricultural year, activities connected with cutting take priority over all others.
>
> Once most of the undergrowth has been slashed, the larger trees must be felled or killed by girdling (cutting a complete ring of bark) so that unwanted shade will be removed. Some trees, however, are merely trimmed but not killed or cut, both to reduce labor and to leave trees to reseed the swidden during the subsequent fallow period.
>
> The crucial and most important single event in the agricultural cycle is swidden burning. The main firing of a swidden is the culmination of many weeks of preparation in spreading and leveling chopped vegetation, preparing firebreaks to prevent flames escaping into the jungle, and allowing time for the drying process. An ideal burn rapidly consumes every bit of litter; in no more than an hour or an hour and a half, only smoldering remains are left.
>
> Swidden farmers note the following as the benefits of a good burn: (1) removal of unwanted vegetation, resulting in a cleared swidden; (2) extermination of many animal and some weed pests; (3) preparation of the soil for dibble (any small hand tool or stick to make a hole) planting by making it softer and more friable; (4) provision of an evenly distributed cover of wood ashes, good for young crop plants and protective of newly planted grain seed. Within the first year of the swidden cycle, an average of between 40 and 50 different types of crop plants have been planted and harvested.
>
> The most critical feature of swidden agriculture is the maintenance of soil fertility and structure. The solution is to pursue a system of rotation of one to three years in crop and ten to twenty in woody or bush fallow regeneration. When population pressures mandate a reduction in the length of fallow period, productivity tends to drop as soil fertility is lowered, marginal land is utilized, and environmental degradation occurs. The balance is delicate.

*Source:* Adapted from *Hanunoo Agriculture*, by Harold C. Conklin (FAO Forestry Development Paper No. 12), 1957.

### Intensive Subsistence Agriculture

Some 45% of the people of the world are engaged in intensive subsistence agriculture, which predominates in the areas shown in Figure 10.8. As a descriptive term, *intensive subsistence* is no longer fully applicable to a changing way of life and economy in which the distinction between subsistence and commercial is decreasingly valid. Although families may still be fed primarily with the produce of their individual plots, the exchange of farm commodities within the system is considerable. Production of foodstuffs for sale in rapidly growing urban markets is increasingly vital for the rural economies of subsistence farming areas and for the sustenance of the growing proportion of national and regional populations no longer themselves engaged in farming. Nevertheless, hundreds of millions of Indians, Chinese, Pakistanis, Bangladeshis, and Indonesians, plus further millions in other Asian, African, and Latin American countries remain small-plot, mainly subsistence producers of rice, wheat, maize, millet, or pulses (peas, beans, and other legumes). Most live in monsoon Asia, and we will devote our attention to that area.

Intensive subsistence farmers are concentrated in such major river valleys and deltas as the Ganges and the Chang Jiang (Yangtze), as well as in smaller valleys close to coasts—level areas with fertile alluvial soils. These warm, moist districts are well suited to the production of rice, a crop that under ideal conditions can provide large amounts of food per unit of land. Rice also requires a great deal of time and attention, for planting rice shoots by hand in standing fresh water is a tedious art (Figure 10.10). In the cooler and drier portions of Asia north of 20°N, wheat is grown intensively, along with millet and, less commonly, upland rice.

Intensive subsistence farming is characterized by large inputs of labor per unit of land, by small plots, by the intensive use of fertilizers, mostly animal manure (see "The Economy of a Chinese Village," p. 352), and by the promise of high yields in good years. For food security and dietary custom, some other products are also grown. Vegetables and some livestock are part of the agricultural system, and fish may be reared in rice paddies and ponds. Cattle are a source of labor and of food. Food animals include swine, ducks, and chickens, but since Muslims eat no pork, hogs are absent in their areas of settlement. Hindus generally eat little meat, mainly goat and lamb but not pork or beef. The large number of cattle in India are vital for labor and are a source of milk and cheese, as well as producers of fertilizer and fuel.

**FIGURE 10.10** Transplanting rice seedlings requires arduous hand labor by all members of the family. The newly flooded diked fields, previously plowed and fertilized, will have their water level maintained until the grain is ripe. This photograph was taken in Indonesia; the scene is repeated wherever subsistence wet-rice agriculture is practiced. *© Sean Sprague.*

Not all of the world's subsistence farming is based in rural areas. Urban agriculture is a rapidly growing activity, with some 800 million city farmers worldwide providing, according to United Nations figures, one-seventh of the world's total food production. Occurring in all regions of the world, developed and underdeveloped, but most prevalent in Asia, urban agricultural activities range from small garden plots, to backyard livestock breeding, to fish raised in ponds and streams. Using the garbage dumps of Jakarta, the rooftops of Mexico City, and meager dirt strips along roadways in Kolkata or Kinshasa, millions of people are feeding their own families and supplying local markets with vegetables, fruit, fish, and even meat—all produced within the cities themselves and all without the expense and spoilage of storage or long-distance transportation.

In Africa, a reported 20% of urban nutritional requirement is produced in the towns and cities; two of three Kenyan and Tanzanian urban families engage in farming, for example, and in Accra, Ghana's capital, urban farming provides the city with 90% of its fresh vegetables. In all parts of the developing world, urban-origin foodstuffs have reduced the incidence of adult and child malnutrition in cities rapidly expanding by their own birth rates and by the growing influx of displaced rural folk.

## Expanding Crop Production

Continuing population pressures on existing resources are a constant spur for seeking ways to expand the food supply available both to the subsistence farmers of the developing economies and to the wider world as a whole. Two paths to promoting increased food production are apparent: (1) expand the land area under cultivation and (2) increase crop yields from existing farmlands.

The first approach—increasing cropland area—is not a promising strategy. Approximately 70% of the world's land area is agriculturally unsuitable, being too cold, too dry, too steep, or totally infertile. Of the remaining 30%, most of the area well suited for farming is already under cultivation, and of that area, millions of hectares annually are being lost

## The Economy of a Chinese Village

The village of Nanching is in subtropical southern China on the Zhu River delta near Guangzhou (Canton). Its subsistence agricultural regime was described by a field investigator, whose account is here condensed. The system is found in its essentials in other rice-oriented societies.

> In this double-crop region, rice was planted in March and August and harvested in late June or July and again in November. March to November was the major farming season. Early in March the earth was turned with an iron-tipped wooden plow pulled by a water buffalo. The very poor who could not afford a buffalo used a large iron-tipped wooden hoe for the same purpose.
>
> The plowed soil was raked smooth, fertilizer was applied, and water was let into the field, which was then ready for the transplanting of rice seedlings. Seedlings were raised in a seedbed, a tiny patch fenced off on the side or corner of the field. Beginning from the middle of March, the transplanting of seedlings took place. The whole family was on the scene. Each took the seedlings by the bunch, ten to fifteen plants, and pushed them into the soft inundated soil. For the first thirty or forty days the emerald green crop demanded little attention except keeping the water at a proper level. But after this period came the first weeding; the second weeding followed a month later. This was done by hand, and everyone old enough for such work participated. With the second weeding went the job of adding fertilizer. The grain was now allowed to stand to "draw starch" to fill the hull of the kernels. When the kernels had "drawn enough starch," water was let out of the field, and both the soil and the stalks were allowed to dry under the hot sun.
>
> Then came the harvest, when all the rice plants were cut off a few inches above the ground with a sickle. Threshing was done on a threshing board. Then the grain and the stalks and leaves were taken home with a carrying pole on the peasant's shoulder. The plant was used as fuel at home.
>
> As soon as the exhausting harvest work was done, no time could be lost before starting the chores of plowing, fertilizing, pumping water into the fields, and transplanting seedlings for the second crop. The slack season of the rice crop was taken up by chores required for the vegetables which demanded continuous attention, since every peasant family devoted a part of the farm to vegetable gardening. In the hot and damp period of late spring and summer, eggplant and several varieties of squash and beans were grown. The green-leafed vegetables thrived in the cooler and drier period of fall, winter, and early spring. Leeks grew the year round.
>
> When one crop of vegetables was harvested, the soil was turned and the clods broken up by a digging hoe and leveled with an iron rake. Fertilizer was applied, and seeds or seedlings of a new crop were planted. Hand weeding was a constant job; watering with the long-handled wooden dipper had to be done an average of three times a day, and in the very hot season when evaporation was rapid, as frequently as six times a day. The soil had to be cultivated with the hoe frequently as the heavy tropical rains packed the earth continuously. Instead of the two applications of fertilizer common with the rice crop, fertilizing was much more frequent for vegetables. Besides the heavy fertilizing of the soil at the beginning of a crop, usually with city garbage, additional fertilizer, usually diluted urine or a mixture of diluted urine and excreta, was given every ten days or so to most vegetables.

Adapted from C. K. Yang, *A Chinese Village in Early Communist Transition*. The MIT Press, Cambridge, Mass. 

through soil erosion, salinization, desertification, and the conversion of farmland to urban, industrial, and transportation uses in all developed and developing countries. In the United States, land is being withdrawn from—rather than added to—the cultivated total in response to governmental "set-aside" programs and individual decisions that the continued farming of some lands is no longer economically attractive. Only the rain forests of Africa and the Amazon Basin of South America retain sizeable areas of potentially farmable land. The soils of those regions, however, are fragile, are low in nutrients, have poor water retention, and are easily eroded or destroyed following deforestation. By most accounts, then, world food output cannot reasonably be increased by simple expansion of cultivated areas.

### Intensification and the Green Revolution

Increased productivity of existing cropland rather than expansion of cultivated area has been the key to the growth of agricultural output over the past few decades. Between 1972 and 2002, world total grain production rose 60%. Despite some 2.5 billion more people in the world, average grain production per capita for the period 1998–2002 was 3% above the 1972 level. The vast majority of that production growth was due to increases in yields rather than expansions in cropland. For Asia as a whole, cereal yields grew by more than 40% between 1980 and the end of the century, largely accounted for by increases in China and India; they increased by more than 35% in South America. Two interrelated approaches to those yield increases mark recent farming practices.

First, throughout much of the developing world production inputs such as water, fertilizer, pesticides, and labor have been increased to expand yields on a relatively constant supply of cultivable land. Irrigated area, for example, nearly doubled between 1960 and 2000 to comprise 18% of the world's cropland. Global consumption of fertilizers has dramatically increased since the 1950s, and inputs of pesticides and herbicides have similarly grown. Traditional practices of leaving land fallow (uncultivated) to renew its fertility have been largely abandoned, and double and triple cropping of land where climate permits has increased in Asia and even in Africa, where marginal land is put to near-continuous use to meet growing food demands.

Many of these intensification practices are part of the second approach, linked to the **Green Revolution**—the shorthand reference to a complex of seed and management improvements adapted to the needs of intensive agriculture and designed to bring larger harvests from a given area of farmland. Genetic improvements in rice and wheat have formed the basis of the Green Revolution. Dwarfed varieties have been developed that respond dramatically to heavy applications of fertilizer, that resist plant diseases, and that can tolerate much shorter growing seasons than traditional native varieties. Adopting the new varieties and applying the irrigation, mechanization, fertilization, and pesticide practices they require have created a new "high-input, high-yield" agriculture. The Green Revolution yield increases and the improved food supplies they represent have been particularly important in densely populated, subsistence farming areas heavily dependent on rice and wheat cultivation. Chinese rice harvests grew by two-thirds, for example, and India's wheat yields doubled between 1970 and the end of the 1990s.

But a price has been paid. The Green Revolution is commercially oriented and demands high inputs of costly hybrid seeds, mechanization, irrigation, fertilizers, and pesticides. As the Green Revolution is adopted, traditional and subsistence agriculture are being displaced. Lost, too, are the food security that distinctive locally adapted native crop varieties provide and the nutritional diversity and balance that multiple-crop intensive gardening assure. Subsistence farming, wherever practiced, is oriented toward risk minimization. Many differentially hardy varieties of a single crop guarantee some yield whatever adverse weather, disease, or pest problems might occur. Commercial agriculture, however, aims at profit maximization, not minimal food security.

There are other costs for Green Revolution successes. For example, irrigation, responsible for an important part of increased crop yields, has destroyed large tracts of land; excessive salinity of soils resulting from poor irrigation practices is estimated to have a serious effect on the productivity of more than 10% of all the world's irrigated land. And the huge amount of water required for Green Revolution irrigation has led to serious groundwater depletion, to conflict between agricultural and growing urban and industrial water needs in developing countries—many of which are in subhumid climates—and to worries about scarcity and future wars over water.

The presumed benefits of the Green Revolution are not available to all subsistence agricultural areas or advantageous to everyone engaged in farming. Most poor farmers on marginal and rain-fed (nonirrigated) lands, for example, have not benefited from the new plant varieties requiring irrigation and high chemical inputs, and women have often been losers in the transition from traditional to the more industrialized farming practices of the Green Revolution (see "Women and the Green Revolution," p. 354). Africa is a case in point. Green Revolution crop improvements have concentrated on wheat, rice, and maize. Of these, only maize is important in Africa, where principal food crops include millet, sorghum, cassava, manioc, yams, cowpeas, and peanuts. Although new varieties of maize resistant to the drought and acidic soils common in Africa were announced in the middle 1990s, belated research efforts directed to other African crops, the continent's great range of growing conditions, and its abundance of yield-destroying pests and viruses have denied it the dramatic regionwide increases in food production experienced elsewhere in the developing world.

In many areas showing greatest past successes, Green Revolution gains are falling off. Recent cereal yields in Asia, for example, are growing at only two-thirds of their 1970s rate; the UN's Food and Agriculture Organization now considers Green Revolution technologies "almost exhausted" of any further productivity gains in Asian rice cultivation. Little prime land and even less water remain to expand farming in many developing countries, and the adverse ecological and social consequences of industrial farming techniques arouse growing resistance. Nor does biotechnology—which many have hailed as a promising new Green Revolution approach—seem likely to fill the gap. Consumer resistance to genetic modification (GM) of crops, fear of the ecological consequences of such modification, and the high cost of and restrictions on the new biotechnologies imposed by their corporate developers all conspire to inhibit the widespread adoption of the new technologies.

## Commercial Agriculture

Few people or areas still retain the isolation and self-containment characteristic of pure subsistence economies. Nearly all have been touched by a modern world of trade and exchange and, in response, have adjusted their traditional economies. Modifications of subsistence agricultural systems have inevitably made them more complex by imparting to them at least some of the diversity and linkages of activity that mark the advanced economic systems of the more-developed world. Farmers in those systems produce not for their own subsistence but primarily for a market off the farm itself. They are part of integrated exchange economies in which agriculture is but one element in a complex structure that includes employment in

## Women and the Green Revolution

Women farmers grow at least half of the world's food and up to 80% in some African countries. They are responsible for an even larger share of food consumed by their own families: 80% in sub-Saharan Africa, 65% in Asia, and 45% in Latin America and the Caribbean. Further, women comprise between one-third and one-half of all agricultural laborers in developing countries. For example, African women perform about 90% of the work of processing food crops and 80% of the work of harvesting and marketing. Women's agricultural dominance in developing states is increasing, in fact, as male family members continue to leave for cities in search of paid urban work. In Mozambique, for example, for every 100 men working in agriculture, there are 153 women. In nearly all other sub-Saharan countries, the female component runs between 120 and 150 per 100 men. The departure of men for near or distant cities means, in addition, that women must assume the effective management of their families' total farm operations.

Despite their fundamental role, however, women do not share equally with men in the rewards from agriculture, nor are they always beneficiaries of presumed improvements in agricultural technologies and practices. Often, they cannot own or inherit the land on which they work and frequently have difficulty in obtaining improved seeds or fertilizers available to male farmers.

As a rule, women farmers work longer hours and have lower incomes than do male farmers. This is not because they are less competent. Rather, it is due to restricting cultural and economic factors. First, most women farmers are involved in subsistence farming and food production for the local market, which yields little cash return. Second, they have far less access than men to credit at bank or government-subsidized rates that would make it possible for them to acquire the Green Revolution technology, such as hybrid seeds and fertilizers. Third, in some cultures women cannot own land and so are excluded from agricultural improvement programs and projects aimed at landowners. For example, many African agricultural development programs are based on the conversion of communal land, to which women have access, to private holdings, from which they are excluded. In Asia, inheritance laws favor male over female heirs, and female-inherited land is managed by husbands; in Latin America, discrimination results from the more-limited status held by women under the law.

At the same time, the Green Revolution and its greater commercialization of crops has generally required an increase in labor per hectare, particularly in tasks typically reserved for women, such as weeding, harvesting, and postharvest work. If women are provided no relief from their other daily tasks, the Green Revolution for them may be more burden than blessing.

mining, manufacturing, processing, and the service activities of the tertiary, quaternary, and quinary sectors. In those economies, farming activities presumably reflect production responses to market demand expressed through price and are related to the consumption requirements of the larger society, rather than to the immediate needs of farmers themselves.

### Production Controls

Agriculture within modern, developed economies is characterized by specialization—by enterprise (farm), by area, and even by country; by *off-farm sale* rather than subsistence production; and by *interdependence* of producers and buyers linked through markets. Farmers in a free market economy supposedly produce the crops that their estimates of market price and production cost indicate will yield the greatest return. Supply, demand, and the market price mechanism are the presumed controls on agricultural production in commercial economies.

Total production costs, eventual harvest yields, and future market prices are among the many uncertainties that individual farmers must face. Beginning in the 1950s in the United States, specialist farmers and corporate purchasers developed strategies for minimizing those uncertainties. Processors sought uniformity of product quality and timing of delivery. Vegetable canners—of tomatoes, sweet corn, and the like—required volume delivery of raw products of uniform size, color, and ingredient content on dates that accorded with cannery and labor schedules. And farmers wanted the support of a guaranteed market at an assured price to minimize the uncertainties of their specialization and to stabilize the return on their investment.

The solution was contractual arrangements or vertical integrations (where production, processing, and sales are all coordinated within one firm) uniting contracted farmer with purchaser-processor. Broiler chickens of specified age and weight, cattle fed to an exact weight and finish, wheat with a minimum protein content, popping corn with prescribed characteristics, potatoes of the kind and quality demanded by particular fast-food chains, and similar product specifications became part of production contracts between farmer and buyer-processor. In the United States, the percentage of total farm output produced under contractual arrangements or by vertical integration rose from 19% in 1960 to well over one-third by the turn of the century. The term *agribusiness* is applied to the growing merging of older, farm-centered crop economy and newer patterns of more integrated production and marketing systems.

But when mechanization is added to the new farming system, women tend to be losers. Frequently, such predominantly female tasks as harvesting or dehusking and polishing of grain—all traditionally done by hand—are given over to machinery, displacing rather than employing women. Even the application of chemical fertilizers ("a man's task") instead of cow dung ("women's work") has reduced the female role in agricultural development programs. The loss of those traditional female wage jobs means that already poor rural women and their families have insufficient income to improve their diets even in the light of substantial increases in food availability through Green Revolution improvements.

If women are to benefit from the Green Revolution, new cultural norms—or culturally acceptable accommodations within traditional household, gender, and customary legal relations—will be required. These must permit or recognize women's land-owning and other legal rights not now clearly theirs, access to credit at favorable rates, and admission on equal footing with males to government assistance programs. Recognition of those realities fostered the Food and Agriculture Organization of the United Nations' "FAO Plan of Action for Women in Development (1996–2001)" and its "Gender and Development Plan (2002–2007)." Both aimed at stimulating and facilitating efforts to enhance the role of women as contributors and beneficiaries of economic, social, and political development. Objectives of the plan include promoting gender-based equity in control of productive resources; enhancing women's participation in decision- and policy-making processes at all levels, local and national; and providing women with access to credit to enable them to engage as creators and owners of small-scale manufacturing, trade, or service businesses.

The model for that credit access is the Grameen Bank, established by a Bangladeshi economist in 1976. Based on the conviction that access to credit should be a basic human right, the bank extends "microcredit"—the average loan is U.S. $160—for "microenterprises," with women the primary borrowers. By 2000, the bank had made more than 2 million loans in 40,000 villages. More than 94% of the borrowers have been women and repayment rates reach 97%. The Grameen concept has spread from its Bangladesh origins to elsewhere in Asia and to Latin American and Africa. By 2002, some 10,000 microfinance institutions worldwide had reached nearly 20 million clients, among the world's poorest people. These represent, however, less than 3% of the estimated 500 to 600 million women who still have virtually no access to credit or to the economic, social, or educational benefits that come from its availability.

Even for family farmers not bound by contractual arrangements to suppliers and purchasers, the older assumption that supply, demand, and the market price mechanism are effective controls on agricultural production is not wholly valid. In reality, those theoretical controls are joined by a number of nonmarket governmental influences that may be as decisive as market forces in shaping farmers' options and spatial production patterns. If there is a glut of wheat on the market, for example, the price per ton will come down and the area sown to it should diminish. It will also diminish regardless of supply if governments, responding to economic or political considerations, impose acreage controls.

Distortions of market control may also be introduced to favor certain crops or commodities through subsidies, price supports, market protections, and the like. The political power of farmers in the European Union, for example, secured generous product subsidies and, for the Union, immense unsold stores of butter, wine, and grains until 1992, when reforms began to reduce the unsold stockpiles even while increasing total farm spending. In Japan, the home market for rice is largely protected and reserved for Japanese rice farmers, even though their production efficiencies are low and their selling price is high by world market standards. In the United States, programs of farm price supports, acreage controls, financial assistance, and other governmental involvements in agriculture have given recurring and equally distorting effects (Figure 10.11).

### A Model of Agricultural Location

Early in the 19th century, before such governmental influences were the norm, Johann Heinrich von Thünen (1783–1850) observed that lands of apparently identical physical properties were used for different agricultural purposes. Around each major urban market center, he noted, developed a set of concentric rings of different farm products (Figure 10.12a). The ring closest to the market specialized in perishable commodities that were both expensive to ship and in high demand. The high prices they could command in the urban market made their production an appropriate use of high-valued land near the city. Rings of farmlands farther away from the city were used for less-perishable commodities with lower transport costs, reduced demand, and lower market prices. General farming and grain farming replaced the market gardening of the inner ring. At the outer margins of profitable agriculture, farthest from the single central market, livestock grazing and similar extensive land uses were found.

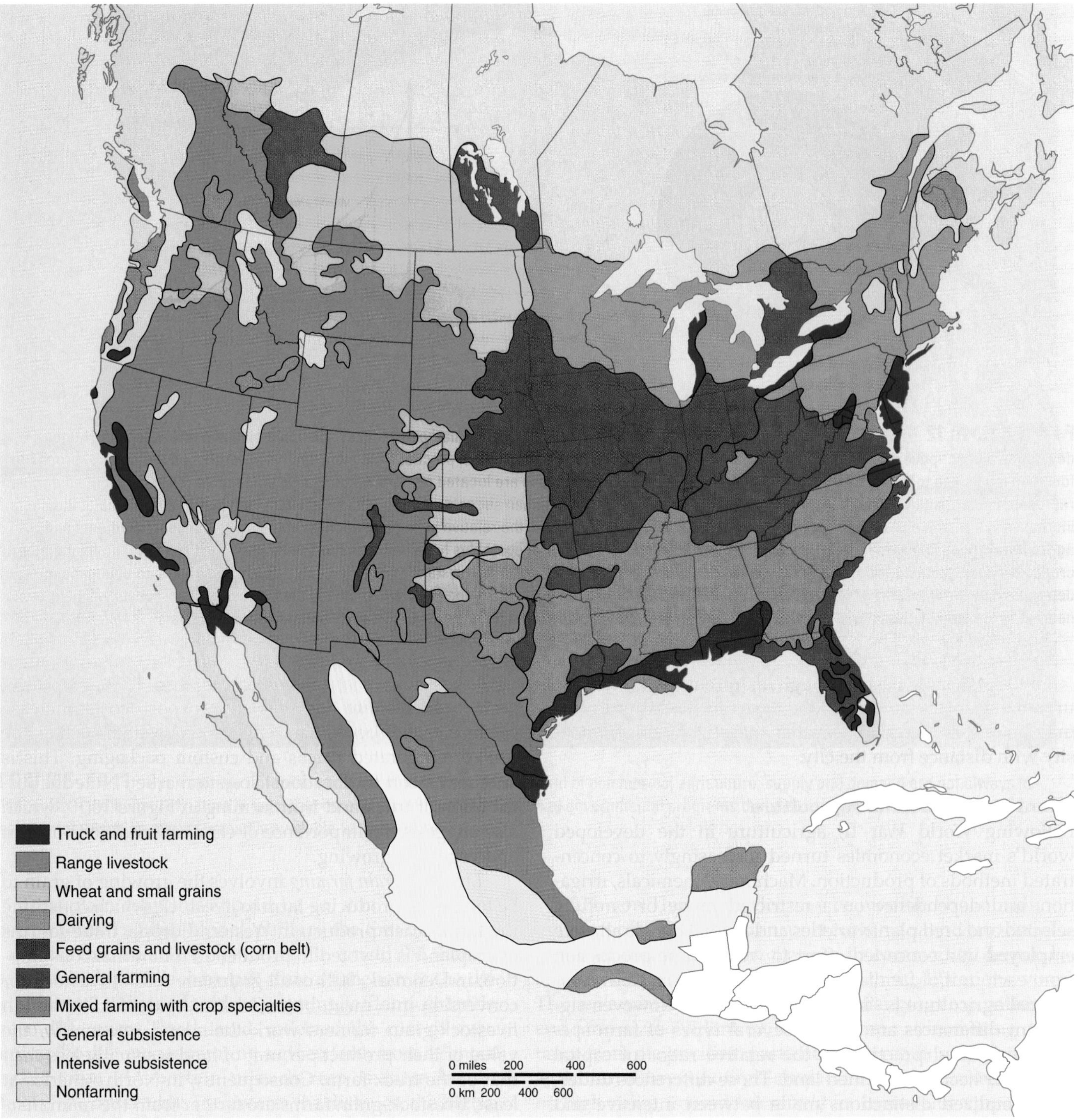


**FIGURE 10.13 Generalized agricultural regions of North America.** Sources: *U.S. Bureau of Agricultural Economics; Agriculture Canada; Mexico, Secretaría de Agricultura y Recursos Hidráulicos.*

*Large-scale wheat farming* requires sizeable capital inputs for planting and harvesting machinery, but the inputs per unit of land are low; wheat farms are very large. Nearly half the farms in Saskatchewan, for example, are more than 400 hectares (1000 acres). The average farm in Kansas is more than 400 hectares; in North Dakota, more than 525 (1300 acres). In North America, the spring wheat (planted in spring, harvested in autumn) region includes the Dakotas, eastern Montana, and the southern parts of the Prairie Provinces of Canada. The winter wheat (planted in fall, harvested in midsummer) belt focuses on Kansas and includes adjacent sections of neighboring states (see Figure 10.13).

Argentina is the only South American country to have comparable large-scale wheat farming. In the Eastern Hemisphere, the system is fully developed only east of the Volga River in northern Kazakhstan and the southern part of West Siberia, and in southeastern and western Australia.

*Livestock ranching* differs significantly from livestock-grain farming and, by its commercial orientation and distribution, from the nomadism it superficially resembles. A product of the 19th-century growth of urban markets for beef and wool in Western Europe and the northeastern United States, ranching has been primarily confined to areas of European settlement. It is found in the western United States and adjacent sections of Mexico and Canada (see Figure 10.13); the grasslands of Argentina, Brazil, Uruguay, and Venezuela; the interior of Australia; the uplands of South Island, New Zealand; and the Karoo and adjacent areas of South Africa (Figure 10.14). All except New Zealand and the humid pampas of South America have semiarid climates. All, even the most remote from markets, were a product of improvements in transportation by land and sea, the refrigeration of carriers, and meat-canning technology. In all, introduced beef cattle or sheep replaced original native fauna, such as bison on North America's Great Plains, almost always with eventual severe environmental deterioration. More recently, the midlatitude demand for beef has been blamed for expanded cattle production and the extensive destruction of tropical rain forests in Central America and the Amazon Basin, though in recent years Amazon Basin deforestation has reflected more the expansion of soybean farming than of beef production.

In all of the ranching regions, livestock range (and the area exclusively in ranching) has been reduced as crop farming has encroached on its more humid margins, as pasture improvement has replaced less-nutritious native grasses, and as grain fattening has supplemented traditional grazing. Since ranching can be an economic activity only where alternative land uses are nonexistent and land quality is low, ranching regions of the world characteristically have low population densities, low capitalization per land unit, and relatively low labor requirements.

### Special Crops

Proximity to the market does not guarantee the intensive production of high-value crops, should terrain or climatic circumstances hinder it. Nor does great distance from the market inevitably determine that extensive farming on low-priced land will be the sole agricultural option. Special circumstances, most often climatic, make some places far from markets intensively developed agricultural areas. Two special cases are agriculture in Mediterranean climates and in plantation areas (see Figure 10.14).

Most of the arable land in the Mediterranean Basin itself is planted to grains, and much of the agricultural area is used for grazing. *Mediterranean agriculture* as a specialized farming economy, however, is known for grapes, olives, oranges, figs, vegetables, and similar commodities. These crops need

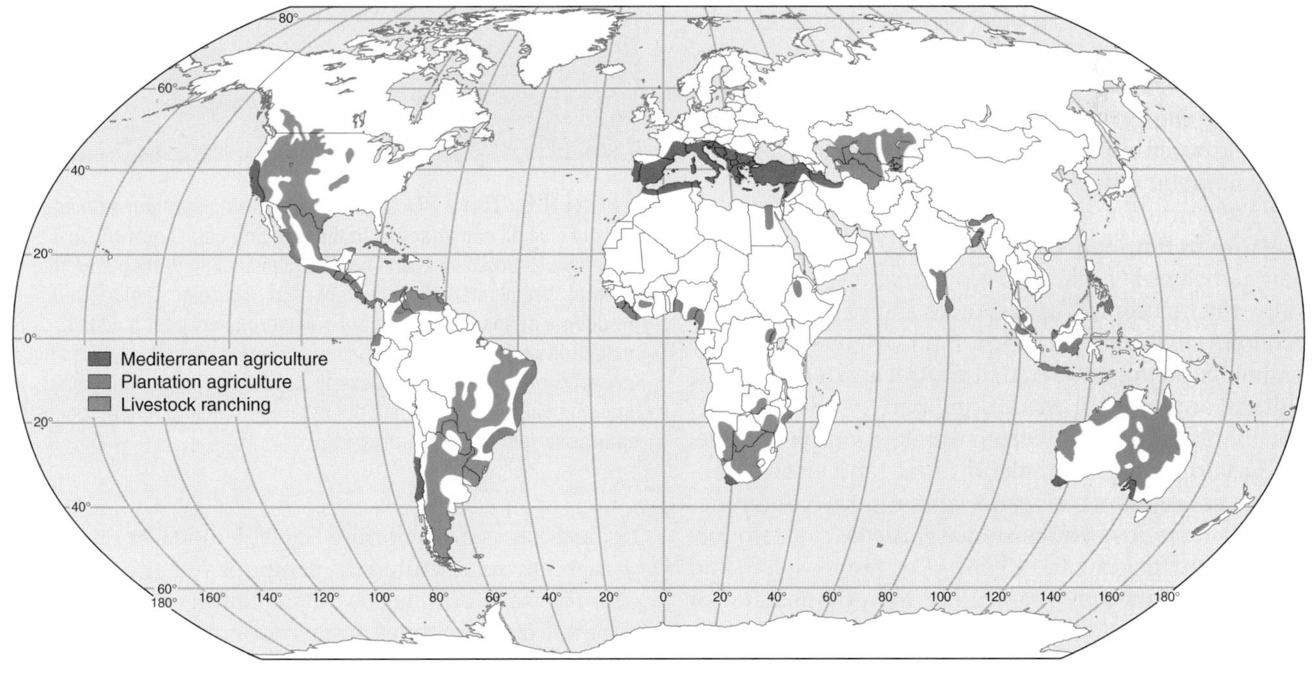


**FIGURE 10.14 Livestock ranching and special crop agriculture.** Livestock ranching is primarily a midlatitude enterprise catering to the urban markets of industrialized countries. Mediterranean and plantation agriculture are similarly oriented to the markets provided by the advanced economies of Western Europe and North America. Areas of Mediterranean agriculture—all of roughly comparable climatic conditions—specialize in similar commodities, such as grapes, oranges, olives, peaches, and vegetables. The specialized crops of plantation agriculture are influenced by both physical geographic conditions and present or, particularly, former colonial control of areas.

warm temperatures all year round and a great deal of sunshine in the summer. The Mediterranean agricultural lands indicated in Figure 10.14 are among the most productive agricultural lands in the world. The precipitation regime of Mediterranean climate areas—winter rain and summer drought—lends itself to the controlled use of water. Of course, much capital must be spent for the irrigation systems, another reason for the intensive use of the land for high-value crops that are, for the most part, destined for export to industrialized countries or areas outside the Mediterranean climatic zone and even, in the case of Southern Hemisphere locations, to markets north of the equator.

Climate is also considered the vital element in the production of what are commonly but imprecisely known as *plantation crops*. The implication of **plantation** is the introduction of a foreign element—investment, management, and marketing—into an indigenous culture and economy, often employing an introduced alien labor force. The plantation itself is an estate whose resident workers produce one or two specialized crops. Those crops, although native to the tropics, were frequently foreign to the areas of plantation establishment: African coffee and Asian sugar in the Western Hemisphere and American cacao, tobacco, and rubber in Southeast Asia and Africa are examples (Figure 10.15). Entrepreneurs in Western countries such as England, France, the Netherlands, and the United States became interested in the tropics partly because they afforded them the opportunity to satisfy a demand in temperate lands for agricultural commodities not producible in the market areas. Custom and convenience usually retain the term *plantation*, even where native producers of local crops dominate, as they do in cola nut production in Guinea, spice growing in India and Sri Lanka, or sisal production in the Yucatán. As Figure 10.14 suggested, for ease of access to shipping most plantation crops are cultivated along or near coasts, since production for export rather than for local consumption is the rule.

**FIGURE 10.15** An Indonesian rubber plantation worker collects latex in a small cup attached to the tree and cuts a new tap just above the previous one. The scene typifies classical plantation agriculture in general. The plantation was established by foreign capital (Dutch) to produce a nonnative (American) commercial crop for a distant, midlatitude market using nonnative (Chinese) labor supervised by foreign (Dutch) managers. Present-day ownership, management, and labor may have changed, but the nature and market orientation of the enterprise remain. *© Wolfgang Kaehler.*

### Agriculture in Planned Economies

As their name implies, planned economies have a degree of centrally directed control of resources and of key sectors of the economy that permits the pursuit of governmentally determined objectives. When that control is extended to the agricultural sector—as it was during particularly the latter part of the 20th century in communist-controlled Soviet Union, Eastern Europe, mainland China, and elsewhere—state and collective farms and agricultural communes replace private farms or subsistence gardens, crop production is divorced from market control or family need, and prices are established by plan rather than by demand or production cost.

Such extremes of rural control have, in recent years, been relaxed or abandoned in most formerly strictly planned economies. Wherever past centralized control of agriculture had been imposed and long endured, however, traditional rural landscapes were altered and the organization of rural society was disrupted. The programs decreed by Stalin and his successors in the former Soviet Union, for example, fundamentally restructured the geography of agriculture of that country, transforming the Soviet countryside from millions of small farm holdings to a consolidated pattern of fewer than 50,000 centrally controlled operating units. Reestablishment of private agriculture was undertaken quickly in Russia following USSR collapse in late 1991, but not until 2002 was a legal framework established for the sale and purchase of farmland. However, the absence of adequate farm regulatory and boundary description and lack of clear ownership rights hindered the outright sale of land, and even in that year the remnants of the old Soviet collective farm system still operated more than 90% of the country's farmland.

A different progression from private and peasant agriculture, through collectivization, and back to what is virtually a private farming system took place in the planned economy of the People's Republic of China. After its assumption of power in 1949, the communist regime redistributed all farmlands to some 350 million peasants in inefficiently small (0.2-hectare, or 0.5-acre) subsistence holdings that were totally inadequate for the growing food needs of the country. By the end of 1957, some 90% of peasant households had been collectivized into about 700,000 communes, a number further reduced in the 1970s to 50,000 communes averaging some 13,000 members.

After the death of Chairman Mao in 1976, what became effectively a private farming system was reintroduced when 180 million new farms were allocated for unrestricted use to peasant families under rent-free leases. Farmland remains owned and controlled by the state and merely leased to farmers, and most staple crops are still sold under enforced contracts at fixed prices to governmental purchasers. Increasingly, however, vegetables and meat are sold on the free market (see Figure 10.4) and per capita food production and availability have increased dramatically.

# Other Primary Activities

In addition to agriculture, primary economic activities include fishing, forestry, and the mining and quarrying of minerals. These industries involve the direct exploitation of natural resources that are unequally available in the environment and differentially evaluated by different societies. Their development, therefore, depends on the occurrence of perceived resources, the technology to exploit their natural availability, and the cultural awareness of their value. (The definition, perception, and utilization of resources are explored in depth in Chapter 5.)

Two of them—fishing and forestry—are **gathering industries** based on harvesting the natural bounty of renewable resources, though ones in serious danger of depletion through overexploitation. Mining and quarrying are **extractive industries,** removing nonrenewable metallic and nonmetallic minerals, including the mineral fuels, from the earth's crust. They are the initial raw material phase of modern industrial economies.

## Fishing

Although fish and shellfish account for less than 20% of all human protein consumption, an estimated 1 billion people—primarily in the developing countries of eastern and southeastern Asia, Africa, and parts of Latin America—depend on fish as their primary source of protein. Fish are also very important in the diets of most advanced states, both those with and without major domestic fishing fleets. Although about 75% of the world annual fish harvest is consumed by humans, up to 25% is processed into fish meal to be fed to livestock or used as fertilizer. Those two quite different markets have increased both the demand for and annual harvest of fish. Indeed, so rapidly have pressures on the world's fish stocks increased that evidence is unmistakable that, at least locally, their *maximum sustainable yield* is actually or potentially being exceeded. The **maximum sustainable yield** of a resource is the largest volume or rate of use that will not impair its ability to be renewed or to maintain the same future productivity. For fishing, that level is marked by a catch or harvest equal to the net growth of the replacement stock.

The annual fish supply comes from three sources:

1. the *inland catch,* from ponds, lakes, and rivers;
2. *fish farming,* where fish are produced in a controlled and contained environment;
3. the *marine catch,* all wild fish harvested in coastal waters or on the high seas.

Inland waters supply less than 10% of the global fish catch; fish farming (aquaculture) accounts for some 30%. The other 60% or more of the harvest comes from the world's oceans (Figure 10.16), and most of that catch is made in coastal wetlands, estuaries, and the relatively shallow coastal waters above the continental shelf. Near shore, shallow embayments and marshes provide spawning grounds and river waters supply nutrients to an environment highly productive of fish. Increasingly, these are also

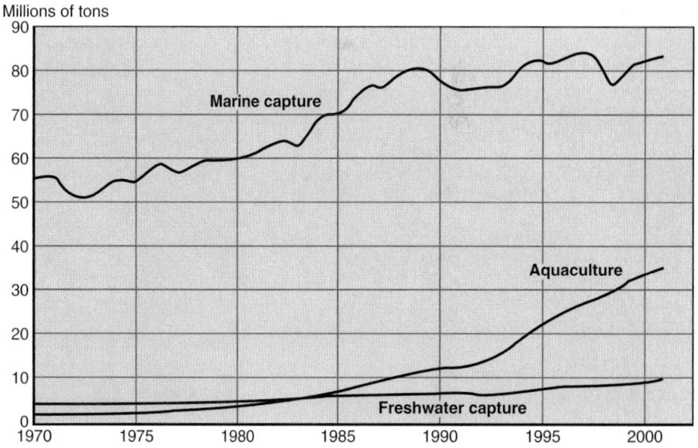


**FIGURE 10.16 Officially recorded annual fish harvests, 1970–2001,** rose irregularly from 64 million tons in 1970 to 131 million tons in 2001. On the basis of individual country reports, the FAO recorded fluctuating but slowly growing harvest totals up to 2001; the 1993 and 1998 dips are associated with El Niño-produced ocean temperature changes. A Chinese admission of regular overreporting of, particularly, their marine capture suggests that world marine catch and composite harvest totals actually registered an irregular downward trend each year since 1988. A compensating adjustment to this graph would reduce 2001 total harvest and marine catch figures by about 9 million tons. The FAO estimates that 20 to 40 million tons per year of unintended marine capture of juvenile or undersized fish and nontarget species are discarded each year. Source: *Food and Agriculture Organization (FAO).*

areas seriously affected by pollution from runoff and ocean dumping, an environmental assault so devastating in some areas that fish and shellfish stocks have been destroyed with little hope of revival.

Commercial marine fishing is largely concentrated in northern waters, where warm and cold currents join and mix and where such familiar food species as herring, cod, mackerel, haddock, and flounder congregate, or "school," on the broad continental shelves (Figure 10.17). Two of the most heavily fished regions are the Northeast Pacific and Northwest Atlantic, which together yield about 40% of the marine catch total. Tropical fish species tend not to school and, because of their high oil content and unfamiliarity, are less acceptable in the commercial market. They are, however, of great importance for local consumption. Traditional, or "artisan," fishermen, nearly all working in inshore waters within sight of land, are estimated to number between 8 and 10 million worldwide. Their annual harvest of some 24 million tons of fish and shellfish is usually not included in world fishery totals. Only a very small percentage of total marine catch comes from the open seas that make up more than 90% of the world's oceans.

Modern technology and more aggressive fishing fleets of more countries greatly increased annual marine capture in the years after 1950. That technology included the use of sonar, radar, helicopters, and satellite communications to locate schools of fish; more efficient nets and tackle; and factory trawlers to follow fishing fleets to prepare and freeze the catch. In addition, more nations granted ever-larger subsidies to expand and reward their marine trawler operations. The rapid rate of increase led to inflated projections of continuing or growing fisheries productivity and to optimism that the resources of the oceans were inexhaustible.

Quite the opposite has proved to be true. In fact, in recent years, the productivity of marine fisheries has declined as *overfishing* (catches above reproduction rates) and the pollution of coastal waters have seriously endangered the supplies of traditional and desirable food species. Adjusted world total catch figures indicate that, rather than the steady increase in capture rates shown in Figure 10.16, there has really been a decline of more than 660,000 tons per year since the late 1980s. Coupled with a growing world population, there has been a serious decline in the average per capita marine catch. The UN reports that all 17 of the world's major oceanic fishing areas are being fished at or beyond capacity; 13 are in decline. The plundering of Anglo American coastal waters has imperiled a number of the most desirable fish species; in 1993, Canada shut down its cod industry to allow stocks to recover, and U.S. authorities report that 67 North American species are overfished and 61 harvested to capacity.

Overfishing is partly the result of the accepted view that the world's oceans are common property, a resource open to anyone's use, with no one responsible for its maintenance, protection, or improvement. The result of this "open seas" principle is but one expression of the so-called **tragedy of the commons**[1]—the economic reality that, when a resource

[1]The *commons* refers to undivided land available for the use of everyone; usually, it meant the open land of a village that all used as pasture. The *Boston Common* originally had this meaning.

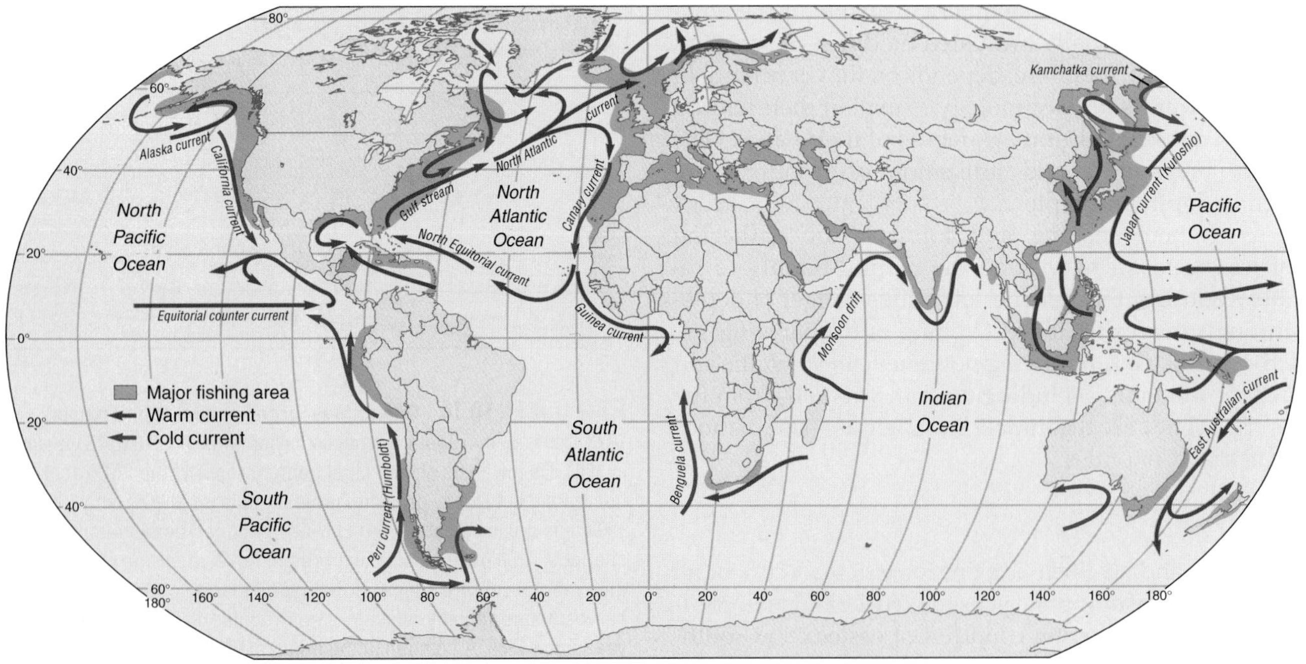


**FIGURE 10.17 The major commercial marine fisheries of the world.** The waters within 325 kilometers (200 mi) of the United States coastline account for almost one-fifth of the world's annual fish and shellfish harvests. Overfishing, urban development, and the contamination of bays, estuaries, and wetlands have contributed to the depletion of fish stocks in those coastal waters.

is available to all, each user, in the absence of collective controls, thinks he or she is best served by exploiting the resource to the maximum, even though this means its eventual depletion. In 1995, more than 100 countries adopted a treaty—to become legally binding when ratified by 30 nations—to regulate fishing on the open oceans outside territorial waters. Applying to such species as cod, pollock, and tuna—that is, to migratory and high-seas species—the treaty requires fishermen to report the size of their catches to regional organizations that would set quotas and subject vessels to boarding to check for violations. These and other fishing control measures could provide the framework for the future sustainability of important food fish stocks.

One approach to increasing the fish supply is through *fish farming,* or **aquaculture,** the breeding of fish in freshwater ponds, lakes, and canals or in fenced-off coastal bays and estuaries or enclosures. Aquaculture production has totaled about 30% of the total fish harvest in recent years; its contribution to the human food supply is even greater than raw production figures suggest. Whereas one-third of the conventional fish catch is used to make fishmeal and fish oil, virtually all farmed fish are used as human food. Fish farming has long been practiced in Asia, where fish are a major source of protein, but now takes place on every continent. Its rapid and continuing production increase makes aquaculture the fastest-growing sector of the world food economy, with the promise of overtaking cattle ranching as a human food source by 2010.

## Forestry

Forests, like fish, are a heavily exploited renewable resource. Even after millennia of land clearance for agriculture and, more recently, commercial lumbering, cattle ranching, and fuelwood gathering, forests still cover nearly a third of the world's land area. Livelihoods based on forest resources are spatially widespread and parts of both subsistence and commercial economies. A full treatment of forest resources and their nature, distribution, and exploitation appears in Chapter 5.

## Mining and Quarrying

Societies at all stages of economic development can and do engage in agriculture, fishing, and forestry. The extractive industries—mining and drilling for nonrenewable mineral wealth—emerged only when cultural advancement and economic necessity made possible a broader understanding of the earth's resources. Now those extractive industries provide the raw material and energy base for the way of life experienced by people in the advanced economies and are the basis for an important part of the international trade connecting the developed and developing countries of the world.

Extractive industries depend on the exploitation of minerals unevenly distributed in amounts and concentrations determined by past geologic events, not by contemporary market demand. Because usable deposits are the result of geologic accident, there is no necessary relationship between the size and wealth of a country and the resources its territory contains. Although larger states are more likely to be the beneficiaries of such "accidents," many smaller developing countries are major sources of one or more critical raw materials and, therefore, are important participants in the growing international trade in minerals.

Transportation costs play a major role in determining where *low-value minerals* will be mined. Materials such as gravel, limestone for cement, and aggregate are found in such abundance that they have value only when they are near the site where they are to be used. For example, gravel for road building has value if it is at the road-building site, not otherwise. Transporting gravel hundreds of miles is an unprofitable activity.

The production of other minerals—especially *metallic minerals* such as copper, lead, and iron ore—is affected by a balance of three forces: the quantity available, the richness of the ore, and the distance to markets. Another factor, land acquisition and royalty costs, may equal or exceed other considerations in mine development decisions (see "Public Land, Private Profit," p. 366). Even if these conditions are favorable, mines may not be developed or even remain operating if supplies from competing sources are more cheaply available in the market. In the 1980s, more than 25 million tons of iron ore–producing capacity was permanently shut down in the United States and Canada as a result of such price competition. Similar declines occurred in North American copper, nickel, zinc, lead, and molybdenum mining as market prices fell below domestic production costs. Beginning in the early 1990s, as a result of both resource depletion and the availability of low-cost imports, the United States became a net importer of nonfuel minerals for the first time. Of course, increases in mineral prices may be reflected in opening or reopening mines that, at lower returns, were deemed unprofitable. However, the developed industrial countries of commercial economies, whatever their former or even present mineral endowment, frequently find themselves at a competitive disadvantage against producers in developing countries with lower-cost labor and state-owned mines with abundant, rich reserves.

When the ore is rich in metallic content, it is profitable (in the case of iron and aluminum ores) to ship it directly to the market for refining. But, of course, the highest-grade ores tend to be mined first. Consequently, the demand for low-grade ores has been increasing in recent years as richer deposits have been depleted. Low-grade ores are often upgraded by various types of separation treatments at the mine site to avoid the cost of transporting waste materials not wanted at the market. Concentration of copper is nearly always mine-oriented (Figure 10.18); refining takes place near areas of consumption.

The large amount of waste in copper (98% to 99% or more of the ore) and in most other industrially significant ores should not be considered the mark of an unattractive

**FIGURE 10.18** Copper ore concentrating and smelting facilities at the Phelps-Dodge mine in Morenci, Arizona. Concentrating mills crush the ore, separating copper-bearing material from the rocky mass containing it. The great volume of waste material removed assures that most concentrating operations are found near the ore bodies. Smelters separate concentrated copper from other, unwanted minerals such as oxygen and sulfur. Because smelting is also a "weight-reducing" activity, it is frequently—though not invariably—located close to the mine as well. *© Cameramann International Ltd.*

deposit. Indeed, the opposite may be true. Many higher-content ores are left unexploited—because of the cost of extraction or the smallness of the reserves—in favor of the use of large deposits of even very low grade ore. The attraction of the latter is a size of reserve sufficient to justify the long-term commitment of development capital and, simultaneously, to assure a long-term source of supply. At one time, high-grade magnetite iron ore was mined and shipped from the Mesabi area of Minnesota. Those deposits are now exhausted. However, immense amounts of capital have been invested in the mining and processing into high-grade iron ore pellets of the virtually unlimited supplies of low-grade iron-bearing rock (taconite) still remaining.

Such investments do not assure the profitable exploitation of the resource. The metals market is highly volatile. Rapidly and widely fluctuating prices can quickly change profitable mining and refining ventures to losing undertakings. Marginal gold and silver deposits are opened or closed in reaction to trends in precious metals prices. Taconite beneficiation (waste material removal) in the Lake Superior region has virtually ceased in response to the decline of the U.S. steel industry and the price advantage of imported ores. In market economies, cost and market controls dominate economic decisions. In planned economies, cost may be a less important consideration than are such other concerns as goals of national development and resource independence.

The advanced economies have reached that status through their control and use of energy. Domestic supplies of mineral fuels, therefore, are often considered basic to national strength and independence. When those supplies are absent, developed countries are concerned with the availability and price of coal, oil, and natural gas in international trade. Mineral fuels and other energy sources are given extended discussion in Chapter 5.

## Trade in Primary Products

International trade expanded by about 6% a year between 1965 and the end of the century; it grew by nearly 13% in 2000 alone, to account for about 17% of all economic activity, though the sluggish world economy through 2003 temporarily reversed that growth. Primary commodities—agricultural goods, minerals, and fuels—made up one-quarter of the total dollar value of the 2000 international trade flows. During the first half of the 20th century, the world distribution of supply and demand for those items in general resulted in an understandable pattern of commodity flow: from raw material producers located within less-developed countries to the processors, manufacturers, and consumers in the more-developed ones (Figure 10.19). The reverse flow carried manufactured goods processed in the industrialized states back to the developing countries. That two-way trade presumably benefited the developed states by providing access to a continuing supply of industrial raw materials and foodstuffs not available domestically and gave less-developed countries needed capital to invest in their own development or to expend on the importation of the manufactured goods, food supplies, or commodities—such as petroleum—they did not themselves produce.

At the end of the century, however, world trade flows and export patterns of the emerging economies radically changed. Raw materials greatly decreased and manufactured goods correspondingly increased in the export flows from developing states as a group. In 1990, nonmanufactured (unprocessed) goods accounted for 60% of their

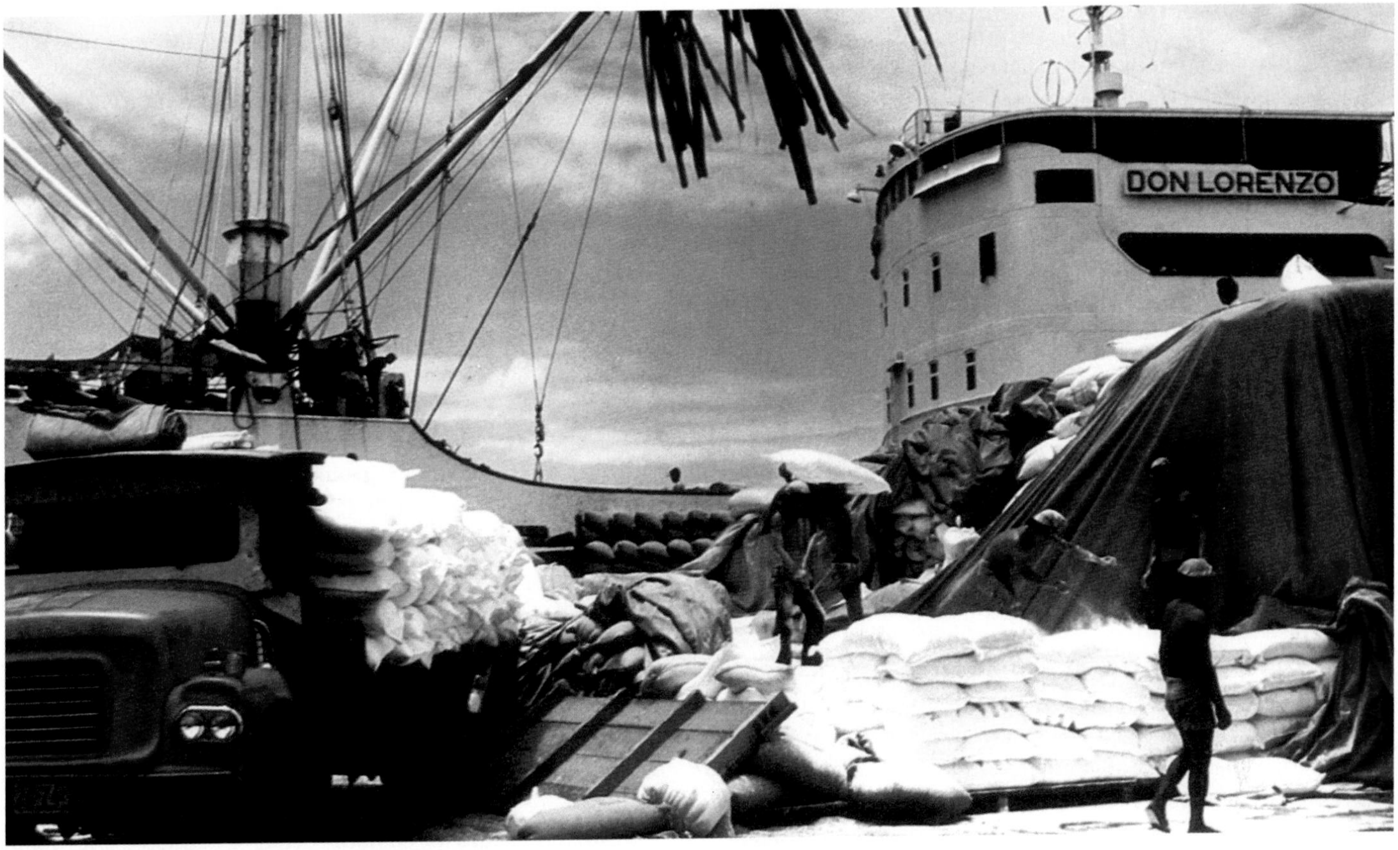


**FIGURE 10.19** Sugar being loaded for export at the port of Cebu in the Philippines. Much of the developing world depends on exports of mineral and agricultural products to the developed countries for the major portion of its income. Fluctuations in market demand and the price of some of those commodities can have serious and unexpected consequences for the economies of exporting states. *© UN/DPI Photo.*

exports; by 2000, that share had been cut in half and, in a reversal, manufactured goods made up 60% of the export flows from the developing to the industrialized world. Even with that earlier overall decline in raw material exports, however, trade in unprocessed goods remains dominant in the economic well-being of many of the world's poorer economies; increasingly, the terms of the traditional trade flows on which they depend on have been criticized as unequal and damaging to commodity-exporting countries.

Commodity prices are volatile; they may rise sharply in periods of product shortage or international economic growth and swiftly decline when the world economy slows. Commodity price movements during the late 1990s and early 21st century were downward, to the great detriment of material-exporting economies. Prices for agricultural raw materials, for example, dropped by 30% between 1975 and 2000, and those for metals and minerals decreased by almost 40%. Such price declines cut deeply into the export earnings of many emerging economies. Many of those same economies, however, benefited from the sudden increase—beginning in early 2004—in world price levels for fuels, metals, and other minerals and for basic food and feed commodities.

Whatever the current world prices of raw materials may be, as a group, raw material exporting states have long expressed resentment at what they perceive as commodity price manipulation by rich countries and corporations to ensure low-cost supplies. Although collusive price fixing has not been demonstrated, other price-depressing agencies are evident. Technology, for example, has provided industries in advanced countries with a vast array of materials which now can and do substitute for the ores and metals produced by developing states. Glass fibers replace copper wire in telecommunications applications; synthetic rubber replaces natural rubber; glass and carbon fibers provide the raw material for rods, tubes, and sheet panels, and other products superior in performance and strength to the metals they replace; and a vast and enlarging array of plastics are the accepted raw materials for commodities and uses for which natural rivals are not even considered. That is, even as the world industrial economy expands, demands and prices for traditional raw materials remain depressed.

While prices paid for developing country commodities tend to be low, prices charged for the manufactured goods offered in exchange by the developed countries tend to be high. To capture processing and manufacturing profits for themselves, some developing states have placed restrictions on the export of unprocessed commodities. Malaysia, the

GEOGRAPHY & Public Policy

## Public Land, Private Profit

When President Ulysses S. Grant signed the Mining Act of 1872, the presidential and congressional goal was to encourage Western settlement and development by allowing any "hard-rock" miners (including prospectors for silver, gold, copper, and other metals) to mine federally owned land without royalty payment. It further permitted mining companies to gain clear title to publicly owned land and all subsurface minerals for no more than $12 a hectare ($5 an acre). Under those liberal provisions, mining firms have bought 1.3 million hectares (3.2 million acres) of federal land since 1872 and each year remove some $1.2 billion worth of minerals from government property. In contrast to the royalty-free extraction privileges granted to metal miners, oil, gas, and coal companies pay royalties of as much as 12.5% of their gross revenues for exploiting federal lands.

Whatever the merits of the 1872 law in encouraging the economic development of lands otherwise unattractive to homesteaders, modern-day mining companies throughout the Western states have secured enormous actual and potential profits from the law's generous provisions. In Montana, a company claim to 810 hectares (2000 acres) of land would cost it less than $10,000 for an estimated $4 billion worth of platinum and palladium; in California, a gold mining company in 1994 sought title to 93 hectares (230 acres) of federal land containing a potential of $320 million of gold for less than $1200. Foreign as well as domestic firms may be beneficiaries of the 1872 law. In 1994, a South African firm arranged to buy 411 hectares (1016 acres) of Nevada land with a prospective $1.1 billion in gold from the government for $5100. And a Canadian firm in 1994 received title to 800 hectares (nearly 2000 acres) near Elko, Nevada, that cover a likely $10 billion worth of gold—a transfer that Interior Secretary Bruce Babbitt dubbed "the biggest gold heist since the days of Butch Cassidy." And in 1995, Mr. Babbitt conveyed about $1 billion worth of travertine (a mineral used in whitening paper) under 45 hectares (110 acres) of Idaho to a Danish-owned company for $275.

The "gold heist" characterization summarized a growing administration and congressional feeling that what was good in 1872 and today for metal mining companies was not necessarily beneficial to the American public, who own the land. In part, that feeling results from the fact that mining companies commit environmental sins that require public funding to repair or public tolerance to accept. The mining firms may destroy whole mountains to gain access to low-grade ores and leave toxic mine tailings, surface water contamination, and open-pit scarring of the landscape as they move on or disappear. The projected public costs of cleaning up more than 50,000 abandoned mine sites, thousands of miles of damaged or dead streams, and several billion tons of contaminated waste are estimated at a minimum of $35 billion.

A congressional proposal introduced, but defeated, in 1993 would have required mining companies to pay royalties of 8% on gross revenues for all hard-rock ores extracted and would have prohibited them from outright purchase of federal land.

Philippines, and Cameroon, for example, have limited the export of logs in favor of increased domestic processing of sawlogs and exports of lumber. Some developing countries have also encouraged domestic manufacturing to reduce imports and to diversify their exports. Frequently, however, such exports meet with tariffs and quotas protecting the home markets of the industrialized states.

Many developing regions heavily dependent on commodity sales saw their share of global trade fall materially between 1970 and the early 21st century: sub-Saharan Africa from 3.8% to 1%, Latin America from 5.6% to 3.3%, and the least-developed states as a group from 0.8% to 0.4%. Those relative declines are understandable in the light of greatly expanding international trade in manufactured goods from China, Korea, Mexico, and other rapidly industrializing states and from the expansion of trade in both manufactured goods and primary products between the industrialized countries themselves within newly established regional free-trade zones. For example, the developed countries acquire some three-quarters by value of their agricultural imports and 70% of their industrial raw materials from one another, diminishing the prospects for developing country exports.

In 1964, in reaction to the whole range of perceived trade inequities, developing states promoted the establishment of the United Nations Conference on Trade and Development (UNCTAD). Its central constituency—the "Group of 77," expanded by 2004 to 135 developing states—continues to press for a new world economic order based, in part, on an increase in the prices and values of exports from developing countries, a system of import preferences for their manufactured goods, and a restructuring of international cooperation to stress trade promotion and recognition of the special needs of poor countries. The World Trade Organization, established in 1995 (and discussed in detail in Chapter 9) was designed, in part, to reduce trade barriers and inequities. It has, however, been judged by its detractors as ineffective on issues of importance to developing countries. Chief among the complaints is the continuing failure of the industrial countries significantly (or at all) to reduce protections for their own agricultural and mineral industries. The World Bank has calculated, for example, that agricultural trade barriers and subsidies in rich countries reduce incomes in developing countries by at least $20 billion a year. Solutions are yet to be achieved to debated questions of the fair treatment of international trade in primary products.

The royalty provision alone would have yielded nearly $100 million annually at 1994 levels of company income. Mining firms claim that an imposition of royalties might well destroy America's mining industry. They stress both the high levels of investment they must make to extract and process frequently low-grade ores and the large number of high-wage jobs they provide as their sufficient contribution to the nation. The Canadian company involved in the Elko site, for example, reports that, since it acquired the claims in 1987 from their previous owner, it has expended more than $1 billion, plus has made additional donations for town sewer lines and schools and has created 1700 jobs. The American Mining Congress estimates that the proposed 8% royalty charge would have cost 47,000 jobs out of 140,000, and even the U.S. Bureau of Mines assumes a loss of 1100 jobs.

Public resistance to Western mining activities are taking their toll. State and federal regulatory procedures, many dragging on for a decade or more, have discouraged opening new mines; newly enacted environmental regulations restricting current mining operations (for example, banning the use of cyanide in gold and silver refining) reduce their economic viability. In consequence, both investment and employment in U.S. mining is in steady decline, eroding the economic base of many Western communities.

**QUESTIONS TO CONSIDER**

1. Do you believe the 1872 Mining Law should be repealed or amended? If not, what are your reasons for arguing for retention? If so, would you advocate the imposition of royalties on mining company revenues? At what levels, if any, should royalties be assessed? Should hard-rock and energy companies be treated equally for access to public land resources? Why or why not?
2. Would you propose to prohibit outright land sales to mining companies? If not, should sales prices be determined by the surface value of the land or by the estimated (but unrealized) value of mineral deposits it contains?
3. Do you think that cleanup and other charges now borne by the public are acceptable in view of the capital investments and job creation of hard-rock companies? Do you accept the industry's claim that an imposition of royalties would destroy American metal mining? Why or why not?
4. Do you favor continued state and federal restrictions on mining operations, even at the cost of jobs and community economies? Why or why not?

# Secondary Activities: Manufacturing

Although primary industries are locationally tied to the natural resources they gather or exploit, secondary and later stages of economic activity are less concerned with conditions of the physical environment. For them, enterprise location is more closely related to cultural and economic than to physical circumstances. They are movable rather than spatially tied, and are assumed to respond to recurring locational requirements and controls.

Those controls are rooted in observations about human spatial behavior in general and economic behavior in particular. We have already explored some of those assumptions in earlier discussions. We noted, for example, that the intensity of spatial interaction decreases with increasing separation of points—distance decay, we called it. Recall that von Thünen's model of agricultural land use was rooted in conjectures about transportation cost and land value relationships.

Such simplifying assumptions help us understand an accepted common set of controls and motivations guiding human economic behavior. We assume, for example, that people are *economically rational;* that is, given the information at their disposal, they make locational, production, or purchasing decisions in light of a perception of what is most cost-effective and advantageous. From the standpoint of producers or sellers of goods or services, it is further assumed that each is intent on *maximizing profit.* To reach that objective, a host of production and marketing costs and political, competitive, and other limiting factors may be considered, but the ultimate goal of profit-seeking remains clear. Finally, in commercial economies it has been accepted that the best measure of the correctness of economic decisions is afforded by the market mechanism and the equilibrium between supply and demand that market prices establish (Figure 10.20). Although increasingly challenged as too rigid and unrealistic as explanations of actual human behavior and economic decision making, those assumptions still underlie most current analyses of the spatial patterns of industry.

## Industrial Locational Models

When market principles are controlling, entrepreneurs seek to maximize profits by locating manufacturing activities at sites of lowest total input costs (and high revenue yields). In

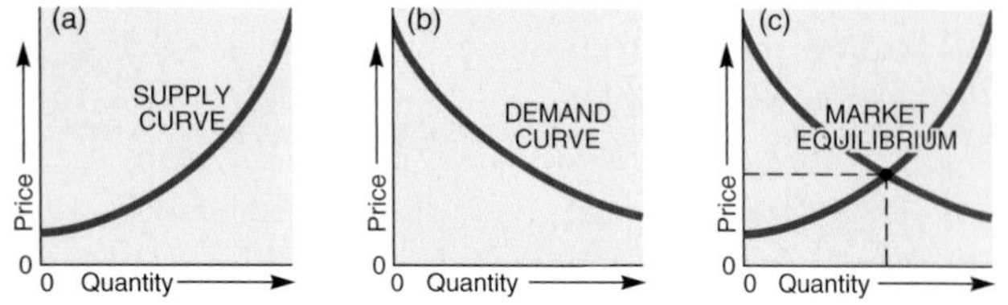


**FIGURE 10.20 Supply, demand, and market equilibrium.** The regulating mechanism of the market can be visualized graphically. (a) The *supply curve* tells us that as the price of a good increases, more of that good will be made available for sale. Countering any tendency for prices to rise to infinity is the market reality that the higher the price, the smaller the demand as potential customers find other purchases or products more cost-effective. (b) The *demand curve* shows how the market will expand as prices drop and the good becomes more affordable and attractive to more customers. (c) *Market equilibrium* is marked by the point of intersection of the supply and demand curves and determines the price of the good, the total demand, and the quantity bought and sold.

order to assess the advantages of one location over another, industrialists must evaluate the most important **variable costs.** They subdivide their total costs into categories and note how each cost will vary from place to place. In different industries, transportation charges, labor rates, power costs, plant construction or operation expenses, the interest rate of money, or the price of raw materials may be the major variable cost. The industrialist must look at each of these and, by a process of elimination, eventually select the lowest-cost site. If the producer then determines that a large enough market can be reached cheaply enough, the location promises to be profitable.

In the economic world, nothing remains constant. Because of a changing mix of input costs, production techniques, and marketing activities, many initially profitable locations do not remain advantageous. Migrations of population, technological advances, and changes in the demand for products affect industrialists and industrial locations greatly. The abandoned mills and factories of New England or of the steel towns of Pennsylvania, even the "deindustrialization" of America itself in the face of changing domestic economy and foreign industrial cost advantages and competition, are testimony to the impermanence of the "best" locations. Similarly, the spread of manufacturing from the more-developed to the less-developed regions of the world since World War II testifies to changing perceptions of manufacturing costs and locational advantages.

The concern with variable costs as a determinant in industrial location decisions has inspired an extensive theoretical literature. Much of it is based on industrial patterns and economic assessments seen as controlling during the later 19th and early 20th centuries and extends the **least-cost theory** proposed by German location economist Alfred Weber (1868–1958), and sometimes called *Weberian analysis.* Weber explained the optimum location of a manufacturing establishment in terms of the minimization of three basic expenses: relative transport costs, labor costs, and agglomeration costs. **Agglomeration** refers to the clustering of productive activities and people for mutual advantage. Such clustering can produce *agglomeration economies* through shared facilities and services. Diseconomies such as higher rents or wage levels resulting from competition for these localized resources may also occur.

Weber concluded that transport costs were the major consideration determining location. That is, the optimum location would be found where the costs of transporting raw materials to the factory and finished goods to the market were at their lowest (Figure 10.21). He noted, however, that, if variations in labor or agglomeration costs were sufficiently great, a location determined solely on the basis of transportation costs might not, in fact, be the optimum one.

Assuming, however, that transportation costs determine the "balance point," optimum location will depend on distances, the respective weights of the raw material inputs, and the final weight of the finished product. It may be either *material-oriented* or *market-oriented.* Material orientation reflects a sizeable weight loss during the production process; market orientation indicates a weight gain (Figure 10.22).

For many theorists, Weber's least-cost analysis is unnecessarily rigid and restrictive. They propose instead a *substitution principle* which recognizes that in many industrial processes it is possible to replace a declining amount of one input (e.g., labor) with an increase in another (e.g., capital for automated equipment) or to increase transportation

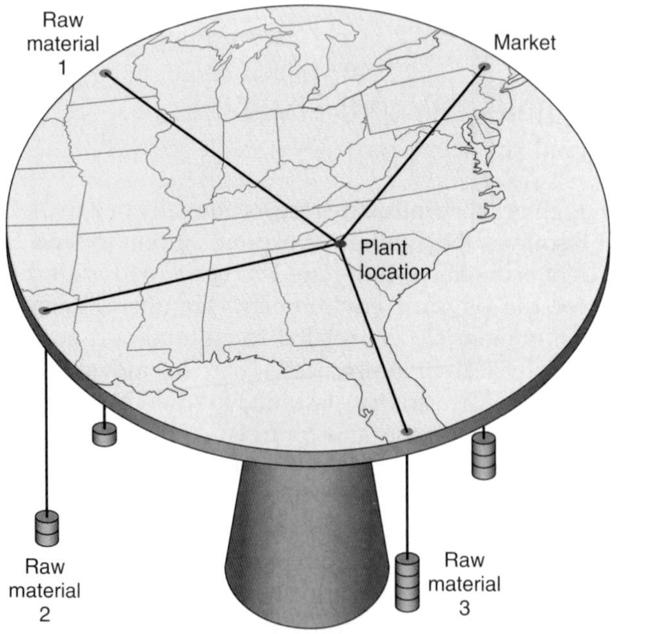


**FIGURE 10.21 Plane table solution to a plant location problem.** This mechanical model, suggested by Alfred Weber, uses weights to demonstrate the least transport cost point where there are several sources of raw materials. When a weight is allowed to represent the "pull" of raw material and market locations, an equilibrium point is found on the plane table. That point is the location at which all forces balance each other and represents the least-cost plant location.

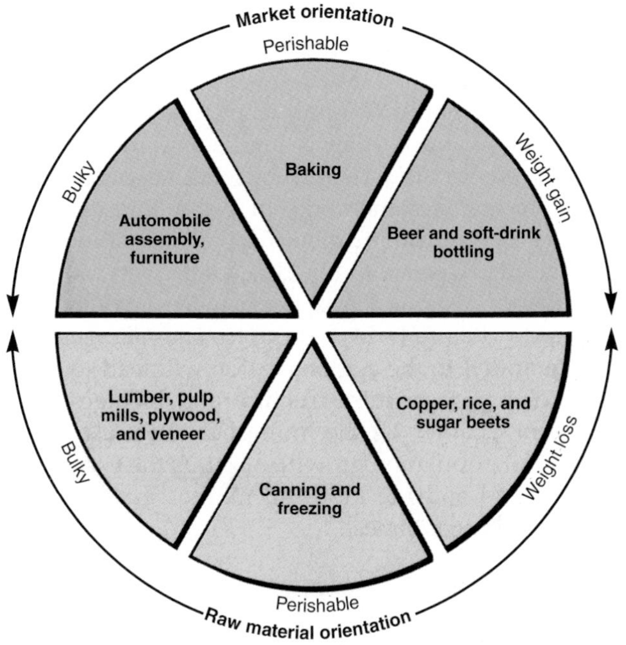


**FIGURE 10.22 Spatial orientation tendencies.** *Raw material orientation* is presumed to exist when there are limited alternative material sources, when the material is perishable, or when, in its natural state, the material contains a large proportion of impurities or non-marketable components. *Market orientation* represents the least-cost solution when manufacturing uses commonly available materials that add weight to the finished product, when the manufacturing process produces a commodity much bulkier or more expensive to ship than its separate components, or when the perishable nature of the product demands processing at individual market points. *Redrawn with permission from Truman A. Hartshorn,* Interpreting the City. *Copyright © 1980 John Wiley & Sons, Inc., New York, N.Y.*

costs while simultaneously reducing land rent. With substitution, a number of different points may be optimal manufacturing locations. Further, they suggest, a whole series of points may exist where total revenue of an enterprise just equals its total cost of producing a given output. These points, connected, mark the *spatial margin of profitability* and define the area within which profitable operation is possible (Figure 10.23). Location anywhere within the margin assures some profit and tolerates both imperfect knowledge and personal (rather than economic) considerations.

## Other Locational Considerations

The behavior of individual firms seeking specific production sites under competitive conditions forms the basis of most classical industrial location theory. But such theory no longer fully explains world or regional patterns of industrial localization or specialization, nor does it account for locational behavior that is uncontrolled by objective "factors," that is influenced by new production technologies and corporate structures, or that is directed by noncapitalistic planning goals.

Traditional theories (including many variants not reviewed here) sought to explain locational decisions for plants engaged in mass production for mass markets where transportation lines were fixed and transport costs were relatively high. Both conditions began to change significantly during the last years of the 20th century. Assembly line production of identical commodities by a rigidly controlled and specialized labor force for generalized mass markets—known as **Fordism** to recognize Henry Ford's pioneering development of the system—became less realistic in both market and technology terms. In its place, post-Fordist *flexible manufacturing* processes based on smaller production runs of a greater variety of goods aimed at smaller, niche markets than were catered to by traditional manufacturing have become common. At the same time, information technology applied to machines and operations, increasing flexibility of labor, and declining costs for transportation services that were increasingly viewed from a cost-time rather than a cost-distance standpoint have materially altered underlying assumptions of the classical theories.

### Transport Characteristics

For example, within both national and international economies, the type and efficiency of transport media, as well as transportation costs, have been central to the spatial patterning of production, explaining the location of a large variety of economic activities. Waterborne transportation is nearly always cheaper than any other mode of conveyance, and the enormous amount of commercial activity that takes place on coasts or on navigable rivers leading to coasts is an indication of that cost advantage. When railroads were

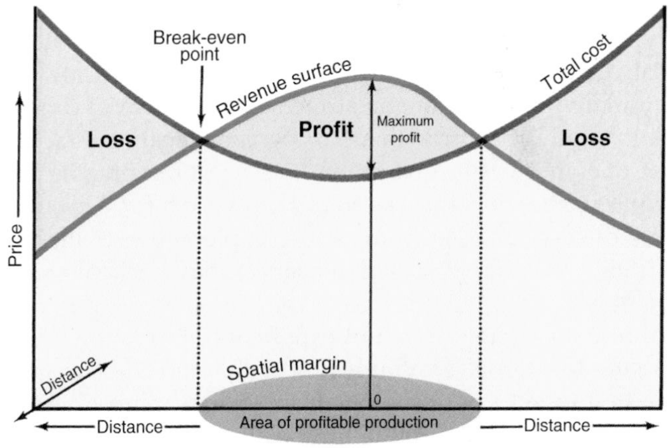


**FIGURE 10.23 The spatial margin of profitability.** In the diagram, 0 is the single optimal profit-maximizing location, but location anywhere within the area defined by the intersects of the total cost and total revenue surfaces will permit profitable operation. Some industries will have wide margins; others will be more spatially constricted. Skilled entrepreneurs may be able to expand the margins farther than less able industrialists. Importantly, a *satisficing* location may be selected by reasonable estimate even in the absence of the totality of information required for an *optimal* decision.

developed and the commercial exploitation of inland areas could begin, coastal sites continued to be important as more and more goods were transferred there between low-cost water and land media. The advent of highway transportation vastly increased the number of potential "satisficing" manufacturing locations by freeing the locational decision from fixed-route production sites. Every change in carrier mode, efficiency, or cost structure has direct implications for locations of economic activity.

In the rare instance when transportation costs become a negligible factor in production and marketing, an economic activity is said to be *footloose.* Some manufacturing is located without reference to raw materials; for example, the raw materials for electronic products such as computers are so valuable, light, and compact that transportation costs have little bearing on where production takes place. Others are inseparable from the markets they serve and are so widely distributed that they are known as *ubiquitous industries.* Newspaper publishing, bakeries, and dairies, all of which produce a highly perishable commodity designed for immediate consumption, are examples.

Overall, transportation costs have been declining and media efficiencies have been increasing. The advent of near-universal commercial jet aircraft service, the development of large ocean-going superfreighters, and the introduction of containerization and intermodal transfers in both water and land shipments of goods have reduced the costs and increased the speed of freight services. As those costs have decreased, manufacturing location has become more influenced by nontransport locational factors. To that extent, Weberian location theories have reduced applicability to the modern global space economy.

### Agglomeration Economies

The geographic concentration of economic, including industrial, activities is the norm at the local or regional scale. The cumulative and reinforcing attractions of industrial concentration and urban growth are recognized locational factors, but ones not easily quantified. Both cost-minimizing and profit-maximizing theories make provision for *agglomeration,* the spatial concentration of people and activities for mutual benefit. That is, both recognize that the areal grouping of industrial activities may produce benefits for individual firms that they could not experience in isolation. Those benefits—**external economies** or *agglomeration economies*—accrue in the form of savings from shared transport facilities, social services, public utilities, communication facilities, and the like. Collectively, these and other installations and services needed to facilitate industrial and other forms of economic development are called **infrastructure.** Areal concentration may also create pools of skilled and ordinary labor, of capital, ancillary business services, and, of course, a market built of other industries and urban populations. New firms, particularly, may find significant advantages in locating near other firms engaged in the same activity, for labor specializations and support services specific to that activity are already in place. Some may find profit in being near other firms with which they are linked either as customers or suppliers.

A concentration of capital, labor, management skills, customer base, and all that is implied by the term *infrastructure* tends to attract still more industries from other locations to the agglomeration. In Weber's terms, that is, economies of association distort or alter locational decisions that otherwise would be based solely on transportation and labor costs, and once in existence, agglomerations tend to grow (Figure 10.24). Through a *multiplier effect,* each new firm added to the agglomeration will lead to the further development of infrastructure and linkages. As we shall see in Chapter 11, the multiplier effect also implies total (urban) population growth and thus the expansion of the labor pool and the localized market that are part of agglomeration economies.

### Just-in-Time and Flexible Production

Agglomeration economies and tendencies are also encouraged by newer manufacturing policies practiced by both older, established industries and by newer post-Fordist plants.

Traditional Fordist industries required the on-site storage of large lots of materials and supplies ordered and delivered well in advance of their actual need in production. That practice permitted cost savings through infrequent ordering and reduced transportation charges, and made allowances for delayed deliveries and for the inspection of received goods and components. The assurance of supplies on hand for long production runs of standardized outputs was achieved at high inventory and storage costs.

*Just-in-Time (JIT)* manufacturing, in contrast, seeks to reduce inventories through the production process by purchasing inputs for arrival just in time to use and producing output just in time to sell. Rather than the costly accumulation and storage of supplies, JIT requires frequent ordering of small lots of goods for precisely timed arrival and immediate deployment to the factory floor. Such "lean manufacturing" based on the frequent purchasing of immediately needed goods demands rapid delivery by suppliers and encourages them to locate near the buyer. Recent manufacturing innovations thus reinforce and augment the spatial agglomeration tendencies evident in the older industrial landscape and deemphasize the applicability of older single-plant location theories.

JIT is one expression of a transition from mass production Fordism to more *flexible production systems.* That flexibility is designed to allow producers to shift quickly and easily between different levels of output and, importantly, to move from one factory process or product to another as market demand dictates. Flexibility of that type is made possible by new technologies of easily reprogrammed computerized machine tools and by computer-aided design and computer-aided manufacturing systems. These technologies permit small-batch, Just-in-Time production and distri-

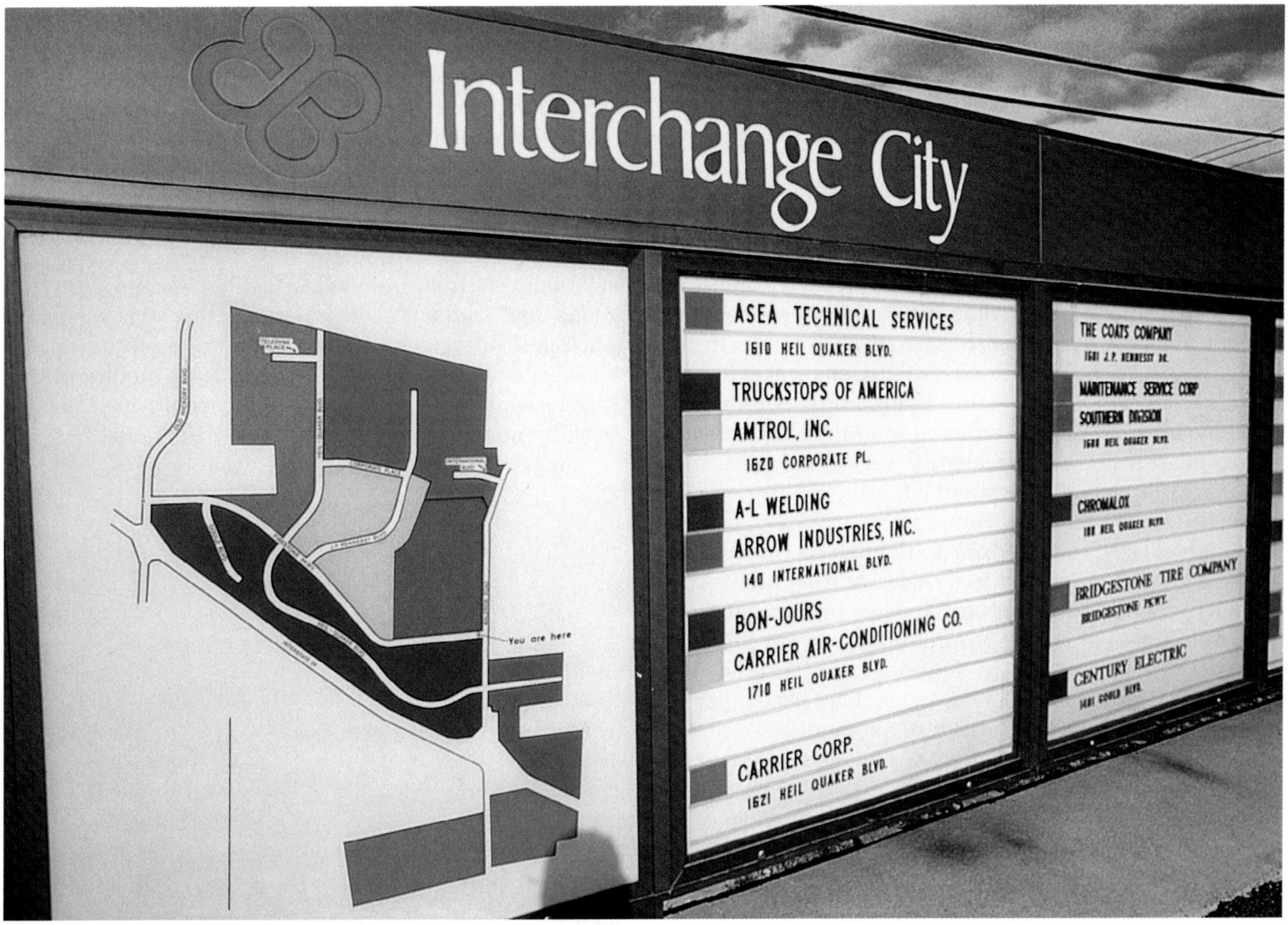


**FIGURE 10.24** On a small scale, a planned industrial park furnishes its tenants external agglomeration economies similar to those offered by large urban concentrations to industry in general. An industrial park provides a subdivided tract of land developed according to a comprehensive plan for the use of (frequently) otherwise unconnected firms. Because the park developers, whether private companies or public agencies, supply the basic infrastructure of streets, water, sewage disposal, power, transport facilities, and perhaps private police and fire protection, park tenants are spared the additional cost of providing these services themselves. In some instances, factory buildings are available for rent, still further reducing firm development outlays. Counterparts of industrial parks for manufacturers are office parks, research parks, science parks, and the like for "high-tech" firms and for enterprises in tertiary and quaternary services. *© Cameramann International Ltd.*

bution responsive to current market demand as monitored by computer-based information systems.

Flexible production, to a large extent, requires significant acquisition of components and services from outside suppliers rather than from in-house production. For example, modular assembly, in which many subsystems of a complex final product enter the plant already assembled, reduces factory space and worker requirements. The premium that flexibility places on proximity to component suppliers adds still another dimension to industrial agglomeration tendencies. "Flexible production regions" have, according to some observers, emerged in response to the new flexible production strategies and interfirm dependencies. Those regions, it is claimed, are usually some distance—spatially or socially—from established concentrations of Fordist industrialization.

### Comparative Advantage

The principle of **comparative advantage** is of growing international importance in industrial location and specialization decisions. It tells us that areas tend to specialize in the production of those items for which they have the greatest relative advantage over other areas, or for which they have the least relative disadvantage, as long as free trade exists. This principle, basic to the understanding of regional specializations, applies as long as areas have different relative advantages for two or more goods.

Assume that two countries have a need for, and are able to produce domestically, two commodities. Further assume that there is no transport cost consideration. No matter what its cost of production of either commodity, Country A will choose to specialize in only one of them if, by that specialization and through exchange with Country B for the other, Country A stands to gain more than it loses. The key to comparative advantage is the use of resources in such a fashion as to gain, by specialization, a volume of production and a selling price that permit exchange for a needed commodity at a cost level that is below that of the domestic production of both.

At first glance, the concept of comparative advantage may seem to defy logic. For example, Japan may be able to produce airplanes and home appliances more cheaply than the United States, thereby giving it an apparent advantage in both goods. But it benefits both countries if they specialize in the good in which they have a comparative advantage. In this instance, Japan's manufacturing cost structure makes it more profitable for Japan to specialize in the volume production of appliances and to buy airplanes from the United States, where large civilian and military markets encourage aircraft manufacturing specialization and efficiency.

When other countries' comparative advantages reflect lower labor, land, raw material and capital costs, manufacturing activities may voluntarily relocate from higher-cost market locations to lower-cost foreign production sites. Such voluntary **outsourcing**—producing parts or products abroad for domestic sale—by American manufacturers has employment and areal economic consequences no different from those resulting from successful competition by foreign companies or from industrial locational decisions favoring one section of the country over others.[2] When comparative advantage is exploited by individual corporations, one expression of flexible production systems is evident in the erosion of the rigid spatial concentration of production assumed by classical location theory.

A clear example of that impact is evident in the changing nature of automobile manufacturing. Formerly, motor vehicle companies were largely self-contained production entities controlling raw material inputs through their own steel and glass plants and producing themselves all the parts and components required in the assembly of their products. Since the late 1990s, that self-containment has been abandoned as car companies have divested themselves of raw material production facilities and have in large part sold off their in-house parts production. Increasingly, they are purchasing parts and subassemblies from independent, often distant suppliers. In fact, some observers of the changing vehicle production scene predict that established automobile companies will eventually convert themselves into "vehicle brand owners," retaining for themselves only such essential tasks as vehicle design, engineering, and marketing. All else, including final vehicle assembly, is projected to be done through outsourcing to parts suppliers. Similar trends are already evident in consumer electronics, in which a third of manufacturing in 2004 was estimated to be outsourced.

[2]*Outsourcing* has also assumed the meaning of subcontracting production work to outside domestic companies.

A distinctive regional impact of more diversified industrial deconcentration through outsourcing is found along the northern border of Mexico. In the 1960s, Mexico enacted legislation permitting foreign (specifically, American) companies to establish "sister" plants, called *maquiladoras,* within 20 kilometers (12 mi) of the U.S. border for the duty-free assembly of products destined for reexport. By 2000, more than 3000 such assembly and manufacturing plants had been established to produce a diversity of goods, including electronic products, textiles, furniture, leather goods, toys, and automotive parts. In that year, the plants generated direct employment for more than 1.2 million Mexican workers (Figure 10.25) and a large number of U.S. citizens, employees of growing numbers of American-side *maquila* suppliers and of diverse service-oriented business

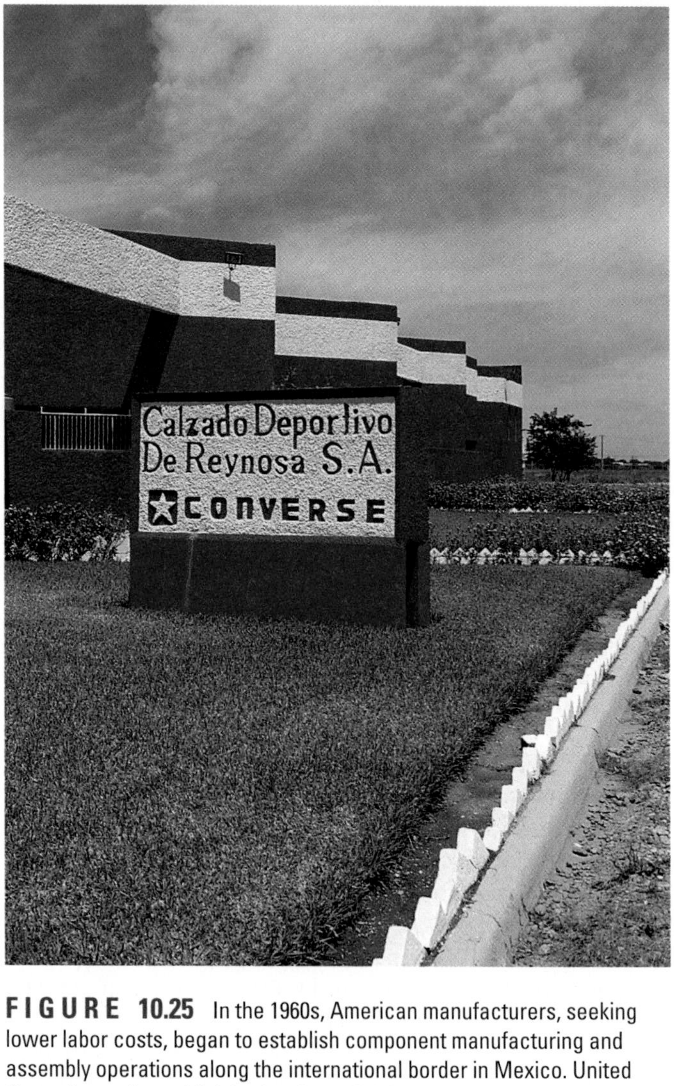


**FIGURE 10.25** In the 1960s, American manufacturers, seeking lower labor costs, began to establish component manufacturing and assembly operations along the international border in Mexico. United States laws allowed finished and semifinished products to be brought into the country duty-free, as they are from the Converse Sport Shoe factory at Reynosa. *Outsourcing* has moved a large proportion of American electronics, small appliance, and garment industries to offshore subsidiaries and contractors in Asia and Latin America.
Source: *Converse, Inc.*

spawned by the multiplier effect. The North American Free Trade Agreement (NAFTA) creating a single Canadian–United States–Mexican production and marketing community promises to turn outsourcing in the North American context from a search abroad for low-cost production sites to a review of best locations within a broadened unified economic environment.

The United States also benefits from outsourcing by other countries. Japanese and European companies have established automobile and other manufacturing plants here in part to take advantage of lower American production and labor costs and in part in response to American protectionist policies and automobile import quotas. A strong yen in relation to the value of the dollar during the 1980s, by raising the relative cost of imported cars, also encouraged Japanese auto plant location in the United States, followed by Japanese auto parts manufacturers. At least part of the products of both automobile assemblers and parts suppliers were available for reexport to other national markets.

Comparative advantage is not an unchanging relationship. One notable outcome of globalization is technology transfer from economically advanced to underdeveloped economies. That transfer is transforming the world economy by introducing a new *international division of labor.* In the 19th and the first half of the 20th centuries, that division invariably involved exports of manufactured goods from the "industrial" countries and of raw materials from the "colonial" or "undeveloped" economies. Roles have now altered. Manufacturing no longer is the mainstay of the economy and employment structure of Europe or Anglo America and, as the NAFTA example showed, the world pattern of industrial production is shifting to reflect the growing dominance of countries formerly regarded as subsistence peasant societies but now emerging as the source areas for manufactured goods of all types.

### Imposed Considerations

Location theories dictate that in a pure, competitive economy, the costs of material, transportation, labor, and plant should be dominant in locational decisions. Obviously, neither in the United States nor in any other market economy do the idealized conditions exist. Other constraints—some representing cost considerations, others political or social impositions—also affect, perhaps decisively, the locational decision process. Land use and zoning controls, environmental quality standards, governmental area development inducements, local tax abatement provisions or developmental bond authorizations, noneconomic pressures on quasi-governmental corporations, and other considerations constitute attractions or repulsions for industry outside of the context and consideration of pure theory (see "Contests and Bribery," p. 374). If these noneconomic forces become compelling, the assumptions of the commercial economy classification no longer obtain, and locational controls reminiscent of those enforced by current or former centrally planned economies become determining.

No other imposed considerations were as pervasive as those governing industrial location in planned economies. The theoretical controls on plant location decisions that apply in commercial economies were not, by definition, determinant in the centrally planned Marxist economies of Eastern Europe and the former Soviet Union. In those economies, plant locational decisions were made by governmental agencies rather than by individual firms.

Bureaucratic rather than company decision making did not mean that locational assessments based on factor cost were ignored; it meant that central planners were more concerned with other than purely economic considerations in the creation of new industrial plants and concentrations. Important in the former Soviet Union, for example, was a controlling policy of the *rationalization* of industry through full development of the resources of the country wherever they were found and without regard to the cost or competitiveness of such development. Inevitably, although the factors of industrial production are identical in capitalist and noncapitalist economies, the philosophies and patterns of industrial location and areal development will differ between them. Because major capital investments are relatively permanent additions to the landscape, the results of their often noneconomic political or philosophical decisions are fixed and will long remain to influence industrial regionalism and competitive efficiencies into the postcommunist present and future. Those same decisions and rigidities continue to inhibit the transition by the formerly fully planned economies to modern capitalist industrial techniques and flexibilities.

## Transnational Corporations (TNCs)

Outsourcing is but one small expression of the growing international structure of modern manufacturing and service enterprises. Business and industry are increasingly stateless and economies borderless as giant **transnational corporations (TNCs)**—private firms that have established branch operations in nations foreign to their headquarter's country—become ever-more important in the globalizing world space economy. Early in the 21st century, there were some 65,000 transnational (or *multinational*) companies (briefly introduced in Chapter 8) controlling at least 850,000 foreign affiliates employing about 54 million workers. Excluding the parent companies themselves, the TNC affiliates accounted for almost $19 trillion in sales, one-tenth of world GDP, and one-third of world exports.

Measured by value added (not total sales), 29 of the world's top 100 economies in 2003 were corporations, not countries. The great majority of them, all TNCs, are engaged in secondary industries. That is, except for a few resource-based firms, they are principally involved in producing and selling manufactured goods. Although tertiary and quaternary activities have also become international in scope and transnational in corporate structure, the locational and operational advantages of multicountry operation were first discerned and exploited by manufacturers.

GEOGRAPHY & Public Policy

## Contests and Bribery

In 1985, it cost Kentucky more than $140 million in incentives—about $47,000 a job—to induce Toyota to locate an automobile assembly plant in Georgetown, Kentucky. That was cheap. By 1993, Alabama had spent $169,000 per job to lure Mercedes-Benz to that state; in 2001, Mississippi agreed to $400 million in spending and tax rebates to Nissan; and Kentucky bid $350,000 per job in tax credits to bring a Canadian steel mill there.

The spirited auction for jobs is not confined to manufacturing. A University of Minnesota economist calculated that his state will have spent $500,000 for each of the 1500 or more permanent jobs created by Northwest Airlines at two new maintenance facilities. Illinois gave $240 million in incentives ($44,000 per job) to keep 5400 Sears, Roebuck employees within the state, and New York City awarded $184 million to the New York Mercantile Exchange and more than $30 million each to financial firms Morgan Stanley and Kidder, Peabody to induce them to stay in the city. For some, the bidding among states and locales to attract new employers and employment gets too fierce. Kentucky withdrew from competition for a United Airlines maintenance facility, letting Indianapolis have it when Indiana's offered package exceeded $450 million. By 2004, following a slowdown in air travel, United walked away from a fully completed operational facility, leaving the city and state with $320 million of bonded debt and a complex of empty hangars and office buildings.

Inducements to lure companies are not just in cash and loans—though both figure in some offers. For manufacturers, incentives may include workforce training, property tax abatement, subsidized costs of land and building or their outright gifts, below-market financing of bonds, and the like. Similar offers are regularly made by states, counties, and cities to wholesalers, retailers, and major office-worker and other service activity employers. The total annual loss of city and state tax revenue through abatements, subsidies, grants, and the like to benefit retained or attracted firms has been estimated at $30 billion to $40 billion. The objective, of course, is not just to secure the new jobs represented by the attracted firm but to benefit from the general economic stimulus and employment growth that those jobs—and their companies—generate. Auto parts manufacturers are presumably attracted to new assembly plant locations; cities grow and service industries of all kinds—doctors, department stores, restaurants, food stores—prosper from the investments made to attract new basic employment.

Not everyone is convinced that those investments are wise, however. A poll of Minnesotans showed that a majority opposed the generous offer made by the state to Northwest Airlines. In the late 1980s, the governor of Indiana, a candidate for Kentucky's governorship, and the mayor of Flat Rock, Michigan, were all defeated by challengers who charged that too much had been spent in luring the Suburu-Isuzu, Toyota, and Mazda plants, respectively. Established businesses resent what often seems to be neglect of their interests in favor of spending their tax money on favors to newcomers. The Council for Urban Economic Development, surveying the escalating bidding wars, has actively lobbied against incentives, and many academic observers note that industrial attraction amounts to a zero sum game: unless the attracted newcomer is a foreign firm, whatever one state achieves in attracting an expanding U.S. company comes at the expense of another state.

Some doubt that inducements matter much anyway. Although, sensibly, companies seeking new locations will shop around and solicit the lowest-cost, best deal possible, their site choices are apt to be determined by more realistic business considerations: access to labor, suppliers, and markets; transportation and utility costs; weather; the nature of the workforce; and overall costs of living. Only when two or more similarly attractive locations have essentially equal cost structures might such special inducements as tax reductions or abatements be determining in a locational decision.

**QUESTIONS TO CONSIDER**

1. As a citizen and taxpayer, do you think it is appropriate to spend public money to attract new employment to your state or community? If not, why not? If yes, what kinds of inducements and what total amount offered per job seem appropriate to you? What reasons support your opinion?
2. If you believe that "best locations" for the economy as a whole are those determined by pure location theory, what arguments would you propose to discourage locales and states from making financial offers designed to circumvent decisions clearly justified on abstract theoretical grounds?

TNCs are increasingly international in origin and administrative home, based in a growing number of economically advanced countries. In the early 21st century, about 90% of the world's 100 largest TNCs have home offices in the European Union, the United States, and Japan; only two developing country firms were on the list. Because of their outsourced purchases of raw materials, parts and components, and services, the total number of worldwide jobs associated with TNCs in 2003 reached 150 million or more.

The foreign impact, however, is limited to relatively few countries and regions. Less than 20% (in 2002) of inflows of foreign direct investment (FDI) went to developing countries, and the majority of that was concentrated in 10 to 15 states, mainly in South, Southeast, and East Asia (China is the largest developing country recipient) and in

Latin America and the Caribbean. The 50 least-developed countries as a group—including nearly all African states—received only 0.3% of world foreign direct investment. Despite poor countries' hopes for foreign investment to spur their economic growth, the vast majority of FDI flows not to the poor or developing worlds but to the rich.

The advanced country destination of those capital flows is understandable: TNCs are actively engaged in merging with or purchasing competitive established firms in already developed foreign market areas, and cross-border mergers and acquisitions have been the main stimuli behind FDI. Between 1980 and 2002, some 225,000 mergers or purchases were announced worldwide. Because most transnational corporations operate in only a few industries—computers, electronics, petroleum and mining, motor vehicles, chemicals, and pharmaceuticals—the worldwide impact of their consolidations is significant. Some dominate the marketing and distribution of basic and specialized commodities. In raw materials, a few TNCs account for 85% or more of world trade in wheat, maize, coffee, cotton, iron ore, and timber, for example. In manufactures, the highly concentrated world pharmaceutical industry is dominated by just 6 firms, and the world's 15 major automobile producers at the start of the 21st century, it has been predicted, will have fallen to 5 or 10 by 2015.

Because they are international in operation with multiple markets, plants, and raw material sources, TNCs actively exploit the principle of comparative advantage. In manufacturing, they have internationalized the plant-siting decision process and have multiplied the number of locationally separated operations that must be assessed. TNCs produce in that country or region where the costs of materials, labor, or other production inputs are minimized, while maintaining operational control and declaring taxes in localities where the economic climate is most favorable. Research and development, accounting, and other corporate activities are placed wherever economical and convenient.

TNCs have become global entities because global communications make it possible. Many have lost their original national identities and are no longer closely associated with or controlled by the cultures, societies, and legal systems of a nominal home country. At the same time, their multiplication of economic activities has reduced any earlier identifications with single products or processes and has given rise to "transnational integral conglomerates" spanning a large spectrum of both service and industrial sectors.

## World Manufacturing Patterns and Trends

Whether locational decisions are made by private entrepreneurs or by central planners—and on whatever considerations those decisions are based—the results over many years have produced a distinctive world pattern of manufacturing. Figure 10.26 suggests the striking prominence of a relatively small number of major industrial concentrations localized within relatively few countries primarily but not exclusively parts of the "industrialized" or "developed" world. These can be roughly grouped into four commonly recognized major manufacturing regions: *Eastern Anglo America, Western and Central Europe, Eastern Europe,* and *Eastern Asia.* Together, the industrial plants within

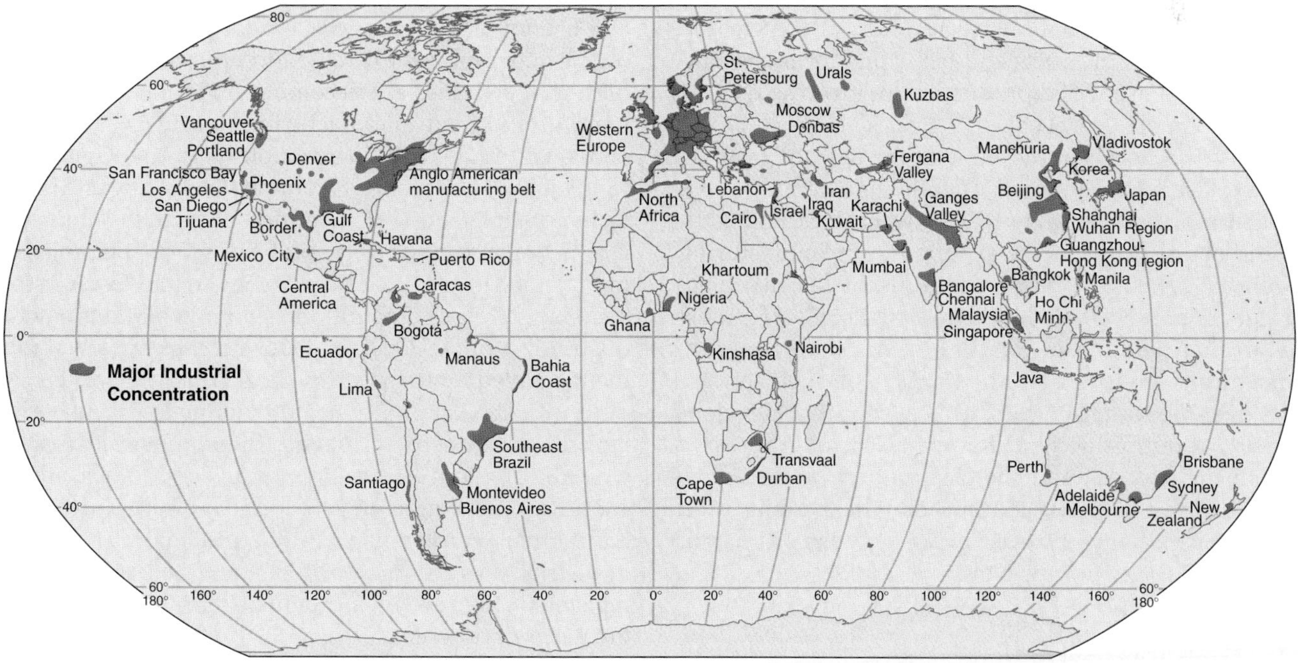


**FIGURE 10.26 World industrial regions.** Industrial districts are not as continuous or "solid" as the map suggests. Manufacturing is a relatively minor user of land even in areas of its greatest concentration.

these established concentrations account for an estimated three-fifths of the world's manufacturing output by volume and value.

Their continuing dominance is by no means assured. The first three—those of Anglo America and Europe—were the beneficiaries of an earlier phase in the development and spread of manufacturing following the Industrial Revolution of the 18th century and lasting until after World War II. The countries within them now are increasingly developing postindustrial economies in which traditional manufacturing and processing are of declining relative importance.

The fourth—the East Asian industrial region—is a part and forerunner of the wider, new pattern of world industrialization that has emerged in recent years, the result of international cultural convergence and technology transfers in the latter half of the 20th century, as briefly reviewed in Chapter 7. The older, rigid industrial "North–South" split between the developed and developing worlds has rapidly weakened as the full range of industrial activities from primary metal processing (e.g., the iron and steel industries) through advanced electronics assembly has been dispersed from, or separately established within, an ever-expanding list of countries.

Such states as Mexico, Brazil, China, and others of the developing world have created industrial regions of international significance, and the contribution to world manufacturing activity of the smaller, newly industrializing countries (NICs) has been growing significantly. Even economies that until recently were overwhelmingly subsistence or dominated by agricultural or mineral exports have become important players in the changing world manufacturing scene. Foreign branch plant investment in low-wage Asian, African, and Latin American states has not only created there an industrial infrastructure but also increased their gross national products and per capita incomes sufficiently to permit expanded production for growing domestic—not just export—markets.

In Malaysia, for example—one of the rapidly growing Southeast Asian economies—agriculture's share of gross domestic product was cut by 60% between 1981 and 2001, and the share of merchandise exports from manufacturing rose from 19% to 80%. For Pakistan in South Asia, manufactured goods jumped from less than half to 85% of export values in that period, and for Bolivia in South America, manufacturing grew from 3% to 22% of merchandise exports over the same 20 years. Collectively, the "developing world" countries account for some 21% of *gross world product* in the first years of the 21st century; increasingly, their share of that output is comprised of the manufactured goods that formerly were nearly exclusively the product of the "advanced world" of the North.

## High-Tech Patterns

Major industrial districts of the world developed over time as entrepreneurs and planners established traditional secondary industries according to the pulls and orientations predicted by classical location theories. Those theories are less applicable in explaining the location of the latest generation of manufacturing activities: the high-technology—or *high-tech*—processing and production that is increasingly part of the advanced economies. For these firms, new and different patterns of locational orientation and advantage have emerged, based on other than the traditional regional and site attractions.

High technology is more a concept than a precise definition. It probably is best understood as the application of intensive research and development efforts to the creation and manufacture of new products of an advanced scientific and engineering character. Professional—"white collar"—workers make up a large share of the total workforce. They include research scientists, engineers, and skilled technicians. When these high-skill specialists are added to administrative, supervisory, marketing, and other professional staffs, they may greatly outnumber actual production workers in a firm's employment structure. In the world of high tech, that is, the distinction between secondary (manufacturing) and quaternary (knowledge) activities and workers is increasingly blurred.

Although only a few types of industrial activity are generally reckoned as exclusively high tech—electronics, communications, computers, software, pharmaceuticals and biotechnology, aerospace, and the like—advanced technology is increasingly a part of the structure and processes of all forms of industry. Robotics on the assembly line, computer-aided design and manufacturing, electronic controls of smelting and refining processes, and the constant development of new products of the chemical and pharmaceutical industries are cases in point.

The impact of high-tech industries on patterns of economic geography is expressed in at least three ways. First, high-tech activities are becoming major factors in employment growth and manufacturing output in the advanced and newly industrializing economies. In the United States, for example, between 1986 and 2000, the largest five high-tech industry groups alone added more than 8 million new jobs to the secondary sector of the economy, helping replace many thousands of other workers who had lost jobs to outsourcing, foreign competition, changing markets, and deindustrialization. In 2000, total U.S. high-technology employment equaled more than 16% of all nonfarm workers, while even workers in more strictly defined "intensive high-technology industries" totaled 7% of all employment. The United Kingdom, Germany, Japan, and other advanced countries—although not yet those of Eastern Europe—have had similar employment shifts, and many of the newly industrializing economies of East and Southeast Asia have registered high-tech employment growth of similar or greater proportions.

Many of the confident generalizations concerning, particularly, the computer and software segments of high-technology industry and employment and the prospects for their continuing expansion were called into question by the abrupt dot-com bubble collapse of the first years of the 21st century. Tens of thousands of software jobs lost; scores of innovative, visionary, or simply hopeful software and hardware ventures bankrupt; billions of dollars of venture capital and stock valuations erased; and lavish offices and plants left vacant marked the end of a period of exuberant growth and expectations that shaped high-tech locational patterns and bolstered national economies throughout the world. Although details of the software and computer hardware distributions, specializations, and employment structures of the end of the 20th century were severely disturbed or destroyed, the new century saw both a continuation of high-technology applications to established general manufacturing activities and a gradual revival of investment and employment in software development. Locational concentrations and specializations may change, but high-tech's fundamental alteration of older economic structures and patterns endures.

The products of high-technology activity represent an increasing share of total industrial output of individual countries and of the trade between them. As early as 1995, high-tech manufactures represented 15% of manufacturing output in both the United States and Japan, 14% in the United Kingdom, and about 10% in Germany and France; among the emerging Asian countries, 12.5% of China's and 15% of South Korea's total output came from high-tech manufacturing. In all these and other countries, high-tech production has increased in volume and percentage. The 2000 U.S. high-tech share increased to near 20%. Global data are incomplete, but Figure 10.27 suggests the great disparity between countries in the importance of high-tech products in their exports of manufactured goods.

A second impact is more clearly spatial. High-tech industries have tended to become regionally concentrated in their countries of development, and within those regions they frequently form self-sustaining, highly specialized agglomerations. California, for example, has a share of U.S. high-tech employment far in excess of its share of American population. Along with California, the Pacific Northwest (including British Columbia), New England, New Jersey, Texas, and Colorado all have proportions of their workers in high-tech industries above the national average. And within these and other states or regions of high-tech concentration, specific locales have achieved prominence: Silicon Valley of Santa Clara County near San Francisco; Irvine and Orange County south of Los Angeles; the Silicon Forest near Seattle; North Carolina's Research triangle; Utah's Software Valley; Routes 128 and 495 around Boston; Silicon Swamp of the Washington, D.C., area; Ottawa, Canada's, Silicon Valley North; and the Canadian Technology triangle west of Toronto are familiar Anglo American examples.

Within such concentration, specialization is often the rule: medical technologies in Minneapolis and Philadelphia; biotechnology around San Antonio; computers and semiconductors in eastern Virginia and at Austin, Texas; biotechnology and telecommunications in New Jersey's Princeton Corridor; and telecommunications and Internet

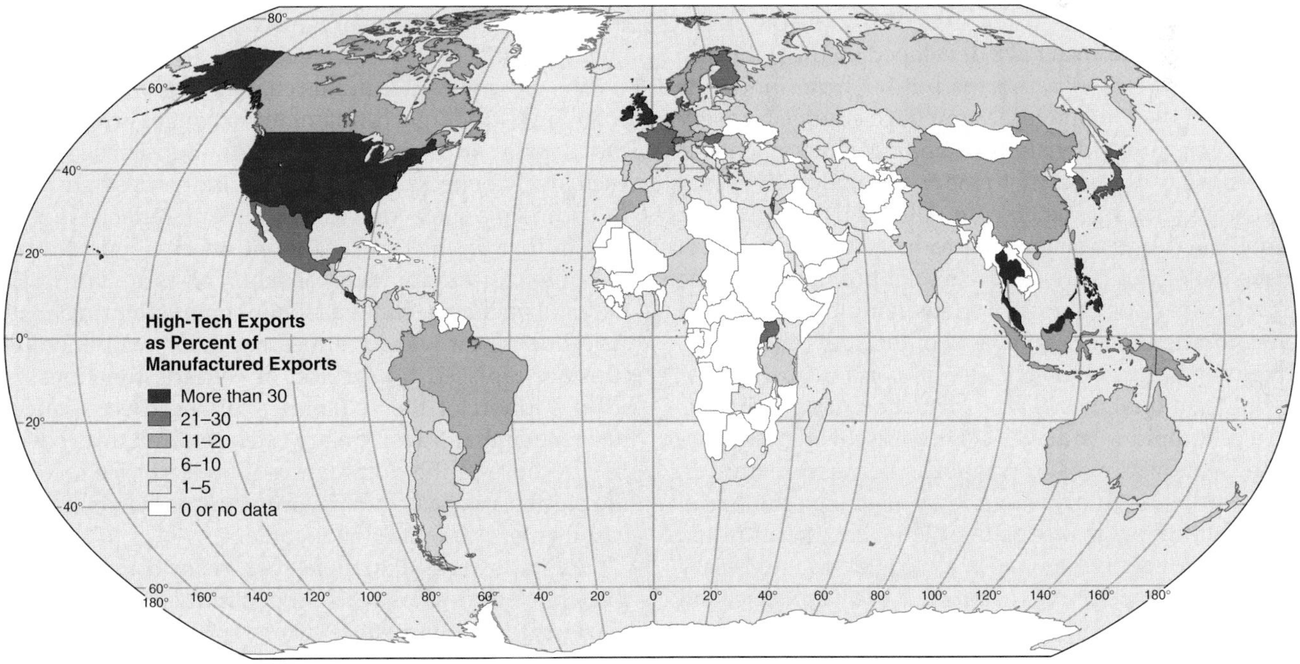


**FIGURE 10.27** The map clearly suggests the importance of the industrialized countries—particularly the United States and Western Europe—in high-tech manufacturing and exports in 2001. Less evident is the relative role of high tech in the manufactured exports of a few smaller, developing states: 70% for the Philippines, 60% for Singapore, and 57% for Malaysia early in the 21st century. The map, of course, does not report a country's ranking in the volumes or values of high-tech manufactured goods exports. Source: *The World Bank.*

industries near Washington, D.C. Elsewhere, Scotland's Silicon Glen, England's Sunrise Strip and Silicon Fen, Wireless Valley in Stockholm, Zhong Guancum in suburban Beijing, and Bangalore, India, are other examples of industrial landscapes characterized by low, modern, dispersed office-plant-laboratory buildings rather than by massive factories, mills, or assembly structures, freight facilities, and storage areas. Planned business parks catering to the needs of smaller companies are increasingly a part of regional and local economic planning. Irvine, California's, Spectrum, for example, at its height housed 44,000 employees and 2200 companies, most of them high-tech start-ups.

The older distributional patterns of high-tech industries suggest that they respond to different localizing forces than those controlling traditional manufacturing industries. At least five locational tendencies have been recognized: (1) proximity to major universities or research facilities and to a large pool of scientific and technical labor skills; (2) the avoidance of areas with strong labor unionization where contract rigidities might slow process innovation and workforce flexibility; (3) locally available venture capital and entrepreneurial daring; (4) location in regions and major metropolitan areas with favorable "quality of life" reputations—climate, scenery, recreation, good universities, and an employment base sufficiently large to supply needed workers and provide job opportunities for professionally trained spouses; (5) the availability of first-quality communication and transportation facilities to unite separated stages of research, development, and manufacturing and to connect the firm with suppliers, markets, finances, and the governmental agencies so important in supporting research. Essentially all the major high-tech agglomerations have developed on the semirural peripheries of metropolitan areas but far from inner-city problems and disadvantages. Many have emerged as self-sufficient areas of subdivisions, shopping centers, schools, and parks in close proximity to the company locations and business parks that form their core. Although the New York metropolitan area is a major high-tech concentration, most of the technology jobs are suburban, not in Manhattan; the periphery's share of computer-related employment in the region amounted to 80% at the end of the 20th century.

Agglomerating forces are also important in this new industrial locational model. The formation of new firms is frequent and rapid in industries where discoveries are constant and innovation is continuous. Because many are "spin-off" firms founded by employees leaving established local companies, areas of existing high-tech concentration tend to spawn new entrants and to provide the necessary labor skills. Agglomeration, therefore, is both a product and a cause of spatial associations.

Not all phases of high-tech production must be concentrated, however. The spatial attractions affecting the professional, scientific, and knowledge-intensive aspects of high tech have little meaning for many of the component manufacturing and assembly operations, which may be highly automated or require little in the way of labor skills. These tasks, in our earlier locational terminology, are "footloose"; they require highly mobile capital and technology investments but may be advantageously performed by young women in low-wage areas at home or—more likely—in countries such as Taiwan, Hong Kong, Malaysia, China, and Mexico. Contract manufacturers totally divorced spatially and managerially from the companies whose products they produce accounted in 2002 for an estimated 15% to 20% of the output of all electronics hardware. Most often, the same factory produces similar or identical products under a number of different brand names. Through manufacturing transfers of technology and outsourcing, therefore, high-tech activities are spread to newly industrializing countries. This globalization through areal transfer and dispersion represents the third impact of high-tech activities on world economic geographic patterns already undergoing significant but variable change in response to the new technologies.

# Tertiary and Beyond

*Primary* activities, you will recall, gather, extract, or grow things. *Secondary* industries give *form utility* to the products of primary industry through manufacturing and processing efforts. A major and growing segment of both domestic and international economic activity, however, involves *services* rather than the production of commodities. These *tertiary* activities consist of those business and labor specializations that provide services to the primary and secondary sectors, to the general community, and to individuals. They imply pursuits other than the actual production of tangible commodities.

As we saw earlier in this chapter, regional and national economies undergo fundamental changes in emphasis in the course of their development. Subsistence societies exclusively dependent on primary industries may progress to secondary-stage processing and manufacturing activities. In that progression, the importance of agriculture, for example, as an employer of labor or as a contributor to national income declines as that of manufacturing expands. As economic growth continues, secondary activities in their turn are replaced by service, or tertiary, functions as the main support of the economy. Advanced economies that have made that transition are often referred to as *postindustrial* because of the dominance of their service sectors and the significant decline of manufacturing as a generator of employment and national income.

Perhaps more than any other major country economy, the United States has reached postindustrial status. Its primary sector component fell from 66% of the labor force in 1850 to 3% in 2003, and the service sector rose from 18% to 81% (Figure 10.28). Of the 13 million new jobs created in the United States between 1993 and 2003, nearly all of them, after discounting for job losses in other employment sectors, occurred in services. Comparable changes are found in

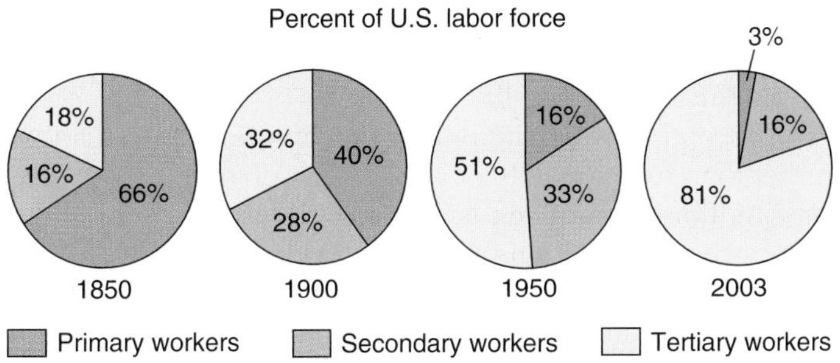


**FIGURE 10.28** The changing sectoral allocation of the U.S. labor force is a measure of the economic development of the country. Its progression from largely agricultural to postindustrial status is clearly evident.

**TABLE 10.1 Contribution of the Service Sector to Gross Domestic Product**

| Country Group | Percentage of GDP 1960 | 1980 | 2001 |
|---|---|---|---|
| Low-income | 32 | 30 | 45 |
| Middle-income | 47 | 46 | 54 |
| High-income | 54 | 59 | 70 |
| United States | 58 | 63 | 73 |
| World | | 55 | 66 |

*Source:* Data from World Bank.

other countries. At the start of the 21st century, between 65% and 80% of jobs in such developed economies as Japan, Canada, Australia, Israel, and all major Western European countries were also in the service sector; Russia and Eastern Europe averaged rather less.

The significance of tertiary activities to national economies and the contrast between more-developed and less-developed states are made clear not just by employment but also by the differential contribution of services to the gross domestic products of states. The relative importance of services displayed in Figure 10.29 shows a marked contrast between advanced and subsistence societies. The greater the service share of an economy, the greater is the integration and interdependence of that society. That share has grown over time among most regions and all national income categories because all economies have shared to some degree in world developmental growth (Table 10.1). Indeed, the expansion of the tertiary sector in modernizing East Asia, South Asia, and the Pacific was three times the world average in the 1990s. In Latin America and the Caribbean, services accounted for more than 60% of total output in 2001.

*Tertiary* and *service,* however, are broad and imprecise terms that cover a range of activities from neighborhood barber to World Bank president. The designations are equally applicable to traditional, low-order personal and retail activities and to higher-order, knowledge-based professional services performed primarily for other businesses, not for individual consumption.

Logically, the composite service category should be subdivided to distinguish between those activities answer-

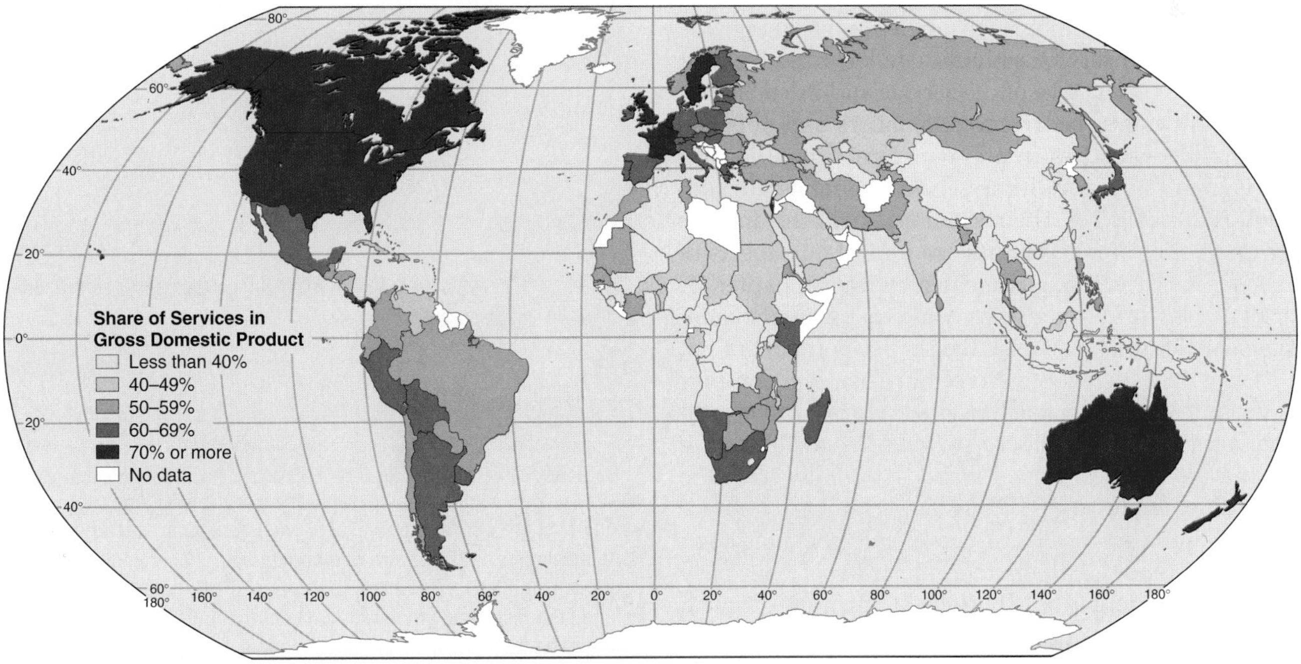


**FIGURE 10.29** Services accounted for more than two-thirds of global GDP early in the 21st century, up sharply from about 50% 20 years earlier. As the map documents, the contribution of services to individual national economies varies greatly. Source: *World Bank,* World Development Indicators.

ing to the daily living and support needs of individuals and local communities and those involving professional, administrative, or financial management tasks on regional, national, and international scales. Those different levels of activity and scope represent different locational principles and quite different roles in their contribution to domestic and world economies.

To recognize such fundamental contrasts, we can usefully restrict the term *tertiary* specifically to the lower-level activities largely related to the day-to-day needs of people and to the usual range of functions found in smaller towns and cities worldwide. We can then assign higher-level, more specialized information, research, and management activities to distinctive *quaternary* and *quinary* categories (see Figure 10.2) with quite different and distinctive characteristics and significance.

## Tertiary Services

Some essential services are concerned with the wholesaling or retailing of goods, providing what economists call *place utility* to items produced elsewhere. They fulfill the exchange function of advanced economies and provide the market transactions necessary in highly interdependent societies. In commercial economies, tertiary activities also provide vitally needed information about market demand without which economically justifiable production decisions are impossible.

Most tertiary activities, however, are concerned with personal and business services performed in shops and offices that, typically, cluster in cities large and small. The supply of those kinds of low-level services of necessity must be identical to the spatial distribution of *effective demand*—that is, wants made meaningful through purchasing power. Retail and personal service activities are localized by their markets, for the production of the service and its consumption are simultaneous occurrences. Retailers and personal service providers tend to locate, therefore, where market density is greatest and multiple service demands are concentrated (Figure 10.30). Their locational patterns and the employment support they imply are important aspects of urban economic structure and are dealt with in Chapter 11.

In all of the world's increasingly interdependent postindustrial societies, the growth of the service component not only reflects the development of increasingly complex social, economic, and administrative structures; it also indicates changes made possible by growing personal incomes or possible and necessary by alterations in family structure and individual lifestyle. For example, in subsistence economies families produce, prepare, and consume food within the household. Urbanizing industrial societies have increasing dependence on specialized farmers growing food and wholesalers and retailers selling food to households that largely prepare and consume it at home. Postindustrial America increasingly opts to purchase prepared foods in restaurants, fastfood, or carry-out establishments with accelerating growth of the tertiary foodservice workers that change demands. People are still fed, but the employment structure has changed.

Part of the growth in the tertiary component is statistical, rather than functional. In our discussion of modern industry, we saw that outsourcing is increasingly used as a device to reduce costs and enhance manufacturing and assembly efficiencies. In the same way, outsourcing of services formerly provided in-house is also characteristic of current business practice. Cleaning and maintenance of factories, shops, and offices—formerly done by the company itself as part of internal operations—now is subcontracted to specialized service providers. The job is still done, perhaps even by the same personnel, but worker status has changed from "secondary" (as employees of a manufacturing plant, for example) to "tertiary" (as employees of a service company).

Special note should be made of *tourism*—travel undertaken for the purposes of recreation rather than business. It has become not only the most important single tertiary sector activity but also the world's largest industry in jobs and total value generated. On a worldwide basis, tourism accounts for some 250 million recorded jobs and untold additional numbers in the informal economy. Altogether, 15% or more of the world's workforce is engaged in providing services to recreational travelers, and in 2001, the total economic value of tourism goods and services reached about $4 trillion, or 13% of the world's gross domestic product. In

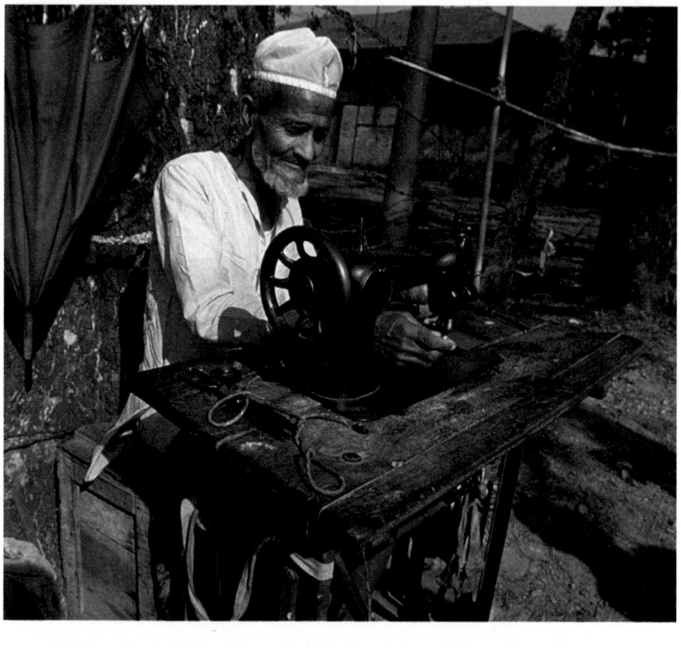

**FIGURE 10.30** Low-level services are most efficiently and effectively performed where demand and purchasing power are concentrated, as this garment repairman in a Bangladesh city marketplace demonstrates. Such informal sector employment—street vendors, odd-job handymen, open-air dispensers of such personal services as barbering, shoe shining, clothes mending, and the like—usually escapes governmental registration and is not included in official service employment totals. *© Laurence Fordyce; Eye Ubiquitous/Corbis Images.*

middle- and high-income countries, tourism supports a diversified share of domestic expenditures through transportation-related costs, roadside services, entertainment, national park visits, and the like. International tourism, on the other hand, generates new income and jobs of growing importance in developing states. Altogether, worldwide international tourist visits numbered about 1 billion annually at the start of the 21st century, more than a quarter of them destined for the less-developed low- and middle-income countries. That inbound flow produced some 8% of all foreign earnings of developing states in 2000 and—in the form of goods and services, such as meals, lodging, and transport consumed by foreign travelers—constituted 44% of their total "service exports." For half of the world's 50 poorest countries, by 2000, tourism had become the leading service export sector.

Whatever the origins of tertiary employment growth, the social and structural consequences are comparable. The process of development leads to increasing labor specialization and economic interdependence within a country. That was true during the latter part of the 20th century for all economies, as Table 10.1 attests. Carried to the postindustrial stage of advanced technology-based economies and high per capita income, the service component of both the employed labor force and the gross domestic product rises to dominance.

### Beyond Tertiary

Available statistics, unfortunately, do not always permit a clear distinction between *tertiary* service employment that is a reflection of daily lifestyle or corporate structural changes and the more-specialized, higher-level *quaternary* and *quinary* activities.

The *quaternary* sector can be seen realistically as an advanced form of services involving specialized knowledge, technical skills, communication ability, or administrative competence. These are the tasks carried on in office buildings, elementary and university classrooms, hospitals and doctors' offices, theaters, accounting and brokerage firms, and the like. With the explosive growth in the demand for and consumption of information-based services—mutual fund managers, tax consultants, Internet and software developers, statisticians, and more—the quaternary sector in the most highly developed economies has unmistakably replaced all primary and secondary employment as the basis for economic growth. In fact, over half of all workers in rich economies are in the "knowledge sector" alone—in the production, storage, retrieval, or distribution of information.

Quaternary activities performed for other business organizations often embody the "externalization" of specialized services similar to the outsourcing of low-level tertiary functions. The distinction between them lies in the fact that knowledge and skill-based free-standing quaternary service establishments can be spatially divorced from their clients; they are not tied to resources, affected by the environment, or necessarily localized by market. They can realize cost reductions through serving multiple clients in highly technical areas and permit client firms to use specialized skills and efficiencies to achieve competitive advantage without the expense of adding to their own labor force.

Often, of course, when high-level personal contacts are required, the close functional association of client and service firms within a country encourages quaternary establishment locational and employment patterns similar to those of the headquarters distribution of the primary and secondary industries served. But the transportability of quaternary services also means that many quaternary activities can be spatially isolated from their client base. In the United States, at least, these combined trends have resulted both in the concentration of certain specialized services—merchant banking or bond underwriting, for example—in major metropolitan areas and in a regional diffusion of the quaternary sector to accompany the growing regional deconcentration of the client firm base. Similar locational tendencies have been noted even for the spatially more-restricted advanced economies of, for example, England and France.

Information, administration, and the "knowledge" activities in their broadest sense are dependent on communication. Their spatial dispersion, therefore, has been abetted by the underlying technological base of most quaternary activities: electronic digital processing and telecommunications transfer of data. That technology permits many "back-office" tasks to be spatially far distant from the home office locations of either the service or the client firms. Insurance claims, credit card charges, mutual fund and stock market transactions, and the like are more efficiently and economically recorded or processed in low-rent, low labor cost locations—often in suburbs or small towns and in rural states—than in the financial districts of major cities. The production and consumption of such services can be spatially separated in a way not feasible for tertiary, face-to-face activities.

Finally, there are the *quinary* activities, another separately recognized subdivision of the tertiary sector representing the special and highly paid skills of top business executives, government officials, research scientists, financial and legal consultants, and the like. These people find their place of business in major metropolitan centers, in and near major universities and research parks, at first-rank medical centers, and in cabinet and department-level offices of political capitals. Within their cities of concentration, they may be highly localized by prestigious street addresses (Park Avenue, Wall Street) or post offices (Princeton, New Jersey) or by notable "signature" office buildings (Transamerica Building, Seagram Building). Their importance in the structure of advanced economies far outweighs their numbers.

## Services in World Trade

Just as service activities have been major engines of national economic growth within the most-advanced states, so, too, have they become an increasing factor in international trade flows and economic interdependence. Between 1980 and 2000, services increased from 15% of total world trade to nearly 25%. The fastest-growing segment of that increase was in such private services as financial, brokerage, and leasing activities, which had grown to 50% of all commercial services trade by early in the 21st century. As in the domestic arena, rapid advances in information technology and electronic data transmission have been central elements in the internationalization of services as wired and wireless communication costs have been reduced to negligible levels. Many services considered nontradable even late in the 1990s are now actively traded at long distance.

Developing countries have been particular beneficiaries of the new technologies. Their exports of services—valued at $250 billion in 2001—grew at an annual 15% rate in the 1990s, twice as fast as service exports from industrial regions. The increasing tradability of services has expanded the international comparative advantage of developing states in relatively labor-intensive long-distance service activities such as mass data processing, computer software development, and the like. At the same time, they have benefited from increased access to efficient, state-of-the-art equipment and techniques transferred from advanced economies.

The concentration of computer software development around Bangalore and Hyderabad has made India a major world player in software innovation, for example, whereas elsewhere in that country increasing volumes of back-office work for Western insurance companies and airlines is being performed. Claims processing for life and health insurance firms have become concentrated in English-speaking Caribbean states to take advantage of lower wages and the availability of a large pool of educated workers there. In all such cases, the result is an acceleration in the transfer rate of technology in such expanding areas as information and telecommunications services and an increase in the rate of developing country integration in the world economy.

That integration has increasingly moved to higher levels of economic and professional services. The cost and efficiency advantages of outsourcing such skilled functions as paralegal and legal services, accountancy, medical analysis and technical services, and research and development work in a nearly unlimited range of businesses are now widely understood and appreciated. Wired and wireless transmission of data, documents, medical and technical records, charts, X rays, and the like make distant quaternary and higher-level services immediately and efficiently accessible. Further, many higher-level services are easily subdivided and performable in sequence or simultaneously in multiple locations. The well-known "follow the sun" practices of software developers who finish a day's tasks only to pass on work to colleagues elsewhere in the world are now increasingly used by professionals in many other fields. When the practice involves highly educated and talented specialists receiving developing world compensation levels, the cost attractions for developed country companies are irresistible. The necessary levels of education and technical expertise are, to an ever-greater extent, more apt to have been acquired not by expensive training in European or North American universities but, rather, through distant learning programs and professional school contacts through the Internet.

Many of the current developing country gains in international quaternary services are the result of increased foreign direct investment (FDI) in the services sector. Those flows accounted for three-fifths of all FDI at the start of the 21st century. The majority of such investment, however, is transferred within the advanced countries themselves rather than between industrial and developing states. In either case, as transnational corporations use mainframe computers around the clock for data processing, they can exploit or eliminate time zone differences between home office countries and host countries of their affiliates. Such cross-border intrafirm service transactions are not usually recorded in balance of payment or trade statistics but materially increase the volume of international services flows.

Despite the increasing share of global services trade held by developing countries, world trade—imports plus exports—in services is still overwhelmingly dominated by a very few of the most-advanced states (Table 10.2). The country and category contrasts are great, as a comparison of the "high-income" and "low-income" groups documents. At a different level, in 2001, the single small island state of Singapore had a greater share (1.6%) of world services trade than did all of sub-Saharan Africa (1.1%).

**TABLE 10.2 Shares of World Trade in Services (Exports plus Imports, 2001)**

| Country or Category | % of World |
|---|---|
| United States | 15.7 |
| Germany | 7.7 |
| United Kingdom | 6.9 |
| Japan | 5.9 |
| France | 4.9 |
| China[a] | 4.8 |
| Italy | 3.9 |
| Netherlands | 3.7 |
| Belgium/Luxembourg | 3.2 |
| Spain | 3.1 |
| Canada | 2.7 |
| Korea (Rep.) | 2.2 |
| Austria | 2.1 |
| **Total** | **66.9** |
| High-income states | 81.5 |
| Low-income states | 3.3 |
| Sub-Saharan Africa | 1.1 |
| European Union | 33.8 |

[a]Includes Hong Kong.

*Source:* Data from International Monetary Fund and World Bank.

The same cost and skill advantages that enhance the growth and service range of quaternary firms and quinary activities on the domestic scene also operate internationally. The principal banks of all the advanced countries have established foreign branches, and the world's leading banks have become major presences in the primary financial capitals. In turn, a relatively few world cities have emerged as international business and financial centers whose operations and influences are continuous and borderless, while a host of offshore banking havens have emerged to exploit gaps in regulatory controls and tax laws (Figure 10.31). Accounting firms, advertising agencies, management consulting companies,

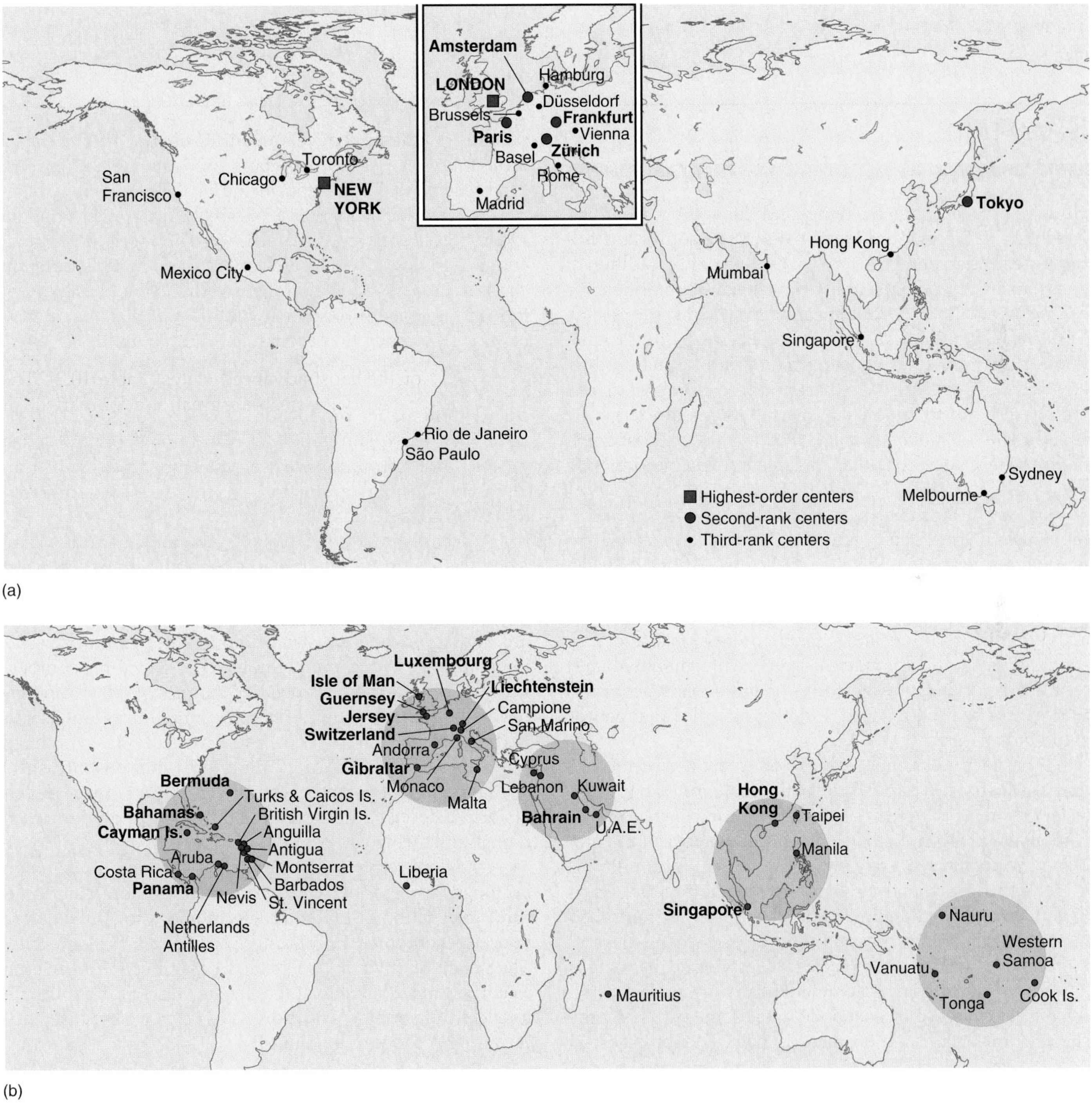


**FIGURE 10.31 The hierarchy of international financial centers,** topped by New York and London, indicates the tendency of highest-order quaternary activities to concentrate in a few world and national centers. (b) At the same time, the multiplication of off-shore locations where "furtive money" avoiding regulatory control and national taxes finds refuge suggests that dispersed convenience sites also serve the international financial community. Source: *(a & b) Peter Dicken,* Global Shift, *4th ed. New York: Guilford Press, 2003, Figs. 13.8 and 13.10.*

and similar quaternary sector establishments of North American or European origin primarily have increasingly established their international presence, with main branches located in principal business centers worldwide.

The list of tertiary, quaternary, and quinary employment is long. Its diversity and familiarity remind us of the complexity of modern life and of how far removed we are from the subsistence economies. As societies advance economically, the shares of employment and national income generated by the primary, secondary, and composite tertiary sectors continually change, and spatial patterns of human activity reflect those changes. The shift is steadily away from primary production and secondary processing and toward the trade, personal, and professional services of the tertiary sector and the information and control activities of the quaternary and quinary.

## SUMMARY

How people earn their living and how the diversified resources of the earth are used by different peoples and cultures are fundamental concerns of the locational tradition of geography. In seeking spatial and activity regularities, we can recognize three types of economic systems: *subsistence, commercial,* and *planned.* The first is concerned with production for the immediate consumption of individual producers and family members. In the second, economic decisions ideally respond to impersonal market forces and reasoned assessments of monetary gain. In the third, at least some nonmonetary social or political goals influence production decisions.

We can further classify economic activities according to the stages of production and the degree of specialization they represent. In that way, we make distinctions between *primary* activities (food and raw material production), *secondary* industries (processing and manufacturing), *tertiary* activities (distribution and general professional and personal service), and the administrative, informational, and technical specializations (*quaternary* and *quinary* activities) that mark highly advanced societies of either planned or commercial systems.

Agriculture, the most extensively practiced of the primary industries, is part of the spatial economy of both subsistence and advanced societies. In the first instance, it is responsive to the immediate consumption needs of the producer group and reflective of the environmental conditions under which it is practiced. In the second, agriculture reacts to consumer demand expressed through free or controlled markets. Its spatial expression reflects assessments of profitability and the dictates of social and economic planning.

Manufacturing is the dominant form of secondary activity, and is evidence of economic advancement beyond the subsistence level. Location theories help explain observed patterns of industrial development. Those theories are based on simplifying assumptions about fixed and variable costs of production and distribution, including the costs of raw materials, power, labor, market accessibility, and transportation. *Weberian analysis* argues that least-cost locations are optimal and are strongly or exclusively influenced by transportation charges. Less rigid location theory admits the possibility of multiple acceptable locations within a *spatial margin of profitability.* Agglomeration economies and the multiplier effect may make attractive locations not otherwise predicted for individual firms, while comparative advantage may influence entrepreneurs' production decisions. Location concepts developed to explain industrial distributions under Fordist production constraints have been challenged, as new Just-in-Time and flexible production systems introduce different locational considerations.

A dominant share of global secondary manufacturing activity is found within a relatively small number of major industrial concentrations and multinational regions. The most-advanced countries within those regions, however, are undergoing deindustrialization as newly industrializing countries with more favorable cost structures compete for markets. In the advanced economies, tertiary, quaternary, and quinary activities become more important as secondary sector employment and share of gross national product declines. The new high-tech and postindustrial spatial patterns are not necessarily identical to those developed in response to theoretical and practical determinants of manufacturing success.

One final reminder is needed: an economy in isolation no longer exists. The world pattern of economic and cultural integration is too complete to allow totally separate national economies. Events affecting one ramify to affect all. The potato blight in an isolated corner of Europe more than a century and a half ago still holds its message. Despite differences in language, culture, or ideology, we are inseparably a single people economically.

# Key Words

agglomeration 368
agriculture 346
aquaculture 363
commercial economy 344
comparative advantage 371
economic geography 342
extensive commercial agriculture 357
extensive subsistence agriculture 348
external economies 370
extractive industries 361
Fordism 369
gathering industries 361
Green Revolution 353
infrastructure 370
intensive commercial agriculture 357
intensive subsistence agriculture 348
least-cost theory 368
maximum sustainable yield 361
nomadic herding 348
outsourcing 372
planned economy 344
plantation 360
primary activity 343
quaternary activity 344
quinary activity 344
secondary activity 344
shifting cultivation 349
subsistence economy 344
tertiary activity 344
tragedy of the commons 362
transnational corporation (TNC) 373
variable costs 368
von Thünen rings 356

# For Review & Consideration

1. What are the distinguishing characteristics of the economic systems labeled *subsistence, commercial,* and *planned?* Are they mutually exclusive, or can they coexist within a single political unit?
2. How is *intensive subsistence* agriculture distinguished from *extensive subsistence* cropping? Why, in your opinion, have such different land use forms developed in separate areas of the warm, moist tropics?
3. Briefly summarize the assumptions and dictates of von Thünen's agricultural model. How might the land use patterns predicted by the model be altered by an increase in the market price of a single crop? A decrease in the transportation costs of one crop but not of all crops?
4. What economic or ecological problems can you cite that do or might affect the *gathering industries* of forestry and fishing? What is *maximum sustainable yield?* Is that concept related to the problems you discerned?
5. What simplifying assumptions did Weber make in his theory of plant location? In what ways does the Weberian search for the *least-cost location* differ from the recognition of the *spatial margin of profitability?*
6. How, in your opinion, do the concepts or practices of *comparative advantage* and *outsourcing* affect the industrial structure of advanced and developing countries?
7. As high-tech industries and *quaternary* and *quinary* employment become more important in the economic structure of advanced countries, what consequences for economic geographic patterns do you anticipate? Explain.
8 What have been the motivations and rewards of the outsourcing of quaternary services by developed country firms? In what ways, do you think, has that outsourcing favorably or unfavorably affected the home country economies of the outsourcing firms?

# Selected References

**WEBSITES**

The World Wide Web has a tremendous variety of sites pertaining to geography. Websites relevant to the subject matter of this chapter appear in the "Web Links" section of the Online Learning Center associated with this book. Access it at www.mhhe.com/getis10e/.

Atkins, Peter J., and Ian Bowler. *Food in Society: Economy, Culture, Geography.* New York: Oxford University Press, 2001.

Berry, Brian J. L., Edgar C. Conkling, and D. Michael Ray. *The Global Economy in Transition.* Upper Saddle River, N.J.: Prentice Hall, 1997.

Chang, Claudia, and Harold A. Kostner. *Pastoralists at the Periphery: Herders in a Capitalist World.* Tucson: University of Arizona Press, 1994.

Chapman, Keith, and David Walker. *Industrial Location.* 2d ed. Cambridge, Mass.: Basil Blackwell, 1991.

Clark, Gordon L., Maryann P. Feldman, and Meric S. Gertler, eds. *The Oxford Handbook of Economic Geography.* New York: Oxford University Press, 2001.

Daniels, Peter, John Bryson, and Barney Warf. *Service Industries in the New Economy.* New York: Routledge, 2003.

Dicken, Peter. *Global Shift: Transforming the World Economy.* 4th ed. New York: Guilford Press, 2003.

Grigg, David. *An Introduction to Agricultural Geography.* 2d ed. New York: Routledge, 1995.

Hanink, Dean M. *Principles and Applications of Economic Geography.* New York: John Wiley & Sons, 1997.

Hardin, Garrett. "The Tragedy of the Commons." *Science* 162 (1968): 1243–48.

Hudman, Lloyd, and Richard Jackson. *Geography of Travel and Tourism.* 4th ed. Albany, N.Y.: Delmar Learning, 2002.

International Bank for Reconstruction and Development/The World Bank. *World Development Indicators.* Washington, D.C.: The World Bank, annual.

Knox, Paul, John Agnew, and Linda McCarthy. *The Geography of the World Economy.* 4th rev. ed. New York: Oxford University Press, 2003.

Mastny, Lisa. *Traveling Light: New Paths for International Tourism.* Worldwatch Paper 159. Washington, D.C.: Worldwatch Institute, 2001.

McGinn, Anne Platt. *Rocking the Boat: Conserving Fisheries and Protecting Jobs.* Worldwatch Paper 142. Washington, D.C.: Worldwatch Institute, 1998.

Peters, William J., and Leon F. Neuenschwander. *Slash and Burn: Farming in the Third World Forest.* Moscow: University of Idaho Press, 1988.

Smil, Vaclav. *Feeding the World: A Challenge for the Twenty-first Century.* Cambridge, Mass.: MIT Press, 2000.

Stutz, Frederick P., and Barney Warf. *The World Economy: Resources, Location, Trade and Development.* 4th ed. Upper Saddle River, N.J.: Prentice-Hall, 2005.

Wheeler, James O., Peter Muller, Grant Thrall, and Timothy Fik. *Economic Geography.* 3d ed. New York: John Wiley & Sons, 1998.

Young, John E. *Mining the Earth.* Worldwatch Paper 109. Washington, D.C.: Worldwatch Institute, 1992.

Cairo was a world-class city in the 14th century. Situated at the crossroads of Africa, Asia, and Europe, it dominated trade on the Mediterranean Sea. By the early 1300s, it had a population of half a million or more, with 10- to 14-story buildings crowding the city center. Cairo's chronicler of the period, Taqui-al-Din-Al-Maqrizi, recorded the construction of a huge building with shops on the first floors and apartments housing 4000 people. One Florentine visitor estimated that more people lived on a single Cairo street than in all of Florence. Travelers from all over Europe and Asia made their way through Cairo, and the shipping at the port of Bulaq outdistanced those of Venice and Genoa combined. There were more than 12,000 shops, some of which specialized in luxury goods from all over the world—Siberian sable, chain mail, musical instruments, cloth, and songbirds. Travelers marveled at the size, density, and variety of Cairo, comparing it favorably with cities like Venice, Paris, and Baghdad.

Today, Cairo is a vast, sprawling metropolis representative of several recent trends in urbanization in developing countries where population growth far outstrips economic development. The population of Egypt was measured at 35 million in 1970. In recent decades, better health care has led to a big drop in infant mortality, and the result is a country of 70 million people. Some 9 million reside in the Cairo greater metropolitan area. Greater Cairo now extends more than 450 square kilometers (175 sq mi) and has a population density of about 30,900 per square kilometer (80,000 per sq mi). Population trends are expected to continue, with a predicted population of around 11.5 million by 2015.

A steady stream of migrants arrives in Cairo daily because it is the place where people think opportunities are available, where life will be better and brighter than in the crowded countryside. Cairo is the symbol of modern Egypt, a place where young people are willing to undergo deprivation for the chance to "make it." But real opportunities continue to be scarce. The poor, of whom there are millions, crowd into row after row of apartment houses, many of them poorly contructed. Tens of thousands more live in rooftop sheds or small boats on the Nile; a half million find shelter in the Northern and Southern Cemeteries—known as the Cities of the Dead—on Cairo's eastern edge. On occasion, buildings collapse; the earthquake of October 12, 1992, measured 5.9 on the Richter scale but caused enormous damage, leveling thousands of buildings. Yet the city continues to grow, spreading onto valued farmland, thus decreasing food availability for the country's increasing population.

One's first impression when arriving in central Cairo is of opulence, a stark contrast to what lies outside the center. High-rise apartments, regional headquarters buildings of multinational corporations, and modern hotels stand amid clogged streets, symbols of the new Egypt (Figure 11.1). The well-to-do can eat at McDonald's, Pizza Hut, Taco Bell, or Chili's; new suburban developments and exclusive residential communities create enclaves for the wealthy. The plush apartments and expensive cars, however, are only a short distance from the slums that are home to masses of underemployed people, perhaps as much as 20% of Cairo's population. The contrasts evident in Cairo between wealth and poverty, between modern structures and slums, between space and overcrowding, are profound. Like a number of large cities, Cairo has witnessed an urban explosion, one that sees an increasing proportion of the world's population housed in cities without the structures to support them all.

Because of its rapid growth, Cairo has not been able to plan and create an adequate infrastructure for its inhabitants. The 1992 earthquake prompted the city's adoption of its first development plan. Traffic is always a problem in Cairo, despite recent improvements, such as a ring road around the city and the construction of a metro. Vehicles idle in traffic, consuming large amounts of fuel; 10% of Cairenes suffer respiratory illnesses due to air pollution. In fact, Cairo's air pollution is worse than that of Mexico City, long thought to be the world's worst. The traffic and the crowding contribute to noise pollution, in which noise levels often reach 80 decibels or more. Water pollution both in the Nile and in treated drinking water shows dangerous levels of lead and cadmium, whose health effects will be borne for years to come. These problems of size, population, poverty, planning, and infrastructure face all cities to one degree or another and are some of the topics we consider in this chapter.

**FIGURE 11.1 Cairo, Egypt.** The population growth in the greater metropolitan area—from some 3 million in 1970 to 9 million today—has been mirrored in many developing countries. The rapid growth of urban areas brings with it housing shortages, inadequate transportation and other infrastructure development, unemployment, poverty, and environmental deterioration. *© Scott Gilchrist/Masterfile.*

# AN URBANIZING WORLD

Figure 11.2 shows that the growth of major metropolitan areas has been astounding since 1900. In that year, 13 cities had a population of more than 1 million people; in 2000, there were 375 such cities. By 2015, there are projected to be 564 cities with populations over 1 million.

Nineteen metropolises had populations of more than 10 million people in 2000 (Table 11.1 and Figure 11.3). In 1900, none were of that size. The United Nations calls these *megacities.* Of course, as we saw in Chapter 6, it follows that, since the world's population has greatly increased, we would expect the urban component to have increased. Urbanization and metropolitanization have increased more rapidly than the growth of total population, however. In the year 1800, only 3% of the world's population lived in cities; now, the figure is about 45%. The amount of urban growth differs from continent to continent and from country to country (Figure 11.4), but all countries have one thing in common: the proportion of the people living in cities is rising.

Table 11.2 shows world urban population by region. Note that the most industrialized parts of the world, North America and Western Europe, are the most urbanized in terms of percentage of people living in cities, whereas Asia and Africa have lower proportions of urban population. Industrialization fosters urbanization, but in developing countries, urbanization has resulted only partly from industrialization. People flock to the cities, seeking a better life than they can find in rural areas, but they often do not find it. The cities of sub-Saharan Africa are growing at a rapid rate, largely due to rural-to-urban migration. However, the urban growth is beyond the capability of the economic system to create employment, housing, and social services. Shantytowns and squatter settlements, in addition to unemployment and underemployment, are characteristic of cities such as Lagos, Nigeria, and Dakar, Senegal.

It is interesting to note in Table 11.2 that China, Southeast Asia (Vietnam, Indonesia, etc.), and South Asia (India, Pakistan, and Bangladesh) have relatively low proportions of people in urban regions. Their absolute number of people in urban areas, however, is among the highest in the world. Given the huge populations in Asia, and the relatively heavy emphasis on agriculture (excluding Japan and Korea), it sometimes escapes us that many large cities exist throughout parts of the world where most people are still engaged in subsistence agriculture.

In this chapter, our first objective is to consider the major factors responsible for the size and location of urban areas. The second goal is to identify the nature of land use patterns within those areas. Third, we attempt to differentiate cities around the world by a review of the factors that help explain their special nature.

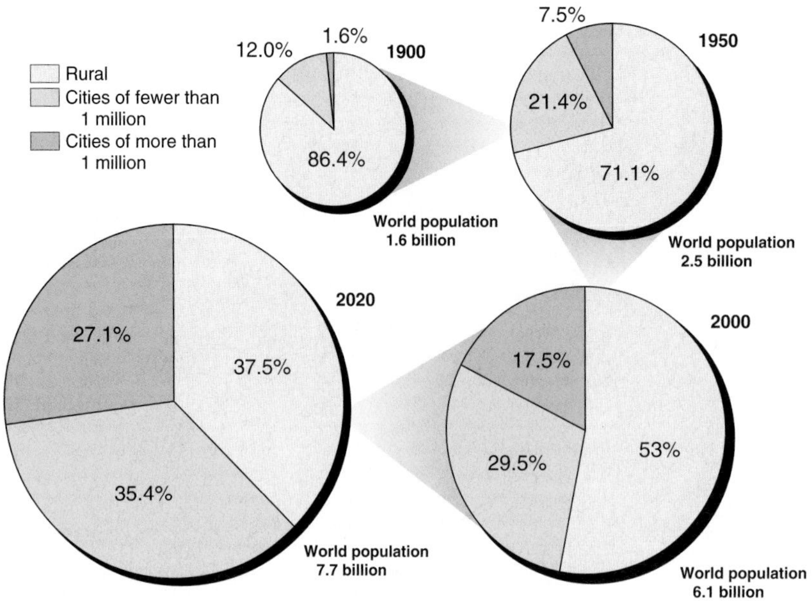


**FIGURE 11.2 Trends in world urbanization.** Note the steady decline in the proportion of people living in rural areas. The United Nations predicts that virtually all of the population growth during 2000–2020 will be concentrated in the urban areas of the world. Sources: *Estimates and projections from Population Reference Bureau and other sources.*

**TABLE 11.1 | Cities with 10 Million or More Inhabitants (Megacities), 2000 and 2015 (Population in Millions)**

| Rank | City in 2000 | Population | Rank | City in 2015 | Projected Population |
|---|---|---|---|---|---|
| 1 | Tokyo, Japan | 26.4 | 1 | Tokyo, Japan | 27.2 |
| 2 | Mexico City, Mexico | 18.1 | 2 | Dhaka, Bangladesh | 22.8 |
| 3 | Mumbai (Bombay), India | 18.1 | 3 | Mumbai (Bombay), India | 22.6 |
| 4 | São Paulo, Brazil | 17.8 | 4 | São Paulo, Brazil | 21.2 |
| 5 | Shanghai, China | 17.0 | 5 | Delhi, India | 20.9 |
| 6 | New York, United States | 16.6 | 6 | Mexico City, Mexico | 20.4 |
| 7 | Lagos, Nigeria | 13.4 | 7 | New York, United States | 17.9 |
| 8 | Los Angeles, United States | 13.1 | 8 | Jakarta, Indonesia | 17.3 |
| 9 | Kolkata (Calcutta), India | 12.9 | 9 | Kolkata (Calcutta), India | 16.7 |
| 10 | Buenos Aires, Argentina | 12.6 | 10 | Karachi, Pakistan | 16.2 |
| 11 | Dhaka, Bangladesh | 12.3 | 11 | Lagos, Nigeria | 16.0 |
| 12 | Karachi, Pakistan | 11.8 | 12 | Los Angeles, United States | 14.5 |
| 13 | Delhi, India | 11.7 | 13 | Shanghai, China | 13.6 |
| 14 | Jakarta, Indonesia | 11.0 | 14 | Buenos Aires, Argentina | 13.2 |
| 15 | Osaka, Japan | 11.0 | 15 | Manila, Philippines | 12.6 |
| 16 | Manila, Philippines | 10.9 | 16 | Beijing, China | 11.7 |
| 17 | Beijing, China | 10.8 | 17 | Rio de Janeiro, Brazil | 11.5 |
| 18 | Rio de Janeiro, Brazil | 10.6 | 18 | Cairo, Egypt | 11.5 |
| 19 | Cairo, Egypt | 10.6 | 19 | Istanbul, Turkey | 11.4 |
| | | | 20 | Osaka, Japan | 11.0 |
| | | | 21 | Tianjin, China | 10.3 |

*Source:* United Nations Population Division, 2000.

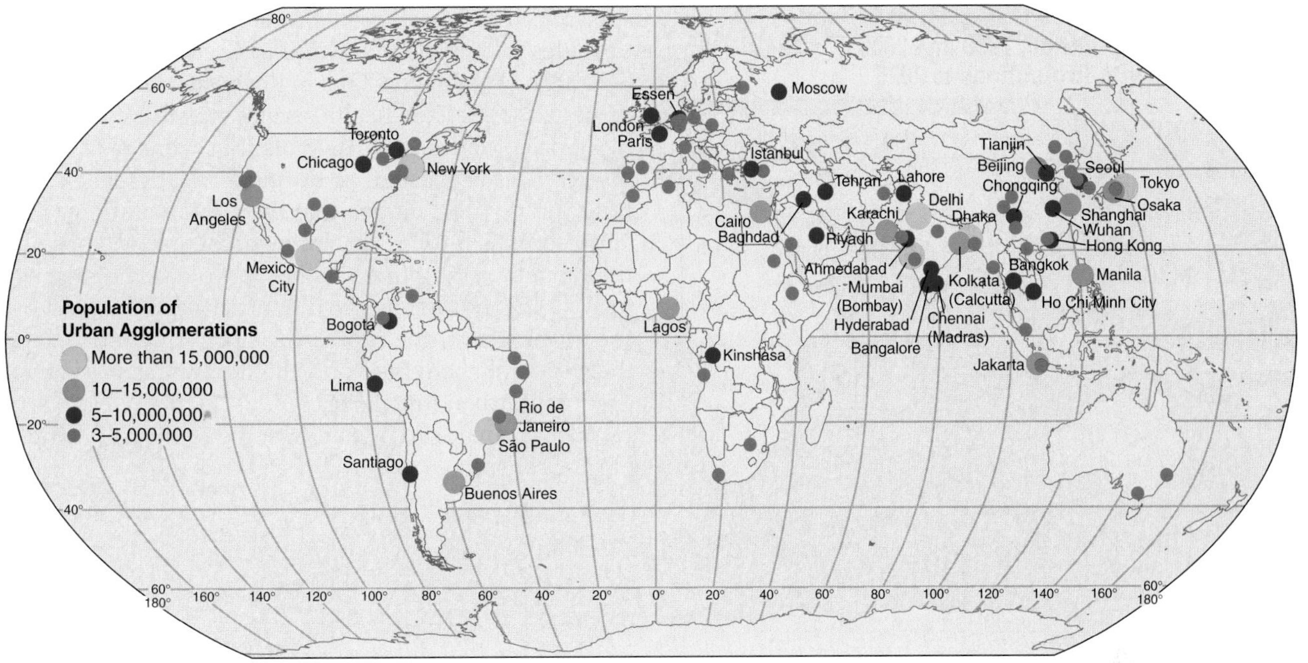


**FIGURE 11.3 Metropolitan areas of 3 million or more in 2005.** Only metropolitan areas with a population of 5 million or more are named. Massive urbanized districts are no longer characteristic only of the industrialized, developed countries. Note the clusters of large metropolitan areas in developing countries, such as Rio de Janeiro–São Paulo in Brazil, Beijing-Tianjin in China, and the Mumbai (Bombay) region of India. Source: *Population projections from United Nations Populations Division.*

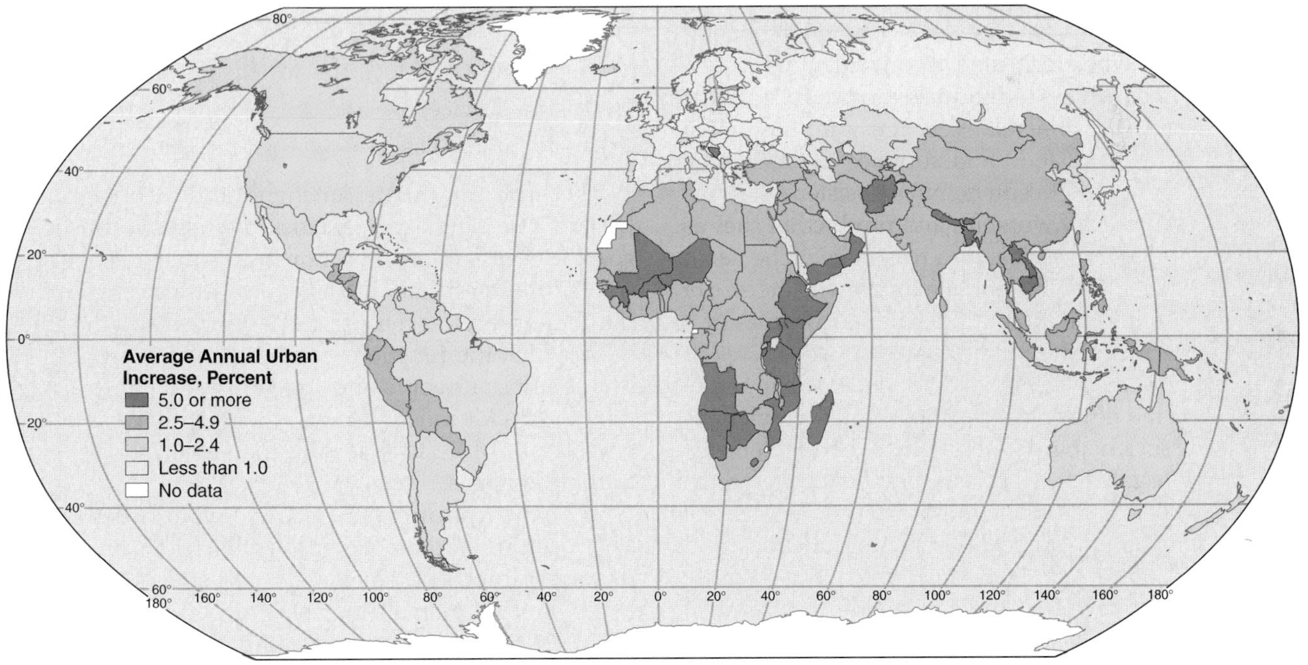


**FIGURE 11.4 Average annual urban population growth rates, 1995–2000.** In general, developing countries show the highest percentage increases in their urban populations, and the already highly urbanized and industrialized countries have the lowest—less than 1% per year in most of Europe. Demographers predict that population increase in cities in developing countries will be the distinguishing demographic trend of the 21st century, with the most explosive growth expected in Africa and Asia. An urban growth rate of 5% means that a city's population will double in just 14 years. Source: *Data from United Nations Population Division.*

**TABLE 11.2 Estimated Urban Share of Total Population, 1950 and 2000, with Projections to 2025**

| Region | 1950 | 2000 | 2025 |
|---|---|---|---|
| North America | 64% | 75% | 85% |
| Europe | 56 | 73 | 80 |
| Russia | 45 | 73 | 82 |
| East Asia (except China) | 43 | 75 | 87 |
| China | 12 | 31 | 55 |
| Southeast Asia | — | 36 | 49 |
| South Asia | 15 | 29 | 49 |
| Latin America | 41 | 74 | 84 |
| Africa | 15 | 33 | 53 |
| Oceania | 61 | 70 | 78 |
| World | 29 | 45 | 61 |

*Source:* United Nations.

# The Functions of Urban Areas

People gather together to form couples, families, groups, organizations, towns, and so forth. This desire to be near one another, however, is more than just a function of the need to socialize. Our human support systems are based on the flow of information, goods and services, and cooperation among people who are located at convenient places relative to one another. Unless individuals can produce all that they need themselves, and relatively few can, they must depend on shipments of food and supplies to their home place or convenient outlet centers. Nonsubsistence groups establish stores, places of worship, and production sites as close to their homes as possible and reasonable. The result is the establishment of towns. These may grow to the size of a Tokyo metropolitan area (about 26 million people) or a Mexico City area (about 18 million).

Whether they are villages, towns, or cities, urban settlements exist for the efficient performance of the functions required by the society that creates them. Urban settlements are points of specialized activities. They have three main functions: (1) central place functions, or providing general services for a surrounding area; (2) transport functions; and (3) special functions, or singular functions not necessarily geared to the local area. All towns provide the first two functions, but not necessarily the third. Detroit, for example, grew to a population of 1,849,568 at its height in 1950 as it provided a point of general merchandise services, convenient transportation on the Great Lakes and the web of railroad networks crisscrossing the Midwest, and a specialized function of manufacturing automobiles. Now, there is only one automobile assembly plant in Detroit, but the city's metropolitan area continues to grow. Other examples of specialized functions are those of Washington, D.C., which provides government services to the whole of the United States, and the Mayo Clinic in Rochester, Minnesota, which provides specialized health services to all who seek them.

Ancient cities grew in spots that were easy to defend, but they also reflected, as do modern cities, the saving of time, energy, and money that the agglomeration of people and activities implies. The more accessible the producer to the consumer, the worker to the workplace, the citizen to the town hall, the worshiper to the church, or the lawyer or doctor to the client, the more efficient is the performance of their separate activities, and the more effective is the integration of urban functions.

Urban areas provide all or some of the following types of functions:

- retailing;
- wholesaling;
- manufacturing;
- business service;
- entertainment;
- political and official administration;
- military defensive needs;
- social and religious service;
- public service, including sanitation and police;
- educational service;
- transportation and communication service;
- meeting place activity;
- recreation;
- visitor service;
- residential areas.

Because all urban functions and people cannot be located at a single point, cities themselves must take up space. Because interconnection is essential, the nature of the transportation system has an enormous bearing on the total number of services that can be performed and the efficiency with which the functions can be carried out (Figure 11.5). The totality of people and functions of a city constitutes a distinctive cultural landscape whose similarities and differences from place to place are the subjects for urban geographic analysis.

Urban areas are not of a single type, structure, or size. Their common characteristic is that they are nucleated, nonagricultural settlements. At one end of the size scale, urban areas are small towns with perhaps a single main street of shops; at the opposite end, they are complex, multifunctional metropolitan areas or megacities (Figure 11.6). The word *urban* is often used to describe such places as a town, city, suburb, or metropolitan area, but it is a general term, not used to specify a particular type of settlement. Not everyone uses common terms the same way. What a resident of rural Vermont or West Virginia calls a city may not at all be afforded that name and status by an inhabitant of

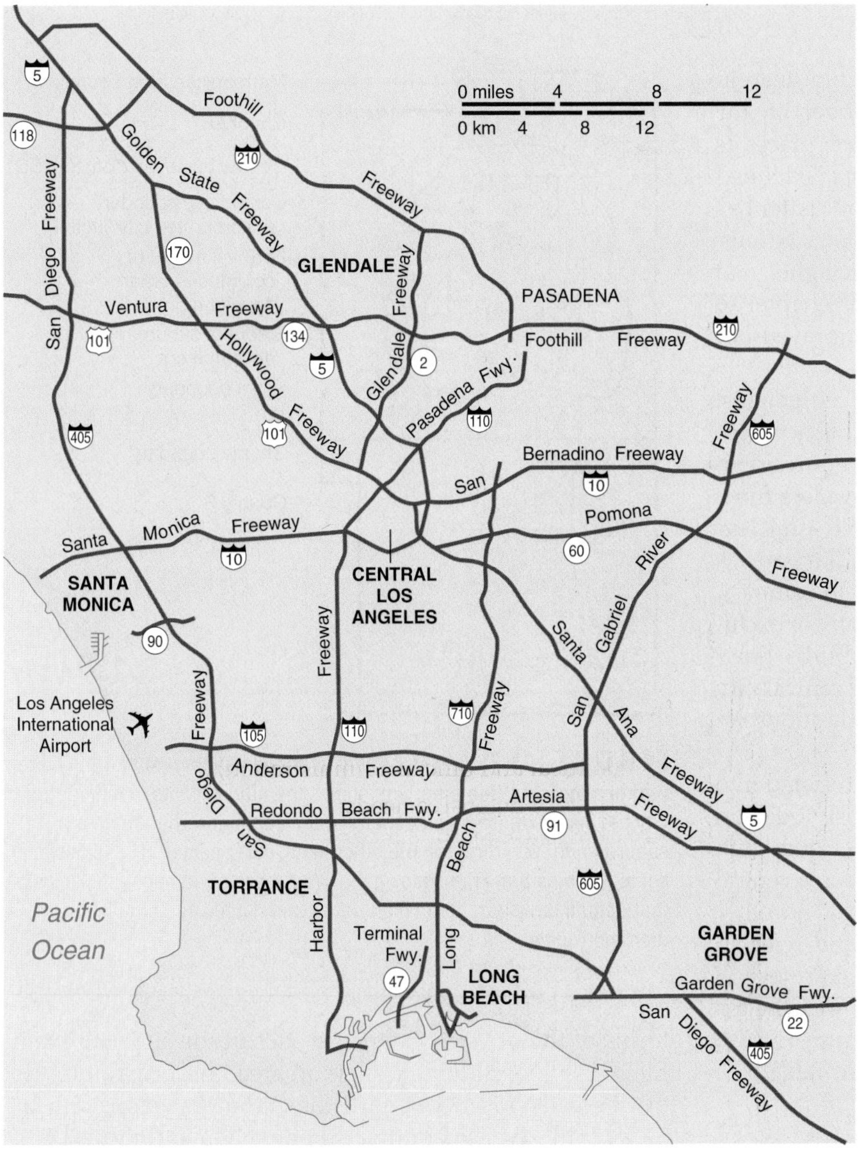


**FIGURE 11.5 The Los Angeles freeway system.** This is the world's most extensive system of high-speed highways for a metropolitan area. It fostered the growth of the Los Angeles metropolitan area, but in recent years the system has become congested.

(a) (b)

**FIGURE 11.6** The differences in size, density, and land use complexity are immediately apparent between (a) New York City and (b) a small town (Amityville, NY). One is a city, one is a town, but both are urban areas. *(a) © Comstock Royalty Free/Getty Images. (b) © Dennis O'Clair/Getty Images.*

California or New Jersey. In addition, one should keep in mind that the term *urban* differs the world over: in the United States, the Census Bureau describes *urban* places as having 2500 or more inhabitants; in Greece *urban* is defined as municipalities in which the largest population center has 10,000 or more inhabitants; in Nicaragua it denotes administrative centers of localities with streets, electric lights, and 1000 or more inhabitants. It is necessary in this chapter to agree on the meanings of terms commonly employed but interpreted in different ways.

The words **city** and **town** denote nucleated settlements, multifunctional in character, including an established central business district and both residential and nonresidential land uses. Towns are smaller in size and have less functional complexity than cities, but they still have a nuclear business concentration. **Suburb** denotes a subsidiary area, a functionally specialized segment of a large urban complex, dependent on an urban area. It may be dominantly or exclusively residential, industrial, or commercial. Suburbs, however, can be independent political entities. The **central city** is the part of the urban area contained within the suburban ring; it usually has official boundaries.

Some or all of these urban types may be associated into larger units. An **urbanized area** is a continuously built-up landscape defined by building and population densities with no reference to political boundaries. It may contain a central city and many contiguous cities, towns, suburbs, and other urban tracts. A **metropolitan area,** on the other hand, refers to a large-scale *functional* entity, perhaps containing several urbanized areas, discontinuously built up but nonetheless operating as an integrated economic whole (Figure 11.7). A list of the large U.S. metropolitan areas is given in Table 11.3.

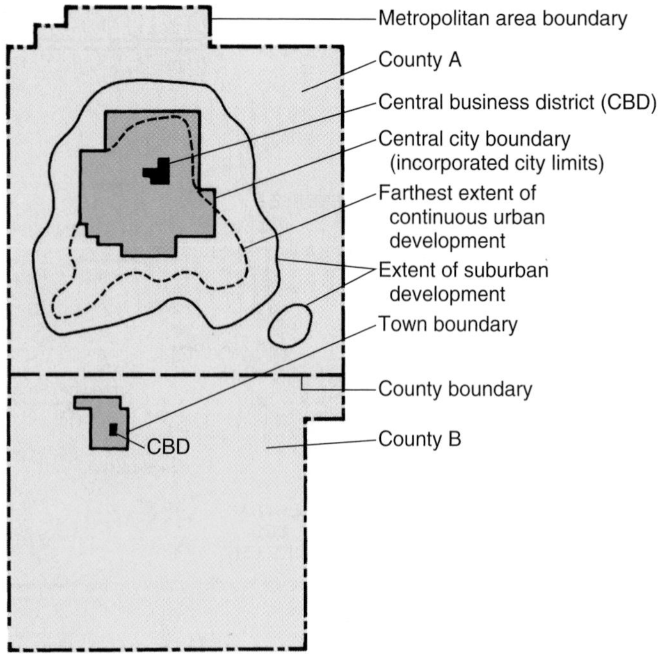


**FIGURE 11.7 A hypothetical spatial arrangement of urban units within a metropolitan area.** Sometimes the official limits of the central city are very extensive and contain areas commonly thought of as suburban or even rural. On the other hand, older eastern U.S. cities and some, such as San Francisco, in the West more often have official limits that contain only part of the high-density land uses and populations of their metropolitan areas.

# THE LOCATION OF URBAN SETTLEMENTS

Urban centers are functionally connected to other cities and rural areas. In fact, the reason for the existence of an urban center is to provide services not only for itself but also for others outside of it. The urban area is a consumer of food, a processor of materials, and an accumulator and a dispenser of goods and services, but it must rely on outside areas for its supplies and as a market for its products and activities.

In order to adequately perform the tasks that support it and to add new functions as demanded by the larger economy, the urban unit must be efficiently located. Efficiency may derive from its centrality, from the physical characteristics of its site, or from its location relative to the resources, the productive regions, and a transportation network connecting it to its markets.

In discussing urban settlement location, geographers frequently differentiate between site and situation, concepts already introduced in Chapter 1. You will recall that **site** refers to the exact location of a settlement and can be described either in terms of latitude and longitude or in terms of the physical characteristics of the site. For example, the site of Philadelphia is an area bordering and west of the Delaware River north of the intersection with the Schuylkill River in southeast Pennsylvania (Figure 11.8).

The description can be more or less exhaustive, depending on the purpose it is meant to serve. In the Philadelphia case, the fact that the city is partly on the Atlantic coastal plain, partly in the piedmont (foothills), and is served by a navigable river is important if one is interested in the development of the city during the Industrial Revolution. As Figure 11.9 suggests, water transportation was an important localizing factor when the major American cities were established.

Classifications of cities according to site characteristics have been proposed, recognizing special placement circumstances. These include *break-of-bulk* locations, such as river crossing points where cargoes and people must interrupt a journey; *head-of-navigation* or *bay head* locations where the limits of water transportation are reached; and *railhead* locations where a railroad ends. In Europe, security and defense—island locations or elevated sites—were considerations in earlier settlement locations.

Whereas *site* suggests absolute location, **situation** indicates relative location; it places a settlement in relation to the physical and cultural characteristics of the surrounding areas. Very often, it is important to know what kinds of possibilities and activities exist in the area near a settlement,

**TABLE 11.3 | United States Metropolitan Area Populations with More Than 1 Million, 2000**

| Rank | Metropolitan Areas (Principal Cities) | Population (in Thousands) | Rank | Metropolitan Areas (Principal Cities) | Population (in Thousands) |
|---|---|---|---|---|---|
| 1 | New York | 21,199 | 26 | Milwaukee | 1,690 |
| 2 | Los Angeles | 16,374 | 27 | Orlando | 1,645 |
| 3 | Chicago | 9,158 | 28 | Indianapolis | 1,607 |
| 4 | Washington-Baltimore | 7,608 | 29 | San Antonio | 1,592 |
| 5 | San Francisco–Oakland | 7,039 | 30 | Norfolk | 1,570 |
| 6 | Philadelphia | 6,188 | 31 | Las Vegas | 1,563 |
| 7 | Boston | 5,819 | 32 | Columbus | 1,540 |
| 8 | Detroit | 5,456 | 33 | Charlotte | 1,499 |
| 9 | Dallas–Fort Worth | 5,222 | 34 | New Orleans | 1,338 |
| 10 | Houston | 4,670 | 35 | Salt Lake City | 1,334 |
| 11 | Atlanta | 4,112 | 36 | Greensboro | 1,252 |
| 12 | Miami | 3,876 | 37 | Austin | 1,250 |
| 13 | Seattle | 3,555 | 38 | Nashville | 1,231 |
| 14 | Phoenix | 3,252 | 39 | Providence | 1,189 |
| 15 | Minneapolis–St. Paul | 2,969 | 40 | Raleigh-Durham | 1,188 |
| 16 | Cleveland | 2,946 | 41 | Hartford | 1,183 |
| 17 | San Diego | 2,814 | 42 | Buffalo | 1,170 |
| 18 | St. Louis | 2,604 | 43 | Memphis | 1,136 |
| 19 | Denver | 2,582 | 44 | West Palm Beach | 1,131 |
| 20 | Tampa | 2,396 | 45 | Jacksonville | 1,100 |
| 21 | Pittsburgh | 2,359 | 46 | Rochester | 1,098 |
| 22 | Portland | 2,265 | 47 | Grand Rapids | 1,088 |
| 23 | Cincinnati | 1,979 | 48 | Oklahoma City | 1,083 |
| 24 | Sacramento | 1,797 | 49 | Louisville | 1,025 |
| 25 | Kansas City | 1,776 | | | |

*Source:* U.S. Bureau of the Census.

such as the distribution of raw materials, market areas, agricultural regions, mountains, and oceans.

The site of central Chicago is 41°52′ N, 87°40′ W, on a lake plain. More important, however, is its situation close to the deepest penetration of the Great Lakes system into the interior of the country, astride the Great Lakes–Mississippi waterways, and near the western margin of the manufacturing belt, the northern boundary of the Corn Belt, and the southeastern reaches of a major dairy region. References to railroads, coal deposits, and ore fields would amplify its situational characteristics (Figure 11.10). As a gateway to the West from the East and vice versa, Chicago's O'Hare International Airport is one of the busiest in the United States. From this description of Chicago's situation, implications relating to market, to raw materials, and to transportation centrality can be drawn.

The site or situation that originally gave rise to an urban unit may not remain the essential ingredient for its growth and development for very long. The already existing markets, labor force, and urban facilities of a successful city may attract people and activities totally unrelated to the initial localizing forces. For instance, although coal mining had

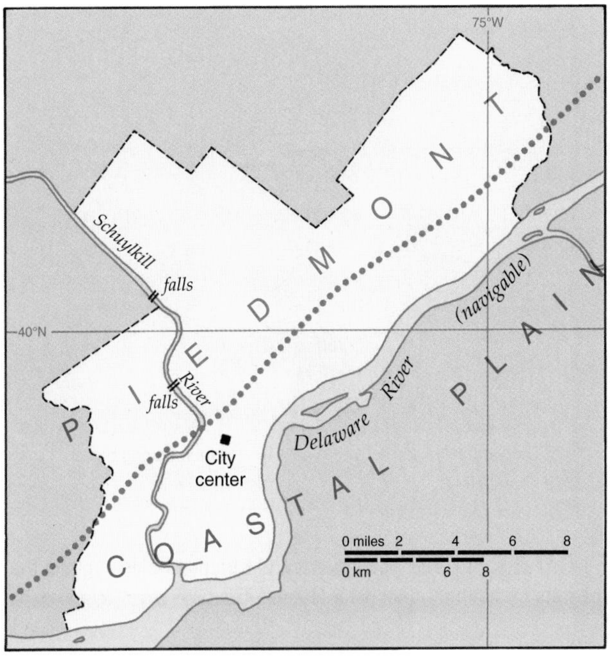


**FIGURE 11.8** The site of Philadelphia.

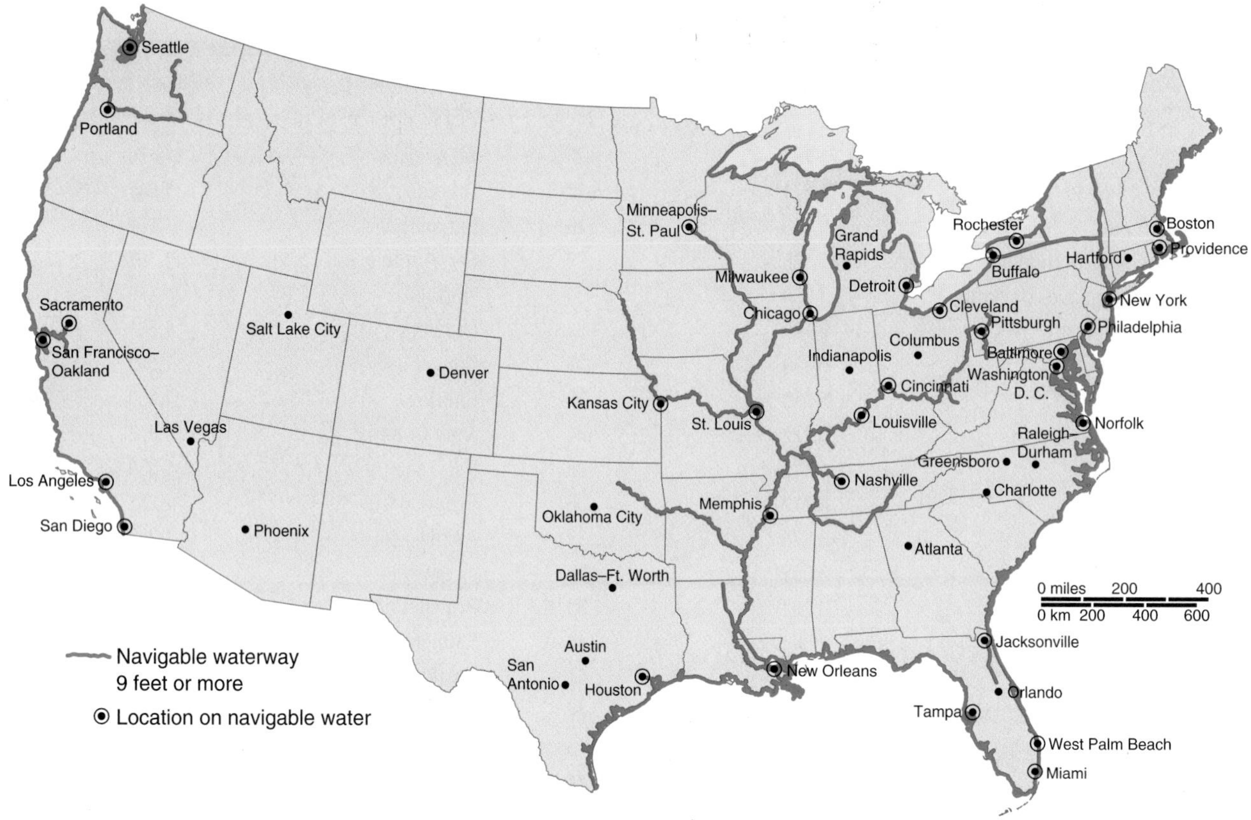


**FIGURE 11.9 Metropolitan areas in the United States with 1 million or more residents in 2000.** Notice the association of principal cities and navigable water. Before the advent of railroads in the middle of the 19th century, all major cities were associated with waterways.

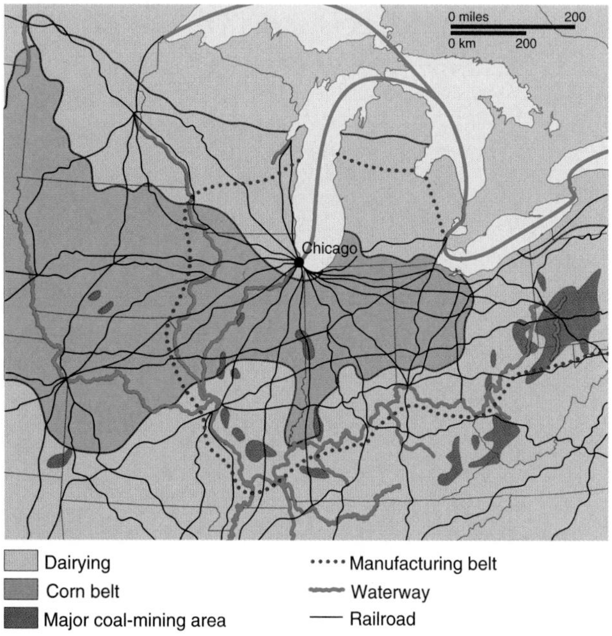


**FIGURE 11.10 The situation of Chicago** helps suggest the reasons for its functional diversity and size.

much to do with Pittsburgh's early growth, other factors are more important to its well-being today.

# The Economic Base

The **economic base** of an urban area can be described in terms of basic and nonbasic sectors; changes in one sector often effect change in the other. Understanding the growth and decline of cities hinges on comprehending the relationship between the two sectors.

Part of the employed population of an urban unit is engaged in either the production of goods or the performance of services for areas and people outside that urban area. They are workers engaged in "export" activities, whose efforts result in money flowing into the community. Collectively, they constitute the **basic sector** of the urban area's total economic structure.

Other workers support themselves by producing goods or services for residents of the urban unit itself. Their efforts, necessary to the well-being and the successful operation of the settlement, do not generate new money for it but constitute a **nonbasic sector** of its economy. These people are responsible for the internal functioning of the urban unit. They are crucial to the continued operation of its stores, offices, city government, local transit, and school systems.

The total economic structure of an urban area equals the sum of its basic and nonbasic activities. In actuality, it is difficult to classify work as belonging exclusively to one sector or the other, as most work involves financial interaction with the residents of other areas. Doctors, for example, may have mainly local patients, and thus are members of the nonbasic sector, but the moment they treat someone from outside the community, they bring new money into the city and become part of the basic sector.

Most cities, especially large ones, perform many export functions. Nonetheless, even in cities with a diversified economic base, one or a very small number of export activities tend to dominate the structure of the community and to identify its operational purpose within a system of cities. Such functional specialization permits the classification of cities into categories: manufacturing, retailing, wholesaling, transportation, government, and so on. Figure 11.11 indicates the functional specializations of some large U.S. metropolitan areas.

Assuming it were possible to divide with complete accuracy the employed population of an urban area into totally separate basic and nonbasic components, a ratio between the

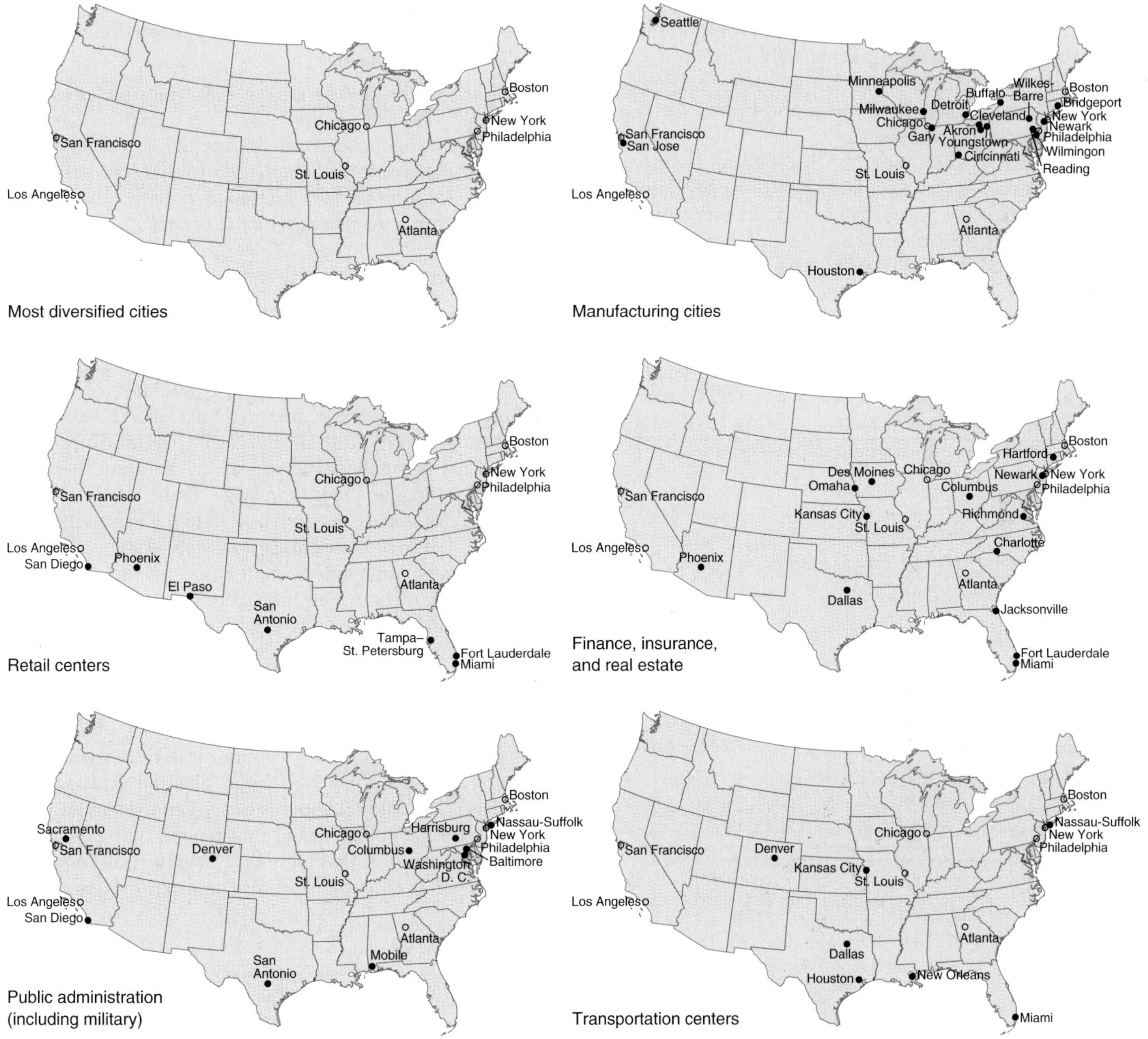


**FIGURE 11.11 Functional specialization of selected U.S. metropolitan areas.** Five categories of employment were chosen to show patterns of specialization for some U.S. metropolitan areas. In addition, the category "most diversified" includes representative examples of cities with a generally balanced employment distribution. Since their "balance" implies performance of a variety of functions, the diversified cities are included as open circles on the specialization maps. Note that the most diversified urban areas tend to be the largest.

two employment groups could be established. The basic/nonbasic ratio shown in Figure 11.12 indicates that as a settlement increases in size, the number of nonbasic personnel grows faster than the number of new basic workers. In cities with a population of 1 million, the ratio is about 2 nonbasic workers for every basic worker. This means that adding 10 new basic employees expands the labor force by 30 (10 basic, 20 nonbasic). The increase in the total population equals the added workers plus their dependents. A **multiplier effect** associated with economic growth thus exists. Recall that the term *multiplier effect* means that a city's employment and population grow with the addition of nonbasic workers and dependents as a supplement of new basic employment. The size of the effect is determined by the community's *basic/nonbasic ratio.* The more basic workers an urban area has, the more nonbasic workers are needed to support them, and the application of the multiplier effect becomes obvious.

The growth of cities may be self-generating—"circular and cumulative"—in a way related not to the development of industries that specialize in the production of material objects for export, such as automobiles and paper products, but to the attraction of what would be classified as *service* activity. Banking and legal services, a sizable market, a diversified labor force, extensive public services, and the like may generate basic and nonbasic additions to the labor force. In recent years, service industries have developed to a point at which new service activities serve older ones. For example, computer systems firms aid banks in developing more efficient computer-driven banking systems.

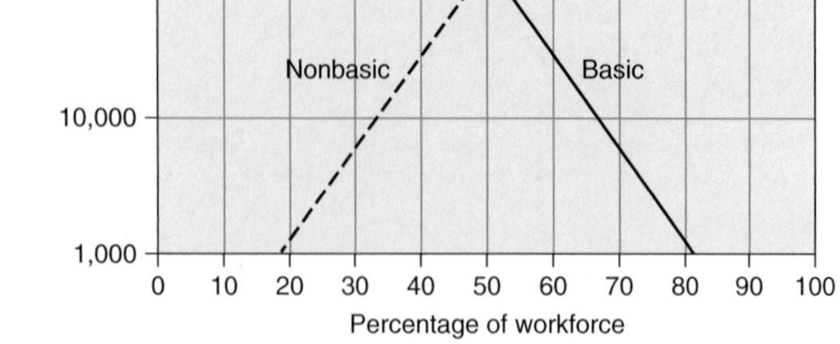

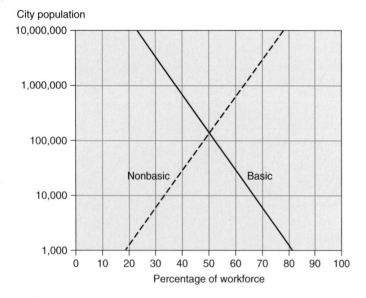

**FIGURE 11.12 A generalized representation of the proportion of the workforce engaged in basic and nonbasic activities.** As settlements become larger, a greater proportion of the workforce is employed in nonbasic activities. A city of 10 million will have about one-quarter of the workforce engaged in basic activities, whereas one with only 100,000 residents will have more than half the workforce so employed.

Just as settlements grow in size and complexity, so do they decline. When the demand for the goods and services of an urban unit falls, fewer workers are needed, and thus both the basic and the nonbasic components of a settlement system are affected. There is, however, a resistance to decline that impedes the process and delays its impact. Whereas settlements can grow rapidly as migrants respond quickly to the need for more workers, under conditions of decline, many of those who have developed roots in the community are hesitant to leave or may be financially unable to move to another locale. Figure 11.13 and Table 11.4 show that, in recent years, urban areas in the South and West of the United States have been growing, while the Northeast and the North Central regions have grown more slowly or have even declined.

## Systems of Urban Settlements

The various functions that an individual urban area performs are reflected not only in the size of that settlement but also in its location and its relationship with other urban units in the larger system of which it is part.

### The Urban Hierarchy

Perhaps the most effective way to recognize how systems of cities are organized is to consider the **urban hierarchy,** a ranking of cities based on their size and functional complexity. One can measure the numbers and kinds of services each city or metropolitan area provides. The hierarchy is then like a pyramid; a few large and complex cities are at the top, and many smaller ones are at the bottom. There are always more smaller cities than larger ones.

When a spatial dimesion is added to the hierarchy, as in Figure 11.14, it becomes clear that an areal system of metropolitan centers, large cities, small cities, and towns exists. Goods, services, communications, and people flow up and down the hierarchy. The few high-level metropolitan areas provide specialized functions for large regions, while the smaller cities serve smaller districts. The separate centers interact with the areas around them, but because cities of the same level provide roughly the same services, those of the same size tend not to serve each other unless they provide a specialized service, such as a major hospital or a research university. Thus, the settlements of a given level in the hierarchy are not independent but interrelated with communities of other levels in that hierarchy. Together, all centers at all levels in the kierarchy constitute an urban system.

### World Cities

Standing at the top of national systems of cities are a relatively few centers that may be called **world cities.** These large urban centers are control points for international production, marketing, and finance. They have been called the "control and command centers" of the global economy.

London, New York, and Tokyo are generally recognized as the three dominant world cities. They contain the highest

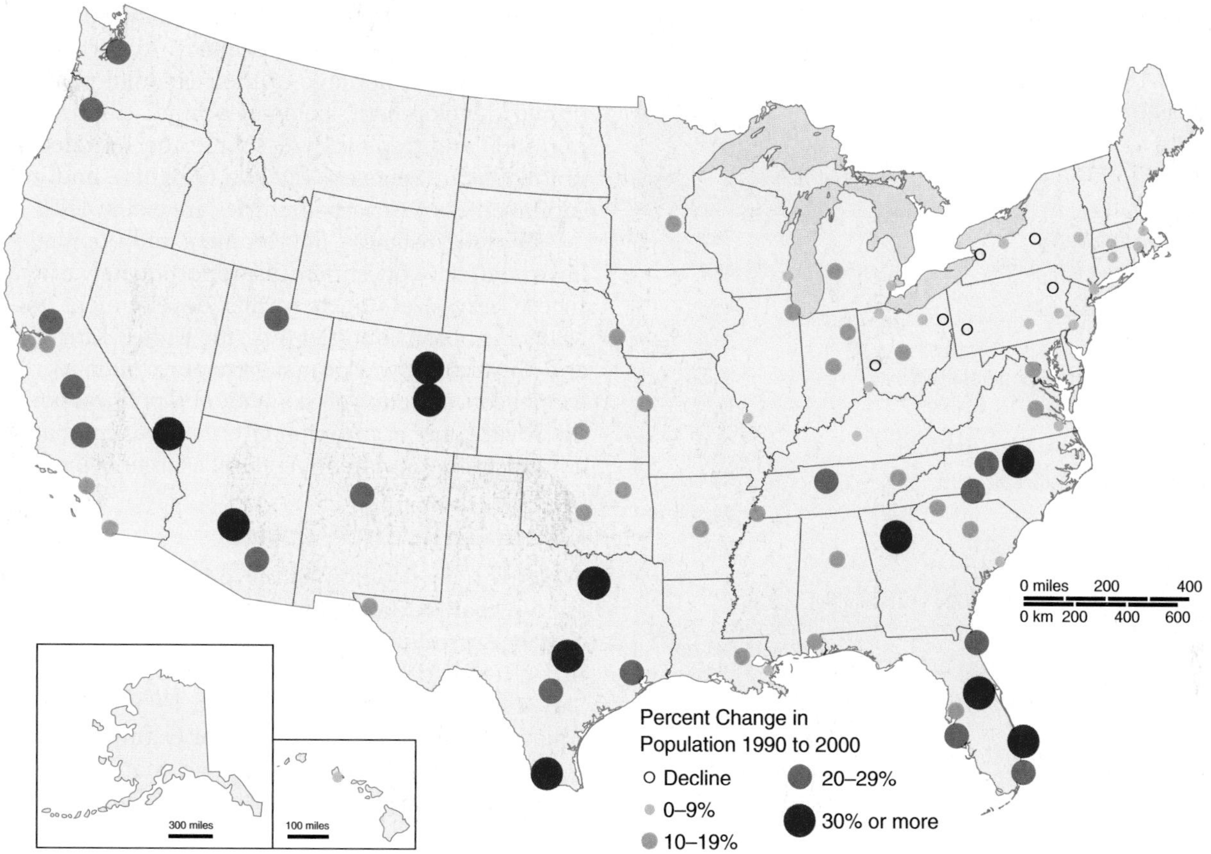


**FIGURE 11.13 The pattern of metropolitan growth and decline in the United States, 1990–2000.** Shown are metropolitan areas with 600,000 or more people in 2000. The cities of the southern and southwestern Sunbelt showed the greatest relative growth. Only modest growth, stability, or decline generally marked the Northeast and Midwest. Source: *Data from U.S. Bureau of the Census.*

number of transnational service offices and the headquarters of multinational corporations, and they dominate commerce in their respective parts of the world. Each is directly linked to a number of other world cities. All are bound together in complex networks that control the organization and management of the global system of finance, manufacturing, and trade. Figure 11.15 shows the links among the dominant centers and the suggested major and secondary world cities, which include Paris, Hong Kong, Singapore, Milan, Toronto, Chicago, and Los Angeles. These cities are all interconnected by advanced communications systems among governments, major corporations, stock and futures exchanges, securities and commodity markets, major banks, and international organizations.

Major international corporations spur world city development and dominance. The growing size and complexity of transnational corporations dictate their need to outsource central managerial functions to specialized service firms to minimize the complexity of control over dispersed operations. Those specialized service agencies—legal, accounting, financial, and so on—in their turn need to draw on the very large pools of expertise, information, and talent available only in world cities.

## Rank-Size and Primacy

The development of city systems on a global scale raises the question of the organization of city systems within regions or countries. In some countries, especially those with complex economies and a long urban history, the **rank-size rule** describes the urban system. It tells us that the *n*th largest city of a national system of cities will be $1/n$ the size of the largest city. That is, the second-largest settlement will be half the size of the largest, the tenth-biggest will be $1/10$ the size of the first ranked city, and so on. Although no national urban system exactly meets the requirement of the rank-size rule, those of Russia and the United States closely approximate it.

The rank-size ordering is less applicable to countries with developing economies and those in which the urban system is dominated by a **primate city,** one that is far more than twice the size of the second ranked city. In fact, there may be no obvious "second city" at all, for a characteristic of a primate city hierarchy is one very large city, few or no intermediate-size cities, and many subordinate smaller settlements. For example, metropolitan Seoul (with 9.8 million in 2000) contains more than 40% of the total population of South Korea. Bangkok is home to more than half of the urban residents of Thailand.

**TABLE 11.4 U.S. Metropolitan Statistical Area[a] (with more than 1 million population) Change, 1990–2000**

| a) Fastest Growing | | |
|---|---|---|
| **Rank** | **MSA** | **% Growth** |
| 1 | Las Vegas, NV–AZ | 83.3 |
| 2 | Austin–San Marcos, TX | 48.7 |
| 3 | Phoenix–Mesa, AZ | 45.3 |
| 4 | Atlanta, GA | 38.9 |
| 5 | Raleigh-Durham–Chapel Hill, NC | 38.9 |
| 6 | Orlando, FL | 34.3 |
| 7 | West Palm Beach–Boca Raton, FL | 31.0 |
| 8 | Denver–Boulder–Greeley, CO | 30.4 |
| 9 | Dallas–Fort Worth, TX | 29.3 |
| 10 | Charlotte, NC | 29.0 |

| b) Slowest Growing or Declining | | |
|---|---|---|
| **Rank** | **MSA** | **% Growth** |
| 1 | Buffalo–Niagara Falls, NY | –1.6 |
| 2 | Pittsburgh, PA | –1.5 |
| 3 | Hartford, CT | 2.2 |
| 4 | Cleveland, OH | 3.0 |
| 5 | Rochester, NY | 3.4 |
| 6 | New Orleans, LA | 4.1 |
| 7 | St. Louis, MO | 4.5 |
| 8 | Providence–Fall River, RI–MA | 4.8 |
| 9 | Philadelphia–Wilmington, PA–DE–NJ | 5.0 |
| 10 | Milwaukee, WI | 5.1 |

| c) Biggest Population Gains | | |
|---|---|---|
| **Rank** | **MSA/CMSA[b]** | **Gain (in thousands)** |
| 1 | Los Angeles–Riverside–Orange County | 1842 |
| 2 | New York–Northern New Jersey–Long Island | 1650 |
| 3 | Dallas–Fort Worth | 1185 |
| 4 | Atlanta | 1152 |
| 5 | Phoenix-Mesa | 1013 |
| 6 | Houston-Galveston-Brazoria | 938 |
| 7 | Chicago-Gary-Kenosha | 918 |
| 8 | Washington-Baltimore | 881 |
| 9 | San Francisco–Oakland–San Jose | 786 |
| 10 | Las Vegas | 711 |

[a]The U.S. Census Bureau defines a Metropolitan Statistical Area (MSA) as an economically integrated urbanized area in one or more contiguous counties.

[b]The U.S. Census Bureau defines a Consolidated Metropolitan Statistical Area (CMSA) as an area composed of more than one Metropolitan Statistical Area and containing more than 1 million people.

*Source:* U.S. Bureau of the Census.

The capital cities of many developing countries display this kind of overwhelming primacy. In part, their primate city pattern is a heritage of their colonial past, when economic development, colonial administration, and transportation and trade activities were concentrated at a single point. Dakar (Senegal), Luanda (Angola), and many other capital cities of African countries are examples.

In other instances, development and population growth have tended to concentrate disproportionately in a capital city whose very size attracts further development and growth. Many European countries (e.g., the United Kingdom, France, and Austria) show a primate structure, often ascribed to the former concentration of economic and political power around the royal court in a capital city that was, perhaps, also the administrative and trade center of a larger colonial empire.

## Urban Influence Zones

Whatever its position in its particular urban hierarchy, every urban settlement exerts an influence upon its immediately surrounding area. The sphere of influence of an urban unit is usually proportional to its size.

A small city may influence a local region of, say, 65 square kilometers (25 sq mi) if, for example, its newspaper is delivered to that region. Beyond that area, another city may be the dominant influence. **Urban influence zones** are the areas outside of a city that are still affected by it. As the distance away from a city increases, its influence on the surrounding countryside decreases (recall the idea of distance decay, discussed in Chapter 8).

A large city located 100 kilometers (62 mi) away from a small city may influence that city and other small cities through its banking services, TV stations, and large shopping malls. There is an overlapping hierarchical arrangement, and the influence of the largest cities is felt over the widest areas.

Intricate relationships and hierarchies are common. Consider Grand Forks, North Dakota, which for local market purposes dominates the rural area immediately surrounding it. However, Grand Forks is influenced by political decisions made in the state capital, Bismarck. For a variety of cultural, commercial, and banking activities, Grand Forks is influenced by Minneapolis. As a center of wheat production, Grand Forks and Minneapolis are subordinate to the grain market in Chicago. Of course, the pervasive agricultural and other political controls exerted from Washington, D.C., on Grand Forks, Minneapolis, and Chicago indicate how large and complex are the urban zones of influence.

## Network Cities

In recent years, a new kind of urban spatial pattern has begun to appear. A **network city** evolves when two or more previously independent nearby cities, potentially complementary in function, strive to cooperate by developing between them high-speed transportation corridors and a communications infrastructure.

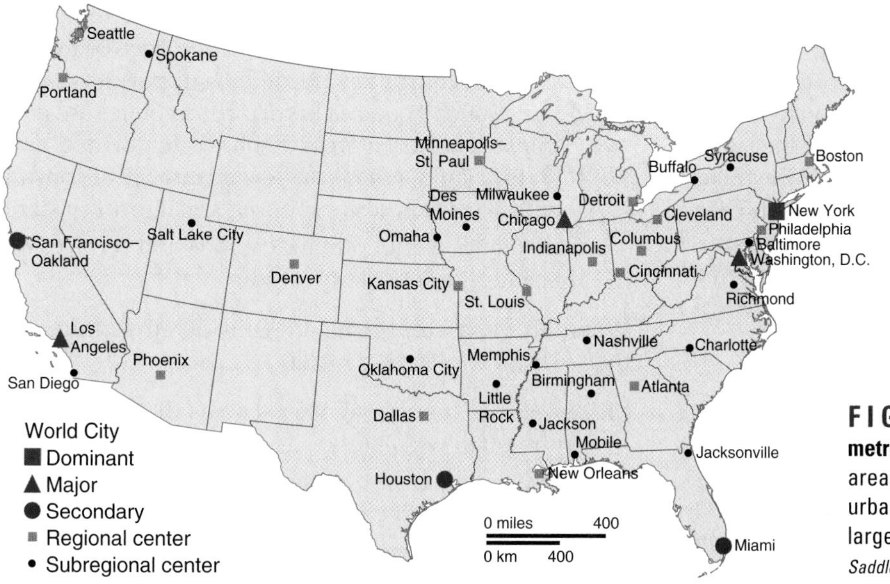


**FIGURE 11.14 A functional hierarchy of U.S. metropolitan regions.** Only the major metropolitan areas are shown. The hierarchy includes smaller urban areas (not shown) that depend on or serve the larger centers. *Redrawn from P. L. Knox, ed.,* Urbanization. *Upper Saddle River, N.J.: Prentice-Hall, 1994, p. 64.*

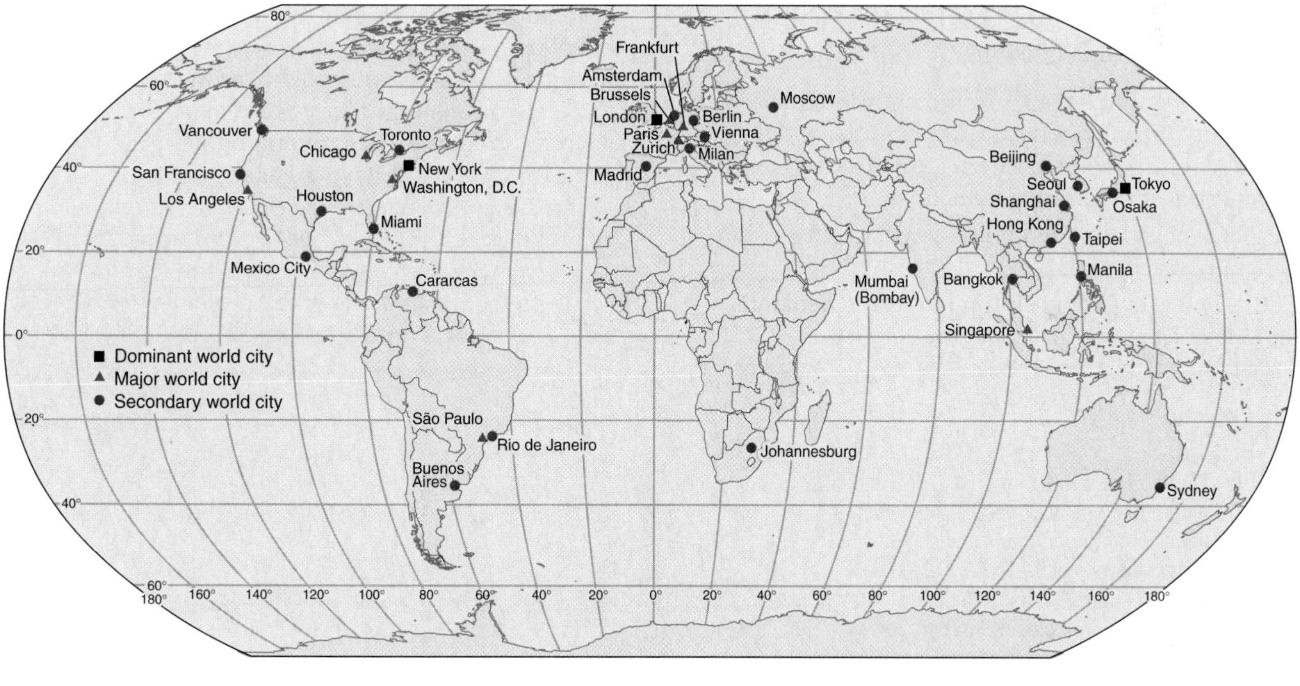


**FIGURE 11.15 A classification of world cities.** World cities are centers of international production, marketing, and finance. They are bound together in complex networks, and all are interconnected in many different ways.

For example, since the reunion of Hong Kong with China proper in 1997, an infrastructure of highway and rail lines and of communications improvements has been developed to help integrate Hong Kong with Guangzhou, the huge, rapidly growing, industrial and economic hub on the mainland of China. In Japan, three distinctive, nearby cities—Kyoto, Osaka, and Kobe—are joining together to compete with the Tokyo region as a major center of commerce. With its temples and artistic treasures, Kyoto is the cultural capital of Japan. Osaka is a major commercial and industrial center, and Kobe is a leading port. Their complementary functional strengths are reinforced by high-speed rail transport connecting the cities and a new airport (Kansai) designed to serve the entire region as a gateway to international business.

In Europe, the major cities of Amsterdam, Rotterdam, and The Hague, together with intermediate cities, such as Delft, Utrecht, and Zaanstad, are connected by high-speed rail lines and a major airport. Each of these cities has special functions not duplicated in the others, and planners have no intention of developing competition among them. This region—called the Randstad—is second only to London in its popularity for international head offices, putting it in a strong position to compete for dominant world city status.

No similar network city has yet developed in the United States. The New York–Philadelphia, the Chicago-Milwaukee, the San Francisco–Oakland, and the Los Angeles–San Diego pairings do not yet qualify for network city status, because there is no concerted effort to bring their competing interests together into a single structure of complementary activities.

## Towns in Agricultural Areas

An effective way to realize how towns in agricultural areas are interrelated is to consider urban settlements as **central places**—that is, as centers for the distribution of economic goods and services to surrounding rural populations. In 1933, the German geographer Walter Christaller attempted to explain the size and location of settlements. He developed a framework, called **central place theory,** for understanding town interdependence. Christaller recognized that his theory would best be developed in rather idealized circumstances, such as the following.

1. Towns that provide the surrounding countryside with such fundamental goods as groceries and clothing would develop where farmers specialized in commercial agricultural production.
2. The farm population would be dispersed in a generally even pattern.
3. The people would possess similar tastes, demands, and incomes.
4. Each kind of product or service available to the population would have its own *threshold,* or minimum number of consumers needed to support its supply. Because such goods as luxury automobiles are either expensive or not in great demand, they would have a high threshold, whereas a fewer number of consumers would be required to support a small grocery store.
5. Consumers would purchase goods and services from the nearest opportunity (store).

When all the assumptions are considered simultaneously, they yield the following results.

1. A series of hexagonal market areas that cover the entire plain will emerge, as shown in Figure 11.16.
2. There will be a central place at the center of each of the hexagonal market areas.
3. The size of the market area of a central place will be proportional to the number of goods and services offered from that place.

In addition, Christaller reached two important conclusions. First, towns of the same size will be roughly evenly spaced, and larger towns will be farther apart than smaller ones. This means that many more small than large towns will exist. In Figure 11.16, the ratio of the number of small towns to towns of the next larger size is 3 to 1. This distinct, steplike series of towns in size classes differentiated by both size and function is called a *hierarchy of central places.*

Second, the system of towns is interdependent. If one town were eliminated, the entire system would have to readjust. Consumers need a variety of products, each of which

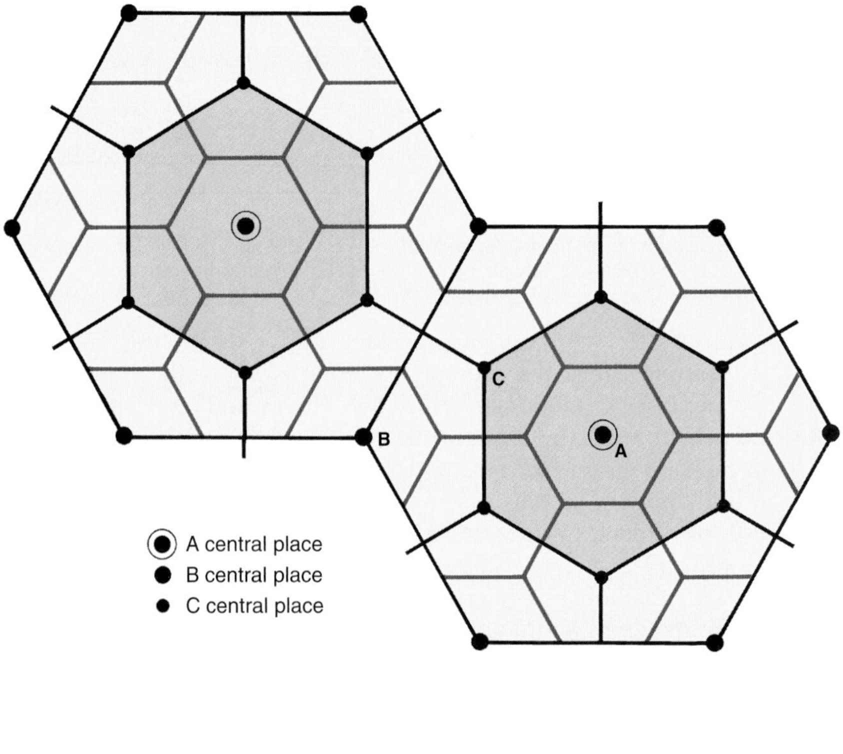


**FIGURE 11.16** The two A central places are the largest on this diagram of one of Christaller's models. The B central places offer fewer goods and services for sale and serve only the areas of the intermediate-size hexagons. The many C central places, which are considerably smaller and more closely spaced, serve still smaller market areas. The goods offered in the C places are also offered in the A and B places, but the latter offer considerably more and more specialized goods. Notice that the places of the same size are equally spaced. *From Arthur Getis and Judith Getis, "Christaller's Central Place Theory,"* Journal of Geography, *1966. Used with permission of the National Council for Geographic Education, Indiana, PA.*

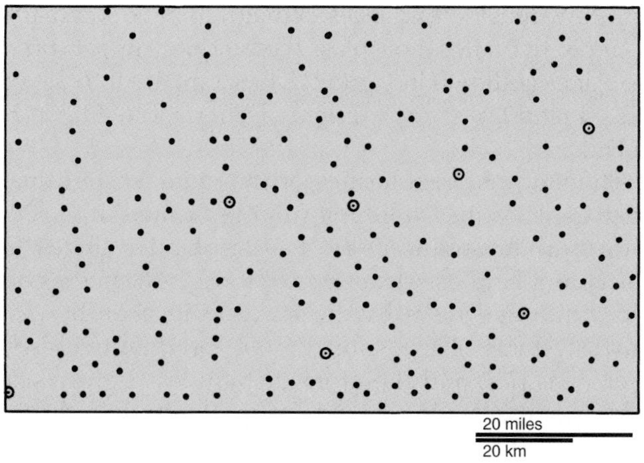


**FIGURE 11.17 The pattern of hamlets, towns, and cities in a portion of Indiana.** This map represents an area 132 by 82 kilometers (82 by 51 mi) just north of Indianapolis. Cities containing more than 10,000 people are circled. Notice that the pattern is remarkably even and includes a number of linear arrangements that correspond to highways and railroads.

has a different minimum number of customers required to support it. The towns containing many goods and services become regional retailing centers, and the small central places serve just the people immediately in their vicinity. The higher the threshold of a desired product, the farther, on average, the consumer must travel to purchase it.

These conclusions have been shown to be generally valid in widely differing areas of the commercial world. When varying incomes, cultures, landscapes, and transportation systems are taken into consideration, the results, although altered to some extent, hold up rather well. They are particularly applicable to agricultural areas, especially with regard to the size and spacing of cities and towns, as Figure 11.17 suggests. If we combine a Christaller-type approach with the ideas that help us understand industrial location and transportation alignments (see Chapter 10), we have a fairly good understanding of the location of the majority of cities and towns.

# INSIDE THE CITY

The structure, patterns, and spatial interactions of systems of cities make up only half the story of urban settlements. The other half involves the distinctive cultural landscapes that are the cities themselves. An understanding of the nature of cities is incomplete without a knowledge of their internal characteristics. So far, we have explored the location, size, and growth and decline tendencies of cities within hierarchical urban systems. Now we look into the city itself to understand better how land uses are distributed, how social areas are formed, and how institutional controls, such as zoning regulations, affect its structure. This discussion will primarily relate to cities in the United States, although most cities of the world have been formed in a somewhat similar manner.

A recurring pattern of land use arrangements and population densities exists within urban areas. There is a certain sameness to the way cities are internally organized, especially within one particular culture sphere, such as North America or Western Europe. The major variables shaping internal land use patterns are accessibility, a competitive market in land, and the transportation technologies available during the periods of urban growth.

## Competitive Bidding for Land

For its effective operation, the city requires a close spatial association of its functions and people. In the past, as long as these functions were few and the population small, pedestrian movement and pack-animal haulage were sufficient for the effective integration of the urban community. With the addition of large-scale manufacturing and the accelerated urbanization of the economy during the 19th century, however, functions and populations—and therefore city area—grew beyond the interaction capabilities of pedestrian movement alone. Cities installed mass-transit systems, which were costly but efficient. Even with their introduction, however, only land within walking distance of the mass-transit routes could be incorporated successfully into the expanding urban structure.

Usable land, therefore, was a scarce commodity, and by its scarcity, it assumed high market value and demanded intensive, high-density utilization. Because of its limited supply of usable land, the industrial city of the mass-transit era (the late 19th and early 20th centuries) was compact, was characterized by high residential and structural densities, and showed a sharp break on its margins between urban and nonurban uses. The older central cities of the northeastern United States and southeastern Canada display that pattern.

Within the city, parcels of land were allocated among alternative potential users on the basis of the relative ability of those users to outbid their competitors for a chosen site. There was a continuous open auction in land, in which users located, relocated, or were displaced in accordance with "rent-paying" ability. The attractiveness of a parcel, and therefore the price that it could command, was a function of its accessibility.

Because uses must arrange themselves spatially, the attractiveness of a parcel is rated by its relative accessibility to all other land uses of the city. Store owners wish to locate where they can be reached easily by potential customers; factories need a convenient meeting point for their workers and materials; residents desire easy connection with jobs, stores, and schools; and so forth. Within the older central city, the radiating mass-transit lines established the elements of the urban land use structure by freezing in the landscape a clear-cut pattern of differential accessibility.

The convergence of that system on the city core gave that location the highest accessibility, the highest desirability, and, hence, the highest land values of the entire built-up area. Similarly, transit junction points were more accessible to larger segments of the city than locations along single traffic routes; the latter were more desirable than parcels lying between the radiating lines (Figure 11.18).

Society deems certain functions desirable without regard to their economic competitiveness. Schools, parks, and public buildings are assigned space without being participants in the auction for land. Other uses, through the process of that auction, are assigned spaces by market forces. The merchants with the highest-order goods and the largest threshold requirements bid most for, and occupy, parcels within the **central business district (CBD),** which is localized at the convergence of mass-transit lines. The successful bidders for slightly less accessible CBD parcels are the developers of the tall office buildings (*skyscrapers*) of major cities, the principal hotels, and similar land uses that help produce the *skyline* of the commercial city.

Comparable, but lower-order, commercial aggregations develop at the outlying intersections—transfer points—of the mass-transit system. Industry takes control of parcels adjacent to essential cargo routes: rail lines, waterfronts, rivers, or canals. Strings of stores, light industries, and high-density apartment structures can afford and benefit from location along high-volume transit routes. The least-accessible locations within the city are left for the least-competitive users: low-density residences. A diagrammatic summary of this repetitive allocation of space among competitors for urban sites is shown in Figure 11.19. Compare it to the generalized land use map of Calgary, Alberta, Canada, in Figure 11.20.

The land use regularities of the older, eastern mass-transit central cities were not fully replicated in the 20th-century urban centers of the western United States. The density and land use structures of those newer cities have been influenced more by the automobile than by mass-transit systems. They spread more readily, evolved at lower densities, and therefore display less tightly structured and standardized land use patterns than do their eastern predecessors.

## Land Values and Population Density

Theoretically, the open land auction should yield two separate, although related, distance-decay patterns, one related to land values and the other to population density (as distance increases away from the CBD, population density decreases). If one were to view the land value surface of the central city as a topographic map (Figure 11.21), with hills representing high valuations and depressions showing low prices, a series of peaks, ridges, and valleys would reflect the differentials in accessibility marked by the pattern of mass-transit lines, their intersections, and the unserved interstitial areas.

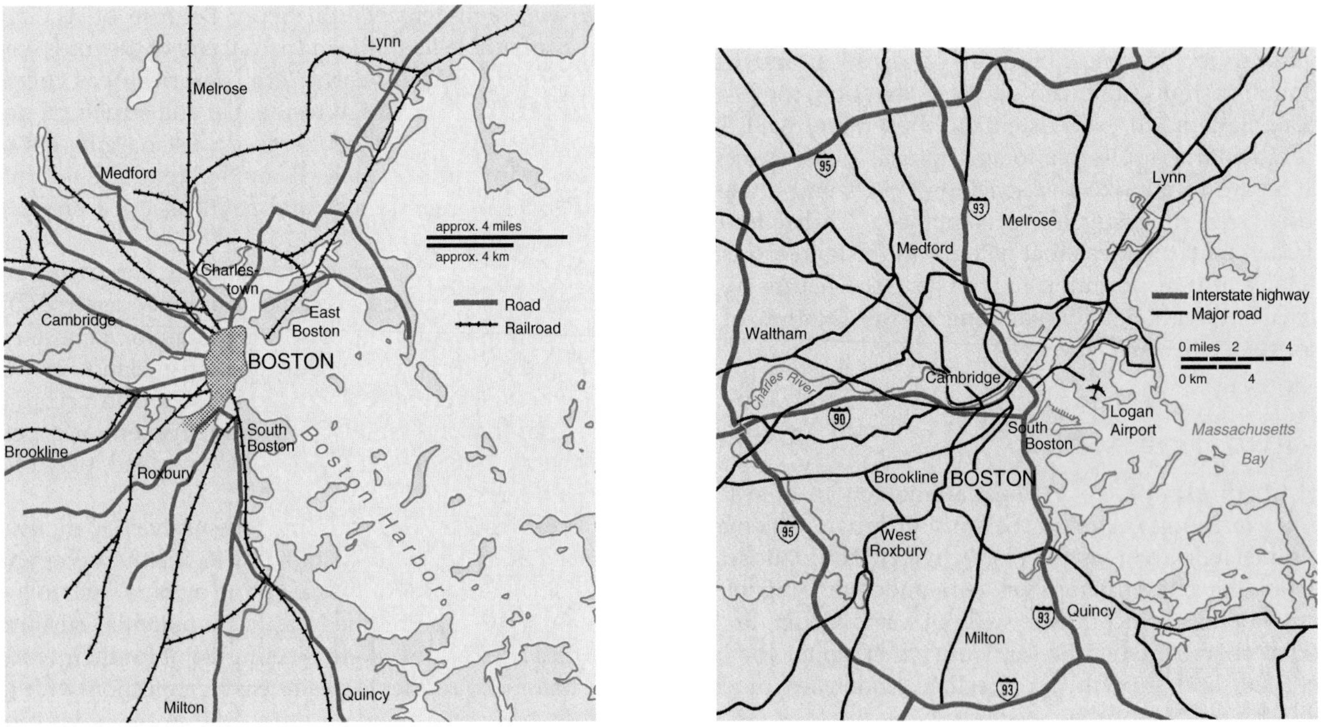


**FIGURE 11.18 Major transit lines in Boston in 1872 and 1994.** (a) Notice how the lines converge on the city center. Compare this map with the late 20th-century freeway patterns shown in (b), the Boston of 1994 and Los Angeles (see Figure 11.5).

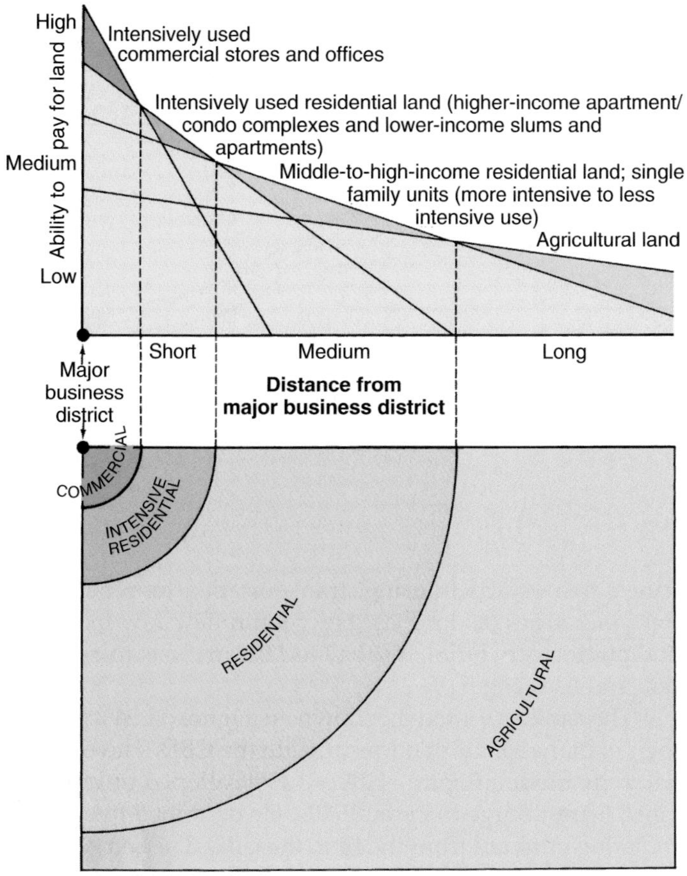


**FIGURE 11.19 A generalized urban land use pattern.** The model depicts the location of various land uses in an idealized urban area where the highest bidder gets the most accessible land.

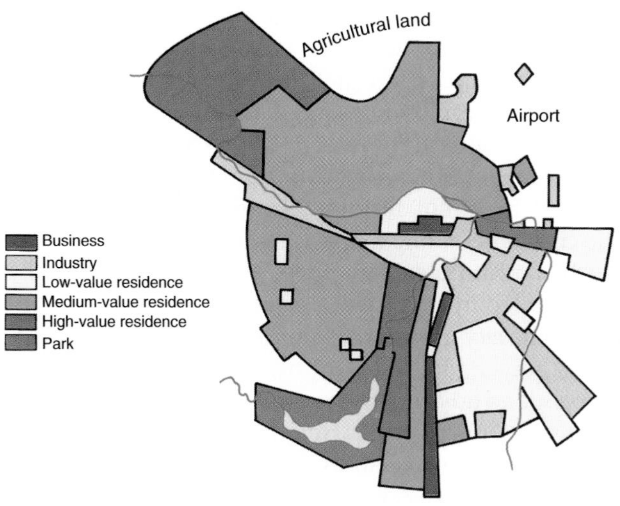


**FIGURE 11.20 The land use pattern in and around Calgary, Alberta, Canada, in 1981.** Physical and cultural barriers and the evolution of urban areas over time tend to result in a sectoral pattern of similar land uses. Calgary's central business district is the focus for many of the sectors. *Revised and redrawn with permission from P. J. Smith, "Calgary: A Study in Urban Patterns,"* Economic Geography, *38, no. 4, p. 328. Copyright © 1962 Clark University, Worcester, MA.*

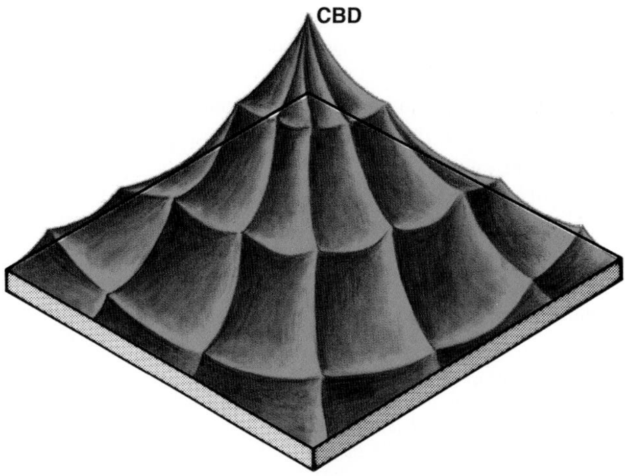


**FIGURE 11.21 A hypothetical pattern of land values.** The "topography" represented on this diagram shows the major land value peak of the CBD. The ridges radiating from the CBD represent the higher land values along transit lines and major business thoroughfares. Regularly occurring minor peaks indicate commercial agglomerations at major transit intersections. *Redrawn with permission from B. J. L. Berry,* Commercial Structure and Commercial Blight, *Research Paper 85, Department of Geography Research Series, The University of Chicago, 1963.*

Dominating these local variations, however, is an overall decline of valuations with increasing distance away from the *peak land value intersection,* the most accessible and costly parcel of the central business district. As would be expected in a distance-decay pattern, the drop in valuation is precipitous within a short linear distance from that point, and then the valuation declines at a lesser rate to the margins of that built-up area.

With one important variation, the population density pattern of the central city shows a comparable distance-decay arrangement, as suggested by Figure 11.22. The exception is the tendency to form a hollow *at the center,* the CBD. The city, always changing, has experienced both declines

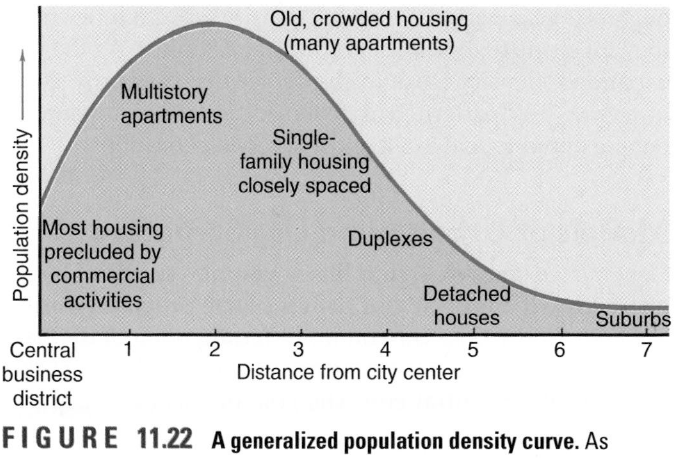


**FIGURE 11.22 A generalized population density curve.** As distance from the area of multistory apartment buildings increases, the population density declines.

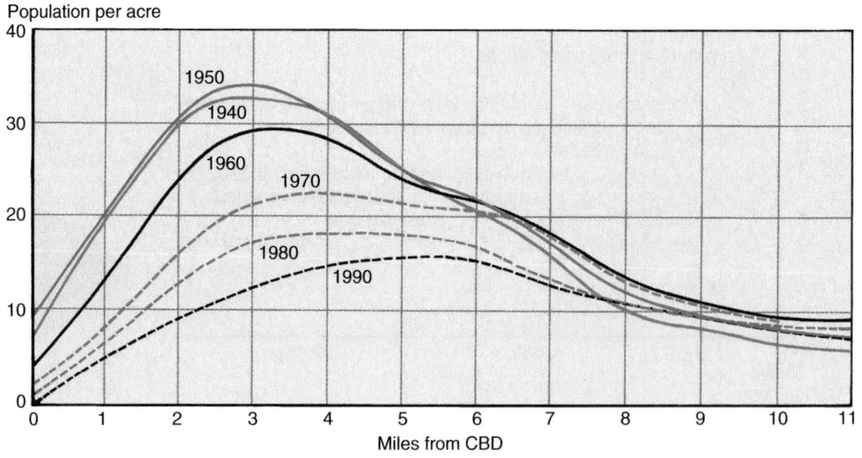


**FIGURE 11.23 Population density gradients for Cleveland, Ohio, 1940–1990.** The progressive depopulation of the central core and flattening of the density gradient to the margin of the city are clearly seen, as Cleveland passed from mass-transit to automobile domination. The Cleveland pattern is consistent with conclusions drawn from other studies of urban density: density gradients tend to flatten over time, and the larger the city, the flatter the gradient.
Source: *Anupa Mukhopadhyay and Ashok K. Dutt, "Population Density Gradient Changes for a Postindustrial City—Cleveland, Ohio 1940–1990,"* GeoJournal *34:517, no. 4, 1994. Redrawn by permission of Kluwer Academic Publishers and Ashok K. Dutt.*

and increases in population at the center. For many years, as cities experienced industrialization, city centers were filled with retail functions, offices, and some housing for the well-to-do. As transportation improvements made lower-priced and larger lots available on the fringes of the cities, the wealthy and middle-class left the center city, leaving the poorest residents behind in obsolescent slum-tenements near the city center. In recent years, however, some middle-class and upper-class urbanites are returning to the city center, attracted by the convenience of the central location, cultural opportunities, and redeveloped upscale apartment buildings. Many cities are fostering this return by creating new parks, building stadiums and sports arenas, and trying to reinvigorate their downtowns. Columbus, Ohio, for example, built a hockey arena downtown and a soccer stadium north of the CBD, promoted rehabilitation of several neighborhoods near downtown, and aided the development of a stretch of art galleries, which dominate the northern part of the main street of the downtown.

These generalizations about population density change over time are represented in Figure 11.23, a time series graph of population density patterns for Cleveland, Ohio, over a 50-year period. The peak density was 2.8 miles from the CBD in 1940, but in 1990 it was at 5.8 miles. As the city expanded, density close to the center decreased. By 2000, however, this pattern had reversed, as affluent younger people opted to live in the most accessible locations.

## Models of Urban Land Use Structure

Generalized models—simplified graphic summaries—of urban growth and land use patterns were proposed during the 1920s and 1930s. The common starting point of the classic models is the distinctive central business district found in every older central city. The core of this area displays intensive land use development: tall buildings, many stores and offices, and crowded streets. Framing the core is a fringe area of warehousing, transportation terminals, and light industries (as long as they require few raw materials and pollute very little). Just beyond the urban core, residential land uses begin.

The land use models shown in Figure 11.24 differ in their explanation of patterns outside the CBD. The **concentric zone model** (Figure 11.24a) was developed by a sociologist, Ernest Burgess, in the 1920s. He described five zones, radiating outward from the first, the CBD. The second ring, the zone of transition, is characterized by stagnation and deterioration and contains high-density, low-income slums, rooming houses, and perhaps ethnic ghettoes. The third is a zone of workers' homes, usually smaller, older homes on modest lots; the fourth and fifth zones are areas of middle-class and well-to-do houses and apartments.

The concentric zone model is dynamic. Each type of land use and each group tends to move outward into the next outer zone. The movement is part of a ceaseless process of invasion and succession that yields a restructured land use pattern and population segregation by income level.

The **sector model** (Figure 11.24b), developed after the concentric zone model, focuses on transportation arterials. It states that high-rent residential areas expand outward from the city center along major transportation routes, such as suburban commuter rail lines. New housing for the wealthy is added in an outward extension of existing high-rent axes as the city grows. Middle-income housing clusters around the housing for the wealthy, and low-income housing occupies land adjacent to the areas of industry and associated transportation, such as freight railroad lines.

The sector model is also dynamic, as there tends to be a filtering-down process as older areas are abandoned by the outward movement of their original inhabitants, with the lowest-income populations (closest to the center of the city and farthest from the current location of the wealthy) becoming the dubious beneficiaries of the least-desirable vacated areas. The accordance of the sector model with the

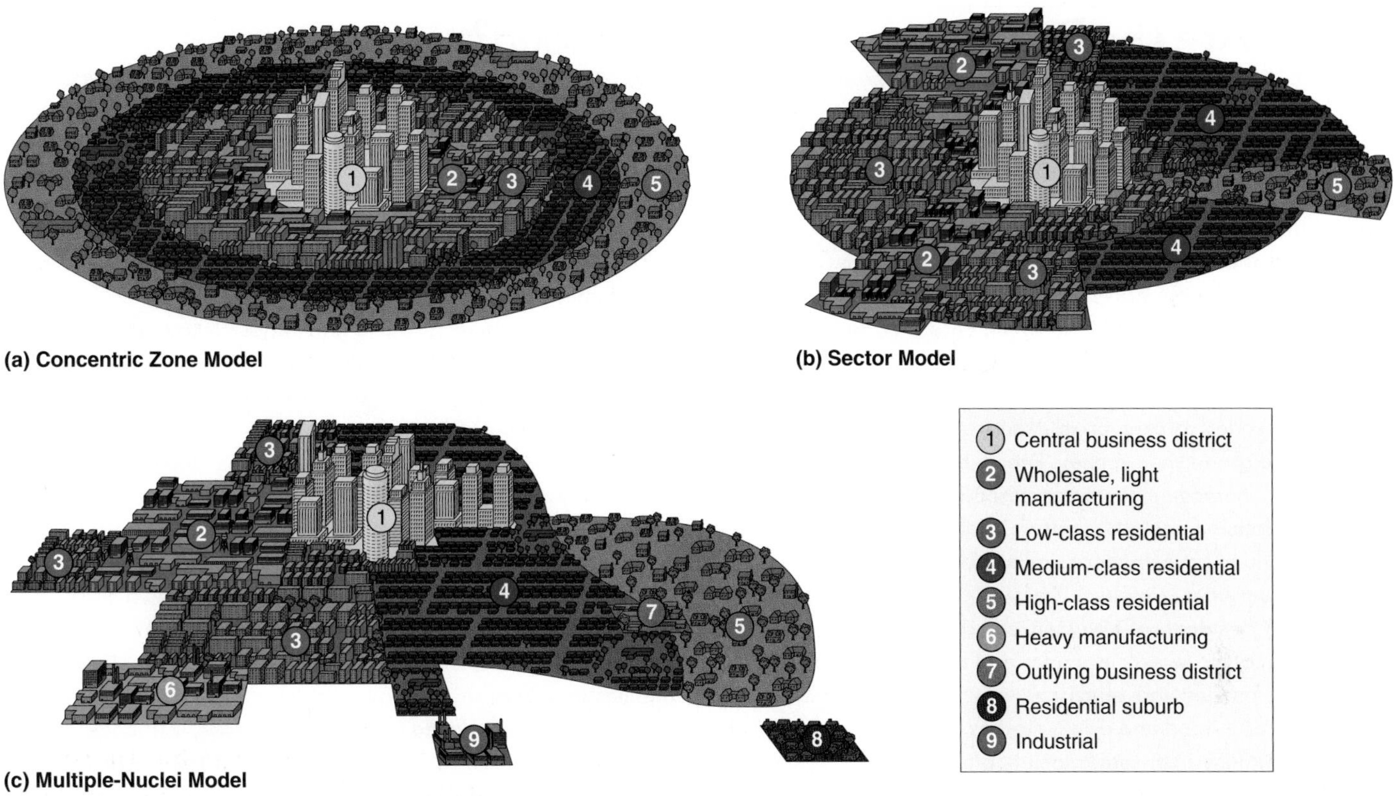


**FIGURE 11.24 Three classic models of the internal structure of cities:** (a) **the concentric zone model,** (b) **the sector model,** and (c) **the multiple-nuclei model.** *Redrawn from "The Nature of Cities" by C. D. Harris and E. L. Ullman in volume no. 242 of* The Annals of the American Academy of Political and Social Science. *Copyright © 1945 The American Academy of Political and Social Science, Philadelphia, PA. Used by permission of the publisher and authors.*

actual pattern that developed in Calgary, Canada, was suggested in Figure 11.20.

The third model of urban land use patterns, the **multiple-nuclei model** (Figure 11.24c), counters the central assumption of the concentric zone and sector models—that urban growth and development spread outward from a single central core. This model states that large cities develop by peripheral spread from several nodes of growth, not just one. Certain activities are limited to specific sites based on their needs: the retail district needs accessibility, whereas a port needs a waterfront location, for example. Peripheral expansion of the separate nuclei eventually leads to coalescence and the meeting of incompatible land uses along the lines of juncture. The urban land use pattern, therefore, is not regularly structured from a single center in a sequence of circles or a series of sectors but is based on separately expanding clusters of contrasting activities.

Although the culture, society, economy, and technology these three models summarized have now been superseded, the physical patterns they explained remain as vestiges and controls on the current landscape. The models still aid in understanding of urban structure and development, but it must be stressed that a model is not a map and that many cities have some of the characteristics of several models. Some geographers suggest that we discard the models and promote new ways of thinking about the city. Modern interpretations of urban land use patterns are based on the observed social segregation within urban areas.

## Social Areas of Cities

The larger and more economically and socially complex cities are, the stronger is the tendency for city residents to segregate themselves into groups on the basis of *social status, family status,* and *ethnicity.* In a large metropolitan region with a diversified population, this territorial behavior may be a defense against the unknown or unwanted; a desire to be among similar kinds of people; a response to income constraints; or a result of social and institutional barriers. Most people feel more at ease when they are near those with whom they can easily identify. In cities, people tend to group according to income or occupation (social status), stage in the life cycle (family status), and language or race (ethnic characteristics) (see "Who Are the People in Your Neighborhood?" p. 410).

Many of these groupings are fostered by the size and value of available housing. Land developers, especially in cities, produce homes of similar quality in specific areas. Of

## Who Are the People in Your Neighborhood?

How does a McDonald's or Burger King decide which one dollar menu items to promote at a certain site, or whether it can profitably offer salads at that franchise? Are there enough families with children to justify building a play area? On what basis does a Starbucks or an Easy Lube determine in what neighborhood to open a new store?

Many businesses, large and small, base their decisions on a marketing analysis system developed by the San Diego–based Claritas, Inc. The system is based on year 2000 census data and uses ZIP codes to categorize Americans by where they live, work, and spend their money. People tend to cluster together in roughly homogeneous areas based on social status, family status, ethnicity, and other cultural markers. People in any one cluster tend to have or adopt similar lifestyles—as Claritas puts it, "You are where you live," because "birds of a feather flock together." Residents of a cluster read the same kinds of books, subscribe to the same magazines and newspapers, and watch the same movies and television shows. They exhibit similar preferences in food and drink, clothes, furniture, cars, and all the other goods a consumer society offers.

Claritas uses a number of variables to classify areas of the country: household density per square mile; area type (city, suburb, town, farm); degree of ethnic diversity; family type (married with children, single, and so on); predominant age group; extent of education; type of employment; housing type; and neighborhood quality. After analyzing the data, the firm characterizes each ZIP code as belonging to from 1 to 5 of 62 possible neighborhood lifestyle categories. Claritas has given catchy names to these clusters, ranging from "Blueblood Estates" (elite, super–rich families) to "Hard Scrabble" (poor, rural families). Some of the others: "Winner's Circle" (executive suburban families); "Pools and Patios" (established empty nesters); "Upward Bound" (dual-income, computer-literate, white-collar families), and "Big City Blend" (middle income immigrant families).

Claritas realizes that the designations don't define the tastes and habits of every person in a community, but they do identify the behavior that most people are apt to exhibit. In the "Towns and Gowns" (college-town singles aged 18–34) area, for example, residents are likely to play soccer, own software purchased in a bookstore, watch Comedy Central, and read *Modern Bride.* But in the "Money and Brains" (dual income families who own their own homes in upscale neighborhoods, predominantly employed in white-collar jobs) ZIP code, residents are most likely to contribute to Public Broadcasting, own or lease a new European luxury automobile, and read *Atlantic Monthly.*

If you want to know how Claritas marketing analysts have categorized your neighborhood, go to *www.claritas.com,* click on "You Are Where You Live" and enter your ZIP code.

course, as time elapses, there is a change in the quality of houses. Land uses may change and new groups may replace old groups, leading to the evolution of new neighborhoods of similar social characteristics.

### Social Status

The social status of an individual or a family is determined by income, education, occupation, and home value, although it may be measured differently in different cultures. In the United States, high income, a college education, a professional or managerial position, and high home value constitute high status. High home value can mean an expensive rented apartment or a large house with extensive grounds.

A good housing indicator of social status is people per room. A low number of people per room tends to indicate high status. Low status characterizes people with low-income jobs living in low-value housing. There are many levels of status, and people tend to filter out into neighborhoods where most of the heads of households are of similar rank.

Social-status patterning agrees with the sector model. In most cities, people of similar social status are grouped in sectors that fan out from the innermost urban residential areas (Figure 11.25). The pattern in Chicago is illustrated in Figure 11.26. If the number of people within a given social group increases, they tend to move away from the central city along an arterial connecting them with the old neighborhood. Major transport routes leading to the city center are the usual migration routes out from the center.

### Family Status

As the distance from the city center increases, the average age of the head of the household declines, or the size of the family increases, or both. Wealthy older people whose children do not live with them and young professionals without families tend to live close to the city center. The young families seek space for child rearing, and older people covet more access to the cultural and business life of the city. Where inner-city life is unpleasant, there is a tendency for older people to migrate to the suburbs or to retirement communities.

Within lower-status populations, the same pattern tends to emerge. Transients and single people are housed in the inner city, and families, if they find it possible or desirable, live farther from the center. The arrangement that emerges is a concentric-circle patterning according to family status (see Figure 11.25). In general, inner-city areas house older people and outer-city areas house younger people.

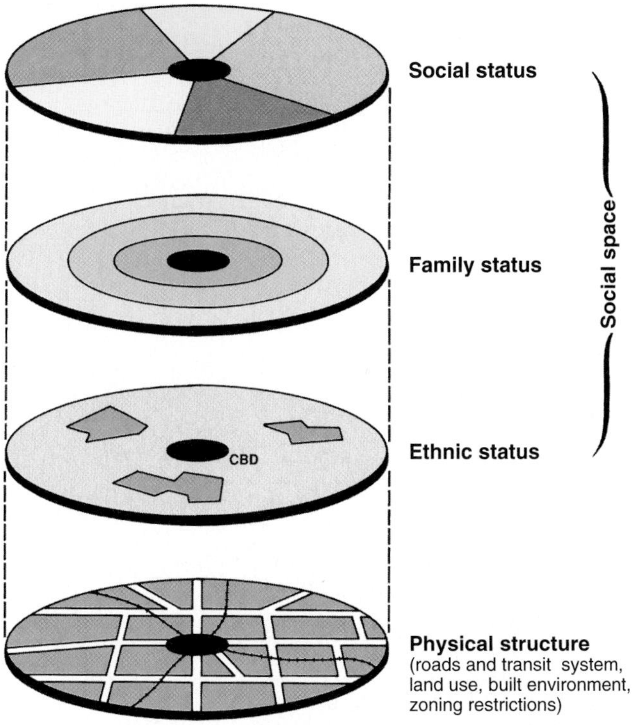


**FIGURE 11.25 The social geography of American and Canadian urban areas.** *Redrawn with permission from Robert A. Murdie,* "Factorial Ecology of Metropolitan Toronto," *Research Paper 116, Department of Geography Research Series, University of Chicago, 1969.*

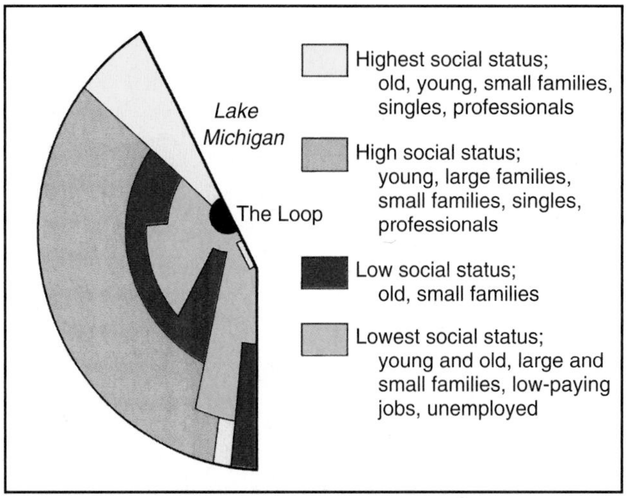


**FIGURE 11.26 A diagrammatic representation of the major social areas of the Chicago region.** The central business district is known as The Loop. *Redrawn with permission from Philip Rees,* "The Factorial Ecology of Metropolitan Chicago," *M. A. Thesis, University of Chicago, 1968.*

### Ethnicity

For some groups, ethnicity is a more important residential locational determinant than social or family status. Areas of homogeneous ethnic identification appear in the social geography of cities as separate clusters or nuclei reminiscent of the multiple-nuclei concept of urban structure. For some ethnic groups, cultural segregation is both sought and vigorously defended, even in the face of pressures for neighborhood change exerted by potential competitors for housing space. The durability of "Little Italys" and "Chinatowns" and of Polish, Greek, Armenian, and other ethnic neighborhoods in many American cities is evidence of the persistence of self-maintained segregation.

Certain ethnic or racial groups, especially blacks, have been segregated in nuclear communities. Every city in the United States has one or more black areas that, in many respects, may be considered cities within a city. Figure 11.27 illustrates the concentration of blacks, Hispanics, and other racial/ethnic groups in distinct neighborhoods in Los Angeles. The social and economic barriers to movement outside the area have always been high. In many American cities, the poorest residents are the blacks, who are often relegated to the lowest-quality housing in the least-desirable areas of the city. Similar restrictions have been placed on Hispanics and other non-English-speaking minorities.

Of the three patterns depicted in Figure 11.25, family status has undergone the most widespread change in recent years. Today, the suburbs house large numbers of singles and childless couples, as well as two-parent families. Areas near the central business district have become popular for young professionals. Gay couples and families often choose to live in urban centers as well. The city structure is constantly changing, reflecting changes in family and employment structure. There are now large numbers of new jobs for professionals in the suburbs and the central business districts, but not in between. With more women in the workforce than ever before, and as a result of multiple-earner families, residential site selection has become a more complex undertaking.

## Institutional Controls

The governments—local and national—of most Western urbanized societies have instituted innumerable laws to control all aspects of urban life, including the rules for using streets, the provision of sanitary services, and the use of land. In this section, we touch only upon land use.

Institutional and governmental controls have strongly influenced the land use arrangements and growth patterns of most cities in the world. Cities have adopted land use plans and enacted subdivision control regulations and zoning ordinances to realize those plans. They have adopted building, health, and safety codes to assure legally acceptable urban development and maintenance. All such controls are based on broad applications of the police powers of municipalities and their rights to assure public health, safety, and well-being even when private-property rights are infringed upon.

These nonmarket controls on land use are designed to minimize incompatibilities (residences adjacent to heavy industry, for example), provide for the creation in appropriate locations of public uses (the transportation system,

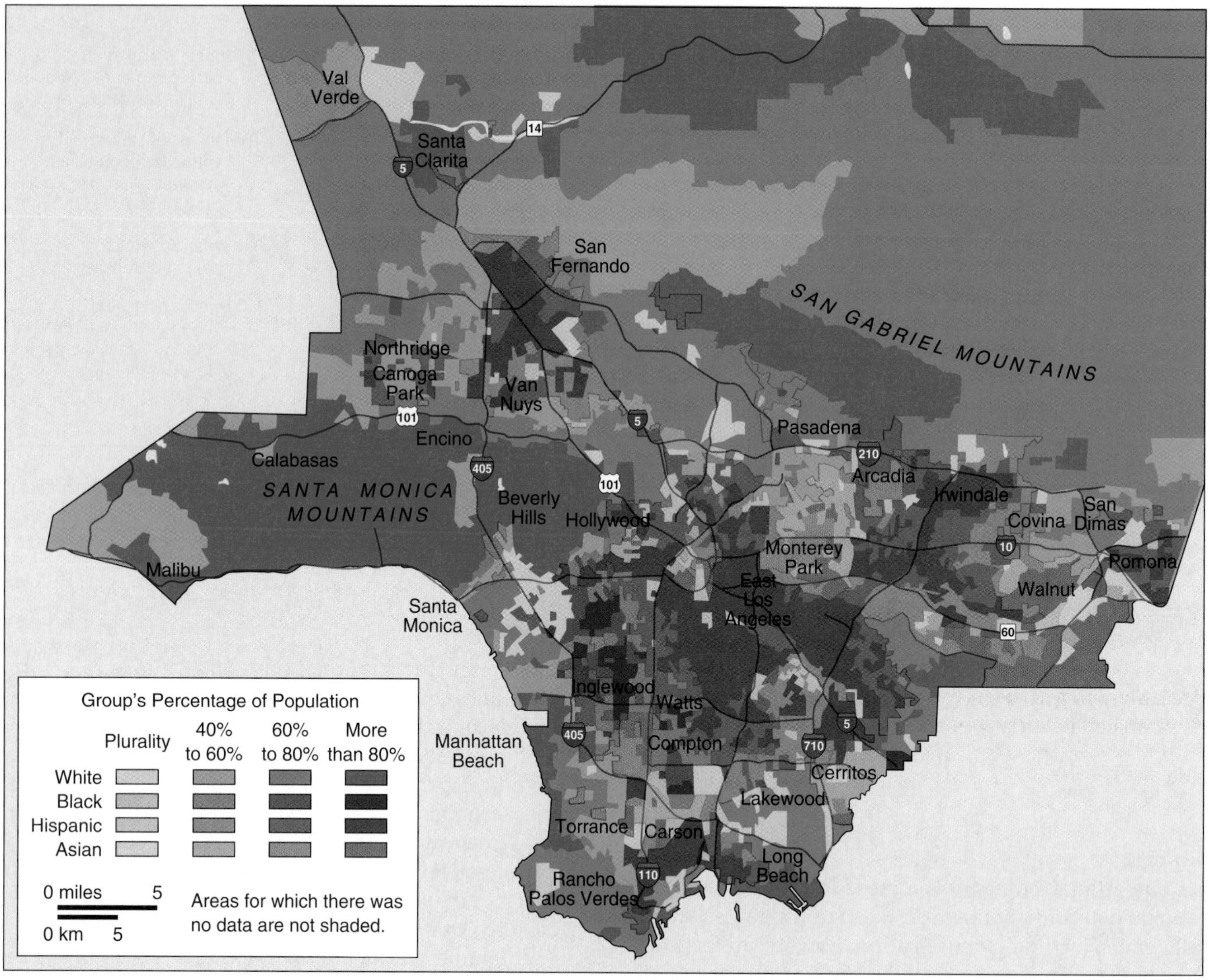


**FIGURE 11.27 Racial/ethnic patterns in Los Angeles, 2000.** Although Los Angeles County has an extremely diverse population, people tend to cluster in distinct neighborhoods by race and ethnicity. Between 1990 and 2000, the Hispanic population of Los Angeles County increased from 38% to 45%, and the proportion of Asians grew from 10% to 12%. The proportion of blacks in the population decreased from 11% to 9%, while the proportion of whites declined from 41% to 31%. Source: *U.S. Census Bureau as published in* The New York Times, *March 30, 2001, p. A16.*

waste disposal, government buildings, prisons, and parks) and private uses (colleges, shopping centers, and housing) needed for and conducive to a balanced, orderly community. In theory, such careful planning should preclude the emergence of slums, so often the result of undesirable adjacent uses, and should stabilize neighborhoods by reducing market-induced pressures for land use change.

In the United States, zoning ordinances and subdivision regulations have been used to exclude "undesirable" uses—apartments, special housing for the aged, low-income housing, half-way homes—from upper-income areas. Bitter court battles have been waged over "exclusionary" zoning practices: while proponents contend that the undesirable uses will harm the city and, more importantly, property values, opponents argue that such uses are important to the city as a whole. In recent years, some newly incorporated suburbs have sought to reduce the costs of growth—like developing infrastructure, building schools and playgrounds, and so on—by zoning out housing that would bring in children.

In most of Asia there is no zoning, and it is quite common to have small-scale industrial activities operating in residential areas. Even in Japan, a house may contain several people doing piecework for a local industry. In both Europe and Japan, neighborhoods have been built and rebuilt gradually over time, and it is usual to have a wide variety of building types from various eras mixed together on the same street. In the United States and Canada, such mixing is much rarer and is often viewed as a temporary condition as areas are in transition toward total redevelopment. Perhaps the only exception to this in a large city is Houston, Texas, which has no zoning regulations.

## Suburbanization in the United States

The 20 years before World War II saw the creation of a technological, physical, and institutional structure that resulted, after that war, in a sudden and massive alteration of past urban forms. The improvement of the automobile increased its reliability, range, and convenience, freeing its owner from dependence on fixed-route public transit for access to home, work, or shopping. The new transport flexibility opened up vast new acreages of nonurban land to urban development. The acceptance of a maximum 40-hour work week in 1938 guaranteed millions of Americans the time for a commuting journey not possible when workdays of 10 or more hours were common.

After World War II, the United States witnessed a significant increase in the number of Americans owning their own homes, from below 50% in 1945 to 60% in 1960. The government stimulated this boom by authorizing increased spending for home loans by the Federal Housing Administration (FHA) and the Veterans Administration (VA). These agencies offered much more generous terms than private bankers had before the war, when buyers had to put down large down payments (sometimes 50% or more) and repay their loans in a short time, often 10 years. The FHA and VA revolutionized home-buying, offering mortgages of up to 90% of a home's value and up to 30 years to pay off loans. The VA permitted many veterans to purchase homes with virtually no down payment.

Demands for housing, pent up by years of economic depression and wartime restrictions, were loosed in a flood after 1945, and a massive suburbanization altered the existing pattern of urban America. Between 1950 and 1970, the two most prominent patterns of population growth were the *metropolitanization* of people and, within metropolitan areas, their *suburbanization.* During the 1970s, the interstate highway system was substantially completed and major metropolitan expressways were put in place, allowing sites 30 to 45 kilometers (20 to 30 mi) or more from workplaces to be within commuting distance from home. The major metropolitan areas rapidly expanded in area and population. Growth patterns for the Chicago area are shown in Figure 11.28. The high energy prices of the 1970s slowed the rush to the suburbs, but in the 1980s suburbanization again proceeded apace.

Residential land uses led the rush to the suburbs. Typically, uniform but spatially discontinuous housing developments were built beyond the boundaries of most older central cities. The new design was an unfocused sprawl because it was not tied to mass-transit lines. It also represented a massive relocation of purchasing power to which retail merchants were quick to respond. The planned major regional shopping center became the suburban counterpart of higher-order central places and the outlying commercial districts of the central city. Smaller shopping malls and strip shopping centers gradually completed the retailing hierarchy.

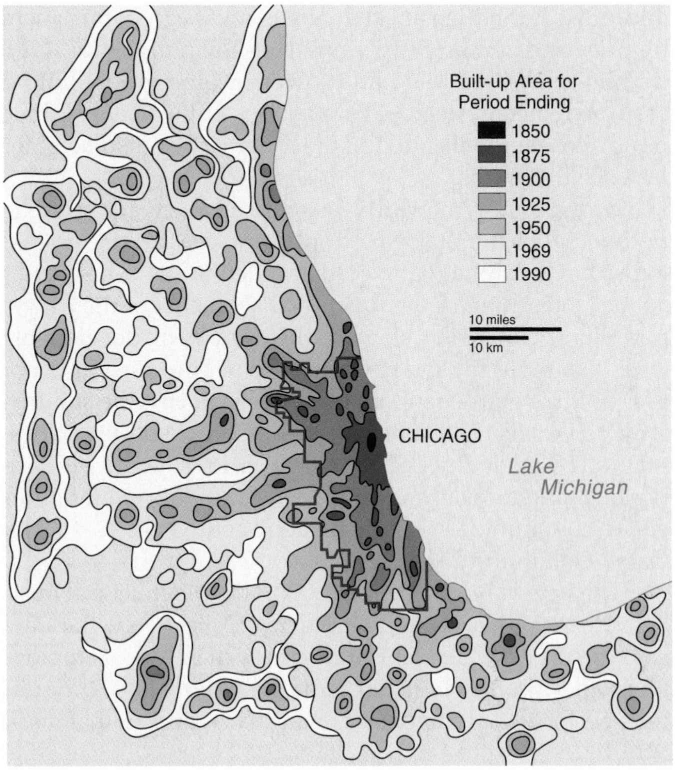


**FIGURE 11.28 A history of urban sprawl.** In Chicago, as in most larger and older U.S. cities, the slow peripheral expansion recorded during the late 19th and early 20th centuries suddenly accelerated as the automobile suburbs began developing after 1945. The red line indicates the Chicago city boundaries. *Revised with permission from B. J. L. Berry,* Chicago: Transformation of an Urban System, *Ballinger Publishing Co., Cambridge, Mass. 1976, with additions from other sources.*

Faced with a newly suburbanized labor force, industry followed the outward move, attracted as well by the economies derived from modern single-story plants with plenty of parking space for employees. Industries no longer needed to locate near railway facilities; freeways presented new opportunities for lower-cost, more flexible truck transportation. Service industries were also attracted by the purchasing power and large, well-educated labor force now living in the suburbs, and complexes of office buildings developed, like the shopping malls, at freeway intersections and along freeway frontage roads and major connecting highways.

In time, in the United States, an established social and functional pattern of suburban land use emerged, giving evidence of a lower-density, more extensive repetition of the models of land use developed to describe the structure of the central city. Multiple nuclei of specialized land uses developed, expanded, and coalesced. Sectors of high-income residential use continued their outward extension beyond the central-city limits, usurping the most scenic and most desirable suburban areas and segregating them by price and zoning restrictions. As shown in Figure 11.29, middle-, lower-middle-, and lower-income groups found their own income-segregated portions of the fringe. Ethnic minorities were relegated to the inner city and some older industrial suburbs.

By the 1990s, a new urban America had emerged on the perimeters of the major metropolitan areas. With increasing sprawl and the rising costs implicit in the ever-greater spatial separation of the functional segments of the fringe, peripheral expansion slowed, the supply of developable land was reduced, and the intensity of land development grew. No longer dependent on the central city, the suburbs were reborn as vast, collectively self-sufficient outer cities, marked by landscapes of industrial parks, skyscraper office parks, massive retailing complexes, and a proliferation of gated communities and apartment complexes (see "The Gated Community," p. 416).

The new suburbia began to rival the older central business district in size and complexity. Collectively, the new centers surpassed the central cities as generators of employment and income. Together with the older CBDs, the suburbs perform the tertiary and quaternary services that mark

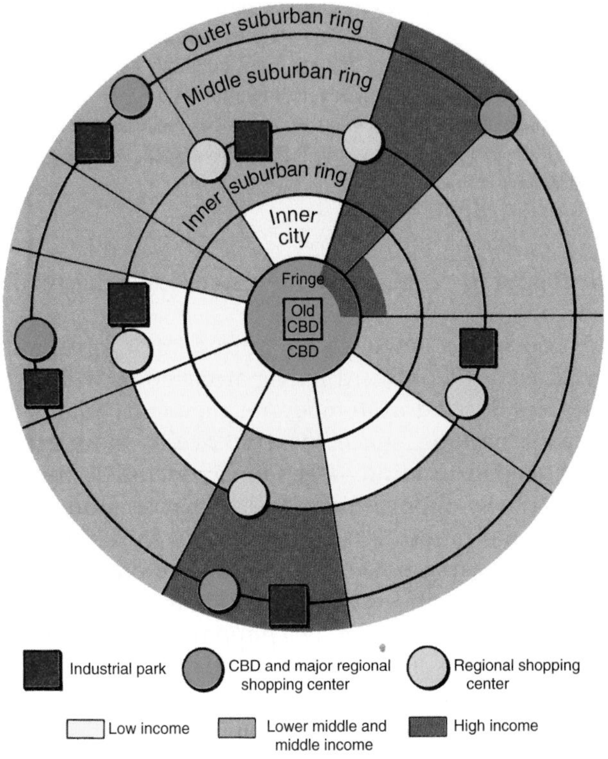


**FIGURE 11.29 A diagram of the present-day United States metropolitan area.** Note that aspects of the concentric zone, sector, and multiple-nuclei patterns are evident and carried out into the suburban fringe. The "major regional shopping centers" of this earlier, mid-1970s model are increasingly the cores of newly developing "outer cities."
*Figure 4.10 (redrawn) from* The North American City, *4th ed. by Maurice Yeates. Copyright © 1990 by Harper & Row, Publishers, Inc.*

the postindustrial metropolis. During the 1980s, more office space was created in the suburbs than in the central cities of America. Tysons Corner, Virginia (between Arlington and Reston), for example, became the ninth-largest central business district in the United States. Regional and national headquarters of leading corporations, banking, professional services of all kinds, major hotel complexes, and recreational centers became parts of the new outer cities, sometimes called **edge cities:** large nodes of office and retail activities at the margin of an urban area.

Edge cities now exist in all regions of urbanized North America. The South Coast Metro Center in Orange County, California; the City Post Oak–Galleria center on Houston's west side; King of Prussia and the Route 202 corridor northwest of Philadelphia; the Meadowlands, New Jersey, west of New York City; and Schaumburg, Illinois, in the western Chicago suburbs are a few examples of the new urban forms. There is even a suburb of Atlanta named Periphery Center. The metropolis has become polynucleated, and urban regions are increasingly "galactic"—that is, galaxies of economic activity nodes organized primarily around the freeway systems (Figure 11.30).

In recent years, suburban areas have expanded to the point where metropolitan areas are coalescing. The Boston-to-Washington corridor, often called a **megalopolis,** is a continuously built-up region with many new centers that compete with the business districts of Boston, Providence, New York, Philadelphia, Baltimore, and Washington, D.C. One summary of the new pattern is shown in Figure 11.31. A further analysis of the northeastern U.S. urban corridor appears in "Megalopolis" in Chapter 13.

# Central City Change

Continuing urbanization and metropolitanization of America at the end of the 20th century led to two contrasting sets of central city patterns and problems. One is characteristic of older, eastern cities and of their older generation of suburbs, unable to expand and absorb the new growth areas on their margins and to maintain their original economic base. The other set is more characteristic of western U.S. cities that developed in the automobile, rather than mass-transit, era. Although they are able to incorporate within their political boundaries the new growth areas on their margins, they are faced with the problem of providing infrastructure, services, and environmental protection to an ever-more-distant residential and functional base.

## Constricted Central Cities

Suburbanization has grievously damaged the economic base and the financial stability of central cities unable to absorb new growth, especially in the eastern United States. In earlier periods of growth, as new settlement areas developed beyond the political margins of the city, annexation absorbed new growth within the corporate boundaries of

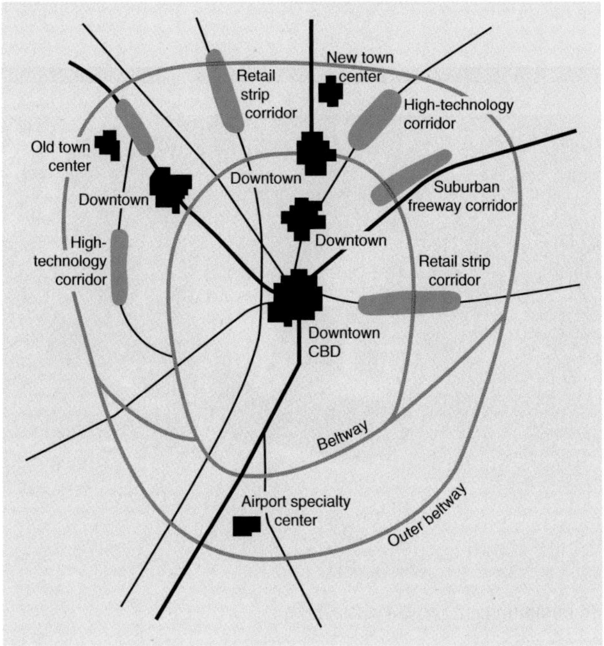


**FIGURE 11.30** The galactic city's multiple downtowns and special function nodes and corridors are linked by the metropolitan expressway systems in this conceptualization. *Redrawn with permission from V. H. Winston and Son, Inc., Truman Hartshorn, "Suburban Downtowns and the Transformation of Atlanta's Business Landscape,"* Urban Geography, *10:382. Silver Springs, Md.: V. H. Winston and Son, Inc., 1989.*

the expanding older city. The additional tax base and employment centers became part of the municipal whole. But in states that recognized the right of separate incorporation of the new growth area, the ability of the city to continue to expand was restricted. Where possible, suburbanites opted for a separation from the central city and for aloofness from the costs, deterioration, and adversities associated with it. Septic tanks and private wells substituted for city sewers and water mains; local dumps substituted for costly garbage pickup. Suburbanites' homes, jobs, shopping, schools, and recreation all existed outside the confines of the city from which they had divorced themselves.

The redistribution of population caused by suburbanization resulted not only in the spatial but also in the political segregation of social groups of the metropolitan area. The upwardly mobile residents of the city—younger, wealthier, and better educated—took advantage of the automobile and the freeway to leave the central city. The poorer, older, least-advantaged urbanites were left behind (Figure 11.32). The central cities and suburbs became increasingly differentiated. Large areas within the cities now contain only the poor and minority groups, including women (see "Women in the City," p. 417), a population barely able to pay the rising costs of social services that their numbers, neighborhoods, and condition require.

The services needed to support the poor include welfare payments, social workers, police and fire protection,

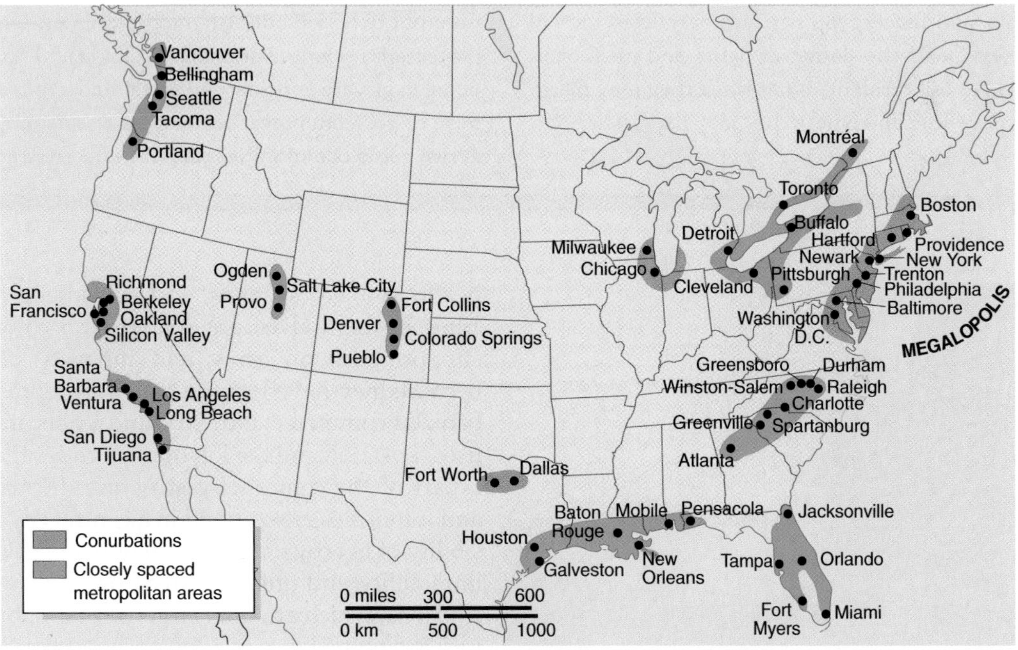


**FIGURE 11.31 Megalopolis and other Anglo American conurbations.** A conurbation is an aggregation or continuous network of urban areas. The northeastern U.S. Boston-to-Norfolk urban corridor comprises the original and largest *megalopolis* and contains the economic, political, and administrative core of the United States. A Canadian counterpart core region, anchored by Montreal and Toronto, connects with U.S. conurbations through Buffalo, New York, and Detroit, Michigan. For some of their extent, conurbations fulfill their classic definition of continuous built-up urban areas. In other portions, they are more statistical than landscape entities, marked by counties that qualify as "urban" or "metropolitan" even though land uses may appear dominantly rural.

## The Gated Community

Approximately one in six Americans—some 48 million people—lives in a *master-planned community.* Particularly characteristic of the fastest growing parts of the country, most of these communities are in the South and West, but they occur everywhere. Master-planned communities date back to the 1960s, when Irvine, California, and Sun City, Arizona, were built.

A subset of the master-planned community is the **gated community,** a fenced or walled residential area where access is limited to designated individuals. By 2002, 9 million Americans were living in these middle- and high-income developments. Entry to these communities within communities is restricted; gates are manned by security guards or are accessible only by computer key card or telephone. Some of the communities hire private security forces to patrol the streets. Surveillance systems monitor common recreational areas, such as community swimming pools, tennis courts, and health clubs. Houses are commonly equipped with security systems. Troubled by the high crime rates, drug abuse, gangs, and drive-by shootings that characterize many urban areas, people seek safety within their walled enclaves.

The typical gated community has been built by a developer following a master plan. To preserve the upscale nature of the development and protect land values, self-governing community associations enact conditions and restrictions. Pervasive and detailed, they specify such things as the size, construction, and color of walls and fences, the size and permitted uses of rear and side yards, and the design of lights and mailboxes. Some go so far as to tell residents what trees they may plant, what pets they may raise, and where they may park their boats or recreational vehicles.

Gated community. *Courtesy of Arthur Getis.*

Gated and sheltered communities are not just an American phenomenon but are increasingly found in all parts of the world. More and more guarded residential enclaves have been sited in such stable Western European states as Spain, Portugal, and France.

Elsewhere, as in Argentina or Venezuela in South America or Lebanon in the Middle East, with little urban planning, unstable city administration, and inadequate police protection, not only rich but also middle-class citizens are opting for protected residential districts. In China and Russia, the sudden boom in private and guarded settlements reflects in part a new form of postcommunist social class distinction, while in South Africa gated communities serve as effective racial barriers.

**FIGURE 11.32 A derelict slum in Philadelphia.** Some areas of large cities have been abandoned and now house only the poor. *© AP/Wide World Photos.*

health delivery systems, and subsidized housing. Central cities, by themselves, are unable to raise the taxes needed to support such an array and intensity of social services because they have lost the tax bases represented by suburbanized commerce, industry, and upper-income residential uses. Lost, too, are the job opportunities that were formerly a part of the central-city structure. Increasingly, the poor and minorities are trapped in a central city without the possibility of nearby employment and are isolated by distance, immobility, and unfamiliarity from the few remaining low-skill jobs, which are now largely in the suburbs. This unfortunate circumstance is often called a *spatial mismatch.*

The population shift and abandonment of the central city by retail and industrial functions have nearly destroyed the traditional active, open auction of urban land. In the vacuum left by the departure of private investors, the federal government, particularly since the landmark Housing Act of 1949, has initiated urban renewal programs with or

## Women in the City

At 6:48 on a recent evening, Isabelle Cuello, a woman of Hispanic heritage in her fifties, climbed aboard the bus taking her from her job in a southwestern suburb of Detroit to her home in the central city. She had left the telecommunications company where she worked in customer service at 6:00 P.M., darted across four lanes of traffic, and stood outside, waiting in the open in the absence of a bus shelter or a bench. She was just starting her journey: three buses would transport her to her home near downtown in just over 3 hours, a journey that would take 45 minutes by car.

Cuello cannot afford a car, having recently left the unemployment rolls. Like many other large industrial cities in the United States, Detroit (the "Motor City") has no subway or rail system Cuello can use. A 3-hour commute, Cuello explains, is hard. It leaves her exhausted at the end of the day, with only enough energy to feed her cats and fall into bed, as she has to be up at 4:00 the next morning to get ready before leaving for the 5:30 A.M. bus. She needs to take pain medications for her rheumatoid arthritis, wait until they begin to take effect, iron her clothes, and shower. Sometimes buses don't come, they are late, or they break down; often they are dirty, smelly, or loud. The journey to work leaves her little time for anything else: she has little time to prepare meals, shop, or socialize.

Cuello is not unusual. Recent studies of the feminization of poverty have shown that

- The central cities of the United States have more households headed by women than men;
- Women constitute the majority of poor people;
- The vast majority of poor women live in central cities;
- Downtowns also house more elderly, among whom women predominate;
- Women depend more than do men on public transportation (and minorities are more likely to use public transportation than whites).

As the structure of American cities has changed over the past half-century, more and more jobs have moved from central cities to the suburbs. This has had long-reaching effects on those—especially women—living in the central cities. Minorities, such as Isabelle Cuello, have significantly longer commutes to work in the suburbs than whites. Work opportunities, as a consequence, are more limited for those living in the central cities and are even more constrained for minority women.

For women in the cities, access to jobs depends not only on where the job is but also on the location of other necessary services. For many women, for example, the presence of child-care facilities at or near the workplace makes some jobs more feasible than others.

Access to jobs is also determined by other restrictions on an individual's life. Studies have found that women experience a lower level of accessibility to jobs because they face more constraints on their nonwork hours than men. For many women, being at home when children return from school limits them in their job opportunities and in the length of time they can commit to a commute. In fact, if Isabelle Cuello, like many of her neighbors, were caring for children or grandchildren, accepting a suburban job would be complicated by questions of child care before and after school, transportation to school, supervision of children's activities, attendance at school functions, and the ability to respond quickly in case of an emergency.

If Cuello could afford to live in the suburb in which she works, she could cut her commuting time. As a low-wage worker, however, she cannot afford the rent there. Besides, she says, people in most suburbs are not kind to minorities and she would rather live near other people like her. Finally, given the instability of jobs in many sectors of the economy, she doesn't want to gamble that this one will last long.

without provisions for a partnership with private housing and redevelopment investment. Under a wide array of programs instituted and funded since the late 1940s, slum areas were cleared, public housing was built, cultural complexes and industrial parks were created, and city centers were reconstructed (Figure 11.33).

With the continuing erosion of the urban economic base and the disadvantageous restructuring of the central-city population mix, however, the hard-fought governmental battle to maintain or revive the central city is frequently judged to be a losing one. Cities such as Detroit, Michigan, Toledo, Ohio, and Bridgeport, Connecticut, witnessed multiple failed attempts at urban renewal. In recent years, the central city has been the destination of thousands of homeless people (see "The Homeless," p. 418). Many live in public parks, in doorways, by street-level warm-air exhausts of subway trains, and in subway stations. Central-city economies, with their high land and housing values, limited job opportunities for the unskilled, and inadequate resources for social services, appeared to many observers to offer few or no prospects for change.

That pessimistic outlook, however, began to change dramatically in the 1990s. Although their death was widely reported, central cities by 2000 were showing many positive signs of revival and renewed centrality both in the expanded metropolitan districts they anchored and in the

(a) (b)

**FIGURE 11.34 Revitalized central cities.** a) Entrepreneurs in New York City. Many Latin American and Asian immigrants have established their own business, fixing, making, or selling things, adding to the vitality of central cities. Some work out of sidewalk stalls, others have stores. According to the U.S. Census Bureau, 36% of New York City residents were born abroad. b) **Gentrified housing in the Georgetown section of Washington, D.C.** Gentrification is especially noticeable in the major urban centers of the eastern United States, from Boston south along the Atlantic Coast to Charleston, South Carolina, and Savannah, Georgia. It is also increasingly a part of the regeneration of older, deteriorated first-generation residential districts in major central cities across the country. *(a) © David Grossman/Image Works. (b) © Carl Purcell.*

Individual home buyers and rehabbers opened the way; commercial developers followed, greatly increasing the stock of quality housing in downtown areas—but often only after the local, state, or federal government made the first investments in slum clearance, park development, cultural center construction, and the like. Milwaukee built a riverside walk and attracted $50 million in private investment, for example. Indianapolis city officials are emptying housing projects in the Chatham Arch neighborhood and selling them to developers for conversion into apartments and condominiums. Renovation of an old cotton mill in the Cabbagetown district of Atlanta has produced 500 new apartments in a building recycling project common to many older cities. And as whole areas are gentrified or redeveloped residentially, other investment flows into nearby commercial activities. For example, Denver's LoDo district, once a skid row, has been wholly transformed into a thriving area of shops, restaurants, and sports bars along with residential lofts.

Gentrification produces tension between long-time residents and newcomers. Each group has different expectations for the area and sometimes the conflicts can be bitter. Once an area is completely gentrified, former residents often find themselves searching for other low-income housing in the city, if it is available. This dynamic also contributes to the continually changing nature of U.S. cities.

## Expanding Central Cities

During the latter part of the 20th century, the most dynamic U.S. urban growth rates were in the 13 states of the Mountain and Pacific West regions. In 1940, little more than half of all westerners lived in cities; by 2000, nearly 90% were urbanites. Arizona, California, Nevada, and Utah all have a higher percentage of city dwellers than New York, and 6 of the 10 fastest-growing U.S. metropolitan areas are in the West (refer back to Table 11.4).

For the most part, these newer, automobile-oriented metropolises were able to expand physically to keep the new growth areas on their peripheries within the central-city boundaries. Nearly without exception they placed few restrictions on physical expansion. That unrestricted growth has often resulted in the coalescence of separate cities into ever-larger metropolitan complexes.

The speed and volume of growth has spawned a complex of concerns, some reminiscent of older eastern cities and others specific to areas of rapid urban expansion in the West. As in the East, the oldest parts of western central cities tend to be pockets of poverty, racial conflict, and abandonment. In addition, western central-city governments face all the economic, social, and environmental consequences of unrestricted marginal expansion. Scottsdale, Arizona, for example, covered 2.6 square kilometers (1 sq mi) in 1950; by 2000 it had grown to nearly 500 square kilometers (about 200 sq mi), four times the physical size of San Francisco. Phoenix, with which Scottsdale has now coalesced, surpasses in sprawl Los Angeles, which has three times as many people. The phenomenal growth of Las Vegas, Nevada, has similarly converted vast areas of desert landscape to low-density urban use (Figure 11.35).

Such unrestricted central-city expansion has introduced its own fiscal crises. In many instances, limited by state law from raising taxes, central cities have been unable to provide the infrastructure improvements and social services

**FIGURE 11.35 Urban sprawl in the Las Vegas, Nevada, metropolitan area.** Like many western cities, Las Vegas has expanded over great expanses of desert in order to keep pace with a rapidly growing population. The fastest growing metropolitan area in the United States in the 1990s, Las Vegas increased from a little more than 850,000 people in 1990 to more than 1,560,000 by the end of the decade, an increase of 83%. *© Robert Cameron/Getty Images.*

their far-flung populations require. Schools remain unbuilt and underfunded; water supplies are increasingly difficult and expensive to obtain; open-space requirements for parkland are ignored; street and highway improvements and repairs are inadequate even as demand for them increases. In short, each additional unit of unrestricted growth costs the municipality more than the additional development generates in tax revenue.

Increasingly, central cities and metropolitan areas of both the East and the West are seeking to restrain rather than encourage physical growth. Portland, Oregon, drew a "do not pass" line around itself in the late 1970s, prohibiting urban conversion of surrounding forests, farmlands, and open space. Rather than losing people and functions, it has added both while preserving and increasing parklands and other urban amenities.

Other cities, metropolitan areas, and states are also beginning to resist and restrict urban expansion. "Smart growth" programs have been adopted by such states as Colorado, Delaware, Minnesota, and Washington. Cities of both the West and the East are beginning to tighten controls on unrestricted growth.

# World Urban Diversity

The city, as Figure 11.3 reminds us, is a global phenomenon. Its structure, form, and functions differ from one region to the next and even within regions, reflecting its historical and cultural heritage. The models and descriptions of the United States city do not generally apply to cities in other parts of the world. Those cities have developed different functional and structural patterns, some so radically different from our United States model that we would find them unfamiliar and uncharted landscapes indeed. The city is universal; its characteristics are cultural and regional.

## The Anglo American City

Even within the seemingly homogeneous culture realm of the United States and Canada, the city shows subtle, but significant, differences—not only between older eastern and newer western U.S. cities but also between cities of Canada and those of the United States. Although the urban form is similar in the two countries, it is not identical. The Canadian city, for example, is more compact than its U.S. counterpart of equal population size, with a higher density of buildings and people and a lesser degree of suburbanization of populations and functions (Figure 11.36).

Space-saving and multiple-family housing units are more the rule in Canada, so a similar population size is housed on a smaller land area with much higher densities within the central area of cities. The Canadian city is better served by and more dependent on mass transportation than is the U.S. city. Because Canadian metropolitan areas have only one-quarter the number of miles of expressway lanes per capita as U.S. metropolitan areas, suburbanization of peoples and functions is less extensive north of the border than south.

The differences are cultural, as well. Although cities in both countries are ethnically diverse (Canadian communities, in fact, have a higher proportion of foreign born), U.S. central cities exhibit far greater internal distinctions in race, income, and social status and more pronounced contrasts between central-city and suburban residents. That is, there has been much less "flight to the suburbs" by middle-income Canadians. As a result, the Canadian city shows greater social stability, higher per capita average income, more retention of shopping facilities, and more employment opportunities and urban amenities than its U.S. central-city counterpart. In particular, it does not have the rivalry from the well-defined competitive edge cities of suburbia that so spread and fragment United States metropolitan complexes.

## The West European City

Western European cities differ even more from the U.S. city. The political history of France, for example, has given Paris an overwhelmingly primate position in its system of cities. Political, economic, and colonial history has done the same for London in the United Kingdom. On the other hand, Germany and Italy came late to nationhood, and no overwhelmingly dominant cities developed in their systems.

Nonetheless, a generally common heritage of medieval origins, Renaissance restructurings, and industrial period growth has given the cities of Western Europe distinct features. Despite wartime destructions and postwar redevelopments, many still bear the impress of past occupants and

**FIGURE 11.36 The central and outlying business districts of Toronto, easily visible in this photo, are still rooted firmly by mass-transit convergence and mass-transit usage.** On average, Canadian metropolitan areas are almost twice as densely populated as metropolitan areas in the United States. On a per capita basis, Canadian urbanites are two and a half times more dependent on public transportation than are American city dwellers. That reliance gives form, structure, and coherence to the Canadian central city, qualities now irretrievably lost in the sprawled and fragmented U.S. metropolis. *© Thomas Kitchin/Tom Stack & Associates.*

technologies, even as far back as Roman times. An irregular system of narrow streets may be retained from the random street pattern developed in medieval times of pedestrian and pack-animal movement. Main streets radiating from the city center and cut by circumferential "ring roads" tell us the location of high roads leading into town through the gates in city walls now gone and replaced by circular boulevards. Broad thoroughfares, public parks, and plazas mark Renaissance ideals of city beautification and the aesthetic need felt for processional avenues and promenades.

Although each is unique historically and culturally, West European cities as a group share certain features. Cities of Western Europe have, for example, a much more compact form and occupy less total area than American cities of comparable population; most of their residents are apartment dwellers. Residential streets of the older sections tend to be narrow, and front, side, or rear yards or gardens are rare.

European cities were developed for pedestrians and still retain the compactness appropriate to walking distances. The sprawl of American peripheral or suburban zones is generally absent. At the same time, compactness and high density do not mean skyscraper skylines. Much of urban Europe predates the steel-frame building and the elevator. City skylines tend to be low, three to five stories in height, sometimes (as in central Paris) held down by building ordinance or by prohibitions on private structures exceeding the height of a major public building, often the central cathedral (Figure 11.37). Where those restrictions have been relaxed, however, tall office buildings have been erected.

Compactness, high densities, and apartment dwelling encouraged the development and continued importance of public transportation, including well-developed subway systems. The private automobile has become much more common of late, though most central-city areas have not yet been significantly restructured with wider streets and parking facilities to accommodate it. The automobile is not the universal need in Europe that it has become in American cities, as home and work are generally more closely spaced.

A generalized model of the social geography of the West European city has been proposed (Figure 11.38). Its exact counterpart can be found nowhere, but many of its general features are seen in most major European cities. In the historic core, now increasingly gentrified, residential units for the middle class, the self-employed, and the older generation of skilled artisans share limited space with preserved historic buildings, monuments, and tourist attractions.

The old city fortifications may mark the boundary between the core and the surrounding transitional zone of

**FIGURE 11.37** Even in their central areas, many European urban centers show a low profile, like that of Paris seen here. Although taller buildings—20, 30, and even 50 or more stories in height—have become more common in major urban areas since World War II, they are not the universal mark of the central business districts they have become in the United States, nor the generally welcomed symbols of progress and pride. *© Scott Gilchrist/Masterfile.*

substandard housing, 19th-century industry, and recent immigrants. The waterfront also has older industry; newer plants are found on the periphery. Public housing and some immigrant concentrations may be near newer industry.

The West European city is not characterized by inner-city deterioration and out-migration. Its core areas tend to be stable in population and attract, rather than repel, the successful middle class and the upwardly mobile, conditions far different from comparable sections of older American central cities.

## The East European City

Cities of Eastern Europe and the former European republics of the Soviet Union make up a separate urban class—the East European city. It is an urban form that shares many of the traditions and practices of West European cities, but it differs from them in the centrally administered planning principles that were, in the communist period (1945–1990), designed to shape and control both new and older settlements. The governments' goals were, first, to limit city size to avoid metropolitan sprawl; second, to assure an internal structure of neighborhood equality and self-sufficiency; and third, to segregate land uses. The planned East European city fully achieved none of these objectives, but by attempting them it has emerged as a distinctive urban form.

The city is compact, with relatively high building and population densities reflecting the nearly universal apartment dwelling, and with a sharp break between urban and rural land uses on its periphery. Like the West European city, the East European city depends nearly exclusively on public transportation.

During the communist period, it differed from its Western counterpart in its purely governmental, rather than market, control of land use and functional patterns. This control dictated that the central area of cities should be reserved for public use, not occupied by retail establishments or office buildings on the Western, capitalist model. Eastern European governments preferred a large central square ringed by administrative and cultural buildings with space nearby for a large recreational and commemorative park. Residential areas were expected to be largely self-contained in the provision of low-order goods and services, minimizing the need for a journey to centralized shopping locations.

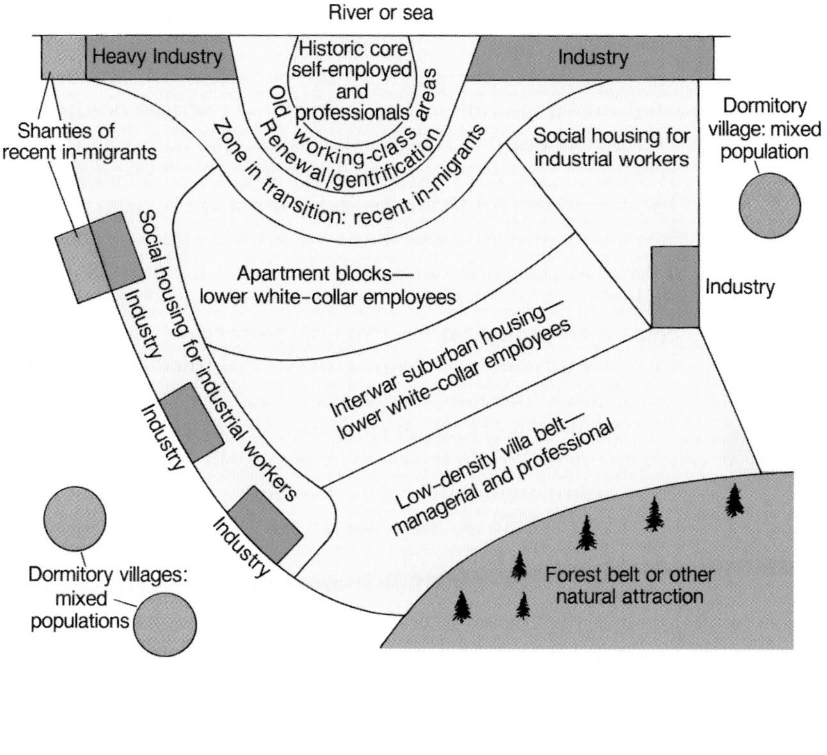


**FIGURE 11.38 A diagrammatic representation of the West European city.** *Redrawn from Paul White,* The West European City: A Social Geography. *Copyright © 1984 Longman Group UK Limited. Used by permission.*

Residential areas are made up of *microdistricts,* assemblages of uniform apartment blocks housing perhaps 10,000 to 15,000 people, surrounded by broad boulevards, and containing centrally sited nursery and grade schools, grocery and department stores, theaters, clinics, and similar neighborhood necessities and amenities (Figure 11.39). Plans called for effective separation of residential quarters from industrial districts by landscaped buffer zones, but in practice many microdistricts were built by factories for their own workers and were located immediately adjacent to the workplace. Because microdistricts were most easily and rapidly constructed on open land at the margins of expanding cities, high residential densities have been carried to the outskirts of town.

This pattern will change in the decades to come as capitalist principles of land allocation are adopted. Now that private interests can own land and buildings, the urban areas may take on the basic form of the West European city. Currently, the trend is to construct more spacious, privately owned apartments and single-family houses for the newly rich.

## Cities in the Developing World

Still farther removed from the United States urban model are the cities of the Middle East, Africa, Asia, and Latin America. Industrialization has come to them only recently. Modern technologies in transportation and public facilities are sparsely available and the structures of cities and the cultures of their inhabitants are far different from the urban world familiar to North Americans. The developing world is vast in extent and diverse in its physical and cultural landscapes.

The backgrounds, histories, and current economies and administrations of developing-world cities vary so greatly that it is impossible in a textbook such as this to generalize about their internal structure. Some are ancient, having been established as early as (or even earlier than) the 21st century B.C., not A.D. Some are still preindustrial, with only a modest central commercial core; they lack industrial districts, public transportation, or any meaningful degree of land use separation. Others, though increasingly Western in form, are only beginning to industrialize.

**FIGURE 11.39** This scene from Bucharest, Romania, clearly shows important recurring characteristics of the East European city: mass transit service to boulevard-bordered "superblocks" of self-contained apartment house microdistricts that contain their own shopping areas, schools, and other facilities. *© Aubrey Diem/Valan Photos.*

Some are the product of colonialism, established as ports or outposts of administration and exploitation, built by Europeans on a Western model. For example, the British built Kolkata (Calcutta), New Delhi, and Mumbai (Bombay) in India and Nairobi and Harare in Africa. The French developed Hanoi and Ho Chi Minh City (Saigon) in Vietnam, Dakar in Senegal, and Bangui in the Central African Republic. The Dutch used Jakarta as their main outpost, Belgium had Kinshasa (formerly Leopoldville) in what is now the Democratic Republic of the Congo, and the Portuguese founded a number of cities in Angola and Mozambique.

Just as socialist planning principles affected cities in Eastern Europe in the latter half of the 20th century, a number of Asian cities also bear their imprint. Examples include Pyongyang (North Korea), Phnom Penh (Cambodia), and Hanoi in Vietnam. Finally, some cities in the developing world are relatively new, having been built to serve as capital cities or to house industrial or transport centers. Among them are Brasília (Brazil), Abuja (Nigeria), and Islamabad (Pakistan).

Urban structure is a function not just of the time when a city was founded, and by whom, but also of the role a city plays in its own cultural milieu. Land use patterns in capital cities reflect the centralization of government functions and the concentration of wealth and power in a single city of a country (Figure 11.40a). The physical layout of a religious center, or sacred city, is conditioned by the religion it serves, whether it is Hinduism, Buddhism, Islam, Christianity, or another faith. Typically, a monumental structure—such as a temple, mosque, or cathedral—and associated buildings rather than government offices occupy the city center (Figure 11.40b). Traditional market centers for a wide area (Timbuktu in Mali and Lahore in Pakistan) or cultural capitals (Addis Ababa in Ethiopia and Cuzco in Peru) have land use patterns that reflect their special functions. Likewise, port cities such as Dubai (UAE), Haifa (Israel), or Shanghai (China) have a land use structure different than that of an industrial or a mining center such as Johannesburg (South Africa). Adding to the complexity is the fact that cities with a long history reflect the changes wrought by successive rulers and/or colonial powers, and that as some of the megacities in the developing world have grown, they have engulfed nearby towns and cities.

Even within a single culture realm, then, urban forms display significant variation. For South Asia, which includes Afghanistan, Pakistan, India, Nepal, Bhutan, and Bangladesh, geographers have developed different models to characterize colonial cities, traditional bazaar cities, planned cities, and

(a)

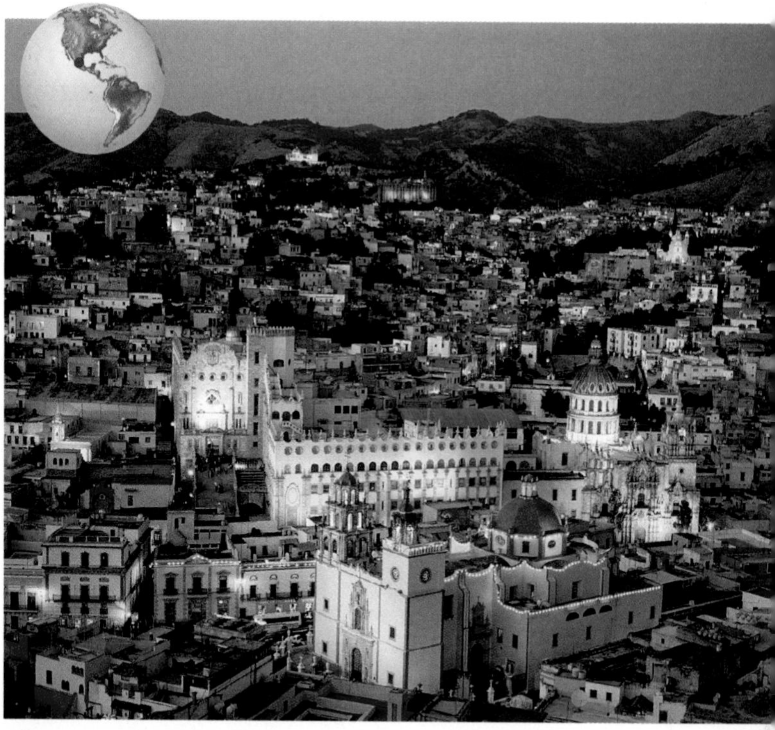

(b)

**FIGURE 11.40 Developing-world cities.** Compare the monumental government buildings in a) the capital city of Brasília, Brazil with b) a state capital, Guanajuato, Mexico, which is dominated by religious structures.
*(a) © Julia Waterlow/Corbis Images. (b) © Jose Fuste Raga/Corbis Images.*

cities that are mixtures of these three. Yet other models describe land use patterns in smaller cities that developed initially as hill stations, railway colonies, or cantonments (military encampments).

Despite these differences in cities in the developing world, we can identify some features common to most of them. Wherever the automobile or modern transport systems are an integral part of the city, the metropolis begins to take on Western characteristics. All of the large cities have modern centers of commerce, not unlike their Western counterparts (Figure 11.41). In places where the public transport system is limited, however, the result has been overcrowded cities centered on a single major business district in the old tradition. Examples include Mumbai (India), Lagos (Nigeria), Jakarta (Indonesia), and Bangkok (Thailand).

The developing countries have experienced disproportionate population concentrations in their national and regional capitals. Few developing countries have mature, functionally complex, small and medium-size centers. The primate city dominates their urban systems. More than a quarter of all Nicaraguans live in Managua, and Libreville contains one-third of the population of Gabon.

All of the countries have experienced massive in-migrations from rural areas as vast numbers of surplus, low-income populations have migrated to cities with the hope of finding jobs and improving their socioeconomic condition. That hope is rarely realized because the cities have populations greater than their formal functions and employment bases can support. In all of them, large numbers of people support themselves in the "informal" sector—as food vendors, peddlers of cigarettes or trinkets, street-side barbers or tailors, errand runners or package carriers outside the usual forms of wage labor. Most of the new urbanites have little choice but to pack themselves into shantytowns and squatter settlements on the fringes of the city, isolated from sanitary facilities, public utilities, and job opportunities found only at the center (see "The Informal Housing Problem," p. 426).

**FIGURE 11.41 High-rise buildings in São Paulo, Brazil.** Contrast this photo with those of *favelas* in Rio de Janeiro on pages 208 and 426.
*© Yann Arthus-Bertrand/Corbis Images.*

## The Informal Housing Problem

Between one-third and two-thirds of the population of most developing world cities is crowded into shantytowns and squatter settlements built by the inhabitants, often in defiance of officialdom. These unofficial communities usually have little or no access to publicly provided services, such as water, sewers and drains, paved roads, and garbage removal. In such megacities as Rio de Janeiro, São Paulo, Mexico City, Bangkok, Chennai (Madras), Cairo, or Lagos, millions find refuge in the shacks and slums of the "informal housing sector." Crumbling tenements house additional tens of thousands, many of whom are eventually forced into shantytowns by the conversion of their residences into commercial property or high-income apartments.

No more than 20% of the new housing in developing countries' urban areas is produced by the formal housing sector; the rest develops informally, ignoring building codes, zoning restrictions, property rights, and infrastructure standards. Informal settlements house varying percentages of these populations, but for low-income developing countries as a whole during the 1980s and 1990s, only one formal housing unit was added for every nine new households, and between 70% and 90% of all new households found shelter in shanties or slums. Peripheral squatter settlements, though densely built, may provide adequate household space and even water, sewers, and defined traffic lanes through the efforts of the residents. More usually, however, overcrowding transforms these settlements into vast zones of disease and squalor subject to constant danger from landslides, fire, and flooding. The informality and often illegality of the squatter housing solution means that those who improvise and build their own shelters lack registration and recognized ownership of their homes or the land on which they stand. Without such legal documentation, no capital accumulation based on housing assets is possible, and no collateral for home improvement loans or other purposes is created.

A *favela* in Rio de Janeiro. These homes are built out of scrap wood and metal and have no city services, such as sewers, running water, or street cleaning. *© Luiz Claudio Margio/Peter Arnold, Inc.*

Impoverished squatter districts exist around most major cities in Africa, Asia, and Latin America. At the beginning of this chapter, we mentioned that one-half million people are estimated to live in Cairo's Northern and Southern Cemeteries. In the sprawling slum district of Nairobi, Kenya, called Mathare Valley, a quarter-million people are squeezed into 15 square kilometers (6 sq mi). Similar shantytowns and squatter settlements house approximately one-fourth of the population of such Asian cities as Bangkok (Thailand), Kuala Lumpur (Malaysia), and Jakarta (Indonesia). The percentage is even higher in a number of Indian cities, including Chennai (Madras) and Kolkata (Calcutta).

## Summary

The urban area is an essential activity center that marks any society advanced beyond the level of subsistence. Although cities are among the oldest forms of civilization, only recently have urban areas become home to the majority of people in industrialized countries and the commercial crossroads for uncounted millions in the developing world. At the global scale, four major world urban regions have emerged: Western Europe, South Asia, East Asia, and Anglo America.

All settlements growing beyond their village origins take on functions uniting them to the countryside and to a larger system of settlements. As they grow, they become functionally complex. Their economic base, composed of both *basic* and *nonbasic* activities, may become diverse. Basic activities are the functions performed for the larger economy and urban system, whereas nonbasic activities satisfy the needs of the urban residents themselves. Functional classifications distinguish the economic roles of urban centers. Hierarchies of cities, central place theory, and network cities help us realize how urban areas are interrelated.

As Anglo American urban centers expanded in population size and diversity, they developed structured land use and social patterns based on market allocations of urban space, channelization of traffic, and socioeconomic aggregation. The observed regularity of land use arrangements has been summarized for U.S. cities by the concentric zone, sector, and multiple-nuclei models. Social area counterparts of land use specializations are based on social status, family status, and ethnicity. Since 1945, land use patterns and social areas have been modified by the suburbanization of people and functions that has led to the creation of new and complex outer urban areas and to the deterioration of the older central city itself. Recent economic trends and gentrification have, however, enhanced the employment and residential importance of central city downtown areas.

Urbanization is a global phenomenon, and the Anglo American models of city systems, land use, and social area patterns are not necessarily or usually applicable to other cultural contexts. In Europe, stringent land use regulations have brought about a compact urban form ringed by greenbelts. Although changing rapidly, the East European urban area still shows a pattern of density and land use reflecting recent communist principles of city structure. Cities in the developing world are currently growing faster than it is possible to provide employment, housing, safe water, sanitation, and other minimally essential services and facilities.

## Key Words

basic sector 398
central business district (CBD) 406
central city 396
central place 404
central place theory 404
city 396
concentric zone model 408
economic base 398
edge city 414
gated community 416
gentrification 419
megalopolis 414
metropolitan area 396
multiple-nuclei model 409
multiplier effect 400
network city 402
nonbasic sector 398
primate city 401
rank-size rule 401
sector model 408
site 396
situation 396
suburb 396
town 396
urban hierarchy 400
urban influence zone 402
urbanized area 396
world city 400

## For Review & Consideration

1. Consider the city or town in which you live or attend school or with which you are most familiar. In a brief paragraph, discuss that community's site and situation; point out the connection, if any, between its site and situation and the basic functions that it performed earlier or now performs.
2. Describe the *multiplier effect* as it relates to the population growth of urban units.
3. What area does a *central place* serve and what kinds of functions does it perform? If an urban system were composed solely of central places, what summary statements could you make about the spatial distribution and the urban size hierarchy of that system?

4. Is there a hierarchy of retailing activities in the community with which you are most familiar? Of how many and of what kinds of levels is that hierarchy composed? What localizing forces affect the distributional pattern of retailing within that community?
5. Briefly describe the urban land use patterns predicted by the *concentric zone,* the *sector,* and the *multiple-nuclei* models of urban development. Which one, if any, best corresponds to the growth and land use pattern of the community most familiar to you?
6. In what ways do social status, family status, and ethnicity affect the residential choices of households? What expected distributional patterns of urban social areas are associated with each? Does the social geography of your community conform to the predicted pattern?
7. How has suburbanization damaged the economic base and the financial stability of the central city?
8. In what ways does the Canadian city differ from the pattern of its United States counterpart?
9. Why are metropolitan areas in developing countries expected to grow larger than many Western metropolises by the year 2015? What do you expect the population density profile for Mexico City to look like in the year 2015?
10. What are *primate* cities? Why are primate cities in the developing countries overburdened? What can be done to alleviate the difficulties?
11. What are the significant differences in the generalized pattern of land uses of Anglo American, West European, and East European cities?

# Selected References

**WEBSITES**

The World Wide Web has a tremendous variety of sites pertaining to geography. Websites relevant to the subject matter of this chapter appear in the "Web Links" section of the Online Learning Center associated with this book. Access it at www.mhhe.com/getis10e/.


Brunn, Stanley D., Jack F. Williams, and Donald Ziegler, eds. *Cities of the World: World Regional Urban Development.* 3d rev. ed. Lanham, Md.: Rowman and Littlefield, 2003.

Carter, Harold. *The Study of Urban Geography.* 4th ed. London: Edward Arnold, 1995.

Chapman, Graham P., and Ashok K. Dutt, eds. *Urban Growth and Development in Asia: Living in the Cities.* Aldershot, Hampshire: Ashgate, 1999.

Drakakis-Smith, David. *Third World Cities.* 2d rev. ed. New York: Routledge, 2000.

Ford, Larry R. *America's New Downtowns.* Baltimore: Johns Hopkins University Press, 2003.

Ford, Larry R. *Cities and Buildings: Skyscrapers, Skid Rows, and Suburbs.* Baltimore: Johns Hopkins University Press, 1994.

Hall, Tim. *Urban Geography.* 2d ed. London: Routledge, 2001.

Hanson, Susan, ed. *The Geography of Urban Transportation.* 2d ed. New York: Guilford, 1995.

Hartshorn, Truman. *Interpreting the City.* 3d ed. New York: Wiley, 1998.

Kaplan, David, James O. Wheeler, and Steven Holloway. *Urban Geography.* New York: Wiley, 2003.

King, Leslie J. *Central Place Theory.* Newbury Park, Calif.: Sage, 1984.

Knox, Paul L., and Steven Pinch. *Urban Social Geography: An Introduction.* 4th ed. Upper Saddle River, N.J.: Prentice Hall, 2000.

Knox, Paul L., and Peter J. Taylor, eds. *World Cities in a World System.* Cambridge: Cambridge University Press, 1995.

Ley, David. *The New Middle Class and the Remaking of the Central City.* Oxford: Clarendon Press, 1996.

Marcuse, Peter, and Ronald van Kampen, eds. *Globalizing Cities: A New Spatial Order?* Oxford: Blackwell, 2000.

Massey, Doreen, John Allen, and Steve Pile, eds. *City Worlds* (in the *Understanding Cities* series). New York: Routledge, 2000.

Pacione, Michael. *Urban Geography: A Global Perspective.* London: Routledge, 2001.

Sassen, Saskia, ed. *Global Networks: Linked Cities.* New York: Routledge, 2002.

Scott, Allen J. *Metropolis: From the Division of Labor to Urban Form.* Berkeley: University of California Press, 1988.

Sheehan, Molly O'Meara. *City Limits: Putting the Brakes on Urban Sprawl.* Worldwatch paper 156. Washington, D.C.: Worldwatch Institute, 2001.

Smith, David W. *Third World Cities.* 2d ed. New York: Routledge, 2000.

Smith, Neil. *The New Urban Frontier: Gentrification and the Revanchist City.* New York: Routledge, 1996.

Stimpson, C. R., E. Dixler, M. J. Nelson, and K. B. Yatrakis. *Women and the American City.* Chicago: University of Chicago Press, 1981.

United Nations Center for Human Settlements (Habitat). *An Urbanizing World: Global Report on Human Settlements 1996.* New York: Oxford University Press, 1996.

Yeates, Maurice. *The North American City.* 5th ed. New York: Longman, 1997.

A-25
H-27

# Human Impact on the Environment

**Ecosystems**

**Impact on Water**

- Availability of Water
- Modification of Streams
- Water Quality
- Agricultural Sources of Water Pollution
  - *Fertilizers*
  - *Biocides*
  - *Animal Wastes*
- Other Sources of Water Pollution
  - *Industry*
  - *Mining*
  - *Municipalities and Residences*
- Controlling Water Pollution

**Impact on Air and Climate**

- Air Pollutants
- Factors Affecting Air Pollution
- Acid Rain
- Photochemical Smog
- Depletion of the Ozone Layer
- Controlling Air Pollution

**Impact on Landforms**

- Landforms Produced by Excavation
- Landforms Produced by Dumping
- Formation of Surface Depressions

**Impact on Plants and Animals**

- Habitat Disruption
- Hunting and Commercial Exploitation
- Exotic Species
- Poisoning and Contamination

**Solid-Waste Disposal**

- Municipal Waste
  - *Landfills*
  - *Incineration*
  - *Source Reduction and Recycling*
- Hazardous Waste

**Summary**

**Key Words**

**For Review & Consideration**

**Selected References**

**Steel cylinders containing spent nuclear reactor fuel at a government facility in Idaho.** © Peter Essick/Aurora.

Times Beach, Missouri, used to be a tight-knit, working-class community along the Meramec River southwest of St. Louis. Now the people are gone. Empty houses and stores, abandoned cars, and rusting refrigerators stand silent. "Thanks for coming," says the sign on the Easy Living Laundromat to nonexistent customers. Wildflowers bloom on overgrown lawns. Flies buzz, squirrels chase each other up trees, and an occasional coyote prowls the streets. Were they allowed, visitors might think they were seeing an uncannily realistic Hollywood stage set. Instead, the ghost town is a symbol of what happens when hazardous wastes contaminate a community.

The trouble began in the summer of 1971, when Times Beach hired a man to spread oil on the town's 16 kilometers (10 mi) of dirt roads to keep down the dust. Unfortunately, his chief occupation was removing used oil and other waste from a downstate chemical factory, and it was a mixture of these products that he sprayed all over town that summer and the next. The tens of thousands of gallons of purple sludge were contaminated with dioxin, one of the most toxic substances ever manufactured.

Effects of the dioxin exposure appeared almost immediately in two riding arenas that had been sprayed. Within days, hundreds of birds and small animals died, kittens were stillborn, and many horses became ill and died. Then residents began reporting a variety of medical disorders, including miscarriages, seizure disorders, liver impairment, and kidney cancer. Not until 1982 did the Environmental Protection Agency (EPA), alerted to high levels of dioxin in wastes stored at the chemical plant, make a thorough investigation of Times Beach. They found levels of the dioxin compound TCDD as high as 300 parts per billion in some of the soil samples; 1 part per billion is the maximum concentration deemed safe. The EPA purchased every piece of property in town and ordered its evacuation (Figure 12.1).

story of Times Beach is one of many that could have ı selected to illustrate how people can affect the quality ıe water, air, and soil on which their existence depends. Terrestrial features and ocean basins, elements of ıther and characteristics of climate, flora, and fauna conıte the building blocks of that complex mosaic called the *ronment,* or the totality of things that in any way may ct an organism. Plants and animals, landforms, soils and ients, weather and climate all comprise an organism's ironment. The study of how organisms interact with one ther and with their physical environment is called **ecol-**. It is critically important in understanding environmenproblems, which usually arise from a disturbance of the ıral systems that make up our world.

Humans exist within a natural environment that they e modified by their individual and collective actions. ests have been cleared, grasslands plowed, dams built, cities constructed. On the natural environment, then, been erected a cultural environment, modifying, alteror destroying the balance of nature that existed before ıan impact was expressed. This chapter is concerned ı the interrelation between humans and the natural ironment that they have so greatly altered.

Since the beginning of agriculture, humans have nged the face of the earth, have distorted delicate bales and interplays of nature, and, in the process, have ı enhanced and endangered the societies and the ıomies that they have erected. The essentials of the natural balance and the ways in which humans have altered it are not only our topics here but also matters of social concern that rank among the principal domestic and international issues of our times. As we shall see, the fuels we consume, the raw materials we use, the products we create, and the wastes we discard all contribute to the harmful alteration of the **biosphere,** the thin film of air, water, and earth within which we live.

**FIGURE 12.1 Times Beach, Missouri.** Most new highway maps no longer mark the location of this former community. *© Joe Sohm/The Image Works.*

# ECOSYSTEMS

The biosphere is composed of three interrelated parts:

1. the *troposphere,* the lowest layer of the earth's atmosphere, extending about 9.5–11.25 kilometers (6–7 mi) above the ground;
2. the *hydrosphere,* including surface and subsurface waters in oceans, streams, lakes, glaciers, or groundwater—much of it locked in ice or earth and not immediately available for use; and
3. the upper reaches of the earth's *crust,* a few thousand feet at most, containing the soils that support plant life, the minerals that plants and animals require to exist, and the fossil fuels and ores that humans exploit.

The biosphere is an intricately interlocked system, containing all that is needed for life, all that is available for living things to use, and, presumably, all that ever will be available. The ingredients of the biosphere must be, and are, constantly recycled and renewed in nature. Plants purify the air, the air helps purify the water, the water and the minerals are used by plants and animals and are returned for reuse.

The biosphere, therefore, consists of two intertwined components: (1) a nonliving outside energy source—the sun—and requisite chemicals and (2) a living world of plants and animals. In turn, the biosphere may be subdivided into specific **ecosystems,** self-sustaining units that consist of all the organisms (plants and animals) and physical features (air, water, soil, and chemicals) existing together in a particular area. The most important principle concerning all ecosystems is that everything is interconnected. Any intrusion or interruption in the balance that has been naturally achieved inevitably results in undesirable effects elsewhere in the system. Each organism occupies a specific *niche,* or place, within an ecosystem. In the energy exchange system, each organism plays a definite role; individual organisms survive because of other organisms that also live in that environment. The problem lies not in recognizing the niches but in anticipating the chain of causation and the readjustments of the system consequent on disturbing the occupants of a particular niche.

Life depends on the energy and nutrients flowing through an ecosystem. The transfer of energy and materials from one organism to another is one link in a **food chain,** a sequence of organisms, such as green plants, herbivores, and carnivores, through which energy and materials move within an ecosystem (Figure 12.2). Most food chains have three or four links, although some have only two—for example, when human beings eat rice. Because the ecosystem in nature is in a continuous cycle of integrated operation, there is no start or end to a food chain. There are, simply, nutritional transfer stages in which each lower level in the food chain transfers part of its contained energy to the next higher-level consumer.

The *decomposers* pictured in Figure 12.2 are essential in maintaining food chains and the cycle of life. They cause the disintegration of organic matter—animal carcasses and droppings, dead vegetation, waste paper, and so on. In the process of decomposition, the chemical nature of the material is changed, and the nutrients contained within it become available for reuse by plants or animals. *Nutrients,* the minerals and other elements that organisms need for growth, are never destroyed; they keep moving from living to nonliving things and back again. Our bodies contain nutrients that were once part of other organisms, perhaps a hare, a hawk, or an oak tree.

Ecosystems change constantly whether people are present or not, but humans have affected them more than has any other species. The impact of humans on ecosystems was small at first, with low population size, energy consumption, and technological levels. It has increased so rapidly and pervasively as to present us with widely recognized and varied ecological crises. Some of the effects of humans on the natural environment are the topic of the remainder of this chapter.

# IMPACT ON WATER

The supply of water is constant. The system by which it continuously circulates through the biosphere is called the **hydrologic cycle** (see Figure 12.3). In that cycle, water may change form and composition, but under natural environmental circumstances, it is purified in the recycling process

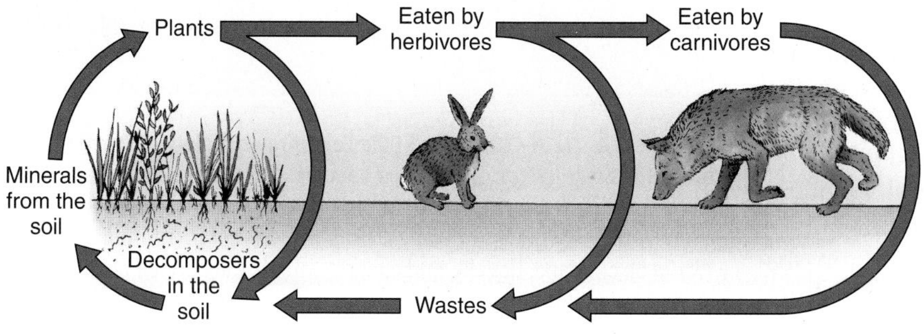


**FIGURE 12.2** The supply of food in an ecosystem is a hierarchy of "who eats what"—a hierarchy that creates a food chain. In this simplified example, green plants are the *producers* (autotrophs), using nutrients and energy from the sun to make their own food. Herbivorous rabbits (*primary consumers*) feed directly on the plants, and carnivorous foxes (*secondary consumers*) feed on the rabbits. A food chain is one thread in a complex *food web,* all the feeding relationships that exist in an ecosystem. For example, a mouse might feed on the plants shown here and then be eaten by a hawk, another food chain in this food web.

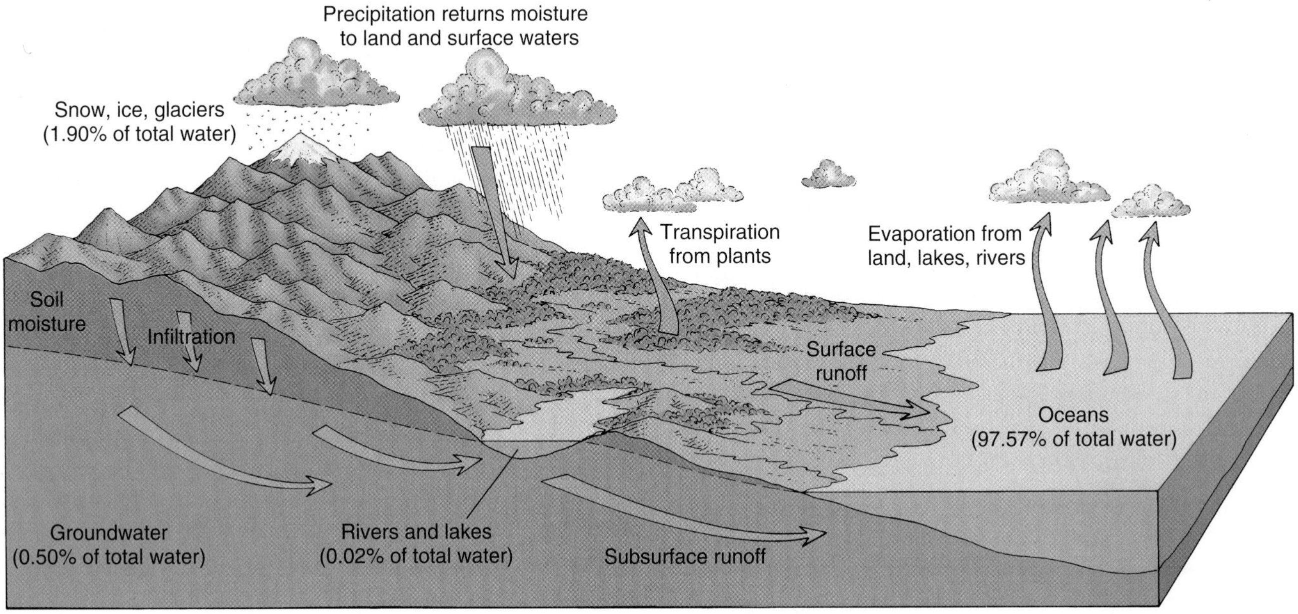


**FIGURE 12.3 The hydrologic cycle.** The sun provides energy for the evaporation of fresh and ocean water. The water is held as vapor until the air becomes supersaturated. Atmospheric moisture is returned to the earth's surface as solid or liquid precipitation to complete the cycle. Because precipitation is not uniformly distributed, moisture is not necessarily returned to areas in the same quantity as it has evaporated from them. The continents receive more water than they lose. The excess returns to the seas as surface water or groundwater. A global water balance, however, is always maintained.

and is again made available with appropriate properties to the ecosystems of the earth. *Evaporation, transpiration* (the emission of water vapor from plants), and *precipitation* are the mechanisms by which water is redistributed. Water vapor collects in clouds, condenses, and then falls again to the earth. There it is reevaporated and retranspired, only to fall once more as precipitation.

People's dependence on water has long led to efforts to control its supply. Such manipulation has altered the quantity and quality of water in rivers and streams.

## Availability of Water

Globally, fresh water is abundant. Enough rain and snow fall on the continents each year to cover the earth's total land area with 83 centimeters (33 in.) of water. It is usually reckoned that the volume of fresh water annually renewed by the hydrologic cycle could meet the needs of a world population 5 to 10 times its present size.

In many parts of the world, however, water supplies are inadequate and dwindling. The problem is not with the global amount of water but with its distribution (the average amount of precipitation an area receives) and reliability (the variability of precipitation from year to year). Regional water sufficiency is also a function of the size of the population using the water and the demands it places on the resource. For the world as a whole, irrigated agriculture accounts for about two-thirds of freshwater use; in the poorest countries, the proportion is 90% (Figure 12.4). Industry uses about one-fifth (21%), and households and municipal-

**FIGURE 12.4 Irrigation in the Lake Argyle project of Western Australia.** Since 1950, irrigation has been extended to between 5 and 6 million additional hectares (12 to 15 million acres) annually. Irrigation agriculture now produces about 40% of the world's harvest from about 17% of its cropland. Usually, much more water is transported and applied to fields than crops actually require, using unnecessary amounts of a scarce resource. "Drip irrigation," which delivers water directly to plant roots through small perforated tubes laid across the field, is one method of reducing water consumption. Source: *Robert Frerck/Odyssey Productions/Chicago.*

ities account for the remainder. Since 1950, humanity has more than tripled its withdrawals of fresh water from streams, lakes, aquifers, and other sources. (An *aquifer* is a zone of water-saturated sands and gravels beneath the earth's surface; the water it contains is called *groundwater,* in contrast to surface waters such as rivers and lakes.)

*Scarcity* is the word increasingly used to describe water supplies in parts of both the developed and developing world (Figure 12.5). Insufficient water for irrigation periodically endangers crops and threatens famine; permanent streams have become intermittent in flow; lakes are shrinking; and from throughout the world come reports of rapidly falling water tables and wells that have gone dry. According to the World Bank, chronic water shortages that threaten to limit food production, economic development, sanitation, and environmental protection already plague 80 countries. Ten countries in North Africa and the Middle East actually run a *water deficit:* they consume more than their annual renewable supply, usually by pumping groundwater faster than it is replenished by rainfall.

National data can mask water scarcity problems at the local level. A number of countries have major crop-producing regions where groundwater overpumping and aquifer depletion have led to serious water shortages and restricted supplies.

- In northern China, for example, irrigated farmland and urban and industrial growth are so depleting water supplies that for much of the year the Huang He (Yellow River) runs dry in its lower reaches before arriving at the Yellow Sea. Some 400 of China's large cities, most of them in the north, already have serious water shortages. The Chinese cabinet in 2002 approved the construction of an enormous project to move water over three 1600-kilometer-long (1000-mi) pathways of canals and aqueducts from Chang Jiang River in the south to the Huang He in the north, where it will serve industries and cities such as Beijing and Tianjin. The project is expected to take 8 years to complete.
- In India, the world's second most populous country, the overall rate of groundwater withdrawal is twice the rate of recharge. In several states, including Haryana and Punjab, the country's "breadbasket," excessive water use is causing water tables to fall and wells to dry up.
- Lake Chapala, the largest body of fresh water in Mexico, has lost about 80% of its intake water since the mid-1970s. The lake is fed primarily by the Rio Lerma, but almost all of the river's flow is now diverted for irrigation and industry. In Mexico City, groundwater is pumped at rates 40% faster than natural recharge; the city responded to those withdrawals by sinking 9 meters (30 ft) during the 20th century.

In the United States, too, groundwater supplies are being depleted faster than they can be renewed in portions

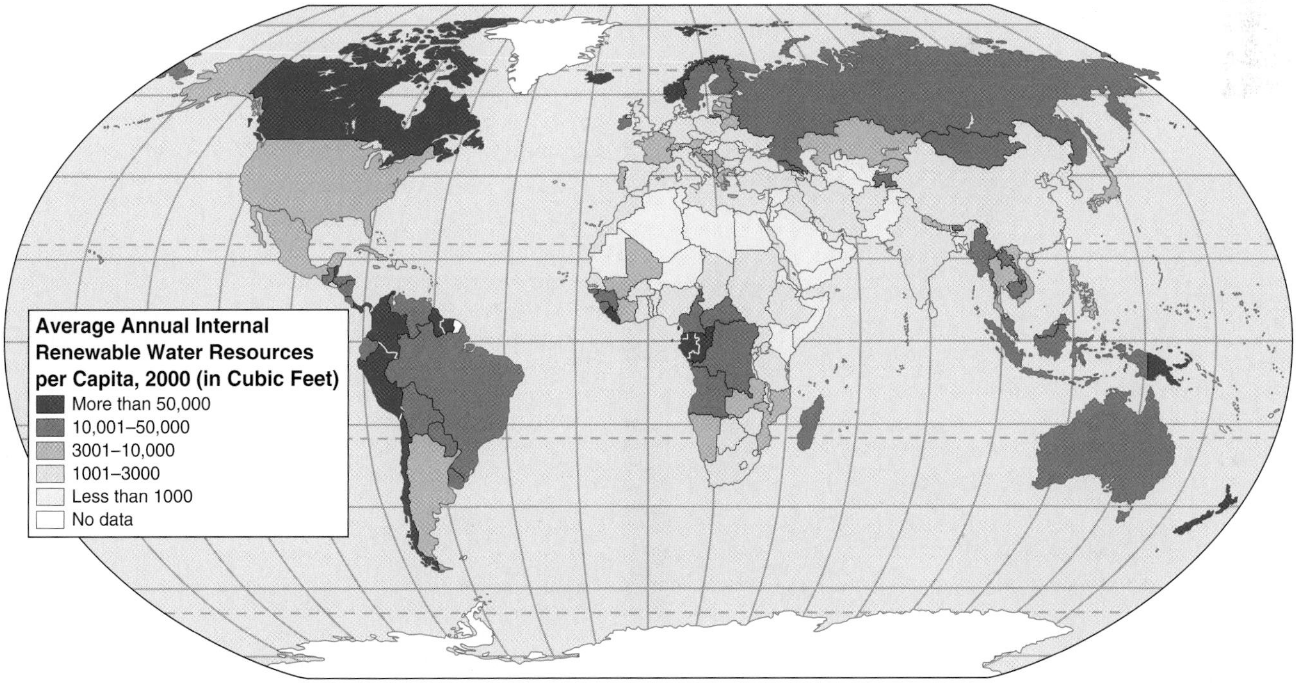


**FIGURE 12.5 Availability of renewable water per capita.** Renewable water resources are those available from streams, lakes, and aquifers. Although the United States has tremendous quantities of fresh water, it also uses tremendous quantities. Water consumption rises with population growth, increases in the standard of living, expansion of irrigation farming, and the enlarged demands of industry and municipalities that come with development. In a growing number of world areas, water is scarce, constraining sustainable development and requiring hard choices in the allocation of water among competing users. Source: Student Atlas of World Geography, *3/e, John Allen, Map 50, p. 63. McGraw-Hill/Dushkin, 2003.*

of Arizona, New Mexico, California, and Nevada. Demands upon the only significant source of surface water in the southwestern United States, the Colorado River, are so great that the stream often reaches the ocean (at the Gulf of California) as little more than a briny trickle. Aqueducts and irrigation canals siphon off its water for use in seven western states and northern Mexico. A number of western cities, including Denver and Santa Fe, have imposed mandatory restrictions on water use.

The country's largest underground water reserve, the Ogallala aquifer, which supplies about 25% of all U.S. irrigated land, is being depleted eight times faster than it is being replenished (Figure 12.6). Stretching from South Dakota to west Texas, the aquifer supports nearly half of the country's cattle industry, a fourth of its cotton crop, and a great deal of its corn and wheat. More than 150,000 wells now puncture the aquifer, pumping water for irrigation, industry, and domestic use.

The UN estimates that by 2015, nearly 3 billion people, or 40% of the projected world population, will live in water-stressed countries where it is difficult or impossible to get enough water to satisfy basic needs. Many analysts believe that regional shortages of this indispensable resource could trigger water wars in the not too distant future. Tensions will be exacerbated in areas where two or more countries share a river system. Turkey, for instance, has by building dams and irrigating fields taken much of the water of the Euphrates River, reducing the amount of water available in downstream Syria and Iraq.

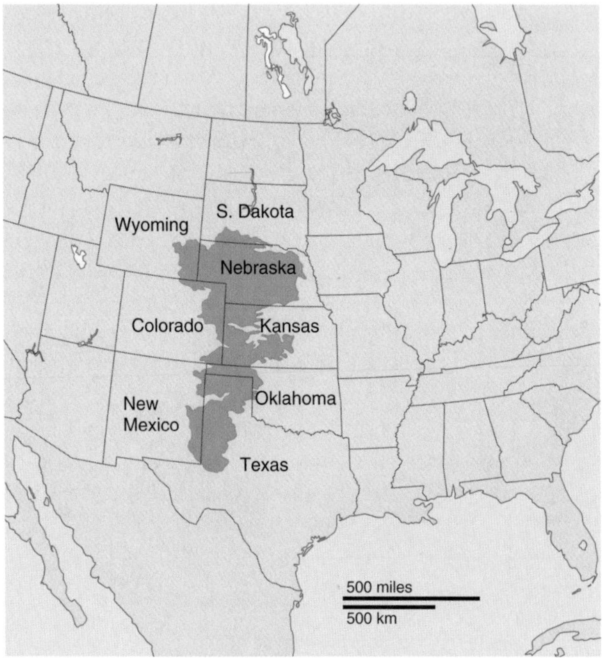


**FIGURE 12.6 The giant Ogallala aquifer,** the largest underground water supply in the United States, provides about one-fourth of all groundwater used for irrigation in the country. Depletion of the aquifer is most severe in its southern portion, where the Ogallala gets little replenishment from rainfall. It is of particular concern in Texas, whose population is expected to double from its year 2000 population of 21 million, by 2050.

Because the oceans contain a never-ending supply of water, some have looked to *desalinization* (cleansing sea water of its salt and mineral content) to provide a technological solution to water shortfall. Because the process is economically inefficient, it is unlikely to supply more than a small portion of humanity's needs. If in the future the costs fall far enough, desalinization could help augment domestic water supplies; it is doubtful that it could ever become cheap enough to provide water for agriculture.

## Modification of Streams

To prevent flooding, to regulate the water supply for agriculture and urban settlements, or to generate power, people have for thousands of years manipulated rivers by constructing dams, canals, and reservoirs. Although they generally have achieved their purposes, these structures can have unintended environmental consequences, as discussed in chapter 5. These include reduction in the sediment load downstream, followed by a reduction in the amount of the nutrients available for crops and fish; an increase in the salinity of the soil; and subsidence (see "Blueprint for Disaster: Stream Diversion and the Aral Sea").

**Channelization,** another method of modifying river flow, is the construction of embankments and dikes and the straightening, widening, and/or deepening of channels to control floodwaters or to improve navigation. Many of the great rivers of the world, including the Mississippi, the Nile, and the Huang He, are lined by embankment systems. Like dams, these systems can have unforeseen consequences. They reduce the natural storage of floodwaters, can aggravate flood peaks downstream, and can cause excessive erosion.

Until 1960, the Kissimmee River in Florida twisted and turned as it traveled through a floodplain between Lake Kissimmee and Lake Okeechobee (Figure 12.7). The habitat supported hundreds of species of birds, reptiles, mammals, and fish. At the request of developers, ranchers, and farmers who had moved onto the wetlands and were disturbed by the tendency of the river to flood, the Army Corps of Engineers dredged the stream, turning 166 kilometers (103 mi) of meandering river into a dirt-lined canal only 90 kilometers (56 mi) long. After the completion of the canal in 1971, the wetlands disappeared, alien plant species moved in, fish populations declined drastically, and 90% of the waterfowl, including several endangered species, disappeared.

Channelization and dam construction are deliberate attempts to modify river regimes, but other types of human action also affect river flow. Urbanization, for example, has significant hydrologic impacts, including a lowering of the water table, pollution, and increased flood runoff. Likewise, the removal of forest cover increases runoff, promotes flash floods, lowers the water table, and hastens erosion.

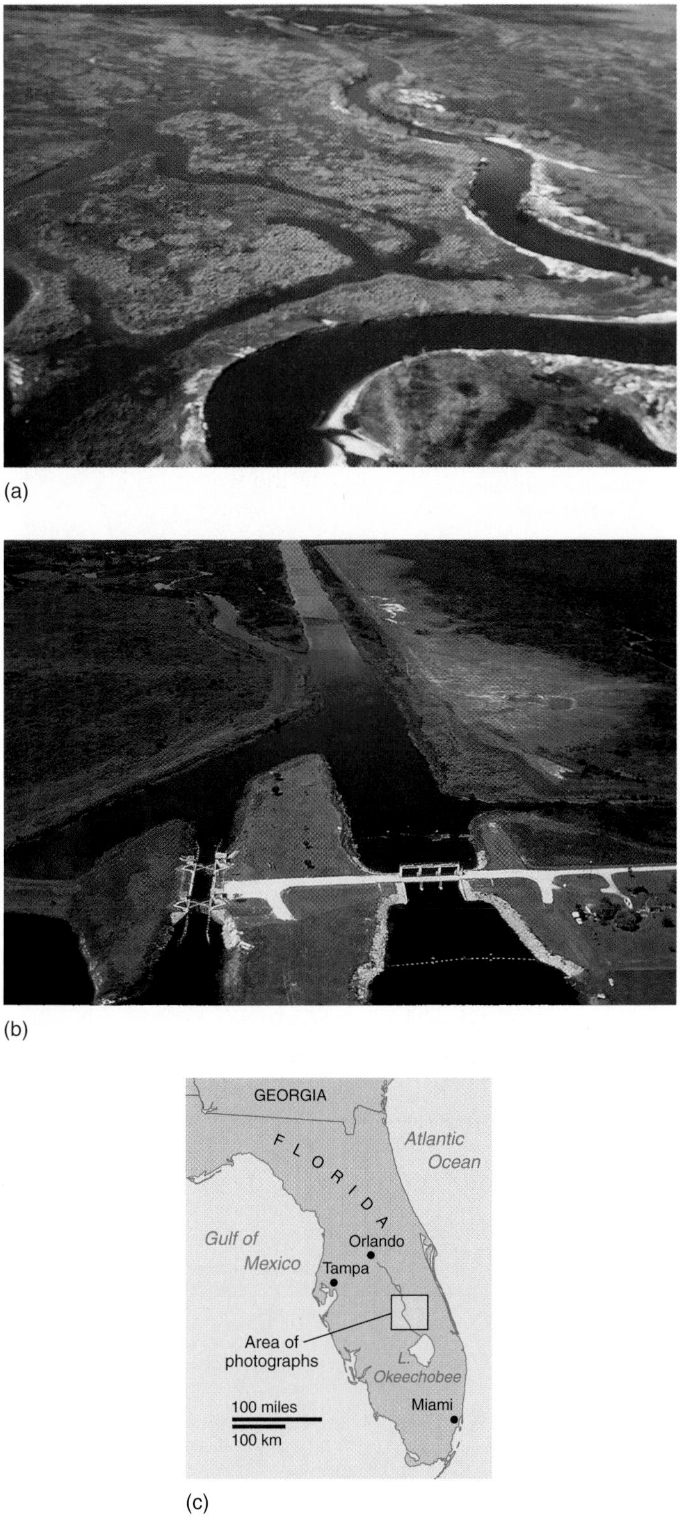


**FIGURE 12.7 The Kissimmee River** in Florida (a) before and (b) after the Army Corps of Engineers turned it into a straight canal. The habitat once supported thousands of fish, waterfowl, and such wading birds as the wood stork, snowy egret, and great blue heron. The channelization contributed to the deterioration of Lake Okeechobee and the Everglades. The federal government and the state of Florida have embarked on a long-term project to return the Kissimmee to a meandering path. *(a), (b) Courtesy of South Florida Water Management District.*

Nevertheless, the primary adverse human impact on water is felt in the area of water quality. People withdraw water from lakes, rivers, or underground deposits to use for drinking, bathing, agriculture, industry, and many other purposes. Although the water that is withdrawn returns to the water cycle, it is not always returned in the same condition as it was at the time of withdrawal. Water, like other segments of the ecosystem, is subject to serious problems of pollution.

## Water Quality

As a general definition, **environmental pollution** by humans means the introduction into the biosphere of wastes that, because of their volume, their composition, or both, cannot be readily disposed of by natural recycling processes. In the case of water, the central idea is that pollution exists when water composition has been so modified by the presence of one or more substances that either it cannot be used for a specific purpose or it is less suitable for that use than it was in its natural state. Pollution is brought about by the discharge into water of substances that cause unfavorable changes in its chemical or physical nature or in the quantity and quality of the organisms living in the water. *Pollution* is a relative term. Water that is not suitable for drinking may be completely satisfactory for cleaning streets. Water that is too polluted for fish may provide an acceptable environment for certain water plants.

Human activity is not the only cause of water pollution. Leaves that fall from trees and decay, animal wastes, oil seepages, and other natural phenomena may affect water quality. There are natural processes, however, to take care of such pollution. Organisms in water are able to degrade, assimilate, and disperse such substances in the amounts in which they naturally occur. Only in rare instances do natural pollutants overwhelm the cleansing abilities of the recipient waters. What is happening now is that the quantities of wastes discharged by humans often exceed the ability of a given body of water to purify itself. In addition, humans are introducing pollutants, such as metals or inorganic substances, that take a very long time to break down or cannot be broken down at all by natural mechanisms.

As long as there are people on Earth, there will be pollution. Thus, the problem is one not of eliminating pollution but of controlling it. Such control can be a matter of life or death. Waterborne pathogens and pollution kill millions of people (mostly children) each year. The people die from diarrhea and water-related diseases, such as cholera and typhoid. One column in Appendix 2 at the back of this book shows, by country, the percent of the population with access to safe water.

The four major contributors to water pollution are agriculture, industry, mining, and municipalities and residences. It is helpful to distinguish between "point" and "nonpoint" sources of pollution. As the name implies, *point sources* enter the environment at specific sites, such as a sewage treatment facility or an industrial discharge pipe. *Nonpoint sources* are

# Blueprint for Disaster: Stream Diversion and the Aral Sea

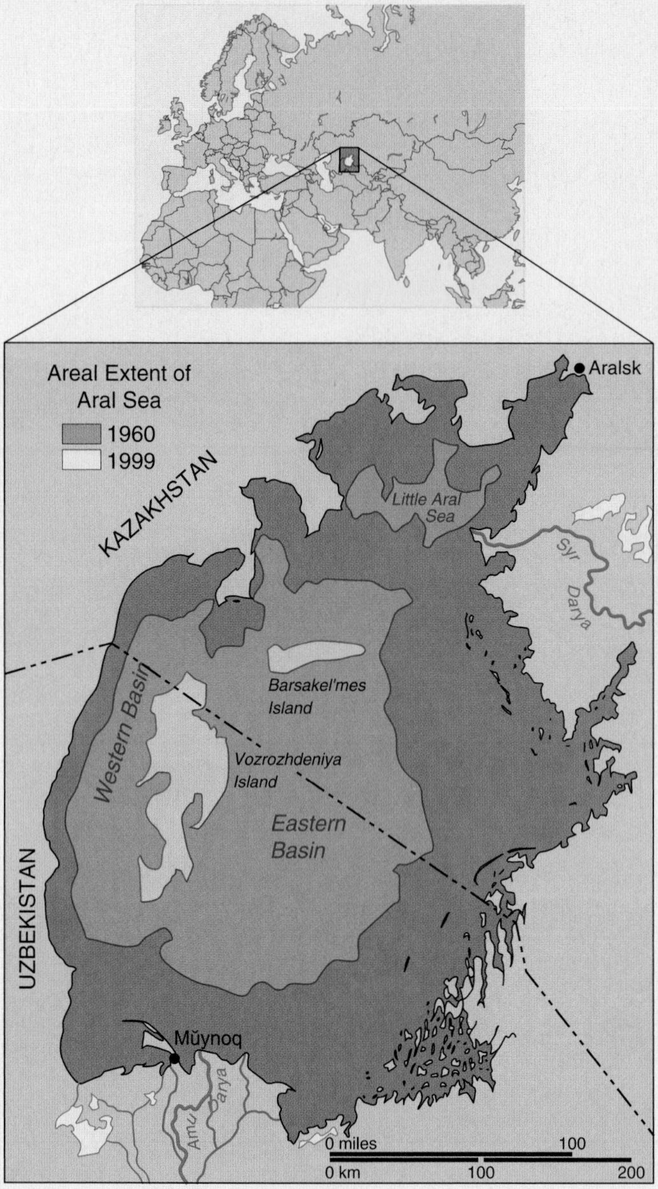


It used to be the fourth largest lake in the world, covering an area of 69,900 square kilometers (27,000 sq mi), larger than the state of West Virginia. Now, the Aral Sea is only the eighth largest lake, and it may break up into three smaller lakes by the year 2010. The level of the sea has dropped more than 18 meters (59 ft) since 1960, and the volume of the water is only 20% of what it used to be.

The shrinkage of the lake is just one consequence of the former Soviet Union diverting nearly all the water from the lake's primary sources, the Amu Darya and the Syr Darya, to irrigate the agricultural fields of central Asia. These are some of the others:

- As the water receded, it exposed 36,000 square kilometers (13,900 sq mi) of seabed. The high concentrations of salt, fertilizers, and pesticides are slowly turning the Aral into a dead sea. Because the water is too salty for most fish species to endure, almost all the fish have died out, and the commercial fishing industry has collapsed. Once a fishing port and popular resort, Mŭynoq is now 70 kilometers (43 mi) from the sea. Aralsk was once the main northern port for a sea that is now 80 kilometers (50 mi) away—a sea that is now called the Little Aral Sea.
- In addition, wind storms whip up salty grit and toxic chemicals from the dried-up seabed and the land around it and deposit it on cropland hundreds of kilometers away, reducing the soil's fertility. Thus, the agricultural crops (cotton, rice, fruit, and vegetables) for which the Aral Sea was sacrificed are themselves at risk.
- Forests and wetlands upon which the region's animal life depend have been decimated. Three-quarters of the animal species in the basin are estimated to have vanished.
- Finally, the rivers that feed the Aral Sea have become sluggish sewers, contaminated by industrial and agricultural wastes and sewage. Although the rivers teem with viruses, bacteria, and other organisms causing dysentery, typhoid, and other diseases, millions of people depend on them for drinking water. Over 65,000 cases of hepatitis have been reported in the region in the last 15 years, and respiratory system diseases (asthma, bronchitis) have increased significantly, as have cancers of the stomach and esophagus. Child mortality rates are among the highest in the world. The newly independent states of central Asia are in a poor position to reduce irrigation or undertake major health improvement projects in the Aral region.

The drastic consequences of river diversion in the Aral Sea basin illustrate why it has been called one of the world's greatest environmental tragedies.

more diffuse and, therefore, more difficult to control; examples include runoff from agricultural fields and road salts.

## Agricultural Sources of Water Pollution

On a worldwide basis, agriculture probably contributes more to water pollution than does any other single activity. In the United States, agriculture is estimated to be responsible for about two-thirds of stream pollution. Agricultural runoff carries three main types of pollutants: fertilizers, biocides, and animal wastes.

### Fertilizers

Agriculture is a chief contributor of *excess nutrients* to water bodies. Pollution occurs when nitrates and phosphates that have been used in fertilizers and that are present in animal manure drain into streams and rivers, eventually accumulating in ponds, lakes, and estuaries. The nutrients hasten the process of **eutrophication,** or the enrichment of waters by nutrients. Eutrophication occurs naturally when nutrients in the surrounding area are washed into the water, but when the sources of enrichment are artificial, as is true of commercial fertilizers, the body of water may become overloaded with nutrients. Algae and other plants are stimulated to grow abundantly, blocking the sunlight that other organisms need. When they die and decompose, the level of dissolved oxygen in the water decreases. Fish and plants that cannot tolerate the poorly oxygenated water are eliminated.

Worldwide, this *accelerated* eutrophication is estimated to affect approximately half of all lakes and reservoirs in North America, Europe, and Southeast Asia, some 40% in South America, and 28% in Africa. Symptoms of a eutrophic lake are prolific weed growth, large masses of algae, fish kills, rapid accumulation of sediments on the lake bottom, and water that has a foul taste and odor.

High levels of nutrients from agricultural runoff in the Mississippi River watershed have helped create a "dead zone," an area of oxygen depletion in which fish, crabs, and other aquatic creatures cannot survive, in the Gulf of Mexico off the coast of Louisiana (Figure 12.8). The zone, whose size varies from year to year, reaches its peak during the summer months, as the water grows warmer and solar radiation increases, causing algal populations to bloom.

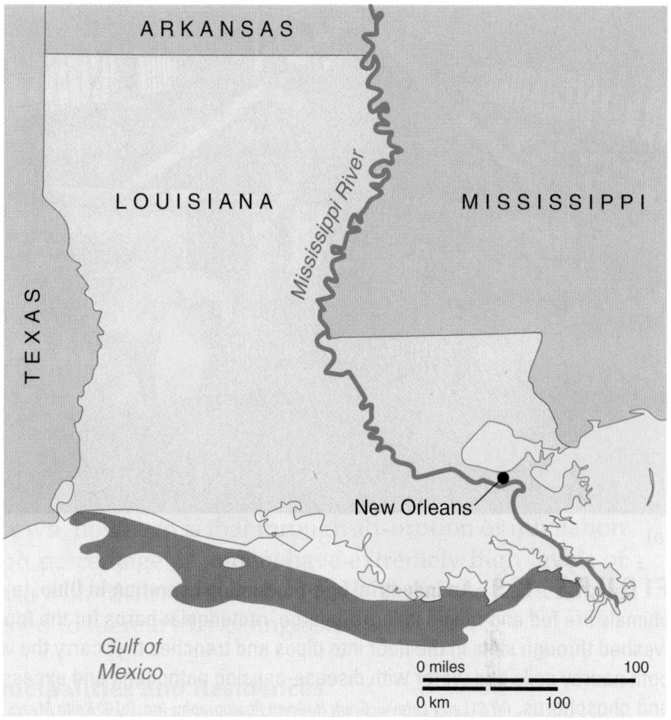


**FIGURE 12.8 The "dead zone" in the Gulf of Mexico** in 2001 covered roughly 23,000 square kilometers (9000 sq mi), an area the size of Massachusetts. Nutrients from agricultural runoff are thought to be a critical factor in creating this area of oxygen depletion. The nutrients feed algal populations that bloom during the summer. After the algae die and sink to the bottom of the Gulf, they decompose, depleting the oxygen near the ocean floor and leaving too little oxygen to sustain life.

### Biocides

The herbicides and pesticides used in agriculture are another source of the chemical pollution of water bodies. Runoff from farms where such *biocides* have been applied contaminates both ground- and surface waters. One of the problems connected with the use of biocides is that the long-term effects of such usage are not always immediately known. DDT (dichlorodiphenyl trichloroethane), for example, was used for many years before people discovered its effect on birds, fish, and water plant life. Another problem is that thousands of these products, containing more than 600 active ingredients, are now in wide use, yet very few have been reviewed for safety by the Environmental Protection Agency (EPA). Finally, biocides that leach into aquifers can remain there long after the chemicals are no longer used. Thus DDT is still found in U.S. waters even though its use was banned in the late 1960s.

### Animal Wastes

A final agricultural source of chemical pollution is animal wastes, especially in countries where animals are raised intensively. This is a problem both in feedlots, where animals are crowded together at maximum densities to be fattened before slaughter, and on the factory-like farms where beef, hog, and poultry production is increasingly concentrated (Figure 12.9). American farms and large feedlots produce vast quantities of manure—nearly 3 billion pounds a day—but usually lack sewage treatment facilities. The main method for disposing of the waste is to put it into open lagoons (containment ponds) and then spray it onto surrounding fields, from which it can leach into the water table and rivers.

The water pollution that occurs from spreading manure on land is suspected by some to be responsible for recent outbreaks of the microorganism *Pfiesteria piscicida* in the

waters in India are polluted, in large part because only about 200 of its more than 3000 cities have full or partial sewage-collection and -treatment facilities. Of Taiwan's 22 million people, fewer than 1 million are served by sewers. Raw sewage from 3.6 million inhabitants of Hong Kong flows into Victoria Harbor.

Although sewage-treatment plants are common in the United States, only half of the American population lives in communities with facilities that meet the minimum goals set by the federal Clean Water Act. Aged sewage systems in 1100 cities still discharge poorly treated sewage into streams, lakes, and oceans. When the sewer system of Dade County, Florida, ruptures, as it does periodically, millions of gallons of raw sewage pour into the Miami River, which empties into Biscayne Bay in downtown Miami.

In many communities, special problems arise after heavy rains, when storm water filled with animal wastes, street debris, and lawn chemicals floods the sewers. As treatment plants become overloaded, both runoff and raw sewage are diverted into rivers, bays, and oceans. New York City alone has more than 500 storm water outlets that overflow in heavy rains, pouring some 65 billion gallons of untreated sewage (about 10% of the city's total sewage) into the Hudson River and Long Island Sound each year. But 1200 other cities in the East, Midwest, and Northwest also have sewage networks that are combined with the overflow systems for storm water.

### Controlling Water Pollution

In recent decades, concern over increased levels of pollution has brought about major improvements in the quality of some surface waters, both in the United States and abroad. The U.S. federal government in 1972 took the lead in regulating water pollution with the enactment of the Clean Water Act. Its objective was "to restore and maintain the chemical, physical, and biological integrity of the nation's waters." Congress established uniform nationwide controls for each category of major polluting industry and directed the government to pay most of the cost of new sewage-treatment plants. Since 1972, such plants have been built to serve over 80 million Americans, and industries have spent billions of dollars to comply with the Clean Water Act by reducing organic waste discharges.

The gains have been impressive. Many rivers and lakes that were ecologically dead or dying are now thriving. Once dumping grounds for all kinds of human and industrial waste, the Hudson, Potomac, Cuyahoga, and Trinity rivers are cleaner, more inviting, and more productive than before, and they now support fishing, swimming, and recreational boating. Similarly, Seattle's Lake Washington and the Great Lakes are healthier than they were two decades ago. Recently, authorities have announced ambitious plans to clean up the waters of Chesapeake Bay, the country's largest estuary, and to undo much of the damage that has been inflicted on Florida's Everglades by improving the water quality of the Kissimmee River and Lake Okeechobee.

Environmental awareness in other countries has also prompted legislation and action. For example, the river Thames in southern England, which had become seriously contaminated by the dumping of sewage and industrial wastes, is now cleaner than it has been in centuries. The enforcement of stringent pollution-control standards has halted the downward trend in quality. Algae and seaweed, fish and wildfowl have returned to the river in abundance.

Even the Mediterranean Sea is on its way to gradual recovery. When the 18 countries that border the sea signed the Convention for the Protection of the Mediterranean Sea against Pollution in 1976, all coastal cities were dumping their untreated sewage into the sea, tankers were spewing oily wastes into it, and tons upon tons of phosphorus, detergents, lead, and other substances were contaminating the waters. Now, many cities have built or are building sewage-treatment plants, ships are prohibited from indiscriminate dumping, and some national governments are beginning to enforce control of pollution from land-based sources.

Such gains should not mislead us. Although some of the most severe problems have been attacked, according to the EPA, in the United States, 50% of estuaries, 45% of lakes, and 39% of streams are polluted. The solution to water pollution lies in the effective treatment of municipal and industrial wastes; the regulation of chemical runoff from agriculture, mining, and forestry; and the development of less-polluting technologies. Although pollution-control projects are expensive, the long-term costs of pollution are even higher.

## Impact on Air and Climate

The **troposphere,** the thin layer of air just above the earth's surface, contains all the air that we breathe. Every day thousands of tons of pollutants are discharged into the air by cars and incinerators, factories and airplanes. Air is polluted when it contains substances in sufficient concentrations to have a harmful effect on living things.

### Air Pollutants

Truly clean air has probably never existed. Just as there are natural sources of water pollution, so are there substances that pollute the air without the aid of humans. Ash from volcanic eruptions, marsh gases, smoke from forest fires, and windblown dust are natural sources of air pollution.

Normally these pollutants are of low volume and are widely dispersed throughout the atmosphere. On occasion, a major volcanic eruption may produce so much dust that the atmosphere is temporarily altered. In general, however, the natural sources of air pollution do not have a significant, long-term effect on air, which, like water, is able to cleanse itself.

Far more important than naturally occurring pollutants are the substances that people discharge into the air. These pollutants result primarily from burning fossil fuels (coal, gas, and oil) and other materials. Fossil fuels are burned in power plants that generate electricity, in many industrial plants, in home furnaces, and in cars, trucks, buses, and airplanes. Scientists estimate that about three-quarters of all air pollutants come from burning fossil fuels. The remaining pollutants largely result from industrial processes other than fuel burning, the incineration of solid wastes, forest and agricultural fires, and the evaporation of solvents. Figure 12.10 depicts the major sources of air pollutants. Table 12.1 summarizes the major sources of six *primary* pollutants emitted in large quantities. Once they are in the atmosphere, these may react with other primary pollutants or with normal atmospheric constituents such as water vapor to form *secondary* pollutants.

Air pollution is a global problem. A recent study by the World Health Organization (WHO) concluded that more than 1.1 billion people live in urban areas with unhealthful air. Sulfur dioxide levels are considered unacceptable for some 625 million people in developing countries alone, and levels of smoke, dust, and other particulates are unacceptable for 1.25 billion people. Particularly at risk of breathing bad air are residents of such megacities as Mexico City, Cairo, Delhi, Seoul, Beijing, and Jakarta.

Like polluted water, dirty air kills people. According to the WHO, at least 3 million people die prematurely each year from illnesses caused by air pollution.

## Factors Affecting Air Pollution

Many factors affect the type and the degree of air pollution found at a given place. Those over which people have relatively little control are climate, weather, wind patterns, and topography. These determine whether pollutants will be blown away or are likely to accumulate. Thus a city on a plain is less likely to experience a buildup than is a city in a valley.

Unusual weather can alter the normal patterns of pollutant dispersal. A *temperature inversion* magnifies the effects of air pollution. Under normal circumstances, air temperature decreases away from the earth's surface. A stationary layer of warm, dry air over a region, however, will prevent the normal rising and cooling of air from below. As described in Chapter 4 (see Figure 4.10 and "The Donora Tragedy"), the air becomes stagnant during an inversion. Pollutants accumulate in the lowest layer instead of being blown away, so that the air becomes more and more contaminated. Normally, inversions last for only a few hours, although certain areas experience them much of the time. Temperature inversions occur often in Los Angeles in the fall and Denver in the winter. If an inversion lingers long enough, over several days, it can contribute to the accumulation of air pollutants to levels that seriously affect human health.

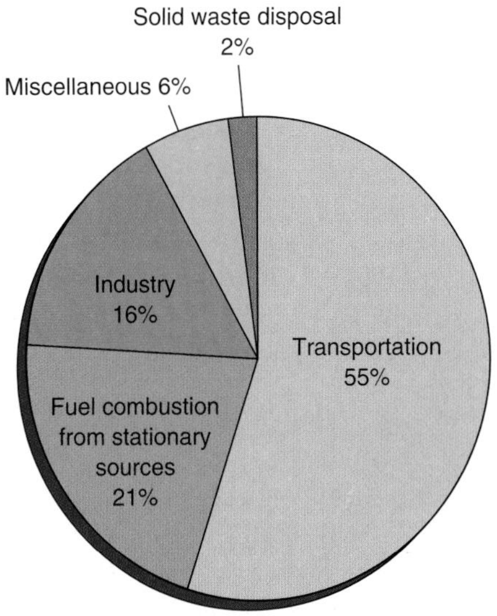


**FIGURE 12.10 Sources of primary air pollutants in the United States.** Primary air pollutants are materials that are emitted directly into the atmosphere in sufficient quantities to exert an adverse effect on human health or the environment. Transportation is the single largest source of human-caused air pollution. The second largest source is fuel combustion in stationary sources such as power plants and factories. *Redrawn from* Biosphere 2000, *by Donald G. Kaufman and Cecilia M. Franz (NY: HarperCollins College Publishers, 1993, Fig. 14.3, p. 251).*

**TABLE 12.1 | Major Sources of Primary Air Pollutants**

| Type of Pollutant | Symbol | Major Sources |
|---|---|---|
| Carbon dioxide | $CO_2$ | Combustion of fossil fuels |
| Carbon monoxide | CO | Incomplete combustion of fossil fuels, mostly in vehicles |
| Hydrocarbons | HC | Combustion of fossil fuels; petroleum refineries; petrochemical plants |
| Nitrogen oxides | $NO_x$ | Transportation vehicles; power plants |
| Particulates | — | Automobile exhaust; oil refining; coal-fired power plants; farming and construction operations |
| Sulfur oxides | $SO_x$ | Combustion of sulfur-containing fuels, especially coal |

The air pollutants generated in one place may have their most serious effect in areas hundreds of kilometers away. Thus, the worst effects of the air pollution that originates in New York City are felt in Connecticut and parts of Massachusetts. The chemical reaction that produces smog takes a few hours, and by that time air currents have carried the pollutants away from New York. In a similar fashion, New York is the recipient of pollutants produced in other places. Much of the acid rain that affects New England and eastern Canada originates in the coal-fired power plants along the lower Great Lakes and in the Ohio Valley that use extremely high smokestacks to disperse sulfurous emissions. And the coal-based industries in Russia and Europe produce sulfate, carbon, and other pollutants that are transported by air currents to the land north of the Arctic Circle, where they result in a contamination known as Arctic haze.

Other factors that affect the type and the degree of air pollution at a given place are the levels of urbanization and industrialization. Population densities, traffic densities, the type and density of industries, and home-heating practices all help determine the kinds of substances discharged into the air at a single point. In general, the more urbanized and industrialized a place is, the more responsible it is for pollution. The United States may contribute as much as one-third of the world's air pollution, a figure roughly equivalent to the proportion of the world's fossil fuel and mineral resources consumed in the country.

The sources of pollution are so many and varied that we cannot begin to discuss them all in this chapter. Instead, we will examine three types of air pollution and their associated effects.

## Acid Rain

Although acid *precipitation* is a more precise description, **acid rain** is the term generally used for pollutants, chiefly oxides of sulfur and nitrogen, that are created by burning fossil fuels and that change chemically as they are transported through the atmosphere and fall back to Earth as acidic rain, snow, fog, or dust. The main sources of these pollutants are vehicles, industries, power plants, and ore-smelting facilities. When sulfur dioxide is absorbed into water vapor in the atmosphere, it becomes sulfuric acid, which is highly corrosive. Sulfur dioxide contributes about two-thirds of the acids in the rain. About one-third comes from nitrogen oxides, transformed into nitric acid in the atmosphere.

Once the pollutants are airborne, winds can carry them hundreds of kilometers, depositing them far from their source. In North America, most of the prevailing winds are westerlies, which means that much of the acid rain that falls on the eastern seaboard and eastern Canada originates in 10 states in the central and upper Midwest (Figure 12.11). Similarly, airborne pollutants from Great Britain, France, and Germany cause acidification problems in Scandinavia.

Acid rain has three kinds of effects: *terrestrial, aquatic,* and *material.* The acids change the *pH factor* (the measure of acidity/alkalinity on a scale of 1 to 14) of both soil and water, setting off a chain of chemical and biological reactions (Figure 12.12). It is important to note that the pH scale is logarithmic, which means every step on the scale represents a factor of 10. Thus, 4.0 is 10 times more acidic than 5.0, and it is 100 times more acidic than 6.0. The average pH of normal rainfall is 5.6, categorized as slightly acidic, but acid rainfalls with a pH of 1.5 (far more acidic than vinegar or lemon juice) have been recorded.

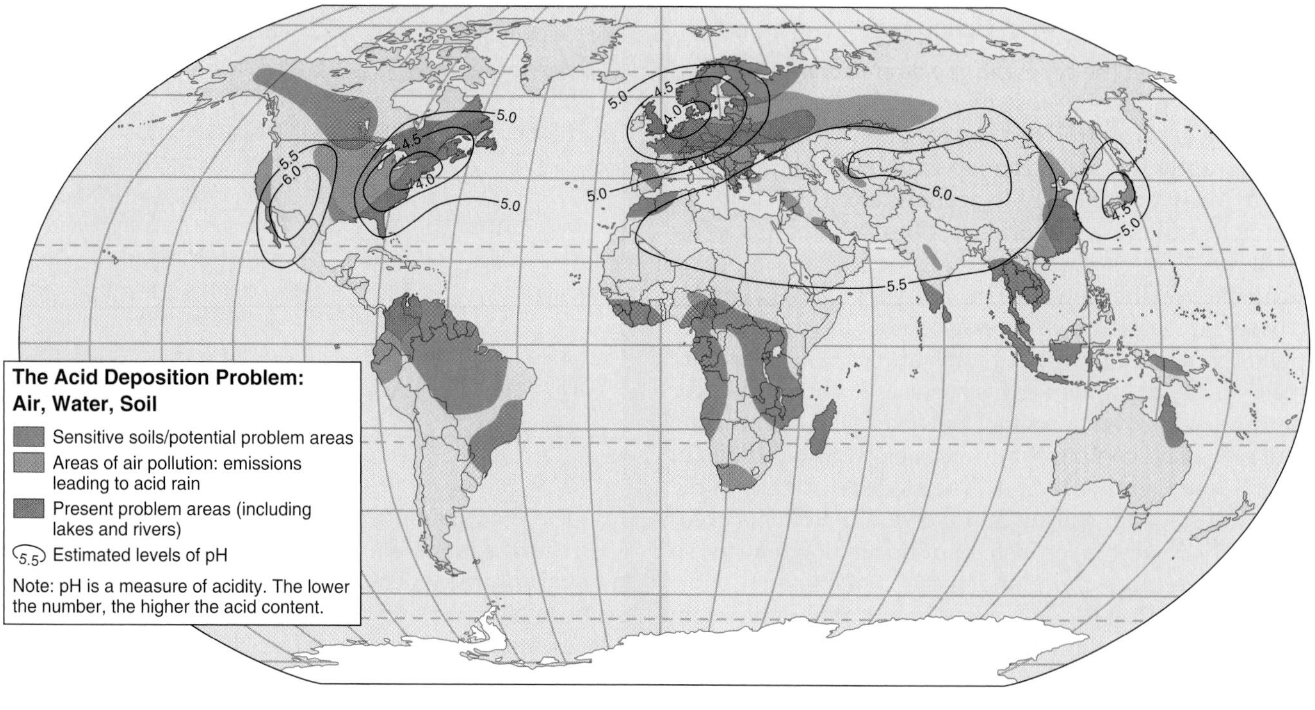


**FIGURE 12.11 Acid precipitation: points of origin and current problem areas.** Prevailing winds can deposit acid precipitation far from its area of origin. The acids in the precipitation harm or destroy soils, vegetation, aquatic life, and buildings. Source: Student Atlas of World Geography, *3d ed., John Allen, Map 46, p. 59. McGraw-Hill/Dushkin, 2003.*

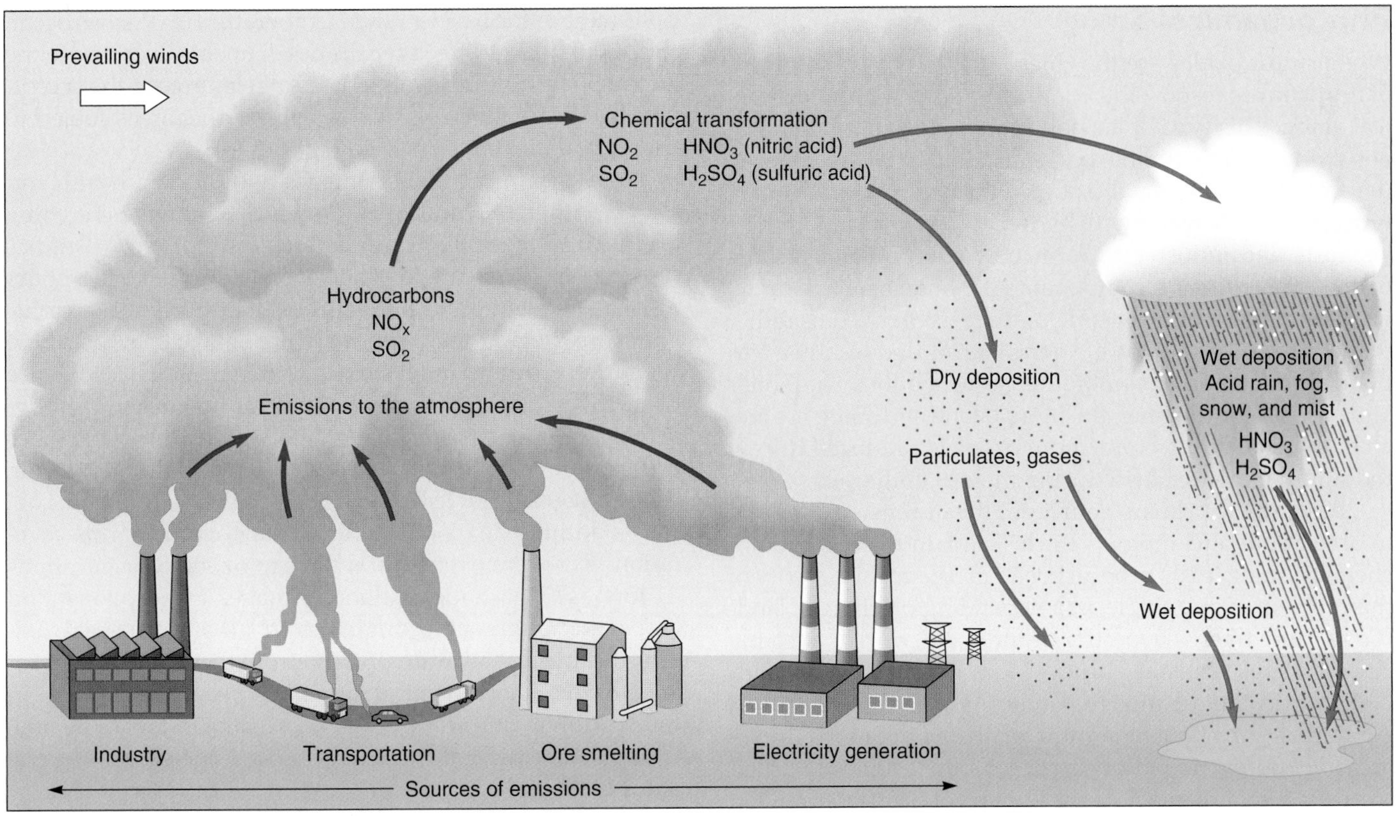


**FIGURE 12.12 The formation of acid precipitation.** Sulfur dioxide and nitrogen oxides produced by the combustion of fossil fuels are transformed into sulfate and nitrate particles; when the particles react with water vapor, they form sulfuric and nitric acids, which then fall to Earth. *Redrawn from* Biosphere 2000, *by Donald G. Kaufman and Cecilia M. Franz (NY: HarperCollins College Publishers, 1993, Fig. 14.5, p. 259).*

Acid deposition harms soils and vegetation by acidifying soils and by coating the ground with particles of aluminum and toxic heavy metals, such as cadmium and lead. It kills microorganisms in the soil that break down organic matter and recycle nutrients through the ecosystem. Significant forest damage has occurred in the eastern United States, northern and western Europe, Russia, and China.

The aquatic effects of acid rain are manifold. The acidity of a lake or stream need not increase much before it begins to interfere with the early reproductive stages of fish. Also, the food chain is disrupted as acidification kills the plants and insects upon which fish feed. Acid rains have been linked to the disappearance of fish in thousands of lakes and streams in the United States, Canada, and Scandinavia, and to a decline of fish populations elsewhere. On the verge of suffering from chronically high acidity are high-elevation streams, lakes, and ponds in the Sierra Nevada, the Cascade Range, the Rocky Mountains, and the Adirondacks.

The material effects of atmospheric acid are evident in damage to buildings and monuments. The acid etches and corrodes many building materials, including marble, limestone, steel, and bronze (Figure 12.13). Worldwide, tens of thousands of structures are slowly being dissolved by acid precipitation.

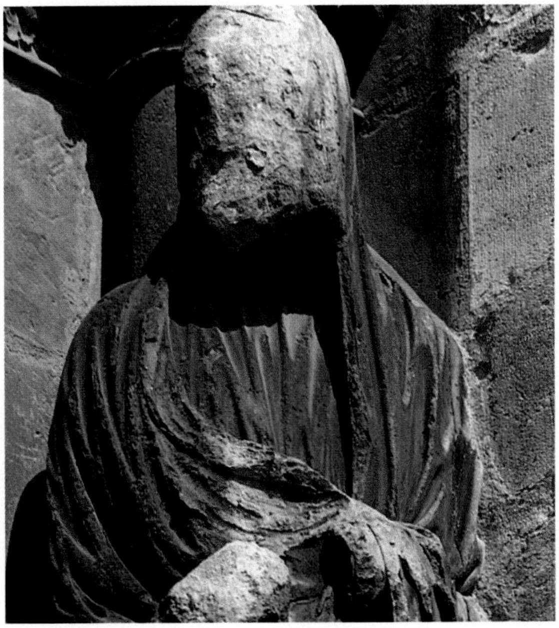

**FIGURE 12.13 The destructive effect of acid rain** is evident on this limestone statue at the cathedral in Reims, France. Sulfuric acid converts limestone to gypsum, which is water-soluble and is washed away over many years of contact with rain. *© William E. Ferguson.*

## Photochemical Smog

While sulfur oxides are the chief cause of acid rain, oxides of nitrogen are responsible for the formation of **photochemical smog.** This type of air pollution is created when nitrogen oxides react with the oxygen present in water vapor in the air to form nitrogen dioxide. In the presence of sunlight, nitrogen dioxide reacts with hydrocarbons from automobile exhausts and industry to form new compounds, such as **ozone.** The primary component of photochemical smog, ozone is a molecule consisting of three oxygen atoms rather than the two of normal oxygen. Warm, dry weather and poor air circulation promote ozone formation. The hotter and sunnier the weather, the more ozone and smog are created. In general, therefore, more ozone is produced during the summer months than during the rest of the year.

Because the primary sources of the nitrogen oxides and hydrocarbons are motor vehicles and industries, photochemical smog tends to be an urban problem, the severity of the problem in any single area depending on climate, landforms, and traffic. It occurs around the world, affecting cities such as Ankara, Turkey; New Delhi, India; Mexico City, Mexico; and Santiago, Chile. As many as 90 million Americans, or about one-third of the population, live in areas where ozone pollution exceeds the limit established by the federal Clean Air Act, 0.12 parts of ozone per million parts of air. Health advisories are issued when the level reaches 0.15 and a stage 1 alert at 0.20.

The warm, sunny climate and topography of California are particularly conducive to ozone pollution (Figure 12.14). Its valleys are encircled by mountains, which help hold air pollutants in the basins. When temperature inversions occur, the pollutants are effectively trapped, unable to escape to the stratosphere. Ozone levels in Los Angeles exceed the acceptable level an average of 70 days a year. Other areas subject to ozone pollution are shown in Figure 12.15.

Photochemical smog damages both human health and vegetation. It has long been known to aggravate coughing and breathing problems for asthmatics, but recent studies indicate that exposure to ozone over a period of 6 to 8 hours also causes respiratory problems in healthy people. Chronic exposure to smog causes permanent damage to lungs, aging them prematurely, and is believed to increase the incidence of such respiratory ailments as pneumonia and emphysema. Because children have smaller breathing passages and less-developed immune systems than adults, they are especially susceptible to damage from the polluted air.

In addition to its effects on humans, ozone harms vegetation. Exposure over several days to ozone concentrations as low as .1 parts per million damages trees, plants, and crops. Although smog originates in urban industrial centers, it can affect areas downwind from them. Damage associated with photochemical smog has been documented in forests downwind from Tokyo and Osaka in Japan; Beijing, China; Karachi, Pakistan; and Los Angeles, California, among other places.

## Depletion of the Ozone Layer

Ozone, the same chemical that is a noxious pollutant near the ground, is essential in the stratosphere. There, approximately 10 to 24 kilometers (6 to 15 mi) above the ground,

(a)

(b)

**FIGURE 12.14 Los Angeles (a) on a clear day and (b) when it is cloaked by photochemical smog.** When air over the city remains stagnant, it can accumulate increasing amounts of automobile and industrial exhausts, reducing afternoon sunlight to a dull haze and sharply lifting ozone levels. Mandates of the Clean Air Act and more stringent restrictions on automobile emissions have resulted in reducing peak levels of ozone to a quarter of their 1955 levels. *(a) © Gregory Mancuso/Stock Boston. (b) © Robert Landau/Corbis Images.*

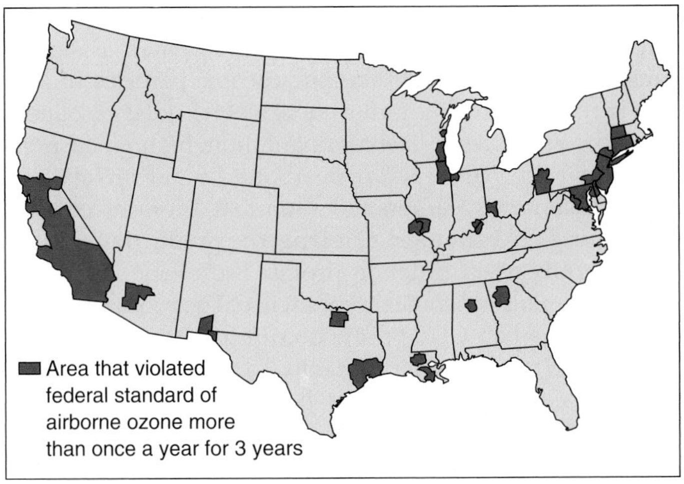


**FIGURE 12.15 Counties where ozone levels violated the federal standard as of June 2000.** Despite significantly reducing harmful emissions in recent years, the Los Angeles basin still has the worst ozone pollution in the United States in terms of the *number* of "bad air" days per year. However some of the worst smog days in the country occur in Houston, Dallas–Fort Worth, and Galveston, Texas, when 1-hour peak concentrations of ozone exceed 0.15 parts per million of air.
Source: *Environmental Protection Agency.*

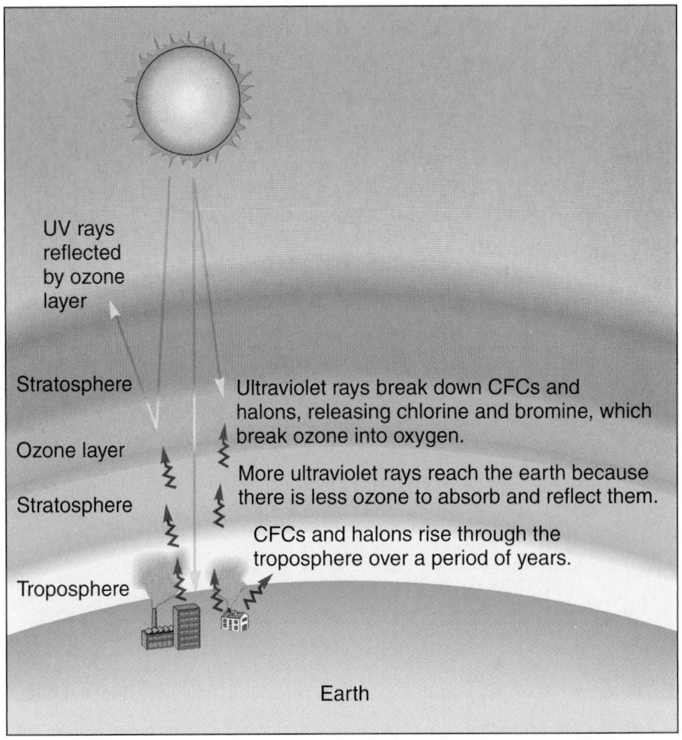


**FIGURE 12.16 How ozone is lost.** CFCs and halons released into the air rise through the troposphere without breaking down (as most pollutants do) and eventually enter the stratosphere. Once they reach the ozone layer, UV rays break them down, releasing chlorine (from CFCs) and bromine. These elements, in turn, disrupt ozone molecules, breaking them up into molecular oxygen, and thus deplete the ozone layer.

ozone forms a protective blanket called the **ozone layer,** which shields all forms of life on Earth from overexposure to lethal ultraviolet (UV) radiation from the sun. Mounting evidence indicates that emissions from a variety of chemicals are destroying the ozone layer. Most important is a family of synthetic chemicals developed in 1931 and known as **chlorofluorocarbons (CFCs).** CFCs are found in hundreds of products. They are used as coolants for refrigerators and air conditioners; as aerosol spray propellants; and as a component in foam packaging, home insulation, and upholstery. In liquefied form, they are used to sterilize surgical equipment and to clean computer chips and other microelectronic equipment.

Also implicated in the depletion of the ozone layer are *halons,* used in fire extinguishers; carbon tetrachloride and methyl chloroform, used as solvents and cleaning agents; and methyl bromide, a pesticide used to sterilize soils and grain silos and to fumigate shipments of perishable goods. CFCs, however, are by far the most important.

After the gases are released into the air, they rise through the lower atmosphere and, after a period of 7 to 15 years, reach the stratosphere (Figure 12.16). There, UV radiation breaks the molecules apart, producing free chlorine and bromine atoms. Over time, a single one of these atoms can destroy tens of thousands, if not a potentially infinite number, of ozone molecules.

Each year, beginning in July, the atmosphere over the Antarctic starts to lose more and more ozone. In 1985, researchers discovered what is popularly termed a "hole" (actually, a broad area of low-ozone concentrations) as big as the continental United States in the ozone layer over Antarctica, extending northward as far as populated areas of South America (Figure 12.17). The ozone depletion intensifies during August and September before tapering off in October as temperatures rise, the winds change, and the ozone-deficient air mixes with the surrounding atmosphere. A less dramatic but still serious depletion of the ozone shield occurs over the North Pole.

A depleted ozone layer allows more UV radiation to reach the earth's surface. Although the exact consequences of that increase will not be known for years, it is almost certain to cause a dramatic rise in the incidence of skin cancers and eye cataracts. Some fear it may also damage the immune system of humans and other animals by impairing the cells that fight viral infections and parasitic disease. Because UV radiation also causes cell and tissue damage in plants, it is likely to reduce agricultural production. The most serious damage may occur in oceans. Increased amounts of UV radiation affect the photosynthesis and metabolism of the microscopic plants called phytoplankton that flourish just below the surface of the Antarctic Ocean. Phytoplankton form the base of the oceanic food chain and also play a central role in the earth's $CO_2$ cycle.

The production of CFCs and other ozone-depleting substances is being phased out under the *Montreal Protocol on the Depletion of the Ozone Layer* of 1987, an international

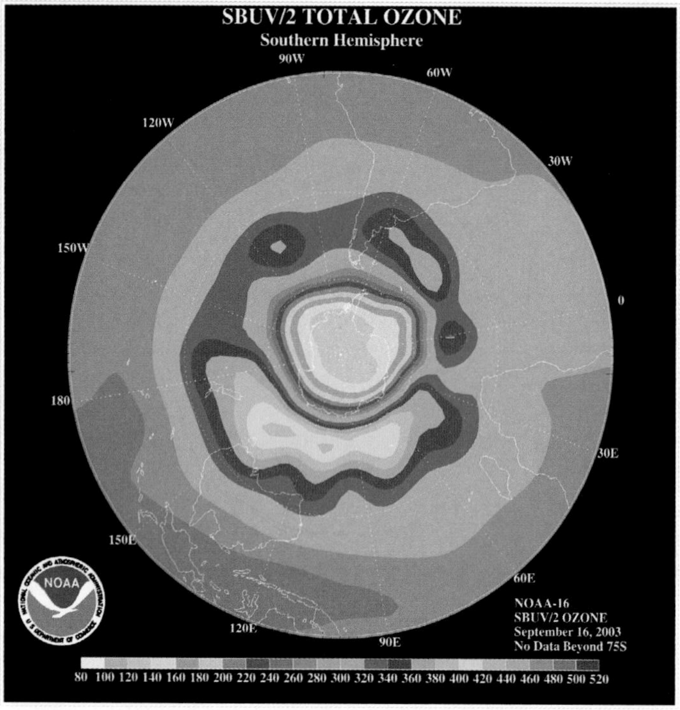


**FIGURE 12.17 Ozone depletion over the Southern Hemisphere on September 16, 2003.** The color scale below the image shows the total ozone levels. The pinks and purples indicate the areas of greatest ozone depletion. The area of significant depletion varies in size from year to year. In 2003, the hole grew to near record size, 29 million square kilometers (11.1 million sq mi). Although depletion of the ozone layer has been particularly severe over Antarctica, a decline in stratospheric ozone has also been observed over other parts of the world. *© National Centers for Environmental Prediction, NOAA.*

agreement that has been endorsed by 160 countries. The treaty required developed countries to stop production and consumption of CFCs, carbon tetrachloride, halons, and methyl chloroform by January 1, 1996; developing states have until 2010 to cease production. The Montreal Protocol spurred a rapid decline in CFC output. By 1998, CFC production had already fallen about 90% from its peak a decade earlier. Concentrations of bromine in the lower atmosphere declined 5% between 1998 and 2003.

Even if all countries eventually comply with the Montreal Protocol, however, past emissions will continue to cause ozone degradation for years to come. The two most widely used forms of CFCs stay in the stratosphere, breaking down ozone molecules, for up to 120 years. Thus, full recovery of the ozone layer cannot be expected until the end of the 21st century.

## Controlling Air Pollution

A number of developments in recent years have given rise to the hope that people can reverse the decline in air quality. The total amount of lead added to gasoline has dropped worldwide by 75% since 1970. Several countries, both industrialized and developing, eliminated leaded gasoline from their markets. Many others reduced the lead content and/or introduced unleaded gasoline. The development is significant because exposure to the microscopic particles of lead that are emitted into the atmosphere when leaded gasoline is burned contributes to mental retardation, high blood pressure, and an increased risk of heart attacks and strokes.

As discussed earlier, the Montreal Protocol of 1987 called for global efforts to eliminate the production of chlorofluorocarbons in order to protect the ozone layer. The treaty has made a significant difference. The protocol and its amendments ban CFC production in industrial countries, and by 1999, total world production of CFCs had fallen 88% from its peak in 1988. The drop has been most pronounced in the European Union and the United States; consumption of CFCs has actually risen in China, India, and other developing economies during the last decade.

Another successful international accord is the 1979 Convention on Long-Range Transboundary Air Pollution, signed by 33 countries in Europe and North America and intended to reduce the emissions of nitrous oxides and sulfur dioxides. During the 1980s, air pollution in Europe was reduced as, for example, Austria, West Germany, Sweden, and Norway cut their $SO_2$ emissions by more than 50%. Emissions of nitrous oxides have proved more difficult to control.

The United States has made significant progress in cleaning up its air in the last 30 years. A series of Clean Air Acts (1963, 1965, 1970, 1977) and amendments identified major pollutants and established national air-quality standards. After many years of debate, Congress in 1990 passed a Clean Air Act that represented the most sweeping legislation to date. It set forth goals to protect public health and the environment by reducing the amount of air pollutants that can be released, and it established a time table for reaching those goals. Major provisions call for

- reducing urban smog by setting limits on allowable levels of particulates and the concentration of ozone in the air;
- using cleaner-burning fuels in the most polluted cities;
- lowering nitrogen oxide and hydrocarbon emissions from motor vehicles; and
- requiring utilities to reduce their emissions of $NO_x$ and $SO_2$.

The country's air is cleaner now than it was when the first of the Clean Air Acts were enacted, despite the fact that the population, economy, and number of vehicles have grown. Since 1985, emissions of lead have dropped 98%, of sulfur dioxide by 50%, and of carbon monoxide by 32%. Nevertheless, the air in many parts of the country still violates the public health standards.

Reaching the goals of the Clean Air Act will require reducing the type and volume of air pollutants from both stationary and nonstationary sources. A number of strategies can be employed to clean up stationary sources. Technological options include switching to cleaner-burning fuels; coal washing, which removes much of the sulfur in coal before it is burned; and removing pollutants from effluent gases in the smokestack by using scrubbers, precipitators, and filters. Another approach is to reduce energy consumption by using more efficient appliances, installing weatherstripping and insulation, and strengthening energy performance standards in the building codes for new buildings.

Reducing emissions from nonstationary sources—mainly motor vehicles of all types—can be accomplished in a number of ways. These include conforming to tighter tailpipe emission standards by retiring older automobiles, driving fuel-efficient automobiles, phasing out leaded gas, and implementing rigorous vehicle inspection programs. Catalytic converters have sharply reduced smog from vehicles. An increase in the price of gasoline would reduce gas consumption. Travel can be made more energy efficient if a community is committed to rewarding those who carpool or use alternative means of transport, such as bicycles or mass transit.

# Impact on Landforms

People have affected the earth wherever they have lived. Whatever we do, or have done in the past, to satisfy our basic needs has had an impact on the landscape (Figure 12.18). To provide food, clothing, shelter, transportation, and defense, we have cleared the land and replanted it, rechanneled waterways, and built roads, fortresses, and cities. We have mined the earth's resources, logged entire forests, terraced mountainsides, even reclaimed land from the sea. The nature of the changes made in any single area depends on what was there to begin with and how people have used the land.

## Landforms Produced by Excavation

Although we tend to think of landforms as "givens," created by natural processes over millions of years, people have played and continue to play a significant role in shaping local physical landscapes. Some features are created deliberately, others unknowingly or indirectly. Pits, ponds, ridges and trenches, subsidence depressions, canals, and reservoirs are the chief landform features resulting from excavations. Some date back to neolithic times, when people dug into chalk pits to obtain flint for toolmaking. Excavation has had its greatest impact within the last two centuries, however, as earth-moving operations have been undertaken for mining; for building construction and agriculture; and for the construction of transport facilities such as railways, ship canals, and highways.

Surface mining, which involves the removal of vegetation, topsoil, and rocks from the earth's surface in order to get at the resources underneath, has perhaps had the greatest environmental impact. Open-pit mining and strip mining are the most commonly used methods of surface mining.

*Open-pit* mining is used primarily to obtain iron, copper, sand, gravel, and stone. As Figure 12.19a indicates, an

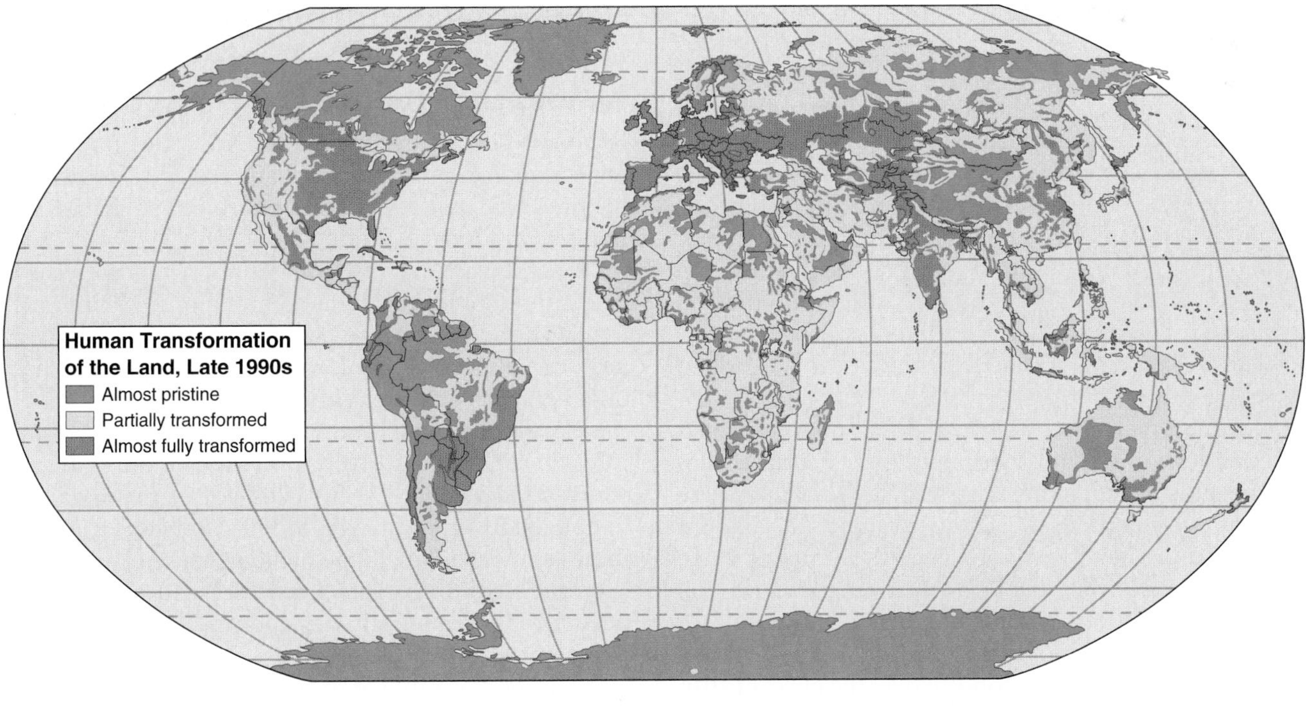


**FIGURE 12.18 Human transformation of the land.** Humans have altered much of the earth's surface in some way. The "almost pristine" areas, covered with original vegetation, tend to be too high, dry, cold, or otherwise unsuitable for human habitation in large numbers. They generally have very low population densities. "Partially transformed" describes areas of secondary vegetation, grown after removal of original vegetation. Most are used for agriculture or livestock grazing. "Almost fully transformed" areas are those of permanent and intensive agriculture and urban settlement; little or no original vegetation remains. Source: *Student Atlas of World Geography, 3d ed., John Allen, Map 61, p. 74. McGraw-Hill/Dushkin, 2003.*

(a)

(b)

**FIGURE 12.19** (a) **An aerial view of the Bingham Canyon open-pit copper mine in Utah, said to be the largest human-made excavation in the world.** Since mining began at the site, approximately 15 billion tons of material have been removed, creating a pit more than 800 meters (2600 ft) deep and 4 kilometers (2.5 mi) wide at the top. The mine has yielded vast quantities of copper, gold, silver, and molybdenum. (b) About 400 square kilometers (154 sq mi) of land surface in the United States are lost each year to the strip mining of coal and other resources; far more is disrupted worldwide. On flat or rolling terrain, strip mining leaves a landscape of parallel ridges and trenches, the result of stripping away the unwanted surface material, or *overburden.* The material taken from one trench to reach the underlying mineral is placed in an adjacent one, leaving the wavelike terrain shown here. Besides altering the topography, strip mining interrupts surface and subsurface drainage patterns, destroys vegetation, and places sterile and often highly acidic subsoil and rock on top of the new ground surface. *(a) © Bettmann/Corbis Images. (b) © Jim Richardson/Corbis.*

enormous pit remains after the mining has been completed because most of the material has been removed for processing. *Strip* mining is increasingly being employed in the United States as a source of coal; more coal per year now comes from strip mines than from underground mines. Phosphate is also mined in this way. A trench is dug, the material is excavated, and another trench is dug, the soil and waste rock being deposited in the first trench, and so on. Unless reclamation is practiced, the result is a ridged landscape (Figure 12.19b).

Landscapes marred by vast open pits or unevenly filled trenches are one of the most visible results of surface mining. Thousands of square miles of land have been affected, with the prospect of thousands more to come as the amount of surface mining increases. Damage to the aesthetic value of an area is not the only liability of surface mining. If the area is large, wildlife habitats are disrupted, and surface and subsurface drainage patterns are disturbed. In the United States in recent years, concern over the effect of strip mining has prompted federal and state legislation to increase regulation and to stop the worst abuses. Strip-mining companies are now expected to level the ridges and regrade the area, restore the soil, and replant grass or other vegetation.

## Landforms Produced by Dumping

Excavation in one area often leads to the creation, via dumping, of landforms nearby. Both surface and subsurface mining produce tons of waste and enormous spoil piles. In fact, in terms of tonnage, mining is the single greatest contributor to solid wastes, with about 2 billion tons per year left to be disposed of in the United States alone. The normal custom is to dump waste rocks and mill tailings in huge heaps near the mine sites. Unfortunately, this practice has secondary effects on the environment.

Carried by wind and water, dust from the wastes pollutes the air, and dissolved minerals pollute nearby water sources. Occasionally the wastes cause greater damage, as happened in Wales in 1966, when slag heaps from the coal mines slid onto the village of Aberfan, burying more than 140 schoolchildren. Such tragedies call attention to the need for less potentially destructive ways of disposing of mine wastes.

Another example of the combined effect of excavating and filling on the landscape is the agricultural terrace characteristic of parts of Asia. In order to retain water and increase the amount of arable land, terraces are cut into the slopes of hills and mountains. Low walls protect the patches of level land.

Human impact on land has been particularly strong in areas where land and water meet. Dredging and filling operations undertaken for purposes of water control create landscape features such as embankments and dikes. In many places, the actual shape of the shoreline has been altered, as builders in need of additional land have dumped solid wastes into landfills. In the Netherlands, millions of acres of land have been reclaimed from the sea by the building of dikes to enclose polders and canals to drain them. Farming practices in river valleys have had significant effects on deltas. For example, increased sedimentation has often extended the area of land into the sea.

### Formation of Surface Depressions

The extraction of material from beneath the ground can lead to **subsidence,** the settling or sinking of a portion of the land surface. Many of the world's great cities are sinking because of the removal of *fluids* (groundwater, oil, and gas) from beneath them. Cities threatened by such subsidence are located on unconsolidated sediments (New Orleans and Bangkok), coastal marshes (Venice and Tokyo), or lake beds (Mexico City). When the fluids are removed, the sediments compact and the land surface sinks. Because many of the cities are on coasts or estuaries and are often only a few feet above sea level, subsidence makes them more vulnerable to flooding from the sea.

Groundwater abstraction has created serious subsidence in many places. Recent evidence indicates that the withdrawal of trillions of gallons of water from a large area of Arizona has resulted in widespread ground subsidence and the formation of more than a hundred earth fissures, jagged ground cracks that can be as much as 14.5 kilometers (9 mi) long and 120 meters (400 ft) deep.

**FIGURE 12.20 Subsidence in Mexico City** is caused by the pumping out of the aquifer underneath the metropolitan area to supply water. The portion of Line 2 of the city Metro system shown here was horizontal when it was built in the 1960s. Parts of the city have sunk as much as 9.15 meters (30 ft). The subsidence has damaged or destroyed hundreds of buildings, particularly in the colonial-era city center, and continues to rupture water pipes, sewer lines, and subway tunnels. *© Gerardo Magallón/The New York Times.*

The removal of *solids* (such as coal, salt, and gold) by underground mining may result in the collapse of land over the mine. *Sinkholes,* or *pits* (circular, steep-walled depressions) and *sags* (larger and shallower depressions) are two types of landscape features produced by such collapse. If surface drainage patterns are disrupted, subsidence *lakes* may form in the depressions. Subsidence has become a more serious problem as towns and cities have expanded over mined-out areas.

As one might expect, subsidence damages structures built on the land, including buildings, roads, and sewage lines. A dramatic example occurred in Los Angeles in 1963, when subsidence caused the dam at the Baldwin Hills Reservoir to crack. In less than 2 hours, the water emptied into the city, resulting in millions of dollars' worth of property damage. The withdrawal of groundwater from beneath Mexico City has led to severe though differential subsidence (Figure 12.20). One of the reasons the 1985 earthquake in that city was so damaging was that subsidence had weakened building structures.

## IMPACT ON PLANTS AND ANIMALS

People have affected plant and animal life on the earth in several ways. When human impact is severe enough, a species can become *extinct;* that is, it no longer exists. Although fossil records show that extinction is a natural feature of life on Earth, scientists estimate that in recent history the rate of extinction has increased exponentially—an increase due to human activity. *Endangered* species are those that are present in such low numbers that they are in immediate jeopardy of becoming extinct in the wild if the causes of endangerment continue, while *vulnerable* species have decreasing populations and are likely to become endangered within the foreseeable future. Both endangered and vulnerable species are termed *threatened.*

Certain kinds of species have a greater probability than others of becoming extinct. They usually exhibit one or more of the following characteristics. They exist in small populations of dispersed individuals; have low reproductive rates, especially as compared with the species that prey upon them; live in a small geographic area; and/or are specialized organisms that rely for their survival on a few key factors in their environment.

The World Conservation Union assessed the state of wildlife on every continent and in 2000 reported that 24% of all known mammal species are considered threatened; of these, 4% are critically endangered. Among the various orders of mammals, listed as threatened are:

- 11 species of hoofed animals (rhinoceroses, zebras, etc.);
- 65 species of carnivores (wild cats and dogs, bears, etc.);

- 70 species of ungulates (hippopotamuses, deer, sheep, etc.);
- 96 species of primates (monkeys and apes).

The countries with the largest numbers of threatened mammal species are China, India, Indonesia, and Brazil. Together they contain more than 40% of the world's human population, which is putting increasing pressure on critical habitats.

The assessment of bird species concluded that as of 2000, of the approximately 9900 bird species in the world, one in eight is at risk of extinction. The countries with the highest numbers of threatened bird species are Indonesia, China, Brazil, and Colombia.

Although endangered mammals and birds have attracted public attention, many species of plants are also in jeopardy. According to a report released by the Worldwatch Institute in 2000, more than 30,000 of the world's 270,000 known plant species are at risk of extinction. The United States ranks first among countries in the number of plant species at risk (4669, or 29%, of its 16,108 plant species are imperiled), followed by Australia and South Africa, but this may be due in part to the fact that plants have been better surveyed in those countries than elsewhere.

In this section, we examine some of the ways people have modified plant and animal life.

## Habitat Disruption

One of the main causes of extinction has been the loss or alteration of habitats for wildlife. About two-thirds of all threatened species are affected by habitat degradation or loss. Agricultural activities (crop and livestock farming), extractive activities (logging and mining), and various forms of development (e.g., buildings, roads, dams, and canals) all modify or destroy the habitats in which plants and animals have lived (Figure 12.21). The destruction of the world's rain forests, the most biologically diverse places on Earth, is by some estimates causing the extinction of hundreds of plant and animal species every year.

Many people fear that as countries in Africa and South America become more industrialized and more urbanized and expand their areas under cultivation, there will be an increasingly negative impact on wildlife. It is already known that in Africa, wild animals are vanishing fast, in part the victims of habitat destruction. In Botswana, for example, 250,000 antelope and zebra died in a decade, disoriented by fences erected to protect cattle. As selected animal species decline in numbers, balances among species are upset and entire ecosystems are disrupted.

Habitat disruption characterizes developed as well as developing countries. Tidal marshes in the United States, for example, have been dredged and filled for residential and industrial development. The loss of such areas reduces the essential habitat of fish, crustaceans, and mollusks. The whooping crane has been virtually eliminated in the United States because the marshes where it nested were drained, and roads and canals brought intruders into its habitat. Its comeback, sought by breeding programs in the United States and Canada, is still uncertain.

## Hunting and Commercial Exploitation

Another way in which people have affected plants and animals is through their deliberate destruction. We have over-

(a)

(b)

**FIGURE 12.21 Two victims of habitat destruction and the demands an expanding population puts on land.** Among the animals facing extinction in the wild are (a) the silvery gibbon and (b) the orangutan. The silvery gibbon exists only on the Indonesian island of Java, which has lost most of its forests to logging. Experts estimate that since 1975 the number of silvery gibbons has plummeted from about 20,000 to between 400 and 2000. Logging is also the chief reason for the decline in the number of orangutans in Indonesia. The population is believed to have been halved during the 1990s. Both the silvery gibbon and the orangutan have been declared endangered species. *(a) © Gerard Lacz/Animals Animals. (b) © R. Lynn/Photo Researchers.*

hunted and overfished, for food, fur, hides, jewelry, and trophies. In the past, unregulated hunting harmed wildlife all over the world and was responsible for the destruction of many populations and species. Beavers, sea otters, alligators, and buffalo are among the species brought to the edge of extinction in the United States by thoughtless exploitation. Under protective legislation, their populations are now increasing, but hunting in developing countries still poses a threat to a number of species.

Three African animals whose existence is threatened by hunting, most of it illegal, are the elephant, rhinoceros, and mountain gorilla. Prized for its ivory tusks, the African elephant has been ruthlessly slaughtered. Between 5 and 10 million of the elephants are estimated to have been alive in the 1930s, a number that had declined to 1.3 million by 1979. Now, only 300,000 to 500,000 remain. The black rhinoceros, killed for its horn, is now an endangered species. The population has declined from nearly a million in sub-Saharan Africa a century ago to about 2500 today. All three subspecies of gorillas are endangered; the rarest of all is the mountain gorilla. Only 650 mountain gorillas are believed to exist, most of them in national parks in the Democratic Republic of Congo, Uganda, and Rwanda (Figure 12.22).

The decline in the productivity of marine fisheries, discussed in Chapter 10, is due in large part to the widespread use of modern fishing technology, which has made hunting easier and more efficient. The technology includes the use of sonar, radar, helicopters, and Global Positioning Systems to locate schools of fish; more efficient nets and tackle; and factory trawlers to follow fishing fleets to prepare and freeze the catch.

According to the UN Food and Agriculture Organization, all 17 major ocean fishing areas in the world are now being fished at or above sustainable levels; 13 are in decline.

**FIGURE 12.22** The mountain gorilla (*Gorilla gorilla beringei*) is one of the world's most endangered primates. Although the gorillas live in national parks, their survival is not guaranteed. Both habitat destruction—the felling of trees for fuel and shelter—and illegal poaching have reduced their numbers. *© Digital Vision/Getty Images.*

The plundering of United States coastal waters has imperiled a number of the most desirable fish species, including haddock, yellowtail flounder, and cod in New England waters; Spanish mackerel, grouper, and red snapper off the Gulf of Mexico; halibut and striped bass off California; and salmon and steelhead in the Pacific Northwest.

Commercial fishermen from Japan, South Korea, Taiwan, and other countries use drift nets to capture squid, tuna, and salmon. The nets, which can stretch up to 65 kilometers (40 mi), have been called "curtains of death" because they are devastatingly effective, essentially scooping up all life in their path—not just the target species but also millions of nontarget fish, sea birds, turtles, and marine mammals. The World Wildlife Fund estimates that at least 60,000 dolphins, porpoises, and whales drown each year after becoming entangled in fishing nets and other equipment. These accidental captures by crews fishing for other species are known as *bycatch;* they have significantly depleted the populations of many of the 80 species of the cetacean (fishlike) groups of sea mammals, critically endangering some of them.

## Exotic Species

A plant or an animal that has been released into an ecosystem in which it did not evolve is a nonindigenous (nonnative) or **exotic species.** The deliberate or inadvertent introduction of a species into an area where it did not previously exist can have damaging and unforeseen consequences. Introduced species have often left behind their natural enemies—predators and diseases—giving them an advantage over native species that are held in check by biological controls. The rabbit, for example, was purposely introduced into Australia in 1859. The original dozen pairs multiplied to a population in the thousands in only a few years and, despite programs of control, to an estimated 1 billion by 1950. Inasmuch as five rabbits eat about as much as one sheep, a national problem had been created. Rabbits became an economic burden and environmental menace, competing with sheep for grazing lands and stimulating soil erosion.

Invasions of exotic (also called *alien*) species have multiplied as the speed and range of world trade and travel have increased. In the United States alone, hundreds of harmful invaders have been discovered in recent years—harmful because they consume or outcompete native species (see "On Becoming Mussel-Bound," p. 454). They include:

- Africanized honeybees, more aggressive and venomous than European bees; they reached the southwestern United States in the 1990s after escaping from an experimental station in Brazil in 1957;
- the Asian gypsy moth, a voracious eater of trees and some crops, which arrived in Oregon, Washington, and British Columbia in 1991;
- the Asian long-horned beetle, which probably arrived in Brooklyn, New York, in wood pallets from China in 1996 and has since appeared in the Chicago area; it threatens species such as maple, elm, willow, and birch;

## On Becoming Mussel-Bound

Zebra mussels appear to be harmless, even attractive creatures, with their tiny, striped shells. But ecologists view their recent entry into North America as nothing but a disaster.

Native to the Caspian and Black seas, zebra mussels (*Dreissena polymorpha*) were first detected in North America in 1988. They are believed to have been stowaways on an East European freighter that dumped ballast water into Lake St. Clair, near Detroit.

In just a few years, the mussels had spread to the rest of the Great Lakes. The floods of 1993 carried the mussel larvae into the Mississippi River system, extending their range as far south as New Orleans.

The rapid advance of the mussels is due both to a scarcity of natural predators, such as diving ducks and crawfish, and to the mussels' prodigious reproductive capacity. Adult females produce from 30,000 to 40,000 eggs a year, and males contribute a like amount of sperm.

The mussels have an unfortunate tendency to attach themselves to anything hard, including water intake pipes. Colonies of mussels now clog the underwater intake pipes of power plants, water treatment plants, and industrial plants in the Great Lakes. A square meter (about 10 sq ft) of wall at one utility plant was found to contain over 700,000 mussels, and a single intake pipe at an Ontario water plant was clogged by 30 tons of the shellfish. The zebra mussels also endanger the Great Lakes fisheries, the largest freshwater fisheries in the world. Because they devour phytoplankton, the microscopic plants at the base of the aquatic food chain, the mussels compete with algae-eating fish for both food and oxygen. Accidentally introduced into the region, the zebra mussel might significantly alter the ecosystem of the entire Great Lakes.

**Severed section of freshwater intake pipe clogged with zebra mussels.** *© Peter Yates/Getty Images.*

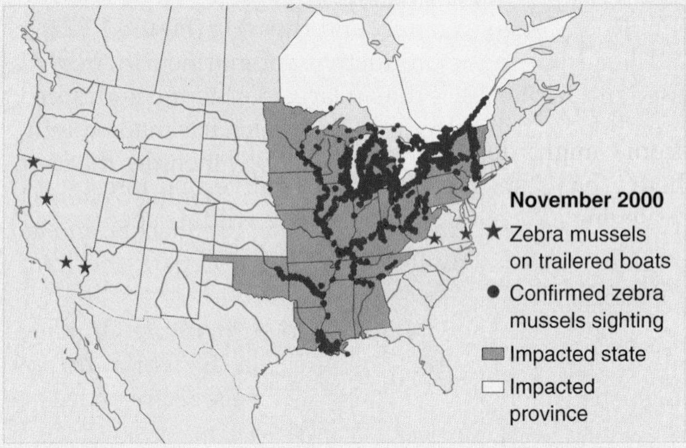


*Map courtesy of USGS.*

- the Asian tiger mosquito, which was discovered in 1985 in a Japanese container of tires headed for a recapping plant in Texas and has since spread to 25 states; it carries a number of tropical viruses, including yellow fever, dengue fever, and various forms of encephalitis;
- the round goby, an aggressive fish that reached U.S. waters in 1990 in ballast chambers of ships from the Black and Caspian seas and threatens to disrupt the Great Lakes' ecosystem.

In the United States, exotic species are second only to habitat loss as a cause of endangerment. About half of the species threatened with extinction in the country are in trouble at least partly because of them.

Plants and animals found on islands are particularly vulnerable to exotics and prone to extinction (Figure 12.23). Native island plants and animals evolved in isolation, with few diseases or predators. Furthermore, because island species often occur on just one or a few islands, the loss of only a few individuals can be devastating to small populations. The Hawaiian islands are plagued by a number of alien species, including feral pigs, mongooses, eucalyptus, ginger, and gorse. Miconia, a large-leaved plant, was introduced as a tropical ornamental. Outside the confines of a flowerpot, it grows to 50 feet or more, and its mammoth leaves (1 meter, or 3 ft, across) cast dense shade, killing native vegetation beneath it, promoting water runoff and erosion. Small colonies of these plants are scattered throughout the islands, and there is a massive eradication program

**FIGURE 12.23 The Galápagos giant tortoise, an endangered species.** Ecuador's Galápagos Islands have embarked on a project to combat scores of exotic plant and animal species that threaten native species. The invaders include feral populations of goats and pigs, which destroy native vegetation; wild dogs, cats, and rats, which feed on lizards and the eggs of birds, tortoises, and snakes; insects that attack native plants; and plants such as kudzu, water hyacinth, and the quinine tree. *© Tui De Roy/Minden Pictures.*

underway in an attempt to curb its spread. The fear is that miconia might do to Hawaii what it has done in Tahiti, where it has displaced 70% of the native rain forest and is threatening 25% of the island's native wildlife species.

As this discussion indicates, introduced plants, as well as animals, can alter vegetative patterns. Some 300 species of invasive plants now threaten native ecosystems in the mainland United States and Canada. At least half were deliberately imported, including purple loosestrife, the melaleuca tree, Norway maple, and water hyacinth. These and other imports have arrived without their natural enemies and have spread uncontrolled, driving out native species.

The Asiatic chestnut blight has destroyed most of the native American chestnut trees in the United States, trees with significant commercial as well as aesthetic value. The cause was the importation of some chestnut trees from China to the United States. They carried a fungus that was fatal to the American chestnut tree but not to the Asiatic variety, which is largely immune to it.

An aquatic vine, hydrilla, imported into Florida from Sri Lanka for use in aquariums, was dumped into a canal in Tampa in 1951. Also known as water thyme, it has overgrown more than 40% of Florida's rivers and lakes and continues to spread rapidly. Transported on motorboat propellers and boat trailers, hydrilla has spread through the lakes and rivers of the South, as far west as California and as far north as Maine and the state of Washington. The plant creates dense canopies at the surface, interfering with swimming, fishing, and boating. The mats clog boat propellers and water intake pipes and reduce the amount of sunlight reaching the water bottom. By monopolizing the dissolved oxygen that fish and aquatic plants require to thrive, hydrilla reduces native diversity.

In an attempt at biological control, Florida officials imported tilapia, a weed-eating fish, to solve the problems created by hydrilla. While the fish have done little to clear Florida waterways, they have driven out many types of native fish, especially large-mouth bass.

These are just a few of the many examples that illustrate an often ignored ecological truth: plant and animal life are so interrelated that when people introduce a new species to a region, whether by choice or by chance, there may be unforeseen and far-reaching consequences.

It is important to note, however, that not all introduced species are harmful. They are of little concern if they assimilate well and coexist with the native stock. Further, an import may become a problem in one area but not be highly invasive elsewhere.

## Poisoning and Contamination

Humans have also affected plant and animal life by poisoning and contamination. In the last several years, we have become acutely conscious of the effect of insecticides, rodenticides, and herbicides, known collectively as *biocides*.

Some of the side effects of these biocides have been well enough documented for scientists to question their indiscriminate use. Once used, a biocide settles into the soil, where it may remain or may be washed into a body of water. In either case, it is absorbed by organisms living in the water, soil, or mud. If an organism is unable to excrete the biocide, its concentration within the animal continues to increase, a process called *bioaccumulation*. Toxins present in very small amounts in the environment can reach dangerous levels inside cells and tissues.

Food webs magnify the effect of toxins in the environment. When a predator eats a large number of plants or animals at a lower trophic level, the predator collects and concentrates the toxins from its prey. **Biological magnification** (also called **biomagnification**) is the accumulation of a chemical in the fatty tissue of an organism and its concentration at progressively higher levels in the food chain. Zooplankton and small fish, for example, collect and retain the toxins from water, sediments, and organic debris. They are eaten in turn by small fish, shrimp, and clams, which build up higher concentrations of the toxins, as shown in Figure 12.24. The higher the level of a plant or animal in the food chain, the greater the concentration of the poisons will be. The carnivores at the top of the food chain, large fish, fish-eating birds, and humans, can accumulate such high levels of the biocide that it adversely affects their health and reproduction. DDT, for example, was shown to cause a decrease in the thickness of the eggshells of some of the larger birds, so that a greater number of eggs broke than normally would. Peregrine falcons, bald eagles, and brown pelicans were among the birds nearly made extinct by this disruption of the reproductive process.

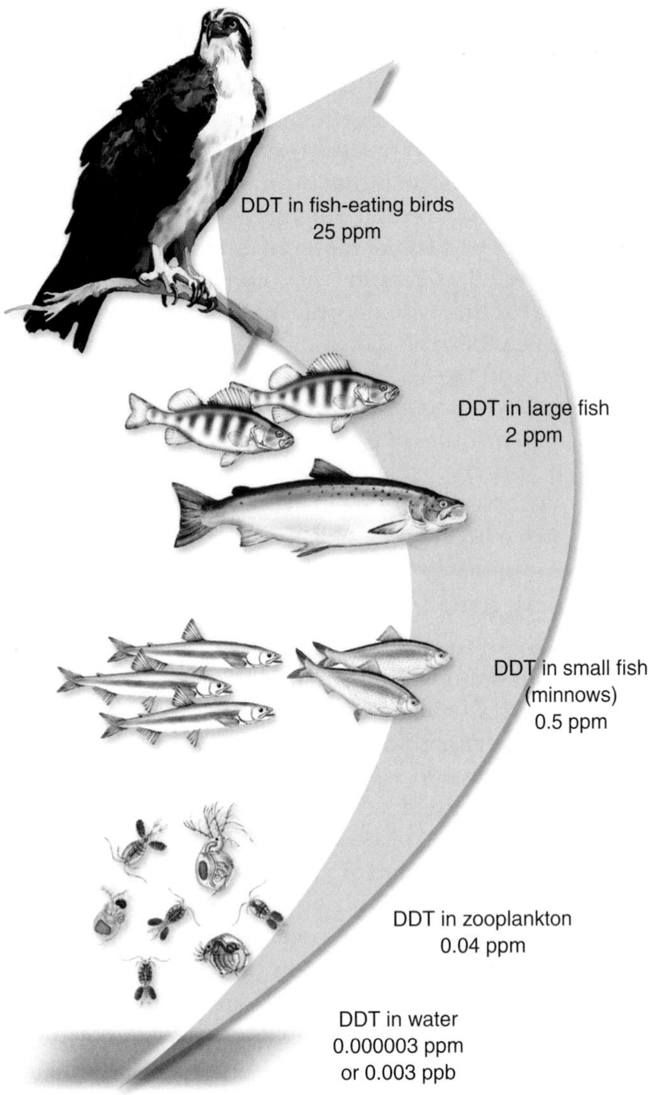


**FIGURE 12.24 The bioaccumulation and biomagnification of DDT.** The numbers are parts per million (ppm). Although the level of DDT in the water may be low, notice how the amount of DDT in the bodies of fish and birds increases as we go up the food chain. In this simplified example, birds at the top of the chain have concentrations of residues 50 times greater than small fish. Mercury, aldrin, chlordane, and other chlorinated hydrocarbons, such as PCBs, undergo magnification in the food chain. Source: *Cunningham, Cunningham, and Saigo,* Environmental Science, *7th ed. Boston: McGraw-Hill, 2003.*

One of the oldest and most dangerous pesticides is DDT. Hailed as a chemical miracle, DDT was first used during World War II as a delousing agent and to clear malaria-bearing mosquitoes from the paths of Allied troops. After the war, one of the first major undertakings of the World Health Organization was the Global Malaria Eradication Program, launched in 1955; its principal tool was DDT. Soon, tons of the pesticide were being used, both to combat disease and to increase agricultural yields, but within a decade two significant problems had emerged: (1) biomagnification, or the increasing concentration of the chemical at higher levels of the food chain; and (2) insecticide resistance—the reproduction within a few years of mosquitoes that were immune to the poison.

Although the use of DDT declined as its effects became apparent, a number of countries, most of them in Africa, still use it in the fight against malaria. Meanwhile, other chlorinated hydrocarbon compounds have been developed and are in wide use. Indeed, the use of pesticides in the United States has more than doubled in the last 25 years. More than 900 million kilograms (2 billion lbs) containing more than 600 active ingredients are applied each year. The pesticides pollute water supplies, often contaminate the crops they are meant to protect, and sometimes sicken the farm workers who apply them. In addition, too often the pesticides are only temporarily effective.

Biocides may in fact exacerbate the problem their use is designed to eradicate. By altering the natural processes that determine which pests (insects, rodents, weeds) in a population will survive, biocides spur the development of resistant species. If all but 5% of the mosquito population in an area are killed by an insecticide, the ones that survive are the most resistant individuals, and they are the ones that will produce the succeeding generations.

There are now insects whose total resistance to certain pesticides has led some scientists to conclude that the entire process of insecticide development may be self-defeating. Despite the enormous growth in the use of biocides, crop loss to insect and weed pests has actually grown. According to Department of Agriculture figures, 32% of crops were lost to pests in 1945; 40 years later, such losses had increased to 37%.

Biocides have also been known to increase problems by destroying the natural enemies of the intended target, leaving it to breed unchallenged. Examples include the tobacco budworm and the brown planthopper, which were relatively minor pests before intensive crop spraying destroyed rival pests.

## Solid-Waste Disposal

Modern technologies and the societies that have developed them produce enormous amounts of solid wastes that must be disposed of. The rubbish heaps of past cultures suggest that humankind has always been faced with the problem of ridding itself of materials it no longer needs. The problem for advanced societies, with their ever-greater variety, amount, and durability of refuse, is even more serious: how to dispose of the solid wastes produced by residential, commercial, and industrial processes. Although these account for much less tonnage than the wastes produced by mining or agriculture, they are everywhere, a problem with which each individual and each municipality must deal.

### Municipal Waste

The wastes that communities must somehow dispose of include newspapers and beer cans, toothpaste tubes and old television sets, broken refrigerators, cars, and tires (Figure 12.25). American communities are now facing twin crises in disposing of these wastes: the sheer volume of trash and the toxic nature of much of it. Solid-waste disposal is a greater problem in the United States than in any other country, because Americans throw away more trash per person than any other country in the world. Solid-waste disposal costs are now the second largest expenditure of most local governments. Americans generate more than twice as much waste per person as do Japanese and Europeans, four times as much as Pakistanis or Indonesians.

This volume of trash is the result of three factors—affluence, packaging, and open space. Craving convenience, Americans rely on disposable goods that they throw away after very limited use. Thus, although readily available substitutes are more economical, Americans annually throw out 20 billion baby diapers, 2 billion razors, 1.7 billion pens, and 1 million tons of paper towels and napkins. People in less affluent countries repair and recycle a far greater proportion of domestic products. In addition, nearly all consumer goods are encased in some sort of wrapping, whether it be paper, cardboard, plastic, or foam. An astounding one-third of the yearly volume of trash consists of these packaging materials. Finally, the United States has traditionally had ample space in which to dump unwanted materials. Countries that ran short of such space decades ago have made greater progress in reducing the volume of waste.

**FIGURE 12.25** Some of the 240 million tires Americans replace each year. Most used tires are dumped, legally or illegally. Some are retreaded, some are shredded and burned to generate electricity, and some are exported abroad. Although many states have enacted scrap-tire management programs, financed by a levy consumers pay when buying new tires, most scrap tires remain unused and unwanted in growing dumps, which are fire hazards and breeding grounds for rats and mosquitoes and the diseases they carry. Source: *Don Kohlbauer/San Diego Union-Tribune.*

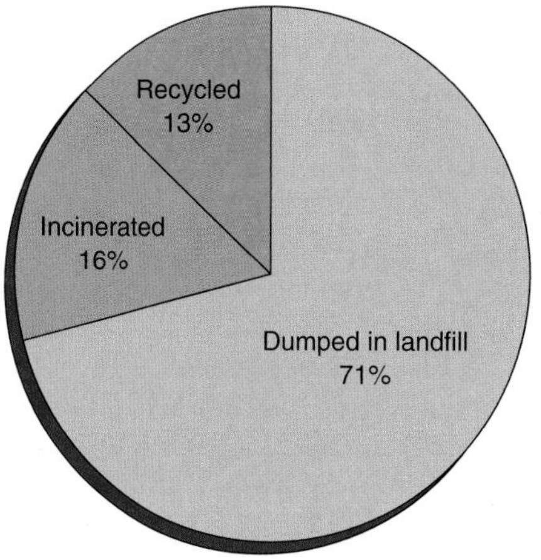


**FIGURE 12.26 Methods of solid-waste disposal in the United States.**

Although ordinary household trash does not meet the governmental designation of **hazardous waste**—defined as discarded material that may pose a substantial threat to human health or to the environment when improperly stored, transported, or disposed of—much of it is hazardous nonetheless. Products containing toxic chemicals include paint thinners and removers, furniture polishes, bleaches, oven and drain cleaners, used motor oil, and garden weed killers and pesticides.

Countries use various methods of disposing of solid wastes; each has its own impact on the environment. Loading wastes onto barges and dumping them in the sea, long a practice for coastal communities, inevitably pollutes oceans. Open dumps on land are a menace to public health, for they harbor disease-carrying rats and insects. Burning combustibles discharges chemicals and particulates into the air. In the United States, three methods of solid-waste disposal are employed: landfills, incineration, and recycling (Figure 12.26).

### Landfills

An estimated 71% of U.S. municipal solid waste is deposited in *sanitary landfills,* where each day's waste is compacted and covered by a layer of soil. "Sanitary" is a deceptive word. Until recently, there were no federal standards to which local landfills had to adhere, and while some communities and states regulated the environmental impact of dumps, many did not. Even if no commercial or industrial waste has been dumped at the site, most landfills eventually produce *leachate* (chemically contaminated drainage), liquids that pollute the groundwater. The EPA estimates that only 15% of the country's landfills have liners to prevent contaminants from seeping into groundwater supplies. Indeed, more than two-thirds of the dumps lack any type of system to monitor groundwater quality. Leachate forms when precipitation entering the landfill interacts with the decomposing materials. Thus, heavy metals are leached from batteries and old electrical parts, while vinyl chlorides come from the plastic in household products. Typical leachate contains more than 40 organic chemicals, many of them poisonous.

New York City's largest landfill, Fresh Kills on Staten Island, illustrates the problem. For more than half a century, trucks and barges daily carried approximately 11,000 tons (22 million lbs) of waste to the site. Opened in 1948 as a temporary facility, Fresh Kills was not constructed to hold its contents securely. Located in an ecologically sensitive wetland area and adjacent to residential communities, it became a malodorous 2200 acres of decomposing garbage. Every day, thousands of gallons of leachate seeped into the groundwater beneath the landfill. Fresh Kills finally closed as a depository for municipal solid waste in 2001, although it temporarily reopened the same year to accept debris from the towers of the World Trade Center. Plans call for turning the site into the largest park in New York City, with sports fields, concert arenas, wildlife sanctuaries, and the like. The conversion will be technically complicated and take years. Construction cannot be completed until the garbage decomposes, the mountains of trash settle, and the polluted water dissipates.

The closing of Fresh Kills made disposal of New York City's trash more complex and expensive. Sanitation trucks, tractor trailers, and railroads now carry the waste to landfills and incinerators in New Jersey, Pennsylvania, Virginia, and Ohio. Many communities face similar problems in disposing of their solid waste. In 2003, for example, Toronto, Canada, began sending all of its trash to a town near Detroit, Michigan, after its own landfill closed and the city was unable to find a town in Ontario willing to be the site of a new landfill.

The number of municipal landfills in the United States shrank significantly during the 1990s, falling from about 6500 in 1990 to about 2000 ten years later. They closed either because they were full or because they posed a threat to the environment. Public opposition has made it increasingly difficult to site new landfills, forcing communities to look for alternative methods of waste disposal. In contrast to the United States, many Western European countries and Japan dispose of little of their municipal solid waste in landfills, relying instead on incineration and recycling.

### Incineration

The quickest way to reduce the volume of trash is to burn it, a practice that was common at open dumps in the United States until it was halted by the Clean Air Act of 1970. Concern over air pollution also forced the closure of old, inefficient *incinerators* (facilities designed to burn waste), providing an impetus to designing a new generation of the plants. Municipal incinerators in the United States, mostly in the Northeast, burn about 16% of the national total of trash; Canada incinerates about 8%. Most municipal incinerators are of the waste-to-energy type, which use extra-high (980°C; 1800°F) temperatures to reduce trash to ash and simultaneously generate electricity or steam, which is sold to help pay operating costs (Figure 12.27).

**FIGURE 12.27 A waste-to-energy incinerator at Peekskill, New York,** one of the new generation of plants originally expected to convert more than one-quarter of the country's municipal waste to energy by the year 2000. A Supreme Court ruling that the ash such plants produce must be tested for hazardous toxicity and appropriately disposed of in protected landfills, growing public rejection and lawsuits, and increasingly stringent controls on the amount and kind of airborne vapors they emit have in many instances raised the operating costs of present incinerators far higher than landfill costs and have altered the economic assessments of new construction. Incineration becomes more cost-effective, however, when cities are unable to dispose of their trash locally and must haul it to distant sites. *© Mark Antman/The Image Works.*

A decade ago, incinerators were hailed as the ideal solution to overflowing landfills, but it has become apparent that they pose environmental problems of their own by generating toxic pollutants in both air emissions and ash. Air emissions from incinerator stacks have been found to contain an alphabet soup of highly toxic elements, ranging from *a* (arsenic) to *z* (zinc), and including, among others, cadmium, dioxins, lead, and mercury, as well as significant amounts of such gases as carbon monoxide, sulfur dioxide, and nitrogen oxides. Emissions can be kept to acceptably low limits by installing electrostatic precipitators, filters, and scrubbers to capture pollutants before they are released into the outside air, although the devices add significantly to the cost of the plant.

A greater problem is created by the concentration of toxins, particularly lead and cadmium, in the ash residue of burning. Incinerators typically reduce trash by 90%; one-tenth remains as ash, which then must be buried in a landfill. In 1994, the U.S. Supreme Court ruled that the ash must be tested for toxicity and handled as hazardous waste if it exceeds federal safety standards. This means it must be disposed of in licensed hazardous waste landfills, which have double plastic linings, moisture collection systems, and tighter operating procedures than do ordinary municipal landfills.

Although the likelihood of pollution from incinerator byproducts has sparked strong protest to their construction in the United States, they have become more accepted abroad. The seriousness of the air pollution and toxic ash problems, however, has aroused concern everywhere. In Japan, where more than three-fifths of municipal waste is incinerated, high atmospheric dioxin levels led the Ministry of Health in 1997 to strengthen earlier emission guidelines. Some European countries called at least temporary halts to incinerator construction while their safety was reconsidered, and increasingly landfills are refusing to take their residue.

### Source Reduction and Recycling

Awareness of the problems associated with landfills and incinerators has spurred interest in two alternative waste-management strategies: source reduction and recycling. By *source reduction,* we mean producing less waste in the first place so as to shrink the volume of the waste stream and lower the monetary and environmental costs associated with landfills and incinerators. Manufacturers can reduce the amount of paper, plastic, glass, and metal they use to package food and consumer products. Since 1977, for example, the weight of plastic soft-drink bottles and of aluminum beverage cans has been reduced by 20% to 30%. Some products, such as detergents and beverages, can be produced in concentrated form and packaged in smaller containers.

Another way to reduce the amount of waste needing disposal is by **recycling,** the recovery and reprocessing or reuse of previously used material into new products for the same or another purpose. Aluminum beverage cans usually are recast into new cans, for example, and glass bottles are crushed, melted, and made into new bottles. Old tires can be shredded and turned into rubberized road surfacing or used as mulch around playground equipment (Figure 12.28). Some communities collect leaves and other yard waste, which accounts for about 20% of the typical municipal waste stream, and turn it into compost. The recycling of plastics is more difficult because there are many types of plastic containers in common use, and they must be separated before being recycled. Recycled plastic can be converted into fiber for carpeting, playground equipment, insulation for clothing, and other products.

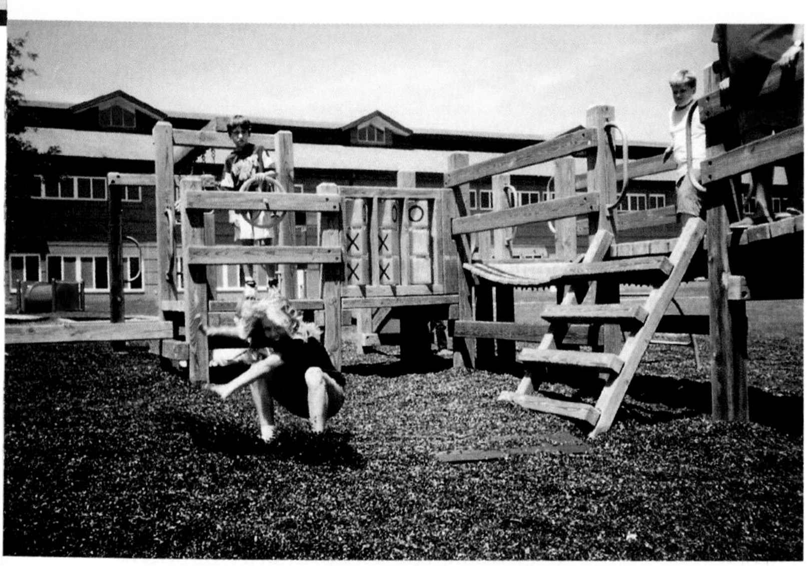

**FIGURE 12.28 Recycled tires used as mulch.** Shredded tire bits can be used as mulch in gardens to help hold moisture in the ground. It is clean and doesn't rot or attract termites, as wood mulch does. It can also be used around playground equipment, as shown here, because it doesn't accumulate dust or compact with use. *Courtesy of American Rubber Technologies, Inc.*

The widespread adoption of municipal recycling programs in the United States now diverts an estimated 25% to 30% of trash away from landfills and incinerators. The country currently recycles some 60% of the aluminum beverage cans, 40% of the paper, and 25% of the glass entering the waste stream. Compared with other methods of solid-waste disposal, recycling has a relatively benign impact on the environment, despite the fact that it takes water, energy, and other resources to recover, process, and convert the materials into new products. Recycling saves natural resources by making it possible to cut fewer trees, burn less oil, and mine less oil. Because it typically takes less energy to make things out of recycled material than out of virgin materials, recycling saves energy. It also reduces the pollution of air, water, and land that stems from the manufacture of new materials and from other methods of waste disposal, and it saves increasingly scarce landfill space for materials that cannot be recycled.

With all of these advantages, why isn't recycling more widely practiced in the United States? Japan and a number of Western European countries recycle a far greater proportion of their waste stream. Analysts have offered a number of explanations: (1) the cost of collecting the goods to be recycled, (2) fluctuations in market prices for commodities, and (3) the lack of a ready market for products manufactured from recycled material. Perhaps the most important factor, however, is that in the United States, the price of energy historically has been low and the supply abundant, and the true monetary and environmental costs of making things from raw materials rather than recycled ones are hidden.

In less affluent countries, where manufactured products are expensive but labor is cheap, recycling plays an important role in reducing the amount of solid waste. In many cities in developing countries, such as Manila (the Philippines), Phnom Penh (Cambodia), Cairo (Egypt), and Mexico City, thousands of poor people make a living sifting through the city's trash, looking for recyclable goods—tin cans, copper, wood, electronics, clothing, and the like—that they can sell to middlemen for businesses and industries (Figure 12.29). The wastepickers, or scavengers, play a vital role in reducing the amount of trash that must be compacted and covered with a fresh layer of dirt each day. It is estimated that in Indonesian cities, for example, wastepickers reduce total urban refuse by one-third.

Unfortunately, the environmental conditions under which wastepickers work are not good, and their life expectancy is considerably less than that of the general population. The health risks include injuries from accidents, infections from disease-carrying organisms, and exposure to gases that seep out of the layers of fermenting trash and to hazardous wastes such as dioxin and heavy metals.

The questions geographers ask, as we saw in Chapter 1 of this book, ultimately focus on matters of location and character of place. We asked how things are distributed over the surface of the earth; how physical and cultural features of areas are alike or different from place to place; how the varying content of different places came about; and what all these differences and similarities mean for people.

## The Nature of Regions

In the earlier chapters, we examined some of the physical and cultural content of areas and some spatially important aspects of human behavior. We looked at physical earth processes that lead to differences from place to place in the environment. We studied ways in which humans organize their actions in earth space—through political institutions, by economic systems and practices, and by cultural and social processes that influence spatial behavior and interaction. Population and settlement patterns and areal differences in human use and misuse of earth resources all were considered as part of the mission of geography.

For each of the topics we studied, from landforms to cities, we found spatial regularities. We discovered that things are not irrationally distributed over the surface of the earth, but reflect an underlying spatial order based on understandable physical and cultural processes. We found, in short, that although no two places are *exactly* the same, it is possible and useful to recognize segments of the total world that are internally similar in some important characteristic and distinct in that feature from surrounding areas.

These regions of significant uniformity of content are the geographer's spatial equivalent of the historian's "eras" or "ages," and are assigned brief summary names to indicate they are different in some important way from adjacent or distant territories. The **region,** then, is a device of areal generalization. It is an attempt to separate into recognizable component parts the otherwise overwhelming diversity and complexity of the earth's surface.

All of us have a general idea of the meaning of region, and all of us refer to regions in everyday speech and action. We visit "the old neighborhood" or "go downtown"; we plan to vacation or retire in the "Sunbelt"; or we speculate on the effects of weather conditions in the "Northern Plains" or the "Corn Belt" on grain supplies or next year's food prices. In each instance we have mental images of the areas mentioned. Those images are based on place characteristics and areal generalizations that seem useful to us and recognizable to our listeners. We have, in short, engaged in an informal place classification to pass along quite complex spatial, organizational, or content ideas. We have applied the **regional concept** to bring order to the immense diversity of the earth's surface.

What we do informally as individuals, geography attempts to do formally as a discipline—define and explain regions (Figure 13.1). The purpose is clear: to make the infinitely varying world around us understandable through spatial summaries. That world is only rarely subdivided into neat, unmistakable "packages" of uniformity. Neither the environment nor human areal actions present us with a compartmentalized order, any more than the sweep of human history has predetermined "eras," or all plant specimens come labeled in nature with species names. We all must classify to understand, and the geographer classifies in regional terms.

Regions are spatial expressions of ideas or summaries useful to the analysis of the problem at hand. Although as many possible regions exist as there are physical, cultural, or organizational attributes of area, the geographer selects for study only those areal variables that contribute to the understanding of a specific topical or spatial problem. All others are disregarded as irrelevant.

In our reference to the Corn Belt, we delimit a portion of the United States showing common characteristics of farming economy and marketing practices. We dismiss—at that level of generalization—differences within the region based on slope, soil type, state borders, or characteristics of population. The boundaries of the Corn Belt are assumed to be marked where the region's internal unifying characteristics change so materially that different agricultural economies become dominant and different regional summaries are required. The content of the region suggests its definition and determines the basis of its delimitation.

Although regions may vary greatly, they all share certain common characteristics related to earth space:

- **Regions have** ***location,*** often expressed in the regional name selected, such as the Midwest, the Near East, North Africa, and the like. This form of regional name underscores the importance of *relative location* (see Chapter 1, p. 9).
- **Regions have** ***spatial extent,*** recognized by territories over which defined characteristics or generalizations about physical, cultural, or organizational content are constant.
- **Regions have** ***boundaries*** based on the areal spread of the features selected for study. Since regions are the recognition of the features defining them, their boundaries are drawn where those features no longer occur or dominate. Regional boundaries are rarely as sharply defined as those suggested by Figure 13.2 or by the regional maps in this and other geography texts. More frequently, there exist broad zones of transition from one distinctive core area to another as the dominance of the defining regional features gradually diminishes outward from the core to the regional periphery. Linear boundaries are arbitrary divisions made necessary by the scale of world regional maps and by the summary character of most regional discussions.
- **Regions may be either** ***formal or functional,*** as we saw in Chapter 1 (p. 17).

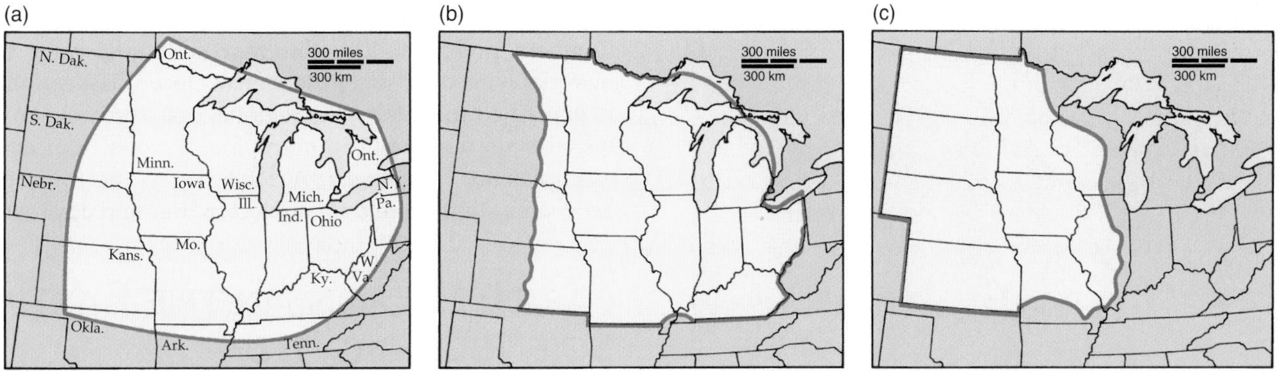


**FIGURE 13.1 The Midwest as seen by different professional geographers.** Agreement on the need to recognize spatial order and to define regional units does not imply unanimity in the selection of boundary criteria. All of the sources concur in the significance of the Midwest as a regional entity in the spatial structure of the United States and agree on its core area. They differ, however, in their assessments of its limiting characteristics.
Sources: *(a) John H. Garland, ed.,* The North American Midwest. *New York: Wiley, 1955; (b) John R. Borchert and Jane McGuigan,* Geography of the New World. *Chicago: Rand McNally, 1961; and (c) Otis P. Starkey and J. Lewis Robinson,* The Anglo-American Realm. *New York: McGraw-Hill, 1969.*

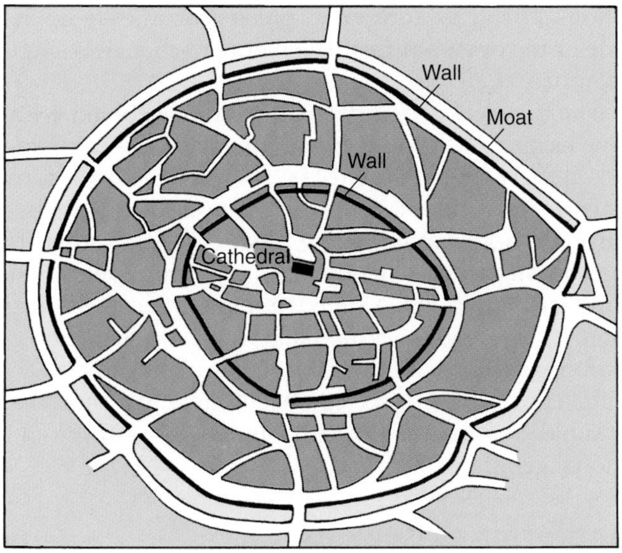


**FIGURE 13.2 Aachen, Germany, in 1649.** The acceptance of regional extent implies the recognition of regional boundaries. At some defined point, *urban* is replaced by *nonurban,* the Midwest ends and the Plains begin, or the rain forest ceases and the savanna emerges. Regional boundaries are, of course, rarely as precisely and visibly marked as were the limits of the walled medieval city. Its sprawling modern counterpart may be more difficult to define, but the boundary significance of the concept of *urban* remains.

**Formal regions** are areas of essential uniformity throughout in one or a limited combination of physical or cultural features. In previous chapters, we encountered such formal physical regions as the humid subtropical climate zone and the Sahel region of Africa, as well as formal (homogeneous) cultural regions in which standardized characteristics of language, religion, ethnicity, or livelihood existed. The frontsheet maps of countries and topographical regions show other formal regional patterns. Whatever the basis of its definition, the formal region is the largest area over which a valid generalization of attribute uniformity may be made. Whatever is stated about one part of it holds true for its remainder.

The **functional region,** in contrast, is a spatial system defined by the interactions and connections that give it a dynamic, organizational basis. Its boundaries remain constant only as long as the interchanges establishing it remain unaltered. The commuting region shown in Figure 1.10 retained its shape and size only as long as the road pattern and residential communities on which it was based remained unchanged.

- **Regions are *hierarchically arranged.*** Although regions vary in scale, type, and degree of generalization, none stands alone as the ultimate key to areal understanding. Each defines only a part of spatial reality.

On a formal regional scale of size progression, the Delmarva Peninsula of the eastern United States may be seen as part of the Atlantic Coastal Plain, which is in turn a portion of the eastern North American humid continental climatic region, a hierarchy that changes the basis of regional recognition as the level and purpose of generalization alter (Figure 13.3). The central business district of Chicago is one land use complex in the functional regional hierarchy that describes the spatial influences of the city of Chicago and the metropolitan region of which it is the core. Each recognized regional entity in such progressions may stand alone and at the same time exist as a part of a larger, equally valid, territorial unit.

Such progressions may also reflect gradations in the intensity in spatial dominance of the phenomenon giving definition to the region. With particular attention to regions defined by the distribution of culture groups—but with application to regions delimited by other criteria—Donald Meinig suggested the term *core* to mean a centralized zone of concentration and greatest homogeneity of regional character. Areas in which the culture (or other defining feature)

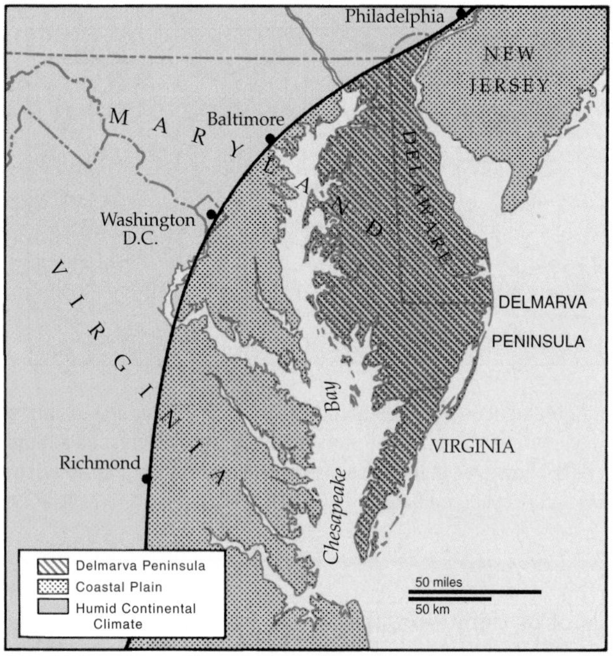


**FIGURE 13.3 A hierarchy of regions.** One possible nesting of regions within a *regional hierarchy* defined by differing criteria. Each regional unit has internal coherence. The recognition of its constituent parts aids in understanding the composite areal unit.

is dominant, but with less intensity and totality of development, he labeled *domains*. Finally, Meinig proposed the term *sphere* to recognize the zone of broadest but least-intensive expression of the regional character, where some of its defining traits are encountered but where it is no longer spatially dominant.

These generalizations about the nature of regions and the regional concept are meant to instill firmly the understanding that regions are human intellectual creations designed to serve a purpose. Regions focus our attention on spatial uniformities. They bring clarity to the seeming confusion of the observable physical and cultural features of the world we inhabit. Regions provide the framework for the purposeful organization of spatial data.

# THE STRUCTURE OF THIS CHAPTER

The remainder of the chapter contains examples of how geographers have organized regionally their observations about the physical and cultural world. Each study or vignette explores a different aspect of regional reality. Each organizes its data in ways appropriate to its subject matter and objective, but in each, some or all of the common characteristics of regional delimitation and structure may be recognized. Each of the regional examples is based on the content of one of the earlier chapters of this book, and the page numbers preceding the different selections refer to the material within those chapters most closely associated with, or explained further by, the regional case study. As an aid to visualizing the application of the regional concept, the examples are introduced by reference to the traditions of geography that unite their subject matter and approach.

# PART I: REGIONS IN THE EARTH SCIENCE TRADITION

The simplest of all regions to define, and generally the easiest to recognize, is the formal region based on a single readily apparent component or characteristic. The island is land, not water, and its unmistakable boundary is naturally given where the one element passes to the other. The terminal moraine may mark the transition from the rich, black soils of recent formation to the parti-colored clays of earlier generations. The dense forest may break dramatically upon the glade or the open prairies. The nature of change is singular and apparent.

The physical geographer, although concerned with all of the earth sciences that explain the natural environment, deals at the outset with *single factor* formal regions. Many of the earth features of concern to physical geography, of course, do not exist in simple, clearly defined units. They must be arbitrarily "regionalized" by the application of boundary definitions. A stated amount of received precipitation, the presence of certain important soil characteristics, the dominance in nature of particular plant associations—all must be decided on as regional limits, and all such limits are subject to change through time or by purpose of the regional geographer.

## Landforms as Regions

(*See "Stream Landscapes in Humid Areas," p. 72*)

The landform region exists in a more sharply defined fashion than such transitional physical features as soil, climate, or vegetation. For these latter areas, the boundaries depend on definitional decisions made (and defended) by the researcher. The landform region, on the other hand, arises—visibly and apparently unarguably—from nature itself, independent of human influence and unaffected by time on the human scale. Landforms constitute basic, naturally defined regions of physical geographic concern. The existence of major landform regions—mountains, lowlands, plateaus—is unquestioned in popular recognition or scientific definition. Their influence on climates, vegetational patterns, even on the primary economies of subsistence populations has been noted in earlier portions of this book. The following discussion of a distinct landform region, describing its constitution and its relationships to other physical and cultural features of the landscape, is adapted from a classic study by Wallace W. Atwood.

## The Black Hills Province[1]

*The Black Hills rise abruptly from the surrounding plains [Figure 13.4]. The break in topography at the margin of this province is obvious to the most casual observer who visits that part of our country. Thus the boundaries of this area, based on contrasts in topography, are readily determined.*

*One who looks more deeply into the study of the natural environment may recognize that in the neighboring plains the rock formations lie in a nearly horizontal position. They are sandstones, shales, conglomerates, and limestones. In the foothills those same sedimentary formations are bent upward and at places stand in a nearly vertical position. Precisely where the change in topography occurs, we find a notable change in the geologic structure and thus discover an explanation for the variation in relief.*

*The Black Hills are due to a distinct upwarping, or doming, of the crustal portion of the earth. Subsequent removal by stream erosion of the higher portions of that dome and the dissection of the core rocks have produced the present relief features. As erosion has proceeded, more and more of a complex series of ancient metamorphic rocks has been uncovered. Associated with the very old rocks of the core and, at places, with the sedimentary strata, there are a number of later intrusions which have cooled and formed solid rock. They have produced minor domes about the northern margin of the Black Hills.*

*With the elevation of this part of our country there came an increase in rainfall in the area, and with the increase in elevation and rainfall came contrasts in relief, in soils, and in vegetation.*

*As we pass from the neighboring plains, where the surface is monotonously level, and climb into the Black Hills area, we enter a landscape having great variety in the relief. In the foothill belt, at the southwest, south, and east, there are hogback ridges interrupted in places by* water gaps, *or gateways, that have been cut by streams radiating from the core of the Hills. Between the ridges there are roughly concentric valley lowlands. On the west side of the range, where the sedimentary mantle has not been removed, there is a plateaulike surface; hogback ridges are absent. Here erosion has not proceeded far enough to produce the landforms common to the east margin. In the heart of the range we find deep canyons, rugged intercanyon ridges, bold mountain forms, craggy knobs, and other picturesque features [Figure 13.5]. The range has passed through several periods of mountain growth and several stages, or cycles, of erosion.*

*The rainfall of the Black Hills area is somewhat greater than that of the brown, seared, semiarid plains regions, and evergreen trees survive among the hills. We leave a land of sagebrush and grasses to enter one of forests. The dark-colored evergreen trees suggested to early settlers the name Black Hills. As we enter the area, we pass from a land of cattle ranches and some seminomadic shepherds to a land where forestry, mining, general farming, and recreational activities give character to the life of the people. In color and form, in topography, climate, vegetation, and economic opportunities, the Black Hills stand out conspicuously as a distinct geographic unit.*

[1]Adapted from *The Physiographic Provinces of North America* by Wallace W. Atwood. © Copyright, 1940, by Ginn and Company. Used by permission of Silver, Burdett & Ginn, Inc.

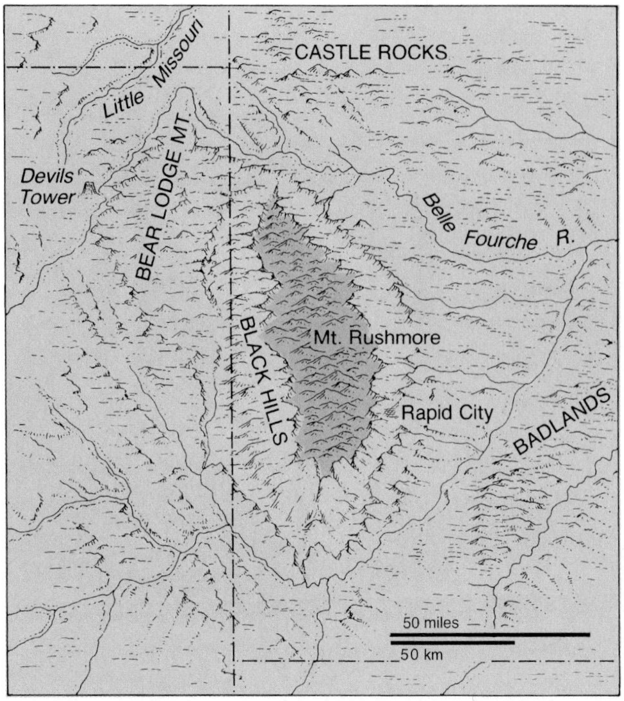


**FIGURE 13.4 The Black Hills landform province.**

## Dynamic Regions in Weather and Climate

*(See "Air Masses," p. 102)*

The unmistakable clarity and durability of the Black Hills landform region and the precision with which its boundaries may be drawn are rarely echoed in other types of formal physical regions. Most of the natural environment, despite its appearance of permanence and certainty, is dynamic in nature. Vegetations, soils, and climates change through time by natural process or by the action of humans. Boundaries shift, perhaps abruptly, as witness the recent migration southward of the Sahara. The core characteristics of whole provinces change as marshes are drained or forests are replaced by cultivated fields.

That complex of physical conditions we recognize locally and briefly as weather and summarize as climate displays particularly clearly the temporary nature of much of the natural environment that surrounds us. Yet even in the turbulent change of the atmosphere, distinct regional entities exist with definable boundaries and internally consistent horizontal and vertical properties. "Air masses" and the consequences of their encounters constitute a major portion of contemporary weather analysis and prediction. Air masses further fulfill all the criteria of *multifactor formal regions,* though their dynamic quality and their patterns of movement obviously mark them as being of a nature distinctly different from such stable physical entities as landform regions, as the following extract from *Climatology and the World's Climates* by George R. Rumney makes clear.

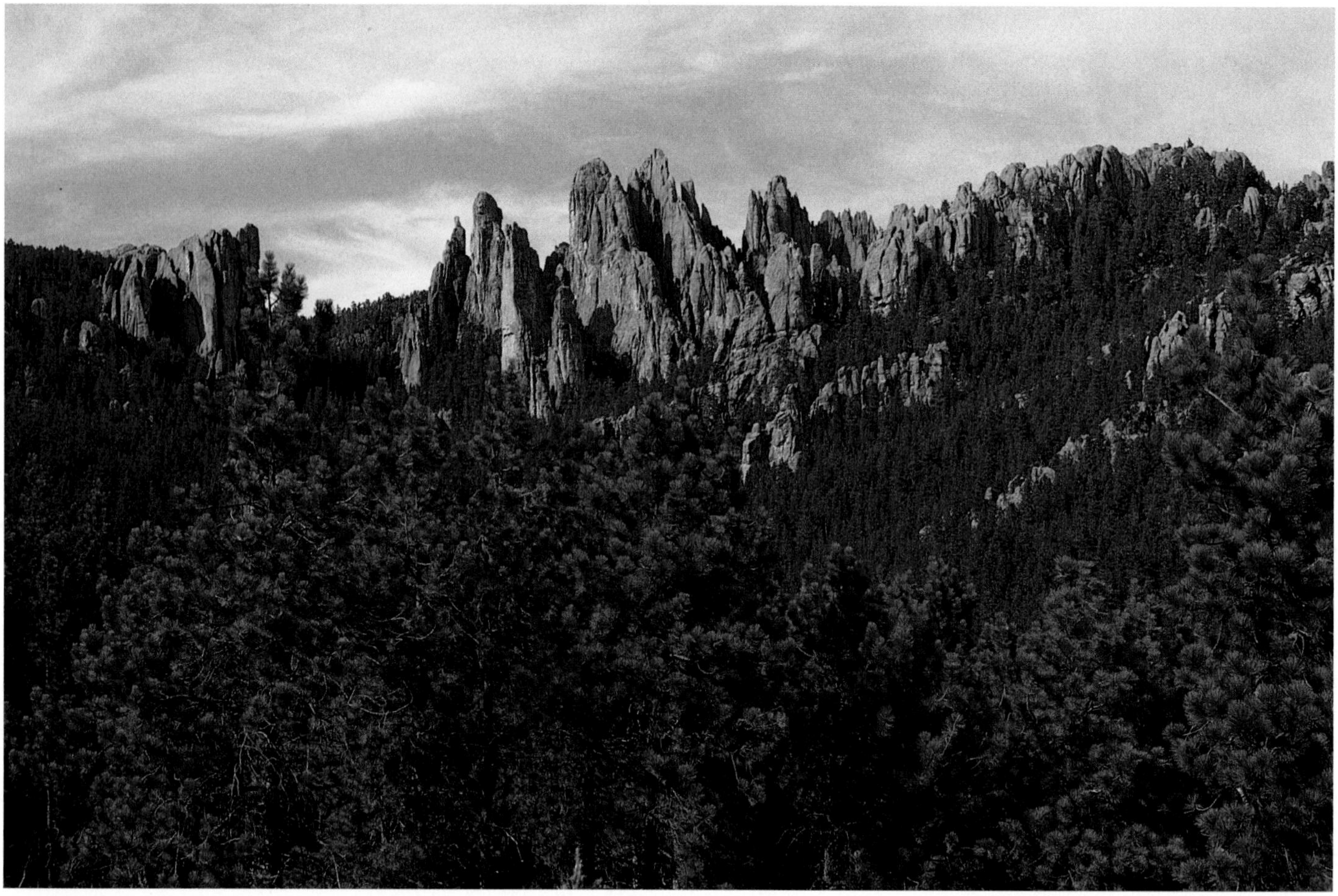

**FIGURE 13.5** The "Needles" of the Black Hills result from erosion along vertical cracks and crevices in granite. *© B. F. Molina/Terraphotographics/BPS.*

## Air Masses[2]

*An* air mass *is a portion of the atmosphere having a uniform horizontal distribution of certain physical characteristics, especially of temperature and humidity. These qualities are acquired when a mass of air stagnates or moves very slowly over a large and relatively unvaried surface of land or sea. Under these circumstances surface air gradually takes on properties of temperature and moisture approaching those of the underlying surface, and there then follows a steady, progressive transmission of properties to greater heights, resulting finally in a fairly clearly marked vertical transition of characteristics. Those parts of the earth where air masses acquire their distinguishing qualities are called* source regions.

*The height to which an air mass is modified depends upon the length of time it remains in its source region and also upon the difference between the initial properties of the air when it first arrived and those of the underlying surface. If, for example, an invading flow of air is cooler than the surface beneath as it comes to virtual rest over a source region, it is warmed from below, and convective currents are formed, rapidly bearing aloft new characteristics of temperature and moisture to considerable heights. If, on the other hand, it is warmer than the surface of the source region, cooling of its surface layers takes place, vertical thermal currents do not develop, and the air is modified only in its lower portions. The process of modification may be accomplished in just a few days of slow horizontal drift, although it often takes longer, sometimes several weeks. Radiation, convection, turbulence, and advection are the chief means of bringing it about.*

*The prerequisite conditions for these developments are very slowly migrating, outward spreading, and diverging air and a very extensive surface beneath that is fairly uniform in nature. Light winds and relatively high barometric pressure characteristically prevail. Hence, most masses form within the great semipermanent anticyclonic regions of the general circulation, where calms, light variable winds, and overall subsidence of the atmosphere are typical.*

*Four major types of source regions are recognized: continental polar, maritime polar, continental tropical, and maritime tropical. Polar air masses are continental when they develop over land or ice surfaces in high latitudes; these are cold and dry. They are maritime when they form over the oceans in high latitudes. An air mass from these sources is cold and moist. Similarly, tropical air is continental when it originates along the Tropics of Cancer and Capricorn over northern Africa and northern Australia and is therefore warm and dry. It is maritime when it forms along the Tropics over the oceans, where it develops as a mass of warm, moist air. A single air mass usually covers thousands of square miles of the earth's surface when fully formed.*

[2]George R. Rumney, *Climatology and the World's Climates.* New York: Macmillan, 1968. Used by permission.

*An air mass is recognizable chiefly because of the uniformity of its primary properties—temperature and humidity—and the vertical distribution of these. Secondary qualities, such as cloud types, precipitation, and visibility, are also taken into account. These qualities are retained for a remarkably long time, often for several weeks, after an air mass has traveled far from its source region, and they are thus the means of distinguishing it from other masses of air.*

The principal air masses of the Americas, their source regions, and paths of movement are shown in Figure 13.6.

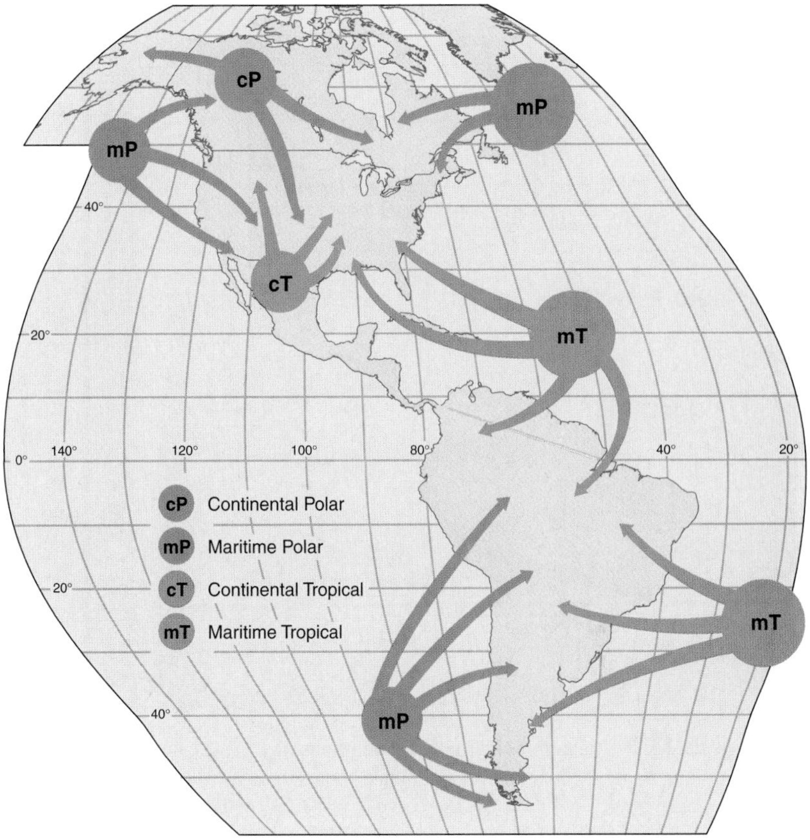


**FIGURE 13.6 Air masses of North and South America,** their source regions, and paths of movement.

## Natural Resource Regions

*(See "Coal," p. 141)*

The unevenly distributed resources upon which people depend for existence are logical topics of interest within the earth science of geography. Resource regions are mapped, and raw material qualities and quantities are discussed. Areal relationships to industrial concentrations and the impacts of material extraction upon alternate uses of areas are typical interests in resource geography and in the definition of resource regions.

Those regions, however, are usually treated as if they were expressions of observable surface phenomena; as if, somehow, an oil field were as exposed and two-dimensional as a soil region or a forested district. What is ignored is that most mineral resources are three-dimensional regions beneath the ground. In addition to the characteristics of an area that may form the basis for regional delimitations and descriptions of surface phenomena, regions beneath the surface add their own particularities to the problem of regional definition. They have, for example, upper and lower boundaries in addition to the circumferential bounds of surface features. They may have an internal topography divorced from the visible landscape. Subsurface relationships—for example, mineral distribution and accessibility in relation to its enclosing rock or to groundwater amounts and movement—may be critical in understanding these specialized, but real, regions. An illustration drawn from the Schuylkill field of the anthracite region of northeastern Pennsylvania helps illustrate the nature of regions beneath the surface.

### The Schuylkill Anthracite Region[3]

*Nothing in the wild surface terrain of the anthracite country suggested the existence of an equally rugged subterranean topography of coal beds and interstratified rock, slate, and fire clays forming a total vertical depth of 900 meters (3000 ft) at greatest development. Yet the creation of the surface landscape was an essential determinant of the areal extent of the Schuylkill district, of the contortions of its bedding, and of the nature of its coal content. A county history of the area reports, "The physical features of the anthracite country are wild. Its area exhibits an extraordinary series of parallel ridges and deep valleys, like long, rolling lines of surf which break upon a flat shore." Both the surface and the subsurface topographies reflect the strong folding of strata after the coal seams were deposited; the anthracite (hard) coal resulted from metamorphic carbonization of the original bituminous beds. Subsequent river and glacial erosion removed as much as 95% of the original anthracite deposits and gave to those that remained discontinuous existence in sharply bounded fields such as the Schuylkill (Figure 13.7), a discrete areal entity of 470 square kilometers (181 sq mi).*

*The irregular topography of the underground Schuylkill region means that the interbedded coal seams, the most steeply inclined of all the anthracite regions (Figure 13.8), outcrop visibly at the surface on hillsides and along stream valleys. The outcrops made the presence of coal known as early as 1770, but not until 1795 did Schuylkill anthracite find its first use by local blacksmiths. Reviled as "stone coal" or "black stone" that would not ignite, anthracite found no ready commercial market, although it was used in wire and rolling mills located along the Schuylkill River before 1815 and to generate steam in the same area by 1830.*

*The resources of the subterranean Schuylkill region affected human patterns of surface regions only after the Schuylkill Navigation Canal was completed in 1825 (Figure 13.9), providing a passage to rapidly expanding external markets for the fuel and for the output of industry newly located atop the region. Growing demand induced a boom in coal exploitation, an exhaustion of the easily available outcrop coal, and the beginning of the more arduous and dangerous underground mining.*

[3]By Jerome Fellmann.

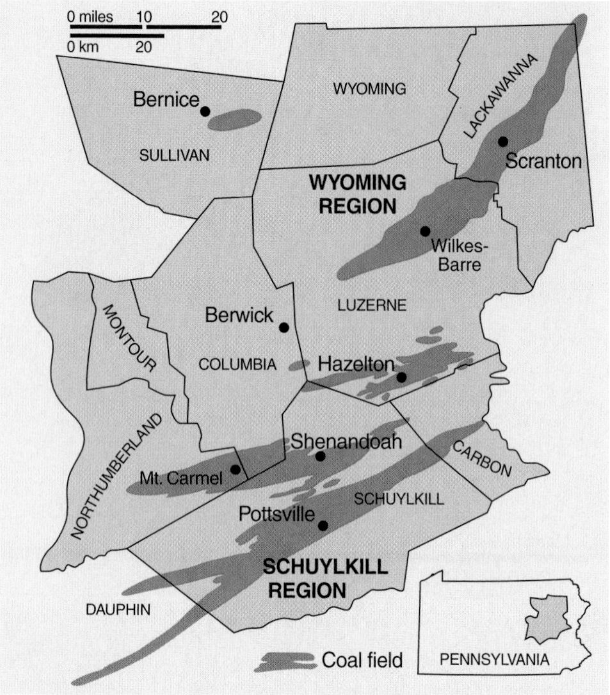


**FIGURE 13.7 The anthracite regions of northeastern Pennsylvania** are sharply defined by the geologic events that created them.

*The early methods of mining were simple: merely quarrying the coal from exposed outcrops, usually driving on a slight incline to permit natural drainage. Deep shafts were unnecessary and, indeed, not thought of, since the presence of anthracite at depth was not suspected. Later, when it was no longer possible to secure coal from a given outcrop, a small pit was sunk to a depth of 9 to 12 meters (30 to 40 ft); when the coal and the water that accumulated in the pit could no longer safely be brought to the surface by windlass, the pit was abandoned and a new one was started. Shaft mining, in which a vertical opening from the surface provides penetration to one or several coal beds, eventually became a necessity; with it came awareness of the complex interrelationships between seam thickness, the nature of interstratified rock and clays, the presence of gases, and the movement of subsurface water.*

*The subterranean Schuylkill region has a three-dimensional pattern of use. The configuration and the variable thickness of the seams demand concentration of mining activities. Minable coal is not uniformly available along any possible vertical or horizontal cross section because of the interstratification and the extreme folding of the beds. Mining is concentrated further by the location of shafts and the construction of passages, in their turn determined by both patterns of ownership and thickness of seam. In general, no seam less than 0.6 meters (2 ft) is worked, and the absolute thickness—15 meters (50 ft)—is found only in the Mammoth seam of the Schuylkill region.*

*Friable interstratified rock increases the danger of coal extraction and raises the costs of cave-in prevention. Although the Schuylkill mines are not gassy, the possibility of gas release from the collapse of coal pillars left as mine supports makes necessary systems of ventilation even more elaborate than those minimally required to provide adequate air to miners. Water is ever-present in the anthracite workings, and constant pumping or draining is necessary for mine operation. The collapse of strata underlying a river may result in sudden disastrous flooding.*

The Schuylkill subterranean anthracite region presents a pattern of complexity in distribution of physical and cultural features and of interrelationships between phenomena every bit as great and inviting of geographic analysis as any purely surficial region.

## Part II: Regions in the Culture–Environment Tradition

The earth science tradition of geography imposes certain distinctive limits on area analysis. However defined, the regions that may be drawn are based on nature and do not result from human action. The culture–environment tradition, however, introduces to regional geography the infinite variations of human occupation and organization of space. There is a corresponding multiplication of recognized regional types and of regional boundary decisions.

Despite the differing interests of physical and cultural geographers, one element of study is common to their con-

**FIGURE 13.8** The deep folding of the Schuylkill coal seams made them costly to exploit. The Mammoth seam runs as deep as 450 to 600 meters (1500 to 2000 ft) below the surface.

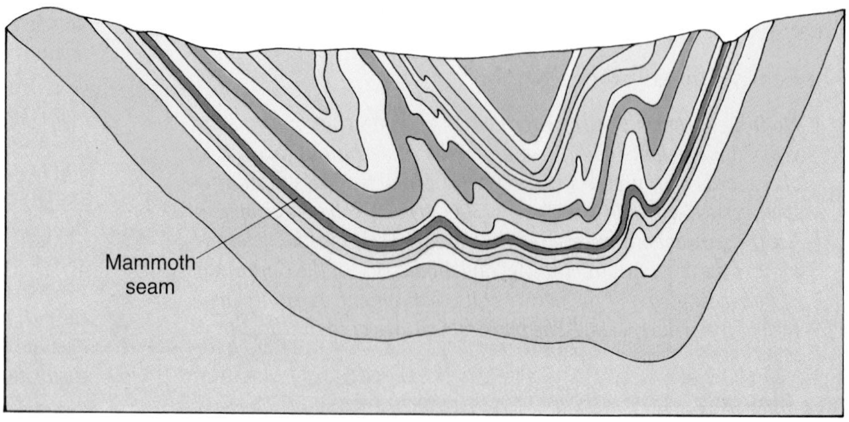

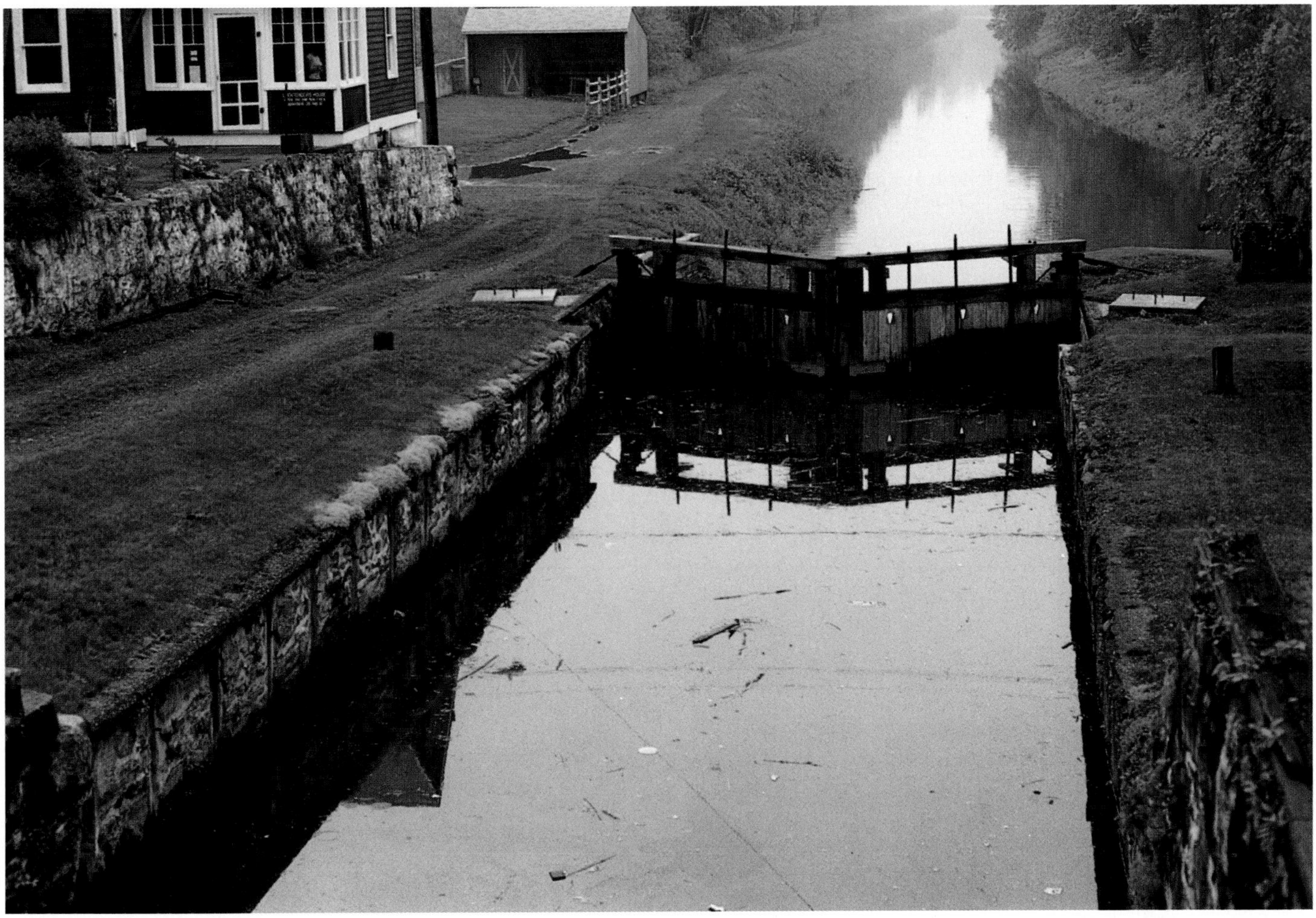

**FIGURE 13.9** The Schuylkill Navigation Canal, like its counterpart Lehigh Canal shown here, provided an outlet to market for the anthracite region's coal resources after 1825. *© Elizabeth J. Leppman.*

cerns: that of process. The "becoming" of an ecosystem, of a cultural landscape, or of the pattern of exchanges in an economic system is an important open or implied part of nearly all geographic study. Evidence of the past as an aid to understanding of the present is involved in much geographical investigation, for present-day distributional patterns or qualities of regions mark a merely temporary stage in a continuing process of change.

## Population as Regional Focus

(*See "World Population Distribution," p. 200*)

In no phase of geography are process and change more basic to regional understanding than in population studies. The human condition is dynamic and patterns of settlement are ever-changing. Although these spatial distributions are related to the ways that people utilize the physical environment in which they are located, they are also conditioned by the purposes, patterns, and solutions of those who went before. In the following extract taken from the work of Glenn T. Trewartha, a dean of American population geographers, notice how population regionalization—used as a focal theme—ties together a number of threads of regional description and understanding. The aspirations of colonist-conquerors, past and contemporary transportation patterns, physical geographic conditions, political separatism, and the history and practice of agriculture and rural land holdings are all introduced to give understanding to population from a regional perspective.

Population Patterns of Latin America[4]

*A distinctive feature of the spatial arrangement of population in Latin America is its strongly nucleated character; the pattern is one of striking clusters. Most of the population clusters remain distinct and are separated from other clusters by sparsely occupied territory. Such a pattern of isolated nodes of settlement is common in many pioneer regions; indeed, it was characteristic of early settlement in*

[4]From Glenn T. Trewartha, *The Less Developed Realm: A Geography of Its Population.* Copyright © 1972 by John Wiley & Sons, Inc. Reprinted by permission of John Wiley & Sons, Inc.

*both Europe and eastern North America. In those regions, as population expanded, the scantily occupied areas between individual clusters gradually filled in with settlers, and the nodes merged. But in Latin America such an evolution generally did not occur, and so the nucleated pattern persists. Expectably, the individual population clusters show considerable variations in density.*

*The origin of the nucleated pattern of settlement is partly to be sought in the gold and missionary fever that imbued the Spanish colonists. Their settlements were characteristically located with some care, since only areas with precious metals for exploitation and large Indian populations to be Christianized and to provide laborers could satisfy their dual hungers. A clustered pattern was also fostered by the isolation and localism that prevailed in the separate territories and settlement areas of Latin America.*

*Almost invariably each of the distinct population clusters has a conspicuous urban nucleus. To an unusual degree the economic, political, and social life within a regional cluster centers on a single large primate city, which is also the focus of the local lines of transport.*

*The prevailing nucleated pattern of population distribution also bears a relationship to political boundaries. In some countries . . . a single population cluster represents the core area of the nation. In more instances, however, a population cluster forms the core of a major political subdivision of a nation-state, so that a country may contain more than one cluster. A consequence of this simple population distribution pattern and its relation to administrative subdivisions is that political boundaries ordinarily fall within the sparsely occupied territory separating individual clusters. In Latin America few national or provincial boundaries pass through nodes of relatively dense settlement [Figure 13.10].*

*Another feature arising out of the cluster pattern of population arrangement is that the* total national territory *of a country is often very different from the* effective national territory, *since the latter includes only those populated parts that contribute to the country's economic support.*

*A further consequence of the nucleated pattern is found in the nature of the transport routes and systems. Overland routes between population clusters are usually poorly developed, while a more efficient network ordinarily exists* within *each cluster, with each such regional network joined by an overland route to the nearest port. Thus, the chief lines of transport connecting individual population clusters are often sea lanes rather than land routes. Gradually, as highways are developed and improved, intercluster overland traffic tends to increase.*

*. . . The spectacular rates at which population numbers are currently soaring in Latin America are not matched by equivalent changes in their spatial redistributions. Any population map of Latin America reveals extensive areas of unused and underutilized land. Part of such land is highland and plagued by steep slopes, but by far the larger share of it is characterized by a moist tropical climate, either tropical wet or tropical wet and dry. Such a climatic environment, with its associated wild vegetation, soils, and drainage, admittedly presents many discouraging elements to the new settler of virgin lands. . . .* [T]*ropical climate alone is scarcely a sufficient explanation for the abundance of near-empty lands south of the Rio Grande. Cultural factors are involved as much as, if not more than, physical ones. One of the former is the unfortunate land-holding system that has been fastened on the continent, under which vast areas of potentially cultivable land are held out of active use by a small number of absentee landlords, who not only themselves make ineffective use of the land, but at the same time refuse to permit its cultivation by small operators. Because of the land-holding system, peasant proprietors are unable to secure their own lands, a situation that discourages new rural settlement. . . .*

*The changes now in progress in the spatial distributions accompanying the vast increase in numbers of people do not appear to involve any large-scale push of rural settlement into virgin territory. Only to a rather limited extent is new agricultural settlement taking place. Intercluster regions are not filling rapidly. The overwhelming tendency is for people to continue to pile up in the old centers of settlement in and around their cities, rather than to expand the frontier into new pioneer-settlement areas. . . .*

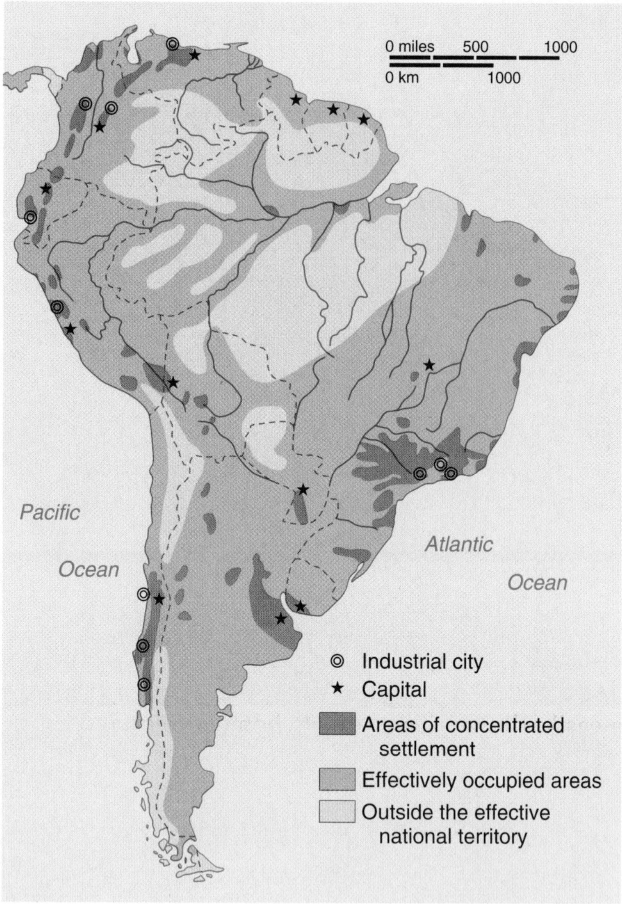


**FIGURE 13.10 Basic settlement patterns of South America.** Population clusters focused on urban cores and separated by sparsely populated rural areas were the traditional pattern in Latin American countries. *Used with permission of Simon & Schuster, Inc. from the Macmillan College Text* Introduction to Latin America, *by Preston E. James. Copyright © 1964 by Macmillan College Publishing Company, Inc.*

## Language as Region

*(See "Language," p. 236)*

The great culture realms of the world (outlined in Figure 7.3) are historically based composites of peoples. They are not closely identified with nation, language, religion, or technology but with all these and more in varying combinations. Culture realms are therefore *multifactor regions* that obscure more than they clarify the distinctions between peoples that are so fundamental to the human mosaic of the earth. Basic to cultural geography is the recognition of small regions of single-factor homogeneity that give character to their areas of occurrence and that collectively provide a needed balance to the sweeping generalizations of the culture realm.

Language provides an example of such small area variation, one partially explored in Chapter 7. The language families shown in Figure 7.19 conceal the identity of and the distinctions among the different official tongues of separate countries. These, in turn, ignore or submerge the language forms of minority populations, who may base their own sense of proud identity upon their regional linguistic separateness. In scale and recognition even below these ethnically identified regional languages are those local speech variants frequently denied status as identifiable languages and cited as proof of the ignorance and the cultural deprivation of their speakers. However, such a limited-area, limited-population tongue contains all the elements of the classic culturally based region. Its area is defined; its boundaries are easily drawn; it represents homogeneity and majority behavior among its members; and it summarizes, by a single cultural trait, a collection of areally distinctive outlooks.

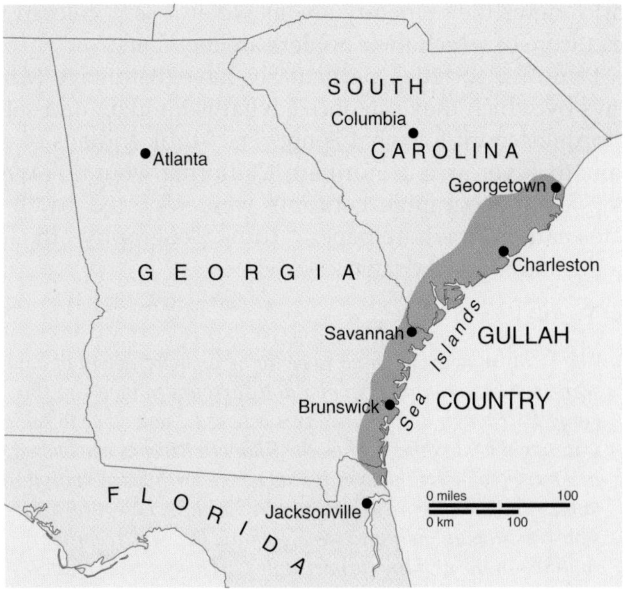


**FIGURE 13.11** Gullah speakers are concentrated on the Sea Islands and the coastal mainland of South Carolina and Georgia. The isolation that promoted their linguistic distinction is now being eroded.

Gullah as Language[5]

*Isolation is a key element in the retention or the creation of distinctive and even externally unintelligible languages. The isolation of the ancestors of the some quarter-million present-day speakers of Gullah—themselves called Gullahs—was nearly complete. Held by the hundreds as slaves on the offshore islands and in the nearly equally remote low country along the southeastern United States coast from South Carolina to the Florida border (Figure 13.11), they retained both the speech patterns of the African languages—Ewe, Fanti, Bambara, Twi, Wolof, Ibo, Malinke, Yoruba, Efik—native to the slave groups and over 4000 words drawn from them. Folk tales told in the Gullah creole can be heard and understood by Krio-speaking audiences in Sierra Leone today.*

*Forced to use English words for minimal communication with their white overseers, but modifying, distorting, and interjecting African-based substitute words into that unfamiliar language, the Gullahs kept intonations and word and idea order in their spoken common speech that made it unintelligible to white masters or to more completely integrated mainland slaves. Because the language was not understood, its speakers were considered ignorant, unable to master the niceties of English. Because ignorance was ascribed to them, the Gullahs learned to be ashamed of themselves, of their culture, and of their tongue, which even they themselves did not recognize as a highly structured and sophisticated separate language.*

*In common with many linguistic minorities, the Gullahs are losing their former sense of inferiority and gaining pride in their cultural heritage and in the distinctive tongue that represents it. Out of economic necessity, Standard English is being taught to their schoolchildren, but an increasing scholarly and popular interest in the structure of their language and in the nature of their culture has caused Gullah to be rendered as a written language, studied as a second language, and translated into English.*

*In both the written and the spoken versions, Gullah betrays its African syntax patterns, particularly in its employment of terminal locator words: "Where you goin' at?" The same African origins are revealed by the absence in English translation of distinctive tenses: "I be tired" conveys the concept that "I have been tired for a period of time." Though tenses exist in the African root languages, they are noted more by inflection than by special words and structures.*

*"He en gut no morratater fer mak no pie wid" may be poor English, but it is good Gullah. Its translation—"He has no more sweet potatoes for making pie"—renders it intelligible to ears attuned to English but loses the musical lilt of original speech and, more important, obscures the cultural identity of the speaker, a member of a regionally compact group of distinctive Americans whose territorial extent is clearly defined by its linguistic dominance.*

[5]By Jerome Fellmann.

## Mental Regions

*(See "Mental Maps," p. 272)*

The regional units so far used as examples, and the methods of regionalization they demonstrate, have a concrete reality. They are formal or functional regions of specified, measurable content. They have boundaries drawn by some objective measures of change or alteration of content, and they have location on an accurately measured global grid.

Individuals and whole cultures may operate, and operate successfully, with a much less formalized and less precise picture of the nature of the world and of the structure of its parts. The mental maps discussed in Chapter 8 represent personal views of regions and regionalization. The private world views they embody are, as we also saw, colored by the culture of which their holders are members.

Primitive societies, particularly, have distinctive world views by which they categorize what is familiar, and satisfactorily account for what is not. The Yurok Indians of the Klamath River area of northern California were no exception. Their geographic concepts were reported by T. T. Waterman, from whose paper, "Yurok Geography," the following summary is drawn.

The Yurok World View[6]

*The Yurok imagines himself to be living on a flat extent of landscape, which is roughly circular and surrounded by ocean. By going far enough up the river, it is believed, "you come to salt water again." In other words, the Klamath River is considered, in a sense, to bisect the world. This whole earth mass, with its forests and mountains, its rivers and sea cliffs, is regarded as slowly rising and falling, with a gigantic but imperceptible rhythm, on the heaving, primeval flood. The vast size of the "earth" causes you not to notice this quiet heaving and settling. This earth, therefore, is not merely surrounded by the ocean but floats upon it. At about the central point of the "world" lies a place which the Yurok call qe'nek, on the southern bank of the Klamath, a few miles below the point where the Trinity River comes in from the south. In the Indian concept, this point seems to be accepted as the center of the world.*

*At this locality also the sky was made. Above the solid sky there is a sky-country, wo'noiyik, about the topography of which the Yurok's ideas are almost as definite as are his ideas of southern Mendocino County, for instance. Downstream from qe'nek, at a place called qe'nek-pul ("qe'nek-downstream"), is an invisible ladder leading up to the sky-country. The ladder is still thought to be there, though no one to my knowledge has been up it recently. The sky-vault is a very definite item in the Yurok's cosmic scheme. The structure consisting of the sky dome and the flat expanse of landscape and waters that it encloses is known to the Yurok as ki-we'-sona (literally "that which exists"). This sky, then, together with its flooring of landscape, constitutes "our world." I used to be puzzled at the Yurok confusing earth and sky, telling me, for example, that a certain gigantic redwood tree "held up the world." Their ideas are of course perfectly logical, for the sky is as much a part of the "world" in their sense as the ground is.*

*The Yurok believe that passing under the sky edge and voyaging still outward you come again to solid land. This is not our world, and mortals ordinarily do not go there; but it is good, solid land. What are breakers over here are just little ripples over there. Yonder lie several regions. To the north (in our sense) lies pu'lekūk, downstream at the north end of creation. South of pu'lekūk lies tsī'k-tsīk-ol ("money lives") where the dentalium-shell, medium of exchange, has its mythical abode. Again, to the south there is a place called kowe'tsik, the mythical home of the salmon, where also all have a "house." About due west of the mouth of the Klamath lies rkrgr', where lives the culture-hero wo'xpa-ku-mä ("across-the-ocean that widower").*

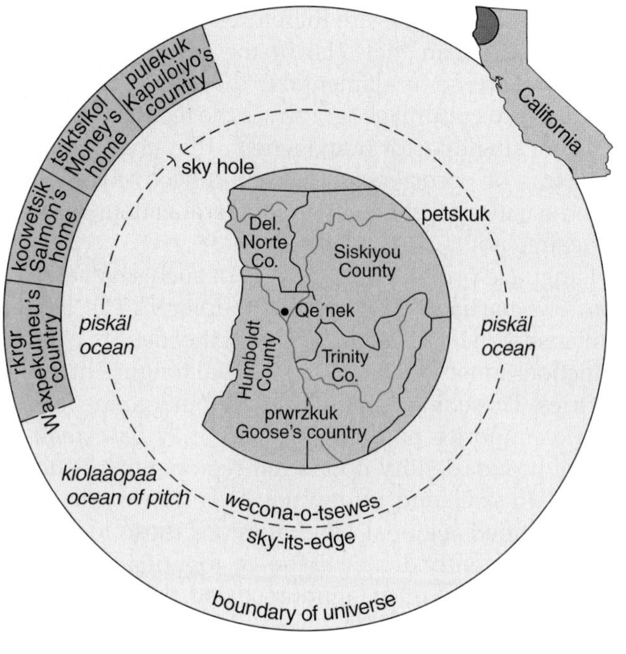


**FIGURE 13.12 The world view of the Yurok** as pieced together by T. T. Waterman during his anthropological study of the tribe. Qe'nek, in the center of the diagram, marks the center of the world in Indian belief. Source: *T. T. Waterman, "Yurok Geography,"* University of California Publications in American Archaeology and Ethnology *16, no. 5: 189–93, 1920.*

*Still to the south of rkrgr' there lies a broad sea, kiolaaopa'a, which is half pitch—an Algonkian myth idea, by the way. All of these solid lands just mentioned lie on the margin, the absolute rim of things. Beyond them the Yurok does not go even in imagination. In the opposite direction, he names a place pe'tskuk ("up-river-at"), which is the upper "end" of the river but still in this world. He does not seem to concern himself much with the topography there.*

The Yurok's conception of the world he lives in may be summed up in the accompanying diagram (Figure 13.12).

## Political Regions

*(See "Boundaries: The Limits of the State," p. 310).*

The most rigorously defined formal culture region is the national state. Its boundaries are presumably carefully surveyed and are, perhaps, marked by fences and guard posts. There is no question of an arbitrarily divided transition zone or of lessening toward the borders of the basic quality of the regional core. This rigidity of a country's boundaries, its unmistakable placement in space, and the trappings—flag, anthem, government, army—that are uniquely its own give to the state an appearance of permanence and immutability not common in other, more fluid culture regions. But its stability is often more imagined than real. Political boundaries are not necessarily permanent. They are subject to change, sometimes violent change, as a result of internal and external pressures. The Indian subcontinent illustrates the point.

[6]*Source:* T. T. Waterman, "Yurok Geography," *University of California Publications in American Archaeology and Ethnology* 16, no. 5 (1920): 189–93.

## Political Regions in the Indian Subcontinent[7]

*The history of the subcontinent since about 400 B.C. has been one of the alternating creation and dissolution of empires, of the extension of central control based upon the Ganges Basin, and of resistance to that centralization by the marginal territories of the peninsula. British India, created largely unintentionally by 1858, was only the last, though perhaps the most successful, attempt to bring under unified control the vast territory of incredibly complex and often implacably opposed racial, religious, and linguistic groupings.*

*A common desire for independence and freedom from British rule united the subcontinent's disparate populations at the end of World War II. That common desire, however, was countered by the mutual religious antipathies felt by Muslims and Hindus, each dominant in separate regions of the colony and each unwilling to be affiliated with or subordinated to the other. When the British surrendered control of the subcontinent in 1947, they recognized these apparently irreconcilable religious differences and partitioned the subcontinent into the second and seventh most populous countries on Earth. The independent state of India was created out of the largely Hindu areas constituting the bulk of the former colony. Separate sovereignty was granted to most of the Muslim-majority area under the name of Pakistan. Even so, the partition left boundaries, notably in the Vale of Kashmir, dangerously undefined or in dispute.*

*An estimated 1 million people died in the religious riots that accompanied the partition decision. In perhaps the largest short-term mass migration in history, some 10 million Hindus moved from Pakistan to India, and 7.5 million Muslims left India for Pakistan, "The Land of the Pure."*

*Unfortunately, the purity resided only in common religious belief, not in spatial coherence or in shared language, ethnicity, customs, food, or economy. During its 23 years of existence as originally conceived, Pakistan was a sorely divided country. The partition decision created an eastern and a western component separated by more than 1600 kilometers (1000 mi) of foreign territory and united only by a common belief in Allah (Figure 13.13). West Pakistan, as large as Texas and Oklahoma combined, held 55 million largely light-skinned Punjabis with Urdu language and strong Middle Eastern cultural ties. Some 70 million Bengali speakers, making up East Pakistan, were crammed into an Iowa-sized portion of the delta of the Ganges and Brahmaputra rivers. The western segment of the country was part of the semiarid world of western Asia; the eastern portion of Pakistan was joined to humid, rice-producing Southeast Asia.*

*Beyond the affinity of religion, little else united the awkwardly separated country. East Pakistan felt itself exploited by a dominating western minority that sought to impose its language and its economic development, administrative objectives, and military control. Rightly or wrongly, East Pakistanis saw themselves as aggrieved and abused. They complained of a per capita income level far below that of their western compatriots, claimed discrimination in the allocation of investment capital, found disparities in the pricing of imported foods, and asserted that their exports of raw materials—particularly jute—were supporting a national economy in which they did not share proportionally. They argued that their demands for regional autonomy, voiced since nationhood, had been denied.*

*When, in November of 1970, East Pakistan was struck by a cyclone and tidal wave that took an estimated 500,000 lives, the limit of eastern patience was reached. Resentful over what they saw as a totally inadequate West Pakistani effort of aid in the natural disaster that had befallen them, the East Pakistanis were further incensed by the refusal of the central government to convene on schedule a national assembly to which they had won an absolute majority of delegates. Civil war resulted, and the separate new state of Bangladesh was created. The sequence of political change in the subcontinent is traced in Figure 13.13.*

*The country and nation-state so ingrained in our consciousness and so firmly defined, as displayed on the map inside the cover of this book, is both a recent and an ephemeral creation of the cultural regional landscape. It rests upon a claim, more or less effectively enforced, of a monopoly of power and allegiance*

[7]By Jerome Fellmann.

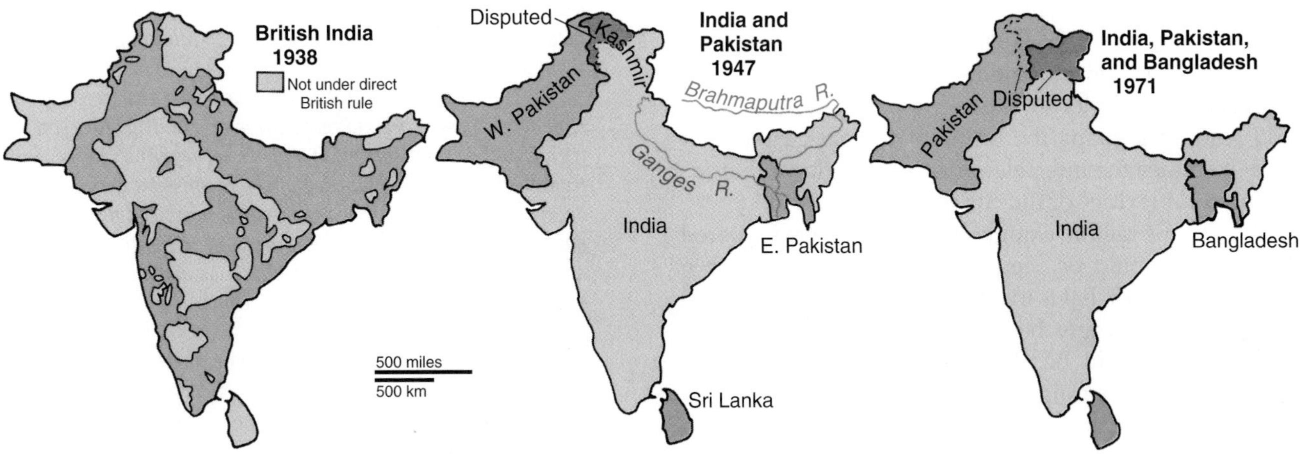


**FIGURE 13.13 The sequence of political change on the Indian subcontinent.** British India was transformed in 1947 to the countries of India and Pakistan, the latter a Muslim state with a western and an eastern component. In 1971 Pakistan was torn by civil war based on ethnic and political contrasts, and the eastern segment became the new country of Bangladesh.

*resident in a government and superior to the communal, linguistic, ethnic, or religious affiliations that preceded it or that claim loyalties overriding it. As the violent recent history of the Indian subcontinent demonstrates, nationalism may be sought, but its maintenance is not assured by the initiating motivations.*

# Part III: Regions in the Locational Tradition

While location, as we have seen, is a primary attribute of all regions, regionalization in the locational tradition of geography implies far more than a named delimitation of earth space. The central concern is with the distribution of human activities and the impact of those activities on the physical and cultural landscape.

In this sense, world regionalization of agriculture and of the soils and climates with which it is related is within the locational tradition. For practical and accepted reasons, however, such underlying physical patterns have been included under the earth science tradition. But the point is made: the locational tradition emphasizes the "doing" in human affairs, and "doing" is not an abstract thing but an interrelation of life and the environment on which life depends.

The locational tradition, therefore, encourages the recognition and the definition of a far wider array of regional types than do the earth science or culture–environment traditions. Any single pattern of economic activity or of cultural interaction invites the recognition of definable *formal regions.* The interchange of commodities, the control of urban market areas, the flows of capital, or the collection and distribution activities of ports are just a few examples of the infinite number of analytically useful *functional regions* that one may recognize.

## Economic Regions

*(See p. 342)*

Economic regionalization is among the most frequent, familiar, and useful employments of the regional method. Through economic regions the geographer identifies activities and resources, maps the limits of their occurrence or use, and examines the interrelationships and flows that are part of the complexities of the contemporary world.

The economic region, examples of which were explored in Chapter 10, should be seen as potentially more than a device for recording what is in either a formal or a functional sense. It has increasingly become a device for examining what might or should be. The concept of the economic region as a tool for planning and a framework for the manipulation of the people, resources, and economic structure of a composite region first took root in the United States during the Great Depression years of the 1930s. The key element in the planning region is the public recognition of a major territorial unit in which economic change or decline is seen as the cause of a variety of interrelated problems including, for example, population out-migration, regional isolation, cultural deprivation, underdevelopment, and poverty.

### Appalachia[8]

*Until the early 1960s, "Appalachia" was for most a loose reference to the complex physiographic province of the eastern United States associated with the Appalachian mountain chain. If thought of at all, it was apt to be visualized as rural, isolated, and tree-covered; as an area of coal mining, hillbillies, and folksongs (Figure 13.14).*

*During the 1950s, however, the economic stagnation and the functional decline of the area became increasingly noticeable in the national context of economic growth, rising personal incomes, and growing concern with the elimination of the poverty and deprivation of every group of citizens. Less dramatically but just as decisively as the Dust Bowl or the Tennessee Valley of an earlier era, Appalachia became simultaneously a popularly recognized economic and cultural region and a governmentally determined planning region.*

*Evidences of poverty, underdevelopment, and social crisis were obvious to a country committed to recognize and eradicate such conditions within its own borders. By 1960, per capita income within Appalachia was $1400 when the national average was $1900. In the decade of the 1950s, mine employment fell 60% and farm jobs declined 52%; the rest of the country lost only 1% of mining jobs and 35% of agricultural employment. Rail employment fell with the drop in coal mining. Massive out-migration occurred among young adults, with such cities as Chicago, Detroit, Dayton, Cleveland, and Gary the targets. Even with these departures, unemployment among those who remained in Appalachia averaged 50% higher than the national rate. Because of the departures, the remnant population—only 47% of whom lived in or near cities in 1960, as against 70% for the entire United States—was distorted in age structure. The very young and the old were disproportionately represented; the productive working-age groups were, at least temporarily, emigrants.*

*When these and other socioeconomic indicators were plotted by counties and by state economic areas, an elongated but regionally coherent and clearly bounded Appalachia as newly understood was revealed by maps (Figure 13.15). It extended through 13 states from Mississippi to New York, covered some 505,000 square kilometers (195,000 sq mi) and contained 18 million people, 93% of them white.*

*By 1963, awareness of the problems of the area at federal and state levels passed beyond recognition of a multifactor economic region to the establishment of a planning region. A joint federal–state Appalachian Regional Commission was created to develop a program designed to meet the perceived needs of the entire area. The approach chosen was one of limited investment in a restricted number of highly localized developments, with the expectation that these would spark economic growth supported by private funds.*

*In outline, the plan was (1) to ignore those areas of poverty and unemployment that were in isolated, inaccessible "hollows" throughout the region; (2) to designate "growth centers" where development potential was greatest and concentrate all spending for economic expansion there; the regional growth potential in tar-*

[8]By Jerome Fellmann.

**FIGURE 13.14** This deserted eastern Kentucky cabin is a mute reminder of the economic and social changes in Appalachia. Increases in per capita income, expansion of urban job opportunities, and improved highway networks have reduced the poverty and isolation that has so long been associated with this region. *© Jean-Yves Rabeuf/Image Works.*

*geted expenditure was deemed sufficient to overcome charges of aiding the prosperous and depriving the poor; and (3) to create a new network of roads so that the isolated jobless could commute to the new jobs expected to form in and near the favored growth centers. Road construction would also, of course, open inaccessible areas to tourism and strengthen the economic base of the entire planning region.*

*In the years since the Appalachian Regional Commission was established, and in ways not anticipated, the economic prospects of Appalachia have altered. During the 1980s, new industrial jobs multiplied as manufacturing relocated to, or was newly developed in, the Appalachia portion of the Sunbelt. The new factory employment opportunities initially exceeded local labor pools, and former out-migrants returned home from cities outside the region, reestablishing a more balanced population pyramid. At the same time, however, coal-mining jobs in the region plummeted and unemployment still remained above the national average. By the early 1990s, the Appalachian Regional Commission had committed more than \$6 billion to the area, and over \$10 billion more had come from other sources.*

*Those investments fundamentally altered the 1960s picture of unrelieved depression. The percentage of people living in poverty within the region declined from 31% in 1960 to 15% in 1990, and per capita income, which had been 79% of the national average in 1960, rose to 85% by the early 1990s. Population flows stabilized, with the number of people coming into Appalachia about equaling those leaving. More than 3500 kilometers (2200 mi) of road had been laid by the mid-1990s; the network continued to grow, reaching nearly 3700 kilometers (2300 mi) by 2002—though its construction did not give the total regional access that had been hoped. Smaller towns and their hinterlands were bypassed, remained as isolated as ever, and lost population. Some basic social services had, for the first time however, reached essentially everyone in the region. For example, each of the 410 counties of the Commission territory had been supplied with a clinic or a doctor by the start of the 21st century.*

*Fortunes again reversed during the latter 1990s and into the 21st century with the widespread loss of jobs in textile and garment industries, particularly to lower wage foreign competitors. Even the acquisition of subsidized foreign automobile parts and assembly plants and growth in service sector employment could not completely preserve the advances of the earlier decades. Once again per capita income slipped well below U.S. averages as the region disproportionately suffered from national economic reversals beginning in 2001; even in counties where manufacturing remained a dominant industry, declining real wages eroded the economic base.*

*The mixed trends of progress and regression shown in Appalachia left much of the region still economicallly distressed in the early 21st century. Its manufacturing base remained largely dependent on primary and fabricated metals, wood products, textiles, and apparel—all eroded by imports replacing domestic products. Mechanization of coal mining and conflict over the extent and application of environmental protection laws held down employment in that sector, more than half of the region's 121 counties (in 2003) remaining officially "distressed" still had a high dependence on tobacco production, and Appalachia's unemployment rate continued well above the national average.*

*Nevertheless, the Appalachian Region's economy had become significantly more diversified since the creation of the Regional Commission in 1963 and after 40 years was more dependent on relatively stable jobs in service industries, retailing, and government. The proportion of its residents living in poverty had been cut in half from the one-in-three level recorded in 1960. Overall, Appalachia has been transformed from a region of almost uniform economic and social distress to a mixed pattern of subregional contrasts. Many counties and communities have successfully diversified and stabilized their economies; some continue to adjust with uncertain success to structural changes in declining employment sectors; and still other counties and districts remain in need of such basic infrastructure as water and sewer systems.*

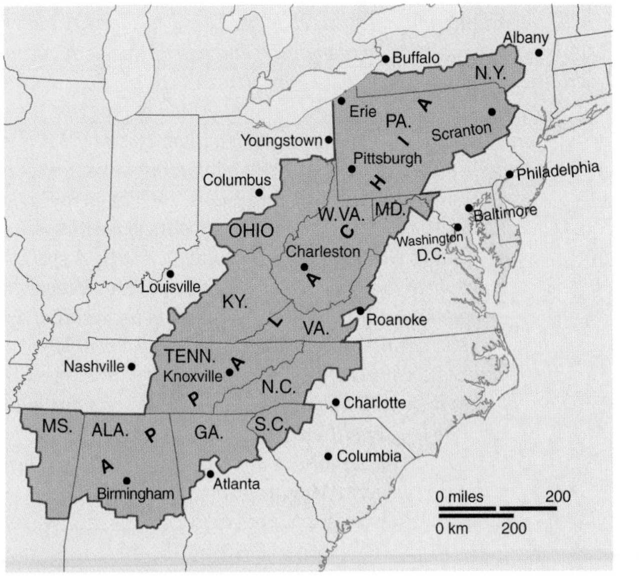


**FIGURE 13.15** **The boundaries of "Appalachia"** as defined by the Appalachian Regional Commission were based on social and economic conditions influenced by political considerations, not topography.

## Urban Regions

(*See "Megalopolis," p. 414*)

Urban geography occupies a center stage position in the locational tradition of geography. Modern integrated, interdependent society on a world basis is urban-centered. Cities are the indispensable functional focuses of production, exchange, and administration. They exist individually as essential elements in interlocked hierarchical systems of cities. Internally, they display complex but repetitive spatial patterns of land uses and functions.

Because of the many-sided character of urbanism, cities are particularly good subjects for regional study. They are themselves, of course, formal regions. In the aggregate, their distributions give substance to formal regions of urban concentration. Cities are also the cores of functional regions of varying types and hierarchical orders. Their internal diversity of functional, land use, and socioeconomic patterns invite regional analysis. The employment of that approach in both its formal and functional modes is clearly displayed by Jean Gottmann, who, in examining the data and landscapes of the eastern United States at mid-20th century, recognized and analyzed Megalopolis. The following is taken from his study.

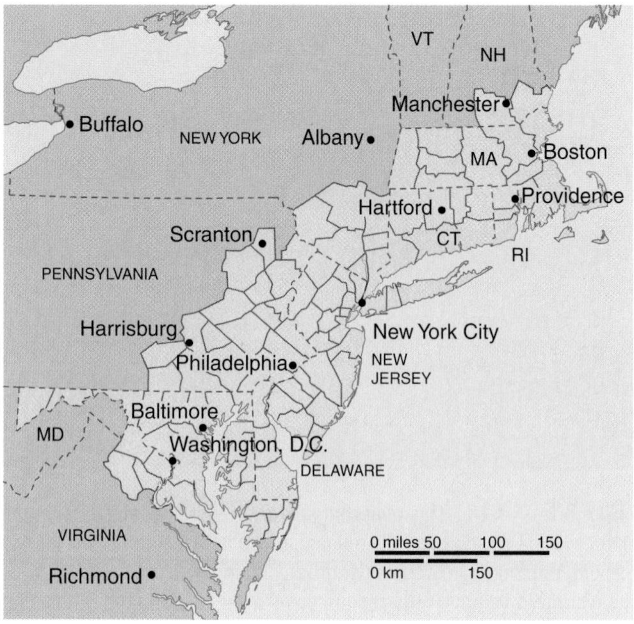


**FIGURE 13.16 Megalopolis in 1960.** The region was then composed of counties that, by United States census definition, were "urban" in population and economic characteristics. Much of the area is still distinctly "rural" in land use.

### Megalopolis[9]

*The northeastern seaboard of the United States is today the site of a remarkable development—an almost continuous stretch of urban and suburban areas from southern New Hampshire to northern Virginia and from the Atlantic shore to the Appalachian foothills [Figure 13.16]. The processes of urbanization, rooted deep in the American past, have worked steadily here, endowing the region with unique ways of life and of land use. No other section of the United States has such a large concentration of population, with such a high average density, spread over such a large area. And no other section has a comparable role within the country or a comparable importance in the world. Here has been developed a kind of supremacy, in politics, in economics, and possibly even in cultural activities, seldom before attained by an area of this size.*

*Great, then, is the importance of this section of the United States and of the processes now at work within it. And yet it is difficult to single this area out from the surrounding areas, for its limits cut across established historical divisions, such as New England and the Middle Atlantic states, and across political entities, since it includes some states entirely and others only partially. A special name is needed, therefore, to identify this special geographical area.*

*This particular type of region is new, but it is the result of age-old processes, such as the growth of cities, the division of labor within a civilized society, the development of world resources. The name applied to it should, therefore, be new as a place name but old as a symbol of the long tradition of human aspirations and endeavor underlying the situations and problems now found here. Hence the choice of the term* ***Megalopolis,*** *used in this study.*

*As one follows the main highways or railroads between Boston and Washington, D.C., one hardly loses sight of built-up areas, tightly woven residential communities, or powerful concentrations of manufacturing plants. Flying this same route one discovers, on the other hand, that behind the ribbons of densely occupied land along the principal arteries of traffic, and in between the clusters of suburbs around the old urban centers, there still remain large areas covered with woods and brush alternating with some carefully cultivated patches of farmland [Figure 13.17]. These green spaces, however, when inspected at closer range, appear stuffed with a loose but immense scattering of buildings, most of them residential but some of industrial character. That is, many of these sections that look rural actually function largely as suburbs in the orbit of some city's downtown. Even the farms, which occupy the larger tilled patches, are seldom worked by people whose only occupation and income are properly agricultural.*

*Thus the old distinctions between rural and urban do not apply here anymore. Even a quick look at the vast area of Megalopolis reveals a revolution in land use. Most of the people living in the so-called rural areas, and still classified as "rural population" by recent censuses, have very little, if anything, to do with agriculture. In terms of their interests and work they are what used to be classified as "city folks," but their way of life and the landscapes around their residences do not fit the old meaning of urban.*

*In this area, then, we must abandon the idea of the city as a tightly settled and organized unit in which people, activities, and riches are crowded into a very small area clearly separated from its nonurban surroundings. Every city in this region spreads out far and wide around its original nucleus; it grows amidst an irregularly colloidal mixture of rural and suburban landscapes; it melds on broad fronts with other mixtures, of somewhat similar though different texture, belonging to the suburban neighborhoods of other cities.*

[9]Jean Gottmann, *Megalopolis: The Urbanized Northeastern Seaboard of the United States.* Copyright © 1961. Twentieth Century Fund, New York. Reprinted with permission.

*Thus an almost continuous system of deeply interwoven urban and suburban areas, with a total population of about 37 million people in 1960, has been erected along the Northeastern Atlantic seaboard. It straddles state boundaries, stretches across wide estuaries and bays, and encompasses many regional differences. In fact, the landscapes of Megalopolis offer such variety that the average observer may well doubt the unity of the region. And it may seem to him that the main urban nuclei of the seaboard are little related to one another. Six of its great cities would be great individual metropolises in their own right if they were located elsewhere. This region indeed reminds one of Aristotle's saying that cities such as Babylon had "the compass of a nation rather than a city."*

The description of Megalopolis in 1960 represents, as does any regional study, a captured moment on the continuum of areal change. When that moment is lucidly described and the threads of areal character clearly delineated, the regional study serves as both a summary of the present and an augury of a future whose roots may be discerned in it. In the years since Gottmann described it, Megalopolis continued to develop along the lines he summarized. Urbanization proceeded, physically as well as functionally, to encroach upon the rural landscapes without regard for state boundaries or even the metropolitan cores that dominated in 1960. As then, new growth centers—becoming equivalents of central cities in their own right—were linked to the expanding transportation corridors, though now the lines of importance are increasingly expressways rather than railroads.

In the south, the corridors around Washington, D.C., are the Capital Beltway in Fairfax County and an extension from it westward to Loudoun County and Dulles Airport in Virginia and, to the north into Montgomery County, Maryland, the I-270 corridor. The Virginia suburbs focusing on Tysons Corner specialized in defense-related industries, but vast office building complexes and commercial centers are rapidly converting rural land to general urban uses. The Maryland suburbs emphasized health, space, and communications interests in their own new office, industrial, and commercial "parks" and complexes. Still farther to the north during the 1980s, in the "Princeton Corridor," a 42-kilometer (26-mi) stretch along Route 1 from New Brunswick to Trenton, New Jersey, huge corporate parks provided office and research space to companies relocating from the New York area and to new technology firms attracted by proximity to Princeton University. Morristown, New Jersey, and White Plains, in Westchester County, New York, were other similar concentrations of commercial, industrial, and office developments north and west of New York City. To the east, Stamford, Connecticut, with 150,000 daily in-commuters became by the late 1980s the headquarters center for major national corporations and had emerged as a truly major central city new on the scene in its present form since Megalopolis was first described.

**FIGURE 13.17** The "Pine Barrens" of New Jersey is still a largely undisturbed natural enclave in the heart of Megalopolis. Its preservation is a subject of dispute between environmentalists and developers. *© Elizabeth J. Leppman.*

## Ecosystems as Regions

(*See "Ecosystems," p. 433*)

A traditional though oversimplified definition of *geography* as "the study of the areal variation of the surface of the earth" suggested that the discipline centered on the classification of areas and on the subdivision of the earth into its constituent regional parts. Considerations of organization and function were secondary and even unnecessary to the implied main purpose of regional study: the definition of cores and boundaries of areas uniform in physical or cultural properties. In our chapter reviews of the earth science, culture–environment, and locational traditions of geography, however, we have stressed that a more dynamic view of the discipline recognizes that humans constantly exert discernible, changeable, and frequently adverse impacts on the physical environments they occupy. The landscape evidences of those human impacts are also subjects of regional study, though often their investigation requires or benefits from techniques different from the descriptive methodologies often associated with regional studies.

Newer research philosophies stress the need for the study of spatial relationships—particularly of human and environmental interactions—from the standpoint of *systems analysis.* That approach emphasizes the organization, structure, and functional dynamics within an area and provides for the quantification of the linkages between the things in space. The *ecosystem* or *biome,* introduced in Chapter 12, provides a systems-analytic concept of great flexibility that permits examination of the relationship between the environment and the biological realm, a relationship often impacted by human interventions. Since that relationship is structured, structure rather than spatial uniformity attracts attention and leads to new understanding of the region. The ecosystem concept, particularly, provides a point of view for investigating the complex consequences of human impact on the natural environment.

Note how, in the following description drawn from an article by William J. Schneider, the ecosystem concept is employed in the recognition and analysis of regions and subregions of varying size, complexity, and nature. Introduced, too, is the concept of *ecotone,* or zone of ecological stress—in this case, induced by human pressures on the natural system.

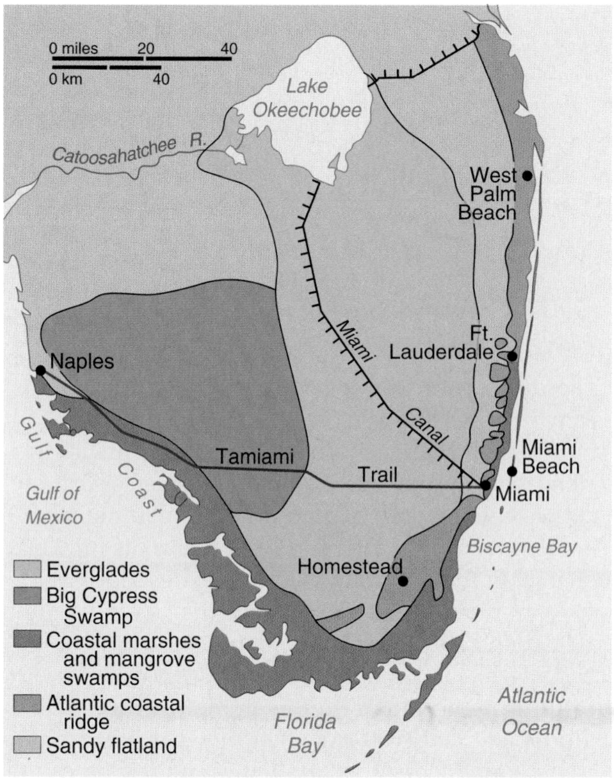


**FIGURE 13.18 The Everglades** is part of a complex of ecosystems stretching southward in Florida from Lake Okeechobee to the sea. Drainage and water-control systems have altered its natural condition.

## The Everglades[10]

*The Everglades is a river. Like the Hudson or the Mississippi, it is a channel through which water drains from higher to lower ground as it moves to the sea. It extends in a broad, sweeping arc from the southern end of Lake Okeechobee in central Florida to the tidal estuaries of the Gulf Coast and Florida Bay. As much as 70 miles (113 km) wide, but generally averaging 40 miles (64 km), it is a large, shallow slough that weaves tortuously through acres of saw grass and past "islands" of trees, its waters, even in the wet season, rarely deeper than 2 feet (0.6 meters). But, again like the Hudson or the Mississippi, the Everglades bears the significant imprint of civilization.*

*Since the close of the Pleistocene, 10,000 years ago, the Everglades has been the natural drainage course for the periodically abundant overflow of Lake Okeechobee. As the lake filled during the wet summer months or as hurricane winds blew and literally scooped the water out of the lake basin, excess water spilled over the lake's southern rim. This overflow, together with rainfall collected en route, drained slowly southward between Big Cypress Swamp and the sandy flatlands to the west and the Atlantic coastal ridge to the east, sliding finally into the brackish water of the coastal marshes [Figure 13.18].*

*Water has always been the key factor in the life of the Everglades. Three-fourths of the annual average 55 inches (140 cm) of rainfall occurs in the wet season, June through October, when water levels rise to cover 90% of the land area of the Everglades. In normal dry seasons in the past, water covered no more than 10% of the land surface. Throughout much of the Everglades, and prior to recent engineering activities, this seasonal rain cycle caused fluctuations in water levels that averaged 3 feet (0.9 meters). Both occasional severe flooding and prolonged drought accompanied by fire imposed periodic stress upon the ecosystem. It may be that randomly occurring ecologic trauma is vital to the character of the Everglades.*

*Three dominant biological communities—open water, saw grass, and woody vegetation—reflect small, but consistent, differences in the surface elevation of peat soils that cover the Everglades [Figure 13.19]. The open-water areas occur at the lower soil elevations; inundated much of the year, they contain both sparse, scattered marsh grasses and a mat of algae. The saw grass communities develop on a soil base only a few inches higher than that in the surrounding open glades. The soil base is thickest under the tree islands. The few inches' difference in soil depth apparently governs the species composition of these three communities.*

*Today, the Everglades is no longer precisely a natural river. Much of it has been altered by an extensive program of water management, including drainage, canalization, and the building of locks and dams. Large withdrawals of groundwater for municipal and industrial use have depleted the underlying aquifer and permitted the landward penetration of seawater through the aquifer and through the surface canals. Thousands of individual water-supply wells have been contaminated by encroaching saline water; large biotic changes have taken place in the former freshwater marshes south of Miami. Mangroves—indicators of salinity—have extended their habitat inland, and fires rage across areas that were formerly much wetter. The ecotone—the zone of stress between dissimilar adjacent ecosystems—is altering as a consequence of these human-induced modifications of the Everglades ecosystem.*

*The organization, structure, and functional dynamics of the Everglades ecosystems are thus undergoing change. The structured relationships of its components—in nature affected and formed by stress—are being subjected to distortions by humans in ways not yet fully comprehended.*

[10]With permission from *Natural History,* November, 1966. © Copyright American Museum of Natural History, 1966.

**FIGURE 13.19** Open water, saw grasses, and tree islands, all visible in this scene, constitute separate *biomes* of the Everglades, which is home also of a teeming animal life. *© Reinhard Eisele/CORBIS.*

## Summary

The region is a mental construct, a created entity whose sole function is the purposeful organization of spatial data. The scheme of that organization, the selection of data to be analyzed, and the region resulting from these decisions are reflections of the intellectual problem posed.

This chapter has not attempted to explore all aspects of regionalism and of the regional method. It has tried, by example, only to document its basic theme: the geographer's regions are arbitrary but deliberately conceived devices for the isolation of things, patterns, interrelations, and flows that invite geographic analysis. In this sense, all geographers are regional geographers, and the regional examples of this chapter may logically complete our survey of the four traditions of geography.

## Key Words

formal region 471
functional region 471
region 470
regional concept 470

## For Review & Consideration

1. What do geographers seek to achieve when they recognize or define regions? On what basis are regional boundaries drawn? Are regions concrete entities whose dimensions and characteristics are agreed upon by all who study the same general segment of earth space? Ask three fellow students who are not participants in this course for their definition of the "South." If their answers differed, what implicit or explicit criteria of regional delimitation were they employing?
2. What are the spatial elements or the identifying qualities shared by all regions?
3. What is the identifying characteristic of a *formal region?* How are its boundaries determined? Name three different examples of formal regions drawn from any earlier chapters of this book. How was each defined and what was the purpose of its recognition?
4. How are *functional regions* defined? What is the nature of their bounding criteria? Give three or four examples of functional regions that were defined earlier in the text.
5. The ecosystem was suggested as a viable device for regional delimitation. What regional geographic concepts are suggested in ecosystem recognition? Is an ecosystem identical to a formal region? Why or why not?
6. National, linguistic, historical, planning, and other regions have been recognized in this chapter. With what other regional entities do you have acquaintance in your daily affairs? Are fire protection districts, police or voting precincts, or zoning districts regional units identifiable with geographers' regions as discussed in this chapter or elsewhere in this book? How influenced are you in your private life by your, or others', regional delimitations?

## Selected References

**WEBSITES**

The World Wide Web has a tremendous variety of sites pertaining to geography. Websites relevant to the subject matter of this chapter appear in the "Web Links" section of the Online Learning Center associated with this book. Access it at www.mhhe.com/getis10e/.


Borchert, John R. *Megalopolis: Washington, D.C., to Boston.* Series: *Touring North America.* New Brunswick, N.J.: Rutgers University Press, 1992.

Freeman, T. W. *A Hundred Years of Geography.* Chapter 2, "The Regional Approach." Chicago: Aldine, 1961.

Gottmann, Jean. *Megalopolis Revisited: 25 Years Later.* College Park: Institute for Urban Studies, University of Maryland, 1987.

Johnston, R. J., J. Hauer, and G. A. Koekveld, eds. *Regional Geography: Current Developments and Future Prospects.* New York: Routledge, 1990.

McDonald, James R. "The Region: Its Conception, Design and Limitations." *Annals of the Association of American Geographers* 56 (1966): 516–28.

Meinig, Donald. "The Mormon Culture Region: Strategies and Patterns in the Geography of the American West, 1847–1964." *Annals of the Association of American Geographers* 55 (1965): 191–220 (esp. 213–16).

Minshull, Roger. *Regional Geography.* Chicago: Aldine, 1967.

Moore, Tyrel G. "Core-Periphery Models, Regional Planning Theory, and Appalachian Development." *Professional Geographer* 46, no. 3 (1994): 317–31.

Murphey, Rhoads. *The Scope of Geography.* 3d ed. Chapter 2, "The Region." London and New York: Methuen, 1982.

Wheeler, James O. "Notes on the Rise of the Area Studies Tradition in U.S. Geography, 1920–1929." *Professional Geographer* 38 (1986):53–61.

Whittlesey, Derwent. "The Regional Concept and the Regional Method." In *American Geography: Inventory and Prospect,* ed. Preston E. James and Clarence F. Jones. Syracuse, N.Y.: Syracuse University Press (for the Association of American Geographers), 1954.

# Appendix 1

## Map Projections

A map projection is a system for displaying the curved surface of the earth on a flat sheet of paper. No matter how one tries to "flatten" the world, it can never be done in such a way as to show all earth details in their correct relative size, shapes, distances, or directions. Something is always wrong, and the cartographer's task is to select and preserve those earth relationships important for the purpose at hand and to minimize or accept those distortions that are inevitable but unimportant.

If we look at a globe directly, only the front is visible; the back is hidden. To make a world map, we must decide on a way to flatten the globe's curved surface on the hemisphere we can see. Then we have to cut the globe map down the middle of its hidden hemisphere and place the two back quarters on their respective sides of the already visible front half. In simple terms, we have to "peel" the map from the globe and flatten it in the same way we might try to peel an orange and flatten the skin (Figure A.1). Inevitably, the peeling and flattening process will produce a resulting map that either shows tears or breaks in the surface or is subject to uneven stretching or shrinking to make it lie flat.

Of course, mapmakers do not physically engage in cutting, peeling, flattening, or stretching operations. Their task, rather, is to construct or project on a flat surface the network of parallels and meridians of the globe grid, or *graticule*. This can be done in a number of ways. Before discussing them, it is important to note that two types of circles appear on a globe's spherical grid. A **great circle** is formed on the surface of a sphere by a plane that passes through the center of the sphere. Thus, the equator is a great circle, and each meridian is half of a great circle. Every great circle bisects the globe, dividing it equally into hemispheres. An arc segment of the great circle joining them is the shortest distance between any two points on the earth's surface. A **small circle** is the line created by the intersection of a spherical surface with a plane that does *not* pass through its center. Except for the equator, all parallels of latitude are small circles. Different projections will represent great and small circles in different ways.

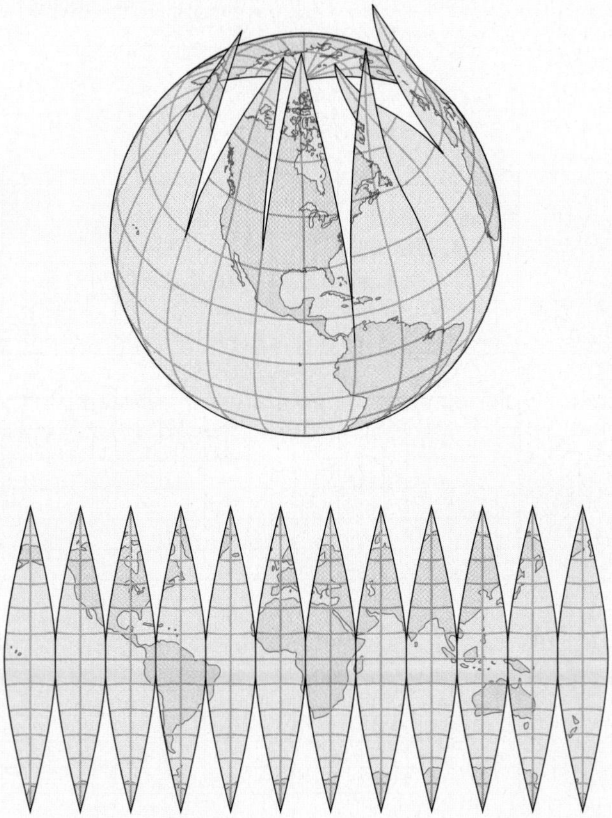

(a)

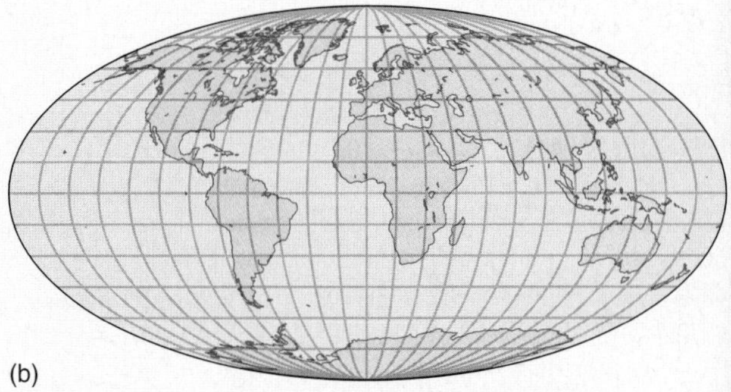

(b)

**FIGURE A.1** (a) A careful "peeling" of the map from the globe yields a set of tapered gores, which, although individually not showing much stretching or shrinking, do not collectively result in a very useful world map. (b) It is usually considered desirable to avoid or reduce the number of interruptions by depicting the entire global surface as a single flat circular, oval, or rectangular shape. That continuity of area, however, can be achieved only at the cost of considerable alteration of true shapes, distances, directions, and/or areas. Although the Mollweide projection shown here depicts the size of areas correctly, it distorts shapes.

# GEOMETRIC PROJECTIONS

Although all projections can be described mathematically, some can be thought of as being constructed by geometric techniques rather than by mathematical formulas. In geometric projections, the grid system is, in theory, transferred from the globe to a geometric figure, such as a cylinder or a cone, which, in turn, can be cut and then spread out flat (or *developed*) without any stretching or tearing (Figure A.2). The surfaces of cylinders, cones, and planes are said to be **developable surfaces**—cylinders and cones can be cut and laid flat without distortion, and planes are flat to begin with. In actuality, geometric projections are constructed not by tracing shadows but by applying geometry and using lines, circles, arcs, and angles drawn on paper.

Imagine a transparent globe with a light source located either inside or outside the globe. Lines of latitude and longitude (or of coastlines or any other features) drawn on that globe will cast shadows on any nearby surface. A tracing of that shadow globe grid would represent a geometric map projection. As Figure A.3 shows, the location of the light source in relation to the globe surface causes significant variation in the projection of the graticule on the developable geometric surface. An *orthographic* projection results from the placement of the theoretical light source at infinity. A *gnomonic* projection is produced when the light source is at the center of the globe. Placing the light at the antipode—the point exactly opposite the point of tangency, or point of contact between globe and map—produces a *stereographic* projection.

## Cylindrical Projections

Suppose that we roll a piece of paper around a transparent globe so that it is tangent to (touching) the sphere at the equator. The line of tangency is called the *standard line* (or *standard parallel,* if it is a parallel of latitude); along it, the map has no distortion. Instead of the paper being the same

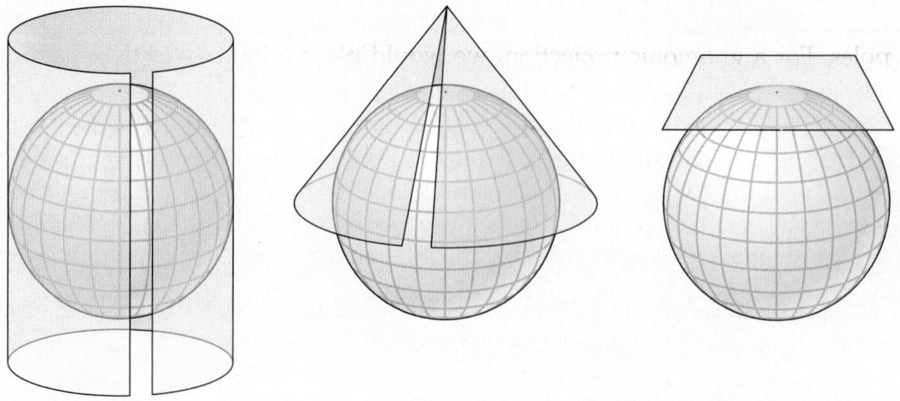

**FIGURE A.2 Geometric projections.** The three common geometric forms used in such projections are the cylinder, the cone, and the plane. One can think of making a map projection by imagining a transparent globe with a light source inside it and a sheet of paper touching the globe in one of the ways shown here. The globe grid and the outline of the continents would be silhouetted on the paper to form the map.

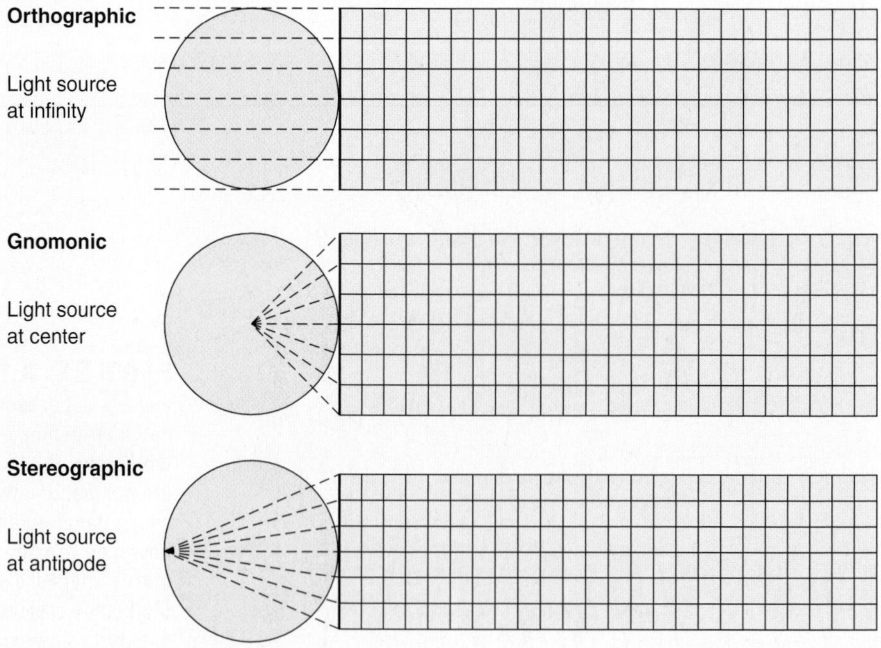


**FIGURE A.3 The effect of the light source location on *planar* surface projections.** Note the variations in the spacing of the lines of latitude that occur when the light source is moved. Completely different map grids would result from using a cylinder or cone as the developable surface.

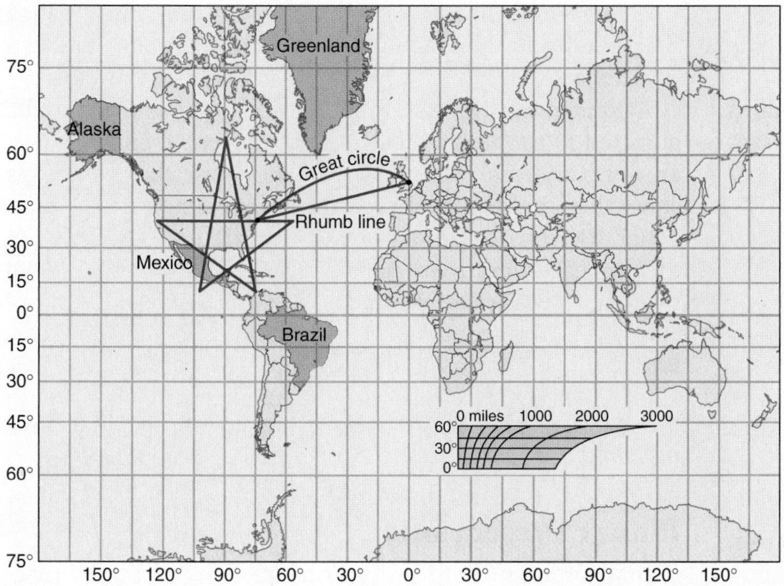


**FIGURE A.4 Distortion on the Mercator projection.** A perfect five-pointed star was drawn on a globe, and the latitude and longitude of the points of the star were transferred to the Mercator map shown here. The manner in which the star is distorted reflects the way the projection distorts land areas. The enlargement of areas with increasing latitude is so great that a Mercator map should not be published without a diagram showing the varying scale of distance at different latitudes. The most significant property of the Mercator projection is that it is the only one on which any straight line is a line of constant compass bearing, or *rhumb line.* Although a rhumb line usually is not the shortest distance between two points, navigators can draw a series of straight lines between the starting point and the destination that will approximate the great circle route.

height as the globe, however, it extends far beyond the poles. For a gnomonic projection, we would place a light source at the center of the globe, and the light would project a shadow map upon the cylinder of paper. The result is one of many **cylindrical projections,** all of which are developed geometrically or mathematically from a cylinder wrapped around the globe.

Note the variance between the grid we have just projected and the true properties of the globe grid. The grid lines of latitude and longitude intersect each other at right angles, as they do on the globe, and they are all straight north-south or east-west lines. But the meridians do not converge at the poles, as they do on a globe. Instead, they are equally spaced, parallel, vertical lines. Because the meridians are everywhere equally far apart, the parallels of latitude have all become the same length. Although there is no scale distortion along the line of tangency, the equator, distortion increases with increasing distance away from it. The polar regions are stretched north and south as well as east and west, and their sizes are enormously exaggerated. The poles themselves can never be shown on a cylindrical projection tangent at the equator with a light source at the center of the globe.

The mathematically derived **Mercator projection** was inspired by the idea of a cylinder tangent at the equator. It is one of the most commonly used (and misused) cylindrical projections. Invented in 1569 by Gerardus Mercator, the Mercator projection was created to serve as a navigational chart during a time when European exploration of other parts of the world was at its height. The Mercator is the standard projection used by navigators because of a peculiarly useful property: a straight line drawn anywhere on the map is a line of constant compass bearing. If such a line, called a **rhumb line,** is followed, a ship's or a plane's compass will show that the course is always at a constant angle with respect to geographic north (Figure A.4). On no other projection is a rhumb line both straight and true as a direction.

Although it is an excellent navigational aid, the Mercator projection frequently has been misused in book or on wall maps as a general-purpose world map—misused because it gives grossly exaggerated impressions of the size of land areas away from the tropics. Notice on Figure A.4 that Greenland appears many times bigger than Mexico, when in fact it is only slightly larger, and that Alaska and Brazil appear to be about the same size; in actuality, Brazil is more than five times as large.

A number of cylindrical projections that are neither equal area nor conformal, such as the *Miller cylindrical projection* shown in Figure A.5, are often used as bases for

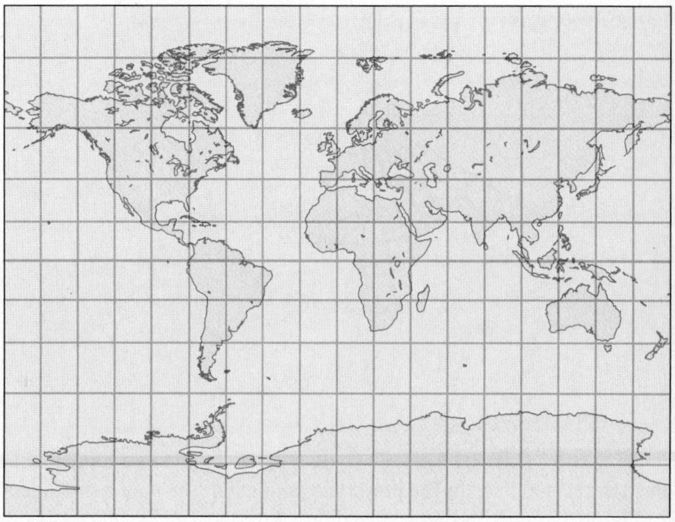

**FIGURE A.5 The Miller cylindrical projection** is mathematically derived.

world maps. The spacing between parallels of latitude on the Miller projection does not increase as rapidly toward the poles as it does on the Mercator, so there is less distortion of the size of areas in the higher latitudes. Although it retains no globe qualities, the Miller cylindrical is used in atlases and wall maps.

## Conic Projections

Of the three developable geometric forms—cylinder, cone, and plane—the cone is the closest in form to one-half of a globe. **Conic projections,** therefore, often are used to depict hemispheres or smaller parts of the earth.

A useful projection in this category, and the easiest to visualize, is the *simple conic* projection. Imagine that a cone is laid over half the globe, as in Figure A.6a, tangent to the globe at the 30th parallel. Distances are true only along this standard parallel. When the cone is developed, of course, the standard parallel becomes an arc of a circle, and all other parallels become arcs of concentric circles. With a central light source, the parallels become increasingly farther apart as they approach the pole, and distortion is accordingly exaggerated.

One can lessen the amount of distortion by shortening the length of the central meridian, spacing the parallels of latitude at equal distances on that meridian, and making the 90th parallel (the pole) an arc rather than a point. Most of the conic projections that are in general use employ such mathematical adjustments. When more than one standard parallel is used, a *polyconic* projection results (Figure A.6b).

Conic projections are widely used because they can be adjusted to minimize distortions and become either equal-area or conformal. By their nature, however, they can never show the whole globe. In fact, they are most useful for and generally restricted to maps of midlatitude regions of greater east-west than north-south extent. Many official map series use types of conic projections. The U.S. Geological Survey, for example, selected the *Albers equal-area conic projection,* shown in Figure A.7, for its *National Atlas of the United States of America.* It is an equivalent projection with very little distortion of shape even in an area as large as the United States.

## Planar Projections

**Planar** (or **azimuthal**) **projections** are constructed by placing a plane surface tangent to the globe at a single point. Although the plane may touch the globe anywhere the cartographer wishes, the polar case with the plane centered on either the North Pole or the South Pole is easiest to visualize (Figure A.8a).

This *equidistant* projection is useful because it can be centered anywhere, facilitating the correct measurement of distances from that point to all others. For this reason, it is

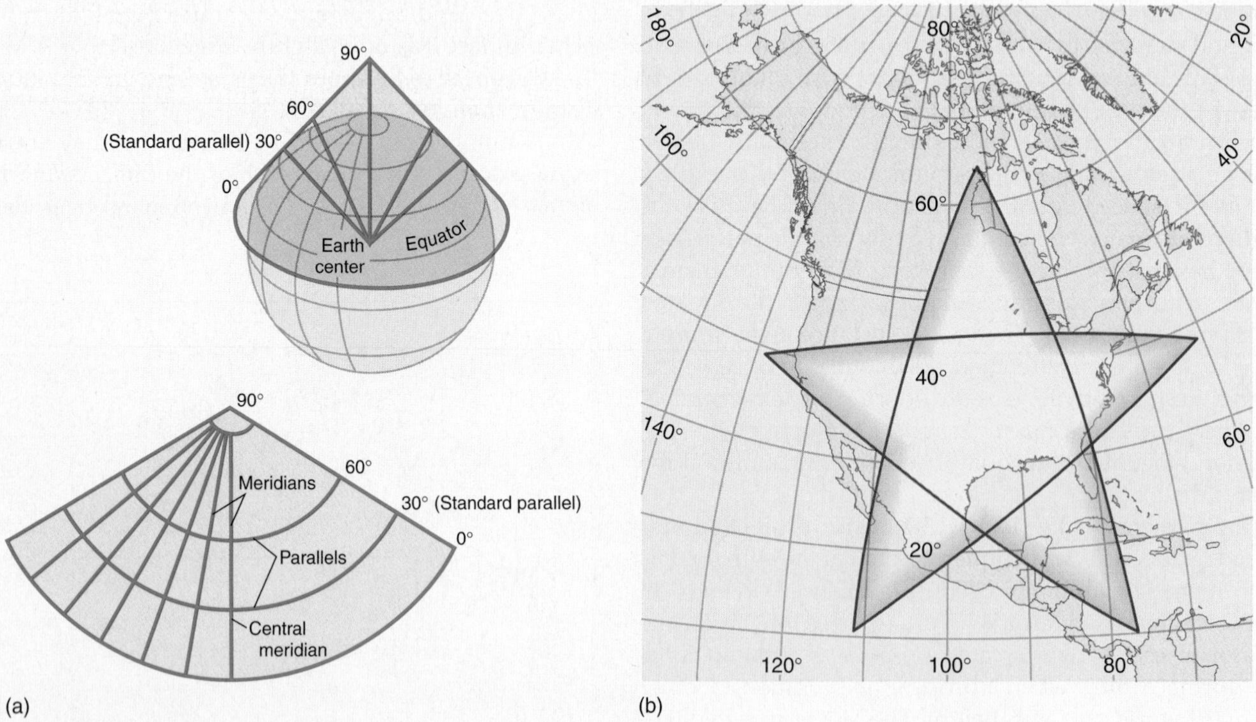


**FIGURE A.6** **(a) A simple conic projection with one standard parallel.** Most conics are adjusted so that the parallels are spaced evenly along the central meridian. (b) **The polyconic projection.** The map is produced by bringing together east-west strips from a series of cones, each tangent at a different parallel. This projection differs from the simple conic in that the parallels of latitude are *not* arcs of concentric circles and the meridians are curved rather than straight lines. Although neither equivalent nor conformal, the projection portrays shape well. Note how closely the star resembles a perfect five-pointed star, unlike the star shown in Figure A.4.

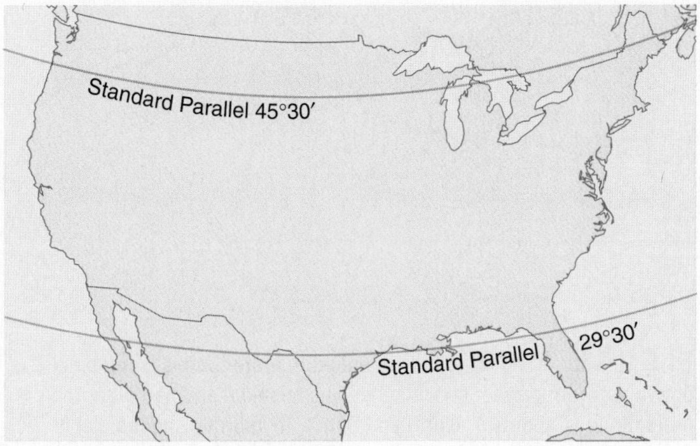


**FIGURE A.7 The Albers equal-area conic projection,** used for many official U.S. maps, has two standard parallels. All parallels of latitude are concentric arcs of circles, the meridians are straight lines, and parallels and meridians intersect at right angles. The projection is best suited for regions of considerably greater east-west than north-south extent.

often used to show air navigation routes that originate from a single place. When the plane is centered on places other than the poles, the meridians and the parallels become curiously curved, as is evident in Figure A.8b.

Because they are particularly well suited for showing the arrangement of polar landmasses, planar maps are commonly used in atlases. Depending on the particular projection used, true shape, equivalency, or a compromise between them can be depicted. In addition, one of the planar projections is widely used for navigation and telecommunications. The *gnomonic planar projection,* shown in Figure A.9, is the only one on which all great circles (or parts thereof) appear as straight lines. Because great circles are the shortest distances between two points, navigators need only connect the points with a straight line to find the shortest route.

# MATHEMATICAL PROJECTIONS

The geometric projections we have just discussed can all be thought of as developed from the projection of the globe grid onto a cylinder, cone, or plane. Many projections, however, cannot be classified in terms of simple geometric shapes. They are derived from mathematical formulas and usually have been developed to display the world or a portion thereof in a fashion that is visually acceptable. Ovals are most common, but hearts, trapezoids, stars, butterflies, and other—sometimes bizarre—forms have been devised for special purposes.

One such projection is *Goode's Homolosine,* developed by the geographer J. Paul Goode as an equal-area projection for statistical mapping. Usually shown in its interrupted form, as in Figure A.10, it is actually a product of fitting together the least distorted portions of two different

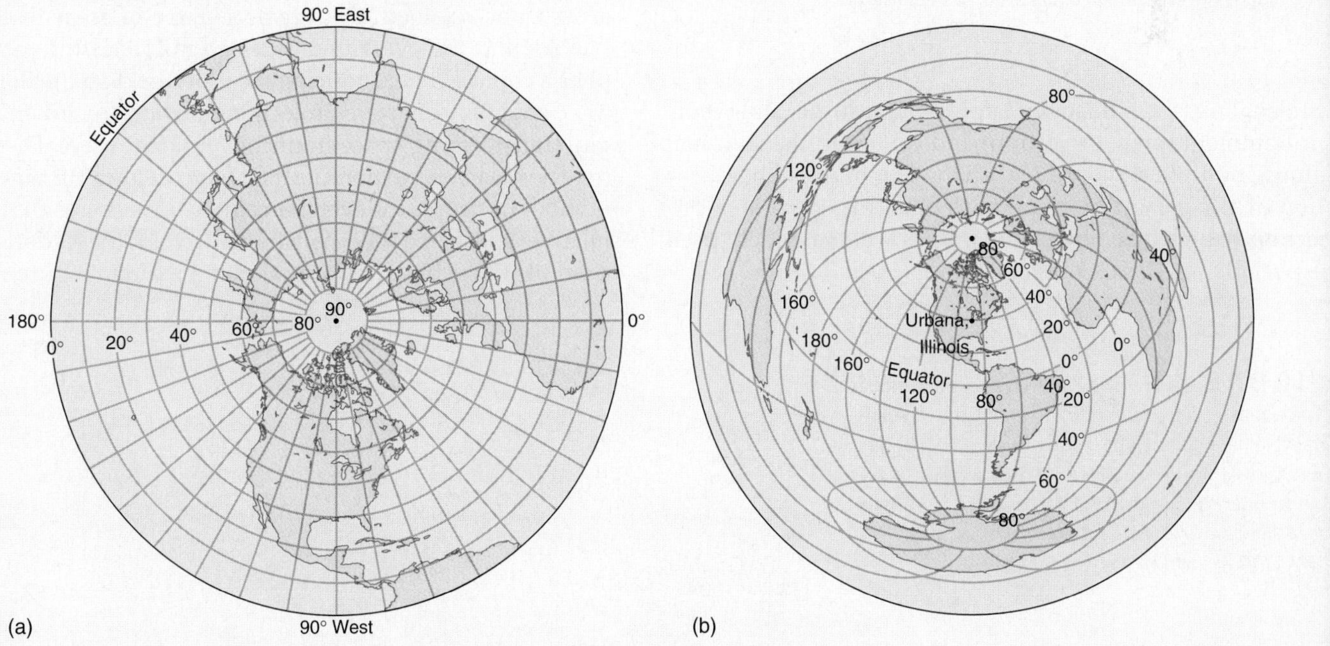


**FIGURE A.8** (a) **The planar equidistant projection.** Parallels of latitude are circles equally spaced on the meridians, which are straight lines. This projection is particularly useful because distances from the center to any other point are true. If the grid is extended to show the Southern Hemisphere, the South Pole is represented as a circle instead of a point. (b) **A planar equidistant projection centered on Urbana, Illinois.** The scale of miles applies only to distances from Urbana or on a line through it. The scale on the rim of the map, representing the antipode of Urbana, is infinitely stretched. *(b) Copyright 1977, Brooks and Roberts; with permission.*

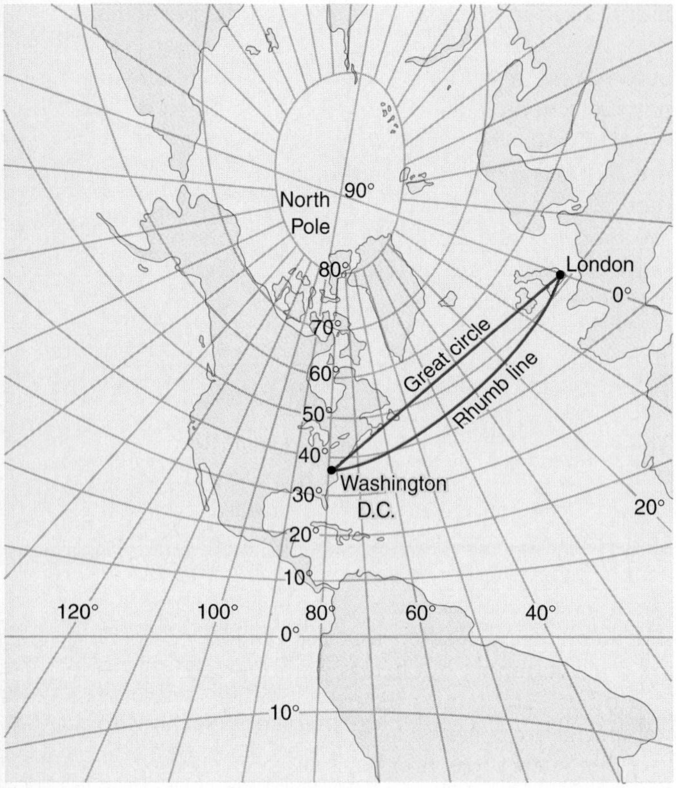


**FIGURE A.9** **The gnomonic projection** is the only one on which all great circles appear as straight lines. Rhumb lines are curved. In this sense, it is the opposite of the Mercator projection, on which rhumb lines are straight and great circles are curved (see Figure A.4). Note that the distortion of shapes and areas increases away from the center point. The map is not conformal, equal area, or equidistant.

projections (the sinusoidal projection and the Mollweide, or homolographic, projection) and centering the split map along multiple standard meridians to minimize the distortion of either land or ocean surfaces. This equal-area projection, which also represents shapes well, is widely used, especially in *Goode's World Atlas.*

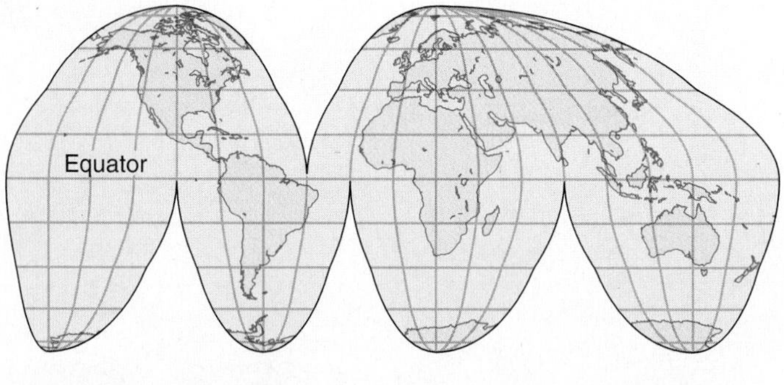


**FIGURE A.10** **Goode's Homolosine projection** is a combination of two different projections. It joins the sinusoidal and homolographic projections at about 40° North and South. To improve shapes, each continent is placed on the middle of a lobe approximately centered on its own central meridian. This projection can also interrupt the continents to display the ocean areas intact. *Copyright by the Committee on Geographic Studies, University of Chicago. Used by permission.*

R. Buckminster Fuller, an architect and a designer perhaps best known as the inventor of the geodesic dome, produced the *Fuller dymaxion projection* (Figure A.11). It consists of 20 equilateral triangles, which can be hinged along different boundaries to show interesting earth relationships. The projection minimizes distortion of the sizes and shapes of the world's landmasses.

Goode's and Fuller's projections show how projections can be manipulated or adjusted to achieve desired objectives. Since most projections are based on a mathematically consistent rendering of the actual globe grid, the possibilities for such manipulation are nearly unlimited. The map properties to be retained, the size and shape of areas to be displayed, and the overall map design influence the cartographer's choices in reproducing the globe grid on the flat map.

Some very effective projections are non-Euclidian in origin, transforming space in unconventional ways. Distances may be measured in nonlinear fashion (in terms of time, cost, number of people, or even perception), and maps that show relative space may be constructed from these data. One example of such a transformation is shown in Figure A.12.

**FIGURE A.11** **The Fuller dymaxion projection.** The equilateral triangles can be folded into a solid approximating the globe. Source: *Buckminster Fuller Institute and Dymaxion Map Design, Santa Barbara, CA. The word* Dymaxion *and the Fuller Projection Dymaxion Map design are trademarks of the Buckminster Fuller Institute, Santa Barbara, California, © 1938, 1967 & 1992. All rights reserved.*

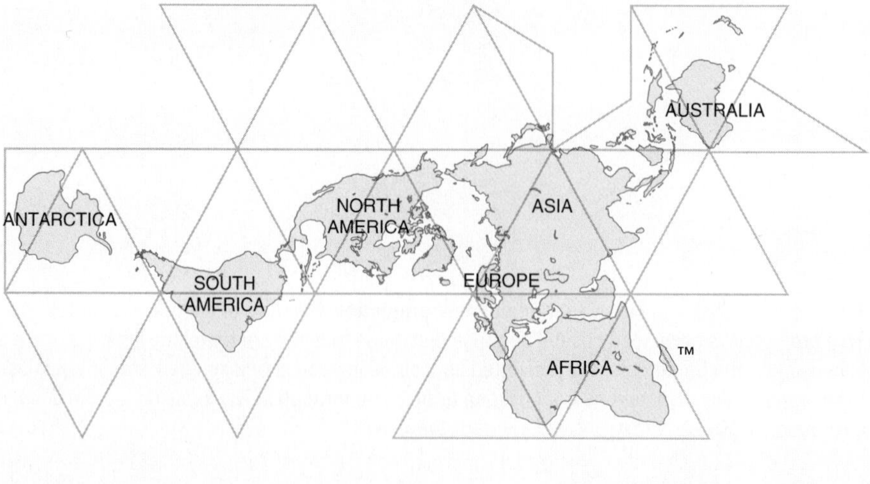

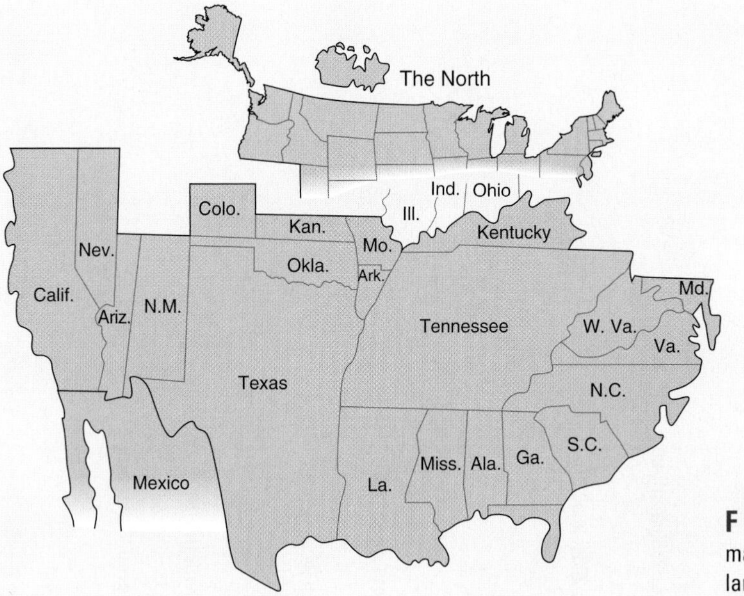

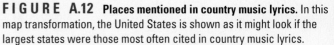

**FIGURE A.12 Places mentioned in country music lyrics.** In this map transformation, the United States is shown as it might look if the largest states were those most often cited in country music lyrics.

## Summary

It is impossible to transform the globe into a flat map without distortion. Cartographers have devised hundreds of possible geometric and mathematical projections to display to their best advantage the variety of the world's features and relationships they wish to emphasize. Some projections are highly specialized and restricted to a single, limited purpose; others achieve a more general acceptability and utility.

## Key Words

azimuthal projection 492
conic projection 492
cylindrical projection 491
developable surface 490
great circle 489
Mercator projection 491
planar projection 492
rhumb line 491
small circle 489

# Appendix 2

## 2004 World Population Data

| | Population Mid-2004 (millions) | Births per 1,000 Pop. | Deaths per 1,000 Pop. | Rate of Natural Increase (%) | Projected Population for 2025 (millions) | Projected Population for 2050 (millions) | Projected Pop. Change 2004–2050 (%) | Infant Mortality Rate[a] | Total Fertility Rate[b] | Percent of Population of Age <15/65+ | Life Expectancy at Birth (years) Total | Percent Urban | Percent of Pop. 15–49 with HIV/AIDS End 2003 | % with Access to Improved Water Source | Per Capita GNI 2002 (US$) |
|---|---|---|---|---|---|---|---|---|---|---|---|---|---|---|---|
| **WORLD** | 6,396 | 21 | 9 | 1.3 | 7,934 | 9,276 | 45 | 56 | 2.8 | 30/7 | 67 | 48 | 1.1 | 82 | 5,073 |
| **MORE DEVELOPED** | 1,206 | 11 | 10 | 0.1 | 1,257 | 1,257 | 4 | 7 | 1.6 | 17/15 | 76 | 76 | 0.5 | + | 26,214 |
| **LESS DEVELOPED** | 5,190 | 24 | 8 | 1.5 | 6,677 | 8,019 | 55 | 62 | 3.1 | 33/5 | 65 | 41 | 1.4 | 78 | 1,154 |
| **LESS DEVELOPED** (Excl. China) | 3,890 | 27 | 9 | 1.8 | 5,201 | 6,582 | 69 | 66 | 3.5 | 36/4 | 63 | 41 | 1.9 | — | — |
| **AFRICA** | 885 | 38 | 14 | 2.4 | 1,323 | 1,941 | 119 | 90 | 5.1 | 42/3 | 52 | 35 | 6.2 | — | — |
| **SUB-SAHARAN AFRICA** | 733 | 41 | 16 | 2.5 | 1,120 | 1,701 | 132 | 96 | 5.6 | 44/3 | 49 | 31 | 7.5 | 57 | 460 |
| **NORTHERN AFRICA** | 191 | 26 | 7 | 2.0 | 265 | 324 | 70 | 49 | 3.4 | 36/4 | 67 | 46 | 0.5 | — | — |
| Algeria | 32.3 | 20 | 4 | 1.5 | 40.5 | 44.3 | 37 | 54 | 2.5 | 34/4 | 73 | 49 | 0.1 | 89 | 1,720 |
| Egypt | 73.4 | 26 | 6 | 2.0 | 103.2 | 127.4 | 74 | 38 | 3.2 | 36/5 | 68 | 43 | z | 97 | 1,470 |
| Libya | 5.6 | 28 | 4 | 2.4 | 8.3 | 10.8 | 92 | 28 | 3.6 | 35/4 | 76 | 86 | 0.3 | 72 | 5,540 |
| Morocco | 30.6 | 21 | 6 | 1.5 | 39.2 | 45.0 | 47 | 40 | 2.5 | 31/5 | 70 | 57 | 0.1 | 80 | 1,190 |
| Sudan | 39.1 | 38 | 10 | 2.8 | 61.3 | 84.2 | 115 | 69 | 5.4 | 45/2 | 57 | 31 | 2.3 | 75 | 350 |
| Tunisia | 10.0 | 17 | 6 | 1.1 | 11.6 | 12.2 | 22 | 22 | 2.0 | 28/6 | 73 | 63 | z | 80 | 2,000 |
| Western Sahara | 0.3 | 29 | 8 | 2.1 | 0.5 | 0.6 | 103 | 59 | 4.1 | —/— | 62 | — | — | — | — |
| **WESTERN AFRICA** | 263 | 42 | 15 | 2.8 | 410 | 633 | 141 | 100 | 5.8 | 44/3 | 51 | 36 | 4.3 | — | — |
| Benin | 7.3 | 41 | 14 | 2.7 | 11.8 | 18.0 | 148 | 89 | 5.6 | 46/3 | 51 | 40 | 1.9 | 63 | 380 |
| Burkina Faso | 13.6 | 45 | 19 | 2.6 | 22.5 | 39.5 | 191 | 83 | 6.2 | 46/3 | 45 | 15 | 1.8 | 42 | 220 |
| Cape Verde | 0.5 | 29 | 7 | 2.3 | 0.7 | 0.8 | 74 | 31 | 4.0 | 42/6 | 69 | 53 | — | 74 | 1,290 |
| Côte d'Ivoire | 16.9 | 39 | 19 | 2.0 | 22.1 | 27.6 | 63 | 102 | 5.2 | 43/3 | 42 | 46 | 7.0 | 81 | 610 |
| Gambia | 1.5 | 41 | 13 | 2.9 | 2.7 | 4.2 | 169 | 78 | 5.6 | 45/3 | 54 | 26 | 1.2 | 62 | 280 |
| Ghana | 21.4 | 33 | 10 | 2.2 | 30.6 | 39.5 | 85 | 64 | 4.4 | 40/3 | 58 | 44 | 2.2 | 73 | 270 |
| Guinea | 9.2 | 43 | 16 | 2.7 | 16.2 | 30.6 | 231 | 98 | 6.0 | 45/3 | 49 | 33 | 3.2 | 48 | 410 |
| Guinea-Bissau | 1.5 | 50 | 20 | 3.0 | 2.8 | 4.7 | 207 | 125 | 7.1 | 47/3 | 45 | 32 | — | 56 | 150 |
| Liberia | 3.5 | 50 | 22 | 2.9 | 6.1 | 9.8 | 182 | 150 | 6.8 | 47/2 | 42 | 45 | 5.9 | — | 150 |
| Mali | 13.4 | 50 | 17 | 3.3 | 25.7 | 46.0 | 243 | 123 | 7.0 | 49/2 | 48 | 30 | 1.9 | 65 | 240 |
| Mauritania | 3.0 | 42 | 15 | 2.7 | 5.0 | 7.5 | 152 | 102 | 5.9 | 43/3 | 54 | 40 | 0.6 | 37 | 410 |

| | Population Mid-2004 (millions) | Births per 1,000 Pop. | Deaths per 1,000 Pop. | Rate of Natural Increase (%) | Projected Population for 2025 (millions) | Projected Population for 2050 (millions) | Projected Pop. Change 2004–2050 (%) | Infant Mortality Rate[a] | Total Fertility Rate[b] | Percent of Population of Age <15/65+ | Life Expectancy at Birth (years) Total | Percent Urban | Percent of Pop. 15–49 with HIV/AIDS End 2003 | % with Access to Improved Water Source | Per Capita GNI 2002 (US$) |
|---|---|---|---|---|---|---|---|---|---|---|---|---|---|---|---|
| **WESTERN AFRICA (continued)** | | | | | | | | | | | | | | | |
| Niger | 12.4 | 55 | 20 | 3.5 | 25.7 | 53.0 | 327 | 123 | 8.0 | 50/2 | 45 | 21 | 1.2 | 59 | 170 |
| Nigeria | 137.3 | 42 | 13 | 2.9 | 206.4 | 307.4 | 124 | 100 | 5.7 | 44/3 | 52 | 36 | 5.4 | 62 | 290 |
| Senegal | 10.9 | 37 | 11 | 2.6 | 17.1 | 24.6 | 126 | 64 | 5.1 | 44/3 | 56 | 43 | 0.8 | 78 | 470 |
| Sierra Leone | 5.2 | 50 | 29 | 2.1 | 7.6 | 10.3 | 100 | 180 | 6.5 | 44/3 | 35 | 37 | — | 57 | 140 |
| Togo | 5.6 | 38 | 11 | 2.7 | 7.6 | 9.7 | 74 | 72 | 5.5 | 46/2 | 54 | 33 | 4.1 | 54 | 270 |
| **EASTERN AFRICA** | 270 | 41 | 18 | 2.3 | 415 | 632 | 134 | 98 | 5.7 | 45/3 | 46 | 22 | 7.6 | — | — |
| Burundi | 6.2 | 40 | 18 | 2.2 | 10.1 | 15.4 | 147 | 74 | 6.2 | 47/3 | 43 | 8 | 6.0 | 78 | 100 |
| Comoros | 0.7 | 47 | 12 | 3.5 | 1.1 | 1.8 | 181 | 84 | 6.8 | 47/2 | 56 | 33 | — | 96 | 390 |
| Djibouti | 0.7 | 41 | 17 | 2.3 | 1.0 | 1.4 | 96 | 106 | 5.9 | 43/3 | 46 | 82 | 2.9 | 100 | 900 |
| Eritrea | 4.4 | 39 | 13 | 2.6 | 7.0 | 10.5 | 137 | 76 | 5.7 | 45/3 | 53 | 19 | 2.7 | 46 | 160 |
| Ethiopia | 72.4 | 41 | 18 | 2.4 | 117.6 | 173.3 | 139 | 105 | 5.9 | 44/3 | 46 | 15 | 4.4 | 24 | 100 |
| Kenya | 32.4 | 38 | 15 | 2.3 | 39.9 | 49.9 | 54 | 78 | 5.0 | 44/4 | 51 | 36 | 6.7 | 57 | 360 |
| Madagascar | 17.5 | 43 | 12 | 3.0 | 33.0 | 65.5 | 274 | 84 | 5.8 | 45/3 | 55 | 26 | 1.7 | 47 | 240 |
| Malawi | 11.9 | 51 | 21 | 3.1 | 23.8 | 47.2 | 296 | 121 | 6.6 | 46/3 | 44 | 14 | 14.2 | 57 | 160 |
| Mauritius | 1.2 | 16 | 7 | 1.0 | 1.4 | 1.5 | 22 | 13.2 | 1.9 | 25/7 | 72 | 42 | — | 100 | 3,850 |
| Mayotte | 0.2 | 41 | 9 | 3.2 | 0.3 | 0.6 | 219 | — | 5.6 | 42/2 | 60 | 28 | — | — | — |
| Mozambique | 19.2 | 40 | 23 | 1.7 | 25.4 | 31.3 | 63 | 127 | 5.5 | 44/3 | 40 | 29 | 12.2 | 57 | 210 |
| Reunion | 0.8 | 20 | 5 | 1.4 | 1.0 | 1.1 | 37 | 27 | 2.5 | 27/7 | 75 | 89 | — | — | — |
| Rwanda | 8.4 | 40 | 21 | 1.9 | 11.7 | 17.2 | 104 | 107 | 5.8 | 43/3 | 40 | 17 | 5.1 | 41 | 230 |
| Seychelles | 0.1 | 18 | 8 | 1.0 | 0.1 | 0.1 | 11 | *18* | 2.0 | 26/7 | 71 | 50 | — | — | 6,530 |
| Somalia | 8.3 | 47 | 18 | 2.9 | 14.9 | 25.5 | 207 | 124 | 7.1 | 45/3 | 47 | 33 | — | — | 130 |
| Tanzania | 36.1 | 40 | 17 | 2.3 | 52.1 | 74.0 | 105 | 105 | 5.3 | 45/3 | 45 | 22 | 8.8 | 68 | 280 |
| Uganda | 26.1 | 47 | 17 | 3.0 | 47.5 | 82.6 | 217 | 88 | 6.9 | 51/2 | 45 | 12 | 4.1 | 52 | 250 |
| Zambia | 10.9 | 42 | 24 | 1.8 | 14.4 | 18.5 | 70 | 95 | 5.6 | 46/3 | 35 | 35 | 16.5 | 64 | 330 |
| Zimbabwe | 12.7 | 32 | 20 | 1.2 | 12.8 | 14.6 | 15 | 65 | 4.0 | 42/3 | 41 | 32 | 24.6 | 83 | 470 |
| **MIDDLE AFRICA** | 107 | 45 | 17 | 2.8 | 182 | 303 | 184 | 103 | 6.4 | 47/3 | 47 | 35 | 5.0 | — | — |
| Angola | 13.3 | 49 | 24 | 2.6 | 23.8 | 40.7 | 206 | 145 | 6.8 | 44/3 | 40 | 33 | 3.9 | 38 | 660 |
| Cameroon | 16.1 | 37 | 15 | 2.2 | 22.4 | 30.9 | 92 | 77 | 4.9 | 43/3 | 48 | 48 | 6.9 | 58 | 560 |
| Central African Republic | 3.7 | 37 | 19 | 1.7 | 4.8 | 6.2 | 65 | 96 | 4.9 | 44/3 | 42 | 39 | 13.5 | 70 | 260 |
| Chad | 9.5 | 49 | 16 | 3.2 | 16.7 | 29.2 | 206 | 103 | 6.6 | 48/3 | 49 | 24 | 4.8 | 27 | 220 |
| Congo | 3.8 | 44 | 15 | 2.9 | 6.8 | 10.6 | 179 | 84 | 6.3 | 47/3 | 48 | 52 | 4.9 | 51 | 700 |
| Congo, Dem. Rep. Of | 58.3 | 46 | 15 | 3.1 | 104.9 | 181.3 | 211 | 100 | 6.8 | 48/3 | 49 | 30 | 4.2 | 45 | 90 |
| Equatorial Guinea | 0.5 | 43 | 17 | 2.6 | 0.8 | 1.2 | 132 | 105 | 5.9 | 44/4 | 49 | 45 | — | 44 | 700 |
| Gabon | 1.4 | 33 | 12 | 2.1 | 1.9 | 2.5 | 84 | 57 | 4.3 | 42/5 | 57 | 73 | 8.1 | 86 | 3120 |
| Sao Tome and Principe | 0.2 | 34 | 6 | 2.8 | 0.3 | 0.3 | 112 | 34 | 4.3 | 41/5 | 69 | 38 | — | — | 290 |

| | Population Mid-2004 (millions) | Births per 1,000 Pop. | Deaths per 1,000 Pop. | Rate of Natural Increase (%) | Projected Population for 2025 (millions) | Projected Population for 2050 (millions) | Projected Pop. Change 2004–2050 (%) | Infant Mortality Rate[a] | Total Fertility Rate[b] | Percent of Population of Age <15/65+ | Life Expectancy at Birth (years) Total | Percent Urban | Percent of Pop. 15–49 with HIV/AIDS End 2003 | % with Access to Improved Water Source | Per Capita GNI 2002 (US$) |
|---|---|---|---|---|---|---|---|---|---|---|---|---|---|---|---|
| **SOUTHERN AFRICA** | 53 | 25 | 14 | 1.0 | 51 | 49 | –9 | 51 | 2.9 | 35/4 | 52 | 50 | 22.6 | — | — |
| Botswana | 1.7 | 27 | 26 | 0.1 | 1.1 | 1.0 | –43 | 62 | 3.5 | 40/4 | 36 | 54 | 37.3 | 95 | 2,980 |
| Lesotho | 1.8 | 33 | 22 | 1.1 | 2.1 | 2.2 | 23 | 90 | 4.4 | 43/5 | 38 | 17 | 28.9 | 78 | 470 |
| Namibia | 1.9 | 31 | 15 | 1.6 | 2.1 | 2.6 | 35 | 38 | 4.2 | 42/4 | 47 | 33 | 21.3 | 77 | 1,780 |
| South Africa | 46.9 | 24 | 13 | 1.0 | 44.6 | 41.7 | –11 | 48 | 2.8 | 34/4 | 53 | 53 | 21.5 | 86 | 2,600 |
| Swaziland | 1.2 | 36 | 16 | 2.0 | 1.1 | 1.1 | –2 | 78 | 4.5 | 43/3 | 43 | 25 | 38.8 | — | 1,180 |
| **NORTH AMERICA** | 326 | 14 | 8 | 0.5 | 386 | 457 | 40 | 7 | 2.0 | 21/12 | 78 | 79 | 0.6 | + | — |
| Canada | 31.9 | 11 | 7 | 0.3 | 36.0 | 36.9 | 16 | 5.2 | 1.5 | 18/13 | 79 | 79 | 0.3 | + | 22,300 |
| United States | 293.6 | 14 | 8 | 0.6 | 349.4 | 419.9 | 43 | 6.7 | 2.0 | 21/12 | 77 | 79 | 0.6 | + | 35,060 |
| **LATIN AMERICA AND THE CARIBBEAN** | 549 | 22 | 6 | 1.6 | 685 | 778 | 42 | 29 | 2.6 | 32/6 | 72 | 75 | 0.7 | 86 | 3,362 |
| **CENTRAL AMERICA** | 146 | 26 | 5 | 2.1 | 189 | 224 | 54 | 27 | 3.0 | 36/4 | 74 | 68 | 0.5 | — | — |
| Belize | 0.3 | 28 | 5 | 2.3 | 0.4 | 0.6 | 102 | 20 | 3.4 | 41/4 | 70 | 49 | 2.4 | 92 | 2,960 |
| Costa Rica | 4.2 | 18 | 4 | 1.4 | 5.6 | 6.3 | 49 | 10 | 2.1 | 30/6 | 79 | 59 | 0.6 | 95 | 4,100 |
| El Salvador | 6.7 | 26 | 6 | 2.0 | 8.5 | 9.9 | 48 | 25 | 3.0 | 36/5 | 70 | 58 | 0.7 | 77 | 2,080 |
| Guatemala | 12.7 | 34 | 7 | 2.8 | 19.8 | 27.2 | 115 | 39 | 4.4 | 44/4 | 66 | 39 | 1.1 | 92 | 1,750 |
| Honduras | 7.0 | 33 | 5 | 2.8 | 10.7 | 14.7 | 109 | 34 | 4.1 | 41/4 | 71 | 47 | 1.8 | 88 | 920 |
| Mexico | 106.2 | 25 | 5 | 2.1 | 131.7 | 149.7 | 41 | 25 | 2.8 | 35/5 | 75 | 75 | 0.3 | 88 | 5,910 |
| Nicaragua | 5.6 | 32 | 5 | 2.7 | 8.3 | 10.9 | 93 | 31 | 3.8 | 43/3 | 69 | 58 | 0.2 | 77 | 370 |
| Panama | 3.2 | 23 | 5 | 1.8 | 4.2 | 5.0 | 58 | 21 | 2.7 | 31/6 | 75 | 62 | 0.9 | 90 | 4,020 |
| **CARIBBEAN** | 39 | 20 | 8 | 1.2 | 46 | 52 | 35 | 41 | 2.7 | 30/7 | 69 | 62 | 2.1 | — | — |
| Antigua and Barbuda | 0.1 | 24 | 6 | 1.7 | 0.1 | 0.1 | 1 | *17* | 2.7 | 26/8 | 71 | 37 | — | 91 | 9,390 |
| Bahamas | 0.3 | 18 | 5 | 1.3 | 0.3 | 0.3 | 8 | 15.8 | 2.1 | 30/5 | 72 | 89 | 3.0 | 97 | 14,860 |
| Barbados | 0.3 | 15 | 8 | 0.6 | 0.3 | 0.2 | –3 | *13.2* | 1.7 | 22/12 | 72 | 50 | 1.5 | 100 | 9,750 |
| Cuba | 11.3 | 11 | 7 | 0.5 | 11.8 | 11.1 | –2 | 7 | 1.6 | 21/10 | 76 | 75 | 0.1 | 91 | 1,170 |
| Dominica | 0.1 | 17 | 7 | 1.0 | 0.1 | 0.1 | 19 | *16.1* | 1.9 | 33/9 | 74 | 71 | — | 97 | 3,180 |
| Dominican Republic | 8.8 | 25 | 6 | 1.9 | 11.1 | 13.4 | 52 | 31 | 3.0 | 34/5 | 69 | 64 | 1.7 | 86 | 2,320 |
| Grenada | 0.1 | 19 | 7 | 1.2 | 0.1 | 0.1 | –3 | *17* | 2.1 | 35/8 | 71 | 39 | — | 95 | 3,500 |
| Guadeloupe | 0.4 | 17 | 7 | 1.0 | 0.5 | 0.5 | 19 | 7.6 | 2.2 | 25/9 | 78 | 100 | — | — | — |
| Haiti | 8.1 | 33 | 14 | 1.9 | 11.7 | 16.0 | 97 | 80 | 4.7 | 43/4 | 51 | 36 | 5.6 | 46 | 440 |
| Jamaica | 2.6 | 20 | 7 | 1.4 | 3.3 | 3.7 | 39 | 24 | 2.4 | 31/7 | 75 | 52 | 1.2 | 92 | 2,820 |
| Martinique | 0.4 | 14 | 8 | 0.7 | 0.4 | 0.4 | 5 | *8* | 2.0 | 24/12 | 79 | 95 | — | — | — |
| Netherlands Antilles | 0.2 | 15 | 7 | 0.8 | 0.2 | 0.2 | 11 | *6* | 2.1 | 24/10 | 76 | 69 | — | — | — |
| Puerto Rico | 3.9 | 14 | 7 | 0.7 | 4.1 | 3.8 | –2 | 9.6 | 1.8 | 23/12 | 77 | 71 | — | — | — |
| St. Kitts-Nevis | 0.05 | 17 | 8 | 1.0 | 0.1 | 0.1 | 32 | *28* | 2.1 | 31/9 | 70 | 33 | — | 98 | 6,370 |

| | Population Mid-2004 (millions) | Births per 1,000 Pop. | Deaths per 1,000 Pop. | Rate of Natural Increase (%) | Projected Population for 2025 (millions) | Projected Population for 2050 (millions) | Projected Pop. Change 2004–2050 (%) | Infant Mortality Rate[a] | Total Fertility Rate[b] | Percent of Population of Age <15/65+ | Life Expectancy at Birth (years) Total | Percent Urban | Percent of Pop. 15–49 with HIV/AIDS End 2003 | % with Access to Improved Water Source | Per Capita GNI 2002 (US$) |
|---|---|---|---|---|---|---|---|---|---|---|---|---|---|---|---|
| **CARIBBEAN (continued)** | | | | | | | | | | | | | | | |
| Saint Lucia | 0.2 | 17 | 6 | 1.1 | 0.2 | 0.2 | 43 | *13.6* | 2.2 | 31/5 | 72 | 30 | — | 98 | 3,840 |
| St. Vincent and the Grenadines | 0.1 | 18 | 7 | 1.1 | 0.1 | 0.1 | –21 | *19.3* | 2.1 | 37/7 | 72 | 44 | — | 93 | 2,820 |
| Trinidad and Tobago | 1.3 | 13 | 7 | 0.6 | 1.3 | 1.2 | –7 | 18.6 | 1.6 | 25/7 | 71 | 74 | 3.2 | 90 | 6,490 |
| **SOUTH AMERICA** | 365 | 21 | 6 | 1.5 | 450 | 502 | 38 | 29 | 2.5 | 31/6 | 71 | 79 | 0.6 | — | — |
| Argentina | 37.9 | 19 | 8 | 1.1 | 45.9 | 53.1 | 40 | 16.3 | 2.4 | 28/9 | 74 | 89 | 0.7 | — | 4,060 |
| Bolivia | 8.8 | 28 | 9 | 1.9 | 12.2 | 15.4 | 75 | 54 | 3.8 | 39/5 | 63 | 63 | 0.1 | 83 | 900 |
| Brazil | 179.1 | 20 | 7 | 1.3 | 211.2 | 221.4 | 24 | 33 | 2.2 | 30/6 | 71 | 81 | 0.7 | 87 | 2,850 |
| Chile | 16.0 | 17 | 5 | 1.2 | 19.5 | 22.2 | 39 | 8.3 | 2.4 | 26/7 | 76 | 87 | 0.3 | 93 | 4,260 |
| Colombia | 45.3 | 23 | 6 | 1.7 | 58.1 | 67.3 | 48 | 26 | 2.6 | 32/5 | 72 | 71 | 0.7 | 91 | 1,830 |
| Ecuador | 13.4 | 25 | 4 | 2.1 | 17.3 | 20.6 | 54 | 30 | 3.0 | 36/5 | 71 | 61 | 0.3 | 85 | 1,450 |
| French Guiana | 0.2 | 31 | 4 | 2.6 | 0.3 | 0.4 | 95 | 12 | 3.9 | 35/4 | 75 | 75 | — | — | — |
| Guyana | 0.8 | 23 | 9 | 1.4 | 0.7 | 0.5 | –34 | 53 | 2.4 | 30/4 | 63 | 36 | 2.5 | 94 | 840 |
| Paraguay | 6.0 | 30 | 5 | 2.5 | 9.2 | 12.1 | 101 | 37 | 3.8 | 39/6 | 71 | 54 | 0.5 | 78 | 1,170 |
| Peru | 27.5 | 23 | 6 | 1.7 | 35.7 | 42.8 | 55 | 33 | 2.8 | 34/5 | 69 | 72 | 0.5 | 80 | 2,050 |
| Suriname | 0.4 | 23 | 7 | 1.5 | 0.4 | 0.4 | –19 | 27 | 2.5 | 32/6 | 70 | 69 | 1.7 | 82 | 1,960 |
| Uruguay | 3.4 | 16 | 9 | 0.6 | 3.8 | 4.2 | 24 | 13.5 | 2.2 | 24/13 | 75 | 93 | 0.3 | 98 | 4,370 |
| Venezuela | 26.2 | 24 | 5 | 1.9 | 35.3 | 41.7 | 59 | 19.6 | 2.8 | 34/4 | 73 | 87 | 0.7 | 83 | 4,090 |
| **ASIA** | 3,875 | 20 | 7 | 1.3 | 4,778 | 5,385 | 39 | 54 | 2.6 | 30/6 | 67 | 39 | 0.4 | — | — |
| **ASIA (Excl. China)** | 2,575 | 24 | 8 | 1.6 | 3,302 | 3,948 | 53 | 59 | 3.0 | 33/5 | 65 | 39 | 0.6 | — | — |
| **WESTERN ASIA** | 209 | 27 | 6 | 2.0 | 302 | 395 | 89 | 48 | 3.7 | 35/5 | 68 | 63 | — | — | — |
| Armenia | 3.2 | 10 | 8 | 0.2 | 3.0 | 2.5 | –24 | 36 | 1.2 | 23/10 | 73 | 64 | 0.1 | — | 790 |
| Azerbaijan | 8.3 | 14 | 6 | 0.8 | 9.7 | 11.6 | 40 | 13 | 1.8 | 29/6 | 72 | 51 | z | 78 | 710 |
| Bahrain | 0.7 | 20 | 3 | 1.7 | 1.0 | 1.3 | 76 | 7 | 2.7 | 28/3 | 74 | 87 | 0.2 | — | 11,130 |
| Cyprus | 0.9 | 12 | 7 | 0.5 | 1.1 | 1.1 | 14 | 6 | 1.6 | 21/11 | 78 | 65 | — | 100 | 12,320 |
| Georgia | 4.5 | 11 | 11 | 0.0 | 4.0 | 3.1 | –32 | 24 | 1.4 | 20/13 | 72 | 52 | 0.1 | 79 | 650 |
| Iraq | 25.9 | 36 | 9 | 2.7 | 41.7 | 57.9 | 124 | 102 | 5.0 | 42/3 | 60 | 68 | z | 85 | 2,170 |
| Israel | 6.8 | 22 | 6 | 1.6 | 9.3 | 10.6 | 56 | 5.3 | 2.9 | 28/10 | 79 | 92 | 0.1 | + | 16,710 |
| Jordan | 5.6 | 29 | 5 | 2.4 | 8.1 | 10.2 | 80 | 22 | 3.7 | 38/4 | 72 | 79 | z | 96 | 1,760 |
| Kuwait | 2.5 | 18 | 2 | 1.7 | 4.6 | 7.0 | 182 | 10 | 4.0 | 26/2 | 78 | 100 | — | — | 18,270 |
| Lebanon | 4.5 | 23 | 7 | 1.7 | 5.7 | 6.9 | 53 | 27 | 3.2 | 28/7 | 73 | 87 | 0.1 | 100 | 3,990 |
| Oman | 2.7 | 26 | 4 | 2.2 | 4.0 | 5.1 | 93 | 16 | 4.1 | 34/2 | 74 | 76 | 0.1 | 39 | 7,720 |
| Palestinian Territory | 3.8 | 39 | 4 | 3.5 | 7.4 | 11.9 | 211 | 26 | 5.7 | 46/3 | 72 | 57 | — | 86 | 930 |
| Qatar | 0.7 | 20 | 4 | 1.6 | 1.0 | 1.2 | 67 | 12 | 4.0 | 27/1 | 72 | 92 | — | — | 12,000 |
| Saudi Arabia | 25.1 | 32 | 3 | 3.0 | 40.1 | 55.2 | 120 | 25 | 4.8 | 40/3 | 72 | 86 | — | 95 | 8,460 |

| | Population Mid-2004 (millions) | Births per 1,000 Pop. | Deaths per 1,000 Pop. | Rate of Natural Increase (%) | Projected Population for 2025 (millions) | Projected Population for 2050 (millions) | Projected Pop. Change 2004–2050 (%) | Infant Mortality Rate[a] | Total Fertility Rate[b] | Percent of Population of Age <15/65+ | Life Expectancy at Birth (years) Total | Percent Urban | Percent of Pop. 15–49 with HIV/AIDS End 2003 | % with Access to Improved Water Source | Per Capita GNI 2002 (US$) |
|---|---|---|---|---|---|---|---|---|---|---|---|---|---|---|---|
| **WESTERN ASIA (continued)** | | | | | | | | | | | | | | | |
| Syria | 18.0 | 28 | 5 | 2.4 | 27.6 | 35.0 | 95 | 18 | 3.8 | 40/4 | 70 | 50 | z | 80 | 1,130 |
| Turkey | 71.3 | 21 | 7 | 1.4 | 88.9 | 97.5 | 37 | 39 | 2.5 | 30/6 | 69 | 59 | — | 82 | 2,500 |
| United Arab Emirates | 4.2 | 16 | 2 | 1.4 | 5.4 | 5.7 | 35 | 8 | 2.5 | 25/1 | 74 | 78 | — | — | 18,060 |
| Yemen | 20.0 | 43 | 10 | 3.3 | 39.6 | 71.1 | 255 | 75 | 7.0 | 48/3 | 60 | 26 | 0.1 | 69 | 490 |
| **SOUTH CENTRAL ASIA** | 1,587 | 26 | 8 | 1.8 | 2,068 | 2,547 | 60 | 69 | 3.3 | 37/4 | 62 | 30 | 0.7 | — | — |
| Afghanistan | 28.5 | 48 | 21 | 2.7 | 50.3 | 81.9 | 187 | 165 | 6.8 | 45/2 | 43 | 22 | — | 13 | 250 |
| Bangladesh | 141.3 | 30 | 9 | 2.1 | 204.5 | 280.0 | 98 | 66 | 3.3 | 37/3 | 60 | 23 | — | 97 | 360 |
| Bhutan | 1.0 | 34 | 9 | 2.5 | 1.5 | 2.1 | 113 | 61 | 4.7 | 42/4 | 66 | 21 | — | 62 | 590 |
| India | 1,086.6 | 25 | 8 | 1.7 | 1,363.0 | 1,628.0 | 50 | 64 | 3.1 | 36/4 | 62 | 28 | 0.9 | 84 | 480 |
| Iran | 67.4 | 18 | 6 | 1.2 | 84.7 | 96.5 | 43 | 32 | 2.5 | 33/5 | 69 | 67 | 0.1 | 92 | 1,710 |
| Kazakhstan | 15.0 | 17 | 11 | 0.6 | 15.8 | 14.8 | –1 | 52 | 2.0 | 27/8 | 64 | 57 | 0.2 | 91 | 1,510 |
| Kyrgyzstan | 5.1 | 21 | 8 | 1.4 | 6.7 | 8.2 | 62 | 42 | 2.6 | 35/6 | 68 | 35 | 0.1 | 77 | 290 |
| Maldives | 0.3 | 18 | 4 | 1.4 | 0.4 | 0.5 | 69 | 18 | 3.7 | 39/4 | 73 | 27 | — | 100 | 2,090 |
| Nepal | 24.7 | 34 | 10 | 2.3 | 37.8 | 50.8 | 105 | 64 | 4.1 | 39/4 | 59 | 14 | 0.5 | 88 | 230 |
| Pakistan | 159.2 | 34 | 10 | 2.4 | 228.8 | 295.0 | 85 | 85 | 4.8 | 42/4 | 61 | 34 | 0.1 | 90 | 410 |
| Sri Lanka | 19.6 | 19 | 6 | 1.3 | 21.9 | 21.6 | 10 | 10 | 2.0 | 27/7 | 72 | 30 | z | 77 | 840 |
| Tajikistan | 6.6 | 25 | 6 | 1.9 | 8.6 | 10.0 | 52 | 50 | 3.1 | 42/4 | 68 | 27 | z | 60 | 180 |
| Turkmenistan | 5.7 | 25 | 9 | 1.6 | 7.6 | 8.7 | 53 | 74 | 2.9 | 38/4 | 67 | 47 | z | — | 1,200 |
| Uzbekistan | 26.4 | 24 | 8 | 1.6 | 36.9 | 48.5 | 84 | 62 | 2.9 | 38/4 | 70 | 37 | 0.1 | 85 | 450 |
| **SOUTHEAST ASIA** | 548 | 22 | 7 | 1.5 | 698 | 800 | 46 | 41 | 2.7 | 31/5 | 68 | 38 | 0.5 | — | — |
| Brunei | 0.4 | 22 | 3 | 1.9 | 0.5 | 0.7 | 85 | 7 | 2.3 | 31/3 | 76 | 74 | z | — | 24,100 |
| Cambodia | 13.1 | 32 | 10 | 2.2 | 19.8 | 26.8 | 104 | 95 | 4.5 | 42/5 | 57 | 16 | 2.6 | 30 | 280 |
| East Timor | 0.8 | 26 | 13 | 1.3 | 1.2 | 1.4 | 75 | 129 | 4.1 | 44/3 | 49 | 8 | — | — | 270 |
| Indonesia | 218.7 | 22 | 6 | 1.6 | 275.5 | 308.5 | 41 | 46 | 2.6 | 30/5 | 68 | 42 | 0.1 | 78 | 710 |
| Laos | 5.8 | 36 | 13 | 2.3 | 8.6 | 11.4 | 98 | 104 | 4.9 | 43/4 | 54 | 19 | 0.1 | 37 | 310 |
| Malaysia | 25.6 | 26 | 4 | 2.1 | 36.0 | 46.9 | 83 | 11 | 3.3 | 34/4 | 73 | 62 | 0.4 | — | 3,540 |
| Myanmar | 50.1 | 25 | 11 | 1.4 | 59.8 | 64.5 | 29 | 87 | 3.1 | 33/5 | 57 | 28 | 1.2 | 72 | 220 |
| Philippines | 83.7 | 26 | 6 | 2.0 | 118.4 | 147.3 | 76 | 29 | 3.5 | 37/4 | 70 | 48 | z | 86 | 1,020 |
| Singapore | 4.2 | 10 | 4 | 0.6 | 4.8 | 4.4 | 6 | 2.2 | 1.3 | 21/8 | 79 | 100 | 0.2 | 100 | 20,690 |
| Thailand | 63.8 | 14 | 7 | 0.8 | 70.2 | 73.2 | 15 | 20 | 1.7 | 23/7 | 71 | 31 | 1.5 | 84 | 1,980 |
| Vietnam | 81.5 | 18 | 6 | 1.2 | 102.9 | 115.1 | 41 | 21 | 2.1 | 29/7 | 72 | 25 | 0.4 | 77 | 430 |
| **EAST ASIA** | 1,531 | 12 | 7 | 0.6 | 1,709 | 1,643 | 7 | 30 | 1.6 | 21/9 | 72 | 46 | 0.1 | — | — |
| China | 1,300.1 | 12 | 6 | 0.6 | 1,476.0 | 1,437.0 | 11 | 32 | 1.7 | 22/7 | 71 | 41 | 0.1 | 75 | 940 |
| China, Hong Kong SAR[c] | 6.8 | 7 | 5 | 0.1 | 8.4 | 9.4 | 38 | 2.4 | 0.9 | 15/12 | 81 | 100 | 0.1 | — | — |

| | Population Mid-2004 (millions) | Births per 1,000 Pop. | Deaths per 1,000 Pop. | Rate of Natural Increase (%) | Projected Population for 2025 (millions) | Projected Population for 2050 (millions) | Projected Pop. Change 2004–2050 (%) | Infant Mortality Rate[a] | Total Fertility Rate[b] | Percent of Population of Age <15/65+ | Life Expectancy at Birth (years) Total | Percent Urban | Percent of Pop. 15–49 with HIV/AIDS End 2003 | % with Access to Improved Water Source | Per Capita GNI 2002 (US$) |
|---|---|---|---|---|---|---|---|---|---|---|---|---|---|---|---|
| **EAST ASIA (continued)** | | | | | | | | | | | | | | | |
| China, Macao SAR[c] | 0.4 | 7 | 3 | 0.4 | 0.5 | 0.6 | 24 | *3* | 0.8 | 20/8 | 77 | 99 | — | — | — |
| Japan | 127.6 | 9 | 8 | 0.1 | 121.1 | 100.6 | –21 | 3.0 | 1.3 | 14/19 | 82 | 78 | z | + | 33,550 |
| Korea, North | 22.8 | 17 | 11 | 0.7 | 24.7 | 25.0 | 10 | 45 | 2.0 | 27/6 | 63 | 60 | — | 100 | — |
| Korea, South | 48.2 | 10 | 5 | 0.5 | 50.6 | 44.3 | –8 | 8 | 1.2 | 20/8 | 77 | 80 | z | 92 | 9,930 |
| Mongolia | 2.5 | 18 | 6 | 1.2 | 3.4 | 4.3 | 72 | 30 | 2.7 | 36/5 | 65 | 57 | z | 60 | 440 |
| Taiwan | 22.6 | 10 | 6 | 0.4 | 24.4 | 22.1 | –3 | 6.0 | 1.2 | 20/9 | 76 | 78 | — | — | — |
| **EUROPE** | 728 | 10 | 12 | –0.2 | 722 | 668 | –8 | 7 | 1.4 | 17/15 | 74 | 74 | 0.5 | — | — |
| **NORTHERN EUROPE** | 96 | 12 | 10 | 0.1 | 102 | 103 | 8 | 5 | 1.7 | 18/16 | 78 | 82 | 0.2 | + | — |
| Channel Islands | 0.2 | 10 | 10 | 0.1 | 0.2 | 0.1 | –3 | *2.8* | 1.5 | 17/15 | 78 | 31 | — | + | — |
| Denmark | 5.4 | 12 | 11 | 0.1 | 5.4 | 5.3 | –3 | 4.4 | 1.8 | 19/15 | 77 | 72 | 0.2 | + | 30,290 |
| Estonia | 1.3 | 10 | 13 | –0.4 | 1.2 | 1.0 | –23 | 6 | 1.4 | 17/16 | 71 | 69 | 1.1 | + | 4,130 |
| Finland | 5.2 | 11 | 9 | 0.2 | 5.3 | 4.8 | –8 | 3.2 | 1.8 | 18/16 | 79 | 62 | 0.1 | + | 23,510 |
| Iceland | 0.3 | 14 | 6 | 0.8 | 0.3 | 0.4 | 22 | *2.4* | 2.0 | 23/12 | 81 | 94 | 0.2 | + | 27,970 |
| Ireland | 4.1 | 16 | 7 | 0.8 | 4.5 | 4.7 | 16 | 5.1 | 2.0 | 21/11 | 77 | 60 | 0.1 | + | 23,870 |
| Latvia | 2.3 | 9 | 14 | –0.5 | 2.2 | 1.8 | –24 | 9 | 1.3 | 16/16 | 72 | 68 | 0.6 | + | 3,480 |
| Lithuania | 3.4 | 9 | 12 | –0.3 | 3.5 | 3.1 | –9 | 7 | 1.3 | 18/15 | 72 | 67 | 0.1 | + | 3,660 |
| Norway | 4.6 | 12 | 9 | 0.3 | 5.1 | 5.6 | 22 | 3.4 | 1.8 | 20/15 | 80 | 78 | 0.1 | + | 37,850 |
| Sweden | 9.0 | 11 | 10 | 0.1 | 9.9 | 10.6 | 18 | 2.8 | 1.7 | 17/18 | 80 | 84 | 0.1 | + | 24,820 |
| United Kingdom | 59.7 | 12 | 10 | 0.1 | 64.0 | 65.4 | 10 | 5.3 | 1.7 | 19/16 | 78 | 89 | 0.2 | + | 25,250 |
| **WESTERN EUROPE** | 185 | 11 | 10 | 0.1 | 190 | 184 | –1 | 4 | 1.6 | 17/17 | 79 | 79 | 0.2 | + | — |
| Austria | 8.1 | 9 | 9 | –0.0 | 8.4 | 8.2 | 1 | 4.5 | 1.4 | 17/15 | 79 | 54 | 0.3 | + | 23,390 |
| Belgium | 10.4 | 11 | 10 | 0.1 | 10.8 | 11.0 | 5 | 4.4 | 1.6 | 17/17 | 79 | 97 | 0.2 | + | 23,250 |
| France | 60.0 | 13 | 9 | 0.4 | 63.4 | 64.0 | 7 | 4.1 | 1.9 | 19/16 | 79 | 74 | 0.4 | + | 22,010 |
| Germany | 82.6 | 9 | 10 | –0.2 | 82.0 | 75.1 | –9 | 4.1 | 1.3 | 15/17 | 78 | 88 | 0.1 | + | 22,670 |
| Liechtenstein | 0.03 | 12 | 6 | 0.5 | 0.04 | 0.04 | 18 | *4.1* | 1.5 | 18/11 | 80 | 21 | — | + | — |
| Luxembourg | 0.5 | 12 | 9 | 0.3 | 0.6 | 0.7 | 56 | *4.9* | 1.6 | 19/14 | 78 | 91 | 0.2 | + | 38,830 |
| Monaco | 0.03 | 23 | 16 | 0.6 | 0.04 | 0.04 | 27 | — | — | 13/22 | — | 100 | — | + | — |
| Netherlands | 16.3 | 12 | 9 | 0.4 | 17.4 | 17.6 | 8 | 4.8 | 1.8 | 19/14 | 79 | 62 | 0.2 | + | 23,960 |
| Switzerland | 7.4 | 10 | 9 | 0.1 | 7.4 | 7.2 | –3 | 4.4 | 1.4 | 17/16 | 80 | 68 | 0.4 | + | 37,930 |
| **EASTERN EUROPE** | 299 | 10 | 15 | –0.5 | 281 | 243 | –19 | 12 | 1.3 | 16/14 | 68 | 68 | 0.8 | — | — |
| Belarus | 9.8 | 9 | 15 | –0.6 | 9.4 | 8.5 | –13 | 8 | 1.2 | 16/14 | 69 | 72 | — | + | 1,360 |
| Bulgaria | 7.8 | 9 | 14 | –0.6 | 6.5 | 4.8 | –38 | 12.3 | 1.2 | 15/17 | 72 | 70 | z | + | 1,790 |
| Czech Republic | 10.2 | 9 | 11 | –0.2 | 10.1 | 9.2 | –10 | 3.9 | 1.2 | 16/14 | 75 | 77 | 0.1 | — | 5,560 |
| Hungary | 10.1 | 9 | 13 | –0.4 | 8.9 | 7.6 | –25 | 7.3 | 1.3 | 16/15 | 73 | 65 | 0.1 | + | 5,280 |

| | Population Mid-2004 (millions) | Births per 1,000 Pop. | Deaths per 1,000 Pop. | Rate of Natural Increase (%) | Projected Population for 2025 (millions) | Projected Population for 2050 (millions) | Projected Pop. Change 2004–2050 (%) | Infant Mortality Rate[a] | Total Fertility Rate[b] | Percent of Population of Age <15/65+ | Life Expectancy at Birth (years) Total | Percent Urban | Percent of Pop. 15–49 with HIV/AIDS End 2003 | % with Access to Improved Water Source | Per Capita GNI 2002 (US$) |
|---|---|---|---|---|---|---|---|---|---|---|---|---|---|---|---|
| **EASTERN EUROPE (continued)** | | | | | | | | | | | | | | | |
| Moldova | 4.2 | 10 | 12 | –0.1 | 3.9 | 3.0 | –28 | 18 | 1.2 | 22/10 | 68 | 45 | 0.2 | 92 | 460 |
| Poland | 38.2 | 9 | 9 | –0.0 | 36.6 | 32.4 | –15 | 7.5 | 1.2 | 18/13 | 75 | 62 | 0.1 | + | 4,570 |
| Romania | 21.7 | 10 | 12 | –0.3 | 18.1 | 15.7 | –27 | 16.7 | 1.2 | 17/14 | 71 | 53 | z | 58 | 1,850 |
| Russia | 144.1 | 10 | 17 | –0.6 | 136.9 | 119.1 | –17 | 13 | 1.4 | 16/13 | 65 | 73 | 1.1 | 99 | 2,140 |
| Slovakia | 5.4 | 10 | 10 | –0.0 | 5.2 | 4.7 | –13 | 7.6 | 1.2 | 18/12 | 74 | 56 | z | + | 3,950 |
| Ukraine | 47.4 | 9 | 16 | –0.8 | 45.1 | 38.4 | –19 | 10 | 1.2 | 16/15 | 68 | 68 | 1.4 | 98 | 770 |
| **SOUTHERN EUROPE** | 149 | 10 | 10 | 0.1 | 149 | 138 | –7 | 6 | 1.3 | 15/17 | 78 | 74 | 0.5 | — | — |
| Albania | 3.2 | 17 | 5 | 1.2 | 3.7 | 3.7 | 15 | 11 | 2.1 | 29/8 | 74 | 42 | — | 97 | 1,380 |
| Andorra | 0.1 | 11 | 3 | 0.8 | 0.1 | 0.1 | –3 | *3.9* | 1.3 | 15/13 | — | 92 | — | + | — |
| Bosnia-Herzegovina | 3.9 | 10 | 8 | 0.1 | 3.9 | 3.3 | –15 | 9 | 1.2 | 18/12 | 74 | 43 | z | + | 1,270 |
| Croatia | 4.4 | 9 | 11 | –0.2 | 4.3 | 3.8 | –14 | 7.0 | 1.3 | 17/16 | 75 | 56 | z | + | 4,640 |
| Greece | 11.0 | 9 | 9 | 0.0 | 10.4 | 9.7 | –12 | 5.9 | 1.3 | 15/17 | 78 | 60 | 0.2 | + | 11,660 |
| Italy | 57.8 | 10 | 10 | –0.1 | 57.6 | 52.3 | –10 | 4.8 | 1.3 | 14/19 | 80 | 90 | 0.5 | + | 18,960 |
| Macedonia | 2.0 | 14 | 9 | 0.5 | 2.2 | 2.1 | 3 | 11.9 | 1.7 | 22/10 | 73 | 59 | z | — | — |
| Malta | 0.4 | 10 | 8 | 0.2 | 0.4 | 0.4 | –9 | *7.2* | 1.5 | 19/13 | 78 | 91 | 0.2 | + | 9,200 |
| Portugal | 10.5 | 11 | 10 | 0.0 | 10.4 | 9.3 | –11 | 5.0 | 1.4 | 16/17 | 77 | 53 | 0.4 | + | 10,840 |
| San Marino | 0.03 | 10 | 7 | 0.3 | 0.03 | 0.03 | 17 | *6.7* | 1.2 | 15/16 | 80 | 84 | — | + | — |
| Serbia and Montenegro | 10.7 | 12 | 11 | 0.2 | 10.7 | 10.2 | –4 | 13 | 1.7 | 19/14 | 73 | 52 | 0.2 | 98 | 1,400 |
| Slovenia | 2.0 | 9 | 10 | –0.1 | 2.0 | 1.7 | –15 | 3.8 | 1.2 | 15/15 | 76 | 51 | z | + | 9,810 |
| Spain | 42.5 | 10 | 9 | 0.1 | 43.5 | 41.3 | –3 | 3.7 | 1.3 | 14/17 | 79 | 76 | 0.7 | + | 14,430 |
| **OCEANIA** | 33 | 17 | 7 | 1.0 | 41 | 47 | 43 | 26 | 2.1 | 25/10 | 75 | 72 | 0.2 | — | — |
| Australia | 20.1 | 13 | 7 | 0.6 | 24.2 | 26.3 | 31 | 4.7 | 1.7 | 20/13 | 80 | 91 | 0.1 | + | 19,740 |
| Federated States of Micronesia | 0.1 | 28 | 7 | 2.1 | 0.1 | 0.2 | 46 | 40 | 4.4 | 40/4 | 67 | 22 | — | — | 1,980 |
| Fiji | 0.8 | 25 | 6 | 1.9 | 1.0 | 1.0 | 18 | 22 | 3.3 | 32/4 | 67 | 39 | 0.1 | 47 | 2,160 |
| French Polynesia | 0.3 | 20 | 5 | 1.5 | 0.3 | 0.4 | 40 | *6* | 2.5 | 31/4 | 72 | 53 | — | — | — |
| Guam | 0.2 | 20 | 4 | 1.6 | 0.2 | 0.2 | 46 | *6.2* | 2.6 | 30/5 | 78 | 93 | — | — | — |
| Kiribati | 0.1 | 26 | 8 | 1.8 | 0.1 | 0.2 | 133 | 43 | 4.3 | 40/3 | 63 | 43 | — | 48 | 810 |
| Marshall Islands | 0.1 | 42 | 5 | 3.7 | 0.1 | 0.1 | 81 | 37 | 4.7 | 42/2 | 69 | 68 | — | — | 2,270 |
| Nauru | 0.01 | 23 | 5 | 1.8 | 0.02 | 0.02 | 92 | *25* | 4.4 | 41/2 | 61 | 100 | — | — | — |

| | Population Mid-2004 (millions) | Births per 1,000 Pop. | Deaths per 1,000 Pop. | Rate of Natural Increase (%) | Projected Population for 2025 (millions) | Projected Population for 2050 (millions) | Projected Pop. Change 2004–2050 (%) | Infant Mortality Rate[a] | Total Fertility Rate[b] | Percent of Population of Age <15/65+ | Life Expectancy at Birth (years) Total | Percent Urban | Percent of Pop. 15–49 with HIV/AIDS End 2003 | % with Access to Improved Water Source | Per Capita GNI 2002 (US$) |
|---|---|---|---|---|---|---|---|---|---|---|---|---|---|---|---|
| **OCEANIA (continued)** | | | | | | | | | | | | | | | |
| New Caledonia | 0.2 | 22 | 5 | 1.7 | 0.3 | 0.4 | 60 | *5* | 2.6 | 30/5 | 73 | 71 | — | — | — |
| New Zealand | 4.1 | 14 | 7 | 0.7 | 4.7 | 5.1 | 26 | 5.6 | 2.0 | 22/12 | 78 | 78 | 0.1 | + | 13,710 |
| Palau | 0.02 | 14 | 7 | 0.8 | 0.02 | 0.03 | 24 | *17* | 1.6 | 24/5 | 70 | 70 | — | — | 6,780 |
| Papua New Guinea | 5.7 | 32 | 10 | 2.2 | 8.2 | 10.8 | 90 | 60 | 4.1 | 40/2 | 57 | 15 | 0.6 | 42 | 530 |
| Samoa | 0.2 | 29 | 6 | 2.4 | 0.2 | 0.2 | 34 | 18 | 4.4 | 41/4 | 73 | 22 | — | 99 | 1,420 |
| Solomon Islands | 0.5 | 36 | 9 | 2.7 | 0.7 | 1.0 | 112 | 66 | 4.8 | 44/3 | 61 | 16 | — | 71 | 570 |
| Tonga | 0.1 | 25 | 7 | 1.8 | 0.1 | 0.1 | 20 | *13* | 3.4 | —/— | 71 | 32 | — | 100 | 1,410 |
| Tuvalu | 0.01 | 27 | 10 | 1.7 | 0.02 | 0.02 | 122 | *34* | 3.8 | 36/6 | — | 47 | — | 100 | — |
| Vanuatu | 0.2 | 28 | 6 | 2.2 | 0.4 | 0.5 | 124 | 27 | 4.8 | 42/3 | 67 | 21 | — | 88 | 1,080 |

[a]Infant deaths per 1,000 live births. Rates shown with decimals indicate national statistics reported as completely registered, while those without are estimates. Rates shown in *italics* are based upon less than 50 annual infant deaths.

[b]Average number of children born to a woman during her lifetime.

[c]Special Administrative Region.

z = Less than 0.05 percent.

A dash (—) indicates data unavailable or inapplicable.

A plus sign (+) indicates essentially 100% access to improved water supplies for the countries of the industrial "North."

Urban population data are the percentage of total population living in areas termed urban by that country.

Table modified from the 2004 World Population Data Sheet of the Population Reference Bureau. Data for access to improved water supply are from UNICEF and the World Health Organization. GNI per capita data are from the World Bank.

# Glossary

## A

**absolute direction** Direction with respect to cardinal east, west, north, and south reference points

**absolute location** (*syn:* mathematical location) The exact position of an object or a place stated in spatial coordinates of a grid system designed for locational purposes. In geography, the reference system is the global grid of parallels of latitude north or south of the equator and of meridians of longitude east or west of a prime meridian.

**accelerated eutrophication** The overnourishment of a water body with nutrients stemming from human activities such as agriculture, industry, and urbanization.

**accessibility** The relative ease with which a destination may be reached from other locations; the relative opportunity for spatial interaction. May be measured in geometric, social, or economic terms.

**acculturation** The cultural modification or change resulting from one culture group or individual adopting traits of a more advanced or dominant society; cultural development through "borrowing."

**acid rain** Precipitation that is unusually acidic; created when oxides of sulfur and nitrogen change chemically as they dissolve in water vapor in the atmosphere and return to Earth as acidic rain, snow, fog, or dry particles.

**activity space** The area within which people move freely on their rounds of regular activity.

**adaptation** A presumed modification of heritable traits through response to environmental stimuli.

**agglomeration** The spatial grouping of people or activities for mutual benefit.

**agglomeration economies** (*syn:* external economies) The savings to an individual enterprise that result from spatial association with other, similar economic activities.

**agricultural density** The number of rural residents per unit of agriculturally productive land. The measure excludes a region's urban population and its nonarable land from the density calculation.

**agriculture** Cultivating the soil, producing crops, and raising livestock; farming.

**air mass** A large body of air with little horizontal variation in temperature, pressure, and humidity.

**air pressure** The weight of the atmosphere as measured at a point on the earth's surface.

**alluvial fan** A fan-shaped accumulation of alluvium deposited by a stream at the base of a hill or mountain.

**alluvium** The sediment carried by a stream and deposited in a floodplain or delta.

**amalgamation theory** In human geography, the concept that multiethnic societies become a merger of the culture traits of their member groups.

**anaerobic digestion** The process by which organic waste is decomposed in an oxygen-free environment to produce methane gas (biogas).

**anecumene** *See* nonecumene.

**animism** A belief that natural objects may be the abode of dead people, spirits, or gods who occasionally give the objects the appearance of life.

**antecedent boundary** A boundary line established before the area in question is well populated.

**aquaculture** Producing and harvesting of fish and shellfish in freshwater ponds, lakes, and canals or in fenced-off coastal bays and estuaries; fish farming.

**aquifer** Underground porous and permeable rock that is capable of holding groundwater, especially rock that supplies economically significant quantities of water to wells and springs.

**arable land** Land that is or can be cultivated.

**Arctic haze** Air pollution resulting from the transport by air currents of combustion-based pollutants to the area north of the Arctic Circle.

**area analysis tradition** One of the four traditions of geography, that of regional geography.

**area cartogram** A type of map in which the areas of the units are proportional to the data they represent; value-by-area map.

**arithmetic density** *See* crude density.

**arroyo** A steep-sided, flat-bottomed gully, usually dry, carved out of desert land by rapidly flowing water.

**artifacts** The material manifestations of culture, including tools, housing, systems of land use, clothing, and the like. Elements in the technological subsystem of culture.

**artificial boundary** *See* geometric boundary.

**assimilation** The social process of merging into a composite culture, losing separate ethnic or social identity and becoming culturally homogenized.

**asthenosphere** A partially molten, plastic layer above the core and lower mantle of the earth.

**atmosphere** The gaseous mass surrounding the earth.

**atoll** A near-circular low coral reef formed in shallow water enclosing a central lagoon; most common in the central and western Pacific Ocean.

**azimuthal projection** *See* planar projection.

## B

**barchan** A crescent-shaped sand dune; the horns of the crescent point downwind.

**basic sector** Those products or services of an urban economy that are exported outside the city itself, earning income for the community.

**bench mark** A surveyor's mark indicating the position and elevation of some stationary object; used as a reference point in surveying and mapping.

**bioaccumulation** The buildup of a material in the body of an organism.

**biocide** A chemical used to kill plant and animal pests and disease organisms. *See also* herbicide, pesticide.

**biological magnification** The accumulation of a chemical in the fatty tissue of an organism and its concentration at progressively higher levels in the food chain; biomagnification.

**biomagnification** *See* biological magnification.

**biomass** Living matter, plant and animal, in any form.

**biomass fuel** Any organic material produced by plants, animals, or microorganisms that can be used as a source of energy through either direct burning or conversion into a liquid or gas.

**biome** The total assemblage of living organisms in a single major ecological region.

**biosphere** (*syn:* ecosphere) The thin film of air, water, and earth within which we live, including the atmosphere, surrounding and subsurface waters, and the upper reaches of the earth's crust.

**birth rate** (*syn:* crude birth rate) The ratio of the number of live births during 1 year to the total population, usually at the midpoint of the same year, expressed as the number of births per year per 1000 population.

**blizzard** A heavy snowstorm accompanied by high winds.

**boundary** A line separating one political unit from another.

**boundary definition** A general agreement between two states about the allocation of territory between them.

**boundary delimitation** The plotting of a boundary line on maps or aerial photographs.

**boundary demarcation** The actual marking of a boundary line on the ground; the final stage in boundary development.

**butte** A small, flat-topped, isolated hill with steep sides, common in dry climate regions.

# C

**carcinogen** A substance that produces or incites cancerous growth.

**carrying capacity** The numbers of any population that can be adequately supported by the available resources upon which that population subsists; for humans, the numbers supportable by the known and utilized resources—usually agricultural—of an area.

**cartogram** A map that has been simplified to present data in a diagrammatic way; the base normally is not true to scale.

**cartography** The art, science, and technology of making maps.

**caste** One of the hereditary social classes in Hinduism that determines one's occupation and position in society.

**central business district (CBD)** The center, or "downtown," of an urban unit, where retail stores, offices, and cultural activities are concentrated and where land values are high.

**central city** That part of the metropolitan area contained within the boundaries of the main city around which suburbs have developed.

**central place** A nodal point for the distribution of goods and services to a surrounding hinterland population.

**central place theory** A deductive theory formulated by Walter Christaller (1893–1969) to explain the size and distribution of settlements through reference to competitive supply of goods and services to dispersed rural populations.

**centrifugal force** In political geography, a force that disrupts and destabilizes a state, threatening its unity.

**centripetal force** In political geography, a force that promotes unity and national identity.

**CFCs** *See* chlorofluorocarbons.

**chain migration** The process by which migration movements from a common home area to a specific destination are sustained by links of friendship or kinship between first movers and later followers.

**channelization** The modification of a stream channel; specifically, the straightening of meanders or dredging of the stream channel to deepen it.

**channelized migration** The tendency for migration to flow between areas that are socially and economically allied by past migration patterns, by economic trade considerations, or by some other affinity.

**chemical weathering** The decomposition of earth materials because of chemical reactions that include oxidation, hydration, and carbonation.

**chlorofluorocarbons (CFCs)** A family of synthetic chemicals that have significant commercial applications but whose emissions are contributing to the depletion of the ozone layer.

**choropleth map** A map that depicts quantities for areal units by varying pattern and/or color.

**circumpolar vortex** High-altitude winds circling the poles from west to east.

**city** A multifunctional nucleated settlement with a central business district and both residential and nonresidential land uses.

**climate** The long-term average weather conditions in a place or region.

**climax community** An association of grasses, shrubs, and/or trees that is in equilibrium with the climate and soil of the site; the last stage of an ecological succession.

**climograph** A bar and line graph used to depict average monthly temperatures and precipitation.

**cogeneration** The simultaneous use of a single fuel for the generation of electricity and low-grade central heat.

**cognition** The process by which an individual gives mental meaning to information.

**cohort** A population group unified by a specific common characteristic, such as age, and who are treated as a statistical unit during their lifetimes.

**commercial economy** The production of goods and services for exchange in competitive markets where price and availability are determined by supply and demand forces.

**commercial energy** Commercially traded fuels such as coal, oil, or natural gas and excluding wood, vegetable or animal wastes, or other biomass.

**Common Market** *See* European Union.

**compact state** A state whose territory is nearly circular.

**comparative advantage** A region's profit potential for a productive activity compared with alternative areas of production of the same good or with alternative uses of the region's resources.

**concentric zone model** A model describing urban land uses as a series of circular belts or rings around a core central business district, each ring housing a distinct type of land use.

**conformal projection** A map projection on which the shapes of small areas are accurately portrayed.

**conic projection** A map projection based on the projection of the grid system onto a cone as the presumed developable surface.

**connectivity** The directness of routes linking pairs of places; all of the tangible and intangible means of connection and communication between places.

**consequent boundary** (*syn:* ethnographic boundary) A boundary line that coincides with some cultural divide, such as religion or language.

**conservation** The wise use or preservation of natural resources so as to maintain supplies and qualities at levels sufficient to meet present and future needs.

**contagious diffusion** The spread of a concept, a practice, or an article from one area to others through contact and/or the exchange of information.

**continental drift** The hypothesis that an original single landmass (Pangaea) broke apart and that the continents have moved very slowly over the asthenosphere to their present locations.

**contour interval** The vertical distance separating two adjacent contour lines.

**contour line** A map line along which all points are of equal elevation above or below a datum plane, usually mean sea level.

**conurbation** An extended urban area formed by the coalescence of two or more formerly separate cities.

**convection** The circulatory movement of rising warm air and descending cool air.

**convectional precipitation** Rain produced when heated, moisture-laden air rises and then cools below the dew point.

**Convention on the Law of the Sea** *See* Law of the Sea Convention.

**coral reef** A rocklike landform in shallow tropical water composed chiefly of compacted coral and other organic material.

**core** The nucleus of a region or country, the main center of its industry, commerce, population, political, and intellectual life; in urban geography, that part of the central business district characterized by intensive land development.

**core area** The nucleus of a state, containing its most developed area, greatest wealth, densest populations, and clearest national identity.

**Coriolis effect** A fictitious force used to describe motion relative to a rotating earth; specifically, the force that tends to deflect a moving object or fluid to the right (clockwise) in the Northern Hemisphere and to the left (counterclockwise) in the Southern Hemisphere.

**countermigration** *See* return migration.

**country** *See* state.

**creole** A language developed from a pidgin to become the native tongue of a society.

**critical distance** The distance beyond which cost, effort, and/or means play an overriding role in the willingness of people to travel.

**crude birth rate (CBR)** *See* birth rate.

**crude death rate (CDR)** *See* death rate.

**crude density** (*syn:* arithmetic density, population density) The number of people per unit area of land.

**crude oil** A mixture of hydrocarbons that exists in a liquid state in underground reservoirs; petroleum as it occurs naturally, as it comes from an oil well, or after extraneous substances have been removed.

**cultural convergence** The tendency for cultures to become more alike as they increasingly share technology and organization structures in a modern world united by improved transportation and communication.

**cultural divergence** The likelihood or tendency for isolated cultures to become increasingly dissimilar with the passage of time.

**cultural ecology** The study of the interactions between societies and the natural environments they occupy.

**cultural integration** The interconnectedness of all aspects of a culture; no part can be altered without impact upon other culture traits.

**cultural lag** The retention of established culture traits despite changing circumstances rendering them inappropriate.

**cultural landscape** The natural landscape as modified by human activities and bearing the imprint of a culture group or society; the built environment.

**culture** A society's collective beliefs, symbols, values, forms of behavior, and social organizations, together with its tools, structures, and artifacts; transmitted as a heritage to succeeding generations and undergoing adoptions, modifications, and changes in the process.

**culture complex** An integrated assemblage of culture traits descriptive of one aspect of a society's behavior or activity.

**culture-environment tradition** One of the four traditions of geography; in this text, identified with population, cultural, political, and behavioral geography.

**culture hearth** A nuclear area within which an advanced and distinctive set of culture traits develops and from which there is diffusion of distinctive technologies and ways of life.

**culture realm** A collective of culture regions sharing related culture systems; a major world area having sufficient distinctiveness to be perceived as set apart from other realms in its cultural characteristics and complexes.

**culture region** A formal or functional region within which common cultural characteristics prevail. It may be based on single culture traits, on culture complexes, or on political, social, or economic integration.

**culture system** A generalization suggesting shared, identifying traits uniting two or more culture complexes.

**culture trait** A single distinguishing feature of regular occurrence within a culture, such as the use of chopsticks or the observance of a particular caste system; a single element of learned behavior.

**cyclone** A type of atmospheric disturbance in which masses of air circulate rapidly about a region of low atmospheric pressure.

**cyclonic precipitation** (*syn:* frontal precipitation) The rain or snow that is produced when moist air of one air mass is forced to rise over the edge of another air mass.

**cylindrical projection** A map projection based on the projection of the globe grid onto a cylinder as the presumed developable surface.

# D

**database** *See* geographic database.

**DDT** A chlorinated hydrocarbon that is among the most persistent of the biocides in general use.

**death rate** (*syn:* crude death rate, mortality rate) A mortality index usually calculated as the number of deaths per year per 1000 population.

**decomposers** Microorganisms and bacteria that feed on dead organisms, causing their chemical disintegration.

**deforestation** The clearing of land through total removal of forest cover.

**delta** A triangular-shaped deposit of mud, silt, or gravel created by a stream where it flows into a body of standing water.

**demographic equation** A mathematical expression that summarizes the contribution of different demographic processes to the population change of a given area during a specified time period: $P_2 = P_1 + B_{1-2} - D_{1-2} + IM_{1-2} - OM_{1-2}$, where $P_2$ is population at time 2; $P_1$ is population at beginning date; $B_{1-2}$ is the number of births between times 1 and 2; $D_{1-2}$ is the number of deaths during that period; $IM_{1-2}$ is the number of in-migrants and $OM_{1-2}$ the number of out-migrants between times 1 and 2.

**demographic momentum** *See* population momentum.

**demographic transition** A model of the effect of economic development on population growth. The first stage involves both high birth and death rates; the second phase displays high birth rates and falling mortality rates and population increases. Phase three shows reduction in population growth as birth rates decline to the level of death rates. The final, fourth, stage implies again a population stable in size but larger in numbers than at the start of the transition cycle.

**demography** The scientific study of population, with particular emphasis upon quantitative aspects.

**density of population** *See* population density.

**dependency ratio** The number of dependents, old or young, that each 100 persons in the productive years must, on average, support.

**deposition** The process by which silt, sand, and rock particles accumulate and create landforms such as stream deltas and talus slopes.

**desertification** The conversion of arid and semiarid lands into deserts as a result of climatic change or human activities such as overgrazing or deforestation.

**developable surface** A geometric form, such as a cylinder or cone, that may be flattened without distortion.

**devolution** The transfer of certain powers from the state central government to separate political subdivisions within the state's territory; decentralization of political control.

**dew point** The temperature at which condensation forms, if the air is cooled sufficiently.

**dialect** A regional or socioeconomic variation of a more widely spoken language.

**diastrophism** The earth force that folds, faults, twists, and compresses rock.

**dibble** Any small hand tool or stick used to make a hole for planting.

**diffusion** *See* spatial diffusion.

**distance** The amount of separation between two objects, areas, or points; an extent of areal or linear measure.

**distance decay** The exponential decline of an activity or a function with increasing distance from its point of origin.

**domestication** The successful transformation of plant or animal species from a wild state to a condition of dependency upon human management, usually with distinct physical change from wild forebears.

**doubling time** The time period required for any beginning total, experiencing a compounding growth, to double in size.

**dune** A wavelike desert landform created by windblown sand.

# E

**earthquake** The movement of the earth along a geologic fault or at some other point of weakness at or near the earth's surface.

**earth science tradition** One of the four traditions of geography, identified with physical geography in general.

**ecology** The scientific study of how living creatures affect each other and what determines their distribution and abundance.

**economic base** The manufacturing and service activities performed by the basic sector of the labor force of a city to satisfy demands both inside and outside the city and earn income to support the urban population.

**economic geography** The study of how people earn a living, how livelihood systems vary by area, and how economic activities are spatially interrelated and linked.

**ecosphere** *See* biosphere.

**ecosystem** A population of organisms existing together in a particular area, together with the energy, air, water, soil, and chemicals upon which it depends.

**ecumene** The permanently inhabited areas of the earth. *See also* nonecumene.

**edge city** An urban development at the margin of a metropolitan area with a large concentration of industrial parks, warehouses, office buildings, and shopping malls and more jobs than residents; characteristic of the postindustrial service and information economy of urbanized Anglo America.

**electoral geography** The study of the delineation of voting districts and the spatial patterns of election results.

**electromagnetic spectrum** The entire range of radiation, including the shortest as well as the longest wavelengths.

**El Niño** The periodic (every 3 to 7 or 8 years) buildup of warm water along the west coast of South America; replacing the cold Humboldt current off the Peruvian coast, El Niño is associated with both a fall in plankton levels (and decreased fish supply) and short-term, widespread weather modification.

**elongated state** A state whose territory is long and narrow.

**enclave** A territory that is surrounded by, but is not part of, a state.

**endangered species** Species that are present in such small numbers that they are in imminent danger of extinction.

**energy** The ability to do work. *See also* kinetic energy, potential energy.

**energy efficiency** The ratio of the output of useful energy from a conversion process to the total energy inputs.

**environment** Surroundings; the totality of things that in any way may affect an organism, including both physical and cultural conditions; a region characterized by a certain set of physical conditions.

**environmental determinism** The view that the physical environment, particularly climate, molds human behavior and conditions cultural development.

**environmental perception** The way people observe and interpret, as well as the ideas they have about near or distant places.

**environmental pollution** *See* pollution.

**epidemiologic transition** Long-term shifts in health and disease patterns as mortality moves from high to low levels.

**equal-area projection** *See* equivalent projection.

**equator** An imaginary line that encircles the globe halfway between the North and South Poles.

**equidistant projection** A map projection on which true distances in all directions can be measured from one or two central points.

**equivalent projection** A map projection on which the areas of regions are represented in correct or constant proportions to earth reality; also called equal-area projection.

**erosion** The result of processes that loosen, dissolve, wear away, and remove earth and rock material. Those processes include weathering, solution, abrasion, and transportation.

**erosional agents** The forces of wind, moving water, glaciers, waves, and ocean currents that carve, wear away, and remove rock and soil particles.

**estuarine zone** The relatively narrow area of wetlands along coastlines where salt water and fresh water mix.

**estuary** The lower course or mouth of a river where tides cause fresh water and salt water from the sea to mix.

**ethnic cleansing** The killing or forcible relocation of one traditional or ethnic group by a more powerful one.

**ethnicity** The social status afforded to, usually, a minority group within a national population. Recognition is based primarily upon culture traits such as religion, distinctive customs, or native or ancestral national origin.

**ethnic religion** A religion identified with a particular ethnic group and largely exclusive to it.

**ethnocentrism** The belief that one's own ethnic group is superior to all others.

**ethnographic boundary** *See* consequent boundary.

**European Union (EU)** An economic association established in 1957 of a number of Western European states that promotes free trade among member countries; often called the Common Market.

**eutrophication** The enrichment of a water body by the addition of nutrients received through erosion and runoff from the watershed. *See also* accelerated eutrophication.

**evapotranspiration** The return of water from the land to the atmosphere through evaporation from the soil surface and transpiration from plants.

**exclave** A portion of a state that is separated from the main territory and surrounded by another country.

**exclusive economic zone (EEZ)** As established in the United Nations Convention on the Law of the Sea, a zone of exploitation extending 200 nautical miles seaward from a coastal state that has exclusive mineral and fishing rights over it.

**exotic species** A plant or an animal that has been deliberately or inadvertently introduced into an ecosystem in which it did not evolve; a nonindigenous species.

**extensive agriculture** A crop or livestock system in which land quality or extent is more important than capital or labor inputs in determining output. It may be part of either a commercial or a subsistence economy.

**extensive commerical agriculture** Extensive farming in a commercial economy; examples include large-scale wheat farming and livestock ranching.

**extensive subsistence agriculture** Extensive farming in a subsistence economy; examples include nomadic herding and shifting cultivation.

**external economies** *See* agglomeration economies.

**extinction** The elimination of all of the individuals of a particular species.

**extractive industries** Primary activities involving the mining and quarrying of nonrenewable metallic and nonmetallic mineral resources.

**extrusive rock** Rock solidified from molten material that has issued out onto the earth's surface.

# F

**false-color image** A remotely sensed image whose colors do not appear natural to the human eye.

**fast breeder reactor** A nuclear reactor that uses uranium-235 to release energy from the more abundant uranium-238.

**faults** Breaks or fractures in rock produced by stress or the movement of lithospheric plates.

**fault escarpment** A steep slope formed by the vertical movement of the earth along a fault.

**fiord** A glacial trough whose lower end is filled with seawater.

**floodplain** A valley area bordering a stream that is subject to inundation by flooding.

**flow-line map** A map used to portray linear movement between places; may be qualitative or quantitative.

**folds** Rock layers that have buckled under pressure by the movement of lithospheric plates.

**folk culture** The body of institutions, customs, dress, artifacts, collective wisdoms, and traditions of a homogeneous, isolated, largely self-sufficient, and relatively static social group.

**food chain** A sequence of organisms through which energy and materials move within an ecosystem.

**footloose** A descriptive term applied to manufacturing activities for which the cost of transporting material or product is not important in determining location of production.

**Fordism** The manufacturing economy and system derived from assembly line mass production and the mass consumption of standardized goods. Named after Henry Ford, who innovated many of its production techniques.

**formal region** A region distinguished by uniformity of one or more characteristics that can serve as the basis for an areal generalization and of contrast with adjacent areas.

**form utility** A value-increasing change in the form—and therefore in the utility—of a raw material or commodity.

**forward-thrust capital** A capital city deliberately sited in a state's frontier zone.

**fossil fuel** Any of the fuels derived from decayed organic material converted by earth processes; especially, coal, petroleum, and natural gas but also including tar sands and oil shales.

**fragmented state** A state whose territory contains isolated parts, separated and discontinuous.

**frictional effect** In climatology, the slowing of wind movement due to the frictional drag of the earth's surface.

**friction of distance** A measure of the retarding effect of distance on spatial interaction. Generally, the greater the distance, the greater the "friction" and the less the interaction or exchange, or the greater the cost of achieving the exchange.

**front** The line or zone of separation between two air masses of different temperatures and humidities.

**frontal precipitation** *See* cyclonic precipitation.

**frontier** That portion of a country adjacent to its boundaries and fronting another political unit.

**frontier zone** A belt lying between two states or between settled and uninhabited or sparsely settled areas.

**Fujita scale** A scale for categorizing tornado intensity.

**functional region** A region differentiated by what occurs within it rather than by a homogeneity of physical or cultural phenomena; an earth area recognized as an operational unit based on defined organizational criteria.

# G

**gated community** A restricted-access subdivision or neighborhood, often surrounded by a barrier, with entry permitted only for residents and their guests; usually totally planned in land use and design.

**gathering industries** Primary activities involving the harvesting of renewable natural resources of land or water; commercial gathering usually implies forestry and fishing industries.

**gender** The socially created, not biologically based, distinctions between femininity and masculinity.

**gender empowerment measure (GEM)** A statistic summarizing the extent of economic, political, and professional participation of women in the society of which they are members; a measure of relative gender equality.

**gene flow** The passage of genes characteristic of one breeding population into the gene pool of another by interbreeding.

**genetic drift** A chance modification of gene composition occurring in an isolated population and becoming accentuated through inbreeding.

**gentrification** The process by which middle- and high-income groups refurbish and rehabilitate housing in deteriorated inner-city areas, thereby displacing low-income populations.

**geodetic control data** The information specifying the horizontal and vertical positions of a place.

**geographic database** In cartography, a digital record of geographic information.

**geographic information system (GIS)** A configuration of computer hardware and software for assembling, storing, manipulating, analyzing, and displaying geographically referenced information.

**geometric boundary** (*syn:* artificial boundary) A boundary without obvious physical geographic basis; often a section of a parallel of latitude or a meridian of longitude.

**geomorphology** The scientific study of landform origins, characteristics, and evolutions and their processes.

**geothermal energy** Energy that is generated by harnessing the naturally occurring steam and hot water produced by contact with heated rocks in the earth's crust.

**gerrymandering** Dividing an area into voting districts in such a way as to give one political party an unfair advantage in elections, to fragment voting blocks, or to achieve other nondemocratic objectives.

**GIS** *See* geographic information system.

**glacial till** The deposits of rocks, silt, and sand left by a glacier after it has receded.

**glacial trough** A deep, U-shaped valley or trench formed by glacial erosion.

**glacier** A huge mass of slowly moving land ice.

**globalization** The increasing interconnection of all parts of the world as the full range of social, cultural, political, economic, and environmental processes and patterns of change becomes international in scale and effect.

**Global Positioning System (GPS)** A method of using satellite observations for the determination of extremely accurate locational information.

**global warming** A rise in surface temperatures on Earth, a process believed by some to be caused by human activities that increase the concentration of greenhouse gases in the atmosphere, magnifying the greenhouse effect.

**globe grid** *See* grid system.

**globe properties** The characteristics of the grid system of longitude and latitude on a globe.

**GPS** *See* Global Positioning System.

**gradational processes** The processes of weathering, gravity transfer, and erosion that are responsible for the reduction of the land surface.

**grade (of coal)** A classification of coals based on their content of waste materials.

**graphic scale** A graduated line included in a map legend by means of which distances on the map may be measured in terms of ground distances.

**great circle** A circle formed by the intersection of the surface of a globe with a plane passing through the center of the globe. The equator is a great circle; meridians are one-half a great circle.

**greenbelt** A ring of parks, farmland, or undeveloped land around a community.

**greenhouse effect** The heating of the earth's surface as shortwave solar energy passes through the atmosphere, which is transparent to it but opaque to reradiated longwave terrestrial energy. Also refers to increasing the opacity of the atmosphere through the addition of increased amounts of carbon dioxide, nitrous oxides, methane, and chlorofluorocarbons.

**greenhouse gases** Heat-trapping gases added to the atmosphere by human activities; carbon dioxide, chlorofluorocarbons, methane gas, and nitrous oxide.

**Green Revolution** The term suggesting the great increases in food production, primarily in subtropical areas, accomplished through the introduction of very high yielding grain crops, particularly wheat and rice.

**Greenwich mean time (GMT)** Local time at the prime meridian (zero degrees longitude), which passes through the observatory at Greenwich, England.

**grid system** The set of imaginary lines of latitude and longitude that intersect at right angles to form a system of reference for locating points on the surface of the earth.

**gross national income (GNI)** *See* gross national product (GNP).

**gross national product (GNP)** The total value of all goods and services produced by a country per year; also called gross national income (GNI).

**groundwater** Underground water that accumulates in aquifers below the water table in the pores and cracks of rock and soil.

# H

**half-life** The time required for one-half the atomic nuclei of an isotope to decay.

**hazardous waste** Discarded solid, liquid, or gaseous material that may pose a substantial threat to human health or the environment when it is improperly disposed of, stored, or transported.

**herbicide** A chemical that kills plants, especially weeds. *See also* biocide, pesticide.

**hierarchical diffusion** The process by which contacts between people and the resulting diffusion of things or ideas occurs first among those at the same level of a hierarchy and then among elements at a lower level of the hierarchy (e.g., small-town residents acquire ideas or articles after they are common in large cities).

**hierarchical migration** The tendency for individuals to move from small places to larger ones.

**hierarchy of central places** The steplike series of urban units in classes differentiated by both size and function.

**high-level waste** Nuclear waste that can remain radioactive for thousands of years, produced principally by the generation of nuclear power and the manufacture of nuclear weapons.

**hinterland** An outlying region that furnishes raw materials or agricultural products to the heartland; the market area or region served by a town or city.

**homeostatic plateau** The equilibrium level of population that can be supported adequately by available resources; equivalent to carrying capacity.

**humid continental climate** A climate of east coast and continental interiors of midlatitudes, displaying large annual temperature ranges resulting from cold winters and hot summers; precipitation at all seasons.

**humid subtropical climate** A climate of the east coast of continents in lower middle latitudes, characterized by hot summers with convectional precipitation and cool winters with cyclonic precipitation.

**humus** Dark brown or black decomposed organic matter in soils.

**hunting-gathering** An economic and social system based primarily or exclusively on the hunting of wild animals and the gathering of food, fiber, and other materials from uncultivated plants.

**hurricane** A severe tropical cyclone with winds exceeding 120 kilometers per hour (75 mph) originating in the tropical region of the Atlantic Ocean, Caribbean Sea, or Gulf of Mexico.

**hydrologic cycle** The system by which water is continuously circulated through the biosphere by evaporation, condensation, and precipitation.

**hydropower** The kinetic energy of moving water converted into electrical power by a power plant whose turbines are driven by flowing water.

**hydrosphere** All water at or near the earth's surface that is not chemically bound in rocks; includes the oceans, surface waters, groundwater, and water held in the atmosphere.

# I

**iconography** In political geography, the study of symbols that unite a country.

**ideological subsystem** The complex of ideas, beliefs, knowledge, and means of their communication that characterize a culture.

**igneous rocks** Rocks formed as molten earth materials cool and harden either above or below the earth's surface.

**incinerator** A facility designed to burn waste.

**inclination** The tilt of the earth's axis about $23^1/_2°$ away from the perpendicular.

**Industrial Revolution** The term applied to the rapid economic and social changes in agriculture and manufacturing that followed the introduction of the factory system to the textile industry of England in the last quarter of the 18th century.

**infant mortality rate** A refinement of the death rate to specify the ratio of deaths of infants age 1 year or less per 1000 live births.

**infrared** Electromagnetic radiation having wavelengths greater than those of visible light.

**infrastructure** The basic structure of services, installations, and facilities needed to support industrial, agricultural, and other economic development.

**innovation** Introduction into an area of new ideas, practices, or objects; an alteration of custom or culture that originates within the social group itself.

**insolation** The solar radiation received at the earth's surface.

**intensive agriculture** The application of large amounts of capital and/or labor per unit of cultivated land to increase output; may be part of either a commercial or a subsistence economy.

**intensive commercial agriculture** Intensive farming in a commercial economy; crops have high yields and high market value.

**intensive subsistence agriculture** Intensive farming in a subsistence economy; the cultivation of small landholdings through the expenditure of great amounts of labor.

**International Date Line** By international agreement, the designated line where each new day begins; generally following the 180th meridian.

**intrusive rock** Rock resulting from the hardening of magma beneath the earth's surface.

**irredentism** The desire of a state to gain or regain territory inhabited by people who have historic or cultural links to the country but who now live in a neighboring state.

**isochrone** A line connecting points that are equidistant in travel time from a common origin.

**isoline** A map line connecting points of equal value, such as a contour line or an isobar.

**isotropic plain** A hypothetical portion of the earth's surface assumed to be an unbounded, uniformly flat plain with uniform distribution of population, purchasing power, transport costs, accessibility, and the like.

# J

**J-curve** A curve shaped like the letter J, depicting exponential or geometric growth (1, 2, 4, 8, 16 . . .).

**jet stream** A meandering belt of strong winds in the upper atmosphere; significant because it guides the movement of weather systems.

# K

**karst topography** A limestone region marked by sinkholes, caverns, and underground streams.

**kerogen** A waxy, organic material occurring in oil shales that can be converted into crude oil by distillation.

**kinetic energy** The energy that results from the motion of a particle or body.

# L

**land breeze** Airflow from the land toward the sea, resulting from a nighttime pressure gradient that moves winds from the cooler land surface to the warmer sea surface.

**landform region** A large section of the earth's surface characterized by a great deal of homogeneity among types of landforms.

**landlocked state** A state that lacks a seacoast.

**Landsat satellite** One of a series of continuously orbiting satellites that carry scanning instruments to measure reflected light in both the visible and near-infrared portions of the spectrum.

**landscape** The appearance of an area and the items comprising that appearance. A distinction is often made between "physical landscapes" confined to landforms, natural vegetation, soils, etc., and "cultural landscapes" (q.v.).

**language** An organized system of speech by which people communicate with each other with mutual comprehension.

**language family** A group of languages thought to have descended from a single, common ancestral tongue.

**La Niña** The cool ebb in low-latitude Pacific Ocean surface temperatures that occurs between El Niño peaks of sea-surface warming.

**lapse rate** The rate of change of temperature with altitude in the troposphere; the average lapse rate is about 6.4°C per 1000 meters (3.5°F per 1000 ft).

**large-scale map** A representation of a small land area, usually with a representative fraction of 1:75,000 or less.

**latitude** The angular distance north or south of the equator, measured in degrees ranging from 0° (the equator) to 90° (the North and South Poles).

**lava** Molten material that has emerged onto the earth's surface.

**Law of the Sea Convention** A code of sea law approved by the United Nations in 1982 that authorizes, among other provisions, territorial waters extending 12 nautical miles from shore and 200-nautical-mile-wide exclusive economic zones; generally referred to as UNCLOS.

**leachate** The contaminated liquid discharged from a sanitary landfill to either surface or subsurface land or water.

**leaching** The downward movement of water through the soil layer, resulting in the removal of soluble minerals from the upper soil horizons.

**least-cost theory** (*syn:* Weberian analysis) The view that the optimum location of a manufacturing establishment is at the place where the costs of transport and labor and the advantages of agglomeration or dispersion are most favorable.

**levee** In agriculture, a continuous embankment surrounding areas to be flooded. *See also* natural levee.

**lingua franca** Any of the various auxiliary languages used as common tongues among people of an area where several languages are spoken.

**liquefied natural gas (LNG)** Methane gas that has been liquefied by refrigeration for storage or transportation.

**lithosphere** The outermost layer of the earth, composed of the crust and upper mantle.

**loam** Agriculturally productive soil containing roughly equal parts of sand, silt, and clay.

**locational tradition** One of the four traditions of geography; in this text, identified with economic, urban, and environmental geography.

**loess** A deposit of windblown silt.

**longitude** The angular distance east or west of the prime (zero) meridian, measured in degrees ranging from 0° to 180°.

**longshore current** A current that moves roughly parallel to the shore and transports the sand that forms beaches and sand spits.

**low-level waste** Hazardous material whose radioactivity will decay to safe levels in 100 years or less, produced principally by industries and nuclear power plants.

# M

**magma** Underground molten material.

**malnutrition** Food intake insufficient in quantity or deficient in quality to sustain life at optimal conditions of health.

**Malthus** Thomas R. Malthus (1766–1834), English economist, demographer, and cleric, who suggested that, unless checked by self-control, war, or natural disaster, population will inevitably increase faster than will the food supplies needed to sustain it.

**mantle** The layer of earth between the crust and the core.

**map projection** A method of transferring the grid system from the earth's curved surface to the flat surface of a map.

**map scale** *See* scale.

**marine west coast climate** A regional climate found on the west coast of continents in upper midlatitudes, rainy all seasons with relatively cool summers and relatively mild winters.

**mass movement** (*syn:* mass wasting) The downslope movement of earth materials due to gravity.

**mass wasting** *See* mass movement.

**material culture** The tangible, physical items produced and used by members of a specific culture group and reflective of their traditions, lifestyles, and technologies.

**maximum sustainable yield** The maximum rate at which a renewable resource can be exploited without impairing its ability to be renewed or replenished.

**mechanical weathering** The physical disintegration of earth materials, commonly by frost action, root action, or the development of salt crystals.

**Mediterranean climate** A climate of lower midlatitudes characterized by mild, wet winters and hot, dry, sunny summers.

**megalopolis** An extensive, heavily populated urban complex with contained open, nonurban land, created through the spread and merging of separate metropolitan areas; (*cap.*) the name applied to the continuous functionally urban area of the northeastern seaboard of the United States from Maine to Virginia.

**megawatt** A unit of power equal to 1 million watts (1000 kilowatts) of electricity.

**mental map** The maplike image of the world, country, region, or other area a person carries in his or her mind; includes knowledge of actual locations and spatial relationships and is colored by personal perceptions and preferences related to place.

**mentifacts** The central, enduring elements of a culture that express its values and beliefs, including language, religion, folklore, artistic traditions, and the like. Elements in the ideological subsystem of culture.

**Mercator projection** A true conformal cylindrical projection first published in 1569, useful for navigation.

**meridian** A north-south line of longitude; on the globe, all meridians are of equal length and converge at the poles.

**mesa** An extensive, flat-topped elevated tableland with horizontal strata, a resistant cap rock, and one or more steep sides; a large butte.

**metamorphic rocks** Rocks transformed from igneous and sedimentary rocks by earth forces that generate heat, pressure, or chemical reaction.

**metropolitan area** A large functional entity, perhaps containing several urbanized areas, discontinuously built up but operating as a coherent economic whole.

**migration** The permanent (or relatively permanent) relocation of an individual or a group to a new, usually distant, place of residence.

**migration field** An area that sends major migration flows to or receives major flows from a given place.

**mineral** A natural inorganic substance that has a definite chemical composition and characteristic crystal structure, hardness, and density.

**ministate** An imprecise term for a state or territory small in both population and area. An informal definition accepted by the United Nations suggests a maximum of 1 million people combined with a territory of less than 700 square kilometers (270 sq mi).

**monoculture** An agricultural system dominated by a single crop.

**monotheism** The belief that there is only one God.

**monsoon** A wind system that reverses direction seasonally, producing wet and dry seasons; used especially to describe the wind system of South, Southeast, and East Asia.

**moraine** Any of several types of landforms composed of debris transported and deposited by a glacier.

**mortality rate** *See* death rate.

**mountain breeze** The downward flow of heavy, cool air at night from mountainsides to lower valley locations.

**multiple-nuclei model** The idea that large cities develop by peripheral spread, not from one central business district but from several nodes of growth, each of specialized use.

**multiplier effect** The expected addition of nonbasic workers and dependents to a city's total employment and population that accompanies new basic sector employment.

# N

**nation** A culturally distinctive group of people occupying a particular region and bound together by a sense of unity arising from shared ethnicity, beliefs, and customs.

**nationalism** A sense of unity binding the people of a state together; devotion to the interests of a particular nation; an identification with the state and an acceptance of national goals.

**nation-state** A state whose territory is identical to that occupied by a particular nation.

**natural boundary** (*syn:* physical boundary) A boundary line based on recognizable physiographic features, such as mountains, rivers, or deserts.

**natural gas** A mixture of hydrocarbons and small quantities of nonhydrocarbons existing in a gaseous state or in solution with crude oil in natural reservoirs.

**natural increase** The growth of a population through excess of births over deaths, excluding the effects of immigration or emigration.

**natural landscape** The physical *environment* unaffected by human activities. The duration and near totality of human occupation of the earth's surface assure that little or no "natural landscape" so defined remains intact. Opposed to *cultural landscape.*

**natural levee** An embankment on the sides of a meandering river formed by deposition of silt during floods.

**natural resource** A physically occurring item that a population perceives to be necessary and useful to its maintenance and well-being.

**natural selection** The process of survival and reproductive success of individuals or groups best adjusted to their environment, leading to the perpetuation of those genetic qualities most suited to that environment.

**natural vegetation** The plant life that would exist in an area if humans did not interfere with its development.

**neo-Malthusianism** The advocacy of population control programs to preserve and improve general national prosperity and well-being.

**neritic zone** The relatively shallow part of the sea that lies above the continental shelf.

**net migration** The difference between in-migration and out-migration of an area.

**network city** One of two or more nearby cities, potentially complementary in function, that cooperate by developing transportation links and communications infrastructure.

**niche** The place an organism or a species occupies in an ecosystem.

**nomadic herding** The migratory but controlled movement of livestock solely dependent upon natural forage.

**nonbasic sector** Those economic activities of an urban unit that supply the resident population with goods and services.

**nonecumene** (*syn:* anecumene) The portion of the earth's surface that is uninhabited or only temporarily or intermittently inhabited. *See also* ecumene.

**nonfuel mineral resource** A mineral used for purposes other than providing a source of energy.

**nongovernmental organization (NGO)** A group of people acting outside government and major commercial agencies and advocating or lobbying for particular causes.

**nonmaterial culture** The oral traditions, songs, and stories of a culture group along with its beliefs and customary behaviors.

**nonpoint source of pollution** Pollution from a broad area, such as one of fertilizer or pesticide application, rather than from a discrete source.

**nonrenewable resource** A natural resource that is not replenished or replaced by natural processes or is used at a rate that exceeds its replacement rate.

**North and South Poles** The end points of the axis about which the earth spins.

**North Atlantic drift** The massive movement of warm water in the Atlantic Ocean from the Caribbean Sea and Gulf of Mexico in a northeasterly direction to the British Isles and the Scandinavian peninsula.

**nuclear fission** The controlled splitting of an atom to release energy.

**nuclear fusion** The combining of two atoms of deuterium into a single atom of helium in order to release energy.

**nuclear power** Electricity generated by a power plant whose turbines are driven by steam produced by the fissioning of nuclear fuel in a reactor.

**nutrient** A mineral or another element an organism requires for normal growth and development.

# O

**oil shale** Sedimentary rock containing solid organic material (kerogen) that can be extracted and converted into a crude oil by distillation.

**ore** A mineral deposit that can be extracted at a profit.

**organic** Derived from living organisms; plant or animal life.

**Organization of Petroleum Exporting Countries (OPEC)** An international cartel composed of 11 countries that aims at pursuing common oil marketing and pricing policies.

**orographic precipitation** The rain or snow caused when warm, moisture-laden air is forced to rise over hills or mountains in its path and is thereby cooled.

**orthophotomap** A multicolored, distortion-free aerial photographic image to which certain supplementary information has been added.

**outsourcing** (1) Producing abroad parts or products for domestic use or sale; (2) subcontracting production or services rather than performing those activities "in house."

**outwash plain** A gently sloping area in front of a glacier composed of neatly stratified glacial till carried out of the glacier by meltwater streams.

**overburden** Soil and rock of little or no value that overlies a deposit of economic value, such as coal.

**overpopulation** A value judgment that the resources of an area are insufficient to sustain adequately its present population numbers.

**oxbow lake** A crescent-shaped lake contained in an abandoned meander of a river.

**ozone** A gas molecule consisting of three atoms of oxygen ($O_3$) formed when diatomic oxygen ($O_2$) is exposed to ultraviolet radiation. In the lower atmosphere, it constitutes a damaging component of photochemical smog; in the upper atmosphere, it forms a normally continuous, thin layer that blocks ultraviolet light.

**ozone layer** A layer of ozone in the high atmosphere that protects life on Earth by absorbing ultraviolet radiation from the sun.

# P

**parallel of latitude** An east-west line indicating the distance north or south of the equator.

**PCBs** Polychlorinated biphenyls; compounds containing chlorine that can be biologically magnified in the food chain.

**peak value intersection** The most accessible and costly parcel of land in the central business district and, therefore, in the entire urbanized area.

**perforated state** A state whose territory is interrupted ("perforated") by a separate, independent state totally contained within its borders.

**permafrost** Permanently frozen subsoil.

**perpetual resource** A resource that comes from an inexhaustible source, such as the sun, wind, and tides.

**pesticide** A chemical that kills insects, rodents, fungi, weeds, and other pests. *See also* biocide, herbicide.

**petroleum** Oil and oil products in all forms, such as crude oil and unfinished oils.

**pH factor** The measure of the acidity/alkalinity of soil or water, on a scale of 0 to 14, rising with increasing alkalinity.

**photochemical smog** A form of air pollution produced by the interaction of hydrocarbons and oxides of nitrogen in the presence of sunlight.

**photovoltaic (PV) cell** A device that converts solar energy directly into electrical energy. *See also* solar power.

**physical boundary** *See* natural boundary.

**physiological density** The number of persons per unit area of agricultural land. *See also* population density.

**pidgin** An auxiliary language derived, with reduction of vocabulary and simplification of structure, from other languages. Not a native tongue, it is employed to provide a mutually intelligible vehicle for limited transactions of trade or administration.

**pixel** An extremely small sensed unit of a digital image.

**place utility** (1) The perceived attractiveness of a place in its social, economic, or environmental attributes; (2) the value imparted to goods or services by tertiary activities that provide things needed in specific markets.

**planar projection** A map projection based on the projection of the globe grid onto a plane as the presumed developable surface.

**planned economy** A system of the production of goods and services, usually consumed or distributed by a governmental agency, in quantities and at prices determined by governmental programs.

**plantation** A large agricultural holding, frequently foreign-owned, devoted to the production of a single export crop.

**plate tectonics** The theory that the lithosphere is divided into plates that slide or drift very slowly over the asthenosphere.

**playa** A temporary lake or lake bed found in a desert environment.

**Pleistocene** The geologic epoch dating from 2 million to about 10,000 years ago during which four stages of continental glaciation occurred.

**point source of pollution** Pollution originating from a discrete source, such as a smokestack or the outflow from a pipe.

**political geography** A branch of human geography concerned with the spatial analysis of political phenomena.

**pollution** The presence in the biosphere of substances that, because of their quantity, chemical nature, or temperature, have a negative impact on the ecosystem or that cannot be readily disposed of by natural recycling processes.

**polychlorinated biphenyls** *See* PCBs.

**polytheism** The belief in or worship of many gods.

**popular culture** The constantly changing mix of material and nonmaterial elements available through mass production and the mass media to an urbanized, heterogeneous, nontraditional society.

**population density** (*syn:* crude density) A measurement of the numbers of persons per unit area of land within predetermined limits, usually political or census boundaries. *See also* physiological density.

**population geography** The branch of human geography dealing with the number, composition, and distribution of humans in relation to variations in earth-space conditions.

**population momentum** (*syn:* demographic momentum) The tendency for population growth to continue despite stringent family planning programs because of a relatively high concentration of people in the childbearing years.

**population projection** A report of future size, age, and sex composition of a population based on assumptions applied to current data.

**population pyramid** A graphic depiction of the age and sex composition of a (usually national) population.

**possibilism** The philosophical viewpoint that the physical environment offers human beings a set of opportunities from which (within limits) people may choose according to their cultural needs and technological awareness.

**potential energy** The energy stored in a particle or body.

**potentially renewable resource** A resource that can last indefinitely if its natural replacement rate is not exceeded; examples include forests, groundwater, and soil.

**precipitation** All moisture, solid and liquid, that falls to the earth's surface from the atmosphere.

**pressure gradient force** Differences in air pressure between areas that induce air to flow from areas of high pressure to areas of low pressure.

**primary activity** The part of the economy involved in making natural resources available for use or further processing; includes mining, agriculture, forestry, fishing or hunting, grazing.

**primary air pollutant** Material, such as particulates or sulfur dioxide, that is emitted directly into the atmosphere in sufficient quantities to adversely affect human health or the environment.

**primate city** A country's leading city, much larger and functionally more complex than any other; usually the capital city and a center of wealth and power.

**prime meridian** An imaginary line passing through the Royal Observatory at Greenwich, England, serving by agreement as the zero degree line of longitude.

**projection** An estimate of future conditions based on current trends. *See also* map projection.

**prorupt state** A state of basically compact form that has one or more narrow extensions of territory.

**proto-language** An assumed, reconstructed, or recorded language ancestral to one or more contemporary languages or dialects.

**proved (usable) reserves** The portion of a natural resource that has been identified and can be extracted profitably with current technology.

**psychological distance** The way an individual perceives distance.

**pull factor** A characteristic of a region that acts as an attractive force, drawing migrants from other regions.

**purchasing power parity (PPP)** A monetary measurement that takes account of what money actually buys in each country.

**push factor** A characteristic of a region that contributes to the dissatisfaction of residents and impels their migration.

# Q

**quaternary activity** That employment concerned with research, with the gathering or disseminating of information, and with administration, including administration of the other economic activity levels.

**quinary activity** A sometimes separately recognized subsection of tertiary activity management functions involving highest-level decision making in all types of large organizations. Also deemed the most advanced form of the quaternary subsector.

# R

**race** A subset of human population whose members share certain distinctive, inherited biological characteristics.

**radar** A device for detecting distant objects by analysis of very high frequency radio waves beamed at and reflected from their surfaces.

**rank (of coal)** A classification of coals based on their age and energy content; those of higher rank are more mature and richer in energy.

**rank-size rule** An observed regularity in the city-size distribution of some countries. In a rank-size hierarchy, the population of any given town will be inversely proportional to its rank in the hierarchy; that is, the *n*th-ranked city will be $1/n$ the size of the largest city.

**rate** The frequency of occurrence of an event during a specified time period.

**rate of natural increase** The birth rate minus the death rate, suggesting the annual rate of population growth without considering net migration.

**recycling** The reuse of disposed materials after they have passed through some form of treatment (e.g., melting down glass bottles to produce new bottles).

**redistricting** The drawing of new electoral district boundary lines in response to changing patterns of population or changing legal requirements.

**reflection** The process of returning to outer space some of the earth's received insolation.

**region** In geography, the term applied to an area of the earth that displays a distinctive grouping of physical or cultural phenomena or is functionally united as a single organizational unit.

**regional autonomy** A measure of self-governance for a subdivision of a country.

**regional concept** The view that physical and cultural phenomena on the surface of the earth are rationally arranged by complex but comprehensible spatial processes.

**regionalism** In political geography, minority group identification with a particular region of a state rather than with the state as a whole.

**relative direction** (*syn:* relational direction) A culturally based locational reference, such as the Far West, the Old South, or the Middle East.

**relative distance** A transformation of *absolute distance* into such relative measures as time or monetary costs. Such measures yield different explanations of human spatial behavior than do linear distances alone. Distances between places are constant by absolute terms, but relative distances may vary with improvements in transportation or communication technology or with different psychological perceptions of space.

**relative humidity** A measure of the moisture content of the air, expressed as the amount of water vapor present relative to the maximum that can exist at the current temperature.

**relative location** The position of a place or an activity in relation to other places or activities.

**relic boundary** A former boundary line that is still discernible and marked by a cultural landscape feature.

**religion** A value system that involves formal or informal worship and faith in the sacred and divine.

**relocation diffusion** The transfer of ideas, behaviors, or articles from one place to another through the migration of those possessing the feature transported; also, spatial relocation in which a phenomenon leaves an area of origin as it is transported to a new location.

**remote sensing** Any of several techniques of obtaining images of an area without having the sensor in direct physical contact with it, as by air photography or satellite sensors.

**renewable resource** A naturally occurring material that is potentially inexhaustible, either because it flows continuously (such as solar radiation or wind) or is renewed within a short period of time (such as biomass). *See also* sustained yield.

**replacement level** The number of children per family just sufficient to keep total population constant. Depending on mortality conditions, replacement level is usually calculated to be between 2.1 and 2.5 children.

**representative fraction (RF)** The scale of a map expressed as a ratio of a unit of distance on the map to distance measured in the same unit on the ground (e.g., 1:250,000).

**reradiation** A process by which the earth returns solar energy to space; some of the shortwave solar energy that is absorbed into the land and water is returned to the atmosphere in the form of longwave terrestrial radiation.

**resource** *See* natural resource.

**return migration** The return of migrants to the region from which they had earlier emigrated.

**rhumb line** A line of constant compass bearing; it cuts all meridians at the same angle.

**Richter scale** A logarithmic scale used to express the magnitude of an earthquake.

# S

**Sahel** The semiarid zone between the Sahara Desert and the savanna area to the south in West Africa; district of recurring drought, famine, and environmental degradation.

**salinization** The concentration of salts in the topsoil as a result of the evaporation of surface water; occurs in poorly drained soils in dry climates, often as a result of improper irrigation.

**sandbar** An off-shore shoal of sand created by the backwash of waves.

**sanitary landfill** The disposal of solid wastes by spreading them in layers covered with enough soil or ashes to control odors, rats, and flies.

**savanna** A tropical grassland characterized by widely dispersed trees and experiencing pronounced yearly wet and dry seasons.

**scale** In cartography, the ratio between length or size of an area on a map and the actual length or size of that same area on the earth's surface; map scale may be represented verbally, graphically, or as a fraction. In more general terms, *scale* refers to the size of the area studied, from local to global.

**S-curve** The horizontal bending, or leveling, of an exponential J-curve.

**sea breeze** Airflow from the sea toward the land, resulting from a daytime pressure gradient that moves winds from the cooler sea surface onto the warmer land surface.

**secondary activity** The part of the economy involved in the processing of raw materials derived from primary activities; includes manufacturing, construction, power generation.

**sector model** A description of urban land uses as wedge-shaped sectors radiating outward from the central business district along transportation corridors. The radial access routes attract particular uses to certain sectors.

**secularism** An indifference to or rejection of religion and religious belief.

**sedimentary rocks** Rocks formed by the accumulation of particles of gravel, sand, silt, and clay that were eroded from already existing rocks.

**seismic waves** Vibrations within the earth set off by earthquakes.

**self-determination** The concept that nationalities have the right to govern themselves in their own state or territory, a right to self-rule.

**shaded relief** A method of representing the three-dimensional quality of an area by use of continuous graded tone to simulate the appearance of sunlight and shadows.

**shale oil** The crude oil resulting from the distillation of kerogen in oil shales.

**shamanism** A form of tribal religion based on belief in a hidden world of gods, ancestral spirits, and demons responsive only to a shaman, or interceding priest.

**shifting cultivation** (*syn:* slash-and-burn agriculture, swidden agriculture) Crop production of forest clearings kept in cultivation until their quickly declining fertility is lost. Cleared plots are then abandoned and new sites are prepared.

**sinkhole** A deep surface depression formed when ground collapses into a subterranean cavern.

**site** The place where something is located; the immediate surroundings and their attributes.

**situation** The location of something in relation to the physical and human characteristics of a larger region.

**slash-and-burn agriculture** *See* shifting cultivation.

**small circle** The line created by the intersection of a spherical surface with a plane that does not pass through its center.

**small-scale map** A representation of a large land area on which small features (e.g., highways, buildings) cannot be shown true to scale.

**sociofacts** The institutions and links between individuals and groups that unite a culture, including family structure and political, educational, and religious institutions; components of the sociological subsystem of culture.

**sociological subsystem** The totality of expected and accepted patterns of interpersonal relations common to a culture or subculture.

**soil** The complex mixture of loose material, including minerals, organic and inorganic compounds, living organisms, air, and water, found at the earth's surface and capable of supporting plant life.

**soil depletion** The loss of some or all of the vital nutrients from soil.

**soil erosion** The wearing away and removal of soil particles from exposed surfaces by agents such as moving water, wind, or ice.

**soil horizon** A layer of soil distinguished from other soil zones by color, texture, and other characteristics resulting from soil-forming processes.

**soil order** A general grouping of soils with broadly similar composition, horizons, and weathering and leaching processes.

**soil profile** A vertical cross section of soil horizons.

**soil properties** The characteristics that distinguish types of soil from one another, including organic and inorganic matter, texture, structure, and nutrients.

**solar energy** Radiation from the sun, which is transformed into heat primarily at the earth's surface and secondarily in the atmosphere.

**solar power** The radiant energy generated by the sun; sun's energy captured and directly converted for human use. *See also* photovoltaic cell.

**solid waste** The unwanted materials generated in production or consumption processes that are solid rather than liquid or gaseous in form.

**source region** In climatology, a large area of uniform surface and relatively consistent temperatures where an air mass forms.

**southern oscillation** The atmospheric conditions occurring periodically near Australia that create the El Niño condition off the coast of South America.

**spatial diffusion** The outward spread of a substance, a concept, a practice or a population from its point of origin to other areas.

**spatial distribution** The arrangement of things on the earth's surface.

**spatial interaction** The movement (e.g., of people, goods, information) between different places; an indication of interdependence between areas.

**spatial margin of profitability** The set of points delimiting the area within which a firm's profitable operation is possible.

**special-purpose map** *See* thematic map.

**spring wheat** Wheat sown in spring for ripening during the summer or autumn.

**stage in life** Membership in a specific age group.

**standard language** A language substantially uniform with respect to spelling, grammar, pronunciation, and vocabulary and representing the approved community norm of the tongue.

**standard parallel** The tangent circle, usually a parallel of latitude, in a conic projection; along the standard line, the scale is as stated on the map.

**state** (*syn:* country) An independent political unit occupying a defined, permanently populated territory and having full sovereign control over its internal and foreign affairs.

**step (stepwise) migration** A *migration* in which an eventual long-distance relocation is undertaken in stages as, for example, from farm to village to small town to city.

**steppe** The name applied to treeless midlatitude grasslands.

**stratosphere** The layer of the atmosphere that lies above the troposphere and extends outward to about 56 kilometers (35 mi).

**stream load** The eroded material carried by a stream in one of three ways, depending on the size and composition of the particles: (1) in dissolved form, (2) suspended by water, or (3) rolled along the stream bed.

**subduction** The process by which one lithospheric plate is forced down into the asthenosphere as a result of a collision with another plate.

**subnationalism** The feeling that one owes primary allegiance to a traditional group or nation rather than to the state.

**subsequent boundary** A boundary line that is established after the area in question has been settled and that considers the cultural characteristics of the bounded area.

**subsidence** The settling or sinking of a portion of the land surface, sometimes as a result of the extraction of fluids, such as oil or water, from underground deposits.

**subsistence agriculture** Any of several farm economies in which most crops are grown for food, nearly exclusively for local consumption.

**subsistence economy** A system in which goods and services are created for the use of producers or their immediate families. Market exchanges are limited and of minor importance.

**substitution principle** In industry, the tendency to substitute one factor of production for another in order to achieve optimum plant location and profitability.

**suburb** A functionally specialized segment of a large urban complex located outside the boundaries of the central city.

**succession** A natural process in which an orderly sequence of plant species will occupy a newly established landform or a recently altered landscape.

**superimposed boundary** A boundary line placed over, and ignoring, an existing cultural pattern.

**supranationalism** The acceptance of the interests of more than one state, expressed as associations of states created for mutual benefit and to achieve shared objectives.

**surface water** Water that is on the earth's surface, such as in rivers, streams, reservoirs, lakes, and ponds.

**sustained yield** The practice of balancing harvesting with growth of new stocks in order to avoid depletion of the resource and ensure a perpetual supply.

**swidden agriculture** *See* shifting cultivation.

**syncretism** The development of a new form of, for example, religion or music, through the fusion of distinctive parental elements.

**syntax** The way words are put together in phrases and sentences.

**systems analysis** An approach to the study of large systems through (1) segregation of the entire system into its component parts, (2) investigation of the interactions between system elements, and (3) study of inputs, outputs, flows, interactions, and boundaries within the system.

# T

**talus slope** A landform composed of rock particles that have accumulated at the base of a cliff, hill, or mountain.

**tar sand** Sand and sandstone impregnated with heavy oil.

**technological subsystem** The complex of material objects together with the techniques of their use by means of which people carry out their productive activities.

**technology** An integrated system of knowledge and skills developed within a culture to carry out successfully purposeful and productive tasks.

**tectonic forces** The processes that shape and reshape the earth's crust, the two main types being diastrophic and volcanic.

**temperature inversion** The condition caused by rapid reradiation in which air at lower altitudes is cooler than air aloft.

**territoriality** The persistent attachment of most animals to a specific area; the behavior associated with the defense of the home territory.

**territorial production complex** In the economic planning of the former Soviet Union, a design for large regional industrial, mining, and agricultural development leading to regional self-sufficiency and the creation of specialized production for a larger national market.

**terrorism** The calculated use of violence against civilians and other symbolic targets in order to publicize a cause or to diminish people's support for a leader, a government, a policy, or a way of life that the perpetrators of violence find objectionable.

**tertiary activity** The part of the economy that fulfills the exchange function and that provides market availability of commodities; includes wholesale and retail trade and associated transportation, government, and information services.

**thematic map** (*syn:* special-purpose map) A map that shows a specific spatial distribution or category of data.

**thermal pollution** The introduction of heated water into the environment, with consequent adverse effects on aquatic life.

**thermal scanner** A remote sensing device that detects the energy (heat) radiated by objects on Earth.

**Third World** Originally (in the 1950s), designating countries uncommitted to either the "First World" Western capitalist bloc or the Eastern "Second World" communist bloc; subsequently, countries considered not yet fully developed or in a state of underdevelopment in economic and social terms.

**threatened species** A species that has declined significantly in total numbers and may be on the verge of extinction; an endangered or vulnerable species.

**threshold** In economic geography, the minimum market needed to support the supply of a product or service.

**topographic map** A map that portrays the shape and elevation of the terrain, often in great detail.

**toponym** A place-name.

**toponymy** The place-names of a region or, especially, the study of place-names.

**tornado** A small, violent storm characterized by a funnel-shaped cloud of whirling winds that can form beneath a cumulonimbus cloud in proximity to a cold front and that moves at speeds as high as 480 kilometers per hour (300 mph).

**total fertility rate (TFR)** The average number of children that would be born to each woman if, during her childbearing years, she bore children at the current year's rate for women that age.

**town** A nucleated settlement that contains a central business district but that is smaller and less functionally complex than a city.

**traditional religion** *See* tribal religion.

**tragedy of the commons** The observation that in the absence of collective control over the use of a resource available to all, it is to the advantage of all users to maximize their separate shares even though their collective pressures may diminish total yield or destroy the resource altogether.

**transform fault** A break in rocks that occurs when one lithospheric plate slips past another in a horizontal motion.

**transnational corporation (TNC)** A large business organization operating in at least two separate national economies.

**tribal religion** (*syn:* traditional religion) An ethnic religion specific to a small, localized, preindustrial culture group.

**tropical rain forest** The tree cover composed of tall, high-crowned evergreen deciduous species, associated with the continuously wet tropical lowlands.

**tropical rain forest climate** The continuously warm, frost-free climate of tropical and equatorial lowlands, with abundant moisture year-round.

**troposphere** The atmospheric layer closest to the earth, extending outward about 11 to 13 kilometers (7 to 8 mi) at the poles to about 26 kilometers (16 mi) at the equator.

**truck farming** The intensive production of fruits and vegetables for market rather than for processing or canning.

**tsunami** Seismic sea waves generated by an earthquake or a volcanic eruption.

**tundra** The treeless area lying between the tree line of Arctic regions and the permanently ice-covered zone.

**typhoon** A hurricane occurring in the western Pacific Ocean region.

# U

**ubiquitous industry** A market-oriented industry whose establishments are distributed in direct proportion to the distribution of population (market).

**underpopulation** A value statement reflecting the view that an area has too few people in relation to its resources and population-supporting capacity.

**unitary state** A state in which the central government dictates the degree of local or regional autonomy and the nature of local governmental units; a country with few cultural conflicts and a strong sense of national identity.

**United Nations Convention on the Law of the Sea (UNCLOS)** *See* Law of the Sea Convention.

**universalizing religion** A religion that claims global truth and applicability and seeks the conversion of all humankind.

**urban hierarchy** The steplike series of urban units (e.g., hamlets, villages, towns, cities, metropolises) in classes differentiated by size and function.

**urban influence zone** An area outside of a city that is nevertheless affected by the city.

**urbanization** The transformation of a population from rural to urban status; the process of city formation and expansion.

**urbanized area** A continuously built-up urban landscape defined by building and population densities with no reference to the political boundaries of the city; it may contain a central city and many contiguous towns, cities, suburbs, and unincorporated areas.

**usable reserves** *See* proved reserves.

# V

**valley breeze** The flow of air up mountain slopes during the day.

**value-by-area map** *See* area cartogram.

**variable costs** In economic geography, the costs of production inputs that change as the level of production changes. They differ from the costs incurred by agricultural or industrial firms that are fixed and do not change as the amount of production changes.

**verbal scale** A statement of the relationship between units of measure on a map and distance on the ground, as "1 inch represents 1 mile."

**vernacular** (1) The nonstandard indigenous language or dialect of a locality; (2) of or related to indigenous arts and architecture, such as a vernacular house; (3) of or related to the perceptions and understandings of the general population, such as a vernacular region.

**volcanism** The earth force that transports subsurface materials (often heated, sometimes molten) to or toward the surface of the earth.

**von Thünen model** The model developed by Johann H. von Thünen (1783–1850) to explain the forces that control the prices of agricultural commodities and how those variable prices affect patterns of agricultural land utilization.

**von Thünen rings** The concentric zonal pattern of agricultural land use around a single market center proposed in the von Thünen model.

**vulnerable species** Species whose numbers have been so reduced that they could become threatened or endangered.

# W

**warping** The bowing of a large region of the earth's surface due to the movement of continents or the melting of continental glaciers.

**wash** A dry, braided channel in the desert that remains after the rush of rainfall runoff water.

**water table** The upper limit of the saturated zone and therefore of groundwater; the top of the water within an aquifer.

**weather** The state of the atmosphere at a given time and place.

**weathering** The mechanical and chemical processes that fragment and decompose rock materials.

**Weberian analysis** *See* least-cost theory.

**Weber model** The analytical model devised by Alfred Weber (1868–1958) to explain the principles governing the optimum location of industrial establishments.

**wetland** A vegetated inland or coastal area that is either occasionally or permanently covered by standing water or saturated with moisture.

**wind farm** A cluster of wind-powered turbines producing commercial electricity.

**wind power** The kinetic energy of wind converted into mechanical energy by wind turbines that drive generators to produce electricity.

**winter wheat** Wheat planted in autumn for early summer harvesting.

**world city** One of a small number of interconnected, internationally dominant centers (e.g., New York, London, Tokyo) that together control the global systems of finance and commerce.

# Z

**zero population growth (ZPG)** A situation in which a population is not changing in size from year to year, as a result of the combination of births, deaths, and migration.

**zoning** Designating by ordinance areas in a municipality for particular types of land use.

# Index

Page numbers in **boldface** indicate glossary terms. Page numbers followed by italic *f* and *t* indicate figures and tables, respectively.

## A


A climates. *See* Tropical climates
Aachen, Germany, boundaries of, 471*f*
AAG. *See* Association of American Geographers
Aberfan, Wales, slag heap disaster, 450
Abortion
  in Cairo Plan, 202
  and population control, 180, 191, 202
Absolute direction, **10**
Absolute distance, **11**
Absolute location, **9,** 307
Accelerated eutrophication, 439
Accessibility, **14**
Acculturation, **233**–235, 244–245
Acid precipitation, **444**–445, 444*f*
  deforestation and, 169
  effects of, 445, 445*f*
  formation of, 443–444, 445*f*
  sources and problem areas, 140, 444, 444*f*
  and water pollution, 441
Acid rain. *See* Acid precipitation
Acidity
  of acid rain, 444
  of soil, 110
Activity space, **271**–275, 274*f*
  factors affecting, 274–275
  travel to work, 274, 274*f*, 275*f*, 278, 280
Adaptation, **259**
Administrative units, in local and regional government, 330, 333–335, 334*f*
Admixture, 259
Advertising, and spread of innovation, 276–277, 277*f*
Aerial photography, 42
  infrared, 42, 44*f*
Affirmative action, 289
Afghanistan
  ethnicity in, 258*f*
  Islamic fundamentalism in, 319
  Mendawi bazaar, 266*f*
  migration from, 200, 281
  population data for, 500*t*
  as terrorist harbor, 323
Africa
  agriculture in, 231*f*, 348*f*, 349, 349*f*, 351, 352, 353, 354, 355
  biocide use in, 456
  birth and fertility rates in, 182, 187, 496*t*–498*t*
  cities in, 191, 402, 424–426
  coal reserves, 142*f*
  colonialism in, 243, 246*f*, 252, 303, 304*f*, 311
  as culture hearth, 230, 231*f*
  as culture realm, 221*f*
  death rates in, 186, 496*t*–498*t*
  desertification in, 160, 161*f*
  economies in, 345
  endangered animal species in, 453, 453*f*
  endangered plant species in, 452
  energy in, 149
  food production in, 206
  forced migration from, 281
  habitat destruction in, 452
  HIV/AIDS in, 189–190, 194, 496*t*–498*t*
  hunting, illegal, in, 453
  hydropower in, 152*f*
  immigration from, 201*f*
  industry in, 375*f*, 376, 377*f*
  infant mortality rates, 188, 496*t*–498*t*
  language regions of, 39*f*
  languages in, 236, 236*f*, 237, 240*f*, 243, 246*f*
  life expectancy and, 496*t*–498*t*
  maternal mortality rates in, 189
  migration from, 200
  mineral resources, 157, 158
  natural gas reserves, 145*f*
  natural gas trade flows, 144*f*
  oil reserves, 139*f*
  oil shale deposits, 145*f*
  oil trade flows, 408*f*
  place names in, 244
  plateau region, 81
  population
    control of, 212
    data on, 496*t*–498*t*
    density of, 208*f*
    growth of, 183*f*, 195*f*
    projections for, 496*t*–498*t*
  rain forests in, 114, 168
  refugee flows in, 281, 283*f*
  religion in, 253, 255*f*
  rift valleys in, 64*f*
  savannas in, 116
  Sub-Saharan
    birth and fertility rates in, 184, 186*f*, 187, 199, 496*t*
    cities in, 191, 402
    commodities trade and, 366
    disease in, 189–190
    food production in, 206
    HIV/AIDS in, 189–190, 194, 496*t*
    maternal mortality rates in, 189, 189*f*
    population, 190
      data on, 496*t*
      growth of, 181*f*
    population pyramid, 191, 192*f*
    trade, in services, 382*t*, 383
    women in, 260, 261*f*
  toxic waste exports to, 462
  transnational corporations and, 375
  tribal *vs.* national boundaries in, 303, 304*f*, 311
  urbanization in, 391, 394*t*
  water pollution in, 439
  women in, 191, 260, 261, 261*f*, 354
African Americans
  birth and fertility rates, 186
  Black English (Ebonics), 239
  defacto segregation of, 411, 412*f*
  female, bone cancer rates, 17*f*
  and racial gerrymandering, 331, 332–333, 332*f*
Afrikaans, 242
Afro-Asiatic language family, 238*f*–239*f*, 240*f*
Age. *See also* Population pyramids
  and mental maps, 272, 272*f*
  migration and, 284, 286*f*
  of world population, 192*f*, 212–213, 213*f*
    by region and nation, 496*t*–503*t*
Age of Degenerative and Human-Origin Diseases, 197
Age of Pestilence and Famine, 197
Agglomeration, **368**
  high-tech industry and, 378
Agglomeration economies, **368,** 370, 371*f*
Agribusiness, 354
Agricultural areas, towns in, 404–405, 404*f*
Agricultural density of population, **205,** 205*t*
Agricultural Revolution. *See also* Green Revolution
  cultural impact of, 227, 230, 232*f*, 260
  diffusion and, 275–276
  population and, 181, 211*f*, 220*f*
Agricultural societies
  classification of, 347–348
  development of, 137
  gender roles in, 260
Agriculture. *See also* Carrying capacity, of land; Food supply; *specific nations and regions*
  acculturation and, 234
  advanced (modern), 346–348
  air pollution and, 443*t*
  biocides and, 439
  biotechnology and, 353
  commercial, 353–361
    extensive, 356, **357**–359
    farm location, 355–357, 357*f*
    intensive, 356–**357**
    production controls in, 354–355
    special crops, 359–360, 359*f*
    supply and demand in, 355, 356*f*
  contour plowing, 162, 162*f*
  crops
    rain forest as source of, 169
    special, 359–360, 359*f*
  and culture, 219*f*
  definition of, 346
  deforestation and, 159, 159*f*
  desertification and, **159**–160
  in developed countries, 223, 224*f*
  in developing countries, 223, 224*f*
    percent of population employed in, 223, 224*f*, 346
  environmental impact of, 14, 432, 439–440, 440*f*, 452
  erosion and, 159–163, 223
  expansion of yields, 351–353
  fertilizers, and water pollution, 439, 440*f*
  grain production, global, 352
  Green Revolution, 212, 352–**353**
    women and, 353, 354–355
  history of, 230, 231*f*
  in hot deserts, 118
  in humid continental climate, 120
  in humid subtropical climate, 120
  intermediate, 346–347
  irrigation
    in Australia, 434*f*
    drip, 434*f*
    Green Revolution and, 353
    in plantation agriculture, 360
    and salinization of soils, 163
    water use in, 434, 434*f*, 435
  land used for. *See also* Soil(s)
    global distribution of, 346, 347*f*
    loss of, 159–163, 223, 351–352
    maintaining, 162, 162*f*
    population density and, 205
    von Thünen model of, 355–357, 357*f*
  livestock ranching, 357, 359, 359*f*
  livestock-grain farming, 357, 358*f*
  and location, 307
  location model for, 355–357, 357*f*
  loess deposits and, 81, 81*f*
  losses to insects and pests, 456
  marine west coast climate, 122
  Mediterranean, 231*f*, 359–360, 359*f*
  ozone layer depletion and, 447
  in planned economies, 360–361
  plantation, 359–**360,** 359*f*, 360*f*
  as pollution source, 440*f*
  population growth and, 159, 162, 181, 197, 211, 211*f*, 351
  population percentage engaged in, 223, 224*f*
  as primary economic activity, 345–361
  regions, U.S., 357, 358, 358*f*, 359
  religion and, 250
  salinization of soils, **163,** 163*f*
  in savannas, 116
  slash-and-burn, 349–350, 349*f*
  soil pH and, 110
  soil productivity, maintaining, 162, 162*f*
  soil properties and, 109–110
  state size and, 303
  on steppes, 118
  subsistence, 224*f*, 346–351, 348*f*
    decline in, 353
    extensive, **348**–349
    intensive, **348,** 348*f*, 350–351, 351*f*, 358*f*
    nomadic herding, **348**–349, 348*f*
    in North America, 358*f*
    shifting cultivation, 348*f*, 349–350, 349*f*
    swidden, 349–350, 349*f*
  technology and, 224*f*, 347–348, 353
  terrace, 2*f*, 162, 204*f*, 450
  trade barriers and, 366
  traditional (intermediate), 346–348
  in tropical rain forest, 114
  urban, 351
  urbanization and, 207
  in Washington State, 468*f*
  water pollution and, 439–440, 440*f*
  water use in, 434, 434*f*, 435
  wheat farming, large-scale, 357–359, 358*f*
  women and, 353, 354–355
Agriculture Improvement and Reform Act of 1996, 162
A-horizon, 108
AIDS. *See* HIV/AIDS
Air. *See also* Wind
  global circulation pattern, 96–97, 98*f*
  moisture in, 99–104, 99*f*. *See also* Precipitation
    condensation of, 100
    and dew point, **101**
    hydrologic cycle, 434*f*
  relative humidity of, 99, 99*f*, **101**
  supersaturated, 100
  water-carrying capacity of, 99, 99*f*
  weight of, 93–94
Air masses, **102**
  as regions, 474–475, 475*f*
  source regions for, **102,** 103*f*, 474–475, 475*f*
    for Americas, 103*f*, 475*f*

Air pollution, 442–449
and acid rain. *See* Acid precipitation
in Cairo, Egypt, 390
control of, 448–449
deforestation and, 168–169
Donora Tragedy, 95
factors affecting, 443
geothermal power and, 153*f*
global warming and, 126
and greenhouse effect, **126**–128, 127*f*, 128*f*
health effects of, 443
incineration of solid wastes and, 458
in Los Angeles, 92, 95*f*, 443, 446, 446*f*
Mexico City and, 443, 446
mining and, 450
natural, 442
and ozone layer depletion, 446–448, 447*f*, 448*f*
photochemical smog, **446,** 446*f*, 447*f*
formation of, 443
health effects of, 446
temperature inversion and, 92, 94*f*, 95, 95*f*, 96, 443, 446
rain forests and, 168, 170*f*
solid waste disposal and, 458–459
sources of, 442–443, 443*f*, 443*t*, 444, 446
temperature inversions and, 92, 94*f*, 95, 95*f*, 96, 443, 446
valleys and, 96
vegetation and, 446
wind currents and, 443–444
Air pressure, **92**–95
air temperature and, 93
causes of, 92–93
and El Niño, 109
equatorial low pressure, 97, 114
and global circulation pattern, 96–97, 98*f*
measurement of, 93
pressure gradient force, **94**–95
subtropical high pressure, 97, 116
tendency toward equalization, 94–95
and water density, 98
and wind generation, 94–96, 95*f*, 96*f*
Air temperature
and air pressure, 93
altitude and, 92
bodies of water and, 92, 93*f*
and convection, **95**–96, 95*f*, 96*f*
daily variation in, 92
and dew point, **101**
earth inclination and, 89–91, 89*f*, 90*f*
factors affecting, **89**–92
global isotherms, 93*f*
lapse rate, **92,** 94*f*
solar energy and, **91**–92, 92*f*
temperature inversions, **92,** 94*f*, 95, 95*f*, 96, 443, 446
Al Sham, 271*f*
Alabama, industry incentives, 374
Alaska
earthquake (1946), 70
earthquake (1964), 64, 65
Fairbanks, climate in, 124*f*
mapping of, 34
wetlands in, 163
Alaskan pipeline, 79
Albers equal-area conic projection, 492, 493*f*
Alcohol, as fuel, 148
Aldrin, biological magnification of, 456*f*
Aleutian Islands, 66
Alfisols, 110*t*, 111*f*, 115*t*, 120
Algeria
agriculture in, 160
Islamic fundamentalism in, 319
population data, 496*t*
*Ali*, 254
Alien species. *See* Exotic species
Alkalinity of soil, 110
*Allah*, 247*f*, 253
Alliance(s)
economic, 327–329
increase in, 300–301
military and political, 329
regional, 327–329
Alliance of Small Island States (AOSIS), 306
Alluvial fans, 73, 73*f*
Alluvium, **73**
Alps, glaciers in, 127
al-Queda, 323
Altitude
and air temperature, 92
geodetic control data and, 37
and population density, 201
and temperature inversion, 95, 95*f*
and wind patterns, 98*f*
Alumina, U.S. imports of, 158*t*
Aluminum, recycling of, 171, 171*f*, 459
Amalgamation theory, **234**
Amazon River, pollution in, 441
Amenities, 283–284
American Civil Liberties Union, 289
American Mining Congress, 367
Americas. *See also* Anglo-America; North America; South America
colonialism in, 251, 252, 303
as culture hearth, 230, 231*f*
languages in, 236*f*
regional trade alliances, 327–328
Amerindian language family, 238*f*–239*f*
Amerindian race, 259
Amish, 232*f*, 233*f*
Amityville, New York, 395*f*
Amnesty International, 295
Amu Darya River, 438, 438*f*
Anaerobic digestion, 148*f*, 415, 415*f*
Anarchists, 323
Anasazi people, 223
Andean Group, 329*f*
Andes Mountains
agriculture in, 231*f*
as culture hearth, 231*f*
population distribution and, 204
Andisols, 110*t*, 111*f*
Andorra, 306*f*, 502*t*
Anecumene, 204
Anglo-America. *See also* Canada; North America; United States
birth rate in, 182
cities in, 421, 422*f*
colonialism in, 252
as culture realm, 221*f*
death rates in, 186
folk groups in, 233
industrial regions of, 375–376, 375*f*
infant mortality rates, 188
place names in, 244–245
population growth in, 183*f*
Angola
cities in, 402
exclaves, 305
population data for, 497*t*
Animal waste
as fuel, 415, 415*f*
as pollutant, 439–441, 440*f*
Animals and wildlife. *See also* Fishing; Livestock
biocides and, 439
by climate type, 115*t*
domestication of, 227, 230, 231*f*
endangered species, 451–452, 452*f*, 453, 453*f*, 455*f*
exotic species, **453**–455
extinction of, 451
habitat disruption, 452*f*
human impact on, 451–456
hunting and commercial exploitation of, 452–453
on islands, 454–455, 455*f*
mining and, 450
threatened species, 451–452
vulnerable species, 451–452
water and, 438
Animism, 247
Antarctic Treaty of 1959, 301, 301*f*
Antarctica
ozone hole over, 447, 448*f*
political status of, 301, 301*f*, 302
Antecedent boundaries, **311**
Anthracite, 142
Schuylkill Anthracite Region, 475–476, 476*f*
Antibiotics
disease control and, 198
impact on mortality rates, 186
Anticlinal valleys, 63*f*
APEC (Asia Pacific Economic Cooperation) forum, 328
Appalachia, 482–483, 483*f*
boundaries of, 483*f*
Appalachian Regional Commission, 482–483, 483*f*
Aquaculture, 361, 361*f*, 362, **363**
Aquifers, 74
contamination of, 441
Ogallala aquifer, 436, 436*f*
overuse of, 435–436
Arab culture
Palestinian issue in, 295
women in, 261
Arabic language
dialects of, 243
as lingua franca, 239, 242
Arable land
global distribution of, 346, 347*f*
loss of, 159–163, 223, 351–352
maintaining, 162, 162*f*
population density and, 205
Arabs
as part-nation state, 302*f*
place names and, 244
Aral Sea, shrinkage of, 438, 438*f*
Arapoosh (Crow chief), 177
Architecture, and culture, 262
*ArcNews*, 47
Arctic and subarctic climates, 115*t*, 123, 124*f*
and global warming, 127
locations of, 114*f*, 124*f*
precipitation, 124*f*
soil, 123
temperature, 124*f*
vegetation, 123
Arctic haze, 444
Arctic Ocean, 79
Area analysis tradition, 19–**21,** 20*f*, 467
Area cartograms, **39**–40, 40*f*
Area, projections maintaining, 30–31
Area symbols, 39–40, 39*f*
Arêtes, 76, 77*f*, 78*f*
Argentina
agriculture in, 359
Antarctic territorial claims, 301*f*
Bolivia and, 312, 313*f*
boundaries of, 313, 313*f*
Buenos Aires, 308
core area of, 308
Falkland Islands War, 306
Los Glacieres National Park, 54*f*
population
data, 499*t*
density, 205*t*
growth, 193
regional trade alliances, 328
time zone of, 29*f*
Arid areas. *See also* Desert(s)
stream landscapes in, 73–74, 75*f*
wind erosion in, 80–81, 81*f*
Aridisols, 110*t*, 111*f*, 115*t*, 118
Arithmetic density, of population, **205**
Arizona
Anasazi people in, 223
cities in, 420
Four Corners Monument, 331*f*
Grand Canyon, 58*f*
migration to, 287*f*
Phelps-Dodge mine, 364*f*
subsidence in, 451
water supply in, 436
Armed forces, and social cohesion, 317
Armenia
Nagorno-Karabakh conflict and, 309
population data for, 499*t*
Armenian Christian Church, 309
Army Corps of Engineers, U.S., 73, 151, 164, 436, 437*f*
Arroyos, 74
Arsenic, as pollutant, 440
Artifacts, 222, 229*f*
Artificial boundaries, **311**
ASEAN (Association of Southeast Asian Nations), 328
Ashkenazim, 250
Asia. *See also* Asia Pacific region; Eurasia
agriculture in, 231*f*, 348*f*, 350–351, 351*f*, 352, 353, 354, 355, 450
aquaculture in, 363
birth and fertility rates in, 182, 184, 186*f*, 187
cities in, 424–426
coal use in, 142
colonialism in, 303
culture hearths in, 231*f*
as culture realm, 221*f*
death rates in, 186
desertification in, 160
Eastern Asia industrial region, 375–376
economies in, 345, 379
El Niño and, 108
energy in, 149
ethnic religions in, 257
high-tech industries in, 377, 377*f*
industry in, 375–376, 375*f*, 377, 377*f*
land use in, 412
languages in, 236*f*, 243
longitude of, 28
maternal mortality rates in, 189, 189*f*
population
control of, 221
data for, 499*t*–501*t*
density of, 208*f*
distribution of, 211
growth of, 181*f*, 183*f*, 195*f*, 196
rain forest in, 168
religion in, 255*f*
savannas in, 116, 118*f*
toxic waste exports to, 462
transnational corporations and, 374–375
urbanization in, 191, 391, 394*t*
water pollution in, 439
wind patterns in, 97
women in, 191, 261, 354
Asia Pacific Economic Cooperation (APEC) forum, 328
Asia Pacific region
hydropower in, 152*f*
natural gas reserves, 145*f*
natural gas trade flows, 144*f*
oil reserves, 139*f*
oil shale deposits, 145*f*
oil trade flows, 408*f*
Asian Americans
birth and fertility rates, 186
residential patterns in Los Angeles, 412*f*
Asian gypsy moth, 453
Asian long-horned beetle, 453–454
Asian tiger mosquito, 454
Asiatic Chestnut blight, 455
Aspiration level, and immigration, 284
Assimilation, **234**–235
of immigrants, 244–245, 258
Association of American Geographers (AAG), 11
Association of Southeast Asian Nations (ASEAN), 328
Asthenosphere, **58,** 59*f*
Aswan Dam, Egypt, 72, 151, 207
Asymmetric federalism, 308, 320
Athabasca deposit, 144
Atlantic City, New Jersey, 82
Atlantic Ocean
continental drift and, 59
currents in, 98, 99*f*

fisheries locations, 362, 362*f*
map of seafloor, 61*f*
Atolls, formation of, 80
Atwood, Wallace W., 472–473
Aum Shinrikyo, 323
Austral European culture realm, 221*f*
Australia
agriculture in, 303, 434*f*
alien species in, 453
Antarctic territorial claims, 301*f*
birth and fertility rates in, 182, 187, 502*t*
Canberra, 308, 310*f*
capital of, 308, 310*f*
climate of, 120, 121*f*
coal reserves, 142*f*
colonies of, 134
economy of, 379
El Niño and, 108
endangered plant species in, 452
Great Barrier Reef, 80
immigration restrictions, 288
industry in, 375*f*
land use map of, 18*f*
longitude of, 28
mineral production by, 157*f*
mineral resources, 157, 158
nuclear power and, 146
oil shale deposits, 144, 145*f*
population
data on, 502*t*
density, 205*t*
regional trade alliances, 328
settlement of, 252, 281
Sydney, climate, 121*f*
vegetation of, 112
waste disposal in, 462
wetlands in, 164
Australian language family, 238*f*–239*f*
Australoids, 259
Austria
air pollution in, 448
exclaves, 305
population data for, 501*t*
population pyramid, 190*f*, 191
regional trade alliances, 327
trade, in services, 382*t*
Austro-Asiatic language family, 238*f*–239*f*
Automated mapping, 44, 45–59
Automobiles. *See* Motor vehicles
Autotrophs, 433*f*
Avalanches, 70
Azerbaijan
Nagorno-Karabakh conflict, 309
population data for, 499*t*
Azimuthal projections, **31, 492**–493, 493*f*

## B

B climates. *See* Dryland climates
Baarle-Hertog, 305
Babbitt, Bruce, 366
Baby boom, 193*f*
Babylonians, ancient, 27
Baha'i faith, 256
Bahrain
population data for, 499*t*
religion in, 254
terrorism and, 323
Bali
agriculture in, 224*f*
religion in, 254
Bangkok, Thailand
as primate city, 401
slums in, 426
Bangladesh
agriculture in, 350
birth and fertility rates in, 184, 500*t*
energy in, 148
establishment of, 305, 481, 481*f*
microcredit initiative in, 355
mineral resources, 157
population, 203
control, 185
data for, 500*t*
density of, 205*t*, 206
services in, 380*f*
women in, 261
Bank(s)
functional regions of, 18*f*
international, 292, 383*f*, 384
off-shore locations, 383*f*, 384
Banqiao Dam collapse, 4
Bantu languages, 237, 240*f*
Baptists, U.S. distribution of, 253*f*
Barchan, 80
Barite, U.S. imports of, 158*t*
Barometers, 93
Barometric pressure, 93
of hurricanes, 105*t*
Barrier islands, and erosion, 82
Basalt, 56, 57*f*
Baseball, as cultural import, 235*f*
Basic sector
ratio to nonbasic sector, 400, 400*f*
of urban areas, **398**–400
Basque language family, 238*f*–239*f*
Basques, 236*f*
regional government of, 308
separatist movement, 314, 314*f*, 321, 321*f*, 322, 323
as stateless nation, 303
territory, 321*f*
Bauxite
reserves of, 156*t*
U.S. imports of, 158*t*
Bay head locations, 396
Bazaar Malay, 242
Beaches
formation of, 79, 80*f*
human impact on, 82
protection and replenishment of, 82
Beijing, China
air pollution in, 443, 446
population of, 392*t*
UN conferences in, 262
Beijing Conference (1995), 262
Beijing Plus Five Conference (2000), 262
Beirut, Lebanon, religious regions of, 248, 250*f*
Belgium
energy in, 154
exclaves, 305
language in, 243
nationalist movements in, 320, 321*f*
nuclear power and, 146
population data for, 501*t*
regional trade alliances, 327
trade, in services, 382*t*
Benares, India, 216*f*
Bench marks
resurvey of, 37
on topographic maps, 36, 37
Beneficiation, 364
Benelux countries, 327
Bengali language, number of speakers, 240*t*
Benin, population data for, 496*t*
Berlin, division of, 305
Berlin Wall, as relic boundary, 312
B-horizon, 108
Bhutan
economy, 358
population data for, 500*t*
population growth, 196
Big Dig (Boston), 338*f*
Bilingual education, 244
Bilingual Education Act of 1974 (U.S.), 244
Bilingualism, 243, 244
Billion, illustration of, 180–181
bin Laden, Osama, 323
Bioaccumulation, 455, 456*f*
Biocides
and environmental contamination, 455–456
Green Revolution and, 353
impact on mortality rates, 186
resistance to, 456
as water pollutant, 439
Biogas, 148
Biological diversity, in tropical rain forest, 169
Biological factor, in soil formation, 106–107
Biological magnification, **455**–456, 456*f*
Biomagnification. *See* Biological magnification
Biomass fuels, **148,** 171
Biomes, as regions, 485–486, 486*f*
Biosphere, **432**
components of, 433
hydrologic cycle in, 434*f*
renewal cycles in, 433, 433*f*
Biotechnology, 353
Birth cohorts, 182*f*
Birth rate
crude, **182**–183, 183*f*
in demographic transition model, 194–200, 196*f*
in developed nations, 182–183, 207*f*, 496*t*
in developing nations, 180, 182–183, 199, 207*f*, 496*t*
Neo-Malthusianism and, 211–212
by region and nation, 496*t*–503*t*
transitional, 182
Bitumen, 144
Bituminous coal, 57, 142
Black Death, 190
Black English (Ebonics), 239
Black Hills of South Dakota
formation of, 67
as region, 473, 473*f*, 474*f*
Blizzard(s), **103**–104, 106*f*
Blue Mosque, Istanbul, 255*f*
Bolivia
and Chile, 307, 312, 313*f*
industry in, 376
location of, 307
mineral production by, 157*f*
and Peru, 312, 313*f*
population data for, 499*t*
port access, 312, 313*f*
regional trade alliances, 328
religion in, 248
Bone cancer distributions, 17*f*
Bosnia
ethnic war in, 258, 283, 295, 322
population data for, 502*t*
UN peacekeeping in, 295, 324
Boston, Massachusetts
Big Dig, 338*f*
map scale and, 33, 34*f*
mass transit lines in, 406*f*
population of, 397*t*
Botswana
habitat destruction in, 452
life expectancy, 190
mineral production by, 157*f*
population data for, 498*t*
population pyramid, 191, 192*f*
Boundaries, 310–316
in Africa, tribal *vs.* national, 303, 304*f*, 311
antecedent, **311**
artificial, **311**
classification of, 310–312
as conflict source, 311, 311*f*, 312–316, 312*f*
consequent, **311**–312
of country, 480–482, 481*f*
ethnographic, 311–312
Four Corners Monument, 331*f*
geometric, **311**
international, 310–316
maritime, 325–326, 325*f*, 326*f*
natural, **311,** 311*f*
physical, 311, 311*f*
of regions, 471, 471*f*
relic, 312
subsequent, **311**
superimposed, 303, 304*f*, **311**
three-dimensional quality of, 310
transportation and, 317–318, 320*f*
of urban areas, 396*f*
Brahmins, 218
Brandt line, 226, 226*f*
*Brandt Report* (1980), 226, 226*f*
Brasilia, Brazil, 310, 424, 425*f*
Brazil
agriculture in, 359
birth and fertility rates in, 184
Brasilia, 310, 424, 425*f*
capital of, 310
cities in, 424, 425*f*, 426
climate of, 120
deforestation in, 159*f*
endangered species in, 452
energy in, 148, 149
industry in, 376
mineral production by, 157*f*
mineral resources, 157
oil shale deposits, 144, 145*f*
population
control of, 211*f*
data, 499*t*
power in, 149
rain forest in, 167, 170*f*
regional trade alliances, 328
Rio de Janeiro
population of, 392*t*
slums of, 208*f*, 426, 426*f*
UN conference in (1992), 202
Breakers (waves), formation of, 76, 80*f*
Break-of-bulk locations, 396
Breakwaters, 82
Breeder reactors, 147
Breton separatist movement, 321
British Commonwealth, 329
British Petroleum Corp., 294*f*
British Phosphate Commission, 134
Brittany, 321*f*
Broad warping, **62**
Brunswick, Georgia, map of, 43*f*
Buddha, 255–256
Buddhism, 255–257
diffusion routes, 250*f*, 256, 256*f*, 257
distribution of, global, 248*f*
as state religion, 317
stupas and pagodas, 256–257, 257*f*
vehicles, 256, 256*f*
Buenos Aires, Argentina, as core area, 308
Built landscape, defined, 14
Burkina Faso
economy of, 358
population data for, 496*t*
Burundi
economy of, 358
population data for, 497*t*
Business centers, U.S., 399, 399*f*
Business district. *See* Central business district
Business parks, 378
Buttes, 74
Bycatch, 453

## C

C climates. *See* Humid midlatitude climates
Cabinda, Angola, 305
Cabot, John, 300
Cadmium, as pollutant, 390, 458
CAFE. *See* Corporate Average Fuel Economy (CAFE) standards
Cairo, Egypt, 391*f*
air pollution in, 443
climate, 119*f*
growth of, 390
history, 390
population of, 392*t*
slums in, 426

Cairo Plan, 202–203
Calgary, Canada, land use pattern in, 407*f*, 409
California. *See also* Los Angeles
  air pollution in, 446
  bilingualism in, 244
  cities in, 420
    edge, 414
  climate of, 120
  cultural landscape in, 13, 13*f*
  earthquakes
    locations of, 64*f*
    Northridge earthquake (1994), 65
    Pasa Robles (2003), 62
    San Andreas fault, 60, 60*f*, 62, 64, 64*f*, 65*f*
  El Niño and, 108
  high-tech industry in, 377
  hydroelectric power in, 149, 149*f*
  and immigration, illegal, 288–289, 315*f*
  migration fields, 290*f*
  Proposition 187, 288–289
  Proposition 209, 289
  Proposition 227, 244
  San Francisco Bay area
    desire line map of, 16*f*
    El Niño and, 108
  San Francisco International Airport, 46*f*
  satellite images of, 47*f*
  soil salinization in, 163
  solar power in, 152, 152*f*
  vegetation of, 112
  water supply in, 436
  wind power and, 154, 155*f*
  Yurok Indians of, 480, 480*f*
Caliphs, 254
Cambodia
  cities in, 424
  population data for, 500*t*
  recycling in, 459
  religion in, 246
Cameroon
  export restrictions, 365–366
  population data for, 497*t*
Campos, 116
Canada
  acculturation and, 235
  acid rain and, 443–444
  agriculture in, 163, 303, 307, 358, 358*f*, 359, 359*f*
  air mass source regions for, 103*f*, 475*f*
  birth and fertility rates in, 187, 498*t*
  borders of, 311
  boundaries of, 313–314
  Calgary, land use pattern in, 407*f*, 409
  capital of, 308, 310, 310*f*
  cities in, 421, 422*f*
  climate of, 122, 124*f*
  colonization of, 252
  core regions of, 308, 415*f*
  economy of, 379
  energy in, 149
  fishing industry in, 362
  folk culture in, 233
  French-Canadians, 233, 320
  high-tech industry in, 377
  language in, 243
  location of, 307
  mental maps in, 272–273, 273*f*
  mineral production by, 157*f*
  mineral resources, 157
  mining and quarrying in, 146*f*, 156, 363
  natural gas trade flows, 144*f*
  nuclear power in, 146
  oil consumption, 140
  oil production, 140
  oil reserves, 140
  oil shale deposits, 145*f*
  oil trade flows, 138*f*
  political subdivisions in, 334–335
  population
    data on, 498*t*
    density of, 205*t*
    distribution of, year 2000, 38*f*
  redistricting in, 330
  regional trade alliances, 327, 328
  resource disputes, 315
  return migration to, 290
  secessionist movement in, 320
  solid waste disposal in, 458
  tar sand deposits, 144, 146*f*
  as terrorist target, 323
  topographic maps of, 34
  Toronto, 422*f*
    political subdivisions in, 335
    solid waste disposal, 458
    work *vs.* nonwork trips, 275*f*
    as world city, 401, 403*f*
  trade
    and NAFTA, 373
    in services, 382*t*
  transportation system in, 318, 320*f*
  urban land use in, 412
  Vancouver, climate of, 122*f*
  waste disposal in, 458
  Whistler Mountain ski resort, 6*f*
Canal(s)
  everglades and, 486
  Lehigh Canal, 477*f*
  Schuylkill Navigation Canal, 475, 477*f*
Canberra, Australia, 308, 310*f*
Canyonlands National Park, 75*f*
Cape Town, South Africa, 222*f*
Cape Verdi, population data for, 496*t*
Capillary action
  groundwater and, 74
  and weathering, 68
Capital cities, 308–310
  in developing nations, 424, 425*f*
  forward-thrust, 309–310
  as primate city, 402
Caprivi Strip, 312
Car(s). *See* Motor vehicles
Carbon dioxide, as air pollutant
  and global warming, 126
  rain forests and, 168–169
  sources of, 140, 443*t*
Carbon monoxide, as air pollutant
  reduction efforts, 448
  sources of, 443*t*, 458
Carbon tetrachloride, and ozone layer, 447, 448
Carbonation, in chemical weathering, 69
Careers in geography, 8, 10–11
Caribbean
  agriculture in, 354
  birth and fertility rates, 186, 498*t*–499*t*
  economies of, 379, 382
  GNI in, 498*t*–499*t*
  language in, 243
  maternal mortality rates in, 189*f*
  migration from, 286
  population control in, 212
  population data for, 498*t*–499*t*
  rain forest in, 168
  transnational corporations and, 374–375
Caribbean Community and Common Market (CARICOM), 328, 329*f*
Caribs, 243
CARICOM (Caribbean Community and Common Market), 328, 329*f*
Carrying capacity, of land, **206**–207, 206*f*
  Agricultural Revolution and, 181
  population growth and, 206–207, 206*f*, 210–211, 212
  projections of, 210–211
    in Africa, 206, 206*f*
  technology and, 223
Cartel(s)
  increase in, 300–301
  regional trade agreements as, 329
Carthage, sacking of, 322
Cartograms, area, **39**–40, 40*f*
Cartography, **27.** *See also* Mapping
Caste, in Hinduism, 246, 254
Castro, Fidel, 282
Catalonia, 308, 321*f*
Catalytic converters, 449
Catholic Church
  cathedrals, 254*f*
  distribution of
    global, 248*f*
    United States, 253*f*
  history of, 251–252
  and population control, 183, 202, 212
Caucasian language family, 238*f*–239*f*
Caucasoids, 259
Caves, formation of, 74–75, 76*f*
CBD. *See* Central business district
Cellular phones, 292, 292*f*
Celtic language, 237
Celts, 237
Census Bureau, U.S., 180, 273, 396
Censuses, 209, 209*f*
  and reapportionment, 330
Central Africa Federation, 329
Central African Republic
  cities in, 424
  population data for, 497*t*
Central America. *See also individual countries*
  coal reserves, 142*f*
  earthquakes in, 66
  Hurricane Mitch in (1998), 5*f*
  hydropower in, 152*f*
  migration from, 286
  natural gas trade flows, 144*f*
  oil reserves, 139*f*
  oil trade flows, 408*f*
  population
    data for, 498*t*
    growth, 195*f*
  rain forest in, 168
Central American Common Market, 329*f*
Central business district (CBD), 396, 396*f*
  in developing country cities, 425
  in land use models, **406**–407, 407*f*, 409*f*, 411*f*
  population of, 407–408, 407*f*, 408*f*
  repopulation of, 411
  suburban, 414
  in urban land use models, 408–409, 409*f*
Central city, **396**
  changes in, 414–421
  constricted, 414–420
  expanding, 420–421
  future of, 419
  gentrification of, **419**–420, 420*f*, 422
  renewal of, 416–419
    private efforts, 417–419
    urban renewal programs, 416–417, 419*f*, 420
Central place(s), **404**–405, 404*f*
  hierarchy of, 404–405, 404*f*
Central place theory, **404**–405, 404*f*, 405*f*
Centrifugal forces, in states, **316**, 318–321
Centripetal forces, in states, **316**–318
Ceuta, Spain, 305, 306*f*
CFCs. *See* Chlorofluorocarbons
Chaco Canyon, 223
Chad, 308, 497*t*
Chain migration, **285**–286
Champaign County, Illinois, administrative units in, 334, 334*f*
Chang Jiang (Yangtze) river, 72, 150–151, 350, 435
Change. *See also* Climate, changes in; Culture change
  economic, inevitability of, 368
  geologic, constancy of, 56
  geologic time and, 58
  in place characteristics, 14, 15*f*
Channelization, **436**
Channelized migration, **290**–291, 291*f*
Chaparral, vegetation, 112, 112*f*
Chechnya, separatist movement in, 321
Chemical weathering, **69**–70
Cherokee tribe, forced migration of, 282*f*
Chesapeake Bay
  cleanup efforts, 442
  *Pfiesteria piscicida* in, 439–440
Chestnut blight, 455
Chicago, Illinois
  central business district of, 471
  climate, 122*f*
  latitude and longitude of, 27, 28
  population of, 397*t*, 402*t*
  public housing in, 419*f*
  situation of, 397, 397*f*, 398*f*
  social areas of, 410, 411*f*
  urban sprawl in, 413, 413*f*
  as world city, 401, 403*f*
Chickasaw tribe, forced migration of, 282*f*
Child care, and women's employment, 417
Children. *See also* Infant mortality rate
  air pollution and, 446
  incentives for having, 185, 196, 197, 199–200, 202–203
  and mental maps, 272*f*
Chile
  air pollution in, 446
  Antarctic territorial claims, 301*f*
  Bolivia and, 307, 312, 313*f*
  boundaries of, 313, 313*f*
  climate of, 120
  infant mortality rate, 188*f*
  mineral production by, 157*f*
  population data for, 499*t*
  regional trade alliances, 328
  religion in, 248
  shape of, 305
  time zone of, 29*f*
  Tsunami strike on (1960), 70
  vegetation of, 112
China
  agriculture in, 2*f*, 207, 231*f*, 352, 360, 435
  Banqiao Dam collapse, 4
  Beijing
    air pollution in, 443, 446
    population of, 392*t*
    UN conferences in, 262
  birth and fertility rates in, 182, 183, 184–185, 199
  border disputes, 311, 311*f*, 313
  censorship in, 295
  CFCs use and production, 448
  climate of, 120, 122
  coal production, 142
  coal reserves, 142*f*
  coal use, 142
  commodities trade and, 366
  as culture hearth, 230, 231*f*
  culture of, 176*f*
  deforestation in, 2*f*
  earthquake in (1976), 66
  economy, 345*f*
  endangered species in, 452
  energy in, 148, 154
  fishing industry in, 361*f*
  food production in, 206
  gated communities in, 416
  high-tech industry in, 377, 378
  history of geography in, 7
  Hong Kong and, 180, 403
  hydroelectric power in, 150–151
  immigration from, 201*f*
  industry in, 376, 377*f*
  loess deposits in, 81
  migration from, 284
  mineral production by, 157*f*
  mineral resources, 157
  Mongol invasion of, 234
  Nanching, economy of, 352
  nuclear power and, 146
  oil consumption by, 138*f*
  oil production by, 138*f*
  oil shale deposits, 144, 145*f*
  political activism in, 295
  population, 180, 184–185, 203
    age of, 213*f*
    control of, 182, 184–185, 211–212, 211*f*, 221

data for, 500t–501t
density of, 205t
growth of, 181f, 183f
population pyramid, 192
regional trade alliances, 328
religion in, 257
rice patties in, 2f
rivers in, 72
Three Gorges Dam, 150–151
time zone of, 29f
trade, in services, 382t
transnational corporations and, 374
urbanization in, 191, 207, 394t
village economy in, 352
water supply in, 435
women in, 184, 191, 263f
Yugoslavian embassy in, NATO bombing of, 26
Chinatowns, 258–259, 260f, 411
Chinese religions, global distribution of, 248f
Chinese Workingmen's Party, 244–245
*Chitemene*, 349
Chlordane, biological magnification of, 456f
Chlorinated hydrocarbons, biological magnification of, 456f
Chlorofluorocarbons (CFCs)
and global warming, 126
and ozone layer, **447**–448, 447f
phase-out of, 447–448
Choctaw tribe, forced migration of, 282f
Cholera, 189, 437
C-horizon, 108
Choropleth maps, 39, **39**, 39f
Christaller, Walter, 404, 404f
Christianity, 250–253. *See also* Catholic Church
churches and cathedrals, 252–253, 254f
and cultural landscape, 252–253, 254f
diffusion routes, 250–251, 250f
distribution of
global, 248f
United States, 253f
Eastern and Western Church, 251
Protestant Reformation, 251
Protestantism, 251–252
distribution of
global, 248f
United States, 253f
in U.S., 252–253, 253f, 254f
Chromium
reserves of, 156t
U.S. imports of, 158t
Chronological factor, in soil formation, 107
Church(es)
attendance decline, 248
and cultural landscape, 252–253, 254f
Cirques, 76, 77f, 78f
Cirrus clouds, 100f, 101
Cities. *See also* Capital cities; Metropolitan areas; Urban areas
in Africa, 191, 402, 424–426
annexation power of, 414–415
in Asia, 424–426
in Canada, 421, 422f
as centers of innovation, 230, 270–271, 277–278
central, **396**
changes in, 414–421
constricted, 414–420
expanding, 420–421
future of, 419
gentrification of, **419**–420, 420f, 422
renewal of, 416–419
private efforts, 417–419
urban renewal programs, 416–417, 419f, 420
classification of, by location, 396
decentralization of, 274, 279, 280
declining, 402t
definition of, **396**
in developing countries, 401–402, 424–426
edge, **414**
employment in, 411
and erosion, 72
in Europe, 403–404
eastern, 423–424, 424f
western, 421–423, 423f
flight from, 408–409, 410, 414–415, 421
galactic, 414, 415f
global diversity in, 421–426
growth of
constricted, 414–420
expandable, 420–421
smart growth programs, 421
land use patterns, 355–357, 357f
in Latin America, 208f, 424–426, 477–478, 478f
mass-transit era of, 405–406
medieval, 14
in Middle East, 424–426
migration to, 191, 282, 285, 286, 291, 425
network, **402**–404
in North America, 421, 422f
political subdivisions in, 335
population of
cities over 3 million (2005), 393f
growth in, 390, 391, 392f, 393f, 393t, 394f
land value and, 407–408, 407f, 408f
megacities, 392t
in U.S., 402t
in post-industrial economy, 419
primate, 308, **401**–402, 421
in developing nations, 401–402, 425
in Latin America, 478
rank-size of, 401–402
as regional cores, 484
in Russia, 401, 423–424, 424f
site of, **396**
situation of, **396**–397, 398f
skyline of
in Anglo America, 406, 422
in Europe, 422, 423f
structure of, 405–412
subsidence of, 451, 451f
and suburbanization, impact of, 414–420
transportation and, 394, 396, 405–406, 406f, 407, 407f. *See also* Mass transit systems
in United States. *See* United States
in USSR, 423–424, 424f
world, **400**–401, 403f
Cities of the Dead (Cairo), 390
Civil Rights movement, U.S., 244
Civilization(s)
defining characteristics of, 230
development of, 230
environmental impact of, 14
Claritas, Inc., 410
Clean Air Acts (1963, 1965, 1970, 1977, 1990), 446, 448, 458
Clean Water Act (1972), 164, 442
Clear cutting, 165–166, 166f, 167
Cleveland, Ohio, population density gradients for, 408, 408f
Cliffs, formation of, 76
Climagraphs, 117f
Cairo, Egypt, 119f
Chicago, Illinois, 122f
Fairbanks, Alaska, 124f
Rome, Italy, 121f
Singapore, 117f
Sydney, Australia, 121f
Tehran, Iran, 119f
Vancouver, Canada, 122f
Yangôn, Myanmar, 118f
Climate, **88.** *See also* Weather; *specific climates* (listed in 115t)
and air pollution, 443
of Cairo, Egypt, 119f
changes in, 123–128
causes of, 75, 124–125
cyclical nature of, 125
desertification, **159**–160, 160f, 161f
and disease, 198
human environmental impact, 126–128
scientific investigation of, 123, 124
short term, 125–126
of Chicago, Illinois, 122f
classification of, 113–115, 115t
of Fairbanks, Alaska, 124f
importance of, 81, **89**
Köppen classification system for, 113–115, 115t
little ice age, 125
medieval warming period, 125
of Moscow, Russia, 123f
natural vegetation and, 110–112
regions, 112–123, 114f, 115t, 473–475
of Rome, Italy, 121f
of Singapore, 117f
soils and, 107f
of Sydney, Australia, 121f
of Tehran, Iran, 119f
transition zones between, 116
tropical rain forest, 114, 117f
of Vancouver, Canada, 122f
volcanoes and, 67
weather as basis of, 104
of Yangôn, Myanmar, 118f
Climate factor, in soil formation, 106
*Climatology and the World's Climates* (Rumney), 473–475
Climax community, **110**
Cloropleth maps, **39**, 39f
Clothing, cultural significance of, 228–229, 229f
Clouds
formation of, 100
types of, 100–101, 100f
CMEA (Council of Mutual Economic Assistance), 328
CNN, 295
Coal, 141–143
and air pollution, 443t, 444
anthracite, 142
Schuylkill Anthracite Region, 475–476, 476f
bituminous, 57, 142
consumption of
in Asia, 142
in U.S., 137f, 141
future use of, 171
and global warming, 126
grade of, 142
lignite, 142
mining of, 142, 450
as nonrenewable resource, 135f, 136
production of, 142
from U.S. government lands, 366
reserves of, 141–142, 142f, 156
sulfur in, 142, 448–449
transportation of, 142, 142f, 143f
types (ranks) of, 142
value determinants for, 142
washing of, 448–449
Coast and Geodetic survey, U.S., 70
Coastal areas
and development, 82
wave and current action and, 76–80, 80f, 82, 82f
wetlands, 163
Cobalt
reserves of, 156t
U.S. imports of, 158t
Cohort, **182**, 182f
Cold War, *vs.* globalization, 293
Colombia
population data for, 499t
religion in, 248
Colonialism
acculturation and, 234
in Africa, 243, 246f, 252, 303, 304f, 311
and developing nations cities, 424
European, 303, 304f, 311
on Indian subcontinent, 481, 481f
and language spread, 237, 238
and mini-states, 306
political status of colonies, 302
political structures and, 303, 304f, 311
Color, of soil, 109
Colorado
Denver, air pollution in, 443
Four Corners Monument, 331f
smart growth programs in, 421
urban renewal in, 420
Colorado River, 150, 436
Cols, 77f
Columbia Plateau, 68, 69f
*Columbia* space shuttle disaster, 29
Columbium, U.S. imports of, 158t
Columbus, Ohio, downtown rejuvenation in, 408
Comecon (Council of Mutual Economic Assistance), 328
Commercial agriculture. *See* Agriculture
Commercial economy, **344**
Commercial exploitation of animals and wildlife. *See also* Fishing, commercial
dairy farms, 357, 358f
hunting, 452–453
impact of, 452–453
livestock-grain farming, 357, 358f
ranching, 357, 359, 359f
and water pollution, 439–440, 440f
Commercial forests, 164–166, 165f
Commission on Sustainable Development (U.N.), 169
Commodity prices, 365–366, 365f
Common Market, 327, 328–329
Commons, tragedy of the, **362**–363
Commonwealth of Nations, 329
Communication(s)
distance decay and, 269
and global trade in services, 382
globalization and, 291–292, 382
high-tech industries, 377–378
Internet
globalization and, 292, 382
political activism and, 295
spatial interaction and, 279–280, 280f
state power and, 323
and national cohesion, 317–318, 319
quaternary sector employment, 381
spatial interaction and, 279–280
transnational corporations and, 375
urban land use and, 419
Communication hierarchy, 277–278, 277f
Communism
city planning under, 423–424, 424f
collapse of, migrations following, 281
planned economies in, **344**
agriculture in, 360–361
city planning, 423–424, 424f
industrial location in, 373
mining and quarrying in, 364
Communities, master-planned, 416
Compact states, **305**, 305f
Comparative advantage, **371**–373, 375
Competition theory, assimilation and, 234–235, 244
Composite volcano, 66, 67f
Computer-aided design/manufacturing, 370–371
Concentric zone model of urban land use, **408**, 409f
Concept(s)
basic. *See* Terminology
definition of, 275
Condensation nuclei, 100
Condensation, of atmospheric moisture, 100
Conference for the Codification of International Law (1930), 325
Conference on Environment and Development (U.N.), 202
Conflict, boundaries as source of, 311, 311f, 312–316, 312f
Conformal projections, **31**, 32f
Confucianism, 257
Confucius, 257

Conglomerate, formation of, 57
Congo Corridor, 312
Congress, U.S.
  Black members of House, 332
  and reapportionment, 330–333, 331*f*, 332*f*
Conic projections, **492,** 492*f*
Coniferous forests, 164, 165*f*
  climate and, 123*f*
  locations of, 112*f*
  soil of, 107*f*
  soil profile, 107*f*
  vegetation, 112, 112*f*
Connecticut, urban renewal in, 417
Connectivity, **14**–15, 18*f*
  mapping of, 16*f*
Consequent boundaries, **311**–312
Conservation, of resources, **171**
Conservation Reserve Program, 162
Constantine (Roman Emperor), 250
Construction industry, 344
Consumers, primary and secondary, in food chain, 433*f*
Contagious diffusion, **276**–277, 277*f*, 278*f*, 279
Contiguous zone, 325*f*, 326
Continental climate. *See* Humid continental climate
Continental drift, **58**–62, 59*f*–62*f*, 74
Continental environments, 92
Contour interval, on topographic maps, **36**–37, 37*f*
Contour lines, on topographic maps, **36**–37, 37*f*
Contour plowing, 162, 162*f*
Contraception. *See also* Abortion
  access to, 185
  Catholic Church and, 183, 202, 212
  developing nations and, 211–212, 211*f*
  education in, 208–209
  Islam and, 183, 202, 212
  sterilization as, 180, 184, 211*f*
Conurbations, 415*f*
Convection, **95**–96, 95*f*, 96*f*, 97, 98*f*
Convectional precipitation, **101,** 101*f*
Convention for the Protection of the Mediterranean Sea (1976), 442
Convention on Long-Range Transboundary Air Pollution (1979), 448
Convergence, cultural, 223
Convergent boundaries, 59, 60*f*
Converse factory, Reynose, Mexico, 372*f*
Copper
  mining, 363, 364*f*, 450*f*
  production of, 158, 158*t*
  properties and uses, 157–158
  recycling, 158
  refining of, 363, 364*f*
  reserves, 156*t*, 157–158
  substitutes for, 158–159
Coral reefs, formation of, 80
Core area, of state, **307**–308
Core, of region, 471–472
Core regions
  in Anglo America, 415*f*
  of separatist movement, 322
Coriolis effect, **96**–97, 97*f*, 98*f*
Corn
  Green Revolution and, 353
  storage of, 356*f*
  trade in, 342
Corn Belt, U.S., 357
Cornucopians, 212
Corporate Average Fuel Economy (CAFE) standards, 140–141
Corporations, transnational (TNCs), 292–293, 294*f*, **373**–375
  employment and, 374
  power of, 323
  and taxes, 376
  trade in world services and, 382–383
  world cities and, 401
Corridors, access, 312
Corsica, 321*f*
Cost
  as barrier to spatial interaction, 270
  as factor in migration, 287
  of transportation, 268, 369–370
Costa Rica
  biodiversity in, 172
  population data for, 498*t*
Côte d'Ivoire, population data for, 496*t*
Council for Urban Economic Development, 374
Council of Mutual Economic Assistance (CMEA), 328
Countermigration, **290**–291, 291*f*
Country, 301–302. *See also* State(s)
  boundaries of, 480–482, 481*f*
  numbers of, 303
Country music, Non-Euclidian map of places mentioned in, 495*f*
Credit, microcredit initiatives, 355
Creek tribe, forced migration of, 282*f*
Creole, **242**
Cretaceous period, 124
Crete, as culture hearth, 231*f*
Critical distance, **269,** 269*f*
Croatia
  ethnic cleansing in, 322
  population data for, 502*t*
Croats, 258
Crops. *See* Agriculture
Crow Indians, 177
Crude birth rate (CBR), **182**–183, 183*f*. *See also* Birth rate
Crude death rate (CDR), **186**–190. *See also* Death rate
Crude density, of population, **205,** 205*t*
Crude oil. *See* Petroleum
Crust, of earth, 58, 59*f*. *See also* Plate tectonics
  biosphere and, 433
  elements in, 154, 156*f*
  rock types in, 56–58
Cuba
  birth and fertility rates, 186
  emigration from, 282
  migration from, 200
  population data for, 498*t*
Cult of true womanhood, 260
Cultural ecology, 220
Cultural geography, 218
Cultural lag, 230
Cultural landscape, **13**–14, 13*f*, 15*f*, **221,** 222*f*
  changes in, 15*f*
  cities and, 394
  *vs.* natural landscape, 6*f*, 13
  religion and, 252–253, 254, 254*f*, 255, 255*f*, 256–257, 256*f*, 257*f*
Cultural traits, **219**
Cultural variables, 283
Culture(s), **218**
  attributes, overview of, 13–14
  categorization of, 235
  complexity of, 219, 227–229
  components of, 218–220
  convergence, 223, 376
  diffusion of, 230–232
  divergence, 223
  diversity of, 218, 219*f*, 223, 235–236, 261–262, 421–426
  and economic activity, 342
  environment and, 220–221
  as factor in migration, 287
  folk, **233**
  gender roles, 235–236, 259–261
  ideological subsystem, 222, **228**–229
  indicators of, 262
  integration, 228–229, 295
  language and, 242–245
  material, **233**
  and migration, 300
  and natural resources perceptions, 135, 135*f*
  nonmaterial, **233**
  percentage invented *vs.* acquired, 243
  popular, **233**
  and religion, 246
  social structure of, 219
  sociological subsystem, 222, **226**–228
  subsystems of, 221–229
  technological subsystem, **222**–226, 224*f*
    in advanced *vs.* developing cultures, 223–226, 224*f*
  transmission of, 226, 227*f*
Culture change, 229–235
  acculturation, **233**–235, 244–245
  assimilation, **234**–235, 244–245, 258
  inevitability of, 229–230
  by innovation, **230,** 232*f*. *See also* Innovation
  spatial diffusion, **230**–232
  syncretism in, **232**
Culture complex, **219**
Culture hearth, **230,** 231*f*
  and spread of innovation, 270
Culture realms, **220,** 221*f*
Culture regions, **220,** 478–479
Culture system, **219**–220
Culture-environment tradition, **19**–20, 20*f*, 177
  regions in, 476–482
Cumulonimbus clouds, 100, 100*f*, 101, 102, 103, 114
Cumulus clouds, 100*f*
Currents. *See also* Ocean currents; Stream landscapes
  erosion by, 76–80, 80*f*
Cyclones, **103**
  Pakistan (1970), 481
  typical paths of, 105*f*
Cyclonic precipitation, **102,** 104*f*
Cyclonic storm, 104*f*
Cylindrical projections, **490**–492, 490*f*, 491*f*
Cyprus
  population data for, 499*t*
  regional trade alliances, 327
Czech Republic
  admission to European Union, 298*f*
  NATO and, 329, 330*f*
  population data for, 501*t*
Czechoslovakia
  collapse of
    and ethnic unrest, 321
    states created by, 303, 307
  Nazi invasion of, 314


# D


D climates. *See* Humid midlatitude climates
Dade County, Florida, waste treatment in, 442
Dairy farms, 357, 358*f*
Dams
  at Baldwin Hills Reservoir (California), 451
  impact of, 150–151, 436
  Three Gorges Dam, 150–151
Dark Ages, 251
Day, length of, and earth's orbit, 91
DDT
  impact on wildlife, 455–456, 456*f*
  malaria and, 198, 456
  as water pollutant, 439
Death rate. *See also* Infant mortality rate; Maternal mortality rate
  crude death rate (CDR), **186**–190
  in demographic transition model, 194–200, 196*f*
  by region and nation, 496*t*–503*t*
  world, 199*f*
Death Valley, California, 119*f*
Deccan Plateau, India, 68
Deceptive practices, in maps, 40–42, 41*f*
Deciduous forests
  locations of, 112, 112*f*, 165*f*
  plant succession in, 111*f*
  as resource, 165, 165*f*
  soil profile, 107*f*
  vegetation, 112, 112*f*
Decision-making, rational, in economics, 367
Declaration of Independence (U.S.), 316
Decomposers, 433, 433*f*
Deep-sea trenches, formation of, 60–62
Deforestation
  agriculture and, 159, 159*f*
  and air pollution, 168–169
  in China, 2*f*
  developing countries and, 168
  and disease, 198
  and erosion, 166, 169
  and global warming, 126, 168–169, 170*f*
  impact of, 166, 169, 436
  in national forests, 166–167
  of old-growth forests, 166–167
  and ozone layer, 169, 170*f*
  resource utilization and, 135, 135*f*
  in tropical rain forests, 114, 159, 159*f*, 167–169, 170*f*, 352
Degrees
  of latitude, 27
  of longitude, 27
Delaware, smart growth programs in, 421
Delmarva Peninsula, as region, 471, 472*f*
Deltas, 72, 72*f*, 451
Demand, 367, 368*f*
  effective, 380
Demand curve, 368*f*
Democratic Republic of Congo
  cities in, 424
  Congo Corridor, 312
  core area of, 308
  endangered species in, 453
  mineral production by, 157*f*
  population data for, 497*t*
Demographic equation, **200**
Demographic momentum, **212**–213
Demographic transition model, **194**–200, 196*f*
  Europe and, 196–197, 196*f*
Demography, definition of, **180**
Denmark
  agriculture in, 357
  energy in, 154
  nuclear power and, 146
  population data, 501*t*
  regional trade alliances, 327
Denver, Colorado, air pollution in, 443
Department of Agriculture, U.S., 456
Department of Defense, U.S., 28
Department of Education, U.S., 244
Department of Energy, U.S., 460–461
Department of Homeland Security, U.S., 323
Dependency ratio, **191**–192
Deposition
  erosional agents and, **71**–81
  by glaciers, 75–76, 77*f*, 78*f*
  by running water, 71–74, 72*f*, 73*f*, 150, 162–163, 169, 451
  by waves and currents, 76–80, 80*f*, 82
  by wind, 80–81, 81*f*
  of wind-driven sand, 119*f*
Desalination plants, 436
Desert(s)
  climate of, 115*t*
  coastal, 99
  cold, 118
  hot, 115*t*, 116–118, 119*f*
    precipitation, 116, 119*f*
    soil, 118
    temperature, 119*f*
    vegetation, 116
  locations of, 112*f*, 114*f*, 118*f*
  midlatitude, 118, 119*f*
  soil profile, 107*f*
  vegetation, 112, 112*f*, 116, 118
  wind erosion and, 80–81, 81*f*
Desert pavement, 80

Desertification, **159**–160, 160*f*, 161*f*
Desire line map, 15*f*
Detroit, Michigan
  functions of, 394
  population of, 397*t*
  urban renewal in, 417
Developable surfaces, **490**
Developed countries
  agriculture in, 223, 224*f*
  AIDS infection in, 496*t*
  birth and fertility rates in, 182–183, 184–186, 186*f*, 207*f*, 496*t*
  CFCs production and use, 448
  death rates in, 186, 496*t*
  in demographic transition model, 194–200
  economy of, 379*t*
  gross national income per capita in, 223–224, 225*f*
  habitat destruction in, 452
  hazardous waste exports by, 461–462
  high-tech industry in, 376–378
  immigration restrictions, 288–289
  life expectancy and, 496*t*
  life expectancy of women in, 191
  maternal mortality rates, 189
  mining in, 363
  population
    data on, 496*t*
    growth of, 181*f*, 194, 496*t*
  trade
    in primary products, 364–366, 365*f*
    in services, 382*t*, 383
  transnational corporations and, 375
Developing countries
  agriculture in, 223, 224*f*
    percent of population employed in, 223, 224*f*, 346
  AIDS infection in, 496*t*
  birth and fertility rates in, 180, 182–183, 184–186, 186*f*, 199, 207*f*, 496*t*
  capital cities in, land use patterns, 424, 425*f*
  carrying capacity and, 206, 212
  CFCs production and use, 448
  cities in, 401–402, 424–426
  convergence with developed world, 198–200
  death rates in, 186, 198, 207*f*, 496*t*
  deforestation and, 168
  demographic momentum in, **212**–213
  in demographic transition model, 194–200
  economies of, 379, 379*t*
  elderly in, 213, 213*f*
  energy resources and, 137, 148
  and global economy, integration with, 382–383
  gross national income per capita in, 223–224, 225*f*
  gross national product per capita in, **225***f*
  gross world product and, 376
  habitat destruction in, 452
  hazardous waste exports to, 461–462
  and HIV/AIDS, 189–190, 194
  industry in, 376
  life expectancy and, 496*t*
  mass transit systems in, 425
  maternal mortality rates, 189
  migration from, 200
  migration within, 282
  population
    control of, 180, 182, 184–185, 211–212, 211*f*
    data, 496*t*
      reliability of, 209, 209*f*
    density of, 206, 207, 208*f*
    growth of, 180, 181*f*, 190, 194, 211–212, 390, 496*t*
  primate cities in, 401–402, 425
  and purchasing power parity, **225***f*
  recycling in, 459, 462*f*
  refugee problem in, 281
  secessionist movements in, 320–321
  sustainable development and, 171
  technology in, 223–226, 224*f*
  technology transfer to, 376
  and tourism, 381
  trade
    in primary products, 364–366, 365*f*
    in services, 382–383, 382*t*
  transnational corporations and, 374–375
  urbanization in, 191, 207, 207*f*, 390
  women in, 191, 260, 261*f*
  zoning regulations in, 426
Development
  and erosion, 82
  and habitat destruction, 452
  smart growth programs, 421
  sustainable, **169**–171
Devolution, **320**
Dew, formation of, 101
Dew point, **101**
Dharma, 254
Dialects, **239**–242
  regional, 241, 242*f*
  social, 239–241
Diaspora, 249, 251*f*
Diastrophism, **62**–66
Diet and nutrition. *See also* Food supply; Hunger and malnutrition
  acculturation and, 234
  Green Revolution and, 352–353
  minimum caloric intake standards, 345
  religion and, 350
Diffusion. *See* Spatial diffusion
Digestion, anaerobic, 148, 148*f*
Dinosaurs, extinction of, 124
Dioxin contamination, 432, 432*f*, 458
Direction
  absolute, **10**
  projections preserving, 31–32
  relational, 10
  relative, **10**
  terminology of, 10
Disease. *See also* Malaria
  in Africa, 189
  animal wastes and, 440*f*
  contagious diffusion of, 277*f*
  control of, 198
  dams and, 151
  epidemiologic transition, 197
  human environmental impact and, 198
  illegal immigrants and, 288–289
  life expectancy and, 188, 197
  most deadly, 198
  resistant strains, return of, 198
  sanitation and, 188, 196, 197, 197*f*, 198, 211*f*
  slums and, 426
  solid waste disposal and, 457*f*
  water pollution and, 437, 438
  world death toll from, 198
Distance
  absolute, **11**
  critical, **269**, 269*f*
  equal-area, **31**, 32*f*
  equidistant, **31**, 32*f*
  equivalent, **31**
  friction of, 14, 269, 269*f*
  projections preserving, 31
  psychological, 11, 269, 270*f*, 271*f*, 272*f*
  relative, **11**
  and spatial interaction, 269, 269*f*, 270
  terminology, 10–11
Distance decay, 14, **269**, 276–277
Distant learning programs, 382
Districting, 330–333, 331*f*, 332*f*
Divergence, cultural, 223
Divergent plate boundaries, 59, 60*f*
Diversity
  biological
    deforestation and, 169
    in tropical rain forest, 169
  cultural, 218, 219*f*, 223, 235–236, 261–262, 421–426
  linguistic, 236–237, 244–245, 246*f*–247*f*
Doctors Without Borders, 295
DOE. *See* U.S. Department of Energy
Domains, 471–472
Dome(s), 67
Domestic terrorism, 322
Donora Tragedy, 95
Dot map, 38*f*
Dot-com bubble, 377
Doubling times, **193**–194, 194*t*, 195*f*, 197
Douglas, William O., 244
Dravidian language family, 238*f*–239*f*
Dredging, and habitat disruption, 451
Drift nets, 453
Drip irrigation, 434*f*
Drought, deforestation and, 169
Drug(s)
  illegal, U.S.-Mexican border and, 315
  pharmaceutical, from tropical rain forests, 172
Drumlins, 76, 78*f*
Dryland climates, 115*t*, 116–118, 119*f*
  locations of, 114*f*, 118*f*
  weathering in, 68
Dumping, landforms produced by, 450–451. *See also* Landfills
Dunes, 80, 81*f*, 119*f*
Dust Bowl, 80
Dutch Harbor, Alaska, earthquake (1946), 70
Dysentery, 438


# E


E climates. *See* Arctic and subarctic climates
Earth
  axis of, 89, 89*f*
    tilt of, 89–91, 89*f*, 90*f*, 125
  circumference of, 27
  crust, 58, 59*f*. *See also* Plate tectonics
    biosphere and, 433
    elements in, 154, 156*f*
    rock types in, 56–58
  geologic change, constancy of, 56
  inclination of, and air temperature, 89–91, 89*f*, 90*f*
  layers of, 58, 59*f*
  orbit of, 89–91, 90*f*, 91*f*, 125
  population of, 180
  rock types in, 56–58
  rotation on axis, 89, 89*f*
  types of rock, 56–58
Earth grid, **28***f*
Earth science tradition, 19–**20**, 20*f*, 29*f*, 53
  regions in, 472–476
Earth Summit (1992), 169
Earthquakes
  Alaska (1946), 70
  Alaska (1964), 64, 65
  cause of, 60, 64
  China (1976), 66
  damage from, 64, 66*f*, 66*t*
  Egypt (1992), 390
  epicenter, 62, 65
  frequency of, 66
  human disregard of, 60
  intensity of, 65
  Iran (2003), 66, 66*f*
  Little Skull Mountain (1992), 461
  locations of
    in California, 64*f*
    global, 60, 61*f*, 66
  magnitude of, 64, 65, 66, 66*t*
  Mexico (1985), 451
  moment magnitude of, 65
  Northridge (1994), 65
  Pasa Robles, California (2003), 62
  radioactive waste storage and, 460–461
  Richter Scale for, 64, 65, 66*t*
  San Andreas fault, 60, 60*f*, 62, 64, 64*f*, 65*f*
  seismic waves, 64
  and tsunami, 64
*Easley* v. *Cromartie* (2001), 333
East Asia zone, 203
East Germany
  and Berlin, 305
  and Berlin Wall, 312
Eastern Anglo-America industrial region, 375–376, 375*f*
Eastern Asia industrial region, 375–376
Eastern Europe
  cities, 423–424, 424*f*
  economies in, 379
  hazardous waste exports to, 462
  industrial location in, 373
  as industrial region, 375–376, 375*f*
  Islam in, 251
  nationalism in, 321
  population data for, 501*t*–502*t*
  population growth incentives, 187
  regional economic alliances, 328
  Warsaw Treaty Organization, 329
Eastern Orthodox Church
  distribution of, global, 248*f*
  history of, 251
Ebonics, 239
Ecology, 432
  cultural, 220
Economic activity
  classification of, 343–344, 343*f*
  factors affecting, 342–343
  inevitability of change in, 368
  primary, **343**, 343*f*, 344*f*
    agriculture, 345–361
    forestry, 363
    hunting and gathering, 345, 346
    mining and quarrying, 363–364, 364*f*
  quaternary, 343*f*, **344**, 381
    global trade in, 382–383, 382*t*
    international trade in, 384
  quinary, 343*f*, **344**, 381–382
    international trade in, 384
  religion and, 246
  secondary, 343*f*, **344**, 344*f*, 367–378
  supranatural regulation of, 345
  tertiary, 343*f*, **344**, 378–382
    characteristics of, 380–381
    GDP and, 379, 379*f*, 379*t*
    global trade in, 382–383, 382*t*
    location considerations for, 380, 380*f*
    outsourcing and, 380–381
    subdivision of, 380–381
  thresholds for, 404
Economic alliances, regional, 327–329
Economic barriers to migration, 287
Economic base, of urban areas, **398**–400
Economic Community of West African States (ECOWAS), 328
Economic development
  and Brandt North-South line, 226, 226*f*
  characterizations of, 224
  technology and, 223–224, 224*f*
Economic geography, **342**
Economic integration, globalization and, 292–293
Economic motivations for migration, 282
Economic planning
  planned economies, **344.** *See also* Planned economies
  in United States, 482–483
Economic rationality assumption, 367
Economic regions, 482–483
Economic systems
  commercial, **344**
  environmentally-sustainable, 169–171
  globalization of, 292–293
  planned economies, **344**. *See also* Planned economies
  postindustrial, 378–382
    city's role in, 419
    and employment, 378–379, 379*f*
    movement toward, 376
    in United States, 293, 342, 368, 376, 378, 379*t*
  subsistence, **344**, 345
  and transportation, 345, 346*f*
  types of, 344–345

Economy, global, oil prices and, 139
Ecosystems, **433,** 433*f*
as regions, 485, 486*f*
Ecotone, 486
ECOWAS (Economic Community of West African States), 328
Ecuador, 308, 499*t*
Ecumene, **203**–204
Edge cities, **414**
Education
bilingual, 244
distant learning programs, 382
and national unity, 316, 316*f*
and social cohesion, 316, 316*f*
in urban areas, 421
of women, and fertility rates, 185
EEC. *See* European Economic Community
Effective demand, 380
Efficiency, energy, 136, 171
EFTA. *See* European Free Trade Association
Egypt
agriculture in, 163, 207
Aswan Dam, 72, 151, 207
Cairo, 391*f*
air pollution in, 443
climate, 119*f*
growth of, 390
history, 390
population of, 392*t*
slums in, 426
as culture hearth, 230, 231*f*
earthquake (1992), 390
food production in, 206
infant mortality rates, 188*f*, 496*t*
Islamic fundamentalism in, 319
population, 390
data on, 496*t*
density of, 205*t*
distribution of, 203
growth, 496*t*
recycling in, 459
urbanization in, 207
Egyptian Arabic, number of speakers, 240*t*
E-horizon, 108
Eire (Ireland), borders of, 311
El Niño, **108**–109, 108*f*, 109*f*, 126, 361*f*
El Salvador
erosion in, 160
migration from, 200
population data for, 498*t*
Elderly, increase in, 213, 213*f*
Electoral geography, **331**
Electricity
generation of, 149
and air pollution, 443, 443*f*, 443*t*
and water pollution, 441
transmission of, 149
Electromagnetic spectrum, 42, 44*f*
Element(s), in earth's crust, 154, 156*f*
Elongated states, **305,** 305*f*
Eluviation, 108
Embankment systems, 436
Emigration, 284. *See also* Migration
Employment
in agriculture, 223, 224*f*, 346
in Appalachia, 482–483
40-hour work week, 413
in geography, 8, 10–11
GIS skills and, 49
in high-tech industries, 376–377
postindustrial economy and, 378–379, 379*f*
suburbanization and, 416, 417
tourism and, 381
transnational corporations and, 374
transportation to
in European cities
Eastern, 424
Western, 422
in U.S., 417
travel to, 274, 274*f*, 275*f*, 278, 280, 417
in urban areas, 411, 419
in developing nations, 425
in U.S.
in high-tech industry, 378
by sector, 378–379, 379*f*
women and, 275, 417
Enclaves, **305,** 306*f*
ethnic, 258–259, 258*f*, 260*f*
politics of, 309
Endangered species, 451–452, 452*f*, 453, 453*f*, 455*f*
Energy, **136.** *See also* Electricity; Geothermal power; Hydroelectric power; Nuclear energy; Power plants
biomass fuels, **148,** 171
cleaner fuels, 448–449
conservation of, 171
consumption of
and economic growth, 137
reducing, 171
in U.S. by source, 137–138, 137*f*
conversion of, 136, 137
in ecosystems, 433
history of use, 137, 137*f*
industrialization and, 137
kinetic, 136
as master resource, 137, 364
nonrenewable, 137–147
potential, 136
renewable, 148–154
solar, **150**–152, 152*f*
Energy efficiency, 136, 171
England. *See* United Kingdom
English language
in Africa, 246*f*
changes in, 237
dialects of, 239, 242*f*, 243
diffusion and acceptance of, 237, 238, 241*f*, 243, 295
number of speakers, 240*t*
as official U.S. language, 244–245
place names in, 244–245
as UN working language, 241*f*
vocabulary imports and exports, 237–238
World Englishes, 243
Entisols, 110*t*, 111*f*
Environment. *See also* Human environmental impact; Natural resources
change over time in, 14
cleansing abilities of, 437, 442
components of, 432
as cultural control, 220–221
Environmental determinism, **220**–221
Environmental pollution, **437.** *See also* Air pollution; Biocides; Human environmental impact; Thermal pollution; Water pollution
point and non-point sources of, 437–439
Environmental Protection Agency (EPA)
biocides and, 439
on hazardous waste sites, 460
and PCBs, 440
purpose of, 13
Superfund sites, 460
and Times Beach, Missouri, 432, 432*f*
on water pollution, 442
Environmental refugees, 294
Environmental Systems Research Institute, Inc., 47
EPA. *See* Environmental Protection Agency
Epicenter, 62, 65
Epidemic of epidemics, 198
Epidemiologic transition, 197
Equador, census, 209*f*
Equal-area projections, **31,** 32*f*
Equator, 27
Equatorial low pressure, 97, 114
Equidistant projections, **31,** 32*f*, 492–493, 493*f*
Equinox, 91, 91*f*
Equivalent projections, **31**
Eratosthenes, 6
Eritrea
population data for, 497*t*
and UN intervention, 324*f*
Erosion, **71**–81
balance with soil creation, 159
beaches and, 82
Black Hills and, 473, 474*f*
deforestation and, 166, 169
desertification and, 159–160
embankment systems and, 436
by glaciers, 75–76, 77*f*, 78*f*
by groundwater, 74–75
by precipitation, 71–72, 71*f*
reduction techniques, 162, 162*f*
by running water, 71–74, 71*f*, 73*f*, 150, 162–163, 169, 451
of topsoil, 159–163, 223
by waves and currents, 76–80, 80*f*, 82
by wind, 80–81, 81*f*
Escarpments, 63, 63*f*
Eskers, 76, 78*f*
Eskimo-Aleut language family, 238*f*–239*f*
Estuarine zone, 62, 163, 163*f*
ETA. *See* Euskadi ta Askatasuna
Ethiopia
agriculture in, 160
energy in, 148
maternal mortality rates in, 189
population data for, 209, 497*t*
and UN intervention, 324*f*
Ethnic cleansing, **322**
Ethnic enclaves, 258–259, 258*f*, 260*f*
Ethnic group
boundary disputes and, 314
defined, 258
Ethnic niche businesses, 286
Ethnic religions, 246–**247**
in East Asia, 257
Ethnicity, **257**–259. *See also* Race
in Afghanistan, 258*f*
border conflicts and, 314
as cultural difference, 235–236
definition of, 257–258
and state boundaries, in Africa, 303, 304*f*, 311
and territorial segregation, 258, 411, 412*f*
and urban residence patterns, 409, 411, 411*f*, 412*f*
in Yugoslavia, 258*f*
Ethnocentrism, 258
Ethnographic boundaries, 311–312
EU. *See* European Union
Eurasia
coal reserves, 142*f*
hydropower in, 152*f*
industry in, 375*f*, 377*f*
natural gas reserves, 145*f*
natural gas trade flows, 144*f*
oil reserves, 139*f*
oil shale deposits, 145*f*
oil trade flows, 408*f*
Euro, 291, 327
Europe
acid rain in, 444
agriculture in, 357
air pollution in, 444, 448
assimilation of minorities, 235
birth and fertility rates in, 182, 187, 501*t*–502*t*
cities in, 403–404
eastern, 423–424, 424*f*
western, 421–423, 423*f*
climate of, 121–122
coal production, 142
coal reserves, 142*f*
colonialism. *See* Colonialism
culture of, 234
as culture realm, 221*f*
death rates in, 186, 501*t*–502*t*
Eastern
cities, 423–424, 424*f*
economies in, 379
hazardous waste exports to, 462
industrial location in, 373
as industrial region, 375–376, 375*f*
Islam in, 251
nationalism in, 321
population data for, 501*t*–502*t*
population growth incentives, 187
regional economic alliances, 328
Warsaw Treaty Organization, 329
economic restructuring in, 345
economy of, 379
energy in, 154
fuel economy standards, 141
HIV/AIDS in, 501*t*–502*t*
hydropower in, 152*f*
immigration from, 201*f*
industry in, 375–376, 375*f*, 377*f*
location of, 423, 423*f*
infant mortality rates, 188, 501*t*–502*t*
influenza pandemic in (1781), 277*f*
land use in, 412
languages in, 236, 236*f*
life expectancy in, 188, 501*t*–502*t*
marriage customs in, 197
migration from, 200
migration to, 281
mineral resources, 158
modern state evolution in, 303
natural gas reserves, 145*f*
natural gas trade flows, 144*f*
oil imports, 139
oil reserves, 139*f*
oil shale deposits, 145*f*
oil trade flows, 408*f*
outsourcing and, 373
population
age of, 213*f*
data on, 501*t*–502*t*
decline in, 196–197
demographic transition in, 196–197, 196*f*
distribution of, 200
growth of, 195*f*
population pyramid, 187, 191, 192*f*
recycling in, 459
secessionist movements in, 320, 321*f*
solid waste disposal in, 458
as terrorist target, 323
tobacco's spread in, 279
transnational corporations and, 374
urbanization in, 391, 394*t*
waste disposal in, 458, 459
water pollution in, 439
Western and Central Europe industrial region, 375–376
wetlands in, 164
women in, 318
European Coal and Steel Community, 327
European Economic Community (EEC), 327
European Free Trade Association (EFTA), 327
European Union (EU)
agriculture in, 355
birth and fertility rates, 186
CFCs use and production, 448
Chech Republic's admission to, 298*f*
future of, 329
globalization and, 291
history of, **327**
membership, 327, 328*f*
migration to, 186
population of, 186
regulation of economic activity by, 345
supranational institutions of, 327
terrorism and, 323
trade, in services, 382*t*
Euskadi language, 314
Euskadi ta Askatasuna (ETA), 314, 314*f*, 323

Euskara, 236*f*
Eutrophication, **439**
Evaporation
  in hydrologic cycle, 434, 434*f*
  and salinization of soils, 163
Everglades
  cleanup efforts, 442
  as ecosystem, 486, 486*f*, 487*f*
  habitat destruction in, 436, 437*f*, 486
  as wetland, 163
Excavation, landforms and, 449–450, 450*f*
Excess vote gerrymandering, 331, 332*f*
Exclaves, **305**
Exclusive economic zone (EEZ), **325**–326, 325*f*, 326*f*
Exotic species, **453**–455
External (agglomeration) economies, **368, 370,** 371*f*
Extinction, 451
Extractive industries, **361**
Extrusive igneous rock, 56
*Exxon Valdez* oil spill, 441
Eye, of hurricane, 103, 105*f*

# F

Fairbanks, Alaska, climate, 124*f*
Falkland Islands
  British-Argentine war over, 306
  population density, 206
Falls, formation of, 72
False-color images, **42**
Family
  activity space of, 271, 274*f*
  in Confucianism, 257
  migration and, 287
  post-industrial economy and, 380
  size, determinants of, 185, 196, 197, 199–200, 202–203
Family planning programs, 202–203, 211–212
Family status, and urban residence patterns, 409–411, 411*f*
Fanagalo, 252
FAO. *See* Food and Agriculture Organization
FAS. *See* Federation of American Scientists
Fault(s), **60,** 63–65, 63*f*–65*f*
  Ghost Dance Fault, 461
  San Andreas fault, 60, 60*f*, 62, 64, 64*f*, 65*f*
  transform, 63*f*, 64, 65*f*
Fault escarpment, 63, 63*f*
Fault steps, 63*f*
Fault-block mountain, 63*f*
FDI. *See* Foreign direct investment
Federal Housing Administration (FHA), 413
Federal states, 308
Federalism, asymmetric, 308, 320
Federation of American Scientists (FAS), 46*f*
Federation of Malaysia and Singapore, 329
Federation of the West Indies, 329
Feminist movement, 260–261, 318
Fertility rates. *See* Total fertility rate
Fertilizer, and water pollution, 439, 440*f*
FHA. *See* Federal Housing Administration
Financial centers, international and off-shore, 383*f*, 384
Finland
  Laplanders and, 315
  population data, 501*t*
  regional trade alliances, 327
Fiords, 76
First Law of Geography, 14
Fish farms (aquaculture), 361, 361*f*, 362, 363
Fishing
  commercial, 361–363
    aquaculture, 361, 361*f*, 362, **363**
    catches, 361, 361*f*, 362
    disputes over, 315
    drift nets and, 453
    environmental impact of, 453
    fisheries locations, 362, 362*f*
    overfishing, 361, 362, 362*f*, 453
  water pollution and, 440
Fission, nuclear, **146**–147
Five Civilized Tribes, forced migration of, 282*f*
Five fundamental themes of geography, 17
Flanders, 321*f*
Flexible manufacturing, 369, 370–371
Flexible production regions, 371
Flood(s)
  Banqiao Dam collapse, 4
  damage caused by, 72
  dams and, 150
  deforestation and, 169
  destructiveness of, 5*f*
  embankment systems and, 436
  erosion and, 72, 163
  global warming and, 127, 128*f*
  and Hurricane Mitch, 5*f*
  India (1978), 97
  Indonesia (2002), 5*f*
Floodplains, **72**
Florida
  Dade County waste treatment, 442
  Everglades
    cleanup efforts, 442
    as ecosystem, 486, 486*f*, 487*f*
    habitat destruction in, 436, 437*f*, 486
    as wetland, 163
  exotic species in, 455
  irrigation in, 340*f*
  Karst topography in, 75, 76*f*
  Kissimmee River modification, 436, 437*f*
  migration fields, 290*f*
  redistricting and, 332*f*
Flow-line maps, qualitative and quantitative, 40, 41*f*
Folding, **62,** 62*f*, 63*f*, 476*f*
Folk culture, **233**
"Follow the sun" practice, 382
Food and Agriculture Organization (FAO), 206, 324, 326, 345, 353, 355, 453
Food chains, **433,** 433*f*
  and acid rain, 445
  and biological magnification of toxins, 455–456, 456*f*
  ozone layer depletion and, 447
  water pollution and, 440, 441
Food supply. *See also* Agriculture; Fishing; Hunger and malnutrition
  and carrying capacity of land, 206–207, 206*f*
  expansion of, 351–353
  Green Revolution and, 352–353
  population growth and, 162, 181, 197, 198, 211, 211*f*, 345–346
Food webs, 433*f*
Footloose economic activities, 370, 378
Forced migration, 281, 282*f*
Ford, Henry, 369
Ford Motor Co., 294*f*
Fordism, **369,** 370
Foreign direct investment (FDI)
  globalization and, 291–292
  trade in world services and, 382
  transnational corporations and, 374–375
Forest(s). *See also* Coniferous forests; Deciduous forests; Deforestation; Logging; Tropical rain forest
  acid rain and, 445
  commercial, 164–166, 165*f*
  functions performed by, 165
  hardwood, 164–165, 165*f*
  management of, 165–167, 171
  national (U.S.), 166–167, 167*f*
  as natural resource, 164–169
  old-growth, 166–167
  percent of earth covered by, 164
  vegetation, 112, 112*f*
Forest Service, U.S., 166
Forestry, as economic activity, 363
Form utility, 344
Formal regions, **17,** 18*f*, 470–**471**
  multifactor, 473, 479
Forward-thrust capital cities, 309–310
Fossil fuels. *See also* Coal; Natural gas; Oil
  and air pollution, 443–444, 443*f*, 443*t*, 445*f*
  dominance of, 137–138
  and global warming, 126
  industrial revolution and, 137
  as nonrenewable resource, 135*f*, 136
Four Corners Monument, 331*f*
Four Noble Truths, 255
Fourth World Conference on Women (1995), 262
Fragmented states, **305,** 305*f*
France
  Antarctic territorial claims, 301*f*
  anti-English-language movement in, 295
  cities in, 421, 423*f*
  colonization of, 252
  core area of, 308
  devolution in, 320
  enclaves, 305, 306*f*
  energy in, 152
  gated communities in, 416
  high-tech industries in, 377
  immigration restrictions, 288
  infant mortality rate, 188*f*
  mineral resources, 157
  nationalist movements in, 320, 321*f*
  nuclear power in, 146–147, 147*f*
  Paris, 196*f*
    as core area, 308
    design of, 422, 423*f*
    Notre Dame Cathedral, 253*f*
    as primate city, 308, 421
    as world city, 401, 403*f*
  population data for, 501*t*
  population growth incentives, 187
  regional trade alliances, 327
  Reims Cathedral, 445*f*
  road system in, 318
  separatist movements in, 321, 321*f*
  trade, in services, 382*t*
  women in, 318
Free trade
  globalization and, 291, 295
  regional trade alliances, 327–329
Freedom to Farm, 162, 162*f*
French Canadians, 243
French language
  in Africa, 246*f*
  as lingua franca, 242
  number of speakers, 240*t*
  as UN working language, 241*f*
French Revolution, 303
Fresh Kills landfill, Staten Island, 458
Fresh water zone, 163*f*
Friction of distance, 14, 269, 269*f*
Frictional effect, in wind, **97**
Friedman, Thomas, 293
Frobisher, Martin, 79
Front(s), precipitation and, **102**–104, 103*f*, 104*f*
Front Liberation du Quebec (FLQ), 323
Frontal precipitation, **102**
Frontier zone, 311
Frost action, 68
Fuel(s). *See also* Fossil fuels
  alcohol, 148
  animal waste, 415, 415*f*
  biomass, **148,** 171
  cleaner, 448–449
  methane, 143, 148
  nuclear
    disposal of, 430*f*, 460–461
    as nonrenewable resource, 136
  waste, 148, 148*f*
  wood, 137, 137*f*, 148, 164–165, 165*f*
Fuel economy of motor vehicles, 140–141, 171
Fuel rods, nuclear, disposal of, 430*f*, 460–461
Fujita scale of tornado intensity, 104
Fuller dymaxion projection, 494, 494*f*
Fuller, R. Buckminster, 494
Functional regions, **17,** 18*f*, 470–**471**
Fundamentalism, militant, 249, 319
*The Fundamentals: A Testimony of the Truth,* 249

# G

Gabon
  cities in, 425
  population data for, 497*t*
Galactic cities, 414, 415*f*
Galápagos giant tortoise, 455*f*
Gambia, population data for, 496*t*
Ganges River, 216*f*, 314, 350
*Garimpeiros,* 441
Gasoline, lead in, 448
Gated communities, 414, **416,** 416*f*
Gathering industries, **361**
GATT. *See* General Agreement on Tariffs and Trade
Gays, urban preference of, 411
Geddes, Patrick, 339
GEM. *See* Gender empowerment measure
Gender, **259**–261. *See also* Women
  cultural differences in, 235–236, 259–261
  definition of, **259**
  gender empowerment measure (GEM), 261, 263*f*
  and language use, 242–243
  and migration, 200, 285
Gender empowerment measure (GEM), 261, 263*f*
Gender gap, 318–319
Gene flow, **259**
General Agreement on Tariffs and Trade (GATT), 326
General purpose map, definition of, 33
Generalized map, definition of, 33
Genetic drift, **259**
Gentrification, **419**–420, 420*f*, 422
Geodetic control data, 37
Geographic characteristics of states, 303–310
Geographic dialects, 241
Geographic information system (GIS), **45**–49
  applications, 47–49
  components of, 47
  output of, 47, 48*f*
  value of familiarity with, 11
Geography
  branches of, 7–8
  careers in, 8, 10–11
  and change, 14, 15*f*
  core concepts, 8–17
  cultural, 218
  in daily life, 4–6
  definition of, 485
  dominant interests of, 8
  economic, **342**
  electoral, **331**
  etymology of term, 4, 6
  five fundamental themes of, 17
  history of discipline, 6–7
  human, 8
  importance of understanding, 8
  job opportunities in, 8
  linguistic, 241
  locational tradition, 339
  National Geography Standards, 17–19
  origin of term, 6
  physical, 8
  political, 300–301
  population, **180**
  process in, 476–477
  regional, 8, 467
  standards of, 17–19
  subjects of study in, 4–6
  systematic, 8
  traditions in, 19–21, 20*f*

*Geography* (Strabo), 21
Geologic time, 58
Geological factor, in soil formation, 106
Geometric boundaries, **311**
Geometric projections, 490–493, 490*f*
Geomorphology, 56
Georgia (Asian nation), 499*t*
Georgia (U.S.)
Gullah language in, 479, 479*f*
maps of, 43*f*
redistricting and, 332*f*, 333
urban renewal in, 420
Geothermal heat pumps, 154
Geothermal power, 152, **152**–154, 153*f*
future use of, 171
nations using, 152, 153*f*
German language, number of speakers, 240*t*
Germanic languages, 237, 238*f*–239*f*
Germany
Aachen, boundaries of, 471*f*
air pollution in, 448
Berlin, division of, 305, 312
cities in, 421
East, and division of Berlin, 305, 312
enclaves, 305
energy in, 154
Hanseatic League, 324
high-tech industries in, 377
immigration restrictions, 288
migration to U.S., 284*f*
mineral imports by, 158
nuclear power and, 146, 147*f*
oil consumption by, 138*f*
population data for, 501*t*
regional trade alliances, 327
reunification of, 312, 321
trade, in services, 382*t*
waste disposal in, 462
in World War II, 314
Gerry, Elbridge, 331*f*
Gerrymandering, **331**–333, 331*f*, 332*f*
racial, 331, 332–333, 332*f*
Ghana
agriculture in, 351
mineral production by, 157*f*
population data for, 496*t*
Ghost Dance Fault, 461
Gibraltar, 306*f*
Gifford Pinchot National Forest, 166*f*
GIS. *See* Geographic information system
Glacial till, 76, 78*f*
Glacial trough, 76, 77*f*
Glacier(s), 75–76, 77*f*. *See also* Ice age
erosion and deposition by, 75–76, 77*f*, 78*f*
extent of, 75, 77*f*
greenhouse effect and, 76, 127, 127*f*
Glacier Peak Wilderness Area, 78*f*
Global Malaria Eradication Program, 456
Global marketing, 293, 294*f*
Global Positioning System (GPS), **28**–30
applications of, 29–30, 49*f*
operation of, 28–29
USGS benchmarks and, 37
Global Positioning System receiver, 30*f*
Global warming, **126**–128, 127*f*, 128*f*
alternative theories of, 127–128
dams and, 150
deforestation and, 126, 168–169, 170*f*
and disease, 198
and glaciers, 76, 127, 127*f*
impact of, 127–128, 127*f*
motor vehicles and, 140
rain forests and, 168–169, 170*f*
Globalization, **15**, **291**–295. *See also* Transnational corporations
*vs.* Cold War, 293
communication and, 291–292, 382
and cultural integration, 295
and economic integration, 292–293
and free trade, 295
outsourcing and, 373
and political integration, 295
and standard of living, 293
technology and, 291–292, 293, 382
time and, 382–383
and transnational corporations, 292–293, 294*f*
Globe(s)
map projections from. *See* Projections
properties of, 30
relative location on, 9*f*
GMT. *See* Greenwich Mean Time
Gneiss, 57*f*, 58
GNI. *See* Gross national income
Gnomic projections, 490, 490*f*, 491, 493, 494*f*
*Goals 2000: Educate America Act* (1994), 10, 19
Gobi Desert, 80
Golan Heights, 325–326
Gold, mining of, 450*f*
Gondwana, 59*f*
Good Friday Agreement (1998), 323
Good Friday earthquake (Alaska, 1964), 64, 65
Goode, J. Paul, 493
Goode's Homolosine, 493–494, 494*f*
*Goode's World Atlas*, 494
Goods and services, as tertiary economic activities, 344
Gorbachev, Mikhail, 309
Gottman, Jean, 484
Government. *See also* Social services
agricultural controls, 355
efficiency of, and loyalty, 317, 319
industry location and, 373, 374
local and regional, 329–335
administrative units in, 330, 333–335, 334*f*
importance of, 329–330
waste disposal costs, 457
representation and, 330–333, 331*f*, 332*f*
urban renewal programs, 416–417, 419*f*, 420
zoning regulations, 335, 373
in developing nations, 426
objections to, 412
purpose of, 411–412
GPS. *See* Global Positioning System
Gradational processes, **68**–81
erosion and deposition, **71**–81
glaciers, 75–76, 77*f*, 78*f*
groundwater, 74–75
running water, 71–74, 71*f*, 73*f*, 150, 162–163, 169, 451
waves and currents, 76–80, 80*f*, 82
wind, 80–81, 81*f*
weathering, **68**–70
chemical, **69**–70
mechanical, **68**–70
Grain production
global output, 352
livestock-grain farming, 357, 358*f*
wheat farming, large-scale, 357–359, 358*f*
Grameen Bank, 355
Grand Canyon, 58*f*
Grand Coulee Dam, 150
Grand Forks, North Dakota, 458*f*
Granite, formation of, 56
Grant, Ulysses S., 366
Graphic scale, 33, 33*f*
Grassland soil, profile, 107*f*
Graticule, **28***f*, 489
Gravel, reserves of, 156
Gravity, landforms and, 70
Great Barrier Reef, 80
Great circle, **489**
Great Lakes, alien species in, 454, 454*f*
Greece
migration to U.S., 286
nuclear power and, 146
population data for, 502*t*
regional trade alliances, 327
urban areas in, 396
women in, 318
Greeks, ancient
alliances, 324
environmental impact of, 14
geographical interests, 6–7, 7*f*, 20
habitability terms of, 203–204
mapping by, 27
Green Line (Beirut, Lebanon), 248, 250*f*
Green Revolution, 212, 352–**353**. *See also* Agricultural Revolution
women and, 353, 354–355
Green River Formation, 144
Greenhouse effect, **126**–128, 127*f*, 128*f*
and glaciers, 76, 127, 127*f*
impact of, 127–128, 127*f*
motor vehicles and, 140
natural gas power and, 143
rain forests and, 168–169, 170*f*
Greenpeace, 295
Greenwich, England, 27
Greenwich Mean Time (GMT), 28
Grid system, **27**–28, 28*f*
Gross domestic product (GDP)
tertiary economic activities in, 379, 379*f*, 379*t*
in urban areas, 419
Gross national income (GNI)
in developed *vs.* developing countries, 223–224, 225*f*
per capita, 225*f*
by region and nation, 496*t*–503*t*
Gross world product, developing countries and, 376
Ground source heat pumps, 154
Groundwater. *See also* Aquifers
contamination of, 441, 457–458, 461, 462*f*, 486
erosion by, 74–75
in hydrologic cycle, 434*f*
overuse of, 435–436
in Everglades, 486
and subsidence, 451, 451*f*
Group of 77, 366
Growth. *See* Development
Guanajuato, Mexico, 425*f*
Guatemala
erosion in, 160
Guatemala City, advertising in, 277*f*
population data for, 498*t*
Guinea
mineral production by, 157*f*
population data for, 496*t*
Guinea-Bissau, population data, 180, 496*t*
Gujarati language, number of speakers, 240*t*
Gulf of Mexico
dead zone in, 439, 439*f*
pollution in, 441
water pollution in, 439, 439*f*
Gulf War. *See* Persian Gulf War
Gullah language, 479, 479*f*
Gullying, 71*f*
Guthrie, Woody, 150
Gypsies, 303

# H

H climates. *See* Highlands climates
Habitat destruction, 452, 452*f*
in Everglades, 436, 437*f*, 486
Haggett, Peter, 312*f*
Haiti
erosion in, 160
language of, 242
migration from, 200
population data for, 498*t*
religion in, 232
Haitian Creole, 242
Haitians, migration of, 283
*Hajj*, 247*f*
Halons, and ozone layer, 447, 447*f*, 448
Hanford, Washington, radioactive waste storage site, 462*f*
Hanseatic League, 324
Hanunóo People of Philippines, 350
Hardwood forests, 164–165, 165*f*
Hasidic Jews, 232
Hausa language, 242
Hawaiian Islands
alien species in, 454–455
Hilo tsunami strike, 64, 70
new island forming in, 56
Tsunami warning system, 70
volcanoes of, 61*f*, 66, 67*f*
Hayakawa, S. I., 245
Hazardous waste, **457**
biological magnification of, **455**–456, 456*f*
civilian, 460–461
disposal of, 460–462, 462*f*
incinerator ash as, 458
low *vs.* high-level, 460
military, 460–461
Times Beach, Missouri, contamination, 140*f*, 432, 432*f*
Head-of-navigation locations, 396
Health. *See* Human health
Heap-leach gold mining, 441
Hebgen Lake, Montana, 24*f*
*Hegira*, 253
Herbicides
Green Revolution and, 353
as water pollutant, 439
Herding, nomadic, **348**–349, 348*f*
Herodotus, 6
Hierarchical diffusion, **277**–278, 277*f*, 278*f*
Hierarchical migration, **291**
Hierarchy
of central places, 404–405, 404*f*
of regions, 471, 472*f*
urban, 400, **400**, 403*f*
High seas, 325*f*, 326
Highlands climates, 114*f*, 115*t*
High-tech industries, 376–378
and agglomeration, 378
dot-com bubble in, 377
outsourcing by, 378
as percent of all industry, 377, 377*f*
Highways
in Canada, 421
suburbanization and, 412, 413
transportation of goods by, 370
Hillary, Edmund, 74
Hilo, Hawaii, tsunami strike, 64, 70
Hindi, 243
as lingua franca, 242
number of speakers, 240*t*
Hinduism, 246, 254–255, 256*f*
and diet, 350
diffusion routes, 250*f*, 254
distribution of, global, 248*f*
and India, partitioning of, 481
as state religion, 317
temples, 254–255, 256*f*
Hispanic Americans
birth and fertility rates, 186
defacto segregation of, 411, 412*f*
Histosols, 110*t*, 111*f*, 123
Hitler, Adolf, 314
HIV/AIDS
deaths from, 198
developing nations and, 189–190, 194, 496*t*
infection with, 198
by region and nation, 496*t*–503*t*
population growth and, 197
Holland, enclaves of, 305
Home rule, 320
Homelessness, 417, 418
Homeostatic plateau, 211, 211*f*
Homolosine, Goode's, 493–494, 494*f*
Honeybees, Africanized, 453
Hong Kong
Chinese takeover of, 180, 403
language in, 243
latitude and longitude of, 28, 30*f*
as network city, 403

population data for, 500*t*
women in, 191
as world city, 401, 403*f*
Horizons, of soil, **107**–108, 107*f*
Horns, 77*f*
Hot spots, 62, 66, 67*f*
House of Representatives, U.S., Black membership, 332
Households, water use by, 434–435
Housing Act of 1949, 416
Houston, Texas
air pollution in, 447*f*
edge cities of, 414
gentrification in, 419
population of, 397*t*, 402*t*
zoning and, 412
Huang He (Yellow) River, 72, 435, 436
Hudson River, 442
Human environmental impact, 13–14, 13*f*. *See also* Air pollution; Biocides; Hazardous waste; Thermal pollution; Water pollution
agriculture and, 14, 432, 439–440, 440*f*, 452
on air and climate, 442–449
on beaches, 82
in Chaco Canyon, 223
climate change and, 126–128
contamination and poisoning, 455–456
cultural landscapes and, 221, 222*f*
dams, 150–151, 436
and disease, 198
fuel economy standards and, 140–141, 171
as fundamental theme in geography, 17
habitat destruction, 452, 452*f*
in Everglades, 436, 437*f*, 486
history of, 14, 432
of hunting and gathering cultures, 222–223
increase in, 433
industry and, 440–441
on landforms, 449–451, 449*f*
in mining and quarrying, 134, 134*f*, 143, 144, 146*f*, 366, 441, 449–450, 450*f*, 452
and natural disasters, 4
on plants and animals, 451–456
rice patties, 2*f*
on soils, 159–163
solid waste disposal, 456–462
technology and, 221
on water, 433–442
Human geography, 8
Human health. *See also* Disease; Life expectancy; Medicine
air pollution and, 443
ozone hole and, 447
smog and, 446
Human rights, and UN intervention, 324
Humid areas, stream landscapes in, 72–73
Humid continental climate, 115*t*, 120, **122**–123, 122*f*
locations of, 114*f*
precipitation, 122, 122*f*, 123*f*
soil, 120, 122
temperature, 122*f*, 123*f*
vegetation, 122, 123*f*
Humid midlatitude climates, 114*f*, 115*t*, 118–123
Humid subtropical climates, 115*t*, **120,** 121*f*
locations of, 114*f*
soil, 120
Humidity, relative, 99, 99*f*, **101**
Humus, 106, 107*f*, 108, 109
Hungary
land claims, 314
migration from, 282
NATO and, 329, 330*f*
population data for, 501*t*
women in, 318
Hunger and malnutrition
causes of, 345–346
global areas of, 345–346, 347*f*
Green Revolution and, 352–353
urban agriculture and, 351
Hunter-gatherers, 230
domestication of, 226–227, 228*f*
gender roles among, 260
as primary economic activity, 345, 346
technology of, 222–223
Hunting, impact on wildlife, 452–453
Hurricane(s)
characteristics of, 105*f*
eye of, 103, 105*f*
force and destructiveness of, 5*f*, 82*f*, 88, 103, 105*t*
formation of, **103**
Isabel (2003), 82*f*, 88, 88*f*
Mitch (1998), 5*f*
typical tracks of, 105*f*
Hussein, Saddam, 322
Hutu, 258, 283
Huxley, Julian, 222
Hydrilla, 455
Hydrocarbons, as air pollutant, 140
reduction efforts, 448
Hydroelectric power
in China, 150–151
transmission of, 149
in United States, 149, 149*f*, 150
Hydrofluorocarbons, and global warming, 126
Hydrologic cycle, **433**–434, 434*f*
Hydrolysis, in chemical weathering, 69
Hydropower, **148**–149
consumption
global, 149, 152*f*
U.S., 137, 137*f*, 149, 149*f*
Hydrosphere, 433

# I

IBM Corp., 294*f*
Ice age
causes of, 75
cyclical nature of, 124
end of, 14
glaciation in, 77*f*
"little," 125
new, from global warming, 128
Ice cap climate, 115*t*
locations of, 114*f*, 124*f*
Ice storm, Northeastern U.S. (1998), 86*f*
Iceland
energy in, 152
location of, 307
as nation-state, 302
population data, 501*t*
time zone of, 29*f*
Iconography, 316
Idaho
hydroelectric power in, 149, 149*f*
nuclear fuel storage in, 430*f*
radioactive waste disposal in, 461
Ideological subsystem, 222, **228**–229, 246
Igneous rock, **56**–57
Ikonos 1, 46
Illinois. *See also* Chicago, Illinois
Champaign County administrative units, 334, 334*f*
edge cities in, 414
industry incentives, 374
population density, by County, 12*f*
and weather, sudden changes in, 53
Illuviation zones, 108
ILO (International Labor Organization), 326
Immigration. *See also* Migration
assimilation, **234**–235, 244–245, 258
to United States, 200, 283, 288, 342
backlash against, 288–289
central cities, 419, 420*f*
by Mexicans, 285, 286, 287*f*, 315, 315*f*
backlash against, 288–289
motivations for, 282
restrictions on, 288–289
return migration, 290
Immigration and Naturalization Service, U.S., 288
Inceptisols, 110*t*, 111*f*, 115*t*
Incineration of solid wastes, 458–459, 458*f*
Income
and activity space, 275
and tertiary activity, 380
India
agriculture in, 163, 350, 352, 466*f*
air pollution in, 446
birth and fertility rates in, 180, 182
border disputes, 311, 311*f*
caste system in, 246, 254
CFCs use and production, 448
cities in, 392*t*, 424, 425, 426
coal reserves, 142*f*
colonization of, 252
culture of, 218
dams in, 150
death rates in, 198
economy of, 382
endangered species in, 452
energy in, 148, 154
erosion in, 160
establishment of, 283, 311–312
ethnic niche businesses in, 286
fertility rates in, 199
flood in (1978), 97
high-tech industry in, 378
immigration from, 201*f*
infant mortality rate, 188*f*
language in, 243
migration in, 286
migration to U.S., 286
mineral production by, 157*f*
monsoon season in, 97
population, 203
control of, 211–212, 211*f*
data, 500*t*
reliability of, 209
density of, 205*t*
growth of, 183*f*
rain forests in, 168
religion in. *See* Hinduism
Varanasi (Benares), 216*f*
water pollution in, 441–442
water supply in, 435
wind currents in, 99
women in, 191
Indian subcontinent, political regions in, 481, 481*f*
Indiana
Indianapolis, urban renewal in, 420
industry incentives, 374
settlement pattern in, 405*f*
Indic culture realm, 221*f*
Indo-European language family, 237, 238*f*–239*f*, 240*f*
Indonesia
agriculture in, 350
air pollution in, 443
cities in, 424, 426
earthquakes in, 66
El Niño and, 108
endangered species in, 452, 452*f*
flood damage in, 5*f*
language of, 242
logging in, 165
migration in, 281
mineral production by, 157*f*
mining in, and water pollution, 441
population
data, 500*t*
reliability of, 209
distribution of, 203
rain forests in, 167, 168
religion in, 246, 248*f*, 253, 255*f*
shape of, 305
women in, 261
Indus River
as culture hearth, 230, 231*f*
water use and, 314
Industrial parks, 371*f*, 378, 417
Industrial Revolution
coal and, 141
and cultural convergence, 223
and demand for minerals, 156
energy and, 137
gender roles and, 260
and global warming, 126
impact of, 227, 232*f*
manufacturing regions and, 376
and migration to cities, 286
and population, 195*f*, 196–197, 211*f*
Industrialization
energy resources and, 137
and pollution, 444
Industry. *See also* Manufacturing industries; Secondary economic activities
decentralization of, 370–371
energy use by, 171
environmental impact, 440–441, 442, 443, 443*f*, 443*t*, 444, 446
in Europe, 375–376, 375*f*, 377*f*
location of, 423, 423*f*
extractive, **361**
gathering, **361**
high-tech, 376–378
and agglomeration, 378
dot-com bubble in, 377
outsourcing by, 378
as percent of all industry, 377, 377*f*
location of
comparative advantage and, 371–373
in Europe, 423, 423*f*
government influence on, 373, 374
high-tech industries, 376–378
models for, 367–368, 368*f*, 369*f*
quaternary, 381
trends in, 375–376
regions of, global, 375–376, 375*f*
suburbanization and, 413, 416, 417
ubiquitous, 370
water use by, 434, 440
Infant mortality rate, 186–188, 188*f*
environmental pollution and, 438
by region and nation, 496*t*–503*t*
Infanticide, Chinese population controls and, 184
Influence zones, urban, **402**
Information services, employment in, 381
Information, state control of, 323
Information technology, and manufacturing, 369
Infrared aerial photography, 42, 44*f*
Infrastructure, **370,** 371*f*
Inheritance laws, women and, 354
Inland wetlands, 163
Inlets, formation of, 79
Inner Six, 327, 327*f*
Innovation, **230**
cities as centers of, 230, 270–271, 277–278
conditions required for, 270–271
definition of, 275
diffusion of, 270, **275**–276, 279
history of, 232*f*
motivation for, 230
parallel, 232
rate of, historical, 230, 232*f*
Inorganic matter in soil, 104, 108–109
Insolation, **89**
Institutions, national unity and, 316–317
Insular Oceanic culture realm, 221*f*
Intensive subsistence agriculture, **348,** 348*f*, 350–351, 351*f*, 358*f*
Interbreeding, and race, 259
Intercontinental migration, 281
International banking, 292
International boundaries, 310–316
International Coffee Agreement, 329

International Conference on Population and Development (U.N.), 202–203
International Court of Justice, 324
International Date Line, **28**
International Labor Organization (ILO), 326
International law, 324
International law of the sea, 325–326, 325*f*, 326*f*
International political systems. *See* Regional alliances; United Nations (U.N.)
International terrorism, 322–323
International Tin Agreement, 329
Internet
  globalization and, 292, 382
  political activism and, 295
  spatial interaction and, 279–280, 280*f*
  state power and, 323
Interregional migration, 281
Interrelations between places, terminology of, 14–15
Intracontinental migration, 281
Intrusive igneous rock, 56
Involuntary migration, 281, 282*f*
Inwashing, 108
Iran
  agriculture in, 163
  censorship in, 295
  earthquake (2003), 66, 66*f*
  fall of Shah, 319
  oil production, 138*f*
  population
    control of, 212
    data for, 500*t*
    density of, 205*t*, 206
  refugee problem in, 281
  religion in, 246, 248, 254, 259
  Tehran, climate, 119*f*
Iraq
  agriculture in, 160, 163
  invasion of Kuwait, 306
  migration from, 281
  Persian Gulf War, 315, 315*f*, 324
  population
    data on, 499*t*
    urban, 38*f*
  religion in, 254
  state terrorism by, 322
  water supply in, 436
Ireland. *See also* Northern Ireland
  Eire, borders of, 311
  population data for, 501*t*
  potato famine in, 342
Iron
  mining of, 363
  reserves of, 156*t*
Irredentism, **314**
Irrigation, 340*f*
  in Australia, 434*f*
  drip, 434*f*
  Green Revolution and, 353
  in plantation agriculture, 360
  and salinization of soils, 163
  water use in, 434, 434*f*, 435
Islam, 247*f*, 253–254
  birth control and, 183, 202, 212
  as culture realm, 221*f*
  diffusion routes, 250*f*, 253, 255*f*
  distribution of, global, 248*f*
  in eastern Europe, 251
  five pillars of faith, 247*f*, 253
  food restrictions in, 342, 350
  fundamentalism, 249, 319
  history of geography and, 7
  and India, partitioning of, 481
  Mecca, 247*f*
  mosques, 254, 255*f*
  as state religion, 317
  Sunni *vs.* Shi'ites, 248*f*, 254
  women in, 261
Islands, wildlife on, 454–455, 455*f*
Islas Malvinas, 306
Isobars, 40
Isochrones, 12*f*
Isohyets, 40
Isolines, 40
Isotherms, 40, 93*f*
Israel
  border disputes, 315–316
  economy of, 379
  establishment of, 249
  Palestinian conflict, 295, 295*f*
  population data for, 499*t*
  separatist movements in, 321
  solar power in, 151
  state religion in, 317
Italian language, number of speakers, 240*t*
Italy
  cities in, 421
  devolution in, 320
  infant mortality rate, 188*f*
  Milan, as world city, 401, 403*f*
  nationalist movements in, 320, 321*f*
  population data, 187, 502*t*
  population growth incentives, 187
  regional trade alliances, 327
  Rome, climate, 121*f*
  separatist movements in, 321*f*
  trade, in services, 382*t*
Itaupu hydroelectric power plant, 149

## J

Jackson, Alan, 322
Jamaica
  mineral production by, 157*f*
  population data for, 498*t*
Japan
  agriculture in, 204*f*, 355
  air pollution in, 458–459
  birth and fertility rates in, 182, 187
  car exports, 343*f*
  commercial fishing by, 453
  culture of, 234, 235*f*, 271
  earthquakes in, 66
  economy of, 379
  energy in, 152, 154
  food production in, 206–207
  fuel economy standards, 141
  high-tech industries in, 377
  immigration from, 201*f*
  industry in, 171
  infant mortality rate, 188*f*
  land use in, 412
  mercury poisoning in, 440
  mineral imports by, 158
  mineral resources, 157
  nationalism in, 316
  network cities in, 403
  Nikko, shrine at, 257*f*
  nuclear power in, 146–147, 147*f*
  oil consumption by, 138*f*
  oil imports, 139
  outsourcing and, 373
  population
    data on, 501*t*
    decline in, 187, 197
    density of, 205*t*, 207
  recycling in, 459
  regional trade alliances, 328
  religion in. *See* Shinto
  rotary clubs diffusion in, 277, 278*f*
  solar power in, 151
  solid waste disposal in, 458
  as terrorism target, 323
  Tokyo
    air pollution in, 446
    population, 392*t*
    subsidence in, 451
    as world city, 400–401, 403*f*
  trade, in services, 382*t*
  transnational corporations and, 374
  tsunami in (1896), 70
  waste disposal in, 458–459
Japanese language, 242
  number of speakers, 240*t*
Japanese-Korean language family, 238*f*–239*f*
Java, migration in, 281
Javanese language, number of speakers, 240*t*
J-curve, in population growth, **194**, 195*f*, 211, 211*f*
Jerusalem, 316
Jesus, 250, 253
Jet streams, **97**, 103
Jews. *See also* Israel; Judaism
  history of, 249–250, 251*f*
Jinyu language, 240*t*
Joints, 64
Jordan
  agriculture in, 160
  population data for, 499*t*
Jordan River, water use and, 314
Jovanovic, Maja, 317
Judaism, 248–250
  distribution of, global, 248*f*
  as state religion, 317
Julius Caesar, 467
Just-in-time (JIT) manufacturing, 370–371

## K

Ka'bab, 247*f*
Kalimantan, rain forest in, 172
Kames, 77*f*, 78*f*
Kampuchea, rain forests in, 168
Karst topography, **75**, 76*f*
Kentucky
  industry incentives, 374
  Mammoth Cave region, 75
Kenya
  language of, 242
  population data for, 497*t*
  population growth in, 193
Kerogen, 144
Kettles, 78*f*
Khajuraho temple complex, India, 256*f*
Khoisan language family, 238*f*–239*f*, 240*f*
Kinetic energy, 136
Kissimmee River, 436, 437*f*
Kitty Hawk, North Carolina, 82*f*
Kleinwalsertal, 305
Knowledge sector, 381
Köppen system, 113–115, 115*t*
Koran, 247*f*, 253
Korea. *See* North Korea; South Korea
Korean language, number of speakers, 240*t*
Kosovo, ethnic cleansing in, 322
Krakatoa eruption (1883), 125
Kuala Lumpur, Malaysia, 388*f*
Kunming, China, market in, 345*f*
Kurds, 302–303, 302*f*, 324
Kuwait
  invasion by Iraq, 306
  Persian Gulf War, 315, 315*f*, 324
  population data for, 499*t*

## L

Labor costs, in industry location model, 368–369
Labor unions, high-tech industry and, 378
*Ladang*, 349
Lagoons, formation of, 79
Lake(s)
  oxbow, 73, 73*f*
  subsidence, 451
Lake Argyle, Australia, 434*f*
Land
  for agriculture
    global distribution of, 346, 347*f*
    loss of, 159–163, 223, 351–352
    maintaining, 162, 162*f*
    population density and, 205
    von Thünen model of, 355–357, 357*f*
  carrying capacity of, **206**–207, 206*f*
    Agricultural Revolution and, 181
    population growth and, 206–207, 206*f*, 210–211, 212
    projections of, 210–211
      in Africa, 206, 206*f*
    technology and, 230
  as natural resource, 159–169
  reclamation of, in Netherlands, 451
  women's ownership of, 354–355
Land breezes, **95**
Land use
  agricultural
    global distribution of, 346, 347*f*
    von Thünen model of, 355–357, 357*f*
  in Latin America, 478
  in planned economies, 423–424, 424*f*
  suburban, 413–414, 413*f*, 414*f*
  urban
    competitive bidding for, 405–408, 407*f*, 416
    gentrification, 419–420, 420*f*, 422
    institutional controls on, 411–412
    models of, 408–409, 409*f*
    and motor vehicles, 420, 425
    residence patterns
      in eastern Europe, 423–424, 424*f*
      ethnicity and, 409, 411, 411*f*, 412*f*
      family status and, 409–411, 411*f*
      gentrification, 419–420, 420*f*, 422
      models of, 408–409, 409*f*
      patterns of, 407–408, 407*f*, 408*f*
      social areas in, 409–411
      social status and, 409, 410, 411*f*
      zoning and, 411–412
Land value, urban, competitive bidding and, 405–408, 407*f*, 416
Landfills, 457–458
  for hazardous waste, 458
  hazardous waste in, 460
Landform(s)
  defined, 56
  forces forming, 56, 60
    gradational processes, **68**–81
    plate tectonics, **58**–62, 59*f*–62*f*, 74
    tectonic forces, **62**–68
  human impact on, 449–451, 449*f*
    dumping, 450–451
    excavation, 449–450, 450*f*
    surface depression formation, 451, 451*f*
  as regions, 81, 472–473
  study of, 56
Landform regions, 81, 472–473
Landlocked states, 307, 308*f*, 312
Landsat satellites, 44, **45**
Landscape
  cultural. *See* Cultural landscape
  physical, 222*f*
Landslides, 70, 76
Language(s), 236–245. *See also* English language
  in Africa, 39*f*, 236, 236*f*, 237, 240*f*, 243, 246*f*
  in Americas, 236*f*
  in Asia, 236*f*
  bilingualism, 243, 244
  birth and death of, 236*f*, 241–242
  creole, **242**
  as cultural difference, 235–236
  and culture, 242–245
  definition of, 236
  dialects, **239**–242, 242*f*
  diversity of, 236–237, 244–245, 246*f*–247*f*
  in Europe, 236, 236*f*
  families and subfamilies, 237
  and gender, 242–243
  Gullah, 479, 479*f*
  Lingua Franca, **242**
  major, 237, 240*t*
  official, 246*f*
  in Pacific region, 236*f*
  pidgin, 241–242
  place names, 244–245

as region, 478–479
spread and change of, 237–238
standard, **238**–242
world distribution of, 236, 236*f*, 238*f*–239*f*
Language families, **237**, 238*f*–239*f*
in Africa, 237, 240*f*
Indo-European, 237, 238*f*–239*f*, 240*f*
Lao Tsu, 257
Laos
population data for, 500*t*
religion in, 246
Laplanders, 315
Lapps (Saami), 349
Lapse rate, **92**, 94*f*
Large-scale maps, 33
Las Vegas, Nevada
population of, 397*t*, 402*t*
treatment plant, 197*f*
urban sprawl and, 420, 421*f*
Lateral moraines, 77*f*
Latin America
agriculture in, 348*f*, 350, 354, 355
birth and fertility rates in, 182, 184, 186*f*, 498*t*
cities in, 208*f*, 424–426, 477–478, 478*f*
colonialism in, 251
commodities trade and, 366
as culture realm, 221*f*
death rates in, 186, 498*t*
desertification in, 160
economies in, 345, 379
HIV/AIDS in, 498*t*
industry in, 376
land-holding system in, 478
maternal mortality rates in, 189*f*
population
control of, resistance to, 202
data on, 498*t*
growth of, 181*f*, 183*f*, 196
patterns of, 477–478, 478*f*
rain forest in, 114, 170*f*
toxic waste exports to, 462
transnational corporations and, 374–375
urbanization in, 394*t*
women in, 191, 260, 261, 354
Latin language, 237, 242
Latitude, 9, **27**
history of concept, 6, 7, 7*f*
parallels of, 27, 28*f*
and temperature, 93*f*
Laurasia, 59*f*
Lava, 56
oozing flows, 67–68, 69*f*
Law of the Sea, 325–326, 325*f*, 326*f*
Leachate, 457–458, 461, 462*f*
Leached soil, 109
Leaching, 349
Lead
in gasoline, 448
mining of, 363
as pollutant, 390, 458
reduction efforts, 448
reserves of, 156*t*
League of Nations, 134, 324
Conference for the Codification of International Law (1930), 325
Least cost theory, **368**
Lebanon
agriculture in, 160
population data for, 499*t*
religious regions of, 248, 250*f*
Lee Kuan Yew, 180
Leeward side, 101
LeHigh Canal, 477*f*
Lesotho, 305, 305*f*, 498*t*
Less-advanced culture, as term, 224
Levees, 73
natural, 73
wetlands and, 164
Libya, population data for, 496*t*
LIDAR (Light Detection and Ranging data) imagery, 49*f*
Life, dependence on ecosystems, 433
Life expectancy
and HIV/AIDS, 189–190
increases in, 186, 188–189, 196
medical advances and, 188, 196
of men, 191
by region and nation, 496*t*–503*t*
sanitation and, 188, 196, 197, 197*f*, 198, 211*f*
of wastepickers, 459
of women, 191
Lignite, 142
Limestone, 57*f*
and erosion, 74–75, 76*f*
formation of, 57
Line symbols, 40
Lingua Franca, **242**
Linguistic families. *See* Language families
Linguistic geography, 241
Linton, Ralph, 234
Liquefied natural gas (LNG), **143**
Lithosphere, **58**, 59*f*, 159
Little Skull Mountain earthquakes (1992, 2002), 461
Livestock
commercial raising of, 439–440, 440*f*
dairy farms, 357, 358*f*
impact of, 452
livestock-grain farming, 357, 358*f*
ranching, 357, 359, 359*f*
and water pollution, 439–440, 440*f*
nomadic herding of, **348**–349, 348*f*
overgrazing, impact of, 160
in subsistence agriculture, 350
Llanos, 116
Llivia, Spain, 305, 306*f*
LNG. *See* Liquefied natural gas
Load, of stream, 72
Loans, microcredit, 355
Local and regional government, 329–335
administrative units in, 330, 333–335, 334*f*
importance of, 329–330
waste disposal costs, 457
Location
absolute, **9**, 307
of agriculture, models for, 355–357, 357*f*
as fundamental theme of geography, 17
of industry
comparative advantage and, 371–373
in Europe, 423, 423*f*
government impact on, 373, 374
high-tech industries, 376–378
models for, 367–368, 368*f*, 369*f*
quaternary, 381
trends in, 375–376
mathematical, 9
relative, **9**–10, 9*f*, 307
satisficing, 369*f*, 370
of states, 305–307
terminology, 9–10
of tertiary economic activities, 380, 380*f*
of urban areas, 396–398
within urban areas, competitive bidding for, 405–406, 407*f*, 416
Location map, definition of, 33
Locational tradition, **19**–21, 20*f*, 339
Loess, **81**, 81*f*
Logashkino, Russia, 41, 41*f*
Logging, 164–167, 165*f*, 166*f*. *See also* Deforestation
clear cutting, 165–166, 166*f*, 167
environmental impact of, 452, 452*f*
in national forests, 166–167
selective cutting, 165–166, 166*f*
U.S. subsidies for, 167
London, England
as financial center, 383*f*
growth of, 339
as primate city, 308, 421
as world city, 400–401, 403*f*
Long Island, New York, 15*f*
Longitude, 9, **27**–28
history of concept, 6, 7, 7*f*
meridians of, 27, 28*f*
Longshore current, 79, 80*f*, 82
Los Angeles, California
air pollution in, 92, 95*f*, 443, 446, 446*f*
defacto segregation in, 411, 412*f*
freeway system, 395*f*
population of, 392*t*, 397*t*, 402*t*
radar image of, 45*f*
residential patterns in, 412*f*
subsidence in, 451
as world city, 401, 403*f*
Los Glacieres National Park, Argentina, 54*f*
Louisiana
New Orleans
subsidence of, 451
topographical map of, 35*f*
redistricting in, 332*f*
water pollution in, 439, 439*f*
wetlands in, 164*f*
Lutherans, U.S. distribution of, 253*f*
Luxembourg
population data for, 501*t*
regional trade alliances, 327
trade, in services, 382*t*

# M

Maasai people, 219, 220*f*
Macao, population density, 206
Magma, 56
Magna Charta, 316
Magnesium, reserves of, 156
Maize. *See* Corn
Malaria
dams and, 151
deaths from, 198
in developing nations, 189
Global Malaria Eradication Program, 456
population growth and, 197
Malay-Indonesian language, number of speakers, 240*t*
Malayo-Polynesian language family, 238*f*–239*f*
Malaysia
cities in, 388*f*, 426
economy of, 376
export restrictions, 365–366
food production and, 207
language in, 243
population data for, 500*t*
rain forests in, 168
raw materials trade and, 366
Mali
agriculture in, 160
desertification in, 141*f*, 161*f*
population data, 496*t*
Malnutrition. *See* Hunger and malnutrition
Malta
population data for, 502*t*
regional trade alliances, 327
Malthus, Thomas Robert, **210**–211
Mammoth Cave region, 75
Mammoth seam, 476, 476*f*
Mandarin
as lingua franca, 242
number of speakers, 240*t*
Mandelbaum, Michael, 293
Manganese
reserves of, 156*t*
U.S. imports of, 158*t*
Manufacturing industries, 344, 367–378. *See also* Secondary economic activities
centers of, U.S., 399, 399*f*
flexible manufacturing, 369, 370–371
footloose activities, 370, 378
Fordism, **369**, 370
high-tech, 376–378
and agglomeration, 378
dot-com bubble in, 377
outsourcing by, 378
as percent of all industry, 377, 377*f*
just-in-time (JIT), 370–371
market *vs.* material orientation in, 368, 369*f*
post-Fordist, 369
transnational corporations and, 373–375
ubiquitous, 370
world patterns and trends, 375–376, 375*f*
Manure. *See* Animal waste
Mao Zedong, 184, 212, 360
Map(s). *See also* Mental map(s); Projections; Scale
cloropleth, **39**, 39*f*
deceptive practices in, 40–42, 41*f*
distortions in, 27, **39**, 39*f*
dot, 38*f*
flow-line, 40, 41*f*
general purpose, definition of, 33
generalized, defined, 33
grid system, **27**–28, 28*f*
history of, 6, 7, 7*f*
importance of, 26
location, 33
orthophotomaps, **42**, 43*f*
patterns and symbols
area symbols, 39–40, 39*f*
line symbols, 40
point symbols, 38–39, 38*f*–39*f*
proportional, 38–39, 38*f*
three-dimensional, 39
on topographic maps, 36, 36*f*
and psychological distance, 269, 270*f*, 271*f*
of Ptolemy, 7*f*
qualitative, 38
quantitative, 38
reference, 33
remote sensing and, **42**–45
of spatial data, 16*f*, 17*f*
special-purpose, 34
thematic, 34, 38–40, 38*f*–40*f*
topographic, **34**–38, 35*f*
Brunswick, Georgia, 43*f*
Hebgen Lake, Montana, 24*f*
remote sensing and, 43*f*
symbols, 36, 36*f*
updating of, 43*f*
U.S. Geological Survey and, 34–36, 35*f*
uses of, 38
types of, 33–42
U.S. Geological Survey and, 34–36, 35*f*, 492, 493*f*
uses of, 26
value-by-area, **39**–40, 40*f*
Mapping
automation of, 44, 45–49
history of, 27
*Maquiladoras*, 372, 372*f*
*Maquilas*, 372
Marathi, number of speakers, 240*t*
Marble, formation of, 58
Marine environments, 92
Marine west coast climate, 115*t*, **121**–122, 122*f*
locations of, 114*f*
precipitation, 121–122, 122*f*
soil, 122
temperature, 122*f*
vegetation, 122
Maritime boundaries, 325–326, 325*f*, 326*f*
Market(s)
for mineral products, 365–366
periodic, 261*f*
supply and demand in, 367, 368*f*
Market equilibrium, 367, 368*f*
Market orientation, 368, 369*f*
Marketing
global, 293, 294*f*
and spread of innovation, 276–277, 277*f*
targeted, 410
Marshes, salt, formation of, 79–80
Marxism, 246. *See also* Communism

Mass movement, **70**
Mass transit systems, 405–406, 407*f*
in Canada, 421, 422*f*
in developing nations, 425
in Eastern European cities, 423
in European cities, 422
inadequacy of, 417
Mass wasting, 70
Massachusetts. *See* Boston
Material culture, **233**
Material orientation, 368, 369*f*
Maternal mortality rate, 186, 189, 189*f*
Mathematical location, 9
Mathematical projections, 493–494
Mauritius, 306, 497*t*
Maximum sustainable yield, **361**
overfishing and, 361, 362, 362*f*
Mayflower compact, 300, 300*f*
Mayo Clinic, 394
Measles, 198
Mecca, 247*f*
Mechanical weathering, **68**
Medial moraines, 77*f*
Medicine
disease control, 198
high-tech industries, 377
illegal immigrants and, 288–289
and life expectancy, 188, 196, 198, 390
and population, 195*f*, 211*f*
in tropical rain forests, 172
Mediterranean agriculture, 231*f*, 359–360, 359*f*
Mediterranean climate, 115*t*, **119**–120, 121*f*
locations of, 112, 112*f*, 114*f*
precipitation, 121*f*, 360
temperature, 121*f*
vegetation, 112, 112*f*, 120, 121*f*
Mediterranean Sea, cleanup efforts, 442
Megacities, 391, 392*t*
air pollution and, 443
Megalopolis, **414**, 415*f*, 484–485, 484*f*, 485*f*
Meinig, Donald, 471–472
Melilla, Spain, 305, 306*f*
Melting pot theory, 234
Men, life expectancy of, 197
Mendawi bazaar, Afghanistan, 266*f*
Mental map(s), 269, 270*f*, 271*f*, **272**–273, 272*f*, 273*f*, 479–480, 480*f*
Mental regions, 479–480, 480*f*
Mentifacts, 222, 228, 229*f*, 235
Mercator, Gerardus, 491
Mercator projection, **491**, 491*f*
MERCOSUR (Southern Cone Community Market), 328, 329*f*
Mercury
biological magnification of, 456*f*
poisoning in Minamata, Japan by, 440
as pollutant, 440, 441, 458
reserves of, 156, 156*t*
Meridians of longitude, 6, 9, 27, 28*f*, 30
Mesas, 74, 75*f*
Meso-America, agriculture in, 231*f*
Mesopotamia, as culture hearth, 230, 231*f*
Metallic minerals
mining of, 363–364, 364*f*
substitutes for, 158–159, 171
Metallurgy, invention of, 230
Metamorphic rock, **58**
Methane
as fuel, 143, 148
and global warming, 126
as greenhouse gas, 150
sources of, 169
Methodists, U.S. distribution of, 253*f*
Methyl bromide, and ozone layer, 447, 448
Methyl chloroform, and ozone layer, 447, 448
Metropolitan areas, 396, 396*f*. *See also* Cities; Urban areas
in United States, 397*t*, 398*f*, 402*t*, 420
expansion of, post-World War II, 413–414, 413*f*, 414*f*
Metropolitan Statistical Areas (MSAs), 402*t*
Metropolitanization, 413
Mexican American Legal Defense and Education Fund, 289
Mexico
agriculture in, 359, 359*f*
boundaries of, 313
commodities trade and, 366
cultural landscape in, 13, 13*f*
earthquake in (1985), 451
El Niño and, 108
emigration to U.S., 285, 286, 287*f*, 315, 315*f*
backlash against, 288–289
energy in, 152
Guanajuato, 425*f*
industry in, 376
mineral production by, 157*f*
natural gas trade flows, 144*f*
oil consumption, 140
oil production, 138*f*, 140
oil reserves, 140
oil trade flows, 138*f*
plateaus of, 204
population data for, 498*t*
regional trade alliances, 327
religion in, 248
resources disputes, 315
trade, and NAFTA, 373
U.S. outsourcing and, 372–373, 372*f*
water supply, 435
Mexico City
air pollution in, 443, 446
population, 392*t*
as primate city, 308
recycling in, 459
slums in, 426
subsidence of, 435, 451, 451*f*
water supply, 435
Miami Beach, Florida, 82
Mica, U.S. imports of, 158*t*
Michigan
Detroit
functions of, 394
population of, 397*t*
urban renewal in, 417
Pictured Rocks National Lakeshore, 52*f*
Microcredit initiatives, 355
Microdistricts, in eastern European cities, 424, 424*f*
Middle East
cities in, 424–426
coal reserves, 142*f*
economies in, 345
maternal mortality rates in, 189*f*
natural gas reserves, 143, 145*f*
natural gas trade flows, 144*f*
oil and, 138*f*, 139
oil reserves, 139*f*
oil trade flows, 408*f*
population control in, 212
women's migration to, 285
Midlatitude climates. *See* Desert(s), midlatitude; Humid midlatitude climates
Midwestern U.S.
boundaries of, 471*f*
population density in, 12*f*
Migration, **281**–291. *See also* Immigration
acculturation and, **233**–235, 244–245
age and, 284, 286*f*
barriers to, 286–289
chain, **285**–286
from cities, 408–409, 410, 414–415, 421
to cities, 191, 282, 285, 286, 291, 425
countermigration, **290**–291, 291*f*
cultural mixing and, 234
cultural variables, 283
decision to migrate, 282–286
for modern Americans, 284
and place utility, **284**–285
and diffusion of innovation, 232
gender and, 200, 285
hierarchical, **291**
from less-developed nations, 186
patterns of, 289–291
and place utility, **284**–285
population and, 200
in demographic equation, 200
principal, of recent centuries, 201*f*
pull factors, **282**, 283, 283*f*
push factors, **282**, 283*f*
refugees, 281–282, 283*f*
return, **290**–291, 291*f*
step, **285**
types of, 281–282
in United States
in 1950s, 41*f*
in 1990s, 283*f*
channelized, 291*f*
of Germans settlers, 284*f*
migration fields, 290*f*
of modern Americans, 284
Migration fields, **289**–290, 290*f*
channelized, **290**–291, 291*f*
in U.S., 290*f*
Milan, Italy, as world city, 401, 403*f*
Military alliances, regional, 329
Military applications
of civilian spy satellites, 46
of Global Positioning System, 29
of maps, 26
of remote sensing, 45
Military hazardous waste, 460–461
Miller cylindrical projection, 491–492, 491*f*
Million, illustration of, 180–181
*Milpa*, 349
Min language, number of speakers, 240*t*
Minamata, Japan, mercury poisoning, 440
Mineral(s), **56.** *See also specific types*
demand for, 156
distribution of, 363
exploitation of, steps in, 154
metallic
mining of, 363–364, 364*f*
substitutes for, 158–159, 171
nonfuel, 154–159
distribution of, 156–157
production of, 157, 157*f*
as nonrenewable resource, 135*f*, 136
reserves of, 156, 156*t*
nation size and, 303, 363
as region, 475–476
reuse of, 136, 171
substitutes for, 158–159, 171, 365
trade in, 364–366
Mining Act of 1872, 366
Mining and quarrying
and air pollution, 450
beneficiation, 364
coal, 142, 450
copper, 158, 158*t*
dangers of, 476
distribution of resources, 156–157
early methods of, 476
economics of, 154–156, 363–364, 364*f*
environmental impact of, 134, 134*f*, 143, 144, 146*f*, 366, 441, 449–450, 450*f*, 452
invention of, 230
landforms impact of, 449–450, 450*f*
in planned economies, 364
Schuylkill Anthracite Region, 475–476, 476*f*
shaft mining, 476
site selection, 363–364
and surface depression formation, 451
transportation and, 363, 364*f*
types of, 449–450, 450*f*
on U.S. government land, 366–367
and water pollution, 441, 450
Ministates, 306, 307*f*
Minnesota
gentrification in, 419
industry incentives, 374
smart growth programs in, 421
Minorities. *See also specific groups*
assimilation of, 234–235
boundary disputes and, 314
defacto segregation of, 411, 412*f*
employment and, 417
ethnic cleansing of, 322
political representation in U.S., 331, 332–333, 332*f*
state terrorism against, 322
in urban areas, 417
Minutes, 27
Mississippi River
alien species in, 454
delta of, 72*f*
floodplain of, 73, 73*f*
modification of, 436
water pollution in, 439, 439*f*
Mobility, and activity space, 275
Mohammed (prophet), 253
Moisture, in atmosphere. *See* Air, moisture in
Mojave Desert, solar power in, 152, 152*f*
Mollisols, 110*t*, 111*f*, 115*t*, 118
Mollweide projection, 489*f*
Molybdenum, mining of, 363, 450*f*
Moment magnitude, 65
Monaco
population data for, 501*t*
population density, 206
Mongolia
as multicore state, 308
population data for, 501*t*
population density, 206
Mongoloids, 270
Monotheism, 246
Monsoon, 97, 116
Montreal Protocol on the Depletion of the Ozone Layer (1987), 447–448
Moore's Law, 293
Moraines, 76, 77*f*
Morenci, Arizona, smelting facility, 364*f*
Mormons, U.S. distribution of, 253*f*
Morocco
elderly in, 213*f*
enclaves, 305, 306*f*
mineral production by, 157*f*
mineral resources, 157
population data for, 496*t*
Moros, separatist movement, 321
Mortality rate, **186**–190. *See also* Death rate
infants, 186–188, 188*f*
by region and nation, 496*t*–503*t*
for mothers, 186, 189, 189*f*
of women, 191
Moscow, Russia
climate, 123*f*
as core area, 308
Mosques, 254, 255*f*
Motor vehicles
activity space and, 274–275, 274*f*, 275*f*
and air pollution, 443*t*, 444, 446, 446*f*
employment and, 417
energy use by, 171
environmental impact of, 448, 449, 457*f*
European cities and, 422
fuel efficiency of, 140–141, 171
GPS navigation systems in, 29–30
inspection programs, 449
international trade in, 343*f*
manufacture of, 369*f*, 372
spatial interaction and, 274–275, 278–279
suburbanization and, 412, 413
tires
disposal of, 457, 457*f*
recycling of, 459, 459*f*
and urban land use, 420, 425
Mount Everest, 74
Mount Kenya, 127
Mount Kilimanjaro, 127
Mount Pelée eruption (1902), 53
Mount Pinatubo eruption (1991), 126
Mount St. Helens eruption (1980), 62, 66, 68*f*

Mountain(s)
as boundaries, 311, 311*f*
dome, 67
fault-block, 63*f*
formation of, 60, **62,** 62*f,* 63*f,* 64
as landform region, 81
soil of, 111*f*
Mountain breezes, **96,** 96*f*
Movement, as fundamental theme in geography, 17
Mozambique
agriculture in, 354
birth and fertility rates in, 184
population data for, 497*t*
MSAs. *See* Metropolitan Statistical Areas
Mud flows, 70
Multicore states, 308
Multicultural societies, centripetal forces in, 316
Multifactor formal regions, 473, 479
Multilingualism, 243
Multinational corporations. *See* Transnational Corporations
Multinational state, 302, 302*f*
Multiple Use Sustained Yield Act of 1960, 166–167
Multiple-nuclei model, of urban land use, **409,** 409*f*
Multiplier effect
in agglomerations, 370
in city growth, 400
Mumbai (Bombay), India, population, 392*t*
Municipalities
as pollution source, 441–442
waste disposal, 456–459, 457*f*
water use by, 434–435
Music, and culture, 262
Myanmar
logging in, 165, 165*f*
population data for, 500*t*
rain forests in, 168
religion in, 246
shape of, 305
Yangôn
climate of, 118*f*
stupa in, 257*f*
Myers, Richard B., 46


# N


NAFTA. *See* North American Free Trade Agreement
Nagorno-Karabakh, 309, 309*f*
Nakhichevan, 309, 309*f*
Namibia
Caprivi Strip, 312
population data for, 498*t*
Nanching, China, economy of, 352
Nation(s), 301–**302**
stateless, 302–303, 302*f*
National Aeronautics and Space Administration (NASA), 45
*National Atlas of the United States of America,* 492
National Council for Geographic Education, 17
National Flood Insurance Program, 82
National forests (U.S.), logging in, 166–167, 167*f*
*National Geographic,* 134
National Geography Standards, 17–19
National Mapping Program, 42
National Oceanic and Atmospheric Administration (NOAA), 37, 61*f*
National political systems, 301–321
National security, in nuclear age, 323
Nationalism
as centrifugal force, 319–320
as centripetal force, 316
religion and, 317
sports and, 317
Nation-building, 303
Nation-state, **302,** 302*f*
desire for, 319–322, 321*f*
Native Americans
acculturation and, 234
and Anglo-American place names, 245
Crow Indians, 177
innovation and, 230
languages of, 237–238
natural resources and, 135
Yurok Indians, 480, 480*f*
NATO. *See* North Atlantic Treaty Organization
Natural boundaries, **311,** 311*f*
Natural disasters, human environmental impact and, 4
Natural gas, **143**–144
consumption of, in U.S., 137, 137*f,* 143
and global warming, 126
liquefied, **143**
as nonrenewable resource, 135*f,* 136
reserves of, 143, 145*f*
trade flows in, 143, 144*f*
Natural increase, rate of, **193**
by region and nation, 496*t*–503*t*
U.S., 194
world, 195*f,* 207*f*
Natural landscape, **13**
*vs.* cultural landscape, 6*f,* 13
Natural resources. *See also specific types*
and carrying capacity of land, 181
conditions required for exploitation of, 361
definition of, **135**
energy
nonrenewable, 137–147
renewable, 148–154
forests, 164–169
land, 159–169
limited nature of, 134
management of, 169–171
nonrenewable, **135**–136, 135*f,* 137–147
perception of, 135, 135*f*
perpetual, **135,** 135*f*
potentially renewable, **135,** 135*f*
regions of, 475–476
renewable, **135,** 135*f,* 148–154
reserves of, 136–137, 137*f*
reusable, 136
rights to, undersea, 325–326, 325*f*
terminology of, 135–137, 135*f*
Natural selection, **259**
Natural vegetation, **110**–112
regions, 110–112, 112*f*
succession in, **110,** 111*f*
Nauru, 307*f*
population data for, 502*t*
mining on, 134, 134*f*
as mini-state, 306
size of, 303
Navigation, maps for, 491, 491*f*
Navigation systems, 28–30
Nazi Germany
mapping practices, 41
state terrorism by, 322
Negroids, 259
Neo-Malthusianism, **211**–212
Nepal
energy in, 148*f*
population data for, 500*t*
religion in, 317
Neritic zone, 163–164, 163*f*
Nestlé Corp., 294*f*
Netherlands
energy in, 154
land reclamation in, 451
network cities in, 403–404
population data for, 501*t*
regional trade alliances, 327
trade, in services, 382*t*
waste disposal in, 462
Network(s), 15
Network cities, **402**–404
Nevada
cities in, 420
Las Vegas
population of, 397*t,* 402*t*
treatment plant, 197*f*
urban sprawl and, 420, 421*f*
water supply in, 436
Yucca Mountain, radioactive waste disposal at, 460–461, 460*f*
New Caledonia, 157, 157*f,* 503*t*
New England
air pollution and, 443–444
high-tech industry in, 377
year without a summer, 125
New Guinea. *See* Papua, New Guinea
New Jersey
edge cities in, 414
high-tech industry in, 377
Pine Barrens, 485*f*
population density of, 205
New Mexico
Anasazi people in, 223
Four Corners Monument, 331*f*
water supply in, 436
New Orleans, Louisiana
subsidence of, 451
topographical map of, 35*f*
New York City, 395*f*
air pollution and, 443
Chinatown, 260*f*
as financial center, 383*f*
as functional area, 17
industry in, 378
industry incentives, 374
landfills, 458
location of, 9
population of, 392*t,* 397*t,* 402*t*
view of U.S. from, 270*f*
as world city, 400–401, 403*f*
New York Mercantile Exchange, industry incentives and, 374
New York State
Peekskill incinerator, 458*f*
radioactive waste disposal in, 461
*New York Times, The,* 103, 317
New Zealand
agriculture in, 359
Antarctic territorial claims, 301*f*
birth and fertility rates in, 182, 187
colonies of, 134
colonization of, 252
energy in, 149, 152, 153*f*
nuclear power and, 146
population data, 503*t*
wetlands in, 164
Newly industrializing countries (NICs), 376, 378
News, globalization of, 295
NGOs. *See* Nongovernmental organizations
Nicaragua
infant mortality rate, 188*f*
population data for, 498*t*
population growth in, 196
urban areas in, 396
Niches, in ecosystems, 433
Nickel
mining of, 363
reserves of, 156*t*
U.S. imports of, 158*t*
NICs (newly industrializing countries), 376, 378
Niger
agriculture in, 160
mineral production by, 157*f*
population data for, 497*t*
Niger-Congo language family, 238*f*–239*f*
Nigeria
capital of, 310
cities in, 424, 425
core area of, 308
language in, 243
mineral resources, 157
Oshodi Market in, 179*f*
population
data for, 209, 497*t*
density, 205*t*
Night, length of, and earth's orbit, 91
Nikko, Honshu Island, Japan, shrine at, 257*f*
Nile River
delta of, 72
modification of, 436
water pollution in, 390
water use and, 314
Nimbostratus clouds, 102
La Niña, 108*f,* **109,** 109*f*
El Niño, **108**–109, 108*f,* 109*f,* 126, 361*f*
Nirvana, 257
Nissan, industry incentives and, 374
Nitrates, as pollutant, 439
Nitrogen, as pollutant, 440*f,* 444
Nitrogen oxides, as pollutant, 140, 443*t,* 446, 458
reduction efforts, 448
Nitrous oxides
and global warming, 126
reduction of, 448
sources of, 169
NOAA. *See* National Oceanic and Atmospheric Administration
Nokia Corp., 294*f*
Nomadic herding, **348**–349, 348*f*
Nonbasic sector, in urban areas, **398**–400
ratio to basic sector, 400, 400*f*
Nonecumene, **204**
Non-Euclidian projections, 494, 495*f*
Nongovernmental organizations (NGOs), 295, 295*f*
Nonindigenous species. *See* Exotic species
Nonindustrial, as term, 224
Nonmaterial culture, **233**
Non-photographic imagery, 42–45
Nonpoint sources of pollution, 437–439
Norgay, Tenzing, 74
Normal fault, 63*f*
Normandy, 321*f*
North America. *See also* Anglo-America; Canada; Mexico; United States
agriculture in, 358, 358*f*
air masses and source regions of, 103*f,* 475*f*
cities in, 421, 422*f*
coal reserves, 142*f*
as culture hearth, 230, 231*f*
hydropower in, 152*f*
industry in, 375–376, 375*f,* 377*f*
latitude of, 27
location *vs.* Russia, 9*f*
longitude of, 27
maternal mortality rates in, 189
natural gas reserves, 145*f*
oil reserves, 139*f*
oil shale deposits, 145*f*
population data for, 498*t*
urbanization in, 391, 394*t*
water pollution in, 439
North American Free Trade Agreement (NAFTA), 295, 327, 329*f,* 373
North Atlantic drift, **98**
North Atlantic Treaty Organization (NATO)
bombing of Chinese embassy in Yugoslavia, 26
membership of, 329, 330*f*
peacekeeping by, 295, 322
role of, 329
North Carolina
high-tech industry in, 377
Kitty Hawk, 82*f*
Outer Banks, 80, 88
redistricting in, 332–333, 332*f*
North, in Brandt development model, 226, 226*f*

North Korea
censorship in, 270
cities in, 424
as nation-state, 302
population data for, 501*t*
satellite imaging and, 46
North Pole
grid system and, 27
ozone hole over, 447
Northeast trades, 97
Northern Ireland
borders of, 311
Good Friday Agreement, 323
government in, 308
religion in, 248
Northridge, California, earthquake (1994), 65
Norway
air pollution in, 448
Antarctic territorial claims, 301*f*
energy in, 149
oil production, 138*f*
population data, 501*t*
regional trade alliances, 327
shape of, 305, 305*f*
No-till farming, 162, 162*f*
Notre Dame Cathedral, 253*f*
Nottingham, England, nuclear power plant, 132*f*
NRCAN. *See* Survey and Mapping and Remote Sensing, Natural Resources
Nuclear arms, and national security, 323
Nuclear energy, 146–147
consumption of, in U.S., 137, 137*f*
from fission, **146**–147
from fusion, 147
production and consumption of, 146–147, 147*f*
and radioactive waste, 460–461, 462*f*
Nuclear fuels
disposal of, 430*f*, 460–461
as nonrenewable resource, 136
Nuclear power plants
breeder reactors, 147
lifespan of, 147
Nottingham, England, 132*f*
in United States, 146–147, 147*f*
water pollution from, 440
Nuclear Regulator Commission, 461
Nutrients
and food chain, 433, 433*f*
in soil, 109
and water pollution, 439–440, 440*f*
Nutrition. *See* Diet and nutrition
Nyerere, Julius, 303

# O

OAS (Organization of American States), 329
Obsidian, formation of, 57
Ocean(s). *See also* Atlantic Ocean; Pacific Ocean
deep-sea trench formation, 60–62
as fresh water source, 436
in hydrologic cycle, 434*f*
Law of the Sea, 325–326, 325*f*, 326*f*
radioactive waste dumping in, 461
state ownership rights in, 325–326, 325*f*, 326*f*
water density, 98
Ocean City, New Jersey, 82
Ocean currents, 98–99, 99*f*
causes of, 98
climate and, 126
El Niño and, 108–109, 108*f*
erosion and, 76–80, 80*f*, 82
global warming and, 128
and precipitation, 98–99
Ocean temperature, El Niño and, 108–109, 108*f*
Oceania
population
data on, 502*t*–503*t*
growth of, 181*f*, 183*f*
urbanization in, 394*t*
Ogallala aquifer, 436, 436*f*
O'Hare Airport, 397, 397*f*
Ohio
Cleveland, population density gradients for, 408, 408*f*
Columbus, downtown rejuvenation in, 408
Toledo, urban renewal in, 417
O-horizon, 107
Oil. *See also* Petroleum (crude oil)
Alaskan pipeline, 79
consumption of, 138*f*, 139*f*
in U.S., 137, 137*f*, 138*f*, 139*f*, 140
crude. *See* Petroleum
flow, by sea, 138*f*, 139
from oil shale, 144
price of, 139
production of, 138*f*, 139, 140
from U.S. government lands, 366
reserves of, 139–140, 139*f*
substitutions for, 171
from tar sands, 144
trade in, 138*f*, 139
undersea, ownership of, 325
United States and
consumption, 137, 137*f*, 138*f*, 139*f*, 140
production, 138*f*, 139, 140, 366
reserves, 140
trade flows, 138*f*, 139, 140
water pollution and, 441
Oil shale, **144**–146
future use of, 171
reserves, 144, 145*f*
Okefenokee Swamp, 163
Oklahoma City bombers, 322
Oklahoma City tornado (1999), 106*f*
Old Order Amish, 232*f*, 233
Old-growth forests, 166–167
Olympic Games, Summer 1976, 317
OPEC. *See* Organization of Petroleum Exporting Countries
Open-pit mining, 449–450, 450*f*
Operation Desert Shield/Storm. *See* Persian Gulf War
Opportunity, and activity space, 275
Orders, soil, **110,** 110*t*
Ore, **154.** *See also* Mineral(s)
grade of, 363–364
Oregon
growth restrictions in, 421
hydroelectric power in, 149, 149*f*
Organic matter in soil, 104, 107–109
Organization for European Cooperation, 327
Organization of African Unity, 462
Organization of American States (OAS), 329
Organization of Petroleum Exporting Countries (OPEC), 138*f*, 139, 329
Orientation
market, 368, 369*f*
material, 368, 369*f*
Orographic precipitation, **101,** 101*f*
Orthographic projections, 490, 490*f*
Orthophotomaps, **42,** 43*f*
Oshodi Market, Nigeria, 179*f*
Outer Banks of North Carolina, 88
Outer Seven, 327, 327*f*
Outsourcing, **372**–373, 378
and tertiary activity, 380–381
Outwash plains, 76, 78*f*
Outwashing, 108
Overburden, 450*f*
Overpopulation, **205**–207
Oxbow lakes, 73, 73*f*
Oxidation, in chemical weathering, 69
Oxisols, 110*t*, 111*f*, 114, 115*t*
Oxygen
in atmosphere, rain forests and, 168
in water, 439
Ozone, as pollutant, 140, **446,** 446*f*
Ozone layer, **447**
deforestation and, 169, 170*f*
depletion of, 446–448, 447*f*, 448*f*

# P

Pacific Ocean
atolls in, 80
continental drift and, 62
currents in, 98, 99*f*
fisheries locations, 362, 362*f*
Pacific region, languages in, 236*f*
Pacific rim, earthquakes and, 62, 66
Packaging, waste and, 457, 459
Pagodas, Buddhist, 256–257, 257*f*
Pakistan
agriculture in, 163, 350
air pollution in, 446
break-up of, 481, 481*f*
capital of, 309
cities in, 424
cyclone in (1970), 481
economy of, 376
energy in, 148
establishment of, 283, 311–312, 481, 481*f*
population, 203
data on, 500*t*
reliability of, 209
religion in, 246, 259, 317
shape of, 305
women in, 191
Paleo-Asiatic language family, 238*f*–239*f*
Palestinians
Jerusalem and, 316
media attention of, 295
population data for, 499*t*
separatist movement, 321
as stateless nation, 303
and terrorism, 322
Palisades, New York, 67
Pancasila, 246
Pangaea, 58, 59*f*
Panjabi, number of speakers, 240*t*
Papua, New Guinea
population data, 503*t*
Tsunami strike on (1998), 70
women in, 191
Papuan language family, 238*f*–239*f*
Parabolic trough reflectors, 152*f*
Paraguay
population data for, 499*t*
power in, 149
regional trade alliances, 328
Parallel, standard, 490
Parallels of latitude, 6, 9, 27, 28*f*, 30
Paris, France, 196*f*
as core area, 308
design of, 422, 423*f*
Notre Dame Cathedral, 253*f*
as primate city, 308, 421
as world city, 401, 403*f*
Particulates, as air pollutant, 443*t*
Part-nation state, 302, 302*f*
Pasa Robles, California, earthquake (2003), 62
Pattison, William D., 19
PCBs (polychlorinated biphenyls), as pollutant, **440**
biological magnification of, 456*f*
Peak land value intersection, 407
Peekskill, New York, incinerator, 458*f*
Pennsylvania
coal deposits in, 142
edge cities in, 414
Philadelphia
population of, 397*t*, 402*t*
site of, 396, 397*f*
slums in, 416*f*
Quecreek Mine disaster, 26
Ridge and Valley region of, 63*f*
Schuylkill Anthracite Region, 475–476, 476*f*
Pentagon, terrorist attack on (2001). *See* September 11th terrorist attacks
Perception, of natural resources, 135, 135*f*
*Perestroika*, 309
Perfluorocarbons, and global warming, 126
Perforated states, **305,** 305*f*
Permafrost, 75, 79, 79*f*, 81
Perpetual natural resources, **135,** 135*f*
Persian Gulf War
Global Positioning System and, 29
as resource dispute, 315, 315*f*
and U.N. intervention, 324
Peru
Bolivia and, 312, 313*f*
mineral production by, 157*f*
mining in, and water pollution, 441
population
data, 499*t*
growth in, 193
rain forest in, 172
regional trade alliances, 328
Pesticides. *See* Biocides
Petroleum (crude oil), **138**–141. *See also* Oil
extraction of, 140–141
flow by sea, 138*f*, 139
formation of, 57
and global warming, 126
as nonrenewable resource, 135*f*, 136
percent of total energy supplied by, 138
products derived from, 138–139
and water pollution, 441
Petronas Twin Towers, 388*f*
*Pfiesteria piscicida*, in Chesapeake Bay, 439–440
pH
of acid rain, 444
of soil, and agriculture, 110
Pharmaceutical industry, 375
Phelps-Dodge mine, Morenci, Arizona, 364*f*
Philadelphia
population of, 397*t*, 402*t*
site of, 396, 397*f*
slums in, 416*f*
Philippines
agriculture in, 350
colonization of, 252
commodities trade and, 365*f*
energy in, 152
export restrictions, 365–366
Hanunóo People of, 350
migration to U.S., 286
Mount Pinatubo, 126
nuclear power and, 146
pollution in, 13*f*
population data, 500*t*
reliability of, 209
rain forests in, 167, 168
recycling in, 459, 462*f*
separatist movements in, 321
shape of, 305*f*
wastepickers in, 462*f*
Phosphate
mining of, 134, 134*f*, 450
as pollutant, 439
reserves of, 156*t*
Phosphorus
mining of, 450
as pollutant, 440*f*
Photochemical smog, **446,** 446*f*, 447*f*
formation of, 443
health effects of, 446
reduction efforts, 448
temperature inversion and, 92, 95, 95*f*, 446
Photovoltaic (PV) cells, **152**
Physical attributes, overview, 13–14

Physical barriers to migration, 287
Physical boundaries, 311, 311*f*
Physical environment, and economic activity, 342
Physical geography, 8
Physical landscape, 222*f*
Physiological density of population, **205,** 205*t*
Pictured Rocks National Lakeshore, Michigan, 52*f*
Pidgin, 241–**242**
Pilgrims, 300, 300*f*
Pine Barrens (New Jersey), 485*f*
Pioneer plant community, 110
Pits, formation of, 451
Place(s)
 characteristics of, 8–9
 as fundamental theme of geography, 17
 names of, 244–245
Place utility, **284**–285, 380
Plain(s), as landform region, 81
Planar projection, 490*f*, **492**–493, 493*f*
Planned economies, **344**
 agriculture in, 360–361
 city planning under, 423–424, 424*f*
 industrial location in, 373
 mining and quarrying in, 364
Planning regions, 482–483
Plantation agriculture, 359–**360,** 359*f*, 360*f*
Plants. *See also* Vegetation
 alien species, impact of, 454–455
 domestication of, 227, 230, 231*f*
 endangered species, 452
 human impact on, 451–456
Plate boundaries
 convergent, 59, 60*f*
 divergent, 59, 60*f*
 transform, 59, 60*f*
Plate tectonics, **58**–62, 59*f*–62*f*, 74. *See also* Tectonic forces
Plateaus, as landform region, 81
Platinum, U.S. imports of, 158*t*
Pledge of Allegiance, U.S., 316*f*
Plutonium, as waste, 460, 461
Point sources, of pollution, 437
Point symbols, 38–39, 38*f*–39*f*
Poisoning of animals and plants, 455–456
Pol Pot regime, 322
Poland
 as nation-state, 302, 302*f*
 NATO and, 329, 330*f*
 population data for, 502*t*
 shape of, 305
Polar easterlies, 97
Polar highs, 97
Polish language, number of speakers, 240*t*
Political alliances, 329
Political barriers to migration, 287–289
Political beliefs, and birth rate, 183
Political control, decentralization of, 308, 321
Political geography, 300–301
Political incentives for migration, 282–283, 284*f*
Political integration, globalization and, 295
Political organization
 and economic activity, 342
 international
  global. *See* United Nations
  regional alliances, 327–329
 local and regional, 329–335
 national, 301–321. *See also* State(s)
Political regions, 480–482, 481*f*
Politics
 activism, Internet and, 295
 women in, 318–319, 319*f*
Pollution. *See* Air pollution; Biocides; Human environmental impact; Thermal pollution; Water pollution
Polychlorinated biphenyls. *See* PCBs
Polyconic projection, 492, 492*f*
Polytheism, 246
Pony Express, 268, 268*f*
Poor people. *See also* Poverty; Slums
 defacto segregation in cities, 415, 416
 services needed by, 415–416, 417
Popular culture, **233**
 spread of, 295
Population. *See also other* Population *entries; specific nations and regions*
 age and, by region and nation, 496*t*–503*t*
 Agricultural Revolution and, 181, 211*f*, 220*f*
 censuses, 209, 209*f*
 of cities
  cities over 3 million (2005), 393*f*
  growth in, 390, 391, 392*f*, 393*f*, 393*t*, 394*f*
  land value and, 407–408, 407*f*, 408*f*
  megacities, 392*t*
  in U.S., 402*t*
 data
  by region and nation, 496*t*–503*t*
  reliability of, 209, 209*f*
  sources of, 207–209
 definitions, 182–194
 demographic equation and, 200
 distribution of
  global, 200–204, 201*f*, 202*f*
  in Latin America, 477–478, 478*f*
  in United States, 38*f*, 203
 and food supply, 345–346
 Industrial Revolution and, 195*f*, 196–197, 211*f*
 and migration, impact of, 200
 natural increase in, **193**
  United States, 194
  world, 195*f*, 207*f*
 as regional focus, 477–478
 shifts in, and reapportionment, 330
 world, 180
  age of, 192*f*, 212–213, 213*f*
  distribution of, 200–204, 201*f*, 202*f*
  projections for, 180, 181*f*, 186, 187, 193, 194, 194*t*, 195*f*, 199, **209,** 212–213
  by region and nation, 496*t*–503*t*
Population and Family Planning Board (Singapore), 212
Population density, 200–204, 201*f*, 202*f*, **204**–207
 agricultural, **205,** 205*t*
 and altitude, 201
 arithmetic, **205**
 crude, **205,** 205*t*
 in developing world, 206, 207, 208*f*
 in Illinois, 12*f*
 in Midwestern U.S., 12*f*
 by nation, 205*t*
 overpopulation, **205**–207
 physiological, 205*t*
 physiological density, **205,** 205*t*
 and scale, 12*f*
 underpopulation, 206
 in urban areas, 207, 208*f*, 407–408, 407*f*, 408*f*
Population explosion, 207*f*
Population geography, **180,** 477–478
Population growth, 180–182. *See also* Birth rate; Total fertility rate (TFR)
 8000 B.C. to A.D. 2000, 195*f*
 and agriculture, 159, 162, 181, 197, 211, 211*f*, 351
 baby boom, 193*f*
 control of
  Cairo Plan for, 202–203
  Catholic Church and, 183, 202, 212
  in China, 182, 184–185
  in developing countries, 180, 182, 184–185, 211–212, 211*f*
  Islam and, 183, 202, 212
  natural, 194–200, 210–211, 211*f*
  Neo-Malthusianism and, 211–212
  resistance to, 202, 212
 demographic momentum, **212**–213
 demographic transition model of, **194**–200, 196*f*
 in developed countries, 181*f*, 194, 496*t*
 in developing countries, 180, 181*f*, 190, 194, 211–212, 390, 496*t*
 doubling times, **193**–194, 194*t*, 195*f*, 197
 fertility rates. *See* Total fertility rate
 homeostatic plateau in, 211, 211*f*
 J-curve in, **194,** 195*f*, 211, 211*f*
 limits on, 181–182
 predictions of, 194
 projections of, 180, 181*f*, 186, 187, 193, 193*f*, 194, 194*t*, 195*f*, 199, **209,** 212–213
  accuracy of, 209, 210
  *vs.* predictions, 209
  by region and nation, 181*f*, 183*f*, 496*t*–503*t*
 replacement level, 184, **187**
 S-curve in, 211, 211*f*
 world
  data by region and nation, 496*t*–503*t*
  doubling times, 194, 194*t*, 195*f*, 197
  history of, 194*t*, 195*f*
Population momentum, **212**–213
Population pressure
 carrying capacity and, **206**–207, 206*f*, 210–211, 212, 230
 physiological density and, 205
Population pyramids, **190**–192, 190*f*, 192*f*, 193*f*
 immigration and, 200
Population Reference Bureau, 180, 207–209, 207*f*
Portland, Oregon, growth restrictions in, 421
Portugal
 colonization by, 251–252
 devolution in, 320
 gated communities in, 416
 population data for, 502*t*
 regional trade alliances, 327
 time zone of, 29*f*
Portuguese language, 242
 in Africa, 246*f*
 number of speakers, 240*t*
Possibilism, **221**
Post-Fordist manufacturing, 369
Postindustrial economy, 378–382
 cities in, 419
 and employment, 378–379
 movement toward, 376
 United States as, 293, 342, 368, 376, 378, 379*t*
Potash
 reserves of, 156, 156*t*
 U.S. imports of, 158*t*
Potential energy, 136
Potential support ratio (PSR), 213
Potentially renewable resources, **135,** 135*f*
Poverty. *See also* Poor people
 as migration motivation, 282
 women and, 417
Power plants
 geothermal, 153*f*
 nuclear
  breeder reactors, 147
  lifespan of, 147
  and radioactive waste, 460–461
  in United States, 146–147, 147*f*
 as secondary economic activity, 344
 solar-thermal, 152*f*
 waste-to-energy, 458, 458*f*
 wind, 154, 155*f*
Practice, definition of, 275
Prairie
 locations of, 112, 112*f*
 vegetation, 112, 112*f*
Precipitation, **99**–104. *See also* Acid precipitation; Air, moisture in
 acidity of, 444
 cause of, 99–100, 100*f*
 climate and, 112
 by climate type, 115*t*
 convectional, **101,** 101*f*
 cyclonic (frontal), **102,** 104*f*
 desertification and, 160
 deserts
  hot, 116, 119*f*
  midlatitude, 119*f*
 El Niño and, 108–109
 erosion by, 71–72, 71*f*
 frontal, **102**–104, 104*f*
 global, by season, 113*f*
 global variation in, 125*f*
 global warming and, 127, 128*f*
 humid continental climate, 122, 122*f*, 123*f*
 humid subtropical climate, 120, 121*f*
 in hydrologic cycle, 434, 434*f*
 marine west coast climate, 121–122, 122*f*
 Mediterranean climate, 121*f*, 360
 ocean currents and, 98–99
 orographic, **101,** 101*f*
 savanna, 116, 118*f*
 seasonal variation in, 113*f*
 subarctic climate, 124*f*
 in tropical rain forest, 114, 117*f*
 types of, 101–102
 in Washington State, 103*f*
 wastewater pollution and, 442
Pregnancy, health risks in, 186, 189, 189*f*
Pressure gradient force, **94**–95
Primary economic activities, **343,** 343*f*, 344*f*
 agriculture, 345–361
 forestry, 363
 hunting and gathering, 345, 346
 mining and quarrying, 363–364, 364*f*
Primary products, trade in, 364–366, 365*f*
Primate cities, 308, **401**–402, 421
 in developing nations, 401–402, 425
 in Latin America, 478
Prime meridian, **27**
Princeton Corridor, 485
Private property, development of, 227
Process, in geographical studies, 476–477
Processing industries, 344. *See also* Secondary economic activities
Producers, in food chain, 433*f*
Production controls, in agriculture, 354–355
Profiles, soil, **107**–108, 107*f*
Profit, in industry location decision, 367
 spatial margin of profitability, 369, 369*f*
Profit maximization assumption, in economics, 367
Program for Monitoring Emerging Diseases (ProMED), 198
Projections, **489**–495
 Albers equal-area conic, 492, 493*f*
 area-preserving, 30–31
 azimuthal, **31, 492**–493, 493*f*
 conformal, **31,** 32*f*
 conic, **492,** 492*f*
 cylindrical, **490**–492, 490*f*, 491*f*
 definition of, **30**
 direction-preserving, 31–32
 distance-preserving, **31**
 distortion in, 30–31, 31*f*
 equal-area, **31,** 32*f*
 equidistant, **31,** 32*f*, 492–493, 493*f*
 equivalent, **31**
 Fuller dymaxion, 494, 494*f*
 geometric, 490–493, 490*f*
 gnomic, 490, 490*f*, 491, 493, 494*f*
 mathematical, 493–494
 mercator, **491,** 491*f*
 Miller cylindrical, 491–492, 491*f*
 non-Euclidian, 494, 495*f*
 orthographic, 490, 490*f*
 planar, 490*f*, **492**–493, 493*f*
 polyconic, 492, 492*f*
 properties of, 30–32
 Robinson, **31,** 32*f*
 selection of, 31–32
 simple cone, 492, 492*f*
 stereographic, 490, 490*f*

ProMED (Program for Monitoring Emerging Diseases), 198
Promised Land dump, Philippines, 462*f*
Proportional symbols, 38–39, 38*f*
Proposition 187 (California), 288–289
Proposition 209 (California), 289
Proposition 227 (California), 244
Prorupt states, **305,** 305*f*
Protectorates, political status of, 302
Protestant Reformation, 251
Protestantism, 251–252
  distribution of
    global, 248*f*
    United States, 253*f*
Proto-Indo-European language family, 237
Proved reserves, 136*f*, **137**
Psychological barriers to interaction, 270
Psychological distance, 11, 269, 270*f*, 271*f*, 272*f*
Ptolemy, 7*f*, 27
Public administration centers, U.S., 399, 399*f*
Public housing, in Chicago, Illinois, 419*f*
Public Law 103-227. *See Goals 2000: Educate America Act*
Public services. *See* Social services
Public transportation
  in Eastern European cities, 423
  motor vehicles and, 278–279
Pueblos, 223
Pull factors, **282,** 283, 283*f*
Pumice, formation of, 57
Purchasing Power Parity (PPP), by nation, **225***f*
Push factors, **282,** 283*f*
PV (photovoltaic) cells, **152**

## Q

Quadrangle, defined, 34
Qualitative flow-line map, 40
Qualitative maps, 38
Quantitative flow-line map, 40, 41*f*
Quantitative maps, 38
Quarrying. *See* Mining and quarrying
Quartz, U.S. imports of, 158*t*
Quaternary economic activity, 343*f*, **344,** 381
  global trade in, 382–383, 382*t*
  international trade in, 384
Quebec, secessionist movement in, 320
Quecreek Mine disaster, 26
Quinary economic activity, 343*f*, **344,** 381–382
  international trade in, 384
Quotas, controversy over, 318

## R

Race. *See also* Ethnicity
  as classification, 258, **259**
  and voting rights, 331, 332–333, 332*f*
Racial gerrymandering, 331, 332–333, 332*f*
Radar imagery, **42**–44, 45*f*
Radioactive waste
  disposal of, 460–462, 462*f*
  low *vs.* high-level, 460
  water pollution by, 440
Railhead locations, 396
Railroad, transportation of goods by, 369–370
Rain. *See* Precipitation
Rain forests, tropical. *See* Tropical rain forest
Raleigh, Sir Walter, 279
Ramadan, 247*f*
Randstad, 403–404
Rank-size rule of urban areas, **401**–402
Rapids, formation of, 72
Rate, definition of, **182**
Rate of natural increase, **193**
  by region and nation, 496*t*–503*t*
  United States, 194
  world, 195*f*, 207*f*
Rational behavior, in economics, 367
Rationalization of industry, 373
Raynolds, Captain, 177
Reapportionment, 330–333, 331*f*, 332*f*
Recessional moraines, 77*f*
Recycling, **459,** 459*f*
  of aluminum, 171, 171*f*, 459
  of copper, 158
  in developing countries, 459, 462*f*
  energy conservation and, 171, 171*f*
  of minerals, 136, 171
  of tires, 459, 459*f*
Redistricting, 330–333, 331*f*, 332*f*
Reduced tillage farming, 162
Reefs, formation of, 80
Reference map, definition of, 33
Reflection of solar energy, **91**–92
Refugee(s), numbers of, 281–282
Region(s), **16**–17, **470**–472, 471*f*
  boundaries of, 471, 471*f*
  characteristics of, 470–472, 471*f*
  climate, 112–123, 114*f*, 115*t*, 473–475
  core of, 471–472
  in culture-environment tradition, 476–482
  definition of, 470
  domains, 471–472
  in earth science tradition, 472–476
  economic, 482–483
  ecosystems as, 485, 486*f*
  formal, **17,** 18*f*, 470–**471**
    multifactor, 473, 479
  functional, **17,** 18*f*, 470–**471**
  as fundamental theme in geography, 17
  hierarchy of, 471, 472*f*
  industrial, 375–376, 375*f*
  landforms as, 472–473
  language as, 478–479
  mental, 479–480, 480*f*
  multifactor formal, 473, 479
  natural resource, 475–476
  political, 480–482, 481*f*
  population as focus of, 477–478
  spheres, 472
  subsurface, 475
  types of, 17, 470–472
  urban areas as, 484–485
  weather as, 473–475
Regional alliances, 327–329
  economic, 327–329, 329*f*
  military and political, 329
Regional autonomy, desire for, 320–322
Regional concept, **470**
Regional dialects, 241, 242*f*
Regional economic alliances, 327–329, 329*f*
Regional geography, 8, 467
Regional government, 329–335. *See also* Asymmetric federalism
  administrative units in, 333–335, 334*f*
Regional studies, in chapter 13, 21*t*
Regional trade agreements, 326, 327–329
Regional tradition, 19–21, 20*f*
Regionalism, **320**
Reims Cathedral, France, 445*f*
Relational direction, 10
Relative direction, **10**
Relative distance, **11**
Relative humidity, 99, 99*f*, **101**
Relative location, **9**–10, 9*f*, 307
Relic boundary, 312
Religion(s), 246–257. *See also* Buddhism; Christianity; Hinduism; Islam; Judaism
  birth control and, 183, 202, 212
  and birth rate, 183
  classification of, 246–247, 248*f*
  and cultural landscape, 252–253, 254, 254*f*, 255, 255*f*, 256–257, 256*f*, 257*f*
  culture and, 235–236, 246
  decline in church attendance, 248
  definition of, 246
  distribution of, global, 247–248, 248*f*
  diversity of, in U.S., 252–253, 254*f*
  dynamic nature of, 247
  economic impact of, 246
  ethnic, 246–**247,** 249, 254
    in East Asia, 257
  principal faiths, 248–257
    diffusion routes of, 250*f*
  Shinto, 257, 257*f*
    distribution of, global, 248*f*
  as source of conflict, 246
  state cohesion and, 317, 319
  Taoism, 257
  tribal, 246–**247**
    distribution of, global, 248*f*
  universalizing, 246–**247,** 255
  in USSR, 319
Reluctant migration, 281
Remote sensing, **42**–45
  aerial photography, 42
  applications of, 44*f*, 45, 47*f*, 49*f*
  non-photographic imagery, 42–45
    radar imagery, **42**–44, 45*f*
    satellite imagery, 44, 46, 46*f*
    thermal imaging, 42, 44*f*, 49*f*
  satellite imagery, 44, 46, 46*f*, 47*f*
  uses of, 46
Renewable resources, **135,** 135*f*
  future use of, 171
Replacement level, 184, **187**
Representation, political, 330–333, 331*f*, 332*f*
Representative fraction (RF) scale, 33, 33*f*
Reradiation of solar energy, **91**–92, 92*f*
  greenhouse effect and, 126, **126**
Research Triangle, 377
Reserves, 136–137, 137*f*
  coal, 141–142, 142*f*, 156
  copper, 156*t*, 157–158
  estimation of, 136–137
  mineral, 156, 156*t*
    nation size and, 303, 363
    as region, 475–476
  natural gas, 143, 145*f*
  oil, 139–140, 139*f*
  oil shale, 144, 145*f*
  proved, 136*f*, **137**
  subeconomic, 136, 136*f*
  tar sand, 144, 146*f*
  uranium-235, 147
Reservoirs, impact of, 436
Residence(s)
  as pollution source, 441–442
  urban
    in eastern Europe, 423–424, 424*f*
    ethnicity and, 409, 411, 411*f*, 412*f*
    family status and, 409–411, 411*f*
    gentrification, 419–420, 420*f*, 422
    models of, 408–409, 409*f*
    patterns of, 407–408, 407*f*, 408*f*
    social areas in, 409–411
    social status and, 409, 410, 411*f*
    zoning and, 411–412
Resources, **135.** *See also* Natural resources
  nation size and, 303
  of oceans, ownership of, 325–326, 325*f*
  as source of conflict, 315–316, 315*f*
Retail centers, U.S., 399, 399*f*
Return migration, **290**–291, 291*f*
Reusable natural resources, 136. *See also* Recycling
Reuters News Agency, 323
Rhine River, 311
Rhode Island, gentrification in, 419
R-horizon, 108
Rhumb line, **491,** 491*f*
Rice
  cultivation of, 2*f*, 350, 351*f*, 352
  Japanese price supports for, 355
Richter, C. F., 65
Richter Scale, 64, 65, 66*t*
Ridge and Valley region of Eastern U.S., 62, 63*f*
Rift valley, 63, 63*f*, 64*f*
Ring of fire, 62
Rio de Janeiro, Brazil
  population of, 392*t*
  slums of, 208*f*, 426, 426*f*
  UN conference in (1992), 202
Rio Grande river, 313
River blindness, 198
Rivers and streams. *See also* Stream landscapes
  as boundaries, 311
  modification of, 436–437, 437*f*, 438, 438*f*
Robert Taylor Homes, 419*f*
Robinson, J. Lewis, 19, 20
Robinson projection, **31,** 32*f*
Rock
  forces altering, 58
  gradational processes and, **68**–81
  rock cycle, 58
  types of, 56–58
Rocky Mountains, formation of, 68
Roma (gypsies), 303
Roman Empire
  acculturation and, 234
  Christianity and, 250–251
  extent of, 251*f*
  geographical interests, 6–7
  and language, 237
  road system in, 317–318
  state terrorism and, 322
Romance languages, 237, 246*f*–247*f*
Romania
  microdistricts in, 424*f*
  mining in, and water pollution, 441
  population data for, 502*t*
Rome, Italy, climate, 121*f*
Root action, and weathering, 68
Rotary clubs, diffusion in Japan, 277, 278*f*
Round goby, 454
Rumaila oil field, 315, 315*f*
Rumney, George R., 473–475
Rural-to-urban migration, 286
Russia. *See also* USSR
  agriculture in, 303, 307, 360
  air pollution in, 444
  birth rate in, 182
  Bolshevik revolution in, 228
  cities in, 401, 423–424, 424*f*
  climate, 122, 123*f*
  core area of, 308
  economies in, 379
  gated communities in, 416
  ideology in, 228
  industrial location in, 373
  influenza pandemic in (1781), 277*f*
  Laplanders and, 315
  location of, 9*f*, 307
  mineral production by, 157*f*
  mineral resources, 157
  Moscow
    climate, 123*f*
    as core area, 308
  NATO and, 329, 330*f*
  natural gas reserves, 143, 145*f*
  nomadic herding in, 349
  nuclear power in, 146–147, 147*f*
  oil shale deposits, 144, 145*f*
  population data for, 502*t*
  population pyramid, 190*f*, 191
  size of, 303
  tar sand reserves, 144
  urbanization in, 394*t*
  water pollution in, 441
Russian Federation
  ethnic unrest in, 321
  oil consumption by, 138*f*
  oil production, 138*f*, 141
Russian language, 242
  number of speakers, 240*t*
Rwanda
  civil war in, 258, 283
  endangered species in, 453

maternal mortality rates in, 189
population data for, 497*t*
state terrorism by, 322

# S

Saami (Lapps), 349
Sags, 451
Sahara Desert, 80, 473
Saharan language family, 238*f*–239*f*, 240*f*
Salinization of soils, **163**, 163*f*
dams and, 150
Salt crystals, and weathering, 68
Salt marshes, formation of, 79–80
San (Bushmen), 227–228, 228*f*
San Andreas fault, 60, 60*f*, 62, 64, 64*f*, 65*f*
San Diego, travel times in, 12*f*
San Francisco Bay area
desire line map of, 16*f*
El Niño and, 108
San Francisco International Airport, satellite image of, 46*f*
San Marino, 305, 502*t*
Sand
reserves of, 156
in soil, 109
wind-driven, 80–81, 81*f*, 119*f*
Sandbars, formation of, 79, 80*f*
Sandstone, formation of, 57
Sanitary landfills, 457–458
radioactive waste and, 461
Sanitation, and life expectancy, 188, 196, 197, 197*f*, 198, 211*f*
Satellite imagery, 47*f*
and mapping, 44
spy satellites, 46, 46*f*
and weather forecasting, 44
Satisficing location, 369*f*, 370
Saudi Arabia
Islamic fundamentalism in, 319
as multicore state, 308
oil production, 138*f*
oil reserves, 139*f*
population data for, 499*t*
terrorism and, 323
Savanna, **114**–116
climate, 115*t*, 118*f*
locations of, 112*f*, 114*f*, 116*f*
precipitation in, 116, 118*f*
soils, 116
vegetation, 112, 112*f*, 114–116, 117*f*
Scale, **11**–12, 34*f*
on globe, 30
graphic, 33, 33*f*
importance of, 12*f*
large, defined, 33
representation of, 33, 34*f*
representative fraction (RF), 33, 33*f*
small, defined, 33
verbal, 33, 33*f*
Scandinavia, population growth incentives in, 187
Schistosomiasis, 151, 198
Schneider, William J., 486
Schuylkill Anthracite Region, 475–476, 476*f*
Schuylkill Navigation Canal, 475, 477*f*
Scotland
industry in, 378
semi-autonomy of, 308, 320, 321*f*
S-curve, in population growth, 211, 211*f*
Sea breezes, **95**
Seasons
climate and, 112–113, 113*f*
earth's rotation and, 89–91, 90*f*, 91*f*
and global air circulation pattern, 97
Secessionist movements, 320–321, 321*f*
Second(s), 27
Secondary economic activities, 343*f*, **344**, 344*f*, 367–378. *See also* Manufacturing industries
Sector(s). *See also specific sectors*
basic and nonbasic, in urban areas, **398**–400, 400*f*
of U.S. labor force, 378–379, 379*f*
Sector model of urban land use, **408**–409, 409*f*
Secularism, 248, 248*f*
rejection of, 249
Sedentary intensive agriculture, 348*f*
Sedimentary rock, **57**, 58*f*
Sedimentation, deposition of, 71–81, 72*f*, 150, 162–163, 169, 451
Segregation
ethnic, in urban areas, 411, 412*f*
of poor, 415, 416
territorial, 258
Seismic waves, 64
Seismographs, 66
Selective cutting, 165–166, 166*f*
Self-determination, 320–322
Semidesert climates, 115*t*
locations of, 112*f*, 114*f*
midlatitude, 118
vegetation, 112, 112*f*
Seminole tribe, forced migration of, 282*f*
Senegal
cities in, 402, 424
population data for, 497*t*
Senior citizens. *See* Elderly
Seoul, Korea
air pollution in, 443
as primate city, 401
Separatist movements, 312*f*, 314, 319–322
Sephardim, 249
September 11th terrorist attacks (2001)
disposal of rubble from, 458
impact of, 292, 322
LIDAR image of site, 49*f*
response planning for, 49, 49*f*
thermal images of site, 44*f*
Serbia and Montenegro
ethnic cleansing in, 322
national identity, 317
population data for, 502*t*
and U.N., 324
Serbs, 258, 283
Serengeti National Park, 116
Service(s). *See also* Postindustrial economy; Tertiary economic activities
global trade in, 382–383, 382*t*
low-level, 380–381, 380*f*
as tertiary economic activities, 344
in urban areas, **398**–400, 400*f*
Sewage. *See also* Animal waste; Sanitation
as pollutant, 441–442
Shaded relief, on topographic maps, 37
Shale, 57, 57*f*
Shale oil, 144
Shamans and Shamanism, 247
Shantytowns, in developing nations, 426, 426*f*
Shape
projections preserving, 31
of states, 305, 305*f*
*Shaw* v. *Reno* (1993), 333
Shield volcano, 66, 67*f*
Shifting cultivation, 348*f*, 349–350, 349*f*
Shi'ites, 248*f*, 254
Shinto, 257, 257*f*
distribution of, global, 248*f*
Siberia
International Date Line and, 28
migration to, 200
Siddhartha Gautama, 255–256
Sierra Leone
maternal mortality rates in, 189
population data for, 497*t*
Sierra Nevada mountains, formation of, 64
Sikhism, 255, 321
Silicon Fen, 378
Silicon Forest, 377
Silicon Glen, 378
Silicon Swamp, 377
Silicon Valley, 377
Silicon Valley North, 377
Silt
deposition of, 71–81, 72*f*, 150, 162–163, 169
in soil, 109
Siltstone, formation of, 57
Silver
mining of, 450*f*
reserves of, 156*t*
U.S. imports of, 158*t*
Similarity of place, terminology, 16–17
Simple cone projection, 492, 492*f*
Singapore
birth and fertility rates in, 187, 212
censorship in, 295
climate, 117*f*
economy of, 383
location of, 307
as mini-state, 306
population, 306
age of, 213*f*
control of, 180, 212
data on, 500*t*
size of, 306
as world city, 401, 403*f*
Singlish, 243
Sinkholes, 75, 76*f*
formation of, 451
Sino-Japanese culture realm, 221*f*
Sino-Tibetan language family, 238*f*–239*f*
Site, **9**
of urban area, **396**
Situation, 9–10
of Chicago, 397, 397*f*
of urban areas, **396**–397, 398*f*
Size, and scale, **11**–12
Skyline, of city
in Anglo America, 406, 422
in Europe, 422, 423*f*
Skyscrapers, European cities and, 422
Slash-and-burn agriculture, 349–350, 349*f*
Slate, formation of, 58
Slavery, and forced migration, 281
Slavic culture realm, 221*f*
Sleeping sickness, 198
Slovenia
as nation-state, 302, 302*f*
population data for, 502*t*
Slums
in Cairo, 390
causes of, 415, 416*f*
clearing of, 417, 419*f*
in developing nations, 208*f*
and disease, 426
in Latin American cities, 208*f*
in Malaysia, 388*f*
urban land use and, 408
urban renewal programs and, 416–417, 419*f*, 420
and zoning regulations, 412
Small circle, **489**
Smallpox, 198
Small-scale maps, 33
Smart bombs, 29
Smart growth programs, 421
Smog. *See* Photochemical smog
Snow, cause of, 99–100, 100*f*
Social areas, in cities, 409–411
Social dialects, 239–241
Social geography, of European city, 422–423, 423*f*
Social services
funding for, 417
illegal aliens and, 288–289
urban poor and, 415–416, 417
Social status
and city structure, 409, 410, 411*f*
determinants of, 410
and targeted marketing, 410
and urban residence patterns, 409, 410, 411*f*
Social structure, 219
Sociofacts, 222, 226, 229*f*
Sociological subsystem, 222, **226**–228
Software Valley, 377
Soil(s). *See also* Erosion
acid rain and, 445
acidity and alkalinity of, 110
arctic and subarctic climate, 123
classification of, **110**, 110*t*
and climate, 107*f*, 115*t*
color of, 109
components of, 108–109
creation of, 69–70
creeping by, 70, 71*f*
desert, 118
fertile
locations of, 345, 347*f*
loss of, 159–163, 223, 351–352
maintaining, 162, 162*f*
formation of, 76, 104–107, 110, 159
horizons (layers) of, **107**–108, 107*f*
human impact on, 159–163
humid continental climate, 120, 122
humid subtropical climate, 120
importance of, 104
layers (horizons) of, **107**–108, 107*f*
leached, 109
marine west coast climate, 122
orders, **110**, 110*t*
preservation of, 162, 162*f*
profiles, **107**–108, 107*f*
properties of, **108**–110
salinization of, **163**, 163*f*
savanna, 116
steppe, 118
structure, 109
taxonomy, 110, 110*t*
texture, 109
topsoil, erosion of, 159–163
tropical rain forest, 107*f*, 114
types of, **110**, 110*t*
Soil order, 110, 110*t*
*Soil Taxonomy*, 110, 110*t*
Solar cells, 152
Solar energy, 89
biosphere and, 433
and climate change, 93
and earth's inclination, 89–91, 89*f*, 90*f*, 91*f*
future use of, 171
hydrological cycle and, 434*f*
and insolation, 89
and latitude, 89–90, 91*f*
ozone layer and, 447, 447*f*
as power source, **150**–152, 152*f*
reflection and reradiation of, **91**–92, 92*f*
seasonal variation in, 89–91, 90*f*, 91*f*
as source of all energy, 137, 150
and temperature, 89–92
Solid waste disposal, 456–462
hazardous waste, 460–462, 462*f*
incineration, 458–459, 458*f*
landfills, 457–458
for hazardous waste, 458
radioactive waste and, 461
municipal waste, 456–459, 457*f*
pollution from, 443, 443*f*, 443*t*
recycling, **459**, 459*f*
source reduction, 459
and water pollution, 457
Solstice, 90, 91*f*
Solution process, 74
Somalia
population data for, 209, 497*t*
UN peacekeeping in, 324
Sony Corp., 294*f*
Source reduction for solid waste, 459
Source regions, **102**, 103*f*, 474–475, 475*f*
for Americas, 103*f*, 475*f*

South Africa
agriculture in, 359
Cape Town, 222*f*
climate of, 120
gated communities in, 416
life expectancy, 190
mineral production by, 157*f*
mineral resources, 157
population, 190
data on, 498*t*
shape of, 305, 305*f*
vegetation of, 112
South America
agriculture in, 231*f*, 352
air mass source regions of, 475*f*
coal reserves, 142*f*
as culture hearth, 230, 231*f*
El Niño and, 108
energy in, 149
habitat destruction in, 452
hydropower in, 152*f*
industry in, 375–376, 375*f*, 377*f*
latitude of, 27
longitude of, 27
mining in, 441
natural gas reserves, 145*f*
natural gas trade flows, 144*f*
oil reserves, 139*f*
oil shale deposits, 145*f*
oil trade flows, 408*f*
population control in, 212
population data for, 499*t*
rain forests in, 114, 170*f*
regional trade alliances, 328, 329*f*
religion in, 248
savannas in, 116
water pollution in, 439
South Asia zone, 203
South Carolina
Gullah language in, 479, 479*f*
radioactive waste disposal in, 461
South, in Brandt development model, 226, 226*f*
South Korea
birth and fertility rates in, 187, 212–213
commercial fishing by, 453
energy in, 148
fertility rates in, 199
food production and, 207
high-tech industries in, 377
migration to U.S., 286
as nation-state, 302
nuclear power and, 146, 147*f*
population data for, 501*t*
Seoul
air pollution in, 443
as primate city, 401
trade
in commodities, 366
in services, 382*t*
women in, 191
South Pole, grid system and, 27
South Tirol, 321*f*
Southern Cone Community Market (MERCOSUR), 328, 329*f*
Southern oscillation, 109
Sovereignty, UN and, 324
Soviet Union. *See* USSR
Space Imaging Corporation, 46
Spain
asymmetric federalism in, 308
Ceuta, 305, 306*f*
colonization by, 251–252
devolution in, 320
exclaves, 305, 306*f*
gated communities in, 416
nationalist movements in, 320, 321*f*
place names in, 244
population data for, 187, 502*t*
regional trade alliances, 327
separatist movements in, 321, 321*f*, 323
time zone of, 29*f*
trade, in services, 382*t*
Spanish language
in Africa, 246*f*
number of speakers, 240*t*
Spatial data, mapping of, 17*f*
Spatial, defined, 8
Spatial diffusion, **15**
of Buddhism, 250*f*, 256, 256*f*, 257
of Christianity, 250–251, 250*f*
contagious, **276**–277, 277*f*, 278*f*, 279
of culture, 230–232
documentation of, 279
of English language, 237, 238, 241*f*, 243, 295
hierarchical, **277**–278, 277*f*, 278*f*
of Hinduism, 250*f*, 254
of innovation, 270, **275**–276, 279
of Islam, 250*f*, 253, 255*f*
Spatial distributions, 16–17
Spatial interaction, **14**–15, 16*f*
barriers to, 269–270
definition of, **269**
distance and, 269, 269*f*, 270
factors affecting, 274–275
and innovation, diffusion of, **275**–276
motor vehicles' impact of, 274–275, 278–279
technology and, 268, 269–270, 278–280
telecommunications' impact on, 279–280
in U.S., 268
Spatial margin of profitability, 369, 369*f*
Spatial mismatch, 416
Spatial orientation, in industry location model, 368, 369*f*
Spatial systems, local and regional government as, 330
Spatial tradition, **19**–21, 20*f*
Special-purpose maps, 34
Speech community, 238–239
Spheres, 472
Spits, 79, 80*f*
Spodosols, 110*t*, 111*f*, 115*t*, 122, 123
Sport utility vehicles (SUVs), 140–141
Sports, and national identity, 317
Spot heights, on topographic maps, 36
SPOT satellites, 44
Spy satellites, 46, 46*f*
Squatters settlements, in developing nations, 426, 426*f*
Sri Lanka, 198, 199
population data for, 203, 500*t*
separatist movements in, 321
Stacked gerrymandering, 331, 332*f*
Stage in life, **274**
and activity space, 274–275
Stalactites, 74
Stalagmites, 74–75
Stalin, Josef, 309, 322, 360
Standard line, 490
Standard of living, globalization and, 293
Standard parallel, 490
Starvation, deaths from, 211
State(s), 301–**302**
administration of. *See* Government
boundaries of, 480–482, 481*f*
as conflict source, 311, 311*f*, 312–316, 312*f*
significance of, 480–482, 481*f*
capitals of, 308–310
in developing nations, 424, 425*f*
forward-thrust, 309–310
as primate city, 402
centrifugal forces in, **316,** 318–321
centripetal forces in, **316**–318
cooperation among, 323–329
core area of, **307**–308
federal, 308
geographic characteristics of, 303–310
capitals, 308–310
core areas, **307**–308
location, 305–307
shape, 305, 305*f*
size, 303
mini, 306, 307*f*
modern
declining power of, 323–324
evolution of, 303
multinational, 302, 302*f*
part-nation, 302, 302*f*
sovereignty of, and UN, 324
unitary, 308
State terrorism, 322
Stateless nation, 302–303, 302*f*
Stayers, 284
Step migration, **285**
Steppe(s), **118**
climate, 115*t*
locations of, 112*f*, 114*f*, 118*f*
soil, 118
vegetation, 112, 112*f*, 118, 120*f*
Stereographic projection, 490, 490*f*
Sterilization, as birth control, 180, 184, 211*f*
Stock market
globalization of, 291–292
September 11th terrorist attacks and, 292
Storm(s), 102–104, 104*f*, 105*f*. *See also* Hurricane(s); Precipitation; Tornadoes; Typhoons
erosion and, 82
fronts, **102**–104, 103*f*, 104*f*
runoff, and water pollution, 442
Strabo, 6, 7, 21
Strata, **57,** 58*f*
Strato, 66
Stratosphere, ozone layer in, **446**–448, 447*f*, 448*f*
Stratus clouds, 100–101, 100*f*
Stream landscapes, 72–74
in arid areas, 73–74, 75*f*
in humid areas, 72–73, 73*f*
Stream load, 72
Strip mining, 449–450, 450*f*
Strip-cropping, 162, 162*f*
Strontium, U.S. imports of, 158*t*
Stupas, 256–257, 257*f*
Subarctic climate. *See* Arctic and subarctic climates
Subduction, **62,** 62*f*
Subduction zones, 60*f*, 62
Subeconomic reserves, 136, 136*f*
Subnational terrorism, 322
Subnationalism, **320**
Subpolar lows, 97
Subsequent boundaries, **311**
Subsidence, **451**–452, 451*f*
Subsidence lakes, 451
Subsistence agriculture. *See* Agriculture
Subsistence economy, **344,** 345
Substance, definition of, 275
Substitution, for scare resources, 158–159, 171
Substitution principle, 368
Subtropical climate. *See* Humid subtropical climates
Subtropical high pressure, 97, 116
Suburbs and suburbanization, **396**
in Canada, 421
and employment, 416, 417
impact on central cities, 414–420
and industry location, 413, 416, 417
land use in, 413–414, 414*f*
in United States, 413–414, 413*f*, 414*f*
Succession, in natural vegetation, **110,** 111*f*
Sudan, population data for, 496*t*
Sudanic language family, 238*f*–239*f*, 240*f*
Sudetenland, 314
Sulfur
in coal, 142
as pollutant, 444
Sulfur oxides, as air pollutant, 443*t*, 458
reduction efforts, 448
Summer, climate in, 112–113, 113*f*
Sun. *See also* Solar energy
orbit of Earth around, 89–91, 90*f*, 91*f*, 125
sunspots, and climate change, 126
Sunbelt, migration to, 284
Sunnis, 248*f*, 254
Sunrise Strip, 378
Superfund sites, 460
Superimposed boundaries, 303, 304*f*, **311**
Supersaturated air, 100
Supply and demand
in commercial agriculture, 355, 356*f*
in markets, 367, 368*f*
Supply curve, 368*f*
Supranationalism, 323–**324**
Supreme Court, U.S. *See* U.S. Supreme Court
Surface(s), developable, **490**
Surface depressions, formation of, 451
Surface mining. *See* Open-pit mining; Strip mining
Survey and Mapping and Remote Sensing, Natural Resources (NRCAN), 34
Sustainable development, **169**–171
SUVs (Sport utility vehicles), 140–141
Swahili, 242
as lingua franca, 242
number of speakers, 240*t*
Swamp Land Acts, 164
Swedagon pagoda, 257*f*
Sweden
air pollution in, 448
currency, 327
energy in, 154
high-tech industry in, 378
nationalism in, 316
nuclear power and, 146
population data, 501*t*
population pyramid, 190*f*, 191
regional trade alliances, 327
Swidden agriculture, 349–350, 349*f*
Switzerland
energy in, 149
language in, 243
maternal mortality rates in, 189
as multinational state, 302*f*
nationalism in, 316
population data for, 501*t*
regional trade alliances, 327
and U.N., 324
Sydney, Australia, climate, 121*f*
Symbols
area, 39–40, 39*f*
line, 40
point, 38–39, 38*f*–39*f*
proportional, 38–39, 38*f*
three-dimensional, 39
Synclinal hills, 63*f*
Syncretism, **232**
Syr Darya River, 438, 438*f*
Syria
agriculture in, 163
border disputes, 315–316
population data for, 500*t*
water supply in, 436
Systematic geography, defined, 8

# T

Taglish, 243
Tahiti, alien species in, 455
Taiwan
birth and fertility rates in, 187
commercial fishing by, 453
food production and, 207
population
data for, 501*t*
growth, 187
women in, 191
Talus, **70,** 71*f*
Tambora, eruption of (1815), 125
Tamil language, number of speakers, 240*t*
Tamils, separatist movement, 321
Tantalum, U.S. imports of, 158*t*

Tanzania
  agriculture in, 351
  language of, 242
  mining in, 441
  population data for, 497t
*Tao*, 257
Taoism, 257
Taqui-al-Din-Al-Maqrizi, 390
Tar sands, **144**–146
  mining of, 145, 146f
  reserves, 144, 146f
Tarns, 76, 77f, 78f
Taxes
  industry location and, 375
  offshore banks and, 384
  urban areas and, 415–416, 417, 420–421
Technological subsystem, **222**–226
  in advanced *vs.* developing cultures, 223–226, 224f
  sociological subsystem and, 226
Technology. *See also* Communication(s)
  in agriculture, 224f, 347–348, 353
  and carrying capacity, 212
  and commercial fishing, 453
  and commodity prices, 365
  connectivity and, 15
  and economic activity, 342
  and employment, 11
  globalization and, 291–292, 293, 382
  high-tech industries, 376–378
  and human environmental impact, 221
  innovation, **230**
    cities as centers of, 230, 270–271, 277–278
    conditions favoring, 270–271
    definition of, 275
    diffusion of, 270, **275**–276, 279
    history of, 232f
    motivation for, 230
    parallel, 232
    rate of, historical, 230, 232f
  and manufacturing, 369
  quaternary economic activity and, 381
  and resource reserves, 136–137
  and spatial interaction, 268, 269–270, 278–280
  transfer to less-developed nations, 376
Tectonic forces, **62**–68
  diastrophism, **62**–66
  volcanism. *See* Volcanism
Tehran, Iran, climate, 119f
Telecommunications. *See* Communication(s)
Telecommuting, 280
Telugu language, number of speakers, 240t
Temperate deciduous soil, 107f
  plant succession in, 111f
Temperature. *See also* Air temperature; Climagraphs
  average, change in, 1860–2002, 126f
  climate and, 112
  by climate type, 115t
  deserts
    hot, 119f
    midlatitude, 119f
  global, by season, 113f
  humid continental climate, 122f, 123f
  humid subtropical climate, 121f
  marine west coast climate, 121, 122f
  Mediterranean climate, 121f
  savanna, 118f
  seasonal variation in, 113f
  solar radiation and, 89–92
  subarctic climate, 124f
  tropical rain forest, 117f
Temperature inversion, **92,** 94f, 95, 95f, 96, 443, 446
Temples, Hindu, 255, 256f
Tennessee Valley Authority, 149
Terminal moraines, 77f
Terminology
  interrelationships, 14–15
  location, direction, and distance, 9–11
  physical and cultural attributes, 13–14
  regions, **16**–17
  similarity of place, 16–17
  size and scale, 11–12
Terrace agriculture, 2f, 162, 204f, 450
Terrain representation. *See* Topographic maps
Territorial sea, 325, 325f
Territorial segregation, 258
Territoriality, **271**
Terrorism, **322**–323. *See also* September 11th terrorist attacks
  definition of, 322, 323
  satellite imagery and, 46
  state responses to, 323
  types of, 322
Tertiary economic activities, 343f, **344,** 378–382
  characteristics of, 380–381
  GDP and, 379, 379f, 379t
  global trade in, 382–383, 382t
  location considerations for, 380, 380f
  outsourcing and, 380–381
  subdivision of, 380–381
Tertullian, 210
Texas
  high-tech industry in, 377
  Houston
    air pollution in, 447f
    edge cities of, 414
    gentrification in, 419
    population of, 397t, 402t
    zoning and, 412
  redistricting and, 332f, 333
Texture of soils, 109
Thailand
  Bangkok
    as primate city, 401
    slums in, 426
  energy in, 148
  fertility rates in, 185, 199
  population data for, 500t
  population growth in, 193
  rain forests in, 168
  religion in, 317
  shape of, 305, 305f
  subsidence in, 451
Thematic maps, 34, 38–40, 38f–40f
Thermal imaging, 39f, 42, 44f
Thermal pollution, **441**
Thermal scanners, **42**
Threatened species, 451–452
Three Gorges Dam, 150–151
Three mile limit, 325
Three Teachings, 257
Three-dimensional symbols, 39
Thresholds, for products and services, 404
Tigris River, water use and, 314
Tilapia, 455
Till, glacial, 76, 78f
Time
  geologic, 58
  globalization and, 382–383
Time zones, 28, 29f
Time-distance, distance measurement in, 269
Times Beach, Missouri, contamination, 432, 432f
Timor-Leste, and U.N., 324
Tin, 156, 156t
Tires
  disposal of, 457, 457f
  recycling of, 459, 459f
Tobacco smoking, cultural spread of, 234, 279
Tobler's First Law of Geography, 14
Tokyo, Japan
  air pollution in, 446
  population, 392t
  subsidence in, 451
  as world city, 400–401, 403f
Toledo, Ohio, urban renewal in, 417
Tongass National Forest, 167, 168f
Topographic factor, in soil formation, 106
Topographic maps, **34**–38, 35f
  Brunswick, Georgia, 43f
  Hebgen Lake, Montana, 24f
  remote sensing and, 42, 43f
  symbols, 36, 36f
  updating of, 43f
  U.S. Geological Survey and, 34–36, 35f
  uses of, 38
Topography, and air pollution, 443
Toponymy, **244**–245
Topsoil, erosion of, 159–163, 223
Torah, 249
*Torii*, 257, 257f
Tornadoes, **104,** 106f
  destructiveness of, 104, 106f
  El Niño and, 108
  measurement of, 104
Toronto, Canada, 422f
  political subdivisions in, 335
  solid waste disposal, 458
  work *vs.* nonwork trips, 275f
  as world city, 401, 403f
Total fertility rate (TFR), **183**–186, 185f, 186f
  decline in, 197, 202–203
  for developing nations, 184–186, 186f
  global, 199
  by region and nation, 496t–503t
  replacement level, 184, **187**
Tourism, 381
Towns, **396**
  in agricultural areas, 404–405, 404f
Toxic wastes. *See* Hazardous waste
Toyota, industry incentives and, 374
Trade
  and alien species introduction, 453–454, 454f
  and disease, 198
  in 14th century, 390
  high-tech industry and, 377f
  in natural gas, 143, 144f
  in oil, 138f, 139
  in primary products, 364–366, 365f
  regional agreements on, 326, 327–329
  in services, 382–383, 382t
  as tertiary activity, 344
Trade alliances, regional, 326, 327–329
Traditional agriculture, 346–348
Traditional societies, innovation and, 230
Tragedy of the commons, **362**–363
Trail of Tears, 282f
Trans-Alaska pipeline, 79
Transform boundaries, 59, 60f
Transform fault, 63f, 64, 65f
Transhumance, 349
Transitional birth rate, 182
Transnational corporations (TNCs), 292–293, 294f, **373**–375
  employment and, 374
  power of, 323
  and taxes, 376
  trade in world services and, 382–383
  world cities and, 401
Transpiration, in hydrologic cycle, 434, 434f
Transport gradients, in agriculture, 357f
Transportation
  in agriculture location model, 355–357, 357f
  centers of, U.S., 399, 399f
  cities and, 394, 396, 405–406, 406f, 407, 407f
  of coal, 142, 142f, 143f
  cost of, 268, 369–370
  economic importance of, 345, 346f
  to employment
    in European cities
      Eastern, 424
      Western, 422
    in U.S., 417
  in industry location model, 368–369, 368f
  in Latin America, 478
  mass transit systems, 405–406, 407f
    in Canada, 421, 422f
    in developing nations, 425
    in Eastern European cities, 423
    in European cities, 422
    inadequacy of, 417
  methods of, and cost, 369–370
  in mining and quarrying, 363, 364f
  and national cohesion, 317–318, 319, 320f
  of natural gas, 143, 144f
  pollution from, 443f, 443t, 444, 446, 446f
  post-Fordist manufacturing and, 369
  of radioactive waste, 461
  state control of, 318, 320f
  suburbanization and, 413
  by water
    of goods, 369
    and urban location, 396, 398f
Trans-Siberian Railroad, 79
Transylvania, 314
Trash. *See* Solid waste disposal
Travel time
  isochrones, 12f
  and relative distance, 11, 12f
Treaty of the Pyrenees (1660), 306f
Trewartha, Glenn T., 477–478
Tribal religions, 246–**247**
  global distribution of, 248f
Trinidad
  population data for, 498t
  tar sand reserves, 144
Tropic of Cancer, 90f, 114
Tropic of Capricorn, 90f, 114
Tropical climates, 114–116, 115t, 117f
  locations of, 114f, 116f
Tropical rain forest, **114**
  biodiversity of, 169
  climate, 114, 115t, 117f
  deforestation in, 114, 159, 159f, 167–169, 170f, 352
  destruction of, 114, 170f, 452
  functions performed by, 168–169
  locations of, 112f, 114f, 116f, 168, 169f
  medical resources in, 172
  precipitation, 117f
  soil, 107f, 114
  vegetation, 112, 112f, 114, 117f
Tropopause, 94f
Troposphere, **89**
  air pollution and, 442
  and air temperature, 92
  biosphere and, 433
Trough reflectors, parabolic, 152f
Truck farms, 369
Truman Proclamation of 1945, 325
Tsunami, **64,** 70, 481
Tsunami Early Warning System, 70
Tuberculosis, 197, 198
Tundra, **123**
  climate, 115t
  locations of, 112f, 114f, 124f
  vegetation, 112, 112f, 123, 124f
Tungsten, U.S. imports of, 158t
Tunisia
  Islamic fundamentalism in, 319
  mineral production by, 157f
  population data for, 496t
Turkey
  air pollution in, 446
  Blue Mosque, Istanbul, 255f
  capital of, 309
  dams in, 150
  erosion in, 160
  Nagorno-Karabakh conflict and, 309
  population data for, 500t
  water management in, 436
Turkish language, number of speakers, 240t
Turks, 251
Tutsi, 258, 283
Tuvalu, 307f
  population data for, 503t
  and U.N., 324

Typhoid, 189, 437, 438
Typhoons
  destructiveness of, 4
  typical paths of, 105*f*
Tysons Corner, Virginia, 414, 485

# U

Ubiquitous industries, 370
Uganda
  endangered species in, 453
  fertility rates in, 185
  population data for, 497*t*
  population pyramid, 190*f*, 191
Ukraine
  nuclear power and, 146
  population data for, 502*t*
Ukrainian language, number of speakers, 240*t*
Ultisols, 110*t*, 111*f*, 116
Ultraviolet radiation, ozone layer and, 447, 447*f*
UN. *See* United Nations
UNAIDS, 190
UNCLOS (United Nations Convention on the Law of the Sea), **325**–326, 325*f*, 326*f*
UNCTAD (United Nations Conference on Trade and Development), 226*f*, 366
Underdeveloped, as term, 224
Underpopulation, 205
UNESCO (United Nations Educational, Scientific, and Cultural Organization), 324
UNICEF (United Nations Children's Fund), 326
Unions, high-tech industry and, 378
Unitary states, 308
United Airlines, and industry incentives, 374
United Arab Republic, 329
United Kingdom
  Antarctic territorial claims, 301*f*
  asymmetric federalism in, 308, 320
  birth and fertility rates in, 212–213, 501*t*
  colonies of, 134, 311–312
  currency, 327
  energy in, 154
  Falkland Islands War, 306
  GNI per capita, 501*t*
  high-tech industries in, 377, 378
  HIV/AIDS in, 501*t*
  London
    as financial center, 383*f*
    growth of, 339
    as primate city, 308, 421
    as world city, 400–401, 403*f*
  marriage customs in, 197
  migration from, 200
  nationalism in, 316
  nationalist movements in, 320, 321*f*
  nuclear power and, 147*f*
  place names in, 244
  population
    data, 501*t*
    density of, 205*t*
  regional trade alliances, 327
  religion in, 248
  trade, in services, 382*t*
  waste disposal in, 462
United Nations (U.N.), 324–326
  affiliates, 326
  agencies of, 334
  on agricultural land use, 346
  Alliance of Small Island States (AOSIS), 306
  Cairo Plan, 202–203
  charter of, 262
  Children's Fund (UNICEF), 326
  Commission on Sustainable Development, 169
  Conference on Environment and Development, 202
  Conference on Trade and Development (UNCTAD), 226*f*, 366
  Convention on the Law of the Sea (UNCLOS), **325**–326, 325*f*, 326*f*
  on desertification, 160
  Development Fund, 293
  Economic and Social Council, 226*f*
  on economic development, 226*f*
  Educational, Scientific, and Cultural Organization (UNESCO), 324
  Food and Agriculture Organization (FAO), 206, 324, 326, 345, 353, 355, 453
  headquarters building, 301*f*
  International Conference on Population and Development, 202–203
  interventionism of, 324, 324*f*
  maritime boundaries and, 325–326
  national sovereignty and, 324
  and news networks, 295
  peacekeeping deployments, 324, 324*f*
  population distribution estimates, 203
  Population Division, 180
  population projections, 180, 181*f*, 186, 187, 194, 199
    accuracy of, 209, 210
  on rain forest deforestation, 168
  small state definition by, 306
  Statistical office, 207
  on toxic waste exporters, 462
  UNAIDS, 190
  on urban growth, 391, 392*f*
  on urban living conditions, 208*f*
  on urbanization, 207
  voting rights in, 306
  on water supply, 436
  working languages of, 241*f*
  World Health Organization (WHO), 189, 189*f*, 190, 324, 326, 443, 456
  World Trade Organization (WTO). *See* World Trade Organization
  Year of Safe Motherhood, 189, 189*f*
United States. *See also agencies under* U.S.; *specific cities and states*
  agriculture in
    agricultural regions, 357, 358, 358*f*, 359
    locational model for, 355–357, 356*f*
    production controls, 354–355
    soil loss, 160–162, 161*f*
    water pollution and, 439–440, 440*f*
  air pollution, 443–444
    cleanup efforts, 448
    photochemical smog, 446, 446*f*, 447*f*
    sources of, 443, 443*f*, 443*t*
  assimilation of minorities, 235
  biocide use in, 456
  birth and fertility rates, 186, 498*t*
  borders of, 311, 313–314
  capital of, 308, 309
  CFCs use and production, 448
  cities in, 421
    central cities, immigration to, 419, 420*f*
    changes in, 414–421
    Eastern, 414–420
    edge, **414**
    ethnicity and, 466
    functional hierarchy, 400, 403*f*
    functional specialization, 399, 399*f*
    galactic, 414, 415*f*
    gated communities in, 416
    growth and decline, 400, 401*f*, 402*t*
    land use in
      competitive bidding and, 405–408, 407*f*, 416
      models of, 408–409, 409*f*
    Megalopolis, **414**, 415*f*, 484–485, 484*f*, 485*f*
    network cities, 404
    over 1 million residents, 398*f*
    percent of population in, 498*t*
    population of, 402*t*
    and rank-size rule, 401
    residential patterns, 407–408, 407*f*, 408*f*
      ethnicity and, 409, 411, 411*f*, 412*f*
      family status and, 409–411, 411*f*
      social status and, 409, 410, 411*f*
    western, 420–421
    zoning regulations in, 411–412
  Civil War, 314
  climate, 120, 122
  conurbations, 415*f*
  core area of, 308
  core regions, 415*f*
  as cultural system, 220
  culture
    complexity of, 219
    folk groups in, 233
    origins of, 234
    post-industrial life and, 380
  dams in, 150
  dialect in, 241, 242*f*
  Eastern
    Anglo-America industrial region, 375–376, 375*f*
    geological formations, 62, 63*f*
    ice storm in (1998), 86*f*
  economic planning in, 482–483
  economy
    classification of, 345
    postindustrial status, 293, 342, 368, 376, 378, 379*t*
    as service economy, 293
    tertiary activities in, 378–379, 379*f*, 379*t*
  employment in
    in high-tech industry, 378
    by sector, 378–379, 379*f*, 379*t*
  energy
    coal
      production, 142
      reserves, 141–142
    consumption, by source, 137–138, 137*f*
    geothermal, 152
    natural gas, 137, 137*f*, 143–144, 144*f*
    nuclear power, 146–147, 147*f*
    oil
      consumption, 137, 137*f*, 138*f*, 139*f*, 140
      production, 138*f*, 139, 140, 366
      reserves, 140
      trade flows, 138*f*, 139, 140
    oil shale deposits, 144, 145*f*
    solar power, 152, 152*f*
    wind power, 154, 155*f*
  erosion in, 160–162, 161*f*
  ethnic niche businesses in, 286
  exotic species in, 453–455, 454*f*
  fertility rates, 187, 498*t*
  fisheries, 453
  gender roles in, 260–261
  GNI per capita, 498*t*
  Gullah language in, 479, 479*f*
  habitat destruction in, 452
  high-tech industries in, 376–378
  HIV/AIDS in, 498*t*
  homelessness in, 417, 418
  hydroelectric power in, 149, 149*f*, 150
  immigration to, 200, 283, 288, 342
    backlash against, 288–289
    central cities and, 419, 420*f*
    by Mexicans, 285, 286, 287*f*, 315, 315*f*
      backlash against, 288–289
    motivations for, 282
    restrictions on, 288–289
    return migration, 290
  industrial regions, 375–376, 375*f*
  infant mortality rates, 188*f*, 498*t*
  as Islamic fundamentalist target, 249
  landfills, 457–458
  language in, 243
    diversity of, 244–245
    official, 244–245
  life expectancy in, 498*t*
  local and regional government, 329–335
    administrative units in, 330, 333–335, 334*f*
    importance of, 329–330
    waste disposal costs, 457
  loess deposits in, 81*f*
  logging subsidies, 167
  manufacturing in, 372–373, 372*f*
  as melting pot, 220, 234
  metropolitan areas in, 397*t*, 398*f*
    growing and declining, 402*t*, 420
  and Mexico
    border between, 313
    immigration from, 285, 286, 287*f*, 315, 315*f*
      backlash against, 288–289
    outsourcing to, 372–373, 372*f*
  Midwest
    boundaries of, 471*f*
    population density in, 12*f*
  and migration
    in 1950s, 41*f*
    in 1990s, 283*f*
    channelized, 291*f*
    of Germans settlers, 284*f*
    migration fields, 290*f*
    of modern Americans, 284
  mineral imports by, 156, 157, 158*t*, 363
  mineral production by, 157*f*
  mineral resources, 157, 158
  mining in, 364*f*
    environmental impact of, 441, 450, 450*f*
    production shutdowns, 156, 363
    on public lands, 366–367
    surface mining, 450, 450*f*
  as multicultural society, 316
  national forests, 166–167, 167*f*
  nationalism in, 316, 316*f*
  New Yorker's view of, 270*f*
  nuclear power plants in, 147*f*
  official language of, 244–245
  oil shale deposits, 145*f*
  place names in, 244–245
  plants
    alien species, 454–455
    endangered, 452
  Pledge of Allegiance, 316*f*
  population
    baby boom, 193*f*
    data on, 498*t*
    density of, 205, 205*t*
    distribution of, 38*f*, 203
    projections, 193*f*, 194, 498*t*
    by state, 40*f*
  public lands, mining on, 366–367
  quaternary economic activity, 381
  regional trade alliances, 327, 328
  religion in, 252, 252*f*, 253*f*
  resources disputes, 315
  sewage treatment in, 441–442
  social services funding, 417
  soil erosion in, 160–162, 160*f*
  soil salinization in, 163
  spatial interaction in, 268
  suburbanization in, 413–414, 413*f*, 414*f*
  as terrorist target, 323. *See also* September 11th terrorist attacks
  topographic maps of, 34–36, 35*f*
  trade
    and NAFTA, 373
    in services, 382*t*
  transnational corporations and, 374
  transportation system in, 318, 320*f*
  urban areas in. *See also* cities *on this page*
    definition of, 396
    growth and decline of, 400, 401*f*, 402*t*
    land use, 412

urban renewal programs, 416–417, 419*f*, 420
waste disposal in
municipal waste, 456–459, 457*f*
radioactive waste, 460–462, 462*f*
toxic waste, 462*f*
toxic waste exports, 462
waste generated in, 457
water pollution in, 441–442
cleanup efforts, 442
control measures, 442
industrial sources, 440–441
water supply in, 435–436, 435*f*
wetlands in, 164, 164*f*
wind patterns in, 97
women in, 260–261, 318–319
Universalizing religions, 246–**247,** 255
Uralic-Altaic language family, 238*f*–239*f*
Uranium-235
in fission, 146–147
reserves, 147
Uranium-238, 147
Urban agriculture, 351
Urban areas. *See also* Cities; Metropolitan areas; Towns
boundaries of, 396*f*
as central places, 404–405, 404*f*
declining, 402*t*
definition of, 396
economic base of, **398**–400
education in, 421
employment in, 411, 419
in developing nations, 425
functions of, 394, 396
hierarchy in, 400, 403*f*
gross domestic product in, 419
hierarchy of, **400,** 403*f*
land use in
competitive bidding for, 405–408, 407*f*, 416
gentrification, 419–420, 420*f*, 422
institutional controls on, 411–412
models of, 408–409, 409*f*
and motor vehicles, 420, 425
residential patterns, 407–408, 407*f*, 408*f*
ethnicity and, 409, 411, 411*f*, 412*f*
family status and, 409–411, 411*f*
social status and, 409, 410, 411*f*
social areas, 409–411
location of, 396–398
megacities, 391, 392*t*
air pollution and, 443
population of
density, 207, 208*f*, 407–408, 407*f*, 408*f*
megacities, 392*t*
rank-size rule for, 401–402
as regions, 484–485
sectors, basic and nonbasic, **398**–400, 400*f*
site of, **396**
situation of, **396**–397, 398*f*
size range of, 394–396, 395*f*
smart growth programs, 421
tax base of, 415–416, 417, 420–421
transportation and, 394, 396, 405–406, 406*f*, 407, 407*f*. *See also* Mass transit systems
in United States
definition of, 396
growth and decline of, 400, 401*f*, 402*t*
land use, 412
urban renewal programs, 416–417, 419*f*, 420
women in, 415, 417
Urban hierarchy, **400,** 403*f*
Urban influence zones, **402**
Urban renewal programs, 416–417, 419*f*, 420
Urban sprawl, 420, 421*f*
automobiles and, 274, 279, 280
in Chicago, Illinois, 413, 413*f*
in Western U.S., 420–421, 421*f*
Urbanization, **207,** 207*f*. *See also* Megalopolis
agriculture and, 207
in developing world, 191, 207, 207*f*, 208*f*, 390
and disease, 198
hydrologic impact, 436
level of, by region and nation, 511*t*–517*t*
and pollution, 441–442, 444
by region, 391, 394*t*
trend toward, 390, 391, 392*f*, 393*f*, 393*t*, 394*f*
Urbanized areas, **396**
Urdu, number of speakers, 240*t*
Uruguay
agriculture in, 359
population data for, 499*t*
regional trade alliances, 328
shape of, 305
U.S. Army Corps of Engineers, 73, 151, 164, 436, 437*f*
U.S. Bureau of Mines, 367
U.S. Census Bureau, 180, 273, 396
U.S. Coast and Geodetic survey, 70
U.S. Department of Agriculture, 456
U.S. Department of Defense, 28
U.S. Department of Education, 244
U.S. Department of Energy (DOE), 460–461
U.S. Department of Homeland Security, 323
U.S. Department of the Interior, 42
U.S. English, 245
U.S. Environmental Protection Agency (EPA). *See* Environmental Protection Agency
U.S. Forest Service, 166
U.S. Geological Survey (USGS)
and bench marks, 37
map projections and, 492, 493*f*
topographic maps, 34–36, 35*f*
symbols, 36, 36*f*
updating of, 43*f*
U.S. Immigration and Naturalization Service, 288
U.S. military, civilian spy satellites and, 46
U.S. Patent Office, 271
U.S. Supreme Court
on bilingual education, 244, 245, 255
on environmental standards, 458
and redistricting, 332–333
U.S.-Mexico Border Counties Coalition, 288
Usable reserves, 136*f*, **137**
USGS. *See* U.S. Geological Survey
USSR. *See also* Russia
agriculture in, 360
Aral Sea environmental tragedy, 438, 438*f*
boundary disputes, 313
cities in, 423–424, 424*f*
collapse of, 228
migrations following, 281
nationalism engendered by, 321
states created by, 303, 304*f*, 307, 309
deceptive mapping practices by, 41, 41*f*
forced migration in, 281
former
coal production, 142
oil production, 141
industrial location in, 373
infant mortality rates, 188
nomadic herding in, 349
political structure of, 302
regional economic alliances, 328
religion in, 319
state terrorism by, 322
and Warsaw Treaty Organization, 329
Utah
cities in, 420
Four Corners Monument, 331*f*
high-tech industry in, 377
tar sand reserves, 144
Utisols, 115*t*

# V

Vaccines, impact on mortality rates, 186
Valley breezes, **96,** 96*f*
Value systems, 246
Value-by-area map, **39**–40, 40*f*
Vancouver, Canada, climate of, 122*f*
Varanasi, India, 216*f*
Variable costs, industrial locational models and, **368**
Vatican City, 305
Vegetation. *See also* Deforestation; Desertification; Forest(s); Plants
acid rain and, 445
air pollution and, 446
arctic and subarctic climate, 123
chaparral, 112, 112*f*
by climate type, 115*t*
coniferous forests, 112, 112*f*
deciduous forest, 112, 112*f*
desert, 112, 112*f*
hot, 116
midlatitude, 118
humid continental climate, 122, 123*f*
humid subtropical climate, 120
marine west coast climate, 122
Mediterranean climate, 112, 112*f*, 120, 121*f*
natural, **110**–112
regions, 110–112, 112*f*
succession in, **110,** 111*f*
ozone layer depletion and, 447
prairie, 112, 112*f*
removal of, and erosion, 71, 71*f*, 159–160, 159*f*
savanna, 112, 112*f*, 114–116, 117*f*
semidesert, 112, 112*f*
steppe, 112, 112*f*, 118, 120*f*
succession in, **110,** 111*f*
tropical rain forest, 112, 112*f*, 114, 117*f*
tundra, 112, 112*f*, 123, 124*f*
Venezuela
agriculture in, 359
oil production, 138*f*
population data for, 499*t*
tar sand reserves, 144
Verbal scale, 33, 33*f*
Vernacular, 241
Vertisols, 110*t*, 111*f*, 115*t*, 116
Veterans Administration (VA), 413
Victorian era, gender roles in, 260
Vietnam
cities in, 424
core area of, 308
emigration from, 282
migration to U.S., 286
population data for, 500*t*
rain forests in, 168
Vietnamese language family, 238*f*–239*f*
Vietnamese language, number of speakers, 240*t*
Vinyl chlorides
as leachate, 458
as pollutant, 440
Virginia
high-tech industry in, 377
redistricting and, 332*f*, 333
Tysons Corner, 414, 485
Volcanism, **62,** 66–68
and air pollution, 442
and climate, 67, 125–126
composite volcanoes, 66, 67*f*
distribution of, global, 61*f*, 62
eruption, 66
in Hawaiian Islands, 61*f*, 66, 67*f*
hot spots, 62, 66, 67*f*
Krakatoa eruption (1883), 125
Mount Pelée eruption (1902), 53
Mount Pinatubo eruption (1991), 126
Mount St. Helens eruption (1980), 62, 66, 68*f*
number of active volcanoes, 66
power of, 53
radioactive waste storage and, 460–461
shield volcanoes, 66, 67*f*
Tambora eruption (1815), 125
and tsunami, 64
volcano belt, 66
and weather, 67
Voluntary migration, 282
von Thünen, Johann Heinrich, 355–356
von Thünen rings, **356,** 357*f*
Voting rights
and language, 244
and race, 331, 332–333, 332*f*
Voting Rights Act of 1965, 244, 332–333
Vulnerable species, 451

# W

Wairakei, New Zealand geothermal plant, 153*f*
Waitsfield, Vermont, church, 254*f*
Wales, 321*f*
government of, 320
language in, 243
slag heap disaster in, 450
Wallonia, 321*f*
Wal-Mart, 279, 279*f*
Warping, **62,** 473
Warsaw Treaty Organization, 329
Washes, 74
Washington, D.C.
establishment of, 308
functions of, 394
gentrification in, 420*f*
high-tech industry in, 377–378
location of, 309
in Megalopolis, 484*f*, 485
population of, 402*t*
smart growth programs in, 421
Washington, George, 233
Washington State
agriculture in, 468*f*
gentrification in, 419
Gifford Pinchot National Forest, 166*f*
hydroelectric power in, 149, 149*f*
pollution control efforts, 442
precipitation in, 102*f*
radioactive waste disposal in, 461, 462*f*
Waste, as fuel, 148, 148*f*
Waste disposal. *See* Solid waste disposal
Wasted vote gerrymandering, 331, 332*f*
Wastepickers, 459, 462*f*
Waste-to-energy power plants, 458, 458*f*
Wastewater treatment, 441–442
Water
and air temperature, impact on, 92, 93*f*
as boundary, 313–314, 313*f*
fresh
access to, by region and nation, 496*t*–503*t*
agricultural use of, 434, 434*f*, 435
supply of, 434–436, 435*f*
ground. *See also* Aquifers
contamination of, 441, 457–458, 461, 462*f*, 486
erosion by, 74–75
in hydrologic cycle, 434*f*
overuse of, 435–436
in Everglades, 486
and subsidence, 451, 451*f*
human impact on, 433–442
hydrologic cycle, **433**–434, 434*f*
quality of, human impact on, 437–439
restrictions on use of, 436
running, erosion by, 71–74, 71*f*, 73*f*, 150, 162–163, 169, 451
transportation by
of goods, 369
and urban location, 396, 398*f*
uses of, 434–435, 434*f*
waves and currents, erosion by, 76–80, 80*f*, 82
wildlife and, 438

Water pollution
acid rain and, 445
biocides and, 456
control of, 442
deforestation and, 166
and disease, 437, 438
leachate, 457–458
natural sources of, 437
in Nile River, 390
solid waste disposal and, 457
sources and types of, 437–442
agricultural sources, 439–440, 440*f*
hazardous waste dump sites, 461
industrial sources, 440–441
landfills, 457–458
mining, 441, 450
municipalities, 441–442
residences, 441–442
sewerage, 441–442
thermal, **441**
wetlands and, 163, 164
Water rights, as conflict source, 314, 315
Water table, **74**, 75*f*
falling of, 435–436
Water temperature, El Niño and, 108–109, 108*f*, 109*f*
Waterman, T. T., 480, 480*f*
Watersheds, as boundaries, 313
Wavelengths of electromagnetic spectrum, 44*f*
Waves
breakers, formation of, 76, 80*f*
erosion by, 76–80, 80*f*, 82
Weather, **88.** *See also* Air temperature; Precipitation; Wind
and air pollution, 443
and air pressure, **92**–95
as basis of climate, 104
convection system and, **95**–96, 95*f*, 96*f*, 97, 98*f*
effect of wind on, 94
factors in, 112
forecasting of, 44, **88**
and global air circulation pattern, 96–97, 98*f*
global warming and, 127
ocean currents and, 98–99
as region, 473–475
sudden changes in (Illinois, 1836), 53
volcanoes and, 67
Weathering, **68**–70, 106
chemical, **69**–70
mechanical, **68**
Weber, Alfred, 368, 368*f*
Weberian analysis, 368, 368*f*
Wegener, Alfred, 58
Wells, contamination of, 441
Westerlies, 97
Western and Central Europe industrial region, 375–376
Western culture
rejection of, 295
role of women in, 260–261
Western United States, desert in, 80
Wetlands, **163**–164, 163*f*, 164*f*
draining of, and disease, 198
loss of, 164
types of, 163
Wheat farming, large-scale, 357–359, 358*f*
Whistler Mountain ski resort, 6*f*
White, Leslie, 221–222
WHO. *See* World Health Organization
Wildlife. *See* Animals and wildlife
Williams, Jody, 295
Wind
air pollution and, 443–444
causes of, 94–96, 95*f*, 96*f*
and convection, **95**–96, 95*f*, 96*f*, 97, 98*f*
Coriolis effect, **96**–97, 97*f*
effect on weather, 94
erosion and deposition by, 80–81, 81*f*
frictional effect, **97**
global circulation pattern, 96–97, 98*f*
land breezes, **95**
mountain breezes, **96**, 96*f*
La Niña, 108*f*, **109**, 109*f*
El Niño, **108**–109, 108*f*, 109*f*, 126, 361*f*
and ocean currents, 99–100
sand deposition by, 119*f*
sea breezes, **95**
seasonal shift in, 97
valley breezes, **96**, 96*f*
Wind farms, 154, 155*f*
Wind power, 154, 155*f*
Windward side, 101
Winter, climate and, 112–113, 113*f*
Wisconsin, urban renewal in, 420
Women. *See also* Gender
abortion and, 180, 184, 202
activity space of, 275
and agriculture, 353, 354–355
in Cairo Plan on population, 202–203
control of childbearing by, 197, 202–203
and credit, 355
education of, and fertility rate, 185
and employment, 275, 417
feminist movement and, 260–261, 318
gender gap and, 318–319
inheritance laws and, 354
land ownership by, 354–355
life expectancy, 191
lower number of, in cultures preferring males, 184, 191
maternal mortality rates, 186, 189, 189*f*
mortality rates, 191
in politics, 318–319, 319*f*
poverty and, 417
rights of, 262
role and status of, 259–261
in Africa, 191, 260, 261, 261*f*, 364
in Arab cultures, 261
in Asia, 191, 261, 354
in Bangladesh, 261
in Caribbean, 191
in China, 184, 191, 263*f*
in developing nations, 191, 260, 261*f*
in Europe, 318
gender empowerment measure (GEM), 261, 263*f*
Green Revolution and, 353, 354–355
history of, 260–261
in Hong Kong, 191
in Hungary, 318
in India, 191
in Indonesia, 261
in Latin America, 191, 260, 261, 354
in Muslim cultures, 261
in New Guinea, 191
in Pakistan, 191
in South Korea, 191
in Taiwan, 191
in United States, 260–261, 318–319
in urban areas, 415, 417
Wood. *See also* Logging
as fuel, 137, 137*f*, 148, 164–165, 165*f*
harvesting of, 163–167, 165*f*, 166*f*
Work. *See* Employment
World Bank, population data, 207–209
World cities, **402**
World Conference on Women (1995), 262
World Conservation Union, 451
World Health Organization (WHO), 189, 189*f*, 190, 324, 326, 443, 456
World Hunger Project, 210
World Refugee Survey, 281
World Trade Center attacks (2001). *See* September 11th terrorist attacks
World Trade Organization (WTO)
activism against, 295, 295*f*
history of, 326
impact of, 345
regional trade alliances and, 328
role of, 326, 366
World War I, and political boundaries, 314
World War II
irredentism in, 314
and migration, 281
veterans of, 413
World Wildlife Fund, 453
Worldwatch Institute, 452
WTO. *See* World Trade Organization
Wu language, number of speakers, 240*t*
Wyoming, population, density, 205


# Y


Yangôn, Myanmar
climate of, 118*f*
stupa in, 257*f*
Yangtze (Chang Jiang) river, 72, 150–151, 350, 435
Year of Safe Motherhood, 189, 189*f*
Year without a summer (New England), 125
Yellow (Huang He) River, 72, 435, 436
Yellowstone River, 73*f*
Yemen, religion in, 254
Yucca Mountain, Nevada, radioactive waste disposal at, 460–461, 460*f*
Yue language, number of speakers, 240*t*
Yugoslavia
bombing of Chinese embassy in, 26
collapse of
and ethnic unrest, 321–322
migrations following, 281
states created by, 303, 307
ethnicity in, 258*f*
return migration to, 291*f*
state terrorism by, 322
Yukon Territory, permafrost in, 79
Yurok Indians, world view of, 480, 480*f*


# Z


Zaire, women in, 285
Zebra mussels, 454, 454*f*
Zero population growth (ZPG), **187**
Zimbabwe
life expectancy, 190
population data for, 497*t*
shape of, 305*f*
Zinc
mining of, 363
reserves of, 156*t*
U.S. imports of, 158*t*
Zionism, 249
Zip code, social status and, 410
Zone of illuviation, 108
Zoning regulations, 335, 373
in developing nations, 426
objections to, 412
purpose of, 411–412
ZPG. *See* Zero population growth
Zulu, 243